ALGAE ABSTRACTS

A Guide to the Literature

Volume 2

1970 - 1972

Volume 1: To 1969

Volume 2: 1970-1972

ALGAE ABSTRACTS

A Guide to the Literature

Volume 2

1970 - 1972

Prepared from material supplied by
Water Resources Scientific Information Center
Office of Water Resources Research
Department of the Interior
Washington, D.C.

IFI/PLENUM · NEW YORK - WASHINGTON - LONDON

Library of Congress Catalog Card Number 73-84004
ISBN 978-1-4757-0422-8 ISBN 978-1-4757-0420-4 (eBook)
DOI 10.1007/978-1-4757-0420-4

Published 1973 by IFI/Plenum Data Corporation
Softcover reprint of the hardcover 1st edition 1973
A Subsidiary of Plenum Publishing Corporation
227 West 17th Street, New York, N. Y. 10011

United Kingdom edition published by Plenum Press, London
A Division of Plenum Publishing Company, Ltd.
Davis House (4th Floor), 8 Scrubs Lane, Harlesden, NW10 6SE, London, England

INTRODUCTION

Algae Abstracts is the first in a series of bibliographies on water resources and pollution published by IFI/Plenum Data Corporation in cooperation with the Water Resources Scientific Information Center (WRSIC). It is produced wholly from the information base comprising material abstracted and indexed for *Selected Water Resources Abstracts.*

The bibliography is divided into volumes according to the publication dates of the source documents. Volume 1 contains 569 abstracts covering publication dates up to and including 1969; Volume 2 contains 730 abstracts covering the years 1970 to 1972. The material included in this bibliography represents computer selections based on the presence of a form of the word "alga" somewhere in the referenced citation. Substantively, the material typifies WRSIC's "centers of competence" approach to information support of the Office of Water Resources Research (OWRR) of the Department of the Interior. Most of the references in this bibliography are the work of the center of competence on eutrophication at the University of Wisconsin.

The indexes refer to the WRSIC accession number, which follows each abstract. The *Significant Descriptor Index* is made up of a fraction of the total descriptors and identifiers by which each paper has been indexed. It represents weighted terms that best describe the information content; this status is indicated by the asterisks which precede them. The *General Index* includes all the remaining descriptors and identifiers by which each paper in this bibliography has been indexed. Through permutation, each word in a multiple-word descriptor or identifier is made to file in its normal alphabetical order, giving access to each abstract by all conceivable routes. To use the indexes, scan the middle rank for a few keywords describing your subject matter interest, then note the WRSIC accession numbers at the right margin. These numbers locate the full record in the bibliography, which is arranged in ascending accession number order.

Each volume, including a complete *Author Index,* is a self-contained entity, offering ready access to the pertinent literature for the time period it spans.

CONTENTS

ABSTRACTS

Publication dates 1970 to 1972

The abstracts are presented in order of increasing accession number. This number appears as the last line of each entry.

CHEMISTRY OF NITROGEN AND PHCSPHORUS IN WATER,

P. L. MCCARTY.

JOURNAL AMERICAN WATER WORKS ASSOCIATION, VOL 62, NO 2, P 127-140, FEB 1970. 14 P, 6 FIG, 8 TAB, 64 REF.

DESCRIPTORS:
*NUTRIENTS, *EUTROPHICATION, *WATER QUALITY CONTRCL, *NITROGEN COMPOUNDS, *PHOSPHORUS CCMPOUNDS, ALGAE, NUTRIENT REQUIREMENTS, PRODUCTIVITY, ESSENTIAL NUTRIENTS, WATER QUALITY, WATER CHEMISTRY, CYCLING NUTRIENTS.

IDENTIFIERS:
NITROGEN REQUIREMENTS(ALGAE), PHOSPHORUS REQUIREMENTS(ALGAE), NUTRIENT CHEMISTRY(AQUATIC).

ABSTRACT:
OF THE MAJOR ELEMENTS ESSENTIAL TO ALGAL GROWTH, NITROGEN AND PHOSPHORUS ARE THE ONES MOST LIKELY TO BE OF CRITICALLY LIMITING AVAILABILITY IN NATURAL WATERS. BECAUSE THEY THEREFORE REPRESENT PROMISING WEAK LINKS IN ALGAL LIFE CYCLES, THEIR CHEMICAL STATES AND BEHAVIOR IN WATER ARE EXAMINED TO SEE HOW WATER TREATMENT MIGHT BENEFIT. LARGE SUPPLIES OF NITROGEN ANC PHOSPHORUS ARE PRESENT IN MANY BODIES OF WATER EITHER IN THE SEDIMENTS, THE ATMOSPHERE ABOVE, OR IN THE FORM OF DISSOLVED GAS. THESE FORMS MAY BE AVAILABLE FOR THE GROWTH OF ALGAE AND OTHER AQUATIC PLANTS, BUT THE RATES AT WHICH THEY MAY BECOME AVAILABLE IS SLOW. THESE RATES ARE IMPORTANT, HOWEVER, AS THEY TEND TO CONTROL THE AMOUNT OF VEGETATIVE GROWTH WHICH CAN BE SUPPORTED. SOLUBLE NITROGEN AND PHOSPHORUS CONTAINED IN THE EFFLUENTS FROM WASTE TREATMENT PLANTS, ON THE OTHER HANC, ARE IN A READILY AVAILABLE FORM. IF DISCHARGED TO NATURAL BCDIES OF WATER, THEY CAN STIMULATE GROWTH FAR IN EXCESS OF THAT WHICH WOULD OCCUR NATURALLY. (KNAPP-USGS)

FIELD 05C, 05A

ACCESSION NO. W70-04080

WISCONSIN WATER RESOURCE PRCBLEMS,

WISCONSIN DEPT. OF NATURAL RESOURCES, MACISON.

L. P. VOIGT.

WISCONSIN CONSERVATION BULLETIN, P 3-5, JANUARY-FEBRUARY 1970. 2 PHOTOS.

DESCRIPTORS:
*WATER QUALITY, *WISCONSIN, PHOSPHORUS, PULP AND PAPER INDUSTRY, ALGAE, AQUATIC WEEDS, RECREATION, GROUNDWATER, FISH, FLCOD PLAINS, IRRIGATION, PESTICIDES.

IDENTIFIERS:
FOX RIVER BASIN(WIS), FWPCA.

ABSTRACT:
WISCONSIN'S WATER PROBLEMS RELATE TO WATER QUALITY AND NOT QUANTITY. THE SOUTHEASTERN AND EASTERN PARTS OF THE STATE ARE PARTICULARLY PROBLEMATIC DUE TO PCPULATION AND INDUSTRIAL CONCENTRATIONS, AND SUBQUENTLY INCREASING DEMANDS FOR QUALITY WATER FCR RECREATION USE. THE PULP AND PAPER INDUSTRY REPRESENTS THE LARGEST WASTE SOURCE, WITH FOUR TIMES THE BIOCHEMICAL OXYGEN DEMAND OF MUNICIPAL WASTES. ANY SOLUTION WILL DEMAND THAT THE PRODUCT PRICES REFLECT TOTAL COSTS, INCLUDING ENVIRONMENTAL DAMAGES. FEDERAL ALLOCATION FUNDS FCR INDUSTRIAL AND MUNICIPAL WASTE TREATMENT PLANTS WERE ONLY 1/3 OF THE AUTHORIZED AMOUNT, THUS THE STATE'S EFFECTIVENESS IN POLLUTICN CONTROL IS HAMPERED. IN THE PAST, THE PROBLEMS OF SILT, NUTRIENTS, AND PESTICIDES HAVE BEEN TACKED BY EDUCATION PROGRAMS, VOLUNTARY ACTION, AND COST SHARING BUT INTENSIFIED LAND USE MAY REQUIRE REGULATORY PROGRAMS. ONLY THE SYMPTOMS OF EUTROPHICATION HAVE BEEN TREATED TO DATE VIA WEED HARVESTING AND ALGAE POISONING, BUT THE STATE IS CCMMITTED TO REDUCE PHOSPHORUS LOADINGS FROM MUNICIPAL ANC INDUSTRIAL WASTES IN LAKE MICHIGAN BY 1972. FLOOD-PLAIN MANAGEMENT, FISH CONTROL, AND IRRIGATION ARE OTHER PROBLEM AREAS TO BE RESOLVEC. (POWERS-WISCCNSIN)

FIELD 05G

ACCESSION NO. W70-05103

ALGAL GROWTH AND DECOMPOSITION: EFFECTS ON WATER QUALITY, NUTRIENT UPTAKE AND CHEMICAL COMPOSITION OF ALGAE IN BATCH CULTURE,

KENTUCKY WATER RESOURCES INST., LEXINGTON.

EDWARD G. FOREE, AND JOHN S. TAPP, JR.

AVAILABLE FROM THE CLEARINGHOUSE AS PB-190 801, $3.00 IN PAPER COPY, $0.65 IN MICROFICHE. RESEARCH REPORT NO. 26, WATER RESOURCES INSTITUTE, UNIVERSITY OF KENTUCKY, LEXINGTON, KENTUCKY, MARCH, 1970. 76 P, 6 TAB, 19 FIG, 43 REF. OWRR PROJECT A-021-KY.

DESCRIPTORS:
*EUTROPHICATION, *ALGAE, *NUTRIENT REQUIREMENTS, CHLOROPHYTA, CYANOPHYTA, CYCLING NUTRIENTS, NITROGEN, PHOSPHORUS, CHEMICAL ANALYSIS, PROTEINS, CARBOHYDRATES, LIPIDS.

IDENTIFIERS:
*ALGAL GROWTH, NUTRIENT UPTAKE, CHEMICAL COMPOSITION, ALGAL DECOMPOSITION..

ABSTRACT:
THE CHEMICAL COMPOSITION OF ALGAE GROWN IN BATCH CULTURE DEPENDS MAINLY ON ENVIRONMENTAL CONDITIONS, NUTRIENT AVAILABILITY, PRESENCE OF PREDATORS, CELL AGE, AND SPECIES. THE EFFECTS OF NUTRIENT AVAILABILITY AND CELL AGE ON THE COMPOSITION OF THREE UNIALGAL CULTURES (ALGAE + BACTERIA) AND ONE HETEROGENEOUS CULTURE (ALGAE + BACTERIA + MICROSCOPIC ANIMALS) WERE EVALUATED. THE CULTURES WERE GROWN IN BATCH CULTURE UNDER BOTH NUTRIENT-ABUNDANT AND NUTRIENT-DEFICIENT CONDITIONS AND THE CHANGES IN COMPOSITIONS WERE OBSERVED. LUXURIOUS UPTAKE, WHERE NUTRIENTS ARE INCORPORATED INTO CELLULAR PROTOPLASM AT LEVELS GREATER THAN THOSE NECESSARY FOR GROWTH, AND SUPER-LUXURIOUS UPTAKE, WHERE SOME NUTRIENTS ARE STORED RATHER THAN CONVERTED INTO ALGAL PROTOPLASM, WERE OBSERVED. THE COMMONLY USED MODEL FOR CALCULATING THE WEIGHT PERCENTAGE OF PROTEIN WAS INACCURATE WHEN SUPER-LUXURIOUS UPTAKE OCCURRED. COMPOSITION OF THE CULTURES WAS GENERALLY CHARACTERIZED BY PROTEIN SYNTHESIS DURING THE NUTRIENT-ABUNDANT GROWTH PHASE, BY A FLUCTUATING COMPOSITION DURING TRANSITION FROM NUTRIENT-ABUNDANT TO NUTRIENT-DEFICIENT GROWTH, AND BY LIPID AND/OR CARBOHYDRATE SYNTHESIS AND THE ESTABLISHMENT OF A RELATIVELY CONSTANT COMPOSITION DURING THE NUTRIENT-DEFICIENT GROWTH PHASE.

FIELD 05C

ACCESSION NO. W70-05469

EFFECT OF LIGHT INTENSITY AND THICKNESS OF CULTURE SOLUTION ON OXYGEN PRODUCTION BY ALGAE,

NAVAL RESEARCH LABORATORY, WASHINGTON, D.C. CHEMISTRY DIV.

R. L. SHULER, AND W. A. AFFENS.

APPLIED MICROBIOLOGY, VOL 19, NO 1, P 76-86, 1970. 10 FIG, 3 TAB, 5 REF.

DESCRIPTORS:
*CHLORELLA, *LIGHT, *PHOTOSYNTHETIC OXYGEN, *ALGAE, CULTURES, VOLUME, MATHEMATICAL STUDIES, CORRELATION ANALYSIS.

IDENTIFIERS:
*CHLORELLA PYRENOIDOSA, OXYGEN PRODUCTION, LIGHT INTENSITY, CULTURE THICKNESS, CULTURE VOLUME.

ABSTRACT:
OBJECTIVE WAS TO OPTIMIZE OXYGEN PRODUCTION EFFICIENCY FOR USE IN CLOSED SYSTEMS, SUCH AS SUBMARINES. MATHEMATICAL RELATIONSHIP BETWEEN OXYGEN PRODUCTION RATE, BY CHLORELLA PYRENOIDOSA (SOROKIN STRAIN--OPTIMUM TEMPERATURE 39C), LIGHT INTENSITY, CULTURE THICKNESS, AND VOLUME OF SUSPENSION WERE WORKED OUT USING A SINGLE TYPE CULTURE VESSEL. THE VESSEL WAS AN UPRIGHT CYLINDER WITH THE INCANDESCENT LIGHT SOURCE (ALSO AN UPRIGHT CYLINDER) LOCATED CENTRALLY. THE CULTURE MEDIUM WAS PUMPED THROUGH CONTINUOUSLY. CULTURE CELL DENSITIES OF 0.5 TO 1.55% WET PACKED CELL VOLUMES WERE SUCH THAT EXPERIMENTAL DATA WERE COLLECTED ON LIGHT LIMITED CELLS UNDERGOING LOG GROWTH. LIGHT INTENSITIES OF 12,000-70,000-FOOT CANDLES WERE EMPLOYED. CULTURE THICKNESS WAS VARIED BY THE USE OF SEPARATE ANNULAR COMPARTMENTS AROUND THE SINGLE LIGHT SOURCE. COOLING WATER WAS PUMPED CONTINUOUSLY IN A LAYER BETWEEN LAMP AND CULTURE VESSEL SO AS TO MAINTAIN TEMPERATURE AT 38.5C. THE CONCLUSION WAS THAT OXYGEN EVOLUTION RATE WAS A LOG FUNCTION OF LIGHT INTENSITY AND THAT RATE OF OXYGEN EVOLUTION PER UNIT VOLUME OF CULTURE IS RELATED LINEARLY TO THE RECIPROCAL OF CULTURE THICKNESS. THE RELATIONSHIPS CITED FOR THE CYLINDRICAL VESSEL MAY HAVE WIDE APPLICABILITY. EFFICIENCIES WITH RESPECT TO ELECTRICAL (LIGHT) REQUIREMENTS WERE COMPUTED. (GERHOLD-WISCONSIN)

FIELD 02K, 05C

ACCESSION NO. W70-05547

RELATIVE CONTRIBUTIONS OF NUTRIENTS TO THE POTOMAC RIVER BASIN FROM VARIOUS SOURCES,

FEDERAL WATER POLLUTION CONTROL ADMINISTRATION, ANNAPOLIS, MD.; AND NEW YORK STATE DEPT. OF HEALTH, ALBANY. DIV. OF ENVIRONMENTAL HEALTH SERVICE.

NORBERT A. JAWORSKI, AND LEO J. HETLING.

CHESAPEAKE TECHNICAL SUPPORT LABORATORY TECHNICAL REPORT NO 31, FEDERAL WATER POLLUTION CONTROL ADMINISTRATION, JANUARY 1970. 36 P, 13 FIG, 6 TAB, 8 REF.

DESCRIPTORS:
*NUTRIENTS, *EUTROPHICATION, *WATER POLLUTION SOURCES, *ESTUARIES, RIVERS, PHOSPHATES, NITRATES, HUDSON RIVER, ALGAE, WATER QUALITY, SEWAGE DISPOSAL, WASTE WATER DISPOSAL, SEWAGE EFFLUENTS, FARM WASTES.

IDENTIFIERS:
POTOMAC RIVER BASIN, POTOMAC ESTUARY.

ABSTRACT:
THE UPPER POTOMAC ESTUARY IS HIGHLY EUTROPHIC. DURING THE SUMMER MONTHS, LARGE BLOOMS OF NUISANCE BLUE-GREEN ALGAE, MAINLY MICROCYSTIS, OCCUR IN THE FRESH-WATER PORTION OF THE UPPER ESTUARY. A RELATIONSHIP BETWEEN HIGH NUTRIENT CONTENT AND THE ACCELERATED EUTROPHICATION IN THE POTOMAC ESTUARY HAS BEEN ESTABLISHED. THE ANNUAL AVERAGE CONCENTRATION OF PHOSPHORUS VARIED FROM 0.09 MG/LITER IN THE SOUTH BRANCH TO 1.9 MG/LITER IN THE ANTIETAM WATERSHED. THE ANNUAL AVERAGE CONCENTRATION OF NITROGEN VARIED FROM 0.3 MG/LITER IN THE SOUTH BRANCH TO 2.2 MG/LITER IN OPEQUON CREEK. ABOUT 92,700 LBS/DAY OF TOTAL PHOSPHORUS ENTERED THE POTOMAC IN 1966, 87% FROM WASTEWATER. THE AVERAGE 1966 LOADING OF TOTAL NITROGEN WAS ABOUT 125,000 LBS/DAY, 51% FROM WASTEWATER. DURING LOW FLOW CONDITIONS A SIGNIFICANT PROPORTION OF THE PHOSPHORUS ENTERING THE SURFACE WATER FROM THE VARIOUS SOURCES IN THE UPPER BASIN IS RETAINED IN THE STREAM CHANNEL. AT HIGH STREAM FLOW, IT APPEARS THAT A LARGE PROPORTION OF THIS PHOSPHORUS IS 'FLUSHED' OUT OF THE STREAM CHANNEL AND TRANSPORTED DOWNSTREAM. A COMPARISON OF SOURCES OF NUTRIENTS IN THE HUDSON RIVER BASIN TO THOSE IN THE POTOMAC SUPPORTS THE CONTENTION THAT IN THE MIDDLE ATLANTIC REGION THE MAJOR SOURCE OF NUTRIENTS TO THE AQUATIC ECOSYSTEM IS FROM WASTEWATER DISCHARGES. (KNAPP-USGS)

FIELD 05B, 05C

ACCESSION NO. W70-06509

STABILIZATION OF DAIRY WASTES BY ALGAL-BACTERIAL SYMBIOSIS IN OXIDATION PONDS,

ALEXANDRIA UNIV. (EGYPT). HIGH INST. OF PUBLIC HEALTH.

F. M. EL-SHARKAWI, AND S. K. MOAWAD.

JOURNAL OF THE WATER POLLUTION CONTROL FEDERATION, VOL 42, NO 1, P 115-125, JANUARY 1970. 4 FIG, 5 TAB, 17 REF.

DESCRIPTORS:
*DAIRY INDUSTRY, *OXIDATION LAGOONS, *PILOT PLANTS, ALGAE, BIOCHEMICAL OXYGEN DEMAND, BIOLOGICAL TREATMENT, PHOTOSYNTHESIS, STABILIZATION, *WASTE WATER TREATMENT, *FARM WASTES.

IDENTIFIERS:
*ALEXANDRIA(EGYPT), PANDORINA, SOLUBLE ORGANIC SOLIDS.

ABSTRACT:
A PILOT-PLANT STUDY OF BOD REDUCTION OF MILK PROCESSING WASTES IS REPORTED. A SYNTHETIC DAIRY WASTE OF 750 MG/L BOD WAS FED CONTINUOUSLY TO RECTANGULAR CONCRETE BASINS WITH SLOPING SIDES TO MINIMIZE SLUDGING. THE DETENTION PERIOD WAS 10 DAYS. AN INFLUENT PH OF 9.8 WAS MAINTAINED TO KEEP THE PH AT A LEVEL CONDUCIVE TO ALGAL GROWTH. TANK DEPTH WAS IMPORTANT IN MAINTAINING BALANCE BETWEEN THE ALGAL AND BACTERIAL FRACTIONS OF THE SYSTEM. THE MICROFLORA SHOWED PLASTICITY IN ADAPTING TO ENVIRONMENTAL VARIATIONS. PANDORINA CONSTITUTED A MAJOR MEMBER OF THE FLORA HIGHLY ADAPTABLE TO INTERACTION WITH DAIRY WASTES. PANDORINA COULD TOLERATE WIDE TEMPERATURE VARIATIONS (11 DEG TO 32 DEG C) AT A CONSTANT DEPTH OF 75 CM. OTHER ORGANISMS WERE RESPONSIVE TO SPECIFIC CONDITIONS AND WHEN THE DOMINANT GROUPS SUFFERED A SERIOUS SETBACK, THE SUBDOMINANTS FLOURISHED. BIOCHEMICAL OXYGEN DEMAND (BOD) REDUCTIONS WERE 80 TO 90 PERCENT AT A BOD LOADING RATE OF 220 LBS/ACRE/DAY (246 KG/DIA/DAY). (AGUIRRE-TEXAS)

FIELD 05D

ACCESSION NO. W70-06619

DISPERSAL OF ALGAE, PROTOZOANS, AND FUNGI BY AQUATIC HEMIPTERA, TRICHOPTERA, AND OTHER AQUATIC INSECTS,

NORTH TEXAS STATE UNIV., DENTON. DEPT. OF BIOLOGY.

KENNETH W. STEWART, LARRY E. MILLIGER, AND BERNARD M. SOLON.

ANNALS OF THE ENTOMOLOGICAL SOCIETY OF AMERICA, VOL 63, NO 1, P 139-144, 1970. 1 FIG, 2 TAB, 13 REF.

DESCRIPTORS:
*ALGAE, *PROTOZOA, *FUNGI, *AQUATIC INSECTS, *DISPERSION, CADDISFLIES, STONEFLIES, DOBSONFLIES, MAYFLIES, CYANOPHYTA, CHLORCPHYTA, CHRYSOPHYTA, CHLORELLA, ROTIFERS, CHLAMYDOMONAS, PHYTOPLANKTON, ZOOPLANKTON, PERIPHYTON.

IDENTIFIERS:
*HEMIPTERA, *TRICHOPTERA, CORIXIDS, EUDORINA, NOTCNECTIDAE, BODO, HYDROPSYCHIDAE, PLECOPTERA, MAGALOPTERA, EPHEMERCPTERA, NANNOCHLORIS, CHLOROCCOCCUM, MICROCYSTIS, FUSARIUM, TRICHORIXA, RAMPHOCORIXA, SIGARA, BOENOA, SALDULA, CRYNELLUS, CHEUMATOPSYCHE, TRIAENCDES INUUSTA, OECETIS, LEPTOCELLA, ISONYCHIA, CAENIS, PERLESTA.

ABSTRACT:
OF 16 SPECIES OF AQUATIC INSECTS STUDIES (5 AQUATIC HEMIPTERA, 5 CADDISFLIES, 1 STONEFLY, 2 DOBSONFLIES, ANC 2 MAYFLIES) 15 WERE FOUND TO BE TRANSPORTING 27 GENERA OF VIABLE SMALL AQUATIC ORGANISMS. THREE CORIXIDS CARRIED 5 GENERA OF BLUE-GREEN ALGAE, 11 GREEN ALGAE, 1 YELLOW-GREEN ALGA, 2 PROTOZOANS, AND FUNGI. EUDORINA, REPRESENTING COLONIAL VOLVOCALEAN ALGAE, WAS FOUND FOR THE FIRST TIME ON FIELD-EXPOSED AQUATIC INSECTS. SUITABILITY OF THE CORIXIDAE AND NOTONECTIDAE AS PASSIVE DISPERSAL VEHICLES FOR SMALL ORGANISMS IS DISCUSSED. ONLY ONE OF SIX CADDISFLIES STUDIED CARRIED ALGAE AND PROTOZOANS, BUT THAT SPECIES, HYDROPSYCHE ORRIS ROSS, CARRIED 10 GENERA. ORGANISMS TAKEN IN CASUAL SAMPLINGS OF PLECOPTERA, MEGALOPTERA, AND EPHEMEROPTERA ARE INDICATED. EXCLUDING FUNGI, CHLORELLA WAS THE MOST COMMON VIABLE FORM TRANSPORTED, FOLLOWED BY BODO, NANNOCHLORIS, CHLOROCCOCCUM, AND MICROCYSTIS, IN ORDER OF FREQUENCY OF OCCURRENCE. UNIDENTIFIED FUNGI AND FUSARIUM WERE CARRIED BY MORE THAN 75% OF THE 16 INSECTS SAMPLED AND OCCURRED IN MORE THAN 50% OF THE 112 CULTURE SAMPLES. THESE DATA FURTHER SUBSTANTIATE ACCUMULATING EVIDENCE THAT AQUATIC INSECTS ARE AMONG THE MOST IMPORTANT OVERLAND PASSIVE TRANSPORT VEHICLES FOR SUCH ORGANISMS, CONTRIBUTING TO THEIR MAINTENANCE IN, AND DISTRIBUTION INTO, SUITABLE ISOLATED AQUATIC HABITATS. (JONES-WISCONSIN)

FIELD 05C, 02H

ACCESSION NO. W70-06655

TRANSFER OF RADIOISOTOPES BETWEEN DETRITUS AND BENTHIC MACROINVERTEBRATES IN LABCRATORY MICROECOSYSTEMS,

OAK RIDGE NATIONAL LAB., TENN. RADIATION ECOLOGY SECTION.

JERRY L. WILHM.

HEALTH PHYSICS, VOL 18, NO 3, P 277-284, 1970. 3 TAB, 14 REF.

DESCRIPTORS:
*RADIOISOTOPES, *DETRITUS, *BENTHIC FAUNA, *INVERTEBRATES, *MICROENVIRONMENT, *ECOSYSTEMS, CESIUM, COBALT RADIOISOTOPES, ALGAE, RADIOACTIVITY, TROPHIC LEVEL, LIFE HISTORY STUDIES, TIME, SNAILS, RADIOACTIVE WASTES, TENNESSEE, CRAYFISH, MIDGES, GAMMA RAYS, BLOODWORMS, SLUDGE WORMS.

IDENTIFIERS:
*MACROINVERTEBRATES, RUTHENIUM-106, CESIUM RADIOISOTOPES, SPIROGYRA, LIMNODRILUS HOFFMEISTERI, STICTOCHIRONOMUS ANNULIORUS, PHYSA HETEROSTROPHA, PROCLADIUS, WHITE OAK LAKE(TENN), UPTAKE.

ABSTRACT:
RUTHENIUM-106, CESIUM-137 AND COBALT-60 TRANSFER BETWEEN DETRITUS OR ALGAE AND BENTHIC MACROINVERTEBRATES WAS STUDIED IN LABORATORY MICROCOSMS. DETRITUS OR ALGAE CONTAINING RADIOISOTOPES WAS COLLECTED FROM A LAKE SERVING AS SETTLING BASIN FOR PARTIALLY DECONTAMINATED WASTE WATER; FOUR SPECIES OF DETRITUS FEEDERS AND ONE CARNIVOROUS SPECIE WERE TAKEN FROM A CONSTANT TEMPERATURE SPRING. MEAN INITIAL ACTIVITY IN PICOCURIES PER ASH-FREE GRAMS IN WHOLE DETRITUS WAS 28,800 FOR RUTHENIUM-106, 6630 FOR CESIUM-137 AND 6870 FOR COBALT-60, WHILE VALUES IN SPIROGYRA WERE 135,000, 4930, AND 62,100, RESPECTIVELY. RADIOISOTOPES WERE PRESENT IN INTERMEDIATE CONCENTRATIONS IN PULVERIZED DETRITUS. WHEN WHOLE DETRITUS WAS USED, ACTIVITY OF ALL THREE RADIOISOTOPES WAS GREATER IN LIMNODRILUS HOFFMEISTERI THAN IN STICTOCHIRONOMUS ANNULIORUS; WHEN PULVERIZED DETRITUS OR ALGAE WAS USED AS NOURISHMENT SOURCE, THE REVERSE WAS TRUE. PHYSA HETEROSTROPHA WAS INTERMEDIATE BETWEEN LIMNODRILUS AND STICTOCHIRONOMUS IN ACTIVITY. PEAK CONCENTRATIONS OF RADIOISOTOPES IN DETRITUS FEEDERS WERE REACHED BY DAY 3, WHILE PROCLADIUS HAD DELAYED UPTAKE, INCREASING IN ACTIVITY UNTIL DAY 14. RUTHENIUM-106 WAS DETECTED IN ALL LIFE STAGES AND EXUVIAE OF STICTOCHIRONOMUS; CESIUM-137 AND COBALT-60 WERE DETECTED ONLY IN LARVAE. ACTIVITY OF RUTHENIUM-106 IN ADULTS WAS 1/3 LESS THAN IN LARVAE. (JONES-WISCONSIN)

FIELD 05C

ACCESSION NO. W70-06668

POLYMERS IN WATER TREATMENT,

UNION CARBIDE CORP. CHEMICAL DIV.

B. MANSFIELD.

PROCESS BIOCHEMISTRY, VOL 5, NO 2, P 28-30, 1970. 4 FIG, 3 TAB, 6 REF.

DESCRIPTORS:
*OXIDES, *FLUID FRICTION, SEWERS, SLURRIES, WASTE WATER TREATMENT, SLUDGE TREATMENT, TOXICITY, ALGAE, BACTERIA, FISH, SEWAGE SLUDGE, FLOW, FLOCCULATION, COAGULATION, FILTRATION, VISCOSITY.

IDENTIFIERS:
*HIGH-MOLECULAR-WEIGHT POLYMERS, TURBULANT FRICTIONAL DRAG, ALKYLENE OXIDE, POLYMERIZATION, ETHYLENE OXIDE, WATER CLARIFICATION, SLUDGE DRYING, POLYETHYLENE OXIDE POLYMERS, SEWAGE SETTLING, UNION CARBIDE CORP, POLYOX, CARBOWAX, FLOCCULANT DOSAGES, BAROCC CLAY SOILS, URANIUM ORE.

ABSTRACT:
THE DISCOVERY AND SUBSEQUENT PRODUCTION BY UNION CARBIDE OF HIGH-MOLECULAR-WEIGHT POLYMERS, SUCH AS POLYOX, PERMITS A REDUCTION OF SURCHARGES IN SEWER PIPELINES. INJECTION OF THESE POLYMERS AT CONCENTRATIONS OF 45 TO 200 PPM INCREASES MORE THAN 100% THE SEWAGE FLOW, THEREBY ELIMINATING THE NECESSITY OF PARALLEL PIPES. THE CHEMICALS DO NOT ADVERSELY INFLUENCE BACTERIA OR FISH, OR INCREASE ALGAL GROWTH. THEY HAVE NO MEASURABLE BIOCHEMICAL OXYGEN DEMAND, BUT IMPROVE SEWAGE SETTLING AND SLUDGE DRYING. ADDITION OF THE POLYMERS TO WATER-BASED SLURRY MARKEDLY DECREASES FLOW RESISTANCE. THE POLYMERS ARE HIGHLY EFFECTIVE AGENTS FOR INCREASING SETTLING AND FILTRATION RATES OF DISPERSED SOLIDS AND FACILITATE WATER CLARIFICATION. (WILDE-WISCONSIN)

FIELD 05D

ACCESSION NO. W70-06968

RELATIONSHIPS BETWEEN WATER QUALITY AND WATER RIGHTS,

R. B. ROBIE.

CONTEMPORARY DEVELOPMENTS IN WATER LAW, WATER RESOURCES SYMPOSIUM, VOL 4, P 72-82, 1970. 11 P, 28 REF.

DESCRIPTORS:
*CALIFORNIA, *WATER RIGHTS, *WATER UTILIZATION, *IMPAIRED WATER QUALITY, RIPARIAN RIGHTS, PRIOR APPROPRIATION, GROUNDWATER, WATER QUALITY, COMPETING USES, DIVERSION, RECIRCULATED WATER, WATER REUSE, LEGAL ASPECTS, JUDICIAL DECISIONS, LEGISLATION, ADMINISTRATION, NON-CONSUMPTIVE USE, ALGAE, EVAPCRATION, REASONABLE USE, BRACKISH WATER, DOMESTIC WATER, GEOCHEMISTRY, WATER CEHMISTRY, SALT BALANCE.

ABSTRACT:
A STUDY OF THE INTERRELATIONSHIPS OF WATER QUALITY TO WATER RIGHTS INVOLVES MANY LEGAL ASPECTS--RIPARIAN, APPROPRIATIVE AND PRESCRIPTIVE RIGHTS, RIGHTS TO THE USE CF GROUNDWATER AND REUSE OF WATER. EVERY EXERCISE OF A WATER RIGHT HAS SOME EFFECT ON WATER QUALITY. THE RETURN FLOW OF DIVERTED WATERS MAY CONTAIN A HIGHER CONCENTRATION OF MINERALS AND SALT THAN PREVIOUSLY. THE REDUCED FLOW CAUSED BY DIVERSION MAY HAVE SEVERAL RESULTS INCLUDING: (1) A REDUCTION IN THE ASSIMILATIVE CAPACITY OF A STREAM; (2) LARGER EVAPORATICN LOSSES; (3) INCREASED STRATIFICATIONS; (4) ACCELERATED ALGAE GROWTHS; AND (5) SALINE INTRUSION. COURTS HAVE LONG RECOGNIZED WATER QUALITY AS A WATER RIGHT. DOWNSTREAM RIPARIAN CWNERS, FOR EXAMPLE, HAVE BEEN ABLE TO ENJOIN THE IRRIGATION USE OF AN UPSTREAM RIPARIAN CWNER. JUNIOR APPROPRIATORS HAVE BEEN HELD LIABLE FOR POLLUTION THAT INTERFERES WITH THE RIGHTS OF A SENIOR APPROPRIATOR. IT IS POSSIBLE THAT THE RIGHT TO DEGRADE WATER COULD BE APPORTIONED AMONG USERS. A UNIFORM APPRCACH IS NECESSARY HOWEVER. VARIOUS STATUTORY LAWS, CASE DECISIONS, AND ADMINISTRATIVE RULES AND REGULATIONS AFFECTING THE SUBJECT ARE DISCUSSED, ESPECIALLY AS RELATES TO CALIFORNIA. (MARSEE-FLORIDA)

FIELD 05G, 06E

ACCESSION NO. W70-07100

FACTORS INFLUENCING THE HERBICIDAL EFFICIENCY OF MCPA AND MCPB IN THREE SPECIES OF MICRO-ALGAE,

UNIVERSITY OF STRATHCLYDE, GLASGOW (SCOTLAND). DEPT. OF BIOLOGY.

R. C. KIRKWOOD, AND W. W. FLETCHER.

WEED RESEARCH, VOL 10, NO 1, P 3-1C, 1970. 2 FIG, 3 TAB, 12 REF.

DESCRIPTORS:
*CHLAMYDOMONAS, *ALGAE, *HERBICIDES, PESTICIDES, CHLORELLA, RESPIRATION, PHOSPHORUS, TOXICITY, RADIOISOTOPES, INHIBITION.

IDENTIFIERS:
*CHLORINATED HERBICIDES, MCPB, MCPA, CHLAMYDOMONAS GLOBUSA, CHLORELLA PYRENOIDOSA, STICHOCOCCUS BACILLARIS, UNDISSOCIATED MOLECULES, ALGAL GROWTH.

ABSTRACT:
THE RELATIVE HERBICIDAL EFFECTIVENESS OF MCPA (4-CHLORO-2-METHYLPHENOXYACETIC ACID) AND MCPB (4-(4-CHLORO-2-METHYLPHENOXY) BUTYRIC ACID) WAS APPRAISED BY USING AS TEST PLANTS THREE SPP OF UNICELLULAR ALGAE. THE ALGAE WERE GROWN IN NUTRIENT SOLUTIONS OF PH 6.5 IN 10-1 ASPIRATORS. THE HERBICIDES WERE C-14-CARBOXYL-LABELLED. MCPB INHIBITED GROWTH, RESPIRATION, AND PHOSPHORUS UPTAKE MORE THAN MCPA AND WAS TOXIC AT THE MINIMUM CONCENTRATION OF 0.0025 MOLES. THE ENHANCED HERBICIDAL EFFECT OF MCPB COINCIDED WITH ITS INCREASED ABSORPTION. BOTH HERBICIDES INHIBITED PHOSPHORYLATION. OPTIMUM UPTAKE OF HERBICIDES WAS AT PH VALUES FAVORING MOVEMENT OF UNDISSOCIATED MOLECULES. (WILDE-WISCONSIN)

FIELD 05F, 05B

ACCESSION NO. W70-07255

ENHANCEMENT BY GLYCEROL OF PHOTOTROPHIC GROWTH OF MARINE PLANKTONIC ALGAE AND ITS SIGNIFICANCE TO THE ECOLOGY OF GLYCEROL POLLUTION,

FISHERIES RESEARCH BOARD OF CANADA, VANCOUVER, (BRITISH COLUMBIA). VANCOUVER LAB.

JOSEPH Y. CHENG, AND NAVAL J. ANTIA.

JOURNAL FISHERIES RESEARCH BOARD OF CANADA, VOL 27, NO 2, P 335-346, 1970. 2 FIG, 2 TAB, 16 REF.

DESCRIPTORS:
*POLLUTANTS, *SEA WATER, *ALGAE, *MARINE ALGAE, CULTURES, PHYTOPLANKTON, CYTOLOGICAL STUDIES, CYANOPHYTA, CHLOROPHYTA, CHRYSOPHYTA, RHODOPHYTA, TESTING, ECOLOGY.

IDENTIFIERS:
*GLYCEROL, PHOTOHETEROTROPHIC GROWTH, LUMINARIN-TYPE POLYSACCHARIDES, CRYPTOPHYCEA, DINOPHYCEA, BACCILARIOPHYCEA, PRYMNESIUM PARVUM, PHOTOAUTOTROPHS, CHROOMONAS SALINA.

ABSTRACT:
EIGHTEEN SPECIES OF MARINE PHYTOPLANKTERS WERE GROWN IN AXENIC CULTURE OF SEA WATER ENRICHED IN NITRATE, ORTHOPHOSPHATE, SILICATE, TRACE-METAL IONS, AND VITAMINES, AND BUFFERED AT PH 7.6-7.8. ASIDE FROM PRYMNESIUM PARVUM AND CHROOMONAS SALINA, NONE OF THE SPECIES SHOWED GROWTH ON GLYCEROL IN THE ABSENCE OF LIGHT. UNDER ILLUMINATION, GLYCEROL STIMULATED THE GROWTH OF 16 SPECIES PARTICULARLY THAT CHRYSOPHYCEA, CRYPTOPHYCEAE, PHAEDACTYLUM TRICORNUTUM, PORPHYRIDIUM CRUENTUM, AND NANNOCHLORIS OCULATA. THE GROWTH OF SOME SPECIES WAS INHIBITED BY GLYCEROL AT DIFFERENT CONCENTRATIONS. THE EFFECTS OF COMPOUND SUGGESTED ITS RELATION TO PHOTOHETEROTROPHIC GROWTH. IN SOME CASES GLYCEROL INDUCED OBVIOUS CYTOLOGICAL AND METABOLIC CHANGES. THE ECOLOGICAL IMPLICATIONS OF GLYCEROL POLLUTION OF SEA WATER ARE DISCUSSED. (WILDE-WISCONSIN)

FIELD 05B

ACCESSION NO. W70-07280

THE LANGE-KUENTZEL-KERR THESIS.

CANADIAN RESEARCH AND DEVELOPMENT, MARCH 1970. 8 P, 1 FIG, 1 TAB, 14 REF.

DESCRIPTORS:
*EUTROPHICATION, *PHOSPHORUS, *CARBON, BACTERIA, ALGAE, SYMBIOSIS, CARBON DIOXIDE, NITROGEN, LAKES, ESTUARIES, NUTRIENTS, DETERGENTS, WISCONSIN, LAKE ERIE, LAKE ONTARIO, ST LAWRENCE RIVER, INTERNATIONAL JOINT COMMISSION.

IDENTIFIERS:
*CARBONACEOUS MATERIAL, *CANADIAN PHOSPHATE DETERGENT BAN, WYANDOTTE CHEMICAL CORPORATION, FMC CORPORATION, SOAP AND DETERGENT ASSOCIATION, CARBOY TRIALS, CANADA, NITROGEN: PHOSPHORUS RATIO.

ABSTRACT:
REPORTS BY W LANGE (NATURE VOL. 215, NO. 5107: 1277-1278, SEP 17, 1967), L E KUENTZEL (JOURNAL WATER POLLUTION CONTROL FEDERATION: 1737-1747, OCT 1969), AND P C KERR (UNPUBLISHED) ARE CITED DEFENDING THE THESIS THAT CARBONACEOUS MATERIAL, NOT PHOSPHORUS, IS THE FACTOR CONTROLLING THE PROCESS OF EUTROPHICATION. AS ADDITIONAL PROOF OF THE MINOR ROLE OF PHOSPHORUS AND NITROGEN IN THE GROWTH OF ALGAE, THE REPORT INCLUDES THE RESULT OF AN EXPERIMENT CONDUCTED IN TWO SMALL OLIGOTROPHIC LAKES IN FLORIDA. ADDITION OF PHOSPHATE AND NITROGEN FERTILIZERS TO ONE OF THESE ORGANIC MATTER-FREE LAKES FAILED TO ALTER SIGNIFICANTLY THE TROPHIC STATE OF THE LAKE AND THE DENSITY OF PLANKTONIC ORGANISMS. THE ISSUE IN QUESTION IS OF A FAR-REACHING IMPORTANCE AS IT IS RELEVANT TO THE USA AND CANADIAN GOVERNMENT ACTION RESTRICTING PHOSPHATE ENRICHED EFFLUENTS OF SOAP AND DETERGENTS INTO ERIE AND ONTARIO LAKES AND THE INTERNATIONAL SECTION OF THE ST LAWRENCE RIVER. (WILDE-WISCONSIN)

FIELD 05B, 05C

ACCESSION NO. W70-07283

ALGAL PERIODICITY AND WASTE RECLAMATION IN A STABILIZATION POND ECOSYSTEM,

IOWA STATE UNIV., AMES.

RONALD L. RASCHKE.

JOURNAL OF THE WATER POLLUTION CONTROL FEDERATION, VOL 42, NO 4, P 518-530, APRIL 1970. 2 FIG, 6 TAB, 40 REF.

DESCRIPTORS:
*WASTE WATER TREATMENT, *PONDS, *ALGAE, LAGOONS, BIOCHEMICAL OXYGEN DEMAND, CHEMICAL OXYGEN DEMAND, PHOSPHATES, NITROGEN, STABILIZATION.

IDENTIFIERS:
*WATER RECLAMATION, VARIATIONS, COMPOSITION.

ABSTRACT:
IN AN ATTEMPT TO EVALUATE POSSIBLE PROCEDURES FOR MEETING NEW STANDARDS OF THE FEDERAL WATER QUALITY ACT, THE AMES WATER POLLUTION CONTROL DEPARTMENT BUILT A NUMBER OF EXPERIMENTAL PONDS. ONE OF THEM WAS MADE AVAILABLE FOR THE PRESENT INVESTIGATION IN WHICH THE INTENT HAS BEEN TO STUDY ALGAL COMPOSITION AND PERIODICITY, DISSOLVED OXYGEN, PH, TEMPERATURE, AND THE EXTENT OF REDUCTION IN 5-DAY BOD, COD, SOLIDS, PHOSPHATES, AMMONIA, AND NITRATES THROUGH THE PLANT POND COMPLEX. THE POND IS 0.12 ACRES IN AREA WITH A WATER HOLDING CAPACITY OF 105, 562 GALLONS AND AN AVERAGE DEPTH OF 2.53 FEET. THE POND DETENTION TIME RANGED BETWEEN 3.7 TO 4.2 DAYS. RESULTS SHOWED THAT ALGAL PERIODICITY AND BOD ANALYSIS MAY BE INFLUENCED BY ALGAL INHIBITORS. GREEN FAGELLATES DOMINATED IN WINTER AND SPRING, WHILE COCCOID GREENS DOMINATED IN SUMMER AND FALL. THE MAXIMUM PER CENT REMOVAL AT ANY ONE TIME WAS 78% AND 75% FOR BOD AND FILTERED BOD RESPECTIVELY. THE MAXIMUM COD REMOVAL WAS 63%. TOTAL SUSPENDED SOLIDS RANGED FROM 458 TO 687 WHICH IS LESS THAN A LIMIT OF 1000 MG/L SET BY THE U.S. PUBLIC HEALTH SERVICE STANDARDS. AMMONIA NITROGEN RANGED FROM 0.0 TO 21.0 MG/L AND NITRATE NITROGEN RANGED FROM 0.5 TO 1.8 MG/L. PH RANGED FROM 7.35 TO 9.09 AND TEMPERATURE FROM 0.8 DEG C TO 29.9 DEG C. THE EFFLUENT USUALLY MET THE IOWA WATER POLLUTION CONTROL COMMISSION'S STANDARDS FOR TEMPERATURE, PH, DISSOLVED OXYGEN CONCENTRATION, BUT PHOSPHORUS AND NITROGEN CONTENTS WERE HIGH. (HANCUFF-TEXAS)

FIELD 05D

ACCESSION NO. W70-07838

ENHANCEMENT BY GLYCEROL OF PHOTOTROPHIC GROWTH OF MARINE PLANKTONIC ALGAE AND ITS SIGNIFICANCE TO THE ECOLOGY OF GLYCEROL POLLUTION,

FISHERIES RESEARCH BOARD OF CANADA, VANCOUVER (BRITISH COLUMBIA). VANCOUVER LAB.

JOSEPH Y. CHENG, AND NAVAL J. ANTIA.

JOURNAL OF THE FISHERIES RESEARCH BOARD OF CANADA, VOL 27, P 335-346, 1970. 11 FIG, 2 TAB, 16 REF.

DESCRIPTORS:
ORGANIC WASTES, WATER POLLUTION EFFECTS, *ALGAE, PHYTOPLANKTON, DIATOMS, CHRYSOPHYTA, CHLOROPHYTA, *GROWTH RATES, INHIBITION, *PLANT PHYSIOLOGY, *PLANT GROWTH, PLANT GROWTH SUBSTANCES, PLANT MORPHOLOGY.

IDENTIFIERS:
*GLYCEROL, *PHOTOTROPHIC GROWTH, *MARINE PHYTOPLANKTON(PRYMNESIUM PARVUM)(CHROOMONAS SALINA), GLYCEROL POLLUTION.

ABSTRACT:
THE EFFECTS OF LOW (0.05 M) AND HIGH (0.5-1.0 M) CONCENTRATIONS OF GLYCEROL ON THE GROWTH OF 18 SPECIES OF MARINE PHYTOPLANKTERS WERE STUDIED. THE ALGAE WERE GROWN IN AXENIC CULTURE IN SEA WATER ENRICHED WITH NITRATE, ORTHOPHOSPHATE, SILICATE, TRACE-METAL IONS, VITAMIN B12, THIAMINE, BIOTIN, AND BUFFERED AT PH 7.6-7.8. APART FROM A CHRYSOMONAD (PRYMNESIUM PARVUM AND A CRYPTOMONAD (CHROOMONAS SALINA), NONE OF THE SPECIES SHOWED ANY SIGNIFICANT GROWTH ON GLYCEROL IN THE ABSENCE OF LIGHT. HOWEVER, IN THE PRESENCE OF LIGHT, GLYCEROL ENHANCED THE GROWTH OF 16 SPECIES, IN PARTICULAR MEMBERS OF THE CHRYSOPHYCEAE AND CRYPTOPHYCEAE, ONE DIATOM (PHAEODACTYLUM TRICORNUTUM), ONE RHODOPHYTE (PORPHYRIDIUM CRUENTUM), AND ONE CHLOROPHYTE (NANNOCHLORIS OCULATA). THE ENHANCEMENT EFFECT WAS OBSERVED IN THE GROWTH RATE AND IN THE PEAK POPULATION DENSITY, WHICH IN MANY INSTANCES WAS SEVERAL TIMES THAT FROM NONGLYCERINATED CULTURES AND SUGGESTED PHOTOHETEROTROPHIC GROWTH. SOME SPECIES SHOWED OBVIOUS CYTOLOGICAL AND METABOLIC CHANGES FROM GROWTH ON GLYCEROL. (SJOLSETH-WASHINGTON)

FIELD 05C

ACCESSION NO. W70-07842

INSECTICIDE RESIDUES IN SOME COMPONENTS OF THE ST LAWRENCE RIVER ECOSYSTEM FOLLOWING APPLICATIONS OF DDD,

DEPARTMENT OF AGRICULTURE, SASKATOON (SASKATCHEWAN); AND SAINT DUNSTAN'S UNIV., CHARLOTTETOWN (PRINCE EDWARD ISLAND). DEPT. OF CHEMISTRY.

F. J. H. FREDEEN, AND J. REGIS DUFFY.

PESTICIDES MONITORING JOURNAL, VOL 3, NO 4, P 219-226, 1970. 4 TAB, 14 REF.

DESCRIPTORS:
*INSECTICIDES, *ST LAWRENCE RIVER, WATER POLLUTION SOURCES, WATER POLLUTION EFFECTS, MUD, MOLLUSKS, FISH, INSECTS, DDT, SILTS, SUCKERS, CATFISHES, BASS, PERCHES, PIKES, SAMPLING, SNAILS, ALGAE, DIATOMS, CLAMS, SHELLFISH, AMPHIPODA, WATER BIRDS, WISCONSIN, ANALYTICAL TECHNIQUES.

IDENTIFIERS:
*RESIDUES, *DDD, DDE, TDE, BLACK-FLY LARVAE, CLADOPHORA, CYPRINUS CARPIO, CATOSTOMUS COMMERSONI, AMEIURUS NEBULOSUS, PERCA FLAVESCENS, AMBLOPLITES RUPESTRIS, ESOX LUCIUS, CAMPELOMA, PSIDIUM, SHADFLY, CHIRONOMIDS, CLEAR LAKE(CALIF), RICHELIEU RIVER(QUEBEC), GREEN BAY(WIS), MACKEREL, MONTREAL(CANADA).

ABSTRACT:
DDD (TDE) RESIDUES WERE MEASURED IN WATER, MUD, MOLLUSKS, AND FISH OF THE ST LAWRENCE RIVER IN 1967 DURING AND AFTER APPLICATIONS OF DDD TO CONTROL NUISANCE INSECTS. CONCENTRATIONS DETECTED IN WATER (UP TO 0.0139 PPM) RANGED FROM 1 TO 17% OF THOSE APPLIED TO THE RIVER 10 MILES UPSTREAM. DDD CONCENTRATIONS IN MOLLUSKS 17 MILES UPSTREAM FROM THE POINT OF APPLICATION AND 10 AND 45 MILES DOWNSTREAM, AVERAGED 0.002, 0.101, AND 0.0 PPM, RESPECTIVELY. IN THE SAME SAMPLES, CONCENTRATIONS OF DDT AND DDE COMBINED (FROM UNKNOWN SOURCES) AVERAGED 0.030, 0.225, AND 0.027 PPM. IN EDIBLE FLESH FROM 216 FISH OF 5 SPECIES, DDD RESIDUES AVERAGED 0.156 PPM IN SAMPLES COLLECTED 17 MILES UPSTREAM AND 0.369 IN THE COMBINED SAMPLES FROM POINTS 10 AND 45 MILES DOWNSTREAM. RESIDUES OF DDT PLUS DDE IN THESE SAME SAMPLES AVERAGED 0.224 AND 0.227 PPM, RESPECTIVELY. THE HIGHEST CONCENTRATION OF DDD IN AN INDIVIDUAL FISH WAS 1.81 PPM. THE ULTIMATE FATE OF THE DDD, THAT IS ITS CONVERSION TO MATERIALS WITH REDUCED TOXICITIES, IS POORLY UNDERSTOOD. DDD IS ITSELF A PRODUCT OF THE REDUCTIVE DECHLORINATION OF DDT IN NATURE. (JONES-WISCONSIN)

FIELD 05C

ACCESSION NO. W70-08098

CURRENT WATER QUALITY CONDITIONS AND INVESTIGATIONS IN THE UPPER POTOMAC RIVER TIDAL SYSTEM,

FEDERAL WATER QUALITY ADMINISTRATION, ANNAPOLIS, MD. CHESAPEAKE TECHNICAL SUPPORT LAB.

JOHAN A. AALTO, NORBERT A. JAWORSKI, AND DONALD W. LEAR, JR.

FEDERAL WATER QUALITY ADMINISTRATION, CHESAPEAKE TECHNICAL SUPPORT LABORATORY TECHNICAL REPORT NO 41, MAY 1970. 38 P, 4 FIG, 6 TAB, 17 REF.

DESCRIPTORS:
*ESTUARIES, *WATER POLLUTION EFFECTS, *DISTRICT OF COLUMBIA, TIDAL EFFECTS, NUTRIENTS, EUTROPHICATION, WATER QUALITY, WATER POLLUTION CONTROL, SURVEYS, INVESTIGATIONS, PATH OF POLLUTANTS, SAMPLING, WATER ANALYSIS, ALGAE, COLIFORMS, DISSOLVED OXYGEN.

IDENTIFIERS:
*POTOMAC RIVER.

ABSTRACT:
BASED ON DATA OBTAINED BY FIELD INVESTIGATIONS AND FROM WASTE WATER TREATMENT AGENCIES IN THE WASHINGTON METROPOLITAN AREA, A STATEMENT ON CURRENT WATER CONDITIONS AND INVESTIGATIONS OF THE UPPER POTOMAC RIVER TIDAL SYSTEM WAS PREPARED AND SUMMARIZED. FECAL CCLIFORM DENSITIES ARE LOWER THAN IN 1965 AS A RESULT OF THE INCREASED CHLORINATION OF TREATED WASTE DISCHARGES SINCE 1969. HIGH FECAL COLIFORM DENSITIES WERE PREVALENT AT TIMES OF HIGH STREAM FLOW ABOVE THE MAJOR WASTEWATER DISCHARGES. THESE HIGH DENSITIES CAN BE ATTRIBUTED TO A COMBINATION OF LAND RUNOFF FROM THE UPPER POTOMAC BASIN, URBAN RUNOFF, STORM SEWERS AND COMBINED SEWER OVERFLOWS. TRIBUTARIES ALSO CONTAINED VERY HIGH FECAL COLIFORM DENSITIES AT TIMES. EXTENSIVE PHYTOPLANKTON BLOOMS WERE DETECTED. SINCE THE LATE 1930'S THE AMOUNT OF PHOSPHCRUS ENTERING THE POTOMAC HAS INCREASED ABOUT TENFOLC ANC NITROGEN INCREASED ABOUT FIVEFOLD. THE MAJOR SHIFT TOWARD NUISANCE BLUE-GREEN ALGAL GROWTHS APPEARS TO BE RELATED TO INCREASES IN NITROGEN AND PHCSPHORUS, AND NOT BOD (CARBON). MOST OF THE PHOSPHORUS WHICH ENTERED THE TIDAL SYSTEM FROM THE UPPER BASIN, PLUS SCME FRCM LOCAL WASTE WATER DISCHARGES, WAS ADSORBED AND DEPOSITED IN THE BOTTOM SEDIMENTS OF THE ESTUARY. DYE AND MATHEMATICAL MODEL INVESTIGATIONS INDICATE THAT WASTE WATER ASSIMILATION AND TRANSPORT RATES ARE VERY LOW. (KNAPP-USGS)

FIELD 05B, 02L

ACCESSION NO. W70-08248

PRELIMINARY ANALYSES OF THE WASTE WATER AND ASSIMILATION CAPACITIES OF THE ANACOSTIA TIDAL RIVER SYSTEM,

FEDERAL WATER QUALITY ADMINISTRATION, ANNAPOLIS, MD. CHESAPEAKE TECHNICAL SUPPORT LAB.

NORBERT A. JAWORSKI, LEO J. CLARK, AND KENNETH D. FEIGNER.

FEDERAL WATER QUALITY ADMINISTRATION, CHESAPEAKE TECHNICAL SUPPORT LAB REPORT NO 39, APRIL 1970. 57 P, 16 FIG, 4 TAB, 5 REF, 3 APPEND.

DESCRIPTORS:
*ESTUARIES, *WATER POLLUTION EFFECTS, *DISTRICT OF COLUMBIA, TIDAL EFFECTS, NUTRIENTS, EUTROPHICATION, WATER QUALITY, WATER POLLUTION CONTROL, SURVEYS, INVESTIGATIONS, PATH OF POLLUTANTS, SAMPLING, WATER ANALYSIS, ALGAE, COLIFORMS, DISSOLVED OXYGEN.

IDENTIFIERS:
*ANACOSTIA RIVER.

ABSTRACT:
A PRELIMINARY ANALYSIS OF WASTE WATER ASSIMILATION AND TRANSPORT CAPACITIES WAS MADE OF THE TIDAL PORTION OF THE ANACOSTIA RIVER, WASHINGTON, D.C. THE MEAN STREAMFLOW IN THE BASIN IS ABOUT 122 CFS AND THE 7-DAY LOW FLOW WITH A RECURRENCE INTERVAL OF ONCE-IN-TEN-YEARS IS 8 CFS. BASED ON 1969 SAMPLING DATA, THE WATER QUALITY CONDITIONS IN THE TIDAL RIVER UNDER SUMMER CONDITIONS ARE TYPIFIED BY: (A) LOW DISSOLVED OXYGEN CONCENTRATIONS OFTEN FALLING BELOW 2.0 MG/L, (B) HIGH FECAL COLIFORM DENSITIES OFTEN ABOVE 10,000 MPN/100 ML, (C) HIGH TURBIDITY LEVELS ESPECIALLY DURING PERIODS OF HIGH RUNOFF, AND (D) HIGH NUTRIENT CONCENTRATIONS. THE MOST PRONOUNCED DEGRADATION RESULTS FROM STORM SEWER AND COMBINED SEWER OVERFLOWS AND DEFECTIVE SANITARY SEWER SYSTEMS. MATHEMATICAL MODEL INVESTIGATIONS OF THE ANACOSTIA RIVER INDICATE THAT WASTE WATER ASSIMILATION AND TRANSPORT CAPABILITIES FOR LARGE ADVECTIVE FLOWS ARE MOST SENSITIVE TO THE DECAY RATE OF A POLLUTANT AND NOT TO THE DISPERSION EFFECT OF THE TIDAL SYSTEM. THE SENSITIVITY OF THE DECAY RATE IS A RESULT OF THE LONG DETENTION TIME, WHILE THE EFFECT OF DISPERSION IS DIMINISHED BY THE PRONOUNCED ADVECTIVE MOVEMENT. THE TIDAL SYSTEM CAPABILITY TO ASSIMILATE OXYGEN-DEMANDING WASTE WATER IS CURRENTLY BEING EXCEEDED. IF WASTE WATER WERE SUBJECTED TO HIGH CARBONACEOUS AND NITROGENOUS BOD REMOVAL AND IF THE EFFLUENT IS AERATED TO 6.0 MG/L, THE PRESENT WATER QUALITY OF THE ANACOSTIA RIVER WOULD BE ENHANCED. (KNAPP-USGS)

FIELD 05A, 05B, 02L

ACCESSION NO. W70-08249

BACTERICIDAL EFFECTS OF ALGAE ON ENTERIC ORGANISMS,

TEXAS UNIV., AUSTIN. CENTER FOR RESEARCH IN WATER RESOURCES.

ERNST DAVIS, AND EARNEST GLOYNA.

FWPCA GRANT 18050 DOL. TECHNICAL REPORT EHE-70-06, CRWR-55, 132 P, MAR 1970. 9 FIG, 144 TAB, 60 REF.

DESCRIPTORS:
*ALGAE, *CULTURES, *ENTERIC BACTERIA, *OXIDATION LAGOONS, *PATHOGENIC BACTERIA.

IDENTIFIERS:
*AUTOGOMISTIC EFFECTS, *AXEMIC CULTURES, *DIEOFF RATES, AFTERGROWTH, BLUE-GREEN ALGAE, FIELD STUDIES, GREEN ALGAE, LABORATORY STUDIES.

ABSTRACT:
A SERIES OF EXPERIMENTS INVOLVING THE EFFECTS OF BLUE-GREEN AND GREEN ALGAE ON THE DIEOFF RATES OF SELECTED BACTERIA WERE CONDUCTED. AXEMIC CULTURES OF ANABAENA CYLINDRICA, A. NIDULANS, OSCILLATORIA CHALYBIA, CHLORELLA PYRENOIDOSA AND SCENEDESMUS OBLUGUUS AMONG OTHERS. CULTURES OF ENTERIC BACTERIA SPECIES (ADCALIGEUES FAECALIS, ENTEROBACTER AEROGENES, E. COLI, PROTEUS VULGARIS, PSUDOMONAS AERARGINOSA, AND SERRATIA MARCERCENS) WERE ADDED TO THE OXENIC ALGAL CULTURES DURING DIFFERENT PERIODS OF THE ALGAL LIFE CYCLES. FILTRATE FROM ACTIVELY GROWING ALGAE WAS EXPOSED TO CULTURES OF ENTERICS TO DETERMINE WHETHER ANY ANTIBIOTIC COMPOUNDS WERE IMPARTED TO THE MEDIUM DURING LAG PHASE GROWTH OF ALGAE. TO DETERMINE AFTERGROWTH OF THE ENTERIC SPECIES, THE DURATION OF THE TESTS WAS EXTENDED TO ABOUT 90 DAYS. MIXED CULTURES OF GREEN AND BLUE-GREEN ALGAE WERE EXPOSED TO BOTH SINGLE SPECIES OF ENTERIC BACTERIA AND MIXED CULTURES. MIXED ALGAL CULTURES CAUSE A GREATER DIEOFF AMONG THE ENTERIC BACTERIA THAN DO INDIVIDUAL SPECIES OF ALGAE. THE DIEOFF CHARACTERISTICS OF PATHOGENIC SPECIES, NAMELY SALMONELLA TYPHOSA, S. PARATYPLIN, SHIGELLA DYSENTIRIAE, S. PARADYSENTESIAE, AND VIBRIO COMMA WERE ALSO DETERMINED. THE PATHOGENIC SPECIES DID NOT SURVIVE AS LONG AS THE ENTERIC TEST SPECIES UNDER SIMILAR TEST CONDITIONS. VIRTUALLY NO AFTERGROWTH WAS DETECTED ON THE PART OF THE PATHOGENES. (AGUIRRE-TEXAS)

FIELD 05D

ACCESSION NO. W70-08319

USES OF WASTE HEAT,

OAK RIDGE NATIONAL LAB., TENN.

SAM E. BEALL.

OAK RIDGE NATIONAL LABORATORY REVIEW, VOL 3, NO 4, P 9-14, SPRING 1970.

DESCRIPTORS:
*THERMAL POLLUTION, *GREENHOUSES, WATER TEMPERATURE, POWERPLANTS, DISTILLATION, ALGAE, AQUICULTURE, FISH, SHELLFISH, CRUSTACEANS, OYSTERS, SEASONAL.

IDENTIFIERS:
*WASTE HEAT, THERMAL EFFICIENCY, HEAT CONSUMPTION.

ABSTRACT:
WASTE HEAT FROM POWERPLANTS WHICH WOULD NORMALLY ADD TO THE PROBLEM OF THERMAL POLLUTION CAN BE REUSED FOR A PROFIT. ONE USE OF THIS WASTE HEAT WOULD BE AS A SUBSTITUTE FOR ELECTRICITY IN LOW-TEMPERATURE INDUSTRIAL USES. WASTE HEAT CAN BE USED FOR AIR COOLING IN AMMONIA OR LITHIUM BROMIDE AIR CONDITIONING SYSTEMS DURING THE SUMMER WHEN THERMAL RELEASES NEED TO BE REDUCED THE MOST. ANOTHER USE OF WASTE HEAT WOULD BE IN THE EVAPORATION OF WASTE WATER TO ALLOW REUSE OF THE VALUABLE WATER. SIMILARLY, DISTILLATION OF SALT WATER OR BRACKISH WATER COULD MAKE USE OF WASTE HEAT. THE BEST POSSIBLE USE WOULD BE IN THE HEATING AND COOLING OF GREENHOUSES AND POULTRY HOUSES LOCATED ON THE REACTOR EXCLUSION AREA. HOWEVER, AQUICULTURE POND HEATING PROMISES LARGE PROFITS IN THE HIGH YIELDS OF FISH, SHELLFISH, AND CRUSTACEANS. FURTHER STUDY ON THE SUBJECT OF WASTE HEAT REUSE IS BEING DONE IN THE PACIFIC NORTHWEST WHERE IRRIGATION USES ARE BEING CONSIDERED. (OSBORNE-VANDERBILT)

FIELD 05G, 08A

ACCESSION NO. W70-08832

ALGAL FLOCCULATION WITH ALUMINUM SULPHATE AND POLYELECTRCLYTES,

NEW SOUTH WALES UNIV., KENSINGTCN (AUSTRALIA).

MICHAEL G. MCGARRY.

JOURNAL WATER POLLUTICN CONTROL FEDERATION, PART II, P R191-R207, MAY 1970, VOL 42, NO 5. 6 FIG, 4 TAB, 14 REF.

DESCRIPTORS:
*ALGAE, *FLOCCULATION, EQUIPMENT, COSTS, SEPARATICN TECHNIQUES, COAGULATION, WASTE WATER TREATMENT.

IDENTIFIERS:
*ALUM, POLYELECTROLYTES, FACTORIAL DESIGN.

ABSTRACT:
THE HIGH RATE POND PROCESS IS BEING DEVELOPED TO TREAT WASTE WATER AND TO PROVIDE A NEW SOURCE OF EDIBLE PROTEIN IN THE FCRM OF ALGAE. AS A FOOD PRODUCTION PROCESS, THE YIELD OF ALGAE EXCEEDS THE YIELDS OF ALL CURRENTLY KNOWN GRAIN CROPS. RECENT STUDIES IN THAILAND HAVE INDICATED THAT PROTEIN PRODUCTION RATES OF 25 TONS/YEAR/ACRE ARE QUITE EASILY ACHEIVED ON A PILOT PLANT SCALE. THIS YIELC COMPARES FAVORABLY WITH THAT OF WHEAT, 135 LB/YEAR/ACRE AND THAT OF SOY BEANS 576 LB/YEAR/ACRE. ALUMINUM SULFATE WAS TESTED AS A PRIMARY COAGULANT IN CONJUNCTION WITH A VARIETY OF POLYELECTROLYTES FOR CHEMICAL FLOCCULATION AS A MEANS OF HARVESTING THE DISPERSED ALGAE FROM HEAVILY LADEN POND WATER. THE INDEPENDENT VARIABLES STUDIED WERE ALUM AND POLYELECTROLYTE CONCENTRATIONS, TIME OF POLYELECTRCLYTE ADDITION, FAST AND SLOW MIXING PERIOD ANC DEGREE OF TURBULENCE. DEPENDENT VARIABLES CONSIDERED WERE: SUPERNATANT TRANSMISSION AND SETTLED VOLATILES AS A MEASURE OF HARVEST OF ALGAE, AFTER ONE HOUR OF SETTLEMENT. COST OF ALGAL HAVESTING BY ALUM ALONE AT CONCENTRATIONS OF 30 MG/L WOULD BE PROHIBITIVE. THE OVERALL MINIMUM COST PER UNIT OF ALGAL YIELD WAS ATTAINED WITH ALUM ALCNG AT DOSAGE RANGE OF 75 TC 100 MG/L OF ALUM. (HANCUFF-TEXAS)

FIELD 05D

ACCESSION NO. W70-08904

BIOLCGICAL EFFECTS ON SEDIMENT-WATER NUTRIENT INTERCHANGE,

CALIFORNIA UNIV., BERKELEY. SANITARY ENGINEERING RESEARCH LAB; AND SUNN, LOW, TOM, AND HARA, INC., HONCLULU, HAWAII.

DONALD B. PORCELLA, JAMES S. KUMAGAI, ANC E. JOE MIDDLEBROOKS.

ASCE PROCEEDINGS, JOURNAL CF THE SANITARY ENGINEERING DIVISION, VOL 96, NO SA4, PAPER 7460, P 911-926, AUGUST 1970. 5 FIG, 8 TAB, 10 REF. CONTRACT AT(11-1)-34 - PROJECT 100, AEC.

DESCRIPTORS:
*NUTRIENTS, *BOTTOM SEDIMENTS, *LIMNOLOGY, *ALGAE, *EUTROPHICATION, LAKES, WATER QUALITY, PHCSPHORUS, PHOSPHATES, PRODUCTIVITY, LEACHING.

IDENTIFIERS:
SEDIMENT-WATER NUTRIENT-EXCHANGE.

ABSTRACT:
DIFFERENT TYPES CF SEDIMENTS VARY IN THEIR ABILITY TO SUPPORT ALGAL GROWTH. THIS IS RELATED TO THE AMOUNT OF AVAILABLE PHOSPHORUS MEASURED IN THE SEDIMENTS. ALTHOUGH THE AMCUNT OF PHOSPHORUS RELEASED FROM THE SEDIMENTS VARIED WITH THE TYPE OF SEDIMENT, ALL OF THE AVAILABLE PHOSPHORUS EVENTUALLY SHOULD BE EXTRACTED IN THE 15-CM LAYER OF SEDIMENT STUDIED. THE DEVELOPMENT OF A THICK MAT OF OSCILLATORIA CAUSED AN INCREASE IN PRODUCTIVITY DUE TO THE INCREASED TRANSFER OF PHOSPHORUS FROM THE SEDIMENT; AND, IN GENERAL, THE MORE PROCUCTIVE SYSTEMS HAD SEDIMENTS CONTAINING GREATER AMOUNTS OF PHOSPHORUS. ALSO, THE EQUILIBRATION OF P-32 WITH STABLE PHOSPHATE OCCURRED MORE RAPIDLY IN THE MORE PROCUCTIVE SYSTEMS, AND THIS RATE OF EQUILIBRATION APPEARED TO BE ASSOCIATED WITH THE SEDIMENTS THEMSELVES. HENCE P-32 EXCHANGE WITH STABLE P WAS RELATED TO THE CONCENTRATION OF AVAILABLE PHCSPHORUS IN THE SYSTEM. (KNAPP-USGS)

FIELD 05C, 02H

ACCESSION NO. W70-08944

TRISODIUM NITRILOTRIACETATE AND ALGAE,

ONTARIO WATER RESOURCES COMMISSION, TORONTO.

A. E. CHRISTIE.

WATER AND SEWAGE WORKS, VOL 117, NO 2, P 58-59, FEBRUARY 1970.

DESCRIPTORS:
*ALGAE, *ACTIVATED SLUDGE, *TOXICITY, *LABORATORY TESTS, DETERGENTS, WASTE WATER TREATMENT.

IDENTIFIERS:
*TRISODIUM NITRILOTRIACETATE, *CHLORELLA PYRENOIDOSA, SOURCE OF NITROGEN, NTA.

ABSTRACT:
THE RELATIONSHIPS BETWEEN TRISODIUM NITRILOTRIACETATE, A POTENTIAL DETERGENT BUILDER, AND THE ALGAE CHLORELLA PYRENOIDOSA WERE INVESTIGATED, IN AN ACCLIMATED ACTIVATED SLUDGE SYSTEM, WITH RESPECT TO TOXICITY AND AS A SOURCE OF NITROGEN BOTH BEFORE AND AFTER TREATMENT OF THE MATERIAL. IT WAS REPORTED THAT TRISODIUM NITRILOTRIACETATE IS NON-TOXIC TO FISH AT CONCENTRATIONS UP TO 100 MG/L, TO MAMMALS FED UP TO SEVERAL THOUSAND MILLIGRAMS PER KILOGRAM BODY WEIGHT, AND NON-IRRITATING TO HUMAN SKIN. RESULTS INDICATED THAT NTANA3 AT CONCENTRATIONS UP TO 275 MG/L WAS NOT TOXIC TO THE TEST ORGANISM CHLORELLA PYRENOIDOSA AND COULD ACT AS A SOURCE OF NITROGEN FOR ALGAL GROWTH, ALTHOUGH, NOT TO THE SAME EXTENT AS EQUIVALENT QUANTITIES OF NITROGEN, FED AS NITRATE, UNDER THE SAME TEST CONDITIONS. TREATMENT OF 275 MG/L NTANA3 BY AN ACCLIMATIZED ACTIVATED SLUDGE SYSTEM DID NOT REDUCE THE ALGAL GROWTH POTENTIAL OF THE CHEMICAL. THE FEASIBILITY OF USING BIODEGRADABLE ORGANIC MATERIALS WHICH CONTAIN NEITHER NITROGEN OR PHOSPHORUS AS DETERGENT BUILDERS SHOULD BE THOROUGHLY CONSIDERED. (GALWARDI-TEXAS)

FIELD 05D

ACCESSION NO. W70-08976

THE EFFECT OF BIOLOGICAL LIFE ON THE DISSOLVED OXYGEN CONCENTRATION IN THE DELAWARE RIVER,

DREXEL UNIV., PHILADELPHIA, PA. DEPT. OF CIVIL ENGINEERING.

WILLIAM L. ZEMAITIS, AND GERALDINE V. COX.

PREPARED FOR THE CITY OF PHILADELPHIA WATER DEPARTMENT. DREXEL UNIVERSITY, DEPARTMENT OF CIVIL ENGINEERING, PHILADELPHIA, JULY 1970. 13 P, 10 FIG, 6 TAB, 131 REF. WORK NO. P-244, CONTRACT NO. 88-2239.

DESCRIPTORS:
*SLUDGE WORMS, *DISSOLVED OXYGEN, *EUTROPHICATION, *BIOCHEMICAL OXYGEN DEMAND, *CHEMICAL OXYGEN DEMAND, *WATER POLLUTION SOURCES, *WATER POLLUTION EFFECTS, *ORGANIC MATTER, *OXYGEN SAG, *IMPAIRED WATER QUALITY, *ESTUARIES, *DELAWARE RIVER, VEGETATION EFFECTS, SLUDGE, RIVER BEDS, ALGAE, MATHEMATICAL MODELS.

IDENTIFIERS:
*DELAWARE ESTUARY, *BIOLOGICAL LIFE, *DISSOLVED OXYGEN CONCENTRATION.

ABSTRACT:
DESCRIBES AN INVESTIGATION OF THE EFFECTS OF BIOLOGICAL LIFE ON THE DISSOLVED OXYGEN CONTENT IN THE DELAWARE ESTUARY. BIOLOGICAL, CHEMICAL, AND PHYSICAL TESTS WERE CONDUCTED FOR EIGHTEEN MONTHS. BENTHAL SURVEYS REVEALED THAT TURBIFICIDS (AQUATIC WORMS), ALGAE, ROOTED AQUATICS, AND TERRESTRIAL VEGETATION WERE PRESENT. THE TURBIFICIDS WERE PREDOMINANT INVERTEBRATE SPECIES IN THE SLUDGE DEPOSITS, AND THE POPULATION WAS GREAT WHERE THE SEDIMENT CONTAINED LARGE AMOUNTS OF ORGANIC COMPOUNDS. RESPIRATION OF MIXED COMMUNITIES OF INVERTEBRATES WAS DETERMINED TO BE AS HIGH AS 4.2 MG/L OF OXYGEN PER GRAM OF TURBIFICIDS. ONE DEAD GRAM OF DEAD TURBIFICIDS WAS FOUND TO CREATE A BOD5 OF 200 MG/L. TOXIC POLLUTION, WITH A RESULTING TOTAL KILL OF TURBIFICIDS IN A REGION, COULD CAUSE A 20,000 MG/L/M2 BOD5. THE BOD/COD RATIO INCREASES REPRESENT A VERY SIGNIFICANT SECONDARY POLLUTION OF THE ESTUARY. A SIGNIFICANT AMOUNT OF NATURAL ORGANICS IN THE FORM OF ALGAE AND OTHER VEGETATION WAS FOUND. SETTLEMENT OF THESE ORGANICS IN SLOWER MOVING WATERS CAUSES ENRICHMENT OF THE SEDIMENT AND PRODUCES A DISSOLVED OXYGEN SAG BELOW TRENTON DURING HIGH FLOW PERIODS WHEN BOTTOM SCOURING TAKES PLACE. MACROFAUNA METABOLISM AND ALGAE OXYGENATION AND DEOXYGENATION WERE NOT INCORPORATED INTO MATHEMATICAL MODELS (DEVELOPED BY OTHERS) FOR THE DELAWARE ESTUARY. THE AUTHORS STATE THAT THE WASTE ALLOCATIONS DETERMINED FROM THESE MODELS MIGHT HAVE DIFFERED GREATLY IF ALL PARAMETERS HAD BEEN CONSIDERED. RECOMMENDATIONS FOR FURTHER STUDIES ARE GIVEN. (POERTNER)

FIELD 05C

ACCESSION NO. W70-09189

USES OF WASTE HEAT,

OAK RIDGE NATIONAL LAB., TENN.

S. E. BEALL.

THIS ARTICLE DRAWS UPON SIX STUDIES CONDUCTED BY ORNL, SOME OF WHICH ARE SPONSORED BY THE U.S. DEPARTMENT OF HOUSING AND URBAN DEVELOPMENT. OAK RIDGE NATIONAL LABORATORY REVIEW, SPRING 1970, P 9-14. 4 FIG.

DESCRIPTORS:
*THERMAL POWERPLANTS, *POWERPLANTS, *ENERGY LOSSES, *WATER CONSERVATION, *AIR CONDITIONING, *HEATING, *COOLING, *THERMAL POWER, *MULTIPLE-PURPOSE PROJECTS, *WATER REUSE, *WATER SUPPLY, *POLLUTION ABATEMENT, *WATER POLLUTION CONTROL, *WATER COOLING, *STEAM TURBINES, *THERMAL POLLUTION, LONG TUBE VERTICAL DISTILLATION, EVAPORATORS, AQUATIC ALGAE, HORTICULTURAL CROPS, POULTRY, GREENHOUSES, FISH FARMING, IRRIGATION WATER.

IDENTIFIERS:
*WASTE HEAT, *USES OF WASTE HEAT, *CENTRAL HEATING AND COOLING, *SPACE HEATING, *FARM USES OF WASTE HEAT, STEAM POWERED BUSES.

ABSTRACT:
PRESENTS NEW CONCEPTS FOR BENEFICIAL USE OF A LARGE PART OF THE HEAT DISCHARGED FROM THE EXHAUSTS OF TURBINES OF FOSSIL FUEL OR NUCLEAR POWER PLANTS. BOTH URBAN AND RURAL SITES OFFER OPPORTUNITIES FOR APPLYING LOW TEMPERATURE HEAT TO USEFUL PURPOSES. WASTE STEAM OR HOT WATER FROM POWER PLANTS, BETWEEN 300 DEG F TO 380 DEG F, CAN BE COMPETITIVE WITH PRESENT URBAN SPACE HEATING SOURCES, EVEN IF THE POWER PLANT IS 10 MILES FROM THE HEAT DISTRIBUTION POINT. IN MOST CITIES IN THE UNITED STATES, MORE WASTE HEAT CAN BE USED FOR COOLING, (IN ABSORBTION SYSTEMS) DURING THE CRUCIAL SUMMER PERIOD THAN IS NEEDED IN THE WINTER FOR RESIDENTIAL AND COMMERCIAL HEATING. THE AVERAGE REDUCTION IN WASTE HEAT DISPOSAL THAT COULD BE ATTAINED IN THIS MANNER IN URBAN AREAS IS ESTIMATED TO BE 60 PER CENT. OTHER USES OF THE WASTE HEAT ARE DISCUSSED, INCLUDING HEATING AND COOLING OF GREENHOUSES AND POULTRY HOUSES, AND HEATING OF AQUACULTURE PONDS. EVAPORATION OF WASTE WATER AND DISTILLATION OF BRACKISH OR SALT WATER FOR POTABLE WATER SUPPLY OR IRRIGATION USES IS CONSIDERED FEASIBLE, BUT COSTLY AT PRESENT. STEAM STORAGE TANKS TO PROVIDE POWER FOR BUSES AND SWITCH ENGINES ARE CONSIDERED TECHNICALLY AND ECONOMICALLY FEASIBLE. THE USE OF 104 DEG F IRRIGATION WATER IS BEING STUDIED IN OREGON ON SEVERAL EXPERIMENTAL FARMS. SIMILAR STUDIES ARE UNDERWAY AT WASHINGTON STATE UNIVERSITY. A STUDY OF INSTALLATIONS OF UNDERGROUND PIPES HEATED WITH CONDENSER DISCHARGE WATER TO STIMULATE GROWTH OF CROPS IS UNDERWAY AT OREGON STATE UNIVERSITY. (POERTNER)

FIELD 03C, 08C, 05D

ACCESSION NO. W70-09193

A STATEMENT ON PHOSPHORUS,

MINNESOTA UNIV., MINNEAPOLIS.

JOSEPH SHAPIRO.

JOURNAL OF THE WATER POLLUTION CONTROL FEDERATION, VOL 42, NO 5, PART 1, PAGE 772-775, MAY 1970. 4 REF.

DESCRIPTORS:
*PHOSPHORUS, *DETERGENTS, *TERTIARY TREATMENT, *EUTROPHICATION, WASTE WATER TREATMENT, ALGAE, EFFLUENTS, NUTRIENTS, SEWAGE TREATMENT, LAKES.

ABSTRACT:
A PAPER BY L. E. KUENTZEL PUBLISHED IN THE OCTOBER 1969 ISSUE OF THE JWPCF IS DISCUSSED AND EVALUATED BY SHAPIRO. THE DISCUSSION IS BASICALLY A REFUTATION OF ARGUMENTS PRESENTED BY KUENTZEL IN HIS CHALLENGING THE EFFICACY OF PHOSPHORUS REMOVAL FOR EUTROPHICATION CONTROL. SEVERAL WORKS CITED, PREVIOUS TO KUENTZEL'S, SUPPORTS THE FACT THAT REDUCTION OF PHCRPHORUS IN EFFLUENTS THROUGH CURTAILED USE OF DETERGENTS OR ADVANCE WASTE WATER TREATMENT WILL REDUCE EUTROPHICATION OF LAKES SIGNIFICANTLY AND WILL ALLOW THOSE ALREADY SEVERELY AFFECTED TO BEGIN THEIR RECOVERY THROUGH NATURAL PROCESSES. (HANCUFF-TEXAS)

FIELD 05D, 05C

ACCESSION NO. W70-09325

DETERGENTS, PHOSPHATES, AND WATER POLLUTION,

DEPARTMENT OF ENERGY, MINES AND RESOURCES, OTTAWA (ONTARIO). INLAND WATERS BRANCH.

P. D. GOULDEN, W. J. TRAVERSY, AND G. KERR.

CANADA DEPARTMENT OF ENERGY, MINES AND RESOURCES INLAND WATERS BRANCH TECHNICAL BULLETIN NO 22, 1970. 8 P.

DESCRIPTORS:
*WATER POLLUTION EFFECTS, *PHOSPHATES, *NITRATES, *NUTRIENTS, *EUTROPHICATION, ALGAE, SURFACTANTS, WATER POLLUTION SOURCES, SURFACE WATERS, WATER POLLUTION, DOMESTIC WASTES, MUNICIPAL WASTES.

IDENTIFIERS:
CANADA.

ABSTRACT:
THIS REPORT WAS PREPARED TO PROVIDE THE GENERAL PUBLIC WITH TECHNICAL INFORMATION ON DETERGENTS AND THE EFFECTS OF PHOSPHATES ON THE WATER ENVIRONMENT. IT ANSWERS SOME OF THE MORE COMMONLY-ASKED QUESTIONS REGARDING PHOSPHATE-BASED DETERGENTS--WHAT THEY ARE, WHAT THEY CONTAIN, HOW THEY WORK AND HOW THEY AFFECT THE QUALITY OF OUR WATER RESOURCES. THE PROPERTY OF BIODEGRADABILITY AS IT APPLIES TO DETERGENTS IS EXPLAINED. WHILE PHOSPHATES AND NITRATES DO NOT IN THEMSELVES POSE ANY THREAT TO HEALTH AT THE CONCENTRATIONS INVOLVED AFTER DIFFUSION IN RECEIVING WATERS, THEY DO POSSESS NUTRITIVE PROPERTIES WHICH ENCOURAGE THE EXCESSIVE GROWTH OF ALGAE AND OTHER FORMS OF UNDESIRABLE AQUATIC VEGETATION. WHEN LARGE AMOUNTS OF THIS VEGETATION DECOMPOSE AT THE END OF THE GROWING PERIOD, DEPLETION OF THE VITAL LIFE-SUSTAINING DISSOLVED OXYGEN IN WATER OCCURS. THIS FORM OF POLLUTION, WHICH IS CAUSED BY EXCESS NUTRIENT ENRICHMENT, IS ALSO REFERRED TO AS EUTROPHICATION AND IN ITS EXTREME FORM RESULTS IN THE ACCELERATED AGING OR DYING OF LAKES. (KNAPP-USGS)

FIELD 05C, 05B

ACCESSION NO. W70-09388

TOXICITY STUDIES WITH AN OIL-SPILL EMULSIFIER AND THE GREEN ALGA PRASINOCLADUS MARINUS,

UNIVERSITY COLL. OF WALES, ABERYSTWYTH. DEPT. OF BOTANY.

A. D. BONEY.

JOURNAL OF THE MARINE BIOLOGICAL ASSOCIATION OF THE UNITED KINGDOM, (1970), P 461-473. 14 REF, 4 TAB, 4 FIG.

DESCRIPTORS:
BIOASSAY, TOXICITY, *EMULSIFIERS, DETERGENTS, OIL, *OIL WASTES, OILY WATER, OIL-WATER INTERFACES, WATER POLLUTION TREATMENT, WATER POLLUTION EFFECTS, ENVIRONMENTAL EFFECTS, ALGAL POISONING, ALGAE, SOLVENTS, SURFACTANTS, WATER TEMPERATURE, PLANT PIGMENTS, PLANT PHYSIOLOGY, *PLANT PATHOLOGY, PLANT GROWTH REGULATORS, SALINITY, PLANT POPULATIONS.

IDENTIFIERS:
*BP 1002, *GREEN ALGAE, PRASINOCLADUS MARINUS, TEMPERATURE EFFECTS.

ABSTRACT:
CYST PHASES OF THE GREEN ALGA PRASINOCLADUS MARINUS HAVE BEEN USED IN AN INVESTIGATION OF THE TOXIC PROPERTIES OF AN OIL-SPILL EMULSIFIER BP 1002, AND OF ITS SOLVENT AND SURFACTANT FRACTIONS. VARIOUS ASPECTS OF A REJUVENATION PROCESS HAVE ALL BEEN UTILIZED AS A MEANS OF ASSAY IN ADDITION TO OBSERVATIONS ON CELL VIABILITY. THE 'AGED' CYSTS WERE MORE TOLERANT OF ALL TYPES OF TOXIC AGENTS THAN WERE THE YOUNG NON-MOTILE CELLS. THE SURFACTANT FRACTIONS WERE MORE TOXIC WHEN USED ALONE, AND THE SOLVENT FRACTION ALONE MORE TOXIC THAN THE COMPOUNDED BP 1002. THE APPLICATION OF ANY OF THE TOXIC AGENTS AT LOW TEMPERATURE (4 DEG. C) RESULTED IN A MARKED REDUCTION IN THEIR EFFECTS AT HIGH CONCENTRATIONS (E.G. 500 PPM). THE TOXIC EFFECT WAS APPRECIABLY INCREASED WITH BOTH 'AGED' AND 'YOUNG' CELLS WHEN ACCOMPANIED BY A LOWERING IN SALINITY. AERATION OF THE TOXIC SOLUTIONS CAUSED A SIGNIFICANT LOWERING OF TOXICITY WITH BOTH BP 1002 AND THE SOLVENT FRACTION. CHLOROPLAST PIGMENT REGENERATION IN 'RECOVERING' CYSTS WAS A SENSITIVE MEANS OF ASSAYING TOXIC EFFECTS. (SJOLSETH-WASHINGTON)

FIELD 05C

ACCESSION NO. W70-09429

THERMAL STANDARDS IN THE UNITED STATES OF AMERICA,

FEDERAL WATER POLLUTION CONTROL ADMINISTRATION, DULUTH, MINN. NATIONAL WATER QUALITY LAB.

D. I. MOUNT.

SYMPOSIUM ON ENVIRONMENTAL ASPECTS OF NUCLEAR POWER STATIONS, NEW YORK, AUGUST 10-14, 1970. VOL SM-146, NO 15, PREPRINT, 5 P.

DESCRIPTORS:
*TEMPERATURE CONTROL, *STANDARDS, *WATER TEMPERATURE, SEASONAL, MONTHLY, AQUATIC LIFE, INTERSTATE RIVERS, HYPOLIMNION, VOLUME, MIXING, SAFETY, WATER COOLING, ANAEROBIC CONDITIONS, ALGAL CONTROL, SPAWNING, WARM-WATER FISH.

ABSTRACT:
TEMPERATURE STANDARDS SPECIFYING PERMISSIBLE WATER TEMPERATURES FOR INTERSTATE RIVERS AND LAKES OF THE UNITED STATES HAVE BEEN ESTABLISHED. STANDARDS FOR AQUATIC LIFE MUST INSURE A SEASONAL TEMPERATURE CYCLE, AND SO TWELVE MAXIMUM MONTHLY TEMPERATURES ARE THE MINIMUM TO DESCRIBE DESIRED CONDITIONS. IN STREAMS, A 5F RISE MAY BE PERMITTED AS LONG AS THE RESULTING TEMPERATURE DOES NOT EXCEEC THE APPROPRIATE MONTHLY MAXIMUM. IN LAKES, A 3F RISE MAY BE USED WITH AN ADDITIONAL REQUIREMENT THAT THE HYPOLIMNION CANNOT BE WARMED OR CHANGED IN VOLUME. MIXING ZONES POSE A DIFFICULT REGULATORY PROBLEM. THEY ARE DIFFICULT TO RESTRICT AS TO SIZE AND SHAPE ESPECIALLY BECAUSE THE EFFECTS OF MIXING ZONES ON RECEIVING WATER DEPENDS, IN PART, ON THE NUMBER OF MIXING ZONES AND THE SIZE OF THE LAKE OR RIVER. WHERE COOLING DEVICES ARE REQUIRED, THE STANDARDS DO NOT SPECIFY THE TYPE TO BE USED. THE FREQUENCY THAT DEVICES WILL NOT MEET THE STANDARDS HAS NOT BEEN WELL DEFINED EXCEPT IN A GENERAL WAY. IF NO SAFETY MARGIN IS PERMITTED, STANDARDS MUST BE MORE COMPLEX IN ORDER TO INSURE ACCEPTABLE CONDITIONS. (OSBORNE-VANDERBILT)

FIELD 05G, 06E

ACCESSION NO. W70-09595

POSSIBILITIES FOR BENEFICIAL USES OF HEATED WATER DISCHARGES INTO COOLING RESERVOIRS,

MISSOURI UNIV., COLUMBIA. DEPT. OF ZOOLOGY.

ROBERT S. CAMPBELL, ARTHUR WITT, JR., AND JAMES R. WHITLEY.

FIFTH ANNUAL WATER RESOURCES RESEARCH CONFERENCE, FEB. 3-4, 1970, WASHINGTON, D.C., U.S. DEPT. OF INTERIOR, OFFICE OF WATER RESOURCES RESEARCH, JUNE 1970. P 57-64, 27 REF.

DESCRIPTORS:
*HEATED WATER, *BENEFICIAL USE, *COOLING WATER, *FISHERIES, WATER QUALITY, FISH ATTRACTANTS, FISH HARVEST, PLANKTON, BENTHIC FAUNA, ALGAE, ECOLOGY, MISSOURI.

IDENTIFIERS:
COOLING RESERVOIR, ARTIFICIAL RESERVOIR, MONTROSE LAKE, THOMAS LAKE.

ABSTRACT:
MONTROSE LAKE AND THOMAS LAKE IN MISSOURI WERE STUDIED. BOTH WERE ARTIFICIAL COOLING RESERVOIRS. STUDIES INCLUDED WATER QUALITY, PRIMARY PRODUCTION OF ALGAE, PLANKTON AND BOTTOM FAUNA AND FISH. IMPACT OF HEATED WATER UPON THE ECOLOGY OF A RESERVOIR WILL BE LESSENED AS THE VOLUME OF THE RESERVOIR IS INCREASED. EMPLOYMENT OF A LONG COOLING CANAL RESULTS IN LOSS OF HEAT TO THE ATMOSPHERE PRIOR TO DISCHARGE INTO THE RESERVOIR. UNDER THE CONDITIONS EXISTING IN THOMAS HILL RESERVOIR, NO LETHAL EFFECTS FROM HEATED WATER INFLOW WAS OBSERVED. BENEFICIAL EFFECTS WHICH MAY BE DERIVED FROM CONTROLLED DISCHARGE OF HEATED WATER ARE: (A) MIGRATION OF FISHES MAY RESULT IN A BUILDUP OF FISH DENSITIES. (B) THE INCREASED DENSITY OF FISHES MAY PROVIDE AN EXTENDED HARVEST INTO THE WINTER PERIOD. (C) CONTINUED GROWTH OF GAME FISHES MAY OCCUR DURING THE WINTER PERIOD. (D) AN INCREASED PRODUCTION OF COMMERCIALLY RAISED FISH MAY BE POSSIBLE. (E) THE LOCATION OF THE POWER PLANT MIGHT BE PLANNED TO PROVIDE OPTIMAL CONDITIONS FOR FISH HARVEST AND FOR USE OF HEATED WATER EFFLUENT IN HATCHERY AND REARING POOLS. (HSIEH-VANDERBILT)

FIELD 05C, 06G

ACCESSION NO. W70-09596

THERMAL EFFECTS AND NUCLEAR POWER STATIONS IN THE U.S.A.,

ATOMIC ENERGY COMMISSION, WASHINGTON, D.C.

R. E. NAKATANI, D. MILLER, AND J. V. TOKAR.

SYMPOSIUM ON ENVIRONMENTAL ASPECTS OF NUCLEAR POWER STATIONS, NEW YORK, AUGUST 12, 1970. PAPER IAEA-SM-146/30. 14 P, 1 TAB, 21 REF. PREPRINT.

DESCRIPTORS:
*THERMAL POLLUTION, *NUCLEAR POWER PLANTS, *COOLING WATER, *WATER TEMPERATURE, CONDENSERS, DESIGN, STANDARDS, FISH KILL, ALGAE, COOLING TOWERS, DENSITY, VISCOSITY, FISH BEHAVIOR, FERTILITY, SEASONAL, DIFFUSIVITY.

IDENTIFIERS:
*WASTE HEAT, SUBLETHEL EFFECTS, MIXING ZONES, MODELING.

ABSTRACT:
NUCLEAR POWER PLANTS IN THE U.S. WERE SURVEYED THAT ARE EITHER OPERABLE, BEING BUILT, OR PLANNED AS OF DECEMBER 31, 1969. OVER SIXTY PERCENT OF THE PLANTS USED THE ONCE-THROUGH SYSTEM WHERE THE CONDENSER COOLING WATER WAS TAKEN FROM NEARBY RIVERS, LAKES, ESTUARIES, OR THE OCEAN AND THEN USUALLY RETURNED TO THE SAME SOURCE. THE AVERAGE MAXIMUM TEMPERATURE RISE ACROSS THE NUCLEAR POWER PLANT CONDENSER WAS APPROXIMATELY 10 DEGREE C. GENERAL OBSERVATIONS AND RECOMMENDATIONS BASED ON STATEMENTS DESCRIBING THE PLANNING, DESIGN, AND PROCEDURES FOR WASTE HEAT RELEASE AND INFORMATION OBTAINED FROM PERSONAL PLANT VISITS INCLUDE: (1) THERMAL QUALITY STANDARDS ARE BECOMING INCREASINGLY STRINGENT; (2) FISH KILLS ARE RARE AND THEIR SEVERITY LIMITED; (3) STUDIES OF THE SHORT- AND LONG-RANGE MOVEMENT AND EFFECTS OF HEAT SHOULD BE UNDERTAKEN; (4) THE EFFORT SHOULD BE MADE TO ACQUIRE KNOWLEDGE OF THE SUBLETHEL EFFECTS OF RELEASING HEAT; (5) ALTERNATE METHODS FOR RELEASING WASTE HEAT SHOULD AND ARE BEING EMPLOYED; (6) THE NEED FOR STOCHASTIC AND ANALYTIC MODELS OF PLUME BEHAVIOR EXISTS. (OSBORNE-VANDERBILT)

FIELD 05B, 05C

ACCESSION NO. W70-09612

THERMAL EFFECTS STUDIES IN NEW YORK STATE,

STOLLER (S. M.) ASSOCIATES, NEW YORK; AND NIAGARA MOHAWK POWER CORP., BUFFALO, N.Y.

THOMAS W. PHILBIN, AND HOWARD D. PHILLIPP.

SYMPOSIUM ON ENVIRONMENTAL ASPECTS OF NUCLEAR POWER STATIONS, NEW YORK, AUGUST 12, 1970. PAPER IAEA-SM-146/32. 13 P, 3 TAB, 10 REF. PREPRINT.

DESCRIPTORS:
*POWER PLANTS, *THERMAL POLLUTION, *THERMAL STRATIFICATION, HYDRAULIC MODELS, FISH, ALGAE, FATHOMETERS, BENTHIC FAUNA, BENTHIC FLORA, LAKE ONTARIO, HUDSON RIVER, STANDARDS, SURFACE WATERS, EPILIMNION, INTAKE STRUCTURES, DISCHARGE(WATER), WATER TEMPERATURE.

IDENTIFIERS:
LAKE CAYUGA, CLADOPHORA, ECOLOGICAL STUDIES.

ABSTRACT:
SEVERAL THERMAL EFFECTS STUDIES HAVE BEEN AND ARE PRESENTLY UNDERWAY TO DETERMINE WHAT THE EFFECTS OF THERMAL DISCHARGES ARE ON THE WATER ENVIRONMENT OF NEW YORK STATE. AT PRESENT, THE RESULTS ARE INCOMPLETE; BUT THEY DO YIELD ENOUGH INFORMATION TO ALLOW SOME INITIAL CONCLUSIONS. ON LAKE ONTARIO, THE GINNA AND NINE MILE POINT STUDIES SHOWED THAT FISH TENDED TO GATHER AROUND BOTH INTAKE AND DISCHARGE STRUCTURES, SLIGHTLY ALTERING THE LOCAL FISH AND BENTHIC DISTRIBUTIONS. ON LAKE CAYUGA, THE BELL NUCLEAR STATION HAS BEEN POSTPONED WITH NO IMMEDIATE PLANS FOR RESUMPTION. PHYSICAL EFFECTS STUDIES ON THE LAKE, IF THE BELL STATION PLANT WERE OPERATING, HAVE YIELDED A PROBABLE 0.7 DEGREE F RISE IN THE AVERAGE SURFACE TEMPERATURE ALONG WITH AN EIGHT TO TEN DAY LONGER STRATIFICATION PERIOD. MOREOVER, IN OCTOBER, WATER COULD POSSIBLY BE DRAWN FROM THE EPILIMNION, INCREASING DISCHARGED WATER ANOTHER 5 DEGREE F. HOWEVER, IF NEW YORK STATE DISCHARGE STANDARDS ARE ADHERED TO, NO ACUTE EFFECTS ARE ANTICIPATED EVEN CLOSE TO THE CUTFALL. ON THE HUDSON RIVER, ON THE OTHER HAND, INCREASED WATER TEMPERATURES AS HIGH AS 91 DEGREE F DO NOT APPEAR TO INFLUENCE THE ABUNDANCE OF FISH. (OSBORNE-VANDERBILT)

FIELD 05C, 08C

ACCESSION NO. W70-09614

THE RELATIVE SIGNIFICANCE OF PHOSPHORUS AND NITROGEN AS ALGAL NUTRIENTS,

NORTH CAROLINA WATER RESOURCES RESEARCH INST., RALEIGH.

CHARLES M. WEISS.

AVAILABLE FROM THE NTIS AS PB-194 054, $3.00 IN PAPER COPY, $0.65 IN MICROFICHE. REPORT NO. 34, WATER RESOURCES RESEARCH INSTITUTE OF THE UNIVERSITY OF NORTH CAROLINA, JUNE 1970. 54 P, 27 FIG, 8 REF, 2 APPEND. OWRR PROJECT B-003-NC(2).

DESCRIPTORS:
EUTROPHICATION, *ALGAL BLOOMS, *NUTRIENTS, *PHOSPHORUS, *NITROGEN, *NORTH CAROLINA.

IDENTIFIERS:
NUTRIENT INTERCHANGE, HAW RIVER, NEW HOPE RIVER.

ABSTRACT:
BY THE EXAMINATION OF THE INTERACTION BETWEEN NITROGEN AND PHOSPHORUS SPECIES RELATIVE TO ALGAL GROWTH IN SEVERAL FRESH WATER ENVIRONMENTS OF DIFFERING TROPHIC STATE, IT HAS BEEN POSSIBLE TO ESTABLISH THE RELATIVE SIGNIFICANCE OF THESE ELEMENTS AS ALGAL NUTRIENTS. THE ALGAL ASSAY WAS CARRIED OUT USING MEMBRANE FILTERED SAMPLES DERIVED FROM A SERIES OF OXIDATION PONDS RECEIVING SECONDARY EFFLUENT FROM A TRICKLING FILTER PLANT AND FROM SAMPLES DERIVED FROM SAMPLING POINTS ON THE NEW HOPE AND HAW RIVERS. THE LATTER REPRESENTED A SERIES OF CHANGING RIVER QUALITIES WITH PARTICULAR RESPECT TO THE OXIDATION STATES OF NITROGEN AND PHOSPHORUS. THE ALGAL ASSAY USED PURE CULTURES OF FIVE SPECIES; EUGLENA ROSTIFERA, CHLAMYDOMONAS REINHARDTII, PANDORINA MORUM, SCENEDUSMUS QUADRICAUDA, AND CHLORELLA ELLIPSOIDEA. EACH OF THESE SPECIES HAS BEEN DESCRIBED AS NORMALLY ASSOCIATED WITH POLLUTED WATERS. THE RESULTS OF BOTH CHEMICAL AND BIOLOGICAL ASSAY EXAMINED THROUGH MULTIPLE REGRESSION ANALYSIS OF THE INDEPENDENT VARIABLE INVOLVED IN THE ALGAL ASSAY AS WELL AS A QUADRATIC ANALYSIS OF COVARIANCE OF NH3-N, NO3N AND PO4-P ESTABLISHED THE RELATIVE SIGNIFICANCE OF THE NITROGEN AND PHOSPHORUS SPECIES IN WATER CONTAINING HIGH CONCENTRATIONS OF THESE ELEMENTS, CONCENTRATIONS NORMALLY FOUND IN DISCHARGES FROM BIOLOGICAL WASTE TREATMENT PLANTS AND FOLLOWING DILUTION IN RECEIVING STREAMS. UNDER THESE CIRCUMSTANCES IT HAS BEEN SHOWN THAT THE QUANTITY OF NITROGEN RATHER THAN THAT OF PHOSPHORUS DETERMINES THE BIOMASS OF ALGAE THAT MIGHT BE EXPECTED TO GROW. THIS RESPONSE HOLDS TRUE FOR THE SEVERAL SPECIES OF ALGAL THAT WERE USED. IT WOULD THUS APPEAR THAT THE QUESTION OF ALGAL NUTRIENTS TAKES ON A SOMEWHAT MORE FORMIDABLE DIMENSION DUE TO THE CONSIDERABLY GREATER DIFFICULTY IN REMOVING IN SIGNIFICANT AMOUNTS THE NITROGEN THAT IS NORMALLY FOUND IN MUNICIPAL WASTE WATERS.

FIELD 05C

ACCESSION NO. W70-09669

THE FATE OF NITROGENOUS COMPOUNDS THROUGH SEWAGE TREATMENT PLANTS,

WASHINGTON UNIV., ST. LOUIS, MC. CENTER FOR THE BIOLOGY OF NATURAL SYSTEMS.

NAVA NARKIS.

WASHINGTON UNIV, ST LOUIS, MISSOURI, CENTER FOR THE BIOLOGY OF NATURAL SYSTEMS, ENVIRONMENTAL FIELD PROGRAM, FEBRUARY 1970. 52 P, 1 FIG, 18 TAB.

DESCRIPTORS:
*NITROGEN COMPOUNDS, *SEWAGE TREATMENT, *DOMESTIC WASTES, CHEMICAL ANALYSIS, EFFLUENTS, TRICKLING FILTERS, HYDROLYSIS, BIOLOGICAL TREATMENT, ACTIVATED SLUDGE, TERTIARY TREATMENT, ION EXCHANGE, ELECTROCHEMISTRY, REVERSE OSMOSIS, ELECTRODIALYSIS, DISTILLATION, NITRIFICATION, DENITRIFICATION, SEPTIC TANKS, OXIDATION LAGOONS, LAND APPLICATION, BACTERIA, ALGAE, SYMBIOSIS, PUBLIC HEALTH, CHLORINE, CORROSION.

IDENTIFIERS:
*LITERATURE SURVEY, PRIMARY SEWAGE TREATMENT, SECCNDARY SEWAGE TREATMENT, PHYSICOCHEMICAL METHODS, AMMONIA STRIPPING, GUGGENHEIM PROCESS.

ABSTRACT:
LITERATURE CONCERNING NITROGENOUS COMPOUNDS IN SEWAGE TREATMENT PLANTS IS REVIEWED, ESPECIALLY EMPHASIZING THE MODIFICATION OR REPLACEMENT OF CLASSICAL TREATMENT OPERATIONS TO REMOVE DISSOLVED INORGANIC MATERIALS MORE EFFICIENTLY. IN PRIMARY TREATMENT, NITROGEN REMOVAL IS LOW AND SEDIMENTATION OF SUSPENDED SOLIDS IS THE ONLY PHYSICAL PROCESS CONTRIBUTING TO ITS TOTAL REDUCTION. RETURNING SUPERNATANT LIQUOR FROM SLUDGE DIGESTERS TO PRIMARY CLARIFIER DECREASES NITROGEN REMOVAL EFFICIENCIES. CLARIFIED WASTE WATER FROM PRIMARY TREATMENT MAY RECEIVE SECONDARY TREATMENT, NORMALLY A BIOLOGICAL PROCESS, DESIGNED PRINCIPALLY TO REDUCE AMOUNT OF SUSPENDED SOLIDS, OXYGEN DEMANDING MATERIAL AND BACTERIA. ACTIVATED SLUDGE AND TRICKLING FILTERS REMOVE 20% TO 50% OF THE INCOMING NITROGEN. EFFLUENTS FRCM TRICKLING FILTERS CONTAIN HIGH ORGANIC AND AMMONIA NITROGEN AS WELL AS NITRATES AND NITRITES. IT IS POSSIBLE TC OPERATE ACTIVATED SLUDGE PROCESSES TO EFFECT HIGH NITRATE CONCENTRATIONS, WHILE AMMONIA NITRITE ANC ORGANIC NITROGEN CONCENTRATICNS ARE MINIMAL. IN THE TERTIARY TREATMENT, AMONG THE PHYSICOCHEMICAL METHODS FOR REMOVING NITROGENCUS COMPOUNDS FROM EFFLUENTS, ION EXCHANGE AND AMMONIA STRIPPING ARE THE ONLY SUITABLE METHODS PRESENTLY EFFICIENT. NITROGEN REMOVAL BY MICROBIAL DENITRIFICATION SEEMS ECONOMICALLY FEASIBLE. SEPTIC TANK EFFLUENTS ARE ANALYZED. STABILIZATION PONDS DO NOT REMOVE NITRCGEN COMPOUNDS ADEQUATELY. (JONES-WISCONSIN)

FIELD 05D, 10

ACCESSION NO. W70-09907

APPLIED RESEARCH CN MARINE SEWAGE DISPOSAL FROM THE GOTHENBURG REGION, SWEDEN,

STCCKHOLM UNIV. (SWEDEN). INTERNATIONAL METEOROLOGICAL INST.

L. ARNBORG, AND E. ERIKSSON.

FIFTH INTERNATIONAL WATER POLLUTION RESEARCH CONFERENCE, SAN FRANCISCO, JULY 26-AUGUST 1, 1970, PREPRINT. 7 P, 7 FIG.

DESCRIPTORS:
*RESEARCH AND DEVELOPMENT, *WATER POLLUTION, PHOSPHORUS, THERMAL STRATIFICATION, CURRENTS(WATER), ALGAE, LAKES, RIVERS, SEWAGE CISPOSAL, SEWAGE TREATMENT, WASTE WATER TREATMENT.

IDENTIFIERS:
*MARINE DISPOSAL, *GOTHENBRUG, SWEDEN.

ABSTRACT:
THE CITY CF GOTHENBURG AT THE WEST COAST OF SWEDEN PLANS TO CONSTRUCT A SYSTEM OF INTERCEPTING SEWERS AND ROCK TUNNELS BY WHICH ALL THE SEWAGE FROM THE CITY CAN BE CONVEYED TO A TREATMENT PLANT. AFTER BIOLOGICAL TREATMENT, THE SEWAGE CAN BE DISCHARGEC EITHER INTO THE RIVER MOUTH OR INTO THE SEA OUTSIDE A NEARBY ARCHIPELAGO. IN ORDER TO COMPARE THE DIFFERENT OUTFALL SITES, AN INVESTIGATION OF THE AREA WAS UNDERTAKEN FROM THE MOUTH OF THE RIVER, GOTA ALV, THROUGH THE ARCHIPELAGO INTO THE OPEN SEA. THE MEAN VALUES AND STANDARD DEVIATIONS OF THE VERTICAL DISTRIBUTION OF TEMPERATURE, SALINITY AND CENSITY DURING SUMMER AND WINTER IN THE AREA OUTSIDE THE ARCHIPELAGO ARE PRESENTED GRAPHICALLY. LARGE VARIATIONS OF CURRENT PATTERNS TAKE PLACE DUE TO TIDES, WINDS, AND VARIATIONS IN THE ATMOSPHERIC PRESSURE. THE MAIN DIRECTIONS OF THE CURRENTS ARE NORTH AND SOUTH WITH THE RESULTANT TO THE NORTH. A FIGURE IS PRESENTED SHOWING THE FREQUENCIES SPECTRUM DENSITY. THE STUDY OF THE USE OF A DIFFUSER SYSTEM FOR THE DISPERSAL OF TREATED SEWAGE RESULTED IN THE FOLLOWING DESIGN AND CONCLUSIONS: THE AREA CF RELEASE IS 3 KM WEST OF THE ARCHIPELAGO. THE DIFFUSER PORTHOLES ARE 0.1 M IN DIAMETER AND LOCATED 30 TO 35 M BELOW THE SURFACE. THE INITIAL VELOCITY THROUGH THE PORTHOLES WILL BE 2.5 METERS/SECONC. WITH THIS ARRANGEMENT AND THE EXISTING VERTICAL DENSITY STRATIFICATION THE PLUM WILL RISE TO A LEVEL OF 14 TO 20 METERS BELOW THE SURFACE. (HANCUFF-TEXAS)

FIELD 05B, 05D

ACCESSION NO. W70-09947

FILTRATION OF ALGAL SUSPENSIONS,

BOHNA ENGINEERING AND RESEARCH INC., SAN FRANCISCO, CALIF.

I. J. WRIGHT, AND W. M. LUIZ.

BOHNA ENGINEERING AND RESEARCH, INCORPCRATED, SAN FRANCISCO, CALIFCRNIA, BOHNA INTERNAL REPORT F170, 1970. 31 P, 7 FIG, 10 TAB.

DESCRIPTORS:
*WASTE WATER TREATMENT, *FILTRATION, *ALGAE, *SUSPENSION, HARVESTING, PONDS, CALIFORNIA, SCENEDESMUS, SUSPENDED LOAD, NITRATES, CALCIUM CARBONATE, PROTEINS, FILTERS, ENGINEERS ESTIMATES, ECONOMICS, FEASIBILITY, COSTS, ANNUAL COSTS, PILOT PLANTS.

IDENTIFIERS:
FIREBAUGH(CALIF), VOLATILE SOLIDS, SANBORN FILTER, BACKWASHING.

ABSTRACT:
THE SANBORN FILTER, PROPRIETARY OF BOHNA ENGINEERING AND RESEARCH, INC, A FABRIC AND GRANULAR-MEDIA DEVICE, HAS BEEN EVALUATED FOR USE IN HARVESTING ALGAE FROM PONDS OF THE AGRICULTURAL WASTE WATER TREATMENT CENTER AT FIREBAUGH, CALIFORNIA. REMOVALS OF UP TO 98% HAVE BEEN ACHIEVED WITHOUT THE USE OF FLOCCULANTS OR OTHER CHEMICAL ADDITIVES. THE BACKWASH FROM THE FILTER HAS SUSPENDED SOLIDS CONCENTRATIONS UP TO 1.7% WHICH SHOWS EVIDENCE OF INCREASED COAGULATION, ALLOWING THE CONCENTRATION OF SOLIDS TO A LEVEL OF 8% TO 12% BY SIMPLE SEDIMENTATION. AN ECONOMIC ANALYSIS OF THE USE OF ALGAE FOR NITROGEN STRIPPING OF SAN JOAQUIN VALLEY DRAINAGE WATER WAS MADE. ASSUMING A FLOW SHEET WHICH INCLUDES LINED, MIXED GROWTH PONDS FOLLOWED BY FILTRATION WITH THE SANBORN FILTER, CONCENTRATION CF THE BACKWASH BY SEDIMENTATION OR CENTRIFUGATION AND AIR DRYING OF THE ALGAE, AN APPROXIMATE TOTAL COST OF $37.01 PER MILLION GALLONS TREATED IS PROJECTED, BASED ON AMORTIZATION AND INTEREST CHARGES OF 10% PER YEAR OF THE INITIAL INVESTMENT. CAPACITY OF 700 MILLION GALLONS PER DAY WAS ASSUMED. (JONES-WISCONSIN)

FIELD 05D

ACCESSION NO. W70-10174

THE FULL-SCALE RECLAMATION OF PURIFIED SEWAGE EFFLUENT FOR THE AUGMENTATION OF THE DOMESTIC SUPPLIES OF THE CITY OF WINDHOEK,

NATIONAL INST. FOR WATER RESEARCH, PRETORIA (SOUTH AFRICA).

L. R. J. VAN VUUREN, M. R. HENZEN, G. J. STANDER, AND A. J. CLAYTON.

FIFTH INTERNATIONAL WATER POLLUTION CONFERENCE, SAN FRANCISCO, JULY 26-AUGUST 1, 1970. PREPRINT, PAPER I-32. 9 P, 5 FIG, 4 TAB, 8 REF.

DESCRIPTORS:
*SEWAGE TREATMENT, *WASTE WATER TREATMENT, *TERTIARY TREATMENT, *WATER REUSE, *WATER QUALITY, WATER SUPPLY, ALGAE, ACTIVATED CARBON, ADSORPTION, FILTRATION, CHLORINATION, CHEMICAL PRECIPITATION, FLOCCULATION, FLOTATION.

IDENTIFIERS:
WINDHOEK, SOUTH AFRICA.

ABSTRACT:
IN EARLY SEPTEMBER 1968, A CRITICAL SITUATION IN WATER SUPPLY AND DEMAND DEVELOPED WHICH NECESSITATED THE EXPEDITOUS CONDITIONING OF A FULL-SCALE WATER RENCVATION PLANT. THE RECLAIMABLE SEWAGE EFFLUENT IS DERIVED FROM A CONVENTIONAL SEWAGE TREATMENT PLANT COMPRISED OF PRIMARY SEDIMENTATION BY FILTRATION, SECONDARY SEDIMENTATION FOLLOWED BY FURTHER BIOLOGICAL PURIFICATION IN 9 MATURATION PONDS IN SERIES. TOTAL HYDRAULIC RETENTION TIME IS 14 DAYS. THE RECLAMATION PLANT CONSTITUTES AN INTEGRAL PART OF THE EXISTING WATER TREATMENT WORKS AND IS COMPRISED OF RECARBONATION, ALGAE FLOTATION, FOAM FRACTIONATION, CHEMICAL FLOCCULATION, BREAK POINT CHLORINATION, CLARIFICATION, SAND FILTRATION AND ACTIVATED CARBON. THE RECLAIMED WATER AFTER CARBON ABSORPTION IS MIXED WITH PURIFIED RAW WATER AND THE CLEAR WATER SUPPLY. THE ADD MIXED STRAINS ARE THEN POST-CHLORINATED TO A FREE RESIDUAL OF 0.2 PPM AND THEN PUMPED TO THE CITY'S SERVICE RESERVOIRS. THE TOTAL COST FOR THE PRODUCTION OF RECLAIMED WATER INCLUDING FIXED CAPITOL, CHEMICAL AND RUNNING COSTS, AMOUNT TO 11.9 CU PM, WHILE THE AVERAGE COSTS FOR PURIFIED DAM WATER ARE 10.7 CU PM. THE COST FIGURE FOR CARBON TREATMENT IS BASED ON A SINGLE USE OF THE CARBON MEDIUM WITHOUT REGENERATION RESULTING IN A HIGH COST. IF REGENERATION FACILITIES ARE PROVIDED IT IS ESTIMATED THAT THE TOTAL COST FOR RECLAIMED WATER WILL BE 9.53 CU PM. (HANCUFF-TEXAS)

FIELD 05D, 03D

ACCESSION NO. W70-10381

ASSAYING ALGAL GROWTH WITH RESPECT TO NITRATE CONCENTRATION BY A CONTINUOUS FLOW TURBIDOSTAT,

CALIFORNIA UNIV., BERKELEY; AND MINISTRY OF HEALTH, JERUSALEM (ISRAEL).

G. SHELEF, W. J. OSWALD, AND C. C. GOLUEKE.

FIFTH INTERNATIONAL WATER POLLUTION RESEARCH CONFERENCE, SAN FRANCISCO, JULY 26-AUGUST 1, 1970. PREPRINT. 9 P.

DESCRIPTORS:
*ANALYTICAL TECHNIQUES, *ALGAE, *ASSAY, CHLORELLA, NITRATES, CHEMICAL ANALYSIS, CHLOROPHYTA, DIATOMS, TEMPERATURE, OPTICAL PROPERTIES, GROWTH RATES.

IDENTIFIERS:
*SAN JOAQUIN DELTA(CALIF), *CONTINUOUS FLOW TURBIDOSTAT, *GROWTH KINETICS, CHEMOSTATS, FLAGELLATE ALGAE, OPTICAL DENSITY SENSOR.

ABSTRACT:
THIS STUDY EMPLOYED A CONTINUOUS FLOW TURBIDOSTAT AND A CHEMOSTAT TO APPRAISE THE EFFECT OF NITRATE-NITROGEN CONCENTRATION ON THE GROWTH OF TWO STRAINS OF CHLORELLA PYRENOIDOSA (TX 71105 AND EMERSON). IN ONE TRIAL A MIXTURE OF ALGAE INDIGENOUS TO SAN JOAQUIN DELTA (CALIFORNIA)--GREEN, GREEN FLAGELLATES AND DIATOMS--WAS USED AS TEST ORGANISMS. THE SPECIFIC GROWTH RATE INCREASED WITH TEMPERATURE AND CONCENTRATION OF NITROGEN. THE CONCENTRATION OF NITRATE-NITROGEN DECREASED WITH TEMPERATURE ABOVE 35C. THE GROWTH OPTIMUM WAS NOT NECESSARILY OPTIMUM FOR NITRATE UTILIZATION. THE MIXED CULTURE OF ALGAE SHOWED SUBSTANTIAL SPECIFIC GROWTH RATE AT MUCH LOWER NITROGEN CONCENTRATION THAN C PYRENOIDOSA. RESULTS OF THE STUDY EMPHASIZE THE WIDE DIFFERENCE IN NITROGEN REQUIREMENTS OF VARIOUS ALGAL SPECIES. THE TURBIDOSTATIC METHOD PROVED MORE EXPENSIVE, BUT MORE ADVANTAGEOUS THAN THE CHEMOSTATIC METHOD. (WILDE-WISCONSIN)

FIELD 05A, 07B

ACCESSION NO. W70-10403

THE EXTENT OF NITROGEN AND PHOSPHORUS REGENERATION FROM DECOMPOSING ALGAE,

KENTUCKY UNIV., LEXINGTON. DEPT. OF CIVIL ENGINEERING; TEXAS UNIV., AUSTIN. DEPT. OF CIVIL ENGINEERING; AND STANFORD UNIV., CALIF. DEPT. OF ENVIRONMENTAL ENGINEERING.

E. G. FOREE, W. J. JEWELL, AND P. L. MCCARTY.

FIFTH INTERNATIONAL WATER POLLUTION RESEARCH CONFERENCE, SAN FRANCISCO, JULY 26-AUGUST 1, 1970. PREPRINT. 15 P.

DESCRIPTORS:
*EUTROPHICATION, *ALGAE, *DECOMPOSING ORGANIC MATTER, NUTRIENTS, SURFACE WATERS, NITROGEN, WATER QUALITY, PHOSPHORUS, AEROBIC CONDITIONS, ANAEROBIC CONDITIONS, CHLORELLA, CHLAMYDOMONAS, SCENEDESMUS, EFFLUENTS, OXIDATION LAGOONS.

IDENTIFIERS:
SAN PABLO ESTUARY(CALIF), SACRAMENTO RIVER(CALIF), SEARSVILLE LAKE(CALIF), SAN JOAQUIN RIVER(CALIF), NITZSCHIA.

ABSTRACT:
DECOMPOSITION OF MONOSPECIFIC AND HETEROGENOUS ALGAE AND SUBSEQUENT RELEASE OF NUTRIENTS WERE INVESTIGATED IN DARK-AEROBIC AND DARK-ANAEROBIC CULTURES. UNDER AEROBIC CONDITIONS, 50% ON THE AVERAGE OF THE INITIAL PARTICULATE NITROGEN AND PHOSPHORUS WERE REGENERATED, BUT THE RECOVERY VARIED FROM ZERO TO NEARLY 100%. UNDER ANAEROBIC CONDITIONS, THE AVERAGE REGENERATION OF NITROGEN AND PHOSPHORUS WAS 40% AND 60%, RESPECTIVELY, WITH VARIATION NEARLY AS GREAT AS THAT FOR AEROBIC DECOMPOSITION. A GENERAL AGREEMENT BETWEEN EQUATION-PREDICTED NITROGEN AND PHOSPHORUS REFRACTORY VALUES AND MEASURED VALUES WAS OBTAINED. THE RESULTS SUGGEST THAT CHANCES FOR A SUCCESSFUL CONTROL OF EUTROPHICATION ARE MUCH GREATER BEFORE REFRACTORY REMAINS OF ALGAE HAVE BEEN PERMITTED TO ACCUMULATE. (WILDE-WISCONSIN)

FIELD 05C, 05G

ACCESSION NO. W70-10405

WATER POLLUTION AND THE TEXTILE INDUSTRY,

W. J. COSGROVE.

CANADIAN TEXTILE JOURNAL, VOL 87, NO 1, 1970, P 63-66. 1 TAB.

DESCRIPTORS:
*ST. LAWRENCE RIVER, BIOCHEMICAL OXYGEN DEMAND, HYDROGEN ION CONCENTRATION, ALGAE, *ENTROPHICATION, NEUTRALIZATION, SEDIMENTATION, TRICKLING FILTERS, WASTE WATER TREATMENT.

IDENTIFIERS:
*TEXTILE MILL WASTES, MONTREAL, BALANCING TANKS.

ABSTRACT:
THERE WAS A DRAMATIC REDUCTION IN THE INCIDENCE OF TYPHOID FEVER DUE TO IMPROVEMENTS IN WATER SUPPLIES IN THE YEARS 1906 TO 1923, AT WHICH TIME THE IMPORTANCE OF TYPHOID CARRIERS WAS DISCOVERED. THE CITY OF MONTREAL NOW FACES PROBLEMS OF SEWAGE DISPOSAL AND THE FORMATION OF ALGAL BLOOMS IN NEAR BY RIVERS AND LAKES. THE TEXTILE INDUSTRY IS ONE OF SEVERAL SOURCES OF WASTES WHICH ADDS TO THE PROBLEM. THE BOD'S AND PH'S OF TYPICAL TEXTILE WASTES ARE TABULATED. ABOUT THREE QUARTERS OF THE TEXTILE PLANTS DISCHARGE THEIR WASTES TO SEWERS AND THE REMAINDER TO RIVERS. TREATMENTS USED FOR TEXTILE WASTES DEPEND UPON BALANCING TANKS, NEUTRALIZATION, SEDIMENTATION, TRICKLING FILTERS AND THE USE OF SPECIFIC OPERATIONS AND CHEMICAL TREATMENTS DECTATED BY THE PARTICULAR PROCESSING OPERATIONS AND CHEMICALS INVOLVED. (WORK-NORTH CAROLINA STATE)

FIELD 05D, 05B

ACCESSION NO. W70-10427

HYDROBIOLOGICAL PROPSPECTIVES OF THE MEURTHE: MINERAL POLLUTION AND ALGAL VEGETATIONS (IN FRENCH),

LABORATOIRE DE BIOLOGIE VEGETALE, NANCY (FRANCE).

JEAN-FRANCOIS PIERRE.

COMPTES RENDUS ACADEMIE DES SCIENCES, PARIS, VOL 270, NC 17, SERIE D, P 2101-2102, 1970. 3 REF.

DESCRIPTORS:
*WATER TREATMENT, *CHLORINATION, *ALGAL CONTROL, RIVERS, WATER POLLUTION EFFECTS, ALGAE, DIATOMS.

IDENTIFIERS:
RHINE-MAAS BASIN(FRANCE), MEURTHE RIVER(FRANCE), CLADOPHORA, PHORMIDIUM, OSCILLATCRIA, OEDOGONIUM.

ABSTRACT:
ONE OF THE FLOWING CULTURES CONTAINED WATER CF THE RIVER MEURTHE WITH CHLORINE ION CONCENTRATION IN THE PROXIMITY OF 37 MILLIGRAM/LITER. IN ANOTHER CULTURE THE CONCENTRATION OF CHLORINE WAS INCREASED TO 760 MILLIGRAM/LITER. INOCULA INCLUDEC CLACOPHORA GLOMERATA, OEDOGONIUM SPP, PHORMIDIUM TENUE, OSCILLATCRIA SPP, ANC NUMEROUS DIATOMS. THE INCREASED CONCENTRATION OF CHLORINE FAILED TO INFLUENCE SIGNIFICANTLY THE ALGAL POPULATION THUS DEMONSTRATING CONSIDERABLE ECOLOGICAL PLASTICITY OF THESE ORGANISMS. THE RESULTS SUGGESTED THAT THE CCNCENTRATION OF 800 MILLIGRAM/LITER OF CHLORINE IONS IN THE RIVER MEURTHE WOULD NOT PRODUCE A HARMFUL SUPPRESSION OF ALGAL GROWTH. (WILDE-WISCCNSIN)

FIELD 05F

ACCESSION NO. W71-00099

CONDENSED PHOSPHATES IN LAKE WATER AND WASTE WATER,

TORONTO UNIV. (ONTARIO). DEPT. OF CIVIL ENGINEERING; AND MCMASTER UNIV., HAMILTON (ONTARIO). DEPT. OF CHEMICAL ENGINEERING.

G. W. HEINKE, AND J. D. NORMAN.

FIFTH INTERNATIONAL WATER POLLUTION RESEARCH CONFERENCE, SAN FRANCISCO, JULY 26-AUGUST 2, 1970. PREPRINT. 6 P, 2 FIG, 2 TAB, 12 REF.

DESCRIPTORS:
*PHOSPHATES, *LAKES, *WASTE WATER(POLLUTION), *HYDROLYSIS, DETERGENTS, GREAT LAKES, TEMPERATURE, HYDROGEN ION CONCENTRATION, ACTIVATED SLUDGE, TREATMENT, EUTROPHICATION, SURFACTANTS, ENZYMES, LAKE ONTARIO, SEASONAL, ALGAE, CULTURES, MICROORGANISMS, EFFLUENTS, SETTLING BASINS, WASTE WATER TREATMENT, LABORATORY TESTS.

IDENTIFIERS:
*CONDENSED PHOSPHATES, SODIUM TRIPOLYPHOSPHATE, SCDIUM PYROPHOSPHATE.

ABSTRACT:
HYDROLYSIS RATE OF SODIUM TRIPOLYPHOSPHATE AND SODIUM PYROPHOSPHATE IN GREAT LAKES WATER AND WASTE WATER UNDER CONDITIONS OF TEMPERATURE, PH AND CONCENTRATION OCCURRING IN THE ENVIRONMENT ARE EVALUATED. CONDENSED PHOSPHATES FROM DETERGENTS ARE MAJOR PHOSPHORUS CONTRIBUTORS TO WASTE WATER--ABOUT DOUBLE THE HUMAN WASTE CONTRIBUTION IN THE UNITED STATES AND ABOUT EQUAL TO IT IN CANADA. CONDENSED PHOSPHATES HYDROLYZE EXTENSIVELY DURING ACTIVATED SLUDGE TREATMENT, WITH ABOUT 15% REMAINING IN THE EFFLUENT, HYDROLYZING IN WASTE WATER THREE TIMES FASTER IN SUMMER THAN WINTER. CNCE DISCHARGEC TO SURFACE WATERS, HYDROLYSIS PROCEEDS AT A MUCH SLOWER RATE, CEPENDING ON THE EXTENT OF MICROBIAL ACTIVITY IN THE SURFACE WATER. STUDIES ON THE EFFECT OF PH CHANGES SHOW FASTEST RATES OF HYDRCLYSIS OCCURRING AT THE NATURAL PH OF WASTE WATER OR LAKE WATER. EXPERIMENTAL DATA AT LOW LEVELS OF CONCENTRATION, BELOW 15 MILLIGRAMS/LITER PHOSPHATES, ARE BEST FITTED BY A ZERO-ORDER MODEL. A GRAPHICAL PRESENTATION ALLOWS AN ORDER OF MAGNITUDE PREDICTION OF HYDROLYSIS RATE TO BE EXPECTED IN A PARTICULAR SITUATION. RESULTS OF THIS WORK ARE OF VALUE IN RESEARCH ON PHOSPHORUS REMCVAL METHODS AND EUTROPHICATION STUDIES. (JONES-WISCONSIN)

FIELD 02K, 05C

ACCESSION NO. W71-00116

BIOLOGICAL AND CHEMICAL INVESTIGATIONS ON THE EFFECT OF SEWAGE ON THE EUTROPHICATION OF BAVARIAN LAKES,

H. LIEBMANN.

FIFTH INTERNATIONAL WATER POLLUTION RESEARCH CONFERENCE, SAN FRANCISCO, JULY 26-AUGUST 2, 1970. PREPRINT. 7 P, 5 FIG, 1 TAB, 27 REF.

DESCRIPTORS:
*SEWAGE, *EUTROPHICATION, *LAKES, AQUATIC ENVIRONMENT, CHEMICAL PROPERTIES, PHOSPHORUS, NITROGEN, POTASSIUM, DIVERSION, ALGAE, NITROGEN FIXATION, PLANKTON, PHYTOPLANKTON, BACTERIA, ZOOPLANKTON, SALMONIDS, SEASON, OXYGEN, HYDROGEN SULFIDE, DISTRIBUTION, NANNOPLANKTON, TURBIDITY, HYDROGEN ION CONCENTRATION, ALKALINITY, CARBONATES, HARDNESS(WATER), SULFATES, CONDUCTIVITY, SEDIMENTS, TERTIARY TREATMENT, NUTRIENTS.

IDENTIFIERS:
*BAVARIA, SCHLIERSEE, TEGERNSEE, ANABAENA, CCREGONS, LYNGBYA LIMNETICA, OSCILLATORIA REDECKEI, OSCILLATORIA RUBESCENS, SYNEDRA ACUS, RHODOMONAS, LAKE CONSTANCE, SEWAGE CANALS.

ABSTRACT:
LAKES SCHLIERSEE AND TEGERNSEE IN UPPER BAVARIA WERE ASSESSED FOR EUTROPHICATION BY COMPARING CHEMICAL AND BIOLOGICAL CHARACTERISTICS BEFORE AND AFTER SEWAGE DIVERSION. BY TERTIARY TREATMENT WITH CHEMICAL FLOCCULATION ABOUT 90% PHOSPHORUS REMOVAL IS POSSIBLE, BUT A PHOSPHORUS RESIDUE AND A RELATIVELY HIGH AMOUNT OF NITROGEN REMAIN. THUS, FULL CANALIZATION AROUND THE LAKES AND DISCHARGE OF THE RAPIDLY FLOWING EFFLUENT AFTER BIOLOGICAL TREATMENT APPEARED THE MOST EFFECTIVE METHOD TO ACHIEVE IMPROVEMENT. BECAUSE CLIMATIC CONDITIONS PRECLUDE PUMPING SEWAGE FOR GREAT DISTANCES IN MIDDLE EUROPE, A PROPOSAL WAS MADE THAT LOCAL PURIFIED SEWAGE EFFLUENTS BE PUMPED INTO PLASTIC PIPES, SUBMERGED IN LAKE CONSTANCE FOR THE SHORTEST ROUTING, WHERE A TERTIARY TREATMENT WITH PHOSPHORUS REMOVAL SHOULD BE ESTABLISHED FOR THE WHOLE SEWAGE. TERTIARY TREATMENT MAY BE ESPECIALLY ADVANTAGEOUS IF THERE ARE OTHER LAKES AND IMPOUNDMENTS DOWNSTREAM. DANGER THAT LEAKAGE FROM PLASTIC PIPELINES MAY INTRODUCE PURIFIED SEWAGE INTO THE LAKE SEEMS MINIMAL WHEN COMPARED TO THE FACT THAT OTHER POSSIBLE SOLUTIONS TO THE PROBLEM MAY DEFER THE REMEDY FOR SO LONG THAT THE EUTROPHICATION PROCESSES BECOME ALMOST IRREVERSIBLE. (JONES-WISCONSIN)

FIELD 02H, 05C

ACCESSION NO. W71-00117

SURFACE PROPERTIES AND ION EXCHANGE IN ALGAE REMOVAL,

CALIFORNIA UNIV., RICHMOND.

C. G. GOLUEKE, AND W. J. OSWALD.

JOURNAL OF THE WATER POLLUTION CONTROL FEDERATION, VOL 42, NO 8, PART II, P R304-R314, AUGUST 1970. 4 FIG, 12 REF.

DESCRIPTORS:
*ALGAE, *HARVESTING OF ALGAE, *ION EXCHANGE, DISSOLVED SOLIDS, FLOCCULATION, HEAD LOSS, HYDROGEN ION CONCENTRATION, CHLORELLA, SCENEDESMUS, WASTE WATER TREATMENT.

IDENTIFIERS:
*SURFACE CHARGE, REGENERATION, SUSPENDED SOLIDS.

ABSTRACT:
A FACET OF ION EXCHANGE YET TO BE REPORTED IN THE LITERATURE IS THE USE OF ION EXCHANGE COLUMNS FOR THE REMOVAL OF CELLULAR MATERIAL HAVING A SURFACE CHARGE. IN EXPERIMENTS ON THE REMOVAL OF ALGAE FROM POND EFFLUENT IT WAS FOUND (1) IN THE PRESENCE OF EITHER STRONG OR WEAK CATION ION EXCHANGE RESINS, ALGAL CELLS AGGREGATE INTO TIGHT CLUMPS AND SETTLE IN A CONE-SHAPED MASS ON THE TOP OF EACH RESIN BEAD, (2) THE AGGREGATED CELLS ARE EASILY REMOVED FROM THE COLUMN BY BACKWASHING AND RETAIN THEIR TENDENCY TO AGGREGATE AND SETTLE IN THE DISPLACED BACKWASH, AND (3) THE EXCHANGE EFFICIENCY OF THE RESIN WITH RESPECT TO ALGAL CELLS IS BUT A SMALL FRACTION OF THAT WITH RESPECT TO CATIONS IN THAT THE CAPACITY FOR REMOVING MG+2, CA+2, AND OTHER CATIONS CONTINUES LONG AFTER THE CAPACITY FOR REMOVING ALGAL CELLS IS EXHAUSTED. AGGREGATION WAS ITS STRONGEST WITHIN THE PH RANGE OF 2.8 TO 3.5 AND DROPPED OFF SHARPLY ON EITHER SIDE OF THIS RANGE. CONCENTRATION OF NA+, CA+2, AND MG+2 FROM 0.5 TO 4.0 MILLIMOLES DID NOT AFFECT THE PRECIPITATION OF THE CELLS. DESPITE THE SIMPLICITY OF THE PROCESS, THE PURITY OF THE ALGAL PRODUCT, AND THE HIGH QUALITY OF THE RECLAIMED WATER ATTAINABLE WITH IT, UNTIL A REGENERANT LESS EXPENSIVE THAN H2SO4 OR HCL IS FOUND, THE INTEREST IN THE PROCESS MUST REMAIN ACADEMIC RATHER THAN ECONOMIC. (AGUIRRE-TEXAS)

FIELD 05D

ACCESSION NO. W71-00138

RELATIONSHIP OF CHLOROPHYLL A TO ALGAL CCUNT AND CLASSIFICATION IN OXIDATION PONDS,

LAMAR STATE COLL. OF TECHNCLOGY, BEAUMONT, TEX.

JOSEPH V. DUST, AND ADAM SHINDALA.

JOURNAL OF THE WATER POLLUTION CONTROL FEDERATION, VOL 42, NO 7, P 1362-1369, JULY 1970. 7 FIG, 7 REF.

DESCRIPTORS:
*ALGAE, *CHLOROPHYLL, *CLASSIFICATION, *OXIDATION LAGCCNS, CHLORELLA, OXYGEN DEMAND, PHOTOSYNTHESIS, RESPIRATION, SCENEDESMUS, WASTE WATER TREATMENT.

IDENTIFIERS:
*ENUMERATION, ANACYSTIS.

ABSTRACT:
A STUDY WAS CCNDUCTED AIMEC AT OBTAINING A QUANTITATIVE COMPARISON OF CHLOROPHYLL A WITH THE TOTAL NUMBER OF ORGANISMS IN THE TRINITY RIVER AUTHORITY WASTE WATER TREATMENT PLANT AS WELL AS A COMPARISON WITH THE VARIOUS SPECIES OF ALGAE EXISTING FROM TIME TO TIME IN THESE OXIDATION PONDS. THE LITERATURE INDICATES THAT THE ANALYTICAL DETERMINATION OF CHLOROPHYLL PIGMENTS CAN BE ACCOMPLISHED SPECTROPHOTOMETRICALLY WITH THE FOLLOWING EQUATICN: CHLOROPHYLL A IN MG/L = (14.3) D665, WHERE D655 IS THE LIGHT ABSORBANCE AT A WAVE LENGTH OF 655 MILLIMICRONS AND A SLIT WIDTH OF 0.3 MM. EIGHT MAJOR TYPES OF ALGAE WERE FOUND IN SIGNIFICANT NUMBERS: OOCYSTIS, ANACYSTIS, CHLORELLA, DESMICS, SCENEDESMUS, COELASTRUM, ZYGNEMA, AND CLOSTERIUM. THE ANACYSTIS ORGANISM WAS FOUND MORE FREQUENTLY AND IN GREATER NUMBERS THAN ANY OTHER ORGANISM. THE AVERAGE WEEKLY COUNT OF ANACYSTIS VARIED FROM ABOUT 1 TO 4 MILLION PER ML DURING THE FIRST 15 WEEKS OF THE STUDY COMPRISING THE SUMMER AND EARLY FALL MONTHS OF THE YEAR. CHLORELLA ANC SCENEDESMUS WERE THE PREDOMINANT ORGANISMS DURING THE LATE FALL, WINTER, AND EARLY SPRING MONTHS, WITH CELL CCUNTS OF 10,000 TO 100,000 CELLS/ML. A RELATIONSHIP BETWEEN CHLOROPHYLL A AND TOTAL ALGAL COUNT WAS NOT FOUND. HOWEVER, A DEFINITE RELATIONSHIP EXISTED BETWEEN THE AMCUNT OF CHLOROPHYLL A AND THE NUMBER OF ANACYSTIS ORGANISMS PRESENT. A SIMILAR RELATIONSHIP WAS FOUND WITH ZYGNEMA. (AGUIRRE-TEXAS)

FIELD 05D

ACCESSION NO. W71-00139

AUTOMATED POTENTIOMETRIC TECHNIQUES FOR THE ON-SITE MONITCRING OF ION CONCENTRATIONS IN WATER,

MISSOURI UNIV., COLUMBIA. DEPT. OF CHEMISTRY.

STANLEY E. MANAHAN.

AVAILABLE FROM NTIS AS PB-195 167, $3.00 IN PAPER COPY, $0.95 IN MICROFICHE. OWRR COMPLETION REPORT, 1970. OWRR PROJECT A-024-MO(11).

DESCRIPTORS:
*ANALYTICAL TECHNIQUES, *WATER ANALYSIS, *NITRATES, *ELECTRODES, IONS, FLUORIDES, ALGAE, VOLUMETRIC ANALYSIS, WATER CHEMISTRY.

IDENTIFIERS:
ION-SELECTIVE ELECTRODES, STANDARD ADDITION, PCTENTIOMETRY, INTERFERING IONS.

ABSTRACT:
THE CHARACTERISTICS CF THE NITRATE ELECTRODE - FLUCRIDE ELECTRODE CELL WITHOUT LIQUID JUNCTION WERE INVESTIGATED FOR APPLICABILITY TO NITRATE ION ANALYSIS. IN ADDITION TO ENABLING ACCURATE NITRATE ANALYSIS, THE CELL WAS FOUND TO BE USEFUL FOR EXAMINING THE EFFECTS OF INTERFERING IONS. AN ATTEMPT WAS MADE TO USE THE NITRATE ELECTRODE IN FOLLOWING NITRATE ION UPTAKE BY ALGAL CULTURES. THE ALGAE INTERFERE TO A CERTAIN EXTENT WITH THE ELECTRODE RESPONSE. USING KNOWLEDGE GAINED FROM RESEARCH UNDER THIS GRANT A POTENTIOMETRIC NITRATE ANALYSIS CAPABILITY HAS BEEN SET UP AT THE UNIVERSITY OF MISSOURI ENVIRONMENTAL TRACE SUBSTANCES LABORATORY.

FIELD 05A, 02K

ACCESSION NO. W71-00474

THE INTERRELATION OF CARBON AND PHOSPHORUS IN REGULATING HETEROTROPHIC AND AUTOTROPHIC POPULATIONS IN AQUATIC ECOSYSTEMS,

FEDERAL WATER QUALITY ADMINISTRATION, ATHENS, GA. NATIONAL POLLUTANTS FATE RESEARCH PROGRAM.

PAT C. KERR, DORIS F. PARIS, AND D. L. BROCKWAY.

AVAILABLE FROM NTIS AS PB-195 195, $0.95 IN MICROFICHE. ALSO FOR SALE BY SUPERINTENDENT OF DOCUMENTS, U.S. GOVERNMENT PRINTING OFFICE, WASHINGTON, D.C. 20402, PRICE $0.60, I67.13/4:16050 FGS 07/70. WATER POLLUTION CONTROL RESEARCH SERIES 16050 FGS 07/70, (JULY 1970). 53 P, 4 TAB, 17 FIG, 51 REF. FWQA PROJECT 16050 FGS.

DESCRIPTORS:
CYCLING NUTRIENTS, *CARBON, *PHOSPHORUS, *CYANOPHYTA, AQUATIC ENVIRONMENT, CARBON DIOXIDE, ECOSYSTEMS, SOUTH CAROLINA, ALGAE, TROPHIC LEVEL.

IDENTIFIERS:
*ANTOTROPHIC POPULATIONS, *HETERCTROPHIC POPULATIONS, LAKE HARTWELL(SOUTH CAROLINA), ANACYSTIS NIDULANS, AQUATIC ECOSYSTEMS, ALGAE GROWTH, ALGAL POPULATIONS.

ABSTRACT:
LABORATORY AND FIELD INVESTIGATIONS WERE CONDUCTED ON THE FATE AND CYCLING OF CARBON AND PHOSPHORUS IN SELECTED AQUATIC ECOSYSTEMS. INORGANIC CARBON, AS CO2, SUPPLIED BY BOTH BACTERIAL CULTURES AND CYLINDER GASES, STIMULATED THE GROWTH OF THE BLUE-GREEN ALGA ANACYSTIS NIDULANS. THE CARBON REQUIREMENT (10-5 UG CO2 PER CELL) FOR THIS ALGA WAS DETERMINED FOR A SINGLE SET OF EXPERIMENTAL CONDITIONS. THE ADDITION OF CO2 TO NATURAL WATER LOW IN PHOSPHORUS (5 UG P) AND NITROGEN (5 UG N) IN THE LABORATORY STIMULATED THE GROWTH OF INDIGENOUS ALGAL POPULATIONS. THE LIMITING AND LUXURY CELLULAR CONCENTRATIONS OF PHOSPHORUS FOR STARVED ANACYSTIS NIDULANS WERE FOUND TO BE 0.3 X 10-8 UG P AND 3.0 X 10-8 UG P PER CELL, RESPECTIVELY. DIEL STUDIES OF A STREAM WHICH RECEIVED BIOLOGICALLY-TREATED SEWAGE DEMONSTRATED THAT THE DISSOLVED CO2 AND HCO3 CONTINUALLY PRODUCED IN THE SYSTEM WERE ESSENTIALLY DEPLETED BY THE AUTOTROPHIC ORGANISMS DURING DAYLIGHT HOURS, WHILE THE CONCENTRATION OF PHOSPHORUS (1.3-2.2 MG/L P) REMAINED UNCHANGED. ADDITION OF ORGANIC CARBON AND INORGANIC NITROGEN AND PHOSPHORUS ALONG AND IN COMBINATION TO THE WATERS STUDIED DIRECTLY STIMULATED THE OXIDATIVE METABOLISM OF THE HETERCTROPHIC POPULATION, WHICH RESULTED IN INCREASED DISSOLVED CO2 AND HCO3-. THIS INCREASED AVAILABILITY OF INORGANIC CARBON, RATHER THAN THE DIRECT METABOLIC REMOVAL OF DISSOLVED PHOSPHORUS BY THE ALGAE, APPEARED TO BE DIRECTLY RESPONSIBLE FOR THE GROWTH OF THE ALGAL POPULATIONS IN THE WATERS STUDIED.

FIELD 05C, 02I, 02H

ACCESSION NO. W71-00475

PROCEEDINGS OF THE CONFERENCE WATER RESOURCES RESEARCH - 1970.

CONNECTICUT UNIV., STORRS. INST. OF WATER RESOURCES.

AVAILABLE FROM NTIS AS PB-195 667, $3.00 IN PAPER COPY, $0.95 IN MICROFICHE. WILLIAM C. KENNARD, EDITOR. CONNECTICUT UNIVERSITY INSTITUTE OF WATER RESOURCES REPORT NO 13, SEPTEMBER 1970. 20 P. OWRR PRCJECT A-999-CONN(8).

DESCRIPTORS:
*WATER RESOURCES RESEARCH ACT, CCNNECTICUT, *UNIVERSITIES, *ABSTRACTS, WATER POLLUTION TREATMENT, INSTRUMENTATION, WASTE DISPOSAL, AERATION, EROSION CONTROL, WATER POLLUTICN SOURCES, WATER LAW, ALGAE, DISSOLVED OXYGEN.

IDENTIFIERS:
WATER RESOURCES RESEARCH.

ABSTRACT:
THESE PROCEEDINGS INCLUDE ABSTRACTS OF PAPERS PRESENTED AT THE CONFERENCE HELD BY THE INSTITUTE OF WATER RESOURCES AT THE UNIVERSITY OF CONNECTICUT ON MAY 18, 1970. THE TITLE OF EACH PRESENTATION IS THE SAME AS THAT OF THE RESEARCH PROJECT. ALSO GIVEN FOR EACH ARE THE PERSONNEL, ACADEMIC DEPARTMENTS, PLANNED DURATION AND THE OBJECTIVES OF THE INVESTIGATIONS. EACH ABSTRACT IS A BRIEF SUMMARY OF ACCOMPLISHMENTS TO DATE AND PLANS FOR THE FUTURE. THE PROGRAM CF THE INSTITUTE HAS BECOME INCREASINGLY DIVERSE IN THE APPROXIMATELY SIX YEARS THAT IT HAS BEEN IN EXISTENCE. PROJECTS IN THE AGRICULTURAL, BIOLCGICAL, ENGINEERING, EARTH, SOCIAL AND PHYSICAL SCIENCES HAVE BEEN OR NOW ARE ACTIVE. CONTINUED EXPANSION, BOTH IN SCOPE OF THE RESEARCH AND IN THE NUMBER OF PROJECTS, IS PLANNED. THE RESULTS CF THESE INVESTIGATIONS WILL, IN MANY CASES, HAVE DIRECT APPLICATION TO SOLVING PROBLEMS OF WATER RESOURCES USE AND DEVELOPMENT IN CCNNECTICUT AND WILL RESULT IN IMPORTANT CONTRIBUTICNS TO THE FUND OF SCIENTIFIC INFORMATION ABOUT WATER IN A BROAD RANGE OF DISCIPLINES. (KNAPP-USGS)

FIELD 09A, 06B, 05G

ACCESSION NO. W71-01192

BIOLCGICAL UPTAKE OF PHCSPHORUS BY ACTIVATED SLUDGE,

ARIZONA UNIV., TUCSON. DEPT. OF MICROBIOLOGY AND MEDICAL TECHNOLOGY.

IRVING YALL, WILLIAM H. BOUGHTCN, RICHARC C. KNUDSEN, AND NORVAL A. SINCLAIR.

APPLIED MICROBIOLOGY, VOL 20, NO 1, P 145-150, 1970. 5 FIG, 2 TAB, 10 REF. FWQA SUPPORTED.

DESCRIPTORS:
*PHOSPHORUS, *ACTIVATED SLUDGE, PHCSPHORUS RADIOISOTOPES, SEWAGE, RADIOACTIVITY, PHOSPHATES, CHEMICAL ANALYSIS, ALGAE, MICROORGANISMS, HYDROGEN ION CONCENTRATION, PHENOLS, ARIZONA.

IDENTIFIERS:
*BIOLOGICAL UPTAKE, CALCIUM RADIOISOTOPES, BLOCMS, 2,4 DINITRO PHENOL, TUCSON(ARIZ).

ABSTRACT:
EXPERIMENTS TO REPRODUCE CONDITIONS POSSIBLY EXISTING IN AN OPTIMALLY AERATED, PLUG FLOW ACTIVATED SLUDGE UNIT ARE REPORTED. ABILITY OF ACTIVATED SLUDGE TO REMOVE PHOSPHATES WAS STUDIED BY ADDING CARRIER-FREE PHOSPHORUS-32 TO RAW SEWAGE AND MEASURING INCORPORATION OF RADIOACTIVITY INTO CELLS. RADIOISOTOPE DETERMINATIONS INDICATED THAT 48% OF PHOSPHORUS-32 RADIOACTIVITY WAS REMOVED IN 12 HOURS. CHEMICAL METHODS INDICATED THAT ONLY 30% CF ORTHOPHOSPHATE APPARENTLY DISAPPEARED FROM THE SEWAGE DURING THIS PERICD. EXPERIMENTS WITH SLUDGE RELABELED WITH PHOSPHORUS INCICATEC CONSIDERABLE PHOSPHATE TURNOVER. CELLS RELEASED LARGE AMOUNTS OF RADIOACTIVITY AS THEY WERE INCORPORATING FRESH PHOSPHATES. STARVATION IN ISCTCNIC SALINE FOR 18 HOURS CAUSED SLUDGE TO DUMP PHOSPHATE. WHEN INTRCDUCED INTO FRESH SEWAGE CONTAINING PHCSPHORUS-32, THE STARVED SLUDGE REMOVED ABOUT 60% OF THE RADIO-ACTIVITY IN 6 HOURS WITH LITTLE PHCSPHATE TURNOVER. ABILITY OF SLUDGE TO REMCVE PHOSPHCRUS-32 WAS INHIBITED APPROXIMATELY 83% BY 0.001 MOLAR 2,4-DINITRO PHENOL. THIS INHIBITICN WAS AT THE EXPENSE OF THE CELL FRACTION THAT CONTAINED RIBO-NUCLEIC ACID AND DEOXYRIBONUCLEIC ACID. THE SLUDGE CELLS RELEASED CRTHOPHOSPHATE WHEN EXPOSED TO THE CHEMICAL AGENT. EXPERIMENTS USING CALCIUM-45 INDICATED THAT CALCIUM PHOSPHATE PRECIPATION PLAYS A MINOR ROLE IN PHOSPHATE REMOVEL UNDER EXPERIMENTAL CONDITIONS. (JONES-WISCCNSIN)

FIELD 05D

ACCESSION NO. W71-01474

SOME ECOLOGICAL EFFECTS OF DISCHARGED WASTES ON MARINE LIFE,

CALIFORNIA UNIV., SAN DIEGO; AND SCRIPPS INSTITUTION OF OCEANOGRAPHY, SAN DIEGO, CALIF.

RICHARD W. GRIGG, AND ROBERT S. KIWALA.

CALIFORNIA FISH AND GAME, VOL 56, NO 3, P 145-155, 1970. 2 FIG, 4 TAB, 10 REF.

DESCRIPTORS:
*ECOLOGY, *WASTES, *MARINE ANIMALS, *MARINE PLANTS, DEPTH, CALIFORNIA, SANDS, SEDIMENTS, PRODUCTIVITY, TOXICITY, CRABS, ALGAE, INVERTEBRATES, FISH, POLLUTANTS, TURBIDITY, ORGANIC LOADING, CORAL, MOLLUSKS, WORMS.

IDENTIFIERS:
SAN PEDRO(CALIF), EPIBENTHIC SPECIES, ABALONE, TUNICATES, ARTHROPODS, COELENTERATES, BRYOZOA, ECHINODERMS.

ABSTRACT:
APPROXIMATELY ONE BILLION GALLONS OF SEWAGE ARE DISCHARGED DAILY INTO THE SHALLOW NEARSHORE MARINE ENVIRONMENT OFF SOUTHERN CALIFORNIA. A SURVEY WAS MADE OF THE EFFECT OF EFFLUENT ON THE BIOLOGY OF THE AREA IN 1954. THE RESULTS OF A 1969 SURVEY TO COLLECT COMPARABLE DATA, AND TO EXAMINE THESE TO DETECT POSSIBLE LONG TERM ECOLOGICAL CHANGES ARE DESCRIBED. THE NUMER OF MACROSCOPIC SPECIES PRESENT AT 5 DIVING STATIONS RANGING IN DEPTHS FROM 45 TO 65 FEET, OFF THE PALOS VERDES PENINSULA, NEAR SAN PEDRO, WAS NEGATIVELY CORRELATED TO THE AMOUNT OF FINE GRAIN ORGANIC-LADEN SAND PRESENT IN THE SEDIMENT. ORGANIC RICH SEDIMENTS WERE THICKEST AT STATIONS NEAR THE OUTFALL. ACCUMULATION OF THIS MATERIAL AT THESE DEPTHS APPEARS TO HAVE MODIFIED OR COVERED SUBSTRATES OTHERWISE SUITABLE FOR THE SETTLEMENT OF MANY EPIBENTHIC SPECIES. FISH, KELP, ABALONE AND SPINY LOBSTER ARE PARTICULARLY AFFECTED. SINCE BOTTOM TOPOGRAPHY AT WHITE POINT HAS NOT CHANGED, DECLINE OF FISHES, IF NOT CAUSED DIRECTLY BY TOXIC WASTE PRODUCTS, MAY INDICATE THAT RELIEF IS MORE IMPORTANT AS A SUBSTRATE FOR FOOD RATHER THAN A SOURCE OF SHELTER OR POINT OF ORIENTATION. (JONES-WISCONSIN)

FIELD 05C

ACCESSION NO. W71-01475

UTROPHICATION CONTROL BY PLANT HARVESTING,

FLORIDA STATE BOARD OF HEALTH, WINTER HAVEN. MIDGE RESEARCH LAB.

JAMES L. YOUNT, AND ROY A. CROSSMAN, JR.

JOURNAL WATER POLLUTION CONTROL FEDERATION, VOL 42, NO 5, PART 2, P R173-R183, 1970. 2 FIG, 1 TAB, 28 REF. FWQA GRANT WP-00216.

DESCRIPTORS:
*EUTROPHICATION, *WATER POLLUTION CONTROL, WATER POLLUTION EFFECTS, *AQUATIC PLANTS, HARVESTING, PONDS, WATER HYACINTH, NUTRIENTS, MIDGES, FISH, LAKES, CONDUCTIVITY, ALGAE, STANDING CROP, CHEMICAL ANALYSIS, MOSQUITOES, BACTERIA, FLOATING PLANTS, ON-SITE INVESTIGATIONS, TURTLES.

IDENTIFIERS:
*CHIRONOMIDS, SAWDUST, EICHHORNIA CRASSIPES, SALVINIA ROTUNDIFOLIA, MILORGANITE, GAMBUSIA AFFINIS, PRODUCTIVITY MEASUREMENTS, NUTRIENT REMOVAL.

ABSTRACT:
DATA REPORTED INDICATE THAT GROWING ORGANISMS IN HYPERTROPHIC PONDS AND HARVESTING THEM TO REMOVE NUTRIENTS, REDUCE THE PONDS' PRIMARY PRODUCTIVITY. THIS CONCLUSION IS CONFIRMED BY EXPERIMENTS SHOWING THAT PLANT PRODUCTIVITY WAS REDUCED IN TEST PONDS, WHILE THE CONTROL PONDS, WHERE NUTRIENTS WERE RETURNED, PRODUCTIVITY REMAINED RELATIVELY HIGH. LARGE-SCALE HARVESTING FROM NATURAL WATERS CAN BE EXPECTED TO REDUCE THE PRODUCTIVITY OF THOSE WATERS, AND PROBABLY REVERSE THE TREND TOWARD HYPERTROPHY, ESPECIALLY IN POLLUTED WATERS. THE METHOD OF CONTROLLING THE 'PEST' WATER HYACINTH AND OTHER PLANTS BY CHEMICAL SPRAYS IS RECIRCULATING THEIR NUTRIENTS TO LAKES AND EXACERBATING HYPERTROPHY. GIVEN A CONSTANT INFLOW OF POLLUTANTS, IF TOO MUCH VEGETATION IS REMOVED, THEN AVAILABILITY OF THESE POLLUTANTS TO OTHER ORGANISMS WOULD INCREASE. THE PROBLEM IS RESOLVED BY MANAGING A POPULATION ON A SUSTAINED-YIELD BASIS. HARVESTING HYACINTHS FROM EFFLUENT STREAMS MIGHT PROVE A MORE PRACTICAL METHOD FOR HANDLING EXCESSIVE NUTRIENTS THAN BY HARVESTING FROM LAKES. (JONES-WISCONSIN)

FIELD 05C

ACCESSION NO. W71-01488

EFFECTS OF FOREST CUTTING AND HERBICIDE TREATMENT ON NUTRIENT BUDGETS IN THE HUBBARD BROOK WATERSHED-ECOSYSTEM,

DARTMOUTH COLL., HANOVER, N.H. DEPT. OF BIOLOGICAL SCIENCES; YALE UNIV., NEW HAVEN, CONN. SCHOOL OF FORESTRY; GEOLOGICAL SURVEY, WASHINGTON, D.C.; AND FOREST SERVICE, (USDA), DURHAM, N.H. NORTHEASTERN FOREST EXPERIMENT STATION.

GENE E. LIKENS, F. HERBERT BORMANN, NOYE M. JOHNSON, D. W. FISHER, AND ROBERT S. PIERCE.

ECOLOGICAL MONOGRAPHS, VOL 40, NO 1, P 23-47, 1970. 15 FIG, 8 TAB, 54 REF.

DESCRIPTORS:
*FORESTS, *CUTTING MANAGEMENT, *HERBICIDES, *NUTRIENTS, *ECOSYSTEMS, *WATERSHEDS(BASINS), STREAMFLOW, IONS, NITRATES, SULFATES, DRAINAGE WATER, DISSOLVED SOLIDS, NITROGEN CYCLE, HYDROGEN ION CONCENTRATION, TEMPERATURE, ELECTRICAL CONDUCTANCE, DISSOLVED OXYGEN, AIR POLLUTION, CHEMICAL ANALYSIS, EUTROPHICATION, TRANSPIRATION, PRECIPITATION(ATMOSPHERIC), RUNOFF, EVAPORATION, GEOLOGIC FORMATIONS, DUSTS, TURBIDITY, EROSION, CONDUCTIVITY, CHLORIDES, SILICA, AMMONIA, MICROORGANISMS, ALGAE, CALCIUM, MAGNESIUM, POTASSIUM, SODIUM, ALUMINUM, BICARBONATES, NITRIFICATION, SEASONAL.

IDENTIFIERS:
*HUBBARD BROOK EXPERIMENTAL FOREST, PARTICULATE MATTER, NITROSOMONAS, NITROBACTER, THIOBACILLUS THIOOXIDANS, DESULFOVIBRIO, ULOTHRIZ ZONATA.

ABSTRACT:
QUANTITY AND QUALITY OF DRAINAGE WATERS WERE SIGNIFICANTLY ALTERED FOLLOWING DEFORESTATION OF A NORTHERN HARDWOODS WATERSHED-ECOSYSTEM. ANNUAL WATER RUNOFF EXCEEDED EXPECTED VALUE (BASED ON UNDISTURBED WATERSHED) BY 39% DURING FIRST WATER-YEAR AFTER DEFORESTATION AND 28% DURING THE SECOND. DEFORESTATION RESULTED IN LARGE INCREASES IN CONCENTRATIONS OF ALL MAJOR IONS EXCEPT AMMONIUM, SULFATE, AND CARBONIC ACID, WITH GREATEST INCREASE FOR NITRATE IN STREAMWATER. SULFATE WAS THE ONLY MAJOR ION THAT DECREASED AFTER DEFORESTATION. IN UNDISTURBED WATERSHEDS, STREAMWATER HAS A PH OF ABOUT 5.1 FROM SULFURIC ACID, WHEREAS AFTER DEFORESTATION IT BECAME A NITRIC ACID SOLUTION OF PH 4.3 ENRICHED IN METALLIC IONS AND DISSOLVED SILICA. THE INCREASE IN NITRATE CONCENTRATION IN PRECIPITATION MAY SOMEWHAT INCREASE AIR POLLUTION. GREATLY INCREASED EXPORT OF DISSOLVED NUTRIENTS FROM THE DEFORESTED ECOSYSTEM WAS DUE TO AN ALTERATION OF THE ECOSYSTEM NITROGEN CYCLE. INCREASED AVAILABILITY OF NITRATE AND HYDROGEN IONS RESULTED FROM NITRIFICATION. TOTAL NEW EXPORT OF DISSOLVED INORGANIC SUBSTANCES WAS 14-15 TIMES GREATER THAN FROM NATURAL ECOSYSTEMS. THE DEFORESTATION EXPERIMENT RESULTED IN SIGNIFICANT POLLUTION OF THE DRAINAGE STREAM, WITH NITRATE CONCENTRATION EXCEEDING THE MAXIMUM RECOMMNDED FOR DRINKING WATER. A BLOOM OF ALGAE APPEARED EACH SUMMER. (JONES-WISCONSIN)

FIELD 02A, 05C

ACCESSION NO. W71-01489

BIOLOGICAL AND CHEMICAL ASPECTS OF RHINE WATER IN THE BERENPLAAT RESERVOIR,

DRINKWATERLEIDING DER GEMEENTE ROTTERDAM (NETHERLANDS).

JOHN J. ROOK, AND GIJSBERT OSKAM.

JOURNAL AMERICAN WATER WORKS ASSOCIATION, VOL 62, NO 4, P 249-259, 1970. 10 FIG, 7 TAB, 30 REF.

DESCRIPTORS:
*WATER PURIFICATION, *DETENTION RESERVOIRS, RESERVOIRS, PHOTOSYNTHESIS, DISSOLVED OXYGEN, PHOSPHATES, WATER YIELD IMPROVEMENT, WATER POLLUTION TREATMENT, ALGAE, POTABLE WATER, TASTE, AMMONIA, SILICA.

IDENTIFIERS:
RHINE RIVER, ASTERIONELLA FORMOSA, CRYPTOMONAS, CHLAMIDOMONAS, BERENPLAAT RESERVOIR(ROTTERDAM), THE NETHERLANDS.

ABSTRACT:
A CONSIDERABLE PURIFICATION OF THE HIGHLY POLLUTED WATER OF THE RHINE RIVER IS ACHIEVED BY ITS RELATIVELY BRIEF STORAGE IN A RETENTION RESERVOIR. THE HIGH LEVEL OF RIVER'S FERTILITY INDUCES A HEAVY GROWTH OF ALGAE, PARTICULARLY ASTERIONELLA FORMOSA. THE STORAGE AND RE-AERATION OF THE ALGAL SUSPENSION PROMOTES PHOTOSYNTHESIS AND RESTORES THE CONTENT OF DISSOLVED OXYGEN. AN OCCASIONAL INCREASE IN DISSOLVED ORGANIC MATTER IN THE RESERVOIR HAS FAILED TO PRODUCE UNFAVORABLE EFFECTS ON POTASSIUM PERMANGANATE CONSUMPTION. THE GROWING ALGAE UTILIZE LARGE AMOUNTS OF PHOSPHATES AND SILICA; THE LATTER APPEARS TO ACT AS A LIMITING FACTOR. THE AMMONIA CONTENT IS REDUCED BY LOSS TO THE ATMOSPHERE AND BY NITRIFICATION, THE NITRATES PRESUMABLY CONTRIBUTING TO THE DISSIMILATION OF SETTLED ORGANIC MATTER. THE MOST IMPORTANT BENEFIT OF THE XTORAGE OF RHINE WATER IS THE STRIKING REDUCTION IN THRESHOLD TASTE VALUE. THE ALGAE ARE WITHDRAWN FROM THE RESERVOIR PRIOR TO THEIR COMPLETION OF THE LIFE CYCLE. AN OPINION IS EXPRESSED THAT THE NETHERLANDS WILL BUILD MANY NEW RHINE RESERVOIRS. (WILDE-WISCONSIN)

FIELD 05F

ACCESSION NO. W71-01491

WATER QUALITY CONTROL WITH SYNTHETIC POLYMERIC FLOCCULANTS,

CONNECTICUT UNIV., STORRS. INST. OF WATER RESOURCES.

J. K. DIXON.

AVAILABLE FROM NTIS AS PB-195 983, $3.00 IN PAPER COPY, $0.95 IN MICROFICHE. TECHNICAL COMPLETION REPORT, NOVEMBER 1970. 4 P. OWRR PROJECT A-016-CONN(1).

DESCRIPTORS:
*FLOCCULATION, FILTRATION, ELECTROPHORESIS, SETTLING, *ALGAE, *SILICA, *BACTERIA, WATER POLLUTION CONTROL, WASTE WATER TREATMENT.

IDENTIFIERS:
POLYACRYLAMIDE, POLYETHYLENIMINE.

ABSTRACT:
STUDIES HAVE BEEN MADE OF THE FLOCCULATION OF SILICE, ALGAE AND BACTERIA BY MEANS OF LOW CONCENTRATIONS OF SYNTHETIC POLYMERS. THE EFFECTS OF POLYMER MOLECULAR WEIGHT AND STRUCTURE HAVE BEEN EMPHASIZED, USING THE SILICA, AND PURE CULTURES OF GREEN ALGAE AND E. COLI UNDER A VARIETY OF CONDITIONS OF PH AND COLLOID CONCENTRATION. FILTRATION RATE, LIGHT TRANSMISSION AND ELECTROPHORETIC MOBILITY HAVE BEEN EMPLOYED TO MEASURE DEGREE OF FLOCCULATION. NON-IONIC AND ANIONIC POLYMERS DID NOT PRODUCE FLOCCULATION, BUT CATIONIC POLYETHYLENEIMINES WERE EFFECTIVE AT CONCENTRATIONS OF 0.1 TO 100 PPM, THE OPTIMUM BEING DEPENDENT ON PH, COLLOID CONCENTRATION AND POLYMER MOLECULAR WEIGHTS ABOVE ABOUT 2000. ALGAE REQUIRED SOME 30 TO 500 TIMES HIGHER SOLUTION CONCENTRATION OF POLYMER THAN DID THE BACTERIA AND SILICA, RESPECTIVELY. TWO TYPES OF POLYETHYLENEIMINE POLYMER TAGGED WITH RADIOACTIVE C14 WERE PREPARED WHICH FLOCCULATED THE THREE KINDS OF COLLOIDS IN THE SAME WAY AS UNTAGGED POLYMERS. IN THE REGION WHERE FLOCCULATION WAS OPTIMUM THE CONCENTRATIONS OF POLYMER ADSORBED ON THE COLLOID SURFACES STOOD IN ROUGHLY THE RATIO1:10:200 FOR SILICA, E. COLI AND ALGAE, RESPECTIVELY. POOR FLOCCULATION EFFICIENCY WAS NOT TO BE ASSOCIATED WITH LOW ADSORPTION, BUT TO SPECIFIC POLYMER-SURFACE INTERACTIONS WHICH REDUCED THE EFFICIENCY OF THE ADSORBED POLYMER. THE RESULTS HAVE BEEN SHOWN TO CONFORM WITH CURRENT THEORIES OF FLOCCULATION OF OTHER COLLOIDS. THEY SHOULD ASSIST IN ATTAINING BETTER WATER QUALITY CONTROL BY THE USE OF THE SYNTHETIC POLYMERS IN NUMEROUS PRACTICAL OPERATIONS. (DIXON-CONN)

FIELD 05D, 05G

ACCESSION NO. W71-01899

TEMPERATURE EFFECTS ON THE SORPTION OF RADIONUCLIDS BY FRESHWATER ALGAE,

DU PONT DE NEMOURS (E. I.) AND CO., AIKEN, S.C. SAVANNAH RIVER LAB.

R. S. HARVEY.

HEALTH PHYSICS, PERGAMON PRESS, VOL 19, AUG 1970, P 293-297. 4 FIG, 3 TAB, 7 REF.

DESCRIPTORS:
TEMPERATURE, SORPTION, RADIOISOTOPES, ALGAE, GROWTH RATES, THERMAL POLLUTION.

IDENTIFIERS:
SAVANNAH RIVER PLANT, CULTURE MEDIA, DRY WEIGHT, THERMAL EFFECTS.

ABSTRACT:
THE SPECIES STUDIED WERE COLLECTED FROM THE REACTOR EFFLUENT STREAMS AT SAVANNAH RIVER PLANT. UNIALGAL CULTURES WERE DEVELOPED IN INORGANIC MEDIA. ALL TESTS WERE CONDUCTED USING THE CONTINUOUS FLOW CULTURE SYSTEM DESCRIBED BY WATTS AND HARVEY. WATER TEMPERATURES OF 23, 26, 29 AND 32C HAD NO SIGNIFICANT EFFECT ON THE SORPTION OF 137CS, 85SR, 65ZN, 59FE, 57CO AND 54MN BY THE FILAMENTOUS GREEN ALGA STIGEOCLONIUM LUBRICUM. RADIONUCLIDE CONCENTRATIONS IN THE UNICELLULAR DIATOM NAVICULA SEMINULUM WERE 2-5 TIMES HIGHER AT 32C THAN THOSE OBTAINED AT LOWER TEMPERATURES. WATER TEMPERATURES OF 25, 30, 35, 40C HAD NO SIGNIFICANT EFFECT ON THE SORPTION OF 137CS, 85SR, 65ZN AND 59FE BY THE FILAMENTOUS BLUE-GREEN ALGA PLECTONEMA BORYANUM. HOWEVER, 57CO CONCENTRATIONS IN P. BORYANUM DECREASED WITH TEMPERATURE, AND 54MN CONCENTRATIONS INCREASED FROM 25 TO 35C. GROWTH RATES OF N. SEMINULUM AND P. BORYANUM WERE INHIBITED AT 32 AND 25C, RESPECTIVELY. GROWTH OF S. LUBRICUM WAS NOT INFLUENCED BY THE TEMPERATURES TESTED. THESE DATA SHOW THAT NONLETHAL CHANGES IN WATER TEMPERATURE HAD NO MAJOR INFLUENCE ON THE SORPTION OF ESSENTIALL ELEMENTS BY THE ALGAE STUDIED. (UPADHYAYA-VANDERBILT)

FIELD 05C

ACCESSION NO. W71-02075

PHOSPHORUS, NITROGEN, AND ALGAE IN LAKE WASHINGTON AFTER DIVERSION OF SEWAGE,

WASHINGTON UNIV., SEATTLE. DEPT. OF ZOOLOGY.

W. T. EDMONDSON.

SCIENCE, VOL 169, NO 3946, P 690-691, AUGUST 14, 1970. 2 FIG, 8 REF.

DESCRIPTORS:
SEWAGE EFFLUENTS, *ALGAE, *PHYTOPLANKTON, *AQUATIC PRODUCTIVITY, *PHOSPHATES, *NITRATES, *CHLOROPHYLL, DIVERSION, CARBON DIOXIDE, ALKALINITY, EPILIMNION, WASHINGTON, PHOSPHORUS, NITROGEN, NUTRIENTS.

IDENTIFIERS:
LAKE WASHINGTON(WASHINGTON).

ABSTRACT:
BECAUSE OF THE LARGE AMOUNTS OF PHOSPHORUS AND NITROGEN BEING ADDED TO LAKE WASHINGTON FROM SEWAGE EFFLUENT, A PROGRAM WAS SET UP TO DIVERT ALL SEWAGE FROM THE LAKE. THE FIRST DIVERSION OF 11 TREATMENT PLANTS OCCURRED IN 1963. FROM 1963 TO 1969, PHOSPHATE DECREASED TO 28% OF THE 1963 CONCENTRATION, BUT NITRATE REMAINED AT MORE THAN 80% OF THE 1963 VALUE. FREE CARBON DIOXIDE AND ALKALINITY REMAINED RELATIVELY HIGH. THE AMOUNT OF PHYTOPLANKTONIC CHLOROPHYLL IN THE SUMMER WAS VERY CLOSELY RELATED TO THE MEAN WINTER CONCENTRATION OF PHOSPHATE, BUT NOT TO THAT OF NITRATE OR CARBON DIOXIDE. PHYTOPLANKTON COUNTS HAVE NOT BEEN COMPLETED, BUT DATA ARE AVAILABLE ON THE CHLOROPHYLL CONTENT OF THE PHYTOPLANKTON IN THE EPILIMNION. THE RELATIONSHIP BETWEEN THE SUMMER MEAN AND THE CONCENTRATION OF PHOSPHATE IN THE SURFACE WATER DURING THE PREVIOUS WINTER STRONGLY SUGGESTS THAT PHOSPHORUS IS THE MOST IMPORTANT LIMITING ELEMENT IN LAKE WASHINGTON. TWO INCLUDED GRAPHS SHOW THE MEAN WINTER VALUES OF PHOSPHATE, NITRATE, CHLOROPHYLL, AND SURFACE PHYTOPLANKTON; AND CORRELATION BETWEEN SURFACE VALUES OF PHOSPHATE AND NITRATE DURING THE SPRING INCREASE OF PHYTOPLANKTON WHEN THE CONCENTRATIONS OF NUTRIENTS ARE DECREASING. (LITTLE-BATTELLE)

FIELD 05C, 05G, 02H

ACCESSION NO. W71-02681

CHEMICAL, PHYSICAL AND BIOLOGICAL DYNAMICS OF NORTHERN PRAIRIE LAKES,

SOUTH DAKOTA STATE UNIV., BROOKINGS.

JOHN G. NICKUM.

AVAILABLE FROM NTIS AS PB-196 357, $3.00 IN PAPER COPY, $0.95 IN MICROFICHE. COMPLETION REPORT, WATER RESOURCES INSTITUTE, SOUTH DAKOTA STATE UNIVERSITY, BROOKINGS, DECEMBER 1970. 38 P, 3 TAB, 10 FIG, 23 REF. OWRR PROJECT B-002-S DAK(4).

DESCRIPTORS:
*WATER QUALITY, *LAKES, *EUTROPHICATION, PHYTOPLANKTON, PLANKTON, PRODUCTIVITY, PRIMARY PRODUCTIVITY, ICED LAKES, IONS, SULFATES, CARBONATES, BICARBONATES, DIATONICS, CYANOPHYTA, ALGAE, SOUTH DAKOTA, WATER POLLUTION EFFECTS, WATER POLLUTION SOURCES, CULTIVATION.

IDENTIFIERS:
LAKE HERMAN(S. DAK), ENEMY SWIM LAKE(S. DAK).

ABSTRACT:
WATER SAMPLES AND SAMPLES OF PLANKTON WERE COLLECTED FROM LAKE HERMAN AND ENEMY SWIM LAKE, BOTH LOCATED IN EASTERN SOUTH DAKOTA, FROM JULY, 1967 THROUGH JUNE, 1970. ALL SAMPLES WERE ANALYZED FOR SELECTED CHEMICAL CHARACTERISTICS AND FOR PHYTOPLANKTON POPULATION. RESULTS INDICATED THAT BOTH LAKES ARE EUTROPHIC BUT LAKE HERMAN IS CONSIDERABLY MORE PRODUCTIVE THAN ENEMY SWIM. CONCENTRATIONS OF MOST IONS WERE FROM 50 PER CENT TO 100 PER CENT GREATER IN LAKE HERMAN THAN IN ENEMY SWIM AND PHYTOPLANKTON POPULATIONS WERE OFTEN 100 TIMES MORE DENSE IN LAKE HERMAN. RELATIVE ABUNDANCE OF MAJOR IONS WERE SIMILAR IN BOTH LAKES EXCEPT THAT SULFATE WAS THE DOMINANT ANION IN LAKE HERMAN AND BICARBONATE - CARBONATE IN ENEMY SWIM. BOTH LAKES SHOWED UNIFORM CHEMICAL CONDITIONS FROM ONE SAMPLING STATION TO ANOTHER AND SIMILAR ANNUAL DYNAMICS MARKED BY A SHARP INCREASE TO PEAK LEVELS OF ALL MAJOR IONS UNDER ICE COVER. PHYTOPLANKTON POPULATIONS WERE DOMINATED BY BLUE-GREEN ALGAE IN LAKE HERMAN, AND BY DIATOMS IN ENEMY SWIM. NUMBERS AND KINDS OF ALGAE VARIED FROM SEASON-TO-SEASON AND FROM YEAR-TO-YEAR INDICATING THAT INTENSIVE LONG TERM SAMPLING IS NECESSARY IN ANY STUDY OF PLANKTON DYNAMICS IN THESE LAKES. THE GREATER EUTROPHICATION OF LAKE HERMAN APPEARS RELATED TO THE INTENSIVE CULTIVATION OF ITS WATERSHED.

FIELD 05C, 02H, 05B

ACCESSION NO. W71-02880

LAKE EUTROPHICATION--WATER POLLUTION CAUSES, EFFECTS AND CONTROL.

MINNESOTA UNIV., MINNEAPOLIS. WATER RESOURCES RESEARCH CENTER.

AVAILABLE FROM NTIS AS PB-196 479, $3.00 IN PAPER COPY, $0.95 IN MICROFICHE. 'SAVE THE LAKES SYMPOSIUM' HELD AT DETROIT LAKES, MINNESOTA, AUGUST 18-19, 1969. WATER RESOURCES RESEARCH CENTER, UNIVERSITY OF MINNESOTA, MINNEAPOLIS, BULLETIN NO 22, 1970. 61 P. OWRR PROJECT A-999-MINN(13).

DESCRIPTORS:
*EUTROPHICATION, *LAKES, *WATER POLLUTION, MINNESOTA, WATER QUALITY, NUTRIENTS, SEDIMENTS, COLIFORMS, MUNICIPAL WASTES, AGRICULTURE, FALLOUT, DRAINAGE, MANAGEMENT, WISCONSIN, PRODUCTIVITY, ALGAE, ECONOMICS, POLITICAL ASPECTS, GEOLOGY, TOPOGRAPHY, GROUNDWATER, BIOCHEMICAL OXYGEN DEMAND, PHOSPHORUS, FISH, WEEDS, CYCLES, BACTERIA, CHEMISTRY, PHYSICS, STRATIFICATION, BIOLOGICAL TREATMENT, VIRUSES, CYANOPHYTA, LEGISLATION, CHEMCONTROL.

IDENTIFIERS:
DETROIT LAKES(MINN), COMMUNITY ACTION, LAKE SALLIE(MINN), LAKE MELISSA(MINN), NUTRIENT REMOVAL.

ABSTRACT:
PAPERS PRESENTED AT 'SAVE THE LAKES SYMPOSIUM' HELD AT DETROIT LAKES, MINNESOTA, AUGUST 18-19, 1969, FOCUSED ON LAKE POLLUTION AND EUTROPHICATION: WHAT IT IS, WHAT CAUSES IT, THE TECHNOLOGY NEEDED AND AVAILABLE TO CONTROL IT, AND THE TYPE OF COMMUNITY ACTION THAT CAN AND MUST BE TAKEN FOR EFFECTIVE CONTROL AND IMPROVEMENT. THE PURPOSE IS TO MAKE FACTS AVAILABLE TO THE PUBLIC AND TO CLARIFY THE COOPERATIVE ROLE OF CITIZENS AND GOVERNMENT. THE LIFE CYCLE OF A LAKE IS DESCRIBED. IMPROVED METHODS OF LAKE MANAGEMENT MAY BE UNCOVERED IN FUNDAMENTAL STUDIES DEALING WITH PRESENT SOURCES OF POLLUTION, MUNICIPAL SEWAGE AND DIFFUSE SOURCES, SUCH AS AGRICULTURAL OPERATIONS, FALLOUT AND WASHOUT FROM AIR, DRAINAGE, AND THE TECHNICAL PROBLEMS INVOLVED. THE HISTORY AND DIVERSION OF WASTES AT MADISON, WISCONSIN, LAKE TAHOE AND LAKE WASHINGTON IS DESCRIBED. BESIDES THE SCIENTIFIC AREAS, THE MAJOR PROBLEMS ARE LACK OF POLITICAL LEADERSHIP NECESSARY FOR PROPER EVALUATION OF WATER QUALITY AND PROVIDING FINANCING NECESSARY FOR SOLUTION. THE GEOLOGICAL HISTORY OF THE DETROIT LAKES IS GIVEN. THERE ARE MANY MEANS OF ALLEVIATING EUTROPHICATION: ECOLOGICAL CONTROL, BIOLOGICAL CONTROL, CHEMICAL CONTROL, AND MECHANICAL CONTROL. (JONES-WISCONSIN)

FIELD 05G

ACCESSION NO. W71-03012

THE EFFECTS OF COPPER SULFATE ON CERTAIN ALGAE AND ZOOPLANKTERS IN WINNISQUAM LAKE, NEW HAMPSHIRE,

NEW HAMPSHIRE UNIV., DURHAM. WATER RESOURCES RESEARCH CENTER.

PHILIP J. SAWYER.

AVAILABLE FROM NTIS AS PB-196 481, $3.00 IN PAPER COPY, $0.95 IN MICROFICHE. NEW HAMPSHIRE WATER RESOURCES RESEARCH CENTER, WRR-2, 1970. 28 P, 2 FIG, 5 TAB, 17 REF. OWRR PROJECT A-004-NH(4).

DESCRIPTORS:
*COPPER SULFATE, *ALGAE, *ZOOPLANKTERS, *CHEMCONTROL, NEW HAMPSHIRE, COPEPODS, DAPHNIA, SAMPLING, DEPTH, CYANOPHYTA, CHRYSOPHYTA, CHLOROPHYTA, PRIMARY PRODUCTIVITY, CARBON RADIOISOTOPES, PHOTOSYNTHESIS.

IDENTIFIERS:
*WINNISQUAM LAKE(N H), ANABAENA CIRCINALIS, BOSMINA LONGIROSTRIS, SYNURA UVELLA, DINOBRYON, OSCILLATORIA LIMNETICA, DICTYOSPHAERIUM, CELL COUNTS, FRAGILARIA, ASTERIONELLA, CRYPTOMONAS, SPHAEROCYSTIS, DAPHNIA GALEATA MENDOTAE, DAPHNIA MAGNA.

ABSTRACT:
IN A RELATIVELY UNBUFFERED LAKE SUCH AS WINNISQUAM, CONTROL OF CYANOPHYCEAN BLOOM CAN BE ACCOMPLISHED WITH RELATIVELY SMALL AMOUNTS OF COPPER SULFATE --4 LBS/ACRE. DATA ON 27 ALGAL SPECIES PRESENT BEFORE, DURING, AND AFTER ADDITION OF COPPER SULFATE INDICATE THAT A RAPID SHIFT IN SPECIES RELATIONSHIPS OCCURRED. CARBON FIXATION RATE WAS FOUND TO BE A MORE SENSITIVE MEASURE OF BLOOM POTENTIAL THAN CELL COUNTS. SUSCEPTIBLE SPECIES DECREASED IN NUMBERS; OTHER FORMS INCREASED IN POPULATION; SOME APPEARED UNAFFECTED BY TREATMENT. THE LATTER GROUP MAY HAVE BEEN PROTECTED BY NATURAL RESISTANCE OR BY THEIR POSITION IN THE WATER COLUMN. ZOOPLANKTON REACTED TO COPPER TREATMENT IN VARIOUS WAYS. THE EFFECTS ON MICROCRUSTACEANS SUCH AS COPEPODS AND DAPHNIDS SEEMED TO FOLLOW A PREDICTABLE PATTERN. COPEPODS AND BOSMINA LONGIROSTRIS WERE MORE RESISTANT THAN DAPHNIA. BOSMINA MULTIPLIED TO A CONCENTRATION OF 154/LITER IN THE TOP METER OF THE WATER COLUMN. THE RISE AND FALL OF SPECIE NUMBERS OF PLANTS AND ANIMALS AND THEIR RAPID RESPONSE TO CONDITIONS AFTER COPPER SULFATE TREATMENT ILLUSTRATE THEIR COMPLEX RELATIONSHIPS. THE CRUDE TREATMENT WITH COPPER SULFATE DID PRODUCE CONTROL OF AN OBNOXIOUS BLUE-GREEN ALGA, ANABAENA CIRCINALIS. (JONES-WISCONSIN)

FIELD 05F, 02H

ACCESSION NO. W71-03014

BIOGEOCHEMICAL MODELING OF EUTROPHIC LAKES FOR WATER QUALITY IMPROVEMENT,

NOTRE DAME UNIV., IND. DEPT. OF CIVIL ENGINEERING.

M. W. TENNEY, W. F. ECHELBERGER, JR., P. C. SINGER, F. H. VERHOFF, AND W. A. GARVEY.

FIFTH INTERNATIONAL CONFERENCE ON WATER POLLUTION RESEARCH, SAN FRANCISCO, CALIFORNIA, JULY 26-AUGUST 1, 1970. PREPRINT. 22 P, 4 FIG, 2 TAB, 17 REF.

DESCRIPTORS:
*SYSTEMS ANALYSIS, *WATER QUALITY, *EUTROPHICATION, *MATHEMATICAL MODELS, SIMULATION, ANALYTICAL TECHNIQUES, PHOSPHORUS, ALGAE, LAKES, MICROORGANISMS, TERTIARY TREATMENT, EFFLUENTS, MODEL STUDIES.

IDENTIFIERS:
HETEROTROPHS, AUTOTROPHS.

ABSTRACT:
A MATHEMATICAL SIMULATION OF BIOGEOCHEMICAL EXCHANGES IN A LAKE SYSTEM AND ITS USE IN SYNTHESISING THE EUTROPHICATION HISTORY ARE DESCRIBED. TAKEN AS A SINGLE-STAGE, CONTINUOUS-FLOWING REACTOR SYSTEM WITH VARIABLE INPUTS AND OUTPUTS, GENERAL TRENDS ARE PREDICTABLE. THE CONCENTRATION OF ORGANIC CARBON AND PHOSPHORUS DETERMINE THE EXTENT OF HETEROTROPHIC AND AUTOTROPHIC ACTIVITY IN THE MODEL, AND THE RESULTANT ACTIVITY IS MEASURED AS SUSPENDED SOLIDS. INTERNAL EXCHANGES BETWEEN STATE AND LOCATION OF POLLUTANTS IS CHARACTERIZED BY RATE FUNCTIONS, AND WHEN NECESSARY, A STOCHIOMETRY OF EACH OF THE BIO- AND PHYSICAL REACTIONS IS INCLUDED. A DETERMINISTIC SIMULATION MODEL WAS APPLIED TO STONE LAKE, CASSOPOLIS, MICHIGAN, AND REASONABLE PREDICTIONS WERE MADE WITH REGARD TO ORGANIC MATERIAL, INORGANIC PHOSPHORUS, AND SUSPENDED MICROORGANISMS. THE MODEL IS APPLICABLE TO PREDICTION OF FUTURE WATER QUALITY AND CERTAIN MANIPULATION PROCEDURES TO ENHANCE WATER QUALITY. ECOLOGICAL MODELING DEVELOPMENTS WILL ASSIST POLITICAL DECISIONS LEADING TO DEVELOPMENT OF SUCCESSFUL CONCEPTS IN WATER QUALITY MANAGEMENT. (AUEN-WISCONSIN)

FIELD 05G, 05C, 02H

ACCESSION NO. W71-03021

COPPER IONS AS POISON IN THE SEA AND IN FRESHWATER,

COPENHAGEN UNIV., HILLEROD (DENMARK). FRESHWATER BIOLOGICAL LAB.

E. STEEMANN NIELSEN, AND S. WIUM-ANDERSEN.

MARINE BIOLOGY, VOL 6, NO 2, P 93-97, 1970. 4 FIG, 20 REF.

DESCRIPTORS:
*COPPER, *IONS, *POISONS, *SEA WATER, *FRESH WATER, ALGAE, PHOTOSYNTHESIS, ORGANIC MATTER, PLANKTON, CHELATION, CARBON RADIOISOTOPES, CHLORELLA, GROWTH RATES, DIATOMS, PHYTOPLANKTON, TOXICITY, IRON, LIGHT, HYDROGEN ION CONCENTRATION.

IDENTIFIERS:
CHLORELLA PYRENOIDOSA, NITZSCHIA PALEA, SKELETONEMA COSTATUM.

ABSTRACT:
DURING THE LAST TWO YEARS INTENSIVE EXPERIMENTAL WORK HAS BEEN APPLIED IN INVESTIGATIONS OF THE INFLUENCE OF IONIC COPPER ON PHOTOSYNTHESIS AND GROWTH OF GREEN ALGA, CHLORELLA PYRENOIDOSA AND THE DIATOM, NITZSCHIA PALEA. COPPER IN IONIC FORM IS FOUND TO BE VERY TOXIC TO PHOTOSYNTHESIS AND GROWTH OF UNICELLULAR ALGAE AT CONCENTRATIONS USUALLY FOUND IN NATURAL WATERS, INDICATING THAT COPPER IS NOT ORDINARILY PRESENT IN IONIC FORM BUT IS COMPLEXED BY ORGANIC MATTER SUCH AS POLYPEPTIDES. THE AFFINITY OF COPPER TO DIETHYL-DITHIOCARBAMINATE IS VERY MUCH HIGHER THAN TO THE ORGANIC MATTER WHICH COMPLEXES COPPER IN NATURE, THUS IT IS NOT POSSIBLE TO DISTINGUISH THE TWO FORMS OF COPPER DURING ANALYSIS. COMPLEXED COPPER IS NOT POISONOUS TO ALGAE. IT HAS RECENTLY BEEN SHOWN THAT OCEAN WATER IN THE CENTERS OF UPWELLING BECOMES SUITABLE FOR PLANKTON GROWTH ONLY AFTER THE ADDITION OF A CHELATOR, SUGGESTING THAT A LARGE PART OF THE COPPER FOUND IN SUBSURFACE WATERS OF OCEANS IS PRESENT IN IONIC FORM. SOME MANUFACTURERS OF CARBON-14 AMPOULES HAVE USED ORDINARY DISTILLED WATER WHICH OFTEN HAS A CONTENT OF ABOUT 250 MICROGRAMS COPPER PER LITER; THUS, IT IS PROBABLE THAT SOME PRODUCTIVITY MEASUREMENTS HAVE BEEN INFLUENCED. (JONES-WISCONSIN)

FIELD 05B

ACCESSION NO. W71-03027

NUTRIENTS AND QUALITY IN IMPOUNDED WATER,

ILLINOIS STATE WATER SURVEY, PEORIA. WATER QUALITY SECTION.

WUM-CHENG WANG, AND RALPH L. EVANS.

JOURNAL AMERICAN WATER WORKS ASSOCIATION, VOL 62, NO 8, P 510-514, 1970. 8 FIG, 3 TAB, 15 REF.

DESCRIPTORS:
*NUTRIENTS, *WATER QUALITY, *IMPOUNDED WATERS, STREAMS, FLOW, SEASONAL, STRATIFICATION, NITRATES, SILICA, AMMONIA, PHOSPHATES, RUNOFF, ILLINOIS, HYDROGEN ION CONCENTRATION, ALKALINITY, HARDNESS(WATER), SAMPLING, DIATOMS, NITROGEN, BACTERIA, ALGAE, THERMAL STRATIFICATION, DISSOLVED OXYGEN, HYPOLIMNION.

IDENTIFIERS:
LAKE BLOOMINGTON(ILL), CHEMOSTRATIFICATION.

ABSTRACT:
AN EXTENSIVE STUDY WAS MADE OF IMPOUNDED LAKE BLOOMINGTON (ILLINOIS), AS SIGNIFICANT CHANGES OCCUR IN STREAM WATER QUALITY WHEN THEY ARE IMPOUNDED. WITH FEW EXCEPTIONS CONCENTRATION OF EACH CONSTITUENT DECREASES WITH PASSAGE THROUGH THE IMPOUNDMENT. MAJOR INFLUENCES WITHIN THE SYSTEM OBSERVED WERE STREAM FLOW, SEASONAL PATTERNS, AND STRATIFICATION. NITRATE NITROGEN AND SILICA CONCENTRATION INCREASED WITH INCREASING STREAM FLOW. AMMONIA-NITROGEN AND ORTHOPHOSPHATE CONCENTRATION WERE MINIMAL. SINCE THESE CONSTITUENTS WERE PRINCIPALLY DERIVED FROM AGRICULTURAL RUNOFF THE RELATIVE MOBILITY OF EACH IN SOIL, APPEARS CLEAR. NITRATE NITROGEN AND PHOSPHORUS CONTRIBUTIONS TO THE WATER SYSTEM REPRESENTED 8.3% AND 0.1%, RESPECTIVELY, OF THAT APPLIED TO THE DRAINAGE AREA. DURING WARM WEATHER, ORGANIC NITROGEN CONCENTRATIONS WERE HIGHEST WHILE NITRATE-NITROGEN WAS MINIMAL, SUGGESTING INFLUENCE OF BIOLOGICAL ACTIVITY. DIFFERENCE BETWEEN RATE OF NITRATE-NITROGEN AND SILICA CONCENTRATION REDUCTION, THOUGH OF THE SAME ORIGIN, ALSO SUGGESTED INFLUENCE OF BIOLOGICAL UPTAKE. WITHIN THE IMPOUNDMENT, STRATIFICATION APPEARED THE GOVERNING INFLUENCE, AS EVIDENCED BY CHANGES IN NITRATE NITROGEN AND SILICA CONCENTRATION WITH DEPTH. FURTHER INSIGHT INTO THE INTERRELATIONSHIPS WOULD APPEAR TO PERMIT INTELLIGENT FLEXIBILITY ON THE PART OF WATER WORKS MANAGEMENT. (JONES-WISCONSIN)

FIELD 05C, 02H

ACCESSION NO. W71-03028

SIMULATING THE EFFECT OF SINKING AND VERTICAL MIXING ON ALGAL POPULATION DYNAMICS,

OREGON STATE UNIV., CORVALLIS. DEPT. OF CIVIL ENGINEERING.

DAVID A. BELLA.

JOURNAL WATER POLLUTION CONTROL FEDERATION, VOL 42, NO 5, PART 2, P R140-R152, 1970. 1 FIG, 1 TAB, 17 REF.

DESCRIPTORS:
*MATHEMATICAL MODELS, *ALGAE, *POPULATION, DYNAMICS, LAKES, DISSOLVED OXYGEN, NUTRIENTS, CYANOPHYTA, MIXING, PHYTOPLANKTON.

IDENTIFIERS:
*ALGAL SINKING, *VERTICAL MIXING, EUPHOTIC ZONE, DIFFUSION COEFFICIENT, LAKE SAMMAMISH(WASH).

ABSTRACT:
THE DEVELOPED SIMULATION MODEL DESCRIBES THE COMBINED EFFECT OF GROWTH, SINKING, AND VERTICAL MIXING ON DYNAMICS OF ALGAL POPULATION OF LAKES. SMALL DIFFERENCES IN SINKING VELOCITIES MAY SIGNIFICANTLY INCREASE THE GROWTH OF ALGAE. A LOWERING OF SINKING OFTEN LEADS TO AN INVASION OF A LESS OBJECTIONABLE PHYTOPLANKTON. VERTICAL MIXING INCREASES THE DENSITY OF ALGAE AND MAY RESULT IN TRANSFORMATION OF THE ENTIRE COMMUNITY. ARTIFICAL MIXING MAY SERVE AS AN AID IN CONTROL OF ALGAE, BUT THE HOMOGENIZATION OF ENVIRONMENT IS APT TO REDUCE THE SPECIES DIVERSITY AT ALL TROPHIC LEVELS. (WILDE-WISCONSIN)

FIELD 05C, 07B

ACCESSION NO. W71-03034

POLLUTION OF THE NORTH SEA AND REARING EXPERIMENTS ON MARINE PHYTOFLAGELLATES AS AN INDICATION OF RESULTANT TOXICITY,

H. KAYSER.

FIFTH INTERNATIONAL WATER POLLUTION RESEARCH CONFERENCE, SAN FRANCISCO, JULY 26-AUGUST 1, 1970. PREPRINT, 7 P, 5 FIG, 5 REF.

DESCRIPTORS:
*WATER POLLUTION EFFECTS, WATER POLLUTION SOURCES, INDUSTRIAL WASTES, DOMESTIC WASTES, DINOFLAGELLATES, *LABORATORY TESTS, CULTURES, *GROWTH RATES, PLANT GROWTH REGULATORS, PLANT GROWTH SUBSTANCES, ALGAE, IRON, BIOASSAY, BIOINDICATORS, FOOD CHAINS, TITANIUM, PLANT POPULATIONS, TOXICITY.

IDENTIFIERS:
*NORTH SEA, PROROCENTRUM MICANS, CERATIUM FURCA, TURBIDOSTAT CULTURES.

ABSTRACT:
THE EXTREME POLLUTION OF THE NORTH SEA WAS DESCRIBED. SINCE THIS TREND IS LIKELY TO CONTINUE, THE AUTHOR DEVELOPED METHODS FOR DETERMINATION OF THE POSSIBLE BIOLOGICAL CONSEQUENCES. THIS METHOD INVOLVES THE CULTURE OF PRIMARY PRODUCERS, NAMELY DINOFLAGELLATES, IN WASTE WATERS. MASS CULTURES PROVIDED INFORMATION ABOUT THE GROWTH RATE AND MAXIMUM DENSITY OF POPULATIONS UNDER THE INFLUENCE OF WASTE WATER. THIS METHOD IS ABLE TO INDICATE THE SUBLETHAL INFLUENCES ON CHARACTERISTIC MEMBERS OF THE MARINE FOOD CHAIN. IT IS HOPED THAT FURTHER ADVANCEMENT OF THE TEST METHOD WILL ALLOW CALCULATION OF THE SUBLETHAL LEVEL OF WASTE WATER DISCHARGE INTO THE NORTH SEA. (SJOLSETH-WASHINGTON)

FIELD 05C

ACCESSION NO. W71-03095

DESIGN CONSTRUCTION AND MAINTENANCE OF WASTE STABILIZATION LAGOONS,

FARMERS HOME ADMINISTRATION, CHAMPAIGN, ILL.

DAVID H. STOLTENBERG.

PUBLIC WORKS, VOL 101, NO 9, P 103-106, SEPTEMBER 1970. 4 FIG, 3 TAB, 17 REF.

DESCRIPTORS:
*OXIDATION LAGOONS, *STABILIZATION, *ALGAE, AEROBIC TREATMENT, BACTERIA, MUNICIPAL WASTES, BIOCHEMICAL OXYGEN DEMAND, WASTE WATER TREATMENT.

IDENTIFIERS:
DEPTH, SLOPE, *BOD LOADING, *WASTE STABILIZATION.

ABSTRACT:
THERE ARE SEVERAL IMPORTANT FACTORS INVOLVED IN WASTE STABILIZATION USING LAGOONS. THE FACT THAT OXYGEN IS PRODUCED BY ALGAE DURING THE DAY AND CONSUMED BY ALGAE AT NIGHT IS PRESENTED AS A CRITICAL FACTOR IN LAGOON DESIGN AND MAINTENANCE. THE VARIANCE OF ALLOWABLE BIOCHEMICAL OXYGEN DEMAND LOADING WITH GEOGRAPHICAL LOCATION IS ALSO DISCUSSED. CONSTRUCTION COSTS ARE ALSO PRESENTED. ESTIMATES RANGE FROM $11 TO $18 PER CAPITA FOR INSTALLATIONS SERVING 1000 PEOPLE, AND FROM $30 TO $36 PER CAPITA FOR INSTALLATIONS SERVING 100 PEOPLE. OPERATION AND MAINTENANCE COSTS ARE ESTIMATED TO BE ONE-FOURTH, OR ONE-FIFTH OF THE COST FOR CONVENTIONAL WASTE TREATMENT. (LOWRY-TEXAS)

FIELD 05D

ACCESSION NO. W71-03896

RESPIRATORY RELATIONSHIPS OF A SYMBIOTIC ALGAL-BACTERIAL CULTURE FOR WASTE WATER NUTRIENT REMOVAL,

OHIO STATE UNIV., COLUMBUS. DEPT. OF CIVIL ENGINEERING.

F. J. HUMENIK, AND G. P. HANNA, JR.

BIOTECHNOLOGY AND BIOENGINEERING, VOL 12, P 541-560, 1970. 8 FIG, 3 TAB, 17 REF.

DESCRIPTORS:
*SYMBIOSIS, *ALGAE, *BACTERIA, *RESPIRATION, ACTIVATED SLUDGE, WASTE WATER TREATMENT, NUTIRIENTS, AEROBIC CONDITIONS, AERATION, MICROBIOLOGY, BIOMASS, CHLORELLA.

IDENTIFIERS:
*NUTRIENT REMOVAL, OXYGEN DECLINE, RESPIRATORY RATES, PHOTOSYNTHETIC OXYGENATION, ACTIVATED ALGAE.

ABSTRACT:
RESEARCH WAS CONDUCTED TO INVESTIGATE THE RESPIRATORY RELATIONSHIPS OF A MIXED ALGAL-BACTERIAL CULTURE AND THE EFFECT OF ITS NATURAL ECOLOGICAL ASSOCIATIONS ON THE STABILIZATION OF A DOMESTIC WASTE WATER. IN ORDER TO DO THIS THE AUTHORS DEVELOPED A CONTINUOUS SYMBIOTIC ALGAL-BACTERIAL SYSTEM OF A MIXED CHLORELLA-ACTIVATED SLUDGE CULTURE WHICH WOULD EFFICIENTLY REMOVE NUTRIENTS FROM WASTE WATER UNDER AEROBIC CONDITIONS AND WITHOUT ADDITIONAL AERATION. DAILY HARVESTING OF EXCESS BIOMASS WAS CONDUCTED TO PROVIDE STABLE RELATIVE BIOLOGICAL POPULATIONS AND A DISSOLVED OXYGEN CONCENTRATION OF APPROXIMATELY 2 MILLIGRAMS PER LITER WAS MAINTAINED THROUGHOUT STEADY STATE OPERATIONS. TO PREDICT RESPIRATORY RATES, PHOTOSYNTHETIC OXYGENATION AND STEADY-STATE OXYGEN CONCENTRATIONS OXYGEN DECLINE DATA WERE FITTED TO MATHEMATICAL MODELS. THE RESPIRATORY AND PHYSIOLOGICAL RELATIONSHIPS INDICATED THAT THE CARBON DIOXIDE-OXYGEN BALANCE MAY BE A PRIMARY CONTROL GOVERNING THE STEADY-STATE OPERATION OF A SYMBIOTIC ALGAL-BACTERIAL CULTURE. MICROSCOPIC EXAMINATIONS SHOWED THAT BACTERIA WERE ATTACHED TO ALGAL CELLS AND THE CHLORELLA WERE INTEGRALLY ENMESHED WITHIN THE BACTERIAL MIX WHICH ALLOWED THIS ALGAL-BACTERIAL FLOC TO SETTLE QUITE RAPIDLY AND YIELD A CLEAR SUPERNATANT. (ELLIS-TEXAS)

FIELD 05D

ACCESSION NO. W71-04079

ACCUMULATION OF PHOSPHATES IN WATER,

AGRICULTURAL RESEARCH SERVICE, MORRIS, MINN. NORTH CENTRAL SOIL CONSERVATION RESEARCH CENTER.

ROBERT F. HOLT, DONALD R. TIMMONS, AND JOSEPH J. LATTERELL.

JOURNAL OF AGRICULTURE AND FOOD CHEMISTRY, VOL 18, NO 5, P 781-784, 1970. 1 TAB, 37 REF.

DESCRIPTORS:
*PHOSPHATES, *PATH OF POLLUTANTS, *EUTROPHICATION, *LEACHING NUTRIENTS, FERTILIZERS, PRECIPITATION(ATMOSPHERIC), RUNOFF, ANIMAL WASTES, SURFACE WATERS, SOIL EROSION, NUTRIENTS, ALGAE, LAKE SUPERIOR, WATER POLLUTION SOURCES, MINNESOTA, WASHINGTON, CONNECTICUT.

IDENTIFIERS:
ILLINOIS RIVER, ST. LOUIS RIVER, BLACK RIVER(MINNESOTA), LAKE WASHINGTON(WASHINGTON), LINSLEY POND(CONNECTICUT), LAKE ZOAR(CONNECTICUT), LAKE MINNETONKA(MINNESOTA), BIG STONE LAKE(MINNESOTA), LAKE CRYSTAL(MINNESOTA).

ABSTRACT:
NATURAL AND AGRICULTURAL SOURCES OF PHOSPHORUS TO SURFACE WATERS INCLUDE PERCIPITATION, ANIMAL WASTES, FERTILIZERS, AND LAND RUNOFF. THE ACTUAL CONTRIBUTION FROM THESE SOURCES IS SHOWN TO BE QUITE LOW. HOWEVER, THE CONCENTRATION OF PHOSPHORUS REQUIRED TO SUPPORT PROFUSE ALGAL BLOOMS IS SO LOW THAT THE LIMITED AMOUNTS SUPPLIED ARE SUFFICIENT TO EXCEED THIS REQUIREMENT. ERODED SOIL DELIVERS APPRECIABLE AMOUNTS OF PHOSPHORUS TO SURFACE WATERS, BUT THE SOIL MATERIALS CAPACITY TO SORB PHOSPHORUS RESULTS IN LITTLE TENDENCY FOR RELEASE OF THIS SOURCE INTO THE WATER. BOTTOM SEDIMENTS APPEAR TO BE A SINK FOR DISSOLVED ORTHOPHOSPHATE THAT IS SUPPLIED TO SURFACE WATERS. LEACHING OF VEGETATION CAN SUPPLY RELATIVELY LARGE AMOUNTS OF PHOSPHORUS TO LAKES AND STREAMS. DEEP INCORPORATION OF PHOSPHATIC FERTILIZERS MATERIALLY REDUCES THE CONCENTRATION OF PHOSPHORUS IN RUNOFF WATERS AS COMPARED TO SHALLOW INCORPORATION. PHOSPHORUS CONCENTRATIONS IN SEVERAL LAKES AND STREAMS ARE SHOWN. (MCCANN-BATTELLE)

FIELD 05B

ACCESSION NO. W71-04216

LABORATORY STREAM STUDIES OF A BENTHIC COMMUNITY,

MISSOURI UNIV., COLUMBIA. DEPT. OF SANITARY ENGINEERING.

GALE ALLEN WRIGHT.

MS THESIS, 1970. 70 P, 32 FIG, 27 REF, APPEND. OWRR PROJECT A-011-MO(2).

DESCRIPTORS:
*SELF-PURIFICATION, *ALGAE, *STREAMS, LABORATORY TESTS, DISSOLVED OXYGEN, HYDROGEN ION CONCENTRATION, ALKALINITY, RESPIRATION, PHOTOSYNTHESIS, CARBON DIOXIDE, PRIMARY PRODUCTIVITY, BIODEGRADATION, ORGANIC MATTER, OXYGEN, CHEMICAL OXYGEN DEMAND, BENTHOS.

IDENTIFIERS:
*OXYGEN BALANCE.

ABSTRACT:
THE AIM OF THIS STUDY WAS TO DETERMINE THE ROLE OF ALGAE IN BIODEGRADATION OF ORGANIC WASTES AND SELFPURIFICATION OF STREAMS. THE MONITORING OF THE SYSTEM IN A LABORATORY STREAM MICROCOSM WAS ACCOMPLISHED BY DETERMINATION OF DISSOLVED OXYGEN, PH, AND BICARBONATE-CARBONATE ALKALINITY. THE RESULTS INDICATED A LESSER CONSTANCY OF PRODUCTION PROCESS IN COMPARISON WITH THAT OF RESPIRATION. THE CARBON:OXYGEN RATIO WAS INFLUENCED BY THE AGE OF ALGAL CELLS. THE OXIDATION OF ORGANIC SUBSTANCES OCCURRED DURING PERIODS OF DARKNESS. PHOTOSYNTHETIC PRODUCTION WAS LIMITED BY THE AVAILABLE CARBON DIOXIDE. THE CONTRIBUTION OF ALGAE TO SELFPURIFICATION OF WATER APPEARED TO BE VERY LOW. THE DETERMINATION OF PHOTOSYNTHETICALLY PRODUCED OXYGEN COULD NOT BE RELIABLY ACCOMPLISHED BEFORE THE SYNTHESIZED MATERIAL HAD BEEN OXIDIZED. (WILDE-WISCONSIN)

FIELD 05G

ACCESSION NO. W71-04518

CONTRIBUTION TO THE EPIPHYTIC ALGAL FLORA OF THE LAKE NEUSIEDLER, (IN GERMAN),

VIENNA UNIV. (AUSTRIA). BOTANISCHES INSTITUT UND BOTANISCHER GARTEN.

LOTHAR GEITLER.

OSTERREICHISCHE BOTANISCHE ZEITSCHRIFT, VOL 118, NC 1/2, P 17-29, 1970. 3 FIG, 24 REF.

DESCRIPTORS:
*EPIPHYTOLOGY, *LAKES, *ALGAE, ALKALINE WATER, MAGNESIUM, SODIUM SULFATES, CYANOPHYTA, RHODCPHYTA.

IDENTIFIERS:
*LAKE NEUSIEDLER(AUSTRIA), CHAMAESIPHON SUBAEQUALIS, APISTONEMA EXPANSUM, PORPHYRIDIUM GRISEUM, CHRYSOTRICHALAE.

ABSTRACT:
DETAILED DESCRIPTIONS ARE GIVEN OF THE FOLLOWING NEW SPECIES OF ALGAE REVEALED IN THE ALKALINE LAKE NEUSIEDLER: CHAMAESIPHCN SUBAEQUALIS OF CYANOPHYCEAE, APISTONEMA EXPANSUM OF CHRYSOTRICHALAE, AND PORPHYRIDUM GRISEUM OF RHODOPHYCEA. THE PAPER INCLUDED A RECCRD OF SCME PREVIOUSLY IDENTIFIED ALGAE OF THE LAKE. (WILDE-WISCONSIN)

FIELD 02H, 05C

ACCESSION NO. W71-04526

A REVIEW OF HERBIVOROUS FISH FOR WEED CONTROL,

BUREAU OF SPORT FISHERIES AND WILDLIFE, WARM SPRINGS, GA. SOUTHEASTERN FISH CONTROL LAB.

JOE B. SILLS.

THE PROGRESSIVE FISH-CULTURIST, VOL 32, NO 3, P 158-161, 1970. 12 REF.

DESCRIPTORS:
*FISH, *HERBIVORES, *AQUATIC WEED CONTROL, TILAPIA, CARP, ALGAE, ROOTED AQUATIC PLANTS, PLANKTON, ARKANSAS, INSECTS, TURBIDITY, CHARA, PONDWEEDS.

IDENTIFIERS:
TILAPIA NILOTICA, TILAPIA MOSSAMBICA, TILAPIA MELANOPLEURA, CYPRINUS CARPIO, CTENOPHARYNGCDON IDELLUS.

ABSTRACT:
THERE IS GROWING INTEREST IN BIOLOGICAL PEST CCNTRCL AS A SUBSTITUTE FOR PESTICIDES. FISH RECEIVING MOST ATTENTION FOR AQUATIC WEED CONTROL POSSIBILITIES ARE TILAPIA NILOTICA, T MOSSAMBICA, AND T MELANOPLEURA; THE ISRAELI STRAIN OF CARP (CYPRINUS CARPIO); AND THE CHINESE OR GRASS CARP (CTENOPHARYNGODCN IDELLUS). RESULTS OBTAINED WITH EACH ARE REVIEWED. ALL TESTS IN WHICH TILAPIA WERE STOCKED IN PONDS RESULTED IN OVERCROWDED POPULATIONS AND UNSATISFACTORY WEED CONTROL. ALTHOUGH THE CARP IS EFFECTIVE IN REDUCING OR CONTROLLING AQUATIC PLANTS, DETRIMENTAL ASPECTS OF ITS PRESENCE MAKE ITS USAGE UNDESIRABLE IN RECREATION AND FISHING WATERS. TESTS SHOWED THAT THE ISRAELI STRAIN IS NOT PRIMARILY HERBIVOROUS, BUT ROOTS IN THE BOTTOM MUCH AS WILD CARP DO. OBSERVATIONS ARE CONTINUED TO LEARN WHY THE ISRAELI STRAIN DOES NOT REPRODUCE IN PONDS WITH MIXED POPULATICNS. ALL CBSERVATIONS TO DATE ON THE GRASS CARP ARE MCST FAVORABLE. IT DOES FEED ON AQUATIC VEGETATION. THERE IS NO INDICATICN OF NATURAL REPRODUCTION, BUT THE FISH CAN BE SPAWNED. AS ITS NATIVE HABITAT IS SIMILAR TO SOME OF OUR LARGE RIVER SYSTEMS, THE GRASS CARP MAY ADAPT TO THEM. MORE INFORMATION IS NEEDED FOR ITS USE EXCEPT IN STRICTLY CONTROLLED ENVIRONMENTS. (JONES-WISCONSIN)

FIELD 05G

ACCESSION NO. W71-04528

STABLE CARBON ISOTOPES IN BLUE-GREEN ALGAL MATS,

TEXAS UNIV., PORT ARANSAS. INST. OF MARINE SCIENCE; AND TEXAS UNIV., PORT ARANSAS. DEPT. OF GEOLOGICAL SCIENCES.

E. W. BEHRENS, AND S. A. FRISHMAN.

JOURNAL OF GEOLOGY, VOL 79, NO 1, P 94-100, JANUARY 1971. 7 P, 3 FIG, 15 REF. NSF GRANT GA 911.

DESCRIPTORS:
*ALGAE, *STABLE ISOTOPES, *CARBON, *ANAEROEIC BACTERIA, *TEXAS, BIODEGRADATION, TRACERS, LAGOONS, CYANOPHYTA.

IDENTIFIERS:
BAFFIN BAY(TEX).

ABSTRACT:
BLUE-GREEN ALGAL MATS HAVE ACCUMULATED FOR OVER 1,000 YEARS IN KLEBERG POINT LAGOON ON THE MARGIN OF BAFFIN BAY, TEXAS. ORGANIC CARBON PRODUCED WITHIN THE MATS IS ISOTOPICALLY HEAVIER THAN THE CARBON ASSOCIATED WITH BAY SEDIMENTS. THE DIFFERENCE BETWEEN BAY AND MAT ORGANIC CARBON IS PRESERVED IN THE OLDER, BURIED SEDIMENTS, AND SEEMS TO BE INCREASED BY FRACTIONATION ASSOCIATED WITH ANAEROBIC BACTERIAL DECOMPOSITION OF ALGAL MAT ORGANIC MATTER. (KNAPP-USGS)

FIELD 02L, 02K, 05A

ACCESSION NO. W71-04873

POTENTIAL ALGICIDES FOR THE CONTROL OF ALGAE,

LOS ANGELES COUNTY SANITATION DISTRICT, CALIF.

JAMES C. GRATTEAU.

WATER AND SEWAGE WORKS, 1970 REFERENCE NUMBER, VOL 117, P R24-R61, NOVEMBER 28, 1970. 17 P, 6 TAB, 27 REF.

DESCRIPTORS:
*ALGICIDES, *WATER QUALITY CONTROL, *AQUATIC WEED CONTROL, *DOMESTIC WATER, *SURFACE WATERS, WATER SUPPLY, EUTROPHICATION, ALGAE, NUTRIENTS, WATER POLLUTION, BIOLOGICAL TREATMENT, CHEMICALS, COPPER SULFATE, CHLORINE, WATER RESOURCES, METHODOLOGY, EVALUATION, CHEMICAL ANALYSIS.

IDENTIFIERS:
ROSIN AMINE D ACETATE(RADA).

ABSTRACT:
PROBLEMS CONCERNING ALGAE IN WATER SUPPLIES AND METHODS OF CONTROLLING ALGAE GROWTH ARE DISCUSSED. IN GENERAL, THERE ARE THREE BASIC METHODS OF ALGAE CONTROL: PHYSICAL, ECOLOGICAL AND CHEMICAL. PHYSICAL METHODS OF CONTROL, RAKING, DRAGGING, PULLING, AND UNDERWATER MOWING HAVE BEEN USED MAINLY ON FLOATING MASSES OF ALGAE AND ROOTED AQUATICS. ECOLOGICAL CONTROL OF ALGAE INVOLVES LIMITATION OF ONE OR MORE OF THE FACTORS NECESSARY FOR GROWTH AND REPRODUCTION: LIGHT OR NUTRIENT MATERIAL. ONLY TWO CHEMICALS ARE WIDELY USED TO CONTROL ALGAE IN WATER SUPPLIES, COPPER SULFATE AND CHLORINE. IN EXCESSIVE CONCENTRATIONS COPPER SULFATE MAY POISON FISH AND OTHER AQUATIC LIFE. APPROXIMATELY 300 CHEMICALS WERE SCREENED FOR TOXICITY AND THEIR EFFECTIVENESS COMPARED WITH THAT OF COPPER SULFATE. DATA FOR SOME OF THE MOST PROMISING CHEMICAL COMPOUNDS (FOR EXAMPLE, ROSIN AMINE D ACETATE) AS DETERMINED FROM PRELIMINARY SCREENING TESTS AND TOXICITY STUDIES ARE TABULATED. A BIBLIOGRAPHY WITH 418 REFERENCES IS INCLUDED. (WOODARD-USGS)

FIELD 05G, 04A, 10

ACCESSION NO. W71-05083

BACTERICIDAL EFFECTS OF ALGAE ON ENTERIC ORGANISMS,

TEXAS UNIV., AUSTIN. CENTER FOR RESEARCH IN WATER RESOURCES.

ERNST M. DAVIS, AND EARNEST F. GLOYNA.

COPY AVAILABLE FROM GPO SUP DOC AS I67.13/4:18050 DOL 03/70, $1.25; MICROFICHE FROM NTIS AS PB-197 862, $0.95. WATER POLLUTICN CONTROL RESEARCH SERIES 18050 DOL 03/70, MARCH 1970. 132 P, 114 TAB, 9 FIG, 60 REF, 3 APPEND. FWQA PROGRAM 18050 DOL.

DESCRIPTORS:
*ALGAE, *ENTERIC BACTERIA, *BACTERICIDES, GROWTH STAGES, PATHOGENIC BACTERIA, CYANOPHYTA, CHLOROPHYTA, SCENEDESMUS, ALGAL TOXINS, ANTIBIOTICS(PESTICIDES), DATA PROCESSING, COLIFORMS, MICROORGANISMS, HYDROGEN ION CONCENTRATION, WASTE WATER TREATMENT, WATER POLLUTION EFFECTS.

IDENTIFIERS:
*MIXED CULTURES, DIEOFF RATES, AFTERGROWTH.

ABSTRACT:
VARIOUS ENTERIC ORGANISMS WERE EXPOSED TO AXENIC CULTURES OF GREEN AND BLUE-GREEN ALGAE TO DETERMINE WHAT EFFECT THE ALGAE MIGHT HAVE. TESTS WERE EXTENDED TO 90 DAYS TO DETERMINE WHETHER AFTERGROWTH WAS POSSIBLE. MIXED CULTURES OF BOTH GREEN AND BLUE-GREEN ALGAE WERE EXPOSED TO BOTH SINGLE SPECIES AND MIXED CULTURES OF ENTERIC BACTERIA, AT VARYING STAGES OF THE ALGAE GROWTH PERIODS. RESULTS INDICATED THAT MIXED CULTURES DO HAVE A PRONOUNCED BACTERICIDAL EFFECT ON ENTERIC ORGANISMS, WHILE FOR SINGLE ALGAL SPECIES, THE EFFECT IS LESS PRONOUNCED. ALSO, THE BACTERICIDAL EFFECTS ARE MORE SPECIFIC FOR PATHOGENS, WITH VIRTUALLY NO AFTERGROWTH OF PATHOGENS DETECTED. (LOWRY-TEXAS)

FIELD 05C, 05D

ACCESSION NO. W71-05155

MODELING OF THE NITROGEN AND ALGAL CYCLES IN ESTUARIES,

MANHATTAN COLL., BRONX, N.Y. ENVIRCNMENTAL ENGINEERING AND SCIENCE PROGRAM.

R. V. THOMANN, D. J. O'CONNOR, AND D. M. DITORO.

PROCEEDINGS, 5TH INTERNATIONAL WATER POLLUTION RESEARCH CONFERENCE, PAPER III-9, JULY-AUGUST 1970. 14 P, 9 FIG, 1 TAB, 10 REF.

DESCRIPTORS:
*MATHEMATICAL MODELS, *NITROGEN CYCLE, *OXIDATION, AMMONIA, NITRITES, OXYGEN, PHYTOPLANKTON, BIOCHEMICAL OXYGEN CEMAND, DISSOLVED OXYGEN, ESTUARIES, WATER POLLUTION EFFECTS, WATER POLLUTICN SOURCES, NITRIFICATION, ALGAE.

IDENTIFIERS:
ALGAL CYCLE, DELAWARE ESTUARY(DEL-NJ), SAN JOAQUIN DELTA(CALIF), STEADY-STATE MODELS.

ABSTRACT:
TWO MATHEMATICAL MODELS WERE CONSTRUCTED TO ADDRESS THE PROBLEMS OF NITROGEN AND ALGAL CYCLES. THE STEADY-STATE, MULTI-DEMENSIONAL MODEL WAS USED FOR ANALYZING NITRIFICATION AND ALGAL UTILIZATION OF AVAILABLE NITROGEN. THE DYNAMIC MODEL INCORPCRATED THE GROWTH AND DEATH OF PHYTOPLANKTON AND HERBIVOROUS ZOOPLANKTON AND THE UTILIZATION OF INORGANIC NITROGEN. APPLICATION OF THE FIRST MODEL TO THE DELAWARE ESTUARY INDICATED THE RATE CF AMMONIA OXIDATION CF ABOUT 0.1/DAY AT 20C WITH NITRIFICATION INHIBITION AT DO LESS THAN 1-2 MG/L. A LOWER DO RESULTED FROM NITROGEN OXIDATION. THE SAME MODEL APPLIED TO THE POTOMAC ESTUARY INDICATED ALGAL UTILIZATION OF NITRATES AT 0.1/DAY AT 20C. THE DYNAMIC NON-LINEAR MODEL, APPLIED TO SACRAMENTO-SAN JOAQUIN DELTA, (CALIF) ADEQUATELY DESCRIBED THE ALGAL GROWTH AND UTILIZATION OF NUTRIENTS. THE NET ALGAL GROWTH COEFFICIENT RANGED UP TO 0.3/DAY DURING THE SPRING. (WILD-WISCONSIN)

FIELD 02L, 05C

ACCESSION NO. W71-05390

WATER RESOURCES POTENTIAL OF AN URBAN ESTUARY (SAUGUS RIVER, PINES RIVER AND LYNN HARBOR COMPLEX),

NORTHEASTERN UNIV., BOSTON, MASS. DEPT. OF CIVIL ENGINEERING; AND NORTHEASTERN UNIV., BOSTON, MASS. DEPT. OF ENVIRONMENTAL SCIENCE.

JOHN J. COCHRANE, CONSTANTINE J. GREGORY, AND GERALD L. ARONSON.

AVAILABLE FROM NTIS AS PB-197 991, $3.00 IN PAPER COPY, $0.95 IN MICROFICHE. MASSACHUSETTS WATER RESOURCES RESEARCH REPORT, JUNE 1970. 83 P, 29 FIG, 7 TAB, 44 REF. OWRR PROJECT A-028-MASS(1).

DESCRIPTORS:
*ESTUARIES, *WATER POLLUTION EFFECTS, *WATER RESOURCES DEVELOPMENT, *MASSACHUSETTS, *EUTROPHICATION, NUTRIENTS, ALGAE, PHOSPHATES, NITRATES, SURVEYS, PESTICIDES, WASTE DISPOSAL, PATH OF POLLUTANTS, WATER QUALITY.

IDENTIFIERS:
*BOSTON(MASS), LYNN HARBOR(MASS).

ABSTRACT:
THE WATER RESOURCES POTENTIAL OF THE URBAN ESTUARINE COMPLEX COMPRISED OF THE SAUGUS AND PINES RIVERS AND LYNN HARBOR, MASSACHUSETTS, WAS EVALUATED. IMPAIRMENT OF RECREATIONAL USAGE IS CAUSED BY EUTROPHICATION RESULTING FROM RAW WASTE DISCHARGES AND NUTRIENT RESERVES IN SEDIMENTS. A LABORATORY STUDY OF THE GROWTH OF MARINE ALGAE, ULVA LATISSIMA, MADE IN FLOWING SEA WATER, INDICATES AN OPTIMUM GROWTH AT NITROGEN-PHOSPHOROUS RATIOS OF BETWEEN 40 AND 60 TO 1, AT A PHOSPHATE CONCENTRATION OF 120 MICROGRAMS PER LITER. AVERAGE VALUES FOR NUTRIENTS IN EUTROPHIC AREAS WERE 268 MICROGRAMS/LITER TOTAL ORTHOPHOSPHATE AND 0.513 MG/L NITRATES. SEDIMENTS FROM THESE AREAS AVERAGED 5.98% VOLATILE SOLIDS, 0.532 MG/L EXTRACTED ORTHOPHOSPHATES, AND 2.177 MG/L TOTAL KJELDAHL NITROGEN. IN CONTRAST, OLIGOTROPHIC AREAS, INCLUDING PARTS OF THE PINES RIVER, HAD AVERAGED VALUES OF 164 MICROGRAMS/L TOTAL ORTHOPHOSPHATE, 0.175 MG/L NITRATES, 0.79% VOLATILE SOLIDS, 0.349 MG/L EXTRACTED ORTHOPHOSPHATES AND 0.294 MG/L TOTAL KJELDAHL NITROGEN. (KNAPP-USGS)

FIELD 05C, 02L

ACCESSION NO. W71-05553

SEDIMENT-WATER NUTRIENT RELATIONSHIPS - PARTS 1 AND 2,

FEDERAL WATER QUALITY ADMINISTRATION, CINCINNATI. FIELD INVESTIGATIONS BRANCH.

GERALD D. MCKEE, LOYS P. PARRISH, CARL R. HIRTH, KENNETH M. MACKENTHUM, AND LOWELL E. LEUP.

WATER AND SEWAGE WORKS, VOL 117, NO 6, P 203-206, JUNE 1970. 23 REF; NO 7, P 246-250, JULY 1970. 4 FIG, 27 REF.

DESCRIPTORS:
*NUTRIENTS, *SEDIMENTS, *WATER ANALYSIS, LAKE MORPHOLOGY, SEDIMENTATION, PLANKTON, DIEL MIGRATION, PROTOZOA, HYDROGEN ION CONCENTRATION, WATER TEMPERATURE, BACTERIA, ALGAE, ALKALINITY, IRON, PHOSPHATE, PHOSPHORUS, CARBON, NITROGEN, CURRENTS(WATER), WINDS, NEVADA, CALIFORNIA, NEW YORK, OREGON, WISCONSIN, MAINE.

IDENTIFIERS:
FECAL MATERIAL, LAKE TAHOE, LAKE SEBASTICOCK(MAINE), FAYETTEVILLE GREEN LAKE(NEW YORK), UPPER KLAMATH LAKE(OREGON), LAKE MENDOTA(WISCONSIN).

ABSTRACT:
A THOROUGH REVIEW OF THE MECHANISMS OF NUTRIENT DEPOSITION AND EXCHANGE IN AQUATIC ENVIRONMENTS LED TO THE FOLLOWING CONCLUSIONS: (1) A PORTION OF THE NUTRIENTS, SOLUBLE AND INSOLUBLE, IN A WATER BODY ARE ULTIMATELY DESTINED TO BECOME PART OF THE SEDIMENTS. (2) ORGANISMS ARE PRIMARY CONCENTRATORS OF DISSOLVED NUTRIENTS. (3) THE INITIAL AREAS OF DEPOSITION MAY BE ONLY TEMPORARY, THE NUTRIENT-CONTAINING SEDIMENTS MAY BE RE-DISSOLVED OR MAY BE PHYSICALLY TRANSPORTED. (4) WIND-INDUCED CURRENTS ARE A MAJOR FACTOR THAT DETERMINES THE RATE AND AREA OF FINAL DEPOSITION AND THE CONTACT TIME BETWEEN SUSPENDED SEDIMENTS AND WATER. (5) MORPHOLOGY OF THE WATER BODY AFFECTS THE ULTIMATE AREA OF DEPOSITION. (6) AQUATIC ORGANISMS CONVERT INORGANIC NUTRIENTS TO ORGANIC NUTRIENTS AND VICE VERSA. (7) SUSPENSION OF SEDIMENTS INCREASES THEIR EFFECT ON THE OVERLYING WATER. (8) THE SEDIMENTS ACT AS RESERVOIRS OF NUTRIENTS FOR THE OVERLYING WATER. (9) THE RAPIDITY OF SEDIMENT BUILD-UP WILL AFFECT THE DEGREE OF INFLUENCE ON THE OVERLYING WATER. (LITTLE-BATTELLE)

FIELD 05B, 05C, 02H

ACCESSION NO. W71-05626

SOME ALGAE OF THE UPPER CUYAHOGA RIVER SYSTEM IN OHIO,

KENT STATE UNIV., OHIO. DEPT. OF BIOLOGICAL SCIENCES.

RUSSELL G. RHODES, AND ANTHONY J. TERZIS.

THE OHIO JOURNAL OF SCIENCE, VOL 70, NO 5, P 295-299, SEPTEMBER 1970. 1 FIG, 1 TAB, 16 REF.

DESCRIPTORS:
*ALGAE, OHIO, CHLOROPHYTA, CHRYSOPHYTA, RHODOPHYTA, CYANOPHYTA, EUGLENOPHYTA.

IDENTIFIERS:
*CUYAHOGA RIVER(OHIO), CLADOPHORA, APHANOCHAETE, RHIZOCLONIUM, VAUCHERIA, TRIBONEMA, OSCILLATORIA.

ABSTRACT:
SIXTY-FOUR SPECIES OF ALGAE WERE FOUND IN A QUALITATIVE SURVEY MADE DURING JUNE AND SEPTEMBER, 1967, IN THREE TRIBUTARIES OF THE CUYAHOGA RIVER IN GEAUGA COUNTY: WEST BRANCH, EAST BRANCH, AND TARE CREEK. EIGHT SPECIES WHICH WERE COLLECTED AT THE MAJORITY OF THE 14 STATIONS SAMPLED ARE CLADOPHORA GLOMERATA, APHANOCHAETE REPENS, RHIZOCLONIUM HIEROGLYPHICUM, EUGLENA GRACILIS, VAUCHERIA SESSILIS, TRIBONEMA BOMBYCINUM, OSCILLATORIA NIGRA, AND O. LIMOSA. (LITTLE-BATTELLE)

FIELD 05A, 05C

ACCESSION NO. W71-05629

BANGIA ATROPURPUREA (ROTH) A. IN WESTERN LAKE ERIE,

OHIO STATE UNIV., COLUMBUS. DEPT. OF BOTANY.

JACK KISHLER, AND CLARENCE E. TAFT.

THE OHIO JOURNAL OF SCIENCE, VOL 70, NO 1, P 56-57, JANUARY 1970. 1 FIG, 5 REF.

DESCRIPTORS:
*RHODOPHYTA, *LAKE ERIE, OHIO, ALGAE.

ABSTRACT:
BANGIA ATROPURPUREA WAS COLLECTED 2 MARCH 1969 AT THE STATE HIGHWAY PARK ON THE EAST SHORE OF MARBLEHEAD PENINSULA, OTTAWA COUNTY, OHIO. THIS IS THE FIRST RECORD OF BANGIA IN WESTERN LAKE ERIE. IT APPEARED AS LAX RED-PURPLE TUFTS, 3/4 INCH LONG, COVERING A FLAGSTONE ON THE SHORELINE WHERE THERE WAS AN OPENING IN THE ICE. (LITTLE-BATTELLE)

FIELD 05A, 05C, 02H

ACCESSION NO. W71-05630

TOXICITY OF ZINC, COPPER AND LEAD TO CHLOROPHYTA FROM FLOWING WATERS,

DURHAM UNIV. (ENGLAND). DEPT. OF BOTANY.

B. A. WHITTON.

ARCHIV FUR MIKROBIOLOGIE, VOL 72, P 353-360, 1970. 1 TAB, 7 REF.

DESCRIPTORS:
*INDICATORS, *ALGAE, *HEAVY METALS, *TOXICITY, COPPER, CHLOROPHYTA, WATER POLLUTION, NUTRIENTS, RIVERS, LAKES, FARM PONDS.

IDENTIFIERS:
*ZINC, *LEAD, STIGEOCLONIUM TENUE, CLADOPHORA GLOMERATA, ULOTRICHALES, ZYGNEMALES, OEDOGONIUM, MOUGOTIA, SPOROTETRAS PYRIFORMIS, GONGROSIRA, MICROSPORA, SPIROGYRA, ULOTHRIX.

ABSTRACT:
THE BIOLOGICAL AND ECONOMIC INTEREST OF HEAVY METALS IN THE RELATION TO FRESHWATER ALGAE LED TO A SURVEY OF TOXICITY OF ZINC, COPPER, AND LEAD TO CHLOROPHYTA IN FLOWING WATERS. TWENTY POPULATIONS EACH OF STIGEOCLONIUM TENUE AND CLADOPHORA GLOMERATA WERE TESTED FOR VARIATIONS IN METAL RESISTANCE. THE ONLY INDICATION FOUND WAS A SLIGHT INCREASE IN RESISTANCE TO ZINC OF ONE STIGEOCLONIUM POPULATION FROM A METAL-POLLUTED STREAM. THIRTY-FIVE OTHER ALGAL POPULATIONS, REPRESENTING ABOUT 25 SPECIES, WERE ALSO TESTED. COMPARISON OF THESE POPULATIONS SHOWED THAT IN THE TEST MEDIUM USED CLADOPHORA GLOMERATA WAS THE MOST OR ALMOST THE MOST SENSITIVE TO ALL THREE METALS. IT WAS ABSENT FROM STREAMS KNOWN TO BE POLLUTED BY LEAD OR LEAD AND ZINC. THE ULOTRICHALES AND MOST OF THE ZYGNEMALES WERE RELATIVELY RESISTANT TO ZINC, WHILE ALL THE OEDOGONIUM SPECIES TAKEN FROM THE FIELD WERE SENSITIVE TO ZINC. HOWEVER, AN OEDOGONIUM POPULATION HIGHLY ZINC RESISTANT WAS OBTAINED FROM A ZINC ENRICHED LABORATORY TANK. BECAUSE THE MEDIUM USED IN THE PRESENT EXPERIMENTS IS RELATIVELY RICH IN NUTRIENTS SUCH AS PHOSPHATES IT MAY WELL MASK EFFECTS THAT WOULD SHOW IN A LESS FAVORABLE ENVIRONMENT. (JONES-WISCONSIN)

FIELD 05C

ACCESSION NO. W71-05991

EVALUATION OF SOME STABILIZATION PONDS IN INDIA,

CENTRAL PUBLIC HEALTH ENGINEERING RESEARCH INST., NAGPUR (INDIA).

R. H. SIDDIQI, AND B. K. HANDA.

JOURNAL OF THE SANITARY ENGINEERING DIVISION, AMERICAN SOCIETY OF CIVIL ENGINEERS, VOL 97, NO SA1, P 91-100, FEBRUARY 1971. 7 FIG, 1 TAB, 11 REF.

DESCRIPTORS:
*OXIDATION LAGOONS, *AEROBIC TREATMENT, *ANAEROBIC CONDITIONS, PONDS, CLIMATIC ZONES, TEMPERATURE, TURBULENCE, OXYGENATION, ALGAE, PHOTOSYNTHESIS, MUNICIPAL WASTES, BIOCHEMICAL OXYGEN DEMAND, WASTE WATER TREATMENT, MICROORGANISMS.

IDENTIFIERS:
FACULTATIVE MICROORGANISMS, INDIA.

ABSTRACT:
CLIMATIC CONDITIONS IN INDIA ARE FAVORABLE TO THE OPERATION OF ENGINEERED WASTE STABILIZATION PONDS. CONSEQUENTLY, THE TREATMENT OF WASTES IN SUCH PONDS IS ECONOMICALLY COMPETITIVE WITH CONVENTIONAL BIOLOGICAL TREATMENT. DATA COLLECTED FROM SEVERAL INSTALLATIONS IN INDIA WERE ANALYZED TO DETERMINE WHAT USEFUL PARAMETERS OF OPERATION COULD BE IDENTIFIED. POND LOADING WAS FOUND TO BE BEST EXPRESSED BY A LOAD FACTOR, L SUB F, WHICH IS THE RATIO OF BOD LOAD TO ALGAL PRODUCED OXYGEN. FOR L SUB F BETWEEN .44 AND 8.0, THE PERFORMANCE WAS DETERMINED FROM THE FOLLOWING EQUATION: E = 100 OVER (1 + 0.188 L SUB F OR EXPLANENTIAL TO 0.48). FROM THE PRECEDING INVESTIGATIONS, IT WAS DETERMINED THAT THE MAJORITY OF ORGANIC MATTER DESTROYED IS DESTROYED ANAEROBICALLY. THEREFORE, PONDS OF DEPTH GREATER THAN 5 FEET ARE MORE EFFICIENT IN THEIR OPERATION SINCE THERE IS LESS TURBULENCE AND LESS CHANGE OF THE ANAEROBES BEING EXPOSED TO OXYGEN. IT WAS ALSO DETERMINED THAT SINGLE CELL REACTORS, OR THE FIRST CELL OF A MULTI-CELL ARRANGEMENT MAINTAIN A HIGHER DESTRUCTION RATE CONSTANT, WITH THE RATE OF BOD REDUCTION BEING DESCRIBED AS A FIRST ORDER EQUATION. THE RATE DROPS OFF WITH EACH SUCCESSIVE CELL IN A MULTI-CELL ARRANGEMENT. (LOWRY-TEXAS)

FIELD 05D

ACCESSION NO. W71-06033

ALGAECIDAL EVALUATION AND ENVIRONMENTAL STUDY OF MAT PRODUCING BLUEGREEN ALGAE,

BUREAU OF RECLAMATION, DENVER, COLO. OFFICE OF CHIEF ENGINEER.

NAMAN E. OTTO.

AVAILABLE FROM NTIS AS PB-194 808, $3.00 IN PAPER COPY, $0.95 IN MICROFICHE. BUR OF RECLAM REPORT REC-OCE-70-25, JUNE 1970. 27 P, 13 FIG, 7 TAB, 27 REF.

DESCRIPTORS:
*ALGAE, *AQUATIC WEEDS, *WEED CONTROL, GROWTH, *CULTURES, *ALGICIDES, BIBLIOGRAPHIES, *ECOLOGY, AQUATIC LIFE, IRRIGATION SYSTEMS, WATER ANALYSIS, TEMPERATURE, LIMNOLOGY, CYANOPHYTA, IRRIGATION CANALS, ALGAL CONTROL.

IDENTIFIERS:
*ALGAE CULTURE TECHNIQUE, AQUATIC WEED STUDY FACILITY, PRODUCT EVALUATION, BLUE-GREEN ALGAE.

ABSTRACT:
ALGAECIDAL TESTS OF 74 COMPOUNDS SHOWED THAT ONLY 9 WERE MORE ACTIVE THAN COPPER SULFATE, AND ONE PROVIDED TOTAL CONTROL OF THE BLUE-GREEN ALGAE MAT COLONIES. RESULTS OF PRELIMINARY ALGAECIDAL FIELD TESTS WITH COPPER SULFATE AND ADMIXTURES OF ENHANCE ACTIVITY ARE GIVEN. OBSERVATIONS OF ALGAECIDAL TREATMENTS ON AN IRRIGATION CANAL SUGGEST THAT COPPER SULFATE APPLICATIONS OF 0.48 LB/CFS APPLIED EVERY 2 WK SUPPRESSED BLUE-GREEN MAT GROWTH. WATER TEMPERATURE APPEARS CRITICAL TO TREATMENT SUCCESS. RESULTS OF STUDIES CONDUCTED IN 1967-1969 TO DETERMINE ENVIRONMENTAL PARAMETERS OF IRRIGATION CANALS ARE PRESENTED. DATA SHOW THAT MAT-TYPE, BLUE-GREEN ALGAE GROW IN A WIDE RANGE OF WATER QUALITY AND TEMPERATURE CONDITIONS AND DO NOT REQUIRE ENRICHED CONDITIONS FOR GROWTH. COLORATION SUBSTANCES IN THE WATER MAY INFLUENCE THE AVAILABILITY OF MICRONUTRIENT METALS AND MAY STIMULATE DIRECTLY BLUE-GREEN ALGAE GROWTH.

FIELD 04A, 05C

ACCESSION NO. W71-06102

PHYSICAL PROPERTIES AND PROCESSING CHARACTERISTICS OF MACROPHYTES AS RELATED TO MECHANICAL HARVESTING,

WISCONSIN UNIV., MADISON. WATER RESOURCES CENTER.

H. D. BRUHN, D. F. LIVERMORE, AND F. O. ABOABA.

AVAILABLE FROM NTIS AS PB-198 129, $3.00 IN PAPER COPY, $0.95 IN MICROFICHE. REPRINT, AMERICAN SOCIETY OF AGRICULTURAL ENGINEERS, ST JOSEPH, MICHIGAN 49085, PAPER 70-582, 1970. 17 P, 12 FIG, 30 REF. OWRR PROJECT B-018-WIS(2).

DESCRIPTORS:
*HARVESTING OF ALGAE, *AQUATIC WEED CONTROL, *MECHANICAL CONTROL, *DEWATERING.

IDENTIFIERS:
FARM MACHINERY, INDUSTRIAL MACHINERY, UTILIZATION, MODIFICATION.

ABSTRACT:
SELECTIVE MECHANICAL HARVESTING OF UNDESIRABLE AQUATIC VEGETATION APPEARS TO BE A FEASIBLE AND ECOLOGICALLY SOUND APPROACH TO CONTROL IN RECREATIONAL WATERS. THE TRANSPORTATION FROM THE HARVEST AREA TO THE DISPOSAL SITE IS A MAJOR EXPENSE IN THE OVERALL HARVEST OPERATION. SINCE AQUATIC VEGETATION IS APPROXIMATELY 90 PER CENT WATER PROCESSING AND DEWATERING THIS MATERIAL AS A PART OF THE INITIAL HARVESTING OPERATION GREATLY FACILITATES ITS TRANSPORTATION AND ULTIMATE DISPOSAL OR UTILIZATION. INTENSIVE MECHANICAL DEWATERING BASED ON PROCEDURES DEVELOPED IN THIS RESEARCH RESULTS IN REDUCING THE FIBROUS FRACTION OF AQUATIC VEGETATION TO 16 PER CENT OF ITS ORIGINAL VOLUME AND 32 PER CENT OF ITS ORIGINAL WEIGHT, WHILE REMOVING FROM THE HARVESTED AREA 90 PER CENT OF THE ORIGINAL DRY MATTER, 85 PER CENT OF THE PROTEIN, 60 PER CENT OF THE POTASSIUM, AND 80 PER CENT OF THE PHOSPHORUS PRESENT IN THE GROWING VEGETATION. THE REDUCTION IN FIBER LENGTH RESULTING FROM THE DEWATERING PROCESS INCREASES THE EASE WITH WHICH THE VEGETATIVE MATERIAL CAN BE TRANSFERRED THROUGH CONVEYING SYSTEMS THUS FURTHER FACILITATING HANDLING AND DISPOSAL. THE MAJOR PORTION OF THE EQUIPMENT REQUIRED FOR THE PREPROCESSING AND DEWATERING CAN BE DEVELOPED BY UTILIZATION AND MODIFICATION OF COMPONENTS OF ALL READILY AVAILABLE FARM AND INDUSTRIAL MACHINERY.

FIELD 04A, 05G

ACCESSION NO. W71-06188

ALGICIDES,

WISCONSIN UNIV., MADISON. WATER RESOURCES CENTER.

GEORGE P. FITZGERALD.

AVAILABLE FROM NTIS AS PB-198 130, $3.00 IN PAPER COPY, $0.95 IN MICROFICHE. WISCONSIN UNIV, MADISON, WATER RESOURCES CENTER, EUTROPHICATION INFORMATION PROGRAM, LITERATURE REVIEW NO 2, 1971. 50 P, 1 FIG, 9 TAB, 153 REF. OWRR PROJECT W-117NO 1614)(3).

DESCRIPTORS:
*ALGICIDES, *ALGAE, *ALGAL CONTROL, TESTING, LABORATORY TESTS, CHEMICALS, BACTERIOCIDES, CULTURES, TEST PROCEDURES, PESTICIDE TOXICITY, RESISTANCE, LAKES, RESERVOIRS, SWIMMING POOLS, CHELATION, APPLICATION METHODS, COOLING TOWERS, WATER POLLUTION EFFECTS.

IDENTIFIERS:
DETOXIFICATION, SYNERGISM, FISH PONDS, ALGISTATIC ACTIVITY, ALGICIDAL ACTIVITY.

ABSTRACT:
AN EVALUATION OF ALGICIDES--WHAT THEY ARE, THEIR USAGE, HOW THEY ARE TESTED, THEIR EFFICIENCY, AND METHODS OF APPLICATION--IS PRESENTED. DETAILED ARE APPROPRIATE TEST ORGANISMS, CULTURE AND TEST MEDIA, THE IMPORTANCE OF ALGICIDAL VERSUS ALGESTATIC EFFECTS, AND EFFECTIVE CHEMICAL CONCENTRATIONS REQUIRED TO ACHIEVE POTENTIAL TOXICITY. CHARACTERISTICS OF AN ALGAL PROBLEM, TOGETHER WITH ITS ENVIRONMENT, INFLUENCE THE METHODOLOGY OF APPLICATION PROCESSES AS WELL AS DETOXIFICATION, RESISTANCE, AND SYNERGISM. ALGAL PROBLEMS OF WATER SUPPLY RESERVOIRS AND RECREATIONAL LAKES, FISH PONDS, SWIMMING POOLS, AND COOLING TOWERS ARE SPECIFICALLY DISCUSSED WITH SUGGESTIONS FOR EFFECTIVE APPLICATION OF TOXICANTS FOR THEIR CONTROL. (AUEN-WISCONSIN)

FIELD 05C, 05F

ACCESSION NO. W71-06189

DEVELOPMENT OF PHOSPHATE-FREE HOME LAUNDRY DETERGENTS,

IIT RESEARCH INST., CHICAGO, ILL.

KARL A. ROSEMAN, AND WARNER M. LINFIELD.

COPY AVAILABLE FROM GPO SUP DOC AS EPA-WQO REPORT NO 16080 DVF, DECEMBER 1970, $1.00; MICROFICHE FROM NTIS AS PB-198 222, $0.95. 103 P, 4 TAB, 10 REF. EPA-WQO CONTRACT 14-12-575.

DESCRIPTORS:
*DETERGENTS, *ALGAL CONTROL, *FORMULATION, *SURFACTANTS, *EUTROPHICATION, LINEAR ALKYLATE SULFONATES, CHELATION, PHOSPHATES, ORGANIC COMPOUNDS, WATER POLLUTION CONTROL.

IDENTIFIERS:
PHOSPHATE-FREE DETERGENTS, TRISODIUM NITRILOTRIACETATE, SODIUM CITRATE.

ABSTRACT:
BASIC STUDIES WERE PERFORMED TOWARDS THE DEVELOPMENT OF PHOSPHATE-FREE HOME LAUNDRY DETERGENTS. FIVE SURFACTANTS WERE SYNTHESIZED WITH THE IDEA THAT THEY MIGHT POSSESS HARD ION CHELATING PROPERTIES. THE CLEANING ABILITIES OF THESE MATERIALS WERE COMPARED TO THE WIDELY USED LINEAR ALKYLBENZENE SULFONATE AS INCORPORATED INTO THE SAME FORMULATIONS. THE DETERGENT COMPOSITIONS CONTAINED 2% CARBOXYMETHYLCELLULOSE AND THE SILICATE CONTENT WAS VARIED. SODIUM ACETATE AND SODIUM CARBONATE WERE INVESTIGATED AS POSSIBLE RESERVOIRS OF ALKALINITY. SURFACTANT COMPATIBILITY WITH SODIUM CHLORIDE AND SODIUM SULFATE WAS EXAMINED. OTHER ADDITIVES INCLUDED TRISODIUM NITRILOTRIACETATE AND SODIUM CITRATE AT MODERATE LEVELS. FIFTEEN DETERGENT FORMULATIONS WERE SCREENED AND THE RESULTS LEAVE LITTLE DOUBT THAT ACCEPTABLE PHOSPHATE-FREE HOME LAUNDRY DETERGENTS CAN BE DEVELOPED.

FIELD 05G

ACCESSION NO. W71-06247

SOURCES OF NITROGEN IN WATER SUPPLIES,

GEOLOGICAL SURVEY, DENVER, COLO.

MARVIN C. GOLDBERG.

AGRICULTURAL PRACTICES AND WATER QUALITY, IOWA STATE UNIVERSITY PRESS, AMES, IOWA, 1970, CHAPTER 7, P 94-124. 4 FIG, 8 TAB, 72 REF.

DESCRIPTORS:
*NITROGEN, *NITRATES, GROUNDWATER, AMMONIA, PRECIPITATION, SEDIMENTS, DENITRIFICATION, RUNOFF, UREAS, FERTILIZERS, DRAINAGE WATER, IRRIGATION, RETURN FLOW, WATER SUPPLY, LIVESTOCK, SEWAGE, INFILTRATION, INDUSTRIAL WASTES, ALGAE, PONDS, FARM WASTES.

IDENTIFIERS:
*SURFACE WATERS, GEOLOGICAL SOURCES, MINERALIZATION, NITROGEN SOURCES, WELL WATER, FEEDLOTS.

ABSTRACT:
WATER SUPPLIES CAN BE CATEGORIZED AS SURFACE WATERS OR GROUNDWATERS. THIS PAPER EXAMINES REPRESENTATIVE STUDIES OF NITRATE ENTRANCE TO BOTH TYPES OF WATER SUPPLIES, WITH SUMMARIES OF SOME OF THE MANY LABORATORY AND FIELD STUDIES DESCRIBED IN THE CURRENT LITERATURE. SOME OF THE SOURCES OF NITROGEN ENTRANCE TO WATER SUPPLIES INCLUDE ATMOSPHERIC, GEOLOGIC, RURAL AND URBAN RUNOFF, SEWAGE, IRRIGATION, ANIMAL WASTES, AND INDUSTRIAL WASTES AMONG MANY OTHERS. SOURCES OF MAJOR IMPORTANCE TO BOTH SURFACE AND GROUNDWATER SUPPLIES ARE POINTED OUT AND FIELD OR LABORATORY STUDIES ARE REPORTED. (WHITE-IOWA STATE)

FIELD 05B

ACCESSION NO. W71-06435

EFFECTS OF AGRICULTURAL POLLUTION ON EUTROPHICATION,

WISCONSIN UNIV., MADISON. DEPT. OF SANITARY ENGINEERING, AND WISCONSIN UNIV., MADISON. DEPT. OF WATER CHEMISTRY.

D. E. ARMSTRONG, AND G. A. ROHLICH.

AGRICULTURAL PRACTICES AND WATER QUALITY, IOWA STATE UNIVERSITY PRESS, AMES, IOWA, 1970, CHAPTER 23, P 314-330. 14 TAB, 2 FIG, 26 REF.

DESCRIPTORS:
*EUTROPHICATION, *NITROGEN, *PHOSPHORUS, NUTRIENTS, ALGAE, NITRATES, SURFACE RUNOFF, BASE FLOW, PERCOLATION, LEACHING, DRAINAGE, FARM WASTES, SOIL MANAGEMENT, GROUNDWATER, WISCONSIN, WATER SUPPLY.

IDENTIFIERS:
*AGRICULTURAL DRAINAGE, LAKE METABOLISM, MOBILITY, PARTICULATE FORM, FEEDLOTS, NUTRIENT SOURCES.

ABSTRACT:
THE PAPER DISCUSSES NITROGEN AND PHOSPHORUS TRANSPORT IN AGRICULTURAL DRAINAGE SINCE THESE ARE THE MOST IMPORTANT NTURIENTS INVOLVED IN EUTROPHICATION. IT IS GENERALLY EXPECTED THAT INORGANIC NITROGEN IS TRANSPORTED MAINLY AS NITRATE BY PERCOLATING WATER, ALTHOUGH THE AMOUNTS OF AMMONIUM AND NITRATE CARRIED IN RUNOFF WATERS MAY BE HIGHLY SIGNIFICANT IN TERMS OF THE RECEIVING WATER. SIMILARLY, THE LARGEST AMOUNT OF PHOSPHORUS IS LIKELY TRANSPORTED IN PARTICULATE FORM IN RUNOFF WATERS, BUT THE AMOUNT OF DISSOLVED PHOSPHORUS IN RUNOFF WATER MAY BE OF EQUAL OR GREATER IMPORTANCE EVEN THOUGH LOWER IN QUANTITY. THE CONTRIBUTION OF AGRICULTURAL DRAINAGE TO THE NITROGEN AND PHOSPHORUS STATUS OF WATERS IS NEXT EXAMINED. THE DATA PRESENTED SUGGEST THAT AGRICULTURAL LAND IS AN IMPORTANT CONTRIBUTOR OF NITROGEN AND PHOSPHORUS TO WATER. ABOUT 60% OF THE NITROGEN AND 42% OF THE PHOSPHORUS WERE ESTIMATED TO COME FROM AGRICULTURAL LAND. NUTRIENT BUDGET ESTIMATIONS WERE BASED ON DATA OBTAINED ON A SMALL SCALE AND EXTRAPOLATED AND THUS HAVE A LOW RELIABILITY. NUTRIENT SOURCES ARE NUMEROUS AND GENERALIZATIONS AS TO WHICH SOURCE IS THE MOST IMPORTANT CANNOT BE MADE. THE CONTRIBUTION OF AGRICULTURE SHOULD BE REDUCED BY IMPROVED AND MORE EFFICIENT AGRICULTURAL MANAGEMENT PRACTICES. (WHITE-IOWA STATE)

FIELD 05C, 02H

ACCESSION NO. W71-06443

EFFECTS OF AGRICULTURAL POLLUTANTS ON RECREATIONAL USES OF SURFACE WATERS,

MISSOURI UNIV., COLUMBIA. DEPT. OF ZOOLOGY; AND MISSOURI DEPT. OF CONSERVATION, COLUMBIA.

ROBERT S. CAMPBELL, AND JAMES R. WHITLEY.

AGRICULTURAL PRACTICES AND WATER QUALITY, IOWA STATE UNIVERSITY PRESS, AMES, IOWA, 1970, CHAPTER 24, P 331-343. 3 TAB, 1 FIG, 43 REF.

DESCRIPTORS:
*POLLUTANTS, *RECREATION, LAKES, STREAMS, ALGAE, PESTICIDES, DDT, RETURN FLOW, SEDIMENTS, SOIL EROSION, FERTILIZERS, FARM WASTES, FISH, FISH EGGS, TURBIDITY, NUTRIENTS, DISSOLVED OXYGEN, NITROGEN, PHOSPHORUS, WATER QUALITY ACT, WATER QUALITY.

ABSTRACT:
UNQUESTIONABLY MANY AGRICULTURAL POLLUTANTS AFFECT RECREATION THROUGH ALTERATION OF WATER QUALITY AND DEGRADATION OF FISH AND AQUATIC LIFE. THE MORE SERIOUS POLLUTING AGENTS ARE ERODED SOIL, AGRICULTURAL FERTILIZERS, ANIMAL WASTES, AND PESTICIDES. WHILE THE PROBLEMS RELATING TO AGRICULTURAL POLLUTION ARE COMPLEX, AND THE SOLUTIONS WILL NOT EASILY BE ATTAINED, IT SEEMS REASONABLE THAT IN MANY INSTANCES ALTERNATIVE PROCEDURES CAN BE DEVELOPED. POLLUTION CONTROL MEASURES ARE AVAILABLE WHICH WILL ALLOW CONTINUATION OF AGRICULTURAL PRODUCTION AND ENHANCE AND PROTECT WATER QUALITY AND RECREATION. WHILE THESE PROCEDURES MAY BE COSTLY TO APPLY, THE EXPENDITURE SHOULD BE JUDGED IN LIGHT OF ITS CONTRIBUTION TOWARD THE PRESERVATION OF MAN'S ENVIRONMENT. ESPECIALLY IN THE INSTANCE OF PESTICIDE USE, PROTECTION OF WATER QUALITY MAY BE REQUISITE TO PROTECTION OF THE HEALTH OF MAN FROM UNKNOWN LONG-TERM EFFECTS OF PESTICIDES. REDUCTION AND CONTROL OF AGRICULTURAL POLLUTANTS ARE ESSENTIAL TO DEVELOP AND MAINTAIN A HIGH QUALITY ENVIRONMENT. QUALITY OF LIFE AND QUALITY OF ENVIRONMENT ARE SYNONYMOUS. (WHITE-IOWA STATE)

FIELD 05C

ACCESSION NO. W71-06444

EFFECTS OF SURFACE RUNOFF ON THE FEASIBILITY OF MUNICIPAL ADVANCED WASTE TREATMENT,

IOWA STATE UNIV., AMES. DEPT. OF CIVIL ENGINEERING.

ROBERT E. BAUMANN, AND SHELDON KELMAN.

AGRICULTURAL PRACTICES AND WATER QUALITY, IOWA STATE UNIVERSITY PRESS, AMES, IOWA, 1970, CHAPTER 25, P 344-362. 7 FIG, 1 TAB, 19 REF.

DESCRIPTORS:
*SURFACE RUNOFF, *SEWAGE TREATMENT, TERTIARY TREATMENT, BIOCHEMICAL OXYGEN DEMAND, INDUSTRIAL WASTES, FARM WASTES, POLLUTANTS, FERTILIZERS, NITROGEN, NITRATES, PHOSPHORUS, PHOSPHATES, ALGAE, EUTROPHICATION, WATER QUALITY, RIVERS, IOWA, CORN, DISCHARGE, CHLOROPHYLL.

IDENTIFIERS:
INDUSTRIAL WATER POLLUTION, DES MOINES RIVER, PACKING PLANTS.

ABSTRACT:
THE PROTECTION OF THE QUALITY OF WATER IN IOWA STREAMS REQUIRES THAT ATTENTION BE DIRECTED AT THE VARIOUS CONTRIBUTORS OF THE SIGNIFICANT POLLUTANTS. ATTENTION IS CURRENTLY BEING DIRECTED AT MUNICIPAL AND INDUSTRIAL WASTES DISCHARGES, SINCE THESE ENTER STREAMS THROUGH A POINT SOURCE AND ARE EASILY CONTROLLED. ALL SUCH WASTES MUST BE GIVEN SECONDARY TREATMENT PRIOR TO DISCHARGE TO IOWA'S STREAMS. AS MORE STRINGENT TREATMENT REQUIREMENTS ARE DEMANDED IN THE FUTURE, THERE IS SOME QUESTION AS TO WHETHER NUTRIENT REMOVALS FROM MUNICIPAL AND INDUSTRIAL WASTES WILL BE SUFFICIENT TO PROTECT THE STREAM. THIS STUDY INDICATED THAT DURING PERIODS OF DRY WEATHER WHEN LIGHT AND TURBIDITY CONDITIONS ARE FAVORABLE FOR PHYTOPLANKTON GROWTH, THE PRINCIPAL SOURCE OF THE N AND P REQUIRED TO SUPPORT SUCH GROWTH IS DERIVED FROM MUNICIPAL AND INDUSTRIAL WASTE WATER DISCHARGES. REMOVAL OF N AND P FROM SUCH WASTE WATER DISCHARGES WILL HELP REDUCE PHYTOPLANKTON GROWTH. IN PERIODS OF HIGH STREAM FLOW, WHEN TURBIDITY LEVELS ARE HIGH ENOUGH TO BE UNFAVORABLE TO PHYTOPLANKTON GROWTH, RUNOFF FROM URBAN AND RURAL LANDS AND CHANNEL EROSION ARE PROBABLY THE PRINCIPAL CONTRIBUTORS OF N AND P TO THE STREAM. REMOVAL OF N AND P FROM MUNICIPAL AND INDUSTRIAL WASTES DURING THESE PERIODS WILL NOT REDUCE NUTRIENT LEVELS SIGNIFICANTLY. UNDER THE LATTER CONDITIONS, TERTIARY TREATMENT OF MUNICIPAL AND INDUSTRIAL WASTES WILL BE OF LESS BENEFIT UNTIL RUNOFF CONTRIBUTIONS OF N AND P ARE ALSO CONTROLLED.
(WHITE-IOWA STATE)

FIELD 05D, 05B

ACCESSION NO. W71-06445

DYNAMIC BEHAVIOUR OF OXIDATICN PONDS,

CAPE TOWN UNIV. (SOUTH AFRICA).

G. V. R. MARAIS.

2ND INTERNATIONAL SYMPOSIUM FOR WASTE TREATMENT LAGOONS, JUNE 23-25, 1970, KANSAS CITY, MISSOURI, P 15-46. 28 FIG, 14 REF, APPEND.

DESCRIPTORS:
*OXIDATION LAGOONS, *MATHEMATICAL MODELS, *COMPUTER SIMULATION, MIXING, TURBULENCE, TEMPERATURE, STRATIFICATION, SOLAR RADIATION, ALGAE, PHOTOSYNTHESIS, ANAEROBIC CONDITIONS, AEROBIC CONCITIONS, SLUDGE, WASTE WATER TREATMENT, BIODEGRADATION, BIOCHEMICAL OXYGEN DEMAND.

ABSTRACT:
A COMPUTER MODEL OF OXIDATION PONDS STIMULATED INTEREST IN SEVERAL PARAMETERS WHICH WERE NOT PREVIOUSLY CONSIDERED TC BE IMPORTANT. DATA WAS COLLECTED FROM THE MATERS NORTH PCND IN LUSAKA, ZAMBIA, AND THEN ANALYZED AND USED TO DEVELOP THE COMPUTER MODEL. THIS MODEL CLEARLY DEMONSTRATED THAT (1) MIXING IN THE POND DEMANDS MUCH GREATER ATTENTION AS AN INFLUENTIAL PARAMETER WITH RESPECT TO BOTH THE THEORETICAL AND THE PRACTICAL ASPECTS, (2) ADVENT OF ANAEROBIC CCNDITIONS IN AEROBIC OXIDATION PONDS IS A MUCH MORE CCMPLEX PHENOMENON THAN WAS PREVIOUSLY SUPPOSED, BEING DEPENDENT UPON TEMPERATURE, MIXING, ALGAL GROWTH AND BOD IN THE POND, AND POSSIBLY OTHER PARAMETERS, (3) THE SLUDGE LAYER OCCUPIES AN IMPORTANT POSITION IN THE DEGRADATION PROCESS AND DIRECTS ATTENTION TO THE BENEFITS IN PARTIAL SEPARATION BY ANAEROBIC PRETREATMENT. (LOWRY-TEXAS)

FIELD 05D

ACCESSION NO. W71-07084

CHLORINATION OF WASTE POND EFFLUENTS,

SACRAMENTO STATE COLL., CALIF.

LECNARD W. HOM.

2ND INTERNATIONAL SYMPOSIUM FOR WASTE TREATMENT LAGOONS, JUNE 23-25, 1970, KANSAS CITY, MISSOURI, P 151-159. 7 FIG, 4 TAB, 15 REF. NATIONAL SCIENCE FOUNDATION GRANT GY3799, NATIONAL INSTITUTE OF HEALTH, USPHS, WP00026-RS.

DESCRIPTORS:
*OXIDATION LAGOONS, *DISINFECTION, DEGRADATION, CHLORINE, CONTACT TIME, SAMPLING, COLIFORMS, BACTERIA, OXIDATION, ORGANIC LOADING, ALGAE, WASTE WATER TREATMENT, BIOCHEMICAL OXYGEN DEMAND.

IDENTIFIERS:
*RESIDUAL, MOST PROBABLE NUMBER.

ABSTRACT:
A TWO YEAR SERIES OF TESTS WERE PERFORMED CN EFFLUENT FROM EXPERIMENTAL STABILIZATION LAGOONS OF THE CITY OF CONCORD, CALIFORNIA, IN AN ATTEMPT TO DETERMINE METHODS FOR CONTROLLING ALGAE KILL BY CONTROLLING CHLORINE DOSAGE. THREE DAY COMPOSITE SAMPLES, OF 500 ML VCLUME, WERE TAKEN AT THREE-HOUR INTERVALS, AND THE CHLORINE DEMAND WAS DETERMINED BY THE OT TEST USING A 30 MIN. CONTACT TIME. IN ADDITION, THE SAMPLES WERE ANALYZED FOR BOD LEVELS AND FOR COLIFORM POPULATICN. SELECTIVE CHLORINATION OF STABILIZATION LAGOON EFFLUENTS CAN BE ACCOMPLISHED. CONTROL OF REACTION AND CHLORINE CONCENTRATION IS CRITICAL, SINCE EXCESSIVE CHLORINE CAN RELEASE NUTRIENTS FROM ALGAL CELLS, THEREBY INCREASING THE BOD. THE SOLUTION LIES IN EXPERIMENTAL OPTIMIZATION OF BOTH CHLORINE DOSAGE AND RESIDUAL CHLORINE CONCENTRATIONS. TIME-CONCENTRATION RELATIONSHIPS REPORTED HERE PROVIDE A RATIONAL SET OF PROCESS DESIGN PARAMETERS FOR CHLORINE DISINFECTICN IN ALGAL-BACTERIAL SYSTEMS. (LOWRY-TEXAS)

FIELD 05D, 05F

ACCESSION NO. W71-07096

EFFECT OF LAGOON EFFLUENT ON A RECEIVING STREAM,

MISSOURI UNIV., COLUMBIA.

DARRELL L. KING, ALLEN J. TOLMSOFF, AND MICHAEL J. ATHERTON.

2ND INTERNATIONAL SYMPOSIUM FOR WASTE TREATMENT LAGOONS, JUNE 23-25, 1970, KANSAS CITY, MISSOURI, P 159-167. 5 FIG, 4 TAB, 3 REF.

DESCRIPTORS:
*OXIDATION LAGOONS, *ALGAE, *TRICKLING FILTERS, EUTROPHICATION, PHOTOSYNTHESIS, DISSOLVED OXYGEN, SEDIMENTATION, SAMPLING, CHEMICAL OXYGEN DEMAND, BIOCHEMICAL OXYGEN DEMAND, STABILIZATION PONDS, MICROORGANISMS, BIODEGRADATION, WASTE WATER TREATMENT, WATER POLLUTION EFFECTS.

IDENTIFIERS:
*RIFFLES, *CALCRIC CONTENT.

ABSTRACT:
BEAR CREEK, A SMALL STREAM WHOSE FLOW IS 21.5% LAGOON EFFLUENT AND SEEPAGE, WAS THE SITE OF A 90 DAY STUDY WHICH ATTEMPTED TO DETERMINE WHAT STREAM PARAMETERS OR CHARACTERISTICS WERE ALTERED BY LAGOON EFFLUENT. PARAMETERS MONITORED WERE, BOD, COD, AND ALSO VOLATILE, SUSPENDED, AND OTHER SOLIDS MEASUREMENTS. THE OXYGEN DEMANDS OF THE VARIOUS CONSTITUENTS WERE THEN CALCULATED AND ANALYZED. FROM THESE EXPERIMENTS, IT WAS DETERMINED THAT RECEIVING STREAMS FOR ALGAE-LADEN EFFLUENTS CAN BE SIGNIFICANTLY INFLUENCED FROM SEVERAL FEET TO SEVERAL MILES DOWNSTREAM. THE AMOUNT OF FLOW IN THE STREAM, AND THE RIFFLE-POOL RATIO WERE FOUND TO BE THE FACTORS WHICH DETERMINE THE DISTANCE DOWNSTREAM WHICH A STREAM WILL BE AFFECTED. SMALLER STREAMS MAY BE BROKEN DOWN INTO RIFFLES, WHICH ACT MUCH LIKE TRICKLING FILTERS, AND POOLS, WHICH ARE MERELY SEDIMENTATION AND DIGESTION UNITS. A MAJOR POINT OF THESE EXPERIMENTS WAS DEMONSTRATION OF THE FACT THAT THE RECEIVING STREAM, IN MANY CASES, IS AN INTEGRAL PART OF THE TREATMENT FACILITY AND MUST BE CONSIDERED AS SUCH BEFORE THE TOTAL SYSTEM CAN BE EVALUATED. (LOWRY-TEXAS)

FIELD 05D, 05C

ACCESSION NO. W71-07097

EFFECTS OF OXIDATION POND EFFLUENT ON RECEIVING WATER IN THE SAN JOAQUIN RIVER ESTUARY,

FEDERAL WATER QUALITY ADMINISTRATION, PACIFIC SOUTHWEST REGION, CALIFORNIA/NEVADA BASINS OFFICE.

RICHARD C. BAIN, JR., PERRY L. MCCARTY, JAMES A. ROBERTSON, AND WILLIAM H. PIERCE.

2ND INTERNATIONAL SYMPOSIUM FOR WASTE TREATMENT LAGOONS, JUNE 23-25, 1970, KANSAS CITY, MISSOURI, P 168-180. 8 FIG, 6 TAB, 8 REF.

DESCRIPTORS:
*OXIDATION LAGOONS, *OXYGENATION, *ALGAE, DEPTH, TEMPERATURE, PHOTOSYNTHESIS, RE-AERATION, LIGHT PENETRATION, NUTRIENTS, PHOSPHORUS, NITROGEN, TIDAL WATERS, CHEMICAL OXYGEN DEMAND, BIOCHEMICAL OXYGEN DEMAND, WASTE WATER TREATMENT, WATER POLLUTION EFFECTS.

IDENTIFIERS:
*SAN JOAQUIN RIVER.

ABSTRACT:
OXIDATION POND EFFLUENT ENTERING THE SAN JOAQUIN RIVER ESTUARY WAS ASSUMED TO BE A MAJOR CAUSE OF LOW DISSOLVED OXYGEN LEVELS, AND RESULTANT FISH KILLS. IN AN EFFORT TO PROVE OR DISPROVE THIS ASSUMPTION, FACTORS CONSIDERED RELATED TO DISSOLVED OXYGEN CONCENTRATION WERE STUDIED. THESE FACTORS INCLUDED WATER TEMPERATURE, ALGAL POPULATION, OXYGEN DEMANDS, NUTRIENTS, AND CHANNEL AND FLOW CHARACTERISTICS. SAMPLES WERE TAKEN FROM THE RIVER AT VARIOUS TIMES, FROM INFLUENT AND EFFLUENT FROM THE TREATMENT PLANT, AND FROM THE OXIDATION LAGOON. THEY WERE THEN ANALYZED FOR NH3-NITROGEN, NO3-NITROGEN, ORGANIC NITROGEN, ORGANIC CARBON, CARBONATE ALKALINITY, BICARBONATE ALKALINITY, ORTHOPHOSPHORUS, TOTAL PHOSPHORUS, 5 DAY BOD, 30 DAY BOD, AND COD. FROM THE PRECEDING INVESTIGATIONS, IT WAS DETERMINED THAT DEPRESSED OXYGEN LEVELS WERE THE RESULT OF BOTH PHYSICAL AND BIOLOGICAL FACTORS. PHYSICALLY, THE DEEPENING OF THE CHANNEL REDUCED TIDAL VELOCITY AND THEREBY REDUCED TURBULENCE AND RATE OF NATURAL RE-OXYGENATION, LEADING TO LOWER OXYGEN TRANSFER RATES. BIOLOGICALLY, ALGAE THRIVES IN BOTH THE OXIDATION PONDS AND THE SHALLOW RIVER, BUT IS TRAPPED IN THE DEEPER CHANNEL WHERE LIGHT PENTRATION IS INSUFFICIENT TO SUPPORT IT. THE ALGAE THEN DECOMPOSES AND REQUIRES OXYGEN. THE PROBLEM THEN, IS A COMPLEX COMBINATION OF FACTORS WHICH REQUIRES THE SYSTEM APPROACH, IF A FULLY COMPREHENSIVE SOLUTION IS TO BE OBTAINED. (LOWRY-TEXAS)

FIELD 05D, 05C

ACCESSION NO. W71-07098

MATHEMATICAL SIMULATION OF WASTE STABILIZATION PONDS,

FEDERAL WATER QUALITY ADMINISTRATION, CINCINNATI, OHIO. ADVANCED WASTE TREATMENT LABS.

JOSEPH F. ROESLER, AND HERBERT C. PREUL.

2ND INTERNATIONAL SYMPOSIUM FOR WASTE TREATMENT LAGOONS, JUNE 23-25, 1970, KANSAS CITY, MISSOURI, P 180-185. 20 REF.

DESCRIPTORS:
*OXIDATION LAGOONS, *MATHEMATICAL MODELS, MIXING, WATER RE-USE, LIGHT INTENSITY, EVAPORATION, RE-AERATION, DISSOLVED OXYGEN, ALGAE, ANAEROBIC CONDITIONS, AEROBIC CONDITIONS, BIODEGRADATION, SEEPAGE, SLUDGE, COST ANALYSIS, *BIOCHEMICAL OXYGEN DEMAND, WASTE WATER TREATMENT.

ABSTRACT:
THE MASS BALANCE APPROACH WAS USED TO OBTAIN EQUATIONS RELATING TO OXIDATION LAGOON PROCESSES. THE MODEL WAS BASED ON THE ASSUMPTION THAT THE INFLUENT WAS DOMESTIC SEWAGE, AND THAT AN ANAEROBIC SLUDGE LAYER WAS FORMED AT THE BOTTOM OF THE POND. EQUATIONS WERE DERIVED FOR BOD REMOVAL BY ALGAE, ANAEROBIC DECOMPOSITION, AND OXIDATION POND DEPTH. IN ORDER TO VERIFY EQUATIONS DERIVED FROM THE MASS BALANCE APPROACH, EXPERIMENTAL DATA ON EFFLUENT BOD, DETENTION TIME, SUNLIGHT INTENSITY, TEMPERATURE, EVAPORATION RATE, AND DISSOLVED OXYGEN DEFICIENCY WERE OBTAINED FROM THE LITERATURE. A COMPUTER PROGRAM WAS THEN DESIGNED TO EVALUATE THE CONSTANTS AND THE DISPOSITION OF ALL BOD'S. FROM THE PRECEDING EQUATION AND CALCULATIONS, IT WAS CONCLUDED THAT: (1) OXIDATION LAGOONS ARE ECONOMICAL TO BUILD AND MAINTAIN, (2) THE ROLE OF ALGAE HAS BEEN OVEREMPHASIZED, WITH RE-AERATION AND ANAEROBIC PROCESSES EQUALLY, IF NOT MORE, IMPORTANT, AND (3) SINCE ALGAE OCCASIONALLY INTERFERE WITH RECEIVING STREAM WATER QUALITY, INVESTIGATIONS SHOULD BE CONDUCTED IN HOW TO MAXIMIZE ANAEROBIC DECOMPOSITION AT THE BOTTOM AND MINIMIZE ALGAE PRODUCTION. (LOWRY-TEXAS)

FIELD 05D

ACCESSION NO. W71-07099

DESIGNING WASTE PONDS TO MEET WATER QUALITY CRITERIA,

CALIFORNIA UNIV., BERKELEY. DEPT. OF SANITARY ENGINEERING AND PUBLIC HEALTH.

WILLIAM J. OSWALD, AARON MERON, AND MARIO D. ZABAT.

2ND INTERNATIONAL SYMPOSIUM FOR WASTE TREATMENT LAGOONS, JUNE 23-25, 1970, KANSAS CITY, MISSOURI, P 186-194, 3 FIG. 4 TAB, 23 REF.

DESCRIPTORS:
*OXIDATION LAGOONS, *DESIGN CRITERIA, *WATER QUALITY, ALGAE, HARVESTING, DEPTH, TEMPERATURE, LIGHT INTENSITY, CHEMICAL PRECIPITATION, HYDROGEN ION CONCENTRATION, COLIFORMS, BACTERIA, PHOTOSYNTHESIS, ORGANIC COMPOUNDS, PHOSPHATES, NITRATES, NUTRIENTS, ANAEROBIC CONDITIONS, AEROBIC CONDITIONS, BIOCHEMICAL OXYGEN DEMAND, WASTE WATER TREATMENT.

ABSTRACT:
THE OVERALL PROCESS OF WASTE STABILIZATION IN OXIDATION LAGOONS WAS EXAMINED FROM A LOGICAL DESIGN STANDPOINT. SIGNIFICANT FACTORS WERE EXAMINED FROM THE LITERATURE, AND THESE INCLUDED INFECTIOUS AGENTS, PLANT NUTRIENTS, ORGANIC CHEMICALS, AND LAND EROSION AND SUBSEQUENT ADDITION OF MINERALS, CHEMICALS, AND SILTS, RADIOACTIVE SUBSTANCES, AND HEAT POLLUTION. IN EACH CASE, DESIGN PRACTICES BOTH FROM LITERATURE RESEARCH AND PRACTICAL EXPERIENCE WERE PRESENTED. A SUMMARY OF THE THREE YEARS OF OPERATION OF THE SAINT HELENA WASTE POND SYSTEM IS PRESENTED IN SUPPORT OF THE DESIGN PRACTICES PREVIOUSLY ADVOCATED. WATER QUALITY CONTROL OF A HIGH ORDER IS CONSISTENTLY OBTAINED AT THE SAINT HELENA PLANT, AND THIS CONTROL IS ATTRIBUTED MAINLY TO THESE IMPROVED DESIGN PRACTICES. FURTHER IMPROVEMENT MAY BE POSSIBLE WITH ADDITION OF ALGAE HARVESTING FACILITIES WHICH WILL ALSO REMOVE PHOSPHATES, BUT SINCE THERE IS LITTLE OR NO EFFLUENT FROM THE PONDS NOW THE PRESENT SYSTEM MORE THAN COMPLIES WITH CURRENT STANDARDS. (LOWRY-TEXAS)

FIELD 05D

ACCESSION NO. W71-07100

NEW EXPERIMENTAL POND DATA,

TEXAS UNIV., AUSTIN. ENVIRONMENTAL HEALTH ENGINEERING RESEARCH LAB.

E. F. GLOYNA, AND J. AGUIRRE.

2ND INTERNATIONAL SYMPOSIUM FOR WASTE TREATMENT LAGOONS, JUNE 23-25, 1970, KANSAS CITY, MISSOURI, P 200-210. 3 FIG, 10 TAB, 10 REF.

DESCRIPTORS:
*OXIDATION LAGOONS, *SLUDGE, ANAEROBIC CONDITIONS, AEROBIC CONDITIONS, PILOT PLANTS, ALGAE, HYDROGEN ICN CONCENTRATION, CHEMICAL OXYGEN DEMAND, *BIOCHEMICAL OXYGEN DEMAND, MIXING, SAMPLING, ORGANIC LOADING.

IDENTIFIERS:
LABORATORY SCALE, *TCTAL ORGANIC CARBON.

ABSTRACT:
THREE LABORATORY-SCALE UNITS AND THREE PILOT-SCALE UNITES WERE CONSTRUCTED AND OPERATED FOR A PERIOD OF EIGHT MONTHS AT THE CONVENTIONALLY ACCEPTED LOADING RATES OF 70 LBS BODU/ACRE/DAY AND 65 DAY DETENTION TIME FOR THE PILOT SCALE PLANTS, WITH CNLY 10 LBS BODU/ACRE/DAY AND 90 DAYS DETENTION TIME IN THE LAB SCALE UNITS. THE ONLY DIFFERENCE IN THE THREE UNITS WAS THE LOCATICN OF THE ANAEROBIC SLUDGE DEPOSITION ZONE. ORGANIC REMOVAL EFFICIENCIES WERE DETERMINED BY TESTS ON BOD5, COD, AND TOC. SUSPENDED SOLICS REMCVAL WAS DETERMINED BY MEMBRANE FILTER ANALYSIS, AND GENERAL POND CHARACTERISTICS WERE EVALUATED THROUGH MEASUREMENT OF PH, DO, TEMPERATURE, NITROGEN, PHOSPHORUS, AND ALGAE TYPING. FROM THIS EXPERIMENTAL EVALUATION, IT WAS DETERMINED THAT PILOT SCALE SYSTEMS ARE OF CONSIDERABLE VALUE IN ARRIVING AT DESIGN CRITERIA FOR A PARTICULAR WASTE IN A PARTICULAR LOCALITY. PILOT PLANTS LACK MANY OF THE DISADVANTAGE OF LAB SCALE PLANTS MAINLY DUE TO THE INCREASED SIZE, AND THEREFORE, DECREASED SENSITIVITY OF THE PILOT PLANTS. ALSO, SERIES ARRANGEMENT OF A MULTIPLE POND SYSTEM HAD A MUCH GREATER EFFECT THAN DID CHANGING THE LOCATION OF THE SLUDGE DEPOSITION ZONE. ONE OF THE POND SYSTEMS UTILIZED FOR THIS STUDY OPERATED ON THE SAME PRINCIPAL AS A SERIES POND ARRANGEMENT, AND THE ANAEROBIC PRETREATMENT PORTICN OF THE SYSTEM PROVIDED 50 TO 76% REDUCTION OF ORGANIC MATERIAL IN FROM 3 TO 5 DAYS BETWEEN 12 AND 24C. (LOWRY-TEXAS)

FIELD 05D

ACCESSION NO. W71-07102

A RATIONAL APPROACH TO THE DESIGN OF AERATED LAGOONS,

CORPS OF ENGINEERS, ANCHORAGE, ALASKA. SANITARY ANC CIVIL ENGINEERING SECTION.

EDWARD F. POHL.

2ND INTERNATIONAL SYMPOSIUM FOR WASTE TREATMENT LAGOCNS, JUNE 23-25, 1970, KANSAS CITY, MISSOURI, P 231-243. 9 FIG, 3 TAB, 26 REF.

DESCRIPTORS:
*OXIDATION LAGOONS, *OXYGENATION, DISSOLVED OXYGEN, SATURATION, ALGAE, METHANE BACTERIA, EUTROPHICATION, NITRATES, PHOSPHATES, BIODEGRADATION, ANAEROBIC CONDITIONS, AEROBIC CONDITIONS, SLUDGE, LIGHT INTENSITY, COLIFORMS, BIOCHEMICAL OXYGEN DEMAND, WASTE WATER TREATMENT.

IDENTIFIERS:
*AERATED LAGOONS, SUSPENDED SOLIDS.

ABSTRACT:
THE VARIOUS PARAMETERS INVOLVED IN THE DESIGN OF AERATED LAGOONS WERE INVESTIGATED WITH PARTICULAR REFERENCE TO THE USE CF AIR DIFFUSED INTO THE LAGOON THROUGH LENGTHS OF TUBING. IT WAS DETERMINED THAT A DESIGN FOR AN AERATED LAGOON SHOULD EVOLVE IN THREE STAGES, EACH TO BE EVALUATED FOR SUMMER AND WINTER CONDITIONS. TO DESIGN AN AERATED LAGOON, THE FIRST STAGE INVOLVES EVALUATION OF THE RECEIVING BODY BY BOTH THE WASTE PRODUCERS, AND THE AGENCY RESPONSIBLE FOR THE PROTECTION OF THE QUALITY OF THE RECEIVING WATER. THE REQUIRED EFFLUENT QUALITY DETERMINATION IS BASED ON BCD, SUSPENDED SOLIDS, EFFLUENT COLIFORM ORGANISM COUNT. DETERMINATION OF THE DETENTION TIME REQUIRED TO ACHIEVE THE STAGE I OBJECTIVES IS THE SECOND STAGE, AND THE FINAL STAGE INVOLVES AN EVALUATION OF THE OXYGEN NEEDED TO SUPPLY BOTH THE AERATION TANK MIXED LIQUOR SOLIDS RESPIRATION, AND THE BENTHAL OXYGEN DEMAND. CONSERVATIVE OVERALL DESIGN VALUES ARE RECOMMENDED BECAUSE OF THE GAPS IN THE RESEARCH DATA. UNTIL MORE RESEARCH HAS BEEN DONE TO INVESTIGATE THE CONSTANTS AND THE MECHANISMS OF THE PROCESS, CONSERVATIVE ESTIMATES WILL CONTINUE TO BE RECOMMENDED IN ORDER TO PROVIDE A LARGER MARGIN OF SAFETY. (LOWRY-TEXAS)

FIELD 05D

ACCESSION NO. W71-07106

A COMPARATIVE STUDY OF AERATED LAGOON TREATMENT OF MUNICIPAL WASTE WATERS,

METROPOLITAN CORP. OF GREATER WINNIPEG (MANITOBA). WATER WORKS AND WASTE DISPOSAL DIV.

G. E. BURNS, R. M. GIRLING, A. R. PICK, AND D. W. VAN ES.

2ND INTERNATIONAL SYMPOSIUM FOR WASTE TREATMENT LAGOONS, JUNE 23-25, 1970, KANSAS CITY, MISSOURI, P 258-276. 14 FIG, 3 TAB, 6 REF.

DESCRIPTORS:
*OXIDATION LAGOONS, *CLIMATIC ZONES, *COST ANALYSIS, TEMPERATURE, NUTRIENTS, DISSOLVED OXYGEN, ALGAE, SLUDGE, OXYGENATION, ANAEROBIC CONDITIONS, AEROBIC CONDITIONS, BIOCHEMICAL OXYGEN DEMAND, WASTE WATER TREATMENT.

IDENTIFIERS:
*AERATED LAGOONS, SURFACE AERATORS.

ABSTRACT:
THREE PILOT SCALE AERATED LAGOONS WERE CONSTRUCTED AT WINNIPEG, CANADA TO TEST THE EFFECT OF THE CANADIAN PRAIRIE CLIMATIC CONDITIONS ON LAGOON OPERATION. THE PILOT CELLS WERE OF THE ANAEROBIC-AEROBIC TYPE. EACH LAGOON WAS EQUIPPED WITH A DIFFERENT TYPE OF AERATION SYSTEM, DIFFUSED AIR, SURFACE AERATORS, OR A COMBINATION OF BOTH. TEST WERE CONDUCTED OVER A 20 MONTH PERIOD ON AN INFLUENT AVERAGING 175 MG/L BOD AND 188 MG/L OF SUSPENDED SOLIDS. FROM THE RESULTS OF THIS INVESTIGATION, IT WAS CONCLUDED THAT AERATED LAGOONS CAN SATISFACTORILY PROVIDE SECONDARY TREATMENT UNDER PRAIRIE CLIMATIC CONDITIONS. BOD REMOVAL EFFICIENCY AND DISSOLVED OXYGEN CONCENTRATION BOTH UNDERGO A SEASONAL SUMMER DECLINE DUE TO THE BUILD-UP OF SLUDGE DURING THE SUMMER MONTHS. THE PREVAILING CONSIDERATION FOR ECONOMIC FEASIBILITY OF AN AERATED LAGOON SYSTEM IS THE EXTENT AND COST IMPLICATIONS OF SLUDGE REMOVAL AND DISPOSAL FACILITIES. ALSO THE USE OF SURFACE AERATORS WAS DEMONSTRATED TO BE IMPRACTICAL IN AREAS WHICH SUSTAIN LARGE AMOUNTS OF ICE COVER. RESEARCH IS CONTINUING ON THE ECONOMIC FACTORS INVOLVED IN SLUDGE REMOVAL AND DISPOSAL. (LOWRY-TEXAS)

FIELD 05D

ACCESSION NO. W71-07109

TERTIARY TREATMENT BY AERATED LAGOON,

V. N. WAHBEH, AND L. W. WELLER.

2ND INTERNATIONAL SYMPOSIUM FOR WASTE TREATMENT LAGOONS, JUNE 23-25, 1970, KANSAS CITY, MISSOURI, P 293-299. 11 FIG, 3 TAB, 5 REF.

DESCRIPTORS:
*OXIDATION LAGOONS, *TERTIARY TREATMENT, *DESIGN CRITERIA, TRICKLING FILTERS, ALGAE, ORGANIC LOADING, ODOR, COLIFORMS, TEMPERATURE, DISSOLVED OXYGEN, AERATION, OXYGENATION, NUTRIENTS, SEDIMENTATION, CHLORINATION, SLOPES, BIOCHEMICAL OXYGEN DEMAND, HYDROGEN ION CONCENTRATION, WASTE WATER TREATMENT, KANSAS.

IDENTIFIERS:
*AERATED LAGOONS, SUSPENDED SOLIDS.

ABSTRACT:
AN OXIDATION POND, ORIGINALLY DESIGNED AS THE TREATMENT FACILITY FOR A SMALL COMMUNITY AND LATER SUPERCEDED BY A HIGH-RATE TRICKLING FILTER PLANT, WAS REMODELED AND EQUIPPED WITH DIFFUSED AIR AERATION USING THREE 600 CFM BLOWERS. THIS UNIT PROVIDED TREATMENT MAINLY FOR STORM BYPASS, UNTIL IT WAS UTILIZED AS A TERTIARY STAGE TO FOLLOW THE TRICKLING FILTER. TESTS WERE THEN CONDUCTED, USING AN AUTOMATIC SAMPLING DEVICE, TO DETERMINE THE EFFICIENCY OF SUCH AN ARRANGEMENT DURING SEVERAL SHORT TESTING PERIODS. IT WAS DISCOVERED THAT CONVENTIONAL OXIDATION PONDS LACKING AERATION WERE INEFFECTIVE IN REDUCTION OF BOD, SUSPENDED SOLIDS, AND INORGANIC NUTRIENTS UNLESS SOME METHOD OF ALGAE HARVESTING WAS EMPLOYED. AERATED LAGOONS, HOWEVER, PROVIDED IN EXCESS OF 60% REDUCTION OF BOD AND SUSPENDED SOLIDS, ALTHOUGH NO DETECTABLE REDUCTION IN INORGANIC NUTRIENTS WAS EVIDENCED. SOME CRITICAL PARAMETERS FOR DESIGN OF AERATED LAGOONS ARE (1) GREATER THAN 4:1 AND PREFERABLY A 2:1 SIDE SLOPE TO PREVENT SOLIDS DEPOSITION ON THE SLOPE, AND (2) MINIMIZATION OF WATER SURFACE AREA PREVENTS UNWANTED ALGAL CELL GROWTH WHICH OTHERWISE WOULD DEFEAT THE PURPOSE OF THE USE OF AERATED LAGOONS INSTEAD OF OXIDATION LAGOONS. (LOWRY-TEXAS)

FIELD 05D

ACCESSION NO. W71-07113

CHALLENGE FOR WASTE WATER LAGOONS,

FEDERAL WATER QUALITY ADMINISTRATION, CINCINNATI, OHIO. ADVANCED WASTE TREATMENT RESEARCH LAB.

FRANCIS M. MIDDLETOWN, AND ROBERT L. BUNCH.

2ND INTERNATIONAL SYMPOSIUM FOR WASTE TREATMENT LAGOONS, JUNE 23-25, 1970, KANSAS CITY, MISSOURI, P 364-366.

DESCRIPTORS:
*OXIDATION LAGOONS, *WASTE WATER TREATMENT, *WATER RE-USE, BIOCHEMICAL OXYGEN DEMAND, ORGANIC LOADING, ODORS, PHOTOSYNTHESIS, ALGAE, BACTERIA, CHLORINATION, GROUNDWATER, COST ANALYSIS.

IDENTIFIERS:
SUSPENDED SOLIDS.

ABSTRACT:
INCREASING USE OF WATER HAS NECESSITATED GREATER AND GREATER USAGE OF RECONDITIONED WASTE WATER AS DRINKING WATER. WITH THIS GREATER USAGE HAS COME INCREASINGLY STRICTER LAWS REGARDING THE QUALITY OF THE EFFLUENT WHICH A TREATMENT FACILITY IS PERMITTED TO DISCHARGE TO A RECEIVING WATER. THEREFORE TREATMENT METHODS MUST BECOME MORE SOPHISTICATED AS TIME GOES ON IN ORDER THAT EFFLUENTS NOT ONLY PRESERVE THE QUALITY OF A RECEIVING STREAM, BUT MUST HELP TO RESTORE IT. IN LIGHT OF THESE FACTS LAGOON PERFORMANCE MUST BE EVALUATED WITH RESPECT TO FUTURE CONDITIONS. LAGOONS HAVE SEVERAL DRAW BACKS FROM A WATER POLLUTION STANDPOINT. THESE ARE: (1) ALTHOUGH COLIFORMS MAY BE REDUCED AS MUCH AS 98% IN A LAGOON, THE WATER QUALITY STANDARDS FOR NUMBER OF ORGANISMS PER MILLILITER MAY STILL BE VIOLATED; (2) IF ALGAE IS ALLOWED TO PASS TO THE RECEIVING STREAM, IT MAY BECOME AS MUCH OF A POLLUTIONAL FACTOR AS THE RAW WASTE SINCE IT CONTAINS THE SAME AMOUNT OF ORGANIC MATTER; (3) LAGOON EFFLUENT IS HIGH IN SUSPENDED SOLIDS WHICH ARE BOTH AESTHETICALLY UNPLEASING AND OXYGEN DEMANDING; (4) LAGOONS ARE A THREAT TO GROUNDWATER QUALITY BECAUSE OF THE DIFFICULTY IN SEALING THEM. UNTIL NOW, THE MOST ATTRACTIVE FEATURE OF LAGOONS HAS BEEN THEIR LOW COST. AS TREATMENT MUST BECOME MORE SOPHISTICATED, LAGOONS AS PRESENTLY CONSTRUCTED WILL NEED ADDITIONAL TREATMENT WHICH WILL MORE THAN OFFSET THEIR COST ADVANTAGE. THEREFORE, UNLESS SIGNIFICANT ADVANCES ARE MADE IN LAGOON TECHNOLOGY, IT IS DOUBTFUL IF LAGOONS WILL HAVE A PLACE IN THE FUTURE. (LOWRY-TEXAS)

FIELD 05D

ACCESSION NO. W71-07123

STATE OF THE ART-OXIDATION PONDS,

NORTH DAKOTA UNIV., GRAND FORKS. DEPT. OF MICROBIOLOGY.

JOHN W. VENNES.

2ND INTERNATIONAL SYMPOSIUM FOR WASTE TREATMENT LAGOONS, JUNE 23-25, 1970, KANSAS CITY, MISSOURI, P 366-376. 3 TAB, 67 REF.

DESCRIPTORS:
*OXIDATION LAGOONS, *WASTE WATER TREATMENT, ORGANIC LOADING, BIOCHEMICAL OXYGEN DEMAND, SOLAR RADIATION, NUTRIENTS, TEMPERATURE, COLIFORMS, CHLORINATION, SEEPAGE, PERCOLATION, EVAPORATION, INSECTS, EUTROPHICATION, ALGAE.

ABSTRACT:
AN INTENSIVE LITERATURE SEARCH OF THE WASTE STABILIZATION POND METHOD OF TREATING WASTE WATER WAS CONDUCTED TO DETERMINE THE PRESENT STATUS OF SUCH UNITS WITH REGARD TO DESIGN CRITERIA, OPERATIONAL PARAMETERS, AND COMPLIANCE WITH WATER POLLUTION PREVENTION REGULATIONS. DESIGN CONSIDERATIONS ARRIVED AT WERE: (1) 1 LANGLEY/DAY OF RADIATION IS SUFFICIENT TO TREAT 1 LB BOD/ACRE/DAY; (2) MULTI-STAGE SERIES PONDS PROVIDE THE BEST ORGANIC AND BACTERIAL REMOVALS; (3) SEEPAGE AND EVAPORATION MUST BE CONSIDERED IN LAGOON DESIGN SINCE IN MANY CASES THESE ARE THE ONLY FORMS OF EFFLUENT; (4) CHLORINATION WHILE KILLING THE BACTERIA, WILL ALSO KILL THE ALGAE WHICH WILL THEN EFFECT AN IMMEDIATE ORGANIC LOAD ON THE SYSTEM OR THE RECEIVING STREAM; (5) ALGAE REMOVAL TECHNIQUES MUST BE DEVELOPED SINCE MOST OF THE ORGANIC MATERIAL IS NOT DESTROYED, BUT MERELY CONVERTED TO ALGAE, AS ARE THE INORGANIC NUTRIENTS AS WELL. WITHOUT ALGAE REMOVAL, OXIDATION POND EFFLUENT IS HIGH IN ORGANIC CONTENT WHICH MAY CAUSE DETERIORATION OF THE RECEIVING STREAM QUALITY. MORE STUDY IS NEEDED ON THE MECHANISMS WHICH ARE UTILIZED IN OXIDATION PONDS FOR THE REMOVAL OF ORGANICS, AND ALSO ON THE ACTUAL MEASUREMENT OF THE ORGANIC LOADINGS ON THE FACILITIES BEFORE ORGANIC POLLUTANTS AND MICRO-ORGANISMS PRESENT IN POND EFFLUENTS CAN BE RELIABLY PREDICTED. (LOWRY-TEXAS)

FIELD 05D

ACCESSION NO. W71-07124

A COMPUTER BASED FLORISTIC ANALYSIS OF PAMLICO RIVER PHYTCPLANKTON,

EAST CAROLINA UNIV., GREENVILLE, N.C. DEPT. OF BIOLOGY.

VINCENT J. BELLIS.

COPIES AVAILABLE FROM WATER RESOURCES RESEARCH INST, 124 RIDDICK BLDG, NORTH CAROLINA STATE UNIV, RALEIGH, NC, 27607, PRICE $2.50. NORTH CAROLINA UNIVERSITY WATER RESOURCES RESEARCH INSTITUTE REPORT NO 46, JAN 1971. 28 P, 7 FIG, 5 TAB, 13 REF. OWRR PROJECT A-044-NC(1) AGREEMENT NO 14-31-0001-3033, FY 1970.

DESCRIPTORS:
*ESTURARIES, *PHYTOPLANKTON, *DISTRIBUTION PATTERNS, *NORTH CAROLINA, *COMPUTER PROGRAMS, DATA PROCESSING, DATA COLLECTIONS, VARIABILITY, PROBABILITY, WATER PCLLUTION EFFECTS, ALGAE, AQUATIC HABITATS, AQUATIC ENVIRONMENT.

IDENTIFIERS:
*PAMLICO RIVER(NC).

ABSTRACT:
A COMPUTER BASED TECHNIQUE FOR ANALYZING DISTRIBUTION PATTERNS AMONG ESTUARINE ORGANISMS WAS DEVELOPED AND TESTED ON PHYTOPLANKTON COLLECTIONS FROM PAMLICO RIVER, NORTH CAROLINA. JACCARD'S COEFFICIENT OF SIMILARITY WAS USED TO GENERATE A SIMILARITY MATRIX WITH FINAL PRINTOUT IN DENDROGRAM FORM SHOWING CLUSTERS OF COLLECTIONS HAVING INHERENT SIMILARITY. USE OF A SIMILARITY INDEX BASED SOLELY UPON PRESENCE OR ABSENCE OF SPECIES WITHIN ALL POSSIBLE PAIRS OF DATA COLLECTIONS PRODUCED A DENDROGRAM WHICH DESCRIBED THE SEASONAL PERIODICITY OF PHYTOPLANKTON IN THE CENTRAL PAMLICO RIVER. COMPUTER PROGRAMS DEVELOPED IN THIS PROJECT ARE WRITTEN IN FORTRAN IV AND HAVE BEEN DEPOSITED IN THE PROGRAM LIBRARY OF THE EAST CAROLINA UNIVERSITY COMPUTER CENTER AND CAN BE OBTAINED UPCN REQUEST FROM ITS DIRECTOR. (KNAPP-USGS)

FIELD 02L, 07C, 05C

ACCESSION NO. W71-07337

NOTES ON FRESHWATER ALGAE, (IN GERMAN),

VIENNA UNIV, (AUSTRIA), BOTANISCHES INSTITUT UND BOTANISCHER GARTEN.

LOTHAR GEITLER.

OSTERREICHISCHE BOTANISCHE ZEITSCHRIFT, NO 118, P 306-310, 1970. 16 REF.

DESCRIPTORS:
*ALGAE, *FRESH WATER, DISTRIBUTICN PATTERNS, BEHAVIOR.

IDENTIFIERS:
MOUGEOTIA GENUFLEXA, ANKISTRODESMUS FALCATUS, PODCHEDRA.

ABSTRACT:
THIS PAPER DESCRIBES DISTRIBUTION AND BEHAVIOR OF FRESHWATER ALGAE, INCLUDING MOUGEOTIA GENUFLEXA, ANKISTRODESMUS FALCATUS, VAR STIPITATUS, AND MEMBERS OF THE GENUS PODCHEDRA. (WILDE-WISCONSIN)

FIELD 05C

ACCESSION NO. W71-07360

LIGHT INTENSITY AND THE VERTICAL DISTRIBUTION OF ALGAE IN TERTIARY OXIDATION PONDS,

NORTH CAROLINA UNIV., CHAPEL HILL. DEPT. OF ENVIRONMENTAL SCIENCES AND ENGINEERING.

WILLIAM R. HARTLEY, AND CHARLES M. WEISS.

WATER RESEARCH, VOL 4, NO 11, NOV 1970, P 751-763. 7 FIG, 2 TAB, 18 REF.

DESCRIPTORS:
*ALGAE, *OXIDATION LAGOONS, *LIGHT INTENSITY, PONDS, ROTIFERS, PROTOZOA, PHOTOSYNTHESIS, CHLOROPHYLL, TEMPERATURE, HYDROGEN ION CONCENTRATION, DISSOLVED OXYGEN, STRATIFICATION, SAMPLING, MICROORGANISMS, SLUDGE, TERTIARY TREATMENT, BIOCHEMICAL OXYGEN DEMAND, WASTE WATER TREATMENT, *NORTH CAROLINA.

IDENTIFIERS:
DURHAM(NC).

ABSTRACT:
FIVE OXIDATION PONDS AT THE THIRD FORK TREATMENT PLANT IN DURHAM, NORTH CAROLINA, WERE INVESTIGATED TO DETERMINE THE OPTIMUM LIGHTING CONDITIONS, THE TYPES OF ORGANISMS PRESENT IN DIFFERENT SITUATIONS, AND THEIR VERTICAL DISTRIBUTION IN THE POND LEVELS. SAMPLES IN THE PONDS WERE TAKEN AT 3,6,9,12,15,18,21,24 INCHES, AND THE SLUDGE LAYER. THESE SAMPLES WERE TAKEN EVERY TWO HOURS OVER THE 24 HOUR DIURNAL CYCLE. THE SAMPLES WERE PRESERVED IN 3% FORMALIN, AND THEN TAKEN TO THE LABORATORY FOR ANALYSIS. LIGHT INTENSITY, PH, DISSOLVED OXYGEN, AND TEMPERATURE WERE ALSO RECORDED AT THE TIME OF SAMPLING. THE SAMPLES TAKEN WERE COUNTED FOR NUMBER OF ORGANISMS PRESENT, AND, ALSO CENTRIFUGED FOR IDENTIFICATION OF SPECIES PRESENT IN LESSER NUMBERS. THE SURFACE WEIRS INSTALLED IN THE PONDS AFFECT THE CHEMICAL AND PHYSICAL PARAMETERS OF POND PERFORMANCE WITH PARTICULAR REFERENCE TO THE DISSOLVED OXYGEN CONCENTRATION, IT BEING MUCH LOWER IN THE POND WHICH ALLOWED THE MAT TO FORM AT THE SURFACE. EUGENA ROSTIFERA WILL CHANGE THEIR VERTICAL POSITION IN THE POND TO REACH THE POINT OF SUBMERGENCE WHERE THE LIGHT INTENSITY IS 75 CANDLES/FT2. WHERE THEY WERE NOT ABLE TO MOVE TO RETREAT FROM HIGH LIGHT INTENSITY CYSTS WERE FORMED. (LOWRY-TEXAS)

FIELD 05D

ACCESSION NO. W71-07382

A POLYCHLORINATED BIPHENYL (AROCLOR 1254) IN THE WATER, SEDIMENT, AND BIOTA OF ESCAMBIA BAY, FLORIDA,

BUREAU OF COMMERCIAL FISHERIES, GULF BREEZE, FLA. CENTER FOR ESTUARINE AND MENHADEN RESEARCH.

T. W. DUKE, J. I. LOWE, AND A. J. WILSON, JR.

BULLETIN OF ENVIRONMENTAL CONTAMINATION AND TOXICOLOGY, VOL 5, NO 2, P 171-180, MAR-APR 1970. 2 FIG, 3 TAB, 7 REF.

DESCRIPTORS:
*WATER POLLUTION EFFECTS, *BIOASSAY, *TOXICITY, *CHLORINATED HYDROCARBONS, *BIOINDICATORS, INDUSTRIAL WASTES, FLORIDA, SESSILE ALGAE, ESTUARIES, CHEMICAL WASTES, GAS CHROMATOGRAPHY, SEDIMENTS, OYSTERS, SHRIMP, CRABS, WATER POLLUTION SOURCES, SEA WATER, TROUT, TOXINS.

IDENTIFIERS:
*POLYCHLORINATED BIPHENYLS, AROCLOR 1254, ESCAMBIA RIVER, PINFISH, FLOUNDER, CROAKER, MENHADEN.

ABSTRACT:
AROCLOR 1254, A POLYCHLORINATED BIPHENYL, HAS BEEN DETECTED IN THE BIOTA, SEDIMENT, AND WATER OF ESTUARINE AREAS NEAR PENSACOLA, FLORIDA. ONLY ONE SOURCE OF THE CHEMICAL, AN INDUSTRIAL PLANT ON THE ESCAMBIA RIVER, HAS BEEN FOUND. WATER, SEDIMENT, AND FISH, CRAB, OYSTER, AND SHRIMP SAMPLES WERE COLLECTED FROM APRIL THROUGH OCTOBER 1969 AND ANALYZED USING GAS CHROMATOGRAPHY PROCEDURES. IN ADDITION, BIOASSAYS ON FISH, SHRIMP, AND OYSTERS WERE CONDUCTED UNDER CONTROLLED LABORATORY CONDITIONS TO DETERMINE THE TOXIC EFFECTS OF AROCLOR 1254. THESE STUDIES SHOWED THAT JUVENILE SHRIMP WERE THE MOST SENSITIVE AND WERE KILLED WHEN EXPOSED TO 5.0 PPB OF AROCLOR 1254 IN FLOWING SEA WATER. THE AROCLOR CONTENT IN WATER FROM ESCAMBIA BAY, EVEN NEAR THE MOUTH OF THE RIVER, WAS LESS THAN 1 PPB. SHRIMP COLLECTED FROM THE BAY CONTAINED A MAXIMUM OF 2.5 P.P.M. THUS, SHRIMP IN THE BAY PROBABLY WERE NOT EXPOSED TO LETHAL LEVELS DURING THE SAMPLING PERIOD. HIGHEST CONCENTRATIONS IN THE WATER OCCURRED DURING AUGUST AND DECREASED WHEN LEAKAGE FROM THE PLANT WAS CORRECTED. (MORTLAND-BATTELLE)

FIELD 05C, 05B, 05A

ACCESSION NO. W71-07412

A COMPREHENSIVE APPRAISAL OF THE EFFECTS OF COOLING WATER DISCHARGE ON AQUATIC ECOSYSTEMS,

BATTELLE MEMORIAL INST., COLUMBUS, OHIO. COLUMBUS LABS.

ARTHUR A. LEVIN, THOMAS J. BIRCH, ROBERT E. HILLMAN, AND GILBERT E. RAINES.

PRESENTED AT THE PUBLIC AFFAIRS WORKSHOP 'WHY NUCLEAR POWER', HILTON HEAD ISLAND, SOUTH CAROLINA, SEPT 13-16, 1970. 45 P, 9 FIG, 2 TAB, 137 REF.

DESCRIPTORS:
*POWER PLANTS, *THERMAL POLLUTION, *REVIEWS, WATER TEMPERATURE, DISSOLVED OXYGEN, BIOINDICATORS, E. COLI, AQUATIC ALGAE, SUNFISHES, SNAILS, EELS, CHINOOK SALMON, BASS, TROUT, MINNOWS, CRAYFISH, MOSQUITOES, FOOD CHAIN, HERRING, OYSTERS, CLAMS, COLUMBIA RIVER, CALIFORNIA, WASHINGTON, CONNECTICUT, FLORIDA, DELAWARE RIVER, INDIANA.

IDENTIFIERS:
BARNACLES, FLOUNDER, HUMBOLDT BAY, CONNECTICUT RIVER, SAN JOAQUIN RIVER(CALIFORNIA), BISCAYNE BAY, WHITE RIVER(INDIANA), PATUXENT RIVER(CANADA), ONTARIO.

ABSTRACT:
A SELECTIVE REVIEW WAS MADE OF THE LITERATURE, AND BASED UPON THE INFORMATION ACQUIRED, A COMPREHENSIVE APPRAISAL OF THE EFFECTS OF COOLING WATER DISCHARGE ON FRESHWATER AND MARINE ECOSYSTEMS IS MADE. THE AQUATIC ORGANISMS CONSIDERED INCLUDE FISH, ALGAE, DIATOMS, ENTERIC BACTERIA, PATHOGENS, SNAILS, AND LARVAE. (LITTLE-BATTELLE)

FIELD 05B, 05C

ACCESSION NO. W71-07417

FEEDING VALUE OF ANIMAL WASTES,

AGRICULTURAL RESEARCH SERVICE, BELTSVILLE, MD. ANIMAL SCIENCE RESEARCH.

L. W. SMITH.

IN: ANIMAL WASTE REUSE--NUTRITIVE VALUE AND POTENTIAL PROBLEMS FROM FEED ADDITIVES--A REVIEW. ARS 44-224, P 5-13, FEB 1971. 1 TAB.

DESCRIPTORS:
*FEEDS, *RUMINANTS, ALGAE, CATTLE, POULTRY, HOGS, ANIMAL DISEASES, CATFISH, WASTE TREATMENT, FARM WASTES, DEHYDRATION, FEASIBILITY.

IDENTIFIERS:
*ANIMAL MANURE, *LITERATURE REVIEW, FEEDING VALUE.

ABSTRACT:
THIS PAPER REVIEWS THE LITERATURE CONCERNED WITH FEEDING ANIMAL WASTE TO LIVESTOCK. FIBER IN DIETS FOR RUMINANTS IS NOT DIGESTED TO THE MAXIMUM POSSIBLE EXTENT DURING THE INITIAL PASS THROUGH THE DIGESTIVE TRACT. OTHER NUTRIENTS ALSO ESCAPE DIGESTION. FEEDING FECES IS NOT A NEW CONCEPT. EARLY IN THE 1940'S COW MANURE WAS LOOKED UPON AS A SOURCE OF B-COMPLEX VITAMINS. POULTRY AND CATFISH HAVE BEEN SUCCESSFULLY FED RATIONS CONTAINING FEEDLOT MANURE. THERE HAVE BEEN MANY ARTICLES CONCERNING THE USE OF POULTRY LITTER IN RUMINANT FEEDING PROGRAMS. FEEDING POULTRY FECES TO POULTRY WAS REPORTED TO HAVE NO ADVERSE EFFECT ON BIRD MORTALITY OR EGG TASTE. ALGAE GROWN ON SEWAGE HAS BEEN FED TO RATS. THE AUTHORS INDICATE THAT ALGAE IS A POTENTIALLY VALUABLE LIVESTOCK FEED. (CHRISTENBURY-IOWA STATE)

FIELD 05G

ACCESSION NO. W71-07544

FARM WASTE DISPOSAL SYSTEMS,

PURDUE UNIV., LAFAYETTE, IND. DEPT. OF AGRICULTURAL ENGINEERING.

A. C. DALE.

COOPERATIVE EXTENSION SERVICE, PURDUE UNIVERSITY, AE-8C, FEB 1971. 10 P, 1 TAB, 5 FIG, 34 REF.

DESCRIPTORS:
*FARM WASTES, *WASTE DISPOSAL, *LAGOONS, *OXIDATICN LAGOONS, AEROBIC CONDITIONS, ANAEROBIC CONDITICNS, ALGAE, ORGANIC MATTER, VOLUME, DRYING, ODOR, NITROGEN, AERATION, HOGS, CATTLE, PCULTRY, RESEARCH AND DEVELOPMENT, SOIL, SOIL CONTAMINATION.

IDENTIFIERS:
*DISPOSAL SYSTEMS, *LAND DISPOSAL, OXIDATION DITCHES, AERATED LAGOONS, COMPOSTING, REFEEDING, ANHYDROUS AMMONIA, WASTE CHARACTERIZATION.

ABSTRACT:
IN THIS PUBLICATION THE PRESENT AVAILABLE ALTERNATIVES FOR ANIMAL WASTE DISPOSAL AND CRITERIA FOR SELECTION OF THESE METHCDS ARE PRESENTED. A BRIEF LITERATURE REVIEW TELLS OF RESEARCH BEING DCNE IN ALL AREAS AND ASPECTS OF ANIMAL WASTE DISPOSAL. LAND DISPOSAL STILL REMAINS THE MOST SUITABLE AND MOST WIDELY USED DISPOSAL METHOD. RESEARCH INDICATES THAT APPROXIMATELY 250 POUNDS OF NITROGEN CAN BE ADDED TO EACH ACRE OF SOIL WITHOUT UNDULY POLLUTING IT. OTHER METHODS OF DISPCSAL DISCUSSED ARE AEROBIC, ANAEROBIC, AND MECHANICALLY AERATED LAGCCNS, AS WELL AS OXIDATION DITCHES, COMPOSTING, AND DRYING. RECOMMENDATIONS ARE MADE FOR THE CHEMICAL TREATMENT OF ANIMAL WASTES TO REDUCE ODORS WHILE SPREADING. (WHITE-IOWA STATE)

FIELD 05D

ACCESSION NO. W71-07551

DETOXICATION OF SIMAZINE BY MICROSCOPIC ALGAE,

VSESOYUZNYI NAUCHNO-ISSLEDOVATELSKII INSTITUT SELSKOKHOZYAISTVENNOI MIKROBIOLOGII, LENINGRAD.

YU V. KRUGLOV, AND L. N. PAROMENSKAYA.

TRANS FROM MIKROBIOLOGIYA, VOL 39, NO 1, JAN-FEB 1970. MICROBIOLOGY, VOL 39, NO 1, P 139-142, 1970. 1 FIG, 2 TAB, 10 REF.

DESCRIPTORS:
*HERBICIDES, *PESTICIDES, *ALGAE, *TOXICITY, SOILS, TRIAGINE PESTICIDES, RESISTANCE, CULTURES, METABOLISM.

IDENTIFIERS:
*SIMAZINE, *DETOXICATION, CHLOROSARCINA, ANKISTRCDESMUS BRAUNII, CHLORELLA VULGARIS.

ABSTRACT:
INOCULATION OF SIMAZINE-TREATED SOILS WITH A CULTURE OF CHLOROSARCINA SP SUGGESTED A DECREASE IN TCXICITY OF THE HERBICIDE, THE PROCESS BEING PARTLY INFLUENCED BY DIRECT SOLAR RADIATION. THE CHEMICAL ABSORBED BY CHLOROSARCINA AND ANKISTRODESMUS BRAUNII WAS PARTLY METABOLIZED BY CELLS AND FORMED SOME PHYSIOCOCHEMICAL BOND WITH PROTEIN. AS SHOWN BY C-14 ANALYSES, THE STRENGTH AND NATURE OF THIS BOND VARY WITH THE SPECIES OF ALGAE. CELLS OF CHLOROSARCINA, RESISTANT TO SIMAZINE, RETAINED CONSIDERABLY GREATER AMOUNT OF THE CHEMICAL THAN DIC THE CELLS OF ANKISTRODESMUS, SENSITIVE TO THE HERBICIDE. (WILDE-WISCONSIN)

FIELD 05G

ACCESSION NO. W71-07675

STUDIES OF PRIMARY PRODUCTIVITY IN COASTAL WATERS OF SOUTHERN LONG ISLAND, NEW YORK,

DOW CHEMICAL CO., FREEPORT, TEX; AND PUERTO RICO UNIV., MAYGUEZ. DEPT. OF MARINE SCIENCES; AND TOWN OF HEMPSTEAD, N.Y. DEPT. OF CONSERVATION AND WATERWAYS; AND VIRGIN ISLANDS COLL., ST. THOMAS.

E. F. MANDELLI, P. R. BURKHOLDER, T. E. DOHENY, AND

MARINE BIOLOGY, VOL 7, NO 2, P 153-160, 1970. 9 FIG, 2 TAB, 34 REF.

DESCRIPTORS:
*PRIMARY PRODUCTIVITY, *COASTS, *PHYTOPLANKTON, *ESTUARIES, *STANDING CROP, SEASONAL, DIATOMS, DINOFLAGELLATES, CHLOROPHYLL, PHOTOSYNTHESIS, TEMPERATURE, SAMPLING, SALINITY, DISSOLVED OXYGEN, NUTRIENTS, CARBON RADIOISOTOPES, PHOSPHATES, SILICATES, SOLAR RADIATION, DEPTH, THERMOCLINE, NITRATES, LIGHT INTENSITY, EUTROPHICATION, MARINE ALGAE, SEA WATER, NEW YORK.

IDENTIFIERS:
*COASTAL WATERS, *SOUTHERN LONG ISLAND, SKELETONEMA COSTATUM, THALASSIOSIRA, CHAETOCEROS, RHIZOSOLENIA ALATA, PERIDINIUM DEPRESSUM, PERIDINIUM PALLIDUM, CERATIUM MASSILENSE, CERATIUM FURCA, CERATIUM TRIPOS, CERATIUM MACROCEROS, NOCTILUCA MILIARIS, GYMNODINIUM SPLENDENS, EXUVIELLA MARINA, EUTREPTIA, NITZSCHIA SERIATA, THALASSIONEMA NITZSCHOIDES, ASTERIONELLA JAPONICA, CHLOROPHYLL-A.

ABSTRACT:
IN CONSERVATION OF COASTAL MARINE RESOURCES, MAINTENANCE OF NATURAL PRIMARY PRODUCTIVITY OF COASTAL WATERS IS SIGNIFICANT. DURING 1966, MONTHLY DETERMINATIONS OF PHYTOPLANKTON PRODUCTIVITY OF TIDAL ESTURARIES AND COASTAL WATERS OF SOUTHERN LONG ISLAND WERE MADE. SUSTAINED BLOOMS OF GREEN FLAGELLATES AND DINOFLAGELLATES WERE FOUND DURING SPRING AND SUMMER IN ESTUARINE WATERS; IN COASTAL AREAS ALTERNATING ABUNDANCES OF DIATOMS AND DINOFLAGELLATES DOMINATED THE STANDING CROP DURING LATE WINTER, EARLY FALL, AND SUMMER. THE CHLOROPHYLL-A DISTRIBUTION IN ESTUARIES EXHIBITED TWO PATTERNS, LASTING ABOUT SIX MONTHS EACH, WITH CONCENTRATION RANGING FROM 1.0 TO 27.6 MG/CU M. IN THE COASTAL AREAS ITS DISTRIBUTION SHOWED A REVERSED PATTERN WITH RANGE FROM 1.45 TO 10.15 MG/CU M. VERTICAL DISTRIBUTION OF CHLOROPHYLL-A WITHIN THE COASTAL REGION EUPHOTIC ZONE SHOWED SIMILAR PATTERNS BOTH NEARSHORE AND OFFSHORE. MEAN PHOTOSYNTHESIS RATES PER UNIT OF CHLOROPHYLL-A VARIED FROM 3.1 TO 3.5 MG CARBON/MG CHLOROPHYLL-A/HOUR; AT LIGHT SATURATION, RATIO VARIED WITH WATER TEMPERATURE AND SPECIES COMPOSITION. MEAN PRIMARY PRODUCTIVITY VALUES WERE 0.35, 0.22, 0.16 G CARBON/CU M PER DAY FOR ESTUARIES, NEARSHORE, AND OFFSHORE AREAS, RESPECTIVELY, DECREASING SEAWARD. (JONES-WISCONSIN)

FIELD 02L, 05C

ACCESSION NO. W71-07875

HIGH MOLECULAR WEIGHT ALGAL SUBSTANCES IN THE SEA,

NAVAL UNDERSEA RESEARCH AND DEVELOPMENT CENTER, PASADENA, CALIF.

J. W. HOYT.

MARINE BIOLOGY, VOL 7, NO 2, P 93-99, 1970. 3 FIG, 3 TAB, 19 REF.

DESCRIPTORS:
*MOLECULAR STRUCTURES, *ALGAE, *FLUID FRICTION, *LITTORAL, PLANT GROWTH SUBSTANCES, PHYTOPLANKTON, FLUID MECHANICS, EXUDATION, OCEANS, CHLOROPHYTA, PHAEOPHYTA, RHODOPHYTA, MARINE BACTERIA.

IDENTIFIERS:
*HIGH MOLECULAR WEIGHT COMPOUNDS, *MOLECULAR SIZE, *ALGAL EXUDATES, EXTRACELLULAR MATERIAL, TURBULENT FRICTION, SEAWEED, PORPHYRIDIUM CRUENTUM, PORPHYRA, GIGARTINA.

ABSTRACT:
TURBULENT-FLOW MEASUREMENTS IN A SPECIAL RHEOMETER REVEALED THAT COMMERCIAL SEAWEED EXTRACTS AND CERTAIN MEMBERS OF PHYTOPLANKTON HAVE EXTRACELLULAR MATERIALS OF LARGE MOLECULES WITH WEIGHTS EXCEEDING 50,000. EXTRACTS FROM A VARIETY OF LITTORAL ALGAE, PARTICULARLY OF PORPHYRA AND GIGARTINA GENERA, HAVE CHANGED THE TURBULENT FRICTION THEREBY INDICATING THE PRESENCE OF HIGH-POLYMER SUBSTANCES. PORPHYRIDIUM CRUENTUM PRODUCED FRICTION REDUCTION AS HIGH AS 60%. RESULTS SUGGEST THAT ALGAL EXUDATES ARE THE SOURCE OF HIGH MOLECULAR COMPOUNDS IN THE SEA. (WILDE-WISCONSIN)

FIELD 02L, 05C

ACCESSION NO. W71-07878

MICROBIOLOGY OF STREAMS,

KENTUCKY UNIV., LEXINGTON. DEPT. OF MICROBIOLOGY.

R. H. WEAVER, AND H. D. NASH.

IN: INFLUENCES OF STRIP MINING ON THE HYDROLOGIC ENVIRONMENT OF PARTS OF BEAVER CREEK BASIN, KENTUCKY, 1955-66, GEOLOGICAL SURVEY PROFESSIONAL PAPER 427-C, P C53-C57, 1970. 5 P, 1 FIG, 4 TAB.

DESCRIPTORS:
*AQUATIC MICROORGANISMS, *ACID MINE WATER, *STRIP MINES, *WATER POLLUTION EFFECTS, *KENTUCKY, SULFUR BACTERIA, WATER QUALITY, SULFATES, HYDROGEN ION CONCENTRATION, ALGAE, BACTERIA, WATER POLLUTION SOURCES.

IDENTIFIERS:
*BEAVER CREEK BASIN(KY).

ABSTRACT:
DRAINAGE FROM STRIP-MINED AREAS APPEARS TO HAVE AFFECTED THE MICROFLORA OF CANE BRANCH, KENTUCKY. CHEMICAL OXIDATION OF PYRITIC COMPOUNDS FOUND EXTENSIVELY IN SPOIL BANKS HAS RESULTED IN THE FORMATION OF FERROUS SULFATE AND SULFURIC ACID. THIS APPEARS TO HAVE LED TO THE ESTABLISHMENT IN THE MINED PART OF THE CANE BRANCH STUDY AREA OF FERROBACILLUS FERROOXIDANTS, WHICH CONTRIBUTES TO THE PRODUCTION OF ACID ENTERING THE STREAM. THE LOWERING OF PH HAS ENABLED THIS ORGANISM TO EXIST THROUGHOUT THE STREAM FROM THE VICINITY OF THE SPOIL BANKS DOWNSTREAM TO THE GAGING STATION. STANDARD PLATE COUNTS SHOW A MUCH SMALLER NUMBER OF SAPROPHYTIC BACTERIA IN CANE BRANCH THAN IN HELTON BRANCH. THIS, TOO, CAN BE ATTRIBUTED TO THE LOW PH OF CANE BRANCH. THE FILAMENTOUS FUNGI ARE MORE NUMEROUS AND DIVERSIFIED IN CANE BRANCH THAN IN HELTON BRANCH. IN ADDITION, THE YEAST, RHODOTORULA, WHICH IS ASSOCIATED WITH INCREASED ACID PRODUCTION BY THIOBACILLUS FERROOXIDANS, AND THE ALGA BUMILLERIA WERE ISOLATED ONLY FROM CANE BRANCH. (KNAPP-USGS)

FIELD 05C

ACCESSION NO. W71-07942

DEMONSTRATING THE EFFECTS OF NUTRIENTS IN BIO-OXIDATION POND RECEIVING STREAMS,

OKLAHOMA UNIV., NORMAN. BUREAU OF WATER RESOURCES RESEARCH.

GEORGE W. REID, LEALE E. STREEBIN, AND OLIVER T. LOVE JR.

AVAILABLE FROM THE NATIONAL TECHNICAL INFORMATION SERVICE AS PB-199 269, $3.00 IN PAPER COPY, $0.95 IN MICROFICHE. MAR 1971. 72 P, 16 FIG, 10 TAB, 49 REF, APPEND. ENVIRONMENTAL PROTECTION AGENCY PROGRAM 16010--03/71, DEMONSTRATION GRANT WPD-98-01-66.

DESCRIPTORS:
*EUTROPHICATION, *OXIDATION LAGOONS, *SEWAGE LAGOONS, BIODEGRADATION, *NUTRIENTS, ALGAE, *DEGRADATION(STREAM), WATER POLLUTION EFFECTS, WASTE WATER TREATMENT, CYANOPHYTE, ENGLENOPHYTA, ORGANIC LOADING.

ABSTRACT:
THIS STUDY CONSIDERED THE RECEIVING STREAM AS AN INTEGRAL PART OF THE BIO-OXIDATION POND METHOD OF TREATMENT WITH THE OBJECTIVE BEING TO PROVIDE A BETTER UNDERSTANDING OF THE BIO-OXIDATION POND - RECEIVING STREAM SYSTEM. AS REPRESENTATIVE OF THIS 'REAL WORLD' SITUATION WITH ALL OF ITS VARIABLES, FIVE EXISTING CENTRAL OKLAHOMA BIO-OXIDATION PONDS WHICH HAD DIVERSE LOADINGS AND DESIGNS WERE UTILIZED. BY OBSERVING THESE SYSTEMS UNDER VARYING CLIMATIC CONDITIONS, THE EFFECTS OF THE BIO-OXIDATION POND NUTRIENTS ALONG WITH OTHER POLLUTIONAL PARAMETERS WHICH WERE DISCHARGED INTO INTERMITTENT RECEIVING STREAMS WERE EVALUATED. EXCEPT FOR SCOURING, BIO-OXIDATION PONDS AND BIO-OXIDATION POND RECEIVING STREAMS WERE FOUND TO BEHAVE ESSENTIALLY THE SAME AS THE STREAMS BECAME A CONTINUATION OF THE POND. IN ADDITION TO MAKING BIOCHEMICAL ADJUSTMENTS, THE STREAMS LOST MUCH OF THEIR BIOLOGICAL IDENTITY AND ASSUMED CHARACTERISTICS MORE CLOSELY ASSOCIATED WITH THE BIOLOGICAL LOADINGS FROM THE POND EFFLUENT. THE MOST PERSISTENT ALGAE IN THE SYSTEMS WERE THE FLAGELLATES (EUGLENOPHYTA) AND THE BLUE-GREEN ALGAE (CYANOPHYTA) AS THESE PLANKTERS HAD LITTLE DIFFICULTY MAKING THE TRANSITION FROM THEIR ACCLIMATED LIFE IN THE POND TO THE STREAM.

FIELD 05C, 05D

ACCESSION NO. W71-07973

POPULATION DYNAMICS AND SALINITY TOLERANCE OF HYADESIA FUSCA (LOHMAN) (ACARINA, SARCOPTIFORMES) FROM BRACKISH WATER ROCKPOOLS, WITH NOTES ON THE MICROENVIRONMENT INSIDE ENTEROMORPHA TUBES,

STOCKHOLM UNIV. (SWEDEN). ASKO LAB.; AND STOCKHOLM UNIV. (SWEDEN). DEPT. OF ZOOLOGY.

BJORN GANNING.

OECOLOGIA (BERL), VOL 5, P 127-137, 1970. 5 FIG, 16 REF.

DESCRIPTORS:
*AQUATIC ANIMALS, *AQUATIC MICROBIOLOGY, *ACARICIDES, BRACKISH WATER, MICROENVIRONMENT, LITTORAL, OXYGEN, CHEMICAL REACTIONS, HYDROGEN ION CONCENTRATION, ALGAE, REPRODUCTION, SAMPLING, SEASONAL, BOTTOM SEDIMENTS, ICE, AMPHIPODA.

IDENTIFIERS:
*HYADESIA FUSCA, *SARCOPTIFORMES, *ROCKPOOLS, *ENTEROMORPHA, SWEDEN, TIGRIOPUS BREVICORNIS, GAMMARUS DUEBENI, FUCUS SERRATUS, FUCUS VESICULOSUS, CLADOPHORA GLOMERATA, HETEROCYPRIS SALINUS, GASTEROSTEUS ACULEATUS, BALTIC SEA, NITOCRA SPINIPES.

ABSTRACT:
THE SALT WATER MITE HYADESIA FUSCA HAS BEEN RECORDED IN SCANDINAVIA FOR THE FIRST TIME. IT IS ENTIRELY RESTRICTED TO THE GREEN ALGA ENTEROMORPHA SPP IN LITTORAL ZONES OR IN ROCKPOOLS, AND IS THE DOMINANT FAUNA OF MANY BRACKISH WATER ROCKPOOLS. THE ALGAE SERVE AS SUBSTRATUM, FOOD AND BREEDING ROOM AND GIVE RISE TO VIOLENT FLUCTUATIONS IN ABIOTIC ENVIRONMENTAL PARAMETERS. OXYGEN AVAILABILITY AND HYDROGEN-ION ACTIVITIES DO NOT DIFFER GREATLY INSIDE AND OUTSIDE THE ALGAL THALLI. DURING BREEDING IN JUNE AND JULY MORE THAN 900 ANIMALS, MOSTLY LARVAE, MAY BE FOUND ON 0.1 GRAM ENTEROMORPHA DRY WEIGHT. IN WINTER AND AVERAGE OF 13 HIBERNATING ANIMALS WERE FOUND PER 0.1 GRAM ALGAE. COPULATION FOLLOWS AFTER A LONG PRECOPULA; NEWBORN LARVAE ARE OFTEN FOUND NEAR THE RHIZOMES INSIDE ALGAL TUBES. THE MITE IS VERY TOLERANT OF SALINITY VARIATIONS. HYADESIA FUSCA IS CHARACTERISTIC OF ROCKPOOL ECOSYSTEMS, OFTEN VERY ABUNDANT, ONE OF THE FEW ENTEROMORPHA GRAZERS OF THE ECOSYSTEM AND SERVES AS FOOD FOR FISHES IN THE SYSTEM. TOGETHER WITH SOME BETTER KNOWN CRUSTACEANS THIS MITE, BECAUSE OF ITS TOLERANCE AND SURVIVAL CAPABILITIES, MUST BE CONSIDERED ONE OF THE HARDIENT MEMBERS OF THE UNSTABLE ROCKPOOL ECOSYSTEMS. (JONES-WISCONSIN)

FIELD 05C, 02L

ACCESSION NO. W71-08026

MICROFLORA OF CHLORELLA K DURING A PROLONGED CULTIVATION OF THE ALGAE IN A PERFUSION UNIT WITH A CONTINUAL RECYCLING OF THE CULTURE MEDIUM, (IN RUSSIAN),

MOSCOW STATE UNIV. (USSR). FACULTY OF BIOLOGY AND SOIL SCIENCE.

M. N. PIMENOVA, I. V. MAXIMOVA, G. I. MELESHKO, E. K. LEBEDEVA, AND T. B. GALKINA.

ENGLISH SUMMARY. MIKROBIOLOGIYA, VOL 39, NO 4, P 645-650, 1970. 1 FIG, 3 TAB, 17 REF.

DESCRIPTORS:
*LABORATORY TESTS, *CHLORELLA, *CULTURES, *AQUATIC MICROORGANISMS, SPORES, ALGAE, CYTOLOGICAL STUDIES, MYCOBACTERIUM, PSEUDOMONAS, BACTERIA.

IDENTIFIERS:
PERFUSION UNIT, FLAVOBACTERIUM, MICROCOCCUS.

ABSTRACT:
THE COMPOSITION AND DENSITY OF MICROFLORA ACCOMPANYING CHLORELLA K WERE DETERMINED DURING A PROLONGED CULTIVATION OF THE ALGAE IN A PERFUSION UNIT WITH A CONTINUAL RECYCLING OF THE NUTRIENT MEDIUM. THE PREDOMINANT ORGANISMS BELONGED TO FOUR GENERA: MYCOBACTERIUM, FLAVOBACTERIUM, MICROCOCCUS, AND PSEUDOMONAS. THE LATTER WERE PARTICULARLY NUMEROUS AND DIVERSIFIED. THE SPORE-FORMING CELLS DID NOT EXCEED 0.01% OF THE TOTAL NUMBER OF BACTERIA. THE FLUCTUATION IN THE DENSITY OF ASSOCIATED MICROORGANISMS AND SPORES WERE RELATED TO PERFORMANCE OF ALGAE AND THE RATIO BETWEEN THE PHOTOSYNTHETICALLY ACTIVE CELLS, PARENT CELLS, AND AUTOSPORES. (WILDE-WISCONSIN)

FIELD 05C, 02H

ACCESSION NO. W71-08027

CONCENTRATION OF STRONTIUM-85 AND CESIUM-137 FROM WATER SOLUTIONS BY CLADOPHORA AND PITHOPHORA,

ALABAMA UNIV., UNIVERSITY. DEPT. OF BIOLOGY.

LOUIS G. WILLIAMS.

JOURNAL OF PHYCOLOGY, VOL 6, NO 3, P 314-316, 1970. 1 TAB, 4 REF.

DESCRIPTORS:
*RADIOISOTOPES, *ALGAE, *ABSORPTION, MONITORING, ANALYTICAL TECHNIQUES, LABORATORY TESTS, CESIUM, STRONTIUM RADIOISOTOPES, IONS, RIVERS, CULTURES, ORGANIC MATTER, BACTERIA, WATER ANALYSIS.

IDENTIFIERS:
*CLADOPHORA GLOMERATA, *PITHOPHORA OEDOGONIA.

ABSTRACT:
WITH THE USE OF PACKAGED LIVE AND PRESERVED GREEN ALGAE AND OTHER ORGANIC MATERIALS, RADIONUCLIDES IN NATURAL WATERWAYS, NOT FOUND BY THE USUAL METHODS OF ANALYSIS OF RAW WATER CAN BE DETECTED. LIVE ALGAE CONCENTRATE CESIUM AND STRONTIUM IONS MUCH MORE THAN ION-EXCHANGE RESINS OR DEAD MATERIALS. THIS IS SIGNIFICANT IN THE ECOLOGY OF FOOD WEBS WHERE RADIONUCLIDES MAY BE PASSED ON TO HIGHER TROPHIC LEVELS. THE CONCENTRATION FACTORS OF NONLIVING ORGANIC MATERIALS WERE MUCH LOWER AND VARIED WITH THE MATERIAL, NOT WITH THE CALCIUM/POTASSIUM RATIO. THE EFFECT OF HIGH CALCIUM IONS AND HIGH POTASSIUM IONS WAS TO LOWER THE UPTAKE OF CESIUM MORE THAN STRONTIUM. CLADOPHORA AND PITHOPHORA WERE GROWN IN AQUARIA, PACKAGED LIVE IN PERFORATED POLYETHYLENE BAGS AND TIED SUBMERGED IN VARIOUS RIVERS FOR PERIODS OF 2 TO 7 DAYS, AFTER WHICH THE CONTENTS SHOWED HIGH CONCENTRATIONS OF MANY FISSION PRODUCTS, INCLUDING STRONTIUM AND CESIUM. WHERE AQUATIC ENVIRONMENTS ARE UNFAVORABLE FOR LIVE PITHOPHORA OR CLADOPHORA, THESE ALGAE, FIXED WITH MERTHIOLATE-IODINE PRESERVATIVE, HAVE BEEN SHOWN TO BE ABOUT TWICE AS EFFECTIVE AS MIXED ION-EXCHANGE RESINS. (JONES-WISCONSIN)

FIELD 05A

ACCESSION NO. W71-08032

DETERMINATION OF THE ACTIVITY OF HETEROTROPHIC MICROFLORA IN THE OCEAN USING C-14-CONTAINING ORGANIC MATTER,

AKADEMIYA NAUK SSSR, MOSCOW. INSTITUT BIOLOGII VNUTRENNYKH VOD.

YU. I. SOROKIN.

TRANS FROM MIKROBIOLOGIYA, VOL 3, NO 1, JAN-FEB 1970. MICROBIOLOGY (USSR) VOL 39, NO 1, P 133-138, 1970. 5 FIG, 4 TAB, 22 REF.

DESCRIPTORS:
*OCEANS, *ORGANIC MATTER, *CARBON RADIOISOTOPES, *ANALYTICAL TECHNIQUES, *MARINE MICROORGANISMS, MARINE PLANTS, BACTERIA, POPULATION, HAWAII, ABSORPTION, TROPICAL REGIONS, PACIFIC OCEAN, COLUMNS, BOTTOM SEDIMENTS, ALGAE, CHLORELLA, BENTHIC FLORA, MUD, MICROORGANISMS, PROTEINS, RADIOACTIVITY TECHNIQUES, FOOD CHAINS.

IDENTIFIERS:
*ALGAL HYDROLYSATE, TONGA ISLANDS, TUTUILA ISLAND, HONSHU ISLAND, OPEN OCEAN, SAPROPHYTES, HETEROTROPHIC BACTERIA, C-14.

ABSTRACT:
THE POSSIBILITY OF USING CARBON-LABELED DISSOLVED ORGANIC MATTER (ALGAL HYDROLYSATE) FOR ESTIMATING THE COMPARATIVE ACTIVITY OF NATURAL POPULATIONS OF MICROFLORA IN THE WATER COLUMN AND BOTTOM DEPOSITS WAS INVESTIGATED IN THE TROPICAL ZONE OF THE PACIFIC OCEAN. THE CRITERION FOR MICROFLORAL ACTIVITY WAS THE RATE OF BACTERIAL ASSIMILATION OF CARBON-14 IN EXPERIMENTS OF SHORT DURATION. THE RATE AND UTILIZATION EFFICIENCY OF LABELED ORGANIC MATTER BY SEA WATER MICROFLORA DEPEND ON ITS INITIAL CONCENTRATION. THE MAXIMUM UTILIZATION EFFICIENCY OF DISSOLVED ORGANIC MATTER BY AQUATIC MICROFLORA FOR BIOSYNTHESIS WAS 45%, STAYING QUITE HIGH EVEN WHEN THE INITIAL CONCENTRATION OF LABELED HYDROLYSATE ADDED WAS 1 TO 2 MG/L. SUPPORTED IS THE IDEA THAT BACTERIAL DESTRUCTION OF ORGANIC MATERIAL IN THE OCEAN IS LIMITED TO THE UPPER WATER LAYER TO A DEPTH OF 600 TO 800 METERS, WITH THE MAXIMUM IN THE TROPICAL ZONE AT DEPTHS OF 400-600. AT GREAT OCEAN DEPTHS THE HETEROTROPHIC MICROFLORA IS SCANT AND SLIGHTLY ACTIVE. INFORMATION ABOUT THE EFFICIENCY OF UTILIZATION OF DISSOLVED ORGANIC MATTER BY THE AQUATIC MICROFLORA IS OF PARAMOUNT IMPORTANCE FOR ESTIMATING ITS ROLE AS A LINK IN THE FOOD CHAIN. (JONES-WISCONSIN)

FIELD 05C, 07B

ACCESSION NO. W71-08042

HOW CAN PORK PRODUCERS COMPLY WITH ENVIRONMENTAL QUALITY STANDARDS,

IOWA STATE UNIV., AMES. DEPT. OF AGRICULTURAL ENGINEERING.

J. RONALD MINER.

AMERICAN PORK CONGRESS-PROCEEDINGS, ENVIRONMENTAL QUALITY WORKSHOP, DES MOINES, IOWA, MAR 3, 1971. P 98-102.

DESCRIPTORS:
*FARM WASTES, *HOGS, *ENVIRONMENT, *POLLUTION ABATEMENT, WATER QUALITY, STANDARDS, WATER POLLUTION, AIR POLLUTION, ODOR, CONFINEMENT PENS, ORGANIC MATTER, NUTRIENTS, NITROGEN, PHOSPHORUS, EUTROPHICATION, ALGAE, PATHOGENIC BACTERIA, EFFLUENT, IRRIGATION, STORAGE, WASTE DISPOSAL.

IDENTIFIERS:
*ENVIRONMENTAL QUALITY, AIR CONTAMINANT, STREAM QUALITY, WASTE MANAGEMENT, MANURE COLLECTION, MANURE TRANSPORT.

ABSTRACT:
TO PREVENT WATER AND AIR POLLUTION WHILE MAINTAINING ENVIRONMENTAL QUALITY IS A COMPLEX PROBLEM. IT BECOMES MORE COMPLICATED BY TRYING TO DESIGN WASTE MANAGEMENT SYSTEMS WHICH CONTRIBUTE MATERIALLY TO OUR EFFECTIVENESS AS PORK PRODUCERS WITH POLLUTION CONTROL AS A SIDE BENEFIT. A SWINE MANURE MANAGEMENT SYSTEM MIGHT INCLUDE A COLLECTION DEVICE, A MANURE TRANSPORT SYSTEM, SOME MEANS OF MANURE STORAGE AND/OR TREATMENT, AND FINALLY, A MANURE OR EFFLUENT DISPOSAL SYSTEM. IN SOME CASES MORE THAN ONE OF THESE COMPONENTS MAY BE INCLUDED IN A SINGLE COMPONENT. GIVING INITIAL CONSIDERATION TO THE DISPOSAL SCHEME WILL HELP DETERMINE DECISIONS TO BE MADE CONCERNING THE OTHER ASPECTS OF THE SYSTEM. THERE IS MUCH REMAINING TO BE LEARNED RELATIVE TO THE CONTROL AND MEASUREMENT OF ODORS. VARIOUS ODOR LEVELS CAN BE ACHIEVED BY THE JUDICIOUS SELECTION OF MANURE HANDLING TECHNIQUES. (WHITE-IOWA STATE)

FIELD 05D, 05G

ACCESSION NO. W71-08214

NITRATE REMOVAL FROM AGRICULTURAL WASTE WATER,

FEDERAL WATER POLLUTION CONTROL ADMINISTRATION, FRESNO, CALIF.; AND CALIFORNIA DEPT. OF WATER RESOURCES, FRESNO.

PERCY P. ST. AMANT, AND LOUIS A. BECK.

IN: WATER QUALITY MANAGEMENT PROBLEMS IN ARID REGIONS, WATER POLLUTION CONTROL RESEARCH SERIES, 13030 DYY, 6/69, OCT 1970, USDI, FEDERAL WATER QUALITY ADMINISTRATION, P 1-8. 1 TAB, 1 FIG.

DESCRIPTORS:
*RETURN FLOW, *NITRATES, WATER POLLUTION, DESALINATION, ALGAE, DENITRIFICATION, CALIFORNIA, FILTERS, PARTICLE SIZE, ANAEROBIC CONDITIONS, WASTE WATER TREATMENT.

IDENTIFIERS:
*NITRATE REMOVAL, ALGAE STRIPPING, POND DENITRIFICATION, FILTER DENITRIFICATION, METHANOL, BACTERIAL DENITRIFICATION, SAN JOAQUIN VALLEY.

ABSTRACT:
THE PROBLEM OF DISPOSING OF IRRIGATION WASTE WATER FROM THE SAN JOAQUIN VALLEY OF CALIFORNIA IS A VERY LARGE ONE. THE MOST SERIOUS POTENTIAL POLLUTANT IS NITROGEN IN THE NITRATE FORM. A WASTE WATER TREATMENT CENTER AT FIREBAUGH, CALIFORNIA HAS ORGANIZED AND IS CARRYING OUT RESEARCH IN THE AREAS OF DESALINATION, ALGAE STRIPPING, AND BACTERIAL DENITRIFICATION. ALGAE STRIPPING SIMPLY INVOLVES GROWING A CROP OF ALGAE TO REMOVE NITROGEN FROM THE WATER, AND THEN HARVESTING THE ALGAE. VARIOUS MARKETS HAVE BEEN PROPOSED FOR THE USE OF ALGAE. TWO METHODS OF BACTERIAL DENITRIFICATION BEING EXPLORED ARE POND DENITRIFICATION, AND FILTER DENITRIFICATION. THE THREE DENITRIFICATION METHODS ARE COMPARED AS TO LAND REQUIREMENTS AND PROJECT COSTS. EACH IS NEARLY THE SAME IN COST - AROUND $10 PER ACRE FOOT, HOWEVER THE ALGAE STRIPPING METHOD REQUIRES MUCH MORE LAND. (WHITE-IOWA STATE)

FIELD 05D, 05G

ACCESSION NO. W71-08223

TRANSFER OF TOXIC ALGAL SUBSTANCES IN MARINE GOOD CHAINS,

HAWAII UNIV., HONOLULU. DEPT. OF BOTANY.

MAXWELL S. DOTY, AND GERTRUDES AGUILAR-SANTOS.

PACIFIC SCIENCE, VOL 24, P 351-355, JULY 1970. 1 FIG, 2 TAB, 11 REF. USPHS GRANT NO FD-00101-03, NIH GRANT NO 5-R01-GM-151-98-03, AEC CONTRACT NO AT (04-3)-235.

DESCRIPTORS:
*FOOD WEBS, *FOOD CHAINS, *TOXICITY, *ALGAL TOXINS, *CHLOROPHYTA, *PHYTOTOXINS, ECOLOGY, FISH FOOD ORGANISMS, PATH OF POLLUTANTS, BIOASSAY, TOXINS, POISONS, POISONOUS PLANTS, CORAL, INVERTEBRATES, ALGAE, MARINE ALGAE, ALGAL POISONING, CHROMATOGRAPHY, CHEMICAL ANALYSIS, BIOCHEMISTRY, PLANT PHYSIOLOGY, BIOLOGICAL COMMUNITIES.

IDENTIFIERS:
*CIGUATERA, CAULERPA SP., CAULERPICIN, CAULERPIN, PALMITIC ACID, BETA-SITOSTEROL, THIN-LAYER CHROMATOGRAPHY.

ABSTRACT:
ALCOHOLIC AND ETHER EXTRACTS OF OBLIGATE HERBIVORES, OMNIVORES AND DETRITUS FEEDERS COMMON ON CAULERPA OR IN ITS COMMUNITIES WERE SOMETIMES FOUND BY THIN-LAYER CHROMATOGRAPHY TO CONTAIN VARYING AMOUNTS OF CAULERPICIN, CAULERPIN, PALMITIC ACID, AND BETA-SITOSTEROL. CERITHIUM AND SOFT CORALS, WHICH MAY BE EITHER OMNIVOROUS OR CARNIVOROUS, ON OCCASION CONTAIN CAULERPICIN. THE CRUSTACEAN DETRITUS FEEDERS DID NOT SEEM TO PRESERVE EITHER CAULERPICIN OR CAULERPIN. CAULERPICIN AND CAULERPIN, WHICH, AS PRODUCED BY CAULERPA ARE PHYSIOLOGICALLY ACTIVE AND TOXIC TO RATS AND TO MICE, ARE APPARENTLY TRANSFERRED ALONG THE FOOD CHAINS AND CONCENTRATED IN AT LEAST SOME HERBIVORES. (LEGORE-WASHINGTON)

FIELD 05C

ACCESSION NO. W71-08597

LAKE MINNETONKA: NUTRIENTS, NUTRIENT ABATEMENT, AND THE PHOTOSYNTHETIC SYSTEM OF THE PHTOPLANKTON,

MINNESOTA UNIV., MINNEAPOLIS. LIMNOLOGICAL RESEARCH CENTER.

ROBERT O. MEGARD.

AVAILABLE FROM THE NATIONAL TECHNICAL INFORMATION SERVICE AS PB-199 915, $3.00 IN PAPER COPY, $0.95 IN MICROFICHE. FINAL REPORT, DEC 1970. 210 P, 21 FIG, 10 TAB, 41 REF, 3 APPEND. OWRR PROJECT A-016-MINN(4).

DESCRIPTORS:
*LIMNOLOGY, *LAKES, *PHOTOSYNTHESIS, *PHYTOPLANKTON, *PRODUCTIVITY, ALGAE, NITROGEN, PHOSPHORUS, MATHEMATICAL MODELS, NUTRIENTS, MEASUREMENT, POPULATION, CHLOROPHYLL, WATER POLLUTION EFFECTS, TEMPERATURE, CARBON, DIATOMS, LIGHT, OXYGEN, MINNESOTA.

IDENTIFIERS:
*LAKE MINNETONKA(MINN).

ABSTRACT:
DENSITY OF PLANKTONIC ALGAE, RATES OF PHOTOSYNTHESIS, NUTRIENT CONCENTRATIONS, AND OTHER LIMNOLOGICAL CHARACTERISTICS WERE RECODED DURING 1968-1970 IN DIFFERENT PARTS OF LAKE MINNETONKA, MINN. IN THE COURSE OF THE PAST 30 YEARS, THE DENSITY OF ALGAE INCREASED 2 OR 3 TIMES, AND THE DOMINANCE OF DIATOMS WAS REPLACED BY THAT OF BLUE-GREEN ALGAE. A LINEAR RELATIONSHIP WAS DISCLOSED BETWEEN THE ALGAE DENSITY AND THE CONTENT OF TOTAL PHOSPHORUS WITHIN THE RANGE OF 50 TO 200 MG/CU M. A DECREASE OF PHOSPHORUS BELOW 45 MG/CU M PROMISES A RAPID IMPROVEMENT OF THE LAKE QUALITY. THE NUTRIENT ENRICHMENT OF THE LAKE IS LARGELY DUE TO THE LAND RUNOFF AND STORM DRAINAGE, RATHER THAN THE DISCHARGE OF SEWAGE. A MATHEMATICAL MODEL OF THE PHOTOSYNTHETIC SYSTEM WAS BASED ON THE RELATION BETWEEN THE INTENSITY OF ILLUMINATION AND THE BIOMASS OF ALGAE. THE REPORT INCLUDES A DETAILED RECORD TO TAXA AND DENSITIES OF ALGAE DURING DIFFERENT MONTHS OF 1968 AND 1969. (WILDE-WISCONSIN)

FIELD 05C

ACCESSION NO. W71-08670

METABOLICALLY ACTIVE SPHEROPLASTS OF BLUE-GREEN ALGAE (IN RUSSIAN),

MOSCOW STATE UNIV (USSR). DEPT. OF MICROBIOLOGY.

M. V. GUSEV, K. A. NIKITINA, AND T. G. KORZHENEVSKAYA.

ENGLISH SUMMARY. MIKROBIOLOGIYA, VOL 39, NO 5, P 862-868, 1970. 5 FIG, 2 TAB, 15 REF.

DESCRIPTORS:
*ALGAE, *METABOLISM, *CYTOLOGICAL STUDIES, PHOTOSYNTHESIS, RESPIRATION.

IDENTIFIERS:
*LYSOZYME, *SPHEROPLASTS, BLUE-GREEN ALGAE, ANABAENA VARIABILIS, ANACYSTIS NIDULANS.

ABSTRACT:
THE USE OF LYSOZYME PERMITTED A QUANTITATIVE SEPARATION OF SPHEROPLASTS FROM THE CELLS OF BLUE-GREEN ALGAE ANABAENA VARIABILIS AND ANACYSTIS NIDULANS. THE YIELD OF SPHEROPLASTS WAS CORRELATED WITH THE CONCENTRATION OF LYSOZYME, THE VOLUME OF REACTION MIXTURE, CONDITIONS OF MIXING, AND THE PERIOD OF INCUBATION. THE TOLERANCE OF CELLS TO LYSOZYME WAS DEPENDENT ON THE SPECIES AND THE GROWTH PHASE OF ALGAE. SPHEROPLASTS OF A VARIABILIS EXHIBITED ENDOGENOUS RESPIRATION AT THE RATE SIMILAR TO THAT OF THE INTACT CELLS. A DECREASE IN THE CONTENT OF PHYCOCYANIN IMMOBILIZED THE ABILITY OF SPHEROPLASTS TO EMIT OXYGEN IN LIGHT, THUS CONFIRMING THE CRITICAL ROLE OF PHYCOCYANIN IN PHOTOCHEMICAL REACTIONS. (WILDE-WISCONSIN)

FIELD 05C

ACCESSION NO. W71-08683

STIMULATORY PROPERTIES OF FILTRATE FROM THE GREEN ALGA HORMOTILA BLENNISTA. I DESCRIPTION,

PROVIDENCE COLL., R.I. DEPT. OF BIOLOGY; AND CONNECTICUT UNIV., STORRS. BIOLOGICAL SCIENCES.

THOMAS J. MONAHAN, AND FRANCIS R. TRAINOR.

JOURNAL OF PHYCOLOGY, VOL 6, NO 3, P 263-269, 1970. 3 FIG, 4 TAB, 37 REF. OWRR PROJECT A-014-CONN(3).

DESCRIPTORS:
*ALGAE, *GROWTH RATES, *PRODUCTIVITY, *PLANT GROWTH SUBSTANCES, PHOTOSYNTHESIS, METABOLISM, BACTERIA, HYDROGEN ION CONCENTRATION, LABORATORY TESTS.

IDENTIFIERS:
*GREEN ALGAE, *HORMOTILA BLENNISTA, FILTRATES, GROWTH AUTOSTIMULATION, GROWTH REGULATION, EXTRACELLULAR PRODUCTS, ALGAL EXCRETIONS.

ABSTRACT:
AUTOSTIMULATION OF GROWTH BY FILTRATES OF HORMOTILA BLENNISTA IS ATTRIBUTED TO SECRETION OF ORGANIC METABOLITES. THE MAXIMUM STIMULATION IN EXCESS OF 100% WAS EXERTED BY FILTRATES OBTAINED FROM 1 TO 4 WEEK OLD ACTIVELY GROWING CULTURES. THE FILTRATES SUPPORTED BACTERIAL GROWTH AND STIMULATED THE GROWTH OF SCENEDESMUS AT PH 6.3, BUT NOT AT PH 7.7. STIMULATORY PROPERTIES OF FILTRATES WERE TERMINATED BY AUTOCLAVE TREATMENT. A SUGGESTION IS MADE THAT EXTRACELLULAR SECRETIONS OF H BLENNISTA INFLUENCE THE SURVIVAL OF THE ALGA AND THE GROWTH OF OTHER ORGANISMS. (WILDE-WISCONSIN)

FIELD 05C

ACCESSION NO. W71-08687

WATER POLLUTION FROM PHOSPHATE,

ALICE Q. HOWARD.

ENFO, JANUARY 1971, 4 P, 1 TAB.

DESCRIPTORS:
*PHOSPHATES, *DETERGENTS, *WATER POLLUTION EFFECTS, *WATER POLLUTION CONTROL, *BIODEGRADATION, FISHKILL, EUTROPHICATION, ALGAL BLOOMS, WATER QUALITY.

IDENTIFIERS:
ENZYME PRESOAKS, DISHWATER DETERGENTS, NTA.

ABSTRACT:
AN EVALUATION OF WATER POLLUTION BY PHOSPHATES IN HOUSEHOLD DETERGENTS IS PRESENTED. THE DISTINCTION IS MADE BETWEEN BIODEGRADABILITY, OR THE BREAKDOWN ABILITY OF THE CLEANING AGENT OF DETERGENTS IN WASTE WATER, AND THE PHOSPHATE PROBLEM. PHOSPHATES ARE PLACED IN DETERGENTS TO ACT AS WATER SOFTENERS TO MAKE THE CLEANING AGENT MORE EFFECTIVE. THE CYCLE OF POLLUTION EFFECTS FROM PHOSPHATE-INDUCED PLANT GROWTH INCLUDE EXCESSIVE ALGAE BLOOMS, EXCESSIVE OXYGEN USE AS THE PLANTS LATER DIE, AND FINALLY SUFFOCATION OF ANIMAL LIFE SUCH AS FISHES. THE USDI ESTIMATES THAT 50-70% OF PHOSPHORUS IN CITY SEWAGE COMES FROM DETERGENTS. PHOSPHATE SUBSTITUTES SUCH AS NTA REQUIRE EXTENSIVE RESEARCH TO BE SURE THEY DO NOT CAUSE PROBLEMS OF THEIR OWN. ENZYME PRESOAKS AND AUTOMATIC DISHWATER DETERGENTS HAVE HIGH PHOSPHATE LEVELS. SUGGESTIONS FOR CONSUMER ACTION TO REDUCE PHOSPHATE USE ARE GIVEN, INCLUDING A LIST OF DETERGENTS AND THEIR PHOSPHATE PER LOAD CONTENT. (MCENTYRE-PAI)

FIELD 05C, 05G

ACCESSION NO. W71-08768

NUTRIENT MANAGEMENT IN THE POTOMAC ESTUARY,

FEDERAL WATER QUALITY ADMINISTRATION, ANNAPOLIS, MD. CHESAPEAKE SUPPORT LAB.

NORBERT A. JAWORSKI, DONALD W. LEAR, JR., AND ORTERIO VILLA, JR.

TECHNICAL REPORT 45, JANUARY 1971, 69 P, 21 FIG, 3 TAB.

DESCRIPTORS:
*ESTUARIES, *NUTRIENTS, WATER QUALITY CONTROL, *ECOLOGY, *ALGAE, *EUTROPHICATION, NUTRIENT REQUIREMENTS, MUNICIPAL WASTES.

IDENTIFIERS:
*POTOMAC RIVER ESTUARY.

ABSTRACT:
THE WATER QUALITY OF THE UPPER POTOMAC ESTUARY HAS BEEN DEGRADED AS A RESULT OF MUNICIPAL WASTEWATER FROM THE METROPOLITAN AREA OF WASHINGTON. PAST STUDIES INDICATED HIGH COLIFORM DENSITIES, LOW DISSOLVED OXYGEN CONTENT, AND LARGE POPULATIONS OF BLUE-GREEN ALGAE AS MAJOR WATER QUALITY MANAGEMENT PROBLEMS OF THE UPPER AND MIDDLE AREAS OF THE ESTUARY. STUDIES AND CONCEPTS USED TO FORMULATE A NUTRIENT MANAGEMENT PROGRAM FOR THE POTOMAC ESTUARY ARE PRESENTED. CURRENT WATER QUALITY CONDITIONS AND ECOLOGICAL TRENDS ARE DISCUSSED. THE SOURCES, CONTROLLABILITY, TRANSPORT AND CRITERIA ESTABLISHMENT OF NUTRIENTS ARE DISCUSSED AND SUGGESTIONS FOR WASTEWATER TREATMENT REQUIREMENTS AND A WATER QUALITY MANAGEMENT PROGRAM ARE MADE. (ENSIGN-PAI)

FIELD 05C, 02L

ACCESSION NO. W71-08775

THE SANTA BARBARA OIL SPILLS OF 1969: A POST-SPILL SURVEY CF THE ROCKY INTERTIDAL,

UNIVERSITY OF SOUTHERN CALIFORNIA, LOS ANGELES. ALLAN HANCOCK FOUNDATION.

NANCY L. NICHOLSON, AND ROBERT L. CIMBERG.

IN: BIOLOGICAL AND OCEANOGRAPHICAL SURVEY OF THE SANTA BARBARA CHANNEL OIL SPILL, 1969-1970, VOL 1, BIOLOGY ANC BACTERIOLOGY, P 325-399, 1971. 18 TAB, 25 REF, 9 PLATES.

DESCRIPTORS:
*MARINE PLANTS, *OILY WATER, WATER POLLUTICN EFFECTS, *INTERTIDAL AREAS, *INVERTEBRATES, *ALGAE, CALIFORNIA.

IDENTIFIERS:
*SANTA BARBARA CHANNEL.

ABSTRACT:
DIFFERENCES BETWEEN THE JANUARY 1969 SANTA BARBARA OIL SPILL AND EARLIER SPILLS, AS THE 1957 TAMPICO MARU ACCIDENT AND THE 1969 TORREY CANYON ACCIDENT ARE CCMPARED. IN THE TWO SHIP INCIDENTS, MASSIVE KILLS OF INTERTIDAL ORGANISMS WERE DUE TO TOXIC COMPONENTS IN DIESEL OIL OR DISPERSANTS. SPECTULAR EFFECTS WERE NOT FOUND AT THE SANTA BARBARA INTERTIDAL, EVEN AFTER A YEAR STUDY. COMPARISONS ARE MADE OF MARINE PLANT POPULATIONS FROM MONTHLY SAMPLINGS TAKEN AT TEN ROCKY INTERTIDAL STATIONS IN ANC OUTSIDE OF THE SPILL AREA. SELECTION OF SAMPLING SITES, SAMPLING METHODS USED, AND TREATMENT OF THE CATA ACCOMPANY DETAILED DISCUSSION OF STUDY FINDINGS. MANY SPECIES ARE LISTED IN EXTENSIVE TABLES THAT CHART THE STUDY FINDINGS. (MOE-PAI)

FIELD 05C

ACCESSION NO. W71-08792

WHAT HAS BEEN THE EFFECT OF THE SPILL ON THE ECOLOGY IN THE SANTA BARBARA CHANNEL,

UNIVERSITY OF SOUTHERN CALIFORNIA, LOS ANGELES. ALLAN HANCOCK FOUNDATION.

DALE STRAUGHAN.

IN: BIOLOGICAL AND OCEANOGRAPHICAL SURVEY OF THE SANTA BARBARA CHANNEL OIL SPILL, 1969-1970, VOL 1, BIOLOGY AND BACTERIOLOGY, P 401-426, 1971. 1 FIG, 5 TAB, 11 REF.

DESCRIPTORS:
*OIL, SEEPAGE, WATER POLLUTION EFFECTS, ENVIRONMENTAL EFFECTS, *BENTHIC FAUNA, *INVERTEBRATES, *FISH POPULATICNS, *MORTALITY, *ALGAE, CALIFORNIA.

IDENTIFIERS:
*SANTA BARBARA CHANNEL.

ABSTRACT:
BECAUSE MANY QUESTICNS LACK ANSWERS, THE EFFECT ON ECOLOGY OF THE 1969 SANTA BARBARA CHANNEL OIL SPILL MAY NEVER BE KNOWN. ENVIRONMENTAL FACTORS SUCH AS PRIOR NATURAL SEEPAGE, HEAVY RAINS, INCREASED SEDIMENTATION AND A POSSIBLE INCREASE IN PESTICIDES, CONTRIBUTE TO THE PROBLEMS OF ISOLATING OIL SPILL EFFECTS. EVIDENCE SUGGESTS INCREASED PRODUCTIVITY OF BENTHIC FAUNA IN INSHORE WATERS AFTER THE SPILL, DUE POSSIBLY TO INCREASED NUTRIENTS. STUDIES OF SANDY BEACH FAUNA SHOWED NO DIRECT EFFECTS. INVERTEBRATES WHOSE SHELLS WERE CCVERED WITH OIL APPEARED HEALTHY. FISH SURVEYS INDICATE A STABLE FISH POPULATICN AND THERE IS NO PROOF THAT OIL WAS DIRECTLY RESPONSIBLE FOR MARINE MAMMAL DEATHS. HIGH MORTALITY RATES WERE RECORDED IN PELAGIC BIRD POPULATIONS, MARINE GRASS, BARNACLE SPECIES AND IN MARINE ALGAE. HYPOTHESES ARE PRESENTED AS TO WHY THE SPILL CAUSED SO LITTLE DAMAGE TO THE AREA'S ECOLOGY, ESPECIALLY WHEN COMPARED TO OTHER SPILLS.
(MOE-PAI)

FIELD 05C

ACCESSION NO. W71-08793

WATER RECLAMATION AND ALGAE HARVESTING,

ASIAN INST. OF TECH., BANGKOK (THAILAND).

M. G. MCGARRY, AND C. TONGKASAME.

JOURNAL WATER POLLUTION CONTROL FEDERATION, VOL 43, NO 5, MAY 1971, P 824-835. 9 FIG, 3 TAB, 9 REF.

DESCRIPTORS:
*PONDS, *WATER REUSE, RECLAMATION, ALGAE, PROTEIN, *OXIDATION LAGOONS, CLIMATIC ZONES, PRECIPITATION, COAGULATION, LIME, HYDROGEN ION CONCENTRATION, TEMPERATURE, CHLORINATION, FLOATATION, DEWATERING, DRYING, COST ANALYSES, WASTE WATER TREATMENT.

IDENTIFIERS:
*ALGAE HARVESTING.

ABSTRACT:
APPLICATION OF HIGH RATE OXIDATION PONDS EQUIPPED WITH ALGAE HARVESTING MAY HELP TO AUGMENT THE DWINDLING WATER SUPPLIES OF LARGE METROPOLITAN AREAS BY RECLAIMING WASTE WATER FOR VARIED USES, WHILE AT THE SAME TIME PRODUCING A USABLE ALGAL FEED SUPPLEMENT. CONDITIONS FOR OPTIMAL OPERATION INCLUDE: (1) 200 LB BOD/ACRE/DAY; (2) 17.7 IN. DEPTH; AND (3) 1 DAY DETENTION TIME. PONDS OPERATED IN THIS MANNER PROVIDE AN AVERAGE EFFLUENT BOD (AFTER ALGAE REMOVAL) OF LESS THAN 10 MG/L, AND ONE ACRE OF POND CAN PRODUCE 100,000 LB PER YER OF ALGAE CONTAINING 60% PROTEIN. CHEMICAL COAGULATION AND PRECIPITATION CHEMICALS STUDIED INCLUDED LIME, ALUM, AND 50 DIFFERENT POLYELECTROLYTES. POLYCATIONS WERE FOUND TO BE MOST ECONOMIC, BUT USAGE OF POLYELECTROLYTES AS AIDS CONTRIBUTED A GREATER CHEMICAL COST THAN USAGE OF ALUM ALONG AT PH 6.5. THE DOWNFLOW SOLIDS CONTACT SYSTEM WAS EXAMINED FOR REMOVAL OF THE ALGAE, EITHER BY THE SPLIT FLOW, DISSOLVED AIR, OR SUPERSATURATED OXYGEN PRINCIPLES. THE ALGAL PASTE WAS SUN DRIED ON UNDRAINED FLAT PLATES, TO LESS THAN 10% MOISTURE. AT SOLAR ENERGY LEVELS OF 480 G CAL/SQ CM/DAY, 2800 LBS/DAY OF DRIED ALGAE COULD BE PROCESSED ON ONE ACRE. THE RESULTS OF THE INVESTIGATIONS WERE INCORPORATED INTO AN URBAN MODEL WHICH INCLUDES RECYCLING OF CLARIFIED POND EFFLUENT FOR HOUSEHOLD CLEANING PURPOSES. A 67% REDUCTION IN RAW WATER NEEDS WAS PREDICTED FOR SUCH A SYSTEM, BUT MUCH FURTHER INVESTIGATION AND REFINEMENT IS NECESSARY. (LOWRY-TEXAS)

FIELD 05D

ACCESSION NO. W71-08951

NUTRIENTS AND NUTRIENT BUDGET IN THE BAY OF QUINTE, LAKE ONTARIO,

ONTARIO WATER RESOURCES COMMISSION, TORONTO.

M. G. JOHNSON, AND G. E. OWEN.

JOURNAL WATER POLLUTION CONTROL FEDERATION, VOL 43, NO 5, MAY 1971, P 836-853. 8 FIG, 7 TAB, 37 REF.

DESCRIPTORS:
*NUTRIENTS, *NITROGEN, *PHOSPHORUS, ALGAE, *EUTROPHICATION, AQUATIC ENVIRONMENTS, GREAT LAKES, BAYS, WATERSHEDS(BASINS), SURFACE DRAINAGE, MUNICIPAL WASTES, INDUSTRIAL WASTES, SAMPLING, WATER QUALITY CONTROL, COST ANALYSIS, WASTE WATER TREATMENT, NUTRIENT REQUIREMENTS.

IDENTIFIERS:
BAY OF QUINTE, *LAKE ONTARIO.

ABSTRACT:
ALGAE BLOOMS IN THE BAY OF QUINTE, LAKE ONTARIO, HAVE BEEN INCREASING IN SEVERITY AND LENGTH OVER THE PAST SEVERAL YEARS. EXAMINATION OF AQUATIC LIFE IN THE BAY REVEALED AN OVERWHELMING PREDOMINANCE OF ORGANISMS WHICH READILY FUNCTION IN ORGANICALLY RICH WATERS. INPUT TO THE BAY WAS MONITORED IN 1968. DURING THAT TIME, THE BAY RECEIVED 9.7 MILLION LBS OF NITROGEN AND 731,000 LBS OF PHOSPHATES FROM ALL SOURCES. APPROXIMATELY 90% OF THE ENTERING NITROGEN ENTERED VIA THE RIVERS WHICH EMPTY INTO THE BAY, AND THESE SAME RIVERS ACCOUNTED FOR 60% OF THE PHOSPHORUS INPUT. THE REMAINING 10% OF THE NITROGEN AND 40% OF THE PHOSPHORUS WAS CONTRIBUTED FROM MUNICIPAL-INDUSTRIAL SOURCES. HOWEVER, MUNICIPAL-INDUSTRIAL FLOWS ARE HIGH STRENGTH LOW VOLUME EFFLUENTS WHICH DISPLACE ONLY A SMALL VOLUME OF WATER FROM THE LAKE. ON THE OTHER HAND, RIVER FLOWS ARE EXTREMELY LOW STRENGTH HIGH VOLUME ADDITIONS TO THE BAY WHICH DISPLACE LARGE VOLUMES OF WATER. THEREFORE, IT WAS ESSENTIAL TO CONSIDER THE NET ADDITIONS OF P AND N RATHER THAN THE TOTAL ADDITIONS, THE NET ADDITION BEING THE AMOUNT OF NUTRIENT CONTAINED IN AN INPUT WHICH IS IN EXCESS OF THE AMOUNT OF NUTRIENTS CONTAINED IN THE VOLUME OF WATER DISPLACED AT THE OUTLET. ON THIS BASIS, 50% OF THE NET NITROGEN AND 85% OF THE NET PHOSPHORUS WERE CONTRIBUTED BY MUNICIPAL-INDUSTRIAL SOURCES. PHOSPHORUS REMOVAL FROM HIGH CONCENTRATION, LOW VOLUME INPUTS, AT A COST OF $200,000 PER YEAR WAS RECOMMENDED. (LOWRY-TEXAS)

FIELD 05C

ACCESSION NO. W71-08953

PILGRIM NUCLEAR POWER STATION. PRELIMINARY SAFETY ANALYSIS REPCRT, AMENDMENT 16.

BOSTON EDISON CO., MASS.

AVAILABLE FROM THE NATIONAL TECHNICAL INFORMATION SERVICE AS DOCKET-50293-31, $3.00 IN PAPER COPY, $0.95 IN MICROFICE. REPORT DOCKET-50293-31, OCT 2, 1970. 189 P.

DESCRIPTORS:
*WATER POLLUTION CONTROL, *NUCLEAR POWER PLANTS, *POTABLE WATER, *BIOTA, SAMPLING, MARINE FISH, MARINE ANIMALS, MARINE ALGAE, TRITIUM, IODINE RADIOISOTOPES, STRONTIUM RADIOISOTOPES, ZINC RADIOISOTOPES, COBALT RADIOISOTOPES, ABSORPTION, WATER POLLUTION SOURCES.

IDENTIFIERS:
MANGANESE RADIOISOTOPES, CESIUM RADIOISOTOPES.

ABSTRACT:
THE SURVEILLANCE PROGRAM OUTSIDE THE STATION SECURITY FENCE INCLUDES THE AIR, DOMESTIC AND SEA WATER, MARINE LIFE, SILT, MILK, AND CROPS. WATER PUMPING STATIONS WHICH CONTROL WATER SUPPLY TO MOST AREA RESIDENTS ARE SAMPLED AND ANALYZED FOR GROSS BETA AND GAMMA BIWEEKLY, AND FOR TRITIUM AND STRONTIUM MONTHLY. SEA WATER INTAKE AND DISCHARGE CANALS ARE ANALYZED MCNTHLY FOR GROSS BETA AND GROSS GAMMA (WITH GAMMA SPECTRUM ANALYSIS IF THE GROSS GAMMA INCREASES), AND TRITIUM. QUARTERLY COMPOSITES OF THE MONTHLY SAMPLES ARE ANALYZED FCR MANGANESE-54, ZINC-65, COBALT-58, AND COBALT-60. FISH, LOBSTERS, AND MOLLUSKS ARE SAMPLED QUARTERLY; IRISH SEA MOSS, DURING ITS HARVEST PERIOD. ALL MARINE LIFE SAMPLES ARE TAKEN FROM THE VICINITY OF THE DISCHARGE CANAL OUTFALL AND OFFSHORE CURRENT PATTERNS AND ARE ANALYZED FOR GROSS BETA, GROSS GAMMA, STRONTIUM-90, CESIUM-137, MANGANESE-54, COBALT-58, COBALT-60, ZINC-65, AND IODINE-131. EQUIPMENT FAILURES ARE ANALYZED WITH RESPECT TO RELEASE OF RADIONUCLIDES TO THE ENVIRONMENT. (BOPP-NSIC)

FIELD 05B

ACCESSION NO. W71-09018

SHORELINE ALGAE OF WESTERN LAKE ERIE,

OHIO STATE UNIV., COLUMBUS. GRADUATE STUDIES IN BOTANY.

RACHEL COX DOWNING.

THE OHIO JOURNAL OF SCIENCE, VOL 70, NO 5, P 257-276, 1970. 97 FIG, 37 REF.

DESCRIPTORS:
*LAKE SHORES, *ALGAE, *LAKE ERIE, AQUATIC HABITATS, LAKES, AQUATIC ENVIRONMENT.

IDENTIFIERS:
*ALGAL SPECIES, WESTERN LAKE ERIE, ARNCLDIELLA CCNCHOPHILA MILLER.

ABSTRACT:
IN SPITE OF SOME 70 YEAR INVESTIGATIONS OF ALGAE INHABITING WESTERN LAKE ERIE, ALMOST NOTHING WAS KNOWN PRIOR TO THIS STUDY OF THE SHORELINE AS A SPECIFIC HABITAT OF THESE ORGANISMS. THIS SITE HARBORS 61 TAXA, 39 OF WHICH ARE NEW RECORDS FCR THIS PART OF THE LAKE, AND ONE, ARNOLDIELLA CONCHOPHILA MILLER, WAS PREVIOUSLY REPORTED ONLY FROM CENTRAL RUSSIA. (WILDE-WISCONSIN)

FIELD 05C

ACCESSION NO. W71-09156

FUTURE PROSPECTS OF ALGAE AND MAN,

VERMONT UNIV., BURLINGTON. DEPT. OF BOTANY.

RICHARD M. KLEIN.

ANNALS NEW YORK ACADEMY OF SCIENCES, VOL 175, P 778-781, 1970. 9 REF.

DESCRIPTORS:
*ALGAE, *HUMAN POPULATION, FERTILIZERS, DIATOMACEOUS EARTH, FOOD CHAINS, OXYGEN, ECOLOGY, RHODOPHYTA, PHAEOPHYTA, CHLORELLA, PLANT PATHOLOGY, INDUSTRIES, IONS, CYANOPHYTA, RADIOISOTOPES, TEMPERATURE, ECOSYSTEMS, CALIFORNIA, IRRIGATION, ECONOMIC FEASIBILITY, WATER POLLUTION CONTROL, CALIFCRNIA, ARIZONA.

IDENTIFIERS:
FODDER, GLUE, AGAR, POLYSIPHONIA, ANTIBIOTICS, IODINE, ATOMIC ENERGY, REACTOR COOLING, IMPERIAL VALLEY(CALIF), LOS ANGELES(CALIF), TUCSON(ARIZ).

ABSTRACT:
ECONOMIC CRITERIA OF HOW AND WHERE ALGAE CAN BE USED TO SERVE MAN'S BIOLOGICAL, PHYSICAL, AND ESTHETIC NEECS IS CONSIDERED. ALTHOUGH THE ACKNOWLEDGED BASE OF ALL FOOD CHAINS AND THE PRIMARY SOURCE FOR MOST OF THE EARTH'S OXYGEN, THE ECCNCMIC POTENTIAL OF ALGAE HAS BARELY BEEN EXPLORED. THE LARGER RED AND BROWN ALGAE ARE INCREASINGLY NEEDED FOR FERTILIZERS AND FCDDER; ANTIBIOTICS ARE DERIVED FRCM CHLORELLA, POLYSIPHONIA, AND OTHER SPECIES; IODINE HAS BEEN OBTAINED BY BURNING MARINE ALGAE. THE POTENTIALS OF UTILIZING THE ABILITY OF ALGAE TO CONCENTRATE MINERAL ICNS HUNDREDS OF TIMES THE CONCENTRATION IN WATER, THE CONTROL OF SPECIATION AND ITS EFFECT ON POLLUTION, POSSIBLE UTILIZATION OF BLUE-GREEN ALGAL PARASITES FOR INCREASING CONCENTRATION OF NITRATES AND PHOSPHATES SUFFICIENTLY THAT ALGAE MAY BE HARVESTED FOR COMMERCIAL USAGE; USE OF HALCPHYTIC ALGAE IN LOWERING EXCESSIVE CONCENTRATION OF SODIUM CHLORIDE IN DRINKING WATER. THE QUALITATIVE AND QUANTITATIVE ALTERATION IN SPECTRUM OF ALGAE SPECIES AS THE RESULT OF ATOMIC ENERGY PLANT CPERATIONS AND THEIR EFFECTS CN MAN ARE UNKNOWN, PARTICULARLY WHETHER ALGAE WILL ACCUMULATE RADIOACTIVE IONS EFFECTIVELY. ALGAL TCXONOMISTS, ALGAL PHYSIOLOGISTS, AND ALGAL HISTOCHEMISTS MUST CONSIDER THEMSELVES ECOLOGISTS AND POOL THEIR KNOWLEDGE TO CONTRIBUTE TO HUMANITY'S WELFARE. (AUEN-WISCONSIN)

FIELD 05C, 06B

ACCESSION NO. W71-09157

UTILIZATION OF HERBIVOROUS FISH IN FISH MANAGEMENT AND MELIORATION OF WATER BASINS, (IN RUSSIAN).

VESTNIK AKADEMII NAUK SSSR NO 11, P 26-30, 1970.

DESCRIPTORS:
*WATER QUALITY CONTROL, *ALGAE CONTROL, *FORAGE FISH, BIOCONTROL, REMEDIES, FOODS, FISH, HERBIVORES, FISH MANAGEMENT.

IDENTIFIERS:
USSR.

ABSTRACT:
AN ADDRESS BY G B NIKOLSKI IS REVIEWED WHICH WAS DELIVERED AT THE MEETING OF THE PRESIDIUM OF THE USSR ACADEMY OF SCIENCE. AT THIS TIME, THE CATCH OF OCEAN INHABITING FISH IS INADEQUATE TO MEET HUMAN POPULATION DEMANDS AND AN ARTIFICIAL CULTURE OF HERBIVOROUS FISH CAN PROVIDE AN INEXPENSIVE SOURCE OF PROTEINS. IN ADDITICN TO INCREASING THE FOOD SUPPLY, SUCH CULTURE WILL HELP TO ERADICATE THE DETRIMENTAL GROWTHS OF ALGAE IN IRRIGATION AND DRAINAGE SYSTEMS AND RESERVOIRS OF HYDRO-ELECTRIC STATIONS. HERBIVOROUS FISH OF DESIRABLE PROPERTIES ARE ABSENT IN BOTH EUROPE AND CENTRAL ASIA, BUT WATERS OF FAR-EASTERN SIBERIA HARBOR THREE HERBIVOROUS (COMMCN AND MOTTLED THICKHEADS AND WHITE AMUR), ATTAINING WEIGHTS OVER 20 AND EVEN 30 KG. THE TWO VARIETIES ARE NOT CCMPETITORS AS THEY DERIVE THEIR SUSTENANCE FROM EITHER NEAR-BOTTOM OR SURFACE PHYTCPLANKTON. THE PRESIDIUM APPROVED A BROAD PROGRAM OF INVESTIGATION OF HERBIVOROUS FISH IN ASIA, AFRICA, AND SOUTH AMERICA, AND A SYSTEMATIC ACCLIMATIZATION CF SUITABLE STOCK IN DIFFERENT PARTS OF THE USSR. (WILDE-WISCONSIN)

FIELD 05G, 02H

ACCESSION NO. W71-09163

ON THE GROSS AND NET PRODUCTION OF PERIPHYTON AND PLANKTON ALGAE, (IN RUSSIAN),

IRKUTSKII GOSUDARSTVENYI UNIVERSITET (SSSR).

O. M. KOZHOVA.

DOKLADY AKADEMII NAUK SSSR, VOL 195, NO 4, P 965-968, 1970. 3 TAB, 5 REF.

DESCRIPTORS:
*PERIPHYTON, *PLANKTON, *PRODUCTIVITY, PHOTOSYNTHESIS, BIOMASS, OXYGEN, ALGAE, COMPARATIVE PRODUCTIVITY, EUTROPHICATION.

IDENTIFIERS:
BRATSK RESERVOIR(SIBERIA), ANGARA RIVER(SIBERIA).

ABSTRACT:
THIS STUDY OF THE BRATSK BASIN INCLUDED DETERMINATIONS OF PHOTOSYNTHESIS OF PERIPHYTON BY LIGHT AND DARK BOTTLES, AND PRODUCTION OF PHYTOPLANKTON BY THE VINBERG OXYGEN METHOD. THE REPORTED RESULTS PRESENT AVERAGES FOR VEGETATION SEASONS OF 1964 AND 1967 FOR PHYTOPLANKTON, AND OF 1965, 1967, AND 1968 FOR PERIPHYTON. THE PHYTOPLANKTON CONSISTED LARGELY OF MELOSIRA, STEPHANODISCUS, ASTERIONELLA, AND CRYPTOMONADACEA SP IN THE SPRING, AND OF APHANIZOMENON AND CERATIUM SP IN SUMMER. THE PREDOMINANT MEMBERS OF PERIPHYTON INCLUDED CLADOPHORA, ULOTHRIX, OEDOGONIUM, SPIROGYRA, AND MOUGEOTIA SP. THE DIURNAL GROSS AND NET PRODUCTION OF PERIPHYTONIC ALGAE WAS 44.2 AND 33.8 MG OXYGEN PER G OF FRESH WEIGHT, RESPECTIVELY. THE PRODUCTION TO BIOMASS RATIO (P/B COEFFICIENT) OF PHYTOPLANKTON, RECALCULATED ON CARBON, WAS APPRECIABLY HIGHER THAN THAT OF THE PERIPHYTON. (WILDE-WISCONSIN)

FIELD 05C

ACCESSION NO. W71-09165

INITIAL UPTAKE, DISTRIBUTION AND LOSS OF SOLUBLE RU-106 IN MARINE AND FRESHWATER ORGANISMS IN LABORATORY CONDITIONS,

CENTRE D'ETUDE DE L'ENERGIE NUCLEAIRE, BRUSSELS (BELGIUM). DEPT. OF RADIOBIOLOGY.

O. VAN DER BORGHT, AND S. VAN PUYMBROECK.

HEALTH PHYSICS, VOL 19 (DEC), P 801-811, 1970. 10 FIG, 4 TAB, 15 REF.

DESCRIPTORS:
*ABSORPTION, *DISTRIBUTION, *RADIOISOTOPES, *MARINE ANIMALS, *MARINE PLANTS, AQUATIC ANIMALS, LABORATORY TESTS, NITRATES, MUSSELS, SNAILS, FISH, ALGAE, GASTROPODS, ADSORPTION, TEMPERATURE, METABOLISM, KINETICS, WATER POLLUTION EFFECTS, NUCLEAR WASTES.

IDENTIFIERS:
RUTHENIUM, FIXATION, TISSUES, ALBURNUS LUCIDUS, BELGIUM.

ABSTRACT:
IN WASTES FROM NUCLEAR REPROCESSING PLANTS, RUTHENIUM-106 IS APPARENTLY THE MOST IMPORTANT RADIOISOTOPE CONTAMINATING ALGAE AND FISH. THIS INVESTIGATION WAS CONDUCTED TO UNDERSTAND THE FACTORS, SUCH AS ENVIRONMENTAL CONDITIONS, AND PHYSICOCHEMICAL STATE OF RUTHENIUM ON THE RADIOACTIVE CONTAMINATION OF ORGANISMS, AND THE EARLY KINETICS OF FIXATION, DISTRIBUTION AND RELEASE OF SOLUBLE RUTHENIUM ORGINATING FROM NITRATE NITROSYL COMPLEXES BY SOME MARINE AND FRESHWATER ORGANISMS. THE EARLY UPTAKE, DISTRIBUTION AND LOSS BY MUSSELS, SNAILS, FISH, ALGAE, AND TWO GASTROPODS OF RUTHENIUM-106 WERE STUDIED FOR SHORT PERIODS UNDER LABORATORY CONDITIONS. INITIAL UPTAKE AND DISTRIBUTION OF RUTHENIUM-106 IS SIMILAR IN THE MARINE AND FRESHWATER LAMELLIBRANCH (MYTILUS AND ANODONTA), BUT FIXATION IS MUCH HIGHER IN THE FRESHWATER SNAIL, VIVIPARUS, THAN IN THE FISH, ALBURNUS. IN THE RADIOCONTAMINATION OF ORGANISMS BY SOLUBLE RUTHENIUM, THIN STRUCTURES, SUCH AS BYSSUS, IN DIRECT CONTACT WITH WATER, SHOW HIGH RUTHENIUM CONTENT, INDICATING THE IMPORTANT ROLE OF SURFACE ADSORPTION. TEMPERATURE-DEPENDANCE SUGGESTS METABOLIC INTERFERENCE, ALTHOUGH TEMPERATURE-DEPENDANCE AND GENERAL KINETICS OF THE UPTAKE ARE SIMILAR BOTH IN ALGAE AND IN ANIMALS STUDIED. SLOWER UPTAKE OF ISOTOPES AT LOWER TEMPERATURES INDICATES THE ADVISABILITY TO RELEASE RADIOACTIVE WASTES DURING PERIODS OF LOW METABOLIC ACTIVITY. (JONES-WISCONSIN)

FIELD 05C

ACCESSION NO. W71-09168

CONTEMPORANEOUS DISEQUILIBRIUM, A NEW HYPOTHESIS TO EXPLAIN THE 'PARADOX OF THE PLANKTON',

CALIFORNIA UNIV., DAVIS. DEPT. OF ZOOLOGY; AND CALIFORNIA UNIV., DAVIS. INST. OF ECOLOGY.

PETER RICHERSON, RICHARD ARMSTRONG, AND CHARLES R. GOLDMAN.

PROCEEDINGS OF THE NATIONAL ACADEMY OF SCIENCES, VOL 67, NO 4, P 1710-1714, 1970. 2 FIG, 1 TAB, 10 REF.

DESCRIPTORS:
*PHYTOPLANKTON, *LAKES, EPILIMNION, ZOOPLANKTON, DISTRIBUTION, PRIMARY PRODUCTIVITY, ALGAE, BIOMASS, OLIGOTROPHY, DIATOMS, EUGLENA, DAPHNIA, HABITATS, NICHES, COMPETITION, CALIFORNIA.

IDENTIFIERS:
*CONTEMPORANEOUS DISEQUILIBRIUM, LAKE TAHOE(CALIF), CASTLE LAKE(CALIF), DIVERSITY, PATCHES.

ABSTRACT:
UNEXPECTEDLY HIGH DIVERSITY FOUND IN EVEN SMALL SAMPLES OF LAKE PHYTOPLANKTON HAS BEEN TERMED 'THE PARADOX OF THE PLANKTON.' A SMALL VOLUME OF WATER, FOR EXAMPLE 10 ML, USUALLY SHOWS SOME TENS OF SPECIES WHERE THE COMPETITIVE EXCLUSION PRINCIPLE MIGHT LEAD TO EXPECTATION OF ONLY ONE OR A FEW SPECIES. PATCHES OF WATER MAY EXIST IN WHICH ONE SPECIES COMPETES ADVANTAGEOUSLY RELATIVE TO THE OTHERS. THEY ARE STABLE ENOUGH FOR CONSIDERABLE PATCHINESS AMONG PHYTOPLANKTON, BUT ARE OBLITERATED FREQUENTLY ENOUGH TO PREVENT EXCLUSIVE OCCUPATION OF EACH NICHE BY A SINGLE SPECIES. WITH EPILIMNION MIXING, ONLY ONE OR AT MOST, A FEW NICHES FOR PRIMARY PRODUCERS MIGHT BE EXPECTED. IN SAMPLES FROM CASTLE LAKE, CALIFORNIA, A HIGH DEGREE OF PATCHINESS FOR MANY PHYTOPLANKTON SPECIES WAS FOUND, INDICATING MIXING RATE IS SUFFICIENTLY SLOW IN RELATION TO THE ALGAL PRODUCTIVITY RATE, FOR MANY DIFFERENT NICHES TO EXIST SIMULTANEOUSLY. IN LAKE TAHOE, PRODUCTIVITY PER UNIT BIOMASS RATIOS SHOW THAT TURNOVER TIMES FOR CARBON ARE OFTEN LESS THAN ONE DAY. HIGH DIVERSITY IS ASSOCIATED WITH HIGH PRODUCTIVITY PER UNIT BIOMASS AND HIGH ZOOPLANKTON POPULATIONS. A CONTEMPORANEOUS DISEQUILIBRIUM MODEL IS A PLAUSIBLE EXPLANATION OF THE DIVERSITY. (JONES-WISCONSIN)

FIELD 05C

ACCESSION NO. W71-09171

PHOTOSYNTHESIS IN THE ALGAE,

BRANDEIS UNIV., WALTHAM, MASS. DEPT. OF BIOLOGY.

MARTIN GIBBS, ERWIN LATZKO, MICHAEL J. HARVEY, ZVI PLAUT, AND

ANNALS NEW YORK ACADEMY OF SCIENCES, VOL 175, P 541-554, 1970. 5 FIG, 6 TAB, 40 REF.

DESCRIPTORS:
*BIOCHEMISTRY, *PHOTOSYNTHESIS, *ALGAE, PHYLOGENY, CARBON CYCLE, ENZYMES, METABOLISM, HYDROGEN, EUGLENA, CHLORELLA, SCENEDESMUS, CHLAMYDOMONAS, CYANOPHYTA, CHLOROPHYTA, OXYGEN, LIGHT, CHRYSOPHYTA, OCHROMONAS, RESPIRATION, CHLOROPHYLL, CARBON RADIOISOTOPES, CARBON DIOXIDE, RHODOPHYTA.

IDENTIFIERS:
WARBURG EFFECT, EUGLENA GRACLIS, CALVIN CYCLE, ANACYSTIS NIDULANS.

ABSTRACT:
THE PHOTOSYNTHETIC CARBON REDUCTION CYCLE IN ALGAE IS DISCUSSED FROM EVIDENCE DESCRIBED IN THE LITERATURE; THIS CYCLE IS NOW FIRMLY ESTABLISHED. TWELVE ENZYMICALLY CATALYZED STEPS ARE INVOLVED. AT LEAST RIBULOSE 1,5-DIP CARBOXYLASE AND FRUCTOSE 1,6-DIPHOSPHATASE ARE INVOLVED IN CARBON DIOXIDE FIXATION. THE BULK OF THE EVIDENCE SUGGESTS THAT THE CARBON REDUCTION CYCLE IS THE MAJOR PATHWAY FOR THE ENTRY OF PHOTOSYNTHETICALLY ASSIMILATED CARBON DIOXIDE IN THE MAJORITY OF CHLOROPHYLLOUS ALGAE. ON THE BASIS OF PRESENT KNOWLEDGE OF PHOTOSYNTHETIC CARBON ASSIMILATION IN ALGAE, THE AUTOTROPHIC PHASE CEASES WITH THE LIGHT-CATALYZED CONVERSION OF CARBON DIOXIDE INTO FRUCTOSE 6-P AND GLUCOSE 6P. THE HETEROTROPHIC PHASE COMMENCES WITH THEIR CONVERSION INTO OTHER CELLULAR COMPONENTS, SOMETIMES REFERRED TO AS SECONDARY PRODUCTS. IMPORTANT EVIDENCE FOR METABOLIC REGULATION OF THE PHOTOSYNTHETIC CARBON REDUCTION CYCLE IN ALGAE INDICATES THE CYCLE IS NOT ISOLATED FROM THE REST OF THE CELL METABOLISM. ASSESSMENT OF THE TRUE VALUE OF THESE RESULTS WAS HINDERED BY POSSIBLE INTERCHANGE OF INTERMEDIATES COMMON TO BOTH PHOTOSYNTHESIS AND RESPIRATION. SUCCESSFUL ISOLATION OF ALGAL CHLOROPLASTS FROM THE GIANT CELL OF ACETABULARIA MEDITERRANEA, ABLE TO CARRY OUT NORMAL PHOTOSYNTHESIS, WAS RECENTLY ACCOMPLISHED. (JONES-WISCONSIN)

FIELD 05C

ACCESSION NO. W71-09172

HYPNODINIUM-LIKE ALGAL BLOOMS IN GEORGIA LAKES,

GEORGIA STATE UNIV., ATLANTA. DEPT. OF BIOLOGY; AND GEORGIA WATER QUALITY CONTROL BOARD, ATLANTA.

D. G. AHEARN, EDWARD T. HALL, JR., AND DONALD J. REINHARDT.

BIOSCIENCE, (RESEARCH REPORTS), P 115, FEBRUARY 1, 1971. 1 FIG, 1 REF.

DESCRIPTORS:
*ALGAE, *LAKES, *GEORGIA, SEWAGE EFFLUENTS, WATER POLLUTION EFFECTS, WATER POLLUTION SOURCES, EUTROPHICATION, COLIFORMS, SPORES, LIFE CYCLES, ODOR, RECREATION, SYSTEMATICS.

IDENTIFIERS:
*WHITE ALGAL BLOOMS, HYPNODINIUM, LAKE SIDNEY LANIER(GEORGIA), LAKE JACKSON(GEORGIA), LAKE CARDINAL(GEORGIA), LAKE BUCKHORN(GEORGIA).

ABSTRACT:
LAKES ON DIFFERENT RIVER SYSTEMS IN NORTHWEST GEORGIA PRODUCED A WHITE ALGAL BLOOM OF VARYING INTENSITY IN JUNE, LASTING 7-10 DAYS. AT ITS PEAK, DENSE SURFACE FLOCS WERE FOUND WHERE SEWAGE EFFLUENTS WERE RECEIVED, IN ALL THE 1970 BLOOM SITES. PRIOR TO, DURING, AND AFTER THE BLOOM, HYDROGRAPHIC CONDITIONS IN THE FLAT CREEK AREA OF LAKE LANIER WERE MONITORED WITH OBSERVATIONS ON THERMOCLINE, TEMPERATURE, PH, DISSOLVED OXYGEN. FOUR DAYS PRIOR TO THE APPEARANCE OF THE BLOOM, THE FECAL COLIFORM DENSITY INCREASED FROM LESS THAN 4/100 ML TO MORE THAN 240/100 ML. THE MILKY WHITE SURFACE SCUM FROM ALL LAKES WAS COMPOSED OF BODIES SURROUNDED WITH A CLEAR ENVELOPE CONTAINING LARGE PROTOPLASMIC BODIES. A DARK PROTOPLASTIC BODY MATURED INTO A DARK MACROSPORE WHICH SUBDIVIDED PRODUCING 2 TO 15 OR MORE BI-PORED MICROSPORES. NO EXIT OF PROTOPLASMIC BODIES THROUGH THE PORES WAS OBSERVED, ALTHOUGH BI-FLAGELLATED ZOOSPORES OCCURRED IN LARGE NUMBERS IN THE MICROSCOPE PREPARATIONS. THE BI-PORED CELLS OF THIS ALGA MAKE IT DISTINCT FROM HYPNODINIUM SPHAERICUM KLEBS. IT HAS NOT BEEN DIRECTLY ASSOCIATED WITH FISH KILLS. A DISAGREEABLE ODOR ACCOMPANIES THE BLOOM AND THE THICK SURFACE FLOCS INTERFERE WITH RECREATION. (JONES-WISCONSIN)

FIELD 05C, 02H

ACCESSION NO. W71-09173

INITIAL UPTAKE, DISTRIBUTION AND LOSS OF SOLUBLE RUTHENIUM-106 IN MARINE AND FRESHWATER ORGANISMS IN LABORATORY CONDITIONS,

CENTRE D'ETUDE DE L'ENERGIE NUCLEAIRE, BRUSSELS (BELGIUM). DEPT. OF RADIOBIOLOGY.

O. VAN DER BORGHT, AND S. VAN PUYMBROECK.

HEALTH PHYSICS, VOL. 19, P 801-811, DEC. 1970.

DESCRIPTORS:
*RADIOISOTOPES, *MARINE ANIMALS, *ABSORPTION, AQUARIA, MOLLUSKS, SNAILS, MUSSELS, MARINE ALGAE, AQUATIC ANIMALS, FRESH WATER, FRESHWATER FISH, TEMPERATURE, KINETICS, NUCLEAR WASTES, SOLUBILITY, ADSORPTION, WATER POLLUTION EFFECTS, FOOD CHAINS.

IDENTIFIERS:
*RUTHENIUM RADIOISOTOPES.

ABSTRACT:
RUTHENIUM UPTAKE EXTENDING TO LONG-TERM EQUILIBRIA WERE STUDIED SINCE THIS ELEMENT IS OFTEN THE LIMITING FACTOR IN THE DISPOSAL OF WASTES FROM NUCLEAR REPROCESSING PLANTS. THE GENERAL KINETICS INCLUDING THE TEMPERATURE DEPENDENCE WERE SIMILAR FOR FOUR MARINE ORGANISMS (AN ALGA, A MUSSEL, AND TWO SNAILS); AND THE INITIAL UPTAKE WAS SIMILAR FOR MARINE AND FRESHWATER MUSSELS. A RELATIVELY HIGH CONCENTRATION IN THIN ORGANS IN DIRECT CONTACT WITH THE WATER SUGGESTED THAT SURFACE ABSORPTION PLAYS A ROLE. RELEASE OF WASTES DURING PERIODS OF LOW TEMPERATURE (WINTER) WILL MINIMIZE BIOLOGICAL CONCENTRATION BEFORE DILUTION IN THE BIOSPHERE. (BOPP-NSIC)

FIELD 05C

ACCESSION NO. W71-09182

INTERACTION OF TRACE ELEMENTS WITH THE ORGANIC CONSTITUENTS IN THE MARINE ENVIRONMENT,

BHABHA ATOMIC RESEARCH CENTRE, BOMBAY (INDIA).

M. V. M. DESAI, AND A. K. GANGULY.

AVAILABLE FROM THE NATIONAL TECHNICAL INFORMATION SERVICE AS BARC-488, $3.00 IN PAPER COPY, $0.95 IN MICROFICHE. REPORT BARC-488, 1970. 117 P, 15 FIG, 41 TAB, 47 REF.

DESCRIPTORS:
*RADIOISOTOPES, *ORGANIC COMPOUNDS, *MARINE ALGAE, PHYTOPLANKTON, HUMIC ACIDS, FULVIC ACIDS, TRACE ELEMENTS, CHEMICAL ANALYSIS, ION EXCHANGE, ABSORPTION, STRONTIUM RADIOISOTOPES, ZINC RADIOISOTOPES, COBALT RADIOISOTOPES, URANIUM RADIOISOTOPES, RADIUM RAIODISOTOPES, ALUMINUM, IRON, COPPER, MAGNESIUM, MANAGNESE, CALCIUM, CHROMIUM, POTASSIUM, CHELATION, METABOLISM, NUCLEAR WASTES, WATER POLLUTION EFFECTS.

IDENTIFIERS:
MANGANESE RADIOISOTOPES, CESIUM RADIOISOTOPES, RUTHENIUM RADIOISOTOPES, CERIUM RADIOISOTOPES, THORIUM RADIOISOTOPES, IRON RADIOISOTOPES, ZIRCONIUM RADIOISOTOPES, ZINC.

ABSTRACT:
LOW-LEVEL NUCLEAR WASTES DISHCHARGE INTO THE ENVIRONMENT THE RADIOISOTOPES OF ZINC, MANGANESE, IRON, COBALT, RADIUM, THORIUM, URANIUM, ETC. IN THE PRESENT WORK IT IS SHOWN THAT THE RADIONUCLIDES ARE SOLUBILIZED BY GROWTH PRODUCTS OF MARINE ORGANISMS. COMPLEXATION OF RADIONUCLIDES WAS STUDIED WHEN PHYTOPLANKTON WAS GROWN IN A MEDIUM SPIKED WITH RADIONUCLIDES IN DIFFERENT PHASES. HUMIC AND FULVIC ACIDS WERE EXTRACTED FROM SEA WATER AND ANALYZED FOR TRACE ELEMENTS. THE STUDIES SHOW MECHANISMS BY WHICH THE AVAILABILITY MAY BE INCREASED THROUGH AQUATIC FOOD CHAINS TO MAN. (BOPP-NSIC)

FIELD 05C

ACCESSION NO. W71-09188

PATHWAYS OF TRACE ELEMENTS IN ARCTIC LAKE ECOSYSTEMS. PROGRESS REPORT, APRIL 15, 1970-APRIL 14, 1971,

ALASKA UNIV., COLLEGE. INST. OF MARINE SCIENCES.

R. J. BARSDATE.

AVAILABLE FROM THE NATIONAL TECHNICAL INFORMATION SERVICE AS SAN-310-P-4-10, $3.00 IN PAPER COPY, $0.95 IN MICROFICHE. REPORT SAN-310-P-4-10, JAN. 1971, 151 P. AEC CONTRACT AT(04-3)-310.

DESCRIPTORS:
*ECOSYSTEMS, *ALASKA, *LAKES, ARCTIC, PRODUCTIVITY, EUTROPHICATION, RED TIDE, SILICA, ALGAE, PHYTOPLANKTON, FREEZING, TEMPERATURE, CHELATION, HYDROGEN ION CONCENTRATION, DYSTROPHY, ZINC RADIOISOTOPES, COBALT RADIOISOTOPES, PHOSPHORUS RADIOISOTOPES, SEDIMENTS, TRACE ELEMENTS, VOLCANOES, BENTHIC FLORA, ALKALINITY, PHOTOSYNTHESIS, HYDROLOGIC ASPECTS, MAGMATIC WATER, WATER POLLUTION SOURCES.

IDENTIFIERS:
ZINC, MANGANESE RADIOISOTOPES.

ABSTRACT:
THIS REPORT CONSIDERS FIVE SUBJECTS: THE FIELD PROGRAM AT POINT BARROW; CHARACTER AND RESIDENCE TIME FOR ORGANIC COMPLEXES OF TRACE METALS ASSOCIATED WITH A RED TIDE; A STUDY OF ZINC COMPLEXATION WITH ORGANIC MATERIAL FROM A NATURAL WATER SYSTEM; PHYSICAL LIMNOLOGY, CHEMISTRY AND PLANT PRODUCTIVITY OF A TAIGA LAKE; AND EFFECTS OF VOLCANIC ASHFALLS ON ALASKAN LAKES. (BOPP-NSIC)

FIELD 05C, 05B, 02H

ACCESSION NO. W71-09190

MECHANISMS OF THE ACCUMULATION OF PLUTONIUM-239 AND POLONIUM-210 BY THE BROWN ALGA ASCOPHYLLUM NODOSUM AND MARINE PHYTOPLANKTON,

POLYARNYI NAUCHNO-ISSLEDOVATELSKII I PROEKTNYI INSTITUT MORSKOGO RYBNOGO KHOZYAISTVA I OKEANOGRAFII, MURMANSK (USSR).

V. S. ZLOBIN, AND O. V. MOKANU.

AVAILABLE FROM THE NATIONAL TECHNICAL INFORMATION SERVICE AS AEC-TR-7205, $3.00 IN PAPER COPY, $0.95 IN MICROFICHE. (RADIOBIOLOGY), VOL. 10, NO. 4, P 584-589, JULY-AUGUST 1970, 3 FIG, 4 TAB, 18 REF) AEC-TR-7205 FROM RADIOBIOLOGIYA.

DESCRIPTORS:
*RADIOISOTOPES, *PHYTOPLANKTON, *METABOLISM, *ABSORPTION, MARINE PLANTS, MARINE ALGAE, PHAEPHYTA, AQUARIA, PARTICLE SIZE, COLLOIDS, WATER POLLUTION EFFECTS.

IDENTIFIERS:
*POLONIUM, *PLUTONIUM.

ABSTRACT:
RADIONUCLIDE UPTAKE BY A BROWN ALGA WAS STUDIED IN LABORATORY AQUARIA; BY PHYTOPLANKTON CULTURED FROM SURFACE SEA WATER, IN SHIPBOARD AQUARIA. THE EFFECT OF INHIBITORS OF THE CYTOCHROME SYSTEM (CADMIUM CHLORIDE, SODIUM CYANIDE, AND AMMONIUM NITRATE) INDICATED THAT UPTAKE WAS AFFECTED BY CELL RESPIRATION. THE RADIONUCLIDES WERE SHOWN TO BE IN THE FORM OF COLLOIDS. THE PARTICLE SIZE OF THE PLUTONIUM SALT WAS 0.1-0.3 MICRONS; OF THE POLONIUM SALT, 0.3-15 MICRONS. (BCPP-NSIC)

FIELD 05C

ACCESSION NO. W71-09193

ECOLOGY AND POLLUTION OF THE ENVIRONMENT (IN FRENCH).

BULLETIN D'INFORMATIONS SCIENTIFIQUES ET TECHNIQUES, NO 151, P 3-73 (SEPT 1970).

DESCRIPTORS:
*STRONTIUM RADIOISOTOPES, *WATER POLLUTION EFFECTS, NUCLEAR WASTES, RIVERS, ABSORPTION, FISH, MOLLUSKS, MARINE ANIMALS, COBALT RADIOISOTOPES, FOOD CHAINS, IRRIGATION, FLOW PROFILES, MARINE ALGAE, WINDS, SEA WATER, PHYSICOCHEMICAL PROPERTIES.

IDENTIFIERS:
*CESIUM RADIOISOTOPES, *RUTHENIUM RADIOISOTOPES, ZIRCONIUM RADIOISOTOPES.

ABSTRACT:
A COLLECTION OF ARTICLES WHICH INCLUDES THE FOLLOWING TOPICS: PLANTS AND PARTICULATE ATMOSPHERIC POLLUTION (WIND TUNNEL EXPERIMENTS), UPTAKE OF RADIONUCLIDES BY MARINE BIOTA, IN-SITU CONTAMINATION OF MARINE BIOTA BY RUTHENIUM-106 AND ZIRCONIUM-95, PHYSICOCHEMICAL BEHAVIOR OF RUTHENIUM IN SEA WATER, A CESIUM SOURCE FOR IRRADIATION STUDIES OF MARINE ORGANISMS, SIMULATION OF NATURAL FLOW IN RIVER POLLUTION STUDIES, MIGRATION OF FISSION PRODUCTS THROUGH SOIL, UPTAKE OF RADIONUCLIDES BY VEGETABLES FROM IRRIGATION WATER. (BOPP-NSIC)

FIELD 05C

ACCESSION NO. W71-09225

TEMPERATURE EFFECTS ON THE SORPTION OF RADIONUCLIDES BY FRESHWATER ALGAE,

SAVANNAH RIVER LAB., AIKAN, S.C.

R. S. HARVEY.

HEALTH PHYSICS, VOL 19, NO 2, P 293-297 (AUG 1970). 4 FIG, 3 TAB, 7 REF.

DESCRIPTORS:
*AQUATIC ALGAE, *RADIOISOTOPES, ABSORPTION, TEMPERATURE, STRONTIUM RADIOISOTOPES, ZINC RADIOISOTOPES, COBALT RADIOISOTOPES, WATER POLLUTION EFFECTS, FRESH WATER.

IDENTIFIERS:
*CESIUM RADIOISOTOPES, IRON RADIOISOTOPES, MANGANESE RADIOISOTOPES.

ABSTRACT:
WATER TEMPERATURES OF 23-32 DEG C DID NOT AFFECT THE SORPTION OF CESIUM-137, STRONTIUM-85, ZINC-65, IRON-59, COBALT-57, AND MANGANESE-54 BY THE FILAMENTOUS GREEN ALGA, STIGEOCLONIUM LUBRICUM. RADIONUCLIDE CONCENTRATIONS IN THE UNICELLULAR DIATOM, NAVUCILA SEMINULUM, WERE 2-5 TIMES HIGHER AT 32 DEG C THAN AT 23 DEG C. WATER TEMPERATURES OF 25-40 DEG C HAD NO SIGNIFICANT EFFECT ON THE SORPTION OF CESIUM-137, STRONTIUM-85, ZINC-65, AND IRON-59 BY THE FILMENTOUS BLUE-GREEN ALGA, PLECTONEMA BORYANUM. HOWEVER, COBALT-57 CONCENTRATIONS IN P. BORYANUM DECREASED WITH TEMPERATURE, AND MANGANESE-54 CONCENTRATIONS INCREASED FROM 25-35 DEG C. GROWTH OF S. LUBRICUM WAS NOT AFFECTED BY THE TEMPERATURE. (BOPP-NSIC)

FIFLD 05C

ACCESSION NO. W71-09250

THE EFFECTS OF ENVIRONMENTAL STRESS ON THE COMMUNITY STRUCTURE AND PRODUCTIVITY OF SALT MARSH EPIPHTIC COMMUNITIES,

CITY COLLEGE, NEW YORK.

J. J. LEE,

AVAILABLE FROM NTIS. ATOMIC ENERGY COMMISSION TECHNICAL REPORT NYO-3995-14, JAN 1970. 46 P.

DESCRIPTORS:
*ECOSYSTEMS, *SALT MARSHES, *PRODUCTIVITY, RADIOISOTOPES, STRONTIUM RADIOISOTOPES, ALGAE, PLANT GROWTH, TEMPERATURE, PROTOZOA, CHLOROPHYTA, ABSORPTION.

ABSTRACT:
THE FOLLOWING TOPICS ARE CONSIDERED: REMOVAL OF CALCIUM-45 AND STRONTIUM-89, -90 FROM COASTAL ENVIRONMENTS BY CALCAREOUS FORAMINIFERS (PROTOZOA); GROWTH AND REPRODUCTION OF FORAMIFERS AND EPIPHYTIC ALGAE; EFFECT OF TEMPERATURE AND LIGHT ON ENTEROMORPHA (ALGAE) AND ITS EPIPHYTES; STIMULATION OF GROWTH OF 9 OUT OF 11 REPRESENTATIVE EIPIHYTIC ALGAE BY EITHER ARGININE OR ALANINE; AND PROGRESS IN OTHER SUBPROJECTS. (BOPP-NSIC)

FIELD 05C

ACCESSION NO. W71-09254

SOME OBSERVATIONS ON ALGAE INVADING A CESIUM-137 CONTAMINATED POND,

ATCMIC ENERGY OF CANADA LTD., PINAWA (MANITOBA).

JANET R. DUGLE, ANC J. E. GUTHRIE.

AVAILABLE FROM NTIS. ATOMIC ENERGY OF CANADA TECHNICAL REPORT AECL-3463, (1970). 12 P.

DESCRIPTORS:
*AQUATIC ALGAE, *RADIOACTIVITY EFFECTS, *PONDS, WATER POLLUTION EFFECTS, RADIOISOTOPES, COMPETITION, SPECIATION.

IDENTIFIERS:
*CESIUM RADIOISOTOPES.

ABSTRACT:
A COMPARISON OF THE SPECIES OF ALGAE FOUND AT VARIOUS COLLECTION SITES WITHIN TWO PONDS, ONE CONTAMINATED WITH CESIUM-137, WAS MADE IN OCTOBER 1968. THESE OBSERVATIONS WILL FORM THE BASIS OF A STUDY OF THE SUCCESSION OF ALGAL SPECIES IN POND CCMMUNITIES. (BOPP-NSIC)

FIELD 05C

ACCESSION NO. W71-09256

THE ADHESIVE PROPERTIES OF 'CHLORELLA VULGARIS',

PUERTO RICO UNIV., MAYAGUEZ. DEPT. OF MARINE SCIENCES.

THCMAS R. TOSTESON, AND LUIS R. ALMODOVAR.

AVAILABLE FROM THE NATIONAL TECHNICAL INFORMATION SERVICE AS AD-721 114, $3.00 IN PAPER COPY, $0.95 IN MICROFICHE. ONR TECHNICAL REPORT, NO. TR-1, APR. 1971, 35 P.

DESCRIPTORS:
*CHORELLA, *ALGAE, *FOULING, *NUISANCE ALGAE, FLCATING PLANTS, SEAWATER, PLANKTON, BIODEGRADATION, LIGHT, TEMPERATURE.

IDENTIFIERS:
ACRYLIC RESINS, GLASS, THYMIDINES, CELLS(BIOLOGY), *BIODETERIORATION, DEMOCOLCINE.

ABSTRACT:
AMBIENT SEA WATER CONTAINS MATERIAL THAT PROMOTES THE ADHESION OF THE PLANKTONIC ALGAE CHLORELLA VULGARIS TO PLASTIC SURFACES. THIS ADHESION TAKES PLACE WITHIN 3 TP 6 HOURS, AND IS INHIBITED BY THE ABSENCE OF LIGHT. THE EFFECT OF LIGHT ON THE RESPONSE OF CHLCRELLA TO THE OCEAN TREATED SURFACE IS DUE TO THE EFFECT OF LIGHT ON THE RATE OF GROWTH OF THE ALGAE POPULATION. THE ADHESION OF CHLORELLA TC GLASS SURFACES IS SIGNIFICANTLY INCREASED IN THE PRESENCE OF THYMIDINE AND COLCEMIDE. THYMIDINE ACCELERATES THE RATE OF GROWTH OF THE ALGAL CELLS AND COLCEMIDE BLOCKS THIS GROWTH DURING MITOSIS. THE EFFECT OF OCEAN DEPOSITED MATERIALS ON THE ADHESION OF CHLORELLA TO PLASTIC APPEARS SIMILAR TO THE EFFECT OF AGENTS THAT INCREASE THE RELATIVE NUMBER OF CELLS TO THE G2 PHASE OF THE CELL CYCLE ON THE ACHESION OF THIS CELL TO GLASS.

FIELD 05C

ACCESSION NO. W71-09261

PHOSPHATES IN DETERGENTS AND THE EUTROPHICATION OF AMERICA'S WATERS.

AVAILABLE FROM SUP OF DOC, US GOVERNMENT PRINTING OFFICE, WASH DC 20402 PRICE $0.40. HOUSE REPORT NO 91-1004, TWENTY-THIRD REPORT BY THE COMM ON GOV'T OPERATIONS, 91ST CONG, 2D SESS (1970). 88 P, 2 FIG, 9 TAB, 87 REF, 2 APP.

DESCRIPTORS:
*EUTROPHICATION, *PHOSPHATES, *DETERGENTS, *POLLUTION ABATEMENT, WATER POLLUTION EFFECTS, NUTRIENTS, PHOSPHOROUS, WATER SOFTENING, LEGAL ASPECTS, HARD WATER, PUBLIC HEALTH, COST COMPARISONS, AQUATIC ALGAE, GRANTS, MODEL STUDIES.

ABSTRACT:
PHOSPHOROUS POLLUTION CAUSES EUTROPHICATION, AND OVER HALF OF THE PHOSPHORUS POLLUTION IN THE UNITED STATES IS CAUSED BY DETERGENTS. IN THESE HEARINGS THE DETERGENT INDUSTRY CONTENDED THAT PHOSPHATE DETERGENTS DID NOT HARM AND WERE ESSENTIAL TO MAINTAIN CLEANLINESS AND SANITATION STANDARDS, SINCE THERE WAS NO SUITABLE REPLACEMENT. REBUTTALS TO THESE DEFENSES SUGGEST SEVERAL ALTERNATIVES TO PHOSPHATE DETERGENTS. THE REMOVAL OF NUTRIENTS FROM WASTE WATERS INSTEAD OF PHOSPHATES FROM DETERGENTS AS SUGGESTED BY THE INDUSTRY, WOULD BE COSTLY, SLOW AND CREATE ADDITIONAL POLLUTION PROBLEMS. THE COMMITTEE CONCLUDED THAT NO PROGRESS HAD BEEN MADE IN COMBATING EUTROPHICATION FROM DETERGENTS AND RECOMMENDED A SERIES OF GRADUAL REDUCTIONS IN THE USE OF PHOSPHATE DETERGENTS, WITH A COMPLETE ELIMINATION OF THEIR USE BY THE END OF 1972. ADDITIONALLY, THE COMMITTEE RECOMMENDED: THE MANDATORY LABELING OF PHOSPHATE CONTENT DURING THE INTERIM PERIOD, EXPANDED RESEARCH BY THE FEDERAL WATER QUALITY ADMINISTRATION TO DEVELOP LOW-PHOSPHORUS OR PHOSPHORUS-FREE DETERGENTS, AND THE IMMEDIATE ELIMINATION OF THE INDUSTRY-GOVERNMENT TASK FORCE, WHOSE ACTUAL PURPOSE WAS TO SECURE GOVERNMENT COOPERATION IN RETAINING PHOSPHATE DETERGENTS. (GALLAGHER-FLORIDA)

FIELD 05C, 06E, 05G

ACCESSION NO. W71-09429

NUTRIENT REMOVAL AND ADVANCED WASTE TREATMENT.

PITTSBURG UNIV., PA. GRADUATE SCHOOL OF PUBLIC HEALTH.

PROCEEDINGS, PITTSBURG SANITARY ENGINEERING CONFERENCE, 8, PITTSBURG, PENNSYLVANIA, FEB 1970. 265 P.

DESCRIPTORS:
*EUTROPHICATION, *TERTIARY TREATMENT, *WATER QUALITY CONTROL, AMMONIA, NITROGEN, PHOSPHORUS, ALGAE, FEASIBILITY STUDIES, TECHNICAL FEASIBILITY, ECONOMIC FEASIBILITY, PILOT PLANTS, ION EXCHANGE, ELECTROLYSIS, CHEMICAL PRECIPITATION, NITRIFICATION, DENITRIFICATION, *WASTE WATER TREATMENT.

ABSTRACT:
THE 8TH PITTSBURG SANITARY ENGINEERING CONFERENCE WAS CONDUCTED IN FEBRUARY OF 1970 TO ASSEMBLE CURRENT TECHNIQUES AND METHODS OF NITROGEN AND/OR PHOSPHORUS REMOVAL. A FEW OF THE METHODS PRESENTED ARE: (1) CATION EXCHANGE; (2) ELECTROLYSIS; (3) BREAKPOINT CHLORINATION; (4) AMMONIA STRIPPING; AND (5) MANY OTHERS. REMOVAL EFFICIENCIES FOR AMMONIA, IN PARTICULAR WERE REPORTED AS HIGH AS 93%. SYSTEMS WERE TESTED AND RETESTED USING VARIOUS FEED RATES, SYSTEM LOADINGS, PH RANGES, TEMPERATURE, ETC. MANY OF THE TECHNIQUES REPORTED NOT ONLY REMOVAL EFFICIENCIES BUT COSTS IN TERMS OF CENTS/1000 GALLONS OF THROUGHPUT. FOR EXAMPLE, COLUMN DENITRIFICATION OF WATER CONTAINING 20 MGL1 AS NO3-N, AND USING A 3/1 RATIO OF METHANOL TO NITROGEN, WAS REPORTED AT 2.8 CENTS/1000 GALLONS FOR CHEMICALS. BY GATHERING TOGETHER ALL OF THIS INFORMATION AND DISCUSSING AND REPRODUCING IT, IT WAS HOPED THAT FUTURE LAKE, RIVER, AND STREAM EUTROPHICATION COULD BE PREVENTED BY TAKING PREVENTIVE ACTION BEFORE THE WATERS ARE DESTROYED OR IRREPARABLY DAMAGED. (LOWRY-TEXAS)

FIELD 05D, 05C

ACCESSION NO. W71-09450

REUSE OF CHEMICAL FIBER PLANT WASTE WATER AND COOLING WATER BLOWDOWN,

FIBER INDUSTRIES INC., CHARLOTTE, N.C.

WILLIAM J. DAY.

COPY AVAILABLE FROM GPO SUP DOC AS EP2.10:12090EOX10/70, $0.70; MICROFICHE FROM NTIS AS PB-200 695, $0.95. WATER POLLUTION CONTROL RESEARCH SERIES, OCTOBER 1970. 66 P, 15 FIG, 5 TAB, 15 REF. EPA GRANT NC WPRD-100-01-68, PROGRAM NO 12090 EUX 10/70.

DESCRIPTORS:
*WATER REUSE, *DOMESTIC WASTES, *INDUSTRIAL WASTES, ACTIVATED SLUDGE, TRICKLING FILTERS, LAGOONS, TERTIARY TREATMENT, ACTIVATED CARBON, FLOCCULATION, ALGAE, SLUDGE, CHROMIUM, TOXICITY, NEUTRALIZATION, HYDROGEN ION CONCENTRATION, TEMPERATURE, BIOCHEMICAL OXYGEN DEMAND, CHEMICAL OXYGEN DEMAND, COOLING TOWERS, *WASTE WATER TREATMENT.

IDENTIFIERS:
*MICRO SCREENING.

ABSTRACT:
WASTE WATERS FROM A FORTREL POLYESTER MANUFACTURING PLANT CONSISTED OF ORGANIC CHEMICAL PROCESS WASTES, COOLING SYSTEM BLOWDOWN, AND SANITARY WASTES FROM THE PLANT. A WATER REUSE PROGRAM WAS INSTITUTED WHICH CONSISTED OF: (1) PRETREATMENT OF COOLING WATERS FOR REMOVAL OF HEAVY METALS; (2) IN-PLANT MODIFICATIONS AND ADDITIONS TO THE EXISTING SYSTEM TO INCREASE TREATMENT PLANT CAPACITY; AND (3) A POST TREATMENT SYSTEM FOR EFFLUENT POLISHING PRIOR TO SELECTED REUSE. THE FINAL SYSTEM CONSISTED OF: (1) A CHROMATE REDUCTION UNIT RATED AT 120 GPM FOR CONCENTRATIONS OF UP TO 300 MG/1 CRO4 AND DESIGNED FOR CONTINUOUS OPERATION; (2) EQUALIZATION BASINS HAVING A COMBINED CAPACITY OF 195,000 GALLONS AND CONTAINING SUFFICIENT MIXING CAPACITY TO PREVENT SHORT-CIRCUITING AND STRATIFICATION; (3) A PLASTIC MEDIA ROUGHING FILTER CONSISTING OF TWO TIERS OF POLY-VINYL CHLORIDE MEDIA 10 FEET THICK AND 25 FT IN DIAMETER; (4) AN AERATION BASIN EQUIPPED WITH 175 HP OF AERATION AND MIXING CAPACITY AND USING 100% RECYCLE OF CLARIFIER SLUDGE; (5) A PERIPHERAL FLOW TYPE CLARIFIER; (6) TWO SERIES CONNECTED POLISHING PONDS; (7) A MICRO SCREEN OR ALGAE SCREEN; (8) A FLOCCULANT AND/OR CARBON UNIT; (9) A SLUDGE POND; AND (10) A DIGESTER. CHROMIUM WAS REMOVED FROM THE COOLING TOWER BLOWDOWN FOR $.21 PER POUND OF CHROMATE REMOVED. THE PLASTIC MEDIA TRICKLING FILTER, USED AS A ROUGHING FILTER, PROVIDED 40% BOD REMOVAL OVER A WIDE RANGE OF LOADING RATES. THE 0.33 MGD INDUSTRIAL AND DOMESTIC WASTE WATER WAS TREATED AND REUSED AT A RATE OF 0.10 MGD FOR APPROXIMATELY 40 CENTS/1000 GALLONS. (LOWRY-TEXAS)

FIELD 05D

ACCESSION NO. W71-09524

EFFECTS OF PLANNEC FRESH WATER DIVERSICNS ON THE SAN FRANCISCO BAY AND SACRAMENT=SAN JOAQUIN ESTUARY,

CALIFORNIA UNIV., CAVIS. DEPT. OF CIVIL ENGINEERING.

RAY B. KRONE.

REPORT TO THE SAN FRANCISCO BAY CONSERVATION AND DEVELOPMENT COMMISSION, MAY 21, 1970. 7 P, 1 FIG.

DESCRIPTORS:
*DELTAS, *BAYS, *WATER QUALITY, *WATER POLLUTION CONTROL, *DIVERSION, SEDIMENTS, WATER CIRCULATICN, ABSORPTION, NUTRIENTS, PESTICIDES, ALGAE, DREDGING, ESTUARINE ENVIRONMENT, CALIFCRNIA WATER CONVEYANCE, CANALS, HEAVY METALS.

IDENTIFIERS:
*SAN FRANCISCO BAY-DELTA.

ABSTRACT:
THE EFFECTS ARE DISCUSSED OF THE PLANNED DIVERSICN OF FRESH WATER FLOWS AND SUSPENDED SEDIMENTS FROM THE SAN FRANCISCO BAY AND SACRAMENTO-SAN JOAQUIN ESTUARY. SEDIMENT IS CARRIED INTO THE BAY SYSTEM WITH FRESH WATER INFLOWS. BAY SEDIMENTS HAVE THREE OUTSTANDING EFFECTS: THEY IMPAIR THE PENETRATICN OF LIGHT, THEY ABSORB TREMENDCUS AMOUNTS OF TOXIC MATERIALS, AND CLOG NAVIGABLE WATERWAYS. THE TURBIDITY CAUSED BY SUSPENDED SEDIMENT MATERIALS LIMITS THE DEPTH TO WHICH THERE IS SUFFICIENT LIGHT FOR ALGAE TO MULTIPLY. SINCE NUTRIENTS FOR ALGAE GROWTH ARE PLENTIFUL IN THE BAY, THE LIMITED LIGHT PENETRATION DUE TO SEDIMENT IS THE LIMITING FACTOR TO THE RAPID MULTIPLICATION OF ALGAE IN THE BAY. THE SEDIMENTS ALSO ABSORB TOXIC COMPOUNDS, SUCH AS HEAVY METALS, PESTICIDES, AND RADIOACTIVE FALLOUT. THE CNLY FORESEEABLE BENEFIT OF THE REDUCED SEDIMENTS ARE RECUCED DREDGING COSTS. THE REPORT RECOMMENDS THAT COMPREHENSIVE STUDIES BE MADE OF THE REDUCED WATER AND SEDIMENT INFLOWS; THAT CRITERIA BE ESTABLISHED AS A BASIS FOR REGULATING FRESH WATER DIVERSION; THAT AN ADEQUATE MCNITORING SYSTEM BE MAINTAINED; THAT THE CONSTITUENT LEVELS AND LOCATICNS OF WASTE DISCHARGES TO THE SYSTEM BE REGULATED; THAT NEW METHCDS CF TREATMENT OF WASTE WATERS FOR REMOVAL OF NUTRIENTS AND TOXIC MATERIALS BE DEVELOPED AND SUPPORTED. (POERTNER)

FIELD 06G, 05G

ACCESSION NO. W71-09581

NUTRIENTS AND ALGAL REMOVAL FROM OXIDATICN PONDS EFFLUENTS,

MISSISSIPPI STATE UNIV., STATE COLLEGE. DEPT. OF SANITARY ENGINEERING.

ADNAN SHINDALA.

CONFERENCE HELD APRIL 13-14, 1971, VICKSBURG, WATER RESOURCES RESEARCH INSTITUTE, MISSISSIPPI STATE UNIVERSITY, P 1-7, 1971. 7 P, 1 FIG, 1 TAB, 3 REF.

DESCRIPTORS:
*ALGAE, *WASTE WATER TREATMENT, *TERTIARY TREATMENT, *COAGULATION, *OXIDATION LAGCONS, AEROBIC TREATMENT, OXIDATION, CRGANIC MATTER.

IDENTIFIERS:
NUTRIENT REMOVAL.

ABSTRACT:
CHEMICAL COAGULATION IS AN EFFECTIVE POST TREATMENT PROCESS FOR ALGAL REMOVAL AND FOR IMPRCVING THE QUALITY OF EFFLUENTS FROM STABILIZATION PONDS. OF THE COAGULANTS TESTED, ALUM WAS THE BEST. THE OPTIMUM DOSAGE FOR BEST REMOVAL OF THE PARAMETERS STUDIED WAS IN THE RANGE OF 75 TO 100 MG/LITER. USING THIS DCSAGE, THE SUPERNATANT FROM THE CHEMICAL COAGULATION PROCESS WAS FOUND TO CONTAIN 2.5 MG/LITER BOD, 22.9 MG/LITER COD, 1.5 MG/LITER TOTAL PHOSPHATES, 3.5 MG/LITER TOTAL PHOSPHATES, 3.5 MG/LITER TCTAL NITROGEN, 500 TO 1000 ALGAL CELLS/ML AND APPROXIMATELY 5,000 COLIFORMS/100 ML. THE ALGAE IN THE POND EFFLUENTS CONTRIBUTE HEAVILY TC THE BOD, COD, AND NITROGEN IN THE EFFLUENT, WHILE THE CONTRIBUTION TO THE PHOSPHATES CONCENTRATION WAS LESS IMPORTANT. (KNAPP-USGS)

FIELD 05G, 05D

ACCESSION NO. W71-09629

ON THE SIGNIFICANCE OF METAL COMPLEXING AGENTS IN SECONDARY SEWAGE EFFLUENTS,

MICHIGAN UNIV., ANN ARBOR. DEPT. OF ENVIRONMENTAL HEALTH.

MICHAEL E. BENDER, WAYNE R. MATSON, AND ROBERT A. JORDAN.

ENVIRONMENTAL SCIENCE AND TECHNOLOGY, VOL 4, NO 6, P 520-521, JUNE 1970. 1 FIG, 1 TAB, 11 REF. FWPCA GRANT 1-FL-WP-26-294-01, NIH GRANT 5-501-FR-05447-07.

DESCRIPTORS:
*WATER POLLUTION EFFECTS, *ORGANIC WASTES, *EUTROPHICATION, *CHELATION, *SEWAGE, *ORGANIC COMPOUNDS, WATER POLLUTION TREATMENT, WATER POLLUTION SOURCES, ENVIRONMENTAL SANITATION, SECONDARY TREATMENT, AQUATIC ALGAE, PRIMARY PRODUCTIVITY, TRACE ELEMENTS, PHOSPHATES, NITRATES, LABORATORY TESTS.

IDENTIFIERS:
*CHELATORS, ANODIC STRIPPING, ORGANIC CHELATORS.

ABSTRACT:
SEVERAL COMPONENTS OF SECONDARY SEWAGE OTHER THAN PHOSPHOROUS AND NITROGEN COULD BE ABETTING LAKE EUTROPHICATION. THE CHELATION OF METALS BY ORGANIC COMPOUNDS HAS BEEN DEMONSTRATED TO CAUSE SIGNIFICANT INCREASES IN ALGAL PRODUCTION. THIS STUDY WAS AN ATTEMPT TO FIND SUCH CHELATORS IN SECONDARY SEWAGE EFFLUENTS. TWO DISTINCT METAL-COMPLEXING MOLECULAR WEIGHT FRACTIONS WERE DEMONSTRATED USING RECENT ADVANCES IN ANODIC STRIPPING TECHNIQUES. ONE FRACTION HAS A MOLECULAR WEIGHT SIMILAR TO THAT OF SYNTHETIC CHELATORS AND HAS PROVEN EFFECTIVE IN STIMULATING ALGAL GROWTH. SUBSEQUENT INVESTIGATIONS ARE ATTEMPTING TO RULE OUT THE POSSIBILITY THAT OTHER SUBSTANCES IN SEWAGE FRACTIONS COULD BE RESPONSIBLE FOR THE APPARENT STIMULATION. (LEGORE-WASHINGTON)

FIELD 05B, 05A

ACCESSION NO. W71-09674

A WATER RESOURCE-WATER SUPPLY STUDY OF THE POTOMAC ESTUARY,

ENVIRONMENTAL PROTECTION AGENCY, ANNAPOLIS, MD. CHESAPEAKE TECHNICAL SUPPORT LAB.

NORBERT A. JAWORSKI, LEO J. CLARK, AND KENNETH D. FEIGNER.

ENVIRONMENTAL PROTECTION AGENCY, WATER QUALITY OFFICE, TECHNICAL REPORT 35, APRIL 1971, 263 P, 39 TAB.

DESCRIPTORS:
*ESTUARIES, *WATER QUALITY, *WATER RESOURCES DEVELOPMENT, *WATER SUPPLY, WASTE WATER, RUNOFF, DISSOLVED OXYGEN, ALGAE, NUTRIENTS, BACTERIA, VIRUSES, HEAVY METALS, HUMAN POPULATION, WASTE WATER TREATMENT, WATER QUALITY CONTROL.

IDENTIFIERS:
POTOMAC ESTUARY.

ABSTRACT:
A DETAILED INVESTIGATION OF THE WATER QUALITY AND WATER RESOURCES OF THE POTOMAC ESTUARY WAS CONDUCTED BY THE CHESAPEAKE TECHNICAL SUPPORT LABORATORY. INCLUDED IN THE STUDY WERE AN EVALUATION OF POLLUTION SOURCES INCLUDING NUTRIENTS; THE DEVELOPMENT OF MATHEMATICAL MODELS TO PREDICT POLLUTANT EFFECTS ON WATER QUALITY; THE PROJECTION OF WATER SUPPLY NEEDS AND WASTEWATER LOADINGS; AN EVALUATION AS A POTENTIAL WATER SUPPLY SOURCE; THE DETERMINATION OF THE MAXIMUM POUND LOADINGS BY ZONE FOR CERTAIN POLLUTANTS UNDER DIFFERENT FLOW CONDITIONS; ALTERNATIVE WASTE TREATMENT PLANS AND COST ANALYSIS OF WASTEWATER TREATMENT. (ENSIGN-PAI)

FIELD 05B, 06D, 02L

ACCESSION NO. W71-09788

EVALUATING OIL SPILL CLEANUP AGENTS. DEVELOPMENT OF TESTING PROCEDURES AND CRITERIA,

CALIFORNIA STATE WATER RESOURCES CONTROL BOARD, SACREMENTO.

CHARLES R. HAZEL, FRED KOPPERDAHL, NORMAN MORGAN, AND WALTER THOMSEN.

CALIFORNIA WATER RESOURCES CONTROL BOARD PUBLICATION NO. 43, JULY 6, 1970, 150 P, 67 FIG, 17 TAB, 100 REF.

DESCRIPTORS:
*OIL WASTES, *CLEANING, *DISPERSION, *CHEMICALS, *SURFACTANTS, *TOXICITY, *BIOASSAY, FISH, SHELLFISH, ALGAE, DEGRADATION.

IDENTIFIERS:
*DISPERSANTS, *COLLECTING AGENTS, *SINKING AGENTS, EFFECTIVENESS, TESTING PROCEDURES, CRITERIA, CLASSIFICATIONS.

ABSTRACT:
CRITERIA FOR LICENSING AND REGULATING THE USE OF OIL SPILL CLEANUP AGENTS WAS STUDIED AND DEVELOPED. CRITERIA INCLUDED TOXICITY, PERFORMANCE EFFECTIVENESS AND PHYSICAL-CHEMICAL DESCRIPTIONS OF THE OSCA. BIOASSAYS WERE PERFORMED TO ESTABLISH 96-HOUR TLM VALUES FOR DISPERSANTS, OIL AND COMBINATIONS OF OIL AND DISPERSANTS. BIODEGRADATION OF DISPERSANTS WAS TESTED BY BOD, FOAM CALIBRATION AND TOXICITY DECAY BIOASSAY. MISCIBILITY WITH SEAWATER, PERCENT OF OIL EMULSIFIED, OIL SINKING AND DISPERSION AFTER SEVERAL HOURS WERE THE CRITERIA FOR PERFORMANCE EFFECTIVENESS TESTS. FEASIBILITY OF 'FINGERPRINTING' DISPERSANTS WERE TESTED QUANTITATIVELY. LICENSING CRITERIA AND TEST PROCEDURES FOR PRODUCT FLASH POINT, P.H, TRACE SUBSTANCES AND OCCUPATIONAL HEALTH HAZARDS WERE DETERMINED. (ENSIGN-PAI)

FIELD 05G, 07B

ACCESSION NO. W71-09789

LAMINARIA SACCHARINA AND MARINE POLLUTION IN NORTH-EAST ENGLAND,

LIVERPOOL UNIV. (ENGLAND). DEPT. OF BOTANY.

E. M. BURROWS, AND C. PYBUS.

MARINE POLLUTION BULLETIN, VOL. 2, NO. 4, P 53-56, APRIL 1971.

DESCRIPTORS:
*POLLUTANT IDENTIFICATION, *BIOINDICATORS, *ALGAE, GROWTH RATES, CULTURES, SUSPENDED LOAD, SEWAGE, INDUSTRIAL WASTES, SILTS, WATER POLLUTION EFFECTS.

IDENTIFIERS:
*ENGLAND, *LAMINARIA SACCHARINA.

ABSTRACT:
THE GROWTH RATE OF LAMINARIA SACCHARINA IN BOTH THE LABORATORY AND IN THE FIELD HAS BEEN STUDIED AS AN INDICATOR FOR THE EFFECTS OF POLLUTION ON THE NORTH-EAST COAST OF ENGLAND. THIS SPECIES PROVIDES A CONSTANT SUBSTRATE FOR OTHER MARINE ORGANISMS OVER A LARGE AREA OF THE COAST MAKING IT AN IMPORTANT COMPONENT OF THE COASTAL ECOSYSTEM. LAMINARIA SACCHARINA IS VERY SENSITIVE TO ITS ENVIRONMENT AND THEREFORE IS CAPABLE OF GIVING A GRADED RESPONSE TO DIFFERENT TYPES AND DEGREES OF POLLUTION. THE FACTOR THOUGHT TO INHIBIT GROWTH MOST WAS THE PRESENCE OF SILT; HOWEVER POLLUTION OF THE WATERS THEMSELVES HAVING DELETERIOUS EFFECTS ON GROWTH CANNOT BE RULED OUT. (ENSIGN-PAI)

FIELD 05C

ACCESSION NO. W71-09795

RESEARCH ON ECOLOGICAL STUDIES ON THE DYNAMICS OF PLANKTONIC BLUE-GREEN ALGAE WITH SPECIAL REFERENCE TO THEIR MICROSTRATIFICATION.

MINNESOTA UNIV., MINNEAPOLIS.

COO-1820-2 (1970), 6 P.

DESCRIPTORS:
*AQUATIC ALGAE, *LAKES, *PHYTOPLANKTON, *DENSITY STRATIFICATION, SPECIATION, VERTICAL MIGRATION, AQUARIA, GROWTH RATES, SEDIMENTATION, BUOYANCY, WIND VELOCITY, CONVECTION, WATTER POLLUTION EFFECTS.

ABSTRACT:
SPECIES ASSOCIATIONS AT VARIOUS DEPTHS ARE FORTUITOUS, TRANSIENT, AND DEPENDENT UPON PHYSICAL CONTROL MANIFEST BY THE ABIOTIC ENVIRONMENT, RATHER THAN UPON ANY DIRECT BIOLOGICAL CONTROL. INDIVIDUALS COMPRISING THE PHYTOPLANKTON COMMUNITY ARE NOT BOUND TO A SUBSTRATE, AND ARE ENTIRELY MOBILE. VERTICAL DISTRIBUTION IS INFLUENCED BY DIFFERENTIAL GROWTH RATES AT VARIOUS DEPTHS, SEDIMENTATION, BUOYANCY, WIND, AND THERMAL CONVECTION. CURRENTLY, PROPORTIONATELY MORE TIME IS BEING DEVOTED WITH SEVERAL LABORATORY MODELS OF THE STUDY LAKES. (BOPP-NSIC)

FIELD 05C

ACCESSION NO. W71-09852

SECONDARY EFFECTS OF NITROGEN IN WATER,

MICHIGAN UNIV., ANN ARBOR, DEPT. OF SANITARY ENGINEERING.

JACK A. BORCHARDT.

PROCEEDINGS 12TH SANITARY ENGINEERING CONFERENCE ON NITRATE AND WATER SUPPLY: SOURCE AND CONTROL, FEBRUARY 11-12, 1970, UNIVERSITY OF ILLINOIS, URBANA: ILLINOIS UNIVERSITY, COLLEGE OF ENGINEERING PUBLICATION, P 66-77, 1970. 12 P, 9 FIG, 1 TAB, 10 REF.

DESCRIPTORS:
*WATER POLLUTION EFFECTS, *NITRATES, *NUTRIENTS, *ALGAE, *BACTERIA, AESTHETICS, COLOR, TASTE, CORROSION, DISEASES, EUTROPHICATION, FOULING, TOXICITY, ODOR, OXYGEN SAG, TURBIDITY, WATER QUALITY, WATER TREATMENT, NITROGEN CYCLE.

ABSTRACT:
ONE OF THE PRIMARY EFFECTS OF INORGANIC NITROGEN IN NATURAL WATERS IS ITS INFLUENCE ON THE GROWTH OF FLORA. ALGAE IN WATER SUPPLIES CAN CAUSE OBNOXIOUS CONDITIONS IN STORAGE RESERVOIRS, CLOGGING OF INTAKE SCREENS, CAN SERIOUSLY AFFECT PRECHLORINATION, MAY MAKE ADDITIONAL CHEMICALS NECESSARY FOR PURIFICATION, MAY CAUSE SHORTER FILTER RUNS, AND MAY RESULT IN SERIOUS TASTE AND ODOR PROBLEMS. THESE PROBLEMS ARE DISCUSSED IN SOME DETAIL. (KNAPP-USGS)

FIELD 05B, 05C

ACCESSION NO. W71-09958

ENERGY FLOW IN A WOODLAND STREAM ECOSYSTEM: I. TISSUE SUPPORT TROPHIC STRUCTURE OF THE AUTUMNAL COMMUNITY, (WITH GERMAN SUMMARY),

PITTSBURG UNIV., PA. PYMATUNING LAB. OF ECOLOGY; AND MICHIGAN STATE UNIV., HICKORY CORNERS. W. K. KELLOGG BIOLOGICAL STATION.

WILLIAM P. COFFMAN, KENNETH W. CUMMINS, AND JOHN C. WUYCHECK.

ARCHIV FUR HYDROBIOLOGIE, VOL 68, NO 2, P 232-276, 1971. 8 FIG, 13 TAB, 36 REF. OWRR PROJECT NOS A-032-MICH(1) AND B-008-MICH(1).

DESCRIPTORS:
*STREAMS, *STANDING CROP, *TROPHIC LEVEL, *ENERGY BUDGET, PRODUCTIVITY, HERBIVORES, CARNIVORES, ECOSYSTEMS, AQUATIC LIFE, SAMPLING, ALGAE, LABORATORY TESTS, DETRITUS, BENTHOS, AQUATIC ANIMALS, DIATOMS, INSECTS, BACTERIA, FUNGI, TURNOVERS.

IDENTIFIERS:
*RIFFLE BENTHOS, WOODLAND STREAM, ECOSYSTEM ANALYSIS, MACROCONSUMERS, AUTUMNAL COMMUNITY, CALORIES, DETRIVORES, LINESVILLE CREEK(PA).

ABSTRACT:
TO DETERMINE AN ENERGY BUDGET FOR A LOTIC ECOSYSTEM, A 2-YEAR INTENSIVE INVESTIGATION OF A WOODLAND STREAM WAS CONDUCTED OF THE AUTUMNAL BENTHIC MACROCONSUMER COMMUNITY OF THE RIFFLE AREA. STANDING CROP AND TROPHIC STRUCTURE WERE DETERMINED AND PRODUCTION PARAMETERS ESTIMATED. THE PROCEDURE WAS DEVELOPED TO EXPRESS LEVELS ON THE BASIS OF TISSUE SUPPORT RATHER THAN SPECIES CATEGORIES. TROPHIC ANALYSES WERE MADE FROM GUT CONTENTS, AND THE DATA DETERMINED THE PROPORTION OF EACH SIZE CLASS OF EACH SPECIES TO BE ASSIGNED TO ONE OF THREE TROPHIC LEVELS. ON AN INSTANTANEOUS INGESTION BASIS, 17% TO 21% OF THE STANDING CROP TISSUE WAS SUPPORTED BY ALGAL CALORIES, 2% TO 5% BY DETRIAL CALORIES, AND 71% TO 73% BY ANIMAL CALORIES. THE ASSIMILATIVE EFFICIENCY FOR INGESTED FOOD IS DEPENDENT UPON SPECIES ENZYME SYSTEMS AND VARIES IN RELATION TO DIFFERENCES IN INGESTIVE MECHANISMS. THE HIGH SUPPORT BY ANIMAL CALORIES DEMANDS A RAPID BIOMASS TURNOVER (LESS THAN A MONTH) BY THE HERBIVORE-DETRITIVORE COMPONENT AND A SLOWER TURNOVER (SEVERAL MONTHS) BY THE CARNIVORE TISSUE. THE STANDING CROP AND TROPHIC STRUCTURE RECORDED YIELD A CONSISTENT PICTURE OVER TIME. (JONES-WISCONSIN)

FIELD 05C

ACCESSION NO. W71-10064

A CHECKLIST OF MARINE ALGAE OF EASTERN CANADA,

MEMORIAL UNIV. OF NEWFOUNDLAND, ST JOHN'S. DEPT. OF BIOLOGY; AND LAVAL UNIV. QUEBEC. DEPARTEMENT DE BIOLOGIE.

G. ROBIN SOUTH, AND ANDRE CARDINAL.

CANADIAN JOURNAL OF BOTANY, VOL 48, P 2077-2095, 1970. 1 FIG, 132 REF, INDEX.

DESCRIPTORS:
*MARINE ALGAE, *SYSTEMATICS, *COASTS, BENTHIC FLORA, RHODOPHYTA, PHAEOPHYTA, CHLOROPHYTA, MAINE, SPECIATION.

IDENTIFIERS:
*CHECKLIST, *EASTERN CANADA, CAPE CHIDLEY(LABRADOR), NEW BRUNSWICK, ANTICOSTI ISLAND(QUEBEC), MAGDALEN ISLAND(QUEBEC), SABLE ISLAND(NOVA SCOTIA), ST PIERRE(CANADA), MIQUELON(CANADA), MIKROSYPHAR PORPHYRAE, PROTECTOCARPUS SPECIOSUS.

ABSTRACT:
A COMPREHENSIVE CHECKLIST DOCUMENTING MARINE BENTHIC ALGAE WITHIN THE POLITICAL BOUNDARIES OF EASTERN CANADA IS PRESENTED. TAYLOR'S SECOND EDITION (1957) FORMS THE BASIC REFERENCE FOR THIS LIST WHICH INCLUDES ALL PREVIOUSLY PUBLISHED RECORDS TOGETHER WITH SOME NEW ONES, INCORPORATING NOMENCLATURAL CHANGES THAT HAVE OCCURRED SINCE 1957. IT CONSISTS OF 371 SPECIES, SUBSPECIES, AND VARIETIES FROM EASTERN CANADA, INCLUDING 157 RHODOPHYCEAE, 127 PHAEOPHYCEAE, AND 87 CHLOROPHYCEAE. RECORDS FOR THE ENTIRE COASTLINE FROM CAPE CHIDLEY, LABRADOR, IN THE NORTH TO THE NEW BRUNSWICK-MAINE BORDER IN THE SOUTH ARE INCLUDED, AS WELL AS FROM ANTICOSTI ISLAND, MAGDALEN ISLAND, SABLE ISLAND, AND ST PIERRE AND MIQUELON. MIKROSYPHAR PORPHYRAE KUCK, AND PROTECTOCARPUS SPECIOSUS (BERG) KUCK ARE NEW RECORDS FOR THE AREA. THE COMPLETE AUTHOR CITATION FOR EACH GENUS AND SPECIES IS GIVEN TOGETHER WITH EXTENSIVE EXPLANATORY NOTES GIVING SYNONYMS AND PREVIOUS AUTHOR NOTES, 132 REFERENCES, AND AN INDEX TO THE GENERA. (JONES-WISCONSIN)

FIELD 05C, 07B, 02L

ACCESSION NO. W71-10066

SOLUBILIZATION OF TRICALCIUM PHOSPHATE BY BLUE-GREEN ALGAE,

INDIAN AGRICULTURAL RESEARCH INST., NEW DELHI. DIV. OF MICROBIOLOGY.

PAROMITA BOSE, UJJAL SINGH NAGPAL, G. S. VENKATARAMAN, AND S. K. GOYAL.

CURRENT SCIENCE, VOL 40, NO 7, P 165-166, 1971. 1 TAB, 22 REF.

DESCRIPTORS:
*PHOSPHATES, *SOLUBILITY, *CYANOPHYTA, ALGAE, CULTURES.

IDENTIFIERS:
*TRICALCIUM PHOSPHATES, ANABAENA, NOSTOC, TOLYPOTHRIX TENUIS, AULOSIRA FERTILISSIMA, ANACYSTIS NIDULANS.

ABSTRACT:
TWENTY-SEVEN STRAINS OF NITROGEN-FIXING BLUE-GREEN ALGAE AND ONE NON-NITROGEN-FIXING ALGA, ANACYSTIS NIDULANS, WERE INCUBATED FOR 15 DAYS IN A NUTRIENT MEDIUM CONTAINING TRICALCIUM PHOSPHATE. RESULTING DETERMINATIONS REVEALED THAT 17 STRAINS, BELONGING TO THE GENERA ANABAENA, NOSTOC, TOLYPOTHRIX, AND AULOSIRA, WERE EFFECTIVE IN SOLUBILIZATION OF ROCK PHOSPHATE. THE SOLUBILIZATION EFFICIENCIES INDICATED BY TOLYPOTHRIX TENUIS AND AULOSIRA FERTILISSIMA ARE OF PARTICULAR INTEREST. (WILDE-WISCONSIN)

FIELD 05C

ACCESSION NO. W71-10070

DISSOLVED ORGANIC PHOSPHORUS EXCRETION BY MARINE PHYTOPLANKTON,

NORTH CAROLINA UNIV., CHAPEL HILL. DEPT. OF ENVIRONMENTAL SCIENCES AND ENGINEERING.

EDWARD J. KUENZLER.

JOURNAL OF PHYCOLOGY, VOL 6, NO 1, P 7-13, 1970. 4 FIG, 2 TAB, 33 REF. AEC CONTRACT AT-(40-1)-3549.

DESCRIPTORS:
*ORGANIC COMPOUNDS, *PHOSPHORUS, *MARINE PLANTS, *PHYTOPLANKTON, ALGAE, LIGHT INTENSITY, SALINITY, NUTRIENT REQUIREMENTS, CHLORELLA, GROWTH RATES, TEMPERATURE, SEA WATER, PHOSPHATES, PHOSPHORUS RADIOISOTOPES, DIATOMS, HYDROGEN ION CONCENTRATION, MARINE ANIMALS, BACTERIA, ABSORPTION, CULTURES.

IDENTIFIERS:
*DISSOLVED ORGANIC PHOSPHORUS, *EXCRETION, ALKALINE PHOSPHATASE, REASSIMILATION, GREEN FLAGELLATE, CYCLOTELLA CRYPTICA, THALASSIOSIRA FLUVIATILIS, DUNALIELLA TERTIOLECTA, SYNECHOCOCCUS, RHODOMONAS, COCCOLITHUS HUXLEYI, AUTOLYSIS, PHAEODACTYLUM.

ABSTRACT:
EXPERIMENTS PERFORMED ON LABORATORY CULTURES SHOW THAT ALGAE ELIMINATE DISSOLVED ORGANIC PHOSPHORUS (DOP) COMPOUNDS UNDER VARYING CONDITIONS FOR DIFFERENT SPECIES, BUT CERTAIN PATTERNS EMERGE. THE AMOUNTS OF EXTRACELLULAR DOP PRODUCED BY EIGHT SPECIES OF MARINE PLANKTONIC ALGAE (AXENIC CULTURES) UNDER VARIOUS CONDITIONS OF LIGHT, SALINITY, AND NUTRITION WERE COMPARED. AFTER INITIAL LABELING, P-32 WAS MEASURED IN THE FRACTIONS. DIFFERENT SPECIES TAKE DIFFERENT FRACTIONS OF THE TOTAL EXCRETED DOP. MORE THAN 20% OF TOTAL PHOSPHORUS IN THE SYSTEM WAS EXCRETED BY CYCLOTELLA CRYPTICA, THALASSIOSIRA FLUVIATILIS, DUNALIELLA TERTIOLECTA, AND SYNECHOCOCCUS UNDER ONE OR MORE OF THE EXPERIMENTAL CONDITIONS. EXCRETION OF DOP WAS PROPORTIONAL TO LIGHT INTENSITY IN DUNALIELLA, RHODOMONAS, CHLORELLA, AND COCCOLITHUS HUXLEYI. PHOSPHORUS LIMITATION REDUCED DOP PRODUCTION BY CYCLOTELLA AND THALASSIOSIRA, NITROGEN LIMITATION REDUCED DOP BY PHAEODACTYLUM, DUNALIELLA, AND RHODOMONAS, AND LACK OF IRON REDUCED DOP LEVELS IN CYCLOTELLA CULTURES. SALINITY AFFECTED GROWTH, BUT NO CLEAR RELATIONSHIP TO DOP EXCRETION WAS EVIDENT. THE DOP ELIMINATED DURING GROWTH WAS REASSIMILATED BY THE SPECIES THAT PRODUCED IT AND BY OTHER SPECIES, BUT LACK OF ALKALINE PHOSPHATASE REDUCED THE AMOUNT OF DOP AVAILABLE TO CERTAIN ALGAE. (JONES-WISCONSIN)

FIELD 05C

ACCESSION NO. W71-10083

PRIMARY PRODUCTION: DEPRESSICN OF OXYGEN EVOLUTION IN ALGAL CULTURES BY ORGANOPHOSPHORUS INSECTICIDES,

NORTHWESTERN UNIV., BOSTON, MASS. DEPT. OF BIOLOGY.

SYLVIA B. DERBY, AND ERNEST RUBER.

BULLETIN OF ENVIRONMENTAL CONTAMINATION AND TOXICOLOGY, VOL 5, NO 6, P 553-557, 1971. 1 TAB, 10 REF.

DESCRIPTORS:
*INSECTICIDES, *ALGAE, *PRIMARY PRODUCTIVITY, CHLORINATED HYDROCARBON PESTICIDES, DDT, ORGANOPHOSPHORUS PESTICIDES, PHYTOPLANKTON, TOXICITY, DISSOLVED OXYGEN, CULTRUES, ENDRIN, DIELDRIN, CXYGENATION, CARBAMATE PESTICIDES.

IDENTIFIERS:
ABATE, BAYTEX, BAYGONE, CYCLOTELLA, DUNALIELLA, SKELETONEMA.

ABSTRACT:
THE EFFECT OF DDT, ORGANOPHOSPHATES (BAYTEX, ABATE), AND CARBAMATE (BAYGONE) INSECTICIDES WAS APPRAISED ON THE BASIS OF OXYGEN PRODUCED BY ALGAE IN ACETONE CULTURE MEDIA. THE INSECTICIDES WERE USED IN CONCENTRATIONS OF 1.0, 0.1, AND 0.01 PPM. THE RESULTS VARIED WITH ALGAL SPECIES AND THE NATURE OF ERADICANTS. THE ORDER FROM THE MOST TO LEAST TOXIC COMPOUNDS WAS: BAYTEX, BAYGONE, DDT, AND ABATE. THE VULNERABILITY OF ALGAE TO INSECTICIDES INCREASED IN THE FOLLOWING CRDER: CYCLOTELLA NANA, PHAEODACTYLUM TRICORNUTUM, SKELETONEMA COSTATUM, AND DUNALIELLA EUCHLORA. (WILDE-WISCONSIN)

FIELD 05C

ACCESSION NO. W71-10096

ASSESSMENT OF POLLUTICN EFFECTS BY THE USE OF ALGAE,

LIVERPOOL UNIV. (ENGLAND). HARTLEY BOTANICAL LABS.

ELSIE M. BURROWS.

PROCEEDINGS OF THE ROYAL SOCIETY OF LONDON B, VOL 177, P 295-306, 1971. 5 FIG, 22 REF.

DESCRIPTORS:
*POLLUTANT IDENTIFICATION, *BIOASSAY, *INDICATORS, *WATER POLLUTION EFFECTS, *ALGAE, *MARINE ALGAE, *BIOINDICATORS, *PHOTOSYNTHESIS, *PLANT PHYSIOLOGY, POLLUTANTS, ANALYTICAL TECHNIQUES, WATER POLLUTION SOURCES, TOXINS, WASTES, CHLOROPHYTA, SESSILE ALGAE, WATER ANALYSIS.

IDENTIFIERS:
*LAMINARIA SP., *ULVA SP., MACRCCYSTIS SP.

ABSTRACT:
AN ATTEMPT WAS MADE TO USE THE GROWTH RATE IN CULTURE OF SOME OF THE LARGER ATTACHEC ALGAE FOR INDICATICN OF POLLUTION. ULVA LACTUA HAS A HIGH POTENTIAL FROM THIS POINT OF VIEW BECAUSE OF THE EASE WITH WHICH IT CAN BE CULTURED AND ALSO BECAUSE OF ITS REACTIONS TO POLLUTION BY SEWAGE. THE GROWTH RATE OF LAMINARIA SACCHARINA SHOWS A GRADED RESPONSE TO CHANGES IN TOTAL MEDIUM AND PHYSICAL CONDITIONS AND TO CONCENTRATIONS OF ADDED SINGLE SUBSTANCES. THIS SPECIES COULD BE A USEFUL POLLUTICN INDICATOR, NOT ONLY BECAUSE OF ITS SENSITIVITY TO CHANGES, BUT ALSO BECAUSE OF THE PART IT PLAYS IN THE ECOSYSTEMS OF THE BRITISH COASTS. (LEGORE-WASHINGTON)

FIELD 05A, 05C

ACCESSION NO. W71-10553

PESTICIDE EFFECT OF GROWTH AND C-14 ASSIMILATION IN A FRESHWATER ALGA,

MISSOURI UNIV., COLUMBIA. DIV. OF BIOLOGY; AND BUREAU OF SPORT FISHERIES AND WILDLIFE, COLUMBIA, MO. FISH-PESTICIDE RESEARCH LAB.

LELYN STADNYK, ROBERT S. CAMPBELL, AND B. THOMAS JOHNSON.

BULLETIN OF ENVIRONMENTAL CONTAMINATION AND TOXICOLOGY, VOL 6, NO 1, P 1-8, 1971. 1 FIG, 1 TAB, 16 REF.

DESCRIPTORS:
*PESTICIDES, *CHLORINATED HYDROCARBON PESTICIDES, *CARBAMATE PESTICIDES, *ORGANOPHOSPHOROUS PESTICIDES, *2-4-D, *DDT, *DIELDRIN, *DIAZINON, *WATER POLLUTION EFFECTS, *PHOTOSYNTHESIS, PESTICIDE TOXICITY, PLANT GROWTH REGULATORS, INHIBITION, INHIBITORS, PHYTOPLANKTON, PHYTO-TOXICITY, HERBICIDES, INSECTICIDES, BIOMASS, PRODUCTIVITY, SCENEDESMUS, ALGAL.

IDENTIFIERS:
*DIURON, *CARBARYL, *TOXAPHENE.

ABSTRACT:
THE EFFECTS WERE EVALUATED OF PESTICIDES (DIURON, CARBARYL, 2,4-D, DDT, DIELDRIN, TOXAPHENE, AND DIAZINON) ON LOW DENSITY POPULATIONS OF THE FRESHWATER ALGA, SCENEDESMUS QUADRICAUDATA, IN TERMS OF CHANGES IN GROWTH AND METABOLISM RATHER THAN DEATH. OF THE COMPOUNDS TESTED, THE SOIL STERILANT AND HERBICIDE DIURON WAS THE MOST TOXIC TO THE ALGAL POPULATION, ALTHOUGH APPARENT BACTERIAL DETOXIFICATION OF THIS SUBSTANCE WAS INDICATED. WHILE DDT, TOXAPHENE AND DIELDRIN GENERALLY CAUSED A DECREASE IN CELL NUMBER, BIOMASS AND CARBON ASSIMILATION, CARBARYL AND DIAZINON EITHER HAD NO EFFECT OR STIMULATED CARBON FIXATION AND CELL DIVISION. LONG TERM CHRONIC EFFECTS OF PESTICIDES IN TERMS OF PLANT LIFE AS WELL AS ANIMAL SPECIES MUST BE CONSIDERED. (LEGORE-WASHINGTON)

FIELD 05C

ACCESSION NO. W71-10566

SALTON SEA, CALIFORNIA--WATER QUALITY AND ECOLOGICAL MANAGEMENT CONSIDERATIONS,

FEDERAL WATER QUALITY ADMINISTRATION, SAN FRANCISCO, CALIF. PACIFIC SOUTHWEST REGION.

R. C. BAIN, A. M. CALDWELL, R. H. CLAWSON, H. L. SCOTTEN, AND R. G. WILLS.

FEDERAL WATER QUALITY ADMINISTRATION PACIFIC SOUTHWEST REGION REPORT, JULY 1970. 54 P, 15 FIG, 10 TAB, 16 REF.

DESCRIPTORS:
*SURFACE WATERS, *SEA WATER, *WATER QUALITY, *ECOLOGY, *CALIFORNIA, HYDROLOGIC DATA, CHEMICAL ANALYSIS, DATA COLLECTIONS, NUTRIENTS, SALINITY, SEDIMENTS, EUTROPHICATION, FISH, ALGAE, RECREATION, ENVIRONMENTAL EFFECTS, INVERTEBRATES, WATER RESOURCES DEVELOPMENT.

IDENTIFIERS:
*SALTON SEA(CALIF), ECOLOGICAL MANAGEMENT.

ABSTRACT:
THIS REPORT PRESENTS DATA RELATED TO SALINITY AND NUTRIENT RELATED PROBLEMS AND CONTROL MEASURES TO ALLEVIATE ADVERSE WATER QUALITY CONDITIONS IN THE SALTON SEA. THE SALTON SEA LIES IN A LOW-LYING DESERT SINK AREA APPROXIMATELY 85 MILES EAST OF SAN DIEGO, CALIFORNIA. FORMED IN 1905-06, THE 230,000 ACRE SEA IS THREATENED WITH RAPIDLY RISING SALINITY LEVELS WHICH IF UNCONTROLLED ARE EXPECTED TO ELIMINATE THE CURRENTLY VALUABLE SPORT FISHERY WITHIN THE NEXT DECADE. FLUCTUATING WATER LEVELS AND EUTROPHICATION SYMPTOMS, SUCH AS DISSOLVED OXYGEN DEFICIENCIES IN DEEPER WATERS, DISCOLORATIONS AND TURBIDITY OF THE WATER AND OFFENSIVE ODORS CAUSED BY DENSE PHYTOPLANKTON POPULATIONS, ARE ALSO MAJOR SALTON SEA PROBLEMS. AN ANNUAL INFLOW OF APPROXIMATELY 1.2 MILLION ACRE FEET PRINCIPALLY FROM THE NEW AND ALAMO RIVERS BRINGS SALTS, NUTRIENTS, PESTICIDES AND FECAL BACTERIA TO THE SEA. MUCH OF THIS WATER IS AGRICULTURAL DRAINAGE FROM THE IMPERIAL VALLEY AND SEWAGE FROM VALLEY COMMUNITIES AND FROM MEXICO. EVAPORATION LOSSES WITHIN THE SALTON SEA APPROXIMATE ANNUAL INFLOW. THUS A HYDRODYNAMIC BALANCE EXISTS. RECENT DATA SHOW SALT LEVELS ARE APPROXIMATELY EQUAL TO OCEANIC SALINITY ALTHOUGH IONIC COMPOSITION IS SOMEWHAT DIFFERENT. (WOODARD-USGS)

FIELD 05G, 02H, 05A

ACCESSION NO. W71-10577

BIOLCGY OF CLADOPHORA IN FRESHWATERS,

DURHAM UNIV. (ENGLAND). DEPT. CF BOTANY.

B. A. WHITTON.

WATER RESEARCH, VOL 4, NO 7, P 457-476, JULY 1970. 20 P, 1 TAB, 80 REF.

DESCRIPTORS:
*EUTROPHICATION, *ALGAE, *SURFACE WATERS, *REVIEWS, *WATER POLLUTION SOURCES, NUTRIENTS, PLANT PHYSIOLOGY, LAKES, PONDS, AQUATIC ALGAE, ECOLOGY, BIOLOGICAL PROPERTIES, FRESH WATER, WATER QUALITY.

IDENTIFIERS:
*CLADOPHORA.

ABSTRACT:
THE LITERATURE ON CLADOPHCRA GROWING IN FRESHWATERS IS SUMMARIZED, ESPECIALLY THOSE SITUATIONS WHERE NUTRIENT ENRICHMENT BY MAD HAS LED TO THE PRESENCE OF CONSPICUOUS AND SOMETIMES TROUBLESOME GROWTHS. IT IS REASONABLE TO ASSUME THAT CLADOPHORA PLAYED A RELATIVELY MINOR ROLE IN AQUATIC COMMUNITIES BEFORE THE ACTIVITIES OF MAN LED TO WIDESPREAD NUTRIENT ENRICHMENT. MASSIVE GROWTHS PROBABLY COULD NOT DEVELOP IN FLOWING WATERS WITHOUT MAN'S ACTIVITIES. THEREFORE, IT IS POSSIBLE THAT THE ALGAE ITSELF MAY BE UNDERGOING EVCLUTICNARY CHANGES IN RESPONSE TO THESE NEW ENVIRONMENTS. CLADOPHORA AS A GENUS IS FAVORED BY HIGH LIGHT INTENSITIES, HIGH WATER TURBULENCE, HIGH NUTRIENT LEVELS, HIGH PH VALUES, HARD WATERS. NOT ALL SPECIES SHARE ALL THESE CHARACTERS, BUT C. GLOMERATA, THE SPECIES WHICH HAS INCREASED THE MOST AS A RESPONSE OF MAN'S ACTIVITIES, IS THE ORGANISM WHICH COMBINES ALL THESE CHARACTERS WITHIN A SINGLE SPECIES. (WOODARD-USGS)

FIELD 05C, 05B

ACCESSION NO. W71-10580

NUTRITIONAL ECOLOGY ANC COMMUNITY STRUCTURE OF THE PHYTOPLANKTON OF GREEN BAY,

WISCONSIN UNIV., MADISON. WATER RESOURCES CENTER.

PAUL E. SAGER.

AVAILABLE FROM THE NATIONAL TECHNICAL INVORMATION SERVICE AS PB-201 696, $3.00 IN PAPER COPY, $0.95 IN MICROFICHE. TECHNICAL CCMPLETION REPORT, WATER RESOURCES CENTER, THE UNIVERSITY OF WISCONSIN-MADISON, WISCONSIN, 1971, 31 P, 7 FIG, 11 TAB, 21 REF, OWRR PROJECT A-017-WIS(1).

DESCRIPTORS:
*PHYTOPLANKTON, *AQUATIC ALGAE, *PHOSPHORUS, *CYANOPHYTA, *EUTROPHICATION, BIOMASS, WATER QUALITY, CHLOROPHYLL, NITROGEN, WISCONSIN, NUTRIENTS, WATER POLLUTION EFFECTS, INTERFACES, DISSOLVED OXYGEN, TEMPERATURE, ORGANCPHOSPHOROUS COMPOUNDS, ALGAE.

IDENTIFIERS:
GREEN BAY, FOX RIVER, SPECIFIC CONDUCTANCE, TRANSPARENCY, ORTHOPHOSPHATES, DRY WEIGHT, PHOSPHATE UPTAKE.

ABSTRACT:
THE INFLUENCE OF THE FOX RIVER ON CERTAIN CHEMICAL AND PHYSICAL PARAMETERS OF WATER QUALITY IN LOWER GREEN BAY HAS BEEN STUDIED WITH RESPECT TO RESPONSES IN THE STRUCTURE AND PHCSPHCRUS NUTRITION OF THE PHYTOPLANKTON COMMUNITY. SECCHI DISC TRANSPARENCY, DISSOLVED OXYGEN, SPECIFIC CONDUCTANCE, TEMPERATURE, ORTHOPHOSPHATE, NITRATE, CHLOROPHYLL A, PHYTOPLANKTON DRY WEIGHT, AND EXTRACTABLE PHOSPHATE (LUXURY UPTAKE BY ALGAE) WERE ANALYZED WEEKLY AT NINE STATIONS ALCNG A TRANSECT RUNNING 13.5 MILES UP THE BAY FROM THE MOUTH OF THE FOX RIVER. THE EXISTENCE OF TWO WATER MASSES CAN BE OBSERVED IN THE LOWER BAY, ONE CHARACTERIZED BY FOX RIVER PARAMETERS AND THE OTHER REPRESENTING THE BAY WATER. THE DIFFUSE INTERFACE BETWEEN THE TWO MASSES CAN BE LOCATED IN THE VICINITY OF LCNG TAIL POINT, APPROXIMATELY FIVE MILES FROM THE MOUTH OF THE RIVER. IN THE EXTREME LOWER BAY, THE PHYTOPLANKTON, DOMINATED BY RIVER ALGAL SPECIES, EXHIBIT HIGH BICMASS AND LOW LUXURY UPTAKE OF ORTHCPHOSPHATE IN THE PRESENCE OF LOW CCNCENTRATIONS OF AVAILABLE PHOSPHATE.

FIELD 05C, 02H

ACCESSION NO. W71-10645

EFFECTS OF HYDROSTATIC PRESSURE ON PHOTOSYNTHESIS AND GROWTH OF UNICELLULAR MARINE ALGAE AND DIATOMS,

HAWAII UNIV., HONOLULU. DEPT. OF MICROBIOLOGY.

LESLIE RALPH BERGER.

AVAILABLE FROM THE NATIONAL TECHNICAL INFORMATION SERVICE AS AD-720 401, $3.00 IN PAPER COPY, $0.95 IN MICROFICHE. ONR PROGRESS REPORT, APR 1971. 11 P. ONR PROJECT NO 306820.

DESCRIPTORS:
*ALGAE, *PHOTOSYNTHETIC OXYGEN, *PHOTOSYNTHESIS, INSTRUMENTATION, HYDROSTATIC PRESSURE, DIATOMS.

IDENTIFIERS:
GROWTH, OXYGEN, MONITORS.

ABSTRACT:
LIGHT-DEPENDENT OXYGEN PRODUCTION AND GROWTH OF ALGAL CULTURES HAVE BEEN MEASURED AT 25C AT VARIOUS LIGHT INTENSITIES AND HYDROSTATIC PRESSURES. A DEVICE WHICH MAINTAINS A DESIRED CONCENTRATION OF DISSOLVED OXYGEN DURING GROWTH AND OXYGEN EVOLUTION BY PHOTOSYNTHETIC ORGANISMS IS DESCRIBED. THE SYSTEM USES A MODIFIED RATE-MEASURING OXYGEN ELECTRODE SYSTEM IN CONJUNCTION WITH AN OXYGEN CONCENTRATION MONITORING UNIT.

FIELD 05C

ACCESSION NO. W71-10791

THE ROLE OF INORGANIC IONS IN THE EUTROPHICATION OF FARM PONDS,

AGRICULTURAL RESEARCH SERVICE, UNIVERSITY PARK, PA. NORTHEAST WATERSHED RESEARCH CENTER.

RICHARD W. TERKELTOUB.

TYPESCRIPT; TO BE PUBLISHED IN THE PROCEEDINGS OF THE SYMPOSIUM ON MAN-MADE LAKES, KNOXVILLE, TENNESSEE, MAY 1971. 4 TAB, 11 REF.

DESCRIPTORS:
*EUTROPHICATION, *IONS, *AQUATIC PRODUCTIVITY, FARM PONDS, NUTRIENTS, SURFACES, TEMPERATURE, PHOSPHORUS, NITRATES, CHLORIDES, CALCIUM, MAGNESIUM, SODIUM, POTASSIUM, IRON, MANGANESE, COPPER, ALGAE, SAMPLING, CHLOROPHYTA.

IDENTIFIERS:
*INORGANIC IONS, ZINC, VOLVOX, HYDRODICTYON, SPOCYLOSIUM, SPIROGYRA.

ABSTRACT:
SEVEN EASTERN PENNSYLVANIA FARM PONDS WERE STUDIED TO ASCERTAIN WHETHER ANY PARTICULAR CONCENTRATIONS OR RATIOS OF THE PRINCIPAL AQUEOUS INORGANIC IONS ARE ASSOCIATED WITH EUTROPHICATION. THE SURFACE AND BOTTOM WATERS WERE SAMPLED FROM MAY 7 TO NOVEMBER 18, 1969. TEMPERATURE WAS MEASURED, AND CALCIUM, MAGNESIUM, SODIUM, POTASSIUM, CHLORIDE, NITRATE, ORTHOPHOSPHATE, IRON, MANGANESE, ZINC, AND COPPER CONCENTRATIONS WERE DETERMINED. THE ALGAE FOUND WERE SPODYLOSIUM, SPIROGYRA, AND VOLVOX. WATER TEMPERATURES WERE USUALLY A FEW DEGREES CENTIGRADE LOWER IN THE BOTTOM WATERS THAN IN THE SURFACE WATERS. THERE WERE NO INDICATIONS OF THERMAL STRATIFICATION IN ANY POND. THE CHEMICAL CONDITIONS THAT SPURRED ALGAL GROWTH IN THE THREE PONDS, 1, 2, AND 7, WERE NOT DISTINGUISHABLY DIFFERENT FROM THOSE PRESENT IN THE ALGAE-FREE PONDS. SIMILARLY, THE SUBSEQUENT DECAY AND DISAPPEARANCE OF THE ALGAE WERE NOT RELATABLE TO ANY OF THE CHEMICAL PARAMETERS MEASURED. NO DISCERNIBLE SET OF CHEMICAL CHARACTERISTICS DIFFERENTIATED THE THREE PONDS THAT SUPPORTED ALGAE DURING THE SUMMER FROM THE FOUR PONDS THAT DID NOT. (JONES-WISCONSIN)

FIELD 05C

ACCESSION NO. W71-11001

POLLUTION POTENTIAL OF SALMONID FISH HATCHERIES,

KRAMER, CHIN AND MAYO, SEATTLE, WASH.

PAUL B. LIAO.

WATER AND SEWAGE WORKS, VOL 117, NO 12, P 291-297, 1970. 6 FIG, 3 TAB, 15 REF.

DESCRIPTORS:
*WATER POLLUTION SOURCES, *FISH HATCHERIES, *SALMONIDS, WATER REQUIREMENTS, TEMPERATURE, NUTRIENTS, ALGAE, WEEDS, TASTE, ODOR, PATHOGENIC BACTERIA, ORGANIC WASTES, SOLID WASTES, CHEMICALS, MICHIGAN, CALIFORNIA, TUBIFICIDS, COLORADO, WATER POLLUTION CONTROL, WASHINGTON, CHEMICAL OXYGEN DEMAND, BIOCHEMICAL OXYGEN DEMAND, DISSOLVED OXYGEN, HYDROGEN ION CONCENTRATION, AMMONIA, NITRATES, PHOSPHATES, SUSPENDED LOAD, DISSOLVED SOLIDS, EFFLUENTS.

IDENTIFIERS:
SETTLEABLE SOLIDS, HATCHERY EFFLUENTS, PARASITES, JORDAN RIVER(MICH), SAN JOAQUIN RIVER(CALIF), RIFLE FALLS(COLO), FISH FECAL WASTES, RESIDUAL FOOD, GREEN RIVER(WASH), COWLITZ TROUT HATCHERY(WASH).

ABSTRACT:
WATER POLLUTION PROBLEMS ASSOCIATED WITH SALMONID HATCHERY OPERATIONS INCLUDE NUTRITIONAL ENRICHMENT, ALGAE AND WEED GROWTH, TASTE, ODOR, SETTLEABLE SOLIDS, PATHOGENIC BACTERIA, PARASITES, ORGANIC MATTER, CHEMICALS AND DRUGS. FISH FECAL WASTES AND RESIDUAL FOOD ARE MOST SERIOUS BECAUSE THEY ARE ENCOUNTERED CONTINUOUSLY UNDER NORMAL OPERATING PROCEDURES; AFTER FIELD TESTING, THESE WASTES ARE CLASSIFIED INTO ORGANIC, NUTRIENT AND SOLID POLLUTANTS. THE AVERAGE BIOCHEMICAL OXYGEN DEMAND (BOD) CONCENTRATION OF HATCHERY EFFLUENTS DURING POND CLEANING IS SEVERAL TIMES GREATER THAN DURING NORMAL OPERATION; CLOSELY RELATED TO BOD, IS THE DISSOLVED OXYGEN LEVEL. THE NUTRIENT POLLUTANTS, NITRATE AND PHOSPHATE, ARE END-PRODUCTS OF DECOMPOSITION OF FISH FOOD. THE HATCHERY EFFLUENT TESTED MAY STIMULATE ALGAL GROWTH AND CAUSE ALGAL BLOOMS UNDER CERTAIN CONDITIONS. THE HIGH PERCENTAGE OF SUSPENDED AND SETTLEABLE SOLIDS INDICATES THAT MOST SOLIDS IN THE CLEANING WATER WILL BE DEPOSITED ON THE STREAM BOTTOM BELOW THE HATCHERY. PROPER FEEDING WOULD GREATLY REDUCE RATE OF POLLUTANT PRODUCTION. THE POLLUTION POTENTIAL OF HATCHERY CLEANING WATER IS COMPARABLY TO DOMESTIC SEWAGE WHEN DILUTED WITH INFILTRATION WATER. HATCHERY OPERATING IMPROVEMENTS SHOULD INCLUDE PROPER FISH LOADING TECHNIQUES, PROPER FEEDING PROCEDURES, AND WATER SUPPLY ADJUSTMENTS. (JONES-WISCONSIN)

FIELD 05C, 05B

ACCESSION NO. W71-11006

TOXICITY OF AMMONIA TO MARINE DIATOMS,

ALASKA UNIV., COLLEGE. INST. OF MARINE SCIENCE.

K. V. NATARAJAN.

JOURNAL OF WATER POLLUTION CONTROL FEDERATION, VOL 42, NO 5, PART 2, P R184-R190, 1970. 2 FIG, 2 TAB, 15 REF.

DESCRIPTORS:
*TOXICITY, *AMMONIA, *MARINE ALGAE, *DIATOMS, CHEMICAL ANALYSIS, HYDROGEN ION CONCENTRATION, SEA WATER, PHOTOSYNTHESIS, PHYTOPLANKTON, AMMONIUM COMPOUNDS, NUTRIENTS, PRIMARY PRODUCTIVITY, FERTILIZERS, EFFLUENTS, WATER POLLUTION SOURCES, WATER POLLUTION EFFECTS, SAMPLING, TEMPERATURE, COPEPODS, NITROGEN, BACTERIA, OXIDATION-REDUCTION POTENTIAL, LABORATORY TESTS.

IDENTIFIERS:
COOK INLET(ALASKA), CYCLOTELLA NANA, LEPTOCYLINDRUS, PACIFIC OCEAN, ALEUTIAN ISLANDS, PRYMNESIUM PARVUM.

ABSTRACT:
ALTHOUGH THE WORLD'S FERTILIZER SUPPLY CONSISTS LARGELY OF AMMONIUM COMPOUNDS, THESE COMPOUNDS ARE ALSO KNOWN TO BE TOXIC. AMMONIA CAN BECOME TOXIC AT CERTAIN CONCENTRATIONS IN THE AQUATIC ENVIRONMENT. EFFECT ON MARINE PHYTOPLANKTON OF A FERTILIZER PLANT EFFLUENT CONSISTING ESSENTIALLY OF AMMONIA COMPOUNDS WAS INVESTIGATED WITH BOTH FIELD AND LABORATORY STUDIES. THE RESULTS OF THE FIELD EXPERIMENTS SHOWED A CLOSE PARALLEL WITH THE LABORATORY EXPERIMENTS WITH AXENIC CULTURES OF MARINE DIATOMS. LEVELS OF THE EFFLUENT TOXIC TO THE ENDEMIC PHYTOPLANKTON WERE BETWEEN 0.1 AND 1% (1.1 AND 11 MG/1 AMMONIA). IT IS EVIDENT THAT THE TOXICITY OF AMMONIA AND AMMONIUM SALTS DEPENDS ON THE CONCENTRATION OF MOLECULAR NONIONIZED AMMONIA OR AMMONIA HYDROXIDE, SINCE THE AMMONIUM ION IS COMPARATIVELY NONTOXIC; (DIRECT EXPERIMENTAL OBSERVATIONS HAVE NOT BEEN OBTAINED). TOXIC EFFECTS OF THE AMMONIA EFFLUENT TO PHYTOPLANKTON OF THIS INVESTIGATION CAN BE EXPLAINED BY THIS PHENOMENON. (JONES-WISCONSIN)

FIELD 05C

ACCESSION NO. W71-11008

NUTRIENTS AND NUTRIENT BUDGETS IN THE BAY OF QUINTE, LAKE CNTARIO,

ONTARIO WATER RESOURCES COMMISSION, TORONTO.

M. G. JOHNSON, AND G. E. OWEN.

JOURNAL OF WATER POLLUTION CONTROL FEDERATION, VOL 43, NC 5, P 836-853, 1971.
8 FIG, 7 TAB, 37 REF.

DESCRIPTORS:
*NUTRIENTS, *LAKE CNTARIC, *EUTROPHICATION, LAKE ERIE, ALGAE, NITROGEN, PHOSPHORUS, DRAINAGE, INDUSTRIES, MUNICIPAL WASTES, TURNOVERS, SEDIMENTS, HUMAN POPULATION, TOURISM, WATERSHEDS(BASINS), TROPHIC LEVELS, TURBIDITY, OXYGEN, FISH, RECREATION, INVERTEBRATES, RIVERS, CYANOPHYTA, GEOLOGIC FORMATICNS, BENTHOS, MAYFLIES, TUBIFICIDS, DIATOMS, WATER POLLUTION EFFECTS, WATER POLLUTICN CONTROL, DISSOLVED OXYGEN, CARBON, SAMPLING, EQUATIONS, WATER POLLUTICN SOURCES, RAINFALL, MIDGES, INPUT-OUTPUT ANALYSIS, MUD, BACTERIA.

IDENTIFIERS:
*NUTRIENT BUDGETS, *BAY OF QUINTE(ONTARIO), NUTRIENT INPUTS, CHIRONOMUS PLUMOSUS, CHIRONOMUS ATTENTUATUS, CHIRONOMUS ANTHRACINUS, LIMNODRILUS HOFFMEISTERI, TUBIFEX TUBIFEX, APHANIZOMENON, CLADCPHORA.

ABSTRACT:
ALGAL BLOOMS, TURBIDITY, DEPLETICN OF DEEP-WATER CXYGEN, AND CHANGES IN COMPOSITION OF THE BIOTA ARE INCREASINGLY OBVIOUS IN BAY OF QUINTE, LAKE ONTARIO. CLARIFICATION OF RESPECTIVE SIGNIFICANCE OF NUTRIENT CONTRIBUTIONS FROM TRIBUTARY RIVERS AND FRCM MUNICIPAL-INDUSTRIAL SOURCES ARE DESCRIBED. THE BAY RECEIVED ABOUT 9,700,000 POUNDS OF NITROGEN AND 700,000 PCUNDS OF PHOSPHORUS IN 1968. 89% OF THE NITROGEN AND 60% OF THE PHOSPHORUS WERE ATTRIBUTABLE TO LAND DRAINAGE AND THE REMAINDER TO MUNICIPAL-INDUSTRIAL SOURCES. COMPARISONS BASED ON 'NET INPUTS', THE AMCUNT OF NUTRIENT CONTAINED IN AN INPUT IN EXCESS OF THE AMOUNT OF NUTRIENT IN THE EQUIVALENT VOLUME OF WATER DISPLACED AT THE OUTLET, ARE PROPOSED. ABOUT 50% OF THE 'NET INPUT' OF NITRCGEN AND 85% OF PHOSPHORUS WERE CONTRIBUTED BY MUNICIPAL-INDUSTRIAL SOURCES IN 1968. IT IS RECOMMENDED THAT PHOSPHORUS BE REMOVED FROM THESE SOURCES. THE WATER TURNOVER RATE IN THE BAY, FIVE TIMES ANNUALLY, TRANSLOCATES RESUSPENDED NUTRIENTS IN SEDIMENTS OF THE INNER BAY SHALLOW WATERS TO SEDIMENTS IN DEEPER WATERS OF THE OUTER BAY AND LAKE CNTARIO, THUS IMPROVING WATER QUALITY. ESTIMATED PHOSPHORUS REMCVAL COST, $200,000 DOLLARS/YEAR, IS JUSTIFIED ON THE BASIS CF ECONOMICS INCOME FROM RECREATION AND TOURISM. (JONES-WISCONSIN)

FIELD 05C

ACCESSION NO. W71-11009

THE PERIPHYTON OF THE SUBMERGED MACROPHYTES OF MIKOLAJSKIE LAKE,

WARSAW UNIV. (POLAND). CHAIR OF HYDROBIOLOGY.

LUCYNA JOANNA BOWNIK.

EKOLOGIA POLSKA, VOL 18, NO 24, P 503-520, 1970. 6 FIG, 2 TAB, 16 REF.

DESCRIPTORS:
*PERIPHYTON, *SUBMERGED PLANTS, BIOMASS, NEMATODES, ALGAE, OLIGOCHAETES, LIFE CYCLES, SEASONAL, DENSITY, DIATOMS, LAKES, PROTOZOA, ROTIFERS, CRUSTACEANS, DIPTERA, MAYFLIES, DRAGONFLIES, CADDISFLIES, MOLLUSKS, CYANOPHYTA, CHRYSOPHYTA, CHLOROPHYTA, PYRROPHYTA, INVERTEBRATES, AQUATIC ANIMALS, LITTORAL, WAVES(WATER).

IDENTIFIERS:
*MACROPHYTES, *MIKOLAJSKIE LAKE(POLAND), COLONIZATION, POTAMOGETON LUCENS, POTAMOGETON PERFOLIATUS, POTAMOGETON PECTINATUS, MYRIOPHYLLUM SPECATUM, ELODEA CANADENSIS, PHRAGMITES, FONTINALIS, CHARALES, CERATOPHYLLUM, LEMNA.

ABSTRACT:
AN ANALYSIS WAS MADE OF THE PERIPHYTON COLONIZING POTAMOGETON LUCENS, P PERFOLIATUS, MYRIOPHYLLUM SPICATUM AND ELODEA CANADENSIS IN MIKOLAJSKIE LAKE, POLAND. COMPOSITION AND THE NUMBERS DYNAMICS OF THE DOMINANT PERIPHYTON GROUPS WERE RECORDED INDICATING THAT NEMATODES ARE DOMINANT WITH CHIRONOMIDS AND OLIGOCHAETA NUMEROUS ON SUBMERGED VEGETATION. A GREAT DIFFERENCE WAS NOTICED BETWEEN THE NUMBERS DYNAMICS OF PERIPHYTON FAUNA COLONIZING PLANTS LIVING IN THE LAKE DURING WINTER. THE LOWEST NUMBERS OF PERIPHYTON FAUNA WERE FOUND ON PLANTS WITH A SHORT LIFE CYCLE, IN SPRING, SHOWING A GRADUAL COLONIZATION OF A GROWING PLANT. ON PLANTS WITH A LONGER LIFE CYCLE, PERIPHYTON FAUNA OCCUR IN LARGE NUMBERS ALREADY IN SPRING BUT MAXIMUM IS REACHED IN AUTUMN. THE LARGEST DENSITY OF DOMINANT GROUPS OF PERIPHYTON ORGANISMS WAS FOUND ON POTAMOGETON LUCENS, DECREASING ON MYRICPHYLLUM SPICATUM AND POTAMOGETON PERFOLIATUS; THE SMALLEST ON ELODEA CANADENSIS. IN A SHORE OVERGROWN BY PLANTS WITHERING DURING WINTER, CHANGES OF PLANT BIOMASS HAVE A DECISIVE INFLUENCE ON PERIPHYTON ABUNDANCE. ON PLANTS WITH LONGER LIFE CYCLES, CHANGES OF THE FAUNA DENSITY DETERMINED ABUNDANCE OF PERIPHYTON. ALGAE ARE MORE NUMEROUS THAN ANIMAL COMPONENTS OF THE PERIPHYTON AND DIATOMS ARE THE MOST NUMEROUS ALGAE. (JONES-WISCONSIN)

FIELD 05C

ACCESSION NO. W71-11012

WASTEWATER LOADING GUIDELINES FOR THE GRAND RIVER BASIN.

ONTARIO WATER RESOURCES COMMISSION, TORONTO.

ONTARIO WATER RESOURCES COMMISSION, CANADA, INTERIM REPORT, JANUARY 1971. 1 FIG, 1 TAB, APPEND.

DESCRIPTORS:
*WASTE WATER(POLLUTION), *WASTE ASSIMILATION CAPACITY, *ORGANIC LOADS, NUTRIENTS, DISSOLVED OXYGEN, AQUATIC LIFE, MUNICIPAL WASTES, PHOTOSYNTHESIS, RESPIRATION, WATER QUALITY, STREAMFLOW, AGRICULTURE, INDUSTRIES, WASTE TREATMENT, WATERSHEDS(BASINS), PHOSPHORUS, ALGAE, ORGANIC MATTER, LAKE ERIE, EFFLUENTS, SEWAGE, DETERGENTS, PLANTS, LAND USE, BIOCHEMICAL OXYGEN DEMAND, FISHERIES, RESERVOIRS, PIPELINES, ESTIMATING, ANALYSIS.

IDENTIFIERS:
*LOADING GUIDELINES, *GRAND RIVER BASIN(ONTARIO), CANADA, BUFFER CAPACITY.

ABSTRACT:
IN CONSIDERING WATER QUALITY OF THE GRAND RIVER BASIN, ONTARIO, ACCEPTABLE LOADINGS, BASED ON THE NEW DISSOLVED OXYGEN CRITERIA ADAPTED BY ONTARIO WATER RESOURCES COMMISSION IN 1970, UPGRADE THE MINIMUM DISSOLVED OXYGEN LEVEL FROM 4.0 TO 5.0 MG/L IN ALL STREAMS EXCEPT THOSE SUPPORTING COLDWATER FISHERIES WHERE MINIMUM DISSOLVED OXYGEN CRITERIA IS 6.0 MG/LITER. AS A RESULT, PREVIOUSLY ACCEPTABLE WASTE DISCHARGES NOW EXCEED PRESENT LOADING GUIDELINES. THE PHOSPHORUS INPUT, CONSIDERED THE CONTROLLING NUTRIENT IN ALGAL PRODUCTION, FROM THE MUNICIPAL SEWAGE TREATMENT PLANTS IS ESTIMATED AS 70% TO 80% OF THE TOTAL ANNUAL INPUT OF THIS NUTRIENT INTO THE BASIN. REDUCTION OF NUTRIENT AND ORGANIC LOADINGS ARE REQUIRED FOR WATER QUALITY IMPROVEMENT AND PROTECTION OF LAKE ERIE. ALTERNATIVES, INCLUDING EFFLUENT POLISHING AND STREAMFLOW AUGMENTATION, WHICH CAN BE UTILIZED TO INCREASE THE POTENTIAL VARIOUS RIVER USES, WHILE REDUCING POLLUTION PRESSURES ARE CONSIDERED. IN MAKING ESTIMATES OF ACCEPTABLE ORGANIC LOADINGS, THE WATERSHED WAS DIVIDED INTO EIGHT SUB-BASINS. THE LOADINGS FOR EACH SUB-BASIN, EXPRESSED IN TERMS OF FIVE-DAY BIOCHEMICAL OXYGEN DEMAND FROM TREATMENT SOURCES, WERE DETERMINED ON BASIC ASSUMPTIONS, DISSOLVED OXYGEN CRITERIA, DESIGN STREAMFLOWS, AND EXISTING WASTEWATER LOADINGS. RESULTS ARE TABULATED; DETAILS APPENDED. (JONES-WISCONSIN)

FIELD 05C, 06B

ACCESSION NO. W71-11017

PERIPHYTON OF THE EXPERIMENTAL LAKES AREA, NORTHWESTERN ONTARIO,

FISHERIES RESEARCH BOARD OF CANADA, WINNIPEG (MANITOBA). FRESHWATER INST.

JOHN G. STOCKNER, AND F. A. J. ARMSTRONG.

JOURNAL FISHERIES RESEARCH BOARD OF CANADA, VOL 28, NO 2, P 215-229, 1971. 9 FIG, 2 TAB, 37 REF.

DESCRIPTORS:
*PERIPHYTON, *ALGAE, *LITTORAL, LIGHT PENETRATION, BENTHIC FLORA, LAKES, DEPTH, DIATOMS, CHLOROPHYTA, CYANOPHYTA, BIOMASS, CHEMICAL ANALYSIS, STATISTICAL METHODS, NITROGEN, PHOSPHORUS, CHLOROPHYLL, DISTRIBUTION, CARBON, SEASONAL, CADDISFLIES, GASTROPODS, SANDS, CHRYSOPHYTA.

IDENTIFIERS:
*EXPERIMENTAL LAKES AREA, *NORTHWESTERN ONTARIO, EPILITHIC ALGAE, ACHNANTHES MINUTISSIMA, ACHNANTHES FLEXELLA, EUNCTIA PECTINALIS, DESMIDS.

ABSTRACT:
SINCE THE LITTORAL ZONE OF THE FOUR LAKES STUDIED IN THE EXPERIMENTAL LAKES AREA (ELA) OF NORTHWESTERN ONTARIO WAS COMPOSED CHIEFLY OF LARGE BOULDERS AND ROCK SHELFS, THE EPILITHIC ALGAL ASSEMBLAGE OF PERIPHYTON WAS DOMINANT. DIATOMS WERE THE DOMINANT GROUP WITHIN THE EPILITHIC ASSEMBLAGE. BENTHIC ALGAL GROWTH IN MOST LAKES WAS NEGLIGIBLE AT DEPTHS OVER 10 METERS. FILAMENTOUS GREEN AND BLUE-GREEN ALGAE INCREASE SIGNIFICANTLY IN JULY AND AUGUST BUT NEVER CONSTITUTED MORE THAN 40% OF TOTAL ALGAL BIOMASS. A WELL=DEFINED DIATOM SUCCESSION OCCURRED. ACHNANTHES MINUTISSIMA WAS THE MOST ABUNDANT DIATOM IN LITTORAL ZONES OF ALL LAKES. VERTICAL DISTRIBUTION OF LITTORAL DIATOMS WAS EXAMINED IN LAKE 240 AND SPECIES DIFFERENCES ARE DISCUSSED IN LIGHT OF POSSIBLE REGULATING MECHANISMS. ACHNANTHES FLEXELLA AND EUNCTIA PECTINALIS WERE FOUND ONLY IN THE PSAMMONAL HABITAT OF LAKE 240. STATISTICAL TREATMENT OF CHEMICAL ANALYSES OF NITROGEN, PHOSPHORUS, AND CHLOROPHYLL-A CONTENT OF PERIPHYTON SHOWED NO SIGNIFICANT DIFFERENCE IN AMOUNTS OF TOTAL NITROGEN, BUT DISTRIBUTION OF BOTH TOTAL PHOSPHORUS AND CHLOROPHYLL-A WAS SIGNIFICANTLY DIFFERENT. COMPARISONS AMONG LAKES ARE MADE, AND ROLE OF NITROGEN AND PHOSPHORUS AS REGULATORS OF PERIPHYTON GROWTH IN ELA LAKES IS DISCUSSED. THIS PAPER CONTAINS 37 REFERENCES. (JONES-WISCONSIN)

FIELD 05C

ACCESSION NO. W71-11029

ALGAL GROWTH ASSESSMENT IN NATURAL WATERS,

ILLINOIS STATE WATER SURVEY, PEORIA. WATER QUALITY SECTION.

WUN-CHENG WANG, AND WILLIAM T. SULLIVAN.

TYPESCRIPT; PRESENTED AT 14TH CONFERENCE ON GREAT LAKES RESEARCH, APRIL 19-21, 1971.

DESCRIPTORS:
*FORECASTING, *ALGAE, *PLANT GROWTH, *EUTROPHICATION, GROWTH RATES, SAMPLING, LAKE MICHIGAN, MEASUREMENT, SUSPENDED LOAD, ALKALINITY, HARDNESS(WATER), ANALYTICAL TECHNIQUES, AMMONIA, NITRATES, PHOSPHATES, WATER QUALITY, NITROGEN, NUTRIENTS, TROPHIC LEVEL.

IDENTIFIERS:
ALGAL GROWTH POTENTIAL, ILLINOIS RIVER(ILL), DIAGNOSTIC CRITERION.

ABSTRACT:
METHODS ARE DESCRIBED FOR DIAGNOSTIC AND PREDICTIVE EVALUATION OF NATURAL WATER QUALITY BASED ON ALGAL GROWTH POTENTIAL. TECHNIQUES FOR MEASUREMENT OF ALGAL GROWTH WERE INVESTIGATED AND ALGAL NUTRITION EXAMINED. AFTER REMOVAL OF PLANKTON BY MEMBRANE FILTRATION, SAMPLES WERE INOCULATED WITH A NATURAL, MIXED ALGAL CULTURE; GROWTH WAS SATISFACTORY MEASURED BY INCREASES IN LIGHT ABSORBANCE AND FILTERABLE ORGANIC AND INORGANIC MASS AND BY DECREASES IN ALKALINITY AND HARDNESS. ALGAL PIGMENT FLUORESCENCE WAS NOT COMPARABLE TO THE OTHER PARAMETERS ABOVE OLIGOTROPHIC LEVELS, POSSIBLY DUE TO INADEQUATE EXTRACTION OF FLUORESCENT COMPOUNDS. MAXIMUM DAILY ALGAL GROWTH WAS ATTAINED IN THREE TO FIVE DAYS FOLLOWING INOCULATION; AMMONIA WAS PREFERRED TO NITRATES AS A NITROGEN SOURCE. RATIO OF FILTERABLE INORGANIC MASS TO ORGANIC MASS INCREASED WITH HIGHER INITIAL SAMPLE CONCENTRATIONS OF ALKALINITY AND HARDNESS. 'ALGAL GROWTH POTENTIAL', REPRESENTING THE TROPHIC LEVEL OF A NATURAL WATER SOURCE, WAS BEST REPRESENTED BY WEIGHT OF FILTERABLE ORGANIC MASS PRODUCED AFTER SEVEN DAYS OF INCUBATION. LAKE MICHIGAN AT THE CHICAGO CENTRAL WATER FILTRATION PLANT WAS JUDGED ESSENTIALLY OLIGOTROPHIC, WHILE THE ILLINOIS RIVER NEAR PEORIA IS EUTROPHIC. THIS PAPER CONTAINS 17 REFERENCES. (JONES-WISCONSIN)

FIELD 05C

ACCESSION NO. W71-11033

DETERMINATION OF MERCURY IN BIOLOGICAL AND ENVIRONMENTAL SAMPLES BY NEUTRON ACTIVATION ANALYSIS,

WESTERN NEW YORK NUCLEAR RESEARCH CENTER, INC., BUFFALO.

K. K. SIVASANKARA PILLAY, CHARLES C. THOMAS, JR., JAMES A. SONDEL, AND CAROLYN M. HYCHE.

TPYESCRIPT; PRESENTED AT 161ST NATIONAL MEETING OF THE AMERICAN CHEMICAL SOCIETY AT LOS ANGELES, CALIFORNIA, MARCH 31, 1971.

DESCRIPTORS:
*SAMPLING, *ENVIRONMENT, *NEUTRON ACTIVATION ANALYSIS, RADIOACTIVITY, ADSORPTION, FISH, LAKE ERIE, SEDIMENTS, SILTS, PLANKTON, ALGAE, TRACE ELEMENTS, VOLATILITY, FREEZE DRYING, X-RAYS, ELECTROLYSIS, COMPUTER PROGRAMS, GAMMA RAYS, POLLUTANT IDENTIFICATION, COLORIMETRY, TRACERS.

IDENTIFIERS:
*MERCURY, *BIOLOGICAL TISSUES, MERCURY ISOTOPES, HUMAN BRAIN, LOW-TEMPERATURE OVEN DRYING, ASHING, EXTRACTIVE DIGESTION, NEUROANATOMY, OXYGEN PLASMA ASHING, ATOMIC ABSORPTION, TITRATION.

ABSTRACT:
THE MINUTE QUANTITIES AND VOLATILE NATURE OF MERCURY CREATES PROBLEMS IN SAMPLING AND ANALYSIS. OF VARIOUS MERCURY DETERMINATION PROCEDURES, ONLY A LIMITED NUMBER CAN BE READILY ADAPTED TO BIOLOGICAL AND ENVIRONMENTAL SAMPLING FOR MONITORING. THE FOLLOWING PROCEDURES WERE DEVELOPED FOR THE INVESTIGATION OF THE MERCURY POLLUTION OF LAKE ERIE AND ITS AQUATIC LIFE. PRE-IRRADIATION PROCESSING OF SAMPLES HAS BEEN SYSTEMATICALLY INVESTIGATED. AFTER REACTOR IRRADIATION, THE SAMPLES ARE WET ASHED WITH MERCURY CARRIER UNDER REFLUX CONDITIONS. A PRELIMINARY SULFIDE PRECIPITATION IS FOLLOWED BY FURTHER PURIFICATION AND EVENTUAL ISOLATION OF MERCURY BY ELECTRODEPOSITION OR BY PRECIPITATION AS MERCURIC OXIDE. THE RADIOACTIVITIES FROM MERCURY-197 AND MERCURY-197-M ISOTOPES ARE MEASURED BY SCINTILLATION GAMMA RAY SPECTROMETRY USING A THIN SODIUM IODIDE DETECTOR TO DETERMINE THE MERCURY LEVELS IN VARIOUS SAMPLES. THESE ANALYTICAL PROCEDURE RESULTS ARE COMPARED WITH OTHER TECHNIQUES. TRACER STUDIES INDICATED THAT THE ERRORS OF THIS PROCEDURE WERE LESS THAN 15% AT 0.01 PPM LEVEL AND LESS THAN 5% AT 2 PPM LEVEL OF MERCURY IN BIOLOGICAL TISSUES. ANALYSIS OF FISH SAMPLES AND SEDIMENTS SAMPLES CONTAINING NATURAL FORMS OF MERCURY SHOWED A STANDARD DEVIATION OF LESS THAN 5% AT 5 PPM LEVELS, LESS THAN 7% AT 1.5 PPM LEVELS AND LESS THAN 17% AT 0.01 PPM LEVELS. (JONES-WISCONSIN)

FIELD 05B, 07B

ACCESSION NO. W71-11036

DETERMINATION OF MICROBIAL BIOMASS IN DEEP OCEAN WATER,

CALIFORNIA UNIV., SAN DIEGO, LA JOLLA. INST. OF MARINE RESOURCES.

OSMUND HOLM-HANSEN.

IN: SYMPOSIUM ON ORGANIC MATTER IN NATURAL WATERS, UNIVERSITY OF ALASKA, COLLEGE, SEPT 2-4, 1968: INSTITUTE OF MARINE SCIENCE OCCASIONAL PUBLICATION NO 1, P 287-300, JUNE 1970. 14 P, 5 FIG, 17 REF. USAEC CONTRACT AT(11-1)GEN 10, PA 20.

DESCRIPTORS:
*BIOMASS, *ORGANIC MATTER, *SEA WATER, WATER ANALYSIS, ALGAE, CHEMICAL ANALYSIS, SAMPLING, ANALYTICAL TECHNIQUES, NUTRIENTS, FOOD CHAINS, BIOLUMINESCENCE, POLLUTANT IDENTIFICATION.

IDENTIFIERS:
*ATP ANALYSIS, *ADENOSINE TRIPHOSPHATE.

ABSTRACT:
THE CONTENT OF ATP IN THE PARTICULATE FRACTION WAS DETERMINED IN FOUR PROFILES IN THE PACIFIC OCEAN DOWN TO 3500 M. THE ATP VALUES WERE CONVERTED TO CELLULAR ORGANIC CARBON VALUES AND COMPARED TO THE TOTAL ORGANIC CARBON IN THE PARTICULATE FRACTION. THE MICROBIAL BIOMASS AS ESTIMATED BY THE ATP DATA IS VERY HIGH IN THE EUPHOTIC ZONE AND DECREASES RAPIDLY TO 1 TO 2 MICROGRAMS OF CELLULAR C/LITER AT ABOUT 200 M. AT THE LOWER DEPTHS SAMPLED, THE CALCULATED BIOMASS CONTAINED ABOUT 0.1 MICROGRAM OF ORGANIC C/LITER. MICROSCOPIC EXAMINATION OF DEEP SAMPLES SHOWED THE PRESENCE OF LARGE NUMBERS OF SMALL FLAGELLATED ALGAL-LIKE CELLS. (KNAPP-USGS)

FIELD 05A, 05B

ACCESSION NO. W71-11234

IMPROVEMENT AND APPLICATION OF BENTHIC ALGAL ISOTOPE PRODUCTIVITY MEASURING METHODS,

HAWAII UNIV., HONOLULU. DEPT. OF BOTANY.

MAXWELL S. DOTY.

AVAILABLE FROM THE NATIONAL TECHNICAL INFORMATION SERVICE AS UH-235P4-4, $3.00 IN PAPER COPY, $0.95 IN MICROFICHE, ATOMIC ENERGY COMMISSION CONTRACT REPORT UH-235P4-4, JUNE 1970. 54 P, 9 FIG, 5 TAB, 47 REF.

DESCRIPTORS:
*RADIOACTIVITY TECHNIQUES, *BENTHIC FLORA, *GROWTH RATE, *AQUATIC PRODUCTIVITY, RADIOISOTOPES, TRACERS, MARINE ALGAE, HAWAII, WATER TEMPERATURE, WAVES, LIGHT INTENSITY.

IDENTIFIERS:
PHILIPPINES, ENIWETOK, MICRONESIA, FANNING ISLANDS.

ABSTRACT:
RADIOISOTOPES HAVE BEEN USED AS QUANTITATIVE TRACERS IN MEASURING RATES OF BENTHIC ALGAL PRODUCTION. BASIC ECOLOGICAL STUDIES ON BENTHIC REEF ALGAE HAVE BEEN COMPLETED OR ARE UNDERWAY. THESE STUDIES HAVE BEEN CONDUCTED IN HONOLULU, ENIWETOK, MICRONESIA, THE PHILIPPINES, AND FANNING ISLAND. IT WAS SHOWN THAT THE CROP OF SOME SPECIES VARY SEASONALLY WITH LIGHT, SOME WITH MEAN WAVE SIZE, AND SOME WITH TEMPERATURE. STUDIES AT WAIKIKI INDICATED THAT AMONG BENTHIC ALGAE, STANDING CROPS OF THE CRUSTOSE FORMS VARY THE LEAST. (MORTLAND-BATTELLE)

FIELD 05A, 05C

ACCESSION NO. W71-11486

AIRBORNE AMMONIA EUTROPHIES LAKES.

AGRICULTURAL RESEARCH (USDA), VOL. 19, NO. 2, P 8-9, AUGUST 1970.

DESCRIPTORS:
*AMMONIA, *EUTROPHICATION, *NITROGEN, ALGAE, WATER POLLUTION SOURCES, COLORADO, CATTLE, URINE, PATH OF POLLUTANTS, FARM WASTES.

IDENTIFIERS:
FEEDLOTS.

ABSTRACT:
AMMONIA TRAPS AND RAIN GAGES WERE INSTALLED AT FIVE SITES AND IN TWO CONTROL AREAS IN COLORADO TO DETERMINE THE RATE AT WHICH AMMONIA IS ADSORBED DIRECTLY FROM THE AIR BY WATER SURFACES UNDER DIFFERENT CONDITIONS OF TEMPERATURE AND CLIMATE AT VARIOUS DISTANCES AND DIRECTIONS FROM CATTLE FEEDLOTS. IN ONE NORTHEAST COLORADO LAKE A LITTLE OVER A MILE FROM A LARGE FEEDLOT, THE SURFACE ADSORBED ABOUT 30 POUNDS OF NITROGEN AS AMMONIA PER ACRE PER YEAR. THIS AMOUNT IS SUFFICIENT TO EUTROPHY A LAKE AVERAGING 20 FEET IN DEPTH TO TWO OR THREE TIMES THE CONCENTRATION NEEDED FOR ALGAL BLOOMS. INDICATIONS ARE THAT EVEN SMALL FEEDLOTS MAY RELEASE ENOUGH AMMONIA TO HAVE AN EFFECT ON NEARBY WATER SURFACE AND THAT AIRBORNE AMMONIA FROM FEEDLOTS MAY CONTRIBUTE MORE NITROGEN THAN RUNOFF AND DEEP PERCOLATION FROM THE SAME SOURCES. (MORTLAND-BATTELLE)

FIELD 05B, 05C

ACCESSION NO. W71-11496

PESTICIDES IN WATER: COPPER SULFATE IN FLOODED CRANBERRY BOGS,

MASSACHUSETTS UNIV., EAST WAREHAM. CRANBERRY EXPERIMENT STATION.

KARL H. DEUBERT, AND IRVING E. DEMORANVILLE.

PESTICIDES MONITORING JOURNAL, VOL. 4, NO. 1, P 11-13, JUNE 1970. 1 TAB, 7 REF.

DESCRIPTORS:
*COPPER SULFATE, *ALGAE, *FLOODWATER, *PATH OF POLLUTANTS, DIELDRIN, DDT, ORGANIC MATTER, ANALYTICAL TECHNIQUES, SAMPLING, MASSACHUSETTS, POLLUTANT IDENTIFICATION.

IDENTIFIERS:
*CRANBERRY BOGS.

ABSTRACT:
CRANBERRY BOGS ARE TREATED WITH COPPER SULFATE TO CONTROL ALGAL GROWTH. IN ORDER TO ASSESS POSSIBLE WATER POLLUTION AFTER RELEASE OF TREATED FLOODWATER INTO STREAMS AND PONDS, THE RATE AT WHICH COPPER DISAPPEARED FROM THE WATER AFTER TREATMENT WAS MONITORED IN TWO SEPARATE BOGS. IN BOTH BOGS THE CONCENTRATION OF COPPER 25 HOURS AFTER APPLICATION WAS HIGHER THAN EXPECTED DUE TO SMALLER VOLUMES OF FLOODWATER. DURING THE FIRST 6 DAYS AFTER TREATMENT, COPPER CONCENTRATIONS DECREASED RAPIDLY, AND AFTER 10 DAYS ABOUT 95 PERCENT OF THE COPPER HAD DISAPPEARED. WHEN FLOODWATER WAS RELEASED ABOUT 4 WEEKS AFTER TREATMENT, THE CONCENTRATION OF COPPER WAS AT THE SAME LEVEL FOUND IN THE WATER PRIOR TO TREATMENT. (MCCANN-BATTELLE)

FIELD 05A, 05B, 05C

ACCESSION NO. W71-11498

A VISUAL DEMONSTRATION OF THE BENEFICIAL EFFECTS OF SEWAGE TREATMENT FOR PHOSPHATE REMOVAL ON PARTICULATE MATTER PRODUCTION IN WATERS OF LAKES ERIE AND ONTARIO,

FISHERIES RESEARCH BOARD OF CANADA, WINNIPEG (MANITOBA). FRESHWATER INST.; AND ONTARIO WATER RESOURCES COMMISSION, TORONTO.

J. R. VALLENTYNE, W. E. JOHNSON, AND A. J. HARRIS.

JOURNAL OF THE FISHERIES RESEARCH BOARD OF CANADA, VOL 27, NO 8, P 1493-1496, AUGUST 1970. 1 FIG, 1 TAB, 5 REF.

DESCRIPTORS:
*SEWAGE TREATMENT, *PHOSPHATES, *EUTROPHICATION, *WATER POLLUTION CONTROL, SEWAGE EFFLUENTS, AQUATIC ALGAE, LAKE ERIE, LAKE ONTARIO, NUTRIENTS, NITROGEN, FILTRATION, MICROSCOPY, INDUSTRIAL WASTES, DETERGENTS, AMMONIA, NITRITES, NITRATES.

IDENTIFIERS:
ORTHOPHOSPHATES.

ABSTRACT:
A PROGRAM TO CONTROL EUTHROPHICATION IN LAKES ERIE AND ONTARIO BY DECREASING THE SUPPLY OF PHOSPHORUS COMPOUNDS HAS BEEN DEVELOPED. THIS STUDY WAS CONDUCTED TO DETERMINE THE EFFECT OF REMOVING PHOSPHATE FROM SEWAGE ON ALGAL GROWTH. FILTERED SAMPLES OF RAW SEWAGE, BIOLOGICALLY TREATED SEWAGE, AND SEWAGE TREATED CHEMICALLY FOR PHOSPHATE REMOVAL WERE ADDED TO UNFILTERED WATERS FROM LAKES ERIE AND ONTARIO, AND PARTICULATE RESIDUES (PR) ON MILLIPORE FILTERS PHOTOGRAPHED AFTER INCUBATION IN LIGHT FOR 10 AND 30 DAYS. PR LEVELS IN THE SEWAGE-ENRICHED FLASKS WERE LEAST IN THE CASE OF SEWAGE TREATED FOR REMOVAL OF PHOSPHATES. ADDITION OF PHOSPHATE TO THE PHOSPHATE-DEPLETED EFFLUENT INCREASED ITS PR GENERATING ABILITY TO THAT OF RAW AND BIOLOGICALLY TREATED SEWAGE. THE REMOVAL OF PHOSPHATES FROM SEWAGE WASTES THUS APPEARS TO ELIMINATE THEIR FERTILIZING EFFECT. (MORTLAND-BATTELLE)

FIELD 05C, 05D, 02H

ACCESSION NO. W71-11507

ANALYSIS OF TRACE ELEMENTS IN SEAWEED,

CENTRAL INST. FOR INDUSTRIAL RESEARCH, OSLO (NORWAY).

G. LUNDE.

JOURNAL OF THE SCIENCE OF FOOD AND AGRICULTURE, VOL 21, NO 8, P 416-418, AUGUST 1970. 4 TAB, 18 REF.

DESCRIPTORS:
*AQUATIC ALGAE, *NEUTRON ACTIVATION ANALYSIS, *TRACE ELEMENTS, COPPER, MOLYBDENUM, MANGANESE, COBALT, IRON, INDUSTRIAL WASTES, SEWAGE EFFLUENTS, POLLUTANT IDENTIFICATION.

IDENTIFIERS:
*ATOMIC ABSORPTION, ARSENIC, SELENIUM, ZINC, ANTIMONY, NORWAY, ASHING, BIOLOGICAL MAGNIFICATION, PELVETIA CANALICULATA, FUCUS SERRATUS, FUCUS SPIRALIS, FUCUS VESICULOSUS, LAMINARIA DIGITATA, LAMINARIA HYPERBOREA, ASCOPHYLLUM NODOSUM, GIGARTINA MAMILLOSA, RHODYMENIA PALMATA.

ABSTRACT:
THE FOLLOWING TRACE ELEMENTS HAVE BEEN ANALYZED BY NEUTRON ACTIVATION AND ATOMIC ABSORPTION IN VARIOUS ALGAE SAMPLES: ARSENIC, COPPER, MOLYBDENUM, MANGANESE, ZINC, COBALT, ANTIMONY, SELENIUM AND IRON. THE ASH CONTENT HAS ALSO BEEN DETERMINED. THE SAMPLES WERE COLLECTED FROM TWO DIFFERENT LOCALITIES IN NORWAY, ONE (TRONDHEIMSFJORD) CHARACTERIZED BY A RELATIVELY STRONG INFLUENCE OF RIVER WATER, INDUSTRIAL WASTE ETC.; THE OTHER (REINE IN LOFOTEN) IS FREE FROM THIS TYPE OF CONTAMINATION. IN THE CASE OF TWO SPECIES, ASCOPHYLLUM NODOSUM AND LAMINARIA HYPERBOREA, THE VARIATION IN THE CONTENT OF TRACE ELEMENTS HAS BEEN STUDIED DURING A ONE-YEAR PERIOD, SAMPLES BEING TAKEN EVERY SECOND MONTH. CONSIDERABLE DIFFERENCES IN THE CONTENT OF TRACE ELEMENTS WAS FOUND BETWEEN THE LAMINARIACEAE AND FUCACEAE. THE RESULTS INDICATE THAT THE MARINE ALGAE GENERALLY CONTAIN LARGER CONCENTRATIONS OF THE TRACE ELEMENTS ANALYZED, ESPECIALLY ARSENIC, THAN DO TERRESTRIAL PLANTS. ARSENIC IS ENRICHED IN MARINE ALGAE BY A FACTOR OF 200 TO 500. MORE ATTENTION SHOULD BE FOCUSED ON HOW THIS ELEMENT OCCURS AND WHETHER OR NOT IT HAS ANY PHYSIOLOGICAL ROLE IN MARINE ALGAE. (MORTLAND-BATTELLE)

FIELD 05A, 05B

ACCESSION NO. W71-11515

DETERGENT POLLUTION CONTROL ACT OF 1971 (A BILL TO AMEND THE FEDERAL WATER POLLUTION CONTROL ACT TO ESTABLISH STANDARDS AND PROGRAMS TO ABATE AND CONTROL WATER POLLUTION BY SYNTHETIC DETERGENTS).

HOUSE BILL 7725, 92D CONG, 1ST SESS (1971). 14 P.

DESCRIPTORS:
*WATER POLLUTION CONTROL, *EUTROPHICATION, *DETERGENTS, *PHOSPHATES, LEGISLATION, WATER LAW, FEDERAL GOVERNMENT, WATER POLLUTION, STANDARDS, WATER POLLUTION EFFECTS, ALKYLBENZENE SULFONATES, WATER POLLUTION SOURCES, AQUATIC ALGAE, PUBLIC HEALTH, BIODEGRADATION, TOXICITY, CONTRACTS, GRANTS, RESEARCH AND DEVELOPMENT, INSPECTION, FEDERAL JURISDICTION.

ABSTRACT:
SURFACE AND GROUNDWATERS IN THE UNITED STATES ARE BEING SERIOUSLY POLLUTED AND DEGRADED BY THE DISCHARGE INTO SUCH WATERS OF SYNTHETIC DETERGENTS. IN ORDER TO ABATE SUCH POLLUTION THE FEDERAL WATER POLLUTION CONTROL ACT WOULD BE AMENDED BY THE DETERGENT POLLUTION CONTROL ACT OF 1971. THE FEDERAL GOVERNMENT WOULD PROVIDE ASSISTANCE IN THE DEVELOPMENT OF POLLUTION-FREE DETERGENTS. THE ENVIRONMENTAL PROTECTION AGENCY WOULD ESTABLISH STANDARDS OF WATER EUTROPHICATION ABILITY, BIODEGRADABILITY, TOXICITY, AND OF EFFECTS ON THE PUBLIC HEALTH AND WELFARE WHICH WOULD HAVE TO BE MET BY ALL DETERGENTS. PROCEDURES FOR ESTABLISHING SUCH STANDARDS ARE HEREIN SET FORTH. SUCH STANDARDS WOULD HAVE TO BE MET BY JUNE 30, 1973. DETERGENTS NOT IN COMPLIANCE WOULD BE PROCEEDED AGAINST AND CONDEMNED. FURTHER ENFORCEMENT PROVISIONS AND JUDICIAL PROCEDURES ARE SET FORTH. PROVISION IS ALSO MADE FOR THE INSPECTION OF MANUFACTURING FACILITIES, AND THE SAMPLING OF BOTH DOMESTIC AND IMPORTED DETERGENTS. (ROBINSON-FLORIDA)

FIELD 06E, 05C

ACCESSION NO. W71-11516

BIOLCGICAL ASPECTS OF THERMAL POLLUTION. I: ENTRAINMENT AND DISCHARGE CANAL EFFECTS,

OAK RIDGE NATIONAL LAB., TENN. ECOLOGICAL SCIENCES DIV.; AND MINNESOTA UNIV., MINNEAPOLIS. DEPT. OF ECOLOGY AND BEHAVIORAL BIOLOGY.

CHARLES C. COUTANT, AND ALAN J. BROOK.

CRITICAL REVIEWS IN ENVIRONMENTAL CONTROL, VOL 1, NO 3, P 341-381, NOVEMBER 1970. 16 FIG, 2 TAB, 140 REF.

DESCRIPTORS:
*THERMAL POLLUTION, *BIOLOGICAL COMMUNITIES, *THERMAL POWER PLANTS, *DISCHARGE(WATER), AQUATIC HABITATS, ENVIRONMENTAL EFFECTS, HEATED WATER, NUCLEAR POWER PLANTS, STEAM, ECOLOGY, PHYTCPLANKTON, ZOOPLANKTON, LARVAE, EGGS, CHINOCK SALMON, STRIPEC BASS, MINNOWS, FISHKILL, HERRING, CCPEPODS, CRUSTACEANS, DIATOMS, CYANOPHYTA, PHOTOSYNTHESIS, OYSTERS, CLAMS, TROUT, MUSSELS, MATHEMATICAL MODELS, DISSOLVED OXYGEN, COOLING TOWERS, LAKE MICHIGAN, CCNDENSERS, DIPTERA, PREDATION, ALGAE, NITROGEN FIXATION, FOOC CHAINS, EUTROPHICATION, CATFISHES, BULLHEADS, CARP, PERCHES, METABCLISM, OXYGEN REQUIREMENTS, RESPIRATION, SUCKERS, SEXUAL MATURITY, SPAWNING, RESISTANCE, AQUATIC PRODUCTIVITY, CANALS.

IDENTIFIERS:
*ENTRAINMENT, WHITE RIVER(INDIANA), CONNECTICUT RIVER(CONNECTICUT), GREEN RIVER(KENTUCKY), CAYUGA LAKE(NEW YORK), PATUXENT ESTUARY, BISCAYNE BAY(FLCRIDA), HUMBOLT BAY(CALIFORNIA), PCTOMAC RIVER, COLUMBIA RIVER(WASHINGTON), COCKLES, GOLDFISH, MERRIMACK RIVER(NEW HAMPSHIRE), DELAWARE RIVER, ILLINCIS RIVER, HOLSTON RIVER(TENNESSEE).

ABSTRACT:
THIS REVIEW ATTEMPTS TO CRITICALLY EVALUATE SOME THERMAL EFFECTS SEEN AT OPERATING THERMAL POWER PLANTS, TO GROUP THESE INTO SEVERAL 'PROBLEMS' ASSOCIATED WITH (1) ENTRAINMENT AND (2) DISCHARGE CANALS, AND TO INDICATE PERTINENT FIELD AND LABORATORY EXPERIMENTS THAT CAN ASSIST IN DEVELOPING INFORMATION OF PREDICTIVE UTILITY. MOST POWER PLANT SURVEYS LACK DETAIL OF OBSERVATICN AND DEFINITION CF GOALS SUFFICIENT TO PROVIDE MORE THAN CIRCUMSTANTIAL EVIDENCE FOR ECOLOGICAL PROCESSES. ON THE OTHER HAND, LABCRATORY EXPERIMENTS ARE OFTEN UNREALISTIC SIMULATICNS OF COMPLEX PHENOMENA. TRUE PREDICTABILITY WILL REQUIRE JUDICIOUS APPLICATION OF DATA FROM BOTH SCURCES. UNTIL COMPLETE INFORMATION IS AVAILABLE, CERTAIN LABORATORY TESTS PROVIDE CONSERVATIVE APPROXIMATIONS THAT CAN GUIDE POWER PLANT SITING AND DESIGN SO THAT SAFE ENVIRCNMENTS CAN BE MAINTAINEC FOR AQUATIC LIFE. (MCCANN-BATTELLE)

FIELD 05C, 05B

ACCESSION NO. W71-11517

THE ROLE OF CARBON IN EUTROPHICATION,

MISSOURI UNIV., COLUMBIA. DEPT. OF CIVIL ENGINEERING.

DARRELL L. KING.

JOURNAL WATER POLLUTION CONTROL FEDERATION, VOL 42, NO 12, P 2035-2051, DECEMBER 1970. 8 FIG, 65 REF.

DESCRIPTORS:
*EUTROPHICATION, *CARBON, *PHOTOSYNTHESIS, *ALGAE, CARBON DIOXIDE, ALKALINITY, NUTRIENTS, AMMONIA, HYDROGEN ION CONCENTRATION, CYANOPHYTA, CHLORELLA, CHLOROPHYTA, SCENEDESMUS, EUGLENA, DISSOLVED OXYGEN, DIURNAL DISTRIBUTION, AQUATIC PRODUCTIVITY, NITROGEN, PHOSPHORUS, LAKES.

IDENTIFIERS:
PHORMIDIUM, CHLORELLA PYRENOIDOSA, CHLAMYDOMONAS, FONTINALIS, ASTINASTRUM, ANACYSTIS NIDULANS, MICROCYSTIS.

ABSTRACT:
THIS LITERATURE REVIEW SURVEYS THE ROLE OF CARBON IN LAKE EUTROPHICATION. THE IMPORTANCE OF CARBONATE ALKALINITY AND THE ROLE OF CARBON DIOXIDE IN PHOTOSYNTHESIS ARE DISCUSSED IN DETAIL. IT IS CONCLUDED THAT ALTHOUGH NITROGEN, PHOSPHORUS, AND A VARIETY OF OTHER NUTRIENTS ARE REQUIRED BY ALGAE, EUTROPHICATION SEEMS TO BE ULTIMATELY A CARBON-ACCUMULATION PHENOMENON. INITIAL ALGAL RESPONSE TO ADDITIONS OF NITROGEN AND PHOSPHORUS TO AN OLIGOTROPHIC LAKE INDICATES THAT ONE OF THESE BASIC NUTRIENTS IS THE FACTOR LIMITING PHOTOSYNTHESIS IN THAT LAKE. HOWEVER, IN A LAKE WITH LOW ALKALINITY, BIOLOGICAL INDICATORS OF EUTROPHICATION, SUCH AS MIDSUMMER BLUE-GREEN ALGAL BLOOMS, OFTEN ARE NOTED FOLLOWING ONLY SLIGHT ADDITIONS OF NITROGEN AND PHOSPHORUS. THE AMOUNT OF NITROGEN, PHOSPHORUS, AND OTHER PLANT NUTRIENTS REQUIRED TO PRODUCE MIDSUMMER BLUE-GREEN ALGAL DOMINANCE IN ANY LAKE SEEMS TO BE RELATED DIRECTLY TO THE BICARBONATE-CARBONATE ALKALINITY OF THAT LAKE. ANY ATTEMPT TO MAXIMIZE ALGAL PRODUCTIVITY WHILE MAINTAINING ALGAL SPECIES SUITABLE FOR HUMAN OR ANIMAL FOOD WILL REQUIRE CONTROL OF CARBON DIOXIDE AVAILABILITY IN ADDITION TO THE SUPPLY OF SUFFICIENT QUANTITIES OF OTHER PLANT NUTRIENTS TO ALLOW UNHINDERED ALGAL PHOTOSYNTHESIS. (MORTLAND-BATTELLE)

FIELD 05C

ACCESSION NO. W71-11519

TAXONOMY AND DISTRIBUTION PATTERNS OF PLANKTONIC DIATOMS IN THE EQUATORIAL PACIFIC,

AKADEMIYA NAUK SSSR, MOSCOW. INSTITUT OKEANOLOGII.

T. V. BELYAYEVA.

OCEANOLOGY (USSR), VOL 10, NO 1, P 101-107, 1970. 6 FIG, 1 TAB, 15 REF.

DESCRIPTORS:
*DIATOMS, *PHYTOPLANKTON, *PACIFIC OCEAN, *ALGAE.

IDENTIFIERS:
JUDAY NET, ASTEROLAMPRA MARYLANDICA, ASTEROLAMPRA HEPTACTIS, BACTERIASTRUM COMOSUM, CHAETOCEROS ATLANTICUS VAR. NEAPOLITANA, CHAETOCEROS COARCTATUS, CHAETOCEROS LORENZIANUS, CHAETOCEROS MESSANENSIS, CHAETOCEROS PERUVIANUS, CLADOGRAMMA KOLBEI, CLADOGRAMMA CRENULATUS, CLADOGRAMMA LINEATUS-EXCENTRICUS, CLADOGRAMMA NODULIFER, DACTYLIOSOLEN MEDITERRANEUS, ETHMODISCUS GAZELLAE, ETHMODISCUS REX, HEMIAULUS HAUCKII, HEMIDISCUS CUNEIFORMIS, PLANKTONIELLA SOL, PODOSIRA SP., RHIZOSOLENIA ALATA, RHIZOSOLENIA BERGONII, RHIZOSOLENIA CALCAR-AVIS, RHIZOSOLENIA CASTRACANEI, RHIZOSOLENIA STYLIFORMIS, ROPERIA TESSELLATA, ROPERIA TESSELLATA VAR. OVATA, THALASSIOSIRA DECIPIENS, THALASSIOSIRA SUBTILIS, TRICERATIUM CINNAMOMEUN VAR. MINOR, AMPHORA SP., NITZSCHIA AEQUATORIALIS, NITZSCHIA BICAPITATA, NITZSCHIA BRAARUDII, NITZSCHIA DELICATISSIMA, NITZSCHIA MARINA, NITZSCHIA SICULA, PSEUDOEUNOTIA DOLIOLUS, STIGMAPHORA ROSTRATA, THALASSIONEMA NITZSCHIOIDES VAR. OBTUSA, THALASSIONEMA NITZSCHIOIDES VAR. PARVA, THALASSIOTHTIX GIBBERULA, THALASSIOTHTIX VANHOFFENII, TROPIDONEIS SP.

ABSTRACT:
THE CONTENT AND DISTRIBUTION OF DIATOMS IN THE PLANKTON OF THE EQUATORIAL PART OF THE PACIFIC OCEAN WERE STUDIED ON FOUR MERIDIONAL SECTIONS ALONG 160 E LONG. AND 176, 154 AND 140 W DURING SEPTEMBER-DECEMBER 1961. THE PHYTOPLANKTON COLLECTIONS WERE MADE WITH A JUDAY NET TO A DEPTH OF 100 M. ALL THE DIATOMS FOUND IN THIS REGION ARE LISTED TOGETHER WITH AN INDICATION OF THE FREQUENCY OF OCCURRENCE OF EACH SPECIES ON THE SECTIONS. NINETY-SEVEN PLANKTONIC DIATOMACEOUS ALGAE WERE FOUND IN THE CENTRAL AND EASTERN PARTS OF THE PACIFIC OCEAN: 86 OF THEM WERE IDENTIFIED AS TO SPECIES, 9 ONLY AS TO GENUS, AND TWO OF THE DIATOMS HAVE NOT BEEN IDENTIFIED AS TO GENUS. MOST OF THE DIATOMS IN OUR REGION ARE TROPICAL SPECIES. EIGHTEEN OF THE SPECIES OCCUR NOT ONLY IN THE TROPICAL REGIONS BUT ALSO IN THE BOREAL REGION (12 SPECIES) OR IN THE ANTARCTIC REGION (6 SPECIES). THE WIDELY DIFFUSED (COSMOPOLITAN) SPECIES FOUND AMOUNTED TO 13. (MCCANN-BATTELLE)

FIELD 05A

ACCESSION NO. W71-11527

CALCIUM AND STRONTIUM DISCRIMINATION BY AQUATIC PLANTS,

ATOMIC ENERGY OF CANADA LTD., CHALK RIVER (ONTARIO). CHALK RIVER NUCLEAR LABS.

I. L. OPHEL, AND C. D. FRASER.

ECOLOGY, VOL 51, NO 2, P 324-327, EARLY SPRING 1970. 1 FIG, 1 TAB, 15 REF.

DESCRIPTORS:
*CALCIUM, *FOOD CHAINS, *SPECTROPHOTOMETRY, *POLLUTANT IDENTIFICATION, ONTARIO, ALGAE, MOSSES, AQUATIC PLANTS.

IDENTIFIERS:
*STRONTIUM, *SAMPLE PRESERVATION, *FLAME EMISSION SPECTROPHOTOMETRY, PERCH LAKE(ONTARIO), ASHING, BIOLOGICAL MAGNIFICATION.

ABSTRACT:
AQUATIC PLANTS WERE COLLECTED OVER A 3-YEAR PERIOD FROM PERCH LAKE, ONTARIO. THE PLANTS WERE TAKEN TO THE LABORATORY IN PLASTIC BAGS, WASHED, DRIED TO A CONSTANT WEIGHT AT 110 C, AND ASHED IN PORCELAIN CRUCIBLES. CALCIUM AND STRONTIUM WERE ANALYZED USING A FLAME EMISSION SPECTROPHOTOMETER. SR/CA RATIOS ARE GIVEN FOR 22 SPECIES OF ALGAE, MOSS, AND VASCULAR PLANTS. WHEN COMPARED WITH THE LAKE WATER, 16 SPECIES WERE FOUND TO DISCRIMINATE AGAINST CALCIUM RELATIVE TO STRONTIUM. THESE SPECIES INTRODUCE A STRONTIUM ENRICHMENT STEP INTO ANY FOOD CHAIN IN WHICH THEY PARTICIPATE. ONE SPECIES (NYMPHAEA ODORATA) SHOWED A REMARKABLE DISCRIMINATION AGAINST STRONTIUM. (MORTLAND-BATTELLE)

FIELD 05C, 05A

ACCESSION NO. W71-11529

CERTAIN ASPECTS OF PRODUCTION AND STANDING STOCK OF PARTICULATE MATTER IN THE SURFACE WATERS OF THE NORTHWEST ATLANTIC OCEAN,

BEDFORD INST., DARTMOUTH (NOVA SCOTIA). MARINE ECOLOGY LAB.

W. H. SUTCLIFFE, JR., R. W. SHELDON, AND A. PRAKASH.

JOURNAL FISHERIES RESEARCH BOARD OF CANADA, VOL 27, NO 11, P 1917-1926, NOVEMBER 1970. 7 FIG, 1 TAB, 25 REF.

DESCRIPTORS:
*CARBON, *CHLOROPHYLL, *SEAWATER, *AQUATIC PRODUCTION, *BIOMASS, ORGANIC MATTER, ATLANTIC OCEAN, GROWTH RATES, ALGAE, INCUBATION, PHYTOPLANKTON.

IDENTIFIERS:
*ADENOSINE TRIPHOSPHATE, *DEOXYRIBONUCLEIC ACID, SAMPLE PRESERVATION, COULTER COUNTER.

ABSTRACT:
SAMPLES COLLECTED DURING JANUARY AND OCTOBER 1968 FROM 44 DEGREES TO 19 DEGREES N IN THE WESTERN NORTH ATLANTIC WERE ANALYZED FOR PARTICULATE ORGANIC CARBON, ATP, DNA, CHLOROPHYLL A AND TOTAL PARTICULATE VOLUME BETWEEN 1 AND 40 MICRON DIAMETER. ALTHOUGH THE RATIO OF CHLOROPHYLL TO ATP DECREASED FROM NORTH TO SOUTH, THE RATIO OF PARTICULATE CARBON TO ATP DID NOT SHOW ANY SYSTEMATIC TREND. A GOOD CORRELATION WAS FOUND BETWEEN ATP AND DNA EVEN THOUGH THE DNA CONCENTRATION WAS TOO GREAT TO BE ENTIRELY REPRESENTATIVE OF LIVING MATERIAL. GROWTH OR PRODUCTION OF PARTICULATE MATERIAL AS MEASURED BY THE COULTER COUNTER METHOD WAS HIGHER THAN THAT MEASURED BY C-14 METHOD. A TREND WAS NOTED TOWARDS HIGHER SPECIFIC GROWTH RATE WITH SMALLER STANDING CROP IN SOUTHERN WATERS. (LITTLE-BATTELLE)

FIELD 05C, 05A

ACCESSION NO. W71-11562

OCCURRENCE AND DISTRIBUTION OF DIATOMS AND OTHER ALGAE IN THE UPPER POTOMAC RIVER,

VIRGINIA POLYTECHNIC INST. AND STATE UNIV., BLACKSBURG; KANSAS UNIV., LAWRENCE; AND ACADEMY OF NATURAL SCIENCES OF PHILADELPHIA, PA.

JOHN CAIRNS, JR., ROGER L. KAESLER, AND RUTH PATRICK.

NOTULAE NATURAE OF THE ACADEMY OF NATURAL SCIENCES OF PHILADELPHIA, NO. 436, DEC 22, 1970. 12 P, 3 FIG, 2 TAB, 12 REF.

DESCRIPTORS:
*WATER POLLUTION EFFECTS, *DIATOMS, WATER POLLUTICN SOURCES, INDUSTRIAL WASTES, AQUATIC ALGAE, BICINDICATORS, BIOASSAY, ALGAE, POWERPLANTS, EFFLUENTS, DISCHARGE(WATER), STATISTICAL MODELS.

IDENTIFIERS:
*SPECIES DIVERSITY, *POTOMAC RIVER.

ABSTRACT:
TEN LIMNOLOGICAL SURVEYS WERE MADE OF A PORTION OF THE UPPER POTOMAC RIVER IN CONJUNCTION WITH INSTALLATION AND EXPANSION OF THE DICKERSON POWER STATION OF THE POTOMAC ELECTRIC POWER COMPANY. CLUSTER ANALYSES OF MATRICES OF JACCARD COEFFICIENTS COMPUTED FROM PRESENCE-ABSENCE DATA ON (1) 344 SPECIES OF DIATOMS IN 46 AGGREGATIONS, AND (2) 161 SPECIES OF OTHER ALGAE IN 50 AGGREGATIONS YIELDED RESULTS IN AGREEMENT WITH THOSE OBTAINED IN SIMILAR STUDIES OF PROTOZOANS AND INSECTS AND ALSO WITH RESULTS OF SPECIES DIVERSITY STUDIES. THESE STUDIES INDICATE THAT THE POPULATIONS OF DIATOMS IN THESE AREAS WERE VERY SIMILAR BEFORE AND AFTER PLANT OPERATIONS STARTED, AND THERE WAS NO EVIDENCE FROM THE DIVERSITY OR KINDS OF SPECIES OF ADVERSE EFFECTS DUE TO POWER PLANT DISCHARGE. (LEGORE - WASHINGTON)

FIELD 05C

ACCESSION NO. W71-11583

HOW AERIAL PHOTOGRAPHS CAN IDENTIFY AND MEASURE POLLUTANTS,

CORNELL AERONAUTICAL LAB., BUFFALO, N.Y.

KENNETH R. PIECH.

PUBLIC WORKS, VOL. 101, NO. 3, P 68-71, MARCH 1970. 4 FIG.

DESCRIPTORS:
*POLLUTANT IDENTIFICATION, *AERIAL PHOTOGRAPHY, *MATHEMATICAL METHODS, MUNICIPAL WASTES, INDUSTRIAL WASTES, NEW YORK, ALGAE, WATER POLLUTION.

IDENTIFIERS:
NIAGARA RIVER.

ABSTRACT:
AERIAL PHCTOGRAPHY CAN BE USED TO BOTH MEASURE AND QUANTIFY POLLUTANTS IN WATERWAYS. IT HAS THE POTENTIAL OF PROVIDING PCLLUTION DATA FOR AN ENTIRE WATERWAY RATHER THAN ISOLATED PCINTS AND IS OFTEN ABLE TO PINPOINT THE SOURCE CF A PARTICULAR POLLUTANT. THE METHCD IS BASED ON THE FACT THAT EVERY SUBSTANCE HAS A CHARACTERISTIC SPECTRAL CURVE BY WHICH IT CAN BE IDENTIFIED. BY USING SPECIALLY CCNSTRUCTED SPECTRAL FILTERS, PHOTOGRAPHS CAN BE TAKEN IN WHICH LIGHT CUTSIDE A GIVEN RANGE OF WAVELENGTHS(THE SPECTRAL CURVE OF THE POLLUTANT TC BE IDENTIFIED FALLS WITHIN THE RANGE) IS FILTERED OUT. ON THE DEVELOPED PHOTOGRAPH, THE POLLUTANT THEN SHOWS UP LIGHTER AND ITS QUANTITY CAN BE DETERMINED BY MATHEMATICAL ANALYSIS, IN PARTS PER MILLION, FCR EXAMPLE. IN REALITY, IT WOULD BE IMPRACTICAL TO CONSTRUCT FILTERS FOR EACH POLLUTANT, SINCE MANY SUBSTANCES HAVE VERY SIMILAR SPECTRAL CURVES. THUS, THE PRACTICALITY OF SINGLING OUT A SPECIFIC POLLUTANT IS LARGELY GOVERNED BY ITS TYPE COMPARED WITH THE COMBINATICNS OF OTHER PCLLUTANTS THAT MAY ALSO BE PRESENT. BECAUSE THE TYPE OF POLLUTANT IS OFTEN KNOWN, THE ABILITY TO TRACK IT BY PHCTOGRAPHY IS PRCBABLY OF MORE IMPORTANCE. THE METHOD CAN ALSO BE USED TO MEASURE ALGAL CONCENTRATIONS. (MORTLAND-BATTELLE)

FIELD 05A, 05B, 07B

ACCESSION NO. W71-11685

METHODS OF REMOVING NITRATES FROM WATER,

INTERAGENCY AGRICULTURAL WASTE WATER TREATMENT CENTER, FIREBAUGH, CALIF.

PERCY P. ST. AMANT, AND LOUIS A. BECK.

JOURNAL OF AGRICULTURAL AND FOOD CHEMISTRY, VOL. 18, NO. 5, P 785-788, SEPTEMBER-OCTOBER 1970. 1 TAB, 7 REF.

DESCRIPTORS:
NITRATES, ALGAE, DENTRIFICATION, FILTRATION, DESALINATION, DRAINAGE WATER, AGRICULTURAL CHEMICALS, FARM WASTES, WASTE WATER TREATMENT, CALIFORNIA.

ABSTRACT:
AN INTERAGENCY NITROGEN REMOVAL GROUP COMPOSED OF THREE AGENCIES, THE FEDERAL WATER POLLUTION CONTROL ADMINISTRATION, THE U.S. BUREAU OF RECLAMATION, AND THE CALIFORNIA DEPARTMENT OF WATER RESOURCES, WAS FORMED TO ATTEMPT TO FIND A METHOD OF REMOVING NITRATES FROM AGRICULTURAL WASTE WATERS IN CALIFORNIA. THREE METHODS OF NITROGEN REMOVAL ARE BEING EVALUATED: POND AND FILTER DENITRIFICATION, WHICH DEPEND UPON BACTERIAL ACTION, AND ALGAE STRIPPING. DESALINATION OF TILE DRAINAGE IS ALSO BEING EVALUATED. DETAILS AND PRELIMINARY RESULTS OF THE STUDIES ARE GIVEN. (LITTLE-BATTELLE)

FIELD 05C, 05D, 05F, 05G

ACCESSION NO. W71-11698

TRACE MATERIALS IN WASTES DISPOSED TO COASTAL WATERS-FATES, MECHANISMS AND ECOLOGICAL GUIDANCE AND CONTROL,

FEDERAL WATER QUALITY ADMINISTRATION, CORVALLIS, OREG. NATIONAL COASTAL RESEARCH PROGRAM.

MILTON H. FELDMAN.

AVAILABLE FROM THE NATIONAL TECHNICAL INFORMATION SERVICE AS PB-202 346, $3.00 IN PAPER COPY, $0.95 IN MICROFICHE. WORKING PAPER, NO. 78, JULY 1970. 102 P, 20 FIG, 17 TAB, 137 REF. EPA PROGRAM 16070---07/70.

DESCRIPTORS:
*OCEANS, *COASTS, *WATER QUALITY CONTROL, BIOMASS, TOXICITY, EUTROPHICATION, TRACE ELEMENTS, NITROGEN, PHOSPHORUS, METABOLISM, ALGAE, SLUDGE, ANALYTICAL TECHNIQUES, INDUSTRIAL WASTES, MUNICIPAL WASTES, METALS, HEAVY METALS, WASTE DISPOSAL, WATER POLLUTION EFFECTS.

IDENTIFIERS:
*TRACE ORGANIC CONTAMINANTS, *TRACE MATERIALS, BIOINHIBITION, BIOSTIMULATION.

ABSTRACT:
WASTES CURRENTLY BEING DISCHARGED TO THE COASTAL WATERS OF THE UNITED STATES INCLUDE TRACE ORGANIC CONTAMINANTS (TC), TRACE ELEMENTS (TE), AND OTHER TRACE MATERIALS (TM). THOSE TM DEALT WITH INCLUDED: (1) KNOWN VIOLENTLY NOXIOUS MATERIALS; (2) MATERIALS WHICH ARE BIOSTIMULATORY TO SOME SPECIES (COBALAMIN, IRON CHELATES, THIAMIN, BIOTIN, MN, MG; AND (3) MATERIALS WHICH ARE BIOINHIBITORY FOR AT LEAST SOME SPECIES IN VARIOUS MECHANISMS (DDT, SE, MN, MG). A DIFFERENT VIEW OF THE PROBLEM WAS TAKEN IN WHICH WAYS OF SELECTING THE OPTIMUM ORGANISM TO UTILIZE A PARTICULAR WASTE WERE PREFERRED TO METHODS OF REMOVING IT CONVENTIONALLY. A THOROUGH LITERATURE SEARCH REVEALED A SIGNIFICANT LACK OF LITERATURE IN WASTE BREAKDOWN FOR BOTH QUALITATIVE AND QUANTITATIVE EVALUATION. FURTHER STUDIES SHOULD BE PERFORMED FOR EACH KNOWN TRACE MATERIAL TO DETERMINE: (1) THE MECHANISM WHEREBY IT IS SECLUDED; (2) THE METABOLIC THRESHOLD; (3) THE ACTIVE LEVEL; (4) THE HARMFUL LEVEL; (5) THE SPECIATION REQUIREMENTS; (6) THE ABSOLUTE RATES IN AND OUT OF THE COMPARTMENTS OF WHICH COASTAL WATERS, SEDIMENTS, CHEMICAL SYSTEM PHASES, AND BIOTA MAY BE CONSIDERED AS COMPOSED. EVALUATIONS SUCH AS THESE WERE PERFORMED FOR DDT, AND THEY MUST BE PERFORMED FOR THE KNOWN SET OF WASTE CONSTITUENTS BEFORE RATIONAL ACTION TO PREVENT DAMAGE TO THE OCEANS IS POSSIBLE. (LOWRY-TEXAS)

FIELD 05B, 05C, 02L, 05G

ACCESSION NO. W71-11793

THE ISOLATION AND OCCURRENCE OF HYALCCHLORELLA MARINA,

CALIFORNIA UNIV. BERKELEY. MICROBIOLOGY GROUP; AND CALIFORNIA UNIV; BERKELEY. CEPT. OF BOTANY.

R. O. POYTON.

JOURNAL OF GENERAL MICROBIOLOGY, VOL. 62, PART 2, P 189-194, AUGUST 1970. 2 TAB, 13 REF.

DESCRIPTORS:
*MARINE FUNGI, *ISOLATION, ALGAE, SEPARATION TECHNIQUES, ATLANTIC OCEAN, PACIFIC OCEAN, GULF OF MEXICO.

IDENTIFIERS:
*HYALOCHLORELLA MARINA, RHODOMELA SP., ENDOCLADIA SP., PHYLLOSPADEX SP., MICROCLADIA SP., CLADOPHORA SP., POLYSIPHONIA SP., GELIDIUM SP., CHLORELLA, AGARS.

ABSTRACT:
FIVE METHODS WERE EVALUATED FOR ISOLATION OF HYALOCHLORELLA MARINA FROM A MARINE ENVIRONMENT. THE METHODS WERE: (1) FILAMENTCUS ALGAE COLLECTED FROM INTERTICAL AND EULITTCRAL ZONES WAS PLACED IN ISOLATION MEDIUM CONTAINING 370 UNITS OF STREPTOMYCIN SULFATE/ML AND 825 UNITS PENICILLIN 'G' SODIUM/ML. (2) SEA WATER WAS CENTRIFUGED AND THE CONCENTRATE PLATED ONTO THE ISOLATION MEDIUM. (3) SEA WATER AND MARINE MUD SUSPENSIONS WERE PREFILTERED TC REMOVE LARGE MUD PARTICLES AND DEBRIS, FILTERED THROUGH A STERILE NUCLEOPORE MEMBRANE FILTER, AND THE FILTERS INCUBATED ON THE ISOLATION MEDIUM. (4) SEA WATER WAS SEEDED WITH POLLEN GRAINS AND OBSERVED AFTER 4, 7, AND 10 DAYS FOR HYALOCHLORELLA FILAMENTS. (5) MICROSCOPE SLICES WERE IMMERSED AT VARIOUS DEPTHS IN THE MARINE ENVIRONMENT AND EXAMINED FOR HYALOCHLORELLA GROWTH AT VARIOUS TIME INTERVALS. THE LATTER TWO METHODS PROVIDED NO SATISFACTORY RESULTS IN THE EXPERIMENTS. THE FIRST THREE METHODS PROVED EFFECTIVE FOR ISOLATION OF HYALOCHLCRELLA MARINA. THESE METHODS SHOWED THAT THIS ORGANISM OCCURS IN NATURE BOTH ATTACHED TO ALGAL FILAMENTS AND FREE FLOATING. HOWEVER, THE CCNCENTRATION OF ORGANISMS ON ALGAL SURFACES WAS 300 TO 13,000 TIMES THAT OF FREE-FLOATING ORGANISMS IN AN EQUIVALENT AMOUNT CF SEA WATER. ADDITIONAL INVESTIGATION OF THE EFFECT OF ANTIBICTICS ON THE VIABILITY OF HYACHLORELLA MARINA SHOWED THAT VIABLE COUNTS WERE THE SAME WITH OR WITHOUT PENICILLIN AND/OR STREPTCMYCIN. (LITTLE-BATTELLE)

FIELD 05A, 05C

ACCESSION NO. W71-11851

COMPARISON OF THE POLLUTED HOOGHLY ESTUARY WITH THE UNPOLLUTED MATLAH ESTUARY, INDIA,

CENTRAL INLAND FISHERIES RESEARCH INST., BARRACKPORE (INDIA).

A. K. BASU, B. B. GHOSH, AND R. N. PAL.

JOURNAL WATER POLLUTION CONTROL FEDERATION, VOL 42, NO 10, P 1771-1781, OCTOBER 1970. 7 FIG, 1 TAB, 17 REF.

DESCRIPTORS:
*INDUSTRIAL WASTES, *WATER QUALITY, *DISSOLVED OXYGEN, *PHYTOPLANKTON, ESTUARIES, WATER TEMPERATURE, TURBIDITY, PHOTOSYNTHESIS, HYDROGEN ION CONCENTRATION, NITRITE, AMMONIA, SALINITY, AQUATIC ALGAE, ZOOPLANKTON, BIOCHEMICAL OXYGEN DEMAND, COPEPODS.

IDENTIFIERS:
COSCINODISCUS, NITZSCHIA, OSCILLATORIA, PLEUROSIGMA, NAUPLII, LAMELLIBRANCH, SPIROGYRA, EUDORINA, CYCLOPS, KERATELLA, ASPLANCHNA, LITHODESMIUM, HARPESTICOIDS, NOCTILUCA, CENTROPYXIS, BIDDULPHIA, CHAETOCEROS, THALASSISTHRIX, INDIA.

ABSTRACT:
A SHORT PRELIMINARY STUDY TO ASSESS THE GENERAL PHYSICOCHEMICAL AND BIOLOGICAL CONDITIONS OF THE HOOGHLY ESTUARY, INTO WHICH A LARGE QUANTITY OF INDUSTRIAL POLLUTANTS IS BEING DISCHARGED FROM DIFFERENT FACTORIES, WAS ATTEMPTED, AND A COMPARISON HAS BEEN DRAWN WITH THE CONDITIONS OF THE UNPOLLUTED MATLAH ESTUARY. THE RESULTS REVEALED THE FOLLOWING POINTS: (1) THE GENERAL CONDITION OF THE HOOGHLY WITH RESPECT TO DISSOLVED OXYGEN, OXYGEN CONSUMPTION VALUE, AND TEMPERATURE WAS INFERIOR TO THAT OF THE MATLAH. (2) TURBIDITY OF THE HOOGHLY WAS GENERALLY HIGHER THAN THAT OF THE MATLAH. (3) PH, NITRITE N, AND FREE AMMONIA VALUES WERE MORE OR LESS THE SAME IN THE MIDSTREAM OF BOTH THE ESTUARIES, ALTHOUGH A LOW PH VALUE OF 4.3 WAS OBSERVED NEAR A DISCHARGE POINT IN THE HOOGHLY. (4) THE QUANTUM OF PLANKTON, WHICH SERVES AS A FISH FOOD, WAS LESS IN THE HOOGHLY THAN THE MATLAH. (5) SOME OF THE PHYSICOCHEMICAL CONDITIONS, MAINLY TEMPERATURE, TURBIDITY, PH, NITRITE NITROGEN, FREE AMMONIA, DO, AND CC, NEAR A FEW DISCHARGE OUTLETS IN THE HOOGHLY ESTUARY INDICATED CONDITIONS UNFAVORABLE FOR FISH AND FISHERIES OR AT LEAST TENDING TO MAKE THE ADJOINING AREA UNINHABITABLE. (MORTLAND-BATTELLE)

FIELD 05C, 02L

ACCESSION NO. W71-11876

HYPNODINIUM-LIKE ALGAL BLOOMS IN GEORGIA LAKES,

GEORGIA WATER QUALITY CONTROL BOARD, ATLANTA; AND GEORGIA STATE UNIV., ATLANTA. DEPT. OF BIOLOGY.

EDWARD T. HALL, JR., DONALD G. AHEARN, AND DONALD J. REINHARDT.

BIOSCIENCE, VOL 21, NO 3, P 115, FEBRUARY 1, 1971. 1 FIG, 1 REF.

DESCRIPTORS:
*EUTROPHICATION, *ALGAE, *SEWAGE EFFLUENTS, GEORGIA, THERMOCLINE, HYDROGEN ION CONCENTRATION, WATER TEMPERATURE, DISSOLVED OXYGEN, COLIFORMS, RECREATION, ODOR.

IDENTIFIERS:
*LAKE SIDNEY LANIER(GEORGIA), LAKE JACKSON(GEORGIA), LAKE CARDINAL(GEORGIA), LAKE BUCKHORN(GEORGIA), HYPNODINIUM SPHAERICUM.

ABSTRACT:
AN ALGAL BLOOM IS DESCRIBED SIMILAR TO HYPNODINIUM SPHAERICUM THAT HAS APPEARED EACH JUNE SINCE 1968 IN LAKE SIDNEY LANIER, GEORGIA. THIS BLOOM OF VARYING INTENSITY LASTS 7-10 DAYS. IN 1969, THE SURFACE OF THE LAKE TURNED MILKY WHITE. IN 1970 THE BLOOM OCCURED SHORTLY AFTER THE WATER STRATIFIED WITH A THERMOCLINE NEAR 9 M. THE SURFACE WATER TEMPERATURE WAS 28 C, THE PH 9.1, AND THE DISSOLVED OXYGEN 8.2 MG/LITER. AT THE THERMOCLINE, THE WATER TEMPERATURE WAS ABOUT 17 C, THE PH 8.3, AND THE DISSOLVED OXYGEN 3.4 MG/LITER. FOUR DAYS PRIOR TO THE RECORDING OF THE BLOOM, THE FECAL COLIFORM DENSITY INCREASED FROM LESS THAN 4/100 ML TO GREATER THAN 240/100 ML. FIFTEEN DAYS LATER THE COLIFORM DENSITY WAS LESS THAN 3/100 ML. THE ALGA HAS NOT BEEN DIRECTLY ASSOCIATED WITH ANY FISH KILLS. A DISAGREEABLE ODOR ACCOMPANIES THE BLOOM AND THE THICK SURFACE FLOCS INTERFERE WITH RECREATIONAL ACTIVITIES. THE SIGNIFICANCE AND LONG-TERM EFFECTS OF THIS ALGA ARE UNKNOWN. (MCCANN-BATTELLE)

FIELD 05A, 05B, 05C

ACCESSION NO. W71-11914

DETERMINATION OF ATP IN CHLORELLA WITH THE LUCIFERIN-LUCIFERASE ENZYME SYSTEM,

UNITED STATES DEPARTMENT OF AGRICULTURE, BELTSVILLE, MD. CRCPS RESEARCH DIV.

J. B. ST. JOHN.

ANALYTICAL BIOCHEMISTRY, VCL 37, P 409-416, OCTOBER 197C. 1 FIG, 4 TAB, 18 REF.

DESCRIPTORS:
*CHLORELLA, *BIOASSAYS, ENZYMES, LIGHT INTENSITY, ALGAE.

IDENTIFIERS:
*ADENOSINE TRIPHOSPHATE, *LUCIFERIN-LUCIFERASE ENZYME SYSTEM, PLANT EXTRACTS, CHLORELLA SORCKINIANA.

ABSTRACT:
A METHOD FOR MEASURING ATP BASED ON THE FIREFLY LUCIFERIN-LUCIFERASE REACTION IS DESCRIBED. THE METHCD INVOLVES A 30 SECOND INTEGRATION OF THE LIGHT PRODUCED FCLLOWING MIXING OF THE SAMPLE AND ENZYME. THE RELATIONSHIP BETWEEN LIGHT MEASUREC ANC QUANTITY CF ATP PRESENT IS LINEAR OVER AT LEAST A THOUSANDFOLD CONCENTRATION RANGE, WITH LESS THAN A 2 PERCENT RELATIVE STANDARD DEVIATION. THE METHCD OF LIGHT MEASUREMENT INCLUDES THE PEAK INTENSITY OF INITIAL FLASH AND A PORTION OF THE SIGNAL WHICH FOLLOWS. THEREFORE, THE PEAK INTENSITY OF THE INITIAL FLASH ALONE IS NOT NECESSARILY THE MOST RELIABLE BASIS FOR MEASUREMENT OF ATP. THE METHCD USES A COMMERCIALLY AVAILABLE LUCIFERIN-LUCIFERASE PREPARATION WITHOUT FURTHER PURIFICATION. THE LEVEL OF SENSITIVITY IS EQUAL TO THAT REPORTED FOR PURIFICATION. THE LEVEL OF SENSITIVITY IS EQUAL TO THAT REPORTED FOR PURIFIED LUCIFERIN-LUCIFERASE PREPARATIONS. THE METHOD OF CCNSTANT ADDITION FOR QUANTITATIVE DETERMINATICN OF ATP IN PLANT EXTRACTS IS DESCRIBED. THE LACK OF SPECIFICITY ASSOCIATED WITH THE LUCIFERIN-LUCIFERASE REACTION IS CONTROLLED BY THIS METHOD. (MORTLANC-BATTELLE)

FIELD 05A

ACCESSION NO. W71-11916

EUTROPHICATION OF NORTHEASTERN OHIO LAKES: I INTRODUCTION, MORPHCMETRY, AND CERTAIN PHYSICO-CHEMICAL DATA CF DCLLAR LAKE,

KENT STATE UNIV., OHIO. INST. OF LIMNOLOGY.

G. D. COOKE, AND ROBERT L. KENNEDY.

THE OHIO JOURNAL OF SCIENCES, VOL. 70, NC. 3, P 150-161, MAY 197C. 7 FIG, 1 TAB, 14 REF.

DESCRIPTORS:
*EUTROPHICATION, *LAKES, *WATER POLLUTION EFFECTS, OHIO, NUTRIENTS, WATER POLLUTICN SOURCES, DISSOLVED OXYGEN, LAKE MCRPHOLOGY, AQUATIC ALGAE, MOSSES, WATER TEMPERATURE, WATER CHEMISTRY, CHLCRCPHYLL, HYPSOMETRIC ANALYSIS, HYPCLIMNION, EPILIMNION, CYANOPHYTA, DINOFLOGELLATES, SEASONAL, ZOOPLANKTON, PHYTOPLANKTON.

IDENTIFIERS:
DOLLAR LAKE(OHIO), SPHAGNUM, APHANIZOMENON FLOS-AQUAE, CERATIUM, OSCILLATORIA, TRANSPARENCY.

ABSTRACT:
THREE NORTHEASTERN OHIO LAKES (EAST TWIN, WEST TWIN, AND DOLLAR) ARE BEING STUDIED TO COMPARE LAKES IN DIFFERENT STAGES OF EUTROPHICATION WITHIN THE SAME AREA AND TC EVOLVE LONG-TERM BEFCRE-AND-AFTER COMPARISONS. THIS IS THE FIRST CF A SERIES OF ARTICLES TC BE PREPARED FROM THESE STUDIES. BASIC EUTROPHICATICN DATA CN DOLLAR LAKE ARE PRESENTED. BECAUSE IT IS MCST STRONGLY INFLUENCED BY HUMAN HABITATION AND IS THE SMALLEST OF LAKES, IT WAS SELECTED AS THE MODEL LAKE FRCM WHICH COMPARATIVE OBSERVATIONS CAN BE CRAWN. DOLLAR LAKE HAS A VOLUME OF 86,400 CUBIC METERS AND AN AREA OF 22,212 SQUARE METERS. THE AVERAGE DEPTH IS 3.89 METERS, THE MAXIMUM CEPTH IS 7.5 METERS. MAXIMUM WIDTH OF THE LAKE IS 140 METERS, THE MAXIMUM LENGTH 215 METERS. EXCEPT FOR BRIEF PERIODS IN SPRING AND FALL, DEEP WATERS ARE CEPLETED OF DISSOLVED OXYGEN. SECCHI DISC TRANSPARENCY IS FREQUENTLY BELCW CNE METER. MASSIVE BLOOMS OF ALGAE ARE OFTEN CBSERVED. THE LAKE STRATIFIES THERMALLY IN APRIL, CIRCULATES IN OCTOBER, ANC RESTRATIFIES IN CECEMBER AFTER ICE FORMATION. SPRING CIRCULATION OCCURS IN MARCH. DCLLAR LAKE IS A DIMICTIC, SECOND-CLASS EUTROPHIC LAKE. (MORTLAND-BATTELLE)

FIELD 05C, 05B

ACCESSION NO. W71-11949

TRISODIUM NITRILOTRIACETATE AND ALGAE,

A. E. CHRISTIE.

WATER AND SEWAGE WORKS, VOL 117, NO 2, P 58-59, FEBRUARY 1970. 4 TAB, 6 REF.

DESCRIPTORS:
*ALGAE, *DETERGENTS, *TOXICITY, *ACTIVATED SLUDGE, *NUTRIENTS, NITROGEN, WATER POLLUTION EFFECTS, WATER POLLUTION SOURCES.

IDENTIFIERS:
*CHLORELLA PYRENOIDOSA, *TRISODIUM NITRILOTRIACETATE.

ABSTRACT:
RELATIONSHIPS BETWEEN TRISODIUM NITRILOTRIACETATE, A POTENTIAL DETERGENT BUILDER, AND THE ALGA CHLORELLA PYRENOIDOSA WITH RESPECT TO TOXICITY AND AS A NITROGEN SOURCE WERE EXAMINED. PRELIMINARY EXPERIMENTS ON THE ROLE OF NTANA3 AS A SOURCE OF NITROGEN FOR ALGAE WERE CARRIED OUT BY MEASURING GROWTH IN A MINERAL MEDIA. EQUIVALENT QUANTITIES OF NITROGEN WERE ADDED EITHER AS NTANA3 OR SODIUM NITRATE, BOTH WITH AND WITHOUT THE ADDITION OF A SMALL QUANTITY OF RAW SEWAGE. PRELIMINARY TESTING OF NTANA3 TOXICITY TO ALGAE WAS ACCOMPLISHED BY ADDING THE COMPOUND TO WEEK-OLD CULTURES GROWN IN A COMPLETE MINERAL MEDIUM. EXAMINATION OF NTANA3 AS A NITROGEN SOURCE FOR ALGAE WAS ALSO INVESTIGATED BY CULTURING CHLORELLA IN THE FILTERED SUPERNATANTS OF ALIQUOTS OF ACTIVATED SLUDGE WHICH HAD RECEIVED VARIOUS CONCENTRATIONS OF NTANA3. THE RESULTS INDICATE THAT NTANA3 AT CONCENTRATIONS UP TO 275 MG/L IS NOT TOXIC TO THE TEST ORGANISM AND CAN ACT AS A SOURCE OF NITROGEN FOR ALGAL GROWTH. NO SIGNIFICANT DIFFERENCES WERE FOUND IN ALGAL RESPONSES IN THE SUPERNATANTS OF THE ACTIVATED SLUDGE EXCEPT IN THOSE RELATED TO THE HIGHEST CONCENTRATIONS OF THE NTANA3. THE HIGHEST CONCENTRATION DID NOT REDUCE THE ALGAL GROWTH POTENTIAL OF THE CHEMICAL. SINCE CARBON IS GENERALLY THE NUTRIENT LIMITING BACTERIAL ACTION IN SEWAGE TREATMENT PLANTS, THE FEASIBILITY OF BIODEGRADABLE ORGANIC MATERIALS AS DETERGENT BUILDERS WHICH CONTAIN NEITHER NITROGEN NOR PHOSPHORUS SHOULD BE THOROUGHLY CONSIDERED. (LITTLE-BATTELLE)

FIELD 05C, 05B

ACCESSION NO. W71-11972

COLOR INFRARED PHOTOGRAPHY FROM 60,000 FEET ALTITUDE AS A TOOL FOR DELINEATING AQUATIC VEGETATION,

NATIONAL AERONAUTICS AND SPACE ADMINISTRATION, HOUSTON, TEX. MANNED SPACECRAFT CENTER.

CURTIS C. MASON.

IN: HYDROBIOLOGY, 'BIORESOURCES OF SHALLOW WATER ENVIRONMENTS', PROCEEDINGS OF SYMPOSIUM, MIAMI BEACH, FLORIDA, JUNE 24-27, 1970; AMERICAN WATER RESOURCES ASSOCIATION, URBANA, ILLINOIS, PROCEEDINGS SERIES NO 8, P 317-328, 1970. 12 P, 6 FIG.

DESCRIPTORS:
*REMOTE SENSING, *AERIAL PHOTOGRAPHY, *INFRARED RADIATION, *AQUATIC ALGAE, *SURFACE WATERS, ANALYTICAL TECHNIQUES, PHOTOMETRY, WATER POLLUTION.

IDENTIFIERS:
*INFRARED PHOTOGRAPHY, *AQUATIC VEGETATION.

ABSTRACT:
COLOR INFRARED (IR) PHOTOGRAPHY WITH ITS ABILITY TO EXTEND THE RANGE OF PHOTOGRAPHIC DETECTABILITY INTO THE NEAR IR WAVELENGTH IS AN EXCELLENT TOOL FOR MAPPING AQUATIC VEGETATION. THIS VEGETATION APPEARS IN A BROADER RANGE OF COLOR THAN ON CONVENTIONAL COLOR PHOTOGRAPHY. WATER APPEARS AS A VERY DISTINCT BLUE OR BLACK SHARPLY OUTLINING THE AQUATIC VEGETATION. USING HIGH ALTITUDE AIRCRAFT PERMITS COVERAGE OF LARGE AREAS ON A SINGLE MISSION. (WOODARD-USGS)

FIELD 05B, 07B

ACCESSION NO. W71-12034

THE EFFECT OF CARBON ON ALGAL GROWTH--ITS RELATIONSHIP TO EUTROPHICATION,

UTAH STATE UNIV., LOGAN. UTAH WATER RESEARCH LAB.

JOEL C. GOLDMAN, DONALD B. PORCELLA, E. JOE MIDDLEBROOKS, AND DANIE F. TOERIEN.

COLLEGE OF ENGINEERING, OCCASIONAL PAPER 6, APRIL 1971. 56 P, 7 FIG, 7 TAB, 242 REF.

DESCRIPTORS:
*CARBON, *ALGAE, *EUTROPHICATION, *LIMITING FACTORS, PHOTOSYNTHESIS, CARBON CYCLE, ABSORPTION, DECOMPOSING ORGANIC MATTER, ALKALINITY, HYDROGEN ION CONCENTRATION, NUTRIENTS, AQUATIC ENVIRONMENT, CARBON DIOXIDE.

IDENTIFIERS:
CHEMISTRY OF INORGANIC CARBON, CARBON TRANSFORMATION, CARBON SOURCES, ORGANIC CARBON, CARBON UTILIZATION.

ABSTRACT:
AVAILABILITY OF INORGANIC AND ORGANIC CARBON FOR PHOTOSYNTHESIS OF ALGAL CELLS DEPENDS UPON IONIC COMPOSITION OF THE MEDIUM AND THE POPULATION OF THE AQUATIC SYSTEM. IN SEWAGE LAGOONS, EUTROPHIC LAKES, OR LABORATORY CULTURES CONTAINING AN EXCESS OF NUTRIENTS, CARBON CAN BE THE FACTOR LIMITING THE GROWTH OF ALGAL MASS. HOWEVER, IN MOST LAKES INORGANIC SOURCES, ATMOSPHERE, AND BACTERIAL DEGRADATION PROVIDE AN ADEQUATE SUPPLY OF CARBON FOR THE AMOUNT OF AVAILABLE LIGHT, NITROGEN, AND PHOSPHORUS. IN TERMS OF PRACTICAL TECHNOLOGY, THE CONTROL OF ALGAL GROWTH IS USUALLY RELATED TO THE CONTENT OF PHOSPHORUS. (WILDE-WISCONSIN)

FIELD 05C, 10, 02H

ACCESSION NO. W71-12068

THE RELATION OF ACETYLENE REDUCTION TO HETEROCYST FREQUENCY IN BLUE-GREEN ALGAE,

WESTFIELD COLL., LONDON (ENGLAND). DEPT. OF BOTANY.

WILLIAM J. JEWELL, AND S. A. KULASOORIYA.

JOURNAL OF EXPERIMENTAL BOTANY, VOL 21, NO 69, P 874-880, 1970. 1 FIG, 3 TAB, 13 REF.

DESCRIPTORS:
*CYANOPHYTA, *ALGAE, *NITROGEN FIXATION, ENZYMES, BACTERIA, SPORES, NITROGEN, METABOLISM.

IDENTIFIERS:
*ACETYLENE REDUCTION, *HETEROCYSTS, ANABAENA CYLINDRICA, ANABAENOPSIS, ACETYLENE TEST, CYLINDROSPERMUM LICHENEFORME, MASTIGOCLADUS LAMINOSUS.

ABSTRACT:
IF NITROGEN-FIXING ENZYME SYSTEM IS LOCATED EXCLUSIVELY IN THE HETEROCYSTS, AND, IF ACETYLENE-REDUCTION TECHNIQUE GIVES A QUANTITATIVE MEASURE OF THIS ACTIVITY, IT MAY BE POSSIBLE TO ESTIMATE NITROGEN FIXATION FROM THE FREQUENCY OF HETEROCYSTS AND THE ACETYLENE REDUCTION CAPACITY OF HETEROCYSTS. THIS STUDY ATTEMPTED TO DETERMINE VARIATION, WITH CULTURE AGE, OF ACETYLENE-REDUCTION ACTIVITY AND HETEROCYST FREQUENCY IN FIVE SPECIES OF BLUE-GREEN ALGAE USING BACTERIA-FREE CULTURES OF ANABAENA CYLINDRICA (NONSPORING), ANABAENA CYLINDRICA (SPORING), ANABAENOPSIS, CYLINDROSPERMUM LICHENIFORME, AND MASTIGOCLADUS LAMINOSUS. VARIATION AND AVERAGE HETEROCYST FREQUENCY ARE SUMMARIZED EXCEPT FOR CYLINDROSPERMUM LICHENIFORME WHICH GREW WITH TIGHTLY INTERTWINED FILAMENTS SO THAT IT WAS IMPOSSIBLE TO COUNT A LARGE NUMBER OF SAMPLES. THE HETEROCYST FREQUENCY VARIED BETWEEN 2.5% TO 12.3% OF TOTAL CELLS; AND THE ACETYLENE REDUCTION ACTIVITY VARIED 2.6-FOLD (AVERAGE). CONSISTENCY OF THE VALUES OF THE COEFFICIENT OF ACETYLENE REDUCTION INDIRECTLY SUPPORTS THE INITIAL ASSUMPTION THAT SITE OF NITROGEN FIXATION IS THE HETEROCYST. ESTIMATES OF NITROGEN FIXATION RATES BY BLUE-GREEN ALGAE MAY BE MADE BY IN SITU HETEROCYST COUNT AND ACETYLENE REDUCTION MEASUREMENTS. (JONES-WISCONSIN)

FIELD 05C

ACCESSION NO. W71-12070

TOLERANCE OF BLUE-GREEN ALGAE TO PESTICIDES,

INDIAN AGRICULTURAL RESEARCH INST., NEW DELHI. DIV. OF MICROBIOLOGY.

G. S. VENKATARAMAN, AND B. RAJYALAKSHMI.

CURRENT SCIENCE, VOL 40, NO 6, P 143-144, 1971. 1 TAB, 8 REF.

DESCRIPTORS:
*CYANOPHYTA, *PESTICIDES, *FERTILIZATION, NUTRIENTS, ALGAE, RICE, NITROGEN, NITROGEN FIXATION, 2-4-D, DALAPON, HERBICIDES, CROP PRODUCTION.

IDENTIFIERS:
*TOLERANCE LEVELS, TOLYPOTHRIX TENUIS, AULOSIRA FERTILISSIMA, ANACYSTIS NIDULANS, CERASON, DIATHENE, PROPAZINE, COTORON, CIURON, LINURON.

ABSTRACT:
INFLUENCE OF ALGALIZATION ON HIGH-YIELDING RICE VARIETIES UNDER HIGH LEVELS OF FERTILIZATION HAS PROMPTED A STUDY ON TOLERANCE OF NITROGEN-FIXING BLUE-GREEN ALGAE TO SOME PESTICIDES. TWENTY-EIGHT STRAINS OF BLUE-GREEN ALGAE BELONGING TO FIVE DIFFERENT GENERA WERE TESTED AGAINST VARYING CONCENTRATIONS OF EIGHT PESTICIDES. THE DIFFERENCE IN GROWTH OF TREATED AND UNTREATED CULTURES WAS USED TO ASSESS THE EFFECT OF PESTICIDES. A LARGE PERCENTAGE OF THE STRAINS TESTED SHOWED A HIGH TOLERANCE LIMIT TO ALL PESTICODES APPLIED MANY TIMES MORE THAN THE RECOMMENDED DOSES OF APPLICATION. TOLYPOTHRIX TENUIS, A POWERFUL NITROGEN-FIXER, WIDELY USED IN INOCULATION TRIALS, GREW IN HIGH CONCENTRATIONS OF ALL PESTICIDES, EXCEPT DIURON. THE RELATED AULOSIRA FERTILISSIMA, WHILE TOLERATING HIGH LEVELS OF CERASAN, DIATHENE, 2,4-D AND DALAPON, WAS SENSITIVE TO THE SUBSTITUTED UREA COMPOUNDS LIKE COTORON, DIURON, AND LINURON AS WELL AS TO THE TRIAZINE HERBICIDE, PROPAZINE. THE NON-NITROGEN-FIXING ANACYSTIS NIDULANS TOLERATED HIGH CONCENTRATIONS OF ALL THE PESTICIDES EXAMINED. THE PRESENT INVESTIGATION INDICATES THAT THESE ALGAE CAN BE USED IN PRESENCE OF PESTICIDES AND ARE ABLE TO FUNCTION IN PRESENCE OF HIGH LEVELS OF NITROGENOUS FERTILIZERS. (JONES-WISCONSIN)

FIELD 05C

ACCESSION NO. W71-12071

ROLE OF PHOSPHORUS IN EUTROPHICATION,

WISCONSIN UNIV., MADISON. WATER CHEMISTRY PROGRAM.

G. FRED LEE.

PRESENTED AT THE SYMPOSIUM OF AMERICAN CHEMICAL SOCIETY, DIVISION OF WATER, AIR AND WASTE CHEMISTRY, LOS ANGELES, CALIFORNIA, APRIL 1970.

DESCRIPTORS:
*PHOSPHORUS, *EUTROPHICATION, FERTILIZATION, AQUATIC PLANTS, MATHEMATICAL MODELS, SELF-PURIFICATION, ALGAE, GREAT LAKES, NUTRIENTS, SEDIMENTS, LAKE MICHIGAN, WATER POLLUTION SOURCES, WATER POLLUTION CONTROL, TROPHIC LEVELS, PHOSPHATES, DETERGENTS, BIOASSAY, ANALYTICAL TECHNIQUES, FORECASTING, SILICA.

IDENTIFIERS:
FLUSHING, GREEN BAY(WIS), NITRILOTRIACETIC ACID, ALGAL ASSAY PROCEDURES, ALGAL GROWTH, IN-LAKE NUTRIENT CONTROL.

ABSTRACT:
SIGNIFICANCE OF PHOSPHORUS AS THE KEY ELEMENT IN EXCESSIVE FERTILIZATION OF NATURAL WATERS IS PRESENTED AND ITS ROLE ON PLANT GROWTH IN LAKES. TOOLS TO ASSESS PHOSPHORUS AND OTHER ELEMENTS FERTILIZING NATURAL WATERS ARE MATHEMATICAL MODELS, ENZYMATIC AND TISSUE ASSAY PROCEDURES. APPRAISAL OF NUTRIENT STATUS OF LAKES CAN BE MADE DURING FEBRUARY AND MARCH OF EACH YEAR IN TEMPERATE LAKES TO SHOW THE POTENTIAL PROBLEMS AND STEPS INITIATED TO CONTROL EXCESSIVE DISCHARGE OF NUTRIENTS. SEDIMENTS SERVE AS A SINK FOR PHOSPHORUS WITH THE NET FLUX OF PHOSPHORUS FROM LAKE WATER TO SEDIMENTS. CONTEMPLATED NUTRIENT REMOVAL PROJECTS SHOULD BE ASSOCIATED WITH SOME LABORATORY LEACHING TESTS ON LAKE SEDIMENTS AND RESULTS COMPARED WITH LAKE RECOVERY RATE UPON NUTRIENT REDUCTION. DEVELOPMENT OF MODELS FOR AQUEOUS ENVIRONMENTAL CHEMISTRY OF AQUATIC NUTRIENTS IN NATURAL WATERS IS ESSENTIAL TO IMPROVE THE PREDICTABILITY OF RELATIONSHIPS BETWEEN THE FLUX OF AQUATIC PLANT NUTRIENTS AND GROWTH OF ALGAE AND OTHER AQUATIC PLANTS. IN REPLACING PHOSPHATES IN DETERGENTS, THE PRIMARY PROBLEM IS THE PROCEDURE USED TO EVALUATE THE REPLACEMENT. (JONES-WISCONSIN)

FIELD 05C

ACCESSION NO. W71-12072

EUTROPHICATION AND THE CURRENT CONTROVERSY OVER ITS CAUSES AND CURES,

FISHERIES RESEARCH BOARD OF CANADA, WINNIPEG (MANITOBA). FRESHWATER INST.

A. L. HAMILTON.

PRESENTED AT 22ND ANNUAL CONVENTION, WESTERN CANADA WATER AND SEWAGE CONFERENCE, SEPTEMBER 23-25, 1970, HELD AT WINNIPEG, P 67-71, 1970. 1 FIG, 11 REF.

DESCRIPTORS:
*WATER POLLUTION CONTROL, *EUTROPHICATION, *WATER POLLUTION SOURCES, POPULATION, INDUSTRIES, WASTE DISPOSAL, MONITORING, NUTRIENTS, ALGAE, INSECTICIDES, HERBICIDES, HEAVY METALS, TASTE, ODOR, RECREATION, PHOSPHORUS, CARBON, LAKE ERIE, DETERGENTS, LAKE ONTARIO, TERTIARY TREATMENT, INTERNATIONAL JOINT COMMISSION.

IDENTIFIERS:
MERCURY, PHOSPHORUS REMOVAL, LAKE WASHINGTON(WASH).

ABSTRACT:
ACUTENESS OF THE WASTE DISPOSAL PROBLEM IS DUE TO BOTH POPULATION AND INDUSTRIAL GROWTH. WASTES EMPTIED INTO AQUATIC ENVIRONMENTS BELONG TO TWO BROAD GROUPS: NUTRIENTS WHICH STIMULATE LIFE AND TOXIC SUBSTANCES DEPRESSING LIFE. THE FIRST POLLUTANTS, GROWTH STIMULATORS, CONTRIBUTE TO EUTROPHICATION, THE MAJOR SYMPTOMS OF WHICH ARE ACCUMULATION OF ALGAE, REDUCED TRANSPARENCY, UNPLEASANT TASTE AND ODOR, OXYGEN DEFICITS, AND REDUCTION IN RECREATION POTENTIAL. MAJORITY OF LIMNOLOGISTS SUPPORT THE ARGUMENT THAT PHOSPHORUS IS THE KEY LIMITING NUTRIENT AND ITS INPUT TO LAKES CAN BE CONTROLLED MORE EFFECTIVELY THAN THAT OF OTHER NUTRIENTS. EFFICIENT AND RELATIVELY INEXPENSIVE METHODS CAN REMOVE PHOSPHORUS DURING SEWAGE TREATMENT, AND ELIMINATION OF PHOSPHATES FROM DETERGENTS WOULD LOWER COSTS OF REDUCING PHOSPHORUS LEVELS IN EFFLUENTS. THE SINGLE MOST CONVINCING PROGRAM OF PRACTICAL BENEFITS IN REDUCING PHOSPHORUS IS EXEMPLIFIED BY THE DIVERSION OF SEWAGE AROUND LAKE WASHINGTON, PROVIDING STRONG EVIDENCE THAT PHOSPHATES WERE THE LIMITING NUTRIENT AND A CONCLUSIVE DEMONSTRATION THAT EUTROPHICATION WAS, AT LEAST IN THIS INSTANCE, A REVERSIBLE PROCESS. VARIOUS ARGUMENTS FAVORING REDUCTION OF PHOSPHORUS INPUTS MAY NOT BE ENTIRELY CONCLUSIVE BUT COMBINED THEY CONSTITUTE A VERY STRONG CASE FOR TAKING MEANINGFUL REMEDIAL ACTIONS - NOW. (JONES-WISCONSIN)

FIELD 05C

ACCESSION NO. W71-12091

EFFECTS OF THERMAL DISCHARGES ON THE CHEMICAL PARAMETERS OF WATER QUALITY AND EUTROPHICATION,

WISCONSIN UNIV., MADISON. WATER CHEMISTRY PROGRAM.

FRED G. LEE, AND GILMAN D. VEITH.

PRE-PRINT: PROCEEDINGS INTERNATIONAL SYMPOSIUM ON THE IDENTIFICATION AND MEASUREMENT OF POLLUTANTS IN THE ENVIRONMENT, HELD AT OTTAWA, ONTARIO, CANADA, JUNE 1971. 27 P, 23 REF.

DESCRIPTORS:
*THERMAL POLLUTION, *CHEMICAL REACTIONS, *WATER QUALITY, *CHEMICAL PROPERTIES, EUTROPHICATION, WATER TEMPERATURE, WATER POLLUTION EFFECTS, WATER POLLUTION CONTROL, ENERGY, DIFFUSION, THERMODYNAMICS, SOLUBILITY, SORPTION, PESTICIDES, TOXICITY, DISSOLVED OXYGEN, LAKE MICHIGAN, ALGAE.

IDENTIFIERS:
*THERMAL DISCHARGES.

ABSTRACT:
DEPENDING ON THE HEAT ASSIMILATIVE CAPACITY OF THE RECEIVING WATER, THERMAL DISCHARGES ARE DIVIDED INTO TWO TYPES: THOSE ELEVATING WATER TEMPERATURE BY 5 TO 10C FOR ONLY A FEW HOURS, AND THOSE INCREASING THE TEMPERATURE FOR SEVERAL DAYS. THE LATTER DISCHARGES MAY IN CERTAIN INSTANCES LEAD TO DETERIORATION OF WATER QUALITY--INTENSIFY OBJECTIONABLE TASTE AND ODOR, INCREASE ALGAL BLOOMS, AND REDUCE DO CONCENTRATION. THE ADVERSE EFFECTS OF THERMAL DISCHARGES CAN PARTICULARLY BE EXPECTED WHEN THE TEMPERATURE IS INCREASED TO ABOUT 15 TO 20C IN NUTRIENT-ENRICHED WATERS. (WILDE-WISCONSIN)

FIELD 05C

ACCESSION NO. W71-12092

FACTORS INFLUENCING PHOSPHOROUS REMOVAL BY BIOLOGICAL TREATMENT,

FECERAL WATER QUALITY ADMINISTRATION, CINCINNATI, CHIO.

ROBERT L. BUNCH.

CHEMICAL ENGINEERING PROGRESS, SYMPCSIUM SERIES, VCL 67, NO 107, P 90-94, 1971. 3 TAB, 33 REF.

DESCRIPTORS:
*BIOLOGICAL TREATMENT, *PHOSPHOROUS, *WASTE WATER TREATMENT, NUTRIENTS, SYNTHESIS, SLUDGE, EFFLUENT, ALGAE, SECONDARY TREATMENT, BIOCHEMICAL OXYGEN DEMAND, CHEMICAL OXYGEN DEMAND, CARBON, FERTILIZATION, DISSOLVED OXYGEN, SEWAGE, HYDRCGEN ION CONCENTRATION.

IDENTIFIERS:
SUSPENDED SOLIDS.

ABSTRACT:
IN EWECENT YEARS THE PHOSPHATE CONTENT OF WASTE WATER TREATMENT PLANT EFFLUENTS HAS BEEN RISING, PRESUMABLY TRACEABLE TC THE USE OF PHOSPHATES IN DETERGENTS. CONVENTIONAL BIOLOGICAL TREATMENT DOES NOT EFFECTIVELY REMOVE NUTRIENTS FROM WASTE WATER. THE INFLUENCE OF AN INCREASED USE OF PHOSPHATES, DIGESTER RECYCLE, AND IMBALANCE OF NUTRIENTS IN WASTE WATER WAS RECENTLY STUDIED. PHCSPHOROUS REMOVAL BY SYNTHESIS CAN BE HIGHER THAN NORMALLY ENCOUNTERED IF DIGESTER SUPERNATANT IS NOT RECYCLED AND PRIMARY CLARIFICATION IS ELIMINATED. THE ELIMINATION OF THESE 2 PRACTICES WILL INCREASE THE CARBON TO PHOSPHOROUS RATIO, CORRECTING THE NUTRIENT IMBALANCE. IF PHOSPHOROUS IS TO BE REMOVED FROM WASTE WATER ON A SUSTAINEC BASIS, THEN CHEMICAL OR CHEMICAL-BIOLOGICAL METHODS MUST BE USED AND EFFCRTS MUST BE CONTINUALLY MADE TO IMPROVE THE DEPENDABILITY AND APPLICABILITY OF THE BIOLOGICAL PROCESS. (GUTIERREZ-TEXAS)

FIELD 05D

ACCESSION NO. W71-12188

PHOSPHATE REMOVAL BY FOAM FRACTIONATION,

GARRETT RESEARCH AND DEVELCPMENT CO., INC., LA VERNE, CALIF.

DONALD E. GARRETT.

CHEMICAL ENGINEERING PROGRESS, SYMPOSIUM SERIES, VOL 67, NO 107, P 296-303, 1971. 10 FIG, 4 TAB, 7 REF.

DESCRIPTORS:
*PHOSPHATES, *WASTE WATER TREATMENT, *FOAM FRACTICNATION, LIME, HYDROGEN ION CONCENTRATION, ECONOMICS, ALGAE, SOLUBILITY, SEWAGE, BIOCHEMICAL OXYGEN DEMAND, ICN EXCHANGE, CHEMICAL CXYGEN DEMAND, FERTILIZER, TURBIDITY.

IDENTIFIERS:
*FRACTIONATION-FLOTATION PROCESS.

ABSTRACT:
A PROCESS HAS BEEN DEVELOPED BY WHICH AIR AND WATER ARE CONTACTED AT HIGH PRESSURES, ADDITIVES ARE MIXED WITH THE SOLUTION, AND VERY FINELY DIVIDED BUBBLES THAT ARE FORMED ON THE RELEASE OF PRESSURE ARE SEPARATED. THE PROCESS ALSO INVOLVES A COMBINATION OF A NEW VARIATION OF THE FOAM FRACTIONATION PROCESS COUPLED WITH A PRECIPITATION STEP TO ALLOW EFFECTIVE REMOVAL OF PHOSPHATES IN AN INEXPENSIVE AND SIMPLE MANNER. THE ADDITION OF IRON AND LIME RESULT IN A COMPLETE, INEXPENSIVE SEPARATION OF PHOSPHATE FROM THE WATER TO BE TREATED. THE FOAM FRACTIONATION-FLOTATION PROCESS IS MORE THAN COMPETITIVE WITH CONVENTIONAL TREATMENT SCHEMES. THE FOAMING PROCESS CFFERS THE ADVANTAGE OF REQUIRING LESS LAND THAN CONVENTIONAL TREATMENT, WHICH IS A SIGNIFICANT FACTOR IN LAND-SHORT AREAS IN HELPING ALLEVIATE DIFFICULT PHOSPHATE REMOVAL PRCBLEMS. ITS GREATEST ACVANTAGES ARE ITS SIMPLICITY AND HIGH REMOVAL EFFICIENCY FOR ALL THE CONTAMINANTS, COD, TURBIDITY, AND PHOSPHATES. FLOWSHEETS AND ECONOMIC EVALUATICNS OF THE PROCESS WERE ALSO DISCUSSED IN THIS PAPER. (GUTIERREZ-TEXAS)

FIELD 05D

ACCESSION NO. W71-12200

TREATMENT OF WASTEWATER RESULTING FROM THE PRODUCTION OF POLYHYDRIC ORGANIC COMPOUNDS,

DOW CHEMICAL CO., FREEPORT, TEX.

M. A. ZEITOUN, W. F. MCLLHENNY, AND W. A. TABER.

CHEMICAL ENGINEERING PROGRESS, SYMPOSIUM SERIES, VOL 67, NO 107, P 495-503, 1971. 18 FIG, 2 TAB, 4 REF.

DESCRIPTORS:
*WASTE WATER TREATMENT, *BIOLOGICAL TREATMENT, *ACTIVATED SLUDGE, EFFLUENT, *ORGANIC WASTES, ECONOMICS, CHEMICAL OXYGEN DEMAND, BIOCHEMICAL OXYGEN DEMAND, AERATION, *OXIDATION, HYDROGEN ION CONCENTRATION, MICROORGANISMS, FERMENTATION, ALKALINE HYDROLYSIS, ALGAE BLOOMS, UTAH, GREAT SALT LAKE, ORGANIC COMPOUNDS, BRINES.

IDENTIFIERS:
FUSARIUM NO 83, BACTERIUM NO 52, ENDOGENOUS RESPIRATION, TOTAL OXYGEN DEMAND, GLYCOLS.

ABSTRACT:
UNDER A GRANT FROM THE FEDERAL WATER QUALITY ADMINISTRATION A DEVELOPMENT PROGRAM WAS UNDERTAKEN TO INVESTIGATE THE TREATMENT OF WASTE WATERS RESULTING FROM THE PRODUCTION OF POLYHYDRIC ORGANIC COMPOUNDS. THE DOW CHEMICAL COMPANY AS THE RECIPIENT OF THE GRANT HAS SUBCONTRACTED WITH TEXAS A AND M UNIVERSITY FOR SUPPORT OF INVESTIGATIONS ON BIOLOGICAL OXIDATION OF ORGANIC COMPOUNDS. FOUR OVERALL OBJECTIVES WERE SET UP: (1) TO DEVELOP A METHOD OF TREATMENT OF WASTES FROM A POLYHYDRIC MANUFACTURING PROCESSES; (2) EXAMINE SEVERAL ALTERNATIVE TREATMENT METHODS; (3) DETERMINE TECHNICAL, ENGINEERING AND ECONOMIC FEASIBILITY, AND (4) DEVELOP SUFFICIENT INFORMATION FOR THE CONCEPTUAL DESIGN OF A WASTE TREATMENT FACILITY. A BIOLOGICAL OXIDATION PROCEDURE USING AN ACTIVATED SLUDGE TECHNIQUE HAS BEEN ABLE TO REDUCE ORGANIC CONTENTS OF SODIUM CHLORIDE BRINES TO ACCEPTABLE LEVELS. A PILOT FACILITY OPERATING ON FULL STRENGTH WASTES HAS BEEN DESIGNED AND IS BEING USED TO OBTAIN DESIGN INFORMATION FOR A LARGE SCALE WASTE CONTROL UNIT. THE PROGRAM INVOLVED IS PRESENTLY STILL UNDERWAY. (GUTIERREZ-TEXAS)

FIELD 05D

ACCESSION NO. W71-12223

PROPERTIES OF TILE DRAINAGE WATER,

IOWA STATE UNIV., AMES. DEPT. OF AGRICULTURAL ENGINEERING.

T. L. WILLRICH.

PAPER NO 70-752 PRESENTED AT 1970 WINTER MEETING OF THE AMERICAN SOCIETY OF AGRICULTURAL ENGINEERS, DECEMBER 8-11, 1970, CHICAGO, ILLINOIS. 10 P, 9 TAB. OWRR PROJECT A-013-IA(4).

DESCRIPTORS:
*TILE DRAINAGE, *WATER PROPERTIES, *FERTILIZATION, *CULTIVATED LANDS, *IOWA, WATER QUALITY, FERTILIZERS, NUTRIENTS, NITROGEN COMPOUNDS, PHOSPHORUS COMPOUNDS, POTASH, ALGAE, AQUATIC PLANTS, CHEMICAL ANALYSIS, WATER POLLUTION SOURCES.

ABSTRACT:
A STUDY OF PLANT NUTRIENT CONTENT AND POLLUTIONAL CHARACTERISTICS OF IOWA TILE DRAINAGE WATER WAS CONDUCTED FROM JULY 1965 THROUGH JUNE 1969. PLANT NUTRIENTS AND SOIL AMENDMENTS ARE LEACHED FROM THE SOIL BY PERCOLATING WATER AND ARE CONVEYED TO RECEIVING STREAMS THROUGH TILE DRAINS WHEN DRAINAGE FLOW OCCURS. SOME OF THESE NUTRIENTS HAVE THE POTENTIAL TO DEGRADE WATER QUALITY TO THE EXTENT THAT BENEFICIAL USES OF THE WATER MAY BE IMPAIRED. THE CONCENTRATIONS OF NITROGEN (MEDIAN VALUES RANGED FROM 12.4 TO 27.4 MG/LITER) AND PHOSPHORUS (MEDIAN VALUES OF 0.2 TO 0.3 MG/LITER) FOUND IN THE TILE DRAINAGE WATER SAMPLES WERE SUFFICIENTLY HIGH TO BE CONDUCTIVE TO THE GROWTH OF ALGAE AND OTHER AQUATIC PLANTS. (WOODARD-USGS)

FIELD 05B, 02G, 02K

ACCESSION NO. W71-12252

DDT UPTAKE AND GROWTH OF EUGLENA GRACILIS,

NATIONAL RESEARCH COUNCIL OF CANADA, OTTAWA (ONTARIO). DIV. OF BIOLOGY.

H. W. DE KONING, AND D. C. MORTIMER.

BULLETIN OF ENVIRONMENTAL CONTAMINATION AND TOXICOLOGY, VOL 6, NO 3, P 244-248, 1971. 2 FIG, 2 TAB, 3 REF.

DESCRIPTORS:
*CHLORINATED HYDROCARBON PESTICIDES, *DDT, *PESTICIDE TOXICITY, *WATER POLLUTION EFFECTS, PESTICIDES, AGRICULTURAL CHEMICALS, PESTICIDE RESIDUES, PESTICIDE KINETICS, ANIMAL PHYSIOLOGY, PLANT PHYSIOLOGY, PHYSIOLOGICAL ECOLOGY, BIOASSAY, MICROORGANISMS, *EUGLENA, ALGAE.

ABSTRACT:
THE EFFECT OF DDT ON THE GROWTH RATE OF EUGLENA GRACILIS AND THE RATE OF DDT UPTAKE BY ACTIVELY GROWING E. GRACILIS CULTURES WERE EXAMINED. DDT SUPPRESSED GROWTH WHEN ADDED IN 1.0 ML OF ETHANOL, BUT HAD NO EFFECT WHEN ADDED ALONE OR IN 0.1 ML OF ETHANOL. THE INHIBITED CULTURES RECOVERED AFTER 3 DAYS, PROBABLY AFTER UTILIZATION OF THE ETOH. DDT ACCUMULATION HAS NO OBSERVABLE EFFECT ON CELL DIVISION, AND NO INTRACELLULAR DEGRADATION PRODUCTS OF DDT WERE DETECTED. (LEGORE-WASHINGTON)

FIELD 05C

ACCESSION NO. W71-12303

FUNDAMENTAL VARIATIONS IN THE WATER QUALITY WITH PERCOLATION IN INFILTRATION BASINS,

INSTITUTE FOR WATER RESEARCH LTD., DORTMUND (WEST GERMANY).

W. H. FRANK.

PAPER NO 7 OF ARTIFICIAL GROUNDWATER RECHARGES CONFERENCE, UNIVERSITY OF READING, ENGLAND, SEPTEMBER 21-24, 1970: THE WATER RESEARCH ASSOCIATION, MARLOW, ENGLAND. 22 P, 25 FIG.

DESCRIPTORS:
*ARTIFICIAL RECHARGE, *INDUCED INFILTRATION, *PIT RECHARGE, *WATER TREATMENT, *FILTATION, ALLUVIA CHANNELS, SURFACE-GROUNDWATER RELATIONSHIPS, PERMEABILITY, ALGAE, DISSOLVED OXYGEN.

IDENTIFIERS:
*RUHR VALLEY.

ABSTRACT:
THE DORTMUNDER STADTWERKE AG SUPPLIES ABOUT 100 MILLION CU M OF MUNICIAL AND INDUSTRIAL WATER PER YEAR FROM THE GROUNDWATER COLLECTED IN THE ALLUVIUM OF THE RUHR VALLEY. THE AQUIFER IS 4 TO 5 M DEEP AND IS COVERED BY A LAYER OF MEADOW LOAM 0.5 TO 2 M THICK. THE PERMEABILITY OF THE GRAVEL IS IN GENERAL VERY GOOD, ABOUT .001 TO 0.01 M/SEC. MORE THAN 90% OF THE GROUNDWATER CONSISTS OF VARYING PROPORTIONS OF BANK INFILTRATION AND ARTIFICIALLY RECHARGED GROUNDWATER. IN DRY YEARS WITH LITTLE WATER FLOW, MORE THAN 75% OF THE REQUIRED WATER MUST BE ARTIFICIALLY RECHARGED. THE INFILTRATION TAKES PLACE IN BASINS MEASURING 25 X 200 M. A LAYER OF SAND 50 TO 70 CM THICK IS LAID DOWN AS A FILTER BED. THIS SAND HAS AN EFFECTIVE PARTICLE DIAMETER (D 10%) OF 0.12 MM, AND AT (D 60%) OF 0.3 MM A COEFFICIENT OF NON-UNIFORMITY OF U = 2.5. ITS PERMEABILITY IN MOST AREAS IS LESS THAN THAT OF THE SUBSOIL. BOTH THE BANK INFILTRATION AND THE ARTIFICIALLY RECHARGED GROUNDWATER ARE RE-COLLECTED AT A DEPTH OF 7 M AFTER A PASSAGE OF AT LEAST 50 M THROUGH THE SOIL AND PASSED DIRECTLY INTO THE MAINS WITHOUT FURTHER TREATMENT AFTER A LIGHT PRECAUTIONARY CHLORINATION. SUSPENDED ORGANIC AND INORGANIC MATTER IS LARGELY RETAINED ON THE FILTER SURFACE. TOGETHER WITH THE ALGAE WHICH GROW THERE, THEY CAUSE AN OBSTRUCTION IN THE FILTER. THE DISSOLVED ORGANIC MATERIAL IN THE WATER IS MINERALIZED WITH THE HELP OF THE BACTERIA WHICH INHABIT THE FILTER, OXYGEN BEING CONSUMED AND CARBON DIOXIDE LIBERATED. IN THE WATER ABOVE AN EXPOSED SAND FILTER, ALGAE CAUSE A DISTINCT DROP IN THE PHOSPHATE ION CONCENTRATION. IN THE UPPER MOST LAYERS OF SAND MORE PHOSPHATE IONS ARE LIBERATED BY MICROBIOL DECOMPOSITION THAN THE BACTERIA NEED FOR THEIR METABOLISM. (KNAPP-USGS)

FIELD 04B, 03A, 05F

ACCESSION NO. W71-12410

PHYSICAL ASPECTS OF ELECTROFILTRATION,

JADAVPUR UNIV., CALCUTTA (INDIA). DEPT. OF PHYSICAL CHEMISTRY.

S. P. MOULIK,

ENVIRONMENTAL SCIENCE AND TECHNOLOGY, VOL 5, NO 9, P 771-776, SEPTEMBER 1971.
7 FIG, 2 TAB, 19 REF.

DESCRIPTORS:
*WATER PURIFICATION, *FILTRATION, *ELECTROCHEMISTRY, *WATER POLLUTION TREATMENT, ANALYTICAL TECHNIQUES, SEPARATION TECHNIQUES, SEDIMENTS, OILY WATER, ALGAE.

IDENTIFIERS:
*ELECTROFILTRATION.

ABSTRACT:
THE PRIMARY PHYSICAL PHENOMENA INVOLVED IN ELECTROFILTRATION AND A THEORY FOR THE PROCESS ARE DISCUSSED. ELECTROPHORETIC FILTRATION WAS EFFECTIVE IN SUSPENSION CLARIFICATION, ESPECIALLY FOR WATER AND WASTE PURIFICATION. SYSTEMS SUCH AS BENTONITE CLAY SUSPENSIONS, SUSPENSIONS OF PURE CULTURE OF ALGAE, OIL-IN-WATER EMULSION, RIVER WATER, ETC. WERE CHOSEN AS REPRESENTATIVE MODELS FOR COMMONLY OCCURRING POLLUTED WATER SAMPLES. THE ELECTROFILTRATION PROCESS INCLUDES ELECTROPHORESIS AND FILTRATION. THE FILTER-CAKE BUILDUP ENCOUNTERED IN NORMAL FILTRATION CAN BE COMPLETELY AVOIDED IN THIS METHOD. CONSIDERATIONS OF FILTER MEDIUM ELECTRO-OSMOSIS, FILTER-CAKE ELECTRO-OSMOSIS, AND PARTICLE ELECTROPHORESIS HAVE MODIFIED THE FILTRATION EQUATION OF SPERRY TO PREDICT ELECTROFILTRATION RESULTS. FOR EFFICIENT ELECTROFILTRATION A CRITICAL VOLTAGE MUST BE ATTAINED. THIS VOLTAGE COMPLETELY PREVENTS THE FILTRATION HINDRANCE OF A DEPOSITED FILTER-CAKE. (WOODARD-USGS)

FIELD 05D, 03A

ACCESSION NO. W71-12467

SHORELINE ALGAE OF WESTERN LAKE ERIE,

RACHEL COX DOWNING.

OHIO J SCI. 70(5): 257-276. ILLUS. MAPS. 1970 (RECD. 1971).

DESCRIPTORS:
*ALGAE, *CHLOROPHYTA, *CYANOPHYTA, *LAKES, RHODOPHYTA, SHORES.

IDENTIFIERS:
ARNOLDIELLA-CONCHOPHILA, LAKE ERIE.

ABSTRACT:
THE ALGAE OF WESTERN LAKE ERIE HAVE BEEN EXTENSIVELY STUDIED FOR MORE THAN 70 YR, BUT, UNTIL THIS STUDY, CONDUCTED BETWEEN APRIL AND OCT., 1967; ALMOST NOTHING WAS KNOWN OF THE SHORELINE AS A SPECIFIC ALGAL HABITAT. A TOTAL OF 61 TAXA WERE IDENTIFIED FROM THE SHORELINES. THE (23 CHLOROPHYTA, 37 CYANOPHYTA, 1 RHODOPHYTA) FOUND 39 ARE NEW RECORDS FOR WESTERN LAKE ERIE, AND ONE, ARNOLDIELLA CONCHOPHILA MILLER, APPEARS TO BE A NEW USA RECORD, HAVING BEEN PREVIOUSLY REPORTED ONLY FROM CENTRAL RUSSIA.--COPYRIGHT 1971, BIOLOGICAL ABSTRACTS, INC.

FIELD 02H, 05C

ACCESSION NO. W71-12489

PRIMARY PRODUCTION AND STANDING CROPS OF EPIPSAMMIC AND EPIPELIC ALGAE,

BRISTOL UNIV. (ENGLAND). DEPT. OF BOTANY.

M. HICKMAN, AND F. E. ROUND.

BRITISH PHYCOLOGICAL JOURNAL, VOL 5 NO 2, P 247-255, 1970. 4 FIG, 6 TAB, 14 REF.

DESCRIPTORS:
*PRIMARY PRODUCTIVITY, *STANDING CROP, *ALGAE, DIATOMS, MEASUREMENT, DETRITUS, METHODOLOGY, SEDIMENTS, LAKES, CHLOROPHYLL, HYDROGEN ION CONCENTRATION, CARBON DIOXIDE.

IDENTIFIERS:
*EPIPSAMMIC ALGAE, *EPIPELIC ALGAE, SHEAR WATER(ENGLAND), AMPHORA OVALIS, OPEPHORA MARTYI, CELL COUNTS, NAVICULA, NITZSCHIA, STEPHANODISCUS, MICROCYSTIS, APHANIZOMENON, ANABENA SPIROIDES.

ABSTRACT:
A METHOD DEVISED FOR DETERMINING PRIMARY PRODUCTION INVOLVED AN ADDITIONAL STEP OF REMOVAL OF EPIPSAMMIC ALGAE FROM THE SAND GRAINS. THE EPIPELIC ALGAE WERE ALSO SEPARATED FROM THE SEDIMENT. SINCE THE EXPERIMENTS WERE CARRIED OUT IN THE LABORATORY AND NOT IN SITU, PRIMARY PRODUCTION IS DESIGNATED 'POTENTIAL' PRODUCTION. THE 'POTENTIAL' PRIMARY PRODUCTION OF THE EPIPELON AND EPIPSAMMON OF SHEAR WATER (ENGLAND) HAS BEEN ESTIMATED AND RELATED TO STANDING CROP PARAMETERS OVER A THIRTY MONTH PERIOD. ALL RESULTS ARE EXPRESSED PER UNIT AREA OF HABITAT TO BE COMPARABLE WITH RESULTS OF OTHER WORKERS. THE EPIPSAMMIC ALGAE COMPRISE MAINLY SMALL DIATOMS. THE EPIPSAMMON CAN BE SEPARATED FROM THE EPIPELON AND DETRITUS BY REPEATEDLY SWIRLING THE SEDIMENT WITH LAKE WATER FILTERED THROUGH WHATMAN GLASS FIBRE FILTER PAPER REMOVING PHYTOPLANKTONIC PRIMARY PRODUCERS AND DETRITAL MATERIAL. DATA HAVE ALSO BEEN OBTAINED FOR THE 'POTENTIAL PRODUCTION' OF THE FLORA IN SHORT CORES FROM THE SAME SITE. STANDING CROP AND PRODUCTION OF THE EPIPSAMMON WAS ALWAYS GREATER THAN THAT OF THE EPIPELON. CORRELATION BETWEEN STANDING CROP AND PRODUCTION IS CLOSER FOR THE EPIPELON THAN FOR THE EPIPSAMMON OWING TO RETENTION OF DEAD CELLS IN THE LATTER COMMUNITY. (JONES-WISCONSIN)

FIELD 05C

ACCESSION NO. W71-12855

LAKE EUTROPHICATION: A LABORATORY INVESTIGATION,

ONTARIO WATER RESOURCES COMMISSION, TORONTO. DIV. OF RESEARCH.

S. A. BLACK.

ONTARIO WATER RESOURCES COMMISSION, DIVISION OF RESEARCH PAPER NO 2026, 1970. 26 P, 3 FIG, 5 TAB.

DESCRIPTORS:
*LABORATORY TESTS, *EUTROPHICATION, *NUTRIENTS, *ALGAE, SEWAGE, PRODUCTIVITY, WATER POLLUTION EFFECTS, PHOSPHORUS, NITROGEN, AMMONIA, CARBON, CHLORELLA.

IDENTIFIERS:
*LIMITING NUTRIENTS, CANADA.

ABSTRACT:
WATER FROM AN OLIGOTROPHIC LAKE WAS SEEDED TO NATIVE ALGAE AND PLACED INTO FOUR 35-LITER PLEXIGLASS CONTAINERS. THE CULTURES WERE FED WITH A MIXTURE OF LAKE WATER, DISTILLED WATER, AND VARIOUSLY TREATED SEWAGE, AND SUBSEQUENTLY ENRICHED IN DISODIUM PHOSPHATE AND SODIUM CARBONATE HYDROGEN. ANALYSES OF THE MEDIA AND ALGAL COUNTS INDICATED THAT 0.015 MG/L OF P AND 0.5 MG/L OF N ARE REQUIRED TO PRODUCE GROWTH RESPONSE OF ALGAE IN THE LAKE INVESTIGATED. CARBON BELOW 8 MG/L INHIBITED THE ALGAL GROWTH. AMMONIA WAS USED BY CHLORELLA IN PREFERENCE TO NITRATES. REMOVAL OF PHOSPHORUS FROM SEWAGE WILL GREATLY REDUCE THE GROWTH EFFECT PRODUCED BY CONVENTIONALLY TREATED ACTIVATED SLUDGE. (WILDE-WISCONSIN)

FIELD 05C

ACCESSION NO. W71-12857

GROWTH OF CHLORELLA IN RELATION TO BORON SUPPLY,

ARGONNE NATIONAL LAB., ILL. DIV. OF BIOLOGICAL AND MEDICAL RESEARCH.

LANDY MCBRIDE, WILLIAM CHORNEY, AND JOHN SKOK.

BOTANICAL GAZETTE, VOL 132, NO 1, P 10-13, 1971. 4 TAB, 18 REF.

DESCRIPTORS:
*ALGAE, *CHLORELLA, *BORON, LABORATORY TESTS, CYTOLOGICAL STUDIES, PLANT GROWTH, SCENEDESMUS.

IDENTIFIERS:
NOSTOC, CYLINDROTHECA.

ABSTRACT:
THE GROWTH OF CHLORELLA VULGARIS, COLUMBIA STRAIN, EITHER IN LOW-BORON OR IN PURIFIED MEDIA WAS NOT INFLUENCED BY BORON CONCENTRATION WITHIN THE RANGE OF 0.001 TO 10.0 MG/L. NO CORRELATION WAS DETECTED BETWEEN THE TRACES OF BORON IN THE ALGAL CELLS AND THE CONCENTRATION OF THE ELEMENT IN THE NUTRIENT SOLUTION. (WILDE-WISCONSIN)

FIELD 05C

ACCESSION NO. W71-12858

ASSESSMENT OF POLLUTION EFFECTS BY THE USE OF ALGAE,

LIVERPOOL UNIV. (ENGLAND). HARTLEY BOTANICAL LABS.

ELSIE M. BURROWS.

PROCEEDINGS OF THE ROYAL SOCIETY OF LONDON, B, VOL 177, P 295-306, 1971. 5 FIG, 22 REF.

DESCRIPTORS:
*ALGAE, *WATER POLLUTION EFFECTS, *INDICATORS, *SEA WATER, MARINE PLANTS, ASSESSMENTS, MONITORING, SEWAGE, ESTUARIES, GROWTH RATES, BIOLOGICAL COMMUNITIES.

IDENTIFIERS:
*ULVA, *LAMINARIA, LIVERPOOL BAY(ENGLAND), IRISH SEA.

ABSTRACT:
INCREASES IN THE DIAMETER OF THALLUS DISCS OF ULVA AND IN THE SURFACE AREA OF LAMINARIA ALGAE SERVED AS THE CRITERION FOR APPRAISAL OF SEWAGE POLLUTION. RELATIVELY SIMPLE CULTURE TRIALS WITH SAMPLES OF PURE AND POLLUTED WATER FROM THE IRISH SEA DISCLOSED THAT ULVA LACTUA AND LAMINARIA SACCHARINA HAVE A HIGH POTENTIAL AS INDICATORS OF POLLUTION EFFECTS BECAUSE OF THEIR EASE OF CULTIVATION AND SENSITIVITY TO CHANGES IN WATER COMPOSITION, PARTICULARLY THOSE CAUSED BY SEWAGE DISCHARGES. THE BEHAVIOR OF DIFFERENT ALGAE SPECIES WITH THEIR ALTERNATION OF GENERATION OFFERS WIDE POSSIBILITIES FOR ASSESSING BOTH THE EUTROPHYING AND TOXIC EFFECTS OF URBAN AND INDUSTRIAL POLLUTION. (WILDE-WISCONSIN)

FIELD 05C

ACCESSION NO. W71-12867

STATISTICAL MODEL FOR THE SELECTION OF UNICELLULAR ALGAL STRAINS BY THE RATE OF INCREASE IN NUMBERS (STATISTICHESKAYA MODEL OTBORA SHTAMMOV ODNOKLETOCHNYKH VODOROSLEI PO SKOROSTI UVELICHENIYA CHISLENNOSTI),

AGROFIZICHESKII NAUCHNO-ISSLEDOVATELSKII INSTITUT, LENINGRAD (USSR).

I. A. SHVYTOV.

ENGLISH SUMMARY, ZHURNAL OBSHCHEI BIOLOGII, VOL 32, NO 3, P 260-265, 1971. 2 TAB, 5 REF.

DESCRIPTORS:
*MODEL STUDIES, *STATISTICAL METHODS, *ALGAE, CYTOLOGICAL STUDIES, SELECTIVITY, PRODUCTIVITY.

IDENTIFIERS:
ALGAL GROWTH.

ABSTRACT:
THE GENETIC SELECTION AIMS TO ISOLATE CELLS FROM A LARGE PHENOTYPIC POPULATION THAT BELONG TO A DEFINITE GENOTYPE OF A HIGH PRODUCTIVE POTENTIAL. THE INTRODUCED STATISTICAL MODEL FOR THE COMPETITIVE SELECTION OF TWO-COMPONENT MICROALGAL POPULATIONS PREDICTS THAT DURING A LONG PERIOD ONE OF THE COMPONENTS WILL UNDERGO DEGENERATION. THE GIVEN FORMULAS DELINEATE THE PROBABILITIES OF ELIMINATION OF STRAINS IN ACCORDANCE WITH THEIR SPECIFIC RATE OF GROWTH AND THE INITIAL DENSITIES. A PREDOMINANCE OF CELLS OF INFERIOR STRAINS MAY CAUSE DEGENERATION OF THE RAPIDLY GROWING CELLS. (WILDE-WISCONSIN)

FIELD 07B, 05C

ACCESSION NO. W71-12868

THE PHYTOPLANKTON OF MINNESOTA LAKES - A PRELIMINARY SURVEY,

MINNESOTA UNIV., MINNEAPOLIS. WATER RESOURCES RESEARCH INST.

A. J. BROOK.

MINNESOTA WATER RESOURCES RESEARCH CENTER, BULLETIN 36, JUNE 1971. 12 P, 1 TAB, 16 REF. OWRR PROJECT A-007-MINN(2).

DESCRIPTORS:
*PHYTOPLANKTON, *EUTROPHICATION, *OLIGOTROPHY, MINNESOTA, LAKES, GLACIAL DRIFT, NUTRIENTS, SALANITY, ALGAE, DIATOMS.

IDENTIFIERS:
*DESMIDS, *EUPLANKTON, *LIMNOPLANKTON.

ABSTRACT:
BETWEEN THE YEARS 1965 AND 1967, PHYTOPLANKTON COLLECTIONS WERE TAKEN IN THE SUMMER FROM NEARLY 200 LAKES IN A DIVERSITY OF AREAS THROUGHOUT THE STATE. ABOUT 220 TAXA OF EUPLANKTONIC ALGAE WERE IDENTIFIED. THE ANALYSIS OF THE MINNESOTA LAKE PHYTOPLANKTON INDICATES THERE IS A DIMINUTION IN SPECIES DIVERSITY IN THE COURSE OF THE EVOLUTIONARY PROGRESSION AS LAKES CHANGE IN CHARACTER FROM OLIGOTROPHY TO EUTROPHY. MANY OF THE MARKEDLY EUTROPHIC LAKES HAVE SUFFERED SEVERE DISTURBANCE OF THE NATURAL SYSTEM DUE TO ARTIFICIAL ENRICHMENT. EUTROPHIC LAKES IN MINNESOTA ARE TYPICALLY DOMINATED IN SUMMER AND EARLY FALL BY WATER BLOOMS OF BLUE GREEN ALGAE AS IS USUAL IN MOST PRODUCTIVE LAKES OF TEMPERATE REGIONS. MICROCYSTIS AERUGINOSA, M. WESENBERGII, COELOSPHAERIUM NAEGELIANIUM, ALPHANIZOMENON FLOS AQUAE, NUMEROUS SPECIES OF ANABAENA, LYNGBYA BIRGEI, AND GLOEOTRICHIA ECHINULATA ARE MOST COMMON. (WALTON-MINNESOTA)

FIELD 05C, 05A, 02H

ACCESSION NO. W71-13149

REVIEW OF NATIONAL RESEARCH POLICY ON EUTROPHICATION PROBLEMS,

WATER POLLUTION RESEARCH LAB., STEVENAGE (ENGLAND).

A. L. DOWNING.

JOURNAL OF THE SOCIETY FOR WATER TREATMENT AND EXAMINATION, VOL 19, PART 3, P 223-238, 1970. DISCUSSION.

DESCRIPTORS:
*EUTROPHICATION, *ALGAL CONTROL, FINANCING, PLANT GROWTH, TOXICITY, NUTRIENTS, WATER QUALITY, NITRATES, PLANNING, NITROGEN, PHOSPHORUS, RESERVOIRS, FISHERIES, WATER SUPPLY, COSTS, WATER DEMAND, ECONOMIC JUSTIFICATION, AQUATIC WEED CONTROL, RIVERS, AGRICULTURE, ROOTED AQUATIC PLANTS, DRAINAGE, RUNOFF, FERTILIZERS, LIVESTOCK, ECONOMICS, FISH KILLS, INHIBITORS, WATER POLLUTION SOURCES, HUMAN DISEASES.

IDENTIFIERS:
*RESEARCH POLICY, *UNITED KINGDOM, *FUTURE TRENDS, CLADOPHORA, RESEARCH STRATEGY, THAMES RIVER(ENGLAND), LEE RIVER(ENGLAND).

ABSTRACT:
FOR NEW WATER SUPPLIES AND FOR GREATER RECREATIONAL EXPLOITATION OF NATURAL WATERS IN THE UNITED KINGDOM, APPROPRIATE LEVELS OF ACTIVITY AND LINES OF INQUIRY ARE NEEDED. COST INCURRED AS A RESULT OF EUTROPHICATION WILL PROBABLY NOT INCREASE TO MORE THAN ABOUT DOUBLE THE PRESENT EXPENDITURES BY THIS CENTURY'S END. IF ALGAL AND WEED GROWTH WERE FULLY UNDERSTOOD, PERHAPS SOME COMPARATIVELY SIMPLE PREVENTIVE MEASURES COULD BE APPLIED FOR ELIMINATING THESE NUISANCES. A BALANCED PROGRAM IS REQUIRED FOR BASIC RESEARCH AND FOR EMPIRICAL INVESTIGATIONS, ESPECIALLY ON STATIC WATERS. SINCE FACTORS INFLUENCING ALGAL GROWTH VARY GEOGRAPHICALLY, EXAMINING THE INFLUENCES OF LOCATION OF WATER MAY SHOW THAT ONE METHOD OF CONTROL MAY BE MORE APPROPRIATE THAN ANOTHER. THOUGH IT IS UNLIKELY ELIMINATION OF PHOSPHATES FROM DETERGENTS WOULD MATERIALLY REDUCE ALGAL PROBLEMS, SEARCH FOR TROUBLE-FREE SUBSTITUTES SHOULD CONTINUE. PROCESSES USED IN REMOVING NITROGEN AND PHOSPHORUS FROM EFFLUENTS SHOULD BE EXAMINED FOR SUITABILITY OF REMOVING OTHER SUBSTANCES (CARBON ADSORPTION, OZONIATION), AND ABILITY OF EFFLUENTS TO SUPPORT ALGAL GROWTH BY UNSUSPECTED COMPONENTS IS PROFOUNDLY IMPORTANT. GROWTH OF WEEDS IN RIVERS AND EFFECT OF SEWAGE EFFLUENT ON CLADOPHORA NEEDS INVESTIGATION. INSURING THAT PRESENT EFFORTS ARE WELL COORDINATED TAKES PRECEDENCE OVER EMBARKING ON MANY NEW INITIATIVES. (JONES-WISCONSIN)

FIELD 05C, 06B

ACCESSION NO. W71-13172

PRIMARY PRODUCTION,

FRESHWATER BIOLOGICAL ASSOCIATION, AMBLESIDE (ENGLAND).

J. W. G. LUND.

JOURNAL OF THE SOCIETY FOR WATER TREATMENT AND EXAMINATION, VOL 19, PART 4, P 332-358, 1970. 6 FIG, 1 TAB, 36 REF, DISCUSSION.

DESCRIPTORS:
*WATER QUALITY, *EUTROPHICATION, *PRIMARY PRODUCTIVITY, *ALGAE, NITROGEN, PHOSPHORUS, THERMAL STRATIFICATION, WATER SUPPLY, SUCCESSION, ECONOMICS, NUTRIENTS, NITROGEN FIXATION, FORECASTING, PHOTOSYNTHESIS, BIOMASS, NITRATES, RESERVOIRS, LIGHT PENETRATION, CYANOPHYTA, CHRYSOPHYTA.

IDENTIFIERS:
*UNITED KINGDOM, ESTHWAITE WATER(ENGLAND), BLELHAM TARN(ENGLAND), LUXURY UPTAKE, LOCH LEVEN(SCOTLAND), LAKE WINDERMERE(ENGLAND).

ABSTRACT:
EUTROPHICATION IS NOT YET A MAJOR ECONOMIC PROBLEM IN THE BRITISH WATER INDUSTRY BUT COULD BECOME ONE AND FURTHER SUPPORT FOR ECOLOGICAL INVESTIGATIONS IS NEEDED. THE MAJOR NUTRIENTS, NITROGEN AND PHOSPHORUS, ARE STUDIED WITH REFERENCE TO ECOLOGICAL FACTORS CONTROLLING ALGAL POPULATION SIZE AND PRODUCTION RATES. TYPICALLY, THE DEGREE TO WHICH THE NUTRIENT CONTENT OF A RESERVOIR APPROACHES THAT OF WINTER INFLOWS IS AFFECTED BY RETENTION TIME AND NUMBER OF ALGAE PRESENT. PRODUCTION RATE IS A GUIDE TO HOW TROUBLES MAY INCREASE AND TO DANGERS AHEAD WHEN ALGAE ARE RELATIVELY UNCOMMON. THE STRONG CHEMICAL STRATIFICATION ACCOMPANYING THERMAL STRATIFICATION IN EUTROPHIC LAKES MAKES IT UNCERTAIN HOW GREAT THE NUTRIENT SUPPLY CAN BE TO ALGAE IN THE EPILIMNION OR EUPHOTIC ZONE EVEN IN ABSENCE OF ACTIVE VERTICAL MIGRATION. THERMAL STRATIFICATION MAY BE ADVANTAGEOUS OR DISADVANTAGEOUS TO WATER SUPPLY. BIOASSAY IS A USEFUL METHOD FOR COMPARING POTENTIAL FERTILITIES OF WATERS AND FOR ELUCIDATING SPECIFIC, LIMITED PROBLEMS. OTHER NUTRIENTS ARE IMPORTANT BESIDES NITROGEN AND PHOSPHORUS. FORECASTING DEVELOPMENT OF ALGAL POPULATIONS AND SEASONAL SPECIES' SUCCESSIONS IS POSSIBLE IN GENERAL TERMS BUT DETAILED FORECASTING OVER RELATIVELY LONG PERIODS IS NOT YET POSSIBLE. (JONES-WISCONSIN)

FIELD 05C

ACCESSION NO. W71-13183

THE NUISANCE ALGAE: CURIOSITIES IN THE BIOLOGY OF PLANKTONIC BLUE-GREEN ALGAE,

WESTFIELD COLL., LONDON (ENGLAND). DEPT. OF BOTANY.

A. E. WALSBY.

JOURNAL OF THE SOCIETY FOR WATER TREATMENT AND EXAMINATION, VOL 19, PART 4, P 359-373, 1970. 2 FIG, 1 TAB, 43 REF, DISCUSSION.

DESCRIPTORS:
*NUISANCE ALGAE, *CYANOPHYTA, *ALGAL CONTROL, LABORATORY TESTS, BUOYANCY, EUTROPHICATION, NIRTROGEN FIXATION, BIOCONTROL, PLANT PHYSIOLOGY, MOVEMENT, VIRUSES, PHOTOSYNTHESIS, TURBULENCE, LIGHT INTENSITY.

IDENTIFIERS:
*GAS VACUOLES, MOTILE FORMS, CYANOPHAGES.

ABSTRACT:
BLUE-GREEN ALGAE, QUITE DIFFERENT FROM ALL OTHER ALGAL GROUPS, HAVE CERTAIN CHARACTERISTICS WHICH CONTRIBUTE TO FORMATION OF BLOOMS; ONE IS THE ABILITY TO FIX ATMOSPHERIC NITROGEN. THEY POSSESS GAS FILLED VACUOLES WHICH PROVIDE A BUOYANCY-REGULATING MECHANISM ENABLING THEM TO OCCUPY WATER LAYERS BELOW THE SURFACE WHERE CONDITIONS ARE OPTIMAL FOR GROWTH, BUT ALSO CAUSE THICK ALGAL SCUMS ON THE SURFACE OF LAKES AND RESERVOIRS. ON THE CREDIT SIDE THEY PRODUCE OXYGEN WHICH KEEPS THE WATER IN THE AEROBIC CONDITION ESSENTIAL FOR PREVENTING GROWTH OF ANAEROBIC BACTERIA AND FOR SUPPORTING FISH AND OTHER ANIMAL POPULATIONS. THEY TAKE UP INORGANIC NUTRIENTS IN GROWTH AND CLEANSE WATER OF NITRATE AND OTHER SUBSTANCES, POTENTIALLY TOXIC. ON THE DEBIT SIDE, THEY CONTRIBUTE TO THE SUSPENDED SOLIDS WHICH MUST BE REMOVED BY FILTERING; THEIR BREAKDOWN PRODUCTS, WHICH PASS FILTERING PROCESSES, MAY PRODUCE TASTES AND ODORS IN TREATED WATER. BEFORE BREAKDOWN OCCURS, SOLUBLE EXTRACELLULAR PRODUCTS, SUCH AS MUCOPOLYSACCHARIDES, OCCASIONALLY CAUSE NUISANCE. SOME PRODUCE TOXIC SUBSTANCES. METHODS OF DESTROYING GAS VACUOLES OR THEIR EFFECTS WOULD PROVIDE MEANS OF CONTROLLING ALGAL BLOOMS; ANOTHER METHOD MIGHT BE TO USE CYANOPHAGES, VIRUSES WHICH ATTACK THESE ALGAE SELECTIVELY. (JONES-WISCONSIN)

FIELD 05C

ACCESSION NO. W71-13184

EUTROPHICATION IN RELATICN TO WATER SUPPLIES,

BOLTON INST. OF TECH. (ENGLAND)

E. G. BELLINGER.

JOURNAL OF THE SOCIETY FOR WATER TREATMENT AND EXAMINATION, VOL 19, PART 4, P 400-409, 1970. 1 FIG, 3 TAB, 21 REF, DISCUSSION.

DESCRIPTORS:
*EUTROPHICATION, *WATER SUPPLY, WATER POLLUTION EFFECTS, NUTRIENTS, RESERVOIR OPERATION, WATER REUSE, POTABLE WATER, ALGAE, NUTRIENT REQUIREMENTS, CYANOPHYTA, CHLAMYDOMONAS.

IDENTIFIERS:
*UNITED KINGDOM, ESTHWAITE WATER(ENGLAND), LAKE WINDERMERE(ENGLAND), GREAT OUSE(ENGLAND), ALGAL GROWTH.

ABSTRACT:
THE GREASTEOT INCREASES IN EUTROPHICATION HAVE TAKEN PLACE IN BRITIAN SINCE 1910 AND ARE CORRELATED WITH POPULATION GRCWTH. DISSOLVED SOLIDS AND PHOSPHATE CONCENTRATIONS HAVE ALSO INCREASED SIGNIFICANTLY COUPLED WITH BIOLOGICAL PRODUCTIVITY--WITH ALGAE THE MAIN OFFENDERS. REMOVAL OF NUTRIENTS BY VARIOUS SEWAGE TREATMENT PROCESSES WCULD NOT BE LOW ENOUGH TO PREVENT ALGAL GROWTH, BUT SUFFICIENTLY LOW TO SLOW DOWN ARTIFICIAL EUTROPHICATION. THE EFFECTS OF EUTROPHICATION ANC CONSEQUENT DIATOM GROWTHS AND BLUE-GREEN ALGAE CAUSE PROBLEMS IN WATERWORKS AND UNPLEASANT OLFACTORY RESPONSES TC POTABLE WATER AS WELL AS ANAEROBIC CONDITIONS DELETERIOUS TO WATER SUPPLIES. RESERVCIR MANAGEMENT SHOULD PROVIDE AT THE OUTLET TOWER FOR A NUMBER OF ABSTRACTION POINTS AT DIFFERENT DEPTHS TO PERMIT USE OF THE BEST QUALITY WATER AVAILABLE. CONTROL OF THE THERMCCLINE LEVEL BY PUMPING, THUS INCREASING THE DEPTH OF THE EPILIMNION, WOULD PROVIDE A GREATER DEPTH CF BETTER QUALITY WATER. INASMUCH AS ALGAE IS LIGHT DEPENDENT, PLASTIC SHEETING TO EXCLUDE LIGHT AND INCREASING TURBIDITY BY PUMPING AND JETTING WOULD REDUCE LIGHT PENETRATION. IT SEEMS CERTAIN THAT MCRE AND MCRE WASTEWATERS WILL HAVE TO BE REUSED AND THAT WASTEWATER SHOULD BE CONSIDERED AS AN INTEGRAL PART OF THE NATICN'S RESOURCES.
(JONES-WISCONSIN)

FIELD 05F

ACCESSION NO. W71-13185

CHLORINATION OF ODORANTS FROM ALGAL BLOOMS,

TECHNION - ISRAEL INST. OF TECH., HAIFA. SANITARY ENGINEERING LAB.

M. REBHUN, M. A. FOX, AND J. B. SLESS.

JOURNAL AMERICAN WATER WORKS ASSCCIATION, VOL 63, NO 4, P 219-224, 1971. 8 FIG, 7 TAB, 17 REF.

DESCRIPTORS:
*WATER QUALITY, *CHLORINATION, *WATER QUALITY CONTROL, *ODOR, ALGAE, EUTROPHICATION, TASTE, RESERVOIRS, SEASONAL, ACTINOMYCETES, CYANOPHYTA, PLANKTON, ON-SITE TESTS, LABCRATORY TESTS.

IDENTIFIERS:
*ISRAEL, *LAKE KINNERET(ISRAEL), PERIDINIUM WESTI, OSCILLATORIA PROLIFICA, THRESHOLD ODOR NUMBER.

ABSTRACT:
THE WATER SUPPLIES DISTRIBUTED BY ISRAEL'S NATIONAL WATER CARRIER FREQUENTLY DEVELOP TASTE AND ODOR PROBLEMS. A STUDY WAS CONDUCTED TO DETERMINE THE RELATICNSHIPS BETWEEN THE PLANKTCNIC ALGA PERIDINIUM WESTI AND THE BENTHONIC ALGA OSCILLATORIA TO ISOLATE THE CHARACTERIZE THE ODORANTS AND TO DETERMINE WHETHER A QUANTITATIVE RELATIONSHIP EXISTS BETWEEN ODOR AND CHLORINATION LEVELS. FIELD TESTS EVALUATED INTERRELATIONSHIPS BETWEEN ALGAL NUMBERS AND SPECIES COMPOSITION, AND THRESHOLD ODOR NUMBER AND CHLORINE DEMAND. ODORANTS PRODUCED BY PERIDINIUM AND OSCILLATORIA WERE CONCENTRATED ANC COLLECTED FOR CHARACTERIZATION AND QUANTITATIVE ANALYSIS OF EFFECT OF CHLORINATION ON THRESHOLD ODOR NUMBER IN THE LABORATORY. FLUCTUATIONS IN LEVELS OF ALGAL NUMBERS, CHLORINE DEMAND, AND THRESHCLD CDCR NUMBER OF RESERVOIR WATER WERE SEASCNAL AND INTERRELATED. CHLCRINE DEMAND WAS EXERTED BY SUBSTANCES IN SOLUTION AND NOT BY ALGAL CELLS AT NORMAL CONCENTRATIONS. ODOR CONTROL CAUSED BY PERIDINUM WAS DEPENCENT CN PROPER CHLORINATION OF FREE RESIDUAL. EARTHLY ODORS CAUSEC BY OSCILLATCRIA COULD NOT BE COMPLETELY REMOVED BY CHLORINATION; THE LEVEL CF CDOR REDUCTION WAS LINEARLY DEPENDENT ON INITIAL OCOR LEVEL. A METHCD OF COLLECTING ODORANTS FROM LABORATORY CULTURES CF AQUATIC ORGANISMS WAS DEVELOPED.
(JONES-WISCONSIN)

FIELD 05F, 05G

ACCESSION NO. W71-13187

OBSERVATIONS CONCERNING THE EXPERIMENTAL CONTAMINATIONS AND THE 'IN SITU' CONTAMINATIONS OF MARINE SPECIES BY RUTHENIUM 106, (OBSERVATIONS CONCERNANT LES CONTAMINATIONS EXPERIMENTALES ET LES CONTAMINATIONS 'IN SITU' D'ESPECES MARINES PAR LE'RUTHENIUM 106),

COMMISSARIAT A L'ENERGIE ATOMIQUE, CHERBOURG (FRANCE). CENTRE DE LA HAGUE.

J. ANCELLIN, AND P. BOVARD.

REVUE INTERNATIONALE D'OCEANOGRAPHIE MEDICALE, NICE, FRANCE, VOL 21, P 85-92, 1971. 8 REF.

DESCRIPTORS:
*WATER POLLUTION EFFECTS, *PLANTS, *ANIMALS, *ALGAE, *INVERTEBRATES, *RADIOACTIVITY EFFECTS, *RADIOISOTOPES.

IDENTIFIERS:
*RUTHENIUM 106, SOLUBLE FORMS, INSOLUBLE FORMS, LABORATORY EXPERIMENTS.

ABSTRACT:
CONTAMINATION OF MARINE FLORA AND FAUNA BY RUTHENIUM 106 WAS STUDIED UNDER BOTH LABORATORY AND IN SITU CONDITIONS. ALGAE AND INVERTEBRATES CONTAMINATED BY SOLUBLE AND INSOLUBLE FORMS OF RUTHENIUM 106 INDICATED THAT THE SOLUBLE RU 106 WAS 5-10 CONCENTRATION FACTORS LOWER THAN THE INSOLUBLE RU 106 FORMS. THE HIGH AMOUNTS OF CONCENTRATION FACTORS OBSERVED IN SITU WERE USUALLY COMPARABLE TO THOSE EXPERIMENTALLY DETERMINED FOR THE SOLUBLE FORMS IN BOTH ALGAE AND INVERTEBRATES. (ENSIGN-PAI)

FIELD 05C

ACCESSION NO. W71-13233

THE ROLE OF ALGAE IN CYCLING OF RADIONUCLIDES,

COMITATO NAZIONALE PER L'ENERGIA NUCLEARE, LA SPEZIA (ITALY). LABORATORIO PER LO STUDIO DELLA CONTAMINAZIONE RADIOATTIVE DEL MARE.

A. LATTERA, AND M. BERNHARD.

REVUE INTERNATIONALE D'OCEANOGRAPHIE MEDICALE, NICE, FRANCE, VOL 20, 1970, P 29-52, 67 REF.

DESCRIPTORS:
*RADIOISOTOPES, *ALGAE, *SEA WATER, *FOOD CHAINS, *ABSORPTION, BIOCHEMISTRY, PHYSIOLOGICAL ECOLOGY, CESIUM, STRONTIUM RADIOISOTOPES, ZINC RADIOISOTOPES, MANGANESE, COBALT RADIOISOTOPES, CADMIUM RADIOISOTOPES, METABOLISM, TROPHIC LEVELS.

IDENTIFIERS:
CONCENTRATION, ABUNDANCE, IRON, CHROMIUM, COPPER, NICKEL, RUTHENIUM, CERIUM, ZIRCONIUM.

ABSTRACT:
UPTAKE, LOSS AND STABLE CONTENT FOR ELEMENTS WHICH POSSESS POSSIBLE DANGEROUS RADIOISOTOPES, AND THE FACTORS INFLUENCING THESE ARE DISCUSSED. THE ACCUMULATION OF TRACE METALS BY ALGAE WAS PROPORTIONAL TO THEIR EXTERNAL CONCENTRATIONS. UPTAKE WAS RELATED TO PHYSICO-CHEMICAL FORMS OF RADIO-NUCLIDES. CONCENTRATION FACTORS FOR THE DIFFERENT ELEMENTS WERE FOUND TO VARY, SOMETIMES TO MORE THAN A THOUSAND TIMES OVER THEIR CONCENTRATION IN THE ENVIRONMENT. (ENSIGN-PAI)

FIELD 05C

ACCESSION NO. W71-13243

ECONOMIC ASPECTS OF ALGAE,

LOUISVILLE UNIV., KY. DEPT. OF BIOLOGY.

V. E. WIEDEMAN.

IN: SYMPOSIUM ON HYDROBIOLOGY, 'BIORESOURCES OF SHALLOW WATER ENVIRONMENTS', JUNE 24-27, 1970, MIAMI BEACH, FLORIDA, P 25-33. 2 TAB, 60 REF.

DESCRIPTORS:
*MARINE ALGAE, *ECONOMICS, *PRODUCTIVITY, ALGAE, RESOURCES, FOOD CHAINS, FISH FOOD ORGANISMS, OXYGEN, ECOSYSTEMS.

ABSTRACT:
ALGAE PROVIDE THE MAIN FOOD SUPPLY IN THE SEA AND HAVE GREAT ECONOMIC IMPORTANCE. THE BLUE-GREEN ALGAE (CYANOPHYCOPHYTE) ARE PRIMITIVE ORGANISMS RESEMBLING BACTERIA AND INCORPORATING A NITROGEN FIXING CAPABILITY WHICH CAUSES HIGH BIOMASS PRODUCTION. THEY SHOULD BE FULLY INVESTIGATED FOR ECONOMICALLY USEFUL METABOLITES. THE GREEN ALGAE (CHLOROPHYCOPHYTA) ARE OF GREAT DIVERSITY BUT CERTAIN GENERA SHOW POTENTIAL AS A PROTEIN-RICH FOOD SOURCE, AS SOIL FERTILIZER, AND AS A LEAD TO FURTHER UNDERSTANDING OF THE PHOTOSYNTHETIC PROCESS AND CYTOLOGY. EUGLENOID FLAGELLATES (EUGLENOPHYCOPHYTA) CAN BE USED AS BIOASSAY ORGANISMS FOR VITAMIN B12, AND SOME OCCUR AS A MAJOR COMPONENT IN WASTE-STABILIZATION PONDS. CERTAIN OF THE GOLDEN ALGAE (CHRYSOPHYCOPHYTA) ARE UTILIZED IN VITAMIN ASSAY BUT HAVE AN UNPLEASANT ODOR AND TASTE IN WATER. THE DIATOMS (BACILLARIOPHYCOPHYTA) CAN BE USED FOR INSULATING, SCOURING AND FILTERING MATERIALS, AND SOME SPECIES HAVE ANTIBIOTIC PROPERTIES. THE BROWN ALGAE (PHAEOPHYCOPHYTA), WHICH INCLUDE KELP, ARE A SOURCE OF IODINE, DRIED FOOD, FODDER, FERTILIZER AND ALGIN, WHICH HAS MANY USES IN MODERN PRODUCTS. THE RED ALGAE (RHODOPHYCOPHYTA) HAVE SIMILAR USEFUL PROPERTIES, WHEREAS THE DINOFLAGELLATES (PYRROPHYCOPHYTA) NEED TO BE CONTROLLED BECAUSE OF TOXIC PROPERTIES. ALGAE THUS CONSTITUTE THE BASE OF MOST FOOD CHAINS AND CONTRIBUTE MOST OF THE WORLD'S OXYGEN SUPPLY. (SMITH-PAI)

FIELD 05C, 06B

ACCESSION NO. W71-13246

THE MEASUREMENT OF OUR NATURAL RESOURCES,

NEW ENGLAND AQUARIUM, BOSTON, MASS.

G. C. MCLEOD.

IN: SYMPOSIUM ON HYDROBIOLOGY, 'BIORESOURCES OF SHALLOW WATER ENVIRONMENTS', JUNE 24-27, 1970, MIAMI BEACH, FLORIDA, P 88-98. 2 FIG, 30 REF. N00014-69-C0380.

DESCRIPTORS:
*MARINE ALGAE, *LIGHT, *GROWTH RATES, PHYTOPLANKTON, NATURAL RESOURCES, DEPTH, DISTRIBUTION PATTERNS, MEASUREMENT, PHOTOSYNTHESIS, PLANT PHYSIOLOGY, PIGMENTS.

IDENTIFIERS:
SUBMARINE LIGHT DATA SPHERE.

ABSTRACT:
OVER SHORT PERIODS OF TIME ALGAE PHOTOSYNTHESIZE FASTER THAN THEY GROW. THE LIGHT-INTENSITY CURVE FOR GROWTH OF ALGAE IS AN INTRINSIC CHARACTERISTIC OF THE ALGAE USED AND THE TEMPERATURE, WHEREAS THAT OF PHOTOSYNTHESIS IS NOT. MEASUREMENTS OF PRODUCTIVITY OF ALGAE MUST CONSIDER THE VARIABILITY IN SYNTHESIS OF STORAGE MATERIALS AND IMPORTANT PIGMENTS. THE QUESTION OF LIGHT RESPONSE MAY INFLUENCE THE DOMINANCE OF CERTAIN PHYTOPLANKTON IN SELECTED ENVIRONMENTS. SOME STUDIES OF THE EFFECT OF LIGHT INTENSITY ON THE GROWTH, PIGMENT COMPOSITION AND METABOLISM OF FRESH WATER AND MARINE PHYTOPLANKTON ARE DESCRIBED. THE INTENSITY AND SPECTRAL COMPOSITION OF LIGHT IN THE MARINE ENVIRONMENT IS VARIABLE. THE EFFECTS OF SPECTRUM-LIMITED LIGHT ON GROWTH, PIGMENTATION AND PHOTOSYNTHESIS HAVE ONLY RECENTLY BEEN STUDIED IN CHROMATIC LIGHT AND DARKNESS. THE EFFECTS OF BLUE LIGHT CAN BE REVERSED BY INCREMENTS OF RED LIGHT. THE KINETICS OF THE ADAPTATION TO BLUE LIGHT AND A POSSIBLE MECHANISM OF ACTION ARE CONSIDERED. SPECIES MUST ADAPT TO THE CHANGES IN SPECTRAL COMPOSITION OF SUBMARINE LIGHT, THEREFORE THE DISTRIBUTION OF CERTAIN SPECIES AT CERTAIN TIMES OF YEAR WILL VARY. A SUBMARINE LIGHT DATA SPHERE IS SPECIFIED, WITH SYSTEMS FOR MEASURING TOTAL INTEGRATED INTENSITY OF LIGHT OVER TIME AS A FUNCTION OF DEPTH. (SMITH-PAI)

FIELD 05C, 05A

ACCESSION NO. W71-13252

EUTROPHICATION: A THREAT TO WATER RESOURCES,

ENVIRONMENTAL PROTECTION AGENCY, CORVALLIS, OREG. PACIFIC NORTHWEST WATER LAB.

A. F. BARTSCH.

IN: SYMPOSIUM ON HYDROBIOLOGY, 'BIORESOURCES OF SHALLOW WATER ENVIRONMENTS', JUNE 24-27, 1970, MIAMI BEACH, FLORIDA, P 127-135. 15 REF.

DESCRIPTORS:
*WATER QUALITY CONTROL, *LAKES, *EUTROPHICATION, AQUATIC ALGAE, NUTRIENTS, WATER RESOURCES, WASTE ASSIMILATIVE CAPACITY, RESEARCH AND DEVELOPMENT, WASTE WATER TREATMENT.

IDENTIFIERS:
NUTRIENT CONTROL METHODS.

ABSTRACT:
QUALITY DETERIORATION OF WATER RESOURCES, ESPECIALLY LAKES, THROUGH EUTROPHICATION IS A WORLDWIDE PROBLEM AND REQUIRES IMMEDIATE ACTION. THE NATURAL PROCESS HAS BEEN AGGRAVATED AND SPEEDED UP BY MAN'S ACTIVITIES IN AUGMENTING THE FERTILITY OF WATER. ARTIFICIALLY CREATED LAKES QUICKLY DEVELOP EUTROPHICATION PROBLEMS. INDUSTRIAL WASTES, SEWAGE WASTES AND AGRICULTURAL RUNOFF ALL CONTRIBUTE NUTRIENTS TO WATERCOURSES. THE QUESTION OF LAKE RESTORATION IS BEING STUDIED BY SCIENTIFIC COMMUNITIES AND THE FWQA. ONE RESTORATIVE AND PREVENTIVE APPROACH IS TO CURB NUTRIENT INPUT, WHICH IS ALREADY BEING DONE IN SOME PLACES BY DIVERTING SEWAGE FROM LAKES. A SECOND CONTROL POSSIBILITY IS TO TREAT SEWAGE AND WASTES TO STRIP THEM OF THEIR NUTRIENT CONTENT, ESPECIALLY PHOSPHORUS. HOWEVER, ALL LAKES ARE DIFFERENT AND THERE IS NEED TO BE ABLE TO PREDICT HOW QUICKLY AND HOW MUCH A LAKE WILL CHANGE AFTER RESTORATIVE EFFORTS. DETERGENTS ARE ANOTHER SOURCE OF PHOSPHORUS IN SEWAGE WHICH MUST BE CONTROLLED. IMPROVED HANDLING OF AGRICULTURAL WASTES CAN ALSO MINIMIZE NUTRIENT INPUT TO WATER. STUDIES ARE BEING MADE OF THE USE OF CERTAIN CHEMICALS FOR NUTRIENT INACTIVATION. OTHER IDEAS UNDER CONSIDERATION ARE THE FEEDING AND HARVESTING OF ALGAE, CONTROLLED DREDGING, STIMULATING DISEASES AND PARASITES TO DESTROY PLANT POPULATIONS, AND, AS A LAST RESORT, THE USE OF ALGICIDES AND HERBICIDES. (SMITH-PAI)

FIELD 05C

ACCESSION NO. W71-13256

BACTERIA-ALGAE SYMBIOSIS - A CAUSE OF ALGAL BLOOMS,

WYANDOTTE CHEMICALS CORP., MICH.

L. E. KUENTZEL.

IN: SYMPOSIUM ON HYDROBIOLOGY, 'BIORESOURCES OF SHALLOW WATER ENVIRONMENTS', JUNE 24-27, 1970, MIAMI BEACH, FLORIDA, P 136-145. 23 REF.

DESCRIPTORS:
*ALGAE, BACTERIA, *LAKES, SYMBIOSIS, ALGAL CONTROL, EUTROPHICATION, WATER POLLUTION, BIOCHEMICAL OXYGEN DEMAND, ORGANIC MATTER, PHOSPHORUS, CARBON DIOXIDE.

ABSTRACT:
THE EXISTENCE OF THE MUTUALLY BENEFICIAL SYMBIOTIC RELATIONSHIP BETWEEN BLUE-GREEN ALGAE AND CERTAIN BACTERIA IS WELL ESTABLISHED. THE PRESENCE OF BIODEGRADABLE ORGANIC MATTER WILL ALWAYS CONTRIBUTE TO MASSIVE ALGAL BLOOM GROWTH IF ALL OTHER FACTORS PERMIT, AND THERE HAS BEEN A GENERAL INCLINATION TO TURN TO PHOSPHORUS AS A CONTROLLING FACTOR. THE LOGISTICS OF SUPPLY FOR CO_2 AND P ARE EXAMINED IN THE LIGHT OF THE ESTABLISHED REQUIREMENTS FOR THE GROWTH OF MASSIVE ALGAL BLOOMS IN INCREASINGLY POLLUTED LAKES, AND FINDINGS FROM PREVIOUS AND NEW STUDIES ARE CITED. NUTRIENT CONTROL HAS LITTLE EFFECT ON LAKES WHICH HAVE BEEN SUPPORTING ALGAL GROWTH FOR SEVERAL YEARS BECAUSE OF THE ACCUMULATION OF ORGANIC DEBRIS. IT APPEARS THAT ORGANIC MATTER AND BACTERIAL ACTION PLAY A MAJOR ROLE IN SUPPORTING ALGAL BLOOM DEVELOPMENT. ATTEMPTS TO CONTROL ALGAL GROWTH VIA CONTROL OF CARBON (BOD) OFFER DISTINCT ADVANTAGES, WHEREAS PHOSPHATE REMOVAL DOES NOT SEEM TO HAVE MUCH EFFECT. WASTEWATERS NEED TO RECEIVE OPTIMUM BOD AND AMMONIA N REMOVAL BY TREATMENT PLANTS TO HELP PREVENT EUTROPHICATION OF RIVERS AND LAKES. (SMITH-PAI)

FIELD 05C, 02H

ACCESSION NO. W71-13257

A LABORATORY EVALUATION OF THE PERFORMANCE OF THREE CELL OXIDATION PONDS IN SERIES,

MISSISSIPPI STATE UNIV., STATE COLLEGE.

ROBERT JACK FREEMAN, JR.

MASTER'S THESIS, MAY 1970. 91 P, 36 FIG, 6 TAB, 26 REF.

DESCRIPTORS:
*OXIDATION PONDS, *ALGAE, LABORATORY TESTS, AEROBIC CONDITIONS, ANAEROBIC CONDITIONS, ORGANIC LOADING, *BIOCHEMICAL OXYGEN DEMAND, NITROGEN, PHOSPHOROUS, PHOTOSYNTHESIS, LIGHT INTENSITY, OXYGENATION, DENITRIFICATION, OXIDATION, ADSORPTION, COLIFORMS, BACTERIA, CENTRIFUGATION, *WASTE WATER TREATMENT.

IDENTIFIERS:
*DETENTION TIME.

ABSTRACT:
THE RELATIVE EASE WITH WHICH THE CRITICAL PARAMETERS COULD BE VARIED PROMPTED THE USE OF LABORATORY SCALE EQUIPMENT FOR EVALUATIONS OF 3 CELL OXIDATION PONDS. 2 SETS OF 3 SERIES PONDS WERE USED, ONCE AT A 300 LB BOD5/ACRE/DAY LOADING RATE AND A 2 DAY PER POND DETENTION TIME. SUBSEQUENT RUNS INVOLVED CONSTANT LOADING RATES OF 50 AND 100 LBS BOD5/ACRE/DAYS AND DETENTION TIMES OF 2,3, AND 4 DAYS PER POND. A SYNTHETIC SUBSTRATE, CONTINUOUS FLOW LOADING, AND CONTROLLED LIGHTING WERE USED. BOTH UNCENTRIFUGED AND CENTRIFUGED SAMPLES WERE ANALYZED FOR BOD, TOTAL NITROGEN, ALGAL CONCENTRATION, SOLUBLE ORTHOPHOSPHATE, PH, AND DISSOLVED OXYGEN. FOR UNCENTRIFUGED SAMPLES, BOD, TOTAL NITROGEN, AND SOLUBLE ORTHOPHOSPHATE REMOVALS RANGED FROM 90 TO 95%, 20 TO 50% AND 16 TO 32% RESPECTIVELY. CENTRIFUGATION OF THE EFFLUENT FOR REMOVAL OF THE ALGAE CELLS EFFECTED A 40 TO 70% REDUCTION OF EFFLUENT BOD, A 25% TO 40% REDUCTION OF TOTAL NITROGEN IN THE EFFLUENT, WHILE LITTLE OR NO REDUCTION OF SOLUBLE ORTHOPHOSPHATE WAS OBSERVED. ALSO, AT CONSTANT LOADING RATE, VARYING DETENTION TIMES PRODUCED NO EFFECTS ON REMOVALS. A SIGNIFICANT PERCENTAGE REMOVAL OF COLIFORMS WAS OBTAINED, BUT EFFLUENT COLIFORM CONCENTRATIONS WERE STILL GREATLY IN EXCESS OF ACCEPTABLE STANDARDS DUE TO THE EXTREMELY HIGH INFLUENT CONCENTRATIONS. (LOWRY-TEXAS)

FIELD 05D

ACCESSION NO. W71-13326

PHOSPHOROUS EXTRACTION FROM SOLUTION BY HETEROGENOUS BIOLOGICAL SYSTEMS,

MAINE UNIV., ORONO, DEPT. OF CIVIL ENGINEERING.

RICHARD D. ROX.

MASTER'S THESIS, UNIVERSITY OF MAINE, JANUARY 1970. 67 P, 20 FIG, 43 REF.

DESCRIPTORS:
*EUTROPHICATION, *PHOSPHOROUS, ALGAE, NITROGEN, ACTIVATED SLUDGE, CHEMICAL OXYGEN DEMAND, OXYGEN, AERATION, DISSOLVED OXYGEN, HYDROGEN ION CONCENTRATION, SETTLING, MICROORGANISMS, ACCLIMATION, CALCIUM, SEDIMENTATION, WASTE WATER TREATMENT.

IDENTIFIERS:
*CHEMICAL-BIOLOGICAL TREATMENT, MIXED LIQUOR, SUSPENDED SOLIDS, UPTAKE.

ABSTRACT:
PHOSPHOROUS HAS BECOME THE MAJOR CONTROLLABLE NUTRIENT IN THE FIGHT AGAINST EUTROPHICATION. AN ATTEMPT HAS BEEN MADE TO FIND EFFECTIVE MEANS OF PHOSPHOROUS REMOVAL THROUGH A CHEMICAL-BIOLOGICAL TREATMENT SCHEME. BENCH SCALE ACTIVATED SLUDGE UNITS OF THE PHOSPHOROUS UPTAKE AND RELEASE BY ACTIVATED SLUDGE SYSTEMS. SOLUBLE COD UPTAKE, MLSS GROWTH, OXYGEN UTILIZATION, INITIAL F:M RATIOS AND DIVALENT CATION CONCENTRATIONS WERE STUDIED TO DETERMINE THEIR EFFECT ON REMOVAL. PHOSPHOROUS RELEASE WAS STUDIED UNDER EXTENDED PERIODS OF AERATION, LOW CONCENTRATIONS OF DISSOLVED OXYGEN, QUIESCENT SETTLING CONDITIONS AND LOW LEVELS OF PH. ACTIVATED SLUDGE MICROORGANISMS WERE SHOWN TO PRODUCE REPRODUCEABLE PHOSPHOROUS UPTAKE DATA WHILE BEING ACCLIMATED TO A SYNTHETIC WASTE. A PHOSPHOROUS UPTAKE OF 0.02 POUNDS P/LB.COD/DAY WAS NOTED FOR BACTERIAL CULTURES OPERATING AT SEVERAL F:M RATIOS. A CLOSE CORRELATION OF THE MICROBIAL OXYGEN UTILIZATION AND THE TOTAL SOLUBLE PHOSPHOROUS UPTAKE WAS OBSERVED. PHOSPHORUS RELEASE WAS SHOWN TO BE INDEPENDENT OF DISSOLVED OXYGEN CONCENTRATIONS AND THE PERIOD OF AERATION. LOW PH CONDITIONS, HOWEVER, SHOWED SIGNIFICANT LEVELS OF PHOSPHOROUS SECRETION. CALCIUM ADDED TO THE ACTIVATED SLUDGE MIXED LIQUOR SHOWED VARYING DEGREES OF ENHANCED PHOSPHOROUS REMOVAL WITH NO SIGNIFICANT DETERIORATION IN BIOLOGICAL ACTIVITY. (ATKINS-TEXAS)

FIELD 05D, 05C

ACCESSION NO. W71-13334

SEVERAL METHODS OF ALGAE REMOVAL IN MUNICIPAL OXIDATION PONDS,

KANSAS UNIV., LAWRENCE.

DONALD M. MARTIN.

MASTER'S THESIS, UNIVERSITY OF KANSAS, DECEMBER 1970. 82 P, 22 FIG, 6 TAB, 19 REF.

DESCRIPTORS:
*OXIDATION LAGOONS, *ALGAE, MUNICIPAL WASTES, *FILTRATION, *FILTERS, BIOCHEMICAL OXYGEN DEMAND, CHEMICAL OXYGEN DEMAND, PHOSPHORUS, NITRATES, AMMONIA, COAGULATION, FLOCCULATION, SEDIMENTATION, FILTRATION, LIME, ROCKS, METABOLISM, DESIGN CRITERIA, *WASTE WATER TREATMENT, EFFLUENTS.

IDENTIFIERS:
*SUSPENDED SOLIDS, SODIUM HYDROXIDE, *ROCK FILTERS.

ABSTRACT:
EFFLUENT FROM AN OXIDATION POND WAS COLLECTED AND USED BOTH FOR A JAR TEST AND FOR RUNS THROUGH AN UPFLOW ROCK FILTER. A WIDE VARIETY OF COAGULANTS AND THE NEW POLYELECTROLYTES WERE USED IN THE HOPE OF DISCOVERING SOME ECONOMICAL MEANS OF REMOVING ALGAE FROM OXIDATION POND EFFLUENTS. LIME, ALUM AND SODIUM HYDROXIDE WERE TESTED FOR USE AS FLOCCULENTS, WHILE CALGON'S CAT-FLOC WAS USED AT A CONCENTRATION OF 8 MG/L AS A FLOCCULENT AID. NO COST FIGURES WERE REPORTED, BUT THE ADDITION OF AN EXTENSIVE DOSING, MIXING, AND SEDIMENTATION SYSTEM, PLUS THE FACT THAT AS MUCH AS 120 MG/L OF FLOCCULENT WOULD BE REQUIRED, INDICATED THE PROCESS WOULD DEFEAT THE MAIN PURPOSE OF OXIDATION PONDS, NAMELY LOW COST AND LITTLE OR NO MAINTENANCE. THE ROCK FILTERS, HOWEVER, SHOWED MUCH MORE PROMISE. OPERATED ONLY BY THE HYDRAULIC HEAD OF THE POND LEVEL, THE FILTERS REQUIRED NO MECHANICAL, ELECTRICAL, OR CHEMICAL EQUIPMENT AND LITTLE OR NO MAINTENANCE. SINCE COD WAS SHOWN TO BE PRESENT BOTH IN DISSOLVED AND SUSPENDED FORMS, THEN THE DIFFERENCE BETWEEN SOLUBLE COD AND TOTAL COD WOULD INDICATE THE EFFICIENCY OF REMOVAL OF SUSPENDED MATTER IN THE ROCK FILTER. ON THIS BASIS, ROCK FILTERS WERE SHOWN TO BE CAPABLE OF REMOVING 80 TO 90% OF THE SUSPENDED MATERIAL COD. FURTHER FIELD SCALE TESTS WERE RECOMMENDED FOR THE ESTABLISHMENT OF OPERATIONAL PARAMETERS AND DESIGN CRITERIA. (LOWRY-TEXAS)

FIELD 05D

ACCESSION NO. W71-13341

A LABORATORY EVALUATION OF CHEMICAL COAGULATION AS A METHOD OF TREATING STABILIZATION POND EFFLUENT,

MISSISSIPPI STATE UNIV., STATE COLLEGE. DEPT. OF CIVIL ENGINEERING.

JERRY STEWART.

MASTER'S THESIS, AUGUST 1970. 73 P, 7 FIG, 5 TAB, 14 REF.

DESCRIPTORS:
*COAGULATION, *WASTE WATER TREATMENT, ANALYTICAL TECHNIQUES, COALESCENCE, FLOCCULATION, LIQUID WASTES, SEPARATION TECHNIQUES, WATER PURIFICATION, *OXIDATION LAGOONS, STABILIZATION, INDUSTRIAL WASTES, MUNICIPAL WASTE, BIOCHEMICAL OXYGEN DEMAND, CHEMICAL OXYGEN DEMAND, *PHOSPHATES, NITROGEN, COLIFORMS, MATHEMATICAL MODELS, ALGAE, SAMPLING, STATISTICAL METHODS.

IDENTIFIERS:
*ALUMINUM SULFATE, *FERRIC CHLORIDE, *FERRIC SULFATE, *STABILIZATION PONDS, ALUM, POST-TREATMENT.

ABSTRACT:
A LABORATORY STUDY WAS PERFORMED TO EVALUATE THE SUITABILITY OF CHEMICAL COAGULATION AS A POST TREATMENT METHOD FOR EFFLUENTS FROM WASTE STABILIZATION PONDS. THE DEGREE OF EFFECTIVENESS WAS MEASURED IN TERMS OF BIOCHEMICAL OXYGEN DEMAND, CHEMICAL OXYGEN DEMAND, TOTAL PHOSPHATES, TOTAL NITROGEN, COLIFORM COUNTS, AND ALGAL CELL COUNTS. THREE COAGULANTS: ALUMINUM SULFATE, FERRIC CHLORIDE, AND FERRIC SULFATE WERE EVALUATED. A MATHEMATICAL MODEL WAS CONSTRUCTED WHICH ALLOWED FOR THE COMPUTATION OF THE REQUIRED DOSAGE OF COAGULANT FOR VARIOUS INFLUENT AND EFFLUENT CONCENTRATION OF PHOSPHATES. THE RESULTS INDICATED THAT CHEMICAL COAGULATION WITH ALUM CAN BE EFFECTIVELY USED AS A POST TREATMENT METHOD OF STABILIZATION POND EFFLUENTS. NEARLY COMPLETE ALGAE REMOVAL, OVER 90 PERCENT BOD AND PHOSPHATE REMOVALS, OVER 70 PERCENT COD REMOVALS, AND AN AVERAGE OF 5,000 COLIFORMS/1000 ML., WERE OBTAINED USING AN ALUM DOSAGE OF 1000 MG/L. SETTLING CHARACTERISTICS OF THE ALUM-ALGAE SLUDGE WERE ALSO INVESTIGATED. (ATKINS-TEXAS)

FIELD 05D

ACCESSION NO. W71-13356

OCCURRENCE OF PHOSPHORUS AND NITROGEN IN SALT CREEK AT LINCOLN, NEBRASKA,

GEOLOGICAL SURVEY, LINCOLN, NEBR.

R. A. ENGBERG, AND T. O. RENSCHLER.

GPO, WASHINGTON, D.C. 20402-PRICE $2.75. GEOLOGICAL SURVEY RESEARCH 1971, CHAPTER C, PROFESSIONAL PAPER 750-C, P C223-C227, 1971. 1 FIG, 2 TAB, 11 REF.

DESCRIPTORS:
*WATER POLLUTION SOURCES, *PHOSPHATES, *NITRATES, *NEBRASKA, URBANIZATION, CITIES, STORM RUNOFF, SEWAGE, WASTE WATER DISPOSAL, PATH OF POLLUTANTS, STORM DRAINS, DETERGENTS, EUTROPHICATION, ALGAE, NUTRIENTS.

IDENTIFIERS:
*URBAN RUNOFF, URBAN HYDROLOGY, LINCOLN(NEBR).

ABSTRACT:
CONCENTRATIONS OF PHOSPHORUS AND NITROGEN IN SALT CREEK INCREASE MARKEDLY IN THE 6-MILE REACH OF THE CREEK WITHIN THE CITY OF LINCOLN, NEBR. MOST OF THE INCREASE IS DUE TO INFLOW FROM THE LINCOLN SEWAGE-TREATMENT PLANT AND FROM STORM-SEWER AND OTHER URBAN RUNOFF ENTERING THE REACH. THE CITY CONTRIBUTES AVERAGE AMOUNTS OF 0.94 TONS PHOSPHORUS AND 1.8 TONS NITROGEN PER DAY TO THE STREAM, INDICATING ANNUAL PER CAPITA CONTRIBUTIONS OF PHOSPHORUS AND NITROGEN OF 4.5 AND 8.7 POUNDS PER YEAR, RESPECTIVELY. (KNAPP-USGS)

FIELD 05B, 05C

ACCESSION NO. W71-13466

KINETIC ASSESSMENT OF ALGAL GROWTH,

CALIFORNIA UNIV., BERKELEY.

E. A. PEARSON, E. J. MIDDLEBROOKS, M. TUNZI, A. ADINARAYANA, AND P. H. MCGAUHEY.

CHEMICAL ENGINEERING PROGRESS, SYMPOSIUM SERIES, VOL 67, NO 107, P 5-14, 1971. 8 FIG, 5 TAB, 9 REF.

DESCRIPTORS:
KINETICS, *BIOLOGICAL TREATMENT, PLANKTON, *EUTROPHICATION, *NUTRIENTS, MICROORGANISMS, EFFLUENT, BIOCHEMICAL OXYGEN DEMAND, SATURATION, OXYGEN, METABOLISM, NITROGEN, PHOSPHOROUS, *ALGAL CONTROL, CALIFORNIA.

IDENTIFIERS:
*ALGAL GROWTH, *LAKE TAHOE(CALIF).

ABSTRACT:
IT IS WELL KNOWN THAT NUTRIENTS SUCH AS NITROGEN AND PHOSPHORUS ARE REQUIRED FOR ALGAL METABOLISM. HOWEVER, THE QUESTION REMAINS, IS NITROGEN OR PHOSPHORUS USUALLY OR FREQUENTLY LIMITING THE GROWTH RATE OR STANDING STOCK IN MOST RECEIVING WATERS. TO EFFECTIVELY CONTROL THE RATE OF EUTROPHICATION OF OUR NATURAL WATERS, THIS QUESTION MUST BE ANSWERED. IT IS NOT ENOUGH TO REMOVE NITROGEN AND PHOSPHORUS FROM WASTEWATERS IN HOPES OF REDUCING THE GROWTH RATE OR STANDING STOCK OF PLANKTON IN WATERS CONCERNED. THERE IS AN URGENT NEED TO DEVELOP METHODS OR TECHNIQUES FOR ASSESSING THE BIOSTIMULATORY CHARACTER OF WASTE EFFLUENTS, NATURAL RUNOFF, ETC., ESTIMATING THE LEVEL OF CONSEQUENCES ONE MIGHT EXPECT FOR A GIVEN LEVEL OF RATE-LIMITING NUTRIENT OR SUBSTANCE. IN NO SINGLE CASE IS THIS NEED MORE URGENT THAN IN THE CASE OF LAKE TAHOE. ALTHOUGH ACTION TO REMOVE SEWAGE EFFLUENTS FROM THE BASIN IS WELL UNDERWAY, HUMAN ACTIVITY IN THE AREA IS RAPIDLY INCREASING AND IT IS NOT KNOWN WHETHER OTHER INPUTS TO THE LAKE SHOULD BE CONTROLLED TO LIMIT THE RATE OF EUTROPHICATION OR ENRICHMENT OF THE LAKE. A LABORATORY METHOD OF EVALUATING THE CAPACITY OF AQUATIC ENVIRONMENTS TO GROW ALGAE IS DEFINITELY NEEDED. (GUTIERREZ-TEXAS)

FIELD 05C, 02H

ACCESSION NO. W71-13553

REPORT OF THE FAO TECHNICAL CONFERENCE ON MARINE POLLUTICN AND ITS EFFECTS ON LIVING RESOURCES AND FISHING.

FOOD AND AGRICULTURE ORGANIZATION (UN), NEW YORK. FISHERY RESOURCES DIV.

FAO FISHERIES REPORTS, NO. 99, FIRM/R99 (EN), ROME, ITALY, 9-18 DECEMBER 1970. 188 P. 132 REF.

DESCRIPTORS:
*POLLUTION ABATEMENT, *WATER POLLUTION EFFECTS, *WATER POLLUTION SOURCES, *FISHERIES, OIL WASTES, FISH REPRODUCTION, WASTE DISPOSAL, MARINE MICROORGANISMS, SEDIMENTATION, ALGAE, INVERTEBRATES, RADIATION, ECOSYSTEMS, ESTUARIES, CORAL, REEFS, WATER RESOURCES, SYSTEMS ANALYSIS, WASTE TREATMENT, AIR POLLUTION, LEGISLATION, BENTHIC FLORA, FORECASTING, LARVAE, LAGOONS, DDT, ENZYMES, MOLLUSKS, NUTRIENTS, RADIOISOTOPES, VIRUSES, SEWAGE, EUTROPHICATION, TOXICITY, LEGAL ASPECTS, PHYTOPLANKTON.

ABSTRACT:
THE FAO TECHNICAL CONFERENCE INCLUDED IN PRINCIPLE THE CONSIDERATION OF ALL TYPES OF POLLUTANTS, FROM WHATEVER SOURCE, AND IN ALL SEA AREAS WITH REFERENCE TO THE CONFERENCE'S DEFINITION OF 'MARINE POLLUTION.' GENERAL OBJECTIVES WERE TO CONCENTRATE IN ONE PLACE AS MUCH AS POSSIBLE OF WORLD-WIDE KNOWLEDGE OF MARINE POLLUTION, TO CONSIDER WAYS OF PREVENTING POLLUTION IN MARINE WATERS, TO FOCUS ATTENTION ON SCIENTIFIC PROBLEMS WHERE INTERNATIONAL COOPERATION WAS IMPERATIVE, AND TO PROVIDE SCIENTIFIC AND TECHNICAL GUIDELINES FOR THE INTERNATIONAL CONTROL OF MARINE POLLUTION. (CURRER-PAI)

FIELD 05C, 05B, 05G

ACCESSION NO. W71-13723

THE KENT COAST IN 1970,

BRITISH MUSEUM LONDON (ENGLAND). DEPT. OF BOTANY.

I. TITTLEY.

MARINE POLLUTION BULLETIN, VOL 2, NO 8, AUGUST 1971, P 120-122. 3 FIG, 22 REF.

DESCRIPTORS:
*WATER POLLUTION EFFECTS, *DATA COLLECTIONS, ALGAE, ECOTYPES, INDUSTRIAL PLANTS, INDUSTRIAL WASTES, OIL WASTES, ESTUARIES, MONITORING.

IDENTIFIERS:
*KENT COAST, INDUSTRIAL EXPANSION, ISLE OF THANET, MEDWELL ESTUARY.

ABSTRACT:
A POLLUTION RECORD IS COMPILED FOR THE KENT COAST. THE EFFECTS OF COAST 'PROTECTION' ON THE CLIFF FLORA AND THE PROBLEMS OF INDUSTRIAL EXPANSION ON THE MEDWAY ARE DESCRIBED. (CURRER-PAI)

FIELD 05C

ACCESSION NO. W71-13746

GROWTH INHIBITORS PROCUCED BY THE GREEN ALGAE (VOLVOCACEAE),

KENTUCKY UNIV., LEXINGTON. DEPT. OF BOTANY.

D. O. HARRIS.

ARCHIV FUR MIKROBIOLOGIE, VOL 76, P 47-50, 1971. 2 TAB, 9 REF. OWRR A-018-KY(1).

DESCRIPTORS:
*CHLOROPHYTA, *INHIBITORS, ALGAL CONTROL, PLANT GROWTH REGULATORS, PLANKTON, CULTURES, WATER POLLUTION EFFECTS.

IDENTIFIERS:
HETEROINHIBITORS, AUTOINHIBITORS, VOLVCCACEAE, PANDORINA MORUM, PLATYDORINA CAUDATA, EUDORINA CALIFORNICA, EUDORINA ILLINOISENSIS.

ABSTRACT:
CELLS OF 12 STRAINS CF VOLVOCACEAE GREEN ALGAE WERE REMOVED BY FILTRATION FROM THE NUTRIENT MEDIUM AFTER 21 DAY GROWTH. THE CELL-FREE MEDIA WERE INOCULATED WITH TEST GENERA AND INCUBATED FOR 14 DAYS. THE GROWTH INHIBITORY EFFECT WAS DETERMINED ON THE BASIS OF OPTICAL DENSITY OF MILLIPORE-FILTERED AND AUTOCLAVED CULTURES. CULTURE FILTRATES OF PANDORINA MORUM, VOLVULINA PRINGSHEIMII AND EUDORINA CALIFCRNICA WERE INHIBITORY TO MOST MEMBERS OF THE FAMILY. A HOPE IS EXPRESSED THAT THESE NATURAL INHIBITORY SUBSTANCES WILL EVENTUALLY SERVE TO CONTROL THE UNDESIRABLE GROWTH OF ALGAE IN WATER BASINS. (WILDE-WISCONSIN)

FIELD 05C

ACCESSION NO. W71-13793

ADENOSINE TRIPHOSPHATE CONTENT OF SELENASTRUM CAPRICCRNUTUM,

WISCONSIN UNIV., MADISON. DEPT. OF SOIL SCIENCE; AND WISCONSIN UNIV., MADISON. WATER CHEMISTRY LAB.

C. C. LEE, R. F. HARRIS, J. K. SYERS, AND D. E. ARMSTRCNG.

APPLIED MICROBIOLOGY, VOL 21, NO 5, P 957-958, 1971. 1 TAB, 6 REF.

DESCRIPTORS:
*TROPHIC LEVELS, *BIOASSAY, PHOSPHORUS, NUTRIENTS, MEASUREMENT, BACTERIA, ALGAE, EUTROPHICATION, DATA COLLECTIONS, TEST PROCEDURES, CYCLING NUTRIENTS, ANALYTICAL TECHNIQUES, WATER PCLLUTION EFFECTS, POLLUTANT IDENTIFICATION.

IDENTIFIERS:
*ADENOSINE TRIPHOSPHATE, *SELENASTRUM CAPRICORNUTUM.

ABSTRACT:
TO QUANTIFY THE LIVING BIOMASS OF HETEROGENEOUS MICROBIAL POPULATIONS IN SEEKING ACCURATE EVALUATION AND EVENTUAL UNDERSTANDING AND CONTROL OF NUTRIENT CYCLES AND OTHER ENVIRONMENTAL PHENOMENA MEDIATED BY MICRO-ORGANISMS, A STUDY WAS MADE OF ADENOSINE TRIPHOSPHATE AS A PARAMETER OF LIVING MICROBIAL BIOMASS. THE POTENTIAL OF ATP AS A PARAMETER DEPENDS ON THE ASSUMPTION THAT ATP IS PRESENT IN A RELATIVELY CONSTANT OR PREDICTABLE AMOUNT IN DIVERSE MICROORGANISMS. SELENASTRUM CAPRICORNUTUM WAS CHOSEN AS TEST ORGANISM FOR BIOASSAY; ITS UNICELLULAR CHARACTERISTICS FACILITATED ACCURATE MEASUREMENT OF VIABLE AND TOTAL ALGAL CELLS FOR CRITICAL EVALUATION OF ATP-LIVING BIOMASS RELATIONSHIPS. BASED ON THE USE OF THE 4-DAY BIOMASS VALUES AS VALID REFLECTIONS OF LIVING BIOMASS, S CAPRICORNUTUM CONTAINED 3.4, 3.1 AND 1.4 MICROGRAMS OF ATP/MG (DRY WEIGHT) OF LIVING BIOMASS UNDER PHOSPHORUS-RICH, BALANCED, AND PHOSPHORUS-DEFICIENT CONDITIONS, RESPECTIVELY. THE EFFECT OF GROWTH STAGE ON ATP CELL CONTENT WAS APPARENTLY MUCH LESS PRONOUNCED FOR S CAPRICCRNUTUM THAN FOR BACTERIA. THESE DATA TEND TO SUPPORT THE POTENTIAL OF ATP MEASUREMENTS FOR PROVIDING VALID APPROXIMATIONS OF FLUCTUATIONS IN LIVING MICROBIAL BIOMASS IN SURFACE WATERS WHERE ALGAE ARE MAJOR CONTRIBUTORS TO THE TOTAL MICROBIAL POPULATION. (JONES-WISCONSIN)

FIELD 05C, 05A

ACCESSION NO. W71-13794

BIOLOGICAL PROCESSES FOR NITROGEN REMOVAL--THEORY AND APPLICATION,

STANFORD UNIV., CALIF.

PERRY L. MCCARTY.

AVAILABLE FROM ENGINEERING PUBLICATIONS OFFICE, 112 ENGINEERING HALL, UNIVERSITY OF ILLINOIS 61801 - PRICE $4.50. IN: PROCEEDINGS 12TH SANITARY ENGINEERING CONFERENCE ON NITRATE AND WATER SUPPLY: SOURCE AND CONTROL, FEBRUARY 11-12, 1970, UNIVERSITY OF ILLINOIS, URBANA: ILLINOIS UNIVERSITY, COLLEGE OF ENGINEERING PUBLICATION, P 136-152, 1970. 17 P, 3 FIG, 29 REF.

DESCRIPTORS:
*WASTE WATER TREATMENT, *NITRIFICATION, *DENITRIFICATION, *NITRATES, NITROGEN COMPOUNDS, AMMONIA, ALGAE, SEWAGE TREATMENT, TERTIARY TREATMENT, WATER POLLUTION CONTROL, BIODEGRADATION.

IDENTIFIERS:
*NITROGEN REMOVAL.

ABSTRACT:
NITROGEN CAN BE REMOVED BIOLOGICALLY FROM WASTE WATER BY THREE DIFFERENT PROCESSES: BACTERIAL ASSIMILATION, ALGAE HARVESTING, AND NITRIFICATION-DENITRIFICATION. BACTERIAL NITRIFICATION-DENITRIFICATION IS PERHAPS THE MOST PROMISING OF THE PROCESSES. AMMONIA NITROGEN REMOVAL IS THE RESULT OF TWO STAGES. IN NITRIFICATION, AMMONIA-NITROGEN IS CONVERTED TO NITRATE-NITROGEN BY TWO DIFFERENT GROUPS OF AUTOTROPHIC NITRIFYING BACTERIA. DENITRIFICATION IS THE REDUCTION OF NITRATES AND NITRITES TO NITROGEN GAS BY A WIDE VARIETY OF FACULTATIVE BACTERIA UNDER ANAEROBIC CONDITIONS. AN ADDITIONAL ORGANIC SOURCE IS NECESSARY FOR EFFICIENT CONVERSION TO NITROGEN GAS, AND METHANOL APPEARS TO BE THE LEAST EXPENSIVE. IN GENERAL FOR AGRICULTURAL AND DOMESTIC WASTE WATER, COSTS FOR NITROGEN REMOVAL MAY RANGE FROM 2 TO 10 CENTS PER 1,000 GALLONS. OF THE THREE PROCESSES CONSIDERED, NITRIFICATION-DENITRIFICATION IS PERHAPS THE MOST GENERALLY APPLICABLE BECAUSE OF GOOD RELIABILITY, SUITABILITY TO A VARIETY OF CONDITIONS, LOW AREA REQUIREMENTS AND MODERATE COST. (KNAPP-USGS)

FIELD 05D

ACCESSION NO. W71-13939

COMBINED WASTEWATER COLLECTION AND TREATMENT FACILITY, MOUNT CLEMENS, MICHIGAN,

SPALDING, DEDECKER AND SSOCIATES, INC, MADISON HEIGHTS, MICH.

VIJAYSINH U. MAHIDA.

PREPRINT, 44TH ANNUAL CONFERENCE, WATER POLLUTION CONTROL FEDERATION, SESSION 4, NO 4, OCTOBER 3-8, 1971, SAN FRANCISCO, CAL 11 P.

DESCRIPTORS:
*STORM RUN-OFF, *SUSPENDED SOLIDS, *RECREATION, *TREATMENT FACILITIES, SAMPLING, MONITORING, AUTOMATIC CONTROLS, AERATION, SEDIMENTATION, FILTRATION, ANAEROBIC DIGESTION, CHLORINATION, DISINFECTION, ULTRAVIOLET LIGHT, ALGAE, OXYGENATION, BIOCHEMICAL OXYGEN DEMAND, WATER SPORTS, PARKS, COST ANALYSIS, MICHIGAN, WASTE WATER TREATMENT.

IDENTIFIERS:
*COMBINED SEWERS, *SUSPENDED SOLIDS, BACKWASHING, *MOUNT CLEMENS(MICH).

ABSTRACT:
THE CITY OF MOUNT CLEMENS, MICHIGAN, HAS 50 MILES OF COMBINED SEWERS SERVING 3.06 SQUARE MILES, AND 10 MILES OF SANITARY SEWERS PLUS 6 MILES OF STORM SEWERS SERVING AN ADDITIONAL 0.8 SQUARE MILES. INTERCEPTORS WERE DESIGNED TO COLLECT STORMWATER OVERFLOWS FROM TWO LOCATIONS ON THE COMBINED SEWERS, SERVING 212 ACRES AND PUMP IT TO A SMALL LAKE. THIS FIRST LAKE PROVIDED SETTLING, NATURAL AND MECHANICAL SURFACE AERATION, AND BOTH AEROBIC AND ANAEROBIC DIGESTION. STORM OVERFLOWS COULD THEN BE DRAWN OFF AT A CONTROLLED 1.0 MGD RATE TO A MICROSTRAINER. ULTRAVIOLET RADIATION EQUIPMENT FOR ALGAE CONTROL WAS ALSO INSTALLED WITH THE MICROSTRAINER, SO THE TWO COMBINED PROVIDED MECHANICAL FILTRATION, SUSPENDED SOLIDS AND INCIDENTAL BOD REMOVAL, AND ALGAE REMOVAL. BEFORE ENTERING THE SECOND LAKELET, THE EFFLUENT ALSO IS SUBJECTED TO CHLORINE-CHLORINE DIOXIDE DISINFECTION. LAKELET 2 WAS DESIGNED TO PROVIDE CHLORINE CONTACT TIME, NATURAL SURFACE AERATION, AND PHOTOSYNTHETIC OXYGENATION, WHILE LAKELET 3 PROVIDED MECHANICAL AS WELL AS SURFACE AERATION. A PRESSURE SAND FILTER WAS THEN DESIGNED AS A POLISHING STEP. THE ENTIRE TREATMENT SYSTEM WAS INCLUDED AS PART OF AN INNER-CITY PARK. COST OF THIS SYSTEM WAS $7000 PER ACRE BENEFITED, WHEREAS COST OF SEWER SEPARATION ONLY WAS ESTIMATED AT $15,000 PER ACRE. (LOWRY-TEXAS)

FIELD 05D

ACCESSION NO. W72-00042

EFFECTS OF ZOOPLANKTON ON ALGAE IN WESTHAMPTON LAKE,

RICHMOND UNIV., VA. DEPT. OF BIOLOGY.

JOHN W. BISHOP.

AVAILABLE FROM THE NATIONAL TECHNICAL INFORMATION SERVICE AS PB-203 961, $3.00 IN PAPER COPY, $0.95 IN MICROFICHE. VIRGINIA POLYTECHNIC INSTITUTE, WATER RESOURCES RESEARCH CENTER BULLETIN 43, 1971. 33 P, 17 FIG, 8 TAB, 22 REF. OWRR A-022-VA (2).

DESCRIPTORS:
*ZOOPLANKTON, *ALGAE, *PHOTOSYNTHESIS, RAINFALL, DIURNAL, FLUCTUATIONS, CHLOROPHYTA, EUGLENOPHYTA, CYANOPHYTA, CHRYSOPHYTA, COPEPODS, CRUSTACEANS, LABORATORY TESTS, VIRGINIA.

IDENTIFIERS:
WESTHAMPTON LAKE (VA).

ABSTRACT:
ASSEMBLAGES OF ALGAE WITH DIFFERENT CONCENTRATIONS OF ZOOPLANKTERS WERE KEPT IN CYLINDERS OF POLYETHYLENE AND FLEXIGLASS, SUSPENDED IN THE LAKE. ANALYSES OF WATER AND BOTH GROUPS OF ORGANISMS WERE CONDUCTED INSIDE AND OUTSIDE THESE AQUARIUMS. ROTIFERS EXHIBITING NO CONSISTENT VERTICAL MIGRATION DECREASED THE DAILY PHOTOSYNTHESIS OF ALGAE BY 22%. VERTICALLY MIGRATING COPEPODS AND CLADOCERANS EXERTED NO SIGNIFICANT EFFECT ON THE PHOTOSYNTHESIS. NONE OF THE ANIMAL GROUPS INFLUENCED THE PHOTOSYNTHETIC EFFICIENCY OF ALGAE, THEIR CARBOHYDRATE PRODUCTION PER UNIT OF CHLOROPHYLL. PHOTOSYNTHESIS AND PHOTOSYNTHETIC EFFICIENCY WERE INCREASED SEVERAL DAYS AFTER RAINFALL. (WILDE-WISCONSIN)

FIELD 05C, 02H

ACCESSION NO. W72-00150

TOXICITY OF SELENIUM TO THE BLUE-GREEN ALGAE, ANACYSTIS NIDULANS AND ANABAENA VARIABILIS,

UDAIPUR UNIV. (INDIA). DEPT. OF BOTANY.

H. D. KUMAR, AND G. PRAKASH.

ANNALS OF BOTANY, VOL 35, P 697-705, 1971. 5 FIG, 12 REF.

DESCRIPTORS:
*TOXICITY, *CYANOPHYTA, SULFUR, ALGAE, GRANULES, INHIBITION, METABOLISM, CULTURES, ENVIRONMENTAL EFFECTS, PLANT GROWTH, SULFATE, LABORATORY TESTS, CYTOLOGICAL STUDIES.

IDENTIFIERS:
*SELENIUM, *ANABAENA VARIABILIS, *ANACYSTIS NIDULANS, GROWTH INHIBITION, SELENATE, SELENITE, SELENOMETHIONINE, SELENOPURINE, ANTIMETABOLITES.

ABSTRACT:
COMPARISON WAS MADE OF TOXIC EFFECTS OF SELENATE, SELENITE, SELENOMETHIONINE, AND SELENOPURINE ON DIFFERENT GROWTH PHASES OF ANACYSTIS NIDULANS AND ANABAENA VARIABILIS IN CONSIDERING SELENIUM TOXICITY. A PROTECTIVE ROLE OF SULFUR AGAINST SELENIUM TOXICITY WAS SUGGESTED AS THESE COMPOUNDS WERE LESS TOXIC IN CULTURE MEDIUM CONTAINING SULFATE THAN IN SULFUR-FREE MEDIUM. GROWTH INHIBITORY EFFECTS OF SELENATE AND SELENITE WERE STUDIED ON ALGAE GROWN ON AGAR MEDIUM AND IN LIQUID MINERAL MEDIUM. SELENOMETHIONINE AND SELENOPURINE WERE STUDIED ONLY IN LIQUID CULTURE. GROWTH WAS ESTIMATED BY MEASURING OPTICAL DENSITY OF CULTURE TUBES AND PERCENT SURVIVAL SCORED BY COLONY COUNTING. DOSE-RESPONSE CURVES OF BOTH ALGAE SUGGEST COMPARATIVELY GREATER KILLING BY SELENITE THAN BY SELENATE. ANACYSTIS NIDULANS IS THREE TIMES MORE TOLERANT TO SELENITE AND SELENATE KILLING THAN ANABAENA VARIABILIS. SELENITE IS MORE TOXIC IN AGAR PLATE CULTURES AND LESS TOXIC IN LIQUID CULTURES THAN SELENATE. CELLS GROWN IN THE HIGHEST GROWTH-PERMITTING CONCENTRATION OF SELENITE IN LIQUID MEDIUM FORM A VARIABLE NUMBER OF RED GRANULES. NO SUCH GRANULATION OCCURS IN CELLS TREATED WITH OTHER SELENO-COMPOUNDS, INDICATING THAT THE MODE OF SELENITE ACTION IN BLUE-GREEN ALGAE DIFFERS FROM THAT OF OTHER SELENIUM COMPOUNDS. (JONES-WISCONSIN)

FIELD 05C

ACCESSION NO. W72-00153

SOME ECOLOGICAL EFFECTS ON THE USE OF PARAQUAT FOR THE CONTROL OF WEEDS IN SMALL LAKES,

NATURE CONSERVANCY, ABBOTS RIPTON (ENGLAND). MONKS WOOD EXPERIMENTAL STATION; AND IMPERIAL CHEMICAL INDUSTRIES LTD., JEALOTT'S HILL (ENGLAND). JEALOTT'S HILL RESEARCH STATION; AND NOTTINGHAM UNIV. (ENGLAND). DEPT. OF BOTANY.

J. M. WAY, J. F. NEWMAN, N. W. MOORE, AND F. W. KNAGGS.

THE JOURNAL OF APPLIED ECOLOGY, VOL 8, NO 2, P 509-532, 1971.

DESCRIPTORS:
*HERBICIDES, *AQUATIC WEED CONTROL, *PARAQUAT, *ENVIRONMENTAL EFFECTS, DIQUAT, ALGAE, ALGAL CONTROL, DIATOMS, PERSISTENCE, BACTERIA, BIRDS, FISH, ON-SITE INVESTIGATION, METHODOLOGY, INVERTEBRATES, AMPHIBIANS, TOXICITY, CHARA, SUBMERGED PLANTS, PROTOZOA, EUGLENA, CHLAMYDOMONAS, CHLOROPHYTA, SCENEDESMUS, CYANOPHYTA, MOLLUSKS, MITES, ANNELIDS, CRUSTACEANS.

IDENTIFIERS:
BIOLOGICAL EFFECTS, NOTTINGHAM(ENGLAND), POLYGONUM, TYPHA, ELODEA.

ABSTRACT:
THE RESULTS OF PARAQUAT APPLICATIONS TO A SERIES OF SMALL LAKES WERE STUDIED AND THE BIOLOGICAL EFFECTS AND PARAQUAT RESIDUE DETERMINATIONS DESCRIBED. IN TWO EXPERIMENTS IN LAKES AT OXTON, NOTTINGHAMSHIRE (ENGLAND), 0.5 MG/L PARAQUAT ERADICATED ALL SUBMERGED AND FLOATING PLANTS, EXCEPT POLYGONUM AMPHIBIUM AND CHARA, WITHIN 32 DAYS OF APPLICATION. ONE LAKE REMAINED SUBSTANTIALLY FREE OF VEGETATION FOR TWO YEARS AFTER TREATMENT. NO MATS OF FILAMENTOUS ALGAE DEVELOPED IN THE TREATED LAKES ALTHOUGH THEY WERE WIDESPREAD IN ADJACENT UNTREATED LAKES. CHANGES IN SPECIES AND POPULATIONS OF PLANKTONIC ALGAE AND DIATOMS AND INCREASES IN POPULATION OF BACTERIA SOON AFTER TREATMENT WERE OBSERVED. SURVIVAL OF CAPTIVE POPULATIONS OF ASELLUS, LYMNAEA AND SIALIS LARVAE UP TO FOUR DAYS AFTER TREATMENT AND OBSERVATIONS ON OTHER FREE-LIVING INVERTEBRATES WERE RECORDED. THE CHEMICAL DID NOT APPARENTLY CAUSE ANY IMPORTANT MORTALITY OF INVERTEBRATES. PARAQUAT DISAPPEARED RAPIDLY FROM THE WATER. SIGNIFICANT QUANTITIES OF THE CHEMICAL WERE FOUND IN THE WEEDS WITHIN 24 HOURS OF APPLICATION. THERE WAS A GRADUAL BUILDUP OF PARAQUAT RESIDUES IN BOTTOM DEPOSITS UP TO 32 AND SUBSEQUENT 197 DAYS AFTER APPLICATION WITH A MARKED FALL-OFF AT 364 DAYS. (JONES-WISCONSIN)

FIELD 04A, 05C

ACCESSION NO. W72-00155

THE USE OF RADIOACTIVE CARBON METHOD IN STUDIES OF TROPHIC RELATIONSHIPS OF PLANKTON (PRIMENENIE RADIOUGLERODNOGO METODA DLYA ISUCHENIYA TROFICHESKIKH VZAIMOOTNOSHENII V PLANKTONE),

AKADEMIYA NAUK SSSR, LENINGRAD. ZOOLOGICHESKII INSTITUT.

S. M. VARDAPETYAN, B. L. GUTELMACHER, AND N. G. OZERETSKOVSKAYA.

DOKLADY AKADEMII NAUK SSSR, VOL 197, NO 3, P 705-707, 1971. 1 FIG, 2 TAB, 5 REF.

DESCRIPTORS:
*ANALYTICAL TECHNIQUES, *RADIOACTIVITY TECHNIQUES, *PLANKTON, PRIMARY PRODUCTIVITY, FOOD HABITS, ALGAE, ZOOPLANKTON, CARNIVORES, DAPHNIA.

IDENTIFIERS:
BOSMINA, CERIODAPHNIA, EUDIAPTOMUS.

ABSTRACT:
SAMPLES OF PLANKTON FROM TWO LAKES WERE PLACED IN 20 LITER CONTAINERS WITH LAKE WATER AND C-14 AS BICARBONATE OF SODIUM IN CONCENTRATION OF 10 MICRO/LITER. IN INTERVALS FROM 1 TO 8 DAYS, THE RADIOACTIVITY OF PHYTOPLANKTON WAS DETERMINED ON 300 ML ALIQUOTS. THE REMAINDER WAS FILTERED THROUGH NO 62 SIEVE. THE RETAINED ZOOPLANKTON WAS FIXED IN FORMALIN AND DISTRIBUTED ACCORDING TO SPECIES. AFTER 24 HOURS DRYING IN A DESICCATOR, THE RADIOACTIVITY OF DIFFERENT SPECIES WAS DETERMINED. THE RESULTS PROVIDED A PICTURE OF UTILIZATION OF THE PRIMARY PRODUCTION BY BOSMINA, DAPHNIA, CERIODAPHNIA, AND EUDIAPTOMUS SPECIES. RESULTS OBTAINED WITH NUTRITION OF PREDATORS WERE LESS WELL PRONOUNCED. (WILDE-WISCONSIN)

FIELD 05C, 07A, 05A

ACCESSION NO. W72-00161

ENVIRONMENTAL CONTROL OF GROWTH AND DIFFERENTIATION IN MULTICELLULAR PHCTOSYNTHETICS,

WASHINGTON UNIV., ST. LOUIS, MO.

H. WAYNE NICHOLS.

AVAILABLE FROM THE NATIONAL TECHNICAL INFORMATION SERVICE AS AD-725 497, $3.00 IN PAPER COPY, $0.95 IN MICROFICHE. FINAL REPORT NOV 17, 1970. 6 P. ONR GRANT NR 104-801.

DESCRIPTORS:
*ALGAE, *ECOLOGY, ENVIRONMENT, PHOTOPERIODISM, GROWTH RATES, TEMPERATURE, BIOCHEMISTRY, PHYSIOLOGY, LIGHT.

IDENTIFIERS:
PHOTOSENSITIVITY.

ABSTRACT:
A SUMMARY IS GIVEN OF A RESEARCH PROJECT WHICH CONCERNED ITSELF WITH BASIC DEVELOPMENT AND BEHAVIORAL ASPECTS OF UNICELLULAR AND MULTICELLULAR RED ALGAE. VARIOUS PHYSIOLOGICAL AND BIOCHEMICAL CHARACTERISTICS OF THE ALGAE HAVE BEEN STUDIED, AND TECHNIQUES FOR CONTROL OF CERTAIN DEVELOPMENTAL ASPECTS OF THE CRGANISMS IN THE LABORATORY HAVE BEEN DEVELCPED. TO DATE THE RESEARCH HAS BASICALLY INVOLVED PHYSICAL ASPECTS OF THE ENVIRONMENT. HOWEVER, IT IS NOW CLEARLY EVIDENT THAT CHEMICAL EVENTS OCCURRING WITHIN THE DIFFERNTIATING CELLULAR SYSTEMS MUST BE STUDIED AND RELATED TO SUCH PHYSICAL CONTROL MECHANISMS AS PHOTOPERIODICITY, TEMPERATURE AND LIGHT QUALITY.

FIELD 05B

ACCESSION NO. W72-00377

BIOLCGICAL--GAMMA-RADIATION SYSTEM FOR SEWAGE PROCESSING,

ENERGY SYSTEMS, INC., MELBCURNE, FLA.

L. A. MANN.

ISCTOPES AND RADIATION TECHNOLOGY, VOL 8, NO 4, P 439-444, SUMMER, 1971. 4 FIG, 1 TAB, 4 REF.

DESCRIPTORS:
*SEWAGE EFFLUENTS, *SEWAGE TREATMENT, *ECONOMIC JUSTIFICATION, *COBALT, *RADIATION, COSTS, SEWAGE BACTERIA, ALGAE, CCLIFORMS, INSECTICIDES.

IDENTIFIERS:
*BIOLOGICAL--GAMMA-RADIATION SYSTEM, MODULAR SYSTEM, COBALT-60.

ABSTRACT:
A BIOLOGICAL--GAMMA-RADIATION METHOD FOR PURIFYING SEWAGE WAS DEVELOPED, AND BOTH A PILOT AND COMMERCIAL PLANT WERE DESIGNED, CONSTRUCTED, AND TESTED. THE PLANT DESIGN INCORPCRATES SEVERAL MODULES INCLUDING A WET WALL, A BIOLCGICAL TREATMENT UNIT, AN IRRADIATOR, A PRIMARY FILTER SYSTEM, AND AN ACTIVATED-CHARCOAL (POLISHING) FILTER. AT DIFFERENT FLOW RATES ON NORMAL SEWAGE, TESTS INDICATE THAT ALMOST ALL COLIFORM BACTERIA WERE KILLED, AND BOTH BIODEGRADABLE AND NONBIODEGRADABLE DETERGENTS WERE MORE THAN 90% DESTROYED. LIMITED TESTS ON PARATHION IN WATER RESULTED IN 25-30% DESTRUCTION OF THIS INSECTICIDE. THERE IS SOME INDICATION THAT ALGAE WILL NOT GROW IN THE EFFLUENT. FURTHERMORE, SETTLEABLE-SOLIDS CONCENTRATION AND TURBIDITY WERE DECREASED, AND A SATISFACTORY BOD LEVEL WAS MAINTAINED IN THE TWO PLANTS. THE COST OF BUILDING AND OPERATING A PLANT USING A BIOLOGICAL--GAMMA-RADIATION SYSTEM DEPENDS ON THE QUALITY OF WATER DESIRED AND THE MODULES SELECTED FOR THE PLANT. ESTIMATES INDICATE THAT THE COST OF TREATMENT PLANTS CONSISTING OF A CONVENTIONAL SECONDARY TREATMENT PLUS IRRADIATOR AND PRIMARY FILTRATION MODULES IS LESS THAN THE COST OF CONVENTIONAL ADVANCED-TREATMENT PLANTS. (SETTLE-WISCONSIN)

FIELD 05D

ACCESSION NO. W72-00383

DISINFECTION OF ALGAL LADEN WATERS,

NOTRE DAME UNIV., IND.

WAYNE F. ECHELBERGER, JR., JOSEPH L. PAVONI, PHILIP C. SINGER, AND MARK W. TENNEY.

JOURNAL OF THE SANITARY ENGINEERING DIVISION, PROCEEDING OF ASCE, VOL 97, NO SA 5, OCTOBER, 1971. P 721-730, 9 FIG, 10 REF.

DESCRIPTORS:
*CHLORINATION, *ALGAE, *PHOTOSYNTHESIS, SOLUBILITY, CHEMICAL OXYGEN DEMAND, BIOCHEMICAL OXYGEN DEMAND, FILTRATION, DISINFECTION, FLOCCULATION, *WASTE WATER TREATMENT.

IDENTIFIERS:
*LYSIS, *CONTACT TIME, CELLULAR METABOLITES.

ABSTRACT:
BATCH LABORATORY SCALE EXPERIMENTS WERE USED TO DEMONSTRATE THE EFFECT OF ALGAE ON THE CHLORINATION PROCESS. TOTAL CHLORINE RESIDUALS WERE MEASURED BY THE STANDARD ORTHO TOLIDINE PROCEDURE. STANDARD JAR TEST FLOCCULATION STUDIES USING CHLORINE AND A CATIONIC POLYAMINE AS COAGULANTS WERE ALSO CONDUCTED. RESULTS INDICATED THAT ALGAL CELLS DO EXERT A CHLORINE DEMAND, AND AS THE ALGAL CONCENTRATION INCREASES, THE CHLORINE DOSAGE REQUIRED TO MAINTAIN A CERTAIN FREE RESIDUAL LEVEL INCREASES. CELLULAR CHLOROPHYLL FRACTIONS OF THE ALGAE WERE REDUCED BY CHLORINE DISINFECTION, RESULTING IN IMPAIRMENT OF THE PHOTOSYNTHETIC CAPABILITY OF THE SLUDGE. APPARENT ALGAL CELL LYSING A CELLULAR METABOLITE LEAKAGE FOLLOWING CHLORINATION TO A DESIRABLE RESIDUAL LEVEL SIGNIFICANTLY INCREASED THE SOLUBLE ORGANIC CONCENTRATION OF THE SUSPENDING MEDIUM. FROM THE ALGAE REMOVAL STANDPOINT, CHLORINATION DID ENHANCE FLOCCULATION, PROBABLY BECAUSE SOME OF THE CELLULAR METABOLITES RELEASED WERE NATURAL ALGAL FLOCCULANTS. WIDESPREAD USAGE OF CHLORINATION FOR DISINFECTION MAY SOON REQUIRE ALGAE REMOVAL PRIOR TO DISINFECTION TO INSURE ADEQUATE DISINFECTIONS. (LOWRY-TEXAS)

FIELD 05D

ACCESSION NO. W72-00411

CHLORINATION DYNAMICS IN WASTEWATER EFFLUENTS,

TECHION - ISRAEL INST. OF TECH., HAIFA (ISRAEL). SANITARY ENGINEERING LAB.

YEHUDA KOTT.

JOURNAL OF THE SANITARY ENGINEERING DIVISION, PROCEEDINGS OF ASCE, VOL 97, NO SA 5, OCTOBER 1971, P 647-659. 2 FIG, 10 TAB, 17 REF.

DESCRIPTORS:
*CHLORINATION, *DISINFECTION, *OXIDATION LAGOONS, COLIFORMS, ALGAE, TEMPERATURE, ANALYTICAL TECHNIQUES, WATER QUALITY CONTROL, *WASTE WATER TREATMENT.

IDENTIFIERS:
*CYSTS, ISRAEL.

ABSTRACT:
THE RESPONSE OF COLIFORMS, ENDAMOEBA HISTOLYTICA CYSTS, AND ALGAE TO CHLORINATION WAS INVESTIGATED ON A LABORATORY SCALE. EFFLUENT USED WAS OBTAINED FROM A TRICKLING FILTER PLANT, AND FROM AN ARTIFICIAL OXIDATION POND. APPLICATION OF 8 MG/L OF CHLORINE TO THE TRICKLING FILTER EFFLUENT REDUCED THE COLIFORM COUNT FROM 10 TO THE 7TH POWER ORGANISMS PER 100 ML TO LESS THAN 100 ORGANISMS PER 100 ML. APPLICATION OF 14 MG/L ACTUALLY PRODUCED A GREATER NUMBER OF COLIFORM ORGANISMS IN THE EFFLUENT THAN WAS NOTICED FOR THE 8 MG/L DOSAGE. THE PROBABLE CAUSE WAS ASSUMED TO BE THE BREAK-DOWN OF ORGANIC MATERIAL AND RELEASE OF ENTRAPPED COLIFORM ORGANISMS. APPLICATION OF THE 8 MG/L DOSAGE TO SOLUTIONS CONTAINING THE ENDAMOEBA HISTOLYTICA CYSTS ALSO PRODUCED A RAPID AND THOROUGH KILL, ALTHOUGH THE DEATH RATE WAS QUITE TEMPERATURE DEPENDENT. ALGAE WAS UNAFFECTED FOR A TWO HOUR INTERVAL, EXCEPT FOR THE CESSATION OF GROWTH. SINCE THE CHLORINE DID NOT ATTACK THE ALGAE FOR 2 HOURS, MUCH LONGER THAN THE NORMAL CONTACT TIME, THEN, CHLORINATION SHOULD BE WELL SUITED FOR DISINFECTION OF OXIDATION POND EFFLUENT. (LOWRY-TEXAS)

FIELD 05D

ACCESSION NO. W72-00422

PLANKTON ASSOCIATIONS AND RELATED FACTORS IN A HYPEREUTROPHIC LAKE,

WASHINGTON UNIV., SEATTLE. WATER AND AIR RESOURCES DIV.

RONALD M. BUSH, AND EUGENE B. WELCH.

AVAILABLE FROM NATIONAL TECHNICAL INFORMATION SERVICE AS PB-204 230, $3.00 IN PAPER COPY, $0.95 IN MICROFICHE. PARTIAL PROJECT COMPLETION REPORT, AUGUST 15, 1971. 34 P, 3 FIG, 2 TAB, 37 REF. OWRR A-034-WASH(1).

DESCRIPTORS:
WATER POLLUTION EFFECTS, *EUTROPHICATION, LAKES, *NUTRIENTS, ALGAL CONTROL, *CYANPHYTA, *NUISANCE ALGAE, PLANKTON, WASHINGTON, DIATOMS, CHLOROPHYTA.

IDENTIFIERS:
HYPEREUTROPHIC LAKES, *MOSES LAKE(WASH).

ABSTRACT:
CLUSTER ANALYSIS WAS USED TO GROUP SAMPLES COLLECTED FROM TEN STATIONS IN MOSES LAKE, WASHINGTON, ACCORDING TO THE SIMILARITY OF THEIR CONTAINED ALGAL SPECIES. DURING THE PERIOD 1968 TO 1970, NINE RECURRING DISTINCT SAMPLE GROUPS, OR ALGAL POPULATIONS, WERE IDENTIFIED. OF THE NINE, THREE WERE MOST DISTINCT; THEY CONSISTENTLY RECURRED AT THE SAME STATIONS, AND WERE DOMINATED BY DIATOM, GREEN, AND BLUE-GREEN ALGAE, RESPECTIVELY. OF THE SIX SPECIES OF BLUE-GREENS THAT CHARACTERIZED THAT POPULATION, THE RECREATIONALLY NUISANCE FORMS, APHANISOMENON FLOS-AQUAE AND MICROCYSTIS ACRUGINOSA, WERE DOMINANT. THE BLUE-GREEN POPULATION WAS THE MOST WIDE-SPREAD IN THE LAKE AND OCCURRED IN WATERS THAT WERE WARMEST AND CONTAINED THE LOWEST CONCENTRATIONS OF INORGANIC NUTRIENTS: NITROGEN, PHOSPHORUS, AND CARBON. GREEN ALGAE DOMINATED IN WATERS THAT RECEIVED TREATED SEWAGE EFFLUENT AND CONTAINED RELATIVELY HIGH CONCENTRATIONS OF NUTRIENTS. AS THE NUTRIENT CONTENT DECLINED PROCEEDING AWAY FROM THAT AREA, BLUE-GREEN ALGAE BECAME DOMINANT. TEMPORAL VARIATION IN BIOMASS (CHLOROPHYLL CONTENT) OF THE BLUE-GREEN POPULATION WAS INVERSELY RELATED TO PHOSPHATE CONTENT, BUT NOT TO THE OTHER NUTRIENTS. THESE RESULTS SUPPORT THE HYPOTHESIS THAT NUISANCE BLUE-GREEN ALGAE DOMINATE IN SHALLOW EUTROPHIC LAKES DURING WARM SUMMER MONTHS WHEN AMBIENT NUTRIENT CONTENT IS LOW BECAUSE, UNDER THESE CONDITIONS, THEY APPARENTLY OUT-COMPETE OTHER FORMS FOR NUTRIENTS. (SEE ALSO W72-00799)

FIELD 05C, 05G

ACCESSION NO. W72-00798

DILUTION AS A CONTROL FOR NUISANCE ALGAL BLOOMS,

WASHINGTON UNIV., SEATTLE. WATER AND AIR RESOURCES DIV.

EUGENE B. WELCH, JAMES A. BUCKLEY, AND RONALD M. BUSH.

PARTIAL PROJECT COMPLETION REPORT, AUGUST 31, 1971. 39 P, 14 FIG, 8 TAB, 18 REF. OWRR A-034-WASH(2).

DESCRIPTORS:
*WASTE DILUTION, WATER POLLUTION CONTROL, *NUISANCE ALGAE, *EUTROPHICATION, LAKES, WASHINGTON, *ALGAL CONTROL, NUTRIENTS, BIOMASS, *CYANOPHYTA, PLANT GROWTH, PLANKTON.

IDENTIFIERS:
*MOSES LAKE(WASH), HYPEREUTROPHIC LAKES.

ABSTRACT:
EXPERIMENTS WERE CONDUCTED IN SITU IN HYPEREUTROPHIC MOSES LAKE IN EASTERN WASHINGTON TO DETERMINE THE EFFECT OF LOW-NUTRIENT DILUTION WATER ON GROWTH AND DOMINANCE OF NUISANCE BLUE-GREEN ALGAE, THE MECHANISM OF EFFECT AND SUGGEST POSSIBLE CONTROL MEASURES BY COMPARISON WITH RESULTS FROM A SUCCESSFUL CONTROL PROGRAM IN GREEN LAKE IN SEATTLE. THE ADDITION OF LOW-NUTRIENT COLUMBIA RIVER WATER TO MOSES LAKE WATER REDUCES THE SUBSEQUENT MAXIMUM BIOMASS OF NUISANCE BLUE-GREEN ALGAE ATTAINED IN THE DILUTION WATER-LAKE WATER MIXTURE IN DIRECT PROPORTION TO THE AMOUNT OF DILUTION WATER ADDED. IN CONTRAST TO BIOMASS, THE RELATIONSHIP BETWEEN DILUTION WATER ADDITION AND MAXIMUM GROWTH RATE OF BLUE-GREEN ALGAE IS NON-LINEAR. GROWTH RATE REMAINED HIGH (TWO DOUBLINGS PER DAY) AND RELATIVELY CONSTANT AT LAKE WATER CONCENTRATIONS DOWN TO ABOUT 50% AND THEN DROPPED RAPIDLY TO NEGATIVE RATES AT 0% LAKE WATER. COMPARISON OF RESULTS OF DILUTION WATER ADDITION TO GREEN LAKE IN SEATTLE WITH EXPERIMENTAL AND CALCULATED POTENTIAL EFFECTS OF DILUTION WATER IN PARKER HORN OF MOSES LAKE ALLOWED ESTIMATES OF DILUTION RATES THAT MIGHT OFFER SOME CONTROL OF BLUE-GREEN ALGAE BLOOMS. ADDITION OF LOW-NUTRIENT RIVER WATER AT ABOUT 1% PER DAY WOULD BE EXPECTED TO RESULT IN A LAKE WATER--RIVER WATER MIXTURE THAT WOULD REDUCE MAXIMUM GROWTH RATE BY ONE HALF IN ABOUT 150 DAYS AND REACH A STEADY STATE BIOMASS IN ABOUT 200 DAYS. (SEE ALSO W72-00798)

FIELD 05G, 05C

ACCESSION NO. W72-00799

NITROGEN REMOVAL AND IDENTIFICATION FOR WATER QUALITY CONTROL,

WASHINGTON UNIV., SEATTLE. DEPT. OF CIVIL ENGINEERING.

DALE A. CARLSON.

AVAILABLE FROM NATIONAL TECHINAL INFORMATION SERVICE AS PB-204 231, $3.00 IN PAPER COPY, $0.95 IN MICROFICHE. AUGUST 15, 1971. 52 P, 12 FIG, 8 TAB, 85 REF. OWRR A-040-WASH(1).

DESCRIPTORS:
*NITRIFICATION, *DENITRIFICATION, WATER QUALITY CONTROL, EUTROPHICATION, ACTIVATED SLUDGE, DESIGN, NITROGEN, DIFFUSION, *ALGAL CONTROL, AERATION, *WASTE WATER TREATMENT.

IDENTIFIERS:
*ION-SELECTIVE CATHODES.

ABSTRACT:
REMOVAL OF NITROGEN FROM WASTE WATER SYSTEMS BY BIOLOGICAL NITRIFICATION AND DENITRIFICATION IN AN ACTIVATED SLUDGE SYSTEM IS CONSIDERED. ONE OF THE MAJOR METHODS PROPOSED FOR CONTROLLING EXCESS ALGAL PRODUCTION IN RECEIVING WATERS, THE PROCESS CONSISTS OF AERATING ACTIVATED SLUDGE MIXED LIQUOR SOLIDS FOR DETENTION TIMES SUFFICIENT FOR NITRIFYING BACTERIA TO OXIDIZE AMMONIA TO NITRATE. THE MIXED LIQUOR THEN GOES TO AN ANAEROBIC DENITRIFICATION ZONE WHERE THE NITRATE IS BIOLOGICALLY CONVERTED TO NITROGEN GAS. PILOT PLANT STUDIES PRECEDING THE CURRENT STUDIES ARE DISCUSSED PRIOR TO PRESENTATION OF THE WORK ON PURE CULTURE KINETIC STUDIES AND DEVELOPMENT OF ICN SELECTIVE CATHODES FOR IDENTIFICATION OF NITROGEN FRACTIONS. FOR THE CONDITIONS USED, THE RATE OF DENITRIFICATION BY PSEUDOMONAS DENITRIFICANS FOLLOWED THE MICHAELIS PATTERN, NOT BEING INHIBITED BY HIGH CONCENTRATIONS OF CH3COONA AS THE SUBSTRATE. THE THEORETICAL MAXIMUM RATE OF DENITRIFICATION (V) WAS ABOUT 10 MICRON 1N2/HR/MG BACTERIAL CELL AND THE THEORETICAL VALUE OF KS WAS 5.8 MM WITH A CELL CONCENTRATION OF 3400 MG/L. THE OBTAINED KS VALUE PROMISES A REASONABLE RATE OF DENITRIFICATION UNDER THE GIVEN CONDITIONS. ALTHCUGH THE RATES DENITRIFICATION AT THE VARIOUS KNO3 CONCENTRATIONS DID NOT FOLLOW THE MICHAELIS PATTERN, INHIBITING EFFECTS OF NITRATE WERE NOT OBSERVED AT KNO3 CONCENTRATIONS AS HIGH AS 8.0 MM/L. THE USE CF ION SELECTIVE CATHODES FOR DETECTING IONS SUCH AS THE NITROGENS WAS INVESTIGATED. IT APPEARS THAT THE SELECTIVITY OF ION-SELECTIVE CATHODES CAN BE INCREASED GREATLY BY THE ADDITION OF TIME DEPENDENT RETARDING VOLTAGES.

FIELD 05D

ACCESSION NO. W72-00800

BIOASSAYS TO DETERMINE ALGAL GROWTH POTENTIAL OF MICRONUTRIENTS,

WASHINGTON STATE UNIV., PULLMAN. WATER RESEARCH CENTER.

WILLIAM H. FUNK, RICHARD J. CONDIT, AND WAYNE T. CRANEY.

AVAILABLE FROM NATIONAL TECHNICAL INFORMATION SERVICE AS PB-204 232, $3.00 IN PAPER COPY, $0.95 IN MICROFICHE. JULY 1971. 66 P, 22 FIG, 12 TAB, 45 REF, 3 APPEND. OWRR A-033-WASH(1).

DESCRIPTORS:
WATER POLLUTION EFFECTS, *MICROORGANISMS, ALGAL CONTROL, *POLLUTANT IDENTIFICATION, ANALYTICAL TECHNIQUES, *ALGAE, *NUTRIENTS, LAKES, PLANT GROWTH, *MOLYBDENUM, BIOASSAY, NEUTRON ACTIVATION ANALYSIS.

IDENTIFIERS:
*MICRO-NUTRIENTS, *ALGAL GROWTH.

ABSTRACT:
ADAPTING CONTINUOUS CULTURE TECHNIQUES FOR THE MEASUREMENT OF MICRO-NUTRIENT EFFECTS UPON ALGAE GROWTH HAS BEEN SHOWN TO BE A SATISFACTORY METHOD. SEVERAL REFINEMENTS AND MODIFICATIONS OF PREVIOUSLY UTILIZED EQUIPMENT FOR THE STUDY OF ALGAE MACRONUTRIENT UPTAKE MADE IT POSSIBLE TO TEST THE EFFECTS OF MOLYBDENUM UPON SCENEDESMUS QUADRICAUDA. HOWEVER, WITH THE PROGRESSIVELY REDUCED SUBSTRATE LEVELS OF MOLYBDENUM TWO PROBLEMS BECAME INCREASINGLY APPARENT: (1) MOLYBDENUM CONTAMINATION OF THE MEDIA BY MAGNESIUM SULFATE APPROACHED OR SURPASSED THE EXPERIMENTAL LEVELS IN THE DILUTION CULTURES AND PREVENTED THE ATTAINMENT OF A MOLYBDENUM DEFICIENT CONTROL; AND (2) THE EXTREMELY LOW SUBSTRATE LEVELS OF MOLYBDENUM COULD NOT BE DETECTED BY THE NEUTRON ACTIVATION ANALYSIS DUE TO INTERFERENCE BY THE HIGH PHOSPHATE CONCENTRATIONS IN THE MEDIA. HOWEVER, THE ANALYSES WERE USEFUL WHERE EITHER THE PHOSPHATE CONCENTRATIONS WERE LOW (LAKE WATER AND REAGENT CHEMICALS) OR MOLYBDENUM CONCENTRATIONS WERE RELATIVELY HIGH (TRIAL IV SUBSTRATES). THE CONTAMINATION PROBLEM MAY BE RESOLVED THROUGH THE PURIFICATION OF THE REAGENT CHEMICALS--HOWEVER, GREAT CARE MUST BE EXERCISED, SINCE FURTHER CONTAMINATION CAN RESULT THROUGH CARELESS TECHNIQUES. IMPROVED MOLYBDENUM ANALYSIS BY NEUTRON ACTIVATION WILL ALLOW THE FUTURE USE OF CONTINUOUS FLOW KINETICS ANALYSIS--A VALUABLE TOOL IN ASSESSING THE EFFECTS OF TRACE ELEMENTS IN ALGAL GROWTH.

FIELD 05C, 05A

ACCESSION NO. W72-00801

INFLUENCE OF MEDIUM PH AND CARBON DIOXIDE ON ASSIMILATION OF SOME ORGANIC ACIDS BY SCENEDESMUS QUADRICAUDA,

MOSCOW STATE UNIV. (USSR), DEPT. OF MICROBIOLOGY; AND MOSCOW STATE UNIV. (USSR). DEPT. OF SOIL BIOLOGY.

N. F. PISKUNKOVA, AND M. N. PIMENOVA.

MIKROBIOLOGIA (USSR), VOL 39, NO 6, P 854-857, 1970. 2 TAB, 18 REF.

DESCRIPTORS:
*CYANOPHYTA, *GROWTH RATES, *ORGANIC ACIDS, *HYDROGEN ION CONCENTRATION, METABOLISM, CARBON DIOXIDE, ALGAE, CULTURES, ACIDITY, ORGANIC COMPOUNDS.

IDENTIFIERS:
SCENEDESMUS QUADRICAUDA, ACETATE, PYRUVATE, LACTATE.

ABSTRACT:
THE NUTRIENT MEDIUM FOR A PURE CULTURE OF GREEN ALGA, SCENEDESMUS QUADRICAUDA, WAS SUPPLEMENTED BY ACETIC, PYRUVIC AND LACTIC ACIDS, ADDED AS SODIUM SALTS IN CONCENTRATION OF 0.05%. THE EFFECT OF ACIDS WAS RECORDED IN TERMS OF DRY BIOMASS ACCRETIONS. IN THE PRESENCE OF LIGHT ALL THREE ACIDS STIMULATED THE ALGAL GROWTH, BUT THEIR UTILIZATION OF THE ACIDS WAS INFLUENCED BY THE PH OF THE MEDIUM. IN THE DARK, GOOD GROWTH OF SCENEDESMUS OCCURRED ONLY IN THE PRESENCE OF ACETATE. PASSAGE OF AIR WITH 5% CARBON DIOXIDE STIMULATED THE GROWTH OF TEST PLANTS IN ACID MEDIA (PH 5.0) ENRICHED IN ACETATE AND PYRUVATE, BUT REDUCED THE CONSUMPTION OF ORGANIC ACIDS. THE STUDY SUGGESTS THAT ALGAE IN ACID WATER BASINS DEFICIENT IN CARBON DIOXIDE ADAPT THEMSELVES TO UTILIZING ORGANIC COMPOUNDS. (WILDE-WISCONSIN)

FIELD 05C

ACCESSION NO. W72-00838

MANAGING OUR ENVIRCNMENT.

AGRICULTURAL RESEARCH SERVICE, WASHINGTON, D. C.

DEPT OF AGRICULTURE, WASHINGTON D C, AGRICULTURE INFORMATICN BULLETIN NO 351, APRIL 1971. 48 P.

DESCRIPTORS:
*MANAGEMENT, *ENVIRONMENT, *AGRICULTURE, *WATER POLLUTION CONTROL, SEDIMENTS, FARM WASTES, NUTRIENTS, PHOSPHORUS, ALGAE, NITRATES, WATER REUSE, SALINITY, PESTICIDES, LIVESTOCK, WASTE DISPOSAL, OXIDATION LAGOONS, DEHYDRATION, RUNOFF, RADIOACTIVITY, FALLCUT, BIOCONTROL, INSECT CONTROL, IRRIGATION, PREDATION, PARASITISM, INSECT RESISTANCE, INSECT ATTRACTANTS, PRECIPITATION(ATMOSPHERIC), GENETICS, EROSION CONTROL, AIR POLLUTION, TREES.

IDENTIFIERS:
FEEDLOTS, COMPOSTING, PLANT RESIDUES, RECYCLING FCCD, PROCESSING WASTES, PATHOGENS, BIOENVIRONMENTAL CONTROLS.

ABSTRACT:
SOME OF THE MAJOR PRCBLEMS IN AGRICULTURAL RESEARCH DEALING WITH NEW AND OLDER METHODS OF ENVIRONMENTAL MANAGEMENT ARE DESCRIBED IN AN EFFORT TOWARD INTERESTING THE PUBLIC IN PRESERVATION OF THE QUALITY OF OUR ENVIRONMENT. GENERAL MATERIAL IS PRESENTED UNDER THE SUBJECTS 'PROTECTING LAND, WATER AND WATERWAYS,' 'MANAGEMENT OF FARM WASTES,' 'RECYCLING FOOD PROCESSING WASTES,' 'NEW WAYS TO FIGHT PESTS--ALTERNATIVES TO PESTICIDES,' AND 'A GREEN WORLD--A CLEAN WORLD.' AMONG THE PROBLEMS DISCUSSED ARE PREVENTION OF ANIMAL WASTES REACHING WATERS, PHOSPHORUS FROM HUMAN WASTES AND DETERGENTS, MULTIPLE WATER REUSE, AND SALINITY IN IRRIGATED LANDS OF THE SOUTHWEST. SCIENTISTS ARE TRYING TO PREVENT PESTICIDE RESIDUES IN SOIL AND WATER AND AVOID PESTICIDE OVERUSE. FALLOUT FROM NUCLEAR WEAPON TESTING CALLS FOR VARIOUS DECONTAMINATION TREATMENTS; FOOD PROCESSING WASTE DISPOSAL AND RECYCLING IS DESCRIBED, AND RECOVERY OF POTABLE WATER FROM SEAWATER BY REVERSE OSMOSIS. ALTERNATIVES TO PESTICIDES ARE DESTRUCTION OF INSECTS AND WEEDS BY INTRODUCTION OF PREDATORS, PARASITES, AND PATHOGENS WHICH FEED ON OR INFECT PESTS; RESISTANT VARIETIES, ATTRACTANTS, GENETIC CONTROL, BIOENVIRONMENTAL CONTROLS, ANC HORMONE AND DAYLIGHT MANIPULATION. (JONES-WISCONSIN)

FIELD 05G, 05C

ACCESSION NO. W72-00846

BOTTCM DEPOSITS,

G. W. FOESS, ANC T. H. FENG.

JOURNAL WATER POLLUTICN CONTROL FEDERATION, VOL 43, NO 6, P 1257-1266, 1971. 80 REF.

DESCRIPTORS:
*REVIEWS, *BOTTOM SEDIMENTS, *AQUATIC ENVIRONMENT, SURFACE WATERS, NUTRIENTS, ALGAE, PHCSPHORUS, NITROGEN, EUTROPHICATION, SEDIMENT TRANSPORT, CYCLING NUTRIENTS, DEGRADATION(DECOMPOSITION), SAMPLING, BENTHIC FAUNA, SEDIMENTATION, SEDIMENT DISTRIBUTICN, PHYSICOCHEMICAL PROPERTIES, PESTICIDES, TROPHIC LEVEL, GASES, MUDS, ADSORPTION, COPPER, CLAYS, HUMIC ACIDS, SALINITY, OXIDATION-REDUCTION POTENTIAL, RADIOISOTOPES, DREDGING, ESTUARIES, BIOCHEMICAL CXYGEN DEMAND, DISSOLVED OXYGEN, WATER PCLLUTION SOURCES.

ABSTRACT:
EIGHTY PAPERS ON BOTTOM DEPOSITS ARE REVIEWED. STUDIES, COVERING THE FIELD EXTENSIVELY ARE REPORTED FROM A WIDE RANGE, INCLUDING THE GREAT LAKES, FINLAND, LONG ISLAND SOUND, ITALY, ALASKA, COLUMBIA RIVER, OHIO, AND LAKE GENEVA. BOTTOM DEPOSITS ARE INCREASINGLY RECOGNIZED FOR THEIR EFFECT ON THE QUALITY OF OVERLYING WATERS. DIFFERENT SEDIMENT TYPES, PROBABLY RELATED TO AMOUNT OF AVAILABLE PHOSPHORUS PRESENT, VARY IN THEIR ALGAL SUPPORT. THE SORPTICN OF PHOSPHORUS BY LAKE MUDS UNDER AEROBIC CONDITICNS MAY BE USED TO REMOVE PHOSPHORUS FROM LAKE WATER. SEDIMENT SAMPLING EQUIPMENT SHOULD NOT ONLY PROVICE AN UNDISTURBED SAMPLE BUT ALSO PERMIT RELIABLE SUBSAMPLING OF THE RETRIEVED MATERIAL. THE MOVEMENT OF SEDIMENTS HAS IMPORTANT PRACTICAL RAMIFICATIONS, AND DIVERSE METHODS FOR MEASURING THIS WERE DEVELOPED. THE SURFACE CHEMISTRY OF BOTTOM DEPOSITS RECEIVED ATTENTION FRCM SEVERAL INVESTIGATORS. INTACT SEDIMENT CCRES WERE USED TC MEASURE EPIBENTHIC ALGAL PRODUCTION AND BENTHIC COMMUNITY RESPIRATICN BY FOLLOWING CHANGES IN DO IN THE WATER OVER THE CORES. THE BIOLOGICAL INTERACTIONS OF BOTTOM DEPOSITS WITH PESTICIDES, HEAVY METALS, AND OTHER TRACE CONSTITUENTS IS INCREASINGLY IMPORTANT. TUBIFICID POPULATION HAS BEEN USED AS AN INDICATOR OF AMOUNT CF ORGANIC POLLUTION. (JONES-WISCONSIN)

FIELD 05C, 10

ACCESSION NO. W72-00847

ECOLOGICAL STUDIES OF THE CONNECTICUT RIVER, VERNON, VERMONT.

WEBSTER-MARTIN, INC., SOUTH BURLINGTON, VT.

AVAILABLE FROM NATIONAL TECHNICAL INFORMATION SERVICE AS NP-18953, $3.00 IN PAPER COPY, $0.95 IN MICROFICHE. AEC CONTRACT REPORT NP-18953, JULY 1971. 24 P, 17 FIG, 2 TAB.

DESCRIPTORS:
*NUCLEAR WASTES, *RADIOACTIVITY EFFECTS, *THERMAL POLLUTION, CONNECTICUT, RIVERS, POST-IMPOUNDMENT, NUCLEAR POWERPLANTS, ECOLOGY, RADIOECOLOGY, MONITORING, POLLUTION ABATEMENT, ENVIRONMENTAL EFFECTS, AQUATIC LIFE, FISH POPULATIONS, PHYTOPLANKTON, BENTHIC FAUNA, PERIPHYTON, AQUATIC ALGAE, PLANTS, ENVIRONMENTAL ENGINEERING, COOLING TOWERS, VERMONT.

IDENTIFIERS:
CONNECTICUT RIVER.

ABSTRACT:
ASPECTS OF VERMONT-YANKEE (A NUCLEAR POWERPLANT) OPERATION OF SPECIAL CONCERN TO THE GENERAL PUBLIC ARE POSSIBLE EFFECTS OF WASTE-HEAT DISPOSAL ON THE ECOSYSTEM OF THE RIVER, AND DISCHARGE OF RADIOACTIVE WASTES. THE PLANT DESIGN PERMITS OPERATION OF THE COOLING SYSTEM IN AN OPEN OR A CLOSED-CYCLE MODE OR A HYBRID OF THESE MODES. ALLOWABLE INCREASES IN THE WATER TEMPERATURE OF THE IMPOUNDMENT CREATED BY VERNON DAM HAVE BEEN ESTABLISHED BY THE VERMONT WATER RESOURCES BOARD. A PRE-OPERATIONAL ENVIRONMENTAL SURVEY INCLUDED WATER PROPERTIES (TEMPERATURE, PH, DISSOLVED OXYGEN, AND CONDUCTIVITY), FISH, PHYTOPLANKTON, ALGAL PERIPHYTON, BENTHIC FAUNA, AND VASCULAR PLANTS. SAMPLES OF WATER, MUD, FISH AND VEGETATION ARE COLLECTED ROUTINELY AND ANALYZED FOR RADIOACTIVITY. FISH SAMPLED FROM BOTH ABOVE AND BELOW VERNON DAM INCLUDED 25 SPECIES. (BOPP-NSIC)

FIELD 05C

ACCESSION NO. W72-00939

RADIONUCLIDES AND SELECTED TRACE ELEMENTS IN MARINE PROTEIN CONCENTRATES,

WASHINGTON UNIV., SEATTLE, LAB. OF RADIATION ECOLOGY.

T. M. BEASLEY, T. A. JOKELA, AND R. J. EAGLE.

AVAILABLE FROM NATIONAL TECHNICAL INFORMATION SERVICE AS RLO-2225-T-14-2, $3.00 IN PAPER COPY, $0.95 IN MICROFICHE. AEC CONTRACT REPORT RLO-2225-T-14-2, 1970. 15 P, 4 TAB, 18 REF.

DESCRIPTORS:
*FOOD CHAINS, *TRACE ELEMENTS, *LEAD RADIOISOTOPES, HEAVY METALS, AIR POLLUTION EFFECTS, SURFACE WATERS, PATH OF POLLUTANTS, MARINE ALGAE, MARINE FISH, ABSORPTION, WATER POLLUTION SOURCES, RADIOACTIVITY EFFECTS, RADIOACTIVITY TECHNIQUES.

ABSTRACT:
CONSUMPTION OF 10-30 G/DAY OF CONCENTRATES PREPARED FROM SURFACE FEEDING FISHES WOULD SUBSTANTIALLY INCREASE INTAKE OF PB-210; PO-210; AND STABLE PB, CO, AND AG. THE STABLE PB IS ATTRIBUTED LARGELY TO AUTOMOTIVE EXHAUST POLLUTION OF SURFACE WATERS. IT IS UNCERTAIN WHETHER AG, CO, CD, AND CU ARE DERIVED SOLELY FROM ECOLOGICAL CONCENTRATION PROCESSES SINCE CONTAMINATION MAY OCCUR DURING PROCESSING. SR-90 IN THE CONCENTRATES WAS LESS THAN 0.05 DISINTEGRATIONS PER MINUTE PER GRAM OF DRY WEIGHT. (BOPP-NSIC)

FIELD 05C

ACCESSION NO. W72-00940

EFFECTS OF ENVIRONMENTAL STRESS ON THE COMMUNITY STRUCTURE, PRODUCTIVITY, ENERGY FLOW, AND MINERAL CYCLING IN SALT MARSH EPIPHYTIC COMMUNITIES,

CITY COLL., NEW YORK. DEPT OF BIOLOGY.

JOHN J. LEE, JOHN H. TIETJEN, AND ROBERT J. STONE.

AVAILABLE FROM NATIONAL TECHNICAL INFORMATION SERVICE $3.00 IN PAPER COPY, $0.95 IN MICROFICHE. AEC REPORT NYO-3995-18 (1970). 31 P, 7 FIG, 1 TAB, 21 REF. (CONF 710501-9). AEC CONTRACT AT(30-1)-3995.

DESCRIPTORS:
*WATER POLLUTION EFFECTS, *PROTOZOA, *MARINE ALGAE, *SALT MARSHES, NEW YORK, NEMATODES, EPIPHYTOLOGY, PRIMARY PRODUCTIVITY, PHYSIOLOGICAL ECOLOGY, RADIOECOLOGY, RADIOACTIVITY EFFECTS, ABSORPTION, FOOD WEBS, STRONTIUM RADIOISOTOPES, MARINE BACTERIA, COASTAL MARSHES, MICROORGANISMS.

ABSTRACT:
THE PURPOSE WAS TO GAIN INSIGHT INTO CHANGES IN PRODUCTIVITY AND COMMUNITY STRUCTURE RESULTING FROM RADIONUCLIDE, THERMAL, ORGANIC, PESTICIDE OR HERBICIDE POLLUTION. FORAMINIFERA (PROTOZOA) WERE ISOLATED IN CULTURE AND THEIR PHYSIOLOGICAL ECOLOGY, ENERGY BUDGETS, MINERAL CYCLING, AND LIFE CYCLES STUDIED. ENTEROMORPHA (ALGAE) HAD HIGH PRODUCTIVITY (ABOUT 10% OF ITS DRY WEIGHT/HR AT THE PEAK PRIMARY PRODUCTIVITY DURING THE SUMMER). REMOVAL RATES OF CA-45 AND SR-89, 90 FROM COASTAL MARINE ENVIRONMENTS BY INCORPORATION IN FORAMINIFERAN SHELLS WERE ESTIMATED. RADIONUCLIDES ARE EXCELLENT TOOLS FOR MEASURING COMPETITION, FEEDING, GROWTH, AND MINERAL CYCLING RATES OF FORAMINIFERA AND NEMATODES. (BOPP-NSIC)

FIELD 05C

ACCESSION NO. W72-00948

THE RAIN FOREST PROJECT. ANNUAL REPORT, JUNE 1970,

PUERTO RICO NUCLEAR CENTER, MAYAGUEZ.

RICHARD G. CLEMENTS, GEORGE E. DREWRY, AND ROBERT J. LAVIGNE.

AVAILABLE FROM NATIONAL TECHNICAL INFORMATION CENTER AS PRNC-147, $3.00 IN PAPER COPY, $0.95 IN MICROFICHE. AEC REPORT PRNC-147, ISSUED JUNE 1971. 142 P, 21 FIG, 30 TAB, 98 REF. AEC CONTRACT AT(40-1)-1833.

DESCRIPTORS:
*RADIOACTIVITY EFFECTS, *ECOSYSTEMS, *RAIN FORESTS, CYCLING NUTRIENTS, SOIL-WATER-PLANT RELATIONSHIPS, RADIOISOTOPES, ABSORPTION, RADIOACTIVITY TECHNIQUES, TRACERS, PATH OF POLLUTANTS, ON-SITE INVESTIGATIONS, FOOD WEBS, FOOD CHAINS, FROGS, LICHENS, ALGAE, FUNGI, SMALL ANIMALS(MAMMALS), CYTOLOGICAL STUDIES, SOIL CHEMISTRY, RADIOSENSITIVITY, RADIOECOLOGY, FALLOUT, INSECT, SOIL WATER MOVEMENT.

IDENTIFIERS:
CHEMISTRY OF THE ATMOSPHERE.

ABSTRACT:
CURRENT RESEARCH IN THE TERRESTRIAL ECOLOGY PROGRAM (INTERIM REPORTS AND COMPLETED SHORT TERM STUDIES, 24 IN NUMBER) INCLUDES STUDIES ON INSECT ECOLOGY; MOVEMENT OF ISOTOPES THROUGH THE ANIMAL FOOD WEB; ELEMENT INPUT THROUGH RAINFALL AND SUBSEQUENT DISTRIBUTION THROUGH THE FOREST; RECOVERY IN IRRADIATED AREAS; MOVEMENT AND DISTRIBUTION OF PREVIOUSLY APPLIED RADIOISOTOPES IN SOIL, PLANTS AND ANIMALS; AND RESEARCH IN THE VISITING SCIENTIST PROGRAM. INCREASED EMPHASIS WILL BE PLACED ON THE PHYSICAL AND CHEMICAL PROPERTIES OF FOREST SOILS AND THE MOVEMENT OF MACRO- AND TRACE ELEMENTS VIA SOIL WATER TO THE STREAM. (BOPP-NSIC)

FIELD 05C

ACCESSION NO. W72-00949

PRELIMINARY ECOLOGICAL STUDIES AT JERVIS BAY,

AUSTRALIAN ATOMIC ENERGY COMMISSION RESEARCH ESTABLISHMENT, LUCAS HEIGHTS.

M. S. GILES.

AVAILABLE FROM NATIONAL TECHNICAL INFORMATION SERVICE AS CONF-710511-3, $3.00 IN PAPER COPY, $0.95 IN MICROFICHE. AEC REPORT CONF-710511-3 (FROM FORTY-THIRD CONGRESS OF THE AUSTRALIAN AND NEW ZEALAND ASSOCIATION FOR THE ADVANCEMENT OF SCIENCE; BRISBANE, AUSTRALIA, 24 MAY 1971). 22 P, 6 FIG, 5 TAB, 8 REF.

DESCRIPTORS:
*FOOD CHAINS, *RADIOACTIVITY EFFECTS, *NUCLEAR POWERPLANTS, MONITORING, PATH OF POLLUTANTS, AIR POLLUTION EFFECTS, WATER POLLUTION EFFECTS, BAYS, WATERSHEDS(BASINS), SMALL WATERSHEDS, CROPS, EGGS, FISHERIES, ON-SITE INVESTIGATIONS, ENVIRONMENTAL EFFECTS, ESTUARINE ENVIRONMENT, REGULATION, FALLOUT, OYSTERS, MUSSELS, PLANKTON, MARINE ALGAE.

IDENTIFIERS:
ENVIRONMENTAL STATEMENT, CRITICAL NUCLIDE PATHWAY, ABALONE.

ABSTRACT:
ATMOSPHERIC CRITICAL NUCLIDE PATHWAYS INCLUDE HONEY PRODUCTION, WATER CATCHMENT, FREE-RANGE EGG PRODUCTION, BACKYARD GARDENING, AND COLLECTION OF RAINWATER FOR DRINKING. LIQUID-EFFLUENT PATHWAYS ARE SEAFOODS (ABALONE, ROCK OYSTERS, AND DREDGE OYSTERS), SEAWEED MULCHES, AND AGAR PRODUCTS. THE SHALLOWER WATERS WILL HAVE THE HIGHEST RADIOACTIVITY SINCE THE WARM EFFLUENT HAS RELATIVELY LOW DENSITY. THIS MAKES ABALONE, WHICH SPENDS SEVERAL YEARS IN ONE (KELP) FEEDING AREA, A MAJOR PATHWAY. DREDGE OYSTERS ARE ANOTHER PATHWAY, BUT LARGE BEDS OF ROCK OYSTERS ARE NOT PRESENT. (BOPP-NSIC)

FIELD 05C

ACCESSION NO. W72-00959

AMCHITKA BIOENVIRONMENTAL PROGRAM. AMCHITKA BIOLOGICAL INFORMATION SUMMARY,

BATTELLE MEMORIAL INST., COLUMBUS, OHIO.

R. GLEN FULLER.

AVAILABLE FROM NATIONAL TECHNICAL INFORMATION SERVICE, AS BMI-171-132, $3.00 IN PAPER COPY, $0.95 IN MICROFICHE. AEC CONTRACT REPORT BMI-171-132, MAY 1971. 16 P, 1 FIG, 18 REF. AEC CONTRACT AT(26-1)-171.

DESCRIPTORS:
*RADIOACTIVITY EFFECTS, *ENVIRONMENTAL EFFECTS, *NUCLEAR EXPLOSIONS, BIRDS, SALMON, STICKLEBACKS, CRABS, OTTERS, ZOOPLANKTON, PHYTOPLANKTON, MAMMALS, MARINE ALGAE, FOOD CHAINS, MARINE FISHERIES, LAKES, POLLUTION ABATEMENT.

IDENTIFIERS:
RADIONUCLIDE UPTAKE.

ABSTRACT:
EARLY IN 1967, THE UNITED STATES ATOMIC ENERGY COMMISSION NEVADA OPERATIONS OFFICE BEGAN A DETAILED EVALUATION OF AMCHITKA ISLAND IN THE WESTERN ALEUTIANS AS A POTENTIAL SITE FOR UNDERGROUND NUCLEAR TESTING. IN OCTOBER 1969, AN UNDERGROUND TEST, PROJECT MILROW, INVOLVED THE DETONATION OF A DEVICE OF ABOUT 1 MEGATON YIELD, ABOUT 4,000 FT BELOW THE SURFACE. STUDIES WERE DESIGNED TO: (1) PREDICT, DOCUMENT, AND EVALUATE THE EFFECTS OF AEC TESTING ACTIVITIES ON THE ENVIRONMENT; (2) RECOMMEND MEASURES TO MINIMIZE EFFECTS; AND (3) PREDICT RADIONUCLIDE UPTAKE BY MARINE FOOD CHAINS, SHOULD THERE BE AN INADVERTENT RELEASE. RESULTS ARE REPORTED OF STUDIES OF SOILS, VEGETATION, FRESHWATER LIFE, BIRDS, AND MARINE LIFE. (BOPP-NSIC)

FIELD 05C

ACCESSION NO. W72-00960

THE USE OF A RUBBLE CHIMNEY FOR DENITRIFICATION OF IRRIGATION RETURN WATERS,

STANFORD UNIV., PALO ALTO, CALIF.

ROY B, EVANS, AND PAUL KRUGER.

AVAILABLE FROM THE NATIONAL TECHNICAL INFORMATION SERVICE AS CONF 700101, VOL 2, $6.00/SET, $0.95 MICROFICHE. AEC CONFERENCE PAPER, CONF 700101, VOL 2, P 1222-1245, JAN 14-18, 1970.

DESCRIPTORS:
*NUCLEAR EXPLOSION, *FILTERS, *DIATOMACEOUS EARTH, *POROUS MEDIA, *WASTE WATER TREATMENT, *DENITRIFICATION, *NITRATES, EUTROPHICATION, IRRIGATION WATER, AQUATIC ALGAE, BIOLOGICAL PROPERTIES, WATER POLLUTION EFFECT, WATER POLLUTION SOURCES, COST COMPARISONS, BENEFICIAL USE, WATER QUALITY, CALIFORNIA, BIOCONTROL, COST-BENEFIT THEORY.

IDENTIFIERS:
RUBBLE CHIMNEY, SAN JOAQUIN VALLEY.

ABSTRACT:
THE USE OF WATER FOR IRRIGATION SERIOUSLY REDUCES THE QUALITY OF THE WATER. IRRIGATION RETURN WATER MAY HAVE HIGH CONCENTRATIONS OF SALTS, PESTICIDES, AND PLANT NUTRIENTS, AND DISCHARGE OF THE RETURN FLOWS INTO RIVERS OR OTHER RECEIVING WATERS CAN SERIOUSLY AFFECT THE QUALITY OF THE RECEIVING WATERS. THE INCREASED AVAILABILITY OF CHEAP FERTILIZERS HAS HAD THE UNDESIRABLE SIDE-EFFECT OF INCREASING CONCENTRATIONS OF NITROGEN AND PHOSPHORUS IN IRRIGATION RETURN FLOWS. THE PRESENCE OF THESE ELEMENTS CAN BE TROUBLESOME. BOTH ARE NUTRIENTS NECESSARY FOR THE GROWTH OF ALGAE, AND IN SOME SITUATIONS THEIR PRESENCE MAY RESULT IN SEVERE ALGAL BLOOMS IN THE RECEIVING WATERS. THEIR REMOVAL FROM WASTE WATER IS CONSIDERED DIFFICULT AND EXPENSIVE. BIOLOGICAL DENITRIFICATION HAS BEEN PROPOSED AS A MEANS OF REMOVING NITRATES FROM WASTE WATERS TO CONTROL EUTROPHICATION IN RECEIVING WATERS. A POTENTIAL USE FOR THIS METHOD IS THE TREATMENT OF IRRIGATION RETURN WATERS CONTAINING HIGH CONCENTRATIONS OF NITRATE-NITROGEN, SINCE DIRECT DISCHARGE OF SUCH WASTES MAY CAUSE OBJECTIONABLE ALGAL GROWTH IN THE RECEIVING WATERS. FOR EXAMPLE, THE PROCESS MAY BE USED TO TREAT AGRICULTURAL WASTE WATERS IN THE SAN JOAQUIN VALLEY IN CALIFORNIA, WHERE AN ESTIMATED 580,000 ACRE-FEET/YEAR OF RETURN WATERS, CONTAINING 20 MG/L OF NITRATE-NITROGEN, WILL REQUIRE DISPOSAL BY A.D. 2020. THIS PAPER PRESENTS THE RESULTS OF A PRELIMINARY INVESTIGATION OF THE FEASIBILITY OF USING A RUBBLE CHIMNEY AS A BIOLOGICAL FILTER FOR DENITRIFICATION. (HOUSER-NSIC)

FIELD 08H, 05D

ACCESSION NO. W72-00974

DETERGENT POLLUTION CONTROL ACT OF 1971 (A BILL TO AMEND THE FEDERAL WATER POLLUTION CONTROL ACT TO BAN POLYPHOSPHATES IN DETERGENTS AND TO ESTABLISH STANDARDS AND PROGRAMS TO ABATE AND CONTROL WATER POLLUTION BY SY NTHETIC DETERGENTS).

HOUSE BILL 6180, 92D CONG. 1ST SESS. (1971). 14 P.

DESCRIPTORS:
*UNITED STATES, *DETERGETNTS, *PHOSPHATES, *WATER POLLUTION CONTROL, *WATER POLLUTION SOURCES, EUTROPHICATION, LEGISLATION, LEGAL ASPECTS, REGULATION, POLLUTION ABATEMENT, WASTE DISPOSAL, STANDARDS, PUBLIC HEALTH, INSPECTION, WATER SOFTENING, SOAPS, AQUATIC ALGAE, WATER POLLUTION EFFECTS, WATER QUALITY, CHEMICAL DEGRADATION.

IDENTIFIERS:
WATER POLLUTION CONTROL ACT.

ABSTRACT:
THIS BILL IS INTENDED TO ABATE AND CONTROL THE POLLUTION AND DEGRADATION OF WATER CAUSED BY THE CONTINUING DISCHARGE OF SYNTHETIC DETERGENTS. RESEARCH HAS DEMONSTRATED THAT POLYPHOSPHATES IN DETERGENTS ARE A MAJOR CAUSE OF WATER POLLUTION, ACCELERATING BEYOND CONTROL THE GROWTH OF ALGAE WHICH INTERFERES WITH THE USE OF THE WATER AND DEGRADES WATER QUALITY. THE BILL RECOGNIZES THAT CERTAIN NONPHOSPHORUS BASED INGREDIENTS PERFORM AT LEAST AS WELL AS PRESENT POLYPHOSPHATE INGREDIENTS IN SYNTHETIC DETERGENTS AND SOAPS. THE BILL WOULD MAKE IT UNLAWFUL TO MANUFACTURE OR IMPORT ANY DETERGENT CONTAINING PHOSPHORUS AFTER JUNE 30, 1972. PROVISIONS ARE MADE FOR THE SEIZURE AND DISPOSAL OF SUCH DETERGENTS AND THE ADJUDICATION OF ALLEGED VIOLATIONS. THE SECRETARY OF THE INTERIOR WOULD BE AUTHORIZED TO MAKE APPROPRIATE INSPECTIONS. STANDARDS OF WATER EUTROPHICATION ABILITY, BIODEGRADIBILITY, TOXICITY, AND OF EFFECTS ON THE PUBLIC HEALTH AND WELFARE WOULD BE ESTABLISHED BY THE SECRETARY. THE SECRETARY IS ALSO AUTHORIZED TO INVENTORY EXISTING TECHNOLOGY AND ASSIST IN THE DEVELOPMENT OF POLYPHOSPHATE SUBSTITUTES. (JOHNSON-FLORIDA)

FIELD 06E, 05G

ACCESSION NO. W72-01085

CHICAGO LEADS FIGHT AGAINST PHOSPHATE POLLUTICN--NOW CONGRESS MUST ACT,

HOUSE OF REPRESENTATIVES, WASHINGTON, D.C.

R. C. PUCINSKI.

CONGRESSIONAL RECORD, VOL 117, NO 8, P H326-29 (DAILY ED. FEBRUARY 1, 1971). 4 P.

DESCRIPTORS:
*DETERGENTS, *WATER POLLUTION CONTROL, *PHCSPHATES, *FEDERAL GOVERNMENT, UNITED STATES, ALGAL CONTROL, STANDARCS, LEGISLATICN, LEGAL ASPECTS, WATER POLLUTION, ADJUDICATION PROCECURE, ADMINISTRATIVE AGENCIES, ADMINISTRATION, ALGAE, BIODEGRADATION, LOCAL GOVERNMENTS, WASTE DISPOSAL, WATER POLLUTICN SOURCES, SOAPS, TCXICITY, PUBLIC HEALTH, WATER POLLUTION EFFECTS, FEDERAL JURISDICTION.

IDENTIFIERS:
*WATER POLLUTION CONTROL ACT.

ABSTRACT:
CONGRESSMAN PUCINSKI INTRODUCED FEDERAL LEGISLATICN, AMENDING THE FEDERAL WATER POLLUTION CCNTROL ACT, TO BAN POLYPHCSPHATES IN DETERGENTS AND ESTABLISH STANDARDS AND PROGRAMS TO ABATE AND CONTROL WATER POLLUTION FRCM SYNTHETIC DETERGENTS. PHOSPHATES USED IN MOST DETERGENTS PROMOTE THE GROWTH OF ALGAE; THE UNRESTRICTED GROWTH OF ALGAE, BY CONSUMING AVAILABLE OXYGEN, INEVITABLY SUFFOCATES FISH AND OTHER MARINE LIFE UNTIL WATER LITERALLY BEGINS TC DIE. NOTING A NEW CHICAGO ORDINANCE WHICH BANS DETERGENTS HAVING A SPECIFIC PHOSPHOROUS CONTENT. PUCINSKI EMPHASIZED THAT FEDERAL LEGISLATION IS ESSENTIAL TO EFFECTIVELY COMBAT WATER POLLUTICN. HIS BILL, IN ADDITION TO MAKING THE MANUFACTURING AND IMPORTING OF DETERGENTS CONTAINING PHOSPHORUS UNLAWFUL, IMPOSES IN REM LIABILITY AND ALLOWS CONDEMNATION IN FEDERAL COURT. ADJUDICATORY PROCEDURES FOR THE DISPOSITION OF SUCH DETERGENTS ARE DETAILED IN THE BILL. THE BILL AUTHORIZES THE SECRETARY OF THE TREASURY TO CONDUCT EXAMINATIONS AND INSPECTIONS. ADDITICNALLY, THE SECRETARY MUST ESTABLISH STANDARDS OF WATER AND EUTROPHICATION ABILITY, BIODEGRADABILITY, TOXICITY, AND OF EFFECTS ON PUBLIC HEALTH. THESE STANDARDS MUST BE MET BY ALL SYNTHETIC DETERGENTS. TO ACCELERATE THE DEVELOPMENT OF PHOSPHATE-FREE DETERGENTS THE SECRETARY MUST REPORT EXISTING TECHNOLOGY OR SUBSTITUTES AND IS AUTHORIZED TO MAKE GRANTS AND CONTRACTS FOR RESEARCH AND DEVELOPMENT.

FIELD 06E, 05G

ACCESSION NO. W72-01093

PROCEEDINGS THIRTEENTH CONFERENCE CN GREAT LAKES RESEARCH, PART I AND II.

GREAT LAKES RESEARCH CENTER, DETROIT, MICH.

AVAILABLE FROM TREASURER, P. O. BOX 640, ANN ARBCR, MICH. 48107. PRICE $18.00 A SET. 1970. 1063 P.

DESCRIPTORS:
*GREAT LAKES, *LAKES, EUTROPHICATION, ALGAE, *LAKE SUPERIOR, *LAKE MICHIGAN, *LAKE HURON, *LAKE ONTARIO, *LAKE ERIE, WATER POLLUTION EFFECTS, WATER PCLLUTION SOURCES, LIMNCLCGY.

ABSTRACT:
THE THIRTEENTH CONFERENCE ON GREAT LAKES RESEARCH WAS HELD 1-3 APRIL, 1970 AT BUFFALO, NEW YCRK AND CO-HOSTEC BY CORNELL AERCNAUTICAL LABORATORY INC. AND THE GREAT LAKES LABORATORY OF THE STATE UNIVERSITY COLLEGE AT BUFFALC. (SEE ALSO W72-01095 THRU W72-C1108)

FIELD 02H, 05C

ACCESSION NO. W72-01094

LASER INDUCED FLUORESCENCE IN RHODAMINE B AND ALGAE,

SYRACUSE UNIV. RESEARCH CORP., N.Y.

G. DANIEL HICKMAN, AND RICHARD B . MOORE.

INTERNATIONAL ASSOCIATION FOR GREAT LAKES RESEARCH, PROCEEDINGS 13TH CONFERENCE ON GREAT LAKES RESEARCH, PART 1, P 1-14, 1970. 11 FIG, 2 TAB, 22 REF.

DESCRIPTORS:
*ALGAE, *REMOTE SENSING, *TRACKING TECHNIQUES, MEASUREMENT, CURRENTS(WATER), CHLOROPHYLL, FLUORESCENCE, BATHYMETRY.

IDENTIFIERS:
LASER, RHODAMINE B.

ABSTRACT:
MAPPING OF THREE-DIMENSIONAL CURRENTS AND OF DISTRIBUTION OF NEAR-SURFACE ALGAE WAS INVESTIGATED BY RECORDING FLUORESCENCE STIMULATED BY THE RECENTLY DEVELOPED OCEANIC PULSED LASER. A 3 NSEC PULSED NEON LASER (540 NM) WAS EMPLOYED TO EXCITE THE EMISSION OF RHODAMINE B (580 NM). THE SAME LASER WAS FILLED WITH NITROGEN (337 NM) TO STIMULATE THE EMISSION OF ALGAE. RHODAMINE FLUORESCENCE WAS OBSERVED FOR CONCENTRATIONS AS LOW AS 0.1 PPB, WHEREAS THAT FROM ANACYSTIS AND CHLORELLA SPECIES, FROM 6 TO 12 MG/CU M. AIRBORNE LASER-RECEIVER UNIT SHOULD DETECT THE FLUORESCENCE FROM AN ALTITUDE OF 100 METERS. (SEE ALSO W72-01094) (WILDE-WISCONSIN)

FIELD 05C, 05A, 02H

ACCESSION NO. W72-01095

BLUE-GREEN ALGAE AND HUMIC SUBSTANCES,

CINCINNATI UNIV., OHIO. TANNER'S COUNCIL LAB.

WILLY LANGE.

INTERNATIONAL ASSOCIATION FOR GREAT LAKES RESEARCH, PROCEEDINGS 13TH CONFERENCE ON GREAT LAKES RESEARCH, PART 1, P 58-70, 1970. 4 FIG, 6 TAB, 17 REF.

DESCRIPTORS:
*CYANOPHYTA, *HUMIC ACIDS, ORGANIC MATTER, ALGAE, EUTROPHICATION, CHELATION, DECOMPOSING ORGANIC MATTER, PHYTOPLANKTON, PRODUCTIVITY, IRON.

IDENTIFIERS:
*FULVIC ACID, GROWTH STIMULATION, ANABAENA, GLOEOTRICHIA, MICROCYSTIS, NOSTOC, SLAKED LIME.

ABSTRACT:
A REPLACEMENT OF EDTA + CITRATE IN THE ZEHNDER-GORHAM MEDIA WITH SYNTHETIC FULVIC ACID INCREASED GREATLY THE CELL NUMBERS OF ANABAENA CIRCINALIS, GLOEOTRICHIA ECHINULATA, MICROCYSTIS AERUGINOSA, AND NOSTOC MUSCORUM. THE GROWTH STIMULATION IS ATTRIBUTED TO CHELATING EFFECTS OF FULVIC ACID AND THE RESULTING INCREASED AVAILABILITY OF IRON AND OTHER TRACE ELEMENTS. RESULTS INDICATE THAT SEEPAGE OF FULVIC ACID FROM RAW HUMUS AND PEAT SOILS, OR ITS RELEASE FROM SEWAGE EFFLUENTS PLAY AN IMPORTANT PART IN THE NUISANCE GROWTH OF CYANOPHYTA. SLAKED LIME IS SUGGESTED AS A PARTIALLY EFFICIENT REMEDY FOR CIRCUM-NEUTRAL WATERS. (SEE ALSO W72-01094) (WILDE-WISCONSIN)

FIELD 05C, 02H

ACCESSION NO. W72-01097

NEUTRON ACTIVATION ANALYSIS OF SEDIMENTS IN WESTERN LAKE ERIE,

OHIO STATE UNIV., COLUMBUS. COLL. OF BIOLOGICAL SCIENCES.

PAUL L. ZUBKOFF, AND WALTER E. CAREY.

INTERNATIONAL ASSOCIATION FOR GREAT LAKES RESEARCH, PROCEEDINGS 13TH CONFERENCE ON GREAT LAKES RESEARCH, PART 1, P 319-325, 1970. 2 FIG, 3 TAB, 11 REF.

DESCRIPTORS:
*SEDIMENTS, *ANALYTICAL TECHNIQUES, *NEUTRON ACTIVATION ANALYSIS, *CHEMICAL ANALYSIS, GREAT LAKES, EUTROPHICATION, NUTRIENTS, ALGAE, BACTERIA, LAKE ERIE, ALUMINUM, MANGANESE, SODIUM, CHROMIUM, IRON.

IDENTIFIERS:
VANADIUM, LANTHANUM, SCANDIUM.

ABSTRACT:
THE CENTERS OF 1 CM LATERAL SECTIONS OF 15 CM SEDIMENT CORES WERE WASHED FREE OF INTERSTITIAL WATER AND SUBJECTED TO A 2.0 X 10 CM 11 POWER NEUTRON 0.01 CM 0.1 SEC FLUX IN RESEARCH REACTOR. ANALYSIS OF GAMMA-RAY SPECTRA, OBTAINED WITH A NAI(TL) CRYSTAL, INDICATED A UNIFORM CONCENTRATION OF AL, V, MN, NA, LA, CR, AND SC. THE CONTENTS OF VANADIUM AND CHROMIUM ARE AT LEAST THREE TIMES AS GREAT AS FOUND IN SOILS. (SEE ALSO W72-01094) (WILDE-WISCONSIN)

FIELD 05C, 02H

ACCESSION NO. W72-01101

AQUEOUS PHOSPHATE AND LAKE SEDIMENT INTERACTION,

GREAT LAKES RESEARCH CENTER, DETROIT, MICH.

R. C. GUMERMAN.

INTERNATIONAL ASSOCIATION FOR GREAT LAKES RESEARCH, PROCEEDINGS 13TH CONFERENCE ON GREAT LAKES RESEARCH, PART 2, P 673-682, 1970. 8 FIG, 1 TAB, 8 REF.

DESCRIPTORS:
*PHOSPHATES, *SEDIMENT-WATER INTERFACES, ADSORPTION, LAKES, NUTRIENTS, LAKE ERIE, LAKE SUPERIOR, LABORATORY TESTS, TEMPERATURE, ALGAE, PHOSPHORUS, ION EXCHANGE, OXIDATION-REDUCTION POTENTIAL, HYDROGEN ION CONCENTRATION.

IDENTIFIERS:
SEDIMENTARY PHOSPHORUS RELEASE.

ABSTRACT:
THIS LABORATORY STUDY OF STERILE SEDIMENTS FROM LAKE ERIE AND LAKE SUPERIOR DISCLOSED THAT THE PHOSPHORUS-SEDIMENT COMPLEX IS FORMED UNDER THE INFLUENCE OF BOTH PHYSICAL AND CHEMICAL ADSORPTION. MAXIMUM REMOVAL OF AQUEOUS P OCCURS WITHIN THE PH RANGE OF 4.5 TO 5.5. LOWERING REDOX POTENTIAL TO ZERO MAY OR MAY NOT EFFECT A RELEASE OF P FROM THE SEDIMENT. THE MAXIMUM ADSORBING CAPACITY OF SEDIMENTS IS IN SURFACE LAYERS LESS THAN 3.5 MM DEEP, AND IS ZERO AT A DEPTH EXCEEDING 14 MM BELOW SEDIMENT-WATER INTERFACE. AS LONG AS THE SEDIMENT CONTAINS SOME ADSORBED PHOSPHORUS, ITS RELEASE WILL MAINTAIN A MINIMUM CONCENTRATION OF 0.1 MG/L OF AQUEOUS PHOSPHATE RADICAL. IN TURN, UNDER SUCH CONDITIONS CESSATION OF PHOSPHATE INPUT MAY NOT EFFECT A REDUCTION OF NUISANCE ALGAL GROWTH FOR A LONG TIME. (SEE ALSO W72-01094) (WILDE-WISCONSIN)

FIELD 05C, 02H

ACCESSION NO. W72-01108

PHYTOPLANKTON SPECIES AND POPULATIONS IN THE PAMLICO RIVER ESTUARY OF NORTH CAROLINA,

NORTH CAROLINA STATE UNIV., RALEIGH. DEPT. OF ZOOLOGY.

J. E. HOBBIE.

AVAILABLE FROM THE NATIONAL TECHNICAL INFORMATION SERVICE AS PB-204 489, $3.00 IN PAPER COPY, $0.95 IN MICROFICHE. NORTH CAROLINA WATER RESOURCES RESEARCH INSTITUTE, RALEIGH. REPORT NO 56, SEPTEMBER 1971. 147 P, 37 FIG, 2 TAB, 27 REF, 2 APPEND. OWRR B-004-NC(11).

DESCRIPTORS:
*ESTUARIES, *EUTROPHICATION, *ALGAL BLOOMS, *PHYTOPLANKTON, *NUTRIENTS, WATER POLLUTION EFFECTS, AQUATIC ALGAE, NORTH CAROLINA, *ALGAE, DIATOMS, *DINOFLAGELLATES, *RED TIDE, WATER POLLUTION SOURCES.

IDENTIFIERS:
*PAMLICO RIVER ESTUARY.

ABSTRACT:
THE PAMLICO RIVER ESTUARY EXTENDS SOME 35 MILES FROM WASHINGTON, N. C. TO PAMLICO SOUND. THE PHYTOPLANKTON CYCLE OF THIS ESTUARY IS COMPLETELY DOMINATED BY DINOFLAGELLATES. DIATOMS, HOWEVER, BECOME MORE AND MORE IMPORTANT IN THE LOWER REACHES OF THE RIVER CLOSE TO THE POINT WHERE IT EMPTIES INTO PAMLICO SOUND. THE DOMINANT ORGANISM IS PERIDINIUM TRIQUETRUM, THAT CREATES A RED TIDE DURING JANUARY, FEBRUARY, AND MARCH. THE PERIDINIUM IS ACCOMPANIED BY OTHER DINCGLAGELLÀTES. THIS BLOOM LASTS UNTIL LATE MARCH AND THEN POPULATIONS REMAIN LOW UNTIL A LATE SUMMER PEAK OF ALGAE DOMINATED BY G. AUREOLUM, G. ESTUARIALE, K. ROTUNDATUM, POLYKRIKOS SP., AND CALYCOMONAS OVALIS. THIS INCREASE TAKES PLACE IN LATE AUGUST AND EARLY SEPTEMBER AND IS FOLLOWED BY A FALL LOW THAT LASTS UNTIL THE EARLY SPRING BLOOM BEGINS IN DECEMBER OR EARLY JANUARY. THIS YEARLY CYCLE IS SIMILAR TO THAT FOUND IN SEVERAL OF THE RIVER ESTUARIES THAT ENTER CHESAPEAK BAY. EVEN THE RED TIDE FORMING ORGANISM, P. TRIQUETRUM, IS THE SAME. PERIDINIUM TRIQUETRUM IS AN INDICATOR OF EXTREMELY RICH OR POLLUTED CONDITIONS. IN THE PAMLICO RIVER ESTUARY, THE ENTICHMENT COMES FROM SEWAGE FROM SEVERAL SMALL CITIES AND FROM FARM AND SWAMP RUNOFF. IT IS LIKELY THAT THE RED TIDE IS THE RESULT OF LARGE AMOUNTS OF NITRATE THAT REACH THE MIDDLE PARTS OF THE ESTUARY IN MID-WINTER. PHOSPHORUS IS IN AMPLE SUPPLY THE YEAR AROUND.

FIELD 05C, 02L

ACCESSION NO. W72-01329

BLUE-GREEN ALGAE BLOOMS--A CURRENT HYDROBIOLOGICAL PROBLEM (ZAKWITY SINIC--AKTUALNY PROBLEM HYDROBIOLOGII),

POLISH ACADEMY OF SCIENCES, WARSAW. INST. OF ECOLOGY; AND POLISH ACADEMY OF SCIENCES, WARSAW. DEPT. OF HYDROBIOLOGY.

IRENA SPODNIEWSKA.

WIADOMOSCI EKOLOGICZNE, VOL 17, NO 2, P 157-163, 1971. 3 REF. ENGLISH SUMMARY.

DESCRIPTORS:
*CYANOPHYTA, *NUISANCE ALGAE, *EUTROPHICATION, REVIEWS.

IDENTIFIERS:
USSR.

ABSTRACT:
BLUE-GREEN ALGAE APPEARS MOST FREQUENTLY IN STAGNANT WATERS AND WATERS WITH A SLUGGISH CURRENT, RICH IN ORGANIC SUBSTANCES. THEIR PERSISTENT BLOOMS, WHICH ACCOMPANY PROGRESSIVE EUTROPHICATION, ARE WIDESPREAD PHENOMENON. THREE RUSSIAN LANGUAGE VOLUMES--'ECOLOGY AND PHYSIOLOGY OF BLUE-GREEN ALGAE' EDITED BY L P BRAGINSKII, MOSCOW AND LENINGRAD 1965, 272 PAGES, 'WATER BLOOMS' PART I, EDITED BY A V TOPACHEVSKII, KIEV, 1968, 386 PAGES AND PART II, 1969, 267 PAGES, DESCRIBING RESEARCH CONDUCTED ON BLUE-GREEN ALGAE DURING THE LAST TEN YEARS IN THE SOVIET UNION AND WHICH CONTAIN A DISCUSSION OF WORLD LITERATURE ARE REVIEWED. EMPHASIS WAS ON THE SPECIFIC PROPERTIES OF BLUE-GREEN ALGAE WHICH ENABLE THEM TO OCCUR UNDER CONDITIONS UNFAVORABLE TO DEVELOPMENT OF OTHER PLANT ORGANISMS. (AUEN-WISCONSIN)

FIELD 05C

ACCESSION NO. W72-01358

AN ECOLOGICAL STUDY OF THREE FRESHWATER PONDS OF HYDERABAD-INDIA. I. THE ENVIRONMENT,

OSMANIA UNIV., HYDERABAD (INDIA). DEPT. OF BOTANY; AND OSMANIA UNIV., HYDERABAD (INDIA). HYDROBIOLOGY LAB.

V. S. RAO.

HYDROBIOLOGIA, VOL 38, NO 2, P 213-223, 1971. 2 FIG, 3 TAB, 30 REF.

DESCRIPTORS:
*ECOLOGY, *PONDS, *CHEMICAL ANALYSIS, TEMPERATURE, SEASONAL, IONS, ALGAE, SALINITY, PHYTOPLANKTON, THERMAL PROPERTIES, HYDROGEN ION CONCENTRATION, DISSOLVED OXYGEN, CARBON DIOXIDE, AMMONIA, ORGANIC MATTER.

IDENTIFIERS:
*HYDERABAD(INDIA), EICHHORNIA CRASSIPES, TYPHA ANGUSTATA, IPOMOEA AQUATICA, TOTAL SOLIDS.

ABSTRACT:
THE DIVERSITY AND COMPLEXITY OF THE ECOLOGICAL PHENOMENA OF INDIAN FRESH WATERS CALL FOR MORE DATA, PARTICULARLY ON IONIC COMPOSITION AND THE BEHAVIOR OF BASIC ELEMENTS. THREE PONDS SUBJECT TO DIVERSE POLLUTION NEAR HYDERABAD WERE STUDIED. CHEMICAL ANALYSES WERE MADE OF WATER FROM EACH POND FOR TWO YEARS. DATA WERE COLLECTED ON WATER TEMPERATURE IN RELATION TO THE AIR TEMPERATURE IN SUMMER, RAINY SEASON, AND IN WINTER. THE TEMPERATURE DIFFERENCES WERE, TO A DEGREE, ATTRIBUTED TO WATER QUANTITY BUT THE INTERMEDIATE SIZED POND DID NOT BEAR OUT THIS CONCLUSION; THE HIGH SALT CONTENT AND DENSE ALGAL POPULATION OF THAT POND WAS POSSIBLY RESPONSIBLE FOR KEEPING IT COOLER. THE IONIC COMPOSITION OF THESE PONDS WAS DETERMINED; THEIR PROPORTIONS DIFFERED CONSIDERABLY FROM RODHE'S (1949) STANDARD COMPOSITION. THE IONIC COMPOSITION OF THE SAME BODY OF WATER ALSO DIFFERED FROM SEASON TO SEASON. THE SEASONAL FLUCTUATIONS, WHICH HAVE SECONDARY IMPORTANCE IN LAKES, BECOME MORE SIGNIFICANT IN SMALL FRESHWATER BODIES. IT IS STRESSED THAT THE IONIC COMPOSITION ASSESSED FROM A LIMITED NUMBER OF OBSERVATIONS MAY LEAD TO WRONG CONCLUSIONS. (JONES-WISCONSIN)

FIELD 05C

ACCESSION NO. W72-01363

ON THE LIFE HISTORIES OF SOME BROWN ALGAE FROM EASTERN CANADA,

NATIONAL RESEARCH COUNCIL OF CANADA, HALIFAX (NOVA SCOTIA). ATLANTIC REGIONAL LAB.

T. EDELSTEIN, L C-M CHEN, AND J. MCLACHLAN.

CANADIAN JOURNAL OF BOTANY, VOL 49, NO 7, P 1247-1251, 1971. 16 FIG, 22 REF.

DESCRIPTORS:
*LIFE HISTORY STUDIES, *PHAEOPHYTA, EPIPHYTOLOGY, ALGAE, HABITATS, CULTURES.

IDENTIFIERS:
*EASTERN CANADA, NOVA SCOTIA, ISTHMOPLEA SPHAEROPHORA, MELANOSIPHON INTESTINALIS, HECATONEMA MACULANS, ELACHISTA LUBRICA, RALFSIA VERRUCOSA.

ABSTRACT:
THE COMPLETE LIFE CYCLES OF FIVE PHAEOPHYCEAN SPECIES, THREE FROM THE ECTOCARPALES AND TWO FROM THE DICTYOSIPHONALES, FROM NOVA SCOTIA, CANADA, WERE COMPLETED IN CULTURE. ISTHMOPLEA SPHAEROPHORA, MELANOSIPHON INTESTINALIS, HECATONEMA MACULANS, ELACHISTA LUBRICA, AND RALFSIA VERRUCOSA REPLICATED DIRECTLY THE PARENT PLANTS. DETAILED DESCRIPTIONS OF THE ALGAE ARE GIVEN, BUT CYTOLOGICAL STUDIES WERE NOT MADE. APART FROM HECATONEMA MACULANS, WHERE PLURILOCULAR SPORANGIA ONLY SERVED AS THE INOCULUM, CULTURES WERE ESTABLISHED FROM SINGLE UNILOCULAR SPORANGIA. CULTURES OF MELANOSIPHON WERE DERIVED FROM BOTH UNI- AND PLURILOCULAR SPORANGIA. NO FUSION WAS NOTED AND THE LIFE CYCLE OBTAINED IN CULTURE WAS PROBABLY ASEXUAL. THE POSSIBILITY IS RECOGNIZED THAT UNDER CONDITIONS OF CULTURE A SPECIES MAY UNDERGO AN ABBREVIATED CYCLE BY ASEXUAL REPRODUCTION ONLY. THE ORIGINAL HABITATS ARE RECORDED. THREE OF THE SPECIES ARE RECORDED AS EPIPHYTIC ON OTHER ALGAE. IN CULTURING, THE MOST SATISFACTORY CONDITIONS OF INCUBATION WERE 13C WITH A PHOTOPERIOD OF 10 HRS LIGHT AND AN INTENSITY OF ABOUT 4000 LUX. THE CULTURE MEDIUM WAS SWM-3 OF CHEN ET AL (1969), CHANGED WEEKLY. (JONES-WISCONSIN)

FIELD 05C

ACCESSION NO. W72-01370

BIOLCGICAL ASPECTS OF WATER POLLUTION,

NATIONAL INST. OF WATER RESEARCH, PRETORIA (SOUTH AFRICA).

R. G. NOBLE, W. A. PRETORIUS, AND F. M. CHUTTER.

SOUTH AFRICAN JOURNAL OF SCIENCE, VOL. 67, NO. 3, P. 132-136, MARCH 1971.

DESCRIPTORS:
*WATER POLLUTION EFFECTS, *ORGANIC WASTES, *NUTRIENTS, *POLLUTANT IDENTIFICATION, ALGAE, AQUATIC BACTERIA, SEWAGE BACTERIA, ENVIRONMENTAL EFFECTS, BIOLOGICAL COMMUNITIES, HEAVY METALS, HERBICIDES, TOXINS, WASTE DILUTION, NITROGEN, PHCSPHCRUS, EUTROPHICATION, SURFACE WATERS.

ABSTRACT:
ALTHOUGH ORGANIC WASTES ARE MAINLY NON-TOXIC, THE PROCESSES INVOLVED IN THEIR DECOMPOSITION IN WATER MAY CREATE CONDITIONS SUCH AS ANAEROBIOSIS AND AMMONIA AND SULPHIDE ACCUMULATIONS WHICH ARE HIGHLY TOXIC. SEWAGE, INDUSTRIAL WASTES AND AGRICULTURAL RUNOFF ALSO CONTAIN HIGH LEVELS OF NITRATES AND PHOSPHATES WHICH PROMOTE THE GROWTH CF ALGAE AND HYDROPHYTES, LEADING TO HIGHLY DESTRUCTIVE EFFECTS IN NATURAL WATERS. IN RECENT YEARS TOXIC SUBSTANCES SUCH AS HERBICIDES AND HEAVY METAL IONS HAVE APPEARED IN HIGH QUANTITIES. ALTHOUGH THEY ACT IN DIFFERENT WAYS, POISONS AND NUTRIENTS HAVE STRIKINGLY SIMILAR EFFECTS ON BIOLOGICAL CCMMUNITIES. IN BOTH CASES, LOCAL ENVIRCNMENTS ARE ALTERED RADICALLY ENOUGH SO THAT ONLY RELATIVELY FEW SPECIES SURVIVE. IN AQUATIC EXOSYSTEMS SUCH LOW SPECIES DIVERSITIES RESULT IN LOW ECOSYSTEM STABILITY SO THAT MANY OF THE PROCESSES IN WATER SELF-PURIFICATION ARE NO LONGER OPERATIVE. BACTERIOLOGICAL, ALGOLGICAL AND FAUNAL MEASURES OF POLLUTION ARE DISCUSSED. SINCE THE COMBATTING CF PCLLUTANTS IS EXTREMELY DIFFICULT CNCE THEY HAVE BEEN RELEASED INTO WATERWAYS; THEIR EARLY DETECTION IS IMPORTANT. (CASEY-ARIZONA)

FIELD 05C, 05B

ACCESSION NO. W72-01472

BIOLOGICAL LIFE IN WATER,

NATIONAL INST. FOR WATER RESEARCH, PRETORIA (SOUTH AFRICA).

B. J. CHOLNOKY.

SOUTH AFRICAN JOURNAL OF SCIENCE, VOL 67, NO. 3, P. 128-131, MARCH 1971.

DESCRIPTORS:
*POLLUTANT IDENTIFICATION, *AQUATIC PLANTS, *AQUECUS SOLUTIONS, *BIOLOGICAL PROPERTIES, MOLECULAR STRUCTURE, PERMEABILITY, PLANT POPULATIONS, ALGAE, CHEMICAL PROPERTIES, WATER POLLUTION, PROTEIN, PRIMARY PRODUCTIVITY.

IDENTIFIERS:
*CHEMICAL EVCLUTION, *MACROMOLECULES.

ABSTRACT:
THE BROAD TRENDS OF CHEMICAL EVOLUTION ARE REVIEWED. THE PROCESSES INVOLVED IN MACROMOLECULE SYNTHESIS DEVELOPED AND OPERATE ONLY IN AQUEOUS SOLUTIONS. AS LARGE ELECTRICALLY CHARGED MOLECULES ACCUMULATED, WATER DIPOLES BECAME ORIENTED, CREATING GRCUPS OF PARTICLES OR MICELLES, AND MOLECULARLY DISPERSE SOLUTIONS WERE PREVENTED. MECHANISMS OF 'DUPLICATION' BY THE INTEGRATION OF DISSOLVED MOLECULES OF DIFFERENT TYPES INTO THE STRUCTURES OF NEW MACROMOLECULES WERE DEVELOPED THROUGH THE HARNESSING OF RADIANT ENERGY LEADING TO THE PRIMARY PRODUCERS, THE GREEN PLANTS. EVENTUALLY, OTHER PLANT FORMS EVCLVED WHICH WERE AT LEAST PARTIALLY DEPENDENT ON DISSOLVED ORGANIC MOLECULES FROM PRIMARY PRODUCERS, AND FINALLY, ANIMALS EVOLVED, WHO ARE TOTALLY DEPENDENT ON PRIMARY PRODUCERS. A MAJOR DIFFERENCE BETWEEN AQUATIC PRIMARY PRODUCERS, SUCH AS ALGAE, AND ALL OTHER LIFE FORMS, IS THE NECESSITY FOR EXTREMELY SELECTIVE PERMEABILITY MECHANISMS IN THE PRIMARY PRODUCERS TO FACILITATE SELECTIVE UPTAKE OF THE IMPORTANT INORGANIC MOLECULES THAT SERVE AS BUILDING BLOCKS OF MACROMOLECULES. IT IS THEREFORE ARGUED THAT THIS PROPERTY CONFERS A GREATER RELIABILITY ON GREEN PLANTS AS WATER POLLUTION INDICATORS. WE ARE STILL INCLINED TO OVERESTIMATE THE IMPCRTANCE OF BACTERIA AND TO UNDERESTIMATE THE IMPORTANCE OF BACTERIA AND TO UNDERESTIMATE THE IMPORTANCE OF ALGAE. (CASEY-ARIZONA)

FIELD 05C, 02I

ACCESSION NO. W72-01473

A BILL TO AMEND THE FEDERAL WATER POLLUTION CONTROL ACT AS AMENDED, AND FOR OTHER PURPOSES. REFERRED TO THE COMMITTEE ON PUBLIC WORKS,

SENATE, WASHINGTON, D.C.

W. F. MONDALE.

CONGRESSIONAL RECORD, VOL 117, NO 24, P S 2079-81 (DAILY ED. FEBRUARY 26, 1971). 3 P.

DESCRIPTORS:
*LAKES, *WATER POLLUTION CONTROL, *WATER QUALITY CONTROL, *FEDERAL GOVERNMENT, *GRANTS, EUTROPHICATION, GOVERNMENT FINANCE, ENVIRONMENTAL SANITATION, ALGAE, NUTRIENTS, INDUSTRIAL WASTES, MUNICIPAL WASTES, FARM WASTES, WASTE DISPOSAL, PESTICIDE REMOVAL, OXYGEN SAG, DREDGING, STATE GOVERNMENTS, TREATMENT FACILITIES, ADMINISTRATIVE AGENCIES, STANDARDS, REMEDIES.

ABSTRACT:
MANY OF THE NATION'S FRESH WATER LAKES ARE DETERIORATING. A BILL ENTITLED THE 'CLEAN LAKES ACT OF 1971' AMENDS THE FEDERAL WATER POLLUTION CONTROL ACT TO PROVIDE FOR A COORDINATED REHABILITATION PROGRAM FOR LAKES. THE PROGRAM INVOLVES INCREASED WASTE TREATMENT AND LAKE CLEANSING, UTILIZING THE LATEST TECHNOLOGY. IT HAS FOUR MAJOR POINTS: (1) FEDERAL GRANTS FOR TREATMENT WORKS LOCATED NEAR OR ADJACENT TO A LAKE AND WHICH DISCHARGE TREATED WASTES INTO THE LAKE OR TRIBUTARY WATERS WOULD BE INCREASED TO A MAXIMUM OF 65%, IF THE STATE PAYS AT LEAST 20%; (2) TECHNICAL AND FINANCIAL ASSISTANCE TO STATES AND CITIES WOULD BE PROVIDED, INCLUDING THE USE OF HARMLESS CHEMICALS TO DESTROY ALGAE, THE DREDGING OF LAKE BOTTOMS TO REMOVE DECAYING SLUDGE AND OTHER NOXIOUS POLLUTANTS, AND THE RECOVERY OF OVERGROWTH OF ALGAE FROM THE SURFACE; (3) THE USE OF FEDERAL WATER RESOURCE AGENCIES TO EXECUTE THE PROGRAM UNDER AGREEMENTS WITH THE STATES; AND (4) MEASURES TO ENFORCE WATER QUALITY STANDARDS, INCLUDING PENALITIES AND INJUNCTIVE RELIEF. ANNUAL APPROPRIATIONS FOR TREATMENT WORKS WOULD BE $150 MILLION FOR FISCAL YEARS 1972-1975. (REES-FLORIDA)

FIELD 06E, 05G

ACCESSION NO. W72-01659

STATEMENT IN SUPPORT OF THE CLEAN LAKES ACT OF 1971,

SENATE, WASHINGTON, D.C.

H. W. CANNON.

CONGRESSIONAL RECORD, VOL 117, NO 24, P S 2110 (DAILY ED. FEBRUARY 26, 1971). 1 P.

DESCRIPTORS:
*LAKES, *WATER POLLUTION CONTROL, *ENVIRONMENTAL SANITATION, *LEGISLATION, *FEDERAL GOVERNMENT, EUTROPHICATION, ENVIRONMENTAL ENGINEERING, WATER TREATMENT, WATER QUALITY CONTROL, STANDARDS, WATER POLLUTION SOURCES, WATER POLLUTION EFFECTS, RECREATION DEMAND, POPULATION, COMMUNITY DEVELOPMENT, ENVIRONMENTAL EFFECTS, GOVERNMENT FINANCE, TECHNOLOGY, ZONING, ALGAL CONTROL, SOIL EROSION, CITIES, ADMINISTRATIVE AGENCIES.

IDENTIFIERS:
*LAKE TAHOE.

ABSTRACT:
THE PROPOSED CLEAN LAKES ACT OF 1971 WOULD BE OF GREAT VALUE IN IMPROVING AND MAINTAINING THE UNIQUE PURITY AND CLARITY OF LAKE TAHOE. THIS GLACIER LAKE IS 22 MILES LONG AND 12 1/2 MILES WIDE, BUT ITS DEPTH OF 1645 FEET IS BEING DIMINISHED BY ALGAE AND MUD EXPANDING IN ITS WATERS AS A RESULT OF MAN'S DISTURBANCE OF WATERSHED SOIL. DUE TO THE INCREASING POPULATION AND RECREATION DEMANDS UPON LAKE TAHOE, IT IS IN DANGER OF PREMATURELY AGING--A PROCESS CALLED EUTROPHICATION. IN THE LAST 10 YEARS THE FERTILITY OF THE LAKE HAS INCREASED BY 72%. THIS FERTILIZATION PROCESS HAS BEEN SPEEDED UP BY COMMUNITY DEVELOPMENT PROJECTS, LUMBERING, ROADBUILDING, AND LAND CLEARING. A REGIONAL PLANNING AGENCY IS ALREADY ESTABLISHING STANDARDS AND CONTROLS OVER AIR AND WATER POLLUTION, ZONING, BUILDING AND THE GENERAL DEVELOPMENT OF THE LAKE AND THE SURROUNDING LAND AREA. HOWEVER, THE LAKE NEEDS IMMEDIATE ATTENTION FROM PAST NEGLECT. THE PROPOSED ACT WOULD DIRECT THE ENVIRONMENTAL PROTECTION AGENCY TO PROVIDE TECHNICAL AND FINANCIAL ASSISTANCE TO STATES AND CITIES AND TO CONDUCT A COMPREHENSIVE PROGRAM OF POLLUTION CONTROL AND REDEVELOPMENT OF THE LAKE AND THE SURROUNDING LAND. (REES-FLORIDA)

FIELD 06E, 05C

ACCESSION NO. W72-01660

THE DETERGENT POLLUTICN CONTROL ACT OF 1970,

SENATE, WASHINGTON, D. C.

G. NELSON.

CONGRESSIONAL RECORD, VOL 116, NO 27, P S 2444-47 (DAILY ED FEBRUARY 26, 1970), 4 P.

DESCRIPTORS:
*WATER POLLUTION CONTROL, *PHOSPHATES, *DETERGENTS, *LEGISLATION, CHEMICALS, CHEMICAL WASTES, CLEANING, DOMESTIC WASTES, EUTROPHICATION, WATER POLLUTION, WATER POLLUTION SOURCES, WATER POLLUTION TREATMENT, NUTRIENTS, OXYGEN REQUIREMENTS, AGING(BIOLOGICAL), AQUATIC ALGAE, WASTE ASSIMILATIVE CAPACITY, WATER POLLUTION EFFECTS, POLLUTANTS, POLLUTION ABATEMENT, FEDERAL GOVERNMENT.

IDENTIFIERS:
*WATER POLLUTION CONTROL ACT.

ABSTRACT:
LEGISLATION TO BAN POLYHOSPHATES FROM DETERGENTS IN THE UNITED STATES IS NECESSARY TO HALT WATER POLLUTION. THE DETERGENT INDUSTRY HAS SHOWN NO INTENTION OF CORRECTING PHOSPHATE POLLUTION WITHOUT LEGISLATION. THE DETERGENT POLLUTION CONTROL ACT OF 1970, WHICH WOULD AMEND THE FEDERAL WATER POLLUTION CONTROL ACT, REQUIRES THE ELIMINATION OF POLYHOSPHATES IN DETERGENTS AND THE ESTABLISHMENT OF NATIONAL ENVIRONMENTAL STANDARDS ON DETERGENT INGREDIENTS. DETERGENTS ARE NOT CONTROLLED BY SEWAGE TREATMENT BECAUSE PHOSPHATES PASS THROUGH MOST TREATMENT SYSTEMS. ONCE ENTERING LAKES AND RIVERS, PHOSPHATES EXCESSIVELY ENTICH THEM WITH NUTRIENTS. THIS CAUSES EXCESSIVE ALGAE GROWTH AND EUTROPHICATION. SEVERAL NON-POLLUTION PHOSPHATE SUBSTITUTES WHICH ARE FEASIBLE FOR USE IN INDUSTRIAL AND HOUSEHOLD DETERGENTS HAVE BEEN DEVELOPED. ENZYMES ARE ANOTHER POLLUTANT USED IN MOST HOUSEHOLD DETERGENTS. THE PROPOSED ACT MAKES IT UNLAWFUL TO MANUFACTURE DETERGENTS WITH PHOSPHATE AFTER JUNE 30, 1972. IT ALSO DIRECTS THE ESTABLISHMENT OF STANDARDS OF TOXICITY, ABILITY, AND BIODEGRADABILITY FOR ALL DETERGENTS. A FEDERAL ASSISTANCE PROGRAM WOULD BE USED TO ACCELERATE THE DEVELOPMENT OF EFFECTIVE PHOSPHATE SUBSTITUTES. (HART-FLORIDA)

FIELD 06E, 05G

ACCESSION NO. W72-01685

POTENTIOMETRIC TECHNIQUES FOR MONITORING IONS INVOLVED IN WATER POLLUTION,

MISSOURI UNIV., COLUMBIA. DEPT. OF CHEMISTRY.

S. E. MANAHAN, M. J. SMITH, D. ALEXANDER, AND P. ROBINSON.

AVAILABLE FROM THE NATIONAL TECHNICAL INFORMATION SERVICE AS PB-204 890, $3.00 IN PAPER COPY, $0.95 IN MICROFICHE. MISSOURI WATER RESOURCES RESEARCH CENTER, COLUMBIA, COMPLETION REPORT, AUG 1, 1971. 26 P, 4 FIG, 2 TAB, 4 REF. OWRR-B-040-MO(1).

DESCRIPTORS:
*POLLUTANT IDENTIFICATION, *NITROGEN, ELECTROCHEMISTRY, TRACE ELEMENTS, HEAVY METALS, ELECTRODES, NITRATES, ALGAE, COPPER, IONS, ION TRANSPORT.

IDENTIFIERS:
POTENTIOMETRY, CADMIUM IONS, COPPER IONS, OOCYSTIS.

ABSTRACT:
THE USE OF ION-SELECTIVE ELECTRODES, PARTICULARLY THE NITRATE ELECTRODE, WAS EXPLORED FOR THE ANALYSIS OF IONIC SPECIES IN NATURAL AQUATIC SYSTEMS. ATTEMPTS TO COMPENSATE QUANTITATIVELY FOR THE EFFECTS OF INTERFERING IONS WERE UNSUCCESSFUL. ION-SELECTIVE ELECTRODES SHOULD NOT BE USED, THEREFORE, WHEN SUBSTANTIAL INTERFERENCES ARE PRESENT. STANDARD ADDITION IS GENERALLY THE PREFERRED TECHNIQUE WITH ION-SELECTIVE ELECTRODES. THE CADMIUM ELECTRODE WAS USED TO DETERMINE THE FORMATION CONSTANT OF THE CITRATE COMPLEX OF CADMIUM ION. THE LOG OF THE FORMATION CONSTANT WAS FOUND TO BE 3.76 PLUS OR MINUS 0.04, 95% CONFIDENCE LEVEL. COPPER ION DEFICIENCY IN ALGAL CULTURES WAS STUDIED AND CORRELATED WITH COPPER ION ACTIVITY AS MEASURED BY THE COPPER ELECTRODE. IT WAS FOUND THAT A MINIMUM LEVEL OF APPROXIMATELY 40 PARTS PER BILLION OF COPPER WAS REQUIRED FOR OPTIMUM GROWTH OF A CULTURE OF OOCYSTIS. THE GROWTH OF THE ALGAE AT MINIMUM COPPER LEVELS COULD BE SUPPRESSED BY THE ADDITION OF CHELATING AGENT.

FIELD 05A

ACCESSION NO. W72-01693

PHYTOPLANKTONIC NITROGEN AS AN INDEX OF CULTURAL EUTROPHICATION,

MICHIGAN STATE UNIV., HICKORY CORNERS. W. K. KELLOGG BIOLOGICAL STATION.

R. G. WETZEL, AND BRUCE A. MANNY.

AVAILABLE FROM THE NATIONAL TECHNICAL INFORMATION SERVICE AS PB-204 707, $3.00 IN PAPER COPY, $0.95 IN MICROFICHE. COMPLETION REPORT, (NOVEMBER, 1971). 15 P, 1 FIG, 36 REF. OWRR-B-009-MICH(1)

DESCRIPTORS:
*NITROGEN, PLANKTON, *EUTROPHICATION, POLLUTANT IDENTIFICATION, ALGAE, ANALYTICAL TECHNIQUES, *ULTRAVIOLET RADIATION, DISSOLVED OXYGEN, LAKES, PHYTOPLANKTON, NUTRIENTS.

IDENTIFIERS:
*POLLUTION INDEX.

ABSTRACT:
ALGAE LESS THAN 10 MICRONS IN DIAMETER ARE NITROGEN RICH ON A CELL VOLUME BASIS COMPARED TO ALGAE LARGER THAN 10 MICRONS IN DIAMETER. A PROCEDURE FOR DISSOLVED ORGANIC NITROGEN (DON) DETERMINATION IN NATURAL WATERS WAS DEVELOPED. THE PROCEDURE UTILIZED HIGH INTENSITY ULTRAVIOLET LIGHT TO DESTROY THE DON AND IS 100-FOLD MORE SENSITIVE THAN THE MICRO-KJELDAHL PROCEDURE. THE PROCEDURE CAN DIFFERENTIATE UV-LABILE AND UV-REFRACTORY DON WITHIN THE DON POOL PRESENT IN A WIDE VARIETY OF NATURAL WATERBODIES. ALLOCHTHONOUS DISSOLVED ORGANIC CARBON (DOC) AND NITROGEN (DON) COMPRISE ABOUT HALF THE DOC AND DON LEAVING THE LAKE OUTLET IMPLYING ABOUT HALF THE DOC AND DON LEAVING THE LAKE ORIGINATES WITHIN THE LAKE AS A RESULT OF PLANT PHOTOSYNTHESIS AND DECOMPOSITION PROCESSES. DURING TRANSPORT IN A HARDWATER STREAM, UV-LABILE DON WAS REMOVED FROM THE WATER AND UV-REFRACTORY DON ACCUMULATED IN THE WATER UNTIL A STABLE EQUILIBRIUM RATIO OF ABOUT 7:3 REFRACTORY TO LABILE DON WAS ATTAINED. SECRETION OF DOC AND DON BY TWO AQUATIC MACROPHYTES IN AXENIC CULTURE WAS DIRECTLY PROPORTIONAL TO INCREASING CARBON FIXATION RATES, LIGHT INTENSITY, PH, CATIONIC CONCENTRATION AND ORGANIC CARBON CONCENTRATION. ORGANIC ENRICHMENT MAY ACCELERATE NUTRIENT CYCLES AND EUTROPHICATION RATES IN HARDWATER LAKES BY STIMULATING INCREASED SECRETION OF DOC AND DON BY THE LITTORAL FLORA.

FIELD 05C, 05A

ACCESSION NO. W72-01780

THE DIVERSITY OF PIGMENTS IN LAKE SEDIMENTS AND ITS ECOLOGICAL SIGNIFICANCE,

MINNESOTA UNIV., MINNEAPOLIS. DEPT. OF BOTANY.

JON E. SANGER, AND EVILLE GORHAM.

LIMNOLOGY AND OCEANOGRAPHY, VOL 15, NO 1, P 59-69, 1971. 2 FIG, 3 TAB, 21 REF. NSF G-23309.

DESCRIPTORS:
*PIGMENTS, *LAKES, *SEDIMENTS, *INDICATORS, CHROMATOGRAPHY, CHLOROPHYLL, TROPHIC LEVEL, EUTROPHICATION, ALGAE, DECOMPOSING ORGANIC MATTER, MINNESOTA, CYANOPHYTA, TREES, GRASSES, LEAVES, FOREST SOILS, SAMPLING, MUD, PHYTOPLANKTON, FLUORESCENCE, PHOTOSYNTHETIC BACTERIA.

IDENTIFIERS:
CAROTENOIDS, XANTHOPHYLLS, MINNESOTA LAKES, LUTEIN.

ABSTRACT:
DIVERSITY OF SEDIMENTARY PIGMENTS IS PARTICULARLY A CONSEQUENCE OF DECOMPOSITION AND SOURCE MATERIAL HAS A PRONOUNCED INFLUENCE. CONCENTRATIONS OF CHLOROPHYLL DERIVATIVES IN THE ORGANIC MATTER OF SURFACE SEDIMENTS IN THE ENGLISH LAKE DISTRICT PROVED SENSITIVE INDICES OF LAKE FERTILITY. THIS STUDY INVESTIGATES FURTHER SOURCES OF ORGANIC MATTER IN LAKE SEDIMENTS, BY EXAMINING PLANT PIGMENT DIVERSITY OF WIDELY DIFFERING TROPHIC, MORPHOLOGICAL, AND CHEMICAL CHARACTERISTICS, AND IN TERRESTRIAL AND AQUATIC PLANT MATERIAL IN VARYING STAGES OF DECOMPOSITION. LAKES CHOSEN (18 IN MINNESOTA AND 6 IN ENGLISH LAKE DISTRICT) INCLUDED THOSE WHOSE PRIMARY PRODUCTIVITY AND WATER AND SEDIMENT CHEMISTRY ARE NOW BEING STUDIED. THIN-LAYER CHROMATOGRAPHY SHOWS A LARGE NUMBER OF PIGMENTS (CHLOROPHYLL DERIVATIVES AND CAROTENOIDS) IN PROFUNDAL LAKE SEDIMENTS, DIVERSITY BEING SOMEWHAT GREATER IN EUTROPHIC THAN IN OLIGOTROPHIC LAKES. WHILE QUANTITY OF PIGMENTS PER GRAM ORGANIC MATTER IS MUCH LOWER IN OLIGOTROPHIC THAN EUTROPHIC LAKES, PIGMENT DIVERSITY IS ONLY SLIGHTLY LOWER. SEDIMENTARY PIGMENTS ARE MUCH MORE NUMEROUS (24-27)THAN THOSE OF UPLAND VEGETATION (7-8), AQUATIC MACROPHYTES (12-15), AND PLANKTONIC ALGAE (10-21). ALGAL DECOMPOSITION, WHICH IS ACCOMPANIED BY A MARKED INCREASE IN PIGMENT NUMBER, SEEMS THE MOST LIKELY CAUSE FOR THE EXTREME DIVERSITY OF SEDIMENTARY PIGMENTS. (JONES-WISCONSIN)

FIELD 05C, 02H

ACCESSION NO. W72-01784

AEROBIC DECOMPOSITION OF ALGAE,

STANFORD UNIV., CALIF.

WILLIAM J. JEWELL, AND PERRY L. MCCARTY.

ENVIRONMENTAL SCIENCE AND TECHNOLOGY, VOL 5, NO. 10, OCTOBER 1971, P 1023-1031, 7 FIG, 4 TAB, 40 REF.

DESCRIPTORS:
*OXYGEN REQUIREMENTS, *DECOMPOSING ORGANIC MATTER, AERATION, *ALGAE, *BIODEGRADATION, LABORATORY TESTS, FILTRATION, AMMONIA, HYDROGEN ION CONCENTRATION, TEMPERATURE, LIGHT INTENSITY, SEDIMANTATION, OXIDATION, STORAGE, BIOCHEMICAL OXYGEN DEMAND, CHEMICAL OXYGEN DEMAND, PHOSPHORUS, NITROGEN, EUTROPHICATION, *AEROBIC CONDITIONS, *AEROBIC TREATMENT.

IDENTIFIERS:
REFRACTORY MATERIALS.

ABSTRACT:
5 GAL. WATER SAMPLES FROM VARIOUS LAKE, STREAMS, AND RESERVOIRS, AS WELL AS EFFLUENTS FROM SEVERAL SEWAGE TREATMENT FACILITIES WERE OBTAINED. EACH SAMPLE WAS FILTERED AND AERATED, TO ALLOW OXIDATION OF ALL REDUCED MATERIALS, EXPOSED TO DIURNAL FLOURESCENT LIGHT, AND AERATED WITH A MIXTURE OF 1% CO2 IN AIR. LIGHTING, PH AND TEMPERATURE WERE HELD CONSTANT DURING ANY GIVEN EXPERIMENT. AT PREDETERMINED TIME INTERVALS, SAMPLES WERE TAKEN FROM THE ILLUMINATED VESSELS AND PLACED IN DARK VESSELS AND AERATED WITH THE SAME AERATION MIXTURE. RESULTS OF THESE TESTS INDICATED THAT ALGAE AND ALGAL-DERIVED ORGANIC MATTER CONSISTED OF THREE FRACTIONS. THE FIRST FRACTION CONSISTED OF DEGRADABLE STORAGE PRODUCTS THAT DISAPPEAR WITHIN A FEW HOURS AFTER THE ORGANISMS ARE PLACED IN THE DARK. OXYGEN DEMAND OF THE STORAGE PRODUCTS WAS DEEMED INSIGNIFICANT IN LONG-TERM CONSIDERATIONS, BUT ITS DECOMPOSITION WAS DETERMINED TO BE A POSSIBLE SIGNIFICANT FACTOR IN INFLUENCING DIURNAL OXYGEN VARIATIONS IN NATURAL WATERS. BIODEGRADABLE ORGANIC MATERIAL, THE SECOND FRACTION, COMPRISED SOME 30% OF THE TOTAL MASS OF ORGANIC ALGAL MATERIALS. A FAIRLY CONSTANT, BUT INTERMEDIATE, FIRST ORDER RATE CONSTANT (.01 TO .06 DAY-1) WAS DETERMINED, INDICATING THAT ALL OF THE BIODEGRADABLE ORGANICS SHOULD BE DECOMPOSED WITHIN ONE YEAR. THE REMAINING FRACTION, REFRACTORY MATERIAL WAS FOUND TO DECOMPOSE BY ONLY A FEW PERCENT PER YEAR, HAVING A SIGNIFICANT OXYGEN DEMAND ONLY BY ACCUMULATION. (LOWRY-TEXAS)

FIELD 05D, 05C

ACCESSION NO. W72-01881

IODINE AND ALGAE IN SEDIMENTARY ROCKS ASSOCIATED WITH IODINE-RICH BRINES,

BUREAU OF MINES, BARTLESVILLE, OKLA. PETROLEUM RESEARCH CENTER.

A. G. COLLINS, J. H. BENNETT, AND O. K. MANUEL.

GEOLOGICAL SOCIETY OF AMERICA BULLETIN, VOL 82, NO 9, P 2607-2610, SEPTEMBER 1971. 2 FIG, 2 TAB, 7 REF. NSF GRANT GA-12099.

DESCRIPTORS:
*SEDIMENTARY ROCKS, *BRINES, *ALGAE, *OKLAHOMA, *IODINE RADIOISOTOPES, URANIUM RADIOISOTOPES, ANALYTICAL TECHNIQUES, PALEOZOIC ERA, WATER CHEMISTRY.

IDENTIFIERS:
NEUTRON ACTIVATION ANALYSIS.

ABSTRACT:
NEUTRON ACTIVATION ANALYSES OF IODINE AND URANIUM IN PALEOZOIC SEDIMENTARY ROCKS FROM THE NORTHERN OKLAHOMA PLATFORM OF THE ANADARKO BASIN SHOWED 0.9 TO 12.3 PPM I AND 0.07 TO 8.7 PPM U. THE SAMPLES WERE TAKEN FROM STRATA IN KINGFISHER COUNTY, OKLAHOMA, WHERE ANOMALOUSLY HIGH CONCENTRATIONS OF IODINE WERE FOUND IN ASSOCIATED SUBSURFACE BRINES. MICROPALEONTOLOGICAL EXAMINATIONS REVEALED ALGAL STRANDS IN THE MORE IODINE-RICH ROCKS. (WOODARD-USGS)

FIELD 02K, 02F, 05A

ACCESSION NO. W72-02073

BLUE-GREEN ALGAL EFFECTS ON SOME HYDROLOGIC PROCESSES AT THE SOIL SURFACE,

ARIZONA UNIV., TUCSON. WATER RESOURCES RESEARCH CENTER.

W. F. FAUST.

IN: HYDROLOGY AND WATER RESOURCES IN ARIZONA AND THE SOUTHWEST, PROCEEDINGS, ARIZONA SECTION-AMERICAN WATER RESOURCES ASSOCIATION AND THE HYDROLOGY SECTION-ARIZONA ACADEMY OF SCIENCE, APRIL 22-23, 1971, TEMPE, VOL 1, P 99-105, 1971. 1 FIG, 2 TAB, 4 REF.

DESCRIPTORS:
*SOIL ALGAE, *SOIL SURFACES, *SEDIMENT YIELD, *SIMULATED RAINFALL, *SOIL TYPES, SOIL TEXTURE, CLAYS, LABORATORY TESTS, STATISTICAL METHODS, SOIL MICROORGANISMS, RUNOFF, HYDROLIC DATA.

IDENTIFIERS:
*SUSPENDED SEDIMENT PRODUCTION, *BLUE-GREEN ALGAE.

ABSTRACT:
PREVIOUS STUDIES HAVE INDICATED THAT BLUE-GREEN ALGAE MAY AFFECT RUNOFF, INFILTRATION AND EROSION AT SOIL SURFACES. USING SOIL PLOTS UPON WHICH BLUE-GREEN ALGAE WERE GROWN UNDER AN ARTIFICIAL WETTING REGIME, STUDIES WERE MADE USING SIMULATED RAINFALL. A 30% CLAY CONTENT PIMA SOIL AND A CONTRASTING 8% CLAY CONTENT RIVER-BOTTOM ANTHONY SOIL WERE USED. SCYTONEMA HOFFMANII AND MICROCOLEUS VAGINATUS GREW ON THE PIMA SOIL WHILE SCHIZOTHRIX CALCICOLA DEVELOPED ON THE ANTHONY SOIL. THE RESULTS SHOWED THAT BLUE-GREEN ALGAL GROWTHS SIGNIFICANTLY REDUCED THE AMOUNT OF SUSPENDED SOIL MATERIAL IN RUNOFF WATER AS COMPARED WITH BARE SOILS. DIFFERENCES IN RUNOFF SUSPENDED SEDIMENTS WERE ALSO RELATED TO DIFFERENCES IN SOIL TYPE AND SIMULATED RAINFALL INTENSITY. AN ANALYSIS OF VARIANCE OF THE EFFECTS OF THESE 3 FACTORS AND THEIR INTERACTIONS SHOWED THAT THE SMALLER DIFFERENCES IN SUSPENDED SEDIMENT PRODUCTION ON THE ANTHONY SOIL DUE TO THE MICROVEGETATION TREATMENT WAS VERIFIED BY A HIGHLY SIGNIFICANT SOILS-MICROVEGETATION INTERACTION, PROBABLY BECAUSE THE FINER PIMA SOILS WASH AWAY MORE EASILY WITHOUT STABILIZING MICROVEGETATION. ALSO, LESS VEGETATION SEEMS TO GROW ON THE ANTHONY SOIL. DIFFERENCES IN RUNOFF AND INFILTRATION VOLUMES AND IN SETTLEABLE SEDIMENT AMOUNTS WERE NOT DETECTED.
(CASEY-ARIZONA)

FIELD 02G, 05C, 04A

ACCESSION NO. W72-02218

EVALUATION OF EFFECT OF IMPOUNDMENT ON WATER QUALITY IN CHENEY RESERVOIR,

COLORADO STATE UNIV., FORT COLLINS.

J. C. WARD, AND S. KARAKI.

BUREAU OF RECLAMATION RESEARCH REPORT NO 25, 1971. 69 P, 38 FIG, 18 TAB, 23 REF, APPEND. BUR. RECLAM CONTRACT 14-06-D-6578.

DESCRIPTORS:
*WATER QUALITY, *IMPOUNDED WATERS, *RESERVOIRS, *HYDROLOGIC DATA, *KANSAS, WATER CHEMISTRY, CHEMICAL ANALYSIS, EVAPORATION, SEEPAGE, TURBIDITY, HYDROLOGIC BUDGET, SOLAR RADIATION, WATER TEMPERATURE, SALT BALANCE, DISSOLVED OXYGEN, ALGAE, BACTERIA, ODOR, INFLOW, DISSOLVED SOLIDS, DATA COLLECTIONS.

IDENTIFIERS:
*CHENEY RESERVOIR(KANSAS).

ABSTRACT:
THE EFFECT OF IMPOUNDMENT ON THE QUALITY OF WATER IN CHENEY RESERVOIR NEAR WICHITA, KANS. IS PRESENTED. THE RESERVOIR DID NOT STRATIFY DURING THE PERIOD OF DATA COLLECTION. THE INCREASE IN THE DISSOLVED SOLIDS CONCENTRATION WAS DIRECTLY RELATED TO EVAPORATION. ON AN ANNUAL BASIS, 42% OF THE TOTAL INFLOW WAS EVAPORATED FROM THE RESERVOIR. BYPASSING THE POOREST QUALITY WATERS OF THE STREAM SERVING THE RESERVOIR IS SUGGESTED TO REDUCE THE DISSOLVED SOLIDS CONCENTRATION IN THE RESERVOIR AND IN THE STREAM BELOW THE RESERVOIR. THE BIOLOGICAL ACTIVITY WITHIN THIS RESERVOIR DID NOT SEEM TO AFFECT THE WATER QUALITY MATERIALLY. ODOR APPEARS TO HAVE STABILIZED AT A THRESHOLD ODOR NUMBER OF ABOUT 5. THE EFFECT OF THE INTERACTION BETWEEN THE MICROORGANISMS AND NUTRIENTS WAS CHARACTERIZED IN THE ANALYSIS OF PHOSPHATES, NITRATES, AND SILICA CONCENTRATIONS IN THE RESERVOIR. THE DISSOLVED OXYGEN PERCENT SATURATION DECREASED FROM 100% AT THE WATER SURFACE TO 82% AT A 25-FOOT DEPTH. (WOODARD-USGS)

FIELD 05C, 02H, 04A

ACCESSION NO. W72-02274

A THERMODYNAMIC ANALYSIS OF A PRIMARY WASTE STABILIZATION POND,

UTAH STATE UNIV., LOGAN. UTAH WATER RESEARCH LAB.

D. W. HENDRICKS, W. D. POTE, AND J. G. ANDREW.

AVAILABLE FROM THE NATIONAL TECHNICAL INFORMATION SERVICE AS PB-205 282, $3.00 IN PAPER COPY, $0.95 IN MICROFICHE. REPORT NO PRCWRR 16-1, SEPTEMBER 1970, 63 P, 14 FIG, 7 TAB, 40 REF. OWRR A-006-UTAH(1).

DESCRIPTORS:
*OXIDATION LAGOONS, *ALGAE, *OXIDATION, PHOTOSYNTHESIS, OXYGEN, *THERMODYNAMICS, ANALYTICAL TECHNIQUES, DESIGN CRITERIA, KINETICS, *WASTE WATER TREATMENT.

IDENTIFIERS:
*SOLAR INSOLATION, *STOICHIOMETRY.

ABSTRACT:
A 97.5 ACRE OXIDATION POND WITH AN AVERAGE DEPTH OF 1 1/2 TO 1 2/3 METERS, IN OPERATION SINCE 1967, WAS SAMPLED IN SEPTEMBER 1969 AND IN JUNE 1970. THE PRIMARY OBJECTIVE WAS TO QUANTITATE THE ACTUAL ENERGY TRADE-OFF, IN TERMS OF ALGAE PRODUCED VS. AMOUNT OF WASTE DEGRADED, FOR OXIDATION PONDS. SUCH QUANTITATION WAS ACCOMPLISHED BY: (1) DEFINING THE CHEMICAL REACTIONS INVOLVED-BOTH STOICHIOMETRICALLY AND THERMODYNAMICALLY (THE LATTER IN TERMS OF EQUILIBRIUM CONDITIONS); (2) MEASURING TERMS IN A DAILY MASS BALANCE MODEL OF AN OPERATING PRIMARY POND; AND (3) EVALUATING THE 'ALGAE PRODUCTION POTENTIAL' FOR THE POND STUDIED, BASED UPON AVAILABLE SOLAR INSOLATION. THESE RESULTS DEFINED RESPECTIVELY: (1) THE CALCULATED ABSOLUTE LOWER LIMIT OF DAILY ALGAL SYNTHESIS NECESSARY FOR PRODUCTION OF THE STOICHIOMETRIC OXYGEN TO SATISFY THE DAILY INFLUENT BOD REQUIREMENT; (2) A MEASURED DAILY SYNTHESIS RATE OF ALGAE TO COMPARE WITH THE DAILY INFLUENT TOC, UNDER CONDITIONS OF MAXIMUM SUNSHINE IN THE ANNUAL CYCLE; AND (3) THE CALCULATED ABSOLUTE UPPER LIMIT OF DAILY ALGAL SYNTHESIS, THROUGH THE ANNUAL CYCLE, IF ALL USABLE SOLAR ENERGY WERE UTILIZED. RESULTS ESTABLISHED: (1) ALGAE PRODUCTION IS SIGNIFICANT IN PROPORTION TO WASTE DEGRADED, EVEN IN THE LOWER LIMIT; (2) ACTUAL PRODUCTION WAS OVER 100 TIMES THE STOICHIOMETRIC AMOUNT; AND (3) THE UPPER PRODUCTION LIMIT WAS OVER 3 TIMES THE ACTUAL PRODUCTION. ALL RESULTS INDICATED A VAST ENERGY OVERTURN WITH LITTLE OR NO NET EFFECT. (LOWRY-TEXAS)

FIELD 05D

ACCESSION NO. W72-02363

BIOLOGICAL RESPONSE TO TERTIARY TREATED EFFLUENT IN INDIAN CREEK RESERVOIR,

UTAH STATE UNIV., LOGAN. DEPT. OF ENVIRONMENTAL BIOLOGY.

D. B. PORCELLA, P. H. MCGAUHEY, AND G. L. DUGAN.

PREPRINT, PRESENTED AT 44TH ANNUAL CONFERENCE OF WATER POLLUTION CONTROL FEDERATION, SESSION 25, OCTOBER 3-8, 1971. 31 P, 5 FIG, 9 TAB, 13 REF.

DESCRIPTORS:
*WATER DILUTION, *NITROGEN, *PHOSPHORUS, *DETENTION RESERVOIRS, TERTIARY TREATMENT, ALGAE, PLANTS, TROUT, AQUATIC ENVIRONMENT, EUTROPHICATION, NITRIFICATION, RESERVOIR EVAPORATION, RESERVOIRS, NUTRIENTS, WATER REUSE.

IDENTIFIERS:
SOUTH TAHOE PUBLIC UTILITIES DISTRICT, *INDIAN CREEK RESERVOIR, NUTRIENT BALANCE.

ABSTRACT:
INDIAN CREEK RESERVOIR RECEIVES TERTIARY EFFLUENT FROM THE SOUTH TAHOE PUBLIC UTILITY DISTRICT. AN ON-GOING STUDY IS OBSERVING THE EFFECTS OF NUTRIENT REMOVAL (MAINLY PHOSPHORUS) AND THE POSSIBILITY OF IMPOUNDMENTS AS A PROCESS FOR RECLAIMING WATER FROM SEWAGE. THE RESERVOIR IS CHANGING FROM A LOW DIVERSITY AND HIGH PRODUCTION TO A HIGHER DIVERSITY AND LOWER PRODUCTION AQUATIC ECOSYSTEM, A MORE BALANCED SYSTEM LESS LIKELY TO BE DISTURBED. HIGH LEVELS OF NITROGEN AND PHOSPHORUS EXIST, ALTHOUGH CONSIDERABLE REMOVAL OF BOTH NITROGEN AND PHOSPHORUS OCCURS IN THE RESERVOIR. A BALANCE ACCOUNTED FOR PHOSPHORUS, BUT A DEFICIT WAS FOUND FOR NITROGEN THAT WAS ATTRIBUTED TO NITRIFICATION-DENITRIFICATION. ALGAL GROWTH WAS OBSERVED THAT WAS ASSOCIATED WITH BENTHIC ALGAE. THE DIVERSITY OF BENTHIC ORGANISMS WAS LOWER THAN EXPECTED IN A MATURE RESERVOIR, BUT IS APPARENTLY INCREASING. FISH SURVIVAL DURING THE 1971 SEASON INDICATES THAT THE RESERVOIR CONTINUES TO BE A SUITABLE ENVIRONMENT FOR PLANTED TROUT SPECIES. AQUATIC VASCULAR PLANTS ARE PRESENT IN SOME ABUNDANCE, PARTICULARLY IN THE SHALLOW WATER AREAS. THOUGH SUFFICIENT PHOSPHORUS IS AVAILABLE NO OBJECTIONABLE ALGAL BLOOMS HAVE OCCURRED. WATER QUALITY, BOTH CHEMICALLY AND BIOLOGICALLY, DIFFERS DRAMATICALLY FROM THE TERTIARY EFFLUENT, FROM WHICH IT PREDOMINANTLY DERIVES, SUGGESTS IMPOUNDMENT AS AN ECONOMICAL PROCESS TO REDUCE NITROGEN FOLLOWING PHOSPHORUS REMOVAL. (MORGAN-TEXAS)

FIELD 05C

ACCESSION NO. W72-02412

PHOTOSYNTHETIC RECLAMATION OF AGRICULTURAL SOLID AND LIQUID WASTE,

CALIFORNIA UNIV., BERKELEY. SANITARY ENGINEERING RESEARCH LAB.

G. L. DUGAN, C. G. GOLUEKE, W. J. OSWALD, AND C. E. RIXFORD.

SERL REPORT NO 70-1, MAY 1970. 165 P, 55 TAB, 24 FIG, 51 REF. USPHS 5 R01 U1 00566-03.

DESCRIPTORS:
*FARM WASTE, *POULTRY, *OXIDATION LAGOONS, NUTRIENTS, AQUATIC ALGAE, NITROGEN, ANAEROBIC DIGESTION, AEROBIC, ALKALINITY, ACIDITY, LABORATORY, DIGESTION TANK, PONDAGE, PILOT PLANT, SETTLING BASIN.

IDENTIFIERS:
*DETENTION PERIOD, FLUSHING OPERATION, ACID INJECTION PUMP, ALUM INJECTION PUMP, MANURE GRINDER.

ABSTRACT:
THE RESEARCH PLAN ON WHICH THE GRANT WAS BASED AND REPORTED CALLED FOR LABORATORY AND PILOT PLANT STUDIES TO DEVELOP A PARTIALLY-CLOSED SYSTEM OF ANIMAL WASTE MANAGEMENT BASED ON THE INTEGRATION OF AN ANAEROBIC AND AN AEROBIC PHASE, THE RECYCLING OF WATER, AND THE RECLAMATION OF A USABLE PRODUCT. CONTAINED HEREIN ARE: A REVIEW OF THE LABORATORY AND 'PRE-PILOT PLANT' STUDIES DESCRIBED IN A PROGRESS REPORT (1) ISSUED DURING THE SECOND YEAR OF THE STUDY, AND A FULL ACCOUNT OF PILOT PLANT STUDIES COMPLETED AT THE TIME OF THIS WRITING. THE PILOT PLANT INCLUDES A POULTRY ENCLOSURE, A HYDRAULIC SYSTEM FOR HANDLING THE WASTES, A HEATED ANAEROBIC DIGESTER WITH AUXILIARY EQUIPMENT, AND AN ALGAE-PRODUCTION POND. (BUNDY-IOWA STATE)

FIELD 05D

ACCESSION NO. W72-02850

REMOVAL OF NITRATE BY AN ALGAL SYSTEM,

CALIFORNIA STATE DEPT. OF WATER RESOURCES, FRESNO. SAN JOAQUIN DISTRICT.

RANDALL L. BROWN.

COPY AVAILABLE FROM GPC SUP DOC FOR $1.25; MICROFICHE FROM NTIS AS PB-205 425, $0.95. ENVIRONMENTAL PROTECTION AGENCY - WATER QUALITY OFFICE, WATER POLLUTION CONTROL RESEARCH SERIES, APRIL, 1971, 132 P, 58 FIG, 27 TAB, 59 REF. EPA PROGRAM 13030 ELY.

DESCRIPTORS:
AGRICULTURAL WASTES, WATER POLLUTION CONTROL, *BIOLOGICAL TREATMENT, *NITRATES, TREATMENT FACILITIES, ALGAE, *WASTE WATER TREATMENT, *ALGAL CONTROL, *AQUATIC WEED CONTROL, CALIFORNIA.

IDENTIFIERS:
*ALGAE STRIPPING, SCENEDESMUS, ALGAL GROWTH, ALGAL HARVESTING, *SAN JOAQUIN VALLEY(CALIF).

ABSTRACT:
AN ALGAL SYSTEM CONSISTING OF ALGAE GROWTH, HARVESTING AND DISPOSAL WAS EVALUATED AS A POSSIBLE MEANS OF REMOVING NITRATE-NITROGEN FROM SUBSURFACE AGRICULTURAL DRAINAGE IN THE SAN JOAQUIN VALLEY OF CALIFORNIA. THE STUDY OF THIS ASSIMILATORY NITROGEN REMOVAL PROCESS WAS INITIATED TO DETERMINE OPTIMUM CONDITIONS FOR GROWTH OF THE ALGAL BIOMASS, SEASONAL VARIATIONS IN ASSIMILATION RATES, AND METHODS OF HARVESTING AND DISPOSAL OF THE ALGAL PRODUCT. A SECONDARY OBJECTIVE OF THE STUDY WAS TO OBTAIN PRELIMINARY COST ESTIMATES AND PROCESS DESIGN. THE GROWTH STUDIES SHOWED THAT ABOUT 75 TO 90 PERCENT OF THE 20 MG/L INFLUENT NITROGEN WAS ASSIMILATED BY SHALLOW (12-INCH CULTURE DEPTH) ALGAL CULTURES RECEIVING 2 TO 3 MG/L ADDITIONAL IRON AND PHOSPHORUS AND A MIXTURE OF 5 PERCENT CO2. THEORETICAL HYDRAULIC DETENTION TIMES REQUIRED FOR THESE ASSIMILATION RATES VARIED FROM 5 TO 16 DAYS, DEPENDING ON THE TIME OF THE YEAR. THE TOTAL NITROGEN REMOVAL BY THE ALGAL SYSTEM, ASSUMING 95 PERCENT REMOVAL OF THE ALGAL CELLS, RANGED FROM 70 TO 85 PERCENT OF THE INFLUENT NITROGEN. THE MOST ECONOMICAL AND EFFECTIVE ALGAL HARVESTING SYSTEM TESTED WAS FLOCCULATION AND SEDIMENTATION FOLLOWED BY FILTRATION OF THE SEDIMENT. THE ALGAL CAKE FROM THE VACUUM FILTER, CONTAINING ABOUT 20 PERCENT SOLIDS, WAS THEN AIR- OR FLASH-DRIED TO ABOUT 90 PERCENT SOLIDS. THE MARKET VALUE FOR THIS PRODUCT AS A PROTEIN SUPPLEMENT WAS ESTIMATED TO BE ABOUT $80 TO $100 PER TON. MINER-IOWA STATE)

FIELD 05D

ACCESSION NO. W72-02975

LYSOGENY OF A BLUE-GREEN ALGA, PLECTONEMA BORYANUM,

DELAWARE UNIV., NEWARK. DEPT. OF BIOLOGICAL SCIENCES.

R. E. CANNON, M. S. SHANE, AND VALERIE N. BUSH.

VIROLOGY, VOL. 45, NO. 1, P 149-153, JULY 1971. 1 FIG, 1 TAB, 9 REF. OWRR A-016-DEL(1).

DESCRIPTORS:
*CYANOPHYTA, ALGAE, *VIRUSES.

IDENTIFIERS:
*LYSOGENY, *PLECTONEMA BORYANUM, LPP-1, LPP-1D, PHYCOVIRUS, MITOMYCIN C.

ABSTRACT:
EVIDENCE FOR LYSOGENY OF A FILAMENTOUS BLUE-GREEN ALGA, PLECTONEMA BORYANUM, IS PRESENTED. A SUSPECTED LYSOGENIC STRAIN WHICH CARRIES PHYCOVIRUS DESIGNATED LPP-1D (DELAWARE STRAIN) HAS BEEN SUBCULTURED FOR FOUR YEARS. VIRAL NEUTRALIZATION TESTS SHOW THAT LPP-1D IS ANTIGENICALLY SIMILAR TO LPP-1. INDUCTION EXPERIMENTS OF LYSOGENIC PLECTONEMA WITH AN ANTIBIOTIC, MITOMYCIN C, RESULT IN 100-FOLD INCREASE IN PHYCOVIRUS TITER 4-5 HR AFTER TREATMENT. GROWTH OF LYSOGENIC CULTURES WITH ANTIPHYCOVIRUS SERUM ELIMINATES ALL FREE PHYCOVIRUS IN THE CULTURE. TEN DAYS LATER, PHYCOVIRUS IS AGAIN PRESENT IN THE MEDIUM. LPP-1 APPEARS TO BE A VIRULENT STRAIN WHILE LPP-1D MAY BE A TEMPERATE STRAIN OF THE SAME TYPE OF PHYCOVIRUS WHICH HAS LYSOGENIZED PLECTONEMA BORYANUM.

FIELD 05C

ACCESSION NO. W72-03188

SAMPLING AND MEASUREMENT IN THE AQUATIC ENVIRONMENT,

WASHINGTON STATE UNIV., PULLMAN. DEPT. OF SANITARY ENGINEERING.

SURINDER K. BHAGAT, DONALD E. PROCTOR, AND WILLIAM H. FUNK.

PRESENTED AT THE 25TH PURDUE INDUSTRIAL WASTE CONFERENCE, LAFAYETTE, INDIANA, MAY 5-7, 1970, MIMEO (UNDATED), 22 P. 13 FIG, 1 TAB, 14 REF. 15-12-68 16080 ERO.

DESCRIPTORS:
*EUTROPHICATION, *MEASUREMENT, *SAMPLING, WATER RESOURCES, AREA REDEVELOPMENT, BIOLOGICAL CHARACTERISTICS, ALGAE, ZOOPLANKTON, BACTERIA, NUTRIENTS, SEDIMENTS, ENVIRONMENTAL EFFECTS, TRACE ELEMENTS, ANALYTICAL TECHNIQUES, COLUMBIA RIVER, WASHINGTON, OREGON.

IDENTIFIERS:
*VANCOUVER LAKE(WASH), VANCOUVER(WASH), PORTLAND(ORE), WATER QUALITY ANALYZER.

ABSTRACT:
VANCOUVER LAKE (WASHINGTON) PRESENTLY POLLUTED, WHICH HAS THE POTENTIAL OF BECOMING A USEFUL MULTIPURPOSE RESOURCE, WAS STUDIED TO DETERMINE THE PRESENT WATER QUALITY CONDITIONS, EVALUATE POLLUTION SOURCES, AND EXPLORE WAYS TO IMPROVE ITS USEFULNESS. SUCH SAMPLING AND MEASUREMENTS WHICH WERE CONSIDERED NECESSARY IN ATTAINING VANCOUVER LAKE PROJECT OBJECTIVES ARE DISCUSSED IN DETAIL. MEASUREMENTS WERE MADE OF BIOLOGICAL AND BACTERIOLOGICAL ACTIVITIES, NUTRIENT LEVELS, AND THE EXISTING ENVIRONMENTAL CONDITIONS. THE MOST PROMINENT ALGA IS APHANIZONMENON FLOS-AQUAE, ONE OF THE MORE UNSIGHTLY AND ODORIFEROUS. A BIOLOGICAL INVENTORY OF THE BOTTOM ORGANISMS SHOWED AQUATIC EARTHWORMS CHARACTERISTIC OF SHALLOW AND TURBID WATERS. BACTERIOLOGICAL EXAMINATION INDICATED THAT BACTERIAL CONTAMINATION IS EXCESSIVELY HIGH PRECLUDING RECREATIONAL USE. SPHAEROTILUS, RESPONSIBLE FOR SLIME GROWTHS IN STREAMS AND FOR DESTROYING HABITATS FOR VARIOUS AQUATIC ANIMALS WAS MEASURED IN THE ADJACENT COLUMBIA RIVER. THE NUTRIENT LEVELS IN VANCOUVER LAKE ARE QUITE HIGH AND RESPONSIBLE FOR EXCESSIVE ALGAL POPULATIONS. EQUIPMENT FOR MEASUREMENTS OF ENVIRONMENTAL FACTORS IS DELINEATED. TRACE ELEMENTS WERE MEASURED BY NEUTRON ACTIVATION AND HIGH RESOLUTION SPECTROMETRY. (JONES-WISCONSIN)

FIELD 05C

ACCESSION NO. W72-03215

PRIMARY PRODUCTIVITY, CHEMO-ORGANOTROPHY, AND NUTRITIONAL INTERACTIONS OF EPIPHYTIC ALGAE AND BACTERIA ON MACROPHYTES IN THE LITTORAL OF A LAKE,

MICHIGAN STATE UNIV., HICKORY CORNERS, W. K. KELLOGG BIOLOGICAL STATION.

HAROLD L. ALLEN.

ECOLOGICAL MONOGRAPHS, VOL 41, NO 2, P 97-127, 1971. 33 FIG, 13 TAB, 18 REF. AEC AT (11-1)-1599 NSF GB-6538.

DESCRIPTORS:
*PRIMARY PRODUCTIVITY, *EUTROPHICATION, *NUTRIENTS, *BIOLOGICAL COMMUNITIES, *AQUATIC PLANTS, ALGAE, BACTERIA, LITTORAL, LAKES, PERIPHYTON, ORGANIC MATTER, BIOMASS, METABOLISM, PHYTOPLANKTON, PLANT PHYSIOLOGY, PHYSIOLOGICAL ECOLOGY, ORGANIC COMPOUNDS, ON-SITE TESTS, METHODOLOGY, CHARA.

IDENTIFIERS:
*EPIPHYTIC ALGAE, *EPIPHYTIC BACTERIA, *MACROPHYTE-EPIPHTYE METABOLISM, LAWRENCE LAKE(MICH), SCIRPUS ACUTUS, NAJAS FLEXILIS.

ABSTRACT:
COMMUNITY METABOLISM OF MACROPHYTE-EPIPHYTE SYSTEMS AND NUTRITIONAL RELATIONSHIPS OF EPIPHYTIC ALGAE AND BACTERIA WERE INVESTIGATED BY C-14 TECHNIQUES IN THE LITTORAL ZONE OF A SMALL LAKE. THE STUDY INDICATED THAT EPIPHYTIC ALGAE CONTRIBUTED 31.3% TO THE TOTAL LITTORAL PRODUCTION AND THAT THE ALGAL EPIPHYTES MAY BE AMONG THE DOMINANT PRODUCERS IN SHALLOW-WATER ECOSYSTEMS WITH SUBMERGED MACROPHYTES. THE CHEMO-ORGANOTROPHY OF EPIPHYTIC BACTERIA WAS EVALUATED ON GLUCOSE AND ACETATE SUBSTRATES BY ENZYME KINETIC ANALYSIS. THE NATURE OF EXTRACELLULAR RELEASE SUGGESTS NUTRITIONAL INTERACTIONS IN MACROPHYTE-EPIPHYTE SYSTEMS. THE METABOLISM OF THE TWO GROUPS OF ORGANISMS MAY BE A SOURCE OF DISSOLVED ORGANIC MATTER THAT CONTRIBUTES TO THE PRIMARY PRODUCTION. (WILDE-WISCONSIN)

FIELD 05C

ACCESSION NO. W72-03216

STUDIES ON THE PHYSIOLOGY OF HETEROCYST PRODUCTION IN THE NITROGEN-FIXING BLUE-GREEN ALGA ANABAENA SP L-31 IN CONTINUOUS CULTURE,

BHABHA ATOMIC RESEARCH CENTRE, BOMBAY (INDIA). BIOLOGY DIV.

J. THOMAS, AND K. A. V. DAVID.

JOURNAL OF GENERAL MICROBIOLOGY, VOL 66, NO 1, P 127-131, 1971. 3 FIG, 1 TAB, 17 REF.

DESCRIPTORS:
*NITROGEN FIXATION, *CYANOPHYTA, PLANT PHYSIOLOGY, CYTOLOGICAL STUDIES, CULTURES, AMMONIA, POTASSIUM COMPOUNDS, NITRATES, DENSITY, GROWTH RATES, LABORATORY TESTS, CYANOPHYTO, ALGAE.

IDENTIFIERS:
*HETEROCYSTS, *ANABAENA SP L-31, CONTINUOUS CULTURE, BATCH CULTURES.

ABSTRACT:
COMPOUNDS WHICH INHIBIT HETEROCYST PRODUCTION IN FILAMENTOUS BLUE-GREEN ALGAE INCLUDE NITRATE, NITRITE, AMMONIA, AND OTHER NITROGENEOUS SUBSTANCES, AMMONIA BEING THE MOST EFFECTIVE. MECHANISM OF THIS INHIBITION HAS NOT BEEN CLEARLY UNDERSTOOD. STUDIES IN BATCH AND CONTINUOUS CULTURES OF ANABAENA SP L-31 ARE REPORTED TO ELUCIDATE FURTHER THE PHYSIOLOGY OF INHIBITION OF HETEROCYSTS. DAILY MEASUREMENTS WERE MADE OF TURBIDITY, CELL NUMBER, CELL SIZE, HETEROCYST FREQUENCY, FILAMENT LENGTH, AND EXTRACELLULAR AMMONIA. INDUCTION OF HETEROCYSTS IN THIS ALGA IS TOTALLY INHIBITED BY POTASSIUM NITRATE IN BATCH CULTURES, WHEREAS IN CONTINUOUS CULTURES NO INHIBITION IS OBSERVED AT HIGH DILUTION RATES. WHEN NITRATE IS UTILIZED, AMMONIA ACCUMULATES IN THE GROWTH MEDIUM, THE QUANTITY OF EXTRACELLULAR AMMONIA DECLINING WITH INCREASING DILUTION RATE. THE RATE OF RELEASE OF AMMONIA PER CELL INCREASES WITH DECREASING DENSITY OF ORGANISMS, AND INDUCTION OF HETEROCYSTS IS CONSISTENTLY OBSERVED WHEN AMMONIA RELEASE PER CELL EXCEEDS 2 X 10 TO THE 10TH POWER MICROGRAMS. IT IS INFERRED THAT SUCH EXCESSIVE RELEASE DEPLETES THE LEVEL OF INTRACELLULAR AMMONIA CAUSING THE INDUCTION OF HETEROCYSTS. THE PRESENT RESULTS SUPPORT THE VIEW THAT HETEROCYSTS ARE THE POSSIBLE SITES OF NITROGEN FIXATION IN BLUE-GREEN ALGAE. (JONES-WISCONSIN)

FIELD 05C

ACCESSION NO. W72-03217

ESTIMATING EUTROPHIC POTENTIAL OF POLLUTANTS,

MONSANTA CO., ST LOUIS, MO; AND WASHINGTON UNIV., ST LOUIS, MO. DEPT. OF ENVIRONMENTAL AND SANITARY ENGINEERING.

DEE MITCHELL, AND JAMES C. BUZZELL, JR.

JOURNAL OF SANITARY ENGINEERING DIVISION, PROCEEDINGS OF THE AMERICAN SOCIETY OF CIVIL ENGINEERS, VOL 97, NO SA 4, P 453-465, 1971. 5 FIG, 3 TAB, 22 REF.

DESCRIPTORS:
*EUTROPHICATION, *LABORATORY TESTS, *WATER POLLUTION EFFECTS, *ALGAE, *PLANKTON, WATER POLLUTION CONTROL, METHODOLOGY, MICROENVIRONMENT, MEASUREMENT.

IDENTIFIERS:
*MICROCOSM ALGAL ASSAY PROCEDURE, SPECIES DIVERSITY INDEX.

ABSTRACT:
THIS STUDY IS CONCERNED WITH EFFECTS OF VARIOUS CHEMICALS AND WASTEWATER ON THE COMPOSITION AND GROWTH OF ALGAL MICROCOSMS OF LAKE WATER AND BOTTOM MUD CULTURES. THE BIOASSAY WAS CONDUCTED IN NINE LITER PYREX BOTTLES FILLED WITH 1:7 MUD-LAKE WATER SUSPENSION. THE TREATMENTS INCLUDED 10% SOLUTION OF DOMESTIC WASTEWATER, 10% SECONDARY TREATMENT EFFLUENT, DIFFERENT CONCENTRATIONS OF 23-19-17 FERTILIZER, AND THE ALGISTAT HERBICIDE. EFFECTS OF THESE TREATMENTS WERE RECORDED IN TERMS OF THE NUMBER OF ALGAL GENERA AND THE TOTAL ALGAL CELLS COUNT. NEARLY ALL ENRICHMENTS OF THE MEDIA CAUSED A MARKED REDUCTION OF THE DIVERSITY INDEX OF MICROORGANISMS. THE RESULTS SUGGESTED THAT THE LABORATORY PROCEDURE MAY SERVE FOR AN APPRAISAL OF THE EFFECT OF DIFFERENT POLLUTANTS. (WILDE-WISCONSIN)

FIELD 05C, 07B, 05A

ACCESSION NO. W72-03218

FRICTION REDUCTION BY ALGAL AND BACTERIAL POLYMERS,

NAVAL UNDERSEA RESEARCH AND DEVELOPMENT CENTER, SAN DIEGO, CALIF.

PAUL R. KENIS, AND J. W. HOYT.

AVAILABLE FROM THE NATIONAL TECHNICAL INFORMATION SERVICE AS AD-726 181, $3.00 IN PAPER COPY, $0.95 IN MICROFICHE. REPORT NUC TP 240, JUNE 1971, 34 P, 18 FIG, 6 TAB, 20 REF. ONR NR 137-699.

DESCRIPTORS:
*HYDRODYNAMICS, *FRICTION, *TURBULENT FLOW, ALGAE, BACTERIA, MARINE PLANTS, ENGINEERING, WATER PROPERTIES, PHYTOPLANKTON, CHLOROPHYTA, PHAEOPHYTA, RHODOPHYTA, DIATOMS, PYRROPHYTA, CHLORELLA, PSEUDOMONAS, TESTING, MEASUREMENT, PHEOLOGY.

IDENTIFIERS:
*BIOLOGICAL POLYMERS, *FRICTION REDUCTION, POLYSACCHARIDES, DRAG REDUCTION.

ABSTRACT:
FRICTION REDUCTION PHENOMENA, ATTRIBUTED TO DISSOLVED LONG-CHAIN POLYSACCHARIDES EXUDED BY ALGAE AND BACTERIA, STIMULATED THE INVESTIGATION OF THE ABILITY OF BIOLOGICAL POLYMERS PRODUCED BY SEAWEEDS, MICROSCOPIC ALGAE, AND BACTERIA TO REDUCE FRICTION IN WATER FLOW, INCLUDING THE EFFECT OF POLYMERS OF BIOLOGICAL ORIGIN ON THE REPORTED UNEXPLAINABLE VARIATIONS IN HYDRODYNAMIC TEST FACILITIES, THEIR USABILITY FOR FRICTION-REDUCTION APPLICATION AND FOR MEASUREMENTS TO QUANTITATE AND CHARACTERIZE BIOLOGICAL POLYMERS. FRICTION-REDUCING MATERIALS HAVE BEEN FOUND IN CULTURE MEDIA. ALL WATER SAMPLES TESTED FROM INLAND AND MARINE SOURCES GAINED FRICTION-REDUCTION ABILITY WHEN ENRICHED WITH SUGAR, AS A CONSEQUENCE OF POLYSACCHARIDE SYNTHESIS BY BACTERIA, WHICH LEADS TO THE ASSUMPTION THAT BIOLOGICAL POLYMERS ARE THE PROBABLE CAUSE OF THE UNEXPLAINABLE VARIATIONS IN HYDRODYNAMIC TEST FACILITIES. BACTERIAL POLYSACCHARIDES WERE MORE EFFECTIVE THAN SEAWEED EXTRACT AT LOW CONCENTRATIONS FOR FRICTION REDUCTION, BUT BOTH WERE MUCH LESS EFFECTIVE THAN SYNTHETIC POLYMERS. TURBULENT-FLOW FRICTIONAL MEASUREMENTS WERE FOUND TO BE SENSITIVE FOR DETECTION, MEASUREMENT, AND PARTIAL CHARACTERIZATION OF LONG-CHAIN POLYMERS. THE FRICTION-REDUCTION TECHNIQUE IS A RAPID AND EFFECTIVE PROCEDURE FOR THE DETECTION AND QUANTIFICATION OF LONG-CHAIN POLYMERS. (JONES-WISCONSIN)

FIELD 05A

ACCESSION NO. W72-03220

REGULATION OF NITRATE REDUCTASE IN CHLORELLA VULGARIS,

CORNELL UNIV., ITHACA, N.Y.; AND AGRICULTURAL RESEARCH SERVICE, ITHACA, N.Y. PLANT, SOIL AND NUTRITION LAB.

F. W. SMITH, AND JOHN F. THOMPSON.

PLANT PHYSIOLOGY, VOL 48, P 224-227, 1971. 4 FIG, 2 TAB, 15 REF.

DESCRIPTORS:
*BIOCHEMISTRY, *NITRATES, *CHLORELLA, ENZYMES, PROTEINS, LABORATORY TESTS, AMMONIA, ASSAY, INHIBITION, UREAS, SYNTHESIS, ALGAE.

IDENTIFIERS:
CHLORELLA VULGARIS, NITRATE REDUCTASE, RIBONUCLEIC ACID.

ABSTRACT:
SEVERAL AMINO ACIDS WERE TESTED AS POSSIBLE INHIBITORS OF NITRATE REDUCTASE. THE RESULTS OF ASSAYS AND INDUCTION STUDIES DISCLOSED THAT NITRATE REDUCTASE IS INCREASED IN CHLORELLA VULGARIS BY AN ADDITION OF NITRATE WITH MAXIMUM INDUCTION ABOVE MM. ACTINOMYCIN D, CYCLOHEXIMIDE, AND PUROMYCIN ANNULED THE EFFECT OF NITRATE THUS INDICATING THAT DE NOVO SYNTHESIS OF MESSENGER RNA AND PROTEIN IS REQUIRED. TESTS OF NITROGENOUS COMPOUNDS REVEALED THAT AMMONIUM CHLORIDE, UREA, AND SEVERAL AMINO ACIDS ARRESTED AN INCREASE OF THE REDUCTASE IN VIVO, BUT NOT IN VITRO. THE STUDY SUGGESTED THAT NITRATE REDUCTION IN CHLORELLA IS CONTROLLED BY REPRESSION OF ENZYME SYNTHESIS. (WILDE-WISCONSIN)

FIELD 05C

ACCESSION NO. W72-03222

STUDIES ON FRESHWATER BACTERIA: FACTORS WHICH INFLUENCE THE POPULATION AND ITS ACTIVITY,

FRESHWATER BIOLOGICAL ASSOCIATION, AMBLESIDE (ENGLAND).

J. G. JONES.

THE JOURNAL OF ECOLOGY, VOL 59, NO 2, P 593-613, 1971. 12 FIG, 7 TAB, 42 REF.

DESCRIPTORS:
*AQUATIC BACTERIA, *POPULATION, *ALGAE, *STRATIFICATION, *ENZYMES, THERMOCLINE, EPILIMNION, MUD, PHYSICOCHEMICAL PROPERTIES, HYDROGEN ION CONCENTRATION, TEMPERATURE, DISSOLVED OXYGEN, EUTROPHICATION, OLIGOTROPHY, PHYTOPLANKTON, ANALYTICAL TECHNIQUES, TURBIDITY, NITRATES, PHOSPHATES, WIND VELOCITY, RAINFALL.

IDENTIFIERS:
ESTHWAITE WATER(ENGLAND), LAKE WINDERMERE(ENGLAND), EXOENZYME PRODUCERS, CHLOROPHYLL A.

ABSTRACT:
THE FACTORS WERE STUDIED CONTROLLING BACTERIAL POPULATIONS DURING THE PERIOD OF WATER STRATIFICATION. THE ANALYZED PARAMETERS INCLUDED TEMPERATURE, PH, DISSOLVED OXYGEN, TURBIDITY, PARTICULATE MATTER, PHOSPHATE, NITRATE, RAINFALL, WIND VELOCITY, AND CHLOROPHYLL A. THE DENSITY OF VIABLE BACTERIA, EXOENZYME PRODUCING BACTERIA, AND THE LEVELS OF CERTAIN ENZYMES WERE ALSO ESTIMATED. IN THE EUTROPHIC ESTHWAITE WATER THE MAJOR FACTORS CONTROLLING BACTERIA APPEARED TO BE TEMPERATURE, DO CONCENTRATION, AND PH. IN THE LAKE WINDERMERE A POSITIVE CORRELATION WAS OBSERVED BETWEEN BACTERIAL NUMBERS AND PH, TEMPERATURE, PARTICULATE MATTER, AND RAINFALL. BACTERIAL POPULATION AND ENZYMATIC ACTIVITY OCCASIONALLY INCREASED WITH PHYTOPLANKTON MAXIMA. (WILDE-WISCONSIN)

FIELD 05C

ACCESSION NO. W72-03224

PHYTOPLANKTON PRIMARY PRODUCTION IN SOME FINNISH COASTAL AREAS IN RELATION TO POLLUTION,

INSTITUTE OF MARINE RESEARCH, HELSINKI (FINLAND); AND HELSINKI UNIV. (FINLAND). DEPT. OF LIMNOLOGY.

PAULI BAGGE, AND PASO O. LEHMUSLUOTO.

MERENTUTKIMUSLAIT JULK/HAVSFORSKNINGSINT SKR NO 235, P 3-18, 1971. 7 FIG, 4 TAB, 24 REF.

DESCRIPTORS:
*PRIMARY PRODUCTIVITY, *PHYTOPLANKTON, *SEA WATER, COASTS, BIORTHYMS, ON-SITE TESTS, WATER POLLUTION SOURCES, WATER POLLUTION EFFECTS, FLUCTUATION, COMPARATIVE PRODUCTIVITY, MUNICIPAL WASTES, INDUSTRIAL WASTES, ANNUAL TURNOVER, ENERGY, CYANOPHYTA, ALGAE, EUTROPHICATION, SUMMER, ENERGY, SEASONAL.

IDENTIFIERS:
*FINLAND, *BOTHNIAN BAY, *GULF OF FINLAND, DARK FIXATION, COASTAL WATERS, TROPHOGENIC LAYER.

ABSTRACT:
CARBON-14 ANALYSES DISCLOSED SIGNIFICANT ANNUAL AND SEASONAL FLUCTUATIONS OF PHYTOPLANKTON PRIMARY PRODUCTION IN THE POLLUTED AND NONPOLLUTED COASTAL AREAS OF FINLAND. IN SOME AREAS, RECEIVING WASTEWATER, THE ANNUAL PRODUCTION EXCEEDS 150 G C/SQ M, IN COMPARISON WITH 15 TO 60 G C/SQ M OF UNPOLLUTED WATERS. BECAUSE OF THE SHADING PLANKTON, THE THICKNESS OF THE TROPHOGENIC LAYER DURING THE SUMMER IN SEWAGE POLLUTED WATERS IS USUALLY LESS THAN 2 M; IN WATERS POLLUTED BY PAPER MILL WASTES, THE THICKNESS IS UNDER 1 M. THE EUTROPHIC WATERS ARE DOMINATED BY BLUE-GREEN ALGAE, WHEREAS OLIGOTROPHIC WATERS BY DIATOMS AND DINOPHYCEAE. DARK FIXATION VALUES OF EUTROPHIC WATERS MAY BE 10 TIMES AS HIGH AS THOSE OF UNPOLLUTED WATERS. (WILDE-WISCONSIN)

FIELD 05C, 02L

ACCESSION NO. W72-03228

NEED FOR BETTER INFORMATION OF EFFECTS OF COASTAL REACTORS,

SCRIPPS INSTITUTION OF OCEANOGRAPHY, LA JOLLA, CALIF.

T. R. FOLSOM.

AVAILABLE FROM THE NATIONAL TECHNICAL INFORMATION SERVICE AS TID-25774, $3.00 IN PAPER COPY, $0.95 IN MICROFICHE. REPORT TID-25774 REV JULY 15, 1971. 16 P.

DESCRIPTORS:
*NUCLEAR WASTES, *MONITORING, *PACIFIC COAST REGION, TRITIUM, OCEAN CURRENTS, COBALT RADIOISOTOPES, NUCLEAR POWERPLANTS, SITES, MOLLUSKS, MARINE ALGAE, ON-SITE INVESTIGATIONS, CALIFORNIA, FOOD CHAINS, PUBLIC HEALTH, WATER POLLUTION SOURCES.

ABSTRACT:
UPTAKE BY ORGANISMS IS THE MOST SENSITIVE KNOWN INDICATOR OF RADIONUCLIDE POLLUTION. CO-58 AND PROBABLY CO-60 AND AG-110M WERE DETECTABLE IN A SPECIES OF RED ALGAE AND IN A MOLLUSK THAT FEEDS UPON THE ALGAE, WHICH WERE COLLECTED FROM AN AREA NEAR THE SAN ONOFRE NUCLEAR POWERPLANT. THE TRITIUM DISCHARGE REACHES 11,000 TU AT OUTFALL, AN AMOUNT SUFFICIENT TO SERVE AS A TRACER OF THE EFFLUENT IN THE LA JOLLA-SAN DIEGO COASTAL REGION. (BOPP-ORNL)

FIELD 05C

ACCESSION NO. W72-03324

THE ALGO-AGRO-INDUSTRIAL COMPLEX. AN AGRO-INDUSTRIAL COMPLEX AT NUCLEAR ENERGY CENTERS WITH ASSOCIATED PRODUCTION OF AUTOTROPHIC MICROORGANISMS,

CESKOSLOVENSKA AKADEMIE VED, PRAGUE. INST. OF MICROBIOLOGY; AND CESKOSLOVENSKA AKADEMIE VED, TREBON. LAB. OF ALGOLOGY.

I. MALEK, J. BARTOS, J. SIMMER, AND B. PROKES.

AVAILABLE FROM THE NATIONAL TECHNICAL INFORMATION SERVICE AS A/CONF.49/P/552. $3.00 PER COPY, $0.95 MICROFICHE. REPORT A/CONF.49/P/552, MAY 1971. 20 P, 7 FIG, 2 TAB, 10 REF.

DESCRIPTORS:
*NUCLEAR POWERPLANTS, *DESALINATION PLANTS, *ELECTRIC POWER PRODUCTION, *IRRIGATION PROGRAMS, *AGRICULTURE, *ECONOMICS, CYANOPHYTA, ALGAE, PLANTS, PROTEINS, NUTRIENTS, BIOMASS, HARVESTING OF ALGAE, PRODUCTIVITY.

IDENTIFIERS:
AGRO-INDUSTRIAL COMPLEX, ELECTRIC ENERGY CONSUMPTION, MODEL TESTING.

ABSTRACT:
THIS PAPER CONSIDERS THE PRODUCTION OF THE BIOMASS OF AUTOTROPHIC MICROORGANISMS (AM), FIRST OF ALL THE GREEN AND BLUE-GREEN ALGAE. THE COMPLEX OF PRODUCTION TYPES CONSIDERED HERE WAS HENCE CALLED THE ALGO-AGRO-INDUSTRIAL COMPLEX (AAIC). THE HETEROGENEITY OF THE STRUCTURE OF THE AAIC PRESENTED OBSTACLES THAT COULD NOT BE ALWAYS OVERCOME SUCCESSFULLY; IT IS HARDLY POSSIBLE IN THIS BRIEF REVIEW TO PRESENT MORE THAN THE BASIC INFORMATION ON THE INDIVIDUAL PARTS OF THE AAIC AND ON THEIR MUTUAL RELATIONS. A FULL-LENGTH PUBLICATION ON THE SUBJECT IS BEING PREPARED. THE MOSAIC OF DATA COMPILED HERE IS MEANT AS A BASIS FOR FURTHER DEVELOPMENT. (HOUSER-ORNL)

FIELD 03A, 03F, 05C

ACCESSION NO. W72-03327

SOILS AS COMPONENTS OF ECOSYSTEMS,

OAK RIDGE NATIONAL LAB. TENN. ECOLOGICAL SCIENCES DIV.

M. WITKAMP.

IN: ANNUAL REVIEW OF ECOLOGY AND SYSTEMATICS; ANNUAL REVIEWS INC.; PALO ALTO, CALIFORNIA, P 85-110, 1971.

DESCRIPTORS:
*SOIL-WATER-PLANT RELATIONSHIPS, *FALLOUT, *CYCLING NUTRIENTS, ECOSYSTEMS, RAIN FORESTS, MATHEMATICAL MODELS, LINEAR PROGRAMMING, SYSTEMS ANALYSIS, DECIDUOUS FORESTS, DECOMPOSING ORGANIC MATERIAL, FERTILITY, RHIZOSPHERE, ROOT SYSTEMS, SOIL FUNGI, SOIL ENVIRONMENT, SOIL TEMPERATURE, SOIL CHEMICAL PROPERTIES, SOIL BACTERIA, SOIL ALGAE.

ABSTRACT:
MINERAL CYCLING LITERATURE OF THE LAST 2-3 YEARS IS REVIEWED TO PROVIDE INSIGHT INTO DRIVING FORCES AND RATES OF TRANSFER, AS WELL AS THE IMPORTANCE OF THE VARIOUS PATHWAYS. BIOTA ARE THE PRIME AGENTS IN MAINTAINING PRODUCTIVITY WHICH IS OFTEN LIMITED BY MINERAL AVAILABILITY AND IS DEPENDENT ON RECYCLING THROUGH PLANT AND ANIMAL DEBRIS. MUCH OF THE INTERNAL TURNOVER IN SUBSYSTEMS CAN BE TREATED AS BLACK BOXES BY CONSIDERING ONLY TURNOVER RATES AND POOL SIZE. TURNOVER TIMES IN A TROPICAL RAIN FOREST CALCULATED AS TOTAL POOL/TURNOVER RATE VARY FROM 0.08 YR FOR K TO 133 YR FOR N, AND DECREASE IN THE ORDER MG, CA, H, C, MN, NA AND P. A LARGE P INFLUX STEMS FROM RAIN. INPUTS OF SODIUM AND CHLORINE BY RAIN DEPEND TO A LARGE EXTENT ON OCEANIC STORMS THAT TRANSPORT SPRAY FAR INLAND. SIMILARLY, ABOUT HALF OF THE POTASSIUM INPUT IS IN RAIN. THE POTASSIUM INPUT IS ONLY ABOUT 1% OF THAT RECYCLED. EFFICIENT RECYCLING ACCOUNTS FOR THE LOWER NUTRIENT REQUIREMENTS WITH FORESTS AS COMPARED TO AGRICULTURE. MICROBIAL ACTIVITY, AND CONSEQUENTLY REMINERALIZATION IF TEMPERATURE DEPENDENT. INTERRUPTION OF MINERALIZATION DURING WINTER MAY ALLOW LOSS OF NUTRIENTS BY LEACHING. LOW RATES OF EVAPORATION AND RESULTING ANAEROBIC CONDITIONS ALSO TEND TO INHIBIT MINERALIZATION. (BCPP-ORNL)

FIELD 02I, 05B, 02G

ACCESSION NO. W72-03342

AMCHITKA RADIOBIOLOGICAL PROGRAM,

WASHINGTON UNIV., SEATTLE. COLL. OF FISHERIES.

E. E. HELD.

AVAILABLE FROM NATIONAL TECHNICAL INFORMATION SERVICE, $3.00 IN PAPER COPY, $0.95 IN MICROFICHE. PROGRESS REPORT, JULY 1970-APRIL 1971, NVO-269-11, JULY 1971. 38 P, 5 FIG, 9 TAB, 13 REF. AEC CONTRACT NO. AT(26-1)-269.

DESCRIPTORS:
*NUCLEAR EXPLOSIONS, *WATER POLLUTION EFFECTS, *RADIOACTIVITY EFFECTS, UNDERGROUND, ALASKA, PONDS, WELLS, FISH, BAYS, AIR POLLUTION EFFECTS, AIR POLLUTION, WATER POLLUTION, LICHENS, SOIL-WATER-PLANT RELATIONSHIPS, MARINE ALGAE, MOSSES, CRABS, ALGAE, SNAILS, MUSSELS, ISOPODS, AMPHIPODA, RADIOISOTOPES, ABSORPTION, PACIFIC OCEAN, FOOD CHAINS, FALLOUT, TRITIUM, ECOLOGY, ON-SITE INVESTIGATIONS, MONITORING.

IDENTIFIERS:
RADIONUCLIDE UPTAKE.

ABSTRACT:
ORGANISMS, WATER, AND AIR SAMPLES WERE ANALYZED FOR RADIONUCLIDES TO DETERMINE WHETHER THERE WAS ANY RELEASE AFTER UNDERGROUND NUCLEAR TESTS. EXCEPTING TRITIUM (3600 T.U. IN PONDS AND TEST WELLS) AND SHORT-LIVED SCANDIUM-46 WHICH COULD BE FROM A NUCLEAR-POWERED VESSEL, THE RADIONUCLIDES DETECTED WERE FROM WORLDWIDE FALLOUT. IN FISH, IRON-55 WAS AS GREAT AS 152 PCI/G DRY WEIGHT (OR 254 PCI/MG FE SPECIFIC ACTIVITY) AND OTHER RADIONUCLIDES (EXCEPTING K-40) WERE LESS THAN 5 PCI/G DRY WEIGHT. IN LICHENS, CESIUM-137 WAS 20 PCI/G DRY WEIGHT. IN FRESHWATER PLANTS, ZIRCONIUM-95, NIOBIUM-95 WAS 12 PCI/G DRY WEIGHT. GENERALLY, THE CONCENTRATIONS OF THE RADIONUCLIDES WERE SMALL AND WITHIN THE RANGE OF VALUES REPORTED FOR SIMILAR SAMPLES FROM OTHER PARTS OF THE NORTHERN HEMISPHERE. (BCPP-ORNL)

FIELD 05C

ACCESSION NO. W72-03347

TRITIUM: DISCRIMINATION AND CONCENTRATION IN FRESH WATER MICROCOSMS,

ARGONNE NATIONAL LAB., ILL.

M. L. STEWART, G. M. ROSENTHAL, AND J. R. KLINE.

AVAILABLE FROM NATIONAL TECHNICAL INFORMATION SERVICE AS CONF-710501-26, $3.00 IN PAPER COPY, $0.95 IN MICROFICHE. REPORT CONF-710501-26, MAY 1971. 29 P, 9 FIG, 1 TAB, 10 REF.

DESCRIPTORS:
*SNAILS, *FROGS, *TRITIUM, ABSORPTION, WATER POLLUTION EFFECTS, RADIOACTIVITY EFFECTS, RADIOACTIVITY TECHNIQUES, AQUARIA, ALGAE, MATHEMATICAL MODELS, PATH OF POLLUTANTS, TRACERS, ANIMAL PHYSIOLOGY, NUCLEAR WASTES.

ABSTRACT:
TRITIUM MOBILITY WAS STUDIED IN AN AQUATIC ENVIRONMENT AND IN AQUATIC ANIMALS (TWO SPECIES OF SNAIL AND ONE SPECIES OF TADPOLE). THE TRITIUM MOBILITY IN SNAILS WAS DESCRIBED BY AN EMPIRICAL, COMPARTMENTAL MODEL. COMPARTMENTS AND RESIDENCE TIMES OF TRITIUM ARE: EXTRACELLULAR, UNBOUND WATER (80% OF THE UNBOUND WATER WITHIN THE SNAIL SHELL), 4 MINUTES; CELLULAR, UNBOUND WATER (20% OF THE UNBOUND WATER WITHIN THE SNAIL SHELL), 330 MINUTES; CELLULAR, BOUND TRITIUM (ACCOUNTING FOR 28% OF THE EXCHANGEABLE HYDROGEN), 2,650 MINUTES; EXTRACELLULAR, BOUND TRITIUM (ACCOUNTING FOR 28% OF THE EXCHANGEABLE HYDROGEN), 8 MINUTES; SURFACE, BOUND TRITIUM (WHICH IS DIRECTLY EXPOSED TO ENVIRONMENTAL TRITIUM AND ACCOUNTS FOR 4% OF THE EXCHANGEABLE HYDROGEN), 2 MINUTES. COMPARTMENTAL SIZES AND RATE TRANSFER COEFFICIENTS ARE GIVEN. (BOPP-ORNL)

FIELD 05C

ACCESSION NO. W72-03348

QUALITATIVE COMPOSITION OF PHYTOPLANKTON IN THE VICINITY OF NEWFOUNDLAND,

ALL-UNION RESEARCH INST. OF MARINE FISHERIES AND OCEANOGRAPHY, MOSCOW (USSR).

O. A. MOVCHAN.

OKEANOLOGIYA, VOL. 10, NO. 3, P 381-387, MAY-JUNE 1970. 5 FIG, 16 REF.

DESCRIPTORS:
*PHYTOPLANKTON, *ALGAE, *DIATOMS, *CURRENTS, *DISTRIBUTION PATTERNS, ATLANTIC OCEAN, PACIFIC OCEAN, TROPICAL REGIONS, POLAR REGIONS, CYANOPHYTA, CHRYSOPHYTA, PYRROPHYTA, SAMPLING, ECOLOGY, WATER TEMPERATURE.

IDENTIFIERS:
NEWFOUNDLAND, PERIDINIANS, QUALITATIVE COMPOSITION, COSCINOSIRA POLYCHORDA, THALASSIOSIRA DECIPIENS, THALASSIOSIRA GRAVIDA, THALASSIOSIRA NORDENSKIOLDII, THALASSIOSIRA ROTULA, SCHRODERELLA DELICATULA, DACTYLIOSOLEN MEDITERRANEUS, LEPTOCYLINDRUS DANICUS, RHIZOSOLENIA FRAGILISSIMA, RHIZOSOLENIA HEBETATA F. SEMISPINA, CHAETOCEROS AFFINIA, CHAETOCEROS ATLANTICUS, CHAETOCEROS BOREALIS.

ABSTRACT:
AMONG THE PHYTOPLANKTON COLLECTED IN THE 590 SAMPLES TAKEN BY THE SHIP R/V MIKHAIL LOMOSOV FROM ITS STATIONS OFF NEWFOUNDLAND WERE 131 SPECIES AND VARIETIES OF ALGAE, 25 GENERA AND 67 SPECIES OF DIATOMS, 15 GENERA AND 50 SPECIES OF PERIDINIANS, 6 GENERA OF COCCOLITHINES, 4 SPECIES OF PROTOCOCCUCI, 2 SPECIES OF SILICO FLAZELLATES, 1 SPECIES OF HETEROCONTAE AND 1 SPECIES OF BLUE-GREEN ALGAE. FIFTY PERCENT OF THE TYPES OF DIATOMS ARE MASSIVE (OVER 1 MILLION CELLS/CU M IN THE 0- TO 100-M LAYER). STUDY OF THE ECOLOGY AND THE GEOGRAPHIC DISTRIBUTION OF THE PHYTOPLANKTON SHOWED ABUNDANCE OF TROPICAL SPECIES BUT FEW ARCTIC-BOREAL SPECIES SOUTH OF 42 DEGREES N. NORTH OF THIS WAS A ZONE OF MIXING OF TROPICAL AND ARCTIC-BOREAL SPECIES. THE COSMOPOLITE SPECIES WAS WIDELY SPREAD IN THE NEWFOUNDLAND REGION. SIMILARITIES WERE FOUND IN THE COMPOSITION OF THE PHYTOPLANKTON IN THE REGION OF NEWFOUNDLAND AND IN ADJACENT WATERS IN THE NORTHWESTERN ATLANTIC AND NORTHWESTERN PACIFIC. INVESTIGATION OF THE DISTRIBUTION OF THE NERITIC AND OCEANIC FORMS OF DIATOMS AND A FEW SPECIES OF PERIDINIANS SHOWED THE INFLUENCE OF CURRENTS AND THE MIXING OF WATERS FROM POLAR AND TROPICAL ZONES ON ABUNDANCE. MAPS SHOW THE CONCENTRATION OF SPECIES, THEIR GEOGRAPHICAL LOCATION, AND SEASONAL VARIATION. (LITTLE-BATTELLE)

FIELD 05A, 05C

ACCESSION NO. W72-03543

GROWTH RESPONSE OF BLUE-GREEN ALGAE TO ALDRIN, DIELDRIN ENDRIN AND THEIR METABOLITES,

WISCONSIN UNIV., MADISON. DEPT. OF ENTOMOLOGY.

J. C. BATTERTON, G. M. BOUSH, AND F. MATSUMURA.

BULLETIN OF ENVIRONMENTAL CONTAMINATION AND TOXICOLOGY, VOL. 6, NO. 6, P 589-594, NOVEMBER/DECEMBER 1971. 2 TAB, 17 REF.

DESCRIPTORS:
*CHLORINATED HYDROCARBON PESTICIDES, *ALDRIN, *DIELDRIN, *ENDRIN, *GROWTH RATES, *CYANOPHYTA, *TOXICITY, PESTICIDES, PESTICIDE RESIDUES, ALGAE, DDT, PHYTOPLANKTON, MARINE ALGAE, AQUATIC ALGAE, COLORIMETRY, TRACES.

IDENTIFIERS:
PHOTODIELDRIN, ANACYSTIS NIDULANS, AGMENELLUM QUADRUPLICATUM STRAIN PR-6, PHOTOALDRIN, KETOENDRIN, METABOLITES, SKELETONEMA, ANABAENA CYLINDRICA, DUNALIELLA.

ABSTRACT:
THE EFFECTS OF DIELDRIN, ENDRIN, ALDRIN, AND FIVE CORRESPONDING METABOLITES ON THE GROWTH RATES OF TWO BACTERIA-FREE BLUE-GREEN ALGAL SPECIES WERE TESTED. THE TEST SPECIES WERE ANACYSTIS NIDULANS (TX20), A FRESHWATER ISOLATE, AND AGMENELLUM QUADRUPLICATUM STRAIN PR-6, A MARINE ISOLATE, BOTH OF WHICH WERE GROWN IN DIFFERENT MEDIA AT THE SAME PH, TEMPERATURE, AND LIGHT INTENSITY. A COMPARISON OF THE RESPONSES OF THE TEST ALGAE SHOWED THAT THE MARINE ISOLATE (PR-6) WAS GENERALLY MORE TOLERANT TO THE INSECTICIDE COMPOUNDS THAN THE FRESHWATER ISOLATE (TX20). THERE WAS AN OVERALL DEPRESSION OF GROWTH RATES WITH HIGH CONCENTRATIONS (475 PPB, 950 PPB) OF DIELDRIN AND ITS METABOLITES. OF THE PARENT COMPOUNDS TESTED, ALDRIN HAD THE LEAST EFFECT ON EITHER TEST SPECIES. THE SELECTED LIGHT INTENSITY AND TEMPERATURE FOR OPTIMUM GROWTH RATES REMAINED UNCHANGED IN THE PRESENCE OF THE INSECTICIDE COMPOUNDS. THAT PR-6 WAS GENERALLY THE MORE TOLERANT OF THE TWO TEST SPECIES RAISES THE QUESTION OF THE EFFECT OF MEDIUM COMPOSITION UPON INSECTICIDE TOXICITY. TABULAR DATA SHOW GROWTH RATES FOR EXPERIMENTS USUALLY LASTING 30-36 HOURS. (HOLOMAN-BATTELLE)

FIELD 05C

ACCESSION NO. W72-03545

PRODUCTION AND BIOENERGETIC ROLE OF THE MIDGE GLYPTOTENDIPES BARBIPES (STAEGER) IN A WASTE STABILIZATION LAGOON,

OREGON STATE UNIV., CORVALLIS. DEPT. OF ENTOMOLOGY.

R. A. KIMERLE, AND N. H. ANDERSON.

LIMNOLOGY AND OCEANOGRAPHY, VOL. 16, NO. 4, P 646-659, JULY 1971. 5 FIG, 5 TAB, 27 REF.

DESCRIPTORS:
*ENERGY BUDGET, *SEWAGE, *AQUATIC PRODUCTIVITY, *MIDGES, LAGOONS, BIOMASS, OREGON, SLUDGE, ALGAE, DIPTERA, DISSOLVED OXYGEN, DREDGING, BOTTOM SEDIMENTS, STABILIZATION, PRIMARY PRODUCTIVITY, GROWTH RATES, RESPIRATION.

IDENTIFIERS:
GLYTOTENDIPES BARBIPES, EKMAN DREDGE.

ABSTRACT:
THE BIOENERGETIC ROLE OF A POPULATION OF GLYPTOTENDIPES BARBIPES IN THE PROCESS OF WASTE STABILIZATION IN TWO SEWAGE LAGOONS WAS STUDIED. WEEKLY PRODUCTION RATES OF THE MULTIVOLTINE MIDGE WERE COMPUTED. ANNUAL PRODUCTION OF G. BARBIPES WAS 808 KCAL/SQ METER IN A NARROW BAND NEARSHORE OF THE SECONDARY LAGOON CONTAINING 90 PERCENT OF THE BIOMASS. BIOMASS DATA FROM BOTH LAGOONS IN 1966 AND 1967 WERE USED TO ESTIMATE PRODUCTION USING A TURNOVER RATIO (TR) OF 8.49 (RATIO OF PRODUCTION:MEAN BIOMASS) FROM DEFINITIVE DATA COLLECTED IN 1967. PRODUCTION IN THE SECONDARY LAGOON WAS 459 KCAL/SQ METER IN 1966 AND 37 IN 1967; IN THE PRIMARY LAGOON IT WAS 165 AND 18 RESPECTIVELY. THE FACTORS CAUSING THESE DIFFERENCES IN PRODUCTION WERE PROBABLY THE DISSOLVED OXYGEN CONCENTRATIONS DURING THE GROWING SEASON, PERCENT OF THE TOTAL LAGOON BOTTOM INHABITABLE BY MIDGE LARVAE, AND THE CONDITION OF THE SLUDGE SUBSTRATE. THE TOTAL ENERGY REMOVED BY EMERGENCE AND RESPIRATION OF G. BARBIPES WAS COMPARED WITH THE ENERGY IN BOTH PATHWAYS IN THE LAGOON: IMPORT OF SEWAGE, PRIMARY PRODUCTION, COMMUNITY RESPIRATION, STORAGE, AND EXPORT. IN 1966, G. BARBIPES REMOVED ABOUT 6.6 PERCENT OF THE NET PRIMARY PRODUCTION IN THE SECONDARY LAGOON AND 0.5 PERCENT IN 1967. (MORTLAND-BATTELLE)

FIELD 05C

ACCESSION NO. W72-03556

EFFECT OF WIDE TEMPERATURE FLUCTUATION ON THE BLUE-GREEN ALGAE OF BEAD GEYSER, YELLOWSTONE NATIONAL PARK,

INDIANA UNIV., INDIANAPOLIS. DEPT. OF MICROBIOLOGY.

J. L. MOSSER, AND T. D. BROCK.

LIMNOLOGY AND OCEANOGRAPHY, VOL. 16, NO. 4, P 640-645, JULY 1971. 3 FIG, 2 TAB, 9 REF.

DESCRIPTORS:
*THERMAL POLLUTION, *WATER TEMPERATURE, *CYANOPHYTA, GEYSERS, PHOTOSYNTHESIS, CULTURES, CHLOROPHYLL, PROTEINS, ALGAE, GROWTH RATES, TRACERS.

IDENTIFIERS:
YELLOWSTONE PARK, SYNECHOCOCCUS, FLEXIBACTERIA, PHORMIDIUM, MASTIGOCLADUS, PHEOPHYTIN, GRASSLAND SPRING.

ABSTRACT:
MEASUREMENT OF PHOTOSYNTHESIS AND GROWTH OF CYANOPHYTA IN THE DRAINWAY OF A SMALL GEYSER IN YELLOWSTONE INDICATES THAT FLUCTUATING TEMPERATURES LIMIT THE EXTENT OF ALGAL GROWTH. AN ALGAL MAT OF 'GRASSLAND SPRING' WAS USED IN THESE EXPERIMENTS AS AN EXAMPLE OF A MAT OCCURRING IN A THERMALLY CONSTANT ENVIRONMENT. PHOTOSYNTHESIS WAS MEASURED BY THE CARBON 14-BICARBONATE METHOD. MAT SAMPLES WERE ALSO ANALYZED FOR CHLOROPHYLL, PROTEIN, AND PHEOPHYTIN. ALGAE FROM THE MAT WERE CULTURED IN MEDIUM D. THE MAT WAS EXPOSED TO HIGH TEMPERATURES ONLY DURING THE BRIEF ERUPTIONS OF THE GEYSER. SINCE OPTIMAL TEMPERATURES FOR THE ALGAE WERE HIGHER THAN THE MEAN ENVIRONMENTAL TEMPERATURE, TEMPERATURES OPTIMAL FOR PHOTOSYNTHESIS AND GROWTH OCCURRED ONLY DURING A SMALL FRACTION OF THE TIME. IT WAS INDICATED THAT THESE ALGAE WERE RESISTANT TO THE HIGH TEMPERATURE THAT OCCURRED DURING ERUPTION, WHEREAS ORGANISMS HAVING THE SAME OPTIMAL TEMPERATURE BUT INHABITING A THERMALLY CONSTANT ENVIRONMENT WERE MORE HEAT SENSITIVE. THE DATA SUGGEST THAT ORGANISMS MAY ADAPT TO SUDDEN TEMPERATURE SHOCKS, SUCH AS THOSE EXPERIENCED NEAR STEAM POWER PLANTS, BUT THAT THEY CANNOT EVOLVE THE ABILITY TO GROW WELL THROUGHOUT THE TEMPERATURE CYCLE. SUCH DATA MAY DIFFER, HOWEVER, IN A MORE RANDOM SITUATION, SINCE THIS WAS A GEYSER THAT ERUPTED PREDICTABLY. FURTHER STUDY OF RANDOMLY ERUPTING GEYSERS IS SUGGESTED. (MORTLAND-BATTELLE)

FIELD 05C

ACCESSION NO. W72-03557

ABUNDANCE OF ETHMODISCUS IN PACIFIC PLANKTON,

AKADEMIYA NAUK SSSR, MOSCOW. INSTITUT OKEANOLOGII.

T. V. BELYAYEVA.

OKEANOLOGIYA, VOL 10, NO 5, P 672-675, SEPTEMBER-OCTOBER 1970. 1 FIG, 1 TAB, 11 REF.

DESCRIPTORS:
*PLANKTON, *DIATOMS, *SAMPLING, *DISTRIBUTION PATTERNS, *PHOSPHATES, *AQUATIC POPULATIONS, MAPS, MAPPING, PHYTOPLANKTON, PACIFIC OCEAN, BOTTOM SEDIMENTS, STATIONS, DATA COLLECTIONS, ALGAE.

IDENTIFIERS:
*ETHMODISCUS REX, ETHMODISCUS GAZELLAE, BAJA CALIFORNIA, HEMIAULUS HAUCKII, STIGMAPHORA ROSTRATA, ETHMODISCUS OOZES.

ABSTRACT:
A MAP WAS COMPILED BRINGING TOGETHER ALL THE SCATTERED INFORMATION ON THE DISTRIBUTION OF THE PLANKTON DIATOM ETHMODISCUS REX IN THE PACIFIC OCEAN. THE DATA WERE ACQUIRED MAINLY FROM ANALYSIS OF SAMPLES COLLECTED DURING SEVERAL VOYAGES OF THE R/V VITYAZ' IN EASTERN, CENTRAL, AND WESTERN PARTS OF THE PACIFIC FROM 1955 TO 1967. MORE THAN 400 SAMPLING LOCATIONS WERE USED FOR THE MAP SHOWING THE QUANTITATIVE DISTRIBUTION OF ETHMODISCUS. THE DISTRIBUTION PATTERN IS EXTREMELY MOSAIC WITH THE MAXIMUM POPULATION OCCURRING BETWEEN THE EQUATOR AND 5 DEGREES N. A RELATIONSHIP WAS SHOWN BETWEEN THE DISTRIBUTION OF ETHMODISCUS AND PHOSPHATE CONCENTRATION, THE MAXIMUM POPULATIONS NOT COINCIDING WITH REGIONS OF HIGHEST PHOSPHATE CONCENTRATION AS IS USUALLY THE CASE WITH MOST DIATOMS. THIS UNUSUAL DISTRIBUTION OF ETHMODISCUS IS UNEXPLAINED, BUT ALSO OCCURS WITH THE DIATOMS, HEMIAULUS HAUCKII AND STIGMAPHORA ROSTRATA. THE ETHMODISCUS POPULATION IN PACIFIC PLANKTON IS INSIGNIFICANT, AVERAGING FROM 0.01 TO 0.5 KL/CU M IN THE 0- TO 100-M LAYER. THEREFORE, THE PRESENT DATA IS INCONCLUSIVE ON THE ORIGIN OF THE ETHMODISCUS OOZES, THEIR CONCENTRATION IN THE WESTERN PACIFIC, OR THE MOSAIC PATTERN OF THEIR DISTRIBUTION. PHYTOPLANKTON SAMPLES WERE COLLECTED MOSTLY WITH A DZHOM NET WITH MOUTH OPENING 80 CM IN DIAMETER AND APERTURE SIZE IN THE FILTERING CONE OF ABOUT 180 MICRONS. (JEFFERIS-BATTELLE)

FIELD 05A, 05C

ACCESSION NO. W72-03567

BIOLOGICAL REMOVAL OF PHOSPHATES FROM AQUATIC MEDIA,

ARIZONA UNIV., TUCSON. DEPT. OF BIOLOGICAL SCIENCES.

DOUGLAS EDWIN GREER.

MASTER'S THESIS, MAY 1971. 29 P, 4 FIG, 4 TAB, 23 REF. OWRR A-019-ARIZ(2).

DESCRIPTORS:
*PHOSPHATES, *CLAMS, *ALGAE, CHEMICAL PRECIPITATION, HYDROGEN ION CONCENTRATION, FILTRATION, COLLOIDS, CARBON DIOXIDE, CALCIUM, MAGNESIUM, NUTRIENTS, *WASTE WATER TREATMENT, TERTIARY TREATMENT, EUTROPHICATION.

IDENTIFIERS:
*ANOXIA.

ABSTRACT:
THE PRECIPITATION OF EXCESS PHOSPHATES IN THE FORM OF HYDROXYL-APATITE HAS BEEN ACCOMPLISHED WITHOUT THE ADDITION OF CHEMICAL REAGENTS. THE PROCESS REQUIRES ELEVATION OF THE OH(-) ION CONCENTRATION THROUGH THE REMOVAL OF CO2 FROM THE WATER BY ALGAE, RESULTING IN A HYDROXYL-APATITE ALGAL SUSPENSION. THIS SUSPENSION IS REMOVED FROM THE WATER BY BEING FILTERED THROUGH BEDS OF CLAMS, CORBICULA FLUMINEA. THESE CLAMS ARE ABLE TO SURVIVE IN EUTROPHIC WATER PROVIDED THE WATER IS CONTINUOUSLY RECIRCULATED, AND THE TEMPERATURE IS KEPT BELOW 30C. ESTUARINE PHOSPHATES MAY BE EVEN MORE SUSCEPTIBLE TO 'AUTOMATIC' REMOVAL SINCE SEA WATER CONTAINS AN AVERAGE OF 410 MG/L CA(+2), WHICH SHOULD RESULT IN MORE COMPLETE PRECIPITATION. OVERALL QUALITY OF WATER RECEIVING CLAM-ALGAE PHOSPHATE-REMOVAL TREATMENT IS ENHANCED, SINCE PH ELEVATION WITHOUT LIME ADDITION RESULTS IN CA(+2) AND MG(+2) PRECIPITATION. TERTIARY TREATED WASTEWATER, SUBJECTED TO NUTRIENT REMOVAL AND CONTAINING LESS THAN 0.30 MG/L TOTAL PHOSPHATES, WILL ALLOW HIGH LEVELS OF PRIMARY PRODUCTIVITY WITHOUT THE DANGER OF ANOXIA FROM EXCESSIVE BUILDUPS OF THE ALGAL STANDING CROP. (LOWRY-TEXAS)

FIELD 05D

ACCESSION NO. W72-03613

PLANT-AVAILABLE PHOSPHORUS STATUS OF LAKES,

WISCONSIN UNIV., MADISON. DEPT. OF WATER CHEMISTRY; AND WISCONSIN UNIV., MADISON. DEPT. OF SOIL SCIENCE.

D. E. ARMSTRONG, R. F. HARRIS, AND J. K. SYERS.

AVAILABLE FROM THE NATIONAL TECHNICAL INFORMATION SERVICE AS PB-206 156, $3.00 IN PAPER COPY, $0.95 IN MICROFICHE. TECHNICAL COMPLETION REPORT, WISCONSIN WATER RESOURCES CENTER, MADISON, 1971. 29 P, 2 TAB, 20 REF, 10 APPEND. OWRR B-022-WIS(6).

DESCRIPTORS:
*PHOSPHORUS, *SEDIMENTS, *ALGAE, *EUTROPHICATION, LAKES, AQUATIC PLANTS, IRON, MICROORGANISMS, COLORIMETRY, WISCONSIN.

IDENTIFIERS:
AVAILABLE PHOSPHORUS, EXCHANGEABLE PHOSPHORUS, A*DENOSINE TRIPHOSPHATE(ATP).

ABSTRACT:
THE FACTORS CONTROLLING THE AVAILABLE P STATUS OF LAKES WERE INVESTIGATED. THE INFLUENCE OF SEDIMENT P ON THE P STATUS OF THE OVERLYING WATER WAS GIVEN PRIMARY ATTENTION. SEDIMENTS FROM SEVERAL NONCALCAREOUS AND CALCAREOUS WISCONSIN LAKES WERE INVESTIGATED. SEDIMENTS CONTAINED LARGE AMOUNTS OF INORGANIC AND ORGANIC P; GENERALLY MORE THAN ONE-HALF OF THE INORGANIC P ASSOCIATED WITH A GEL COMPLEX DOMINATED BY SHORT-RANGE ORDER HYDRATED IRON OXIDES AND HYDROUS OXIDES. MEASUREMENTS OF EXCHANGEABILITY USING 32P SHOWED THAT A LARGE PORTION OF THE SEDIMENT INORGANIC P EXCHANGED RAPIDLY WITH P IN THE SURROUNDING WATER. INVESTIGATION OF THE NUTRIENT STATUS OF SEDIMENT MICROORGANISMS (ATP PER CELL RELATIONSHIPS) AND THE ABILITY OF ALGAE INOCULATED INTO SEDIMENT-WATER SYSTEMS TO UTILIZE SEDIMENT P AS THEIR SOLE SOURCE OF P INDICATED THAT A SUBSTANTIAL PORTION OF SEDIMENT P IS AVAILABLE TO ORGANISMS IN CLOSE CONTACT WITH SEDIMENTS. EVIDENCE BASED IN PART ON COMPARISON OF 32P AND INORGANIC PHOSPHATE MEASUREMENTS INDICATED THAT THE MURPHY AND RILEY (1962) COLORIMETRIC METHOD PROVIDED VALID INFORMATION ON THE LEVELS OF DISSOLVED INORGANIC P IN LAKE WATERS.

FIELD 05C, 02H

ACCESSION NO. W72-03742

REVIEW OF STARCH PROBLEMS AS RELATED TO STREAM POLLUTION,

WESTERN MICHIGAN UNIV., KALAMAZOO. DEPT. OF PAPER TECHNOLOGY.

W. J. GILLESPIE, C. A. MAZZOLA, AND D. W. MARSHALL.

PAPER TRADE JOURNAL, VOL. 154, NC. 9, MARCH 2, 1970, P 29-32, 6 FIG, 16 REF.

DESCRIPTORS:
*WASTE WATER TREATMENT, *TURBIDITY, BIOCHEMICAL OXYGEN DEMAND, DISSOLVED OXYGEN, ALGAE, THERMAL STRATIFICATION, COAGULATICN, BENTONITE, LIME, *PULP WASTES, TEOTIARY TREATMENT, EFFLUENTS.

IDENTIFIERS:
*STARCH REMOVAL, STARCH SUBSTITUTION, POYELECTROLYTES.

ABSTRACT:
ALTHOUGH THERE ARE OPERATING LOSSES OF STARCH DURING PRODUCTION, THE PRIMARY ADVERSE EFFECT RESULTS FROM THE REPULPING OF STARCH BEARING BROKE OR WASTE PAPER. THIS EXCESS STARCH, PARTICULARLY OXIDIZED STARCH, IN THE SYSTEM ADVERSELY AFFECTS THE PAPER MAKING PROCESS AND INCREASES TURBIDITY OF THE EFFLUENTS THAT MUST BE TREATED PRIOR TO DISCHARGE. STARCH DISPERSIONS ARE DIFFICULT TO COAGULATE WITH CONVENTIONAL FLOCCULANTS, HOWEVER, BENTONITE CLAYS HAVE BEEN USED SUCCESSFULLY AT DOSAGES OF 200 MG/L. STARCH DISPERSED IN THE MILL DISCHARGES EXERTS TWO DIBILITATING EFFECTS CN THE RECEIVING STREAM, A HIGH BOD AND AN INCREASED TURBIDITY WHICH LIMITS LIGHT PENETRATION. THOUGH STARCH SUBSTITUTION AND PROCESS MODIFICATIONS OFFER CONSIDERABLE PROMISE FOR ELIMINATION OF STARCH INDUCED DISPERSIONS, IT IS PROBABLE THAT UNIVERSAL APPLICATION OF SUCH ALTERNATIVES CANNOT BE ACHIEVED. SECONDARY TREATMENT DOES PROVIDE A MECHANISM FOR STARCH TURBIDITY REMOVAL. HOWEVER, ITS UTILIZATICN FOR ONLY THE AESTHETIC IMPROVEMENT OF WASTE DISCHARGE REPRESENTS A COSTLY APPROACH IN AREAS WHERE EFFLUENT STANDARDS DO NOT OTHERWISE REQUIRE IT. THE TASK THAT REMAINS IS FOR TECHNOLOGY TO DEVISE A MEANS OF STARCH PROBLEM RESOLUTION IN THE PRIMARY CLARIFIER. (GOESSLING-TEXAS)

FIELD 05D

ACCESSION NO. W72-03753

TECHNICAL PHOSPHORUS POSITION PAPER, A RESPONSE TO THE IPCB PROPOSED WATER QUALITY STANDARDS REVISIONS, NO. 71-14.

METROPOLITAN SANITARY DISTRICT OF GREATER CHICAGO, ILL. RESEARCH AND DEVELOPMENT DEPT.

SEPTEMBER 1971. 14 P, 37 REF.

DESCRIPTORS:
*PHOSPHORUS, *WASTE WATER TREATMENT, *ALGAL CONTROL, *AQUATIC ALGAE, *EUTROPHICATION, MUNICIPAL WASTES, CITIES, PHOSPHATES, ILLINOIS, ECOLOGY, TERTIARY TREATMENT, SEWAGE EFFLUENTS, WATER REUSE.

IDENTIFIERS:
*PHOSPHORUS REMOVAL, CHICAGO(ILLINOIS).

ABSTRACT:
THE EXACT RELATIONSHIP BETWEEN PHOSPHORUS CONCENTRATION AND AQUATIC GROWTH IS NOT KNOWN. INVESTIGATORS REPORT QUITE DIVERGENT OPINIONS CONCERNING THE AMOUNT OF PHOSPHORUS NECESSARY FOR AQUATIC GROWTH. SOME HAVE SUGGESTED THAT VERY LOW CONCENTRATIONS (LESS THAN .01 MG/L) ARE LIMITING, AND SOME HAVE SUGGESTED THAT EVEN UNDETECTABLE AMOUNTS WILL CAUSE GROWTH. SOME INVESTIGATORS HAVE SUGGESTED THAT OTHER FACTORS MAY BE LIMITING. PHOSPHATE REMOVAL PROCESSES ARE QUESTIONABLE AND VERY DIFFICULT TO CONTROL. ASSUMING THAT IT MAY BE POSSIBLE TO MEET THE LIMITING PERMISSABLE PHOSPHORUS OF 0.1 MG/L PROPOSED BY THE ILLINOIS POLLUTION CONTROL BOARD (USING COAGULATION-SEDIMENTATION OF SECONDARY EFFLUENT PLUS MICRO-FILTRATION) THE CHEMICAL COSTS ALONE HAVE BEEN ESTIMATED TO BE IN EXCESS OF $18 MILLION PER YEAR FOR THE MSDGC. CAPITAL COSTS FOR REQUIRED DOSING EQUIPMENT AND SETTLING TANKS WOULD EXCEED $50 MILLION. THE INCREASED SLUDGE DISPOSAL COST TO HANDLE THE ADDITIONAL SOLIDS GENERATED BY REMOVING PHOSPHORUS WOULD INVOLVE AN ANNUAL COST OF AT LEAST $9,000,000. CONSEQUENTLY, AN ADDITIONAL ESTIMATED ANNUAL OPERATING COST OF $27,000,000 MUST BE ASSIGNED TO PHOSPHORUS REMOVAL COMPARED TO THE PRESENT TOTAL ANNUAL COST OF TREATMENT OF 10-14 MILLION DOLLARS. THERE IS NO SUBSTANTIAL EVIDENCE TO INDICATE THAT THE LIMITING NUTRIENT IS PHOSPHORUS. PHOSPHORUS REMOVAL AT WASTEWATER TREATMENT PLANTS WHICH DO NOT DISCHARGE TO LAKE MICHIGAN DOES NOT APPEAR AT THIS TIME TO BE JUSTIFIED. (POERTNER)

FIELD 05C

ACCESSION NO. W72-03970

FINAL REPORT, PROVISIONAL ALGAL ASSAY PROCEDURES.

CALIFORNIA UNIV., BERKELEY. SANITARY ENGINEERING RESEARCH LAB.

D. F, TOERIEN, C. H. HUANG, J. RADIMSKY, E. A. PEARSON, AND J. SCHERFIG.

AVAILABLE FROM THE NATIONAL TECHNICAL INFORMATION SERVICE AS PB-206 140, $3.00 IN PAPER COPY, $0.95 IN MICROFICHE. ENVIRONMENTAL PROTECTION AGENCY, WATER POLLUTION CONTROL RESEARCH SERIES, SERL REPORT 71-6, OCTOBER 1971. 211 P, 51 FIG, 37 TAB, 111 REF, 4 APPEND. EPA PROJECT 16010 DQB 10/71.

DESCRIPTORS:
*BIOASSAY, CULTURES, *BICMASS, NUTRIENTS, *NUTRIENT REQUIREMENTS, NUISANCE ALGAS, *ENTROPHICATION, CYTOLCGICAL STUDIES, ANALYTICAL TECHNIQUES, POLLUTANT IDENTIFICATION.

IDENTIFIERS:
*CHEMOSTAL ASSAYS, *ALGAL ASSAY PROCEDURES, SELENASTRUM CAPRICORNUTUM.

ABSTRACT:
BATCH AND CONTINUOUS FLOW (CHEMOSTAT) ASSAYS WERE INVESTIGATED AS PART OF A JOINT INDUSTRY-GOVERNMENT SPONSORED, MULTILABCRATORY EFFORT TO DEVELOP A STANDARDIZED ALGAL ASSAY PROCEDURE FOR NUTRIENT LEVEL ASSESSMENT. ASSAYS WERE CONDUCTED WITH SELENASTRUM CAPRICORNUTUM AS A STANDARD ASSAY CRGANISM. BATCH CULTURE ASSAYS WERE FOUND TO HAVE A LOWER LEVEL CF PRECISION THAN CHEMOSTAT ASSAYS IN THE ASSESSMENT OF GROWTH RESPONSE AS A FUNCTION OF NUTRIENT CONCENTRATION. THE BIOMASS PARAMETER, MAXIMUM CELL CCNCENTRATION, X, OF THE BATCH ASSAY GENERALLY RESPONDED TO THE NUTRIENT CONCENTRATION OF THE SAMPLE; HOWEVER, THE CHEMOSTAT BICMASS PARAMETER, STEADY STATE CELL CCNCENTRATION, X SUB 1, ALWAYS WAS FOUND TO BE PROPORTIONAL TO THE NUTRIENT CONCENTRATION OF THE SAMPLES. THE RESULTS OF SPIKING TESTS WITH BATCH ASSAYS GENERALLY WERE INCONCLUSIVE, WITH RESPECT TO IDENTIFICATION CF THE GROWTH RATE LIMITING NUTRIENT, WHEREAS THE RESULTS OF SPIKING TESTS WITH CHEMOSTATS INDICATED CLEARLY THE GROWTH RATE LIMITING NUTRIENT. IT IS RECOMMENDED THAT BATCH TYPE ALGAL ASSAYS BE USED ONLY FOR CRUDE SCREENING OR ROUTINE MONITORING PURPOSES AND THAT THE CHEMOSTAT SHOULD BE USED FOR THE QUANTITATIVE ASSESSMENT CF THE ALGAL GROWTH SUPPORTING PROPERTIES OF WATERS AS WELL AS FOR THE DEVELOPMENT OF KINETIC DESCRIPTIONS FOR NUISANCE ALGAE AND THE RATE LIMITING NUTRIENTS OF CONCERN. A KINETIC DESCRIPTION OF SELENASTRUM CAPRICORNUTUM INDICATED A LOW HALF SATURATION CONSTANT, K SUB S, (THE CONCENTRATION CF NUTRIENT SUPPORTING ONE-HALF THE MAXIMUM GROWTH RATE) OF ABOUT 5 MICRCGRAM P/L FOR PHOSPHATE PHOSPHORUS AND A YIELD COEFFICIENT, Y, THAT VARIED AS A FUNCTION CF GROWTH RATE. A THEORETICAL MODEL WAS PROPOSED AND EVALUATED WHICH DESCRIBES THE VARYING YIELD COEFFICIENT (THE RESULT OF 'ECESS' UPTAKE) AS A FUNCTION OF THE GROWTH RATE (MEAN CELL RESIDENCE TIME). THE FUNCTION WAS VERIFIED EXPERIMENTALLY AT A VERY HIGH STATISTICAL CONFIDENCE LEVEL. THE SIGNIFICANCE OF THESE FINDINGS AND THEIR APPLICATION TO THE PRACTICAL PROBLEM OF EUTRCPHICATION ASSESSMENT IS PRESENTED. (EPA ABSTRACT)

FIELD 05C, 05A

ACCESSION NC. W72-03985

NUTRIENT LIMITATION,

MINISTRY OF AGRICULTURE, CRUMLIN (NORTHERN IRELAND). FRESHWATER BIOLOGICAL INVESTIGATION UNIT.

C. E. GIBSON.

JOURNAL WATER POLLUTICN CONTROL FEDERATICN, VOL 43, NO 12, DECEMBER 1971, P 2436-2440, 2 TAB, 12 REF.

DESCRIPTORS:
*NUTRIENTS, *ALGAE, *EUTROPHICATION, *PHOSPHORUS, *NITROGEN, EVAPORATION, STORM RUNOFF, BIOASSAYS, BIO-INDICATCRS, STORAGE, FLOW RATES, INHIBITION, ANALYTICAL TECHNIQUES, ENZYMES, COLOR, WATER QUALITY CONTROL, RESEARCH AND DEVELCPMENT.

ABSTRACT:
DETERMINATION OF LIMITING LEVELS OF NITROGEN AND PHOSPHORUS HAS BEEN EXCEEDINGLY DIFFICULT FOR SEVERAL REASONS. INPUTS OF PHOSPHATES AND NITRATES ARE EXTREMELY VARIABLE WITH SEASON OR WITH SHIFTS IN POPULATION. SECONDLY, THE RATIO CF CARBON TO NITRCGEN TO PHOSPHORUS IS JUST AS IMPORTANT AS THE ACTUAL AMOUNTS OF EACH PRESENT. THIRDLY, CERTAIN ALGAE HAVE A CAPACITY FOR STORING PHOSPHATES, AND THRIVING ALGAL GROWTHS MAY APPEAR IN AREAS WHERE PHOSPHATE IS SERIOUSLY DEPLETED. PERHAPS THE GREATEST DIFFICULTY ENCOUNTERED IS THE INABILITY OF RESEARCHERS TO DEVELOP A SYNTHETIC MEDIUM UPON WHICH ALGAE WILL PRODUCE GROWTH PATTERNS SIMILAR TO THOSE OF ALGAE GRCWN IN STREAMS OR LAKES. PHCSPHORUS AND NITROGEN REQUIREC BY ALGAE GROWN ON SYNTHETIC MEDIA FAR EXCEED THE SUPPLY AVAILABLE IN ANY WATERS LESS THAN GROSSLY POLLUTED. HOWEVER, ADVANCEMENTS IN BIOASSAY TECHNIQUES ARE CONTINUING, AND THERE IS NOW A SIMPLE SPCT TEST FOR PHOSPHORUS LIMITATION. PRESENCE OF ALKALINE PHOSPHATES HAS BEEN SHOWN TO BE AN INDICATOR OF PHOSPHATE LIMITATION, AND A RAPID COLOR TEST HAS BEEN DEVELCPED. TEAMS OF RESEARCHERS ARE CONTINUING TO DEVOTE THEIR ATTENTION TO THIS AND OTHER NEW METHOCS FOR OBTAINING INFORMATION ABOUT NUTRIENT LIMITATION. (LOWRY-TEXAS)

FIELD 05C

ACCESSION NO. W72-04070

NITRCGEN REMOVAL FROM WASTEWATERS--STATEMENT OF THE PROBLEM,

FEDERAL WATER QUALITY ADMINISTRATION, CINCINNATI, OHIO. ADVANCED WASTE TREATMENT RESEARCH LAB.

EDWIN F. BARTH, AND ROBERT B. DEAN.

IN: NITROGEN REMOVAL FROM WASTEWATERS, PAPER NO 1, MAY 1970. 6 P. FWQA PROJECT 17010---05/70.

DESCRIPTORS:
*NITROGEN, *AMMONIA, NITRATES, NITRITES, TOXICITY, CCRROSION, PUBLIC HEALTH, ODORS, OXIDATION-REDUCTICN POTENTIAL, BIOCHEMICAL OXYGEN DEMAND, NITRIFICATION, NUTRIENTS, SLUDGE, ALGAE, BACTERIA, *WASTE WATER TREATMENT, EUTROPHICATION.

IDENTIFIERS:
*METHEMOGLOBINEMIA.

ABSTRACT:
RELEASE OF AMMCNIA TO A RECEIVING WATER BODY MAY CAUSE A VARIETY OF UNWANTED RESULTS. THE TOXICITY OF AMMONIA TO ADULT HUMANS IS SO SLIGHT AS TO REQUIRE NO LIMITS IN DRINKING WATER, ALTHOUGH THE PRESENCE OF AMMONIA IS REGARDED AS EVIDENCE CF POLLUTION. HOWEVER, INFANT HUMANS, BECAUSE OF THE LACK CF SUFFICIENT INTESTINAL FLORA, ARE NOT ABLE TO REDUCE NITRITES TO NITROGEN GAS. THE NITRITE THEN COMBINES WITH THE HEMOGLOBIN WHICH RESULTS IN METHEMOGLOBINEMIA. HIGH AMMONIA LEVELS IN DRINKING WATER WILL ALSO DISSOLVE COPPER PLUMBING. FOR THESE REASONS, LIMITS HAVE BEEN SET AT NC MCRE THAN 10 MG OF NITRATE NITROGEN IN WATER TO BE FED TO BABIES, ALTHOUGH NO STANDARDS HAVE BEEN SET FOR AMMONIA. THE OXYGEN DEMAND OF NITROGENOUS MATERIAL HAS SIGNIFICANTLY ACCELERATED EUTROPHICATION IN MANY AREAS, INCLUDING THE POTOMAC ESTUARY, THE GRAND AND CLINTON RIVERS IN MICHIGAN AND KANAWHA RIVER IN WEST VIRGINIA. A NITRIFIED EFFLUENT, HOWEVER, WILL PROVIDE OXYGEN TC SLUDGE BEDS AND PREVENT SEPTIC ODORS, BE MCRE EFFICIENTLY AND EFFECTIVELY DISINFECTED BY CHLORINE, AND WILL CONTAIN LESS SOLUBLE ORGANIC MATTER THAN THE SAME EFFLUENT BEFORE CHLORINATION. (LOWRY-TEXAS)

FIELD 05D

ACCESSION NO. W72-04086

ACCUMULATION AND DISTRIBUTION OF CHLORINATED HYDROCARBONS IN CULTURES OF (CHLOROPHYCEAE),

LUND UNIV. (SWEDEN). DEPT. OF ANIMAL ECOLOGY.

A. SODERGREN.

OIKOS, VOL 22, P 215-220, 1971. 7 FIG, 1 TAB, 20 REF.

DESCRIPTORS:
*CHLORINATED HYDROCARBON PESTICIDES, *PESTICIDE RESIDUES, *ALGAE, CHLOROPHYTA, CHLORELLA, TOXICITY, WATER ANALYSIS, ABSORPTION, CULTURES, DDT, ANALYTICAL TECHNIQUES, BIOMASS, GAS CHROMATOGRAPHY, CENTRIFUGATION, PESTICIDES, BIOASSAY, DISTRIBUTION PATTERNS, CHROMATOGRAPHY, GROWTH RATES, POLLUTANT IDENTIFICATION.

IDENTIFIERS:
*CHLORELLA PYRENOIDOSA, LINDANE, P P'-DDE, CLOPHEN A 50, ELECTRON CAPTURE GAS CHROMATOGRAPHY, POLYCHLORINATED BIPHENYLS, BIOLOGICAL MAGNIFICATION, CHLAMYDOMONAS, METABOLITES, HAEMATOCRIT TUBE.

ABSTRACT:
THE ACCUMULATION AND DISTRIBUTION OF LINDANE, P,P'-DDE AND CLOPHEN A 50 (PCB) HAVE BEEN STUDIED IN CONTINUOUS FLOW CULTURES OF CHLORELLA PYRENOIDOSA CHICK. IN ESTIMATING THE CELL VOLUMES OF ALGAE IN THE CULTURE A TECHNIQUE HAS BEEN DEVELOPED USING CENTRIFUGATION IN A MODIFIED HAEMATOCRIT TUBE. THE NUMBER OF CELLS IN THE CULTURE WAS CALCULATED BY CALIBRATING THE CELL VOLUMES VERSUS THE CELL COUNTS. SAMPLES OF ALGAE, AIR, AND WATER FROM WITHIN THE CULTURE SYSTEM WERE ANALYZED FOR RESIDUES OF THE CHLORINATED HYDROCARBON PESTICIDES ON A VARIAN AEROGRAPH 204 GAS CHROMATOGRAPH WITH AN ELECTRON CAPTURE DETECTOR. IT WAS FOUND THAT: (1) LINDANE WAS ACCUMULATED BY THE ALGAE BUT WAS ALSO DISTRIBUTED IN THE WATER AND AIR OF THE CULTURE. (2) P,P'-DDE AND CLOPHEN A 50 WERE TAKEN UP BY THE ALGAE TO A GREAT EXTENT AND ONLY A SMALL PART REMAINED IN THE WATER. AT THE END OF THE EXPERIMENTS 82 PERCENT AND 88 PERCENT RESPECTIVELY OF P,P'-DDE AND CLOPHEN A 50 WERE ACCUMULATED BY THE ALGAE. (3) IN THE EXPERIEMENTS THERE WERE NO DISTURBANCES OF THE GROWTH RATE OF THE CULTURED ALGA DUE TO THE PRESENCE OF THE TEST SUBSTANCE. THE UPTAKE AND DISTRIBUTION OF THE PESTICIDES WITHIN THE CULTURE SYSTEM AND THE RELATIONSHIP BETWEEN CELL VOLUMES AND CELL COUNTS ARE SHOWN GRAPHICALLY. (HOLOMAN-BATTELLE)

FIELD 05A

ACCESSION NO. W72-04134

PHARMACOLOGICAL TESTING OF BLUE-GREEN ALGAE FOR CONSTITUENTS HAVING THERAPEUTIC VALUE,

WORLD LIFE RESEARCH INST., COLTON, CALIF.

BRUCE W. HALSTEAD.

AVAILABLE FROM SOD EP2.10:16010DQU06/70, $0.95 IN MICROFICHE. ENVIRONMENTAL PROTECTION AGENCY, WATER QUALITY OFFICE, WATER POLLUTION CONTROL RESEARCH SERIES, JUNE 1970. 17 P, 1 TAB, 9 REF. EPA PROGRAM 16010DQU06/70, 14-12-535.

DESCRIPTORS:
*CYANOPHYTA, *LABORATORY TESTS, ALGAL TOXINS, HUMAN PATHOLOGY.

IDENTIFIERS:
*THERAPEUTIC VALUE, *APHANIZOMENON FLOS-AQUAE, *PHARMACOLOGICAL TESTS, ANTIMICROBIOTICS, ESCHERICHIA COLI, BETA STREPTOCOCCUS, PSEUDOMONAS AERUGINOSA, MYCOBACTERIUM FURTUITUM, STAPHYLOCOCCUS AUREUS, CANDIDA ALBICANS.

ABSTRACT:
THE BLUE-GREEN ALGA APHANIZOMENON FLOS-AQUAE (LINNAEUS) RALFS, WAS COLLECTED DURING THE PEAK OF ITS PRIMARY BLOOM AND PREPARED WITH ACETONE AND ETHANOLIC EXTRACTS FOR PHARMALOGICAL AND MICROBIOLOGICAL TESTING. THE FRACTIONS EXHIBITED LITTLE ACUTE OVERTSYMPTOMATOLOGY AND THUS PROBABLY DO NOT CONTAIN PRODUCTS THAT EXERT PHARMACOLOGICAL EFFECTS SIMILAR TO THE TYPE SEEN BY DRUGS WHICH ORDINARILY DIRECTLY PRODUCE PSYCHOMATIC, NEUROLOGICAL, OR CARDIOVASCULAR EFFECTS. GENERALLY THE ACTIVITY OF THE FRACTIONS RESULTED IN DEPRESSION OF FUNCTION, ALERTNESS, AND APPETITE, AND DECREASED BODY WEIGHT IN RATS, CHARACTERISTIC OF AN ANTIMETABOLITE ACTION COMMON TO MANY NATURALLY OCCURRING SUBSTANCES. THERE WAS LOW POTENCY, NO FRACTIONAL LOCALIZATION OF ACTIVITY, NO SIGNIFICANT NEUROLOGICAL, CARDIOVASCULAR, OR ANY OUTSTANDING BIOLOGICAL ACTIVITY. ANTIMICROBIAL TESTS PERFORMED ON ESCHERICHIA COLI, BETA STREPTOCOCCUS, PSEUDOMONAS AERUGINOSA, MYCOBACTERIUM FURTUITUM, STAPHYLOCOCCUS AEREUS, AND CANDIDA ALBICANS. NO ANTIMICROBIAL ACTIVITY WAS OBSERVED, HOWEVER THE ALGAE COULD BE CLASSIFIED AS 'TOXIC' IF SUFFICIENT QUANTITIES ARE INGESTED. A SAMPLE WAS SCREENED, TESTED, AND EVALUATED FOR ANTITUMOR ACTIVITY BUT WAS NOT FOUND TO HAVE ANY SIGNIFICANT ANTITUMOR ACTIVITY. THE TESTS FAILED TO SHOW EVIDENCE OF PHARMACOLOGICAL PROPERTIES HAVING COMMERCIAL PHARMACEUTICAL POTENTIAL. FURTHER INVESTIGATION AS TO THE COMMERCIAL POTENTIAL OF THE PHYTOCHEMICAL CONSTITUENTS OF APHANIZOMENON FLOS-AQUAE DOES NOT APPEAR WARRANTED. (AUEN-WISCONSIN)

FIELD 05C, 05G

ACCESSION NO. W72-04259

WATER QUALITY CONTROL THROUGH FLOW AUGMENTATION,

HEIDELBERG COLL., TIFFIN, OHIO. DEPT. OF BIOLOGY.

DAVID B. BAKER, AND JACK W. KRAMER.

AVAILABLE FROM SOD EP2.10:16080DF001/71, $0.95 IN MICROFICHE. ENVIRONMENTAL PROTECTION AGENCY, WATER QUALITY OFFICE, WATER POLLUTION CONTROL RESEARCH SERIES, JANUARY 1971. 156 P, 27 FIG, 8 TAB, 14 REF, 12 APPEND. EPA PROGRAM 16080DF001/71.

DESCRIPTORS:
*WATER QUALITY CONTROL, *FLOW AUGMENTATION, WATER QUALITY, RESERVOIRS, WATER POLLUTION SOURCES, ALGAE, RIVER FLOW, PHOSPHORUS, NITRATES, POTASSIUM, DISSOLVED OXYGEN, FORECASTING, FLUORIDES, OHIO.

IDENTIFIERS:
SANDUSKY RIVER(OHIO).

ABSTRACT:
A 60-MILE SECTION OF THE SANDUSKY RIVER, OHIO, WAS INVESTIGATED TO EVALUATE THE RELATIONSHIP BETWEEN THE VOLUME AND THE QUALITY OF FLOWING WATER. THE CONTENT OF FLUORIDE, CALCIUM, MAGNESIUM, AND SODIUM WERE DIRECTLY AND THOSE OF TOTAL AND SOLUBLE PHOSPHORUS INDIRECTLY RELATED TO THE FLOW VOLUME. NO CORRELATION WAS OBSERVED BETWEEN THE FLOW AND CONCENTRATION OF EITHER POTASSIUM OR NITRATES. OXYGEN CONTENT WAS HIGH AT AN ABUNDANT FLOW, BUT SHOWED CONSIDERABLE VARIATION, BEING INFLUENCED BY RESPIRATION OF ALGAE. THE STUDY SUGGESTED THAT AN INCREASED CURRENT VELOCITY REDUCES THE DENSITY OF ALGAL POPULATIONS. (WILDE-WISCONSIN)

FIELD 05G

ACCESSION NO. W72-04260

PHOSPHATE NUTRIENT OCCURRENCE AND DISTRIBUTION IN GREAT LAKES SEDIMENTS,

ILLINOIS UNIV., CHICAGO. SOIL MECHANICS LAB.

MARSHALL L. SILVER, AND CHARLES A. MOORE.

SM PUBLICATION NO 12, 1971. 27 P, 7 FIG, 5 TAB, 55 REF. OWRR A-053-ILL(1).

DESCRIPTORS:
*PHOSPHATES, *GREAT LAKES, *SEDIMENTS, *NUTRIENTS, EUTROPHICATION, NITROGEN, LITTORAL, HARBORS, SPATIAL DISTRIBUTION, ALGAE, TRIBUTARIES, CYCLING NUTRIENTS.

IDENTIFIERS:
*NUTRIENT LEVELS, CUYAHOGA RIVER(OHIO).

ABSTRACT:
PUBLISHED AND UNPUBLISHED SOURCES ARE COMBINED TO PROVIDE INSIGHT INTO THE SOURCE AND FATE OF POLLUTANTS IN THE GREAT LAKES. DATA ARE PRESENTED SUMMARIZING THE QUANTITY OF PHOSPHORUS PRESENT IN CENTRAL LAKE WATERS AND CENTRAL LAKE SEDIMENTS AS WELL AS PHOSPHORUS CONTENTS IN SEDIMENTS AND WATERS OF RIVERS AND HARBORS TRIBUTARY TO THE LAKES, AND NUTRIENT CONTENTS IN THE GREAT LAKES MARGINS. AVAILABLE LITERATURE IS SUMMARIZED DESCRIBING THE EFFECT OF PHOSPHORUS ON THE GROWTH RATE OF AQUATIC PLANTS WHOSE ACCELERATED GROWTH IS RESPONSIBLE FOR EUTROPHICATION. AN ATTEMPT IS MADE TO ORGANIZE MECHANISMS CONTRIBUTING TO THE CYCLING OF PHOSPHORUS AND OTHER NUTRIENTS IMPORTANT TO GREAT LAKES STUDIES. GENERALLY, THE MINIMUM PHOSPHORUS CONTENT NEEDED FOR ALGAL GROWTH IS BELOW 0.01 MG/L AND MAY BE LESS THAN 0.001 MG/L. PHOSPHORUS LEVELS IN EXCESS OF 0.1 MG/L OFTEN CAUSE ACCELERATED GROWTH OF VARIOUS PHYTOPLANKTON AND ALGAE. NITROGEN CONTENTS GREATER THAN 0.1 MG/L ARE NECESSARY FOR GROWTH WHILE CONCENTRATIONS GREATER THAN 1 MG/L LEAD TO ACCELERATED PLANT GROWTH. (AUEN-WISCONSIN)

FIELD 05C

ACCESSION NO. W72-04263

DETERGENTS A STATUS REPORT.

NATIONAL INDUSTRIAL POLLUTION CONTROL COUNCIL, WASHINGTON, D.C.

AVAILABLE FROM THE NATIONAL TECHNICAL INFORMATION SERVICE AS COM-71-50084, $3.00 IN PAPER COPY, $0.95 IN MICROFICHE. SUB-COUNCIL REPORT MARCH 1971. 16 P.

DESCRIPTORS:
*DETERGENTS, *EUTROPHICATION, *ECONOMIC EFFICIENCY, *PUBLIC HEALTH, WATER POLLUTION, HAZARDS, NUTRIENTS, ALGAE.

IDENTIFIERS:
NITRILOTRIACETIC ACID, HOUSEHOLD DETERGENTS, SOAP AND DETERGENT INDUSTRY, NTA.

ABSTRACT:
THE USE OF THE MOST PROMISING SUBSTITUTE FOR PHOSPHATES IN DETERGENTS, NITRILOTRIACETIC ACID (NTA) HAS BEEN SUSPENDED PENDING FURTHER STUDY AT THE REQUEST OF THE U S SURGEON GENERAL AND THE ENVIRONMENTAL PROTECTION AGENCY. THE SEARCH FOR PHOSPHATE SUBSTITUTES CONTINUES. EUTROPHICATION IS DEFINED AS SIMPLY THE OVERABUNDANCE OF A NATURAL AND NECESSARY PROCESS--TROUBLESOME INDEED WHEN IT OCCURS, BUT, BY NO STRETCH OF THE IMAGINATION, A FORM OF WATER POLLUTION. THERE IS NO EVIDENCE WHATEVER THAT THE REMOVAL OF PHOSPHATES FROM DETERGENTS WILL STOP ACCELERATED EUTROPHICATION. ADEQUATE TECHNOLOGY EXISTS TO REMOVE PHOSPHORUS FROM SEWAGE BY MEANS OF CHEMICAL TREATMENT. FOR RELATIVELY MODEST COSTS, CHEMICAL SEWAGE TREATMENT PERMITS THE REMOVAL OF PHOSPHORUS FROM ALL SOURCES WITHOUT THE CAPITAL COST OF SECONDARY OR TERTIARY SEWAGE PLANTS AND IT IS HIGHLY FLEXIBLE, IN THAT IT CAN BE EMPLOYED IN ONLY THOSE LOCALITIES WHERE CULTURAL EUTROPHICATION IS A PROBLEM. (JONES-WISCONSIN)

FIELD 05C

ACCESSION NO. W72-04266

LIMNOLOGICAL CHARACTERISTICS OF NORTH AND CENTRAL FLORIDA LAKES,

FLORIDA UNIV., GAINESVILLE. DEPT. OF ENVIRONMENTAL ENGINEERING.

EARL E. SHANNON, AND PATRICK L. BREZONIK.

RESEARCH REPORT, 1971. 36 P, 9 FIG, 3 TAB, 31 REF. OWRR B-004-FLA(5). FWQA DON 16010.

DESCRIPTORS:
*LIMNOLOGY, *FLORIDA, *LAKES, PONDS, SAMPLING, THERMAL STRATIFICATION, HARDNESS(WATER), COLOR, OLIGOTROPHY, EUTROPHICATION, ALGAE, CYANOPHYTA, PHYTOPLANKTON, LAKE MORPHOMETRY, GEOLOGY, LIMESTONE, TEMPERATURE, DEPTH, CHEMICAL PROPERTIES, ACIDITY, ALKALINITY, CONDUCTIVITY, CALCIUM, SCENEDESMUS, WATER HYACINTH, LIGHT PENETRATION, OXYGEN, SILICA, AQUATIC PLANTS.

IDENTIFIERS:
MESOTROPHIC, MICROCYSTIS, ANABAENA, LYNGBYA.

ABSTRACT:
AN EXTENSIVE SURVEY OF THE PHYSICAL, CHEMICAL, AND BIOLOGICAL CHARACTERISTICS OF 55 NORTH-CENTRAL FLORIDA LAKES AND PONDS WAS MADE. RESULTS OF THE ONE-YEAR INVESTIGATION, WITH PARTICULAR REFERENCE TO THEIR TYPOLOGIC AND TROPHIC FEATURES ARE DESCRIBED. THEY ARE TYPICALLY SHALLOW, WITH MEAN DEPTHS RANGING FROM 0.7 TO 8.1 M AND DERIVED FROM SOLUTION OF LIMESTONE FORMATIONS WHICH UNDERLY THE FLORIDA PENINSULA. STABLE THERMAL STRATIFICATION OCCURS IN 13 LAKES. THEY ARE TYPICALLY SOFT WATER, LOW IONIC STRENGTH WATERS. HIGH ORGANIC COLOR, DERIVED PRIMARILY FROM PINE LITTER, IS COMMON. TROPHIC CONDITIONS OBTAIN FROM ULTRAOLIGOTROPHIC TO HYPEREUTROPHIC. MULTI-VARIATE CLUSTER ANALYSIS CONSIDERING SEVEN QUANTITATIVE TROPHIC INDICATORS GROUPED THE CLEAR LAKES INTO THREE READILY INTERPRETABLE GROUPS (OLIGOTROPHIC, MESOTROPHIC, AND EUTROPHIC) AND THE COLORED LAKES INTO FIVE GROUPS. HIGHEST EUTROPHIC CONDITIONS ARE ASSOCIATED WITH ALKALINE (HARDWATER) CLEAR LAKES, WHEREAS THE SOFT-WATER CLEAR LAKES WERE ALL OLIGOTROPHIC; COLOR APPEARS TO DAMPEN THE RANGE OF TROPHIC CONDITIONS. ALGAL BLOOMS CAN OCCUR ALMOST CONTINUOUSLY THROUGHOUT THE YEAR IN EUTROPHIC LAKES BECAUSE OF FAVORABLE GROWTH CONDITIONS; THE BLUE-GREEN BLOOM-FORMING ALGAE, MICROCYSTIS, ANABAENA, AND LYNGBYA ARE THE MOST DOMINANT PHYTOPLANKTON OF EUTROPHIC FLORIDA LAKES. (JONES-WISCONSIN)

FIELD 05C, 02H

ACCESSION NO. W72-04269

PRIMARY PRODUCTIVITY IN FIVE SALT VALLEY RESERVOIRS,

NEBRASKA UNIV., LINCOLN. DEPT. OF ZOOLOGY.

JOHN L. ANDERSEN.

MS THESIS, AUGUST 1971. 54 P, 12 FIG, 5 TAB, 22 REF. OWRR A-014-NEB(1).

DESCRIPTORS:
*PRIMARY PRODUCTIVITY, *RESERVOIRS, *NEBRASKA, *EUTROPHICATION, PLANKTON, CHLOROPHYLL, RECREATION, RUNOFF, MEASUREMENT, TEST PROCEDURES, PHOTOSYNTHESIS, SEASONAL, LIGHT PENETRATION, LIGHT INTENSITY, DISSOLVED SOLIDS, PHYTOPLANKTON, ALGAE, NUTRIENTS.

IDENTIFIERS:
*SALT VALLEY RESERVOIRS(NEB), ULTRA PLANKTON, C-14.

ABSTRACT:
FIVE SALT VALLEY RESERVOIRS, LOCATED WITHIN A 20-MILE RADIUS OF LINCOLN, NEBRASKA, WERE CONSTRUCTED TO PROVIDE FLOOD PROTECTION BUT RECREATIONAL FACILITIES HAVE ALSO BEEN EXTENSIVELY DEVELOPED. THESE RESERVOIRS DEPEND UPON RUNOFF FROM AGRICULTURAL LANDS AS THEIR SOURCE OF WATER AND ARE RICH IN NUTRIENTS. MORPHOMETRY AND PHYSICAL-CHEMICAL PARAMETERS ARE RECORDED. THE IN SITU CARBON-14 TECHNIQUES WERE USED FOR STUDYING PRODUCTIVITY. ABSOLUTE ACTIVITY OF THE CARBON-14 MUST BE DETERMINED SINCE THERE OFTEN EXISTS A WIDE DIFFERENCE BETWEEN THE LABELED AND ACTUAL VALUE OF COMMERCIALLY PREPARED RADIOISOTOPE SOLUTIONS. THE LAKES ARE DISCUSSED INDIVIDUALLY AND RANKED ACCORDING TO PRODUCTIVITY INCLUDING PRODUCTIVITY VALUES, CHLOROPHYLL VALUES, AND THE ULTRA PLANKTON CONTRIBUTION, WHICH RANGED FROM 0 TO 13.3% OF THE TOTAL PRODUCTIVITY. PRIMARY PRODUCTIVITY OCCURRING UNDER THE ICE CONSTITUTES A LARGE PORTION OF THE ANNUAL PRODUCTIVITY AND CANNOT BE IGNORED. IN COMPARING THE PRODUCTIVITY OF THE SALT VALLEY LAKES WITH OTHER LAKES AROUND THE WORLD, IT CAN BE SEEN THAT THE SALT VALLEY LAKES ARE QUITE EUTROPHIC. IF THE RATE OF YEARLY PRODUCTIVITY INCREASES IN THESE RESERVOIRS (500-600 MG/YEAR) CONTINUES, IT MAY BE A SHORT TIME UNTIL THESE LAKES ARE IN THE PRODUCTIVITY RANGE OF SEWAGE LAGOONS. (JONES-WISCONSIN)

FIELD 05C

ACCESSION NO. W72-04270

PHYTOCHEMICAL, PHARMACOLOGICAL AND ANTIMICROBIAL SCREENING OF MINNESOTAN AQUATIC PLANTS,

MINNESOTA UNIV., MINNEAPOLIS.

KWEI LEE SU.

PHD THESIS, JUNE 1971. 158 P, 13 FIG, 23 TAB, 342 REF. OWRR A-025-MINN(1).

DESCRIPTORS:
*AQUATIC PLANTS, *ALGAE, *PUBLIC HEALTH, ANALYTICAL TECHNIQUES, TOXICITY, CHEMICAL PROPERTIES, MINNESOTA, LIPIDS, HUMAN PATHOLOGY, DISEASES.

IDENTIFIERS:
ANTIMICROBIOTICS, PHARMACOLOGICAL PROPERTIES, PHYTOCHEMICAL SCREENINGS, NYMPHAEA TUBEROSA, NUPHAR VARIEGATUM, NUPHAR JAPONICUM, NUPHAR LUTEN, ANTICANCER POTENTIAL, ANTICOAGULANT POTENTIAL, SPARGANIUM FLUCTUANS, ANTINEOPLASTIC ACTIVITY, ALKALOIDS, FLAVONOIDS, STEROIDS, TANNINS, SAPONINS,

ABSTRACT:
SEVENTEEN MONOCOTS, 4 DICOTS, AND 1 ALGA, COLLECTED FROM MINNESOTA LAKES WERE EXTRACTED WITH SKELLYSOLVE F, CHLOROFORM, 80% ETHANOL, AND ACIDIC AND BASIC WATER. AS REVEALED BY THIN-LAYER CHROMATOGRAPHY AND OTHER ANALYSES, MANY PLANTS CONTAIN ALKALOIDS, FLAVONOLES, FLAVONONES, BETA-SITOSTEROL, LIPIDS, TANNINS, AND SAPONINS. THE TOXICITY, INCLUDING LETHAL DOSE, MEDIAN TEST, AND ANTINEOPLASTIC ACTIVITY IN VIRO WERE DETERMINED USING SWISS WEBSTER MICE AND SYRIAN HAMSTERS. SCREENING WAS PERFORMED ON HUMAN EPIDERMOID CARCINOMA AND LYMPHOID LEUKEMIA. ANTIMICROBIAL EFFECTS OF EXTRACTS WERE STUDIED ON BOTH ANIMAL AND PLANT PATHOGENES BY THE DISC DIFFUSION METHOD. SOME EXTRACTS WERE EFFECTIVE AGAINST STAPHYLOCOCCUS AUREUS, MYCOBACTERIUM SMEGMATIS, AND CANDIDA ALBICANS. RECOMMENDATIONS FOR ADDITIONAL STUDIES INCLUDE IDENTIFICATION OF CERTAIN ALKALOIDS, ANTICANCER POTENTIAL AGAINST AMELANOMA MALIGNANCY, ANTICOAGULANT EFFECTS OF SEVERAL AQUATIC PLANTS, AND INHIBITION OF CANDIDA ALBICANS. (WILDE-WISCONSIN)

FIELD 05C

ACCESSION NO. W72-04275

WORKSHOP-CONFERENCE ON RECLAMATION OF MAINE'S DYING LAKES.

MAINE UNIV., BANGOR.

AVAILABLE FROM THE NATIONAL TECHNICAL INFORMATION SERVICE AS PB-206 432, $3.00 IN PAPER COPY, $0.95 IN MICROFICHE. UNIVERSITY OF MAINE AT ORONO, MAINE, JUNE 1971. 113 P, 4 FIG, 1 MAP, 6 PHOTO, 39 REF, 2 APPEND. OWRR A-999-ME(8).

DESCRIPTORS:
*MAINE, *EUTROPHICATION, *WATER QUALITY CONTROL, *NON-STRUCTURAL ALTERNATIVES, AGING(BIOLOGICAL), AQUATIC ALGAE, AQUATIC PRODUCTIVITY, FISH, FISHKILL, LIMNOLOGY, NUTRIENTS, OXYGEN, OXYGEN REQUIREMENTS, OXYGEN SAG, WASTE ASSIMILATIVE CAPACITY, WATER PROPERTIES, WATER POLLUTION, WATER POLLUTION SOURCES, WATER POLLUTION EFFECTS, WATER POLLUTION CONTROL, WATER POLLUTION TREATMENT, WATER QUALITY, WATER QUALITY CONTROL, LAND USE, LAND TENURE, LAND DEVELOPMENT, LAND MANAGEMENT.

ABSTRACT:
METHODS OF CORRECTING AND PREVENTING EUTROPHIC LAKES IN MAINE ARE COMPREHENSIVELY DISCUSSED IN THIS REPORT OF A CONFERENCE CONDUCTED AT THE UNIVERSITY OF MAINE. EUTROPHICATION WAS THE FIRST CONFERENCE TOPIC CONSIDERED; THE EUTROPHICATION PROCESS, EFFECTS, AND REMEDIES WERE DISCUSSED. PARTICULAR EMPHASIS WAS PLACED UPON LAKE FERTILIZATION, CULTURAL EUTROPHICATION AND ITS CAUSES, LAND DRAINAGE, AND WASTEWATER DISCHARGE. GENERALLY, THE TOPICS WERE EXAMINED BY MEANS OF PANEL DISCUSSIONS WHICH SOLICITED QUESTIONS FROM THE AUDIENCE. VARIOUS TECHNIQUES FOR CORRECTING EUTROPHICATION TO ACHIEVE INLAND LAKE RENEWAL WERE DISCUSSED; THE ADDRESS DESCRIBED THE PROBLEM, RECLAMATION PROJECTS, AND DESTRATIFICATION METHODS. LAND USE AND PLANNING CONSIDERATIONS IN CONTROLLING EUTROPHICATION WERE DESCRIBED, WITH EMPHASIS UPON AN INTEGRATED ZONING, PLANNED UNIT DEVELOPMENT, AND SANITATION PROGRAM TO PREVENT EUTROPHICATION. A HIGHLY TECHNICAL ADDRESS CONCERNING LAKE FLUSHING RATES FOR WATER QUALITY CONTROL WAS ALSO PRESENTED. METHODS OF SAMPLING LAKE WATER QUALITY WERE DISCUSSED, ALONG WITH THE FUNCTION OF THE MAINE INLAND FISH AND GAME DEPARTMENT IN WATER QUALITY CONTROL. THE CONCLUSION ADDRESSES THE ROLE OF LAW IN MAINTAINING LAKE WATER QUALITY. (SEE ALSO W72-04280)
(HART-FLORIDA)

FIELD 05C, 06E, 05B, 06F, 05G

ACCESSION NO. W72-04279

EUTROPHICATION - THE PROBLEM - DETECTION - EXTENT,

MAINE UNIV., ORONO. WATER RESOURCES RESEARCH CENTER.

MILLARD W. HALL.

PROCEEDINGS, WORKSHOP--CONFERENCE ON RECLAMATION OF MAINE'S DYING LAKES, HELD AT UNIVERSITY OF MAINE, BANGOR, MARCH 24-25, 1971. P 5-13.

DESCRIPTORS:
*EUTROPHICATION, *LAKES, *MAINE, RECREATION, WATER SUPPLY, WATER POLLUTION EFFECTS, WATER POLLUTION SOURCES, ALGAE, NUTRIENTS.

IDENTIFIERS:
*LAKE ANNABESSACOOK(ME), *KENNEBEC RIVER(ME), CULTURAL EUTROPHICATION.

ABSTRACT:
NATURAL AND CULTURAL EUTROPHICATION ARE DEFINED AND STRESS IS PLACED ON THE LIMITED POSSIBILITIES AVAILABLE TO MAN IN HIS EFFORTS TO REDUCE THE ACCUMULATION OF NUTRIENTS IN WATER BASINS AND TO LIQUIDATE THE RESULTING BIOLOGICAL 'IMBALANCE.' PARTICULAR EMPHASIS IS PLACED ON THE FUTILITY OF ALGAL CONTROL BY COPPER SULFATE AND OTHER CHEMICALS. DIVERSION OF WASTEWATERS MAY PREVENT THE CULTURAL EUTROPHICATION OF CERTAIN LAKES, E.G., LAKE ANNABESSACOOK, BUT ONLY AT THE COST OF POLLUTING A RIVER. THE USE OF INOCULA OF PARASITIC VIRUSES MAY ACHIEVE CONTROL OF BLUE-GREEN ALGAE. (SEE ALSO W72-04279) (WILDE-WISCONSIN)

FIELD 05C

ACCESSION NO. W72-04280

ADENOSINE TRIPHOSPHATE IN LAKE SEDIMENTS,

WISCONSIN UNIV., MADISON. DEPT. OF SOIL SCIENCE.

C. C.LEE.

WISCONSIN WATER RESOURCES CENTER, MADISON, PH.D. THESIS, 1971. 98 P, 5 FIG, 14 TAB, 111 REF. OWRR B-022-WIS(7).

DESCRIPTORS:
*SOIL BACTERIA, *LAKE SOILS, *PHOSPHORUS, *EUTROPHICATION, LAKES, AQUATIC ALGAE, BACTERIA, BIOCHEMISTRY, WISCONSIN.

IDENTIFIERS:
*ADENOSINE TRIPHOSPHATE, *MICROBIAL BIOMASS, *GROWTH STAGE, *DETERMINATION, *ORIGIN, *SIGNIFICANCE, *EXTRACTION, LUCIFERIN-LUCIFERASE METHOD, SEDIMENT-WATER INTERACTIONS, BIOLUMINESCENCE, SELENASTRUM CAPRICONUTUM, AEROBACTER AEROGENES.

ABSTRACT:
A MODIFICATION OF THE LUCIFERIN-LUCIFERASE BIOLUMINESCENCE TECHNIQUE WAS DEVELOPED FOR DETERMINING ADENOSINE TRIPHOSPHATE (ATP) IN SEDIMENTS. ATP RECOVERY RANGED FROM 20 TO 85%, WAS A CHARACTERISTIC, REPRODUCIBLE PROPERTY OF A GIVEN SEDIMENT, BUT WAS NOT RELATED CONSISTENTLY TO ANY OTHER SEDIMENT PROPERTY. GENERALLY, THE DETECTION LIMITS OF THE METHOD WERE ABOUT 0.05 MICROGRAM ATP/G OVEN DRY SEDIMENT. THE ATP CONTENTS OF NINE SEDIMENT SAMPLES OBTAINED FROM DIFFERENT LAKES IN WISCONSIN RANGED FROM 0.34 TO 9.5 MICROGRAM ATP/G SEDIMENT. THE ATP CONTENT OF SELENASTRUM CAPRICONUTUM WAS EVALUATED IN RELATION TO GROWTH STAGE, SUBSTRATE P STATUS, AND THE APPLICABILITY OF ATP AS AN INDEX OF LIVING ALGAL CELL MATERIAL IN BIOASSAY SYSTEMS. GENERALLY, THE ATP CONTENT PER VIABLE CELL IN PAAP MEDIUM WAS MAINTAINED RELATIVELY CONSTANT WITHIN A 4 TO 6.5 X 10 TO THE MINUS 8TH POWER MICROGRAM/CELL RANGE OVER A 32-DAY INCUBATION PERIOD. VIABLE CELLS COMPRISED ONLY 40 TO 50% OF THE TOTAL CELLS FROM 8 TO 32 DAYS. ATP PER LIVING BIOMASS WAS 3 MICROGRAM/MG IN PAAP MEDIUM. IN P-DEFICIENT PAAP MEDIUM THE ATP CONTENT OF S. CAPRICONUTUM DECLINED TO ABOUT 2 X 10 TO THE MINUS 8TH POWER MICROGRAM/VIABLE CELL AND 1.5 MICROGRAM/MG LIVING BIOMASS. NEVERTHELESS, PRECISE CONVERSION OF ATP TO LIVING ALGAL CELL MATERIAL NECESSITATES CONSIDERATION OF THE EFFECT OF NUTRIENT, PARTICULARLY P, DEFICIENCIES ON THE ALGAL ATP POOL.

FIELD 05C, 02H

ACCESSION NO. W72-04292

ENZYMATIC PATTERNS IN TWO DENITRIFYING MICROBIAL POPULATIONS,

NATIONAL INST. FOR WATER RESEARCH, PRETORIA (SOUTH AFRICA).

P. J. DUTOIT, D. F. TOERIEN, AND T. R. DAVIES.

WATER RESEARCH, VOL 4, 1970, P 149-156. 1 FIG, 4 TAB, 23 REF.

DESCRIPTORS:
*WASTE WATER TREATMENT, *DENITRIFICATION, WATER REUSE, *NITROGEN, PHOSPHORUS, CARBON, *SEWAGE BACTERIA, SEWAGE, ALGAE, HYDROGEN ION CONCENTRATION, CARBON DIOXIDE, NITRATES, NITRITES, *ENZYMES.

IDENTIFIERS:
*GLUCOSE, *MALATE, CITRIC ACID CYCLE, CYTOCHROMES, PHOSPHORYLATION, CLYCOLYSIS.

ABSTRACT:
IN VIEW OF THE INCREASED ATTENTION GIVEN TO DENITRIFICATION AS A PROCESS IN WASTEWATER RENOVATION, IT WAS NECESSARY ALSO TO STUDY THE MICROBIOLOGY AND BIOCHEMISTRY OF THE MICROBIAL POPULATIONS ACTIVE IN THIS PROCESS. THE DENITRIFYING BACTERIA WERE OBTAINED FROM A LABORATORY-SCALE, DENITRIFYING UNIT RECEIVING THE SUPERNATANT FLUID OF SETTLED DOMESTIC SEWAGE. TO ENSURE THAT AN ACTIVELY DENITRIFYING POPULATION WOULD BE USED, THE ORIGINAL POPULATION WAS ENRICHED IN A SUITABLE MEDIUM. GLUCOSE AND MALATE WERE DEGRADED BY THE ENRICHED DENITRIFYING CULTURES WITH AN ACCOMPANYING DECREASE IN NITRATE CONTENT. NITRATE ACCUMULATED TOWARDS THE END OF THE EXPERIMENT BUT CONCENTRATIONS WERE LOW COMPARED TO THE AMOUNT OF NITRATE REMOVED IN BOTH CULTURES. BOTH CULTURES WERE STARTED FROM A SMALL INOCULUM OF ONLY 1 ML IN 6.1. WITHIN 36 HOURS THE CULTURES WERE TURBIC AND GROWTH WAS PRONOUNCED AFTER 80 HOURS AT 30 DEG C. CONCLUSIONS WERE: (1) DENITRIFICATION TOOK PLACE WITH EITHER GLUCOSE OR MALATE AS THE SOURCE OF CARBON, (2) THE INTERMEDIARY METABOLIC PATTERNS OF THE TWO CULTURES WERE DIFFERENT, (3) GLUCOSE DEGRADATION OCCURRED VIA GLYCOLYSIS, THE PENTOSE PHOSPHATE SHUNT AND THE CITRIC ACID CYCLE, (4) DEGRADATION OF MALATE OCCURRED MAINLY THROUGH THE CITRIC ACID CYCLE, (5) CYTOCHROMES WERE MORE ABUNDANT IN THE MALATE CULTURES THAN IN THE GLUCOSE CULTURE. SUBSTRATE PHOSPHORYLATIONS AS WELL AS OXIDATIVE PHOSPHORYLATION APPEARED TO OCCUR IN THE GLUCOSE CULTURE, BUT ONLY OXIDATIVE PHOSPHORYLATION WAS DETECTED IN THE MALATE CULTURES. (BIGGS-TEXAS)

FIELD 05D

ACCESSION NO. W72-04309

CAN BOD CONTRIBUTE TO ALGAL MASS,

WASHINGTON STATE UNIV., PULLMAN.

NING-HSI TANG, AND SURINDER K. BHAGAT.

WATER AND SEWAGE WORKS, VOL 118, NO 12, P 396-401, DECEMBER 1971. 9 FIG, 3 TAB, 4 REF.

DESCRIPTORS:
*ALGAE, *BIOCHEMICAL OXYGEN DEMAND, *EUTROPHICATION, PHOSPHOROUS, NITROGEN, CHEMICAL OXYGEN DEMAND, LABORATORY TESTS, ALKALINITY, CARBONATES, BICARBONATES, CARBON DIOXIDE, HYDROGEN ION CONCENTRATION, WATER QUALITY CONTROL, WASTE WATER TREATMENT, CHLOROPHYTA, GROWTH RATES.

IDENTIFIERS:
*ORGANIC CARBON, *ALGAL MASS, ALGAL GROWTH.

ABSTRACT:
THE EFFECT OF ORGANIC CARBON ON THE GROWTH OF ALGAE WAS DEMONSTRATED USING GLUCOSE AS A CARBON SOURCE AND COD AS A MEASURE OF CARBON PRESENT. ALGAE DERIVE CARBON FROM THREE SOURCES, CO2 FROM THE ATMOSPHERE, FROM CARBONATES AND BICARBONATES, AND FROM ORGANIC MATERIAL IN THE WATER. ON A SEMICONTINUOUS FLOW BASIS, ONCE THE ALGAL-BACTERIAL SYMBIOSIS WAS ESTABLISHED, MORE ALGAE GREW IN THE FLASKS WITH ORGANIC CARBON THAN GREW IN THE FLASKS WITHOUT ORGANIC CARBON, EVEN THOUGH ALL FLASKS WERE FED WITH EQUAL, ADEQUATE AMOUNTS OF N AND P. RESULTS INDICATE THAT THE AMOUNT OF EXCESSIVE ALGAL GROWTH IN AN AQUATIC SYSTEM WILL BE ABOUT THE SAME AS THE AMOUNT OF INFLUENT BOD FOR A GIVEN WASTEWATER INFLUENT BOD OF LESS THAN 50 MG/L. FOR AN INFLUENT WASTEWATER BOD IN EXCESS OF 50 MG/L THE AMOUNT OF EXCESSIVE ALGAL GROWTH WILL BE SOMEWHAT LESS THAN THE INCOMING BOD. A GRAPH WAS CONSTRUCTED TO SERVE AS A GUIDE FOR PREDICTING THE AMOUNT OF EXCESSIVE ALGAL GROWTH FOR A GIVEN EFFLUENT BOD. (LOWRY-TEXAS)

FIELD 05C

ACCESSION NO. W72-04431

EVALUATION OF ALGAL ASSAY PROCEDURES--PAAP BATCH TEST,

CALIFORNIA UNIV., IRVINE. SCHOOL OF BIOLOGICAL SCIENCES; AND CALIFORNIA UNIV., IRVINE. SCHOOL OF ENGINEERING.

STEVEN MURRAY, JAN SCHERFIG, AND PETER S. DIXON.

JOURNAL WATER POLLUTION CONTROL FEDERATION, VOL 43, NO 10, P 1991-2003, OCTOBER 1971. 4 FIG, 13 TAB, 3 REF.

DESCRIPTORS:
*BIOASSAY, *LABORATORY TESTS, RESEARCH AND DEVELOPMENT, TRACE ELEMENTS, ALGAE, NUTRIENTS, LIGHT INTENSITY, HYDROGEN ION CONCENTRATION, CARBON DIOXIDE, AERATION, MIXING, *EUTROPHICATION, *ANALYTICAL TECHNIQUES, *ALGAL CONTROL.

IDENTIFIERS:
SELENASTRUM CAPRICORNUTUM.

ABSTRACT:
PRELIMINARY INVESTIGATIONS HAVE DEMONSTRATED FOUR IMPORTANT DIFFICULTIES INVOLVED IN THE PROVISIONAL ALGAL ASSAY PROCEDURE (PAAP) BATCH TEST. THE FOUR DIFFICULTIES WERE: (1) THE EFFECTS OF MEDIUM PREPARATION ON THE FINAL COMPOSITION OF THE MEDIUM; (2) PH CHANGES DURING THE TEST PERIOD IN RELATION TO CARBON DIOXIDE AVAILABILITY; (3) THE METHOD OF GAS ADDITION IN THE BATCH TEST UNITS; AND (4) THE EFFECTS OF LIGHT INTENSITY ON ALGAL GROWTH. BECAUSE OF THESE LIMITATIONS, CERTAIN MODIFICATIONS TO THE STANDARDIZED PROCEDURES WERE DEVELOPED. A NEW CULTURE MEDIA PREPARATION WHICH ELIMINATES THE PRECIPITATION OF ESSENTIAL TRACE ELEMENTS WAS DEVELOPED AND PROVIDES A RELIABLE REFERENCE MEDIUM FOR ALGAL ASSAY WORK. THE ADDITION OF CO2 -ENRICHED AIR REDUCED THE GROWTH-LIMITING EFFECTS OF CO2 AND STABILIZES THE PH AT 7.5 TO 8.0. AERATION WAS SHOWN TO BE MORE EFFECTIVE THAN VENTILATION FOR CO2 ADDITION AND GASEOUS AS A RESULT OF BICARBONATE UTILIZATION. LIGHT INTENSITIES OF 500 FT-C REDUCED GROWTH RATES OF SELENASTRUM CAPRICORNUTUM IN BOTTLE TEST EXPERIMENTATION COMPARED WITH 350 FT-C. (LOWRY-TEXAS)

FIELD 05C, 05G

ACCESSION NO. W72-04433

ECOLOGICAL ASPECTS OF PLUTONIUM DISSEMINATION IN AQUATIC ENVIRONMENTS; WHAT HAS PU DATA TO TELL US ABOUT OTHER TRANSURANICS,

WOODS HOLE OCEANOGRAPHIC INSTITUTION, MASS.

VICTOR E. NOSHKIN.

AVAILABLE FROM THE NATIONAL TECHNICAL INFORMATION SERVICE, SPRINGFIELD, VA. 22151 AS NYO-2174-132, $3.00 IN PAPER COPY, $0.95 IN MICROFICHE. REPORT NYO-2174-132, 1971. 31 P, 2 FIG, 8 TAB, 40 REF. AEC CONTRACT AT(30-1)-2174.

DESCRIPTORS:
*RADIOACTIVITY EFFECTS, *FALLOUT, *FOOD CHAINS, *AQUATIC LIFE, RADIOISOTOPES, ABSORPTION, NUCLEAR WASTES, WATER POLLUTION EFFECTS, MOLLUSKS, MARINE PLANTS, MARINE ALGAE, PUBLIC HEALTH, ESTUARINE ENVIRONMENT, ON-SITE INVESTIGATIONS, REVIEWS.

IDENTIFIERS:
PLUTONIUM.

ABSTRACT:
THIS REVIEW INDICATES THAT FALLOUT PLUTONIUM CONTRIBUTES MORE THAN STRONTIUM OR CESIUM TO THE ARTIFICIAL RADIATION EXPOSURE OF MANY MARINE SPECIES. ORGANISMS AT HIGHER TROPHIC LEVELS AND THOSE WHICH FEED ON SEDIMENTS HAVE MORE PLUTONIUM. BONE AND LIVER HAVE MUCH MORE THAN MUSCLE TISSUE. AS USE OF NUCLEAR POWER INCREASES, SUCH FOOD ITEMS AS ANCHOVIES, SARDINES, CANNED SALMON, ETC. WHICH ARE CONSUMED WHOLE, AND FISH PROTEIN CONCENTRATES WILL REPRESENT IMPORTANT TRANSFER VECTORS FOR PLUTONIUM. THE RELATIVELY UNIFORM PLUTONIUM UPTAKE BY THE MUSSEL (CONCENTRATION FACTOR OF 200-500) SUGGESTS USE AS A BIOLOGICAL INDICATOR. THE MUSSEL INDICATES THE PLUTONIUM MORE RECENTLY INTRODUCED INTO THE WATER IN SUSPENSION RATHER THAN THAT PRESENT IN SEDIMENTS. (BOPP-ORNL)

FIELD 05C

ACCESSION NO. W72-04461

INFLUENCE OF THE PHYSICO-CHEMICAL FORMS OF RADIONUCLIDES AND STABLE TRACE ELEMENTS IN SEAWATER IN RELATION TO UPTAKE BY THE MARINE BIOSPHERE,

BATTELLE-NORTHWEST, RICHLAND, WASH. PACIFIC NORTHWEST LAB.

D. E. ROBERTSON.

AVAILABLE FROM THE NATIONAL TECHNICAL INFORMATION SERVICE, SPRINGFIELD, VA. 22151, AS BNWL-SA-4048, $3.00 IN PAPER COPY, $0.95 IN MICROFICHE. BNWL-SA-4048, SEPTEMBER 8, 1971. 60 P, 105 REF. AEC CONTRACT NO. AT(45-1)-1830.

DESCRIPTORS:
*RADIOACTIVITY EFFECTS, *NUCLEAR WASTES, *FALLOUT, *OCEANS, WATER POLLUTION EFFECTS, FOOD CHAINES, PUBLIC HEALTH, ESTUARINE ENVIRONMENT, ZINC RADIOISOTOPES, PHYSICOCHEMICAL PROPERTIES, FISH, MARINE PLANTS, MARINE ALGAE, HYDROLYSIS, CHELATION, SEDIMENTS, SETTLING VELOCITY, RADIOISOTOPES, ABSORPTION, REVIEWS.

IDENTIFIERS:
PLUTONIUM RADIOISOTOPES, RUTHENIUM RADIOISOTOPES, IRON RADIOISOTOPES.

ABSTRACT:
THIS REVIEW PRESENTS EXAMPLES OF DISSIMILAR BIOLOGICAL UPTAKE BETWEEN IMPORTANT ARTIFICIAL RADIONUCLIDES (FE-55, ZN-65, AND RU-106) AND THEIR STABLE ISOTOPES IN THE MARINE ENVIRONMENT. ALTHOUGH DEFINITIVE WORK IS NEEDED, THERE IS EVIDENCE THAT ZN, FE, AND CU ARE PRESENT AS SOLUBLE METAL CHELATES. RU, FE, AND MN HYDROLYZE TO FORM POLYMERIC SPECIES. MOST PU IS IN PARTICULATE FORM. THE RELATIVELY HIGH UPTAKE OF PU BY SEAWEEDS AND CERTAIN CRUSTACEANS APPEARS TO BE A SURFACE ABSORPTION PROCESS. THE MAXIMUM PERMISSIBLE CONCENTRATION OF PU IN SEAWATER IS ABOUT 6 X 0.01 MICROCI/ML, BASED UPON A SINGLE INTAKE OF 200 GMS OF FISH PER INDIVIDUAL. (BOPP-ORNL)

FIELD 05C

ACCESSION NO. W72-04463

REMOVAL OF TRACE METALS FROM MARINE CULTURE MEDIA,

NATIONAL MARINE WATER QUALITY LAB., WEST KINGSTON, R.I.

EARL W. DAVEY, JOHN H. GENTILE, STANTON J. ERICKSON, AND PETER BETZER.

LIMNOLOGY AND OCEANOGRAPHY, VOL 15, NO 3, P 486-488, MAY 1970. 1 FIG, 5 REF.

DESCRIPTORS:
*CHELATION, CHROMATOGRAPHY, HEAVY METALS, *TRACE ELEMENTS, RADIOCHEMICAL ANALYSIS, RESINS, *SEA WATER, BIOASSAY, COPEPODS, OYSTERS, DINOFLAGELLATES, DIATOMS, ALGAE CULTURES, COPPER, TOXICITY.

IDENTIFIERS:
CHELEX 100, ARTEMIA SALINA, BIO-RAD, CYCLOTELLA NANA.

ABSTRACT:
THE SODIUM FORM OF PURIFIED CHELEX-100 CAN BE USED TO REMOVE TRACE METALS FROM ARTIFICIAL AND NATURAL SEAWATER WITHOUT ALTERING THE MAJOR CATION-ANION COMPOSITION OR CONTRIBUTING ORGANIC TOXICANTS OR CHELATORS. SPECIFIC CONTAMINANT TRACE METALS, OFTEN PRESENT IN ARTIFICIAL SEAWATER AT TOXIC LEVELS, ARE REDUCED TO LEVELS THAT PERMIT THE GROWTH OF SENSITIVE MARINE PHYTOPLANKTON AND THE DEVELOPMENT OF CERTAIN INVERTEBRATES FROM THE EGG THROUGH EARLY LARVAL STAGES. THE CONCENTRATIONS OF ESSENTIAL TRACE METALS ARE REDUCED BELOW THE NUTRITIONAL REQUIREMENTS OF SELECTED SPECIES OF MARINE PHYTOPLANKTON. (EPA ABSTRACT)

FIELD 05A

ACCESSION NO. W72-04708

STRATEGIES FOR CONTROL OF MAN-MADE EUTROPHICATION,

COMMITTEE ON PUBLIC WORKS (U.S. SENATE).

R. D. GRUNDY.

ENVIRONMENTAL SCIENCE AND TECHNOLOGY, VOL 5, NO 12, P 1184-1190, DECEMBER 1971. 6 FIG, 5 REF.

DESCRIPTORS:
*EUTROPHICATION, *PHOSPHATES, *WATER POLLUTION CONTROL, *WASTE WATER TREATMENT, *COST-BENEFIT ANALYSIS, DETERGENTS, PHOSPHORUS, NITROGEN, NUTRIENTS, SEWAGE, WATER TEMPERATURE, LIGHT PENETRATION, CARBON, LAKE ERIE, RUNOFF, EROSION, CHLORELLA, SODIUM COMPOUNDS, ALKALINITY, MINING, BIOCHEMICAL OXYGEN DEMAND, ECONOMICS, ALGAE, AGRICULTURAL CHEMICALS, LEGISLATION.

IDENTIFIERS:
POTOMAC RIVER, ALAFIA RIVER, CHLORELLA PYRENOIDOSA, SODIUM SILICATE, SODIUM METASILICATE, LAKE WASHINGTON, NITRILOTRIACETATE, HUMAN FECES.

ABSTRACT:
SOURCES OF PHOSPHATES, AND OTHER NUTRIENTS IN THE AQUATIC ECOSYSTEM INCLUDE NOT ONLY DETERGENTS, BUT ALSO SEWAGE, EROSION, AND AGRICULTURAL RUNOFF. ON A NATIONAL BASIS, DETERGENTS PROVIDE 30 TO 40 PERCENT OF ALL THE PHOSPHORUS ENTERING THE AQUATIC ENVIRONMENT. HOWEVER, THIS FIGURE IS SHOWN TO VARY ON A REGIONAL BASIS. BECAUSE THERE ARE MANY OTHER SOURCES OF PHOSPHATES, THEIR CONTROL IN DETERGENTS IS NOT IN ITSELF A SUFFICIENT STRATEGY TO CONTROL EUTROPHICATION. ALSO, ANY REGULATION ON PHOSPHATES IN DETERGENTS SHOULD NOT BE UNDERTAKEN WITHOUT CAREFUL CONSIDERATION OF THE PUBLIC HEALTH AND ENVIRONMENTAL IMPLICATIONS OF ALTERNATIVE FORMULATIONS. VALID CONTROL STRATEGIES SHOULD INCLUDE ADVANCED WASTE WATER TREATMENTS, DIVERSION, DILUTION, AND LAND DISPOSAL. WASTE WATER TREATMENT USING CHEMICAL PRECIPITATION PROCESSES, NOT ONLY REMOVES PHOSPHORUS BUT ALSO FACILITATES THE REMOVAL OF BOD, TOXICANTS, AND OTHER NUTRIENTS. ECONOMIC CONSIDERATIONS SHOW THAT 90 PERCENT OF MUNICIPAL WASTE WATER COULD BE TREATED FOR PHOSPHORUS REMOVAL AT LESS COST TO THE CONSUMER THAN THE INCREASED PRODUCT COSTS FOR DETERGENT PHOSPHATE SUBSTITUTES. IT IS ALSO SIGNIFICANT THAT TREATMENT REDUCES TOTAL PHOSPHATE LEVELS WHILE PRODUCT CONTROLS AFFECT DETERGENT PHOSPHATE ALONE. (MORTLAND-BATTELLE)

FIELD 05D, 05G, 05C

ACCESSION NO. W72-04734

DISTRIBUTION OF BLUE-GREEN ALGAL VIRUSES IN VARIOUS TYPES OF NATURAL WATERS,

DELAWARE UNIV., NEWARK. DEPT. OF BIOLOGICAL SCIENCES.

M. S. SHANE.

WATER RESEARCH, VOL 5, NO 9, P 711-716, SEPTEMBER 1971. 1 FIG, 3 TAB, 10 REF.

DESCRIPTORS:
*VIRUSES, *CYANOPHYTA, *DISTRIBUTION PATTERNS, *WATER ANALYSIS, *BIOINDICATORS, ALGAE, RIVERS, STREAMS, DELAWARE RIVER, OXIDATION LAGOONS, PONDS, FARM PONDS, MARYLAND, QUARRIES, OHIO RIVER, HUDSON RIVER, INDICATORS, DELAWARE, PATH OF POLLUTANTS, HOSTS, PENNSYLVANIA, NEW YORK, NEW HAMPSHIRE, OHIO, MICHIGAN.

IDENTIFIERS:
*LPP VIRUS, *BLUE-GREEN ALGAL VIRUSES, *PLECTONEMA BORYANIUM, SUSQUEHANNA RIVER, BRANDYWINE RIVER, RED CLAY RIVER, WHITE CLAY RIVER, CHRISTINA RIVER, BIG ELK RIVER, LITTLE ELK RIVER, OCTORARA RIVER, PLECTONEMA, LYNGBYA, PHORMIDIUM, CYANOPHYCEAE, OSCILLATORIACEAE, SCHIZOTHRIX CALCICOLA, ELKTON POND, RISING SUN POND, INDUSTRIAL STORAGE TANKS.

ABSTRACT:
A SURVEY WAS CONDUCTED IN THE DELAWARE-MARYLAND AREA TO ASCERTAIN THE DISTRIBUTION OF LPP BLUE-GREEN ALGAL VIRUSES IN ALL TYPES OF NATURAL WATERS. THEIR PRESENCE WAS DETERMINED ON THE BASIS OF THE ABILITY OF SAMPLES TO PRODUCE LYSIS IN CULTURES OF THE ALGAL TEST ORGANISM, PLECTONEMA BORYANUM (UI581), AND TO FORM PLAQUES ON THE HOST ALGAL LAWN USING THE SOFT-AGAR TECHNIQUE OF SAFFERMAN AND MORRIS. THE VIRAL CONCENTRATION WAS DETERMINED BY THE DIRECT COUNT OF THE NUMBER OF PLAQUES FORMED PER ML OF WATER SAMPLE. THE VIRAL STRAIN, LPP-1, WAS DETERMINED FOR SELECTED VIRAL CULTURES ISOLATED FROM THE CHRISTINA, RED CLAY, AND WHITE CLAY RIVERS AND THE ELKTON AND RISING SUN OXIDATION PONDS. LPP-VIRUSES EXISTED IN EVERY TYPE OF WATER. THE OXIDATION PONDS GAVE THE HIGHEST PERCENTAGE OF SAMPLES CONTAINING AT LEAST ONE VIRUS AND THE HIGHEST CONCENTRATION OF VIRUSES. ALL RESERVOIRS AND INDUSTRIAL STORAGE TANKS WHICH CONTAINED VIRUSES HAD INFLUENTS WHICH ALSO CONTAINED VIRUSES. TEST DATA ON THE RIVERS SHOWED THESE VIRUSES PRESENT IN MOST STREAMS. THE VIRUSES WERE CONSISTENTLY ABSENT FROM THE HEADWATERS OF THE THREE RIVERS LISTED ABOVE BUT INCREASED AS THESE RIVERS FLOWED THROUGH MORE POPULATED AREAS. THE HEADWATERS FOR ALL THREE ARE IN LOW POPULATION DENSITY AND RURAL ENVIRONMENTS. IT IS SUGGESTED THAT THE ALGAL HOSTS OF THESE VIRUSES THRIVE IN AREAS OF HIGH ORGANIC MATERIAL SUCH AS MAY BE ASSOCIATED WITH POLLUTION. TABLES SHOW THE DISTRIBUTION AND INCIDENCE OF THE LPP-VIRUSES IN THE WATERS. (HOLOMAN-BATTELLE)¬28(U)

FIELD 05B, 05C

ACCESSION NO. W72-04736

EARTH RESOURCE TECHNOLOGY USED IN POLLUTION DETECTION,

B. M. ELSON.

AVIATION WEEK AND SPACE TECHNOLOGY, VOL 95, NO 24, P 46-49, DECEMBER 13, 1971.

DESCRIPTORS:
*REMOTE SENSING, *WATER POLLUTION CONTROL, *SEA WATER, *OILS, *NUTRIENTS, AIRCRAFT, SATELLITES(ARTIFICIAL), PHOSPHORUS, NITROGEN, HEAVY METALS, ORGANOPHOSPHORUS PESTICIDES, INDUSTRIAL WASTES, VIRUSES, ENERIC BACTERIA, TRITIUM, RADIOACTIVE WASTES, MONITORING, PHOTOGRAPHY, ULTRAVIOLET RADIATION, FLUORESCENCE, INFRARED RADIATION, RADAR, CHLOROPHYLL, WATER PROPERTIES, ALGAE.

IDENTIFIERS:
SPACE PLATFORMS, LEAD, MERCURY, LASERS, MICROWAVE RADIOMETRY, VIDEO, SCANNERS.

ABSTRACT:
VARIOUS REMOTE SENSING DEVICES, CARRIED IN AIRCRAFT OR SATELLITES, ARE BEING DEVELOPED TO MONITOR OCEAN POLLUTANTS. PRIMARY EMPHASIS IN THIS PAPER IS ON THE DETECTION, IDENTIFICATION,MAPPING, AND VOLUMETRIC MEASUREMENT OF OIL SPILLS. ONE MEANS OF DETECTING SOME KINDS OF OILS ON WATER IS PHOTOGRAPHY IN THE BLUE AND ULTRAVIOLET REGIONS. FLUORESCENCE UNDER ULTRAVIOLET LIGHT IS ANOTHER MEANS. OIL SPILL BOUNDARIES HAVE BEEN ACCURATELY DEFINED USING INFRARED SCANNERS, BUT CERTAIN ANOMALIES ARE ALSO OCCURRING WHICH LIMIT THE USEFULNESS OF THIS TECHNIQUE. MICROWAVE RADIOMETRY WAS USED TO MEASURE OIL SPILLS OFF THE SOUTHERN CALIFORNIA COAST. THE NAVY HAS USED SYNTHETIC APERTURE RADAR TO PERFORM WIDE-AREA SURVEILLANCE OVER OIL SPILLS. MANY POLLUTANTS OTHER THAN OIL ALSO CAN BE REMOTELY DETECTED. A MULTISPECTRAL RADIOMETRIC SCANNER CAN BE USED TO DETERMINE OPTICAL AND THERMAL PROPERTIES OF WATER AND TO MONITOR ALGAL GROWTHS AND SHORELINE VEGETATION. AN AIRBORNE SYSTEM FOR DETERMINING CHLOROPHYLL CONTENT OF WATER HAS BEEN DEVELOPED. THIS TECHNIQUE UTILIZES A DIFFERENTIAL RADIOMETER THAT MEASURES CHANGES IN SPECTRAL RADIANCE BETWEEN TWO OR MORE WAVELENGTH REGIONS OF THE SPECTRUM. (MORTLAND-BATTELLE)

FIELD 05A, 07B

ACCESSION NO. W72-04742

ISOLATION AND COUNTING OF ATHIORHODACEAE WITH MEMBRANE FILTERS,

PENNSYLVANIA STATE UNIV., UNIVERSITY PARK. DEPT. OF MICROBIOLOGY.

W. C. SWOAGER, AND E. S. LINSTROM.

APPLIED MICROBIOLOGY, VOL 22, NO 4, P 683-687, OCTOBER 1971. 3 TAB, 13 REF.

DESCRIPTORS:
*CULTURES, *ANAEROBIC BACTERIA, *ISOLATION, *MEMBRANE PROCESSES, BACTERIA, SULFUR BACTERIA, PHOTOSYNTHETIC BACTERIA, AQUATIC BACTERIA, LAKES, FILTERS, COLIFORMS, METABOLISM, CARBON, SULFUR, SAMPLING, WATER ANALYSIS, ALGAE, SPECTROPHOTOMETRY, PENNSYLVANIA, INHIBITORS, ECOLOGY, ECOTYPES, FRESH WATER.

IDENTIFIERS:
*ATHIORHODACEAE, WHIPPLE DAM, STONE VALLEY DAM(PA.), ATRAZINE, BIOTYPES, POUR PLATE METHOD, MOST PROBABLE-NUMBER TEST, MEMBRANE FILTRATION, RHODOPSEUDOMONAS GELATINOSA, THIORHODACEAE, CHLOROBACTERIACEAE.

ABSTRACT:
A MEMBRANE FILTER TECHNIQUE WAS DEVELOPED TO GROW, ISOLATE, AND ENUMERATE ATHIROHODACEAE FROM AQUATIC ECOSYSTEMS. WATER SAMPLES WERE TAKEN FROM THE SPILLWAYS OF TWO CENTRAL PENNSYLVANIA ARTIFICIAL LAKES DURING SPRING, SUMMER, AND FALL OF 1969. UPPER LAYER AND DEPTH SAMPLES WERE TAKEN AT 1-M INTERVALS FROM STONE VALLEY DAM IN MAY, JUNE, AND JULY 1970. THE SAMPLES WERE FILTERED THROUGH A 47-MM HA MEMBRANE FILTER AND THE FILTRATE CULTURED ON PERFORATED PETRI DISHES IN A CARBON-BASED MEDIUM TO PROMOTE GROWTH OF ATHIORHODACEAE. ATRAZINE WAS ADDED TO THE BASAL CULTURE MEDIUM TO INHIBIT THE GROWTH OF ALGAE. ABSORPTION SPECTRA WERE PREPARED USING A BECKMAN DK-2A SPECTROPHOTOMETER. PURE CULTURES WERE ISOLATED BY REPEATED STREAKING ON PLATES INCUBATED ANAEROBICALLY IN 200-400 FT-C OF INCANDESCENT LIGHT. IDENTIFICATION OF THE ISOLATES AS ATHIORHODACEAE WAS CORROBORATED BY THE FACT THAT THEY USED NO SULFIDE OR THIOSULFATE IN CULTURE MEDIA AND ALL (40) HAD SPECTRAL CHARACTERISTICS OF ATHIORHODACEAE. QUANTITATIVE DATA AND A WIDE RANGE OF BIOTYPES CAN BE OBTAINED BY THE CULTURING METHOD DESCRIBED. ACCURACY WAS VALIDATED BY PARALLEL COUNTS USING THE POUR PLATE AND MOST PROBABLE NUMBERS METHODS. BY USING A SULFUR-BASED MEDIUM AND MORE ANAEROBIC CONDITIONS, THE METHOD COULD PROBABLY BE USED TO COUNT AND ISOLATE SULFUR PHOTOSYNTHETIC BACTERIA SUCH AS CHLOROBACTERIACEAE AND THIROHADACEAE. (JEFFERIS-BATTELLE)

FIELD 05A

ACCESSION NO. W72-04743

CULTURAL ESTIMATION OF YEASTS ON SEAWEEDS,

RHODE ISLAND UNIV., KINGSTON. NARRAGANSETT MARINE LAB.

R. SESHADRI, AND J. M. SIEBURTH.

APPLIED MICROBIOLOGY, VOL 22, NO 4, P 507-512, OCTOBER 1971. 6 TAB, 31 FIG.

DESCRIPTORS:
*YEASTS, *CULTURES, *MARINE ALGAE, RHODOPHYTA, PHAEOPHYTA, CHLOROPHYTA, INHIBITORS, ANTIBIOTICS, SEA WATER, RHODE ISLAND.

IDENTIFIERS:
*AGARS, EPIPHYTES, CANDIDA PARAPSILOSIS, CHONDRUS CRISPUS, RHODOTORULA GLUTINIS, SARGASSUM NATANS, ASCOPHYLLUM NODOSUM, FUCUS VESICULOSUS, LAMINARIA DIGITATA, LAMINARIA AGARDHII, ULVA LACTUCA, RHODYMENIS PALMATA, POLYSIPHONIA LANOSA, POLYSIPHONIA HARVEYI, RHODOTORULA RUBRA, RHODOTORULA LACTOSA, CANDIDA ZEYLANOIDES, CAMP VARNUM BEACH.

ABSTRACT:
FIVE PERCENT SUSPENSIONS OF FRESHLY HARVESTED SEAWEEDS WERE USED AS AN INOCULUM TO DEVELOP A SELECTIVE MEDIUM FOR EPIPHYTIC YEASTS. CONDITIONS FOR SATISFACTORY YEAST GROWTH AND VISUALIZATION AS RED COLONIES ON MEMBRANE FILTERS WERE OBTAINED BY SUPPLEMENTING A BASAL GLUCOSE-TRYPTICASE-YEAST EXTRACT-AGAR AT PH 7.0 WITH 100 MG EACH OF CHLORAMPHENICOL AND 2,3,5-TRIPHENYL TETRAZOLIUM CHLORIDE PER LITER. MAXIMAL COUNTS WERE OBTAINED BY TRITURATING THE ALGAE IN PRECHILLED (4 C) SEAWATER WITH A BLENDER FOR 2 TO 5 MIN. INHIBITORY PHENOLIC MATERIALS RELEASED FROM PHAEOPHYTES DURING THIS PROCESS WERE REMOVED WITH A MODIFIED CHOLODNY FILTRATION. A PRELIMINARY SURVEY INDICATED THAT YEASTS WERE EPIPHYTIC ON ALL NINE SPECIES OF SEAWEEDS AND THAT MAXIMAL POPULATIONS OCCURRED ON THE CHLOROPHYTES AND RHODOPHYTES ESPECIALLY DURING THE PERIODS OF WARMER WATER. (HOLOMAN-BATTELLE)

FIELD 05A

ACCESSION NO. W72-04748

EUTROPHICATION OF SMALL RESERVOIRS IN THE GREAT PLAINS,

NEBRASKA UNIV., LINCOLN. DEPT. OF CIVIL ENGINEERING; AND NEBRASKA UNIV., LINCOLN. DEPT. OF ZOOLOGY.

MARK J. HAMMER, AND GARY L. HERGENRADER.

NEBRASKA ENGINEER, JUNE 1971. 5 P, 3 FIG, 1 TAB. OWRR A-014-NEB(4).

DESCRIPTORS:
*EUTROPHICATION, *RESERVOIRS, *GREAT PLAINS, NUTRIENTS, RECREATION, LIGHT PENETRATION, NEBRASKA, WASTEWATER TREATMENT, TURBIDITY, RUNOFF, PHOSPHORUS, ALGAE, AQUATIC PLANTS, CLAYS, ODOR, WATER QUALITY, FOOD CHAINS, DISSOLVED OXYGEN, COPPER SULFATE, HERBICIDES, WATER POLLUTION EFFECTS, WATER POLLUTION SOURCES, WATER POLLUTION CONTROL.

IDENTIFIERS:
SALT VALLEY RESERVOIRS(NEB), MADISON(WIS), LAKE WASHINGTON(WASH), DYEING WATER.

ABSTRACT:
RESERVOIRS BUILT WITHIN A 20-MILE RADIUS OF LINCOLN, NEBRASKA, FOR FLOOD CONTROL AND SOIL CONSERVATION WITH RECREATION AS A SECONDARY BENEFIT HAVE DETERIORATED DUE TO EUTROPHICATION. THE RUNOFF WATERS ENTERING ARE PRINCIPALLY FROM CULTIVATED FARMLAND AND CONTAIN SUFFICIENT NUTRIENT SALTS TO SUPPORT ABUNDANT GROWTHS OF AQUATIC PLANTS. THERE IS NO READY SOLUTION FOR REMOVAL OF NUTRIENTS FROM LAND RUNOFF. IT IS DOUBTFUL THAT SOIL AND WATER CONSERVATION PRACTICES CAN REDUCE THE NUTRIENT LEVELS IN RUNOFF SUFFICIENTLY TO PREVENT EUTROPHICATION. THE RAINFALL-RUNOFF PATTERNS OF TYPICAL NEBRASKA WEATHER MAKE NUTRIENT CONTROL IN DRAINAGE WATER IMPRACTICAL. USE OF COPPER SULFATE FOR ALGAL BLOOMS HAS BEEN ABANDONED DUE TO ITS TOXICITY IN BOTTOM MUDS TO AQUATIC LIFE; HERBICIDES MUST BE USED WITH CAUTION TO PREVENT UNWANTED BIOLOGICAL DAMAGE. PLANT GROWTH HARVESTING IN MOST SITUATIONS CAN REMOVE ONLY SMALL QUANTITIES OF NUTRIENTS. A NEW AND NOVEL APPROACH TO CONTROLLING WEED AND ALGAE GROWTH IS BEING EVALUATED--INHIBITING SUNLIGHT PENETRATION THROUGH APPLICATION OF CHEMICAL DYES INTO THE WATER OR ONTO THE SURFACE. COMMERCIALLY AVAILABLE DYES ARE BEING TESTED WITH CULTURES OF BLUE-GREEN ALGAE, COMMONLY BLOOMING IN THE SALT VALLEY RESERVOIRS. (JONES-WISCONSIN)

FIELD 05C, 02H

ACCESSION NO. W72-04759

THE DISTRIBUTION AND NET PRODUCTIVITY OF SUBLITTORAL POPULATICNS OF ATTACHED MACROPHYTIC ALGAE IN AN ESTUARY ON THE ATLANTIC COAST OF SPAIN,

DURHAM UNIV. (ENGLAND). DEPT. OF BOTANY.

D. M. JOHN.

MARINE BIOLOGY, VOL 11, NO 1, P 90-97, 1971. 3 FIG, 3 TAB, 39 REF.

DESCRIPTORS:
*AQUATIC PLANTS, *DISTRIBUTICN, *MARINE ALGAE, *ESTUARIES, PRODUCTIVITY, POPULATICN, ATLANTIC OCEAN, DEPTH, STANDING CROP, CURRENTS(WATER), WAVES(WATER), SAMPLING, BENTHOS, GRAZING, GASTROPODS, LITTORAL, AGE, BIOMASS, LIGHT QUALITY, LIGHT PENETRATION, EFFICIENCES, COMPETITION, GROWTH RATES, PLANT MCRPHOLCGY, WATER CIRCULATION.

IDENTIFIERS:
*SUBLITTORAL BENTHIC MACROPHYTES, RIA CE ALDAN(SPAIN), LAMINARIA OCHROLEUCA, SACCORHIZA POLYSCHIDES.

ABSTRACT:
DISTRIBUTION OF SUBLITTORAL POPULATICNS OF LAMINARIA OCHROLEUCA ANC SACCORHIZA POLYSCHIDES WAS STUDIED AT SEVEN LOCALITIES ON THE NORTHWEST COAST OF SPAIN. POSSIBLE FACTORS CONTRCLLING THE EQUILIBRIUM BETWEEN THE TWO POPULATIONS WERE ASSESSEC AND THE NET PRODUCTIVITY OF THESE POPULATIONS MEASURED BY IN SITU CROPPING USING SCUBA DIVING TECHNIQUES. A NARROW TRANSITION ZCNE BETWEEN THE TWO POPULATICNS WAS FOUND AND THE DEPTH OF THIS ZONE WAS GOVERNED BY AMOUNT OF WATER MOVEMENT. TOTAL STANDING CROP AND PRCDUCTIVITY PER UNIT AREA DECREASED WITH BOTH AN INCREASE IN DEPTH AND WAVE ACTION WHILE THE HIGHEST VALUES OF ALL WERE IN TWO LOCALITIES WHERE THERE WAS CONSIDERABLE CURRENT SURGE. THE MAXIMUM PRODUCTIVITY OF L OCHROLEUCA WAS FOUND IN THE MOST SHELTERED LOCALITY, WHILE THE HIGHEST VALUE FOR S POLYSCHIDES WAS FOUND WHERE CURRENT SURGE WAS GREATEST BUT WAVE ACTION ONLY MCDERATE. THE NET PERCENTAGE EFFICIENCY OF ENERGY FIXATION SHOWS A NEARLY LINEAR RELATIONSHIP WITH DEPTH WHEN BASED ON SURFACE RACIATION. WHEN EFFICIENCY IS BASED ON THE RADIATION REACHING EACH DEPTH, THERE IS A FALL-OFF ABOVE AND BELOW 8.4 METERS; AN INVERSE RELATIONSHIP EXISTS BETWEEN PRODUCTIVITY AND EFFICIENCY DOWN TO THIS DEPTH BUT NOT BELOW. (JONES-WISCONSIN)

FIELD 05C, 02L

ACCESSION NO. W72-04766

ABCS OF CULTURAL EUTROPHICATION AND ITS CONTROL. PART I--CULTURAL CHANGES,

METCALF AND EDDY, INC., BOSTON, MASS.

CLAIR N. SAWYER.

WATER AND SEWAGE WORKS, P 278-281, SEPTEMBER 1971. 5 FIG, 2 TAB, 2 REF.

DESCRIPTORS:
*EUTROPHICATION, *CARBON, *WATER POLLUTION CONTROL, WATER POLLUTION EFFECTS, NUTRIENTS, ALGAE, CYANOPHYTA, PHOSPHORUS, WATER POLLUTION SOURCES, HYDROGEN ION CONCENTRATION, CARBON DIOXIDE, NITROGEN, CHLOROPHYTA, NITROGEN FIXATION, LAND MANAGEMENT, FERTILIZATION, PRIMARY PRODUCTIVITY.

ABSTRACT:
DRASTIC CHANGES IN NUTRIENT INPUT MUST BE ACCOMPLISHED IN ORDER TO CONTROL CULTURAL EUTROPHICATION. ALTHOUGH INCICATIONS ARE THAT GREEN ALGAE AND MANY BLUE-GREEN ALGAE ARE DEPENDENT UPON FIXED FORMS OF NITROGEN, AT LEAST FOUR GENERA OF BLUE-GREENS (ANABAENA, GLOEOTRICHIA, APHANIZOMENON AND NOSTIC) ARE CAPABLE OF FIXING ATMOSPHERIC NITROGEN. SUCCESS OF NITROGEN CONTROL IN SOME AREAS COULD DEPEND TO A GREAT EXTENT ON THE COOPERATION OF FARMERS IN CURTAILING BAD LAND MANAGEMENT AND FERTILIZATION PRACTICES. PHOSPHORUS GAINS ACCESS TO NATURAL WATERS MAINLY THROUGH THE DISCHARGE OF WASTEWATERS AND TO SOME EXTENT THROUGH SURFACE WASH FROM FARMLANDS. WASTEWATER TREATMENT METHODS FOR PHOSPHORUS REMOVAL FROM EFFLUENTS HAVE BEEN KNOWN FOR SEVERAL YEARS, AND A LARGE PART OF PHOSPHORUS IN DOMESTIC WASTEWATERS AND ESSENTIALLY ALL PHOSPHORUS IN SOME INDUSTRIAL WASTES, IS CONTRIBUTED BY SYNTHETIC DETERGENTS. PHOSPHORUS LIMITATION IN LAKES AND STREAMS SEEMS TO BE THE ONLY KNOWN MEANS TO CONTROL NITROGEN-FIXING BLUE-GREEN ALGAE. THUS PHOSPHORUS REMOVAL MUST BE PART OF ANY PLAN TO CONTROL EUTROPHICATION IN ADDITION TO NITROGEN REMOVAL, WHICH IS THE SECOND MOST IMPORTANT NUTRIENT IN PRIMARY PRODUCTION. (JONES-WISCONSIN)

FIELD 05C, 05G

ACCESSION NO. W72-04769

SOME OBSERVATIONS ON THE LIMNOLOGY OF A POND RECEIVING ANIMAL WASTES,

OKLAHOMA STATE UNIV., STILLWATER. DEPT. OF ZOOLOGY.

DALE W. TOETZ.

PROCEEDINGS OKLAHOMA ACADEMY OF SCIENCE, VOL 51, P 30-35, 1971. 4 FIG, 2 TAB, 11 REF.

DESCRIPTORS:
*OXIDATION LAGOONS, *WATER POLLUTION EFFECTS, *FARM WASTES, OKLAHOMA, CYANOPHYTA, CHLORELLA, IONS, PHYTOPLANKTON, NITRATES, NITROGEN FIXATION, ALGAE, CONDUCTIVITY, CHLOROPHYLL, PIGMENTS, DISPERSION, RUNOFF.

IDENTIFIERS:
*FEEDLOTS, LEMNA,

ABSTRACT:
A SMALL POND RECEIVING RUNOFF FROM A HOG YARD WAS DOMINATED BY A LARGE POPULATION OF BLUE-GREEN ALGAE AND PHYTOFLAGELLATES DURING SUMMER AND BY CHLORELLA SP DURING WINTER. HEAVY RAINFALLS DECREASED THE IONIC CONCENTRATION OF WATER AND ALTERED THE PHYTOPLANKTON COMPOSITION. OCCASIONAL CONCENTRATION OF OXYGEN BELOW 5 MG/L AND HIGH AMMONIA CONTENT SUGGESTED THAT THE POND IS UNSUITABLE FOR WARM-WATER FISH CULTURE. (WILDE-WISCONSIN)

FIELD 05C

ACCESSION NO. W72-04773

KINETICS OF ALGAL SYSTEMS IN WASTE TREATMENT--AMMONIA-NITROGEN AS A GROWTHLIMITING FACTOR AND OTHER PERTINENT TOPICS,

CALIFORNIA UNIV., BERKELEY. SANITARY ENGINEERING RESEARCH LAB.

M. SOBSEY, J. E. HARRISON, H. GEE, G. SHELET, AND J. C. GOLDMAN.

AVAILABLE FROM THE NATIONAL TECHNICAL INFORMATION SERVICE AS PB-206 811, $3.00 IN PAPER COPY, $0.95 IN MICROFICHE. FINAL REPORT, PART II, SEPTEMBER 1970, 156 P. 45 FIG, 11 TAB, 102 REF. FWQA PROGRAM 17010 DZQ 09/70.

DESCRIPTORS:
*ALGAE, *KINETICS, *NITROGEN, *NUTRIENT REQUIREMENTS, AMMONIA, *GROWTH RATES, *WASTE ASSIMILATIVE CAPACITY, OXIDATION, DISSOLVED OXYGEN, HYDROGEN ION CONCENTRATION, DESIGN CRITERIA, MATHEMATICAL MODELS, *BIOASSAY, *WASTE DILUTION, LAGOONS, *WASTE WATER TREATMENT, ALGAL CONTROL, EUTROPHICATION, MODEL STUDIES, MATHEMATICAL MODELS.

IDENTIFIERS:
*OSTRACODS, PREDATOR-PREY RELATIONSHIPS, *AMMONIA-NITROGEN, *GROWTH REACTORS, ALGAL PONDS, ALGAL GROWTH RATES, *POTAMCCYPRIS.

ABSTRACT:
4 SMALL OUTDOOR REACTORS, A PREDATOR INFESTED OUTDOOR POND, AND NUMEROUS INDOOR SYSTEMS WERE DESIGNED, CONSTRUCTED AND OPERATED TO PROVIDE DATA DESCRIBING THE KINETICS ASPECTS OF ALGAL SYSTEMS USED AS WASTEWATER TREATMENT SCHEMES. THE STUDIES INCLUDED DETERMINATIONS OF: (1) THE KINETICS OF AMMONIA-NITROGEN AS A GROWTH LIMITING FACTOR; (2) THE EFFECT OF NUTRIENTS IN WASTEWATER EFFLUENTS ON ALGAL GROWTH; (3) SUGGESTED DESIGNS FOR BOTH INDOOR AND OUTDOOR GROWTH REACTORS; AND (4) PREDATOR-PREY RELATIONSHIPS IN OUTDOOR ALGAL SYSTEMS. SPECIFIC GROWTH RATE DATA FOR NITROGEN WERE QUITE CONSISTENT, ALLOWING THE USE OF THE MAXIMUM SPECIFIC NET GROWTH RATE AS AN EXTREMELY RELIABLE PARAMETER OF SYSTEM OPERATION. PARAMETERS KA, KS, AND SN ALL DIRECTLY RELATED TO THE MAXIMUM SPECIFIC NET GROWTH RATE, WERE SHOWN TO BE USEFUL BOTH IN MATHEMATICAL INTERPRETATION OF SYSTEM KINETICS AND IN DETERMINING TO WHAT PERCENT OF CAPACITY AN ALGAL SYSTEM IS OPERATING. OTHER STUDIES DEMONSTRATED THAT PERMISSIBLE DILUTION RATIOS FOR WASTE WATER DISCHARGES INTO RECEIVING WATERS CAN BE FORMULATED EMPIRICALLY. DESIGN AND OPERATIONAL INFORMATION OF VARIOUS SYSTEMS IS PRESENTED, AS WELL AS PRELIMINARY BIOASSAY RESULTS ON THE MICROBIAL POPULATIONS OF ALGAL PONDS. (LOWRY-TEXAS)

FIELD 05D, 05C

ACCESSION NO. W72-04787

KINETICS OF ALGAL SYSTEMS IN WASTE TREATMENT--PHOSPHORUS AS A GROWTH LIMITING FACTOR,

CALIFORNIA UNIV., BERKELEY. SANITARY ENGINEERING RESEARCH LAB.

MARIO D. ZABAT, WILLIAM J. OSWALD, CLARENCE G. GOLUEKE, AND HENRY GEE.

AVAILABLE FROM THE NATIONAL TECHNICAL INFORMATION SERVICE AS PB-206 810, $3.00 IN PAPER COPY, $0.95 IN MICROFICHE. FINAL REPORT PART I, SEPTEMBER 1970, 210 P. 57 FIG, 15 TAB, 210 REF. FWQA PROGRAM 17010 DZQ 09/70.

DESCRIPTORS:
*ALGAE, *GROWTH RATES, *KINETICS, *PHOSPHORUS, *NUTRIENT REQUIREMENTS, TEMPERATURE, HYDROGEN ION CONCENTRATION, EFFICIENCIES, WATER QUALITY CONTROL, EUTROPHICATION, ALGAL CONTROL, *CHLORELLA, *WASTE WATER TREATMENT.

IDENTIFIERS:
*CHEMOSTATIC CONTINUOUS CULTURES, MAXIMUM SPECIFIC GROWTH RATE, SATURATION CONSTANT, DILUTION, CHLORELLA PYRENOIDOSA, MONOD GROWTH MODEL, ALGAL GROWTH RATES.

ABSTRACT:
THE KINETICS AND CHARACTERISTICS OF PHOSPHATE-LIMITED ALGAL CULTURES WERE STUDIED ON A LABORATORY SCALE IN ORDER TO: (1) EVALUATE KINETIC PARAMETERS OF ALGAL GROWTH IN RELATION TO PHOSPHORUS CONCENTRATION; (2) DERIVE AN EXPRESSION DESCRIBING THE GROWTH RATE-LIMITING NUTRIENT RELATIONSHIP; AND (3) DETERMINE DESIRABLE OR OPTIMUM DESIGN AND OPERATING PARAMETERS FOR EFFECTIVE PHOSPHORUS REMOVAL BY ALGAL SYSTEMS. PHOSPHORUS LIMITING ALGAL GROWTH WAS DESCRIBED BY THE MONOD GROWTH MODEL FOR SPECIFIC GROWTH RATE. MAXIMUM SPECIFIC GROWTH RATE FOR THE ALGAE TESTED WAS 4.19/DAY, AND ALL GROWTH CHARACTERISTICS WERE SIGNIFICANTLY AFFECTED BY VARIATIONS IN PH AND TEMPERATURE OF THE CULTURE. REMOVALS OF MODERATE CONCENTRATIONS OF PHOSPHORUS (LESS THAN OR EQUAL TO 7 MG/L AS P) IN EXCESS OF 85% CAN BE ACCOMPLISHED BY CONTINUOUS ALGAL CULTURES. (LOWRY-TEXAS)

FIELD 05D, 05C

ACCESSION NO. W72-04788

KINETICS OF ALGAL SYSTEMS IN WASTE TREATMENT FIELD STUDIES,

CALIFORNIA UNIV., BERKELEY. SANITARY ENGINEERING RESEARCH LAB.

AARON MERON, WILLIAM J. OSWALD, AND HENRY GEE.

AVAILABLE FROM THE NATIONAL TECHNICAL INFORMATION SERVICE AS PB-206 812, $6.00 IN PAPER COPY, $0.95 IN MICROFICHE. FINAL REPORT PART III, MAY 1971, 334 P, 46 FIG, 44 TAB, 122 REF. FWQA PROGRAM 17010 DZQ 05/71.

DESCRIPTORS:
*ALGAE, *KINETICS, *WASTE WATER TREATMENT, *NUTRIENT REQUIREMENTS, TEMPERATURE, *NITROGEN, *PHOSPHATES, ANAEROBIC CONDITIONS, AEROBIC CONDITIONS, METABOLISM, SLUDGE, BIOCHEMICAL OXYGEN DEMAND, CHEMICAL OXYGEN DEMAND, CARBON DIOXIDE, LININGS, DISSOLVED OXYGEN, LIGHT INTENSITY, DESIGN CRITERIA, *OXIDATION LAGOONS, SEWAGE TREATMENT, ALGAL CONTROL, BIOLOGICAL TREATMENT, *GROWTH RATES.

IDENTIFIERS:
ALGAL GROWTH RATES.

ABSTRACT:
THREE DOMESTIC SEWAGE STABILIZATION POND SYSTEMS WERE STUDIED TO EVALUTE THE RELATIONSHIP BETWEEN DESIGN CRITERIA AND PERFORMANCE CRITERIA FOR POND SYSTEMS IN ACCOMPLISHING SPECIFIC WASTE TREATMENT STEPS. SYSTEM 1 CONSISTED OF A LINED HIGH-RATE POND FOLLOWING SEDIMENTATION, SYSTEM 2 WAS COMPRISED OF CONVENTIONAL SECONDARY WASTE TREATMENT FOLLOWED BY A POND SYSTEM, AND SYSTEM 3 CONSISTED SOLELY OF SEVERAL DEEP PONDS IN SERIES. STUDY RESULTS INDICATED THAT SYSTEM 3, CONSISTING OF AN ANEROBIC POND, AND UNLINED HIGH-RATE POND, AND DEEP HIGH-RATE POLISHING POND, WAS EFFECTIVE IN BOTH BOD AND NUTRIENT REMOVALS, WITH MEAN ANNUAL REMOVALS OF 97.3%, 93.2%, 91.6%, AND 64.5% FOR BOD, COD, TOTAL NITROGEN, AND PHOSPHATE. RESULTS CONFIRMED THAT POND FUNCTIONING IS DETERMINED BY POSITION WITH RESPECT TO OTHER PONDS AND UNIT PROCESSES, AS WELL AS THEIR INDIVIDUAL DESIGN CHARACTERISTICS. OTHER STUDY RESULTS DEMONSTRATED THAT: (1) EFFICIENT N AND P REMOVAL DEPENDS UPON CARBON AVAILABILITY; (2) HIGH-RATE PONDS DEVELOP A SLUDGE-SLIME BOTTOM LAYER WHICH PREVENTS EROSION; (3) PROPER NUTRIENT BALANCE, AS MEASURED BY C:N AND C:P IS ESSENTIAL FOR EFFECTIVE ALGAL GROWTH AND NUTRIENT REMOVAL; (4) DEEP ANAEROBIC PONDS SHOULD BE OPERATED AS AN AEROBIC LAYER ABOVE AN ANAEROBIC LAYER; AND (5) NUTRIENT BALANCE AS WELL AS PH CAN BE ADJUSTED IN HIGH RATE PONDS BY CO2 ADDITION. (LOWRY-TEXAS)

FIELD 05D

ACCESSION NO. W72-04789

WATER QUALITY OFFICE.

ENVIRONMENTAL PROTECTION AGENCY, SAN FRANCISCO, CALIF. WATER QUALITY OFFICE.

ENVIRONMENTAL PROTECTION AGENCY, PACIFIC SOUTHWEST REGION WATER QUALITY OFFICE REPORT, DECEMBER 1970. 59 P, 3 FIG, 10 TAB, 11 REF, 3 APPEND.

DESCRIPTORS:
*WATER QUALITY, *WATER POLLUTION EFFECTS, *LAKES, *COLORADO, *RECREATION FACILITIES, WATER TEMPERATURE, CHEMICAL ANALYSIS, ALGAE, NUTRIENTS, WATER ANALYSIS, DOMESTIC WASTES.

IDENTIFIERS:
GRAND LAKE(COLO), SHADOW MOUNTAIN LAKE(COLO), LAKE GRANBY(COLO).

ABSTRACT:
LABORATORY ANALYSES OF SAMPLES COLLECTED FROM GRAND LAKE, SHADOW MOUNTAIN LAKE, AND LAKE GRANBY (ONE OF THE MOST POPULAR RECREATIONAL AREAS IN COLORADO) SHOW THAT THE INORGANIC CONCENTRATIONS WERE UNIFORM THROUGHOUT THE LAKES AND AVERAGE 0.107 MG/L. THIS CONCENTRATION IS APPROXIMATELY ONE-THIRD OF THAT CONSIDERED NECESSARY TO INITIATE AN ALGAE BLOOM. THE ORTHOPHOSPHATE CONCENTRATION IN THE THREE LAKES AVERAGED 0.0255 MG/L. THIS AVERAGE CONCENTRATION EXCEEDS THE MINIMUM LEVEL OF ORTHOPHOSPHATE CONSIDERED AS NECESSARY TO SUSTAIN AN ALGAE 'BLOOM' (0.010 MG/L). LAKE GRANBY AND GRAND LAKE HAD THERMALLY STRATIFIED LAYERS ALTHOUGH THERMOCLINES WERE NOT IN EVIDENCE. SHADOW MOUNTAIN LAKE WAS GENERALLY ISOTHERMAL BECAUSE OF ITS SHALLOW DEPTH. THE DISSOLVED OXYGEN CONCENTRATION IN THE NEAR SURFACE WATERS OF ALL THREE LAKES EXCEEDED THE 6.0 MG/L STANDARD FOR COLD WATER FISHERIES AT ALL STATIONS EXCEPT FOR ONE STATION ON SHADOW MOUNTAIN LAKE. THE WATER QUALITY OF THE THREE LAKES WAS EQUAL TO OR BETTER THAN THE STANDARDS ESTABLISHED FOR THE THREE LAKES EXCEPT THAT THE DISSOLVED OXYGEN CONCENTRATIONS IN THE HYPOLIMNION OF LAKE GRANBY WERE LESS THAN THE ESTABLISHED STANDARDS OF 6.0 MG/L. (WOODARD-USGS)

FIELD 05C, 02H

ACCESSION NO. W72-04791

NATURAL CONDITIONS AND AQUATIC LIFE OF THE BRATSK RESERVOIR (FORMIROVANIYE PRIRODNYKH USLOVIY I ZHIZNI BRATSKOGO VODOKHRANILISHCHA).

IZDATEL ' STVO 'NAUKA', MOSCOW, G. I. GALAZIY, EDITOR, 1970. 280 P.

DESCRIPTORS:
*AQUATIC LIFE, *AQUATIC BACTERIA, *AQUATIC ANIMALS, *AQUATIC PLANTS, *RESERVOIR STAGES, EARLY IMPOUNDMENT, BIOMASS, AQUATIC ALGAE, PHYTOPLANKTON, ZOOPLANKTON, FISH POPULATIONS, WATER TEMPERATURE, THERMAL STRATIFICATION, OXIDATION, ORGANIC MATTER, WATER QUALITY, WATER CHEMISTRY, WATER POLLUTION SOURCES, METEOROLOGICAL DATA, SEASONAL.

IDENTIFIERS:
*USSR, *IRKUTSK OBLAST, *BRATSK RESERVOIR, ANGARA RIVER, OKA RIVER, HYDROBIOLOGY, BACTERIOLOGY, SAPROPHYTES, STURGEON, METALIMNION.

ABSTRACT:
THIS COLLECTION OF 8 PAPERS DEALS PRIMARILY WITH BIOLOGICAL ASPECTS OF AQUATIC LIFE IN THE BRATSK RESERVOIR, EASTERN SIBERIA, (IRKUTSK OBLAST)--THE LARGEST MANMADE LAKE IN THE WORLD. THE PAPERS, MAY BE USED AS A BASIS FOR FORECASTING WATER QUALITY OF THE RESERVOIR FOR DRINKING-WATER SUPPLY AND INDUSTRIAL USE, AND FOR RECLAIMING THE RESERVOIR AS A FISHERY. SUBJECTS EXAMINED INCLUDE: (1) SEASONAL METEOROLOGICAL CONDITIONS IN THE REGION OF THE RESERVOIR; (2) FORMATION OF PHYTOPLANKTON IN THE RESERVOIR; (3) AQUATIC BACTERIA IN THE RESERVOIR (1965); (4) ESTABLISHMENT OF ZOOPLANKTON IN THE RESERVOIR DURING THE FIRST TWO YEARS OF IMPOUNDMENT (1962-63); (5) FISH POPULATION AND DISTRIBUTION IN THE OKA RIVER SECTION OF THE RESERVOIR; (6) STERLET IN THE RESERVOIR AND ANGARA RIVER; (7) LONG-TERM WATER-QUALITY CHARACTERISTICS OF THE RESERVOIR AND FORECAST OF ORGANIC MATTER ACCUMULATION; AND (8) WATER QUALITY OF THE RESERVOIR DURING THE FIRST YEARS OF IMPOUNDMENT (1962-64). THE COLLECTION IS OF PARTICULAR VALUE TO HYDROBIOLOGISTS, ICHTHYOLOGISTS, HYDROLOGISTS, CLIMATOLOGISTS, ZOOLOGISTS AND BOTANISTS. (SEE ALSO W72-04855)
(JOSEFSON-USGS)

FIELD 02H, 05C

ACCESSION NO. W72-04853

FORMATION OF PHYTOPLANKTON IN THE BRATSK RESERVOIR (FORMIROVANIYE FITOPLANKTONA BRATSKOGO VODOKHRANILISHCHA),

LIMNOLOGICHESKII INSTITUT, IRKUTSK (USSR).

O. M. KOZHOVA.

IN: FORMIROVANIYE PRIRODNYKH USLOVIY I ZHIZNI BRATSKOGO VODOKHRANILISHCHA: IZDATEL 'STVO 'NAUKA', MOSCOW, P 26-160, 1970. 24 FIG, 37 TAB, 68 REF.

DESCRIPTORS:
*AQUATIC PLANTS, *AQUATIC ALGAE, *PHYTOPLANKTON, *RESERVOIRS, *RESERVOIR STAGES, EARLY IMPOUNDMENT, BIOMASS, DIATOMS, CHLOROPHYTA, CYANOPHYTA, PYRROPHYTA, CHRYSOPHYTA, EUGLENOPHYTA, WATER TEMPERATURE, THERMAL STRATIFICATION, ICE, EUTROPHICATION, OXYGEN, SEASONAL.

IDENTIFIERS:
*USSR, *IRKUTSK OBLAST, *BRATSK RESERVOIR, *HYDROBIOLOGY, METALIMNION, BACILLARIOPHYTA, VOLVOCINEAE, PROTOCOCCINEAE, ULOTRICHINEAE, DESMIDIALES.

ABSTRACT:
SEASONAL CHANGES IN THE COMPOSITION, PRODUCTION, AND VERTICAL AND HORIZONTAL DISTRIBUTION OF PHYTOPLANKTON IN THE BRATSK RESERVOIR, EASTERN SIBERIA, (AREA--5,500 SQ KM; VOLUME--179 CU KM) DURING ITS EARLY IMPOUNDMENT (1963-65) ARE EXAMINED. THE PHYTOPLANKTON IS REPRESENTED BY 118 ALGAL SPECIES, CONSISTING OF 47 CHLOROPHYTA, 20 CYANOPHYTA, 17 BACILLARIOPHYTA, 14 PYRROPHYTA, 11 CHRYSOPHTYA, AND 9 EUGLENOPHYTA. A TOTAL OF 17 DOMINANT SPECIES WERE IDENTIFIED IN 1964 AND 13 IN 1965. THE MOST MASSIVE ALGAL POPULATIONS ARE APHANIZOMENON FLOS-AQUAE, STEPHANODISCUS HANTZSCHII, MELOSIRA ISLANDICA HELVETICA, AND ASTERIONELLA FORMOSA. NO WATER BLOOMS ARE FOUND UNDER ICE (ALTHOUGH THE BIOMASS IS SEVERAL MG/CU M). BACILLARIOPHYTA ARE PREDOMINANT IN SPRING (SEVERAL G/CU M); AN ANNUAL BIOMASS MAXIMUM OF 320 G OF DOMINANT APHANIZOMENON FLOS-AQUAE/CU M IS OBSERVED IN SUMMER AND FALL. MAXIMUM BIOMASSES INCREASED FROM 12 TO 320 G/CU M BETWEEN 1963 AND 1965; AVERAGE AREAL INDICES FOR THE RESERVOIR DURING THE SUMMER MAXIMUM WERE 55 G/SQ M IN 1964 AND 195 G/SQ M IN 1965. MAXIMUM BIOMASS INDEX WAS 816 G/SQ M AND APPROACHED EUTROPHIC CONDITIONS. TOTAL PHYTOPLANKTON PRODUCTION OF OXYGEN ALSO CHANGED. IN 1964 IT WAS 100 G/SQ M OR 370 KCAL/SQ M DURING THE GROWING PERIOD AND IN 1965-66 IT WAS 389 G/SQ M. RIVER INFLOW CONTAINS ABOUT 13,000 METRIC TONS OF PHYTOPLANKTON. A TOTAL OF 30,000 METRIC TONS IS DISCHARGED FROM THE RESERVOIR. PHYTOPLANKTON COMPOSITION, PRODUCTION, VERTICAL DISTRIBUTION, AND BIOMASS ARE DETERMINED BY THE MORPHOMETRIC HETEROGENEITY OF THE RESERVOIR, ITS ENORMOUS SIZE, AND BY DIFFERENCES IN RESERVOIR DEPTHS, WHICH RANGE FROM 3 TO 100 M. (SEE ALSO W72-04853) (JOSEFSON-USGS)

FIELD 05C, 02H

ACCESSION NO. W72-04855

ALGAL-BACTERIAL SYMBIOSIS FOR REMOVAL AND CONSERVATION OF WASTEWATER NUTRIENTS,

NORTH CAROLINA STATE UNIV., RALEIGH. DEPT. OF BIOLOGICAL AND AGRICULTURAL ENGINEERING; AND NORTH CAROLINA STATE UNIV., RALEIGH. DEPT. OF CIVIL ENGINEERING.

FRANK J. HUMENIK, AND GEORGE P. HANNA, JR.

JOURNAL WATER POLLUTION CONTROL FEDERATION, VOL 43, NO 4, APRIL 1971, P 580-594. 9 FIG, 1 TAB, 27 REF.

DESCRIPTORS:
*ALGAE, *BACTERIA, *SYMBIOSIS, PHOTOSYNTHESIS, OXIDATION, RESPIRATION, OXYGENATION, TEMPERATURE, LIGHT INTENSITY, MIXING, DISSOLVED OXYGEN, METABOLISM, CARBON DIOXIDE, PHOSPHATES, NITRATES, *NUTRIENTS, ALKALINITY, CHEMICAL OXYGEN DEMAND, ANALYTICAL TECHNIQUES, TERTIARY TREATMENT, *WASTE WATER TREATMENT.

ABSTRACT:
A LABORATORY-SCALE CONTINUOUS SYMBIOTIC ALGAL-BACTERIAL SYSTEM WAS DEVELOPED THAT REMOVED NUTRIENTS FROM WASTEWATER AND REMAINED WITHOUT SUPPLEMENTAL AERATION. THE BIOMASS GROWTH UNIT SUPPORTED A MIXED CHLORELLA-ACTIVATED SLUDGE CULTURE. MAXIMUM AND MOST CONSISTENT NUTRIENT REMOVAL AND CONSERVATION OCCURRED DURING UNAERATED OPERATION WITH SOLIDS CONTROL BY DAILY HARVESTING. PHOSPHORUS REMOVALS WERE NOT SIGNIFICANT. THE BIOMASS CULTURE WAS AN INTIMATELY ASSOCIATED MIXTURE OF ALGAE AND ACTIVATED SLUDGE, AND THE RESULTING SYMBIOTIC BENEFITS PRESENTED MORE FAVORABLE CONDITIONS FOR THE REMOVAL AND CONSERVATION OF WASTEWATER NUTRIENTS. THE REESTABLISHMENT OF THE SYMBIOTIC ALGAL-BACTERIA SYSTEM AFTER AN EQUIPMENT FAILURE PROVIDES EVIDENCE THAT SUCH A PROCESS COULD BE RECOVERED AFTER SOME OPERATIONAL MISHAP HAD MADE IT NECESSARY TO REVERT TO SUPPLEMENTAL AERATION. (LOWRY-TEXAS)

FIELD 05D

ACCESSION NO. W72-04983

PREDICTING QUALITY EFFECT OF PUMPED STORAGE,

WATER RESOURCES ENGINEERS, INC., WALNUT CREEK, CALIF.

C. W. CHEN, AND G. T. ORLOB.

PAPER, INTERNATIONAL CONFERENCE ON PUMPED STORAGE DEVELOPMENT AND ENVIRONMENTAL EFFECTS, MILWAUKEE, WIS, SEPT 1971. 21P, 7FIG, 2 TAB, 13 REF.

DESCRIPTORS:
*ECOSYSTEMS, *ECOLOGY, *PUMPED STORAGE, *WATER QUALITY, *MODEL STUDIES, ZOOPLANKTON, BIOTA, BACTERIA, BENTHIC FAUNA, FISH, ENVIRONMENTAL EFFECTS, RESERVOIR STORAGE, NUTRIENTS, TEMPERATURE, ALGAE, HYDROLOGIC DATA, DISSOLVED OXYGEN, WEATHER, RESERVOIR OPERATION, WATER TEMPERATURE, PHYTOPLANKTON, EUTHROPHICATION.

IDENTIFIERS:
SMITH MOUNTAIN PROJECT(VA), WEATHER EFFECT.

ABSTRACT:
WHILE MUCH ATTENTION HAS BEEN PAID TO THE ECONOMIC ASPECTS OF PUMPED STORAGE, THE ENVIRONMENTAL EFFECTS, PARTICULARLY THE WATER QUALITY WITHIN THE RESERVOIR, HAVE SELDOM BEEN FULLY ASSESSED. THE PUMPED STORAGE IMPOUNDMENT IS INVARIABLY A COMPLEX ECOSYSTEM HOSTING A WIDE SPECTRUM OF BIOTA--BACTERIA, PHYTOPLANKTON, ZOOPLANKTON, FISH AND BENTHIC ANIMALS. NUTRIENTS FROM RUNOFF CAN BE CONCENTRATED AND RETAINED IN THE ECOSYSTEM THROUGH ALGAL ASSIMILATION AND SEDIMENTATION. ACCELERATED EUTHROPHICATION OF THE ENVIRONMENT MAY RESULT. A PREDICTIVE TECHNIQUE HAS BEEN DEVELOPED FOR CALCULATING THE ENVIRONMENTAL CHANGES AND EVALUATING THE ALTERNATIVES OF PUMPED STORAGE OPERATION THAT MAY ENHANCE A NORMAL ECOLOGIC SUCCESSION AND FORESTALL ADVERSE ENVIRONMENTAL PROBLEMS. THE MODEL, APPLIED TO SMITH MOUNTAIN RESERVOIR, VA, REVEALS THAT IMPROVED WATER QUALITY CAN BE ACHIEVED BY MODIFYING THE PUMPING-RETURN SCHEDULE. (USBR)

FIELD 05F, 06B

ACCESSION NO. W72-05282

LIQUID WASTE TREATMENT PROCESS,

J. M. VALDESPINO.

U. S. PATENT NO. 3,625,883, 5 P, 3 FIG, 11 REF. OFFICIAL GAZETTE, VOL. 893, NO. 1, P. 254, DECEMBER 7, 1971.

DESCRIPTORS:
*PATENTS, *WASTE WATER TREATMENT, *LIQUID WASTES, SEWAGE TREATMENT, SEPARATION TECHNIQUES, *CHEMICAL WASTES, INDUSTRIAL WASTES, CHLORINATION, PHOSPHATES, NITRATES, FILTRATION, POLLUTION ABATEMENT, ALGAE, NUTRIENT, *BIODEGRADATION, WATER PURIFICATION, *AEROBIC TREATMENT, *OXIDATION LAGOONS.

IDENTIFIERS:
AEROBIC BIOLOGICAL DEGRADATION.

ABSTRACT:
A METHOD IS DESCRIBED FOR THE TREATMENT OF LIQUID WASTE FROM DOMESTIC AND INDUSTRIAL SOURCES. THE LIQUID WASTE IS FIRST FED TO A COMMINUTOR FOR CHOPPING AND MIXING SOLID WASTE WITH THE LIQUID. THE PRODUCT IS THEN FED TO A SURGING TANK FOR PROVIDING AN INTERMITTENT FLOW INTO A CENTRIFUGING STEP. THE WASTE IS CENTRIFUGED THROUGH A FILTER MEDIUM SUCH AS SAND OR DIATOMACEOUS EARTH. THE SEPARATED EFFLUENT IS FED TO AN OXIDATION LAGOON FOR AEROBIC BIOLOGICAL DEGRADATION. THE GENERATION OF ALGAE MAY BE PROVIDED TO REMOVE PHOSPHATES, NITRATES AND OTHER NUTRIENTS. THE EFFLUENT THEN IS FED TO A SECOND CENTRIFUGE FILTER TO SEPARATE REMAINING SOLIDS WHICH ARE BURNED. CHLORINATION PROVISIONS ARE INCLUDED. (SINHA-OEIS)

FIELD 05D

ACCESSION NO. W72-05315

AQUATIC AND MARINE MICROORGANISMS, INTERRELATIONSHIPS IN ENRICHMENT CULTURES,

WASHINGTON UNIV., SEATTLE. DEPT. OF MICROBIOLOGY.

E. J. ORDAL.

AVAILABLE FROM THE NATIONAL TECHNICAL INFORMATION SERVICE AS AD-731 409, $3.00 IN PAPER COPY, $0.95 IN MICROFICHE. FINAL TECHNICAL REPORT 71-1, OCTOBER 14, 1971. 18 P, 2 TAB, 10 REF. PROJ. NC. NR306-487, N00014-67-A-0103-0006.

DESCRIPTORS:
*BACTERIA, *CULTURES, *SEA WATER, *NITRIFICATION, AMMONIA, NITRITE, SPHAEROTILUS, NUTRIENTS, PACIFIC OCEAN, NITRATE, CHEMICAL ANALYSIS, SODIUM CHLORIDE, COLUMBIA RIVER, ORGANIC COMPOUNDS, VITAMINS, YEASTS, SULFUR, FOULING, ALGAE, SULFATES, PERCOLATION, HERRING, SALMON, TROUT, TOXICITY, WASHINGTON, MARINE ALGAE, MARINE BACTERIA.

IDENTIFIERS:
HYPHOMICROBIUM, VIBRIO ANGUILLARUM, ENRICHMENT CULTURES, AGARS, CHEMOSTATS, NITROSOMONAS, NITROBACTER, NITROSOCYSTIS OCEANUS, SPHAEROTILUS NATANS, GLUCOSE, GALACTOSE, SUCROSE, MALTOSE, MANNITAL, SORBITAL, SUCCINATE, FUMARATE, BUTYRATE, BUTANOL, GLYCEROL, LACTATE, PYRUVATE, ACETATE, ETHANOL, BENZOATE, PROPIANATE, PROPANOL, LAURATO, PALMITATE, N-HEPTYLATE, XYLOSE, CITRATE, YEAST EXTRACT, MINERAL SALTS, CAULOBACTER, OLIGOCARBOPHILIC BACTERIA, ULVA, DESULFOVIBRIO, MACROMONAS, THIOSPIRA, VIBRIO PARAHAEMOLYTICUS, NITRIFYING BACTERIA.

ABSTRACT:
THE METHOD OF STEADY-STATE ENRICHMENT CULTURES WAS APPLIED TO THE STUDY OF MARINE NITRIFYING BACTERIA. WHEN LOW CONCENTRATIONS OF AMMONIA OR NITRITE WERE EMPLOYED, THE NITRIFYING BACTERIA WERE FOUND ATTACHED TO THE WALLS OF THE CULTURE VESSELS. WHEN CONCENTRATIONS OF AMMONIA OR NITRITE WERE INCREASED, THE NITRIFYING BACTERIA APPEARED ALSO IN THE CULTURE FIELDS ENABLING THE ISOLATION OF PURE CULTURES IN LIQUID MEDIA. AGAR WAS TOXIC AND THE NITRIFYING BACTERIA WERE FOUND TO REQUIRE AN UNKNOWN COMPONENT IN SEA WATER IN ADDITION TO SODIUM CHLORIDE. CONTROL FLOW-THROUGH CULTURES OF OCEAN WATER ALONE SUPPORTED LARGE POPULATIONS OF THE BUDDING, BRANCHING BACTERIUM HYPHOMICROBIUM, WHICH ALSO APPEARED IN CULTURES WITH LOW CONCENTRATIONS OF AMMONIA OR NITRITE. HYPHOMICROBIUM FED BY SMALL CONCENTRATIONS OF ORGANIC MATTER IN SEA WATER QUICKLY APPEARED ON GLASS, CELLOPHANE AND METAL SURFACES AND MAY BE IMPORTANT AS AN INITIAL STEP IN FOULING OF VESSELS. STEADY-STATE ENRICHMENT CULTURES WERE EMPLOYED TO DETERMINE THE RANGE OF ORGANIC COMPOUNDS USED BY SPHAEROTILUS. CULTURES GROWN ON GLUCOSE PLUS MINERAL SALTS GREW WELL IN ENRICHMENT CULTURES BUT FAILED TO GROW IN PURE CULTURE UNLESS VTAMINS WERE ADDED. IT WAS CONCLUDED THAT ASSOCIATED BACTERIA WERE PROVIDING THE NECESSARY VITAMINS FOR GROWTH OF SPHAEROTILUS. A GROUP OF MARINE VIBRIOS ISOLATED FROM DISEASED SALMON AND HERRING WERE INVESTIGATED BY SEVERAL METHODS. STRAINS OF VIBRIO ANGUILLARUM FROM THE PACIFIC NORTHWEST WERE CLOSELY RELATED TO STRAINS FROM SCOTLAND AND DENMARK. HOWEVER, A NUMBER OF OTHER VIBRIOS WERE CLEARLY DISTINCT, INDICATING THAT THERE ARE SEVERAL KINDS OF MARINE VIBRIOS. (MORTLAND-BATTELLE)

FIELD 05C, 05A

ACCESSION NO. W72-05421

TWO SIMPLE DURABLE EPIFAUNAL COLLECTORS,

LAVAL UNIV., QUEBEC. DEPARTEMENT DE BIOLOGIE.

E. BOURGET, AND G. LACROIX.

JOURNAL OF THE FISHERIES RESEARCH BOARD OF CANADA, VOL. 28, NO. 8, P 1205-1207, AUGUST 1971. 3 FIG, 1 TAB, 5 REF.

DESCRIPTORS:
*BENTHOS, *SAMPLING, *SESSILE ALGAE, *INVERTEBRATES, ST. LAWRENCE RIVER, ALGAE, NEMATODES, GASTROPODS, MOLLUSKS, AMPHIPODA, MECHANICAL EQUIPMENT.

IDENTIFIERS:
*VERTICAL COLLECTORS, *HORIZONTAL COLLECTORS, BRYOZOANS, FORAMINIFERA, HYDROIDS, POLYCHAETA, OSTRACODS, HARPACTICOIDS, CIRRIPEDIA, ACARINA.

ABSTRACT:
TWO TYPES OF EPIFAUNAL SAMPLERS (VERTICAL AND HORIZONTAL) FOR USE IN THE INFRALITTORAL ZONE WERE DEVELOPED AND EVALUATED AT FOUR STATIONS IN THE ST. LAWRENCE ESTUARY. THE VERTICAL SAMPLER CONSISTED OF AN IRON PEDESTAL WITH A PROJECTING ARM DESIGNED TO HOLD 12 PLYWOOD COLLECTION PANELS. THE ARM WAS CONSTRUCTED TO ALLOW IT TO ROTATE TO OFFER MINIMAL RESISTANCE TO THE WATER FLOW AND AVOID TOPPLING THE STRUCTURE. THE HORIZONTAL SAMPLER CONSISTED OF A BOX-LIKE FRAME OF IRON WITH A HINGED TOP TO WHICH 16 COLLECTION PANELS WERE ATTACHED. BOTH SAMPLERS WERE DESIGNED FOR EASY REMOVAL OF THE SAMPLE PANELS BY A SCUBA DIVER. NO BUOYS WERE ATTACHED DIRECTLY TO THE SAMPLERS SO THAT ALGAL COLLECTIONS ON MOORING LINES WOULD NOT INTERFERE WITH THE SAMPLES. EVALUATION OF THE TWO TYPES OF SAMPLERS SHOWED THAT THE VERTICAL SAMPLER IS MORE SUITABLE FOR STUDY OF EPIBENTHIC COMMUNITIES SINCE IT DRAWS A WIDER RANGE OF SPECIES. THE HORIZONTAL SAMPLER, ON THE OTHER HAND, COLLECTED A LARGER NUMBER OF SPECIMENS AND IS CONCLUDED TO BE PREFERABLE IF ONLY SESSILE AND SEDIMENTARY ORGANISMS ARE TO BE STUDIED. FOR COMPLETE SAMPLING OF THE EPIFAUNA, USE OF BOTH SAMPLERS IS RECOMMENDED. (JEFFERIS-BATTELLE)

FIELD 05A, 07B, 02L

ACCESSION NO. W72-05432

BACTERICAL PATHOGENS OF FRESHWATER BLUE-GREEN ALGAE,

DUNDEE UNIV. (SCOTLAND). DEPT. OF BIOLOGICAL SCIENCES.

M. J. DAFT, AND W. D. P. STEWART.

NEW PHYTOLOGIST, VOL 70, NO 5, P 819-829, 1971. 6 FIG, 3 TAB, 2 PLATES, 12 REF.

DESCRIPTORS:
*BIOCONTROL, *PATHOGENIC BACTERIA, *ALGAL CONTROL, FRESH WATER, CYANOPHYTA, MYXOBACTERIA, CULTURES, METABOLISM, SEWAGE, CARBOHYDRATES, HYDROGEN ION CONCENTRATION, TEMPERATURE, ENZYMES, CELLULOSE, VIRUSES, HUMAN PATHOLOGY, SYSTEMATICS.

IDENTIFIERS:
LYTIC BACTERIA, ENGLAND, METABOLIC INHIBITION, HOSTS, MYXOPHYCEAE, SCOTLAND.

ABSTRACT:
EVIDENCE IS PRESENTED THAT HETEROTROPHIC BACTERIA WHICH ARE PATHOGENIC TO CERTAIN BLUE-GREEN ALGAE OCCUR IN BRITISH FRESHWATER HABITATS. FOUR BACTERIAL ISOLATES WERE OBTAINED WHICH CAUSE LYSIS IN LABORATORY CULTURES AND NATURAL POPULATIONS OF BLUE-GREEN ALGAE. THE ORGANISMS ARE TENTATIVELY IDENTIFIED AS MEMBERS OF THE MYXOBACTERALES. THE HOST RANGE OF THESE STRAINS IS WIDE AND SPECIES BELONGING TO ALL ORDERS OF MYXOPHYCEAE ARE SUSCEPTIBLE. THE ISOLATES ALSO LYSE A VARIETY OF GRAM-NEGATIVE AND GRAM POSITIVE BACTERIA AND ARE CAPABLE OF GROWTH FREE FROM HOSTS, AND THEIR INHIBITORY EFFECT IS RAPID (IN SOME INSTANCES WITHIN 2-5 HOURS). SOME BACTERIA SUSCEPTIBLE ARE HUMAN PATHOGENS THUS IT MAY BE THAT THREE OF THE LYTIC BACTERIA OBTAINED FROM SEWAGE PONDS DECREASE BACTERIAL NUMBERS WHICH ARE PATHOGENIC TO MAN. THE FINDINGS, IF EXTRAPOLATED TO NATURAL POPULATIONS, SUGGEST THAT GIVEN SUFFICIENT LYTIC BACTERIA, ALGAL BLOOMS WILL NOT OCCUR, UNLESS RESISTANT STRAINS ARE PRESENT. POSSIBLY ALGAL FORMATION COULD BE REGULATED TO SOME DEGREE BY THE EXTENT TO WHICH ALGAL GROWTH OUTSTRIPS, OR IS OUTSTRIPPED BY, THE GROWTH OF ALGAL PATHOGENS. THE ISOLATES CAN BE PROPAGATED READILY IN THE ABSENCE OF HOSTS. (AUEN-WISCONSIN)

FIELD 05C

ACCESSION NO. W72-05457

PHYTOPLANKTON ENERGETICS IN A SEWAGE-TREATMENT LAGOON,

SOUTHERN ILLINOIS UNIV., CARBONDALE. DEPT. OF BOTANY.

JACOB VERDUIN.

ECOLOGY, VOL 52, NO 4, P 626-631, 1971. 5 FIG, 3 TAB, 20 REF. NSF GE-2649.

DESCRIPTORS:
*PHYTOPLANKTON, *METABOLISM, *SEWAGE TREATMENT, *SEWAGE LAGOONS, OHIO, OXYGEN, DISSOLVED OXYGEN, LIGHT, DIFFUSIVITY, CARBON DIOXIDE, LIGHT INTENSITY, PHOTOSYNTHESIS, EQUATIONS, LIGHT PENETRATION, ALGAE, HYDROGEN ION CONCENTRATION.

IDENTIFIERS:
EXTINCTION COEFFICIENT, LAMBERT-BEER EQUATION, EDDY DIFFUSIVITY, DESHLER(OHIO).

ABSTRACT:
INTENSIVE INVESTIGATIONS WERE CARRIED OUT FROM MARCH 21 THROUGH JUNE 10, 1964 OF A SEWAGE-TREATMENT LAGOON SYSTEM NEAR DESHLER, OHIO--A PERIOD CRITICAL FOR SEWAGE PONDS IN THE TEMPERATE ZONE. THE PONDS WERE WELL MIXED VERTICALLY, EXHIBITED DAYTIME AVERAGE EDDY DIFFUSIVITIES OF ABOUT 1 SQ CM/SEC AND HAD A HIGH TITRATABLE BASE. LOW OXYGEN LEVELS WERE ENCOUNTERED FOR ONLY A SHORT PERIOD IN THE SECOND STAGE OF THE TWO-POND SYSTEM. LIGHT ABSORPTION AVERAGED ABOUT 90%/M. THE PHYTOPLANKTON COMMUNITIES, DOMINATED BY SCENEDESMUS, ROSE TO 50 MICROLITERS/L. RATES OF CARBON DIOXIDE UPTAKE IN THE ORDER OF 1 M/CM PER DAY WERE OBSERVED, BUT OXYGEN PRODUCTION AVERAGED ONLY 1/5 AS HIGH. A GRAPH OF CARBON DIOXIDE UPTAKE VERSUS LIGHT INTENSITY REVEALED A CURVE CLOSELY SIMILAR IN SHAPE TO THE TYPICAL PHOTOSYNTHESIS VERSUS LIGHT CURVE, BUT A SPEARATE SET OF DATA, BASED ON C-14 UPTAKE, SHOWED ANOMALOUSLY HIGH VALUES AT SURFACE LIGHT INTENSITIES. CARBON DIOXIDE UPTAKE BY A PHOTOSYNTHETIC METABOLIC PROCESS THAT DOES NOT PRODUCE OXYGEN IS POSTULATED TO EXPLAIN THE IMBALANCE BETWEEN CARBON DIOXIDE AND OXYGEN BUDGETS. (JONES-WISCONSIN)

FIELD 05D

ACCESSION NO. W72-05459

LAKE KINNERET: THE NUTRIENT CHEMISTRY OF THE SEDIMENTS,

KINNERET LIMNOLOGICAL LAB., TIBERIAS (ISRAEL).

C. SERRUYA.

LIMNOLOGY AND OCEANOGRAPHY, VOL 16, NO 3, P 510-521, 1971. 8 FIG, 7 TAB, 15 REF.

DESCRIPTORS:
*TROPHIC LEVEL, *LAKES, *NUTRIENTS, *CHEMICAL PROPERTIES, *SEDIMENTS, CLAYS, CORES, SEASONAL, ALGAE, MUD-WATER INTERFACES, IRON, MANGANESE, PHOSPHORUS, NITROGEN, ORGANIC MATTER, CALCIUM CARBONATE, DETRITUS, PLANKTON, ADSORPTION, EUTROPHICATION, TEMPERATURE, LIGHT PENETRATION, AMMONIA, PHOSPHATES, SALINE WATER.

IDENTIFIERS:
*LAKE KINNERET(ISRAEL), BOTTOM WATERS, RIVER JORDAN(ISRAEL).

ABSTRACT:
TO UNDERSTAND EUTROPHICATION PROCESSES IN LAKE KINNERET, ISRAEL, IMPORTANT AS A WATER SUPPLY, BOTTOM DEPOSITS WERE STUDIED. THE BOTTOM WATERS WERE SAMPLED ONCE A WEEK FOR TWO YEARS, MUD SAMPLES WERE OBTAINED WITH A DREDGE, AND CORES TAKEN. SINCE THE SEDIMENTS MAY ACTIVELY CONTRIBUTE TO THE NUTRIENT SUPPLY CAUSING WINTER ALGAL BLOOM, THE MUD-WATER EXCHANGES WERE INVESTIGATED. TEMPERATURE WAS RECORDED WITH RELATION TO THE SEASONS, COLDWATER BROUGHT IN BY THE RIVER JORDAN, DISCHARGE OF THE SUBMARINE HOT SPRINGS, AND INTERNAL WAVES. LIGHT TRANSMISSION WAS MEASURED. THE CONCENTRATIONS OF IRON, MANGANESE, PHOSPHORUS, NITROGEN, ORGANIC CARBON, AND CALCIUM CARBONATE IN RECENT SEDIMENTS WERE DETERMINED FOR THE WHOLE LAKE AND THE RESULTS DISCUSSED IN THE LIGHT OF THE CHEMICAL CONDITIONS OF BOTTOM WATERS, THE DETRITAL INPUT, AND THE PLANKTON COMPOSITION. ONLY 4% OF THE TOTAL AUTOTROPHIC CARBON IS WITHDRAWN FROM THE LAKE CYCLE BY THE SEDIMENTS. THE LAKE MUDS CONCENTRATE PHOSPHORUS IN SPITE OF THE PREVAILING REDUCING CONDITIONS, WHICH SUGGESTS AN ACTIVE PHYSICAL ADSORPTION. THE SPECIFIC CLAY SURFACES COULD BE USED AS NUTRIENT FIXERS TO SLOW EUTROPHICATION. (JONES-WISCONSIN)

FIELD 05C, 02H

ACCESSION NO. W72-05469

ABCS OF CULTURAL EUTROPHICATION AND ITS CONTROL: PART 2--WASTEWATERS,

METCALF AND EDDY, INC., BOSTON, MASS.

CLAIR N. SAWYER.

WATER AND SEWAGE WORKS, P 322-327, OCTOBER 1971. 9 FIG, 1 TAB, 4 REF.

DESCRIPTORS:
*EUTROPHICATION, *WATER POLLUTION CONTROL, *WASTE WATER(POLLUTION), VIRGINIA, NUTRIENTS, NITROGEN, PHOSPHORUS, SEWAGE EFFLUENTS, SEWAGE TREATMENT, ALGAE, WATER POLLUTION SOURCES, CHLOROPHYTA, CYANOPHYTA, WASHINGTON, DIVERSION, LAKE ERIE, NITROGEN FIXATION, DETERGENTS, CARBON DIOXIDE, ALKALINITY, BURNING, DOMESTIC WASTES, INDUSTRIAL WASTES, AGRICULTURE, SURFACE RUNOFF, WISCONSIN.

IDENTIFIERS:
*NITROGEN:PHOSPHORUS RATIO, LAKE WAUBESA(WIS), OCCOQUAN RESERVOIR(VA), MADISON(WIS), LAKE MENDOTA(WIS), LAKE MONONA(WIS), LAKE KEGONSA(WIS), LAKE WASHINGTON(WASH), GREEN LAKE(WASH).

ABSTRACT:
WASTEWATERS AS A NUTRIENT SOURCE BECOMES APPARENT IN THE CULTURAL EUTROPHICATION PROBLEM. DATA PRIOR TO THE ADVENT OF SYNTHETIC DETERGENTS CONTAINING PHOSPHATES WERE COLLECTED FROM LAKE WAUBESA, WISCONSIN. OCCOQUAN RESERVOIR, VIRGINIA, INDICATED A LARGER RELATIVE INCREASE IN PHOSPHORUS, AS COMPARED TO NITROGEN THAN THAT ENTERING LAKE WAUBESA, PROBABLY DUE TO THE HIGHER PHOSPHORUS CONTENT OF MODERN SEWAGES DUE TO PHOSPHATE-BEARING SYNTHETIC DETERGENTS. CULTURAL EUTROPHICATION CONTROL DEPENDS UPON LIMITING THE INPUT OF BOTH PHOSPHORUS AND NITROGEN. OF THE MAJOR INORGANIC NUTRIENT SOURCES, DOMESTIC AND INDUSTRIAL WASTEWATERS ARE EASIEST TO CONTAIN AND TREAT WHILE AGRICULTURAL SOURCES ARE PROBABLY THE MOST DIFFICULT TO CONTROL; PHOSPHORUS CONTROL ALONE MAY SUFFICE IN SOME LOCATIONS. LAKES WAUBESA AND KEGONSA (WISCONSIN) AND LAKE WASHINGTON HAVE RECOVERED AFTER DIVERSION OF SEWAGE. THE BASIC PHILOSOPHY IS SCIENTIFICALLY CORRECT THAT, IF THE DEGREE OF CULTURAL EUTROPHICATION IS RELATED TO THE DEGREE OF FERTILIZATION, ANY REDUCTION IN WASTEWATER QUANTITY OR IN EQUIVALENT NUTRIENTS SHOULD REDUCE PRIMARY PRODUCTIVITY. ELIMINATION OF PHOSPHATES IN DETERGENTS WOULD BE BENEFICIAL TO THE GREAT LAKES; REMOVAL OF 80% OF PHOSPHORUS FROM WASTEWATERS MAY OR MAY NOT BE ADEQUATE, DEPENDING UPON LAKE SIZE AND DETENTION TIME. (JONES-WISCONSIN)

FIELD 05C

ACCESSION NO. W72-05473

RELATIONSHIP BETWEEN PHOSPHATES AND ALKALINE PHOSPHATASE OF ANABAENA FLOS-AQUAE IN CONTINUOUS CULTURE,

QUEEN'S UNIV. KINGSTON (ONTARIO). DEPT. OF CHEMICAL ENGINEERING.

DEREK H. BONE.

ARCHIV FUR MIKROBIOLOGIE, VOL 80, P 147-153, 1971. 3 TAB, 16 REF.

DESCRIPTORS:
*ALGAE, *PHOSPHATES, *ENZYMES, *CULTURES, NITRATES, PLANT PHYSIOLOGY, PLANT GROWTH, BIOMASS, GROWTH RATES, LIGHT INTENSITY, CARBON DIOXIDE, NITROGEN, LABORATORY TESTS, CYANOPHYTA, CYTOLOGICAL STUDIES.

IDENTIFIERS:
*ALKALINE PHOSPHATASE, *ANABAENA FLOS-AQUAE, *CONTINUOUS CULTURE, CHEMOSTATS, PHOSPHATASE ACTIVITY.

ABSTRACT:
TO DEFINE INTERRELATIONSHIPS BETWEEN GROWTH AND IMPORTANT PHYSIOLOGICAL CHARACTERISTICS, THE CONTINUOUS CULTIVATION TECHNIQUE WAS APPLIED. ANABAENA FLOS-AQUAE WAS GROWN IN CHEMOSTATS WITH PHOSPHATE-LIMITING GROWTH AND VARIOUS DILUTION RATES TO ASCERTAIN THE INFLUENCE OF DILUTION RATES, PHOSPHATE AND NITRATE CONCENTRATIONS ON THE YIELDS AND ALKALINE PHOSPHATASE ACTIVITY OF THE ALGA. STEADY STATE CONDITIONS WERE ASSUMED TO EXIST WHEN THE DRY WEIGHT ENZYME ACTIVITIES AND CHEMICAL ANALYSIS WERE CONSTANT FOR FOUR CONSECUTIVE DAYS. THE CELL YIELDS WERE DEPENDENT ON DILUTION RATE AND A TWO-FOLD INCREASE OBTAINED BY GROWTH IN THE PRESENCE OF 15 MILLIMOLE POTASSIUM NITRATE. ALKALINE PHOSPHATASE ACTIVITY VARIED 20-FOLD, LOWEST ACTIVITY WITH EXCESS PHOSPHATE LIGHT-LIMITED CELLS AND HIGHEST ACTIVITY WITH CELLS GROWN IN THE PRESENCE OF 15 MILLIMOLE POTASSIUM NITRATE. THERE WAS NO CORRELATION BETWEEN HOT WATER SOLUBLE PHOSPHATE OF CELLS AND ALKALINE PHOSPHATASE ACTIVITY. INCREASING THE INTENSITY OF LIGHT 1.5 FOLD DID NOT AFFECT THE YIELD OF THE ALGAL CULTURES. THE ORTHOPHOSPHATE CONTENT OF CELLS FROM THE CHEMOSTATS, THROUGHOUT THESE EXPERIMENTS WAS IN THE RANGE OF 10-15% OF THE TOTAL PHOSPHATE INPUT, AND ORTHOPHOSPHATE WAS NEVER DETECTED IN THE GROWTH MEDIUM. (JONES-WISCONSIN)

FIELD 05C

ACCESSION NO. W72-05476

INVESTIGATION OF FOAM AS A MEANS OF DETECTING SMALL CONCENTRATIONS OF MINERAL AND ORGANIC SUBSTANCES IN NATURAL WATERS (ISSLEDOVANIYE PENY KAK METOD OBNARUZHENIYA MINERAL'NYKH I ORGANICHESKIKH VESHCHESTV, SODERZHASHCHIKHSYA V PRIRODNYKH VODAKH V MALYKH KOLICHESTVAKH),

AKADEMIYA NAUK SSSR, MOSCOW. INSTITUT BIOLOGII VNUTRENNYKH VOD.

S. M. DRACHEV, AND A. A. BYLINKINA.

IN: KHIMICHESKIYE RESURSY MOREY I OKEANOV; 'NAUKA', MOSCOW, P 202-208, 1970. 6 TAB, 7 REF.

DESCRIPTORS:
*WATER CHEMISTRY, *FOAM SEPARATION, *FOAM FRACTIONATION, *ORGANIC MATTER, *INORGANIC COMPOUNDS, IRON, MANGANESE, PHOSPHORUS, CARBON, NITROGEN, OXIDATION, BIOCHEMICAL OXYGEN DEMAND, RADIOACTIVITY, RADIOACTIVE WASTES, WASTE WATER(POLLUTION), WATER POLLUTION SOURCES, ALGAE, PLANKTON, SEASONAL.

IDENTIFIERS:
*USSR, MOSCOW RIVER, OKA RIVER, FOAM, NATURAL WATERS.

ABSTRACT:
FORMATION OF FOAM ON NATURAL AND POLLUTED WATERS OF THE MOSCOW AND OKA RIVERS IN THE FALL OF 1964 AND SUMMER OF 1965 IS DESCRIBED. THE CONTENT OF ORGANIC MATTER IN DRY FOAM RESIDUE FORMED DURING FOAM BREAKAGE IS USUALLY HIGH AND OFTEN EXCEEDS 50%. ORGANIC CARBON-NITROGEN RATIOS AND CONCENTRATIONS IN A POLLUTED REACH OF THE MOSCOW RIVER DECREASED WITH INCREASING DISTANCE FROM THE SOURCE OF POLLUTION. A HIGH FE, MN, AND P CONTENT WAS OBSERVED IN THE FOAM. THE LEVEL OF RADIOACTIVITY IN FOAM SAMPLES COLLECTED FROM NATURAL WATERS WAS HIGHER THAN THAT IN WATER AND INDICATED CONSIDERABLE CONCENTRATION OF URANIUM FISSION PRODUCTS. HIGHER RADIOACTIVITY IN FOAM IS PRESUMED TO BE RELATED TO ACCUMULATION OF IRON HYDROXIDES AND SURFACE-ACTIVE SUBSTANCES INTRODUCED WITH DOMESTIC AND INDUSTRIAL WASTES. (FOSEFSON-USGS)

FIELD 02K, 05A

ACCESSION NO. W72-05506

LOPEZ WATER SUPPLY PROJECT,

KOEBIG AND KOEBIG, INC., LOS ANGELES, CALIF.

C. H. LAWRANCE, K. G. TRANBARGER, AND R. A. CRAHN.

JOURNAL OF THE AMERICAN WATER WORKS ASSOCIATION, VOL. 63, NO. 11, P 711-727, NOVEMBER 1971. 16 FIG, 7 TAB, 15 REF.

DESCRIPTORS:
*WATER SUPPLY, *WATER RESOURCES DEVELOPMENT, *DISTRIBUTION SYSTEMS, *WATER QUALITY, WATER CONSERVATION, WILDLIFE CONSERVATION, FLOOD CONTROL, RECREATION, HYDROLOGIC DATA, WATER TREATMENT, STREAM GAGES, GROUNDWATER, SAFE YIELD, ALGAE, MINERALOGY, TURBIDITY, COLOR, ALKALINITY, VOLUMETRIC ANALYSIS, FISH, INTAKE STRUCTURES, CARBON DIOXIDE, CALCIUM CARBONATE, CALCIUM, MAGNESIUM, SODIUM, IRON, MANGANESE, SULFATES, CHLORIDES, CHLORINE, IONS, BICARBONATES, CARBONATES, COPPER, NITRATES, FLUORIDES, FLUORINE, CALIFORNIA.

IDENTIFIERS:
*LOPEZ WATER SUPPLY PROJECT, LOPEZ CREEK, ARROYO GRANDE CREEK.

ABSTRACT:
THE LOPEZ WATER SUPPLY PROJECT, A MULTIPURPOSE WATER-SUPPLY SYSTEM, HAS BEEN RECENTLY COMPLETED IN SAN LUIS OBISPO COUNTY, CALIFORNIA AT A COST OF $17,714,000. THIS PROJECT INCORPORATES WATER CONSERVATION, FLOOD CONTROL, DOMESTIC WATER SUPPLY, FISH AND WILDLIFE PRESERVATION AND RECREATION. THE LOPEZ AND ARROYO GRANDE CREEKS ARE THE SOURCES OF WATER UTILIZED BY THIS PROJECT. THEY WERE SAMPLED AND QUALITY CHECKED FOR MINERAL CONTENT, ALKALINITY, HARDNESS, COLOR AND TURBIDITY PRIOR TO AND AFTER CONSTRUCTION OF THE LOPEZ RESERVOIR. THE LABORATORY ANALYTICAL DATA FOR THE CREEKS AND THE RESERVOIR ARE SUMMARIZED IN TABULAR FORM. THE FEATURES OF THE ENTIRE PROJECT ARE SUMMARIZED INCLUDING INFORMATION ON THE INTAKE AND OUTLET FACILITIES (PIPING AND INSTRUMENTATION) OF THE RESERVOIR. ALGAL PROBLEMS AND CONTROL AND THE EFFECTS OF THE PROJECT ON FISH AND WILDLIFE ARE DISCUSSED. (HOLOMAN-BATTELLE)

FIELD 04A, 05A, 02K

ACCESSION NO. W72-05594

ALGAL DISTRIBUTION IN SIX THERMAL SPRING EFFLUENTS,

SOUTHEAST MISSOURI STATE COLL., CAPE GIRARDEAU. DEPT. OF BIOLOGY.

R. G. KULLBERG.

THE AMERICAN MICROSCOPICAL SOCIETY TRANSACTIONS, VOL 90, NO 4, P 412-434, OCTOBER 1971. 9 FIG, 6 TAB, 20 REF.

DESCRIPTORS:
*ALGAE, *DISTRIBUTION PATTERNS, *THERMAL SPRINGS, CYANOPHYTA, ALKALINITY, NUTRIENTS, WATER QUALITY, DIATOMS, CHLOROPHYTA, THERMAL WATER, *MONTANA, PHOSPHATES, NITRATES, NITRITES, IRON, SILICON, SULFATES, SODIUM, POTASSIUM, CALCIUM, MAGNESIUM, CHLORIDES, HYDROGEN ION CONCENTRATION, PERIPHYTON, WATER TEMPERATURE.

IDENTIFIERS:
ALHAMBRA HOT SPRINGS, BOULDER HOT SPRINGS, JACKSON HOT SPRINGS, LOLO HOT SPRINGS, PIPESTONE HOT SPRINGS, SLEEPING CHILD HOT SPRINGS.

ABSTRACT:
THE DISTRIBUTION OF ALGAE WAS STUDIED IN SIX THERMAL SPRINGS OF WESTERN MONTANA. ALHAMBRA (NORTH), 54.4 C; ALHAMBRA (SOUTH) 48.0 C; BOULDER, 61.3 C; JACKSON, 61.5 C; LOLO, 46.0 C; PIPESTONE, 59.5 CP AND SLEEPING CHILD, 52.0 C. THE BLUE-GREEN ALGAE (MYXOPHYCEAE) WERE THE ONLY ALGAE NEAR THE SOURCES OF THE STREAMS. THE MEAN MAXIMUM TEMPERATURE ENDURED BY THE DIATOMS (BACILLARIOPHYCEAE) WAS 43.2 C; BY THE GREEN ALGAE (CHLOROPHYCEAE), 40.9 C. PRESENCE LISTS OF THE ALGAE ALONG A TEMPERATURE GRADIENT INDICATE THE ORDER IN WHICH THE ALGAE APPEARED IN THE STREAM AS THE WATER COOLED. THE PER CENT VOLUMES FOR THE MAJOR AND INTERMEDIATE SPECIES WERE PLOTTED ALONG THE TEMPERATURE GRADIENT TO SHOW THE INTERACTIONS AMONG THE POPULATIONS OF THE CONTINUUM. THE STREAMS CONTAINING THE GREATEST VARIATIONS OF HABITATS RESULTED IN MORE ERRATIC CURVES OF SEVERAL MODES AMONG THE MAJOR POPULATIONS AND SCATTERED OCCURRENCES AMONG THE INTERMEDIATE SPECIES. VARIATIONS IN TYPES OF HABITATS DOWNSTREAM PROMOTED THE RELOCATION OF TRANSLOCATED ALGAL MASSES UNTIL OVERWHELMED BY THE PREVIOUSLY ESTABLISHED POPULATIONS. FIVE NEW TAXA ARE DESCRIBED: CHAMAESIPHON PRESCOTTI N. SP.; PSEUDANABAENA OBLONGA N. SP.; SYNECHOCOCCUS LIVIDUS COPELAND VAR. NANUM N. VAR; OSCILLATORIA GEMINATA COPELAND VAR. FRAGILIS FORMA BREVE N. FORMA; OSCILLATORIA GEMINATA COPELAND VAR. TENELLA FORMA MINOR N. FORMA. (MORTLAND-BATTELLE)

FIELD 05B, 05C

ACCESSION NO. W72-05610

POLYCHLORINATED BIPHENYLS: TOXICITY TO CERTAIN PHYTOPLANKTERS,

STATE UNIV. OF NEW YORK, STONY BROOK. MARINE SCIENCES RESEARCH CENTER.

J. L. MOSSER, N. S. FISHER, T. C. TENG, AND C. F. WURSTER.

SCIENCE, VOL 175, NO 4018, P 191-192, JANUARY 14, 1972. 1 FIG, 18 REF.

DESCRIPTORS:
*POLYCHLORINATED BIPHENYLS, *ALGAE, *WATER POLLUTION EFFECTS, *PESTICIDES, DIATOMS, PHYTOPLANKTON, TOXICITY, CULTURES, DDT, GROWTH RATES.

IDENTIFIERS:
THALASSIOSIRA PSEUDONANA, SKELETONEMA COSTATUM, DUNALIELLA TERTIOLECTA, EUGLENA GRACILIS, CHLAMYDOMANAS REINHARDTII, LETHAL DOSAGE.

ABSTRACT:
POLYCHLORINATED BIPHENYLS (PCBS) AND DDT WERE ADDED TO CULTURES OF FIVE SPECIES OF UNICELLULAR ALGAE TO INVESTIGATE THEIR EFFECTS ON GROWTH RATE. THE GROWTH RATES OF TWO CENTRIC DIATOMS (THALASSIOSIRA PSEUDONANA AND SKELETONEMA COSTATUM) WERE REDUCED BY PCB CONCENTRATIONS AS LOW AS 25 PPB FOR THE FORMER AND 10 PPB FOR THE LATTER. FOR BOTH THESE SPECIES, PCBS WERE MORE TOXIC THAN THE EQUIVALENT AMOUNT OF DDT. BY CONTRAST, A MARINE GREEN ALGAE (DUNALIELLA TERTIOLECTA) AND TWO FRESHWATER ALGAE (EUGLENA GRACILIS AND CHLAMYDOMONAS REINHARDTII) WERE NOT INHIBITED AT THESE, OR HIGHER, CONCENTRATIONS. THIS SENSITIVITY PARALLELED THAT TO DDT. (MORTLAND-BATTELLE)

FIELD 05C

ACCESSION NO. W72-05614

GREENE COUNTY SURFACE WATER RESOURCES,

ILLINOIS DEPT. OF CONSERVATION, SPRINGFIELD. DIV. OF FISHERIES.

R. LOCKART.

ILLINOIS DIVISION OF FISHERIES REPORT, AUGUST 1971. 44 P, 5 FIG, 12 TAB, 12 REF.

DESCRIPTORS:
*SURFACE WATERS, *LAKES, *STREAMS, *REVIEWS, *ILLINOIS, RECREATION, FISHING, SWIMMING, BOATING, DOCUMENTATION, WATER QUALITY, SILTING, AGRICULTURAL CHEMICALS, WATER POLLUTION SOURCES, ALGAE, GEOLOGY, DATA COLLECTIONS.

IDENTIFIERS:
*GREENE COUNTY(ILL).

ABSTRACT:
GREENE COUNTY, ILLINOIS, HAS A TOTAL SURFACE WATER INVENTORY OF 1,136.0 ACRES OF LAKES AND PONDS AND 494.70 ACRES OF NAMED STREAMS. THE LARGEST SINGLE WATER AREA (58.8 ACRES) IN GREENE COUNTY IS GREENFIELD CITY LAKE. THE MAIN DRAINAGE SYSTEMS ARE APPLE CREEK IN THE NORTHERN PART OF THE COUNTY AND MACOUPIN CREEK IN THE SOUTHERN PART OF THE COUNTY. MOST OF THE STREAMS IN GREENE COUNTY ARE SLOW MOVING. THE TWO EXCEPTIONS TO THIS ARE HURRICANE CREEK AND WOLF RUN. FLOODING ON ALL STREAMS IS A YEARLY OCCURRENCE DURING THE SPRING RAINS. ALONG THE ILLINOIS RIVER AND APPLE CREEK FLOODPLAIN DRAINAGE IS A MAJOR PROBLEM. MANY CANALS AND DITCHES HAVE BEEN CONSTRUCTED IN THESE TWO AREAS. SILTATION OCCURS IN ALL OF THE SLOWER MOVING STREAMS AND IS THE MAIN SOURCE OF WATER POLLUTION IN GREENE COUNTY. THE IMPOUNDMENTS AND STREAMS ARE MEDIUM HARD IN ALKALINITY. THE HARDNESS RANGES FROM 125 TO 350 PPM. PH VALUES RANGE FROM 7.4 TO 8.0. MOST OF THE LAKES AND PONDS CONTAIN SOME SPECIES OF SPORT FISH SUCH AS LARGEMOUTH BASS, BLUEGILL, CHANNEL CATFISH, AND BULLHEADS. (WOODARD-USGS)

FIELD 02H, 02E, 05B

ACCESSION NO. W72-05869

MERCURY: ITS OCCURRENCE AND EFFECTS IN THE ECOSYSTEM,

CORNELL UNIV., ITHACA, N.Y. ECOLOGY AND SYSTEMATICS SECTION.

D. B. PEAKALL, AND R. J. LOVETT.

BIOSCIENCE, VOL. 22, NO. 1, P 20-25, JANUARY 1972. 1 FIG, 3 TAB, 69 REF.

DESCRIPTORS:
*WATER POLLUTION, *FOOD CHAINS, *ECOSYSTEMS, ABSORPTION, FISH, INDUSTRIAL WASTES, ALGAE, AQUATIC PLANTS, SEA WATER, TRACE ELEMENTS, TOXICITY, CHROMOSOMES, FUNGICIDES, PESTICIDES, ECOLOGY, BIRD TYPES, BIRDS, AQUATIC ANIMALS, MOSSES, INVERTEBRATES, DIATOMS, PHYTOPLANKTON, PHOTOSYNTHESIS, WATER POLLUTION SOURCES, WATER POLLUTION EFFECTS, DDT, OCEANS, AIR POLLUTION, VOLCANOES, BIOASSAY, DIETS, SOIL MICROORGANISMS, ANIMALS, URINE, GENETICS, PATH OF POLLUTANTS.

IDENTIFIERS:
*MERCURY, *METHYLMERCURY, BRAIN, ORGANOMERCURIAL COMPOUNDS, ESOX LUCIUS, ALKYL MERCURY, EAGLES, SALMO GAIRDNERII, IRIS PSEUDACORUS, LYSIMACHIA NUMMULARIA, BLUE HERON, STERNA HIRUNDO, RATS, PARTRIDGE, DDE, KIDNEYS, LIVER, BLOOD, GASTEROSTEUS ACULEATUS, DAPHNIA MAGNA, DAPHNIA, GASTEROSTEUS.

ABSTRACT:
THE WORLD PRODUCTION AND INDUSTRIAL USES OF MERCURY, ITS NATURAL AND INDUSTRIAL ROUTES OF CONTAMINATION OF THE WATER AND OF THE ATMOSPHERE, AND ITS TOXICITY TO ANIMALS AND MAN ARE REVIEWED. MERCURY CAN FORM STABLE ORGANOMERCURY COMPOUNDS. OF THESE METHYL MERCURY IS OF SPECIAL IMPORTANCE BECAUSE OF ITS TOXICITY, ITS SLOW EXCRETION FROM THE BODY, AND ITS RELATIVELY LONG BIOLOGICAL HALF-LIFE. THE HUMAN HAZARD OF ENVIRONMENTAL MERCURY SEEMS TO BE FOR PERSONS WHOSE MAIN DIET IS FISH WHICH CONTAIN METHYL MERCURY OBTAINED THROUGH THE AQUATIC FOOD CHAIN. MERCURY IN THE ECOSYSTEM APPEARS TO MAINTAIN A BALANCED LEVEL. MERCURY LEVELS IN THE OCEAN HAVE NOT BEEN CHANGED BY MAN'S ACTIVITY. HOWEVER, MAN'S USE OF MERCURY CAN HAVE A SIGNIFICANT EFFECT ON FRESHWATER AND THE LAND ENVIRONMENT. (JEFFERIS-BATTELLE)

FIELD 05B, 05A, 05C

ACCESSION NO. W72-05952

RADIOCHEMICAL DETERMINATION OF PLUTONIUM IN SEA WATER, SEDIMENTS AND MARINE ORGANISMS,

WOODS HOLE OCEANOGRAPHIC INSTITUTION, MASS.

K. M. WONG.

ANALYTICA CHIMICA ACTA, VOL 56, NO 3, P 355-364, OCTOBER 1971. 1 FIG, 5 TAB, 18 REF.

DESCRIPTORS:
*RADIOCHEMICAL ANALYSIS, *SEA WATER, *SEDIMENTS, *MARINE ANIMALS, *MARINE ALGAE, RADIOISOTOPES, FALLOUT, CHEMICAL PRECIPITATION, IRON, BIOASSAY, LEACHING, VOLUMETRIC ANALYSIS, SOIL ANALYSIS, SPECTROMETERS, ZOOPLANKTON, MUSSELS, CLAMS, FISH, SHELLFISH, AQUEOUS SOLUTIONS, PHAEOPHYTA, ATLANTIC OCEAN, SHARKS, RADIOACTIVITY TECHNIQUES.

IDENTIFIERS:
*PLUTONIUM, DETECTION LIMITS, BIOLOGICAL SAMPLES, SARGASSO WEED, STARFISH, ASTERIAS FORBESI, PLUTONIUM 236, PLUTONIUM 238, PLUTONIUM 239, POLYMERIZATION, LIVER, BONE, DETECTION LIMITS, CHEMICAL RECOVERY, SAMPLE PREPARATION.

ABSTRACT:
A RADIOCHEMICAL PROCEDURE FOR PLUTONIUM IS DESCRIBED WHICH HAS A SENSITIVITY OF 0.004 D.P.M. PER 100 LITER OF SEA WATER (FOR A 50-LITER SAMPLE), 0.02 D.P.M. PER KG OF SEDIMENTS (100-G SAMPLE) AND 0.002 D.P.M. PER KG OF MARINE ORGANISMS (1-KG SAMPLE). AN IRON(II) HYDROXIDE COPRECIPITATION METHOD IS USED FOR THE CONCENTRATION OF PLUTONIUM IN SEA WATER. A NITRIC-HYDROCHLORIC ACID LEACHING METHOD IS ADAPTED FOR THE TREATMENT OF SEDIMENTS AND ASHED ORGANISMS. OF 30 SAMPLES OF 5-60 LITERS OF SEA WATER ANALYZED RADIOCHEMICALLY, THE AVERAGE PLUTONIUM RECOVERY WAS 34-70 PERCENT AS COMPARED WITH 11-39 PERCENT RECOVERY ON 38 SAMPLES BY THE IRON(III) HYDROXIDE METHOD. A POSSIBLE REASON FOR THE LOW RECOVERY WITH THE IRON(III) HYDROXIDE PROCEDURE IS THAT PLUTONIUM IS LOST THROUGH THE FORMATION OF PLUTONIUM(IV) POLYMERS. THE AVERAGE RECOVERY OF 75 SAMPLES OF SEDIMENTS AND ASHED ORGANISMS WAS 43-83 PERCENT BY THE LEACHING METHOD. THE RESULTS OF THE VARIOUS REPLICATE ANALYSES ARE TABULATED, AND THE FACTORS INFLUENCING THE RECOVERY, CONTAMINATION, AND BLANK ACTIVITY ARE DISCUSSED. (HOLOMAN-BATTELLE)

FIELD 05A, 02K, 02L

ACCESSION NO. W72-05965

ECOLOGICAL INVESTIGATIONS ON THE PLANKTON OF THE RIVERIS RESERVOIR (OKOLOGISCHE UNTERSUCHUNGEN AM PLANKTON DER RIVERISTALSPERRE),

BONN UNIV. (WEST GERMANY).

H. SCHNITZLER.

ARCHIV FUR HYDROBIOLOGIE, VOL 69, NO 1, P 60-94, AUGUST 1971. 18 FIG, 17 REF.

DESCRIPTORS:
*PLANKTON, *RESERVOIRS, *AQUATIC ALGAE, *AQUATIC ANIMALS, *BIOMASS, *PRIMARY PRODUCTIVITY, *SECONDARY PRODUCTIVITY, PHYTOPLANKTON, ZOOPLANKTON, EUTROPHICATION, OLIGOTROPHY, PROTOZOA, CHLORELLA, CHLAMYDOMONAS, CYANOPHYTA, CHLOROPHYTA, CHRYSOPHYTA, DIATOMS, DINOFLAGELLATES, SCENEDESMUS, ROTIFERS, CRUSTACEANS, COPEPODS, NUTRIENTS, NITROGEN, PHOSPHORUS, ECOLOGY, BIOINDICATORS, WATER QUALITY, NITRATES, PHOSPHATES, SEASONAL, DISTRIBUTION PATTERNS.

ABSTRACT:
FROM NOVEMBER 1966 TO OCTOBER 1968 THE QUANTITY AND QUALITY OF THE ZOOPLANKTON AND PHYTOPLANKTON OF THE RIVERIS RESERVOIR NEAR TRIER WERE INVESTIGATED IN RELATION TO THE ABIOTIC FACTORS IN THE ENVIRONMENT. THE QUANTITY OF INORGANIC NUTRIENTS COMING INTO THE RESERVOIR WAS CALCULATED FROM THE RESULTS OF CHEMICAL ANALYSIS OF THE TRIBUTARIES. THE RESULTS OBTAINED BY MEASURING NITROGENOUS AND PHOSPHOROUS COMPOUNDS AND OXYGEN QUALIFY THE RIVERIS RESERVOIR AS AN OLIGOTROPHIC WATER. THE PHYTOPLANKTON AND ZOOPLANKTON SPECIES FOUND ARE TABULATED AND THE SEASONAL DISTRIBUTION OF THE MORE COMMON SPECIES IS DIAGRAMMED. MOST OF THE PHYTOPLANKTON SPECIES ARE CHARACTERISTIC OF OLIGOTROPHIC WATERS. THE BIOMASS OF THE PHYTOPLANKTON WAS CALCULATED AND OWING TO THE OCCURRENCE OF THE LARGE GYMNODINIUM UBERRIMUM THERE WAS FROM TIME TO TIME IN SUMMER 1967 A COMPARATIVELY HIGH TOTAL BIOMASS WHICH, HOWEVER, SHOULD NOT NECESSARILY BE INTERPRETED AS A SIGN OF INCIPIENT EUTROPHY. (HOLOMAN-BATTELLE)

FIELD 05C, 05A, 02H

ACCESSION NO. W72-05968

MERCURY INHIBITION ON LIPID BIOSYNTHESIS IN FRESHWATER ALGAE,

WESTERN WASHINGTON STATE COLL., BELLINGHAM, WASHINGTON. DEPT. OF CHEMISTRY.

ROBERT S. MATSON, GEORGE E. MUSTOE, AND S. B. CHANG.

ENVIRONMENTAL SCIENCE AND TECHNOLOGY, VOL 6, NO 2, FEBRUARY 1972, P 158-160.
2 FIG, 2 TAB, 12 REF.

DESCRIPTORS:
*ALGAE, *INHIBITION, *LIPIDS, TOXICITY, HEAVY METALS, LABORATORY TESTS, PHOTOSYNTHESIS, CHLOROPHYLL, ENZYMES, PERMEABILITY, WATER QUALITY CONTROL, WATER POLLUTION EFFECTS.

IDENTIFIERS:
*MERCURY, *SYNTHESIS, MERCURIC CHLORIDE, METHYL MERCURIC CHLORIDE, ANKISTRODESMUS BRAUNII, EUGLENA GRACILIS.

ABSTRACT:
WHOLE CELLS OF SPECIMENS OF UNICELLULAR ALGAE, ANKISTRODESMUS BRAUNII AND EUGLENA GRACILIS, WERE EXPOSED TO VARIOUS CONCENTRATIONS OF INORGANIC MERCURIC AND METHYL MERCURIC CHLORIDE TO DETERMINE IF INHIBITION OF LIPID BIOSYNTHESIS WAS ONE OF THE TOXIC EFFECTS OF MERCURY COMPOUNDS. AT A MERCURIC CHLORIDE CONCENTRATION OF 3.5 PPM, CHLOROPHYLL SYNTHESIS WAS 98% INHIBITED AND GALACTOLIPID SYNTHESIS WAS 50% INHIBITED IN A. BRAUNII. HOWEVER, SIGNIFICANT INHIBITION OF BOTH SYNTHESIS PROCESSES WAS DETECTED AT MERCURIC CHLORIDE LEVELS LESS THAN 1.0 PPM. FOR METHYL MERCURIC CHLORIDE, A 2.0 PPM LEVEL INHIBITED 98% OF CHLOROPHYLL SYNTHESIS AND 85% OF GALACTOLIPID SYNTHESIS. THESE MERCURY COMPOUNDS WERE ALSO SHOWN TO SPECIFICALLY INHIBIT THE GALACTOSYL TRANSFERASE ACTIVITY IN EUGLENA CHLOROPLASTS. THESE RESULTS COULD LOGICALLY BE EXTENDED TO OTHER PHOTOSYNTHETIC ORGANISMS, INCLUDING PHYTOPLANKTON AND OTHER SPECIES OF UNICELLULAR ALGAE. (LOWRY-TEXAS)

FIELD 05C

ACCESSION NO. W72-06037

THE ROLE OF BENTHIC PLANTS IN A FERTILIZED ESTUARY,

HARVARD UNIV., CAMBRIDGE, MASS. LAB. OF APPLIED MICROBIOLOGY.

T. WAITE, AND R. MITCHEL.

PREPRINT, PRESENTED AT 44TH ANNUAL CONFERENCE OF WATER POLLUTION CONTROL FEDERATION, SESSION 13, NO. 1, SAN FRANCISCO, CALIFORNIA, OCTOBER 5, 1971, 17 P, 1 FIG, 6 TAB, 17 REF.

DESCRIPTORS:
*NUTRIENTS, *BENTHIC FAUNA, *BENTHIC FLORA, *PHYTOPLANKTON, ALGAE, GROWTH RATES, BIOASSAY, DISSOLVED OXYGEN, CARBON DIOXIDE, AMMONIA, PHOSPHATE, ANALYTICAL TECHNIQUES, *PHOTOSYNTHESIS, CARBON FIXATION, EUTROPHICATION, WATER QUALITY CONTROL, MASSACHUSETTS, *NUTRIENTS, OUTLETS.

IDENTIFIERS:
*WOODS HOLE(MASS).

ABSTRACT:
FIELD PRODUCTIVITY TESTS WERE CONDUCTED USING CARBON-14 TECHNIQUES TO COMPARE PHYTOPLANKTON AND BENTHIC PLANTS. TEST STATIONS WERE CHOSEN AT VARIOUS DISTANCES FROM THE WOODS HOLE, MASSACHUSETTS, SEWAGE OUTFALL. THE BOTTOM FLORA OF THE AREA GENERALLY CONSISTS OF A LITTORAL ZONE, AND THE PRODUCTIVITY DETERMINATIONS WERE RUN ONLY ON THE BENTHIC ALGAL ZONE OR NON-ROOTED PLANTS. ON AN AREA BASIS THE BENTHIC PLANTS IN THE LITTORAL ZONE FIX CARBON AT ABOUT 40 TIMES THE RATE OF PHYTOPLANKTON. ALSO, THE WATER MOVEMENT ASSOCIATED WITH ESTUARINE OR COASTAL AREAS MIXED THE PHYTOPLANKTON DILUTING THE EFFECT OF ENRICHMENT, BUT DID NOT MIX BENTHIC MACROPHYTES WHICH WERE CONTINUALLY AFFECTED BY ENRICHMENT. THE BENTHIC PLANTS ARE BETTER INDICATORS OF NUTRIENT STIMULATION. IT IS PROPOSED THAT THE BAULEP MITSCHERLICH EQUATION BE USED TO PREDICT THE CONTRIBUTION TO PHOTOSYNTHETIC YIELD OF THE BOTTOME FLORA AS A FUNCTION OF NUTRIENT ENRICHMENT, DISCOUNTING THE CONCEPT OF A 'SINGLE' LIMITING NUTRIENT. (MORGAN-TEXAS)

FIELD 05C, 02L, 05B

ACCESSION NO. W72-06046

METHODS IN MICROBIOLOGY-VOLUME 5B.

ACADEMIC PRESS INC., NEW YORK, N.Y. 1971. J. R. NORRIS AND D. W. RIBBONS, EDITORS. 695 P, 21 REF.

DESCRIPTORS:
*SEPARATION TECHNIQUES, *ALGAE, *BACTERIA, *FUNGI, *BACTERIOPHAGE, *VIRUSES, ABSORPTION, ELECTROPHORESIS, DETERGENTS, AMINO ACIDS, SPECTROPHOTOMETRY, MAGNESIUM, SULFATES, SPORES, AZOTOBACTER, CYANOPHYTA, YEASTS, CARBON DIOXIDE, CHLORELLA, CLOSTRIDIUM, E. COLI, EUGLENA, EUGLENOPHYTA, CHLOROPHYTA, FERROBACILLUS, GAS CHROMATOGRAPHY, IONS, CALCIUM, CARBON, SNAILS, LACTOBACILLUS, MAGNESIUM, MANGANESE, MYCOBACTERIUM, HYDROGEN ION CONCENTRATION, PHOSPHATES, POTASSIUM, SALMONELLA, SHIGELLA, PROTOZOA, SODIUM, CENTRIFUGATION, ULTRASONICS, CULTURES, HEAVY METALS, IRON, COLIFORMS, ISOLATION.

ABSTRACT:
THE DISINTEGRATION OF CELLS, THEIR CHEMICAL ANALYSIS, AND THE TECHNIQUES USED TO SEPARATE AND CHARACTERIZE THEIR COMPONENTS ARE DISCUSSED. TOPICS INCLUDED ARE: FREE-FLOW ELECTROPHORESIS, DISC ELECTROPHORESIS, PREPARATIVE ZONAL ELECTROPHORESIS, REFLECTANCE SPECTROPHOTOMETRY, ISOELECTRIC FOCUSING AND SEPARATION OF PROTEINS, CHEMICAL EXTRACTION METHODS OF MICROBIAL CELLS, CHEMICAL ANALYSIS OF MICROBIAL CELLS, CENTRIFUGAL TECHNIQUES FOR ISOLATION AND CHARACTERIZATION OF SUB-CELLULAR COMPONENTS FROM BACTERIA. (HOLOMAN-BATTELLE)

FIELD 05A, 02K

ACCESSION NO. W72-06124

MICROSCOPIC WATER QUALITY AND FILTRATION EFFICIENCY,

NEW YORK STATE DEPT. OF HEALTH, ALBANY. BUREAU OF WATER AND WASTEWATER UTILITIES MANAGEMENT.

S. SYROTYNSKI.

JOURNAL AMERICAN WATER WORKS ASSOCIATION, VOL. 63, NO. 4, P 237-245, APRIL 1971. 8 FIG, 6 TAB, 3 REF.

DESCRIPTORS:
*WATER TREATMENT, *FILTRATION, *MICROORGANISMS, *SURFACE WATERS, *PLANKTON, *ALGAE, TURBIDITY, *NEW YORK, *WATER QUALITY.

IDENTIFIERS:
*MICROSCOPIC COUNTS, *AMORPHOUS MATTER.

ABSTRACT:
HISTORICAL RECORDS OF THE NEW YORK STATE DEPARTMENT OF HEALTH FOR THE PERIOD 1946-1967 WERE STUDIED. IN SURFACE WATERS THE MEAN VALUES FOR TOTAL MICROSCOPIC COUNT, AMORPHOUS MATTER, TOTOAL PLANKTON AND TOTAL ALGAE WERE 2,000, 1,500, 350, AND 140 ASU/ML, RESPECTIVELY. THE RANGE OF TOTAL MICROSCOPIC COUNTS WAS FROM 14 TO 195,900 ASU/ML; VALUES ABOVE 20,000 GENERALLY WERE ASSOCIATED WITH HIGH TURBICY (TO 30 UNITS). OF 172 TURBIDITY VALUES WITHIN THE RANGE 1,500-2,500 ASU/ML, 91/8 PERCENT WERE LESS THAN 10, AND 65.7 PERCENT LESS THAN 5 TURBIDITY UNITS. A MEAN VALUE OF 140 ASU/ML WAS DETERMINED FOR TOTAL ALGAE, THOUGH MAXIMUM VALUES WERE AS HIGH AS 10,700. THE HISTORICAL DATA FOR FILTERED-WATER TOTAL COUNTS SHOW 80 PERCENT EQUAL TO OR LESS THAN 200 ASU/ML. A VALUE OF 200 ASU/ML TOTAL MICROSCOPIC COUNT IS RECOMMENDED AS A LIMIT IN FILTER EFFLUENT QUALITY. OF UNTREATED SURFACE WATERS 3.65 PERCENT CONTAINED LESS THAN 200 ASU/ML AND 7 PERCENT LESS THAN 300. (BEAN-AWWARF)

FIELD 05F, 05A

ACCESSION NO. W72-06191

PHOSPHATE AND TURBIDITY CONTROL BY FLOCCULATION AND FILTRATION,

WAHNBACHTALSPERRENVERBAND, SIEGBURG (WEST GERMANY).

H.BERNHARDT, J. CLASEN, AND H. SCHELL.

JOURNAL OF AMERICAN WATER WORKS ASSOCIATION, VOL. 63, NC. 6, P 355-368, JUNE 1971. 16 FIG, 12 TAB, 40 REF.

DESCRIPTORS:
*WATER TREATMENT, *PHOSPHATES, *TURBIDITY, *FLOCCULATION, *FILTRATION, *ALGAE, ELECTROLYTES, ACTIVATED CARBON, HYDROGEN ION CONCENTRATION, NUTRIENTS.

IDENTIFIERS:
*FERRIC CHLORIDE, MULTI-LAYER FILTERS, ANTHRACITE, *WAHNBACH RESERVOIR(GERMANY), OSCILLATORA RUBESCENS.

ABSTRACT:
HIGH NUTRIENT LOAD ENTERING THE WAHNBACH RESERVOIR HAS LED TO INCREASES IN BIO-PRODUCTION AND A CHANGE IN PREDOMINANT SPECIES FROM DIATOMACEAE TO BLUE-GREEN ALGAE, THE MAIN ALGAE BEING OSCILLATORIA RUBESCENS, WHICH OCCUR IN MASSES. BECAUSE 70 PERCENT OF THE NUTRIENTS ORIGINATE FROM AGRICULTURAL ACTIVITY, EXCESS FERTILIZATION CANNOT BE COMBATTED BY TREATING ONLY THE DOMESTIC SEWAGE OF THE 8,000 INHABITANTS. ALSO IRON COMPOUNDS ARE PRESENT IN THE RUNOFF. THE RIVER WATER WILL BE TREATED AS IT FLOWS INTO THE RESERVOIR BY FLOCCULATION WITH FERRIC IRON AND FILTRATION, SO THAT TOTAL PHOSPHORUS WILL BE REDUCED TO A MAXIMUM OF 10 PPB. ALSO TURBIDITY WILL BE REMOVED. REMOVAL OF PHOSPHATE IS OPTIMUM AT A PH OF UP TO 7.2 USING 40 TIMES THE STOICHIOMETRIC FERRIC IRON REQUIREMENT. USE OF AN ANIONIC FLOCCULANT AID EXTENDED FILTER RUNS TO APPROXIMATELY DOUBLE THE ORIGINAL LENGTH. USE OF A THREE-LAYER FILTER PRODUCED BETTER FILTER RUNS THAN A ONE-LAYER OR TWO-LAYER FILTER AND BETTER USE OF THE FILTER BED. BY COMBINING ACTIVATED CARBON (THE LIGHTEST LAYER), HYDROANTHRACITE (CENTRAL LAYER), AND QUARTZ SAND (BOTTOM) A STABLE, THREE LAYER FILTER WAS CONSTRUCTED WITHOUT USE OF EXPENSIVE HEAVY MATERIALS. (BEAN-AWWARF)

FIELD 05F, 05C

ACCESSION NO. W72-06201

THE BEHAVIOR OF CHLORELLA PYRENOIDOSA IN STEADY STATE CONTINUOUS CULTURE,

CALIFORNIA UNIV., BERKELEY. LAWRENCE RADIATION LAB.

J. N. DABES, C. R. WILKE, AND K. H. SAUER.

AVAILABLE FROM THE NATIONAL TECHNICAL INFORMATION SERVICE AS UCRL-19958, $3.00 IN PAPER COPY, $0.95 IN MICROFICHE. AUGUST 1970. 184 P, 43 FIG, 3 TAB, 138 REF.

DESCRIPTORS:
*ALGAE, *CULTURES, PHOTOSYNTHESIS, BIOMASS, LIGHT INTENSITY, GROWTH RATES, MATHEMATICAL MODELS, WATER TEMPERATURE, WATER QUALITY, CARBON DIOXIDE, HYDROGEN ION CONCENTRATION, CARBON, OXYGEN, NITROGEN, HYDROGEN, PRIMARY PRODUCTIVITY, RESPIRATION, PLANT PHYSIOLOGY, NUTRIENTS, NITRATES, SULFATES, PHOSPHATES, CHLORIDES, BORATES, ABSORBANCE.

IDENTIFIERS:
DNA, CHLOROPHYLL A, CHLORELLA PYRENOIDOSA, SPIROGIRA, CHLORELLA ELLIPSOIDEA, RNA, CHLOROPHYLL B, ASHING, LABORATORY TECHNIQUES.

ABSTRACT:
THE GROWTH OF CHLORELLA PYRENOIDOSA IN STEADY STATE CONTINUOUS CULTURE WAS NEVER LIMITED BY CO2, MINERALS, PH, OR TEMPERATURE. THE EFFECTS OF THE TWO REMAINING INDEPENDENT VARIABLES, SPECIFIC GROWTH RATE AND INCIDENT LIGHT INTENSITY, ON ALGAL BIOMASS PRODUCTIVITY AND ALGAL PHYSIOLOGY WERE EXAMINED. IT WAS FOUND THAT OPTIMUM ALGAL BIOMASS PRODUCTIVITY WAS OBTAINED AT A SPECIFIC GROWTH RATE OF APPROXIMATELY 1.6 DAY, WHEN THE INCIDENT LIGHT INTENSITY WAS 8.05 MW/SQ CM. THIS OPTIMUM SPECIFIC GROWTH RATE IS NOT EXPECTED TO CHANGE SIGNIFICANTLY AS A FUNCTION OF INCIDENT LIGHT INTENSITY. TOTAL CHLOROPHYLL CONTENT, CHLOROPHYLL A/CHLOROPHYLL B RATIO, LIGHT SATURATED RATE OF PHOTOSYNTHESIS, DARK RESPIRATION RATES, AND RNA CONTENT WERE STRONG FUNCTIONS OF SPECIFIC GROWTH RATE. ON THE OTHER HAND, MAXIMUM QUANTUM EFFICIENCY, LIGHT SATURATED RATE AND MAXIMUM QUANTUM EFFICIENCY OF THE QUINONE HILL REACTION, AND DNA CONTENT CHANGED LITTLE, IF AT ALL, AS A FUNCTION OF SPECIFIC GROWTH RATE. A MATHEMATICAL EXPRESSION FOR THE LIGHT RESPONSE CURVE OF PHOTOSYNTHESIS WAS FORMULATED AND A MATHEMATICAL MODEL FOR THE PERFORMANCE OF OPTICALLY DENSE ALGAL SYSTEMS, WHICH ARE OF INTEREST FOR THE MASS CULTURE OF ALGAE, IS PRESENTED. THIS MODEL DIFFERS FROM PREVIOUS MODELS, SINCE IT USES THE ABOVE-MENTIONED LIGHT RESPONSE CURVE TO DESCRIBE THE LOCAL RATE OF PHOTOSYNTHESIS AND ALSO COUNTS FOR CHANGES IN THE PHYSIOLOGY OF THE ALGAE. (MORTLAND-BATTELLE)

FIELD 05C, 05A

ACCESSION NO. W72-06274

IMPROVEMENT AND APPLICATION OF BENTHIC ALGAL ISOTOPE PRODUCTIVITY MEASURING METHODS,

HAWAII UNIV., HONOLULU. DEPT. OF BOTANY.

M. S. DOTY.

AVAILABLE FROM THE NATIONAL TECHNICAL INFORMATION SERVICE AS UH-235-P-4-5, $3.00 IN PAPER COPY, $0.95 IN MICROFICHE. JUNE 1971. 94 P, 12 FIG, 14 TAB, 51 REF. CONT. NO. AT(04-3)-235.

DESCRIPTORS:
*MARINE ALGAE, *BENTHIC FLORA, *PRIMARY PRODUCTIVITY, *ENERGY CONVERSION, STANDING CROP, CHLOROPHYTA, CYANOPHYTA, RHODOPHYTA, PHAEOPHYTA, CARBON RADIOISOTOPES, MARINE FISH, INSTRUMENTATION, ABSORPTION, SEDIMENT DISCHARGE, REGRESSION ANALYSIS, BIOMASS, METHODOLOGY, REEFS, SAMPLING, OXYGEN, TEMPERATURE, LIGHT, WAVES(WATER), PHYTOPLANKTON, LIMESTONES, GASTROPODS, MOLLUSKS, INVERTEBRATES, CRABS, PHOTOSYNTHESIS, MARINE ANIMALS, BENTHIC FAUNA, PACIFIC OCEAN, RESPIRATION, RADIOACTIVITY TECHNIQUES, REPRODUCTION, ABSORPTION, CARBON, TRACERS.

IDENTIFIERS:
SEAWEEDS, APHIDS, SPONGES, CARBON-14.

ABSTRACT:
THERE IS NEED FOR A QUANTIFICATION OF THE DIFFERENT FACTORS THAT CAUSE VARIATION IN BENTHIC ALGAL POPULATIONS AND THE PASSAGE OF ENERGY AND MATERIALS THROUGH THEM. A REVIEW OF THREE YEARS' WORK ON THE ROLES OF BENTHIC ALGAE (PRIMARY PRODUCTIVITY, SEDIMENT DEPOSITION, LIMESTONE PRODUCTION) IN THE TROPICAL PACIFIC IS GIVEN. THE WORK CARRIED OUT IN THE 3-YEAR PERIOD INCLUDED (1) TECHNIQUES FOR ELUCIDATING THE ROLES OF BENTHIC ALGAE; (2) QUANTITATIVE DESCRIPTIONS OF THE ALGAL CROPS, THEIR COMPONENTS, AND COMPOSITION, AS WELL AS PRODUCTIVITY; AND (3) DETERMINATIONS OF STANDING CROP OR BIOMASS(C), PRIMARY PRODUCTION OR PHOTOSYNTHESIS(P), RESPIRATION AND REPRODUCTION(R), AND LOSSES TO THE ALGAL CROP(D). NEW TECHNIQUES HAVE BEEN DEVELOPED AND APPLIED FOR RADIOISOTOPIC ANALYSES (FOR EXAMPLE, RADIOACTIVE TRACER AND CHEMICAL OXYGEN PRODUCTIVITY MEASUREMENT), AND INSTRUMENTATION FOR RADIATION AND OTHER MEASUREMENTS HAS BEEN ACCUMULATED. TABULATED AND GRAPHIC DATA INCLUDE EFFECTS OF TEMPERATURE, LIGHT, AND WAVE ACTION ON STANDING ALGAL CROP, STATISTICS DEALING WITH CROP VARIATIONS, PRIMARY PRODUCTIVITY, REDUCED OXYGEN IN RELATION TO PRIMARY PRODUCTIVITY, UPTAKE AND UTILIZATION OF RADIOCARBON, AND ALGAL-ANIMAL ASSOCIATIONS. SOME OF THE INCLUDED RESEARCH REPORTS ARE 'PHYSICAL FACTORS IN THE PRODUCTION OF TROPICAL BENTHIC MARINE ALGAE', 'ABRASION BY WATER-BORNE SEDIMENTS AS AN ERROR PRODUCING FACTOR IN THE MEASUREMENT OF DIFFUSION GRADIENTS THROUGH DISSOLUTION OF CALCIUM SULFATE', 'THE EFFECT OF OXGEN CONCENTRATION ON PRODUCTIVITY MEASUREMENT', 'BENTHIC ALGAL PRODUCTIVITY MEASUREMENTS CORRELATED WITH STANDING CROP HARVESTS'. (HOLOMAN-BATTELLE)

FIELD 05C, 05A

ACCESSION NO. W72-06283

THE SUSPENSION AND SINKING OF PHYTOPLANKTON IN THE SEA,

RHODE ISLAND UNIV., KINGSTON. GRADUATE SCHOOL OF OCEANOGRAPHY.

THEODORE J. SMAYDA.

OCEANOGR. MAR. BIOL. ANN. REV., VOL. 8, (HAROLD BARNES, EDITOR), 1970, P 353-414. 7 FIG, 9 TAB, 226 REF.

DESCRIPTORS:
*PHYTOPLANKTON, *MARINE ALGAE, *SEDIMENTATION RATES, *REVIEWS, BIOCHEMISTRY, GEOCHEMISTRY, RADIOISOTOPES, PATH OF POLLUTANTS, PHYSIOLOGICAL ECOLOGY, ENVIRONMENTAL EFFECTS, PRODUCTIVITY, RADIOECOLOGY, OCEANOGRAPHY.

ABSTRACT:
FAECEL-PELLET TRANSPORT MAY EXPLAIN RADIONUCLIDE SINKING RATES OBSERVED BY OSTERBERG AND OTHERS (NATURE, 1963, 200, 1276-71). THIS REVIEW CONSIDERS: SUSPENSION MECHANISMS (MORPHOLOGICAL, PHYSIOLOGICAL, PHYSICAL); SINKING-RATES (DETERMINATION METHODS, RESULTS, AND EFFECTS ON POPULATION DISTRIBUTIONS); EFFECTS ON ECOLOGY; BIOGEOCHEMICAL CONSEQUENCES (NON-CONSERVATIVE-SUBSTANCE DISTRIBUTIONS, MATTER TRANSPORT TO DEPTH, PHYTOPLANKTON REMAINS IN SEDIMENTS); ACCELERATED SINKING (DENSITY INVERSION CURRENTS, AGGREGATES, DOWNWELLING, FAECAL PELLETS). (BOPP-ORNL)

FIELD 05B, 05C

ACCESSION NO. W72-06313

LABORATORY STREAM RESEARCH: OBJECTIVES, POSSIBILITIES, AND CONSTRAINTS,

OREGON STATE UNIV., CORVALLIS. DEPT. OF FISHERIES AND WILDLIFE.

CHARLES E. WARREN, AND GERALD E. DAVIS.

IN: ANNUAL REVIEW OF ECOLOGY AND SYSTEMATICS, VOLUME 2; ANNUAL REVIEWS INC.; PALO ALTO, CALIFORNIA. R. F. JOHNSTON, EDITOR, 1971, P 111-144. 8 FIG, 88 REF.

DESCRIPTORS:
*WATER POLLUTION EFFECTS, *MODEL STUDIES, *STREAMS, *ECOSYSTEMS, AQUATIC HABITATS, BIOMASS, AQUATIC ALGAE, AQUATIC ANIMALS, BALANCE OF NATURE, ECOLOGY, TROPHIC LEVEL, HERBIVORES, NATURAL STREAMS, PULP WASTES, PRODUCTIVITY, DIELDRIN, PATH OF POLLUTANTS, REVIEWS, FOOD CHAINS, FORECASTING, CONSTRAINTS, RESEARCH FACILITIES.

ABSTRACT:
RELATIVELY SMALL, CONSTRUCTED CHANNELS HAVING CONTROLLED WATER FLOW, WHICH HAVE BEEN DEVELOPED MAINLY OVER THE PAST 15 YEARS, HAVE POTENTIAL FOR DETAILED EXAMINATION OF THE FUNCTION OF ECOSYSTEM PARTS AS THEY LEAD TO THE BEHAVIOR OF THE WHOLE. FOOD ORGANISM BIOMASS MAY SERVE AS A 'SUFFICIENT' (R. LEVINS, AM. SCI. 54, 421-431(1966)) PARAMETER FOR SOME TROPHIC MODELS. A VERY WIDE RANGE OF CIRCUMSTANCES MAY BE MODELLED WITH REGARD TO POLLUTION. HOWEVER CONSTRAINTS, HAVING TO DO WITH THE SUCCESSION, DIVERSITY, AND STABILITY OF COMMUNITIES MAY BE DIFFERENT FROM NATURAL ECOSYSTEMS, INTENTIONALLY OR NOT. (BOPP-ORNL)

FIELD 05C

ACCESSION NO. W72-06340

TRITIATION OF AQUATIC ANIMALS IN AN EXPERIMENTAL FRESHWATER POOL,

CALIFORNIA UNIV., LIVERMORE. LAWRENCE RADIATION LAB.

FLORENCE L. HARRISON, AND JOHN J. KORANDA.

AVAILABLE FROM THE NATIONAL TECHNICAL INFORMATION SERVICE AS UCRL-72930, $3.00 IN PAPER COPY, $0.95 IN MICROFICHE. AUGUST 3, 1971. 35 P.

DESCRIPTORS:
*FOOD CHAINS, *RADIOACTIVITY EFFECTS, *TRITIUM, *NUCLEAR WASTES, WATER POLLUTION EFFECTS, CATTAILS, AQUATIC ANIMALS, CLAMS, CARP, CRAYFISH, AQUATIC ALGAE, ABSORPTION, AQUARIA, ANIMAL PHYSIOLOGY, WATER BALANCE, TRACERS, PLANT PHYSIOLOGY, AQUATIC PLANTS, RADIOECOLOGY.

ABSTRACT:
FRESHWATER ANIMALS (CLAMS, CRAYFISH, AND GOLDFISH) AND ALGAE SHOWED HALF TIMES OF LESS THAN ONE DAY FOR EQUILIBRATION OF MORE THAN 95% OF TISSUE WATER TO THE SPECIFIC ACTIVITY OF POOL WATER. THE SPECIFIC ACTIVITY OF THE TISSUE WATER OF A PLANT (CATTAILS) REACHED ONLY 60-70% OF THAT OF THE POOL WATER SINCE THERE WAS INTERCHANGE WITH ATMOSPHERIC WATER THROUGH THE LEAVES. RATIOS OF THE SPECIFIC ACTIVITY OF THE ORGANICALLY BOUND TRITIUM TO THE SPECIFIC ACTIVITY OF TISSUE WATER WERE 1.0 FOR ALGAE, 0.62-0.99 FOR CATTAILS, 0.60 FOR THE VISCERAL ORGANS OF CLAMS AND FISHES, AND 0.30 FOR THE MUSCLE TISSUE. CONSUMPTION OF 1 KILOGRAM OF THE ANIMAL TISSUE WOULD GIVE A WHOLE BODY DOSE OF ABOUT 4 MREM, WHEN THE RADIOACTIVITY OF THE POOL WATER WAS ABOUT 0.05 MICROCURIES/ML. (BOPP-ORNL)

FIELD 05C

ACCESSION NO. W72-06342

A PRELIMINARY ECOLOGICAL STUDY OF AREAS TO BE IMPOUNDED IN THE SALT RIVER BASIN OF KENTUCKY,

KENTUCKY WATER RESOURCES INST., LEXINGTON.

LOUIS A. KRUMHOLZ.

AVAILABLE FROM THE NATIONAL TECHNICAL INFORMATION SERVICE AS PB-207 868, $3.00 IN PAPER COPY, $0.95 IN MICROFICHE. RESEARCH REPORT NO. 43, SEPTEMBER 1971, 34 P, 2 FIG, 1 TAB, 8 REF. OWRR B-005-KY(1).

DESCRIPTORS:
*ECOLOGY, *PREIMPOUNDMENTS, *AQUATIC HABITATS, *SAMPLING, *ECOSYSTEMS, *ALGAE, *CHLORAPHYTA, *AQUATIC INSECTS, *CHLORAPHYTA, *AQUATIC INSECTS, *CRUSTACEANS, *MOLLUSKS, *AQUATIC ANIMALS, *BIOTA, *AQUATIC PLANTS, WATER QUALITY, ENVIRONMENTAL EFFECTS, LIMNOLOGY, PLANNING, EUTROPHICATION, EVALUATION, WATER CHEMISTRY, WATER TEMPERATURE, DISSOLVED OXYGEN, TURBIDITY, ALKALINITY, NITRATES, IRON, MANGANESE, ANIONS, CATIONS, DIPTERA, CLAMS, GASTROPODS, BOTTOM FAUNA, KENTUCKY.

IDENTIFIERS:
*SALT RIVER(KENT), VASCULAR PLANTS, NITRATE NITROGEN, ORTHOPHOSPHATES, *OLIGASAPROBIC STREAMS, FINGERNAIL CLAMS, UNIONIDS, ORCONECTES RUSTICUS, LIRCEUS LINEATUS, GAMMARUS.

ABSTRACT:
A SERIES OF 25 SAMPLING STATIONS WAS ESTABLISHED IN THE MAINSTREAM AND TRIBUTARIES OF THE SALT RIVER. SAMPLING FOR WATER CHEMISTRY AND BIOTA WAS CARRIED OUT SEMIMONTHLY. PHYSICAL AND CHEMICAL DATA, ALONG WITH THE FLORA AND FAUNA PRESENT THE CHARACTERISTICS OF A RELATIVELY HEALTHY ECOSYSTEM. WATER TEMPERATURES REFLECT AIR TEMPERATURES CLOSELY AND DISSOLVED OXYGEN VALUES ARE NEAR SATURATION. TURBIDITY INCREASED WITH RUNOFF, THE STREAM FLOW INCREASING RAPIDLY DURING RAINY PERIODS AND FALLING TO A MINIMUM DURING DRY PERIODS. TOTAL ALKALINITIES RANGED FROM 135 TO 210 MG/1 AS CACO3 WITH RANGES IN PH FROM 6.3 TO 8.2. NITRATE NITROGEN RANGED FROM 2.0 TO 11.3 MG/1 AND ORTHOPHOSPHATE FROM 0.25 TO 2.78 MG/1. IRON AND MANGANESE RANGED FROM 0.07 TO 0.46 AND 0.09 TO 0.39 MG/1, RESPECTIVELY. A TOTAL OF 74 SPECIES OF ALGAE REFERABLE TO 35 FAMILIES WERE IDENTIFIED. GREEN ALGAE (CHLOROPHYTA) WERE REPRESENTED BY 38 SPECIES. MORE THAN 200 SPECIES OF VASCULAR PLANTS REFERABLE TO 50 FAMILIES HAVE BEEN COLLECTED FROM THE RIPARIAN VEGETATION. BOTTOM FAUNA INCLUDES 98 SPECIES OF INSECTS REPRESENTING 8 ORDERS AND 42 FAMILIES. PROMINENT AMONG THESE ARE THE 23 SPECIES OF CHIRONOMIDS THAT HAVE BEEN IDENTIFIED TO DATE. THE MOST COMMON CRUSTACEANS ARE ORCONECTES RUSTICUS AND LIRCEUS LINEATUS ALONG WITH SEVERAL SPECIES OF GAMMARUS. MOLLUSCS INCLUDE GASTROPODS, FINGERNAIL CLAMS, AND UNIONIDS. MORE THAN 50 SPECIES OF FISHES HAVE BEN COLLECTED.

FIELD 05A, 05C, 05B

ACCESSION NO. W72-06526

CHEMISTRY OF NITROGEN AND PHOSPHORUS IN WATER.

AMERICAN WATER WORKS ASSOCIATION, NEW YORK. WATER QUALITY DIV.; AND AMERICAN WATER WORKS ASSOCIATION, NEW YORK. COMMITTEE ON NUTRIENTS IN WATER.

JOURNAL OF AMERICAN WATER WORKS ASSOCIATION, VOL. 62, NO. 2, P 127-140, FEB. 1970. 6 FIG, 8 TAB, 64 REF.

DESCRIPTORS:
*WATER QUALITY, *AQUATIC PRODUCTIVITY, *AQUATIC MICRORGANISMS, *ALGAE, *NITROGEN, *PHOSPHORUS, SEDIMENTS, ORGANIC WASTES.

ABSTRACT:
OF THE MAJOR ELEMENTS ESSENTIAL TO ALGAL GROWTH, NITROGEN AND PHOSPHORUS ARE MOST LIKELY TO BE FOUND IN LIMITED AMOUNTS IN NATURAL WATERS. SINCE THEY THEREFORE REPRESENT PROMISING WEAK LINKS IN THE ALGAL LIFE CYCLES, THEIR CHEMICAL STATES AND BEHAVIOR IN WATER ARE EXAMINED TO SEE HOW WATER TREATMENT MIGHT BENEFIT. SUPPLIES OF NITROGEN AND PHOSPHORUS ARE PRESENT IN BODIES OF WATER EITHER IN THE SEDIMENTS, THE ATMOSPHERE ABOVE, OR IN THE FORM OF DISSOLVED GASES. THESE MAY BE AVAILABLE FOR THE GROWTH OF ALGAE AND OTHER AQUATIC PLANTS, BUT THE RATES AT WHICH THEY MAY BECOME AVAILABLE IS SLOW. THESE RATES ARE IMPORTANT AS THEY TEND TO CONTROL THE AMOUNT OF GROWTH WHICH CAN BE SUPPORTED. SOLUBLE NITROGEN AND PHOSPHORUS CONTAINED IN EFFLUENTS FROM WASTE TREATMENT PLANTS ARE IN READILY SOLUBLE FORM AND THEY CAN STIMULATE GROWTH FAR IN EXCESS OF THAT WHICH WOULD OCCUR NATURALLY. IN ASSESSING THE EXTENT OF NUTRIENT-RELATED PROBLEMS AND THEIR CONTROL, THE WATER MANAGER MUST EVALUATE THE SIGNIFICANCE BOTH OF THE READILY AVAILABLE FORMS AND THE FORMS WHICH MAY BE SLOWLY RELEASED FROM SUSPENDED PARTICLES AND FROM SEDIMENTS. (BEAN-AWWARF)

FIELD 05C, 05B, 05G

ACCESSION NO. W72-06532

BIOLCGICAL PROBLEMS ENCOUNTERED IN WATER SUPPLIES,

FEDERAL WATER QUALITY ADMINISTRATION, WASHINGTON, D.C. DIV. OF TECHNICAL SUPPORT.

K. M. MACKENTHUN, AND L. E. KEUP.

JOURNAL OF AMERICAN WATER WORKS ASSOCIATION, VOL. 62, NC. 8, P 520-526, AUGUST 1970. 2 FIG, 5 TAB, 36 REF.

DESCRIPTORS:
*WATER TREATMENT, *MICROORGANISMS, *ALGAE, *ALGICIDES, MICROSTRAINING, AERATION, PHOSPHATES, CLAY, TASTE, ODOR, *SURFACE WATERS, *GROUNDWATERS, SURVEYS, *WATER SUPPLY.

IDENTIFIERS:
*FILTER CLOGGING, IRCN BACTERIA, PONC WEEDS, PERMANGANATE.

ABSTRACT:
THE TYPES OF ORGANISMS GENERALLY RECOGNIZED AS CAUSING DIFFICULTIES IN WATER SUPPLIES AND SYSTEMS, AND THE TYPE OF PROBLEMS ENCCUNTERED ARE DISCUSSED, INCLUDING TASTES AND ODORS, FILTER CLCGGING, ALGAE ON RESERVOIR WALLS, IRON BACTERIA, AND ANIMALS. ALSO, VARIOUS CONTROLS ARE DESCRIBED. THE RESULTS OF A 1969 SURVEY ARE REPORTED. REPLIES WERE RECEIVED FROM 869 MANAGERS REPRESENTING OVER 1,372 MUNICIPAL AND 24 INDUSTRIAL SYSTEMS; THE MUNICIPAL SYSTEMS SERVE 80 MILLICN PEOPLE. ORGANISMS HAVE CREATED PROBLEMS FOR 25 PERCENT OF THE MANAGERS WITHIN THE PAST FIVE YEARS; THE FRECUENCY IN THE LARGER SYSTEMS WAS TWICE THE AVERAGE. THE MOST FREQUENTLY REPORTED PROBLEMS WERE ALGAE AND POND WEEDS IN SURFACE SOURCES AND IRON BACTERIA IN WELLS AND DISTRIBUTION SYSTEMS. WITH SURFACE WATERS, ALGAL PROBLEMS WERE REPORTED 14 TIMES AS OFTEN AS IRON BACTERIA, WHEREAS WITH GROUNDWATERS THE IRON BACTERIA WERE REPORTED FOUR TIMES AS OFTEN, BUT 75 PERCENT OF THESE WERE IN WELLS. NON-CHEMICAL METHODS CF PROBLEM CCNTRCL ARE SCREENING, MICROSTRAINING, MECHANICAL CLEANING, FLUSHING, REDESIGN OF SYSTEM, AERATION OF RESERVOIR, CUTTING POND WEEDS, AND ACJUSTMENT OF FILTER RATES. CHEMICALS USED IN CONTROLLING ORGANISMS ARE CHLORINE, COPPER SULFATE, CARBON, POTASSIUM PERMANGANATE, ACIC, SCCIUM METAPHOSPHATE, AMMONIA, 2,4,-D, AROMATIC SOLVENTS, CLAY, 2,4,5-T, DCW-PON, ANC PITT CHLOR. THE FIRST FOUR WERE MOST FREQUENTLY USED. CARBON IS ALMOST UNIVERSALLY USED TO REMOVE OBJECTICNABLE BY-PRODUCTS THROUGH ABSORPTION. (BEAN-AWWARF)

FIELD 05F

ACCESSION NO. W72-06536

FACTORS REGULATING THE GROWTH OF ALGAE IN CONTINUOUS CULTURE IN DILUTED SECONDARY SEWAGE TREATMENT PLANT EFFLUENT AND SUBSEQUENT BIODEGRADABILITY,

KENTUCKY WATER RESOURCES INST., LEXINGTON.

EDWARD G. FOREE, AND CAROLINE P. WADE.

AVAILABLE FROM THE NATIONAL TECHNICAL INFORMATION SERVICE AS PB-208 030, $3.00 IN PAPER COPY, $0.95 IN MICROFICHE. RESEARCH REPORT NUMBER 45, JANUARY 1972, 56 P, 10 FIG, 5 TAB, 29 REF. OWRR A-023-KY(1).

DESCRIPTORS:
*ALGAE, *NITROGEN, HYDROGEN ION CONCENTRATION, BIOCHEMICAL OXYGEN DEMAND, *SEWAGE EFFLUENTS, *SEWAGE TREATMENT, CULTURE, *WASTE-WATER TREATMENT, *CARBON DIOXIDE, *GROWTH RATES, *AEROBIC CONDITIONS, *GROWTH STAGES, *BIODEGRADATION, *BIOCHEMICAL OXYGEN DEMAND, CHEMICAL OXYGEN DEMAND.

IDENTIFIERS:
*ALGAE GROWTH, *CHEMOSTATS.

ABSTRACT:
HETEROGENEOUS ALGAL CULTURES WERE GROWN IN LABORATORY CONTINUOUS CULTURE IN CONTINUOUS FLOW, COMPLETELY MIXED CHEMOSTATS IN SECONDARY SEWAGE TREATMENT PLANT EFFLUENT DILUTED TO GIVE AN AMMONIA NITROGEN CONCENTRATION OF 10 MG/L. VARIABLES WERE LIGHTING, PH, CARBON DIOXIDE AVAILABILITY, AND HYDRAULIC RESIDENCE TIME. OPTIMUM GROWTH OCCURRED UNDER PH.7.0, EXCESS CO2, AND CONTINUOUS LIGHTING CONDITIONS. THE AVAILABILITY OF ARTIFICIALLY SUPPLIED EXCESS CO2 GREATLY INCREASED THE MASS (STANDING CROP) AT STEADY-STATE OVER THAT PRODUCED UNDER OTHERWISE IDENTICAL CONDITIONS FOR ALL RESIDENCE TIMES STUDIED. FOR THE CASE OF EXCESS CO2 AVAILABILITY, THE NITROGEN CONCENTRATION IN THE ALGAL CELLS REGULATED GROWTH RATHER THAN THE CONCENTRATION OF NUTRIENTS IN SOLUTION. A MATHEMATICAL EXPRESSION WAS HYPOTHESIZED TO DESCRIBE THIS PHENOMENON AND WAS CONFIRMED BY THE EXPERIMENTAL RESULTS. UNDER DARK-AEROBIC CONDITIONS, THE ALGAL CULTURES EXERTED A TWO-STAGE BOD, THE SECOND STAGE APPARENTLY BEGINNING AFTER THE DEATH OF THE ALGAL CELLS. LONGER CHEMOSTAT RESIDENCE TIMES DURING GROWTH PRODUCED CULTURES WITH LOWER PERCENTAGE BIODEGRADABILITY. CARBON DIOXIDE ENRICHED GROWTH CONDITIONS PRODUCED CULTURES WITH LOWER PERCENTAGE BIODEGRADABILITY THAN CULTURES GROWN IN A CARBON DIOXIDE DEFICIENT MEDIUM.

FIELD 05C, 05D

ACCESSION NO. W72-06612

PHOSPHATES IN DETERGENTS: THE CHICAGO-TYPE ORDINANCE AND OTHER REMEDIES,

B. NIEHOFF.

CINCINNATI LAW REVIEW, VOL. 40, NO. 3, P 548-568, 1971. 115 REF.

DESCRIPTORS:
*EUTROPHICATION, *DETERGENTS, *PHOSPHATES, *LOCAL GOVERNMENTS, *WATER POLLUTION CONTROL, WATER QUALITY CONTROL, PHOSPHORUS, AQUATIC ENVIRONMENT, WATER POLLUTION EFFECTS, ALGAE, NUTRIENTS, WATER POLLUTION SOURCES, SEWAGE TREATMENT, PUBLIC HEALTH, WASTE WATER TREATMENT, RESEARCH AND DEVELOPMENT, MUNICIPAL WASTES, LEGISLATION, REGULATION, PHOSPHORUS COMPOUNDS, FEDERAL GOVERNMENT.

ABSTRACT:
PHOSPHATE DETERGENTS ARE A MAJOR CAUSE OF ACCELERATED EUTROPHICATION. THIS PROCESS FOULS LAKES AND RIVERS WITH MATS OF ALGAE AND WEEDS. A NUMBER OF LOCAL GOVERNMENTS HAVE ENACTED CHICAGO-TYPE ORDINANCES TO CONTROL THE USE OF PHOSPHATE DETERGENTS. UNDER THE CHICAGO ORDINANCE IT IS ILLEGAL TO SELL DETERGENTS CONTAINING PHOSPHATES. THE CHICAGO ORDINANCE ALLOWS SOME EXEMPTIONS FOR MACHINES BUILT TO USE ONLY PHOSPHATE DETERGENTS, AND FOR HEALTH AND SANITATION REASONS. MANY ORDINANCES HAVE LABELING REQUIREMENTS. A FEDERAL TRADE COMMISSION PROPOSAL WOULD GO FURTHER AND REQUIRE A WARNING ON THE CONTAINER. THIS PROPOSAL MAY, HOWEVER, BE CONFUSING AND RAISE UNNECESSARY CONSUMER DOUBTS. THE DETERGENT INDUSTRY INSISTS LOCAL ORDINANCES ARE AN UNCONSTITUTIONAL RESTRAINT OF INTERSTATE COMMERCE. HOWEVER, THEY ARE PROBABLY A CONSTITUTIONAL EXERCISE OF LOCAL POLICE POWERS. AT LEAST FOUR PHOSPHATE-DETERGENT CONTROL LAWS ARE NOW BEFORE CONGRESS. THE SOUNDEST OF THESE PROHIBITS THE MANUFACTURING OR IMPORTATION OF ANY DETERGENT CONTAINING PHOSPHATE AFTER JUNE 30, 1973. (BRACKINS-FLORIDA)

FIELD 05G, 06E

ACCESSION NO. W72-06638

LEGISLATIVE RESPONSE TO 'SOFT SOAP' ON DETERGENTS,

CONGRESS, WASHINGTON, D.C.; AND SENATE, WASHINGTON, D.C.

T. J. MCINTYRE.

CONGRESSIONAL RECORD, VOL. 117, NO. 170, S 18015-18023 (DAILY ED.) NOVEMBER 10, 1971.

DESCRIPTORS:
*DETERGENTS, *PHOSPHATES, *WATER POLLUTION SOURCES, *POLLUTION ABATEMENT, EUTROPHICATION, LEGISLATION, PUBLIC HEALTH, WATER QUALITY CONTROL, ALGAL CONTROL, FEDERAL GOVERNMENT, NUTRIENTS, SEWAGE TREATMENT, WATER POLLUTION, WATER POLLUTION EFFECTS, RESEARCH AND DEVELOPMENT, WASTE WATER TREATMENT, BIODEGRADATION, LOCAL GOVERNMENTS, MUNICIPAL WASTES, ADMINISTRATIVE AGENCIES, GREAT LAKES.

ABSTRACT:
PHOSPHATE DETERGENTS ARE A MAJOR CAUSE OF THE DEATH OF MANY LAKES AND RIVERS. ALTHOUGH THE EFFECTS OF PHOSPHATE DETERGENTS ARE WELL KNOWN, THE FEDERAL GOVERNMENT HAS BEEN RECALCITRANT IN TAKING ACTION TO CONTROL THEIR USE. THE MOST OSTENSIBLE REASON IS THE LACK OF A SAFE AND EFFECTIVE SUBSTITUTE. ANOTHER REASON IS THE NORMAL DELAY WHEN THERE ARE POWERFUL INTERESTS OR OPPOSITION TO SUCH ACTION. A PROPOSAL BEFORE THE SENATE COMMERCE COMMITTEE'S ENVIRONMENTAL SUBCOMMITTEE WOULD ACCOMPLISH THE FOLLOWING: (1) LIMIT THE PHOSPHATE CONTENT IN DETERGENTS FOR THE PRESENT, WITH A CLEAR MANDATE FOR THEIR EVENTUAL TOTAL REMOVAL; (2) ESTABLISH TEST PROTOCOLS, STANDARDS, AND REGULATIONS FOR ALL DETERGENT INGREDIENTS; AND (3) ESTABLISH A FEDERAL PROGRAM FOR THE DEVELOPMENT OF SAFE SUBSTITUTES WHICH WILL NOT HARM THE ENVIRONMENT. SOME, INCLUDING THE DETERGENT INDUSTRY, CLAIM ELIMINATION OF PHOSPHATES AT THE SEWAGE TREATMENT STAGE IS THE MOST EFFECTIVE METHOD OF DEALING WITH THE PROBLEM. STRICTER LABELING REQUIREMENTS, DISCLOSING THE AMOUNT OF PHOSPHATE IN A DETERGENT, ARE ALSO OFFERED AS AN IMMEDIATE STEP WHICH COULD BE TAKEN TO PARTIALLY ALLEVIATE THE PROBLEM. (BRACKINS-FLORIDA)

FIELD 06E, 05G

ACCESSION NO. W72-06655

MICROBIOLOGICAL STUDIES OF OXYGEN DEPLETION IN THE LAKE ERIE CENTRAL BASIN,

DEPARTMENT OF NATIONAL HEALTH AND WELFARE, KINGSTON (ONTARIO). DIV. OF PUBLIC HEALTH ENGINEERING.

A. S. MENON, AND A. A. JURKOVIC.

MANUSCRIPT REPORT KR 70-4, 1970. 51 P, 15 FIG, 13 TAB, 18 REFS.

DESCRIPTORS:
*AQUATIC BACTERIA, *SULFUR BACTERIA, *OXYGEN, *SEDIMENT-WATER INTERFACES, *LAKE ERIE, HYPOLIMNION, AQUATIC ALGAE.

ABSTRACT:
THE SIGNIFICANCE OF BACTERIAL ACTIVITY IN THE OVER-ALL PROCESSES OF OXYGEN DEPLETION AND NUTRIENT REGENERATION IN THE CENTRAL BASIN OF LAKE ERIE WAS ASSESSED. MOST INTENSIVE BACTERIAL ACTIVITY OCCURRED AT THE SEDIMENT-WATER INTERFACE. BACTERIAL DECOMPOSITION OF ORGANIC MATTER ACCUMULATING AT THE INTERFACE RESULTED IN THE FORMATION OF REDUCED PRODUCTS OF LOW MOLECULAR WEIGHT AND DEPLETION OF OXYGEN IN THE HYPOLIMNION. THESE COMPOUNDS WERE SUBSEQUENTLY OXIDIZED BY CHEMOAUTOTROPHIC BACTERIA WITH FURTHER LOSS OF O2. REDUCING CONDITIONS ON THE BOTTOM ADVERSELY AFFECTED NITRIFYING BACTERIAL DENSITIES. HOWEVER, ACTIVELY PHOTOSYNTHESIZING ALGAE FRESHLY DEPOSITED ON THE BOTTOM STIMULATED MULTIPLICATION OF NITRIFYING BACTERIA AND NITRIFICATION. LARGE BACTERIAL POPULATIONS WERE ABSENT IN THE THERMOCLINE, SUGGESTING THAT THIS ZONE WAS NOT A SITE FOR INTENSIVE BACTERIAL ACTIVITY. QUANTITATIVE ANALYSIS INDICATED THAT THE HIGH BACTERIAL DENSITIES IN THE HYPOLIMNION, ESPECIALLY AT THE SEDIMENT-WATER INTERFACE, RESPIRING AT THE RATE OF 2.4 X 10-11 MG O2 PER CELL PER HOUR COULD ACCOUNT FOR OXYGEN DEPLETION IN THE LAKE. (CCIW)

FIELD 05C, 02H

ACCESSION NO. W72-06690

EFFECTS OF DETERGENTS ON WATER SUPPLIES,

ONTARIO WATER RESOURCES COMMISSION, TORONTO.

A. J. HARRIS, K. J. ROBERTS, AND A. E. CHRISTIE.

JOURNAL AMERICAN WATER WORKS ASSOCIATION, VOL. 63, NO. 12, P 795-799, DECEMBER 1971. 2 TAB, 56 REF.

DESCRIPTORS:
*DETERGENTS, *PHOSPHATES, *POTABLE WATER, EUTROPHICATICN, ALKYLBENZENE SULFONATES, SURFACTANTS, LINEAR ALKYLATE SULFONATES, SEWAGE, NUTRIENTS, COAGULATION, SEDIMENTATION, TOXICITY, TASTE, ODOR, CORROSION, ENZYMES, CHLEATION, ALGAE, WATER QUALITY, BIODEGRADATION, SEWAGE TREATMENT PLANTS, DIATOMS, PHYTOPLANKTON, HYDORGEN ICN TEMPERATURE, WATER POLLUTION EFFECTS.

IDENTIFIERS:
ARSENIC, NITRILOTRIACETATE, POLYSILICATES, POLYCARBOXYLATES.

ABSTRACT:
THE DEVELOPMENT OF DETERGENTS AND THEIR DIRECT AND INDIRECT EFFECTS ON WATER SUPPLIES ARE REVIEWED. THE MAJORITY OF PRESENT-DAY PROBLEMS ARE ASSOCIATED WITH PHOSPHATES IN THE DETERGENTS. THESE PROBLEMS INCLUDE EUTROPHICATION OF LAKES, WATER QUALITY AND TREATMENT PROBLEMS, AND UNDESIRABLE TASTES AND ODORS ASSOCIATED WITH ALGAL GROWTH. SEVERAL MATERIALS SUCH AS POLYSILICATES, POLYCARBOXYLATES, NTA, AND STARCH DERIVATIVES HAVE BEEN STUDIED AS POSSIBLE REPLACEMENTS FOR PHOSPHATE. LACK OF BIODEGRADABILITY, UNDESIRABLE CHELATING PROPERTIES, AND CORROSION PROBLEMS ARE SOME OF THE FAULTS THAT HAVE BEEN DETECTED IN THESE MATERIALS. NTA HAS BEEN MOST PROMISING BUT HAS BEEN WITHDRAWN BECAUSE OF A POSSIBLE HEALTH HAZARD. BECAUSE IT IS NOW POSSIBLE TO REMOVE PHOSPHATE FROM WASTE WATER BY CHEMICAL PRECIPITATION AND BECAUSE DETERGENT PHOSPHATE IS ONLY ONE PART OF THE TOTAL PHOSPHATE IN SEWAGE EFFLUENTS, REMOVAL FROM THE WASTE WATER RATHER THAN THE DETERGENT PRODUCT SHOULD BE CONSIDERED. (MORTLANC-BATTELLE)

FIELD 05C, 05G

ACCESSION NO. W72-06837

SUPPLEMENTARY AERATION OF LAGOONS IN RIGOROUS CLIMATE AREAS,

WYOMING UNIV., LARAMIE. DEPT. OF CIVIL ENGINEERING.

R. L. CHAMPLIN.

COPY AVAILABLE FROM GPO SUP DOC AS EP2.10:17050 DVO-10/71, $0.75; MICROFICHE FROM NTIS AS PB-2C8 204, $0.95.ENVIRONMENTAL PROTECTION AGENCY WATER POLLUTION CONTROL RESEARCH SERIES, OCTOBER 1971, 73 P. 24 FIG, 15 TAB, 7 REF. EPA PROGRAM 17050DVC--10/71.

DESCRIPTORS:
*ON-SITE INVESTIGATION, *COLD REGIONS, *OXIDATION LAGOCNS, ORGANIC LOADING, AERATION, MIXING, ALGAE, TEMPERATURE, ALTITUDE, CLIMATES, METABOLISM, BIOCHEMICAL OXYGEN DEMAND, PILOT PLANTS, *WASTE WATER TREATMENT, WYOMING, *AERATED LAGOONS.

ABSTRACT:
A PILOT SCALE FIELD INVESTIGATION OF THE EFFECTS OF SUPPLEMENTAL AERATION ON WASTE STABILIZATION LAGOONS WAS CONDUCTED AT LARAMIE, WYOMING, A LOW TEMPERATURE, HIGH ALTITUDE AREA. BOTH BATCH AND COMPLETE MIXED EXPERIMENTS WERE CONDUCTED USING CONSTANT AIR FLOWS. LOADING RATES, BOTH HYDRAULIC AND PROCESS , WERE VARIED FROM 160 LBS BOD5/ACRE/DAY (0.725 LBS/1000 FT3/DAY) TO 900 LBS BOD5/ACRE/DAY (4.08 LBS/1000 FT3/DAY). THE SUPPLEMENTAL AERATION PROVIDED BOTH AERATION AND MIXING, THEREBY INCREASING METABOLIC RATES. BOD REDUCTIONS VARIED FROM 72 TO 85% UNDER THREE DIFFERENT LOADINGS, AT TEMPERATURES OF LESS THAN 12 DEG C. LOADING BELOW 320 LBS/ACRE/DAY AND SECONDARY CELL OPERATION PRODUCED SIGNIFICANT ALGAL GROWTH EVEN AT TEMPERATURES AROUND 6 DEG C. NO SETTLEABLE SOLIDS WERE FOUND IN THE EFFLUENT FROM THE AERATED SYSTEM. SERIES OPERATION WAS DEMONSTRATED TO HAVE THE ADVANTAGES OF DAMPING VARIATIONS IN QUALITY PARAMETERS, PROVIDING FOR SHOCK LOADING, AND REDUCING COLIFORM COUNTS TO MINIMUM LEVELS. SHORTER DETENTION PERIODS ALSO TAKE GREATER ADVANTAGE OF THE WARMER INFLUENT TEMPERATURES IN ORDER TO SATISFY EASILY OXIDIZED ORGANIC MATERIAL. (LOWRY-TEXAS)

FIELD 05D

ACCESSION NO. W72-06838

ECOSYSTEM ALTERNATION BY MOSQUITOFISH (GAMBUSIA AFFINIS) PREDATION,

SAN DIEGO STATE COLL., CALIF. DEPT. OF BIOLOGY.

S. H. HURLBERT, J. ZEDLER, AND D. FAIRBANKS.

SCIENCE, VOL. 175, NO. 4022, P 639-641, FEBRUARY 11, 1972.

DESCRIPTORS:
*ECOSYSTEMS, FOOD CHAINS, *ZOOPLANKTON, *PREDATION, *BIOASSAY, BIOLOGICAL COMMUNITIES, DOMINANT ORGANISMS, AQUATIC ENVIRONMENT, AQUATIC ALGAE, AQUATIC ANIMALS, AQUATIC INSECTS, DAPHNIA, AQUATIC POPULATIONS, FISH DIETS, FRESHWATER FISH, PONDS, LAKES, FOOD HABITS, EUTROPHICATION, LARVAE, ROTIFERS, AQUATIC MICROORGANISMS, DIPTERA, PHOSPHORUS, WATER TEMPERATURE, WATER CHEMISTRY, WATER ANALYSIS, COPEPODS, CYANOPHYTA, MAYFLIES, PHYTOPLANKTON, SESSILE ALGAE, CYCLING NUTRIENTS, BIOMASS, CHARA, LIGHT PENETRATION, BENTHIC FLORA, MODEL STUDIES, AQUATIC LIFE, CRUSTACEANS, FISH, ANALYTICAL TECHNIQUES.

IDENTIFIERS:
*GAMBUSIA, CYCLOPS NAUPLII, COCCOCHLORIS PENIOCYSTIS, SPIROGYRA, EPHEMEROPTERA BAETIDAE, EPHYDRA, CHAETOGASTER, AEOLOSOMA, CHAOBORUS, *GAMBUSIA AFFINS, EPHEMEROPTERA, MENICIA, BRACHYDEUTERA, CHYDORUS SPHAERICUS, CHYDORUS, MONOSTYLA, LEPADELLA, TRICHOCERCA, HAEMATOCOCCUS LACUSTRIS.

ABSTRACT:
A MODEL ECOSYSTEM CONSISTING OF EIGHT PLASTIC POOLS, 2 M IN DIAMETER AND 30 CM DEEP, WAS SET UP OUTDOORS TO DETERMINE THE EFFECT OF FISH ON BIOLOGICAL POPULATIONS AND PHOSPHORUS CYCLE. A 3 CM LAYER OF SAND, TAP WATER TO A DEPTH OF 20 CM, A LITER OF DRY ALFALFA PELLETS AS A SOURCE OF NUTRIENTS, AND INOCULA OF PLANKTON FROM A LAKE AND FROM A LABORATORY COLONY OF DAPHNIA PULEX WERE ADDED TO EACH POOL. MOSQUITOFISH (GAMBUSIA AFFINS) WERE ADDED TO THREE POOLS WITH THE REMAINING POOLS SERVING AS CONTROLS. THE STATUS OF THE SYSTEM WAS DETERMINED BY REGULAR SAMPLING OF PHYTOPLANKTON, ZOOPLANKTON, INSECT POPULATION, BENTHOS, ALGAE, AND BY DETERMINATION OF PHOSPHORUS CONTENT, WATER TEMPERATURE, AND TRANSPARENCY. GAMBUSIA AFFINS GREATLY REDUCED ROTIFER, CRUSTACEAN, AND INSECT POPULATIONS AND THUS PERMITTED EXTRAORDINARY DEVELOPMENT OF PHYTOPLANKTON POPULATIONS (2 X 108 CELLS PER MILLILITER). OTHER EFFECTS INCLUDED DECREASED OPTICAL TRANSMISSIVITY AND INCREASED TEMPERATURE OF THE WATER, DECREASED AMOUNTS OF DISSOLVED INORGANIC PHOSPHORUS, AND INCREASED AMOUNTS OF DISSOLVED ORGANIC PHOSPHORUS, INHIBITION OF SPIROGYRA, AND REPLACEMENT OF ONE ANNELID, CHAETOGASTER, BY ANOTHER, AEOLOSOMA. THE RESULTS OF THE EXPERIMENT ALSO INDICATED FISH CAN BE USED AS A POSSIBLE CONTROLLING AGENT FOR EUTROPHICATION. FURTHER, THE BIOLOGIC CHANGES IN AQUATIC SYSTEMS ARE PIMARILY A RESULT OF CHANGES IN THE AQUATIC FOOD CHAIN. (SNYDER-BATTELLE)

FIELD 05C

ACCESSION NO. W72-07132

COMPARATIVE STUDY OF SPECIES NAUTOCOCCUS MAMMILATUS AND NAUTOCOCCUS PYRIFORMIS (TETRASPORALES),

CESKOSLOVENSKA AKADEMIE VED, TREBON. LAB. OF ALGOLOGY.

J. LUKAVSKY.

ARCHIV FUR HYDROBIOLOGIE, VOL. 39, NO. 4, P 245-258, OCTOBER 1971. 70 FIG, 18 REF.

DESCRIPTORS:
*ALGAE, *SYSTEMATICS, *PLANT MORPHOLOGY, REPRODUCTION, LIFE CYCLES, AQUATIC ALGAE, PERIPHYTON.

IDENTIFIERS:
EPINEUSTONT, NAUTOCOCCUS MAMMILATUS, NAUTOCOCCUS PYRIFORMIS NAUTOCOCCUS, APIOCOCCUS, OSCILLATORIA, NAUTOCOCCUS CAUDATUS, NAUTOCOCCUS CONSTRICTUS, NAUTOCOCCUS EMERSUS.

ABSTRACT:
THE MORPHOLOGICAL VARIABILITY AND THE KIND OF REPRODUCTION OF TETRASPORAL ALGA NAUTOCOCCUS MAMMILATUS KORSH. IN NATURE AND IN THE CULTURE WERE STUDIED AND COMPARED WITH THE CULTURE OF TWO STRAINS OF NAUTOCOCCUS PYRIFORMIS KORSH. THE STABILITY AND TAXONOMICAL VALUE OF SOME FEATURES, NAMELY THE MORPHOLOGY AND THE CELL STRUCTURE, THE KIND OF FLOATING, THE MORPHOLOGY OF THE FLOATING CAP AND THE MORPHOLOGY OF ZOOSPORES WERE STUDIED. FOR THE FIRST TIME THE ISOGAMICAL SEXUAL PROCESS OF ONE STRAIN, NAUTOCOCCUS PYRIFORMIS, WAS OBSERVED. THE SURFACE FILM OF WATER IS CONSIDERED NOT TO BE THE REASON OF FLOATING OF THE CELLS OF NAUTOCOCCUS; THE CELLS ARE LIFTED BY GAS BUBBLES. THE DIAGNOSIS OF THE GENUS NAUTOCOCCUS KORSH. WAS EXTENDED, THE GENUS APIOCOCCUS KORSH. WAS INCLUDED AS A SYNONYM. (HOLCMAN-BATTELLE)

FIELD 05C, 05A, 02I

ACCESSION NO. W72-07141

ION TRANSPORT STUDIES ON PHYTOPLANKTON OF A FISH POND AT ILE-IFE,

IFE UNIV. (NIGERIA). DEPT. OF BIOLOGICAL SCIENCES.

A. M. A. IMEVBORE, AND Z. BOSZORMENYI.

ARCHIV FUR HYDROBIOLOGIE, VOL. 69, NO. 2, P 200-209, OCTOBER 1971. 4 FIG, 3 TAB, 41 REF.

DESCRIPTORS:
*PHYTOPLANKTON, *ION TRANSPORT, *ABSORPTION, *PHOSPHORUS, *PHOSPHORUS RADIOISOTOPES, ALGAE, FRESH WATER, FILTRATION, LEACHING, FUNGI, LIGHT, PHOSPHATES, NUTRIENTS, WATER POLLUTION EFFECTS, RADIOACTIVITY TECHNIQUES.

IDENTIFIERS:
CYANIDES, PHOSPHORUS-32.

ABSTRACT:
THE ION TRANSPORT OF PHOSPHORUS BY PHYTOPLANKTON WAS INVESTIGATED BY EXPOSING MICROORGANISMS TO WATER SAMPLES FROM A FISHPOND AT ILE-IFE (NIGERIA) WHICH WERE LABELED WITH PHOSPHORUS-32. THE ANALYSES WERE RESTRICTED TO THE PROCESSES ACTING BETWEEN THE PHYTOPLANKTON AND THE EXTERNAL INORGANIC PHOSPHATE POOL. THE STUDIES PRODUCED DATA ON THE RATE OF UPTAKE, EFFECT OF PHOSPHORUS CONCENTRATION ON UPTAKE, EFFECT OF WASHING WITH INACTIVE PHOSPHORUS, AND THE EFFECTS OF LIGHT, CYANIDE, AND SHAKING ON UPTAKE. THE QUALITATIVE AND QUANTITATIVE DISSIMILARITY BETWEEN THE ABSORPTION AT THE HIGH (1MM) AND THE LOW (0.01MM) PHOSPHORUS-32 CONCENTRATIONS SEEM TO EXCLUDE THE POSSIBILITY THAT THE ABSORPTION PROCESSES ACTING IN BOTH RANGES ARE THE SAME. THEREFORE THERE ARE MORE LIKELY TWO DIFFERENT MECHANISMS IN ION TRANSPORT. AT THE LOW PHOSPHORUS-32 CONCENTRATIONS, CYANIDE WAS ABLE TO BLOCK THE PHOSPHATE UPTAKE ALMOST COMPLETELY; AT THE HIGH PHOSPHORUS-32 CONCENTRATION THERE WAS LITTLE OR NO INHIBITION. THE OBSERVED UPTAKE AT THE LOW CONCENTRATION WAS IN ACCORDANCE WITH THE GENERAL VIEW THAT PHOSPHATE ABSORPTION WAS DEPENDENT UPON METABOLISM AND THAT IT WAS A BIOLOGICAL TRANSPORT PROCESS. ON THE OTHER HAND, IT WAS PROBABLE THAT THE ACTION OF CYANIDE AT HIGH PHOSPHORUS-32 CONCENTRATION DEPENDED UPON THE DAMAGE IT CAUSED TO THE STRUCTURE OF BIOLOGICAL MEMBRANES. THEN THE INCREASE IN UPTAKE MUST BE DUE ENTIRELY TO A PHYSICAL PERMEATION PROCESS. CONTRARY TO EXPECTATION, SHAKING TENDED TO REDUCE THE ABSORPTION RATE. (HOLOMAN-BATTELLE)

FIELD 05B, 05C

ACCESSION NO. W72-07143

THE MICROCHAMBER CULTIVATION OF ALGAE,

CESKOSLOVENSKA AKADEMIE VED, BRNO. LAB. OF SCIENTIFIC FILM.

J. HRIB, AND V. BREZINA.

ARCHIV FUR HYDROBIOLOGIE, VOL. 39, NO. 4, P 349-354, OCTOBER 1971. 2 FIG, 13 REF.

DESCRIPTORS:
*AQUATIC ALGAE, *CHLOROPHYTA, *GROWTH CHAMBERS, *LIFE CYCLES, CULTURES, WATER QUALITY, SCENEDESMUS, LABORATORY EQUIPMENT, PHOTOGRAPHY.

IDENTIFIERS:
*SYNCHRONOUS CULTURES, SCENEDESMUS SOLI, SCENEDESMUS QUADRICAUDA, MOUGEOTIA, ULOTHRIX AEQUALIS, STEPHANOPYXIS TURRIS.

ABSTRACT:
A CULTIVATION METHOD FOR ALGAE BY THE SYSTEM OF MICROCHAMBER CULTIVATION IS PRESENTED. THE METHOD ENABLES CONTINUAL CULTIVATION OF ALGAL CELLS WITH THE POSSIBILITY OF CINEMATOGRAPHIC REGISTRATION, ESPECIALLY OF TIME LAPSE MICROCINEMATOGRAPHY. THE SYSTEM OF THE CULTIVATION MICROCHAMBER IS DIVIDED INTO FOUR SECTIONS: A CULTIVATION SECTION, A MICROMANIPULATION SECTION, A LIGHT SECTION AND A CINEMATOGRAPHIC SECTION. THE MAIN PART OF THE SYSTEM IS A CULTIVATION CHAMBER WHICH IS OF FLOWING CHARACTER WITH THE POSSIBILITY OF SIMPLE MICROMANIPULATION. THIS SYSTEM PROVED TO BE USEFUL IN STUDYING THE ONTOGENETIC CYCLE OF THE ALGA SCENEDESMUS QUADRICAUDA, STRAIN GREIFSWALD/15. (HOLOMAN-BATTELLE)

FIELD 05C, 05A

ACCESSION NO. W72-07144

OPTICAL DENSITY PROFILES AS AN AID TO THE STUDY OF MICROSTRATIFIED PHYTOPLANKTON POPULATIONS IN LAKES,

MINNESOTA UNIV., MINNEAPOLIS. LIMNOLOGICAL RESEARCH CENTER.

A. L. BAKER, AND A. J. BROOK.

ARCHIV FUR HYDROBIOLOGIE, VOL. 69, NO. 2, P 214-233, OCTOBER 1971. 12 FIG, 1 TAB, 14 REF.

DESCRIPTORS:
*PHYTOPLANKTON, *LAKES, *TURBIDITY, *PROFILES, PHOTOMETER, HYPOLIMNION, EPILIMNION, LAKE MORPHOMETRY, DISTRIBUTION PATTERNS, AQUATIC ALGAE, LIMNOLOGY, CHRYSOPHYTA, EUGLENOPHYTA, PYRROPHYTA, BIOMASS, HYDROGEN ION CONCENTRATION, MINNESOTA, WATER TEMPERATURE, DISSOLVED OXYGEN.

IDENTIFIERS:
*MICROSTRATIFICATION, *OPTICAL DENSITY, MICROAMMETER, OSCILLATORIA AGARDHII VAR ISOTHRIX, LAKE ITASCA, LAKE ELK, LAKE MARY, LAKE JOSEPHINE, LAKE ARCO, LAKE DEMING, LOWER LA SALLE LAKE, LAKE LONG, LAKE SQUAW, LAKE BUDD, OSCILLATORIA REDEKEI, LYNGBYA LIMNETICA, ANABAENA SPP., CHROMATIUM, MERISMOPEDIA TROLLERI, CRYPTOMONADS, STICHOGLOEA DEODERLEINII, APHANIZOMENON FLOS-AQUAE, MALLOMONAS ACAROIDES, OSCILLATORIA ORNATA, TURBIDIMETER, METALIMNION.

ABSTRACT:
TURBIDITY PROFILES OF TEN NORTH-CENTRAL MINNESOTA LAKES HAVE BEEN DETERMINED WITH A SCHENK-TURBIDIMETER. THE LAKES WERE STUDIED AT RELATIVELY CLOSE TIME INTERVALS AND EACH TIME A TURBIDITY PROFILE WAS TAKEN, TEMPERATURE AND DISSOLVED OXYGEN PROFILES AND IN MANY CASES PH AND CONDUCTIVITY PROFILES WERE ALSO DETERMINED. SOME OF THESE DATA ARE GRAPHED WITH THE OPTICAL DENSITY PROFILES FOR EACH LAKE. OBSERVED DIFFERENCES IN TURBIDITY WERE CORRELATED WITH THE MICROSTRATIFICATION OF SPECIFIC POPULATIONS OF PHYTOPLANKTON. THE TURBIDITY PROFILES DEMONSTRATED THE FOLLOWING: (1) MOVEMENT AND DISRUPTION OF MICROSTRATA DURING VERNAL AND AUTUMNAL OVERTURN, THEIR REAPPEARANCE UNDER ICE IN WINTER AND ESPECIALLY IN THE METALIMNION IN EARLY SUMMER; (2) PRONOUNCED MICROSTRATIFICATION ESPECIALLY IN SMALL LAKES WITH RELATIVE DEPTH VALUES GREATER THAN 5.0 METERS; AND (3) INTEGRITY AND STABILITY OF THE MICROSTRATA. THERE IS A DISCUSSION OF THE TURBIDITY PROFILES AS COMPARED WITH THOSE RECORDED BY OTHER INVESTIGATORS AND THE CAUSES OF OPTICAL DENSITY MAXIMA OTHER THAN BY PHYTOPLANKTON. (HOLOMAN-BATTELLE)

FIELD 05C, 05A, 02H

ACCESSION NO. W72-07145

PHOSPHORUS UPTAKE BY PLANKTONIC ALGAE IN THE DARK AND UNDER FAINT LIGHT,

INSTITUTE OF BIOLOGY OF THE SOUTHERN SEAS, SEVASTOPOL (USSR).

D. K. KRUPATKINA.

OKEANOLOGIYA, VOL 11, NO 2, P 221-226, 1971. 3 TAB, 18 REF.

DESCRIPTORS:
*ALGAE, *PHYTOPLANKTON, *ABSORPTION, *AQUATIC ALGAE, *LIGHT INTENSITY, *PHOSPHORUS, PLANKTON, CHLOROPHYTA, PYRROPHYTA, DINOFLAGELLATES, PRIMARY PRODUCTIVITY, MINERALOGY, LIGHT, CULTURES, BIOASSAY.

IDENTIFIERS:
PHAEODACTYLUM TRICORNUTUM, GONYAULAX POLYEDRA, PLATYMONAS VIRIDIS, CHAETOCEROS CURVISETUS, SKELETONEMA COSTATUM, GLENODINIUM FOLIACUM, GYRODINIUM FISUM.

ABSTRACT:
THE UPTAKE OF MINERAL PHOSPHORUS AND ORGANIC PHOSPHORUS IN THE DARK AND UNDER FAINT LIGHT BY PLANKTONIC ALGAE NOT DEFICIENT IN PHOSPHORUS HAS BEEN INVESTIGATED. SIX SPECIES OF PLANKTONIC ALGAE (GREEN, DIATOM, DINOFLAGELLATE) WERE STUDIED. MINERAL PHOSPHORUS WAS TAKEN UP IN THE DARK BY CELLS NOT DEFICIENT IN PHOSPHORUS IN MOST SPECIES (EXCEPT PHAEODACTYLUM TRICORNUTUM AND GONYAULAX POLYEDRA). THE PHOSPHORUS CONTENT INCREASED 1.5- TO 2.5-FOLD WHEN THE CELLS (PERCENT OF DRY WEIGHT) WERE KEPT IN THE DARK FOR 3 DAYS. ORGANIC PHOSPHORUS WAS TAKEN UP BOTH IN THE DARK AND IN THE PRESENCE OF LIGHT. SOME SPECIES (PLATYMONAS VIRIDIS AND CHAETOCEROS CURVISETUS) TAKE UP MINERAL PHOSPHORUS AND ORGANIC PHOSPHORUS SIMULTANEOUSLY BOTH IN THE DARK AND IN THE PRESENCE OF FAINT LIGHT, WHILE OTHERS (PHAEODACTYLUM TRICORNUTUM AND GONYAULAX POLYEDRA) TAKE UP ONLY ORGANIC PHOSPHORUS. THE SOURCES OF ORGANIC PHOSPHORUS IN THE MEDIUM ARE THE INTRAVITAL EXCRETIONS OF CELLS AND DECOMPOSITION PRODUCTS OF DEAD ALGAE. ORGANIC PHOSPHORUS IN THE MEDIUM CONSTITUTES 2 PERCENT OF THE DISSOLVED AND TOTAL P. ALL SUPPORTING DATA ARE TABULATED. (HOLOMAN-BATTELLE)

FIELD 05C, 05A

ACCESSION NO. W72-07166

THE EFFECTS OF HEATED WASTE WATERS ON SOME MICROORGANISMS,

VIRGINIA POLYTECHNIC INST. AND STATE UNIV., BLACKSBURG. WATER RESOURCES RESEARCH CENTER.

J. CAIRNS, AND G. R. LANZA.

AVAILABLE FROM THE NATIONAL TECHNICAL INFORMATION SERVICE AS PB-208 414, $3.00 IN PAPER COPY, $0.95 IN MICROFICHE. WATER RESOURCES RESEARCH CENTER BULLETIN 48, FEBRUARY 1972. 101 P, 18 FIG, 19 TAB, 48 REF. OWRR B-017-VA(6).

DESCRIPTORS:
*THERMAL POLLUTION, *WATER POLLUTION EFFECTS, *AQUATIC MICROORGANISMS, *INVESTIGATIONS, *ANALYTICAL TECHNIQUES, DATA COLLECTIONS, ALGAE, PROTOZA, FUNGI, EQUIPMENT, INSTRUMENTATION, FLUORESCENCE, WATER TEMPERATURES, THERMAL STRESS.

IDENTIFIERS:
RESEARCH, CHYTRIDIACEOUS FUNGI, HEAT SHOCK.

ABSTRACT:
THE EFFECTS OF HEATED WASTE WATER ON AQUATIC MICROORGANISMS WERE STUDIED. NUMEROUS CATEGORIES OF TEMPERATURE SHOCK WERE INVESTIGATED WITH EMPHASIS ON (1) GENERAL SIMULATION OF THE PASSAGE OF WATER THROUGH A CONDENSING SYSTEM OF A STEAM ELECTRIC POWER PLANT TO DETERMINE THE EFFECTS OF THIS EXPOSURE AND (2) THE EFFECTS OF SUCH ENTRAINMENT UPON THE MICROBIAL SYSTEM BELOW THE DISCHARGE POINT. THREE GENERAL GROUPS OF ORGANISMS WERE EXAMINED: PROTOZOANS, ALGAE, AND CHYTRIDIACEOUS FUNGI. NEW EQUIPMENT DEVELOPED SPECIFICALLY TO AID IN THESE STUDIES (A SIMPLE APPARATUS FOR DELIVERING HEAT SHOCK TO MICROORGANISMS) AND A NEW FLUORESCENT SURVEY TECHNIQUE TO CHARACTERIZE STRESS INDUCED CELLULAR ALTERATIONS ARE DESCRIBED IN DETAIL. RESULTS OF THE STUDY ARE SUMMARIZED IN TABLES AND GRAPHS. (WOODARD-USGS)

FIELD 05C

ACCESSION NO. W72-07225

AQUATIC PLANTS FROM MINNESOTA, PART 2 - TOXICITY, ANTI-NEOPLASTIC, AND COAGULANT EFFECTS,

MINNESOTA UNIV., MINNEAPOLIS. WATER RESOURCES RESEARCH CENTER.

K. LEE SU, AND E. JOHN STABA.

AVAILABLE FROM THE NATIONAL TECHNICAL INFORMATION SERVICE AS PB-208 609, $3.00 IN PAPER COPY, $0.95 IN MICROFICHE. MINNESOTA WATER RESOURCES RESEARCH CENTER BULL. 47, FEBRUARY 1972. 24 P, 4 FIG, 5 TAB, 73 REF. OWRR A-025-MINN(3).

DESCRIPTORS:
*AQUATIC PLANTS, *MINNESOTA, *TOXICITY, COAGULATION, WATER POLLUTION EFFECTS, *ALGAL TOXINS.

IDENTIFIERS:
*PROTHROMBIN TIME, *PHARMACOLOGICAL PROPERTIES, ANTICANCER, ANTINEOPLASTICS.

ABSTRACT:
TOXICITY, ANTINEOPLASTIC, COAGULANT AND ANTICOAGULANT EFFECTS OF THE FOLLOWING 22 MINNESOTAN AQUATIC PLANTS WERE EVALUATED IN TERMS OF PHARMACOLOGICAL PROPERTIES: ANACHARIS CANADENSIS, CALLA POLUSTRIS, CAREX LACUSTRIS, CERATOPHYLLUM DEMERSUM, CHARA VULGARIS, ELEOCHARIS SMALLII, LEMNA MINOR, MYRIOPHYLLUM EXALBESCENS, NUPHAR VARIEGATUM, NYMPHAEA TUBEROSA, POTAMOGETON AMPLIFOLIUS, P. NATANS, P. PECTINATUM, P. RICHARDSONU, P. ZOSTERIFORMIS, SAGITTARIA CUNEATA, S. LATIFOLIA, SPARGANIUM EURYCARPUM, S. FLUCTUANS, TYPHA ANGUSTIFOLIA, VALLISNERIA AMERICANA, AND ZIZANIA AQUETICA. TOXICITY OF SKELLYSOLVE F, CHLOROFORM, 80% ETHANOL AND WATER EXTRACTS OF THESE AQUATIC PLANTS WERE EVALUATED IN A NUMBER OF ANIMAL EXPERIMENTS IN SWISS WEBSTER MICE. ANTINEOPLASTIC EXPERIMENTS INVOLVED AMELANOMA TUMOR CELLS. IN VIVO PROTHROMBIN TIME (PT) AND PARTIAL THROMBOPLASTIN TIME (PTT) WERE ASSAYED IN ANTI-COAGULATION EXPERIMENTS. THE TOXICITY OF THE AQUATIC PLANTS IN GENERAL WAS FOUND TO BE RELATIVELY LOW. THE LD50 FOR THE MOST TOXIC ONE, I.E., N. TUBEROSA (STEM), IN MICE WAS 3 GM OF DRY PLANT MATERIAL/KG (CA. 25.4 GM OF WET PLANT MATERIAL/KG). ONLY NUPHAR VARIEGATUM INDICATED AN ANTICANCER POSSIBILITY, THE REMAINING AQUATIC PLANTS HAD NO SIGNIFICANT INHIBITION ACTIVITY AT THE DOSES SELECTED. NORMAL PARTIAL THROMBOPLASTIN TIME FOR MICE WAS 51 SECONDS AND ONLY THE PROLONGATION OF PTT (LONGER THAN 61 SECONDS) WAS OBSERVED IN 50% OF THE AQUATIC PLANTS TESTED. AMONG THESE PLANTS, THE MOST SIGNIFICANT INCREASE OF PTT (MORE THAN 20 MINUTES) WAS OBSERVED IN CAREX LACUSTRIS, MYRIOPHYLUM EXALBESCENS, NUPHAR VARIEGATUM AND NYMPHAEA TUBEROSA.

FIELD 05C

ACCESSION NO. W72-07360

WESTERN AUSTRALIA UNIV., NEDLANDS.

R. C. JENNINGS.

AUSTRALIAN JOURNAL OF BIOLOGICAL SCIENCES, VOL 24, P 1115-1124, 1971. 3 FIG, 4 TAB, 18 REF.

DESCRIPTORS:
*ALGAE, *PLANT GROWTH, RHODOPHYTA, INHIBITORS, PLANT TISSUES, BIOASSAY, PHAEPHYTA.

IDENTIFIERS:
*BIOSYNTHESIS, *GIBBERELLIN, HYPNEA MUSCIFCRMIS, GRACILARIA VERUCOSA, ECKLONIA RADIATA, GROWTH REGULATION.

ABSTRACT:
THAT ENDOGENOUS GIBBERELLINS MAY BE INVOLVED IN GRCWTH REGULATION OF BROWN ALGA, ECKLONIS RADIATA, AND THE GREEN ALGA, ENTEROMORPHA PROLIFERA, PRCMPTED STUDIES OF RED ALGAE, HYPNEA MUSCIFORMIS AND GRACILARIA VERUCOSA. CCC AND AMO-1618, AT RELATIVELY HIGH CONCENTRATIONS ONLY, INHIBITED GROWTH OF EXCISED BRANCH APICES OF H MUSCIFORMIS. NEITHER GA-3 NOR GA-7 STIMULATED GRCWTH OF THE ALGA IN PRESENCE OR ABSENCE OF THESE COMPOUNDS, ANC GIBBERELLIN-LIKE MATERIAL EXTRACTED FROM H MUSCIFCRMIS FAILEC TO STIMULATE GROWTH. BOTH GIBBERELLIS STIMULATED GROWTH OF SLOW-GROWING, BUT NOT FAST-GROWING, BRANCH APICES CF GRACILARIA VERUCOSA. IT IS CONCLUDED THAT ENDOGENOUS GIBBERELLINS MAY NOT REGULATE GRCWTH OF H MUSCIFCRMIS, BUT THIS MAY BE A SPECIES PECULIARITY AND NOT A GENERAL PHENCMENON IN RED ALGAE. CCC INHIBITED GAMETOPHYTE GROWTH OF ECKLONIA RADIATA, AND GA-3 SIGNIFICANTLY OVERCAME THIS INHIBITION IN A MANNER CCMPATIBLE WITH THE CONCEPT OF CCC INHIBITING GIBBERELLIN BISCYNTHESIS. THE COMPLEMENT OF ACIDIC, ETHYL ACETATE-SOLUBLE GIBBERELLINS, EXTRACTED FRCM THOSE REGIONS OF E RADIATA SPOROPHYTES ACTIVE IN CELL DIVISION, WAS CHROMATOGRAPHICALLY SIMILAR BUT CIFFERED FROM THCSE EXTRACTED FROM A RELATIVELY QUIESCENT REGION OF THE ALGA. THESE DATA SUPPCRT CONCLUSIONS THAT ENDOGENOUS GIBBERELLINS ARE INVOLVED IN GROWTH REGULATION OF E RADIATA. (JONES-WISCCNSIN)

FIELD 05C

ACCESSION NO. W72-07496

EXTRUSION OF CARBON ACCOMPANYING UPTAKE OF AMINO ACIDS BY MARINE PHYTOPLANKTERS,

CALIFRONIA UNIV., IRVINE. DEPT. OF DEVELOPMENTAL AND CELL BIOLOGY.

C. G. STEPHENS, AND B. B. NORTH.

LIMNOLOGY AND OCEANOGRAPHY, VOL 16, NO 5, P 752-757, 1971. 4 FIG, 1 TAB, 22 REF.

DESCRIPTORS:
*ANALYTICAL TECHNIQUES, *CARBON, *AMINO ACIDS, *MARINE ALGAE, PHYTOPLANKTON, NITROGEN, ABSORPTION, SEA WATER.

IDENTIFIERS:
PLATYMONAS, NITZSCHIA OVALIS, CARBON EXTRUSION, CARBCN-14, NITROGEN SOURCE.

ABSTRACT:
TO GAIN INFORMATION ABOUT CONCENTRATIONS OF FREE AMINO ACIDS IN THE OCEAN, SHORT-TERM UPTAKE MEASUREMENTS WITH C-14-LABELED AMINO ACIDS OF CELL NITROGEN, DISSOLVED C-14, AND DISSOLVED AMINC NITROGEN WERE MADE. IT IS INDICATED THAT AMBIENT AMINO ACIDS ENTER PHYTOPLANKTON CELLS, THE AMINO NITROGEN IS RETAINED, AND SOME CARBON IS RETURNED TO THE MEDIUM. PLATYMONAS AND NITZSCHIA OVALIS WERE USED THROUGHCUT. THE DISTRIBUTION OF C-14 WAS FOLLOWED BY DETERMINING RACIO-ACTIVITY IN MEDIUM SAMPLES AND IN CELL SAMPLES COLLECTED ON MILLIPORE FILTERS. UPTAKE OF THE NITROGENOUS PORTION OF THE AMINO ACID MOLECULE COULD BE FOLLOWED DIRECTLY IN TWO WAYS. CHRCMATOGRAPHY ANC AUTORADICGRAPHY OF A PORTION OF THE ELUENT PRODUCED A SINGLE SPOT AT THE PROPER POSITION FOR THE AMINO ACID CONCERNED. RESULTS IMPLY THAT RADIOCHEMICAL MEASUREMENTS OF UPTAKE RATES USING C-14-LABELED AMINO ACIDS MAY PRCDUCE SERIOUS UNDERESTIMATES OF ACTUAL UPTAKE RATES, PARTICULARLY WITH LONG INCUBATION TIMES. ALSO, IT IS CONSISTENT WITH THE HYPOTHESIS THAT AMINO ACIDS MAY BE A SIGNIFICANT NITROGEN SOURCE FOR PHYTOPLANKTERS IN THE OCEAN. THESE OBSERVATIONS MAKE IT NECESSARY TO RE-EVALUATE PROCEDURES FOR STUDY OF AMINO ACID UPTAKE BY ALGAE. (JONES-WISCONSIN)

FIELD 05C

ACCESSION NO. W72-07497

PRODUCTION OF BENTHIC MICROALGAE IN THE LITTORAL ZONE OF A EUTROPHIC LAKE,

COPENHAGEN UNIV., HILLERCD (DENMARK). FRESHWATER-BIOLOGICAL LAB.

C. HUNDING.

OIKOS, VOL 22, NO 3, P 389-397, 1971. 8 FIG, 5 REF.

DESCRIPTORS:
*PRIMARY PRODUCTIVITY, *BENTHIC FLORA, MICROORGANISMS, ALGAE, LAKES, EUTROPHICATION, LITTORAL, SANDS, SHALLCW WATER, PERIPHYTON, SEDIMENTS, DIATOMS, LIGHT INTENSITY, PHOTOSYNTHESIS, SESSILE ALGAE, TEMPERATURE, DEPTH, LIGHT PENETRATION.

IDENTIFIERS:
LAKE FURESO (CENMARK), COCCONEIS, RHOICOSPHENIA, ACHANTHES, FRAGILARIA CONSTRUENTS, NAVICULA, NITZSCHIA, EPIPYTIC ALGAE.

ABSTRACT:
EFFECT OF VARYING LIGHT INTENSITY AND TEMPERATURE ON MICROBENTHIC COMMUNITY FLORA WERE STUDIED THRCUGHOUT THE YEAR SIMULTANEOUSLY WITH PRIMARY PRODUCTION, WHICH WAS MEASURED ON BENTHIC MICROALGAE OF THE SANDY LITTORAL ZONE IN DANISH EUTROPHIC LAKE FURESO BY THE C-14 METHOD. ANNUAL POTENTIAL GROSS PRODUCTION WAS CALCULATED SHOWING HIGHEST VALUES IN JULY AND ANNUAL TRUE GROSS PRODUCTICN ESTIMATED. DARK FIXATION WAS RELATIVELY HIGH. ALTHOUGH THE FOUR STATIONS DIFFERED SCMEWHAT IN WAVE ACTION, NO SIGNIFICANT DIFFERENCES BETWEEN PRODUCTION WERE NOTED. DAILY PRODUCTION FOR ALL STATIONS WAS AVERAGED. SESSILE ALGAE ACCOUNTED FOR THE BULK OF PRIMARY PRODUCTICN (70 TO 90%). ALGAE WERE MAINLY PENNATE DIATOMS. DIATCMS WITH INTACT CHRCMATOPHORES WERE FCUND DOWN TO A 10 CM DEPTH IN SEDIMENT. WHEN CELLS FRCM THE ANAEROBIC ZONE WERE BROUGHT INTO LIGHT, PHOTOSYNTHESIS STARTED IMMEDIATELY. DURING SUMMER, LIGHT INTENSITY VALUE WAS 9 TO 10 WHILE THE AUTUMN VALUE WAS 3 TO 5 KLUX. PHOTOINHIBITION WAS DEMONSTRATED FOR THE AUTUMN POPULATICN AT LIGHT INTENSITIES EXCEEDING 25 TO 30 KLUX. IT SEEMS TO BE CHARACTERISTIC FOR BENTHIC ALGAE THAT LIGHT SATURATED PHOTOSYNTHESIS TAKES PLACE OVER A WIDE RANGE OF LIGHT INTENSITIES. BENTHIC SESSILE DIATOM POPULATIONS APPEAR RATHER STABLE. (JONES-WISCONSIN)

FIELD 05C

ACCESSION NO. W72-07501

A PRELIMINARY INVESTIGATION OF A POTENTIAL NEW ALGICIDE,

HATFIELD POLYTECHNIC (ENGLAND). DEPT. OF BIOLOGICAL SCIENCES.

K. H. GOULDING.

PRCCEEDINGS 6TH BRITISH INSECTICIDE AND FUNGICIDE CONFERENCE, P 621-629, 1971. 4 REF.

DESCRIPTORS:
*ALGICIDES, *ALGAL CCNTROL, WATER POLLUTION, EUTRCPHICATION, NUTRIENTS, CHLORELLA, CHLAMYDOMCNAS, DIATOMS, SILICATES, CYANCPHYTA, TOXICITY, COSTS, CARBON DIOXIDE.

IDENTIFIERS:
*DACONIL, *CHLOROPHTHALONIL, ULOTHRIX, ANABAENA, CSCILLATORIA, MICROCYSTIS, ALGISTAT.

ABSTRACT:
SINCE THE PROBLEM OF INCREASED ALGAL GROWTH INVOLVES SUCH A WIDE RANGE OF SPECIES ANY SUCCESSFUL ALGICIDE SHOULD HAVE THE WIDEST RANGE OF EFFECTIVENESS; IT MUST BE PARTICULARLY ACTIVE AGAINST BLUE-GREEN ALGAE. A LARGE NUMBER OF CRITERIA SHOULD BE SATISFIED BEFORE AN ALGICIDE IS USED ON A WIDE SCALE AND SHOULD REALLY ACT MORE AS AN ALGISTATIC AGENT SLOWING DOWN GROWTH AND SPREADING OUT AN ALGAL 'BLOOM' OVER A LONGER TIME PERIOD. IF THIS COULD BE ACHIEVED WATER TREATMENT PLANTS COULD COPE WITH THE ALGAL PROBLEMS. SOME OF THE WORK SC FAR CARRIED OUT IN EXAMINING THE ALGICIDAL PROPERTIES OF THE COMPOUND CHLOROPHTHALONIL INDICATES THE POTENTIAL OF THIS COMPOUND. IT IS EFFECTIVE AGAINST A RANGE OF ALGAE INCLUDING CHLORELLA, CHLAMYCOMONAS, ULOTHRIX, ANABAENA, OSCILLATORIA, AND MICROCYSTIS AT LOW CONCENTRATICNS--OFTEN LESS THAN 0.001 PPM. THE EFFECT IS GENERALLY LESS AFTER 300 HOURS THAN AFTER 150 HOURS AND IS DEPENDENT UPON THE SIZE OF THE INITIAL CELL INOCULUM. THE WORK CARRIED OUT EVALUATES CHLORCPHYTHALONIL WITH A LIST OF CRITERIA FOR AN IDEAL ALGICIDE. (JONES-WISCONSIN)

FIELD 05C

ACCESSION NO. W72-07508

ALGAL GROWTH EXCITERS,

KAPPE ASSOCIATES, INC., ROCKVILLE, MD.

D. S. KAPPE, AND S. E. KAPPE.

WATER AND SEWAGE WORKS, VOL 118, NO 8, P 245-248, 1971. 14 REF.

DESCRIPTORS:
*ALGAL CONTROL, *ESSENTIAL NUTRIENTS, *DEFICIENT NUTRIENTS, *AQUATIC ALGAE, *CARBON DIOXIDE, NUISANCE ALGAE, EUTROPHICATION, PHOSPHATES, PHOSPHORUS, CARBON, ORGANIC MATTER, MANGANESE, IRON, SULFUR, POTASSIUM, MOLYBDENUM, DETERGENTS, SEWAGE TREATMENT, CARBONATES, BICARBONATES, HARDNESS(WATER), ALKALINITY, ENZYMES, TEMPERATURE, PHOTOSYNTHESIS, DECOMPOSING ORGANIC MATTER, LIGHT INTENSITY, CALCIUM, CYANOPHYTA, SCENEDESMUS, SODIUM, MANGANESE, CHLOROPHYLL, POTASSIUM, COPPER, ALGICIDES, COBALT.

IDENTIFIERS:
LAKE TAHOE(CALIF), ZINC, ASCORBIC ACID, PYRIDOXINE, INDAL COMPOUNDS.

ABSTRACT:
NO APPARENT CONSISTENT RELATIONSHIP APPEARS TO EXIST BETWEEN LAKE EUTROPHICATION AND PHOSPHORUS RETENTION CAPACITY AND EUTROPHICATION CANNOT BE EXPLAINED SOLELY ON THE BASIS OF SEDIMENTS' CAPACITY TO RETAIN PHOSPHORUS. CURRENT FINDINGS INDICATE THAT EUTROPHICATION IS MORE DIRECTLY RELATED TO BICARBONATE-CARBONATE ALKALINITIES OF WATER THAN TO PHOSPHATE CONCENTRATIONS AND DISCLOSE THAT WATER TEMPERATURE PLAYS AN IMPORTANT PART IN THE ALGAL CYCLE; THAT THE BICARBONATE-CARBONATE ALKALINITY OF WATER IS ESSENTIAL TO ALGAL GROWTH; THE CONCENTRATIONS OF NUTRIENTS ARE CRITICAL FOR PHOTOSYNTHESIS; THE INTENSITY OF SUNLIGHT, THE CLARITY OF WATER, AND THE ABILITY OF ALGAE TO ADSORB AND ABSORB ENERGY FROM SUNLIGHT HAVE A GREAT EFFECT ON ALGAL GROWTH; AND THAT ALGAL GROWTH IS CONTROLLED AND STIMULATED BY MANY METAL-ACTIVATED ENZYMES. THE CURRENT CONCEPT IS THAT ALTHOUGH NITROGEN AND PHOSPHORUS ARE MAJOR FACTORS IN THE EUTROPHIC CONDITION THAT LEADS TO ALGAL BLOOMS, THEIR PRESENCE IS NOT ALWAYS THE CRITICAL FACTOR. NITROGEN IS THE CRITICAL LIMITING FACTOR TO ALGAL GROWTH AND EUTROPHICATION IN COASTAL MARINE WATERS AND REMOVAL OF PHOSPHATES FROM DETERGENTS IS NOT LIKELY TO SLOW EUTROPHICATION. PHOTOSYNTHETIC BACTERIA, PHOTOCHEMICAL BACTERIA, AND OTHER AUTOTROPHIC AND HETEROTROPHIC BACTERIA COULD BE THE PRIMARY REACTORS THAT PRODUCE NUTRIENTS AND ESSENTIAL ELEMENTS DIRECTLY OR INDIRECTLY IN A FORM RESPONSIBLE FOR ALGAL GROWTH. PHOSPHORUS CANNOT BE THE GROWTH LIMITING NUTRIENT IN ALL WATERS. (AUEN-WISCONSIN)

FIELD 05C

ACCESSION NO. W72-07514

THE EFFECTS OF HEATED WASTE WATERS UPON MICROBIAL COMMUNITIES,

VIRGINIA POLYTECHNIC INST. AND STATE UNIV., BLACKSBURG. WATER RESOURCES RESEARCH CENTER.

J. CAIRNS, JR., AND R. A. PATERSON.

AVAILABLE FROM THE NATIONAL TECHNICAL INFORMATION SERVICE AS PB-208 697, $3.00 IN PAPER COPY, $0.95 IN MICROFICHE. VIRGINIA WATER RESOURCES RESEARCH CENTER, BLACKSBURG, PROJECT COMPLETION REPORT SEPTEMBER 1971 24 P, 2 FIG, 9 REF. OWRR B-017-VA(7).

DESCRIPTORS:
*THERMAL STRESS, *THERMAL POLLUTION, *MICROORGANISMS, *FLUORESCENCE, FOOD CHAINS, EQUIPMENT, ANALYTICAL TECHNIQUES, HEATED WATER, CONDENSERS, PROTOZOA, ALGAE, FUNGI, WATER POLLUTION EFFECTS.

ABSTRACT:
THE EFFECTS OF THERMAL STRESS ON SELECTED AQUATIC MICROORGANISMS WERE INVESTIGATED. NUMEROUS CATEGORIES OF TEMPERATURE SHOCK WERE STUDIED WITH EMPHASIS ON (1) SIMULATION OF THE PASSAGE OF WATER THROUGH A CONDENSING SYSTEM TO DETERMINE THE EFFECTS OF THIS EXPOSURE AND (2) THE EFFECTS OF SUCH ENTRAINMENT UPON THE MICROBIAL SYSTEM BELOW THE DISCHARGE POINT. THREE GENERAL GROUPS OF ORGANISMS WERE EXAMINED: PROTOZOANS, ALGAE, AND CHYTRIDIACEOUS FUNGI. NEW EQUIPMENT DEVELOPED SPECIFICALLY TO AID IN THESE STUDIES (A SIMPLE APPARATUS FOR DELIVERING HEAT SHOCK TO MICROORGANISMS) AND A NEW FLUORESCENT SURVEY TECHNIQUE TO CHARACTERIZE STRESS INDUCED CELLULAR ALTERATIONS ARE DESCRIBED IN DETAIL.

FIELD 05C, 05B

ACCESSION NO. W72-07526

CARBON, NITROGEN, AND PHOSPHORUS AND THE EUTROPHICATION OF FRESHWATER LAKES,

FISHERIES RESEARCH BOARD OF CANADA, WINNIPEG (MANITOBA). FRESHWATER INST.

D. W. SCHINDLER.

JOURNAL OF PHYCOLOGY, VOL. 7, NO. 4, P 321-329, DECEMBER 1971. 1 FIG, 2 TAB, 83 REF.

DESCRIPTORS:
*EUTROPHICATION, *LAKES, *NUTRIENTS, *FRESHWATER, *BIOASSAY, *STANDING CROP, CARBON, NITROGEN, PHOSPHORUS, PHYTOPLANKTON, CARBON DIOXIDE, SESTON, FERTILIZATION, WATER QUALITY, RADON RADIOISOTOPES, PRIMARY PRODUCTIVITY, CULTURES, BIOMASS, CARBON RADIOISOTOPES, LIMITING FACTORS, PHOTOSYNTHESIS, GREAT LAKES, LAKE ERIE, LAKE ONTARIO, HYDROGEN ION CONCENTRATION, TURBIDITY, OLIGOTROPHY, CHLOROPHYTA, ALGAE, CHRYSOPHYTA, CYANOPHYTA, EPILIMNION, OCHROMONAS, LAKES.

IDENTIFIERS:
SECCHI DISC, DISSOLVED NITROGEN, DISSOLVED PHOSPHORUS, OSCILLATORIA, RADON RADIOISOTOPES, CHROMULINA, CRYPTOMONAS, MALLOMONAS, STAURASTRUM, PHACOMYXA, SPONDYLOSIUM, LYNGBYA, PSEUDOANABAENA, DATA INTERPRETATION.

ABSTRACT:
THE QUESTION OF NUTRIENTS RESPONSIBLE FOR EUTROPHICATION OF FRESHWATER LAKES IS REVIEWED, AND RECENT ADDITIONS TO THE LITERATURE ON NUTRIENT LIMITATION ARE DISCUSSED. THE PAPER BY LANGE IS CRITICIZED ON SEVERAL GROUNDS, INCLUDING THE FACTS THAT UTILIZATION OF BICARBONATE BY PHYTOPLANKTON AND THE INVASION OF LAKE WATERS BY ATMOSPHERIC CO2 ARE IGNORED AS SOURCES OF PHOTOSYNTHETIC CARBON. THE PHOSPHORUS AND NITROGEN CONCENTRATIONS USED IN LANGE'S EXPERIMENTS ARE FAR HIGHER THAN VALUES PUBLISHED BY OTHERS FOR LAKES ERIE AND ONTARIO. PRELIMINARY RESULTS OF FERTILIZING A SMALL OLIGOTROPHIC LAKE WITH NITROGEN AND PHOSPHORUS ARE DESCRIBED. THE STANDING CROP OF PHYTOPLANKTON INCREASED BY 30-50 TIMES, WHILE THE P:N:C RATIO IN SESTON DID NOT CHANGE FROM RATIOS FOUND IN UNFERTILIZED LAKES. OTHER EXPERIMENTS DONE IN WATER COLUMNS ISOLATED WITH POLYETHYLENE FILM SHOWED THAT ADDITION OF CARBON DID NOT INCREASE THE PHYTOPLANKTON STANDING CROP. SINCE THE FERTILIZED LAKE WAS INITIALLY LOWER IN TOTAL CO2 THAN ANY OTHER RECORDED IN THE LITERATURE, IT IS CONCLUDED THAT CARBON IS UNLIKELY TO LIMIT THE STANDING CROP OF PHYTOPLANKTON IN ALMOST ANY SITUATION. MEASUREMENTS OF INVASION OF ATMOSPHERIC GASES TO THE FERTILIZED LAKE BY THE RADON-222 TECHNIQUE WERE COMPARED WITH PHYTOPLANKTON PRODUCTION MEASUREMENTS, REVEALING THAT ATMOSPHERIC INVASION OF CO2 IS SUFFICIENT TO SUPPORT THE HIGH PHYTOPLANKTON STANDING CROP IN THE EPILIMNION OF THE LAKE. POSSIBLE ERRORS IN INTERPRETATION OF CULTURE AND BOTTLE-BIOASSAY EXPERIMENTS WITH RESPECT TO EUROPHICATION ARE DISCUSSED. (MORTLAND-BATTELLE)

FIELD 05C, 02H

ACCESSION NO. W72-07648

EFFECT OF MERCURY ON ALGAL GROWTH RATES,

NAVAL RESEARCH LAB., WASHINGTON, D.C.

P. J. HANNAN, AND C. PATOUILLET.

BIOTECHNOLOGY AND BIOENGINEERING, VOL 14, P 93-101, 1972. 3 FIG, 1 TAB, 12 REF.

DESCRIPTORS:
*HEAVY METALS, *ALGAE, *GROWTH RATES, *TOXICITY, *FLUORESCENCE, TRACE ELEMENTS, COPPER, CHLORELLA, CULTURES, NUTRIENTS, RADICISOTOPES, CHLOROPHYLL, PIGMENT, MAGNESIUM, PHOSPHATES, NITROGEN, SEA WATER, FLUOROMETRY, EQUIPMENT, INHIBITION, IONS, ABSORPTION, BIOASSAY, *MERCURY.

IDENTIFIERS:
SILVER, CADMIUM, LEAD, CHLCRELLA PHYRENOIDOSA, PHAEODACTYLUM TRICORNUTUM, CYCLOTELLA NANA, CHAETOCEROS GALVESTONENSIS, MERCURIC CHLORIDE, DIMETHYL MERCURY, FLUOROMICROPHOTOMETER.

ABSTRACT:
COMPARISONS CF THE EFFECTS OF 0.1 PPM OF MERCURY AND OTHER METALLIC IONS (SILVER, CADMIUM, LEAD, COPPER) ON THE GROWTH RATES OF ONE FRESHWATER (CHLORELLA PYRENOIDOSA) AND THREE MARINE ALGAE (PHAEODACTYLUM TRICORNUTUM, CYCLOTELLA NANA, CHAETOCEROS GALVESTONENSIS) HAVE BEEN MADE UNDER CONTROLLED LABORATORY CONDITIONS. GROWTH, DETERMINED BY AN INCREASE IN FLUORESCENCE, WAS MONITORED USING A FLURO MICROPHOTOMETER. FLUORESCENCE WAS MEASURED INITIALLY AND ONCE EACH DAY FOR THREE DAYS. GRAPHIC DATA SHOW THAT MERCURY WAS MORE TOXIC THAN THE OTHER METALS TESTED. MERCURY AS MERCURIC CHLORIDE WAS MORE TOXIC THAN AS DIMETHYL MERCURY. THE TOXICITY OF MERCURY WAS FOUND TO BE COMPARATIVELY IRREVERSIBLE AND TC VARY INVERSELY WITH THE CONCENTRATION OF NUTRIENTS PRESENT IN THE GROWTH MEDIA. JUST HOW MUCH MERCURY ALGAE CAN ABSORB AND STILL GROW IS YET TO BE DETERMINED. (JEFFERIS-BATTELLE)

FIELD 05C, 05A

ACCESSION NO. W72-07660

EFFECTS OF DETERGENT PROTEASE ENZYMES ON SEWAGE OXIDATICN POND PHYTOPLANKTON,

VIRGINIA POLYTECHNIC INST., BLACKSBURG. DEPT. OF BIOLOGY.

B. C. PARKER, G. L. SAMSEL, AND E. K. OBENG-ASAMOA.

BIOSCIENCE, VOL 21, NO 20, P 1035-1042, OCTOBER 15, 1971. 4 FIG, 4 TAB, 15 REF.

DESCRIPTORS:
*ENZYMES, *DETERGENTS, *ALGAE, *PHYTOPLANKTON, *SEWAGE LAGOONS, BIODEGRADATION, SEWAGE TREATMENT, WASTE WATER TREATMENT, PONDS, SCENDESOMUS, CYANOPHYTA, BIOASSAY, SAMPLING, OXIDATICN LAGOONS, EUGLENA, CHLAMYDOMONAS, EUGLENOPHYTA, PROTOZOA, EQUIPMENT, WATER POLLUTION EFFECTS.

IDENTIFIERS:
*PROTEASE ENZYMES, CHLOROGONIUM, OOCYSTIS, OSCILLATORIA, SPIRULINA, PANDORINA, PHOCUS, STAURASTRUM, TRACHELOMONAS.

ABSTRACT:
FIELD INVESTIGATIONS ON THE EFFECTS OF DETERGENT PROTEASE ENZYMES ON ALGAL COMMUNITIES OF SEWAGE CXIDATION PONDS SUGGEST THAT ADDITIONS UP TO 1.0 MG/L CAUSE ONLY RELATIVELY SMALL CHANGES IN PHYTOPLANKTON POPULATIONS AND ALGAL COMMUNITY STRUCTURE. AS CONCENTRATIONS APPROACH 10 MG/L, THE COMMUNITY STRUCTURE MIGHT BE DISTURBED SIGNIFICANTLY IN SOME PONDS, DEPENDING ON THE PARTICULAR ENZYME PREPARATION. ENZYME PREPARATIONS NUMBERS 1 AND 2 (EP-1 AND EP-2) FROM THE SOAP AND DETERGENT ASSOCIATION WERE USED IN CONCENTRATICNS OF 10, 1.0, AND 0.1 MG/L ON SEVERAL BIOASSAY SYSTEMS. EITHER ONE LITER, OPEN POLYETHYLENE BAGS, FLOATING AT THE SURFACE IN POLYSTYRENE FRAMES, OR TRANSPARENT ACRYLIC PLASTIC CYLINDERS, WITH THEIR BASES STUCK IN THE MUD AND TOPS PROTRUDING ABOVE THE WATER LINE, WERE USED IN SAMPLING. EP-2 FAVORED GROWTH OF BLUE-GREEN ALGAE, INDICATING THAT WIDESPREAD USE OF SUCH PREPARATIONS SHOULD BE AVCIDED. (MORTLANC-BATTELLE)

FIELD 05C

ACCESSION NO. W72-07661

COMPOSITION OF PHYTOPLANKTON OFF THE SOUTHEASTERN COAST OF THE UNITED STATES,

OLD DOMINION UNIV., NORFOLK, VA.

H. G. MARSHALL.

BULLETIN OF MARINE SCIENCE, VOL 21, NO 4, P 806-825, DECEMBER 1971. 1 FIG, 12 TAB, 12 REF.

DESCRIPTORS:
*PHYTOPLANKTON, *SYSTEMATICS, *DIATOMS, SEA WATER, SEASONAL, FLUCTUATIONS, CONTINENTAL SHELF, VERTICAL DISTRIBUTION, NORTH CAROLINA, SOUTH CAROLINA, SAMPLING, MICROSCOPY, ELECTRON MICROSCOPY, PYRROPHYTA, DISTRIBUTION PATTERNS, WATER TEMPERATURE, WATER QUALITY, PLANKTON, ANALYTICAL TECHNIQUES, ALGAE.

IDENTIFIERS:
COCCOLITHOPHORES, SILICOFLAGELLATES, SARGASSO SEA, SHELF WATERS, GULF STREAM, CAPE HATTERAS, NANSEN BOTTLE, SPECIES DIVERSITY, SAMPLE PRESERVATION, COUNTING.

ABSTRACT:
PHYTOPLANKTON COMPOSITION OFF THE SOUTHEASTERN COAST OF THE UNITED STATES WAS STUDIED DURING A 42-MONTH PERIOD FROM 1964 TO 1968. COLLECTIONS WERE MADE USING NANSEN BOTTLES AT NINE DEPTHS FROM 0 TO 300 FEET IN CONTINENTAL SHELF WATERS, THE GULF STREAM, AND THE SARGASSO SEA. A 500-ML WATER SAMPLE WAS PRESERVED FOR ANALYSIS IMMEDIATELY AFTER SAMPLING USING NEUTRALIZED FORMALIN. THE SUPERNATANT LIQUID WAS SIPHONED OFF 4-6 WEEKS LATER TO OBTAIN A 10 - 40-ML CONCENTRATE. MICROSCOPIC TECHNIQUES WERE THEN USED TO COUNT PHYTOPLANKTON. CONSIDERABLE SPECIES DIVERSITY AND SEASONAL FLUCTUATIONS WERE NOTED. DIATOMS WERE PREDOMINANT IN SHELF WATERS AND TO THE WESTERN BOUNDARY OF THE GULF STREAM, BUT THEIR NUMBERS DECLINED RAPIDLY INTO THE SARGASSO SEA. SKELETONEMA COSTATUM AND RHIZOSOLENEA ALATA WERE THE MAJOR SPECIES FOUND. COCCOLITHOPHORES, PYRRHOPHYCEANS, AND SILICOFLAGELLATES WERE FOUND IN GREATEST NUMBERS IN THE GULF STREAM. IN THE SARGASSO SEA, COCCOLITHOPHORES AND PYRROHOPHYCEANS PREDOMINATED. HOWEVER, THE TOTAL CONCENTRATION OF PHYTOPLANKTON AND THE NUMBER OF SPECIES WERE LESS IN THE SARGASSO SEA THAN IN THE SHELF WATERS OR GULF STREAM. (MORTLAND-BATTELLE)

FIELD 05B, 05A, 02L

ACCESSION NO. W72-07663

BACTERIAL DIEOFF IN PONDS,

TEXAS UNIV., HOUSTON. SCHOOL OF PUBLIC HEALTH.

E. M. DAVIS, AND E. F. FLCYNA.

JOURNAL OF THE SANITARY ENGINEERING DIVISION, AMERICAN SOCIETY OF CIVIL ENGINEERS, VOL 98, NO SA 1, P 59-69, FEBRUARY 1972. 2 FIG, 5 TAB, 14 REF.

DESCRIPTORS:
*ALGAE, *ENTERIC BACTERIA, *PATHOGENIC BACTERIA, CULTURES, E. COLI, BIOINDICATORS, PONDS, ANAEROBIC CONDITIONS, WASTE WATER TREATMENT, CHLOROPHYTA, CYANOPHYTA, AEROBIC CONDITIONS, BIOASSAY, GROWTH RATES, MORTALITY.

IDENTIFIERS:
STABILIZATION PONDS, ANABAENA CYLINDRICA, ANACYSTIS NIDULANS, GLOEOCAPSA ALPICOLA, OSCILLATORIA CHALYBIA, OSCILLATORIA FORMOSA, PHORMIDIUM FAVEOLARUM, ANKISTRODESMUS BRAUNII, CHLORELLA PYRENOIDOSA, CHLORELLA VULGARIS, SCENEDESMUS OBLIQUUS, ALCALIGENES FAECALIS, ENTEROBACTER AEROGENES, PROTEUS VULGARIS, PSEUDOMONAS AERUGINOSA, SERRATIA MARCESCENS, SALMONELLA PARATYPHI, SALMONELLA TYPHOSA, SHIGELLA PARADYSENTERIAE, SHIGELLA DYSENTERIAE, VIBRIO COMMA.

ABSTRACT:
LABORATORY AND FIELD TESTS WERE CONDUCTED TO EVALUATE SPECIFIC INTERACTIONS BETWEEN ALGAE AND TYPICAL ENTERIC AND SELECTED PATHOGENIC BACTERIA IN WASTE STABILIZATION PONDS. BACTERIAL DIEOFF FROM EXPOSURE TO SINGLE ALGAL SPECIES AND MIXED ALGAL ENVIRONMENTS WAS EVALUATED. RESULTS FROM STUDIES OF BOTH LABORATORY AND PILOT-SCALE STABILIZATION PONDS SHOWED THAT SINGLE ALGAL SPECIES HAD LITTLE EFFECT ON DIEOFF RATES, BUT AS THE COMPLEXITY OF THE ALGAL ENVIRONMENT INCREASED SO DID THE DIEOFF RATE. THIS WAS TRUE FOR BOTH ENTERIC AND PATHOGENIC BACTERIA. ESCHERISCHIA, PSEUDOMONAS, AND SERRATIA EXHIBITED AFTERGROWTH POTENTIAL BUT PROTEUS, ALCALIGENES, ENTEROBACTER, SALMONELLA, SHIGELLA, AND VIBRIO DID NOT. ANAEROBIC PRETREATMENT IN COMBINATION WITH FACULTATIVE AND MATURATION PONDS RESULTED IN HIGHER DIEOFFS OF ENTERIC BACTERIA WHEN COMPARED TO AEROBIC CONDITIONS. (MORTLAND-BATTELLE)

FIELD 05C, 05B

ACCESSION NO. W72-07664

A LIST OF NEW GENERA AND TYPE SPECIES OF FLAGELLATES AND ALGAE PUBLISHED IN 1966-1968. PART IV,

B. V. SKVORTZOV.

HYDROBIOLOGIA, VOL. 39, ISSUE 1, P 1-7, JANUARY 31, 1972. 24 FIG, 9 REF.

DESCRIPTORS:
*ALGAE, *CHLOROPHYTA, *EUGLENOPHYTA, *PYRROPHYTA, *SYSTEMATICS, DINOFLAGELLATES, SPECIATION.

IDENTIFIERS:
KOLBEANA OVOIDEA, PALMERIAMONAS PLANCTONICA, MANCHUDINIUM SINICUM, SINAMONAS STAGNALIS, KOFOIDIELLA UNIFLAGELLATA, TSUMURAIA NUMEROSA, HIROSEIA QUINQUELOBATA, TROITSKIELLA TRIANGULATA, AKIYAMAMONAS TERRESTRIS, BRASILOBIA PISCIFORMIS, PROWSEMONAS TROPICA, STIGMOBODO BRASILIANA, PAVLOVIAMONAS FRUTICOSA, NODAMASTIX SPIROGYRAE, COLEMANIA VERRUCOSA, GORDYMONAS VITALIS, PROCTORMONAS CHARACOLA, IOLYA PLANCTONICA, PROTOACEROMONAS SPINOSA, ANGULOMONAS TRIQUETRA, SWIRENKOIAMONAS UNIFLAGELLATA, PROTOCHROCMONAS GRANULATA, ROTUNDOMASTIX PLUVIALIS.

ABSTRACT:
TWENTY-THREE NEW GENERA AND TYPE SPECIES ARE LISTED. OTHER DATA INCLUDE ILLUSTRATIONS OF THE TYPE SPECIES, SIZE OF THE CELLS AND THEIR HABITATS. (HOLOMAN-BATTELLE)

FIELD 05A, 05C

ACCESSION NO. W72-07683

COMPARATIVE METABOLISM OF DDT, METHYLCHLOR, AND ETHOXYCHLOR IN MOUSE, INSECTS, AND IN A MODEL ECOSYSTEM,

ILLINOIS UNIV., URBANA. DEPT. OF ENTOMOLOGY; AND ILLINOIS UNIV., URBANA. DEPT. OF ZOOLOGY.

I. P. KAPOOR, R. L. METCALF, A. S. HIRWE, P.-Y. LU, AND J. R. COATS.

JOURNAL OF AGRICULTURAL AND FOOD CHEMISTRY, VOL 20, NO 1, P 1-6, JANUARY/FEBRUARY 1971. 5 FIG, 7 TAB, 14 REF.

DESCRIPTORS:
*BIODEGRADATION, *METABOLISM, *DDT, *RADIOACTIVITY TECHNIQUES, PESTICIDES, ECOSYSTEMS, POLLUTION SOURCES, PESTICIDE TOXICITY, FOOD CHAINS, FISH, BIOASSAY, ALGAE, SNAILS, MOSQUITOES, MODEL STUDIES.

IDENTIFIERS:
MICE, INSECTS, *ETHOXYCHLOR, *METHYLCHLOR, METABOLITES, BIOLOGICAL MAGNIFICATION, BIOLOGICAL SAMPLES, LIVER, URINE, OEDOGONIUM, PHYSA, CULEX, GAMBUSIA, FATE OF POLLUTANTS.

ABSTRACT:
METABOLIC PATHWAYS FOR ETHOXYCHLOR AND METHYLCHLOR IN INSECTS, MICE, AND IN A MODEL ECOSYSTEM DEMONSTRATE THAT THESE COMPOUNDS ARE PERSISTENT, INSECT TOXIC, BIODEGRADABLE ANALOGS OF DDT. THE EXPERIMENTAL PROCEDURE INVOLVED RADIO 'TAGGING' THE PESTICIDES AND FOLLOWING THEM AND THEIR REACTION PRODUCTS THROUGH THE METABOLIC PATHWAYS. ANALYSIS SHOWED THAT ETHOXYCHLOR IS OXYGEN DEALKYLATED TO MONO AND BISPHENOLIC PRODUCTS WITH DEHYDROCHLORINATION BEING A MAJOR METABOLIC PATHWAY. METHYLCHLOR IS OXIDIZED TO BENZYL ALCOHOL AND BENZOIC ACID ANALOGS AND CONJUGATES. EXCRETION-RECOVERY RATES INDICATE INITIAL RAPID ELIMINATION OF BOTH ANALOGS OVER DDT, BUT OVER A LONGER PERIOD OF TIME THEIR RETENTION SUPERCEDES THAT OF DDT. METABOLISM BY THE MOUSE ALSO INDICATES DEGRADATION TO ALCOHOLIC AND CARBOXYLIC ACID COMPONENTS. IN THE MODEL ECOSYSTEM, PESTICIDES WERE FOUND IN ALGAE, SNAILS, MOSQUITOES, AND FISH. ETHOXYCHLOR WAS FOUND IN FISH AT 1500 TIMES THE AMOUNT IN WATER, METHYLCHLOR WAS FOUND AT 1400 TIMES THE AMOUNT IN WATER, AND DDT WAS FOUND CONCENTRATED AT 85,000 TIMES THE AMOUNT IN WATER, ILLUSTRATING THE BIODEGRADATIVE CHARACTERISTICS OF THE ANALOGS. (MACKAN-BATTELLE)

FIELD 05C, 05B

ACCESSION NO. W72-07703

GROWTH AND THE PRODUCTION OF EXTRACELLULAR SUBSTANCES BY TWO STRAINS OF PHAEOCYSTIS POUGHETI,

WOODS HOLE OCEANOGRAPHIC INSTITUTION, MASS.

R. R. L. GUILLARD, AND J. A. HELLEBUST.

JOURNAL OF PHYCOLOGY, VOL 7, NO 4, P 330-338, DECEMBER 1971. 2 FIG, 7 TAB, 27 REF.

DESCRIPTORS:
*ALGAE, *CARBOHYDRATES, *ORGANICS ACIDS, *PLANT PHYSIOLOGY, *GROWTH RATES, PLANT MORPHOLOGY, CULTURES, PHOTOSYNTHESIS, CARBON RADIOISOTOPES, CHROMATOGRAPHY, AMINO ACIDS, GAS CHROMATOGRAPHY, PHAEOPHYTA, EUTROPHICATION, PLANKTON, ATLANTIC OCEAN, SEA WATER, SALINITY, PHENOLS, WATER TEMPERATURE, CHEMICAL ANALYSIS, SAMPLING, ECOLOGY, PHYTOPLANKTON, ALCOHOLS, SIEVE ANALYSIS, INHIBITORS, SURFACE WATERS, BIOASSAY, TRACERS.

IDENTIFIERS:
*EXCRETION, DISACCHARIDES, MONOSACCHARIDES, OLIGOSACCHARIDES, CLONES, PHAEOCYSTIS POUCHETI, GLUCOSE, MANNOSE, RHAMNOSE, ACRYLIC ACID, SODIUM, PYRIDINE, DEXTRAM, RAFFINOSE, SUCROSE, SENESCENT CULTURES, XYLOSE, RIBOSE, ARABINOSE, GALACTOSE, HEXURONIC ACID, VINEYARD SOUND, ORGANIC CARBON, ACID-VOLATILE COMPOUNDS, WOODS HOLE, HAPTAPHYCEAE, UNIALGAL CULTURES, GEL FILTRATION, AXENIC CULTURES, AUTORADIOGRAPHY, C-14, SURVIVAL.

ABSTRACT:
THE GROWTH AND RELEASE OF EXTRACELLULAR SUBSTANCES BY COLD-WATER STRAINS OF PHAEOCYSTIS, ISOLATED FROM THE WINTER SURFACE WATERS AT WOODS HOLE, WERE COMPARED WITH A WARM-WATER STRAIN FROM THE TROPICAL ATLANTIC NEAR THE COAST OF SURINAM. THE COLD-WATER STRAINS WERE CULTURED AT 4-8C UNDER ABOUT 4.5 KLUX OF LIGHT FOR 14 HOURS/DAY, AND THE TROPICAL CLONE, AT ABOUT 25 DEGREES C IN 14-HR. LIGHT - 10-HR. DARK CYCLE UNDER 3-4.5 KLUX OF LIGHT. NORTHERN STRAINS SURVIVED ONLY UP TO 14C, WHILE THE TROPICAL STRAIN SURVIVED ONLY AS LOW AS 17C. COLONY SHAPES OF THE NORTHERN AND TROPICAL CLONES DIFFERED SOMEWHAT, BUT THE MOTILE AND NONMOTILE SINGLE CELLS OF BOTH STRAINS SEEMED IDENTICAL IN THE LIGHT MICROSCOPE. BY CURRENT TAXONOMIC CRITERIA BOTH STRAINS BELONG TO THE SPECIES P. POUCHETI (HARIOT) LAGERHEIM. WHEN GROWING IN THE FORM OF COLONIES, BOTH STRAINS EXCRETED 16-64 PERCENT OF THEIR PHOTOASSIMILATED CARBON INTO THE MEDIUM, MAINLY AS CARBOHYDRATES OF VARYING MOLECULAR WEIGHTS. HOWEVER, CULTURES PREDOMINANTLY IN THE FORM OF SINGLE CELLS RELEASED ONLY ABOUT 3 PERCENT OF THEIR PHOTOASSIMILATED CARBON. THE QUALITATIVE COMPOSITION OF THE CARBOHYDRATES RELEASED IS SIMILAR FOR THE 2 STRAINS, CONSISTING OF SOME 8 SUGARS OR SUGAR DERIVATIVES WITH GLUCOSE, MANNOSE, AND RHAMNOSE AS THE DOMINANT COMPONENTS. THE PRODUCTION OF ACRYIC ACID WAS CONFIRMED. IT HAS BEEN ESTIMATED THAT AS MUCH AS 7 MICROGRAMS/LITER OF ACRYLIC ACID, AND AT LEAST 0.3 MG/LITER OF POLYSACCHARIDES CAN BE LIBERATED IN A PHAEOCYSTIS BLOOM. (SNYDER-BATTELLE)

FIELD 05C, 05A

ACCESSION NO. W72-07710

ORGANIC MATERIAL ANALYZER MONITORS SHORELINE SEWAGE DISPOSAL,

CHEMURGIC COUNCIL, NEW YORK.

J. W. TICKNOR.

WATER AND SEWAGE WORKS, VOL 119, NO 3, P 50-51, MARCH 1972. 3 FIG.

DESCRIPTORS:
*CHEMICAL ANALYSIS, *MONITORING, *CYCLING NUTRIENTS, *FOOD CHAINS, *ORGANIC COMPOUNDS, SHELLFISH, INSTRUMENTATION, PHYTOPLANKTON, AUTOMATION, EQUIPMENT, ANALYTICAL TECHNIQUES, SEWAGE, RESEARCH EQUIPMENT, SEWAGE EFFLUENTS, SEWAGE TREATMENT, SEWAGE DISPOSAL, MARINE ALGAE, CLAMS, MUSSELS, OYSTERS, EUTROPHICATION, NITROGEN, PHOSPHOROUS, CARBON, HYDROGEN, MARINE ANIMALS, NUTRIENTS, MOLLUSKS, CULTURES.

IDENTIFIERS:
ELEMENTAL ANALYZERS, MACROINVERTEBRATES.

ABSTRACT:
SCIENTISTS AT THE WOODS HOLE OCEANOGRAPHIC INSTITUTION, MASSACHUSETTS, ARE WORKING ON A PROJECT CONCERNED WITH THE HEALTHFUL DISPOSAL OF TREATED SEWAGE AND PROVIDING NUTRIENTS FOR INCREASED SHELLFISH CULTURE ALONG THE EASTERN SEABOARD. EXPERIMENTATION INVOLVES FEEDING TREATED SEWAGE TO ALGAE AND FEEDING THE ALGAE TO SHELLFISH (OYSTERS, MUSSELS, SCALLOPS) UNDER CONTROLLED CONDITIONS. IN ORDER TO DETERMINE WHAT NUTRITIVE VALUES THE SHELLFISH DERIVE FROM THE SEWAGE-PHYTOPLANKTON-SUNLIGHT SYSTEM, THE FOOD CHAIN IS CONSTANTLY MONITORED AT SELECTED POINTS USING A PERKIN-ELMER MODEL 240 ELEMENTAL ANALYZER. THIS INSTRUMENT AUTOMATICALLY ANALYZES ORGANIC COMPOUNDS FOR CARBON, HYDROGEN, AND NITROGEN CONTENT SIMULTANEOUSLY, PERFORMING A COMPLETE ANALYSIS WITHIN 13 MINUTES. IT IS USED, IN THIS CASE, TO DETERMINE LEVELS AND RATIOS OF N AND P PRESENT IN ALGAE, MOLLUSCS, AND MOLLUSCS' DETRITUS. THE ANALYTICAL METHOD CONSISTS OF: (1) BURNING THE ORGANIC SAMPLE IN AN O2 ATMOSPHERE AIDED BY CHEMICALS SUCH AS SILVER TUNGSTATE AND MAGNESIUM OXIDE, (2) FLUSHING THE GASEOUS COMBUSTION PRODUCTS WITH A HELIUM STREAM THROUGH A REDUCTION TUBE, AND (3) REMOVING H, C, AND N AS THE GASEOUS PRODUCTS MOVE THROUGH A SERIES OF TRAPS. THE ANALYZER CAN BE READILY CONVERTED FOR OXYGEN ANALYSIS. (MACKAN-BATTELLE)

FIELD 05A, 05C, 05E

ACCESSION NO. W72-07715

CLUSTER ANALYSIS OF FISH IN A PORTION OF THE UPPER POTOMAC RIVER,

VIRGINIA POLYTECHNIC INST. AND STATE UNIV., BLACKSBURG. DEPT. OF BIOLOGY.

J. CAIRNS, JR., AND R. L. KAESLER.

TRANSACTIONS OF THE AMERICAN FISHERIES SOCIETY, VOL. 100, NO. 4, P 750-756, OCTOBER 1971. 2 FIG, 3 TAB, 10 REF.

DESCRIPTORS:
*SURVEYS, *STATISTICAL METHODS, *POTOMAC RIVER, DATA COLLECTIONS, EVALUATION, ON-SITE DATA COLLECTIONS, RIVERS, *ELECTRIC POWERPLANTS, WATER POLLUTION EFFECTS, WATER POLLUTION SOURCES, *FISH, DIATOMS, PROTOZOA, AQUATIC INSECTS, AQUATIC ANIMALS, AQUATIC ALGAE.

IDENTIFIERS:
*CLUSTER ANALYSIS(FISH), *DICKERSON PLANT.

ABSTRACT:
A SERIES OF SURVEYS WERE CARRIED OUT FROM 1956 UNTIL 1965 IN THE VICINITY OF THE DICKERSON PLANT OF THE POTOMAC ELECTRIC POWER CO, LOCATED BELOW THE CONFLUENCE OF THE POTOMAC AND MONOCACY RIVERS. THESE SURVEYS WERE MADE AT VARIOUS STAGES OF GROWTH OF THE POWER STATION AND AT BOTH HIGH AND LOW WATER LEVELS OF THE RIVER. CLUSTER ANALYSES WERE MADE OF THE FISH DATA IN ORDER TO PROVIDE A RAPID AND EXPLICIT MEANS OF ANALYZING THE DATA. A SECONDARY PURPOSE WAS TO MEASURE THE REDUNDANCY OF INFORMATION OBTAINED BY ANALYSIS OF OTHER GROUPS, NAMELY ALGAE, PROTOZOANS, AQUATIC INSECTS, AND OTHER INVERTEBRATES. ON THE BASIS OF THE CLUSTER ANALYSIS OF DATA ON OCCURRENCE AND DISTRIBUTION OF FISH, NO CHANGES IN THE ENVIRONMENT OF THE UPPER POTOMAC RIVER CAN BE ASCRIBED TO THE OPERATION OF THE DICKERSON PLANT. THIS CONCLUSION AGREES WITH OTHER RESULTS OBTAINED BY ANALYSIS OF PROTOZOANS, ALGAE, AND INSECTS. JACCARD COEFFICIENTS AMONG AGGREGATES OF SPECIES OF FISH ARE APPRECIABLY HIGHER THAN FOR OTHER GROUPS OF ORGANISMS, INDICATING THE RELATIVE MOBILITY OF FISH. THERE IS MUCH INFORMATION REDUNDANCY WHEN ALL GROUPS OF ORGANISMS ARE STUDIED SIMULTANEOUSLY. THIS REDUNDANCY IS CONSIDERED DESIRABLE FOR CONFIRMATION OF RESULTS. (LEGORE-WASHINGTON)

FIELD 05C

ACCESSION NO. W72-07791

CONCENTRATIONS OF PU, CO AND AG RADIONUCLIDES IN SELECTED PACIFIC SEAWEEDS,

SCRIPPS INSTITUTION OF OCEANOGRAPHY, LA JOLLA, CALIF.

K. M. WONG, V. F. HODGE, AND T. R. FOLSOM.

AVAILABLE FROM THE NATIONAL TECHNICAL INFORMATION SERVICE AS CONF 710817-1 $3.00 IN PAPER COPY, $0.95 IN MICROFICHE. (1971) 11P.

DESCRIPTORS:
*RADIOACTIVITY EFFECTS, *MONITORING, *MARINE ALGAE, RHODOPHYTA, PHAEOPHYTA, *COBALT RADIOISOTOPES, NUCLEAR WASTES, PATH OF POLLUTANTS, ESTUARINE ENVIRONMENT, RADIOACTIVITY TECHNIQUES, ABSORPTION.

IDENTIFIERS:
*PLUTONIUM, *SILVER RADIOISOTOPES.

ABSTRACT:
RED ALGAE NEAR A NUCLEAR PLANT SHOWED EVIDENCE OF ENHANCED CO-58, CO-60, AND AG-110 (TYPICAL RADIOACTIVITIES WERE 2200, 180, AND 260 DPM/KG WET SAMPLE, RESPECTIVELY.), BUT NO ENHANCEMENT OF PU-239. ALL SPECIES OF SEAWEED WERE CONTAMINATED WITH PU-239 FROM (WORLD-WIDE) FALLOUT, BUT FURTHER WORK IS NEEDED TO SHOW THE CORRELATION BETWEEN THE AMOUNT IN THE SEA WATER AND THAT IN THE SEAWEED. (BUPP-ORNL)

FIELD 05C, 05B

ACCESSION NO. W72-07826

SECOND ANNUAL REPORT, ENVIRONMENTAL STUDIES, MAINE YANKEE ATOMIC POWER COMPANY.

MAINE YANKEE ATOMIC POWER CO., WESTBORO, MASS.

AVAILABLE FROM THE NATIONAL TECHNICAL INFORMATION SERVICE AS DOCKET 50309-45, $6.00 IN PAPER COPY, $0.95 IN MICROFICHE. DOCKET 50309-45, 1971. 364 P.

DESCRIPTORS:
*MONITORING, *SURVEYS, *MEASUREMENT, *DATA COLLECTIONS, *ENVIRONMENT, *RADIOACTIVITY, *RADIOACTIVITY EFFECTS, RIVERS, EFFLUENT, RADIOECOLOGY, AQUATIC POPULATIONS, HYDROLOGIC ASPECTS, NUTRIENTS, CHEMICALS, PLANKTON, BENTHOS, ALGAE, FISH, WATER POLLUTION, WATER POLLUTION SOURCES.

IDENTIFIERS:
CONCENTRATION, RADIATION SAFETY, RADIATION CONTROL.

ABSTRACT:
THE ENVIRONMENTAL SURVEY OF THE MONTSWEAG BAY-BACK RIVER AREA WAS INITIATED AS A FULL-SCALE INTEGRATED PROGRAM IN OCTOBER, 1969. PORTIONS OF THE PROGRAM HAD BEEN STARTED AS EARLY AS THE SUMMER OF 1968, BUT SAMPLING WAS NOT CONDUCTED ON A REGULAR SCHEDULE UNTIL THE FALL AND SPRING OF 1969-1970. INITIAL EFFORTS INVOLVED ASSEMBLING AND TESTING EQUIPMENT, EVALUATING SAMPLING PROCEDURES, AND SELECTING PERMANENT SAMPLING STATIONS. ALMOST ALL PORTIONS OF THE PROGRAM HAVE NOW BEEN OPERATIVE FOR A YEAR OR MORE AND SOME CONCLUSIONS CAN BE DRAWN FROM THE FINDINGS. HYDROLOGY AND NUTRIENT CHEMISTRY, PLANTON, BENTHOS, SOME COMMERCIAL INVERTEBRATES, FINFISH, AND MARINE ALGAE ARE DISCUSSED. (HOUSER-ORNL)

FIELD 05A, 05B

ACCESSION NO. W72-07841

ALGAE CONTROL BY MIXING, STAFF REPORT ON KEZAR LAKE IN SUTTON, NEW HAMPSHIRE.

NEW HAMPSHIRE WATER SUPPLY AND POLLUTION CONTROL COMMISSION, CONCORD.

AVAILABLE FROM THE NATIONAL TECHNICAL INFORMATION SERVICE AS COM-71-01087, $3.00 IN PAPER COPY, $0.95 IN MICROFICHE. DECEMBER 1970. 103 P, 28 FIG, 30 TAB, 76 REF.

DESCRIPTORS:
WATER QUALITY CONTROL, *CYANOPHYTA, *WATER ANALYSIS, LAKES, SAMPLING, DIATOMS, TURBIDITY, WATER POLLUTION SOURCES, NUTRIENTS, SEWAGE, WATER QUALITY, AERATION, WATER PROPERTIES, *NEW HAMPSHIRE, ALGAE, COLOR, COPPER SULFATE, NETS, HYDROGEN ION CONCENTRATION, PHOSPHATES, NITROGEN, HARDNESS(WATER), CONDUCTIVITY, ZINC, WATER TEMPERATURE, ALKALINITY, CHLORIDES, COPPER, MANGANESE, DISSOLVED OXYGEN, IRON, LIGHT PENETRATION, CHLOROPHYTA, PIPES, MECHANICAL EQUIPMENT, NITRATES, EUTROPHICATION, CARBONATES, CARBON DIOXIDE, WATER POLLUTION EFFECTS, MIXING.

IDENTIFIERS:
*KEZAR LAKE(N.H.), APHANIZOMENON, ANABAENA SPP, MICROCYSTIS AERUGINOSA, APHANIZOMENON HOLSATICUM, ICTALURUS NEBULOSUS, NOTEMIGONUS CRYSOLEUCAS, CATOSTOMUS COMMERSONII COMMERSONII, ASTERIONELLA SPP, ORTHOPHOSPHATES, SECCHI DISC, NEWFOUND LAKE, SQUAM LAKE, APHANIZOMENON FLOW-AQUAE, DESTRATIFICATION, ANABAENA FLOW-AQUAE, HOMOGENIZING, TRANSPARENCY.

ABSTRACT:
KEZAR LAKE IS A NEW HAMPSHIRE RECREATIONAL LAKE, WHICH IN 1964 WAS OBVIOUSLY SUFFERING FROM OBJECTIONABLE ALGAE BLOOM. AFTER SEVERAL COPPER SULFATE TREATMENTS FAILED TO SOLVE THE PROBLEM, A DESTRATIFICATION PROCESS WAS ATTEMPTED. THIS WAS ACCOMPLISHED BY FORCING COMPRESSED AIR, FROM SHORE-LOCATED COMPRESSORS, THROUGH P.V.C., 2-INCH DIAMETER PIPING TO THE DEEPEST PORTION OF THE LAKE WHERE IT WAS RELEASED THROUGH CERAMIC DIFFUSERS, AND ALLOWED TO BUBBLE UP TO THE LAKE SURFACE THROUGH THE WATER COLUMN. CLARITY OF THE WATER WAS VISIBLY AND MEASURABLY IMPROVED; THE POPULATIONS OF NOXIOUS ALGAE, SO OBJECTIONABLE TO RECREATION INTERESTS, WERE DECREASED IN NUMBER; AND NO HARMFUL EFFECTS WERE DETECTED DURING 1968 AND 1969 SUMMER MIXING OF KEZAR LAKE. MANY IMPROVEMENTS IN VARIOUS WATER QUALITY PARAMETERS WERE ALSO NOTED. OPERATING PRESSURES FOR THE COMPRESSORS ARE LOW. THE EQUIPMENT IS CONVENIENT FOR NECESSARY MAINTENANCE, AND THE OPERATING AND STUDY BUDGET IS MODEST. (MORTLAND-BATTELLE)

FIELD 05G, 05C, 04A

ACCESSION NO. W72-07890

THE ECOLOGY OF THE PLANKTON OFF LA JOLLA, CALIFORNIA, IN THE PERIOD APRIL THROUGH SEPTEMBER, 1967.

CALIFORNIA UNIV., BERKELEY.

AVAILABLE FROM THE NATIONAL TECHNICAL INFORMATION SERVICE AS UCSD-10-P20-54, $3.00 IN PAPER COPY, $0.95 IN MICROFICHE. BULLETIN OF THE SCRIPPS INSTITUTION OF OCEANOGRAPHY OF THE UNIVERSITY OF CALIFCRNIA, LA JOLLA, J.D.H. STRICKLANC, EDITOR, VOL 17, NOV 16, 1970. 103 P, 42 FIG, 17 TAB, 98 REF.

DESCRIPTORS:
*PLANKTON, *BIOMASS, *SAMPLING, *NUTRIENTS, *HYDRCGRAPHY, *PACIFIC OCEAN, ECOLOGY, WATER TEMPERATURE, PHYTOPLANKTON, ZOOPLANKTON, CHLOROPHYLL, SOLAR RADIATION, VITAMIN B, PIGMENTS, GROWTH RATES, ALGAE, FOOD WEBS, COPEPODS, ORGANOPHOSPHORUS COMPOUNDS, PRIMARY PRODUCTIVITY, SALINITY, PROFILES, PHOSPHATES, NITRATES, SILICATES, HYDROGRAPHS, ISOTHERMS, CARBON, NITROGEN, PHOSPHORUS, RED TIDE, PHOTOSYNTHESIS, DEPTH, SYSTEMATICS, CINOFLAGELLATES, DIATOMS, TROPHIC LEVELS, PROTOZOA, CRUSTACEANS, FLUORESCENCE, ANALYTICAL TECHNIQUES, DETRITUS, STANDING CROP, DISTRIBUTION PATTERNS, STATISTICAL METHODS, MECHANICAL EQUIPMENT, SAMPLING, PUMPS, MORTALITY.

IDENTIFIERS:
ORGANIC CARBON, CHLOROPHYLL A, ORGANIC NITROGEN, THIAMINE, BIOTIN, VITAMIN B 12, GONYAULAX POLYEDRA, PERIDINIUM DEPRESSUM, METAZOA, CALANUS HELGOLANDICUS(PACIFICUS), THERMISTCRS, DATA INTERPRETATION, COUNTING.

ABSTRACT:
WEEKLY OBSERVATIONS OF NEARSHORE PLANKTON ANC RELATED HYDROGRAPHIC VARIABLES WERE MADE FROM MID-APRIL TO MIC-SEPTEMBER, 1967, AT THREE STATIONS, 1.4, 4.6, AND 12.1 KM OFFSHORE, JUST NORTH OF LA JOLLA. DAILY WATER-TEMPERATURE MEASUREMENTS WERE OBTAINED FROM THE NAVY ELECTRONIC LABORATORY OCEANOGRAPHIC RESEARCH (NEL) TOWER ANC THE SCRIPPS INSTITUTION OF OCEANCGRAPHY PIER. THE AMCUNT OF INCOMING SOLAR RADIATION WAS MEASURED AT THE SCRIPPS INSTITUTION CF OCEANOGRAPHY. AT EACH WEEKLY STATION, MEASUREMENTS WERE MACE OF TEMPERATURE, SALINITY, SUBMARINE LIGHT ATTENUATION, PHYTOPLANKTON, MICRCZOOPLANKTON, CHLOROPHYLL A, PHOSPHATE, NITRATE, AND SILICATE, THE LAST FOUR USING AUTOMATED METHODS OF ANALYSIS. IN ADDITION, AN ESTIMATE WAS OBTAINED OF THE TOTAL AMOUNT OF PARTICULATE ANC DISSOLVEC ORGANIC CARBON, NITROGEN, AND PHOSPHORUS OVER THE 'PLANT PIGMENT DEPTH'. HYCROGRAPHY AND CHEMISTRY; VITAMIN B12, THIAMINE, AND BIOTIN; ESTIMATES OF PHYTOPLANKTON CROP SIZE, GROWTH RATE, AND PRIMARY PRCDUCTION; RELATIONSHIPS OF PHYTOPLANKTCN SPECIES DISTRIBUTICN TO THE DEPTH DISTRIBUTION OF NITRATE; PHYTOPLANKTON TAXCNOMY ANC STNDING CROP; NUMERICAL ABUNDANCE AND ESTIMATED BIOMASS OF MICRCZOOPLANKTON, PRODUCTION OF THE PLANKTONIC COPEPCD, CALANUS HELGCLANCICUS ARE DISCUSSED. BECAUSE THERE WAS A CLOSE CORRELATICN AMONG EACH OF THE THREE PLANT NUTRIENTS AND TEMPERAUTURE, 'UPWELLING', CAUSEC LAYERS OF HIGH NUTRIENT CONCENTRATICN TC MOVE NEARER THE AREA SURFACE. THE SHOALING OF THE 'TROPHOCLINE' WAS THE FEATURE PROBABLY MCST RESPONSIBLE FOR QUALITATIVE AND QUANTITATIVE CHANGES OF PRODUCTIVITY. ALTHOUGH THE AMOUNT OF DETRITUS IN THE WATER APPEARED TC DEPENC ON THE LEVEL CF PRIMARY PRODUCTION, THE PRODUCTION HAD LITTLE EFFECT ON THE AMOUNT OF DISSOLVED ORGANIC MATERIAL EXCEPT PERHAPS AT THE STATION CLOSEST TO THE COAST, WHERE THE PLANT-CELL CONCENTRATION WAS DENSEST.
(JEFFERIS-BATTELLE)

FIELD 05C, 05A, 02L

ACCESSION NO. W72-07892

A HYDROBIOLOGICAL STUDY OF THE POLLUTED RIVER LIEVE (GHENT, BELGIUM),

RIJKSUNIVERSITAIR CENTRUM ANTWERPEN (BELGIUM). LAB. OF ECOLOGY.

W. H. O. DE SMET, AND F. M. J. C. EVENS.

HYDROBIOLOGIA, VOL 39, ISSUE 1, P 91-154, JANUARY 31, 1972. 3 FIG, 38 TAB, 75 REF.

DESCRIPTORS:
*RIVERS, *BACTERIA, IPLANKTON, PHYSICOCHEMICAL PROPERTIES, *ALGAE, *HYDROBIOLOGY, ENTERIC BACTERIA, SAMPLING, ZOOPLANKTCN, PHYTOPLANKTON, WATER ANALYSIS, POLLUTANTS, AQUATIC ALGAE, AQUATIC MICROORGANISMS, AQUATIC ANIMALS, NUTRIENTS, WATER POLLUTION EFFECTS, WATER QUALITY, SEASONAL, SUSPENDED LCAD, BIOINDICATORS, WATER POLLUTION, MICROBIOLOGY, WATER TEMPERATURE, HYDROGEN ION CONCENTRATION, DISSOLVED OXYGEN, BIOCHEMICAL OXYGEN DEMAND, NITROGEN, NITRATES, NITRITES, CHLORIDES, PHOSPHATES, COLIFORMS, CYANOPHYTA, PYRROPHYTA, EUGLENOPHYTA, CHRYSOPHYTA, CHLOROPHYTA, PROTOZCA, CCPEPCDS, CRUSTACEANS, ROTIFERS, EUGLENA, DIATOMS, SCENEDESMUS, ODOR, AGARS, ODOR-PRODUCING ALGAE, VOLUMETRIC ANALYSIS, COLORIMETRY, PLANKTON NETS, CULTURES, COLOR, CHLAMYDOMONAS, DAPHNIA, PRIMARY PRODUCTIVITY, BICMASS, SECONDARY PRODUCTIVITY, AERATICN, TURBIDITY, SESTON, DOMESTIC WASTES, CHEMICAL WASTES, INDUSTRIAL WASTES, SUMMER, WINTER, AUTUMN, METHODOLOGY.

IDENTIFIERS:
AMOEBOBACTER ROSEUS CHROMATIUM OKENII, CHROMATIUM, MACROMONAS MOBILIS, THIOVULUM MAJUS, THIOSPIRA WINOGRADSKY, ZOOGLEA RAMIGERA, BEGGIATOA ALBA, BEGGIATOA, THICTHRIX NIVEA, THIOTHRIX TENUIS, MERISMOPEDIA GLAUCA, DACTYLOCOCCOPSIS SMITHII, OSCILLATORIA PRINCEPS, OSCILLATORIA AMOENA, OSCILLATORIA TENUIS, OSCILLATORIA CHLORINA, OSCILLATORIA, LYNGBYA, ANABAENA CONSTRICTA, LIEVE RIVER, POTASSIUM PERMANGANATE-CONSUMPTION, SAPROBIC VALENCY, SARCCOINA, RHIZOPOCA, QUANTITATIVE ANALYSIS, SUCTORIANS, HARPACTOIDEA, CCUNTING CHAMBERS, FECES, *BELGIUM, BACILLARICPHYTA, BACTERIAPHYTA, CRYPTOPHYTA, XANTHOPHYTA, CILIATA, SUCTOREA, ROTATORIA, *LIEVE RIVER(BELG.).

ABSTRACT:
PHYSICO-CHEMICAL, BACTERIOLOGICAL, AND PLANKTON LEVELS WERE DETERMINED ABOUT EVERY TWO WEEKS (FROM 14 JULY 1964 TO 15 MARCH 1965) AT SIX SITES ON THE LIEVE RIVER. ALL SAMPLES WERE TAKEN FROM THE UPPER 20 CM OF WATER AT THE MIDDLE CF THE RIVER. THREE OF THE SAMPLING STATIONS WERE LOCATED OUTSIDE AND THREE INSIDE THE AREA OF THE RIVER WHICH WAS AERATED. FROM THE CHEMICAL AND BACTERIOLOGICAL PCINT OF VIEW, THE LIEVE WAS CHARACTERIZED AS HEAVILY POLLUTED. A POLLUTICN GRADIENT CAN BE DEMONSTRATED AS WELL B PHYSICO-CHEMICAL INVESTIGATIONS AS BY THE BIOLOGICAL RESULTS. THE PRESENCE AND THE PRODUCTION OF THE PLANKTON IN THE DIFFERENT PLACES ARE DISCUSSED AND SOME GENERAL CONSIDERATIONS LEADING TO A NEW WORK HYPOTHESIS ARE MADE. (HOLOMAN-BATTELLE)

FIELD 05C, 05A

ACCESSION NO. W72-07896

STUDIES AT OYSTER BAY IN JAMAICA, WEST INDIES. V. QUALITATIVE OBSERVATIONS ON THE PLANKTONIC ALGAE AND PROTOZOA,

JOHNS HOPKINS UNIV., BALTIMORE, MD. MCCOLLUM-PRATT INST.; AND JOHNS HOPKINS UNIV., BALTIMORE, MD. DEPT. OF BIOLOGY.

R. J. BUCHANAN.

BULLETIN OF MARINE SCIENCE, VOL. 21, NO. 4, P 914-937, DECEMBER 1971. 1 FIG, 3 TAB, 45 REF.

DESCRIPTORS:
*ALGAE, *PROTOZOA, *PRIMARY PRODUCTIVITY, INVERTEBRATES, *PLANKTON, *WATER TEMPERATURE, SYSTEMATICS, SAMPLING, NETS, CENTRIFUGATION, MICROSCOPY, DIATOMS, NUTRIENTS, BACTERIA, CHRYSOPHYTA, PHYTOPLANKTON, ZOOPLANKTON, PYRROPHYTA, DINOFLAGELLATES, CYANOPHYTA, CHLOROPHYTA, EUGLENOPHYTA, GRAZING, WATER QUALITY.

IDENTIFIERS:
*OYSTER BAY, *JAMAICA, FALMOUTH HARBOR, *TINTINNIDS, *TYCHOPELAGIC, *TEMPERATURE TOLERANCES, SAMPLE PRESERVATION, COUNTING CHAMBERS, NUTRIENT CYCLING.

ABSTRACT:
PLANKTON SAMPLES WERE COLLECTED FROM OYSTER BAY, JAMAICA OVER A 15-MONTH PERIOD TO MAKE A QUALITATIVE ANALYSIS OF PLANKTONIC ALGAE AND PROTOZOA. ONE HUNDRED FIFTY-TWO ORGANISMS WERE IDENTIFIED FROM THE ENTIRE COLLECTION SERIES. OF THESE, 28 TAXA WERE FOUND TO BE MOST COMMON AND ABUNDANT. MANY TAXA (ALL OF THE SARCODINA AND OVER HALF THE BACILLARIOPHYCEAE) WERE TYCHOPELAGIC AND IT IS SUGGESTED THAT THESE SPECIES ARE IMPORTANT IN THE PRIMARY PRODUCTION OF OYSTER BAY. ALL THE CILIATED PROTOZOANS IDENTIFIED WERE TINTINNIDS, WHICH, BEING PLANKTONIC, ARE PROBABLY SIGNIFICANT GRAZERS ON BACTERIA AND THE SMALLEST ALGAE. THEY MIGHT ALSO PLAY AN IMPORTANT ROLE IN THE REGENERATION OF NUTRIENTS IN OYSTER BAY. EXAMINATION OF THE LITERATURE ON THE KNOWN TEMPERATURE TOLERANCES OF THE 28 MOST COMMON AND ABUNDANT TAXA SHOWED THAT ONLY PYRODINIUM BAHAMENSE AND CERATIUM HIRCUS WERE RESTRICTED TO THE TEMPERATURE RANGE FOUND IN OYSTER BAY. THIS, AND OTHER EVIDENCE, INDICATES THAT OYSTER BAY SELECTS AGAINST HIGHLY SPECIALIZED ORGANISMS AND IN FAVOR OF HIGHLY ADAPTABLE ONES. (MORTLAND-BATTELLE)

FIELD 05B, 05C

ACCESSION NO. W72-07899

BENTHIC ALGAL COMMUNITIES OF THE METOLIUS RIVER,

OREGON STATE UNIV., CORVALLIS.

B. J. SHERMAN, AND H. K. PHINNEY.

JOURNAL OF PHYCOLOGY, VOL. 7, NO. 4, P 269-273, DECEMBER 1971. 1 FIG, 1 TAB, 10 REF.

DESCRIPTORS:
*ALGAE, *BENTHIC FLORA, *OREGON, RIVERS, BIOLOGICAL COMMUNITIES, SEASONAL, DIATOMS, *LIGHT, SOLAR RADIATION, *PHOTOPERIODISM, PERIPHYTON, *BENTHOS.

IDENTIFIERS:
*CLADOPHORA, *ACHNANTHES, *SPIROGYRA, CLODOPHORA GLOMERATA, *METOLIUS RIVER(ORE).

ABSTRACT:
BENTHIC ALGAE IN THE METOLIUS RIVER (OREGON) WERE SAMPLED AT REGULAR INTERVALS OVER A NEARLY 1-YEAR PERIOD TO DETERMINE CAUSAL FACOTRS IN THE OCCURRENCE AND DISTRIBUTION OF ALGAE SPECIES. THIS RIVER IS PRACTICALLY FREE OF SEASONAL CHANGES IN CURRENT, TURBIDITY, DISSOLVED SUBSTANCE, AND TEMPERATURE. THUS PHOTOPERIOD AND TOTAL LIGHT ENERGY ARE THE ONLY VARIABLE ENVIRONMENTAL FACTORS TO BE CONSIDERED. THREE GENERA, CLODOPHORA, ACHNANTHES, AND SPIROGYRA, PREDOMINATED THROUGHOUT THE YEAR. CLADAPHORA GLOMERATA WAS THE MOST ABUNCANT FILAMENTOUS SPECIES. IT DECREASED AT ONE STATION DURING WINTER MONTHS WHEN LESS LIGHT WAS AVAILABLE, INDICATING ITS DEPENDENCE ON ABUNDANT LIGHT. OTHER SPECIES WERE PRESENT ONLY AT CERTAIN TIMES OF THE YEAR, ATTRIBUTED ALSO TO CHANGE IN AVAILABLE SOLAR RADIATION. HOWEVER, OF APPROXIMATELY 60 SPECIES OF ALGAE IDENTIFIED, ONLY 9 SHOWEC A DEFINITE SEASONAL DISTRIBUTION. THE PRESENCE OF RELATIVELY FEW SPECIES WAS ATTRIBUTED TO THE PREDOMINANCE OF CLADOPHORA GLOMERATA. (MORTLAND-BATTELLE)

FIELD 05A

ACCESSION NO. W72-07901

THERMAL DISCHARGES: ECOLOGICAL EFFECTS,

BATTELLE MEMORIAL INST., COLUMBUS, OHIO.

A. A. LEVIN, T. J. BIRCH, R. E. HILLMAN, AND G. E. RAINES.

ENVIRONMENTAL SCIENCE AND TECHNOLOGY, VOL. 6, NO. 3, P 224-230, MARCH 1972. 1 FIG, 1 TAB.

DESCRIPTORS:
*THERMAL POLLUTION, *CRUSTACEANS, *FISH, HEATED WATER, NUCLEAR POWER PLANTS, ELECTRICAL POWER PLANTS, HEAT TRANSFER, WATER POLLUTION EFFECTS, WATER TEMPERATURE, ALGAE, BULLHEADS, SALMON, SHELLFISH, CLAMS, MARINE ANIMALS, AQUATIC LIFE, ECOSYSTEMS, SALMONIDS.

IDENTIFIERS:
CHALK POINT, COLUMBIA RIVER, PATUXENT RIVER, CONTRA COSTA POWER PLANT, SAN JOAQUIN RIVER, MORRO BAY POWER PLANT, HUMBOLDT BAY, CONNETICUT YANKEE NUCLEAR PLANT, CONNECTICUT RIVER, TURKEY POINT, BISCAYNE BAY, FLORIDA, BALANUS, BARNACLES, EPIFAUNA, SAGARTIA, ANEMONE, MOLGULA, TUNICATE, PISMO CLAM, TIVELA STULTORUM, OSTREA LURIDA, CARDIUM CORBIS, COCKLES, PROTOTHACA STAMINAE, LITTLENECK CLAMS, SAXIDOMUS GIGANTEUS, BUTTER CLAMS, TRESUS NUTTALLI, GAPER CLAMS, SPECIES DIVERSITY, MACROINVERTEBRATES, SURVIVAL.

ABSTRACT:
BY USING PROJECTIONS OF BOTH FOSSIL AND UNCLEAR FUELED ELECTRICAL GENERATION CAPACITY, DATA ON THERMAL EFFICIENCY AND WATER WITHDRAWAL AS WELL AS THE QUANTITY OF WASTE HEAT THAT WILL BE DISSIPATED INTO CONDENSER COOLING WATERS BY THE ELECTRICAL UTILITY INDUSTRY CAN BE ESTIMATED. BASED ON STUDIES CONDUCTED AT GENERATOR STATION SITES, DEGRADATION OF AQUATIC ECOSYSTEMS APPEARS TO VARY WITH THE GENERATOR SYSTEM AND ITS OUTPUT AND THE FLORA AND FAUNA OF THE SITE. SOME FISH AND CRUSTACEANS TOLERATE TEMPERATURE CHANGES EVEN TO AN INSTANT INCREASE OF 25 DEGREES F WITH NO MORTALITY. CERTAIN BIVALVES AND SHELLFISH FIND DISCHARGE CANALS SUPPORTIVE OR FAVORABLE. HEATED EFFLUENTS HAVE BEEN SHOWN TO REDUCE THE DIVERSITY AND ABUNDANCE OF PHYTOPLANKTON, ALGAE, AND ANIMALS IN SOME AREAS SUGGESTING THAT AN INCREASED OUTPUT AND EXPANSION OF THE INDUSTRY MAY INCREASE THE DELETERIOUS EFFECTS AND MAKE WASTE MANAGEMENT MORE DIFFICULT. NO INFORMATION IS AVAILABLE YET ON SUBLETHAL EFFECTS OF THERMAL DISCHARGE. ALTHOUGH NO MAJOR DAMAGE HAS BEEN OBSERVED, ECOLOGICAL CHANGES HAVE TAKEN PLACE. RECOMMENDATIONS FOR STANDARDS FOR LIMITING THERMAL LOAD ON AQUATIC ECOSYSTEMS ARE SUGGESTED. (MACKAN-BATTELLE)

FIELD 05C, 05B

ACCESSION NO. W72-07907

OIL POLLUTION DAMAGE OBSERVED IN TROPICAL COMMUNITIES ALONG THE ATLANTIC SEABOARD OF PANAMA,

SMITHSONIAN INSTITUTION, WASHINGTON, D.C. DEPT. OF INVERTEBRATE ZOOLOGY.

K. RUTZLER, AND W. STERRER.

POLLUTION - FOUNDATIONS FOR TODAY, VOL. 2, P 70-73, 8 FIG, 1 TAB, 3 REF, 1971.

DESCRIPTORS:
*WATER POLLUTION EFFECTS, *OIL WASTES, *CRUSTACEANS, OILY WATER, ALGAE, NEMATODES, OYSTERS, MUSSELS, PROTOZOA, MANGROVE SWAMPS, LITTORAL, SHORES, TIDES, BEACHES, SANDS, INTERTIDAL AREAS, ANNELIDS, COPEPODS, CRABS, AQUATIC LIFE, BACTERIA.

IDENTIFIERS:
*OIL SPILLS, S.S. WITWATER, GALETA ISLAND, CANAL ZONE, MACROINVERTEBRATES, DIESEL FUEL, BUNKER C OIL, AVICENNIA, RHIZOPHORA, BOSTRYCHIETUM, FIDDLER CRABS, MESOFAUNA, TURBELLARIA, SPONGES, TUNICATES, BRYOZOANS, SEA TURTLES.

ABSTRACT:
THE RUPTURE OF THE TANKER S.S. WITWATER IN DECEMBER, 1968, SPILLED NEARLY 20,000 BARRELS OF DIESEL AND BUNKER C OIL NEAR THE SHORELINE OF GALETA ISLAND, CANAL ZONE. MUCH OF THE OIL WAS REMOVED BY BURNING AND PUMPING. HOWEVER, SUBSEQUENT INVESTIGATIONS CONDUCTED TO STUDY THE EFFECTS OF THE REMAINING OIL SHOWED A NUMBER OF DETRIMENTAL EFFECTS. IN ROCKY SHORE AREAS, POLLUTED WATER SPRAY KILLED TREES AND SHRUBS. SUPRALITTORAL SPRAY POOLS AND UPPER MESOLITTORAL TIDAL POOLS WHICH WERE COVERED WITH AN OIL LAYER WERE DEVOID OF LIFE. DAMAGE TO GASTROPOD AND BARNACLE POPULATIONS WAS ASSUMED. SUBTIDAL CORAL REEFS AND ASSOCIATED ORGANISMS SHOWED NO EFFECTS SINCE THEY WERE EXPOSED TO THE OIL. BEACH MEIOFAUNA SUCH AS TURBELLARIA, NEMATODES, ANNELIDA, COPEPODS AND OTHERS WERE RADICALLY DESTROYED BECAUSE OF THE COMPLETE PERMEATION OF THE OIL. CRUSTACEANS WERE FIRST TO DISAPPEAR. SMALL CILIATES THRIVED ON THE INCREASING NUMBERS OF OIL DEGRADING BACTERIA. IN THE MANGROVES, TREES WERE SEVERELY DAMAGED OR KILLED, THE FIDDLER CRAB POPULATION WAS REDUCED AND THE INTERTIDAL ALGAE COMMUNITY 'BOSTRYCHIETUM' AND ITS MICROFAUNA WERE PRACTICALLY ELIMINATED, AS WERE OYSTERS, MUSSELS, SPONGES, TUNICATES, BRYOZOANS, SEA TURTLES, AND SEA BIRDS. (MACKAN-BATTELLE)

FIELD 05C

ACCESSION NO. W72-07911

PRELIMINARY NOTES ON CHANGES IN ALGAL PRIMARY PRODUCTIVITY FOLLOWING EXPOSURE TO CRUDE OIL IN THE CANADIAN ARCTIC,

OTTAWA UNIV. (ONTARIO). DEPT. OF BIOLOGY.

MIKE DICKMAN.

CAN FIELD NAT. 85(3): 249-251. 1971.

DESCRIPTORS:
*PRIMARY PRODUCTIVITY, *OIL WASTES, *ARCTIC, *ALGAE.

IDENTIFIERS:
*CANADIAN ARCTIC, *CHLAMYDOMONAS-SPP, *CRYPTOMONAS-SPP.

ABSTRACT:
MACKENZIE VALLEY CRUDE OIL WHICH HAD BEEN EXPOSED FOR 2 MO. TO NATURAL ARCTIC SUMMER CONDITIONS WAS ADDED TO BOTTLES CONTAINING ALGAE TAKEN FROM A MARSH NEAR INUVIK, N. W. T. (NORTHWEST TERRITORIES). C-14 PRIMARY PRODUCTIVITY WAS 10 TIMES LOWER IN THE OIL TREATED SAMPLES (0.59 + 0.30 MGC/M3/HR) THAN IN THE UNTREATED CONTROL SAMPLES (5.12 + 1.2 MGC/M3/HR) AFTER A 4 HR INCUBATION PERIOD. SMALL FLAGELLATES SUCH AS CRYPTOMONAS SPP. AND CHLAMYDOMONAS SPP. COMPRISED NEARLY 80% OF THE PRIMARY PRODUCERS IN THE INUVIK MARSH SAMPLES. SOME IMPLICATIONS OF THE SIGNIFICANCE OF THESE PRELIMINARY FINDINGS ARE DISCUSSED IN VIEW OF THE PROPOSED 800 MI. MACKENZIE VALLEY PIPELINE ROUTE.--COPYRIGHT 1972, BIOLOGICAL ABSTRACTS, INC.

FIELD 05C, 02C

ACCESSION NO. W72-07922

MICROBIOLOGICAL ASPECTS OF THE POLLUTION OF FRESH WATER WITH INORGANIC NUTRIENTS,

WISCONSIN UNIV., MADISON. DEPT. OF SOIL SCIENCE; AND EDINBURGH UNIV. (SCOTLAND). DEPT. OF MICROBIOLOGY.

D. R. KEENEY, R. A. HERBERT, AND A. J. HOLDING.

IN: MICROBIAL ASPECTS OF POLLUTION, G. SYKES AND F. H. SKINNER, EDITORS, SOCIETY FOR APPLIED BACTERIOLOGY SYMPOSIUM SERIES NO 1, P 181-200, 1971. ACADEMIC PRESS, LONDON. 2 FIG, 3 TAB, 99 REF. EPA PROGRAM 16010 EHR.

DESCRIPTORS:
*WATER POLLUTION, MICROORGANISMS, *INORGANIC COMPOUNDS, *NUTRIENTS, FRESH WATER, ALGAE, BACTERIA, NITROGEN, PHOSPHORUS, OXIDATION-REDUCTION POTENTIAL, LAKES, SOILS, EUTROPHICATION, AMMONIFICATION, NITRIFICATION, DENITRIFICATION, NITROGEN FIXATION, SEDIMENTS, NITRATES, WISCONSIN.

IDENTIFIERS:
*LIMITING NUTRIENTS, LAKE MENDOTA(WIS).

ABSTRACT:
THIS LITERATURE REVIEW CONSIDERS BOTH PHOSPHORUS AND NITROGEN BUT EMPHASIZES MICROBIAL PROCESSES GOVERNING NITROGEN AVAILABILITY, PRIMARILY IN LAKE SYSTEMS. HOWEVER, MANY PRINCIPLES ARE APPLICABLE TO RIVER SYSTEMS WHEN COMPARABLE ENVIRONMENTAL CONDITIONS PREVAIL. DUE TO THE MANY COMPLEX COMPETING BIOLOGICAL REACTIONS OCCURRING WITHIN AN ECOSYSTEM IT IS EXTREMELY DIFFICULT TO DETERMINE RELATIVE IMPORTANCE OF EACH INDIVIDUAL MICROBIAL PROCESS. THE NITRATE-NITROGEN BALANCE AT ANY GIVEN TIME IS GOVERNED BY THE RELATIVE RATE OF NITRATE-NITROGEN LOSS (RESULTING FROM DENITRIFICATION AND IMMOBILIZATION) TO THE RATE OF NITRATE-NITROGEN REGENERATION (FROM GROUNDWATER SEEPAGE, STREAM DRAINAGE, AND AMMONIFICATION/NITRIFICATION). ALGAL GROWTH IS LARGELY GOVERNED BY NITROGEN AND PHOSPHORUS AVAILABILITY. NITROGEN FIXATION BY BLUE-GREEN ALGAE PROBABLY CONSTITUTES ONLY 1-2% OF THE INPUT INTO FRESH-WATERS BUT IT MAY BE IMPORTANT TO BIOLOGICAL PRODUCTIVITY WHEN INORGANIC NITROGEN LEVELS BECOME DEPLETED DURING THE SUMMER. THE SMALL BACTERIAL POPULATIONS FOUND IN OLIGOTROPHIC WATERS SUGGEST THEIR ROLE IN NUTRIENT CYCLING IS PERHAPS MINIMAL. IN EUTROPHIC WATERS, OXYGEN DEPLETION BY MICROORGANISMS CAN LEAD TO A SERIES OF EVENTS LARGELY GOVERNED BY THE REDOX POTENTIAL. VALID MODEL SYSTEMS, WHEREBY INDIVIDUAL PARAMETERS CAN BE CONTROLLED ARE INVALUABLE IN DIFFERENTIATING SPECIFIC MICROBIAL PROCESSES TO AID IN PREDICTING FRESH WATER BEHAVIOR IN RESPONSE TO VARYING NUTRIENT LOADS. (JONES-WISCONSIN)

FIELD 05C, 10F

ACCESSION NO. W72-07933

A REVIEW OF THE FACTORS LIMITING THE GROWTH OF NUISANCE ALGAE,

MICHIGAN WATER RESOURCES COMMISSION, LANSING.

ALBERT MASSEY, AND JOHN ROBINSON.

WATER AND SEWAGE WORKS, VOL 118, NO 11, P 352-355, 1971. 38 REF.

DESCRIPTORS:
*ALGAL CONTROL, *NUISANCE ALGAE, EUTROPHICATION, REVIEWS, TROPHIC LEVEL, PHOSPHORUS, PHOSPHATES, CARBON, VITAMIN B, NITROGEN, CYANOPHYTA, LAKE MICHIGAN, NUTRIENTS.

IDENTIFIERS:
*LIMITING FACTORS, SILICON.

ABSTRACT:
ISOLATING THE KEY FACTOR WHICH LIMITS ALGAL GROWTH HAS PROVEN CONFUSING BECAUSE THE AQUATIC ECOSYSTEM IS A MULTIFACTOR SYSTEM IN DYNAMIC EQUILIBRIUM ESTABLISHED BY THE PARTICULAR GEOCHEMICAL CHARACTER AND BIOTA OF THE LAKE. THAT PHOSPHORUS IS THE LIMITING ELEMENT HAS BEEN THEORIZED SINCE LIMNOLOGY'S INFANCY. THROUGH INDEPENDENT RESEARCH NUMEROUS INVESTIGATORS HAVE ESTABLISHED THAT PHOSPHORUS IS THE ELEMENT WHICH USUALLY LIMITS ALGAL GROWTH. SINCE SHORTAGE OF ANY OF 15 ELEMENTS MAY LIMIT ALGAE IF PHOSPHORUS OR NITROGEN IS ADDED, PRIMARY PRODUCTION INCREASES UNTIL SOME OTHER ELEMENT BECOMES LIMITING. CONTINUED DEPLETION OF SILICON WILL FAVOR PROLIFERATION OF PHYTOPLANKTON OTHER THAN DIATOMS AND COULD RESULT IN BLOOMS OF UNDESIRABLE ALGAE. BECAUSE SOME ALGAE FIX ATMOSPHERIC NITROGEN, CONTROL OF CULTURAL EUTROPHICATION BY LIMITING NITROGEN SUPPLY IS HIGHLY QUESTIONABLE. THREE RECENT PAPERS PROPOSING THAT CARBON RATHER THAN PHOSPHORUS IS THE GROWTH-LIMITING FACTOR, CONTAIN NO ORIGINAL RESEARCH. THE CARBON SUPPLY FROM INORGANIC SOURCES, FROM THE ATMOSPHERE, FROM BACTERIAL DEGRADATION, WOULD BE MORE THAN ADEQUATE TO FORCE SOME OTHER FACTOR TO BE LIMITING. IN TERMS OF PRACTICAL TECHNOLOGY ONE WOULD ALMOST HAVE TO ANSWER THAT IT WOULD BE NECESSARY TO CONTROL PHOSPHORUS. WITH THE REDUCTION OF PHOSPHATE INPUTS, CULTURAL EUTROPHICATION OF LAKES MAY BE SLOWED, STOPPED OR EVEN IN SOME CASES REVERSED. (JONES-WISCONSIN)

FIELD 05C, 10F

ACCESSION NO. W72-07937

FOOD QUALITY AND ZOOPLANKTON NUTRITION,

OXFORD UNIV. (ENGLAND). DEPT. OF ZOOLOGY.

JAMES E. SCHINDLER.

JOURNAL OF ANIMAL ECOLOGY, VOL 40, NO 3, P 589-595, 1971. 2 FIG, 3 TAB, 30 REF.

DESCRIPTORS:
*ZOOPLANKTON, *FOODS, DAPHNIA, COPEPODS, ALGAE, NUTRIENT REQUIREMENTS, CYANOPHYTA, CHLOROPHYTA, PHYTOPLANKTON, GRAZING.

IDENTIFIERS:
DAPHNIA LONGISPINA, DIAPTOMUS GRACILIS, CYCLOPS STRENUUS, INGESTION RATES, ASSIMILATION EFFICIENCES, CRYPTOMONAS, ANKISTRODESMUS, ELAKOTOTHRIX, OSCILLATORIA, ANABAENA, MICROCYSTIS, TRIBONEMA, APHANIZOMENON, OOCYSTIS, GLOEOCYSTIS, COELASTRUM, ASTERIONELLA, CRYPTOMONAS.

ABSTRACT:
FOOD SUPPLY HAS BEEN CONSIDERED AMONG THE POSSIBLE FACTORS LEADING TO SEASONAL CHANGES IN ZOOPLANKTON POPULATIONS AND A QUANTITATIVE DETERMINATION OF THE DEGREE OF NUTRITION THAT CAN BE SUPPLIED BY DIFFERENT FOODS PERMITS A VERY RAPID ASSESSMENT OF INGESTION AND ASSIMILATION RATES. IN AN ATTEMPT TO DETERMINE THE AMOUNTS AS WELL AS KIND OF FOOD INGESTED AND ASSIMILATED BY THREE SPECIES OF ZOOPLANKTON, DAPHNIA LONGISPINA, DIAPTOMUS GRACILIS, AND CYCLOPS STRENUUS, C-14 WAS USED TO LABEL 11 DIFFERENT TYPES OF ALGAE. FRESHLY COLLECTED ZOOPLANKTON WAS PRECONDITIONED IN THE DARK AT A CONSTANT TEMPERATURE OF 15C AND WITH THE FOOD TYPE TO BE USED IN THE EXPERIMENTS. ONLY THE LARGER ADULT FEMALES WITHOUT EGGS WERE SELECTED. IN GENERAL THIS WORK TESTS THE RELATIONSHIP BETWEEN SOME CRUDE MEASURE OF PHYTOPLANKTON CONCENTRATION AND SOME MEASURE OF ZOOPLANKTON DENSITY. THE RESULTS SHOW HOW FOOD QUALITY AFFECTS THE ASSIMILATION AND INGESTION RATES OF ZOOPLANKTON. A RELATIONSHIP WAS FOUND BETWEEN THE INGESTION RATES AND ASSIMILATION EFFICIENCIES OF DIAPTOMUS. THE ECOLOGICAL IMPLICATIONS OF THE RELATIONSHIP BETWEEN FOOD QUALITY AND ZOOPLANKTON NUTRITION ARE DISCUSSED. (JONES-WISCONSIN)

FIELD 05C

ACCESSION NO. W72-07940

KINETICS OF SYNTHESIS OF NITROGENASE IN BATCH AND CONTINUCUS CULTURE OF ANABAENA FLOS-AQUAE,

QUEEN'S UNIV. KINGSTON (ONTARIO), DEPT. OF CHEMICAL ENGINEERING.

H. BONE.

ARCHIV FUR MICROBIOLOGIE, VOL 80, NO 3, P 242-251, 1971. 2 FIG, 3 TAB, 28 REF.

DESCRIPTORS:
*NITROGEN FIXATION, *SYNTHESIS, *CULTURES, *NITROGEN, OXYGEN, BIOCHEMISTRY, INHIBITORS, ALGAE, PROTEINS.

IDENTIFIERS:
*ANABAENA FLOS-AQUA, *NITROGENASE ACTIVITY, OXYGEN INACTIVATION.

ABSTRACT:
BACTERIAL NITROGENASES ARE OXYGEN SENSITIVE AND CAN BE SEPARATED INTO TWO FRACTIONS, AN IRON-MOLYBDENUM-PROTEIN AND AN IRON-PROTEIN. FOR A FEW AEROBIC FACULTATIVE NITROGEN-FIXING BACTERIA, THE IRON-PROTEIN IS OXYGEN SENSITIVE. NITROGENASE OF ANABAENA FLOS-AQUAE WAS INACTIVATED BY OXYGEN AND RECOVERY OF ACTIVITY WAS MEASURED IN BATCH AND IRON, PHOSPHATE AND UREA-LIMITED CONTINUOUS CULTURES. A FLOS-AQUAE WAS SHOWN TO ACTIVELY SYNTHESISE NEW NITROGENASE COMPONENTS AFTER OXYGEN INACTIVATION. IN BATCH CULTURE, CANAVANINE, CHLORAMPHENICAOL, METHYLAMINE, PROFLAVINE, PUROMYCIN AND UREA INHIBITED THE RECOVERY PROCESS. A FLOS-AQUAE GROWING AT A DILUTION RATE OF 0.03 PER HOUR HAS 180% THE SPECIFIC ACTIVITY OF NITROGENASE COMPARED TO PHOSPHATE-LIMITED CELLS WHICH SUGGEST THAT PHYTOFLAVIN MIGHT BE ACTIVE IN A FLOS-AQUAE. THE KINETICS OF OXYGEN INACTIVATION OF NITROGENASE OF IRON AND PHOSPHATE-LIMITED A FLOS-AQUAE ARE THE SAME AND ALL THESE PIECES OF EVIDENCE INDICATE THAT THE OXYGEN SENSITIVE SITE IS A NITROGENASE COMPONENT. THE RATE OF RECOVERY OF NITROGENASE ACTIVITY IN CONTINUOUS CULTURES WAS DEPENDENT ON LIGHT INTENSITY, CONCENTRATION OF UREA, AMMONIUM SALTS AND NITRATE, AND INDEPENDENT OF GROWTH RATE. STEADY-STATE NITROGENASE ACTIVITIES SEEM TO BE MAINTANINED BY BALANCING NITROGENASE SYNTHESIS AND INACTIVATION PROCESSES. (JONES-WISCONSIN)

FIELD 05C

ACCESSION NO. W72-07941

ALIGICIDAL NONFRUITING MYXOBACTERIA WITH HIGH G + C RATIOS,

TEXAS UNIV. MEDICAL SCHOOL, SAN ANTONIO. DEPT. OF BIOCHEMISTRY; AND NORTH CAROLINA UNIV., CHAPEL HILL. DEPT. OF BOTANY.

J. R. STEWART, AND R. M. BROWN, JR.

ARCHIV FUR MICROBIOLOGIE, VOL 80, NO 2, P 176-190, 1971. 6 FIG, 2 TAB, 52 REF. FWQA PROGRAM 18050 DBR.

DESCRIPTORS:
*ALGAL TOXINS, *MYXOBACTERIA, CHLOROPHYTA, CYANOPHYTA, SYSTEMATICS, VITAMINS, ANAEROBIC BACTERIA, ELECTRON MICROSCOPY, PIGMENTS, CARBOHYDRATES, PROTEINS, ACTINOMYCETES, CHLORELLA.

IDENTIFIERS:
*G + C RATIOS, *NONFRUITING MYXOBACTERIA, DNA, MOTILITY, CYST FORMATION, ANTIBIOTIC SENSITIVITY, CYTOPHAGA JOHNSONII, SPOROCYTOPHAGA MYXOCOCCOIDES.

ABSTRACT:
BACTERIA WHICH KILL ALGAE HAVE BEEN REPORTED ON GLIDING ORGANISMS IN THE ORDER MYXOBACTERALES OF FRUITING AND NONFRUITING GROUPS BUT ONLY THE LATTER APPEARS TO CONTAIN ORGANISMS KNOWN TO KILL BOTH GREEN AND BLUE-GREEN ALGAE. DEOXYRIBONUCLEIC ACID BASE RATIOS ARE GENERALLY AGREED TO BE IN THE 70'S FOR FRUITING MYXOBACTERIA AND IN THE 30'S FOR NONFRUITING. A FEW NONFRUITING MYXOBACTERIA WITH HIGH G + C RATIOS HAVE BEEN REPORTED. FIVE NONFRUITING MYXOBACTERIA WITH G + C RATIOS RANGING FROM 69-71 MOLE PERCENT, THREE BEING NEW ISOLATES AND TWO OBTAINED ELSEWHERE ARE REPORTED. THESE BASE RATIOS OF THE DEOXYRIBONUCLEIC ACIDS OF THESE ALGICIDAL MYXOBACTERIA WERE DETERMINED BY THERMAL DENATURATION TEMPERATURES; NO UNUSUAL NUCLEIC ACID BASES WERE DETECTED. THESE ORGANISMS ARE DESCRIBED AS AMICRO-CYSTOGENOUS, GLIDING, GRAM-NEGATIVE BACILLI CAPABLE OF DEGRADING GELATIN, CASEIN, STARCH, CELLULOSE, CHITIN, AND ALGINATE. ALL HAVE BEEN SHOWN TO BE ALGICIDAL. POLY-BETA-HYDROXYBUTYRATE IN EACH WAS INDICATED BY CONVERSION TO CROTONIC ACID. ANTIBIOTIC SENSITIVITY WAS SIMILAR TO THAT OF KNOWN NONFRUITING MYXOBACTERIA. FINE STRUCTURE OF ONE, MYXOBACTER 44, REVEALED A TRIPLE-LAYERED CELLULAR ENVELOPE WHOSE MIDDLE LAYER IS LYSOZYME SENSITIVE. RUTHENIUM RED-POSITIVE SLIME MATERIAL ADHERED TO THE OUTER SURFACE. (JONES-WISCONSIN)

FIELD 05C

ACCESSION NO. W72-07951

INTERACTIONS OF DISSOLVED AND PARTICULATE NITROGEN IN LAKE METABOLISM,

MICHIGAN STATE UNIV., HICKORY CORNERS. W. K. KELLOGG BIOLOGICAL STATION; AND MICHIGAN STATE UNIV., HICKORY CORNERS. DEPT. OF BOTANY AND PLANT PATHOLOGY.

BRUCE A. MANNY.

PHD THESIS, 1971. 189 P, 32 FIG, 7 TAB, 268 REF. B-009-MICH(2).

DESCRIPTORS:
*NITROGEN, *LAKES, ANALYTICAL TECHNIQUES, MICHIGAN, SESTON, ALGAE, PHYTOPLANKTON, NANNOPLANKTON, AMMONIA, NITRATES, NITRITES, HARDNESS(WATER), NITROGEN CYCLE, ORGANIC MATTER, PHOTOSYNTHESIS, EUTROPHICATION, PHOSPHORUS, CHLOROPHYTA, CYANOPHYTA, CHRYSOPHYTA, EUGLENOPHYTA, DISSOLVED OXYGEN, CARBON, LAKE MORPHOMETRY, CHEMICAL PROPERTIES, OXYGEN, CHLAMYDOMONAS, AQUATIC PLANTS, ZOOPLANKTON, DAPHNIA, SEDIMENTS, CONDUCTIVITY, BACTERIA, PSEUDOMONAS, NITROGEN COMPOUNDS.

IDENTIFIERS:
*LAKE METABOLISM, *DISSOLVED ORGANIC NITROGEN, *PARTICULATE ORGANIC NITROGEN, CALCAREOUS LAKES, ULTRA-VIOLET LABILE NITROGEN, ULTRA-VIOLET REFRACTORY NITROGEN, NETPLANKTON, LAWRENCE LAKE(MICH), WINTERGREEN LAKE(MICH), CARBONATE PARTICLES, PHOSPHORUS CYCLE.

ABSTRACT:
TO COMPARE RATES OF ALLOCHTHONOUS NITROGEN ENTRY, RATES OF NITROGEN REGENERATION AND ORGANIC NITROGEN PRODUCTION, TWO LAKES, REPRESENTING EXTREMES OF HARDWATER, WERE SAMPLED WEEKLY. MEASUREMENTS OF DISSOLVED AMMONIA, NITRITE, AND NITRATE FOR 22 MONTHS AND ORGANIC NITROGEN FOR 12 MONTHS WERE MADE AND SEASONALLY IN FOUR OTHER LAKES REPRESENTING CHEMICAL AND TROPHIC SPECTRUM IN GLACIATED MICHIGAN. SESTON AT 1 M DEPTH WAS FRACTIONATED WEEKLY FOR 14 MONTHS; NET AND NANNO FRACTIONS ANALYZED. SEASONAL CHANGES IN ALL NITROGENOUS PARAMETERS WERE RELATED TO CHANGES IN 15 OTHER CHEMICAL AND BIOLOGICAL PARAMETERS ASSAYED SIMULTANEOUSLY. NANNOPHYTOPLANKTON CONTAIN MORE NITROGEN PER UNIT CELL VOLUME THAN NETPHYTOPLANKTON. DISSOLVED ORGANIC NITROGEN COMPOUNDS AT NATURAL CONCENTRATIONS WERE MEASURED AND UV-LABILE FRACTION DIFFERENTIATED FROM UV-REFRACTORY FRACTION WITHIN THE NITROGEN POOL IN SIX LAKES. MEASUREMENTS OF ALLOCHTHONOUS DISSOLVED ORGANIC NITROGEN ENTERING THE MOST CALCAREOUS LAKE IN SPRING REVEALED THE TWO FRACTIONS ENTERED THE LAKE. THE MOST CALCAREOUS LAKE REVEALED THE UV-REFRACTORY FRACTION ORIGINATED LARGELY IN THE SURROUNDING WATERSHED, WHEREAS THE UV-LABILE FRACTION ORIGINATED LARGELY WITHIN THE LAKE. PELAGIC DISSOLVED ORGANIC NITROGEN INTERACTIONS SEEM TO EXPLAIN HOW AQUATIC PHOTOSYNTHESIS IN HARDWATER LAKES IS REGULATED. EUTROPHICATION RATES STEMMING FROM VARIOUS COMBINATIONS OF THESE INTERACIONS ARE DISCUSSED. (JONES-WISCONSIN)

FIELD 05C, 02H

ACCESSION NO. W72-08048

EFFECTS OF LOW NUTRIENT DILUTION WATER AND MIXING ON THE GROWTH OF NUISANCE ALGAE,

WASHINGTON UNIV., SEATTLE.

JAMES A. BUCKLEY.

MS THESIS, 1971. 116 P. 54 FIG, 20 TAB, 33 REF, 3 APPEND. OWRR A-034-WASH(3).

DESCRIPTORS:
*ALGAL CONTROL, *DIFFUSION, *NUTRIENTS, *NUISANCE ALGAE, WASHINGTON, PHYTOPLANKTON, CYANOPHYTA, NITRATES, PHOSPHATES, LAKES, STRATIFICATION, MIXING, BOTTOM SEDIMENTS, SUCCESSION, CHLOROPHYTA, DIATOMS, CHLOROPHYLL, ALKALINITY, CARBON DIOXIDE, TEMPERATURE, DISSOLVED OXYGEN, HYDROGEN ION CONCENTRATION, SCENEDESMUS, BIOMASS.

IDENTIFIERS:
*MOSES LAKE(WASH), NUTRIENT DILUTION, LIMITING NUTRIENTS, ALGAL CELL WASHOUT.

ABSTRACT:
THE PURPOSE OF THE DILUTION EXPERIMENTS WAS TO STUDY THE EFFECT OF LOW NUTRIENT DILUTION WATER ON ALGAL GROWTH IN SITU. OF PRIMARY INTEREST WAS THE POSSIBILITY OF CHANGING THE ALGAL COMMUNITY STRUCTURE FROM PREDOMINATELY BLUE-GREEN ALGAE TO A COMMUNITY DOMINATED BY LESS OBNOXIOUS FORMS BY ADDING LOW NUTRIENT WATER TO THE LAKE. ADDITIONAL DILUTION EXPERIMENTS, EMPLOYING NUTRIENT ADDITIONS, WERE DESIGNED TO RELATE OBSERVED ALGAL GROWTH CHANGES TO DILUTION OF NITROGEN OR PHOSPHORUS AT GROWTH LIMITING LEVELS. THE EXPERIMENTS, CONDUCTED IN MOSES LAKE, WASHINGTON, FROM JUNE THROUGH SEPTEMBER 1970, SHOWED THAT DILUTION RESULTED IN REDUCTION OF THE YIELD AND GROWTH RATE OF BLUE-GREEN ALGAE WITHOUT INCREASING OR DECREASING THE ABUNDANCE OF OTHER ALGAL FORMS. NITRATE WAS FOUND TO STIMULATE ONLY BLUE-GREEN ALGAL GROWTH WHEN ADDED TO THE DILUTION WATER WHILE ADDITIONS OF ORTHOPHOSPHATE GENERALLY HAD NO EFFECT. IN SITU EXPERIMENTS WERE ALSO CONDUCTED TO DETERMINE THE EFFECT OF LAKE STRATIFICATION AND MIXING ON ALGAL GROWTH. RESULTS INDICATE WHEN THE LAKE IS STRATIFIED AND CONDITIONS ARE AEROBIC THE WATER NEAR THE BOTTOM HAS THE GREATEST ALGAL GROWTH POTENTIAL. DURING LAKE MIXING RESULTS MAY BE UNPREDICTABLE. (JONES-WISCONSIN)

FIELD 05C

ACCESSION NO. W72-08049

ARE SOME BACTERIA TOXIC FOR MARINE ALGAE,

CENTRE D'OCEANOGRAPHIE, MARSEILLE (FRANCE). STATION MARINE D'ENDOUME.

B. R. BERLAND, D. J. BONIN, AND S. Y. MAESTRINI.

MARINE BIOLOGY, VOL 12, P 189-193, 1972. 4 FIG, 1 TAB, 18 REF.

DESCRIPTORS:
*BACTERIA, *TOXICITY, *MARINE ALGAE, SEA WATER, CHLOROPHYTA, CHLAMYDOMONAS, LABORATORY TESTS, CHRYSOPHYTA, OPTICAL PROPERTIES.

IDENTIFIERS:
*ANTIBIOSIS, OPTICAL DENSITY, PRASINOPHYCEAE, BACCILARIOPHYCEAE, XANTHOPHYCEAE, VIBRIO, ACHROMOBACTER, FLAVOBACTERIA.

ABSTRACT:
ALGAL AND BACTERIAL ENVIRONMENT IN THE SEA IS NOT ONLY RULED BY A SIMPLE TROPHIC RELATIONSHIP, BUT ANTIBIOSIS MAY ALSO PLAY AN IMPORTANT ROLE IN MARINE ECOLOGY. IN VITRO EXPERIMENTS WERE CARRIED ON WITH 13 ALGAL SPECIES, 20 BACTERIAL STRAINS, AND THREE BACTERIAL UNDETERMINED STRAINS ISOLATED FROM ALGAL CULTURES. EXPERIMENTS REVEALED AN OBVIOUS INHIBITION OF SEVERAL MARINE ALGAE BY SOME BACTERIA. PSEUDOMONAS AERUGINOSA AND RELATED STRAINS ARE NOT MARINE ORGANISMS, BUT THEY CAN BE TRANSPORTED TO THE SEA. INSOFAR AS THERE IS NO QUANTITATIVE SIMILARITY BETWEEN IN SITU AND CULTURE DENSITIES OF BACTERIAL POPULATIONS, BACTERIAL POISONS MAY ONLY BE IMPORTANT WHEN CONCENTRATED; FOR EXAMPLE, IN NARROW WATERS RICH IN SUSPENDED OR DISSOLVED ORGANIC MATTER. THE IN VITRO EXPERIMENTS INDICATE THAT THE PIGMENTED POISON OF THE BACTERIUM PSEUDOMONAS AERUGINOSA IS A STRONG GROWTH INHIBITOR OF THE ALGA TETRASELMIS STRIATA. SEVERAL BACTERIA STRAINS FROM DIFFERENT ORIGINS ARE RECOGNIZED TO HAVE THE SAME TOXICITY AGAINST VARIOUS MARINE ALGAE. TAKING INTO ACCOUNT THE VERY GREAT DIFFERENCES BETWEEN IN VITRO EXPERIMENTS AND IN SITU ENVIRONMENTAL CONDITIONS, IT IS NOT AT PRESENT POSSIBLE TO STATE THAT BACTERIAL POISONS REALLY PLAY A ROLE IN THE ALGAE-BACTERIA RELATIONSHIPS IN THE SEA. (JONES-WISCONSIN)

FIELD 05C, 02I

ACCESSION NO. W72-08051

NITROGEN FIXATION IN LAKE ERKEN,

UPPSALA UNIV (SWEDEN). INST. OF LIMNOLOGY.

U. GRANHALL, AND A. LUNDGREN.

LIMNOLOGY AND OCEANOGRAPHY, VOL 16, NO 5, P 711-719, 1971. 8 FIG, 3 TAB, 26 REF.

DESCRIPTORS:
*NITROGEN FIXATION, *ALGAE, MEASUREMENT, CYANOPHYTA, PHYTOPLANKTON, PHOTOSYNTHESIS, PRIMARY PRODUCTIVITY, EUTROPHICATION, SOLAR RADIATION, DISTRIBUTION, DEPTH, DIURNAL, SEASONAL.

IDENTIFIERS:
*LAKE ERKEN(SWEDEN), *DARK NITROGEN FIXATION.

ABSTRACT:
TO DETERMINE SEASONAL NITROGEN FIXATION BY PLANKTONIC ALGAE IN THE WHOLE PELAGIAL OF A LAKE AND ESTIMATE ITS IMPORTANCE IN RELATION TO OTHER SOURCES OF COMBINED NITROGEN, LAKE ERKEN, EAST OF UPPSALA, UNPOLLUTED AND MODERATELY EUTROPHIC, WAS STUDIED. IN SITU FIXATION OF MOLECULAR NITROGEN WAS MEASURED IN 1970 BY THE ACETYLENE REDUCTION TECHNIQUE EVERY TWO WEEKS DURING FIVE MONTHS. SAMPLES WERE TAKEN FROM 20 RANDOMLY DISTRIBUTED STATIONS. SUBSAMPLES WERE WITHDRAWN FOR NITROGEN FIXATION EXPERIMENTS, PRIMARY PRODUCTION MEASUREMENTS, ALGAL COUNTS OR CHEMICAL ANALYSES. THE ALGAL DIURNAL CYCLES AND VERTICAL DISTRIBUTION WERE INVESTIGATED. FIXATION WAS CORRELATED WITH THE PRESENCE OF HETEROCYSTOUS BLUE-GREEN ALGAE, ESPECIALLY APHANIZOMENON, IN THE PHYTOPLANKTON, AND WAS LIGHT DEPENDENT, THOUGH APPRECIABLE DARK FIXATION ALSO OCCURRED, OWING TO ENDOGENOUS UTILIZATION OF PHOTOSYNTHETIC PRODUCTS FORMED DURING PREVIOUS LIGHT PERIODS. ANNUAL CONTRIBUTION OF NITROGEN FIXATION IN THE PELAGIAL WAS AROUND 0.5 G N/SQ M. HIGHEST VALUES WERE OBTAINED BEFORE MASS DEVELOPMENT OF APHANIZOMENON AND MAXIMAL PRIMARY PRODUCTION. HIGHEST NITROGENASE ACTIVITY WAS SHOWN IN THE FIRST ALGAL DEVELOPMENT PERIOD. CONTRIBUTION OF EASILY AVAILABLE COMBINED NITROGEN BY PELAGIC NITROGEN FIXATION INCREASES ANNUAL COMBINED NITROGEN LOADING BY 40% AND IS IMPORTANT. (JONES-WISCONSIN)

FIELD 05C

ACCESSION NO. W72-08054

THE DISTRIBUTION OF UREA IN COASTAL AND OCEANIC WATERS,

WOODS HOLE OCEANOGRAPHIC INSTITUTION, MASS.

C. C. REMSEN.

LIMNOLOGY AND OCEANOGRAPHY, VOL 16, NO 5, P 732-740, 1971. 5 FIG, 3 TAB, 19 REF.

DESCRIPTORS:
*DISTRIBUTION, *UREAS, *COASTS, *OCEANS, SURFACE WATERS, CONTINENTAL SHELF, DEPTH, NITROGEN, PHYTOPLANKTON, HYDROGRAPHY, NITRITES, NITRATES, AMMONIA, ALGAE, CHLORELLA.

IDENTIFIERS:
*COASTAL WATERS, PANAMA, CALLAO(PERU), CAPE COD(MASS), CAPE MAY(N J), SARGASSO SEA.

ABSTRACT:
IT HAS BEEN SUGGESTED THAT UREA SHOULD BE CONSIDERED A PART OF THE NITROGEN RESERVE IN COASTAL WATERS AND PERHAPS IN OCEANIC WATERS AS WELL. THE DISTRIBUTION WAS DETERMINED FOR CERTAIN COASTAL AND OCEANIC WATERS, AS FOLLOWS: UREA-NITROGEN IN SURFACE WATERS OFF THE CONTINENTAL SHELF BETWEEN PANAMA AND CALLAO, PERU, WAS EXTREMELY PATCHY AND VARIED IN CONCENTRATION FROM 0.54 TO 5.00 MICROGRAM-ATOM UREA-N/LITER. HIGHER VALUES WERE GENERALLY FROM SAMPLES COLLECTED WITHIN A FOAM SLICK OR WINDROW. SURFACE WATERS IN NONUPWELLING WATERS NORTH OF CALLAO AVERAGED 1.83 MICROGRAM-ATOM UREA-N/LITER WHILE SURFACE WATERS IN UPWELLING WATERS SOUTH OF CALLAO AVERAGED 3.46. ALONG THE CONTINENTAL SHELF OF THE NORTHEAST UNITED STATES BETWEEN CAPE COD AND CAPE MAY, THE CONCENTRATION OF UREA RANGED FROM 0.25 MICROGRAM-ATOM UREA-N/LITER ON THE 1000 FATHOM (1830 METER) LINE TO A HIGH OF 11.20 WITHIN NEW YORK HARBOR. THE VERTICAL DISTRIBUTION OF UREA IN PERUVIAN WATERS, ALONG THE NORTHEAST UNITED STATES, AND SARGASSO SEA FLUCTUATED CONSIDERABLY WITH DEPTH BUT THERE WERE INDICATIONS OF PEAKS. THE SUGGESTION THAT UREA MAY SERVE AS AN AVAILABLE SOURCE OF NITROGEN FOR PHYTOPLANKTON GROWTH IS SUPPORTED. (JONES-WISCONSIN)

FIELD 05C

ACCESSION NO. W72-08056

DISTRIBUTIONAL PATTERNS IN ASSEMBLAGES OF ATTACHED DIATOMS FROM YAQUINA ESTUARY, OREGON,

OREGON STATE UNIV., CORVALLIS. DEPT. OF BOTANY; AND CREGON STATE UNIV., CORVALLIS. DEPT. OF STATISTICS.

C. D. MC INTIRE, AND W. S. OVERTON.

ECCLOGY, VOL. 52, NO. 5, P 758-777, LATE SUMMER 1971. 9 FIG, 11 TAB, 25 REF.

DESCRIPTORS:
*DIATOMS, *RESEARCH EQUIPMENT, ENVIRONMENTAL EFFECTS, ESTUARIES, *DISTRIBUTION PATTERNS, *AQUATIC ALGAE, STATISTICAL METHODS, MARINE ALGAE, *OREGON, *SEASONAL, CHRYSOPHYTA, SYSTEMATICS, *ESTUARINE ENVIRONMENT, SEA WATER, SALINITY, WATER TEMPERATURE, CHEMICAL ANALYSIS, WATER ANALYSIS, TIDAL EFFECTS, STABILITY, SOLAR RADIATION, PHOTOPERIODISM, FRESH WATER, SAMPLING, ANALYTICAL TECHNIQUES, SPECIES DIVERSITY, *DIATOMS.

IDENTIFIERS:
*YAQUINA BAY(ORE), YAQUINA ESTUARY, DATA INTERPRETATION, MACROALGAE, SPECIES DIVERSITY INDEX, FRAGILARIA CONSTRUENS, COCCONEIS PLACENTULA, NAVICULA SRYTOCEPHALA, ACHNANTHES BREVIPES, ACHNANTHES JAVANICA, LICMOPHORA JURGENSII, AMPHORA OVALIS, SYNEDRA FASCIULATA, EUNOTIA PECTINALIS, LITHOPHYTES, ENTEROMORPHA, INSOLATION, TEXTURE, SUBSTRATES, NAVICULA MUTICA, SYNEDRA FASCICULATA, FRAGILARIA STRIATULA VAR CALIFORNIA, MELOSIRA MONILIFORMIS, NAVICULA DISERTA, NITZSCHIA FRUSTULUM VAR PERPUSILLA, MELOSIRA NUMMULOIDES, NAVICULA SPP, ACHNANTHES SPP, ACHNANTHES LANCEOLATA.

ABSTRACT:
SUMMER AND WINTER DISTRIBUTIONAL PATTERNS OF ATTACHED DIATOMS WERE INVESTIGATED IN YAQUINA BAY AND ESTUARY, OREGON. DIFFERENCES IN SPECIES COMPOSITION AND DIVERSITY OF DIATOM ASSEMBLAGES AT SELECTED STATIONS FROM FRESH WATER JUST BELOW ELK CITY, OREGON, TO THE MARINE WATERS OF LOWER YAQUINA BAY WERE RELATED TO ENVIRONMENTAL GRADIENTS. A TOTAL OF 16,475 DIATOMS FROM 30 SAMPLES WAS SEPARATED INTO 256 SPECIES AND VARIETIES, OF WHICH 97 WERE FOUND IN ONLY ONE SAMPLE, AND 72 WERE REPRESENTED BY A SINGLE INDIVIDUAL. THE MOST ABUNDANT DIATOMS IN THE AUGUST SAMPLES WERE FRAGILARIA STRIATULA VAR. CALIFORNIA, MELOSIRA MONILIFORMIS, MELOSIRA NUMMULOIDES, NAVICULA MUTICA, AND SYNEDRA FASCICULATA, WHILE IN THE FEBRUARY SAMPLES ACHNANTHES NO. 2 AND NO. 4, NAVIVULA DISERTA, NAVICULA MUTICA, AND NITZSCHIA FRUSTULUM VAR. PERPUSILLA WERE DOMINANT. OF THE MOST ABUNDANT TAXA, NAVIVULA NO. 2, NAVICULA DISERTA, NAVICULA GREGARIA, NITZSCHIA FRUSTULUM VAR PERPUSILLA, SYNEDRA FASCICULATA, AND THALASSIONEMA NITZSCHIODES WERE THE MOST EVENLY DISTRIBUTED AMONG THE STATIONS. THE MEAN SPECIES DIVERSITY FOR DIATOM ASSEMLAGES SAMPLE IN FEBRUARY WAS SLIGHTLY HIGHER THAN THAT FOR ASSEMBLAGES COLLECTED IN AUGUST. IN FEBRUARY THE MEAN SPECIFIC DIVERSITY WITHIN A GENUS WAS HIGHER AND THE MEAN GENERIC DIVERSITY SLIGHTLY LOWER THAN IN AUGUST. IN GENERAL, DIFFERENCES IN ASSEMBLAGES WERE CLOSELY RELATED HORIZONTALLY TO THE SALINITY GRADIENT AND VERTICALLY TO THE DESICCATION AND INSOLATION GRADIENTS. HOWEVER, BIOLOGICAL FACTORS WERE NORE IMPORTANT IN ACCOUNTING FOR DIFFERENCES AMONG ASSEMBLAGES IN THE SUMMER THAN IN THE WINTER, AND THESE FACTORS WERE PRIMARILY SPECIES INTERACTIONS BETWEEN DIATOMS AND MACRO-ALGAE. (HOLOMAN-BATTELLE)

FIELD 05B, 05C, 02L

ACCESSION NO. W72-08141

INFESTATION OF BENTHIC CRUSTACES, FISH EGGS, AND TROPICAL ALGAE,

RHCDE ISLAND UNIV., KINGSTCN. DEPT. OF BACTERIOLOGY AND BIOPHYSICS.

P. W. JOHNSON, J. MCN. SIEBURTH, A. SASTRY, C. R. ARNOLD, AND M. S. DOTY.

LIMNOLOGY AND OCEANOGRAPHY, VOL. 16, NO. 6, P 962-969, NOVEMBER 1971. 1 FIG, 3 TAB, 17 REF.

DESCRIPTORS:
*CRUSTACEANS, *MARINE BACTERIA, *WATER QUALITY, BENTHOS, CRABS, COPEPODS, MARINE ALGAE, SHRIMP, FISH EGGS, FISH, BACTERIA, FISH PARASITES, ANTIBIOTICS(PESTICIDES), BACTERICIDES, ATOLLS, CULTURES, SPHAEROTILUS, CLADOPHORA.

IDENTIFIERS:
PLEOPODS, URCPODS, ENRICHMENT, LEUCOTHRIX MUCCR, MAJURO ATOLL, MARCHALL ISLAND, POLYSIPHONIA LANCSA, CHONDRUS CRISPUS, COD, FLCUNDER, CRAB EGGS, PELAGIC EGGS, GADUS MORHUA, GRASS SHRIMP, GREEN CRAB, PENICILLIUM, STREPTOMYCIN, EPIFLCRA, WALLEYE LARVAE, SHAD, MACROINVERTEBRATES, CYANOPHYCEAE.

ABSTRACT:
CHARACTERISTIC FILAMENTS OF THE BACTERIUM LEUCOTHRIX MUCOR ARE OFTEN FOUND ON APPENDAGES AND EGGS OF BENTHIC MARINE CRUSTACEANS AND ON A WIDE VARIETY OF ALGAE. PLANKTONIC CRUSTACEA AND FISH EGGS CAN BECOME INFESTED IN AQUARIA IN THE ABSENCE OF ANTIBIOTICS. DEATH OF COD, FLOUNDER, BENTHIC INVERTEBRATES, AND CRAB EGGS HAS BEEN ATTRIBUTED TO LARGE POPULATICNS OF L. MUCOR. COPEPODS, SHRIMP, AND A VARIETY OF CRABS CAN BE INFECTED. ALTHOUGH L. MUCOR IS NOT PATHOGENIC, IT MAY CAUSE DEATH BY CAUSING PELAGIC EGGS TO SINK BELOW THE SURFACE AND BY INTERFERING WITH THE FILTERING APPARATUS OF CRUSTACEAN LARVAL FORMS, E.G., PLEOPODS AND UROPODS. ANTIBIOTICS SUCH AS PENICILLIN AND STREPTOMYCIN PREVENT L. MUCOR DEVELOPMENT ANC CECREASE ASSOCIATED MORTALITY. EXAMINATICN CF 48 MARINE ALGAE SAMPLES FROM THE LAGOON AT MAJURO ATOLL IN THE MARSHALL ISLANDS SHOWEC THAT 81 PERCENT WAS INFECTED BY THE BACTERIA. EIGHTEEN RANDOM SAMPLES SHOWED 100 PERCENT INFESTATION UPON ENRICHMENT. THESE OBSERVATIONS ARE AT VARIANCE WITH PREVIOUS REPORTS OF L. MUCCR'S ABSENCE OR RARITY IN WARM WATERS. SPHAEROTILUS NATANS IS THE FRESHWATER COEFFICIENT CF THE ORGANISM, DISPLAYING THE SAME EFFECTS CN FLORA AND FAUNA OF THE AREAS INFESTED. (MACKAN-BATTELLE)

FIELD 05C, 05A

ACCESSION NO. W72-08142

GRANULAR CARBON FILTERS FOR TASTE AND ODOR REMCVAL,

WATER PURIFICATION PLANT AND PUMPING STATION, MT. CLEMENS, MICH.

R. E. HANSEN.

JOURNAL OF THE AMERICAN WATER WORKS ASSOCIATION, VOL 64, NO. 3, P 176-181, MARCH 1972, 11 FIG, 2 REF.

DESCRIPTORS:
*TASTE, *ODOR, *POTABLE WATER, ACTIVATED CARBON, ADSORPTION, FILTRATION, ALGAE, PHENOLS, DETERGENTS, PESTICIDES, OPERATION AND MAINTENANCE, *WATER TREATMENT, *MICHIGAN.

IDENTIFIERS:
*CARBON REGENERATION, *BACKWASHING, *MT CLEMENS(MICH).

ABSTRACT:
, THE MT. CLEMENS, MICHIGAN WATER PURIFICATION PLANT EXPERIENCED TASTE AND ODOR PROBLEMS FOR 33 YEARS, STARTING WITH SEPTIC AND PHENOL TASTES AND ODORS AND GRADUALLY MOVING TOWARDS ACTINOMYCETES AND ALGAL PROCUCED TASTES ANC ODORS. ALTHOUGH POWDERED ACTIVATED CARBON SLURRIES WERE SOMEWHAT EFFECTIVE IN KEEPING DOWN THE TASTE AND CDOR PROBLEMS, THERE WAS NOT ENOUGH SLURRY HANDLING CAPACITY TO ELIMINATE THE PROBLEM. EFFORTS TOWARD IMPROVING AND EXPANDING CARBON SLURRY OPERATIONS WERE TERMINATED WHEN IT WAS FCUND THAT GRANULAR ACTIVATED CARBON WITH AN EFFECTIVE SIZE OF 0.55 TO 0.65 MM AND A UNIFORMITY CCEFFICIENT OF 1.4-1.7 COULD BE USED IN THE EXISTING SANDFILTERS. SINCE PLACING THE ACTIVATED CARBON FILTERS ON LINE, NO FURTHER TASTE AND ODOR PROBLEMS HAVE BEEN ENCOUNTERED. BASED ON A 3 YEAR LIFE, A TOTAL INSTALLED COST OF $4.97/MILLION GALLCNS OF WATER TREATED OR $0.C5/MONTH/HOUSEHOLD WAS COMPUTED. IN ADDITION TO REMOVING TASTE AND ODORS, AS WELL AS DOING AN EXCELLENT FILTERING JOB, THE CARBON FILTERS ALSO REMCVE ALL ORGANICS, DETERGENTS, AND PESTICIDES. (LOWRY-TEXAS)

FIELD 05F

ACCESSION NO. W72-08357

KINETICS OF ALGAL GROWTH IN AUSTERE MEDIA,

ALBANY COUNTY SEWER DISTRICT, N.Y.

G. C. MCDONALD, R. D. SPEAR, R. J. LAVIN, AND N. L. CLESCERI.

IN: PROPERTIES AND PRODUCTS OF ALGAE, PLENUM PRESS 1970, P 97-105, 5 FIG, 2 TAB, 5 REF.

DESCRIPTORS:
*ALGAE, *GROWTH RATES, *NUTRIENTS, CULTURES, INHIBITION, OLIGOTROPHY, PHOSPHORUS, NITROGEN, *EUTROPHICATION, LABORATORY TESTS, WATER QUALITY CONTROL, *KINETICS, *CHLOROPHYTA, ALGAL CONTROL, *NEW YORK.

IDENTIFIERS:
*SELENASTRUM CAPRICORNUTUM, *ALGAL GROWTH, *LAKE GEORGE(NY).

ABSTRACT:
SELENASTRUM CAPRICORNUTUM IS A UNI-CELLULAR GREEN ALGA WHICH IS EASY TO CULTURE IN THE LABORATORY AND HAS BEEN DOCUMENTED AS HAVING PRODUCED NUISANCE BLOOMS IN EUROPEAN LAKES. S. CAPRICORNUTUM WAS CULTURED IN LABORATORY USING THE BASIC ASM MEDIUM OF THE PROVISIONAL ALGAL ASSAY PROCEDURE (PAAP) AND A 10 TO 1 DILUTION OF GORHAM'S MEDIUM WITH AN INCREASE IN SODIUM CARBONATE CONCENTRATION TO 50 MG/L AS A PH CONTROL AID. DILUTION WATER FOR BOTH MEDIA PREPARATIONS WAS GLASS DISTILLED WATER AND/OR 0.45 MICRON MEMBRANE FILTERED LAKE GEORGE WATER (AN OLIGOTROPHIC SOFT WATER). AFTER GROWTH RATES HAD BEEN ESTABLISHED IN THE VARIOUS MEDIA, DISTILLED WATER, AND LAKE GEORGE WATER COMBINATIONS, THE CONCENTRATIONS OF NITROGEN AND PHOSPHORUS WERE REDUCED TO ONE HALF AND ONE QUARTER OF THE FULL AMOUNT AND NEW GROWTH RATE LEVELS AT EACH CONCENTRATION WERE ASCERTAINED. RESULTS DEMONSTRATED AN INHIBITORY EFFECT, AS EVIDENCED BY REDUCED GROWTH RATES, WHEN LAKE GEORGE WATER WAS USED AS DILUTION WATER IN EITHER MEDIUM. THE EXTENT AND CAUSATIVE FACTORS FOR SUCH INHIBITION BECAME THE FOCAL POINT FOR FURTHER INVESTIGATION. RESULTS ALSO SHOWED THAT THE NITROGEN CONCENTRATION IN MODIFIED TENTH GORHAM'S MEDIUM MAY BE REDUCED TO ONE-HALF THE POSITED LEVEL WITHOUT ANY SIGNIFICANT GROWTH RATE CHANGES FOR SELENASTRUM CAPRICORNUTUM. (LOWRY-TEXAS)

FIELD 05C, 02H

ACCESSION NO. W72-08376

RECYCLING SYSTEM FOR POULTRY WASTES,

LAKE TAHOE AREA COUNCIL, TAHOE CITY, CALIF.

G. L. DUGAN, C. G. GOLUEKE, AND W. J. OSWALD.

JOURNAL WATER POLLUTION CONTROL FEDERATION, VOL. 44, NO. 3 P 432-440, MARCH 1972, 3 FIG, 2 TAB, 9 REF. EPA GRANT 5R01 U100566-03.

DESCRIPTORS:
*FARM WASTES, POULTRY, NITROGEN, LAGOONS, PUMPING, AEROBIC TREATMENT, BIOCHEMICAL OXYGEN DEMAND, CHEMICAL OXYGEN DEMAND, NUTRIENTS, *WASTE TREATMENT, *ALGAE, COSTS.

IDENTIFIERS:
*OXIDATION DITCH, HYDRAULIC MANURE HANDLING.

ABSTRACT:
AN INTEGRATED WASTE MANAGEMENT SYSTEM WAS DEVELOPED IN WHICH ANIMAL ENCLOSURE SANITATION WAS INTEGRATED WITH WASTE TREATMENT. IT WAS A LARGELY CLOSED HYDRAULIC SYSTEM INVOLVING AN ANAEROBIC PHASE AND AN AEROBIC PHASE IN WHICH OXYGENATION COULD BE ACCOMPLISHED EITHER BY THE PHOTOSYNTHETIC ACTIVITY OF ALGAE OR BY MECHANICAL AERATION. WHEN PHOTOSYNTHETIC OXYGENATION WAS USED, ALGAE WERE HARVESTED. THE RANGE OF APPLICATION OF THE PROCESS IS FROM SMALL-SCALE TO LARGE-SCALE OPERATIONS. ALGAE RECLAMATION WOULD BE PRACTICED IN LARGE-SCALE OPERATIONS AND INDUCED AERATION IN SMALLER ONES. AN IMPORTANT OPERATIONAL FEATURE OF THE SYSTEM IS TO KEEP THE SOLIDS CONTENT OF THE MANURE SLURRY TO LESS THAN 3 PERCENT, WET WEIGHT. AT CONCENTRATIONS OF 3 PERCENT OR LESS, 70 PERCENT OR MORE OF SUSPENDED SOLIDS IN MANURE SLURRIES SETTLE OUT OF SUSPENSION IN LESS THAN 30 MIN. POND DEPTH SHOULD NOT EXCEED 12 IN. (30.5 CM). THE INDICATED POND AREA PER BIRD WAS 2 SQ. FT. (0.19 SQ. M.). AN ECONOMIC EVALUATION BASED ON AN INTEGRATED SYSTEM OF 100,000 EGG LAYERS AND THE APPLICATION OF THE LOW-LOADING, HIGH-COST, AND OVERDESIGNED COMPONENTS USED IN THE RESEARCH INDICATES THAT THE WASTE-HANDLING COSTS OF THE SYSTEM WOULD BE AT THE MOST, $0.02/DOZEN EGGS. IF THE VALUE OF THE ALGAL CROP WERE CREDITED TO THE OPERATION, THE NET WASTE-HANDLING COST WOULD BE ABOUT $0.01/DOZEN EGGS. (BUNDY-IOWA STATE)

FIELD 05D

ACCESSION NO. W72-08396

CHARACTERISTICS AND EFFECTS OF CATTLE FEEDLOT RUNOFF,

ROBERT S. KERR WATER RESEARCH CENTER, ADA, OKLA.

M. R. SCALF, W. R. DUFFER, AND R. D. KREIS.

IN: PROCEEDINGS, INDUSTRIAL WASTE CONFERENCE, 25TH, MAY 5, 6, AND 7, 1970. PURDUE UNIVERSITY, ENGINEERING EXTENSION SERIES NO. 137, PART 2, P 855-864, 10 FIG, 3 TAB, 6 REF.

DESCRIPTORS:
*FARM WASTES, *RUNOFF, *FISHKILL, CATTLE, DISSCLVED CXYGEN, DIVERSION STRUCTURES, SEDIMENTATION, BIOCHEMICAL OXYGEN DEMAND, ALGAE, CONFINEMENT PENS, IMPOUNDMENTS, *WATER PCLLUTICN SOURCES, *AGRICULTURAL RUNOFF, *WATER POLLUTION EFFECTS, *CATTLE, *FEED LCTS.

IDENTIFIERS:
ALGAL BLOCMS.

ABSTRACT:
CATTLE FEEDLOT CAPACITY IN THE UNITED STATES HAS BEEN INCREASING AT ABOUT 10 PERCENT ANNUALLY IN RECENT YEARS. ESSENTIALLY, ALL THIS GROWTH HAS BEEN IN THE FORM OF LARGE SCALE FEEDLOTS OF 5000 TO 100,000 HEAD CAPACITY. AS WITH THE CONCENTRATIONS OF PEOPLE, THE CONCENTRATION OF THOUSANDS OF ANIMALS IN A SMALL AREA PRODUCES MASSIVE ENVIRONMENTAL PROBLEMS. RAINFALL RUNOFF MAY CONTAIN POLLUTANT CCNCENTRATIONS 10 TO 100 TIMES THOSE OF RAW MUNICIPAL SEWAGE, AND UNCCNTRCLLED ACCESS TO STREAMS CAN RESULT IN OXYGEN DEPLETION, FISH KILLS, AND OTHER LONG TERM, UNDESIRABLE ECCLOGICAL CONCITIONS FOR MILES DOWNSTREAM. THIS STUDY WAS DESIGNED TC MEASURE THE QUANTITY OF RAINFALL RUNOFF AND ITS POLLUTIONAL CHARACTERISTICS FROM A COMMERCIAL FEEDLOT AND EVALUATE THE EFFECT OF THIS WASTEWATER CN SMALL IMPOUNDMENTS. LESS THAN TWO WEEKS OF SEDIMENTATION IN RUNOFF COLLECTICN PONDS PROCUCED ON EFFLUENT WITH POLLUTANT CONCENTRATICNS CF 10 TO 30 PER CENT OF THE MEAN DIRECT RUNOFF CONCENTRATIONS. THE NECESSITY OF FURTHER TREATMENT WAS DEMONSTRATED WHEN THE FEEDLOT OPERATOR PUMPED COLLECTION POND EFFLUENT THROUGH AN INADEQUATE TREATMENT SYSTEM INTO A 45-ACRE FLOCD CCNTRCL RESERVOIR. ESSENTIALLY, ALL GAME FISH IN THE RESERVOIR WERE KILLED DUE TO DISSOLVED OXYGEN STRESS AND HIGH AMMONIA CONCENTRATICNS. (DORLAND-IOWA STATE)

FIELD 05B, 05C

ACCESSION NO. W72-08401

POLYCHLORINATEC BIPHENYLS: TOXICITY TO CERTAIN PHYTOPLANKTERS,

STATE UNIV. OF NEW YORK, STONY BROOK. MARINE SCIENCE RESEARCH CENTER.

N. S. FISCHER, J. L. MOSSER, T. C. TENG, AND C. F. WURSTER.

SCIENCE, VOL. 175, P 191-192, 14 JANUARY 1972. 1 FIG, 18 REF.

DESCRIPTORS:
*POLYCHLORINATED BIPHENYLS, *CHLCRNATED HYDROCARBON PESTICIDES, *PESTICIDE TOXICITY, *DDT, PESTICIDES, PHYTOPLANKTCN, DIATOMS, ALGAE, MARINE ALGAE, AQUATIC ALGAE, EUGLENA, CHLAMYDOMONAS.

IDENTIFIERS:
FRESHWATER ALGAE, THALASIOSIRA SPP., SKELETONEMA SPP., DUNALIELLA SPP.

ABSTRACT:
THE GROWTH RATES OF TWO SPECIES OF MARINE CIATOMS WERE REDUCED BY POLYCHLORINATED BIPHENYLS (PCB'S), WIDESPREAD POLLUTANTS OF THE MARINE ENVIRONMENT, AT CONCENTRATIONS AS LOW AS 10-25 MICROGRAMS/1. IN CONTRAST, A MARINE GREEN ALGA AND TWO SPECIES OF FRESHWATER ALGA WERE NOT INHIBITED AT THESE OR HIGHER CCNCENTRATIONS. THE SENSITIVITY OF THESE SPECIES TO PCB'S PARALLELED THEIR SENSITIVITY TC DDT (1,1,1-TRICHLORO-2,2-BIS(P-CHLOROPHENYL)ETHANE). (SVENSSON-WASHINGTON)

FIELD 05C

ACCESSION NO. W72-08436

TOXICITY OF 2,4-D AND PICLORAM TO FRESH WATER ALGAE,

PURDUE UNIV., LAFAYETTE, IND.

J. H. ELDER, C. A. LEMBI, AND D. J. MORRE.

PURDUE UNIV. AND INDIANA HIGHWAY COMM. JOINT PUBLICATION NO. 23, 1970. 10 P, 4 TAB, 7 REF.

DESCRIPTORS:
*PESTICIDE TOXICITY, *WATER POLLUTION EFFECTS, BIOASSAY, PESTICIDES, HERBICIDES, 2-4-D, PLANT GROWTH REGULATORS, ALGAE, AQUATIC ALGAE, CHLORELLA.

IDENTIFIERS:
TORDON, FRESHWATER ALGAE, PICLORAM, PEDIASTRUM, 4-AMINO-3,5,6-TRICHLOROPICOLINIC ACID, 2,4-DICHLOROPHENOXYACETIC ACID.

ABSTRACT:
THE SOLUBILITY OF 2,4-D ACID IN WATER IS APPROXIMATELY 0.0025 M, THAT OF PICLORAM 0.0018 M. THEREFORE, THE HIGHEST CONCENTRATION TESTED WAS 0.001 M. AT THIS CONCENTRATION, NO EFFECT WAS OBSERVED ON MOST OF THE ORGANISMS WITH 2,4-D AND NONE OF THE ORGANISMS WITH PICLORAM. WITH TECHNICAL PICLORAM, MOTILE SPECIES WERE FOUND TO LOSE MOTILITY AT 0.001 M AND 0.005 M, BUT NOT AT 0.0001 M. THE TOXIC PRINCIPLE IS AN IMPURITY IN THE TECHNICAL PICLORAM TENTATIVELY IDENTIFIED AS 2(3,4,5,6-TETRACHLORO-2-PYRIDYL) GUANIDINE. THE RESULTS SHOW THAT THE POTENTIAL HAZARD OF 2,4-D OR PICLORAM TO BOTH FRESHWATER AND MARINE ALGAE FROM TERRESTRIAL RUNOFF WATER OR FROM DIRECT OR INDIRECT CONTAMINATION IS NIL. CERTAIN 2,4-D DERIVATIVES (PARTICULARLY ESTERS) MAY BE SUBSTANTIALLY MORE TOXIC THAN THE PARENT ACID. IN THIS REGARD, IT WILL BE NECESSARY TO EXAMINE A NUMBER OF 2,4-D DERIVATIVES IN DIFFERENT FORMULATIONS TO SEEK THOSE WHICH WILL RESULT IN MINIMUM DAMAGE TO ALGAE, FISH AND OTHER AQUATIC ORGANISMS. THERE IS NO EVIDENCE FOR BIOLOGICAL MAGNIFICATION OF EITHER 2,4-D OR PICLORAM IN ALGAE. (SVENSSON-WASHINGTON)

FIELD 05C

ACCESSION NO. W72-08440

THE FATE OF NITROGEN IN AQUATIC ECOSYSTEMS,

WISCONSIN UNIV., MADISON. WATER RESOURCES CENTER.

DENNIS R. KEENEY.

AVAILABLE FROM THE NATIONAL TECHNICAL INFORMATION SERVICE AS PB-209 217, $3.00 PAPER COPY, $0.95 IN MICROFICHE. WISCONSIN WATER RESOURCES CENTER, MADISON. EUTROPHICATION INFORMATION PROGRAM, LITERATURE REVIEW NO 3, 1972. 59 P. 4 FIG, 13 TAB, 198 REF. OWRR W-117(NO 1614)(4).

DESCRIPTORS:
*NITROGEN, *AQUATIC ENVIRONMENT, PRODUCTIVITY, SEDIMENTS, NITRIFICATION, OXIDATION-REDUCTION POTENTIAL, DENITRIFICATION, NITROGEN FIXATION, NITROGEN CYCLE, AMMONIA, NITRATES, MODEL STUDIES, DISSOLVED OXYGEN, CARBON, DISTRIBUTION, PHOSPHORUS, EUTROPHICATION, ALGAE, AQUATIC WEEDS, PHYTOPLANKTON, OLIGOTROPHY, HARDNESS(WATER), SOILS, AMINO ACIDS, DREDGING, SULFUR, ORGANIC MATTER, HYDROGEN ION CONCENTRATION, BACTERIA, FUNGI, RUNOFF, WATER POLLUTION SOURCES, PRECIPITATION(ATMOSPHERIC), GROUNDWATER, STREAMS, SEEPAGE, DRAINAGE, MARSHES, DIVERSION, WASTE WATER TREATMENT, ION EXCHANGE.

IDENTIFIERS:
LAKE MENDOTA(WIS), BANTAM LAKE(CONN), TROUT LAKE(WIS), SEDIMENT COMPACTION.

ABSTRACT:
THIS REVIEW OF 198 PAPERS CONSIDERS THE MANY ASPECTS OF NITROGEN IN AQUATIC ECOSYSTEMS. EVIDENCE INDICATES THAT ALGAE CAN USE ORGANIC AND INORGANIC FORMS OF NITROGEN AND THAT AMMONIUM IS USED PREFERENTIALLY BY SOME SPECIES. BECAUSE OF THE MANY TRANSFORMATIONS INVOLVED, THE CONCENTRATIONS OF VARIOUS NITROGEN SPECIES IN LAKE WATERS VARY WIDELY AND ARE THE NET RESULT OF NUMEROUS ENVIRONMENTAL FACTORS INFLUENCING RATE OF NITROGEN IMMOBILIZATION, MINERALIZATION, NITRIFICATION, AND DENITRIFICATION. THE LARGE POTENTIAL SOURCE OF NITROGEN TO LAKES FROM SEDIMENTS IS ILLUSTRATED. NITROGEN REQUIREMENTS FOR ALGAL GROWTH MUST BE MET FROM SOURCES OTHER THAN JUST THE NITROGEN IN WATER. THE OXIDATION-REDUCTION SYSTEM IN LAKES IS DISCUSSED EXTENSIVELY. THE RELATIVE ROLES OF VARIOUS BIOLOGICAL ENTITIES (BACTERIA, FAUNA, AND FLORA) RESPONSIBLE FOR NITROGEN ASSIMILATION AND RELEASE WILL VARY WITH THE SYSTEM AND WITH CHANGES IN MORPHOLOGICAL, PHYSICAL AND CHEMICAL VARIABLES WITHIN A GIVEN SYSTEM. NITRIFICATION, DENITRIFICATION, AND NITROGEN FIXATION WITH THE ORGANISMS RELATED TO THEM ARE DETAILED. (JONES-WISCONSIN)

FIELD 05C, 10F

ACCESSION NO. W72-08459

NITROGEN FIXATION BY BLUE-GREEN ALGAE IN YELLOWSTONE THERMAL AREAS,

DUNDEE UNIV. (SCOTLAND). DEPT. OF BIOLOGICAL SCIENCES.

W. D. P. STEWART.

PHYCOLOGIA, VOL 9, NO 3-4, P 261-268, DECEMBER 1970. 3 FIG, 5 TAB, 22 REF. NSF GRANT GB-5258.

DESCRIPTORS:
*NITROGEN FIXATION, *ALGAE, *CYANOPHYTA, *THERMAL WATER, *WYOMING, NATIONAL PARKS, HOT SPRINGS, WATER TEMPERATURES, NOSTOC STREAMS, DATA COLLECTIONS, ANALYTICAL TECHNIQUES, NITROGEN FIXING BACTERIA, THERMAL POLLUTION.

IDENTIFIERS:
*YELLOWSTONE NATIONAL PARK(WYO), BLUE-GREEN ALGAE, MASTIGOCLADUS, CALOTHRIX.

ABSTRACT:
POTENTIAL NITROGEN-FIXING BLUE-GREEN ALGAE WERE COMMON IN THREE HOT SPRING STREAMS IN YELLOWSTONE NATIONAL PARK. IN TWO STREAMS, WHERE THE DOMINANT NITROGEN-FIXING ALGAE WERE SPECIES OF CALOTHRIX, NITROGEN FIXATION, AS MEASURED BY UPTAKE OF N-15 WAS DETECTED IN SITU IN THE TEMPERATURE RANGE 28-46 C. AT HIGHER TEMPERATURES NITROGEN FIXATION WAS NOT DETECTED, ALTHOUGH THE ALGAE MAY HAVE RECEIVED FIXED NITROGEN FROM A GROWTH OF CALOTHRIX, NOSTOC, AND UNICELLULAR ALGAE WHICH OCCURRED AT LOWER TEMPERATURES ON THE SIDES OF THE STREAMS. IN THE THIRD STREAM, WHERE MASTIGOCLADUS WAS ABUNDANT, NITROGEN FIXATION WAS DETECTED AT TEMPERATURES UP TO 54 C, ALTHOUGH THE OPTIMUM FOR FIXATION WAS NEAR 42.5 C. THE OVERALL DATA IMPLY THAT IN SITU NITROGEN FIXATION CONTRIBUTES TO THE PRODUCTIVITY OF YELLOWSTONE HOT SPRINGS REGIONS AND THAT MASTIGOCLADUS AND CALOTHRIX ARE THE MOST IMPORTANT NITROGEN-FIXING BLUE-GREEN ALGAE. (WOODARD-USGS)

FIELD 05C, 05B

ACCESSION NO. W72-08508

A KINETIC MODEL OF PHYTOPLANKTON GROWTH, AND ITS USE IN ALGAL CONTROL BY RESERVOIR MIXING,

MUNICIPAL WATERWORKS OF ROTTERDAM (NETHERLANDS).

F. OSKAM.

PRESENTED AT INTERNATIONAL SYMPOSIUM ON MAN-MADE LAKES, THEIR PROBLEMS AND ENVIRONMENTAL EFFECTS, KNOXVILLE, TENNESSEE, MAY 3-7, 1971. 21 P, 5 FIG, 36 REF.

DESCRIPTORS:
*PHYTOPLANKTON, *MATHEMATICAL MODELS, *ALGAL CONTROL, *RESERVOIRS, DESTRATIFICATION, MIXING, PHOSPHORUS, NITROGEN, TURBULENCE, PHOTOSYNTHESIS, DEPTH, LIGHT PENTRATION, RESPRIATION, NUTRIENTS, TURBIDITY, PRODUCTIVITY, PLANKTON, STANDING CROP, LIMNOLOGY, EUTROPHICATION.

IDENTIFIERS:
RHINE RIVER(THE NETHERLANDS), MEUSE RIVER(THE NETHERLANDS), BIESBOSCH RESERVOIR(THE NETHERLANDS), KINETIC EQUATION.

ABSTRACT:
AN IN-RESERVOIR METHOD OF ALGAL CONTROL BY TURBULENT MIXING IS CONSIDERED. BASED ON THE CONCLUSION THAT LIGHT IS THE BASIC CONTROLLER OF PLANKTON DEVELOPMENT AND BY CONSIDERING NET GROWTH AS THE BALANCE BETWEEN GROSS PHOTOSYNTHESIS AND RESPIRATION, A SIMPLIFIED MATHEMATICAL EXPRESSION WAS DEVELOPED FOR THE RELATION BETWEEN WATER TRANSPARENCY, LIGHT PENTRATION, AND MIXING DEPTH. A MIXED LAYER OF ABOUT 20 M MIGHT SEVERLY RESTRICT ALGAL GROWTH AT HIGH LEVELS OF NATURAL TURBIDITY, WHEREAS PRODUCTIVITY IN A CLEAR WATER RESERVOIR WOULD HARDLY BE AFFECTED. IN THE LATTER CASE NUTRIENT LIMITATION CAN BE EXPECTED TO BE OPERATIVE LONG BEFORE LIGHT WOULD BECOME LIMITING. THIS EXPLAINS THE FACT THAT UNTIL RECENTLY THE CONNECTION BETWEEN TURBULENCE AND ALGAL GROWTH HAS BEEN NEGLECTED IN LIMNOLOGICAL STUDIES. DEEPENING THE MIXED LAYER MAY ALSO BE AT LEAST PARTLY RESPONSIBLE FOR THE DECREASE IN STANDING CROP, OBSERVED IN SEVERAL DESTRATIFICATION EXPERIMENTS. AN EVEN DISTRIBUTION OF ALGAE IS AN ESSENTIAL REQUIREMENT FOR THE MODEL TO BE VALID. MAINTAINING A HOMOGENEOUS STATE ASKS FOR A CONTINUOUS LEVEL OF TURBULENCE, WHICH MUST BE INTRODUCED ARTIFICIALLY. THE CONDITIONS FOR PRARCTICAL REALIZATION OF EFFECTIVE TURBULENT MIXING HAVE TO BE ESTABLISHED BEFORE THE RELIABILITY OF ALGAL CONTROL BY THIS METHOD CAN BE ASCERTAINED. (AUEN-WISCONSIN)

FIELD 05C, 02H, 04A

ACCESSION NO. W72-08559

ALGAL POPULATIONS IN MOSES LAKE, WASHINGTON: TEMPORAL AND SPATIAL DISTRIBUTION AND RELATIONSHIP WITH ENVIRONMENTAL PARAMETERS,

WASHINGTON UNIV., SEATTLE.

R. M. BUSH.

MS THESIS, 1971. 113 P. 9 FIG, 2 PLATES, 8 TAB, 35 REF, 2 APPEND. OWRR A-034-WASH(2).

DESCRIPTORS:
*ALGAE, *POPULATION, *TEMPORAL DISTRIBUTION, *SPATIAL DISTRIBUTION, ENVIRONMENT, SAMPLING, CHEMICAL ANALYSIS, PLANKTCN, COMPUTER PROGRAMS, BIOMASS, TEMPERATURE, CONDUCTIVITY, PHCSPHATES, NITRATES, ALKALINITY, DIATOMS, CYANOPHYTA, CHLOROPHYTA, SEASONAL, NUTRIENTS, THERMAL STRATIFICATION, CARBCN, SEDIMENTS, CARBON CIOXIDE, CHLORPHYLL, DOMESTIC WASTES, EUTROPHICATION, TURBIDITY, SESTON, HYDROGEN ION CONCENTRATION, SCENEDESMUS, DINOFLAGELLATES, CHRYSOPHYTA, PROTOZCA, *WASHINGTON.

IDENTIFIERS:
*MOSES LAKE(WASH).

ABSTRACT:
NATURALLY OCCURRING ALGAL POPULATICNS IN MCSES LAKE, WASHINGTON WERE STUDIED RELATIVE TO ENVIRONMENTAL PARAMETERS INFLUENCING GROWTH AND DISTRIBUTION, ESPECIALLY THOSE LIMITING OR CAUSING NUISANCE GROWTHS. USING CLUSTER ANALYSIS, THE DIFFERENT ALGAL POPULATIONS WERE DELINEATED FROM PLANKTON SAMPLES COLLECTED OVER TWO YEARS. SAMPLES WERE GROUPED ON THE BASIS OF HIGH JOINT OCCURRENCES OF ALGAL SPECIES. THE THREE MOST PREVALENT POPULATIONS WERE (1) COMPOSED PREDOMINANTLY OF DIATOM GENERA AND OFTEN DOMINATED DIATCMS, (2) DOMINATED BY BLUE-GREEN ALGEE, AND (3) COMPOSED PREDCMINATLY OF GREEN ALGAE AND DCMINATED BY GREEN ALGAE. ALGAL POPULATIONS 1 AND 3 WERE FCUND CONSISTENTLY IN SPECIFIC LAKE AREAS AND DEMONSTRATED LITTLE TEMPCRAL VARIATICN. THE REMAINING LAKE AREAS FOLLOWED A SUCCESSIONAL PATTERN IN ALGAL PCPULATIONS RESULTING IN ONE DOMINATED BY BLUE-GREEN ALGAE IN THE SUMMER. MULTIPLE REGRESSION ANALYSIS INDICATED THAT ORTHPHOSPHATE AND CONDUCTIVITY BEST EXPLAINED THE VARIATION IN BLUE-GREEN POPULATION BIOMASS. THIS OCCURRED DESPITE THE FACT THAT NITROGEN SEEMED TO BE THE LIMITING NUTRIENT. AS WITH THE BLUE-GREEN ALGAL POPULATION, THE GREEN ALGAL BIOMASS WAS CORRELATED NEGATIVELY WITH CONDUCTIVITY AND PHOSPHATE-PHOSPHORUS INDICATING AND UPTAKE OF PHOSPHATE-PHOSPHORUS AND DISSOLVED SALTS (REPRESENTED BY CONDUCTIVITY) AS THE BIOMASS INCREASED. (JONES-WISCONSIN)

FIELD 05C, 02H

ACCESSION NO. W72-C8560

SEWAGE FUNGUS IN RIVERS IN THE UNITED KINGDOM: THE SLIME COMMUNITY AND ITS CONSTITUENT ORGANISMS,

WATER POLLUTION RESEARCH LAB., STEVENAGE (ENGLAND).

E. J. C. CURTIS, AND C. R. CURDS.

WATER RESEARCH, VOL 5, P 1147-1159, 1971. 3 FIG, 4 TAB, 27 REF.

DESCRIPTORS:
*SEWAGE, *FUNGI, *SLIME, *BIOLOGICAL COMMUNITIES, RIVERS, BACTERIA, WATER POLLUTION, ACTIVATED SLUDGE, ALGAE, PROTOZOA, DIATOMS, EUGLENA, MATHEMATICAL STUDIES.

IDENTIFIERS:
*UNITED KINGDOM, *SEWAGE FUNGUS, SPHAEROTOTILUS NATANS, ZOOGLOEAL BACTERIA.

ABSTRACT:
BIOLOGICAL COMPOSITION OF SLIMES EXAMINED DURING A SURVEY OF SLIME OUTBREAKS IN THE UNITED KINGDOM IS DESCRIBED AND COMPARED WITH OTHER POLLUTED WATER BIOTA. SLIMES WERE FREQUENTLY DOMINATED BY SPHAEROTILUS NATANS OR ZOOGLOEAL BACTERIA, AND THE INTERRELATIONSHIPS OF THESE AND OTHER SPECIES WERE STUDIED USING CLUSTER ANALYSIS AND DIVERSITY INDICES. COMPARED WITH S NATANS AND ZOOGLOEAL FORMS, ALL OTHER ORGANISMS CONTRIBUTING TO THE ACTUAL SLIME MATRIX OCCURRED ONLY INFREQUENTLY. THE MOST COMMON TO OCCUR AS DOMINANT ORGANISMS WERE THE FUNGI, LEPTOMITUS LACTEUS AND GEOTRICHUM CANDIDUM OCCURRING IN LARGE AMOUNTS IN 3.4 AND 4.5% OF SITES RESPECTIVELY, AND THE GLIDING BACTERIUM BEGGIATOA ALBA (5.6%). OTHER FILAMENTOUS ORGANISMS WERE MORE WIDELY FOUND AS MINOR COMPONENTS OF SLIMES FORMED PRIMARILY BY S NATANS OR ZOOGLOEAL BACTERIA. FILAMENTS OF FLEXIBACTERIUM AND FLAVOBACTERIUM WERE SOMETIMES ATTACHED TO FILAMENTS OF S NATANS OR OTHER MATERIAL WITHOUT SPECIFIC HOLDFAST. FILAMENTOUS ALGAE ONLY RARELY OCCURRED IN SEWAGE FUNGUS SLIMES. MOST SAMPLES CONTAINED PROTOZOA AND ALL CILIATED PROTOZOA. SEVENTY-SEVEN CILIATE SPECIES WERE RECORDED. IN ADDITION TO CLUSTER ANALYSIS, ANOTHER TECHNIQUE DESCRIBING ASSOCIATION BETWEEN ORGANISMS AND GROUPS OF ORGANISMS IS CONSTRUCTION OF A MINIMUM-SPANNING TREE. (JONES-WISCONSIN)

FIELD 05C

ACCESSION NO. W72-08573

POLLUTION CONTROLLED CHANGES IN ALGAL AND PROTOZOAN COMMUNITIES,

VIRGINIA POLYTECHNIC INST., BLACKSBURG. CENTER FOR ENVIRONMENTAL STUDIES; AND VIRGINIA POLYTECHNIC INST., BLACKSBURG. DEPT. OF BIOLOGY.

J. CAIRNS, JR., AND G. R. LANZA.

IN: WATER POLLUTION MICROBIOLOGY, P 245-272. JOHN WILEY AND SONS, INC, NEW YORK, 1972. OWRR B-017-VA(5).

DESCRIPTORS:
*AQUATIC MICROORGANISMS, *WATER POLLUTION EFFECTS, *ALGAE, *PROTOZOA, *BIOLOGICAL COMMUNITIES, ECOLOGICAL DISTRIBUTION, BIOINDICATORS, WATER POLLUTION SOURCES.

IDENTIFIERS:
RIDLEY CREEK(PA), LITITZ CREEK(PA), SPECIES DIVERSITY, DIRECT EFFECTS, INDIRECT EFFECTS.

ABSTRACT:
STRESSES CAUSED BY WASTES AFFECT ALGAL AND PROTOZOAN COMMUNITIES BY REDUCTION IN SPECIES NUMBER, INCREASE IN RANGE OF NUMBERS OF INDIVIDUALS PER SPECIES, REDUCTION IN COLONIZATION RATES, CHANGES IN SELECTIVE PREDATOR OR PARASITE PRESSURE, AND SHIFT IN DOMINANCE WITHIN THE COMMUNITY. A RATHER SIMPLE RESPONSE PATTERN TO STRESS IS CHARACTERISTIC OF BOTH POPULATIONS AND COMMUNITIES AND MAY BE USED TO ASSESS THE STRESS EFFECT. ALGAL AND PROTOZOAN SPECIES FREQUENTLY HAVE COMPLEX REQUIREMENTS THAT MAY RIVAL THOSE OF HIGHER ORGANISMS. IT IS BECOMING INCREASINGLY EVIDENT THAT THE SAME ECOLOGICAL PRINCIPLES APPLICABLE TO HIGHER PLANTS AND ANIMALS ARE ALSO VALID FOR ALGAE AND PROTOZOA. MICROBIAL COMMUNITIES HAVE A STRUCTURE WHICH IS MAINTAINED DESPITE SUCCESSION; OTHER HOMEOSTATIC MECHANISMS ARE ALSO OPERATIVE. GENERAL EFFECTS OF STRESS CAUSED BY INDUSTRIAL AND MUNICIPAL WASTES AND AGRICULTURAL RUNOFF ARE DISCUSSED. THE BASIC CATEGORIES OF POLLUTION ARE: NONTOXIC ORGANIC AND INORGANIC SUBSTANCES, THERMAL CHANGES, TOXIC SUBSTANCES, SUSPENDED SOLIDS, AND RADIOACTIVE MATERIALS. WITH A FINITE ECOLOGICAL BASE AND INCREASED PRESSURES ON THIS BASE, STANDARDS MUST BE DEVELOPED TO PROTECT THE SYSTEM AND KNOWLEDGE OF ITS OPERATIONAL PREREQUISITES MUST BE GAINED TO MANAGE IT WELL. (JONES-WISCONSIN)

FIELD 05C

ACCESSION NO. W72-08579

PHOTCRESPIRATION AND NITROGENASE ACTIVITY IN THE BLUE-GREEN ALGA, ANABAENA CYLINDRICA,

DUNDEE UNIV. (SCOTLAND). DEPT. OF BIOLOGICAL SCIENCES.

M. LEX, W. B. SILVESTER, AND W. D. STEWART.

PROCEEDINGS OF THE ROYAL SOCIETY OF LONDON, B, VCL 180, P 87-102, 1972. 7 FIG, 7 TAB, 41 REF.

DESCRIPTORS:
*LIGHT, *PHOTOSYNTHESIS, *CYANOPHYTA, NITROGEN FIXATICN, OXYGEN, CARBON DIOXIDE, BIOCHEMISTRY, ALGAE, INHIBITICN, LABORATORY TESTS.

IDENTIFIERS:
*PHOTORESPIRATION, *ANABAENA CYLINDRICA, ACETYLENE REDUCTICN, WARBURG EFFECT.

ABSTRACT:
GROWN UNDER ASEPTIC CONDITIONS IN NITROGEN-FREE MEDIUM UNDER CONTINUOUS LIGHT, THE PHOTORESPIRATION OF ANABAENA CYLINDRICA AND ITS EFFECT ON PHOTOSYNTHESIS AND NITROGENASE ACTIVITY WERE STUDIED. IN LIGHT, OXYGEN UPTAKE MAY BE UP TO 20 TIMES DARK RESPIRATION RATE; UPTAKE RATE IN LIGHT INCREASES LINEARLY WITH INCREASING OXYGEN PRESSURE, WHILE DARK RESPIRATION IS SATURATED AT CXYGEN PRESSURE NEAR 0.05 ATMOSPHERES. DCMU, NOT KCN, INHIBITS PHOTORESPIRATION. EXCGENCUSLY SUPPLIED HYDROXYETHANE SULPHONATE, A GLYCCLLATE OXIDASE ACTIVITY INHIBITOR, AND GLYCOLLATE DO NOT AFFECT RESPIRATION, ALTHOUGH C-14-LABELED GLYCOLLATE IS ASSIMILATED IN BOTH LIGHT AND DARK. PHOTORESPIRATICN IS HIGHLY SENSITIVE TO CARBON DIOXIDE PRESSURE AND TO BICARBONATE CONCENTRATION AND APPROACHES TRUE PHOTOSYNTHETIC OXYGEN PRODUCTICN AT THE CARBON DIOXIDE COMPENSATION POINT OF 10 PARTS/1,000,000. CARBON DIOXIDE CONCENTRATION (0.02 ATMOSPHERES) COMPLETELY INHIBITS PHOTORESPIRATION, WHEREAS TRUE PHOTOSYNTHESIS IS SCARCELY AFFECTED. CONDITIONS WHICH STIMULATE PHOTORESPIRATION (LOW PRESSURE CARBON DIOXIDE AND HIGH PRESSURE OXYGEN) PROGRESSIVELY INHIBIT ACETYLENE REDUCTION. IN SHORT-TERM STUDIES DCMU INHIBITS ACETYLENE REDUCTION UNDER CONDITIONS STIMULATING PHOTORESPIRATION BUT HAS LITTLE EFFECT UNDER CONDITIONS INHIBITING PHOTORESPIRATICN. PHCTORESPIRATION AND NITROGENASE ACTIVITY APPARENTLY COMPETE INDIRECTLY FOR REDUCING POWER AND AT LEAST ONE MECHANISM OF OXYGEN INHIBITICN OF NITROGENASE ACTIVITY IS VIA A STIMULATION OF PHOTORESPIRATION. (JONES-WISCONSIN)

FIELD 05C

ACCESSION NO. W72-08584

AQUATIC PLANTS FROM MINNESOTA, PART 3 - ANTIMICROBIAL EFFECTS,

MINNESOTA UNIV., MINNEAPOLIS. WATER RESOURCES RESEARCH CENTER.

K. LEE SU, E. J. STABA AND Y. ABUL-HAJJ.

AVAILABLE FROM THE NATIONAL TECHNICAL INFORMATION SERVICE AS PB-209 530. MINNESOTA WATER RESOURCES RESEARCH CENTER, BULLETIN 48, FEBURARY 1972. 36 P, 6 FIG, 9 TAB, 145 REF. OWRR A-025-MINN(4).

DESCRIPTORS:
*AQUATIC PLANTS, *MINNESOTA, BACTERIA, FUNGI, PROTOZOA, PATHOGENS, MICROORGANISMS, ALGAL TOXINS.

IDENTIFIERS:
*ANTIMICROBIAL ACTIVITY, *SKELLYSOLVE, *FILTER PAPER DISC DIFFUSION METHOD, *ZONES OF INHIBITION, ANTIFUNGAL ACTIVITY.

ABSTRACT:
THE ANTIMICROBIAL ACTIVITY OF THE FOLLOWING 22 MINNESOTAN AQUATIC PLANTS WAS INVESTIGATED: ANACHARIS CANADENSIS, CALLA PCLUSTRIS, CAREX LACUSTRIS, CERATOPHYLLUM DEMERSUM, CHARA VULGARIS, ELECCHARIS SMALLII, LEMNA MINOR, MYRIOPHYLLUM EXALBESCENS, NUPHAR VARIEGATUM, NYMPHAEA TUBEROSA, POTAMOGETON AMPLIFCLUS, P. NATANS, P. PECTINATUS, P. RICHARDSONU, P. ZCSTERIFORMIS, SAGITTARIA CUNEATA, S. LATIFOLIA, SPARGANIUM EURYCARPUM, S. FLUCTUANS, TYPHA ANGUSTIFOLIA, VALLISNERIA AMERICANA, AND ZIZANIA AQUETICA. THE CHEMICAL CONSTITUENTS RESPONSIBLE FOR THE SIGNIFICANT ANTIMICRCBIAL EFFECT WERE ISOLATED AND IDENTIFIED. THE SKELLYSOLVE F, CHLOROFORM, 80% ETHANOL AND FRESH WATER EXTRACTS OF PLANT SPECIES WERE TESTED FOR ANTIMICROBIAL ACTIVITY EMPLOYING THE QUALITATIVE FILTER PAPER DISC DIFFUSION METHOD AND REFERENCE ANTIBIOTIC DISCS. ETHANOL (80%) EXTRACTS OF MYRIOPHYLLUM EXALBESCENS (ACTIVITY RATIO OF 0.34 AS COMPARED TO THE 30 MCG CHLORAMPHENICOL DISCS), NYMPHAEA TUBEROSA (LEAF: 0.40, STEM: 0.38) AND NUPHAR VARIEGATUM COLLECTED IN LAKE MINNETONKA (LEAF: 0.43, STEM: 0.45) WERE MODERATELY ACTIVE AGAINST S. AUREUS. ETHANOL (80%) EXTRACTS CF CAREX LUCUSTRIS (ACTIVITY RATIO OF 0.34 AS CCMPARED TO THE 10 MCG STREPTCMYCIN DISCS), NUMPHAEA TUBEROSA (LEAF: 1.01, STEM: 1.10) AND NUPHAR VARIEGATUM COLLECTED IN LAKE MINNETCNKA (LEAF: 0.73, STEM: 0.58) WERE ACTIVE AGAINST M. SMEGMATIS. ALL EXTRACTS WERE RELATIVELY INACTIVE AGAINST E. COLI EXCEPT THE WATER EXTRACT OF PCTAMOGETON NATANS WHERE A LOW ACTIVITY RATIO OF 0.10 AS COMPARED TO THE 30 MCG CHLORAMPHENICOL DISCS WAS INDICATED. (SEE ALSO W72-05877) (WALTON-MINNESOTA)

FIELD 05C, 05A

ACCESSION NO. W72-08586

BIOLOGICAL EFFECTS OF COOLING TOWER BLOWDOWN,

NATIONAL ENVIRONMENTAL RESEARCH CENTER, CORVALLIS, OREG.

R. R. GARTON.

PREPRINT, OF PAPER PRESENTED AT AMERICAN INSTITUTE OF CHEMICAL ENGINEERS NATIONAL MEETING, 71ST, DALLAS, TEXAS, FEBRUARY 20-23, 1972, 25 P, 5 FIG, 9 TAB, 6 REF.

DESCRIPTORS:
*COOLING TOWERS, *TOXICITY, *BIOASSAYS, LABORATORY TESTS, FISH, ALGAE, CHEMICAL WASTES, ALGICIDES, BACTERICIDES, CHROMIUM, ZINC, TEMPERATURE, HYDROGEN ION CONCENTRATION, CORRCSION, WATER POLLUTICN EFFECTS.

IDENTIFIERS:
*SLIMICIDES.

ABSTRACT:
A SIMULATED COOLING TOWER BLOWDOWN WASTE WAS ASSEMBLED, USING THOSE CHEMICALS AND CONCENTRATIONS PRESENTED IN A WASTE DISCHARGE PERMIT APPLICATION TO THE ENVIRONMENTAL PROTECTION AGENCY AS MAXIMUMS TO BE RELEASED BY A 1130 MWE NUCLEAR POWER PLANT UNDER CONSTRUCTION. THIS SIMULATED BLOWDOWN MIXTURE AND DILUTIONS THEREOF WERE THEN USED IN BIOASSAY EXPERIMENTS TO DETERMINE TOXICITY TO REPRESENTATIVES FROM TWO TROPHIC LEVELS, A GREEN ALGA AND A SALMONID FISH. BIOASSAY RESULTS POINTED OUT ONLY TWO OF THE TOXIC COMPOUNDS, ZINC AND CHROMIUM. ZINC WAS TOXIC TO ALGAE AT 0.064 MG/L AND THE 96 HOUR LC50 TO JUVENILE STEEL HEAD WAS 0.09 MG/L. FOR CHROMATE, NO LETHAL EFFECT ON JUVENILE STEEL HEAD WAS NOTED IN 96 HRS. AT 31 MG/L CRO4, BUT ALGAL PRODUCTIVITY WAS DECREASED AT LEVELS OF 0.14 MG/L CRO4. THESE RESULTS INDICATE: (1) TOXICITY MUST BE DETERMINED FOR EACH INDIVIDUAL SPECIES EXPOSED TO BLOWDOWN MIXTURES; (2) SAFE EXPOSURE LEVELS MAY BE AS LITTLE AS 1% OF THE 96-HOUR LC50; (3) CONCENTRATIONS OF CHEMICALS CF KNOWN TOXICITY MUST BE MAINTAINED WITHIN SAFE LEVELS. (LOWRY-TEXAS)

FIELD 05C

ACCESSION NO. W72-08683

METHODS, MODELS AND INSTRUMENTS FOR STUDIES OF AQUATIC POLLUTION. AN ANNOTATED BIBLIOGRAPHY,

OCEAN ENGINEERING INFORMATION SERVICE, LA JOLLA, CALIF.

E. SINHA.

AVAILABLE FROM OCEAN ENGINEERING INFORMATION SERVICE, LA JOLLA, CAL. 92037, PRICE: $6.00. OCEAN ENGINEERING INFORMATION SERIES VOL 5, 1971. 29 P.

DESCRIPTORS:
*BIBLIOGRAPHIES, *ANALYTICAL TECHNIQUES, *WATER POLLUTION CONTROL, *INSTRUMENTATION, MATHEMATICAL MODELS, WATER ANALYSIS, POLLUTANT IDENTIFICATION, DETERGENTS, NUTRIENTS, PESTICIDES, OIL, METALS, THERMAL POLLUTION, SAMPLING, BACTERIA, TOXICITY, BEHAVIOR, BITUMINOUS MATERIALS, ANIMAL WASTES, FISH, FLUORESCENCE, INFRARED RADIATION, IRRADIATION, SPECTROPHOTOMETRY, SPECTROSCOPY, HEAVY METALS, LIGNINS, FARM WASTES, POLAROGRAPHIC ANALYSIS, SEDIMENTOLOGY, SEDIMENTS, SURVEYS, PHOSPHORUS, DISSOLVED OXYGEN, BIOCHEMICAL OXYGEN DEMAND, CLAMS, CRUSTACEANS, PHOSPHATES, ANION EXCHANGE, LAKES, RIVERS, NITROGEN, GREAT LAKES, VOLUMETRIC ANALYSIS, ENTROPHICATION, E. COLI, BIOASSAY, BOTTOM SEDIMENTS, MEMBRANE PROCESSES, DIGITAL COMPUTERS, ALGAE, RADIOACTIVITY TECHNIQUES, SEPARATION TECHNIQUES.

IDENTIFIERS:
VOLTAMMETRY, RADIOGRAPHY, ION SELECTIVE ELECTRODES, LASERS, ORTHOPHOSPHATES.

ABSTRACT:
THIS BIBLIOGRAPHY CONTAINS 204 ABSTRACTS OF LITERATURE PROVIDING SUBSTANTIAL SCIENTIFIC AND TECHNICAL INFORMATION ON METHODS, MODELS AND INSTRUMENTS USED IN STUDIES OF ACUATIC POLLUTION AND MEANS OF ABATEMENT. THESE DEAL WITH THE DETECTION, IDENTIFICATION AND MEASUREMENT OF THE PARAMETERS OF POLLUTION, BIOTIC CONSTITUENTS, DETERGENTS AND NUTRIENTS, PESTICIDES, OIL, METALS, AND NON-METALLIC TOXICANTS. VARIOUS ASPECTS OF WATER QUALITY MANAGEMENT ARE ENCOMPASSED. PERTINENT PATENTS, A BIBLIOGRAPHY OF BIBLIOGRAPHIES, A SUBJECT OUTLINE, A KEYTERM INDEX, AND AN INDEX CITING ALL AUTHORS AND CO-AUTHORS ARE INCLUDED. (MORTLAND-BATTELLE)

FIELD 05A, 07B, 10B

ACCESSION NO. W72-08790

BENTHIC ALGAE IN POLLUTED ESTUARIES,

DURHAM UNIV. (ENGLAND). DEPT. OF BOTANY.

P. EDWARDS.

MARINE POLLUTION BULLETIN, VOL 3, NO 4, P 55-60, APRIL 1972. 4 FIG, 4 TAB, 13 REF.

DESCRIPTORS:
*ALGAE, *ESTUARIES, *WASTE DISPOSAL, *SEWAGE, *INDUSTRIAL WASTES, WATER POLLUTION SOURCES, ON-SITE INVESTIGATIONS, METHODOLOGY, FOREIGN RESEARCH, MARINE ALGAE, PRODUCTIVITY, POLLUTANTS, CHEMICAL WASTES, SEWAGE EFFLUENTS, WATER POLLUTION EFFECTS.

IDENTIFIERS:
*ENGLAND, *POLLUTED ESTUARIES, DURHAM COUNTY RIVERS.

ABSTRACT:
ON-SITE STUDIES WERE MADE OF THREE ESTUARIES IN NORTHEASTERN ENGLAND IN WHICH DIFFERENT CONDITIONS OF POLLUTION EXIST. THE RIVERS, LOCATED IN COUNTY DURHAM, ARE THE WEAR, A RELATIVELY UNPOLLUTED STREAM, THE TYNE, WHICH RECEIVES A LARGE VOLUME OF UNTREATED SEWAGE, AND THE TEES, WHICH IS MIXED WITH INDUSTRIAL WASTES. THESE SYSTEMS PROVIDE A HUGE NATURAL EXPERIMENT SINCE THE DEGREE OR TYPE OF POLLUTION IS PROBABLY THE ONLY ENVIRONMENTAL FACTOR THAT VARIES SIGNIFICANTLY BETWEEN THE THREE ESTUARIES. THE ALGAL FLORA OF THE ESTUARIES IS COMPARED TO REVEAL THE EFFECTS OF DIFFERENT KINDS OF CONTAMINATION. A TOTAL OF 69 STATIONS AT ABOUT 1 KM INTERVALS REACHING FROM THE MOUTH TO THE TIDAL LIMIT OF EACH ESTUARY WERE USED TO DETERMINE THE VARIOUS SPECIES OF ALGAE. VEGETATION IDENTIFIED CONSISTS OF 54 SPECIES FROM THE THREE ESTUARIES; THESE ARE LISTED. A TABLE ALSO GIVES THE SPECIES OF BENTHIC MARINE ALGAE IN THE TEES ESTUARY FOR 1935. A DECREASE IN ALGAL VEGETATION SINCE THE 1930'S IN THE TEES IS PROBABLY DUE TO GROWTH OF THE CHEMICAL INDUSTRY AND THE ASSOCIATED DISCHARGE OF TOXIC CHEMICAL WASTES. (LANG-USGS)

FIELD 05C, 05B, 02L

ACCESSION NO. W72-08804

EUTROPHICATION FACTORS IN NORTH CENTRAL FLORIDA LAKES,

FLORIDA UNIV., GAINSVILLE. DEPT. OF ENVIRONMENTAL ENGINEERING.

H. D. PUTNAM, W. H. MORGAN, P. L. BREZONIK, E. E. SHANNON, AND P. E. MASLIN.

COPY AVAILABLE FROM GPO SUP DOC, $1.25; MICROFICHE FROM NTIS AS PB-209 863. ENVIRONMENTAL PROTECTION AGENCY WATER POLLUTION CONTROL RESEARCH SERIES, FEBRUARY 1972, 141 P, 36 TAB, 40 FIG, 87 REF. EPA PROGRAM 16010 DON 02/72.

DESCRIPTORS:
*EUTROPHICATION, *MATHEMATICAL MODELS, *ESSENTIAL NUTRIENTS, *PRIMARY PRODUCTIVITY, WATER QUALITY, TROPHIC LEVEL, AQUATIC ALGAE, FISH POPULATIONS, FLORIDA, WATER POLLUTION EFFECTS, NITROGEN, PHOSPHORUS.

IDENTIFIERS:
ANDERSON-CUE LAKE, MELROSE(FLORIDA).

ABSTRACT:
A SMALL FLORIDA LAKE HAS BEEN RECEIVING A REGIMEN OF NUTRIENT ADDITION EQUIVALENT TO 500 MG/CU M-YEAR N AND 43 MG/CU M-YEAR P SINCE 1967. DATA WERE ACCUMULATED THROUGH 1969. THE EFFECT ON THE LACUSTRINE ECOSYSTEM OF VARIOUS BIOGENES INCLUDES PRODUCTION BY PRIMARY PRODUCERS, SPECIES DIVERSITY OF PLANKTON AND CERTAIN PRODUCTION ESTIMATES AT THE SECONDARY TROPHIC LEVEL USING NATURAL POPULATIONS OF PLANKTIVOROUS FISH. PLANKTON PRODUCTION USING ISOTOPIC CARBON IS CA. 58 G/SQ M-YEAR. SPECIES DIVERSITY IS SLOWLY CHANGING TO A MIXED CHLOROPHYCEAN AND YELLOW-GREEN. BIOMASS OF BENTHIC GREEN FILAMENTOUS TYPES HAS INCREASED SLIGHTLY. NUTRIENT ADDITION HAS HAD LITTLE INFLUENCE ON ZOOPLANKTON PRODUCTION. RELATED STUDIES ON 53 OTHER REGIONAL LAKES HAVE BEEN DONE USING A MULTI-DIMENSIONAL HYBRID CONCEPT AS DEFINED BY SEVERAL TROPHIC STATE INDICATORS. THIS TROPHIC STATE INDEX HAS PROVIDED A MEANS FOR RANKING THE LAKES ON AN ARBITRARY SCALE. CLUSTER ANALYSIS UTILIZING PERTINENT CHARACTERISTICS RESULTED IN CLASSIFICATION OF OTHER LAKES. LAND USE PATTERNS AND POPULATION CHARACTERISTICS WERE DETERMINED PHOTOGRAPHICALLY AND N AND P BUDGETS ESTIMATED. USING MULTIPLE REGRESSION AND CANONICAL ANALYSIS, SEVERAL SIGNIFICANT RELATIONSHIPS WERE FOUND BETWEEN LAKE TROPHIC STATE, LAKE BASIN, LAND USE, AND POPULATION CHARACTERISTICS. IN GENERAL, TROPHIC STATE OF LAKES CAN BE EXPRESSED AS A SIMPLE RELATIONSHIP INCORPORATING N AND P INFLUX RATES. (EPA ABSTRACT)

FIELD 05B, 05C

ACCESSION NO. W72-08986

VITAMIN B12 PRODUCTION AND DEPLETION IN A NATURALLY OCCURRING EUTROPHIC LAKE,

OREGON STATE UNIV., CORVALLIS. DEPT. OF MICROBIOLOGY; AND OREGON STATE UNIV., CORVALLIS. DEPT. OF OCEANOGRAPHY.

A. GILLESPIE, AND Y. MORITA.

APPLIED MICROBIOLOGY VOL. 23, NO. 2, P 341-348, 1972, 11 FIG, 3 TAB, 9 REF. EPA PROGRAM NO. 16010 EBB 02/72.

DESCRIPTORS:
*VITAMIN B, *EUTROPHICATION, *AQUATIC MICROBIOLOGY, OREGON, *SEDIMENT-WATER INTERFACES, SEDIMENTS, AQUATIC PRODUCTIVITY, PHYTOPLANKTON, ALGAE.

IDENTIFIERS:
UPPER KLAMATH LAKE(OREGON).

ABSTRACT:
THE DISTRIBUTION OF VITAMIN B12 WITHIN UPPER KLAMATH LAKE WAS SURVEYED AT APPROXIMATELY MONTHLY INTERVALS DURING A PERIOD FROM SEPTEMBER 1968 TO NOVEMBER 1969. HIGH CONCENTRATIONS (UP TO 1.8MICRO G/G OF DRY SEDIMENT) CHARACTERISTICALLY OCCURRED AT THE WATER-SEDIMENT INTERFACE, WITH A SHARP DECLINE BELOW THIS AREA. A HEAVY BLOOM OF APHANIZOMENON FLOS-AQUAE OCCURRED FROM THE LATTER PART OF MAY THROUGH OCTOBER 1969. B12 CONCENTRATIONS OF THE UPPERMOST SEDIMENTS, FROM ALL BUT ONE SAMPLING SITE, INCREASED GRADUALLY THROUGH THE BLOOM FOLLOWED BY A DRASTIC INCREASE DURING THE DIE-OFF PERIOD. B12 IS PROBABLY NOT A LIMITING FACTOR FOR PRIMARY PRODUCTIVITY, SINCE SUFFICIENT LEVELS OF THIS VITAMIN WERE FOUND TO OCCUR THROUGHOUT THE YEAR. OF 42 CULTURES ISOLATED FROM UPPER KLAMATH LAKE WATER AND SEDIMENTS, 20 WERE FOUND CAPABLE OF PRODUCING 50PG OR MORE OF B12/ML OF MEDIUM. PHYTOPLANKTON SAMPLES WERE FOUND TO CONTAIN UP TO 5 MICRO G OF B12/G OF DRY MATERIAL. DEGRADATION OF B12 OCCURRED IN STERILIZED AS WELL AS FRESH SEDIMENT SAMPLES. (EPA ABSTRACT)

FIELD 05C, 02H

ACCESSION NO. W72-08989

SOME ECOLOGICAL EFFECTS OF ARTIFICIAL CIRCULATION ON A SMALL EUTROPHIC NEW HAMPSHIRE LAKE,

NEW HAMPSHIRE UNIV., DURHAM. DEPT. OF ZOOLOGY.

R. C. HAYNES.

PHD THESIS, 1971. 166 P, 13 FIG, 21 TAB, 93 REF. OWRR A-004-NH(6).

DESCRIPTORS:
*AERATION, *EUTROPHICATION, *ECOLOGICAL DISTRIBUTION, *AQUATIC ALGAE, *CYANOPHYTA, *BEGGIATOA, PHYTOPLANKTON, *NEW HAMPSHIRE, MIXING, ENVIRONMENTAL EFFECTS.

IDENTIFIERS:
*ARTIFICIAL CIRCULATION, *KEZAR LAKE(NH).

ABSTRACT:
FOR SEVERAL YEARS ANNUAL NOXIOUS BLUE-GREEN ALGAL BLOOMS PLAGUED KEZAR LAKE, N. H. ARTIFICIAL CIRCULATION WAS TRIED IN AN ATTEMPT TO IMPROVE LAKE CONDITIONS. THIS STUDY WAS MADE TO HELP UNDERSTAND SOME OF THE ECOLOGICAL EFFECTS OF MIXING THE LAKE. SAMPLES COLLECTED WEEKLY WERE ANALYZED BY STANDARD METHODS. ARTIFICIAL CIRCULATION COMPLETELY DESTRATIFIED KEZAR LAKE AND ISOTHERMAL CONDITIONS WERE MAINTAINED THROUGHOUT THE TEST PERIODS OF 1968 AND 1969. THE SUPPLY AND DISTRIBUTION OF OXYGEN, CARBON DIOXIDE, ALKALINITY AND MANY ALGAL NUTRIENTS WERE MEASURED BEFORE AND AFTER LAKE MIXING. CONCENTRATIONS OF THE BLOOM-FORMING ALGAE OCCURRED AFTER LAKE MIXING IN BOTH 1968 AND 1969, BUT SECCHI DISK READINGS IMPROVED. CHLOROPHYLL CONCENTRATION IN RELATION TO BLOOMING OF A. FLOS=AQUAE WAS DETERMINED AND RATE OF CARBON FIXATION BY PHYTOPLANKTON WAS EXPLORED. A DISCUSSION OF THE RESULTS IS INCLUDED AND A COMPARISON MADE WITH THE FINDINGS OF OTHER AUTHORS.

FIELD 05C, 05G, 02H

ACCESSION NO. W72-09061

KAISER REFRACTORIES ENVIRONMENTAL STUDIES,

MOSS LANDING MARINE LABS. CALIF.

J. P. HARVILLE.

AVAILABLE FROM THE NATIONAL TECHNICAL INFORMATION SERVICE AS COM-71-01107, $3.00 IN PAPER COPY, $0.95 IN MICROFICHE. REPORT NO. TP-71-3, NOAA-71112214, JUNE 1, 1971. 201 P, 61 FIG, 33 TAB, 31 REF. GRANT NO. GH-94.

DESCRIPTORS:
*INDUSTRIAL WASTES, *WATER POLLUTION EFFECTS, EFFLUENTS, CALIFORNIA, SEDIMENTS, SAMPLING, CHLOROPHYLL, SEDIMENT TRANSPORT, BENTHIC FAUNA, PLANKTON, BIOINDICATORS, CURRENTS(WATER), TIDES, WIND, LIGHT, WAVES, CLAMS, CRUSTACEANS, WORMS, MOLLUSKS, WATER TEMPERATURE, SALINITY, SEA WATER, WATER QUALITY, ON-SITE TESTS, HYDROGEN ION CONCENTRATION, CALCIUM, MAGNESIUM, ALGAE, CYANOPHYTA, DIATOMS, MUSSELS, PHYTOPLANKTON, LABORATORY STUDIES, MORTALITY, CRABS, BIOASSAY, LITTORAL DRIFT, INVERTEBRATES, CALCITE, SILICON, SODIUM, POTASSIUM, PERIPHYTON, ALUMINUM, IRON, QUARTZ, SEWAGE, ANNELIDS, TOPOGRAPHY, COPEPODS, ISOPODS, NEMATODES, GASTROPODS, ORGANIC MATTER, ZOOPLANKTON, FISH EGGS, AMPHIPODS, PERCHES, CHRYSOPHYTA, PYRROPHYTA, DINOFLAGELLATES, PROTOZOA, ROTIFERS, LARVAE, X-RAY DIFFRACTION.

IDENTIFIERS:
*MOSS LANDING, BARNACLES, LETHAL DOSAGE, CALCIUM ION CONCENTRATION, VERTEBRATES, ABSORBANCE, MAGNESIUM ION CONCENTRATION, BRUCITE, BOTTOM TOPOGRAPHY, PRIONOSPIO, LUMBRINEREIS, GLYCERA, NEPHTYS, CAPITELLA, MAGELONA, ARMANDIA, TELLINA BUTTONI, VENUS, TIVELA STULTORUM, EMERITA ANALOGA, DIASTYLOPSIS, OXYUROSTYLIS, HEMILAMPROPS, LAMPROPS, COLUROSTYLIS, CUMELLA, BATHYCUMA, LEPTOSTYLIS, LEUCON, BRACHYURA, ANOMURA, OLIVELLA, METZGERIA, NASSARIUS, VELUTINA, POLINICES, COLUS, CLINOCARDIUM, ZIRFAEA, COSCINODISCUS.

ABSTRACT:
A LONG-RANGE STUDY WAS CONDUCTED OF POTENTIAL EFFECTS OF THE KAISER REFRACTORIES INDUSTRIAL EFFLUENT ON THE MARINE ENVIRONMENT OF THE MOSS LANDING AREA. THE INVESTIGATIONS FOLLOWED THREE PRIMARY THRUSTS: (1) DETERMINATION OF PHYSICAL CHARACTERISTICS AND DYNAMICS OF THE WATER MASS IN THE PROPOSED OUTFALL AREA, AND ASSESSMENT OF BOTTOM STRUCTURE, SEDIMENTS, AND SEDIMENT TRANSPORT CHARACTERISTICS OF THE AREA. (2) BIOLOGICAL INVESTIGATIONS OF BOTTOM FAUNA OF THE PROPOSED OUTFALL AREA, WITH ANCILLARY STUDIES OF PLANKTON, FISHES, AND INTERTIDAL FAUNA. (3) FIELD AND LABORATORY STUDIES OF THE IMPACT OF VARIOUS DILUTIONS OF KAISER EFFLUENT UPON SELECTED BIOINDICATOR PLANT AND ANIMAL SPECIES. SELECTED CONCLUSIONS ARE: (A) INVESTIGATIONS OF BENTIC COMMUNITIES INDICATE GREAT VARIABILITY IN THE DISTRIBUTION AND POPULATION DENSITIES OF ORGANISMS. THE FOLLOWING GENERALIZATIONS APPEAR SIGNIFICANT: (1) THE PISMO CLAM (TIVELA STULTORUM), THE BUTTON CLAM (TELLINA BUTTONI), AND CERTAIN POLYCHAETE WORMS ARE DOMINANT. (2) CRUSTACEANS APPEAR TO CONSTITUTE AN IMPORTANT COMPONENT OF THE SURFACE FAUNA, AND POLYCHAETE WORMS USUALLY DOMINATE THE SEDIMENTS BELOW THE SURFACE. (3) DUMPING OF DREDGE SPOIL HAS AN IMMEDIATE DEVASTATING EFFECT UPON THE BENTHIC COMMUNITY, BUT THE COMMUNITY APPEARS TO RECOVER IN THE YEAR FOLLOWING. (B) THE KAISER EFFLUENT IS CHARACTERIZED CHEMICALLY BY HIGH PH, HIGH CALCIUM ION CONCENTRATIONS, LOW MAGNESIUM ION CONCENTRATIONS, AND IS CHARGED WITH A MILKY-WHITE PRECIPITATE WHICH IS PRINCIPALLY CALCITE (CACO3) AND BRUCITE (MG(OH)2). (C) THE KAISER EFFLUENT SHARPLY RESTRICTED THE AMOUNT AND VARIABILITY OF BIOTA. (D) THE PHOTOSYNTHETIC RATE OF MIXED PHYTOPLANKTON CULTURES INDICATED A LINEAR DEPRESSION OF THAT RATE BY THE KAISER EFFLUENT. (E) UNDILUTED EFFLUENT (APPROXIMATELY 1700 PPM CA IONS) WAS LETHAL TO COPEPODS TO VARIOUS DILUTIONS OF THE EFFLUENT FOR 24 HOUR PERIODS, BUT GRADUAL ADAPTATION TO UP TO 50 PERCENT EFFLUENT WAS POSSIBLE. (F) THE SIZE AND ABUNDANCE OF MUSSELS ARE DIRECTLY RELATED TO DISTANCES FROM THE KAISER OUTFALL. (G) THE COMMON SAND CRAB, EMERITA ANALOGA, IS SENSITIVE PHYSIOLOGICALLY TO ABNORMALLY HIGH CALCIUM-MAGNESIUM RATIOS. (H) THE PISMO CLAM, TIVELA STULTORUM, SHOWED BEHAVIORAL AND PHYSIOLOGICAL DAMAGE WHEN SUBJECTED TO HIGH CONCENTRATIONS OF KAISER EFFLUENT UNDER LABORATORY CONDITIONS. (MORTLAND-BATTELLE)

FIELD 05C, 05A

ACCESSION NO. W72-09092

RELATIONSHIP BETWEEN LIGHT CARBON DIOXIDE FIXATION AND DARK CARBON DIOXIDE FIXATION BY MARINE ALGAE,

NOVA UNIV., DANIA, FLA. PHYSICAL OCEANOGRAPHIC LAB.

I. MORRIS, C. M. YENTSCH, AND C. S. YENTSCH.

LIMNOLOGY AND OCEANOGRAPHY, VOL. 16, NO. 6, P 854-858, NOVEMBER 1971. 2 FIG, 2 TAB, 6 REF.

DESCRIPTORS:
*LIGHT INTENSITY, *PRIMARY PRODUCTIVITY, *RADIOACTIVITY TECHNIQUES, *PHOTOSYNTHESIS, CARBON DIOXIDE, PHYTOPLANKTON, ALGAE, CULTURES, *FLORIDA, SEA WATER, CHLOROPHYLL, NITRATES, NITRITES, PHOSPHATES, NUTRIENTS.

IDENTIFIERS:
*FLORIDA STRAIT, CARBON-14, DUNALIELLA TERTIOLECTA, PHAEODACTYLUM TRICORNUTUM.

ABSTRACT:
EXPERIMENTS WERE CONDUCTED ON NATURAL POPULATIONS (IN THE FLORIDA STRAIT) AND CULTURED MARINE PHYTOPLANKTON (DUNALIELLA TERTIOLECTA AND PHAEODACTYLUM TRICORNUTUM) TO DETERMINE RATES OF DARK FIXATION OF CARBON DIOXIDE. SUCH INFORMATION IS IMPORTANT TO A BETTER UNDERSTANDING OF THE CARBON-14 TECHNIQUE FOR ASSESSING PRIMARY PRODUCTIVITY. THE RATIO OF CARBON DIOXIDE FIXATION IN LIGHT TO THAT IN DARK INCREASES WITH INCREASING CONCENTRATIONS OF PHYTOPLANKTON. THIS IS TRUE IN BOTH NATURAL POPULATIONS AND CULTURES. POSSIBLE EXPLANATIONS OF THIS ARE PRESENTED AND DISCUSSED AND THE USE OF DARK FIXATION VALUES IN THE CARBON-14 TECHNIQUE IS CONSIDERED. THE C-14 TECHNIQUE IS ONLY AS GOOD AS THE KNOWLEDGE OF HOW THE CARBON IS FIXED. (SEE ALSO W72-09103) (MORTLAND-BATTELLE)

FIELD 05C, 05B

ACCESSION NO. W72-09102

THE PHYSIOLOGICAL STATE WITH RESPECT TO NITROGEN OF PHYTOPLANKTON FROM LOW-NUTRIENT SUBTROPICAL WATER AS MEASURED BY THE EFFECT OF AMMONIUM ION ON DARK CARBON DIOXIDE FIXATION,

NOVA UNIV., DANIA, FLA. PHYSICAL OCEANOGRAPHIC LAB.

I. MORRIS, C. M. YENTSCH, AND C. S. YENTSCH.

LIMNOLOGY AND OCEANOGRAPHY, VOL. 16, NO. 6, P 859-868, NOVEMBER 1971. 4 FIG, 3 TAB, 11 REF.

DESCRIPTORS:
*PHYTOPLANKTON, *NITROGEN, *NUTRIENTS, *PHOTOSYNTHESIS, LIGHT INTENSITY, AMMONIA, CARBON DIOXIDE, FLORIDA, CULTURES, *ALGAE, INCUBATION, CENTRIFUGATION, RADIOACTIVITY TECHNIQUES, CHLOROPHYLL, SEA WATER, IONS, NUTRIENT REQUIREMENTS, AQUATIC POPULATIONS, PRIMARY PRODUCTIVITY.

IDENTIFIERS:
CARBON-14, *FLORIDA STRAIT, DUNALIELLA TERIOLECTA, PHAEODACTYLUM TRICORNUTUM, SKELETONEMA COSTATUM, CARBOY, OSCILLATORIA ERYTHRAEA.

ABSTRACT:
STUDIES CONDUCTED WITH THREE SPECIES OF MARINE ALGAE (DUNALIELLA TERTIOLECTA, PHAEODACTYLUM TRICORNUTUM, AND SKELETONEMA COSTATUM) INDICATE THAT THERE IS NO INCREASE IN DARK FIXATION IN NON-NITROGEN DEFICIENT CULTURES AFTER THE ADDITION OF AMMONIUM ION. THE PRESENCE OF AMMONIUM IONS HAD NO EFFECT ON THE CAPACITY FOR LIGHT FIXATION OF CARBON-14. THE EXPERIMENTS WERE CONDUCTED IN LOW-NITROGEN, SUBTROPICAL WATERS OF THE FLORIDA STRAIT, AND THE RESULTS INDICATED THAT THE PHYTOPLANKTON CELLS WERE NOT PHYSIOLOGICALLY NITROGEN DEFICIENT, ALTHOUGH NITROGEN MAY LIMIT POPULATION SIZE. THE AMMONIUM ION ENHANCEMENT WHICH WAS USED OFFERS THE ADVANTAGE OF ASSESSING THE POPULATION AT TIME ZERO. WHEN ENHANCEMENT WAS NOT NOTED AT TIME ZERO, INCUBATION OF THE WATER IN A CARBOY RESULTED IN DEPLETION OF THE AVAILABLE NITROGEN AND A PHYSIOLOGICAL STATE OF NITROGEN DEFICIENCY SUBSEQUENTLY BECAME MEASURABLE BY THIS METHOD. THE EVENTUAL POSITIVE RESULTS FROM THE TEST THEREFORE CONFIRM THE NEGATIVE RESULTS AT TIME ZERO. (SEE ALSO W72-09102) (MORTLAND-BATTELLE)

FIELD 05C, 05A

ACCESSION NO. W72-09103

THE SEASONAL VARIATION OF DISSOLVED ORGANIC CARBON IN THE INSHORE WATERS OF THE MENAI STRAIT IN RELATION TO PRIMARY PRODUCTION,

UNIVERSITY COLL. OF NORTH WALES, MENAI BRIDGE. MARINE SCIENCE LABS.

A. W. MORRIS, AND P. FOSTER.

LIMNOLOGY AND OCEANOGRAPHY, VOL. 16, NO. 6, P 987-989, NOVEMBER 1971. 1 FIG, 6 REF.

DESCRIPTORS:
*ANALYTICAL TECHNIQUES, *ORGANIC COMPOUNDS, *CARBON, GRAVIMETRIC ANALYSIS, PRODUCTIVITY, SEASONAL, CHLOROPHYLL, ALGAE, PHYTOPLANKTON, WATER ANALYSIS, CARBON CYCLE, ORGANIC MATTER, NUTRIENTS.

IDENTIFIERS:
*DISSOLVED ORGANIC CARBON, *MENAI STRAIT(WALES).

ABSTRACT:
A REGULAR ANNUAL CYCLE FOR DISSOLVED ORGANIC CARBON WAS FOUND IN THE MENAI STRAIT FROM MEASUREMENTS MADE OVER A TWO-YEAR PERIOD. ULTRAVIOLET PHOTOOXIDATION OF 1-LITER SAMPLES OF WATER FOLLOWED BY GRAVIMETRIC ESTIMATION OF THE LIBERATED CARBON DIOXIDE WAS USED FOR DETERMINATIONS. WINTER LEVELS OF ABOUT ONE MG C/LITER ARE FOLLOWED BY A STEADY INCREASE THROUGH SPRING AND SUMMER, REACHING MAXIMUM VALUES OF 3-4 MG C/LITER IN AUTUMN FOLLOWED BY A SHARP RETURN TO WINTER VALUES. STUDIES INDICATED A HIGH RATE OF PRODUCTION IS MAINTAINED THROUGHOUT THE SUMMER PERIOD ALTHOUGH THE STANDING CROP MAY VARY CONSIDERABLY. (SNYDER-BATTELLE)

FIELD 05A, 05B

ACCESSION NO. W72-09108

DISTRIBUTION OF PHYTOPLANKTON IN A POLLUTED SALINE LAKE, ONONDAGA LAKE, NEW YORK,

CORNELL UNIV., ITHACA, N.Y. DIV. OF BIOLOGICAL SCIENCES.

P. SZE, AND J. M. KINGSBURY.

JOURNAL OF PHYCOLOGY, VOL. 8, NO. 1, P 25-37, MARCH 1972. 4 FIG, 2 TAB, 42 REF.

DESCRIPTORS:
*PHYTOPLANKTON, *DISTRIBUTION PATTERNS, *SYSTEMATICS, *SALINE LAKES, *EUTROPHICATION, WATER SAMPLING, WATER POLLUTION EFFECTS, BIOMASS, NUTRIENTS, ALGAE, EUGLENOPHYTA, CHRYSOPHYTA, CHLOROPHYTA, PYRROPHYTA, CYANOPHYTA, MANGANESE, ZINC, METALS, SECCHI DISKS, HEAVY METALS, BIOINDICATORS, SILICA, CHROMIUM, COPPER, *NEW YORK, DIATOMS, PHOSPHORUS, NITROGEN, CHLORELLA, CHLAMYDOMONAS, SCENEDESMUS, DINOFLAGELLATES, EUGLENA, E. COLI, PHOSPHATES, DISSOLVED OXYGEN, BIOCHEMICAL OXYGEN DEMAND, HYDROGEN ION CONCENTRATION, CALCIUM, SODIUM, POTASSIUM, MAGNESIUM, IRON.

IDENTIFIERS:
*ONONDAGA LAKE, OEDOGONIUM, RHIZOCLONIUM, COSMARIUM, PHACUS, CHAETOCEROS, MOUGEOTIA, NAVICULA, PINNULARIA, SURIRELLA, PERIDINIUM, SENECA RIVER, NINEMILE CREEK, VAN DORN SAMPLER, KEMMERER SAMPLER, CHLAMYDOMONAS EPIPHYTICA, CHLAMYDOMONAS SPP, CARTERIA FRITSCHII, PANDORINA MORUM, SPHAEROCYSTIS SCHROETERI, ULOTHRIX SPP, MICROTHAMNION KUETZINGIANUM, MICRACTINIUM PUSILLUM, ERRERELLA BORNHEMIENSIS.

ABSTRACT:
DURING 1969, ONONDAGA LAKE (NEW YORK) WAS REGULARLY SAMPLED AT FIVE SITES IN ORDER TO STUDY THE PHYTOPLANKTON. SAMPLES OBTAINED FOR QUANTITATIVE EXAMINATIONS WERE TAKEN FROM THE SURFACE WATERS AT ALL STATIONS AND AT DEPTHS OF 3, 6, AND 12 M AT TWO STATIONS USING DIPPERS AND VAN DORN OR KEMMERER SAMPLERS. SAMPLES FOR GROSS QUALITATIVE ANALYSES WERE OBTAINED BY MAKING VERTICAL AND HORIZONTAL HAULS WITH A NO. 20 NET AT EACH STATION. AN IBM 1130 COMPUTER WAS USED FOR ANALYZING THE RAW DATA. THE LAKE IS RELATIVELY SALINE AND HAS BEEN FOUND TO SUPPORT AN ALGAL FLORA CHARACTERISTIC OF A EUTROPHIC LAKE WITH AN ADMIXTURE OF SALINE SPECIES. SEASONS IN THE LAKE CAN BE IDENTIFIED BY FLORAL SUCCESSION WITH CERTAIN SPECIES OF ALGAE APPEARING FIRST AT THE OUTFLOW AND THEN SPREADING THROUGH THE LAKE. PHOSPHORUS AND NITROGEN WERE NEVER LIMITING, BUT SILICA DIMINUTION WAS LIMITING FOR FURTHER DIATOM POPULATION GROWTH AND WAS RELATED TO DIATOM BLOOMS. SINCE THE EUTROPHIC LAYER WAS MUCH SHALLOWER THAN THE THERMOCLINE, TURBULENCE AND MIXING WERE THOUGHT TO PLAY AN IMPORTANT ROLE IN CONTROLLING CERTAIN POPULATIONS. NO OBVIOUS CONTROLLING RELATIONSHIP EXISTED BETWEEN HERBIVORES AND PHYTOPLANKTON POPULATIONS. CR AN CU WERE HIGH AS A RESULT OF INDUSTRIAL DISCHARGES AND MAY BE RESPONSIBLE FOR INHIBITING BLOOMS. BIOMASS WAS CALCULATED FOR THE MAJOR PHYTOPLANKTERS. (HOLOMAN-BATTELLE)

FIELD 05C, 02H

ACCESSION NO. W72-09111

A LIST OF NEW GENERA AND TYPE SPECIES OF FLAGELLATES AND ALGAE PUBLISHED IN 1969, PART V,

B. V. SKVORTZOV.

HYDROBIOLOGIA, VOL 39, NO. 2, P 241-245, FEBRUARY 29, 1972. 15 FIG, 4 REF.

DESCRIPTORS:
*PROTOZOA, *SYSTEMATICS, *ALGAE, CHLOROPHYTA, HABITATS, AQUATIC HABITATS, EUGLENOPHYTA, PYRROPHYTA, FRESHWATER ALGAE.

IDENTIFIERS:
PROTOCRYPTOCHRYSIS OBOVATA, GUTTULA BACILLARIOPHAGA, REFRACTOCYSTIS PLUVIALIS, GOMESIAMONAS STAGNALIS, REFRACTODES BRASILIANA, OYEMONAS PULSULAE, REFRACTOMONAS BRASILIANA, HORTOBAGGIAMONAS PLICATA, TRICHOCYANELLA SPIRALIS, STROMIA SUBSPHAERICA, TETRACULAMONAS NATANS, MARINIAMONAS SAUPAULENSIS, ENEIDAMONAS APPLANATA, BICUDOMONAS CYANOPHORA, LIANGIANA TERRESTRIS, CHLOROMONADS, CRYPTOMONADS, BICUDOMONAS, *FLAGELLATES.

ABSTRACT:
A LIST OF 15 NEW GENERA OF FLAGELLATES AND ALGAE IS GIVEN. INCLUDED ARE ILLUSTRATIONS OF THE ORGANISMS, SIZES OF THE CELLS, TYPE SPECIES, AND THE HABITATS IN WHICH THEY MAY BE FOUND. (MACKAN-BATTELLE)

FIELD 05A, 05C

ACCESSION NO. W72-09119

SURVEY OF MACROPHYTE RESOURCES IN THE COASTAL WATER OF ALASKA,

ALASKA UNIV., COLLEGE. INST. OF MARINE SCIENCE.

C. P. MCROY, M. MUELLER, S. STOKER, J. J. GOERING, AND M. T. GOTTSCHALK.

AVAILABLE FROM THE NATIONAL TECHNICAL INFORMATION SERVICE AS COM-71-01141, $3.00 IN PAPER COPY, $0.95 IN MICROFICHE. REPORT NO. 3, R71-6, (NOAA-71101806), MAY 1971. 43 P, 14 FIG, 3 TAB, 40 REF.

DESCRIPTORS:
*MARINE PLANTS, *ALASKA, SAMPLING, *SYSTEMATICS, BENTHOS, SCUBA DIVING, DEPTH, DISTRIBUTION PATTERNS, AERIAL PHOTOGRAPHY, REMOTE SENSING, SATELLITES(ARTIFICIAL), ALGAE, CHLOROPHYTA, PHAEOPHYTA, RHODOPHYTA.

IDENTIFIERS:
*SEA WEEDS, *MACROPHYTES, *SEA GRASSES, TELEVISION, DEPTH SOUNDER, VERTICAL DISTRIBUTION, COLD BAY, BERMERS BAY, ALARIA, NEREOCYSTIS, LAMINARIA, THALASSIOPHYLLUM, FUCUS, SCHIZYMENIA, RHODOMELA, RHODOMENIA, MARMION ISLAND, MORRIA REEF, BEAR BAY, NEVA ISLAND, BAIRD ISLAND, SHOLIN ISLAND, WINIFRED ISLAND, IZEMBEK LAGOON, CHAETOMORPHA CANNABINA, CHAETOMORPHA MELAGONIUM, CLADOPHORA GLAUCESCENS, CLADOPHORA STIMPSONII.

ABSTRACT:
A PROJECT WAS INITIATED TO QUANTITATIVELY ASSESS NATURAL STOCKS OF SEAWEEDS AND SEAGRASSES IN ALASKA COASTAL WATERS. THE FIRST YEAR'S PROGRESS IS REPORTED. A TECHNIQUE UTILIZING SCUBA DIVERS AND SUBMARINE TELEVISION HAS BEEN DEVISED IN WHICH A DIVER MAKES PARALLEL TRANSECTS AT RIGHT ANGLES TO SHORE ON A SLED TOWED BY A SURFACE VESSEL OR WINCH. ON THE SLED IS ALSO MOUNTED THE TV CAMERA THAT MAKES A PERMANENT RECORD OF THE TRANSECT. A DEPTH PROFILE IS ALSO MADE WITH A RECORDING DEPTH SOUNDER MOUNTED ON A BOSTON WHALER. IN ADDITION THIS CAN BE USED TO IDENTIFY CERTAIN TYPES OF KELP BEDS. THIS TYPE OF TRANSECT IS RAPID AND GIVES A GOOD RECORD OF THE VERTICAL DISTRIBUTION OF THE MAJOR SPECIES. FOLLOWING THE VISUAL SURVEY, DIVERS TAKE REPLICATE SAMPLES OF ALL PLANTS WITHIN A QUADRAT. THE SURVEY TO DATE INDICATES THAT SEAWEED SPECIES IN THE FOLLOWING GENERA ATTAIN HIGH ENOUGH STANDING STOCKS IN ALASKA TO SUSTAIN EXPLOITATION: ALARIA, NEREOCYSTIS, LAMINARIA, THALASSIOPHYLLUM, FUCUS, SCHIZYMENIA, RHODOMELA, AND RHODOMENIA. REPRINTS OF JOURNAL ARTICLES COVERING THIS RESEARCH AND A SURVEY OF EELGRASS DISTRIBUTION ON THE ALASKA COAST ARE INCLUDED. (MORTLAND-BATTELLE)

FIELD 02L, 05A

ACCESSION NO. W72-09120

CARBON DIOXIDE AND PRIMARY PRODUCTIVITY IN THE GLACIAL FIORD SYSTEM OF SOUTHEAST ALASKA,

ALASKA UNIV., COLLEGE. INST. OF MARINE SCIENCE.

L. L. LONGERICH, M. BILLINGTON, V. ALEXANDER, J. J. KELLEY, AND D. W. HOOD.

AVAILABLE FROM THE NATIONAL TECHNICAL INFORMATION SERVICE AS AD-734 672, $3.00 IN PAPER COPY, $0.95 IN MICROFICHE. REPORT NO. R71-19, SEPTEMBER 1971. 24 P, 1 FIG, 1 TAB, 5 REF. ONR CONTRACT N00014-67-AC-317-0001AB.

DESCRIPTORS:
*CARBON DIOXIDE, *PRIMARY PRODUCTIVITY, ALASKA, FIORDS, SURFACE WATERS, NITROGEN, SEA WATER, WINDS, SAMPLING, INCUBATION, RUNOFF, SALINITY, DEPTH, ABSORPTION, ALGAE, SEASONAL, WEATHER, WATER ANALYSIS, TRACERS, RADIOACTIVITY TECHNIQUES.

IDENTIFIERS:
AUKE BAY, GLACIAL WATER, BECKMAN ANALYZER, VAN DORN SAMPLERS, C-14, BIOCARBONATES.

ABSTRACT:
EARLY SPRING STUDIES OF CARBON DIOXIDE CONCENTRATIONS AND PRIMARY PRODUCTIVITY WERE MADE IN THE GLACIAL FIORD SYSTEM OF SOUTHEASTERN ALASKA. THE SURFACE WATERS OF AUKE BAY AT THE BEGINNING OF THE SPRING BLOOM HAD A LOWER CARBON DIOXIDE CONCENTRATION AND HIGHER PARTICULATE NITROGEN AND CARBON-14-HCO3 UPTAKE THAN THE MORE OPEN WATERS SURROUNDING IT. SURFACE WATER CARBON DIOXIDE IN A GLACIAL OUTFLOW AREA WAS UNEXPECTEDLY LOW COMPARED TO VALUES FOR SURFACE SEAWATER IN A NORMAL FRESH WATER STREAM OUTFLOW. AN EXAMPLE OF WIND INDUCED UPWELLING WAS NOTED WHICH PRODUCED INCREASED SURFACE CARBON DIOXIDE CONCENTRATIONS OVER THE DURATION OF A STORM. IT WAS EVIDENT FROM PREVIOUS CARBON DIOXIDE STUDIES AND FROM THIS SURVEY THAT SURFACE CARBON DIOXIDE CONCENTRATIONS ARE DEPENDENT ON LOCATION, PRIMARY PRODUCTIVITY, PROXIMITY TO FRESH WATER RUNOFF, THE SEASON, AND THE WEATHER. PAST WORK SHOWS THAT LONG-TERM, ON-SITE MULTIDISCIPLINARY OBSERVATIONS ARE ESSENTIAL TO AN UNDERSTANDING OF HOW CARBON DIOXIDE INTERACTS IN THE MARINE ENVIRONMENT. (MORTLAND-BATTELLE)

FIELD 05C, 05A, 02C

ACCESSION NO. W72-09122

NUTRIENTS AND EUTROPHICATION: THE LIMITING NUTRIENT CONTROVERSY.

AMERICAN SOCIETY OF LIMNOLOGY AND OCEANOGRAPHY SPECIAL SYMPOSIA, VOL I, ALLEN PRESS, LAWRENCE, KANSAS, 1972. 328 P. OWRR X-109(NO. 3415)(1).

DESCRIPTORS:
*EUTROPHICATION, *CARBON, *PHOSPHORUS, NITROGEN, ALGAE.

IDENTIFIERS:
*LIMITING NUTRIENTS.

ABSTRACT:
THE INTENT OF THE AMERICAN SOCIETY OF LIMNOLOGY AND OCEANOGRAPHY TO PROVIDE AN OPEN FORUM FOR COMMUNICATION BETWEEN ACADEME, STATE AND FEDERAL AGENCIES, AND INDUSTRY, CULMINATED IN THIS SYMPOSIUM HELD AT W. K. KELLOGG BIOLOGICAL STATION OF MICHIGAN STATE UNIVERSITY ON FEBRUARY 11-12, 1971. IT WAS THE HOPE TO PROVIDE A CLEAR STATEMENT OF CURRENT IDEAS RELATIVE TO THE IMPORTANCE OF VARIOUS REGULATING OR LIMITING NUTRIENTS IN THE EUTROPHICATION OF AQUATIC ECOSYSTEMS. ONLY ONE FACTOR--NUTRIENT AVAILABILITY--WAS SELECTED AS THE SUBJECT. THE FOCUS WAS ON THE RELATIVE IMPORTANCE OF CARBON AND PHOSPHORUS IN REGULATING EUTROPHICATION. BECAUSE OF THE RECENT CARBON VS. PHOSPHORUS CONTROVERSY (CENTERED ON THE PROPOSAL THAT CARBON RATHER THAN PHOSPHORUS OR NITROGEN LIMITS ALGAL PRODUCTIVITY) THIS SUBJECT CARRIES POLITICAL AND ECONOMIC OVERTONES. SINCE PHOSPHORUS IN DETERGENTS IS LINKED TO CULTURAL EUTROPHICATION, THE CONTROVERSY IS NOW EMOTIONALLY CHARGED FOLLOWING LEGISLATIVE PROPOSALS TO REMOVE PHOSPHORUS FROM DETERGENT FORMULATIONS. THE PROCEEDINGS OF THIS SYMPOSIUM WHERE IDEAS AND DATA WERE OPENLY AND AUTHORITATIVELY DISCUSSED, QUESTIONED, AND DEBATED, ARE INTENDED TO PROVIDE THE PUBLIC AND POLITICIANS WITH SOME USEFUL GUIDELINES TO THIS PROBLEM. IT IS UP TO EACH READER TO EVALUATE THE CONTROVERSY AND TO DETERMINE WHETHER THERE WAS ONE. (SEE W72-09159 THRU W72-09171) (AUEN-WISCONSIN)

FIELD 05C

ACCESSION NO. W72-09155

THE INTERRELATION OF CARBON AND PHOSPHORUS IN REGULATING HETEROTROPHIC AND AUTOTROPHIC POPULATIONS IN AN AQUATIC ECOSYSTEM, SHRINER'S POND,

FEDERAL WATER POLLUTION CONTROL ADMINISTRATION, ATHENS, GA. SOUTHEAST WATER LAB.

P. C. KERR, D. L. BROCKWAY, D. F. PARIS, AND J. T. BARNETT, JR.

IN: NUTRIENTS AND EUTROPHICATION: THE LIMITING NUTRIENT CONTROVERSY, P 41-62, 9 FIG, 58 REF. AMERICAN SOCIETY OF LIMNOLOGY AND OCEANOGRAPHY SPECIAL SYMPOSIA VOL I, ALLEN PRESS, LAWRENCE, KANSAS, 1972.

DESCRIPTORS:
*CARBON, *PHOSPHORUS, *POPULATION, AQUATIC ENVIRONMENT, ECOSYSTEMS, GEORGIA, BIORTHYTHMS, LABORATORY TESTS, ALGAE, CARBON DIOXIDE, CYCLING NUTRIENTS, BICARBONATES, PLANT GROWTH, DURNAL.

IDENTIFIERS:
*LIMITING NUTRIENTS, POTASSIUM CHLORIDE, SHRINER'S POND(GA).

ABSTRACT:
ANY OF THE TYPES OF BIOLOGICAL REGULATION--PHYSICAL, CHEMICAL, GENETIC, AND NUTRITIONAL--CAN REGULATE BIOLOGICAL ACTIVITY OF ORGANISMS AT DIFFERENT TIMES AND PLACES. SHRINER'S POND, GEORGIA WAS STUDIED TO ASCERTAIN CHEMICAL AND BIOLOGICAL CHANGES ASSOCIATED WITH ADDITION OF NITROGEN AND PHOSPHORUS OR CARBON DIOXIDE. WITHIN 12 HOURS AFTER REAGENT-GRADE NITROGEN, PHOSPHORUS, AND POTASSIUM CHLORIDE WERE ADDED, THE BACTERIAL POPULATION INCREASED. ALGAL POPULATION INCREASED 36-48 HOURS AFTER FERTILIZATION. DIEL CYCLING OF CARBON DIOXIDE AND BICARBONATE WAS MEASURED. ALTHOUGH THESE DATA DO NOT CLEARLY INDICATE THAT INCREASED AVAILABILITY OF INORGANIC CARBON WAS RESPONSIBLE FOR INITIATION OF ALGAL GROWTH, CONTINUED BIOLOGICAL PRODUCTION OF CARBON DIOXIDE APPEARED TO PROLONG THE BLOOM DURATION. DATA INDICATE MORE NITROGEN AND PHOSPHORUS WERE REMOVED DURING NIGHT THAN DURING DAY; THE HIGHER REMOVAL WAS ASSOCIATED WITH OXYGEN REMOVAL, DECREASING PH, AND ACCUMULATION OF CARBON DIOXIDE AND BICARBONATE, INDICATING HETEROTROPHIC ACTIVITY. CARBON DIOXIDE WAS REMOVED AND BICARBONATE WAS DEPLETED FROM THE WATER DURING LIGHT HOURS. ALGAL GROWTH WAS STIMULATED BY BUBBLING 5% AND 0.03% CARBON DIOXIDE IN AIR THROUGH WATER. THESE EXPERIMENTS INDICATE IMPORTANCE OF CARBON IN REGULATING ALGAL GROWTH. (SEE ALSO W72-09155) (JONES-WISCONSIN)

FIELD 05C

ACCESSION NO. W72-09159

CARBON LIMITATION IN SEWAGE LAGOONS,

MISSOURI UNIV., COLUMBIA. DEPT. OF CIVIL ENGINEERING.

D. L. KING.

IN: NUTRIENTS AND EUTROPHICATION: THE LIMITING NUTRIENT CONTROVERSY, P 98-110, 8 FIG, 16 REF. AMERICAN SOCIETY OF LIMNOLOGY AND OCEANOGRAPHY SPECIAL SYMPOSIA VOL I, ALLEN PRESS, LAWRENCE, KANSAS, 1972.

DESCRIPTORS:
*CARBON, *SEWAGE LAGOONS, *LIMITING FACTORS, ALGAE, PRODUCTIVITY, PHOTOSYNTHESIS, NUTRIENTS, EUTROPHICATION, CYANOPHYTA, ALKALINITY, DIURNAL, NITROGEN.

IDENTIFIERS:
*LIMITING NUTRIENTS.

ABSTRACT:
OBSERVATIONS OF SEWAGE LAGOONS OFFER SIGNIFICANT INSIGHT INTO THE PROCESSES INVOLVED IN LAKE EUTROPHICATION. IT IS CONCLUDED THAT BOTH THE QUALITATIVE AND QUANTITATIVE ASPECTS OF EUTROPHICATION MUST BE CONSIDERED. ADDITIONS OF REQUIRED ALGAL NUTRIENTS TO A LAKE ALLOW INCREASES IN QUANTITY OF ALGAE AND CAN STRAIN THE CARBON AVAILABILITY TO THE POINT WHERE THERE IS ALSO A CHANGE IN ALGAL QUALITY. ESTABLISHMENT OF SUMMER BLUE-GREEN ALGAL BLOOMS USUALLY IS OF GREATER CONCERN THAN THE PRECURSORY INCREASE IN QUANTITY OF THE MORE DESIRABLE ALGAL FORMS. CALCULATIONS SUGGEST THAT AMOUNTS OF ALGAL NUTRIENTS, OTHER THAN CARBON, REQUIRED TO PROMOTE SUMMER BLOOMS OF BLUE-GREEN ALGAE ARE DETERMINED BY ALKALINITY OF THE WATER IN QUESTION. ATTEMPTS TO CONTROL EUTROPHICATION BY LIMITING JUST CARBON AVAILABILITY PROBABLY WOULD RESULT IN ESTABLISHMENT OF BLUE-GREEN ALGAE DOMINANCE BUT PERHAPS IN LOWER QUANTITIES. THE BLUE-GREEN ALGAE PROBABLY WOULD ACCELERATE AND RATE OF CARBON DIOXIDE GAIN FROM THE ATMOSPHERE. ATTEMPTS TO LIMIT AVAILABLE NITROGEN MAY RESULT IN ESTABLISHMENT OF BLOOMS BLUE-GREENS WHICH CAN FIX ATMOSPHERIC NITROGEN. LIMITATION OF PHOSPHORUS APPEARS TO OFFER THE BEST CHANCE OF CONTROLLING BOTH QUALITATIVE AND QUANTITATIVE ASPECTS OF EUTROPHICATION. (SEE ALSO W72-09155) (JONES-WISCONSIN)

FIELD 05C

ACCESSION NO. W72-09162

CHARACTERIZATION OF PHOSPHORUS-LIMITED PLANKTON ALGAE,

NEW YORK STATE DEPT. OF HEALTH, ALBANY. ENVIRONMENTAL HEALTH CENTER.

G. W. FUHS, S. C. DEMMERLE, E. CANELLI, AND M. CHEN.

IN: NUTRIENTS AND EUTROPHICATION: THE LIMITING NUTRIENT CONTROVERSY, P 113-133, 16 FIG, 3 TAB, 59 REF. AMERICAN SOCIETY OF LIMNOLOGY AND OCEANOGRAPHY SPECIAL SYMPOSIA VOL I, ALLEN PRESS, LAWRENCE, KANSAS, 1972.

DESCRIPTORS:
*PHOSPHORUS, *PLANKTON, *ALGAE, *LIMITING FACTORS, GROWTH RATES, DIATOMS, BACTERIA, LAKES, NITROGEN, ANALYTICAL TECHNIQUES, CARBON, EUTROPHICATION, STANDING CROPS, LABORATORY TESTS, CHEMICAL ANALYSIS, BIOASSAY, CYTOLOGICAL STUDIES, NUTRIENTS, ENZYMES.

IDENTIFIERS:
*LIMITING NUTRIENTS, CANADARAGO LAKE(NY), LAKE GEORGE(NY).

ABSTRACT:
GROWTH RATE OF MICROORGANISMS AS A FUNCTION OF CONCENTRATION OF PHOSPHORUS SOURCE IS CALCULATED FROM CONTINUOUS CULTURE AND PHOSPHORUS UPTAKE EXPERIMENTS. FOR TWO DIATOMS AND THREE BACTERIA, VALUES OF THE MINIMUM PHOSPHORUS CONTENT, GROWTH RATE WITH PHOSPHORUS NONLIMITING, THE MICHAELIS CONSTANT, AND RATE FOR UPTAKE OF THE PHOSPHORUS SOURCE ARE GIVEN. PHOSPHORUS CONTENT OR ORGANISMS DEPENDS ON PROTOPLASMA VOLUME ALTHOUGH MAXIMUM GROWTH RATES VARY WIDELY. MAXIMUM UPTAKE RATES PER UNIT AREA OF CELL SURFACE ARE SIMILAR. BACTERIA WITH MORE FAVORABLE SURFACE-TO-VOLUME RATIO MAY OUTGROW ALGAE EVEN THOUGH THEY SHOW LOWER AFFINITY TOWARD ORTHOPHOSPHATE. PHOSPHORUS-LIMITED DIATOMS SHOW INCREASES IN MEAN CELL VOLUME, CELL CARBON, REFRACTILITY OF CELLS, AND ALKALINE PHOSPHATASE CONTENT, WHEREAS CELL NITROGEN AND CELL PROTEIN DECREASE. THE CARBON:PHOSPHORUS ATOMIC RATIO REFLECTS AVAILABILITY OF PHOSPHORUS AND, ALTHOUGH LESS DRAMATICALLY, SO DOES THE NITROGEN:PHOSPHORUS RATIO. THESE RATIOS ARE IMPORTANT IN DETECTING NUTRIENT LIMITATION. FOR PRESERVATION OF CANADARAGO LAKE AND LAKE GEORGE, NEW YORK FEASIBILITY OF REMOVING LIMITING OR NEAR-LIMITING ELEMENTS FROM TRIBUTARY WATERS SHOULD BE CONSIDERED. KINETIC MODELS AND MEASUREMENT OF SPECIES CONSTANTS UNDER LABORATORY AND FIELD CONDITIONS ARE REQUIRED FOR PREDICTION OF SHORT-TERM CHANGES AND TURNOVER OF LAKE BIOMASS. (SEE ALSO W72-09155) (JONES-WISCONSIN)

FIELD 05C

ACCESSION NO. W72-09163

ALGAL RESPONSES TO NUTRIENT ADDITIONS IN NATURAL WATERS. I. LABORATORY ASSAYS,

ENVIRONMENTAL PROTECTION AGENCY, CORVALLIS, OREG.

T. E. MALONEY, W. E. MILLER, AND T. SHIROYAMA.

IN: NUTRIENTS AND EUTROPHICATION: THE LIMITING NUTRIENT CONTROVERSY, P 134-140, 12 FIG, 2 TAB, 8 REF. AMERICAN SOCIETY OF LIMNOLOGY AND OCEANOGRAPHY SPECIAL SYMPOSIA VOL I, ALLEN PRESS, LAWRENCE, KANSAS, 1972.

DESCRIPTORS:
*ALGAE, *NUTRIENTS, *LABORATORY TESTS, *BIOASSAY, OREGON, LAKES, NITROGEN, PHOSPHORUS, CARBON, GROWTH RATES, OLIGOTROPHY, EUTROPHICATION.

IDENTIFIERS:
*LIMITING NUTRIENTS.

ABSTRACT:
IN THE FALL OF 1970 WATERS OF NINE OREGON LAKES OF VARYING WATER QUALITY WERE STUDIED IN A SERIES OF LABORATORY ALGAL ASSAYS ON A QUARTERLY BASIS TO DETERMINE EFFECTS OF SEASONAL CHANGES AND OF WATER CHEMISTRY UPON ALGAL GROWTH AND TO EVALUATE POTENTIAL EFFECTS ON ALGAL GROWTH OF VARIOUS NUTRIENT ADDITIONS. SELANASTRUM CAPRICORNUTUM WAS USED AS THE TEST SPECIES. ADDITIONS OF NITROGEN, PHOSPHORUS, AND CARBON, SINGLY AND IN COMBINATION, WERE MADE TO THE WATERS AND ALGAL GROWTH RATES WERE DETERMINED. IN FOUR OF THE WATERS ADDITION OF PHOSPHORUS ALONE GREATLY STIMULATED ALGAL GROWTH AND IN TWO OF THE LAKE WATERS ADDITION OF NITROGEN ALONE SLIGHTLY STIMULATED ALGAL GROWTH RATES. THREE OF THE TEST WATERS WERE CAPABLE OF SUPPORTING RELATIVELY HIGH ALGAL GROWTH RATES WITHOUT NUTRIENT ADDITIONS, AND IN ONE HIGHLY OLIGOTROPHIC LAKE WATER THE ADDITION OF NITROGEN, PHOSPHORUS, AND CARBON HAD NO EFFECT ON ALGAL GROWTH RATES. IN ALL CASES ALGAL GROWTH RATES WERE DIRECTLY PROPORTIONAL TO THE AMOUNTS OF DISSOLVED PHOSPHORUS IN THE WATERS, BUT THERE WAS NO OBVIOUS CORRELATION BETWEEN ALGAL GROWTH RATES AND CONCENTRATIONS OF NITROGEN AND CARBON. (SEE ALSO W72-09155) (JONES-WISCONSIN)

FIELD 05C

ACCESSION NO. W72-09164

ALGAL RESPONSES TO NUTRIENT ADDITIONS IN NATURAL WATERS. II. FIELD EXPERIMENTS,

PACIFIC NORTHWEST WATER LAB., CORVALLIS, OREG.; AND SHAGAWA LAKE EUTROPHICATION CONTROL PROJECT, ELY, MINN.

C. F. POWERS, D. W. SCHULTS, K. W. MALUEG, R. M. BRICE, AND M. D. SCHULDT.

IN: NUTRIENTS AND EUTROPHICATION: THE LIMITING NUTRIENT CONTROVERSY, P 141-156, 9 FIG, 5 TAB, 6 REF. AMERICAN SOCIETY OF LIMNOLCGY AND OCEANOGRAPHY SPECIAL SYMPOSIA, VOL I, ALLEN PRESS, LAWRENCE, KANSAS, 1972.

DESCRIPTORS:
*EUTROPHICATION, *ON-SITE TESTS, *NUTRIENTS, ALGAE, LAKES, PHOSPHORUS, CARBON, NITROGEN.

IDENTIFIERS:
*LIMITING NUTRIENTS, SHAGAWA LAKE(MINN), WALDO LAKE(ORE).

ABSTRACT:
RESULTS OF THE IN SITU ASSAY EXPERIMENTS HAVE SOME DEFINITE IMPLICATIONS APPLICABLE TO EUTROPHICATION AND LAKE RESTORATION. WHILE EUTROPHIC SHAGAWA LAKE, MINNESOTA RECEIVES ABOUT 80% OF LCW NUTRIENT CONTENT AND EXCELLENT QUALITY WATER FROM BURNTSIDE RIVER ABOUT 70% OF THE TOTAL PHOSPHORUS AND 25% OF THE TOTAL NITROGEN ORIGINATES FROM DISCHARGES OF THE ELY MUNICIPAL WASTE TREATMENT PLANT. NEITHER PHOSPHORUS NOR NITROGEN ALONE STIMULATEC ALGAL GRCWTH IN BURNTSIDE WATER; CARBON LIKEWISE HAD NO POSITIVE EFFECT. ALL DISSOLVED MATERIALS WERE PRESENT IN EXTREMELY LOW CONCENTRATICNS IN PRISTINE WALDO LAKE, OREGON; A POSITIVE RESPONSE TO PHOSPHORUS ENRICHMENT ON THREE OF THE FOUR TESTING DATES INDICATED THAT INCREASING RATES OF PRODUCITIVITY MIGHT RAPIDLY RESULT FROM RELATIVELY SMALL ADDITICNS OF THIS OBVIOUSLY CRITICAL ELEMENT. AT NO TIME DID THIS LAKE RESPOND TC ADDITIONS OF EITHER NITROGEN OR CARBON. TESTS AT MODERATELY PRODUCTIVE TRIANGLE LAKE, OREGON AND EUTROPHIC CLINE'S PONC SHOWED VARIABLE REACTIONS TO NUTRIENT ENRICHMENT SERVING TO POINT UP THE IMPORTANCE OF NUMEROUS OBSERVATIONS OVER DIFFERENT SEASONS. FREQUENTLY MCRE THAN ONE NUTRIENT ELEMENT APPEARS TO BE LIMITING PHYTOPLANKTON AT THE SAME TIME. EVEN THOUGH POSITIVE GROWTH RESPONSE OCCASIONALLY FOLLCWED ADDITIONS OF OTHER NUTRIENTS, PHCSPHCRUS WAS THE MOST FREQUENTLY LIMITING. (SEE ALSO W72-09155) (AUEN-WISCONSIN)

FIELD 05C

ACCESSION NO. W72-09165

NUTRIENTS AND PHYTOPLANKTON IN LAKE WASHINGTON,

WASHINGTON UNIV., SEATTLE. DEPT. OF ZOOLOGY.

W. T. EDMONDSON.

IN: NUTRIENTS AND EUTROPHICATION: THE LIMITING NUTRIENT CONTROVERSY, P 172-193, 8 FIG, 3 TAB, 34 REF. AMERICAN SOCIETY OF LIMNOLOGY AND OCEANOGRAPHY SPECIAL SYMPOSIA VOL 1, ALLEN PRESS, LAWRENCE, KANSAS, 1972.

DESCRIPTORS:
*NUTRIENTS, *PHYTOPLANKTON, WATER POLLUTION EFFECTS, EUTROPHICATION, PRODUCTIVITY, ALGAE, PHOSPHATES, NITROGEN, SEWAGE, CARBON DIOXIDE, SEASONAL, PHOSPHORUS, WATER POLLUTION SOURCES, BIOCHEMICAL OXYGEN DEMAND, WATER POLLUTION CONTROL, NITRATES, EFFLUENTS, DETERGENTS, SODIUM CHLORIDE, CHEMICAL PROPERTIES, SECCHI DISKS, METHODOLOGY.

IDENTIFIERS:
*LIMITING NUTRIENTS, *LAKE WASHINGTON(WASH).

ABSTRACT:
LAKE WASHINGTON RECEIVED INCREASING AMOUNTS OF RAW SEWAGE UNTIL A DIVERSION SYSTEM TO PUGET SOUND WAS COMPLETED ABOUT 1963. INFORMATION ABOUT THE LAKE'S CHANGES PERTINENT TO THE RELATIVE IMPORTANCE OF DIFFERENT ELEMENTS IN CONTROL OF PRODUCTIVITY IN NATURAL LAKES IS SUMMARIZED. RESULTING FROM ENRICHMENT (1941 TO 1963) DUE TO EFFLUENT FROM SECONDARY SEWAGE TREATMENT PLANTS, PRODUCTION AND ABUNDANCE OF ALGAE INCREASED SEVERAL FOLD. THE WINTER PHOSPHATE CONCENTRATION INCREASED PROPORTIONALLY MUCH MORE THAN DID NITRATE OR CARBON DIOXIDE. AFTER DIVERSION, BY 1969, WINTER PHOSPHATE DECREASED TO 28% OF ITS 1963 VALUE, SUMMER CHLOROPHYLL CONCENTRATIONS DECREASED ABOUT AS MUCH, BUT NITRATE AND CARBON DIOXIDE FLUCTUATED YEAR TO YEAR AT RELATIVELY HIGH VALUES. THUS PHOSPHORUS AND PHYTOPLANKTON SHOWED THE MAJOR CHANGE, NOT NITRATE OR CARBON DIOXIDE. SIMILARLY AT SUMMER'S END, PHOSPHATE CONCENTRATION IN HYPOLIMNION CHANGED MORE THAN DID NITRATE AND CARBON DIOXIDE. THIS IS CONSISTENT WITH SECONDARY EFFLUENT, WHICH CONTAINS MORE PHOSPHORUS THAN NITROGEN OR CARBON. THE BIOLOGICAL OXYGEN DEMAND OF EFFLUENTS REACHING THE HYPOLIMNION IS SMALL RELATIVE TO OBSERVED CHANGES IN HYPOLIMNETIC OXYGEN. PHYTOPLANKTON WAS NOT APPRECIABLY CHANGED BY A SMALL SALT WATER INVASION. EFFECTIVENESS OF A GIVEN PHOSPHATE SUPPLY MAY BE AFFECTED BY NITRATE. (SEE ALSO W72-09155) (JONES-WISCONSIN)

FIELD 05C

ACCESSION NO. W72-09167

NATURAL CARBON SOURCES, RATES OF REPLENISHMENT, AND ALGAL GROWTH,

WARF INST., INC., MADISON, WIS.

S. D. MORTON, R. SERNAU, AND P. H. DERSE.

IN: NUTRIENTS AND EUTROPHICATION: THE LIMITING NUTRIENT CONTROVERSY, P 197-204, 2 FIG, 3 TAB, 15 REF. AMERICAN SOCIETY OF LIMNOLOGY AND OCEANOGRAPHY SPECIAL SYMPOSIA VOL I, ALLEN PRESS, LAWRENCE, KANSAS, 1972.

DESCRIPTORS:
*CARBON, *ALGAE, *EUTROPHICATION, *GROWTH RATES, CHLORELLA, CARBON DIOXIDE, BICARBONATES, ALGAL CONTROL, AERATION.

IDENTIFIERS:
*LIMITING NUTRIENTS, ANABAENA, CARBON SOURCES.

ABSTRACT:
IN RESEARCH ORIGINALLY UNDERTAKEN TO DETERMINE WHETHER EUTROPHICATION IN SMALL AREAS COULD BE CONTROLLED BY SWEEPING CARBON DIOXIDE OUT OF WATER BY AERATION WITH AIR CONTAINING LOW CARBON DIOXIDE AMOUNTS, THREE AREAS WERE STUDIED: THE STEADY STATE (ALGAL GROWTH RATES AT VARIOUS CONSTANT, MAINTAINED, DISSOLVED CARBON DIOXIDE CONCENTRATIONS), THE NONEQUILIBRIUM CASE (NATURAL ATMOSPHERIC REPLENISHMENT THE SOLE CARBON SOURCE), AND ALGAL GROWTH WITH INORGANIC BICARBONATE AS SOLE CARBON SOURCE. CARBON AVAILABILITY WAS STUDIED BY OBSERVATIONS OF GROWTH RATES OF CHLORELLA, MYCROCYSTIS, AND ANABAENA. ALGAE CAN EFFICIENTLY UTILIZE CARBON DIOXDE AT CONCENTRATIONS MUCH LOWER THAN THOSE PRESENT FROM ATMOSPHERIC EQUILIBRIA. IT IS VERY DIFFICULT TO CONTROL GROWTH BY CARBON DIOXIDE CONTROL IN SYSTEMS OPEN TO THE ATMOSPHERE. BICARBONATE IS A GOOD SOURCE OF CARBON AND IS AT LEAST 50% UTILIZED AT GROWTH RATES OF AT LEAST 5 MG/L PER DAY. MAY LAKES CAN HAVE MASSIVE ALGAL BLOOMS, USING NATURALLY PRESENT BICARBONATE AS SOLE CARBON SOURCE. THE ATMOSPHERE, WITHOUT ANY VIGOROUS OR CONTINUOUS WIND MIXING, IS AN ADEQUATE SOURCE OF CARBON DIOXIDE FOR DEPTHS TO AT LEAST 1.7 M, PERMITTING ALGAL GROWTH RATES UP TO 2 MG/L PER DAY. (SEE ALSO W72-09155) (JONES-WISCONSIN)

FIELD 05C

ACCESSION NO. W72-09168

DETERGENTS: NUTRIENT CONSIDERATIONS AND TOTAL ASSESSMENT,

PROCTER AND GAMBLE CO., CINCINNATI, OHIO. ENVIRONMENTAL WATER QUALITY RESEARCH DEPT.

J. R. DUTHIE.

IN: NUTRIENTS AND EUTROPHICATION: THE LIMITING NUTRIENT CONTROVERSY, P 205-216, 2 FIG, 24 REF. AMERICAN SOCIETY OF LIMNOLOGY AND OCEANOGRAPHY SPECIAL SYMPOSIA VOL I, ALLEN PRESS, LAWRENCE, KANSAS, 1972.

DESCRIPTORS:
*DETERGENTS, *PHOSPHATES, *EUTROPHICATION, NUTRIENTS, ASSESSMENTS, ALGAE, FORMULATION, WASTE TREATMENT, INDUSTRIES, WATER QUALITY, TECHNOLOGY, SURFACTANTS, TOXICITY, HAZARDS, BIODEGRADATION.

IDENTIFIERS:
NITRILOTRIACETATE.

ABSTRACT:
PROCTOR AND GAMBLE INTENDS TO TAKE PHOSPHATES OUT OF LAUNDRY DETERGENTS AS RAPIDLY AS POSSIBLE AND IN A THOROUGHLY REASONABLE MANNER. SEARCH FOR AND QUALIFICATION OF A PHOSPHATE REPLACEMENT HAS BEEN AND CONTINUES TO BE THE PRIME TECHNICAL OBJECTIVE OF THEIR INDUSTRY. THE DETERGENT INDUSTRY DOES NOT EXPECT THOSE IN THE LIMNOLOGY FIELD TO SHARE EQUALLY THE BURDEN OF FINDING ALTERNATIVES FOR PHOSPHATES, BUT THERE ARE CERTAIN THINGS THAT SCIENTISTS CAN DO INDIVIDUALLY AND AS A GROUP TO CREATE THE CLIMATE IN WHICH RESPONSIBLE AND PRODUCTIVE CHANGES CAN BE MADE. WHEN TALKING TO LAYMEN, DETERGENT PHOSPHATES CAN BE PUT INTO PROPER PERSPECTIVE MAKING IT CLEAR THAT PHOSPHATE IS AN IMPORTANT DETERGENT INGREDIENT AND, ALTHOUGH AN ALGAL NUTRIENT, IT IS NOT A POLLUTANT OR POISON. ENVIRONMENTAL ASSESSMENTS OF NTA, NOW UNDER WAY, SHOULD BE COMPLETED. NTA STILL REPRESENTS A VERY REAL HOPE AS A PHOSPHATE REPLACEMENT; IT IS THE MOST TESTED REPLACEMENT MATERIAL. IF, AS EXPECTED, FURTHER TESTING CLEARS THE TERATOLOGY QUESTIONS RAISED FROM THE PRELIMINARY TESTING OF EXAGGERATED LEVELS OF HEAVY METALS AND NTA, ITS USE IS SURE TO BE CONSIDERED. REALISTIC PLANNING OF WASTE TREATMENT PROCESSES SHOULD BE CONTINUED AS NUTRIENT REMOVAL IS CERTAIN TO BE A NECESSITY OF THE FUTURE. (SEE ALSO W72-09155) (JONES-WISCONSIN)

FIELD 05C

ACCESSION NO. W72-09169

NUTRIENT MANAGEMENT IN THE POTOMAC ESTUARY,

ENVIRONMENTAL PROTECTION AGENCY, ANNAPOLIS, MD. CHESAPEAKE TECHNICAL SUPPORT LAB.

N. A. JAWORSKI, D. W. LEAR, JR., AND O. WILLA, JR.

IN: NUTRIENTS AND EUTROPHICATION: THE LIMITING-NUTRIENT CONTROVERSY, P 246-273, 11 FIG, 17 TAB, 35 REF. AMERICAN SOCIETY OF LIMNOLOGY AND OCEANOGRAPHY SPECIAL SYMPOSIA, VOL I, ALLEN PRESS, LAWRENCE, KANSAS, 1972.

DESCRIPTORS:
*ESTUARIES, *EUTROPHICATION, *WATER POLLUTION CONTROL, MANAGEMENT, ANALYSIS, WATER QUALITY, CYANOPHYTA, NUISANCE ALGAE, COMPREHENSIVE PLANNING, AQUATIC WEEDS, WATER POLLUTION SOURCES, MATHEMATICAL MODELS, NITROGEN, ABATEMENT, COSTS, ANALYTICAL TECHNIQUES, MUNICIPAL WATER, PHOSPHORUS.

IDENTIFIERS:
*POTOMAC ESTUARY.

ABSTRACT:
WATER QUALITY STUDIES WERE UNDERTAKEN TO DEFINE WASTEWATER TREATMENT REQUIREMENTS OF UPPER POTOMAC ESTUARY SINCE 1965. STUDIES AND CONCEPTS USED TO FORMULATE A NUTRIENT MANAGEMENT PROGRAM ARE PRESENTED. CAUSES AND CONTROL NEEDS WERE STUDIED RELATIVE TO THE CHANGES IN NUTRIENT ENRICHMENT, INCLUDING APPEARANCE OF NUISANCE BLUE-GREEN ALGAE. DATA FROM ALGAL COMPOSITION ANALYSIS, ANNUAL NUTRIENT CYCLES AND LONGITUDINAL PROFILES, BIOASSAY STUDIES, ALGAL MODELING, COMPARISON WITH A NONEUTROPHIC ESTUARY, AND REVIEW OF HISTORICAL MATERIAL WERE USED TO ESTABLISH NUTRIENT CRITERIA. BASED ON A SUBJECTIVE ANALYSIS, DESIRED UPPER LIMITS OF CHLOROPHYLL A CONCENTRATIONS WERE DETERMINED FOR ESTABLISHING DEGREE OF EUTROPHICATION CONTROL REQUIRED TO MINIMIZE DETRIMENTAL EFFECTS ON WATER QUALITY AND WATER USES. ALTHOUGH AT THE PRESENT TIME NO SPECIFIC CRITERIA RELATIVE TO REQUIREMENTS FOR WASTEWATER TREATMENT HAVE BEEN ESTABLISHED FOR THE MESOHALINE PORTION OF THE ESTUARY, SPECIFIC NUTRIENT CRITERIA HAVE BEEN DEVELOPED FOR THE FRESHWATER PORTION. WITH A PROPERLY DESIGNED FACILITY THE DISSOLVED OXYGEN CONCENTRATION MAY BE ENHANCED AND ALGAL GROWTH REDUCED. THE WATER QUALITY MANAGEMENT PROGRAM BEING DEVELOPED WILL NOT ONLY IMPROVE THE WATER QUALITY TO MEET MINIMUM DESIGNATED STANDARDS, BUT WILL RENDER IT A FEASIBLE SOURCE OF MUNICIPAL WATER SUPPLY. (SEE ALSO W72-09155) (JONES-WISCONSIN)

FIELD 05C

ACCESSION NO. W72-09171

PHOSPHORUS IN WASTEWATER EFFLUENTS AND ALGAL GROWTH,

INDIAN INST. OF SCIENCE, BANGALORE. DEPT. OF BIOCHEMISTRY.

E. G. SRINATH, AND S. C. PILLAI.

JOURNAL OF WATER POLLUTION CONTROL FEDERATION, VOL 44, NO 2, P 303-308, 1972. 1 FIG, 4 TAB, 7 REF.

DESCRIPTORS:
*PHOSPHORUS, *SEWAGE EFFLUENTS, *ALGAE, WASTE WATER TREATMENT, ACTIVATED SLUDGE, LIME, AEROBIC TREATMENT, DOMESTIC WASTES.

IDENTIFIERS:
BANGALORE(INDIA).

ABSTRACT:
SAMPLES FROM THE WASTEWATER TREATMENT PLANT AT THE INDIAN INSTITUTE OF SCIENCE AND FROM THREE OUTFALLS AT BANGALORE, INDIA WERE EXAMINED OVER A PERIOD OF 12 YEARS AND EVIDENCE COLLECTED ON THE CONCENTRATION OF PHOSPHORUS IN DOMESTIC WASTEWATER, PHOSPHORUS CONTENTS IN EFFLUENTS FROM DIFFERENT WASTEWATER TREATMENT PROCESSES, EXTENT OF ALGAL GROWTH IN VARIOUS WASTEWATER EFFLUENTS, INFLUENCE OF DIFFERENT CONCENTRATIONS OF PHOSPHORUS ON THE GROWTH OF ALGAE IN WATER, AND METHODS OF REMOVAL OF PHOSPHORUS FROM WASTEWATER EFFLUENTS. PHOSPHORUS IN WASTEWATER AND EFFLUENTS WAS DETERMINED BY THE METHOD RECOMMENDED BY FISKE AND SUBBAROW AND MODIFIED BY KING. THE OTHER ITEMS OF ANALYSIS, EXCEPT TURBIDITY AND ALGAL GROWTH, WERE CARRIED OUT BY STANDARD METHODS. AMONG THE AEROBIC BIOLOGICAL METHODS OF WASTEWATER TREATMENT, THE ACTIVATED SLUDGE PROCESS REMOVES THE MAXIMUM AMOUNT OF PHOSPHORUS; THE REMAINING PHOSPHORUS IS RELATIVELY SMALL, BUT IT MUST BE REMOVED IF ALGAL GROWTH IN THE RECEIVING WATER IS TO BE CONTROLLED. IT CAN PREFERABLY BE ELIMINATED BY THE ADDITION OF CALCULATED AMOUNTS OF A MIXTURE OF ALUM AND LIME. EXPERIMENTS INDICATED THAT THE CONCENTRATION OF PHOSPHORUS IN WATER SHOULD BE LESS THAN 0.05 MGP/L FOR CONTROL OF ALGAL GROWTH. (JONES-WISCONSIN)

FIELD 05C

ACCESSION NO. W72-09176

SPECIES AND INDIVIDUAL PRODUCTIVITY IN PHYTOPLANKTON COMMUNITIES,

TEXAS UNIV., AUSTIN. DEPT. OF ZOOLOGY.

B. MAGUIRE, JR., AND W. E. NEILL.

ECOLOGY, VOL. 52, NO. 5, P 903-907, LATE SUMMER 1971. 1 FIG, 2 TAB.

DESCRIPTORS:
*ALGAE, *PHYTOPLANKTON, *PRIMARY PRODUCTIVITY, MICROSCOPY, CHLOROPHYTA, SCENEDESMUS, *CYANOPHYTA, CHLORELLA, BIOLOGICAL COMMUNITIES, DOMINANT ORGANISMS, AQUATIC PRODUCTIVITY, BIOMASS, CARBON RADIOISOTOPES, SUCCESSION, RADIOACTIVITY TECHNIQUES, CULTURES.

IDENTIFIERS:
CARBON-14, CARTERIA EUGAMETOS, STICHOCOCCUS, CHLOROCOCCUM, HYALOTHECA, LAKE LIVINGSTON, AUTORADIOGRAPHY, SAMPLE PREPARATION, SCINTILLATION COUNTING.

ABSTRACT:
THE RELATIVE PRODUCTIVITY OF INDIVIDUAL CELLS IN A MIXED PHYTOPLANKTON COMMUNITY CAN BE DETERMINED BY C-14 AUTORADIOGRAPHY. IN GENERAL, THE TECHNIQUE INVOLVES PLACING C-14-LABELLED ALGAL CELLS ON MICROSCOPE SLIDES AND COVERING THEM WITH A RADIOSENSITIVE EMULSION. AFTER SUITABLE EXPOSURE AND PHOTOGRAPHIC DEVELOPMENT, THE CELLS AND THE SILVER GRAINS PRODUCED BY THE RADIOACTIVE DECAY OF INCORPORATED C-14 ARE CLEARLY VISIBLE. THE PROPORTION OF GRAINS PRODUCED BY EACH SPECIES IS THEN USED TO PARTITION COMMUNITY-WIDE MEASUREMENTS OF PRIMARY PRODUCTION. ALGAL SAMPLES FROM ARTIFICIAL LAKE WATER IN THE LABORATORY AND LAKE LIVINGSTON WERE STUDIED. LABORATORY DATA SHOWED CARTERIA, BECAUSE OF ITS SIZE (685 CUBIC MICRONS), TO BE MORE PRODUCTIVE PER CELL THAN CHLORELLA (41 CUBIC MICRONS) OR STICHOCOCCUS (27 CUBIC MICRONS) DURING BOTH THE 1-HR AND 24 HR PERIODS TESTED. HOWEVER, IN TERMS OF PRODUCTIVITY PER UNIT VOLUME PER 24 HRS, CHLORELLA WILL BECOME MOST DOMINANT, STICHOCOCCUS SUBDOMINANT, AND CARTERIA PROGRESSIVELY LESS IMPORTANT IN NUMBER AND ECOLOGICAL EFFECT. LAKE LIVINGSTON DATA SHOWED CHLORELLA AND BLUE-GREEN ALGAE PREDOMINATING NUMERICALLY BUT WITH CHLORELLA HAVING THE GREATEST BIOVOLUME. ALTHOUGH CHLORELLA PRODUCED FAR MORE PHOTOSYNTHATE THAN ALL OTHER SPECIES COMBINED, ITS PRODUCTIVITY PER UNIT VOLUME WAS CONSIDERABLY LOWER THAN THAT OF SCENEDESMUS, HYALOTHECA, AND CHLOROCOCCUM. CHLORELLA COULD THUS BE REPLACED FROM THE DOMINANT POSITION AND UNICELLULAR BLUE-GREEN ALGAE ELIMINATED AS A SIGNIFICANT PART OF THE COMMUNITY. (MACKAN-BATTELLE)

FIELD 05C

ACCESSION NO. W72-09239

SOME QUANTITATIVE FEEDING PATTERNS EXHIBITED BY THE COPEPOD CALANUS,

AKADEMIYA NAUK SSSR, KALININGRAD. INSTITUT OKEANOLOGII.

YE. G. PERUYEVA.

OKEANOLOGIYA, VOL. 11, NO. 2, P 232-239, 1971. 5 FIG, 1 TAB, 29 REF.

DESCRIPTORS:
*COPEPODS, CRUSTACEANS, ZOOPLANKTON, ALGAE, FEEDING RATES, *LARVAL GROWTH STAGE, PHYTOPLANKTON, NUTRIENT REQUIREMENTS, INVERTEBRATES, CHLAMYDOMONAS, MATHEMATICAL STUDIES, BIOLOGICAL COMMUNITIES, FOOD CHAIN, DIATOMS, LARVAE, DIETS.

IDENTIFIERS:
*FEEDING PATTERNS, *CALANUS HELGOLANDICUS, COSCINODISCUS, COPEPODIDS, *CALANUS GLACIALIS, DITYLUM, PROROCENTRUM, NITZSCHIA, LAUDERIA, CHAETOCEROS.

ABSTRACT:
A STUDY WAS MADE OF THE FEEDING OF TWO WIDESPREAD SPECIES-CALANUS GLACIALIS AND C HELGOLANDICUS - ON THE LARGE DIATOM COSCINODISCUS SP. THE DIFFERENCES NOTED IN THE SHAPING OF THE DIET OF THE TWO SPECIES - ASYMPTOTIC GROWTH IN STAGE V C. GLACIALIS AND LINEAR GROWTH IN FEMALE C. HELGOLANDICUS IN RELATION TO THE TIME OF FEEDING - ARE CAUSED BY THE DIFFERENT FOOD REQUIREMENTS IN THE VARIOUS STAGES OF CALANUS DEVELOPMENT. SATIATION CONDITIONS (STAGE V C. GLACIALIS) ARE CHARACTERIZED BY THE CONSTANT RATE OF MOVEMENT OF THE CRUSTACEANS IN THE RANGE OF CONCENTRATIONS USED AND BY INCREASED SELECTIVITY WITH INCREASING FOOD CONCENTRATION. WHEN C. HELGOLANDICUS FEMALES FEED IN PLACES OF INSUFFICIENT CONCENTRATIONS, THEIR SPEED OF MOVEMENT CHANGES AT DIFFERENT FOOD CONCENTRATIONS AND SELECTIVITY BECOMES INDEPENDENT OF THE CONCENTRATION. (MACKAN-BATTELLE)

FIELD 05C

ACCESSION NO. W72-09248

ALGAE CONTROL BY MIXING, STAFF REPORT ON KEZAR LAKE IN SUTTON, N.H.

NEW HAMPSHIRE WATER SUPPLY AND POLLUTION CONTROL COMMISSION, CONCORD.

REPORT TO THE NEW ENGLAND REGIONAL COMMISSION, APRIL 1971. 103 P, 28 FIG, 30 TAB, 76 REF.

DESCRIPTORS:
*ALGAL CONTROL, *EUTROPHICATION, *DESTRATIFICATION, *WATER QUALITY CONTROL, *LAKE MORPHOLOGY, AERATION, MIXING, ALGAL BLOOMS, ALGAL TOXINS, WATER CIRCULATION, WATER POLLUTION CONTROL, FISHKILL, NEW HAMPSHIRE, WATER POLLUTION EFFECTS, COPPER SULFATE.

IDENTIFIERS:
*KEZAR LAKE, AIR INJECTION, COMPRESSED AIR.

ABSTRACT:
THE CONTROL OF ALGAE IN KEZAR LAKE BY DESTRATIFICATION WAS DEMONSTRATED IN FIELD EXPERIMENTAL WORK CONDUCTED IN 1968 AND 1969. KEZAR LAKE HAS SUFFERED FROM INCREASINGLY VIRULENT BLOOMS OF ALGAE WHICH APPARENTLY HAVE BEEN INDUCED BY NUTRIENTS CONTAINED IN WASTEWATER REACHING THE LAKE. A GROWING NUMBER OF NEW ENGLAND LAKES SHARES THIS PROBLEM, WHICH THREATENS ONE OF THE REGION'S MAJOR RECREATIONAL RESOURCES. BEYOND THE EUTROPHICATION OF THE LAKES THEMSELVES, THE RESULTS CAN BE MEASURED IN DECLINING PROPERTY VALUES AND DECREASES IN RECREATION BUSINESS WHICH ARE DIRECTLY ATTRIBUTABLE TO THE UNAESTHETIC APPEARANCE OF AN ALGAE-CHOKED LAKE. THE NEW HAMPSHIRE WATER SUPPLY AND POLLUTION CONTROL COMMISSION APPEARS TO HAVE DEVELOPED AN EFFECTIVE AND INEXPENSIVE METHOD OF BRINGING SUCH ALGAL BLOOMS UNDER CONTROL BY MIXING THE WATER IN THE LAKE AND ALLOWING NATURAL PROCESSES TO TAKE OVER. HOMOGENIZING, MIXING OR DESTRATIFYING KEZAR LAKE WAS ACCOMPLISHED BY FORCING COMPRESSED AIR, FROM SHORE-LOCATED COMPRESSORS, THROUGH P.V.C., 2-INCH-DIAMETER PIPING TO THE DEEPEST PORTION OF THE LAKE WHERE IT WAS RELEASED THROUGH CERAMIC DIFFUSERS, AND ALLOWED TO BUBBLE UP TO THE LAKE SURFACE THROUGH THE WATER COLUMN. CLARITY OF THE WATER WAS VISIBLY AND MEASUREABLY IMPROVED; THE POPULATIONS OF NOXIOUS ALGAE, SO OBJECTIONABLE TO RECREATION INTERESTS, WERE DECREASED IN NUMBER; AND NO HARMFUL EFFECTS WERE DETECTED DURING 1968 AND 1969 SUMMER MIXING OF KEZAR LAKE. OPERATING PRESSURES FOR THE COMPRESSORS ARE LOW. THE EQUIPMENT IS CONVENIENT FOR NECESSARY MAINTENANCE, AND THE OPERATING AND STUDY BUDGET IS MODEST. (POERTNER)

FIELD 05G

ACCESSION NO. W72-09304

GAS CHROMATOGRAPHIC ANALYSIS OF THE HIGHER FATTY ACIDS OF THE ALGA CHLORELLA VULGARIS (PYRENOIDOSA),

INSTITUTE FOR RESEARCH, PRODUCTION AND USES OF RADIOISOTOPES, PRAGUE (CZECHOSLOVAKIA).

M. MATUCHA, L. ZILKA, AND K. SVIHEL.

JOURNAL OF CHROMATOGRAPHY, VOL. 65, NO. 2, P 371-376, FEBRUARY 23, 1972. 5 FIG, 1 TAB, 20 REF.

DESCRIPTORS:
CHEMICAL ANALYSIS, *ALGAE, COLUMNS, ORGANIC ACIDS, SEPARATION TECHNIQUES, *CHROMATOGRAPHY.

IDENTIFIERS:
*FATTY ACIDS, *GAS LIQUID CHROMATOGRAPHY, METHYL ESTERS, ESTERS, CHLORELLA VULGARIS, SAMPLE PREPARATION.

ABSTRACT:
LABELED (C-14) FATTY ACIDS WERE PREPARED FROM RADIOACTIVE CHLORELLA VULGARIS FOR GAS-LIQUID CHROMATOGRAPHY BY THE FOLLOWING PROCEDURE. LIPIDS WERE EXTRACTED FROM EXTRACTED AND WASHED CELLS WITH 96 PERCENT ETHANOL AND AN ETHANOL-DIETHYL ETHER MIXTURE; REMAINING SUGARS WERE REMOVED FROM THE LIPID EXTRACT BY WATER EXTRACTION. THE DRIED LIPIDS WERE THEN TRANSESTERIFIED WITH METHANOLIC HYDROCHLORIC ACID. AFTER EXTRACTION OF THE METHYLESTERS BY PETROLEUM ETHER, THEY WERE PURIFIED BY VACUUM SUBLIMATION. TO IDENTIFY CHLORELLA FATTY ACIDS WITH A RELATIVE ABUNDANCE HIGHER THAN 0.5 PERCENT, A POLAR POLYESTER DIETHYLENE GLYCOL SUCCINATE (DEGS) AND A NON-POLAR GREASE, APIEZON-L, WERE CHOSEN AS STATIONARY PHASES. G-L CHROMATOGRAPHY WAS CARRIED OUT ON A COLUMN 2 M LONG AND 4MM I.D. FILLED WITH 15 PERCENT DEGS WITH AN ARGON-IONIZATION DETECTOR AND ON A COLUMN 1.5 M LONG AND 3 MM I.D. FILLED WITH 15 PERCENT APIEZON WITH A FLAME-IONIZATION DETECTOR. A 2 ML GAS PROPORTIONAL COUNTER GAVE A SIMULTANEOUS DETERMINATION OF RADIOACTIVITY. IN ADDITION TO THE MAJOR ACIDS (PALMITIC, OLEIC, LINOLEIC, AND LINOLENIC), EVEN-NUMBERED, STRAIGHT-CHAIN FATTY ACIDS AND ODD-NUMBERED, SATURATED FATTY ACIDS WERE FOUND. (MACKAN-BATTELLE)

FIELD 05A, 05C

ACCESSION NO. W72-09365

CONTROLLED EUTROPHICATION-INCREASING FOOD PRODUCTION FROM THE SEA BY RECYCLING HUMAN WASTES,

WOODS HOLE OCEANOGRAPHIC INSTITUTION, MASS.

J. H. RYTHER, W. M. DUNSTAN, K. R. TENORE, AND J. E. HUGUENIN.

BIOSCIENCE, VOL. 22, NO. 3, P 144-152, MARCH 1972. 4 FIG, 11 REF.

DESCRIPTORS:
*EUTROPHICATION, *RECYCLING, WATER QUALITY CONTROL, SEWAGE TREATMENT, SEWAGE EFFLUENTS, SANITARY ENGINEERING, SECONDARY TREATMENT, MARINE ALGAE, PHYTOPLANKTON, BIOMASS, GROWTH RATES, PRODUCTIVITY, ANIMAL GROWTH, PLANT GROWTH, OYSTERS, SHELLFISH, MUSSELS, SHELLFISH FARMING, AQUACULTURE, COMMERCIAL SHELLFISH, BIOLOGICAL COMMUNITIES, FOODCHAINS, HABITATS, NICHES, TURNOVERS, CULTURES, WASTE WATER TREATMENT, *CYCLING NUTRIENTS, NITROGEN, CARBON DIOXIDE, CHLORELLA, NUTRIENTS, TERTIARY TREATMENT, FLOW RATES, WATER POLLUTION EFFECTS, PRIMARY PRODUCTIVITY.

IDENTIFIERS:
GROWTH KINETICS, BIOENERGETICS, ENRICHMENT, CULTURE MEDIA, MACROINVERTEBRATES.

ABSTRACT:
THE ESSENTIAL FEATURE OF 'CONTROLLED EUTROPHICATION' IS THE PHYSICAL SEPARATION AND COMPARTMENTALIZATION OF THE PRODUCER AND CONSUMER LEVELS OF A BIOLOGICAL COMMUNITY. FOLLOWING THESE GUIDELINES, LABORATORY EXPERIMENTS WERE BEGUN IN THE SUMMER OF 1970 ON THE GROWTH KINETICS OF MARINE PLANKTON ALGAE GROWN IN SEAWATER ENRICHED WITH EFFLUENT FROM A SECONDARY SEWAGE TREATMENT PLANT. IN GENERAL, DILUTED SEWAGE WAS FOUND TO BE AN EXCELLENT CULTURE MEDIUM FOR THE MARINE PHYTOPLANKTON. AT CONCENTRATIONS OF 10 PERCENT SEWAGE, THE YIELD OF ALGAE INCREASED WITH FLOW RATE THROUGH THE SYSTEM UP TO A 'TURNOVER RATE' OF 50 PERCENT OF THE CULTURE PER DAY. THESE ALGAE WERE SUBSEQUENTLY FED TO MONITORED OYSTER CULTURES, THUS COMPLETING THE PRODUCER-CONSUMER FOOD CHAIN. IN ONE SUCH EXPERIMENT, A NATURAL POPULATION OF DIATOMS GROWN ON 10 PERCENT SEWAGE WAS PASSED THROUGH A 3M X 1.5M X .5 IN. TANK CONTAINING SUSPENDED STRINGS OF OYSTER SPAT ATTACHED TO SCALLOP SHELLS. OVER 30 DAYS, AT 7-10C, THE OYSTERS REMOVED 77 PERCENT OF THE ALGAE AND CONVERTED 22 PERCENT OF THE CELLS INTO NEW OYSTER FLESH. THESE AND SIMILAR EXPERIMENTS HAVE PROVIDED BASIC DATA ON THE KINETICS AND BIOENERGETICS OF A SMALL SCALE 'CONTROLLED EUTROPHICATION PROGRAM' AND REVEALED THE VALUE FOR DEVELOPING APPLICATIONS IN ADVANCED SEWAGE TREATMENT AND COMMERCIAL AQUACULTURE. (MACKAN-BATTELLE)

FIELD 05G, 05C

ACCESSION NO. W72-09378

WASTE TREATMENT LAGOONS-STATE OF THE ART.

MISSOURI BASIN ENGINEERING HEALTH COUNCIL, CHEYENNE, WYO.

COPY AVAILABLE FROM GPO SUP DOC, $1.25; MICROFICHE FROM NTIS AS PB-209 937, $0.95, ENVIRONMENTAL PROTECTION AGENCY WATER POLLUTION CONTROL RESEARCH SERIES, JULY 1971. 152 P, 9 TAB, 42 REF. EPA PROGRAM 17090 EHX 07/71.

DESCRIPTORS:
*SEWAGE LAGOONS, *AERATED LAGOONS, *OXIDATION LAGOONS, *DESIGN CRITERIA, BIODEGRADATION, ALGAE, PHOTOSYNTHESIS, SUSPENDED SOLIDS, SEPARATION TECHNIQUES, FILTRATION, MUNICIPAL WASTES, INDUSTRIAL WASTES, OPERATION AND MAINTENANCE, *WASTE WATER TREATMENT, WATER QUALITY CONTROL, *REVIEWS.

IDENTIFIERS:
*FACULTATIVE LAGOONS, *ANAEROBIC LAGOONS.

ABSTRACT:
A REVIEW OF PUBLISHED LITERATURE AND FIELD EVALUATIONS REVEALED THE PRESENCE OF OVER 3500 WASTE TREATMENT LAGOONS CURRENTLY IN OPERATION IN THE UNITED STATES. THE THREE TYPES OF LAGOONS IN USE INCLUDE: (1) OXIDATION LAGOONS; (2) AERATED LAGOONS, AND (3) ANAEROBIC LAGOONS. OXIDATION LAGOONS DEPEND UPON ALGAE TO SUPPLY OXYGEN BY PHOTOSYNTHESIS AND DEGRADE THE WASTE PRODUCTS. EFFLUENT QUALITY IS DETERMINED BY THE QUANTITY OF ALGAE IN THE EFFLUENT AND SEVERAL METHODS OF ALGAE REMOVAL ARE CURRENTLY UNDER INVESTIGATION. AERATED LAGOONS MAY BE MERELY OXIDATION PONDS WITH SUPPLEMENTAL AERATION, PARTIALLY MIXED ACTIVATED SLUDGE (FACULTATIVE AERATED) OR COMPLETE MIX ACTIVATED SLUDGE (CMAS) SYSTEMS. HIGH QUALITY EFFLUENTS FROM AERATED LAGOONS CAN BE ACHIEVED ONLY BE REMOVING EFFLUENT MICROBIAL SOLIDS. ANAEROBIC LAGOONS CAN PROVIDE UP TO 80% BOD REMOVALS, BUT MUST BE FOLLOWED BY SOME TYPE OF AEROBIC TREATMENT TO PRODUCE A HIGH QUALITY EFFLUENT. THIS REVIEW HAS DEMONSTRATED THAT LAGOONS DO HAVE APPLICABILITY TO THE TOTAL WASTE TREATMENT PROBLEM, BUT THE FUTURE OF LAGOONS DEPENDS UPON PROPER DESIGN AND OPERATION IN RELATIONSHIP WITH THE FUNDAMENTAL BIOCHEMISTRY OF THE MICROBES IN THE VARIOUS SYSTEMS. (LOWRY-TEXAS)

FIELD 05D, 10F

ACCESSION NO. W72-09386

FILTRATION OF WATER AND WASTEWATER,

UNIVERSITY COLL., LONDON (ENGLAND).

K. J. IVES.

CRC CRITICAL REVIEWS IN ENVIRONMENTAL CONTROL, VOL 2, NO. 2, AUGUST 1971. P 293-335, 8 FIG, 9 TAB, 99 REF.

DESCRIPTORS:
*FILTRATION, *OPERATION AND MAINTENANCE, *COST ANALYSIS, *FILTERS, HEADLOSS, PRESSURE, TURBIDITY, ALGAE, FLOCCULATION, CLEANING, *WATER TREATMENT, TERTIARY TREATMENT, *WASTE WATER TREATMENT.

IDENTIFIERS:
*MICROSTRAINING, *PRECOAT FILTRATION, BACKWASHING, SPECIFIC RESISTANCE.

ABSTRACT:
A REVIEW OF THE RECENT ADVANCEMENTS IN FILTRATION TECHNOLOGY WAS CONDUCTED, DEALING PRIMARILY WITH SLOW SAND FILTERS, RAPID SAND FILTERS, PRECOAT FILTERS, AND MICROSTRAINERS. APPLICATIONS, MODES OF OPERATION, MAINTENANCE REQUIRED, AND COST FIGURES WERE ASSEMBLED FOR EACH TYPE OF FILTER. SOME GENERAL RULES FOR FILTRATION WHICH WERE SELECTED FROM THE TECHNICAL LITERATURE INCLUDE: (1) HEAVY TURBIDITY LOADS CANNOT BE SUSTAINED BY EITHER SLOW SAND OR PRECOAT FILTERS; (2) ALUM FLOC WILL CLOG SLOW SAND FILTERS WHEREAS MICROSTRAINERS WILL NOT RETAIN IT; (3) SLOW SAND AND PRECOAT FILTERS RETAIN FINE TURBIDITY, RAPID SAND FILTERS RETAIN IT IF IT IS PREVIOUSLY FLOCCULATED, AND MICROSTRAINERS CANNOT RETAIN IT; (4) ALL 4 FILTERS RETAIN PLANKTON, ALTHOUGH SMALLER ALGAE ARE NOT RETAINED BY MICROSTRAINERS AND RAPID FILTERS; (5) ONLY MICROSTRAINERS AND RADIAL FLOW RAPID FILTERS OPERATE CONTINUOUSLY; (6) MICROSTRAINER HEADLOSS IS 0.15 M COMPARED TO 1 TO 20 M FOR THE OTHER FILTER TYPES; AND (7) RAPID FILTERS HAVE THE WIDEST APPLICABILITY TO BOTH WATER AND WASTEWATER TREATMENT. COSTS PRESENTED WERE NOT FOR OPTIMUM FILTERS, SINCE TREATMENT SYSTEMS, NOT TREATMENT COMPONENTS, ARE OPTIMIZED. (LOWRY-TEXAS)

FIELD 05D, 05F, 10F

ACCESSION NO. W72-09393

ALGAL INFLUENCES ON DIEOFF RATES OF INDICATOR BACTERIA,

TEXAS UNIV., AUSTIN. COLL. OF ENGINEERING.

E. M. DAVIS, AND E. F. GLOYNA.

PROCEEDINGS, INDUSTRIAL WASTE CONFERENCE, 25TH, MAY 5-7, 1970, PURDUE UNIVERSITY ENGINEERING EXTENSION SERIES NO. 137, P 266-273, 1 FIG, 4 TAB, 17 REF.

DESCRIPTORS:
*ENTERIC BACTERIA, *WATER QUALITY, *DOMESTIC WASTES, ALGAE, INDUSTRIAL WASTES, METABOLISM, GROWTH RATES, OXIDATION LAGOONS, COLIFORMS, NUTRIENTS, ANALYTICAL TECHNIQUES, MICROBIOLOGY, *WASTE WATER TREATMENT.

IDENTIFIERS:
*AFTERGROWTH.

ABSTRACT:
DESTRUCTION OF COLIFORM BACTERIA IN WASTE TREATMENT FACILITIES HAS BEEN ATTRIBUTED TO A NUMBER OF PHYSICAL, BIOLOGICAL, AND CHEMICAL FACTORS. HOWEVER, SEVERAL SPECIES OF ALGAE APPEAR TO BE CAPABLE OF STIMULATING SPECIFIC COLIFORM ORGANISMS, THROUGH ALGAL METABOLIC EXUDATES WHICH ARE RELEASED DURING LOG GROWTH CONDITIONS AND FROM NUTRIENTS WHICH MAY BE RELEASED AFTER DEATH OF THE ALGAE. INVESTIGATIONS OF BEHAVIORAL PATTERNS OF SELECTED SPECIES OF ENTERIC BACTERIA WHEN IN THE PRESENCE OF SINGLE AXENIC CULTURES AND MIXED CULTURES OF SELECTED BLUE-GREEN AND GREEN ALGAE, AND INVESTIGATIONS OF THE BEHAVIOR OF ENTERIC BACTERIA IN THE PRESENCE OF ALGAE AND KNOWN QUANTITIES OF INDUSTRIAL WASTE WERE CONDUCTED. RESULTS INDICATED THAT ALCALIGENES FAECALIS, ENTEROBACTER AEROGENES, AND PROTEUS VULGARIS EXHIBITED ONLY LIMITED TENDENCIES FOR AFTERGROWTH, WHILE ESCHERICHIA COLI, PSEUDOMONAS, AND SERRATIA MARCESCENS SHOWED DISTINCT AFTERGROWTH CAPABILITIES IN ALL TEST ENVIRONMENTS USED. THESE INVESTIGATIONS DEMONSTRATED THAT THE ORGANISMS WHICH ARE USED AS POLLUTION INDICATORS MAY APPEAR IN NUMBERS SEEMING TO INDICATE DOMESTIC WASTEWATER POLLUTION WHEN NONE IS PRESENT. (LOWRY-TEXAS)

FIELD 05D

ACCESSION NO. W72-09590

PROBLEMS IN WATER TOXICOLOGY (VOPROSY VODNOY TOKSIKOLOGII).

'NAUKA', MMOSCOW, 1970. 224 P.

DESCRIPTORS:
*HYDROBIOLOGY, *PATHOLOGY, *TOXICITY, *WATER QUALITY CONTROL, *WATER POLLUTION EFFECTS, WATER POLLUTION SOURCES, AQUATIC LIFE, AQUATIC ANIMALS, AQUATIC MICROORGANISMS, AQUATIC PLANTS, AQUATIC ALGAE, PESTICIDES, INSECTICIDES, HERBICIDES, PHOSPHORUS, PHENOLS, METALS, POLLUTANTS, INDUSTRIAL WASTES, CONFERENCES.

IDENTIFIERS:
*USSR, *TOXICOLOGY, *TOXICANTS, *HYDROBIONTS, ICHTHYOLOGY, ICHTHYOFAUNA, MACROPHYTES, SAPROPHYTES, CERIUM.

ABSTRACT:
THIS COLLECTION OF 37 PAPERS CONTAINS REPORTS PRESENTED AT THE ALL-UNION CONFERENCE ON PROBLEMS IN WATER TOXICOLOGY, HELD IN MOSCOW JANUARY 30 TO FEBRUARY 2, 1968. THE BOOK IS DIVIDED INTO 5 SECTIONS: (1) GENERAL PROBLEMS IN WATER TOXICOLOGY (10 PAPERS); (2) EFFECTS OF PESTICIDES ON HYDROBIONTS (8 PAPERS); (3) EFFECTS OF PHOSPHORUS AND PHOSPHORUS COMPOUNDS ON HYDROBIONTS (6 PAPERS); (4) EFFECTS OF PHENOLS ON HYDROBIONTS (7 PAPERS); AND (5) EFFECTS OF METALS AND METALLIC COMPOUNDS ON HYDROBIONTS (6 PAPERS). SAMPLES COLLECTED FOR ANALYSIS INCLUDED BACTERIA, ALGAE, AND A VARIETY OF INVERTEBRATES AND FISHES. CHANGES IN BIOLOGICAL PROCESSES IN SEA WATER AND FRESHWATER UNDER INFLUENCE OF DIFFERENT POLLUTANTS AND INDUSTRIAL WASTES WERE EXAMINED TOGETHER WITH METHODS FOR FUTURE INVESTIGATION OF THE TOXICITY OF THE AQUATIC ENVIRONMENT. (SEE W72-09649 AND W72-09650) (JOSEFSON-USGS)

FIELD 05C, 02H, 05G

ACCESSION NO. W72-09646

DETERMINATION OF THE TOXICITY OF CONTAMINATED FRESHWATER WITH RESPECT TO CERTAIN HYDROBIONTS (OPREDELENIYE TOKSICHNOSTI ZAGRYAZNENNYKH PRESNYKH VOD V OTNOSHENII NEKOTORYKH GIDROBIONTOV),

MOSCOW STATE UNIV. (USSR).

G. D. LEBEDEVA.

IN: VOPROSY VODNOY TOKSIKOLOGII; 'NAUKA', MOSCOW, P 57-61, 1970. 2 FIG, 10 REF.

DESCRIPTORS:
*HYDROBIOLOGY, *TOXICITY, *WATER POLLUTION, *FRESHWATER, *AQUATIC LIFE, AQUATIC ANIMALS, CARP, DAPHNIA, AQUATIC PLANTS, AQUATIC ALGAE, SCENEDESMUS, RADIOISOTOPES, STRONTIUM RADIOISOTOPES, PHOSPHORUS RADIOISOTOPES, CESIUM, RADIOACTIVITY EFFECTS, LABORATORY TESTS.

IDENTIFIERS:
*USSR, *HYDROBIONTS, *CERIUM, RADIOMETRY.

ABSTRACT:
THE EFFECTS OF RADIOACTIVE ELEMENTS ON BIOLOGICAL FUNCTIONS OF HYDROBIONTS WERE BASED ON LABORATORY ANALYSES OF FISH (CYPRINUS CARPIO), CRUSTACEANS (DAPHNIA MAGNA), HIGHER AQUATIC PLANTS (LEMNA MINOR AND ELODEA CANADENSIS), AND GREEN ALGAE (SCENEDESMUS QUADRICAUDA). THE RADIOACTIVE ISOTOPES USED WERE SR-90, SR-89, CS-137, CE-144, AND P-32. PLANT ORGANISMS, WHICH CONCENTRATE LARGE AMOUNTS OF RADIOACTIVE STRONTIUM, ARE LESS SENSITIVE TO WATER CONTAMINATION BY THIS ELEMENT. OF THE ANIMAL ORGANISMS EXAMINED, DAPHNIA ARE THE MOST SENSITIVE TO WATER CONTAMINATION BY RADIOSTRONTIUM. THE TOXICITY CRITERIA FOR FISH WERE THE GROWTH RATE AND AMOUNT OF EDIBLE FOOD; FOR DAPHNIA--THE SURVIVAL RATE, REPRODUCTION AND FERTILITY; AND FOR AQUATIC PLANTS--THE POPULATION AND INCREASE IN THE BIOMASS. (SEE ALSO W72-09646) (JOSEFSON-USGS)

FIELD 05C, 02H

ACCESSION NO. W72-09649

PROBLEM OF PESTICIDES IN WATER TOXICOLOGY (PROBLEMA PESTITSIDOV V VODNOY TOKSIKOLOGII),

AKADEMIYA NAUK URSR, KIEV. INSTYTUT HIDROBIOLOGII.

L. P. BRAGINSKIY.

IN: VOPROSY VODNOY TOKSIKOLOGII; 'NAUKA', MOSCOW, P 81-88, 1970. 2 FIG, 3 TAB, 4 REF.

DESCRIPTORS:
*HYDROBIOLOGY, *PESTICIDES, *PESTICIDE RESIDUES, *PESTICIDE TOXICITY, *AQUATIC PLANTS, AQUATIC ALGAE, PHYTOPLANKTON, AQUATIC MICROORGANISMS, AQUATIC ANIMALS, DAPHNIA, APPLICATION METHODS, GRANULES, PLANT GROWTH REGULATORS, INHIBITORS, HERBICIDES, MONURON, PHOTOSYNTHESIS.

IDENTIFIERS:
*USSR, *TOXICOLOGY, *HYDROBIONTS, MACROPHYTES, SAPROPHYTES.

ABSTRACT:
PESTICIDES IN AN AQUATIC ENVIRONMENT AFFECT HYDROBIOLOGICAL PROCESSES BY (1) ALTERING THE HABITATS OF HYDROBIONTS BY INHIBITING PLANT PHOTOSYNTHESIS (OXYGEN DEPLETION) AND BY STIMULATING SAPROPHYTE GROWTH; (2) ACUTE POISONING OF HYDROBIONTS BY TOXIC CONCENTRATIONS OF PESTICIDES; AND (3) PRODUCING A VARIETY OF ACUTE AND CHRONIC EFFECTS ON HYDROBIONTS BY THE PRESENCE OF PESTICIDE RESIDUES ACCUMULATED THROUGH FOOD CHAINS. PROSPECTS FOR THE USE OF PESTICIDES TO REGULATE ALGAL AND MACROPHYTE GROWTH ARE BASED PRIMARILY ON THE APPLICATION OF GRANULES TO PLANT PROTECTION. METHODS ARE GIVEN FOR STUDYING THE EFFECTS OF PESTICIDES ON WATER BODIES AND ON THE BIOLOGICAL PROCESSES OCCURRING IN THEM. (SEE ALSO W72-09646) (JOSEFSON-USGS)

FIELD 05C, 05G

ACCESSION NO. W72-09650

POSSIBLE EFFECTS OF ORGANOCHLORINE PESTICIDES ON PRIMARY PRODUCTIVITY AND SKELETOGENESIS OF NEW ENGLAND ESTUARINE AND COASTAL MARINE ALGAE,

MASSACHUSETTS UNIV., AMHERST. WATER RESOURCES RESEARCH CENTER.

J. P. SEARS, AND C. YENTSCH.

AVAILABLE FROM THE NATIONAL TECHNICAL INFORMATION SERVICE AS PB-210 146, $3.00 IN PAPER COPY, $0.95 IN MICROFICHE. PROJECT COMPLETION REPORT, FEBRUARY 1972. 13 P, 6 FIG, 2 REF. OWRR A-035 MASS.(1).

DESCRIPTORS:
*ANALYTICAL TECHNIQUES, RATES, *PHOTOSYNTHESIS, *MARINE ALGAE, DDT, *PESTICIDES, *PRIMARY PRODUCTIVITY, WATER POLLUTION EFFECTS, ESTUARINE ENVIRONMENT.

IDENTIFIERS:
RHODYMENIA, FUCUS, ULVA.

ABSTRACT:
TECHNIQUES HAVE BEEN DEVELOPED TO ESTIMATE RATES OF PHOTOSYNTHESIS IN MACROSCOPIC MARINE ALGAE BY MONITORING OXYGEN PRODUCTION. ATTEMPTS TO USE 14C TECHNIQUES WERE UNSECCESSFUL BECAUSE OF THE DIFFICULTY OF RECOVERING ASSIMILATED 14C BY MACERATION, ACID HYDROLOSIS AND STRONG BASE SOLUBILIZATION OF THALLI. THE PROBLEM OF 14C RECOVERY WOULD BE SOLVED IF THE PLANT MATERIAL COULD BE COMPLETELY COMBUSTED AND TOTAL CARBON RECOVERED IN ETHANOLAMINE. BASE LINE DATA ON THE INFLUENCE OF TEMPERATURE AND LIGHT INTENSITY ON RATES OF PHOTOSYNTHESIS WERE OBTAINED FOR RHODYMENIA, FUCUS AND ULVA. PHOTOSYNTHESIS BY MACROSCOPIC ALGAE WAS STRONGLY INFLUENCED BY WATER MOVEMENT. UNDER CONDITIONS OF STANDING WATER, OXYGEN PRODUCTION DECREASED TO A FRACTION OF OXYGEN PRODUCTION IN A STIRRED SYSTEM. THE REDUCED RATE OF PHOTOSYNTHESIS WAS PROBABLY DUE TO DEPLETION OF CO2 AT THE THALLUS SURFACE, AND A SUBSEQUENT RISE IN PH. FUTURE ESTIMATES OF PHOTOSYNTHESIS IN MACROALGAE SHOULD INCLUDE SOME MECHANISM FOR WATER AGITATION, ESPECIALLY DURING ESTIMATES OF PRODUCTIVITY. THE EFFECT OF DDT ON PHOTOSYNTHESIS OF RHODYMENIA, FUCUS AND ULVA COULD NOT BE CONSISTENTLY DEMONSTRATED. IF ANY EFFECT DID OCCUR, ITS MAGNITUDE WAS BELOW THE LEVEL OF DETECTION BECAUSE OF THE PHOTOSYNTHETIC VARIABILITY AMONG PLANTS. THESE FINDINGS ARE IN CONTRAST TO THE PRONOUNCED EFFECTS OF DDT ON PHYTOPLANKTON PHOTOSYNTHESIS REPORTED BY MENZEL, 1970; AND WURSTER, 1969.

FIELD 05C, 02L

ACCESSION NO. W72-09653

THE RELATIONSHIPS BETWEEN 32P ACCUMULATICN IN ALGAE, BACTERIA AND TUBIFICIDS,

WESTERN MICHIGAN UNIV., KALAMAZOO, DEPT. OF BIOLOGY.

W. L. STROMBERG, AND C. J. GOODNIGHT.

AVAILABLE FROM THE NATIONAL TECHNICAL INFORMATION SERVICE AS COO-1803-6, $3.00 IN PAPER COPY, $0.95 IN MICROFICHE. FINAL REPORT 1970. 27 P, 8 FIG, 2 TAB, 18 REF. CONTRACT NO. AT(11-1)-1803.

DESCRIPTORS:
*PHOSPHORUS RADIOISOTOPES, *TUBIFICIDS, *FOOD CHAINS, *PATH OF POLLUTANTS, *ABSORPTION, ALGAE, CHORELLA, TRACERS, PHOSPHATES, ECOSYSTEMS, AQUATIC LIFE, ANTIBICTICS(PESTICIDES), ECOLOGY, BENTHOS, INVERTEBRATES, PHYTOPLANKTON, PLANKTON, WATER TEMPERATURE, E. COLI, BENTHIC FAUNA, CELLULOSE, ENZYMES, PULP WASTES, RADIOECOLOGY, WATER POLLUTION, PHOSPHORUS, WATER TEMPERATURE, HYDROGEN ION CONCENTRATION, NUTRIENTS, RADIOISOTOPES, ALGAE, WATER QUALITY, ANNELIDS, ADSORPTION.

IDENTIFIERS:
*CHLORELLA PYRENOIDOSA, *CHLORAMPHENICAL, LIMNODRILUS SPP., TUBIFEX, UNIALGAL CULTURES, OUABAIN, BIOLCGICAL MAGNIFICATION, PHOSPHORUS-32, LIMNODRILUS.

ABSTRACT:
KINETICS OF ACCUMULATION AND RETENTION OF P-32 WERE DETERMINED FOR UNIALGAL CULTURES OF CHLORELLA PYRENOIDOSA IN CARRIER (52 MG NA2HPO4/1.) ANC CARRIER-FREE KNOP'S SOLUTION (MODIFIED) USING 10 MICROCURIES P-32/1 AS TRACER AND A GEIGER-MUELLER COUNTING SYSTEM. THE NUMBER OF COUNTS WAS SUFFICIENT TO ASSURE ACCURACY WITHIN 5 PERCENT. ACCUMULATION IN A CARRIER-FREE SOLUTION WAS EXPONENTIAL AND APPROACHED A 90 PERCENT EFFICIENCY OF REMOVAL WITHIN 30 MINUTES. IN THE CARRIER SOLUTION, ACCUMULATION WAS APPROXIMATELY LINEAR AND DID NOT REACH EQUILIBRIUM WITHIN A 27-HOUR PERIOD. USING OUABAIN AS AN INHIBITOR, IT WAS SHOWN THAT THE OBSERVED SORPTION OF RADIOPHOSPHATE IN CARRIER-FREE SOLUTION BY THIS ALGA WAS INDEPENDENT OF METABOLIC ACTIVITY. MOST LOSS OF THE PHOSPHATE TAKEN UP IN A 27-HOUR INTERVAL OCCURRED WITHIN ONE HOUR IN AGITATED CARRIER AND CARRIER-FREE KNOP'S SOLUTION. ACCUMULATION OF P-32 BY TUBIFICID WORMS WAS ALSO STUDIED USING SIMILAR COUNTING METHODS. UPTAKE BY THESE WORMS WAS APPROXIMATELY EXPONENTIAL IN MEDIA WITH A BACTERIAL SOURCE OF RADIOSPHSPHATE, AND WAS REDUCED BY ADDITION OF CHLORAMPHENICOL AND REMOVAL OF P-32 FROM SOLUTION BY ALGAE. PRELIMINARY RESULTS INDICATED THAT C. PYRENOIDOSA COMPETED EFFECTIVELY WITH THE WORMS FOR INORGANIC PHOSPHATE, AND THAT ACCUMULATION OF P-32 BY INGESTION OF ALGAE AMONG TUBIFICIDS MAY BE SLIGHT. THE ECOLOGICAL IMPLICATIONS OF THESE RESULTS ARE DISCUSSED. (JEFFERIS-BATTELLE)

FIELD 05C, 05B

ACCESSION NO. W72-09668

SUMMER CONFERENCE OF SOCIETY FOR APPLIED BACTERIOLOGY, LIVERPOOL, 13-15 JULY 1971,

OFFICE OF NAVAL RESEARCH, LONDON (ENGLAND).

G. A. HOTTLE.

AVAILABLE FROM THE NATIONAL TECHNICAL INFORMATION SERVICE AS AD731-726, $3.00 IN PAPER COPY, $0.95 IN MICROFICHE. REPORT NO. ONRL-C-19-71, 31 AUGUST 1971. 12 P, 11 REF.

DESCRIPTORS:
*MICROORGANISMS, *WATER POLLUTION CONTROL, *WASTES, *PATHOGENIC BACTERIA, CONFERENCES, *PUBLIC HEALTH, RIVERS, *SEWAGE, BIOCHEMICAL OXYGEN DEMAND, SOLID WASTES, AMMONIA, NITROGEN COMPOUNDS, NITRATES, PHOSPHATES, SLUDGE TREATMENT, DISSOLVED OXYGEN, BACTERIA, FUNGI, PROTOZOA, FERMENTATION, BACTERIOPHAGE, NUTRIENTS, FILTERS EQUIPMENT, OXIDATION, LAKES, WISCONSIN, ANAEROBIC DIGESTIION, ALGAE, EUTROPHICATION, BIODEGRADATION, PLASTICS, PHOSPHORUS COMPOUNDS, SOIL FUNGI, PSEUDOMONAS, DAIRY INDUSTRY, MUNICIPAL WASTES, HERBICIDES, SOIL BACTERIA, PESTICIDES, INDUSTRIAL WASTES, DDT, TRACE ELEMENTS, WATER QUALITY, DETERGENTS, FARM WASTES, URINE, AEROBIC BACTERIA, CARBOHYDRATES, BACTERICIDES, CLOSTRIDIUM, STREPTOCOCCUS, SALMONELLA, WATER PURIFICATION, SHEEP, YEASTS, FOODS, ANAEROBIC BACTERIA, SOIL CONTAMINATION, WASTE TREATMENT, ORGANIC MATTER, DIGESTION, ACTIVATED SLUDGE, ACTINONYCETES, HYDROCARBON PESTICIDES, VIRUSES, WATER POLLUTION SOURCES, WATER POLLUTION EFFECTS, PATH OF POLLUTANTS.

IDENTIFIERS:
BDELLOVIBRID, PELODICTYON, POLIOVIRUSES, PLASTICIZERS, PHTHALATE, THERMOPHILIC FUNGI, POLYETHYLENE, POLYPROPYLENE, BACILLUS, CORYNEBACTERIA, STREPTOTHRIX HYORHINA, CADMIUM, DIMETHYLNITROSAMINE, AMINES, CARCINOGENS, TORULOPSIS SPP., ASPERGILLUS, PENICILLIUM, FUSARIA, VIBRIS.

ABSTRACT:
THE ANNUAL CONFERENCE OF THE SOCIETY FOR APPLIED BACTERIOLOGY, HELD AT THE UNIVERSITY OF LIVERPOOL, 13-15 JULY 1971, INCLUDED A TWO-DAY SYMPOSIUM ON 'MICROBIOL ASPECTS OF POLLUTION' AND A ONE-DAY SESSION DEVOTED TO PAPERS ON INDIVIDUAL RESEARCH. THE FOLLOWING TOPICS WERE DISCUSSED: (1) WATER POLLUTION BY DOMESTIC, AGRICULTURAL AND INDUSTRIAL WASTES, (2) SEWAGE TREATMENT USING COMBINED AEROBIC-ANAEROBIC SYSTEMS, (3) MICROBIAL ECOLOGY OF THE ACTIVATED SLUDGE PROCESS, (4) MICROBIAL ASPECTS OF POLLUTION IN THE FOOD AND DAIRY INDUSTRY, (5) POLLUTION OF FRESHWATERS WITH INORGANIC NUTRIENTS, (6) MICROBIAL DEGRADATION OF PLASTICS, HERBICIDES, AND PESTICIDES, (7) AEROBIC METHODS FOR THE TREATMENT OF FARM WASTES, (8) FACTORS AFFECTING ALGAL BLOOMS, (9) THE ROLE OF OBLIGATE ANAROLS IN THE DIGESTION OF ORGANIC MATERIAL, (10) HEALTH HAZARD OF POLLUTION, AND (11) SEWAGE POLLUTION OF NATURAL WATERS. INDIVIDUAL RESEARCH TOPICS INCLUDED (1) 'AN EVALUATION OF PROCEDURES FOR ENUMERATING BACTERIA IN ACTIVATED SLUDGE', (2) 'THE MICROBIAL POLLUTION OF WATER COURSES AS A RESULT OF THE SEWAGE AND ANIMAL WASTES AND THE APPLICATION OF ANIMAL SLURRY TO LAND', AND (3) 'METHODS FOR ANALYZING THE MICROBIAL DECAY OF SOLID WASTES'. (JEFFERIS-BATTELLE)

FIELD 05C, 05B, 05D

ACCESSION NO. W72-09675

A MICROBIOLOGICAL SURVEY IN LAKE ERIE NEAR CLEVELAND, OHIC,

CONNECTICUT UNIV., STORRS. BIOLOGICAL SCIENCES GROUP.

R. P. COLLINS.

COPY AVAILABLE FROM GPC SUP DOC, $0.50; MICROFICHE FROM NTIS AS PB-210 324, $0.95, ENVIRONMENTAL PROTECTION AGENCY, WATER POLLUTICN CONTROL RESEARCH SERIES, OCTOBER 1971. 31 P, 14 TAB, 12 REF. EPA PROGRAM 16020 GDQ 10/71.

DESCRIPTORS:
*LAKE ERIE, *POTABLE WATER, *TASTE, *ODOR, SURVEYS, MICROBIOLOGY, WATER TREATMENT, PHYTOPLANKTON, FUNGI, BACTERIA, ALGAE.

IDENTIFIERS:
*CLEVELAND(OHIO), CERATIUM, COELOSPHAERIUM, DINOBRYON, FRAGILARIA, PEDIASTRUM, STAURASTRUM, TABELLARIA, MOUGEOTIA.

ABSTRACT:
PERIODIC TASTE AND ODOR AT THE CLEVELAND, OHIC CRCWN WATER TREATMENT PLANT PROMPTED INVESTIGATION OF THE ROLE MICROORGANISMS PLAY IN THIS PROBLEM. FUNGI, BACTERIA, AND ALGAE COLLECTED NEAR THE PLANT INTAKE WERE STUDIED DURING JUNE THROUGH AUGUST 1971. DURING THE THREE MONTHS OF SAMPLING, NO VERTICAL DISTRIBUTION PATTERN WAS NOTED IN QUANTITATIVE ANALYSIS OF THE PHYTOPLANKTON. IN JUNE THE MAXIMUM CONCENTRATION OF PHYTOPLANKTON WAS CONSIDERABLY LOWER THAN THAT OBSERVED IN JULY AND AUGUST. MAXIMUM CONCENTRATION OCCURRED AT THE SURFACE WITH 13,674 ORGANISMS PER LITER; SIMILAR CONCENTRATIONS WERE AT THE 3 AND 9 METER DEPTHS; LCWEST CONCENTRATICNS WERE AT THE 6 ANC 12 METER DEPTHS. STUDIES SHOWED THAT FUNGI AND BACTERIA PLAYED LITTLE, IF ANY, ROLE IN THE PROBLEM. A NUMBER OF ALGAE, REPORTED TO INDUCE TASTE AND ODOR IN WATER, WERE IDENTIFIED. WHATEVER THE SOURCE OF THESE ODORS, THEY WERE NOT DUE TO BENTHIC OR PERIPHYTON ALGAE, BUT COULD HAVE BEEN ASSOCIATED WITH THE PHYTOPLANKTON COMMUNITY AS THE REPORTED 'LAKE ERIE ODOR' COINCIDED WITH PHYTOPLANKTCN INCREASE. THE TASTE AND ODOR PRODUCING ALGAE INCLUDED: COELOSPHAERIUM, DINOBRYON, FRAGILARIA, PEDIASTRUM, STAURASTRUM, TABELLARIA, AND MOUGEOTIA. THERE WAS NO EVIDENCE THAT BENTHIC ORGANISMS PLAYED ANY SIGNIFICANT ROLE. (JCNES-WISCONSIN)

FIELD 05C

ACCESSION NO. W72-10076

STUDY OF THE POSSIBLE ROLE OF POLLUTION IN THE PREVALENCE OF SEA NETTLES IN THE CHESAPEAKE BAY AND THE DEVELOPMENT OF A CENSUS TAKING METHOD.

BICSPHERICS INC., ROCKVILLE, MD.

AVAILABLE FROM THE NATIONAL TECHNICAL INFORMATION SERVICE AS N72-102 73, $3.00 IN PAPER COPY, $0.95 IN MICROFICHE. FINAL REPORT, OCTOBER 15, 1971. 135 P, 29 FIG, 12 TAB, 26 REF. NASW-2115.

DESCRIPTORS:
*SEA NETTLES, *POLLUTANTS, *MAINTENANCE, *PLANT MORPHOLOGY, *PHOSPHATES, *NITRATES, *AMMONIUM COMPOUNDS, SEWAGE EFFLUENTS, TEMPERATURE CONTROL, ASSAY, REMOTE SENSING, ALGAE, ANTIBIOTICS, MERCURY, ARSENIC COMPOUNDS, CHESAPEAKE BAY.

IDENTIFIERS:
*CHRYSAORA QUINQUECIRRHA, *MEDUSAE MAINTENANCE, *POLYP MAINTENANCE, *POLYP METABOLISM, TETRACYCLINE, SULFANILAMIDE, PENICILLIN G, POLYMYXIN B, PODOCYSTS.

ABSTRACT:
THE EFFECT OF POLLUTANTS CN THE POLYP STAGE OF THE SEA NETTLE, CHRYSAORA QUINQUECIRRHA, AND A MEANS TO DETECT THE MEDUSAE FORM BY REMOTE SENSING WERE INVESTIGATED. PHOSPHATE, NITRATE, AMMONIUM, COMBINATIONS OF THESE, PH FROM 6-8, AND SYNTHETIC SEWAGE EFFLUENTS WERE EXAMINED FOR MAINTENANCE AND MORPHOLOGY OF THE POLYPS. PHOSPHATE, NITRATE, ANC THEIR CCMBINATICNS WERE FOUND TO CONTRIBUTE TO THE PROLIFERATION OF POLYPS, PH HAD NO EFFECT WHILE AMMONIUM, AMMONIUM COMBINATIONS AND SEWAGE WERE FOUND TO BE DETRIMENTAL TO POLYPS. PHOSPHATE AND NITRATE SEEMED TO ACT AS A PROTECTICN AGAINST THE LETHAL EFFECTS OF AMMONIUM. AN ASSAY BASED ON THE LABELED RELEASE TECHNIQUE WAS USED TO MEASURE POLYP METABOLISM. THE RESULTS SUPPORT THOSE OBTAINED IN THE MAINTENANCE STUDY SUGGESTING A PCTENTIAL TO PREDICT AND SUPPORT THE MORPHOLOGICAL EFFECTS OF POLLUTANTS ON POLYPS. (ENSIGN-PAI)

FIELD 05C, 02L

ACCESSION NO. W72-10162

BIOCHEMICAL CHANGES IN OXIDATION PONDS,

MAHARAJA SAYAJIRAO UNIV. OF BARODA (INDIA). DEPT. OF BIOCHEMISTRY.

P. M. AMIN, AND S. V. GANAPATI.

JOURNAL WATER POLLUTION CONTROL FEDERATION, VOL 44, NO 2, FEBRUARY 1972, P 183-200, 11 FIG, 5 TAB, 68 REF.

DESCRIPTORS:
*OXIDATION LAGOONS, *ON-SITE INVESTIGATIONS, *BACTERIA, *ALGAE, OXIDATION, METABOLISM, RESPIRATION, LIPIDS, AMMONIA, CARBOHYDRATES, CARBON DIOXIDE, POLYMERS, HYDROGEN ION CONCENTRATION, PROTOZOA, DISSOLVED OXYGEN, SLUDGE, WASTE WATER TREATMENT.

ABSTRACT:
A LABORATORY STUDY OF THE FIRST OPERATIONAL STAGE OF A WASTEWATER LAGOON, NAMELY, THE 28 DAY PERIOD OF NON-CHANGING ENVIRONMENT WHEN ALGAE ARE ALLOWED TO DEVELOP NATURALLY, SHOWED THAT THIS STAGE IS SEPARATED INTO TWO DISTINCT PHASES, THE BACTERIAL PHASE I AND THE ALGAL PHASE II. THE MOST NOTABLE FINDINGS WERE: (1) ABSENCE OF ACIDITY IN BOTH PHASES; (2) ABSENCE OF DISSOLVED OXYGEN IN PHASE I AND ITS ABUNDANCE IN PHASE II; (3) LARGE REDUCTIONS IN COLIFORM DENSITY; (4) LARGE INCREASES AND DECREASES IN BIOCHEMICAL CONSTITUENTS, SUCH AS SUGARS, IN PHASE I AND PHASE II, RESPECTIVELY; (5) INCREASES IN FATTY SUBSTANCES AND CHLOROPHYLLS; (6) LARGE PROTOZOA POPULATION IN PHASE I COMPARED TO LARGE ALGAL POPULATIONS IN PHASE II; AND (7) LACK OF APPRECIABLE SLUDGE FORMATION. CARBOHYDRATE USAGE MOSTLY FOR SYNTHESIS OF FATTY SUBSTANCES AND MUCH LESS FOR POLYMER ACCUMULATIONS WITHIN CELLS, RESULTED IN MORE FAT IN THE CLEAR FINAL EFFLUENT, THUS EXPLAINING THE ABSENCE OF SLUDGE IN THE ECOSYSTEM. (LOWRY-TEXAS)

FIELD 05D, 05G

ACCESSION NO. W72-10233

REMOVAL OF ALGAE FROM WASTE STABILIZATION POND EFFLUENTS--A STATE OF THE ART,

ILLINOIS STATE WATER SURVEY, URBANA.

V. KOTHANDARAMAN, AND R. L. EVANS.

STATE OF ILLINOIS DEPARTMENT OF REGISTRATION AND EDUCATION CIRCULAR NO 108, 1972. 9 P, 3 FIG, 1 TAB, 16 REF.

DESCRIPTORS:
*OXIDATION LAGOONS, *HARVESTING OF ALGAE, *SEPARATION TECHNIQUES, CHEMICAL PRECIPITATION, COAGULATION, SEDIMENTATION, POLYELECTROLYTES, LIME, NEUTRALIZATION, FILTRATION, CENTRIFUGATION, VACUUM DRYING, FLOTATION, WASTEWATER TREATMENT.

IDENTIFIERS:
*AGGLOMERATION.

ABSTRACT:
UNI-ALGAL CELLS HAVE BEEN FOUND TO CARRY A NEGATIVE CHARGE AT PH BETWEEN 2 AND 11. AT PH 2 AND PH 11, THE ALGAL CELLS POSSESS A VERY HIGH CHARGE DENSITY, WHILE AT PH 7 A VERY LOW NEGATIVE CHARGE DENSITY IS EXHIBITED. CHEMICAL PRECIPITATION OF ALGAE HAS BEEN POSTULATED AS BEING DUE TO CHARGE NEUTRALIZATION, AGGLOMERATION, AND SEDIMENTATION. ALGAE HANDLING IN THE REMOVAL PROCESS CONSISTS OF CONCENTRATION OF ALGAL CELLS FROM 200-4000 MG/L TO 1 TO 4% BY WEIGHT, DEWATERING THE 1 TO 4% MIXTURE TO BETWEEN 8 AND 20%, AND DRYING THE 8 TO 20% MIXTURE TO AN 85 TO 90% MIXTURE OF ALGAL CELLS. CHEMICAL COAGULANTS, PARTICULARLY LIME, ALUM, FERRIC SALTS AND CATIONIC POLYMERS, ARE EFFECTIVE IN CAUSING COAGULATION AND SEDIMENTATION OF ALGAL CELLS. DEWATERING AND DRYING OF ALGAL SLURRIES FROM THE CONCENTRATION STEP CAN BE MOST ECONOMICALLY ACHIEVED BY SAND BED APPLICATION. VACUUM FILTRATION HAS BEEN ONLY PARTIALLY SUCCESSFUL, WHILE CENTRIFUGATION HAS BEEN TOO EXPENSIVE. A MARKET FOR SEWAGE-GROWN ALGAE AS LIVESTOCK FEED HAS YET TO DEVELOP, BUT USE AS A SOIL CONDITIONER HAS BEEN SUGGESTED AND OTHER CONVENTIONAL DISPOSAL METHODS ARE BEING INVESTIGATED. (LOWRY-TEXAS)

FIELD 05D

ACCESSION NO. W72-10237

NUTRIENT REMOVAL BY NATURAL GAS FERMENTATION,

BRITISH COLUMBIA RESEARCH COUNCIL, VANCOUVER. DIV. OF APPLIED BIOLOGY.

J. C. MUELLER.

JOURNAL WATER POLLUTION CONTROL FEDERATION, VOL 44, NO 1, JANUARY 1972, P 25-33, 4 FIG, 10 TAB, 13 REF.

DESCRIPTORS:
*NUTRIENTS, *MUNICIPAL WASTES, *SEPARATION TECHNIQUES, FERMENTATION, NITROGEN, PHOSPHORUS, AMMONIA, METHANE, TEMPERATURE, HYDROGEN ION CONCENTRATION, PROTEIN, LABORATORY TESTS, COST ANALYSIS, ALGAE, WASTE WATER TREATMENT.

ABSTRACT:
NITROGEN AND PHOSPHORUS WERE REMOVED FROM SECONDARY WASTE TREATMENT PLANT EFFLUENTS BY FERMENTATION OF GASEOUS HYDROCARBONS TO ASSIMILATE BOTH NITROGEN AND ORTHOPHOSPHATE INTO A BACTERIA CELL MASS. IN LABORATORY TESTS, 93 TO 100% OF THE NITROGEN AND 98 TO 99% OF THE PHOSPHORUS WERE RECOVERED IN THE FORM OF A PROTEIN RICH BIOMASS. THIS BIOMASS WAS SHOWN TO CONTAIN 60 TO 70% PROTEIN, INDICATING THAT ONE POSSIBLE OUTLET FOR THE PRODUCT COULD BE ANIMAL FODDER OR FEED SUPPLEMENTS. AT PRESENT EFFICIENCIES, AS INDICATED IN THE LABORATORY TESTS, PRODUCTION COSTS OF THE BIOMASS PROTEIN WERE ESTIMATED AT 25% MORE THAN THE COST OF COMMERCIALLY AVAILABLE ANIMAL FEED SUPPLEMENTS. (LOWRY-TEXAS)

FIELD 05D

ACCESSION NO. W72-10390

SEQUENTIAL PROCESSING IN WASTEWATER LAGOONS,

ARIZONA STATE UNIV., TEMPE.

J. W. KLOCK.

JOURNAL WATER POLLUTION CONTROL FEDERATION, VOL 44, NO 2, FEBRUARY 1972, P 241-254. 12 FIG, 3 TAB, 5 REF.

DESCRIPTORS:
*OXIDATION LAGOONS, *CHANNELS, *HEAT TRANSFER, *PLASTICS, DOMESTIC WASTES, BACTERIA, ALGAE, NUTRIENTS, OXIDATION-REDUCTION POTENTIAL, ODORS, OPERATION AND MAINTENANCE, WASTE WATER TREATMENT.

ABSTRACT:
A RELIABLE AND SIMPLY OPERATED LAGOON, INCORPORATING THE PRINCIPLES OF HEAT CONSERVATION, SEQUENTIAL-PHASE PROCESSING, AND UTILIZATION OF THIN PLASTIC FILMS TO FORM CHANNEL BARRIERS AND HEAT TRANSFER SURFACES, WAS DEVELOPED. THE CONSTRUCTION OF THE SYSTEM INCLUDED THE USE OF CHANNELS CONSTRUCTED OF BLACK POLYETHYLENE SHEETS. THE INFLUENT TO THE SYSTEM FLOWED INTO THE LAGOON IN THE LOWER PHASE I, AND THE PHASE II CHANNELS OF THE BLACK POLYETHYLENE FLOATED ON THE SURFACE OF THE PHASE I LIQUID. THIS ARRANGEMENT EFFECTIVELY PREVENTED HEAT LOSS DIRECTLY TO THE ATMOSPHERE ABOVE THE PHASE I LIQUID, AND THE THIN PLASTIC MATERIAL ALLOWED HEAT EXCHANGE FROM THE LOWER PHASE TO THE UPPER PHASE. PHASE I WAS DESIGNED AS STRICTLY A BACTERIAL PHASE TO PREVENT PREMATURE ALGAE GROWTH AND SUBSEQUENT NUTRIENT FIXATION, WITH PHASE II INTENDED AS AN ALGAL PHASE FOLLOWED BY LIMITED GROWTH OF CRUSTACEANS AND AQUATIC INSECTS. AVERAGE BOD REDUCTION WAS 76.8%, WITH MINIMAL SURFACE SOLIDS AND NO ODORS. ASSOCIATED OPERATIONAL PROBLEMS AND SOLUTIONS ARE PRESENTED. (LOWRY-TEXAS)

FIELD 05D

ACCESSION NO. W72-10401

LIMNOLOGICAL STUDIES OF LAKE JACOMO, JACKSON COUNTY, MISSOURI. I. WATER QUALITY AND SURFACE PLANKTON, 1970 - 1971,

MISSOURI WATER RESOURCES RESEARCH CENTER, COLUMBIA.

D. H. STERN, AND M. S. STERN.

AVAILABLE FROM THE NATIONAL TECHNICAL INFORMATION SERVICE AS PB-210 587, $3.00 IN PAPER COPY, $0.95 IN MICROFICHE. MISSOURI WATER RESOURCES RESEARCH CENTER COMPLETION REPORT MAY 1972. 61 P, 14 FIG, 4 TAB, 18 REF. PROJECT NO. OWRR A-041-MO(1). 14-31-0001-3225.

DESCRIPTORS:
*ALGAE, *EUTROPHICATION, *NUTRIENTS, *PLANKTON, *RECREATION, *WATER QUALITY, CYANOPHYTA, LAKES, MISSOURI, NITROGEN, PHOSPHORUS.

IDENTIFIERS:
LAKE JACOMO(MO).

ABSTRACT:
WATER QUALITY AND SURFACE PLANKTON IN LAKE JACOMO, THE PRIMARY RECREATIONAL WATER FOR KANSAS CITY, WERE STUDIED SYNOPTICALLY FROM JUNE 1970 THROUGH MAY 1971. THE LAKE, WHICH HAS 2 MILLION VISITORS PER YEAR, HAS BEEN CULTURALLY EUTROPHICATED BY NUTRIENT-RICH RUNOFF FROM WATERSHED LAND CONTAINING FARMS, FEEDLOTS, AND URBAN AREAS. NUISANCE BLOOMS ARE PROMOTED BY ITS DISSECTED SHORELINE AND RELATIVELY SMALL SIZE. AMMONIA NITROGEN CONCENTRATIONS RANGED FROM 0.25 PPM TO 2.64 PPM, NITRATE NITROGEN FROM 0.02 PPM TO 1.00 PPM, AND ORTHOPHOSPHATE FROM 0.02 PPM TO 1.79 PPM. ONE-HUNDRED AND EIGHTY PLANKTON TAXA WERE IDENTIFIED. CYANOPHYTA, IN PARTICULAR APHANIZOMENON FLOS-AQUAE, WERE DOMINANT, EXCEPT IN JUNE AND FROM MID-NOVEMBER THROUGH JANUARY WHEN CHRYSOPHYTA DOMINATED THE PLANKTON. RECOMMENDATIONS FROM THIS STUDY INCLUDE THE FOLLOWING: (1) PURCHASE OR ZONING CONTROL OF THE LAKE'S WATERSHED; (2) LAKE BANK STABILIZATION BY THE PARK DEPARTMENT; (3) ENCOURAGEMENT OF APPROPRIATE POWERBOAT OPERATION TO MINIMIZE WAKES ALONG UNSTABLE BANKS; (4) REDUCTION OF NUTRIENT-RICH RUNOFF FROM AGRIBUSINESSES; (5) EROSION CONTROL BY WISE CONSTRUCTION AND FARMING PRACTICES; (6) REDUCTION OF BLUE-GREEN ALGAE BY REPEATED APPLICATIONS OF COPPER SULFATE OR COPPER-CONTAINING ALGICIDES; AND (7) LIMITED DREDGING OF PRODUCTIVE SHALLOW COVES. AERATION AND DESTRATIFICATION ARE NOT RECOMMENDED AT THIS TIME. (MISSOURI ABSTRACT)

FIELD 05C

ACCESSION NO. W72-10431

THE CARBON DIOXIDE SYSTEM AND EUTROPHICATION,

WARF INST., INC., MADISON, WIS.

S. D. MORTON, P. H. DERSE, AND R. C. SERNAU.

COPY AVAILABLE FROM GPO SUP DOC, $0.75; MICROFICHE FROM NTIS AS PB-210 706, $0.95. ENVIRONMENTAL PROTECTION AGENCY, WATER POLLUTION CONTROL RESEARCH SERIES NOVEMBER 1971. 72 P, 2 FIG, 8 TAB, 15 REF, APPEND. EPA PROGRAM 16010 DXV 11/71.

DESCRIPTORS:
*NUISANCE ALGAE, *CARBON DIOXIDE, *ALGAL CONTROL, CYANOPHYTA, CARBON, CHLORELLA, BICARBONATES, AERATION, LIMITING FACTORS, EUTROPHICATION.

IDENTIFIERS:
*ATMOSPHERIC CARBON, MICROCYSTIS, ANABAENA.

ABSTRACT:
TO DETERMINE THE FEASIBILITY OF EUTROPHICATION CONTROL BY CONTROLLING CARBON, THREE MAJOR AREAS WERE STUDIED: THE STEADY STATE, IN WHICH THE GROWTH RATES OF ALGAE AT VARIOUS CONSTANT, MAINTAINED DISSOLVED CARBON DIOXIDE CONCENTRATIONS WERE DETERMINED; THE NON-EQUILIBRIUM, WHERE NATURAL ATMOSPHERIC REPLENISHMENT WAS THE SOLE CARBON SOURCE; AND ALGAL GROWTH WITH INORGANIC BICARBONATE AS THE SOLE CARBON SOURCE. IN STUDYING GROWTH RATES OF CHLORELLA, MICROCYSTIS, AND ANABAENA WITH RESPECT TO CARBON AVAILABILITY, IT WAS FOUND THAT ALGAE CAN UTILIZE DISSOLVED CONCENTRATIONS OF CARBON DIOXIDE MUCH LOWER THAN THOSE FROM ATMOSPHERIC EQUILIBRIA. ALGAL GROWTH CONTROL BY SWEEPING CARBON DIOXIDE OUT BY AERATION WITH AIR CONTAINING VERY LOW CARBON DIOXIDE CONCENTRATIONS IS DIFFICULT BECAUSE OF ATMOSPHERIC REPLENISHMENT OF CARBON DIOXIDE. BICARBONATE IS AT LEAST 50% UTILIZED AT GROWTH RATES AS HIGH AS 7 MG/L PER DAY. ATMOSPHERIC REPLENISHMENT OF CARBON DIOXIDE, WITHOUT ANY WIND MIXING, CAN SUSTAIN GROWTH RATES OF 1.5-2 MG/L PER DAY FOR DEPTHS OF AT LEAST 1.7 METERS. (JONES-WISCONSIN)

FIELD 05C, 05G

ACCESSION NO. W72-10607

NUTRIENT SOURCES FOR ALGAE AND THEIR CONTROL,

WISCONSIN UNIV., MADISON. WATER RESOURCES CENTER.

G. P. FITZGERALD.

COPY AVAILABLE FROM GPO SUP DOC, $1.00; MICROFICHE FROM NTIS AS PB-210 707, $0.95. ENVIRONMENTAL PROTECTION AGENCY, WATER POLLUTION CONTROL RESEARCH SERIES AUGUST 1971. 77 P, 4 FIG, 26 TAB, 53 REF. EPA PROGRAM 16010 EHR 08/71.

DESCRIPTORS:
*ANALYTICAL TECHNIQUES, *NUTRIENTS, *ALGAE, *BIOASSAY, PHOSPHORUS COMPOUNDS, NITROGEN COMPOUNDS, CYCLING NUTRIENTS, ESSENTIAL NUTRIENTS, LAKES, MUD, WISCONSIN, AQUATIC WEEDS, EUTROPHICATION.

IDENTIFIERS:
*NUTRIENT SOURCES, ACETYLENE REDUCTION, LAKE MENDOTA(WIS), LAKE MONONA(WIS).

ABSTRACT:
BIOASSAYS FOR NUTRIENT AVAILABILITY WERE EVALUATED TO DEFINE CONDITIONS AND LIMITS UNDER WHICH EACH METHOD CAN GIVE MEANINGFUL RESULTS. THE BIOLOGICAL AVAILABILITY OF ALGAL NUTRIENTS IN A WATER SAMPLE AND THE ALGAL RESPONSE TO CHANGES IN THE GROWTH-LIMITING NUTRIENT WERE MEASURED. FACTORS OTHER THAN INSOLUBILITY PREVENT THE NITROGEN OR PHOSPHORUS OF CERTAIN SAMPLES OF AEROBIC LAKE MUDS FROM BEING READILY AVAILABLE FOR ALGAL GROWTH; P-LIMITED SPIROGYRA HAS BEEN FOUND GROWING THROUGH MUD LAYERS CONTAINING 0.1% TOTAL PHOSPHORUS. THE FACTS, THAT LIVE ALGAE AND AQUATIC WEEDS DO NOT SHARE THEIR ADEQUATE OR SURPLUS NUTRIENTS WITH NUTRIENT-LIMITED ALGAE AND THAT LAKE MUDS DO NOT PROVIDE READILY AVAILABLE NITROGEN OR PHOSPHORUS, INDICATE THAT ONCE LAKE WATERS ARE STRIPPED OF AVAILABLE NUTRIENTS BY PLANT PRODUCTION, FURTHER PLANT PRODUCTION WILL DEPEND UPON NUTRIENTS FROM CONTINUOUS SOURCES, SUCH AS WASTEWATER EFFLUENTS. PHOSPHORUS-STARVED CELLS OF ANABAENA RAPIDLY INCREASE THEIR CAPACITY TO REDUCE ACETYLENE TO ETHYLENE WHEN THEY RECEIVE PHOSPHORUS. THIS RESPONSE MAY BE USED AS A BIOASSAY FOR DETECTING AVAILABLE PHOSPHORUS IN AQUATIC ECOSYSTEMS. SENSITIVITY OF THE METHOD COMPARES FAVORABLY WITH CONVENTIONAL METHODS FOR MEASURING DISSOLVED ORTHOPHOSPHATE, AND HAS THE ADVANTAGE THAT IT MEASURES AVAILABLE PHOSPHORUS. (JONES-WISCONSIN)

FIELD 05C, 02H

ACCESSION NO. W72-10608

PLANKTONIC BLUE-GREEN ALGAE: GROWTH AND ODOR-PRODUCTION STUDIES,

NORTH TEXAS STATE UNIV., DENTON. DEPT. OF BIOLOGICAL SCIENCES; AND TELEDYNE BROWN ENGINEERS, HUNTSVILLE, ALA. AND EAST TENNESSEE STATE UNIV., JOHNSON CITY. DEPT. OF BIOLOGY.

J. K. SILVEY, D. E. HENLEY, AND J. T. WYATT.

JOURNAL OF THE AMERICAN WATER WORKS ASSOCIATION, P 35-39, JANUARY 1972. 2 FIG, 47 REF.

DESCRIPTORS:
*DOMESTIC WATER, *CYANOPHYTA, RESERVOIRS, ODOR-PRODUCING ALGAE, SOUTHWEST U.S., TEMPERATURE, OXYGEN, LIGHT INTENSITY, NUTRIENTS, ACTINOMYCETES, TASTE, EUTROPHICATION.

IDENTIFIERS:
ANABAENA CIRCINALIS, ANABAENA CYLINDRICA, APHANIZOMENON FLOS-AQUAE.

ABSTRACT:
PRODUCTION AND CONTROL OF BLUE-GREEN ALGAE, ESPECIALLY THOSE RESPONSIBLE FOR TASTE AND ODOR IN WATER SUPPLIES, ARE SURVEYED. BLUE-GREEN ALGAL BLOOMS IN SOUTHWESTERN RESERVOIRS ALMOST ALWAYS CAUSE TASTE AND ODOR PROBLEMS. MAJOR OFFENDERS ARE THE HETEROCYSTOUS BLUE-GREEN FORMS, PARTICULARLY ANABAENA CIRCINALIS AND APHANIZOMENON FLOS-AQUAE. THE BOTTOM WATERS OF A RESERVOIR SHOULD BE EXAMINED TO DETERMINE THE PHYSICOCHEMICAL AND BIOLOGICAL PARAMETERS INFLUENCING THESE BLOOMS. CONTROL HAS BEEN PRIMARILY LIMITED TO CHEMICAL TREATMENT OF THE WATER IN TREATMENT PLANTS. ACTIVATED CARBON REMOVES SOME ODOR; CHLORINE REMOVES SOME TASTES AND ODORS, BUT IT IS KNOWN TO INTENSIFY OTHERS. THE MOST LOGICAL METHOD OF CONTROLLING TASTES AND ODORS IN WATER SUPPLIES IS TO RESEARCH THEIR SOURCE. NEW APPROACHES OF ECONOMIC IMPORTANCE INCLUDE USE OF WEAK ELECTROLYTES TO CAUSE LYSIS OF ALGAL CELLS, USE OF VIRAL AGENTS AND BACTERIA THAT LYSE BLUE-GREEN ALGAE, AND DEVELOPMENT OF AGENTS THAT WOULD GRADUALLY RELEASE ALGICIDES WHILE FLOATING OR AFTER SINKING TO THE BOTTOM. MANIPULATION OF ECOLOGICAL CONDITIONS MAY PROVE TO BE THE MOST DESIRABLE CONTROL METHOD. EXPERIMENTATION IN THIS AREA HAS INCLUDED FORCED AERATION AND CIRCULATION AND CONTINUOUS LOW-LEVEL SUPPLIES OF NITROGEN COMPOUNDS. (JONES-WISCONSIN)

FIELD 05C, 02H, 05G

ACCESSION NO. W72-10613

A TECHNIQUE FOR BIOASSAY OF FRESHWATER, WITH SPECIAL REFERENCE TO ALGAL ECOLOGY,

FRESHWATER BIOLOGICAL ASSOCIATION, WINDERMERE (ENGLAND).

J. W. G. LUND, G. H. M. JAWORSKI, AND H. BUCKA.

ACTA HYDROBIOLOGICA, VOL. 13, NO. 3, P 235-249, 1971. 3 FIG, 2 REF.

DESCRIPTORS:
*ANALYTICAL TECHNIQUES, *TESTING PROCEDURES, *BIOASSAY, *ALGAE, FRESHWATER, MEASUREMENT, DIATOMS, EUTROPHICATION.

IDENTIFIERS:
BRITISH ISLES.

ABSTRACT:
THE USE OF BIOASSAY ENLARGES KNOWLEDGE OF ALGAL ECOLOGY AS OBTAINED ON THE BASIS OF THEIR GROWTH IN CULTURES. AMONG OTHER USES IT CAN BE APPLIED FOR DETERMINATION OF THE POTENTIAL FERTILITY OF WATERS, AS WELL AS UTILIZED FOR THEIR DIFFERENTIATION. A TECHNIQUE FOR BIOASSAY OF FRESHWATER AND POSSIBILITIES OF ITS PRACTICAL APPLICATION ARE DESCRIBED. THE GROWTH OF THE ALGA IN THE SAMPLES IS A MEASURE OF THE RELATIVE POTENTIAL, QUALITATIVE AND QUANTITATIVE, FERTILITIES OF THESE WATERS. THE METHOD DESCRIBED HAS ALREADY BEEN OF VALUE IN UNPUBLISHED TESTS ON WATER FROM BRITISH LAKES. IT CONSISTS BASICALLY OF THE COLLECTION OF A WATER SAMPLE, FILTRATION OF THE WATER THROUGH GLASS FIBER INTO FLASKS, ADDITION OF A TEST ALGA TO ALL THESE FLASKS AND OF CERTAIN SUBSTANCES, SUCH AS NUTRIENTS, TO SOME OF THE FLASKS. THE WATER IN THE FLASKS IS TRANSFERRED TO EXPERIMENTAL TRANSPARENT OR TRANSLUCENT CONTAINERS WHICH ARE THEN EXPOSED FOR GIVEN TIMES TO CONSTANT CONDITIONS OF LIGHT AND TEMPERATURE. TO REDUCE CHEMICAL CONTAMINATION OR ALTERATION OFTHE WATER SAMPLE TO A MINIMUM, ALL GLASSWARE AND OTHER EQUIPMENT IS CLEANED AND WASHED VERY CAREFULLY. (JONES-WISCONSIN)

FIELD 05C, 05A

ACCESSION NO. W72-10619

EUTROPHICATION,

FRESHWATER BIOLOGICAL ASSOCIATION, AMBLESIDE (ENGLAND).

J. W. G. LUND.

PROCEEDINGS OF THE ROYAL SOCIETY OF LONDON, B, VOL. 180, P 371-382, 1972. 3 FIG, 25 REF.

DESCRIPTORS:
*EUTROPHICATION, *NUTRIENTS, *ALGAE, SEWAGE TREATMENT, WATER POLLUTION SOURCES, WATER POLLUTION EFFECTS, AGRICULTURAL RUNOFF, PHOSPHORUS, NITROGEN, PHOSPHATES, DETERGENTS, POTABLE WATER.

IDENTIFIERS:
ASTERIONELLA FORMOSA, BLELHAM TARN(ENGLAND), WINDERMERE(ENGLAND).

ABSTRACT:
ALTHOUGH THE MAIN CAUSES OF EUTROPHICATION ARE WELL KNOWN, THERE IS LITTLE DETAILED UNDERSTANDING OF THEIR INTERACTION WITH OTHER ENVIRONMENTAL FACTORS; IT IS NOT YET POSSIBLE TO FORECAST THE EXACT CHANGES TO BE EXPECTED IN AQUATIC ECOSYSTEMS. OVER THE LAST 27 YEARS, OBSERVATIONS WERE MADE ON TWO ENGLISH LAKE DISTRICT WATERS--WINDERMERE, RECEIVING THE MAIN SOURCE OF URBAN SEWAGE, AND BLELHAM TARN; PHYTOPLANKTON WAS EXAMINED, USUALLY AT WEEKLY INTERVALS; CONCENTRATIONS OF PHOSPHATES, NITRATES, AND SILICATES WERE ALSO DETERMINED WEEKLY NEARLY ALL THROUGH THIS PERIOD; OTHER CHEMICAL ANALYSES HAVE BEEN CARRIED OUT OVER SHORTER PERIODS. THE MAXIMUM CONCENTRATION OF PHOSPHATES HAS RISEN IN BOTH WATERS. A COMPARISON OF THE CHANGES IN THE ABUNDANCE AND RATE OF GROWTH OF SOME MAJOR ALGAE WAS MADE. RESULTS SHOWED THAT CHANGES IN THE TWO WATERS HAVE NOT FOLLOWED THE SAME PATH. MOREOVER, CERTAIN ASPECTS OF THE CHEMICAL CHANGES IN THE WATER ARE DIFFICULT TO UNDERSTAND. THE INCREASE IN PHOSPHATE OVER THE YEARS HAS FOLLOWED A SIMILAR COURSE IN EACH BODY OF WATER BUT THE AVERAGE NUMBER OF CELLS OF ASTERIONELLA HAS DECREASED IN WINDERMERE AND INCREASED IN BLELHAM TARN. (JONES-WISCONSIN)

FIELD 05C

ACCESSION NO. W72-10625

SURVEY OF TOXAPHENE LEVELS IN GEORGIA ESTUARIES,

GEORGIA UNIV., SAPELO ISLAND. MARINE INST.

R. J. REIMOLD, AND C. J. DURANT.

AVAILABLE FROM THE MARINE RESOURCE CENTER, SAVANNAH, GA. TECHNICAL REPORT SERIES NO. 72-2, FEB. 1972, 51 P, 2 FIG, 25 TAB, 10 REF.

DESCRIPTORS:
*SURVEYS, *DATA COLLECTIONS, *HYDROLOGIC DATA, *MEASUREMENT, *ASSAY, ANALYTICAL TECHNIQUES, EVALUATION, WATER POLLUTION, POLLUTANT IDENTIFICATION, SAMPLING, BIOASSAY, ENVIRONMENT, TOXICITY, POISONS, ALGAL TOXINS, FISH TOXINS, SEDIMENTS, SALT MARSHES, ECOLOGY, PESTICIDES, ESTUARIES, INTERTIDAL AREAS, ESTUARINE FISHERIES.

ABSTRACT:
DATA COLLECTED AND PROCESSED FROM 1 AUGUST 1970 TO 31 JULY 1971 ARE INCLUDED. AN EFFORT IS MADE TO COVER EACH PORTION OF THE RESEARCH WITH EXPLANATIONS OF METHODOLOGY AND RESULTS. THIS PROCEDURE DEVIATES FROM THE NORMAL INTRODUCTION, METHODS, RESULTS AND DISCUSSION TO PROVIDE THE READER WITH A COHERENT SUMMARY OF THE RESEARCH FINDINGS. THE SECTIONS ARE SUBDIVIDED IN THE FOLLOWING BROAD CATEGORIES: (1) ENVIRONMENTAL TOXAPHENE RESIDUE RESULTS (EXCEPT SEDIMENT); (2) SEDIMENT ANALYSIS; (3) BIOASSAYS; (4) TRAWL DATA; AND (5) SUMMARY OF FINDINGS RELATED TO THE PROBLEMS OF TOXAPHENE CONTAMINATION IN THE SALT MARSH. (HOUSER-ORNL)

FIELD 05B, 05C, 02L

ACCESSION NO. W72-10678

RADIUM IN AQUATIC FOOD CHAINS: RADIUM UPTAKE BY FRESH WATER ALGAE,

ATOMIC ENERGY OF CANADA LTD., CHALK RIVER (ONTARIO).

B. HAVLIK.

RADIATION RESEARCH VOL 46, NO. 3, JUNE 1971, P 490-505 10 FIG, 3 TAB, 31 REF.

DESCRIPTORS:
*RADIOISOTOPES, *RADIUM, *ABSORPTION, *PATH OF POLLUTANTS, *WATER POLLUTION, WATER POLLUTION SOURCES, AQUATIC ALGAE, FOOD CHAINS, MEASUREMENT, ADSORPTION.

IDENTIFIERS:
CONCENTRATION, UPTAKE.

ABSTRACT:
ACCUMULATION OF 226RA WAS STUDIED IN FOUR SPECIES OF GREEN ALGAE AND TWO SPECIES OF BLUE-GREEN ALGAE. ALGAE WERE CULTIVATED IN INORGANIC CULTURE. RADIUM WAS ADDED IN AMOUNTS OF 1 MICRO CI/1., 0.1 MICRO CI/1. AND 0.01 MICRO CI/1. AFTER 14 DAYS THE AMOUNT OF RADIUM IN THE MEDIUM IN DEAD AND LIVING ALGAL CELLS, AND IN WASHES FROM THE ALGAE, WAS DETERMINED AT INTERVALS USING A LIQUID SCINTILLATION COUNTER. FROM 50 TO 80% OF THE RADIUM WAS ABSORBED WITHIN THE CELLS AND THE AMOUNT ABSORBED WAS PROPORTIONAL TO THE LENGTH OF EXPOSURE. RADIUM WAS MOSTLY ADSORBED ON THE CELL SURFACE (25-50%) AND ONLY 1-8% WAS PRESENT WITHIN THE CELLS. THE HIGHEST ACCUMULATIVE FACTOR WAS REACHED AFTER 24 HOURS EXPOSURE. THE ACCUMULATIVE FACTOR OF THE RADIUM ABSORBED BY THE ALGAE WAS INVERSELY PROPORTIONAL TO CONCENTRATION OF RADIUM IN THE MEDIUM AND WAS DEPENDENT ON: THE SPECIES OF ALGAE; THE CONCENTRATION RADIUM IN THE MEDIUM; THE GROWTH RATE OF THE ALGAE AND THEIR PHYSIOLOGICAL CONDITION; THE PERIOD OF EXPOSURE. THE FACTORS RESPONSIBLE FOR RADIUM ACCUMULATION WERE ADSORPTION, ABSORPTION AND INCORPORATION, IN THAT ORDER OF IMPORTANCE. (HOUSER-ORNL)

FIELD 05C, 05B

ACCESSION NO. W72-10686

EFFECT OF A MIXTURE OF URANIUM FISSION PRODUCTS ON THE SANITARY CONDITIONS AND HYDROBIONTS OF WEAKLY-MINERALIZED FRESH-WATER BASINS, (IN RUSSIAN),

V. N. GUSKOVA, A. N. BRAGINA, A. A. ZASEDATELEV, B. N. ILYIN, AND V. M. KUPRIYANOVA.

GIDROBIOL. ZH., VOL 6, NO. 4, P 5-11, 1970, 3 TAB, 3 FIG.

DESCRIPTORS:
*RADIOISOTOPES, *RADIOACTIVITY EFFECTS, *RADIOECOLOGY, *RESERVOIRS, *FRESHWATER, HYDROBIOLOGY, ALGAE, FISH, ABSORPTION, BIOASSAY, METABOLISM.

IDENTIFIERS:
CONCENTRATION, MUSCLE, BONE PHYSIOLOGY/METABOLISM.

ABSTRACT:
EXPERIMENTAL INVESTIGATIONS DID NOT ESTABLISH A CONSIDERABLE NEGATIVE EFFECT OF TWO MIXTURES OF FISSION RADIONUCLIDES IN CONCENTRATIONS OF 2.0 X 10-1 AND 1.0 X 10-5 CI/1 ON SANITARY CONDITIONS AND HYDROBIANTS OF A RESERVOIR. A CONCENTRATION OF THE MIXTURES OF 1.0 X 10-3 CI/1 CAUSED INHIBITION OF BIOCHEMICAL CONSUMPTION OF OXYGEN AND DEVELOPMENT OF FISH SPAWN. A DECREASE IN CONCENTRATION OF MIXTURES (RADIONUCLIDES) IN WATER OCCURRED MOST INTENSIVELY (BY 50 TO 60% AS COMPARED WITH THE INITIAL ONE) DUE TO PROTOCOCCUS ALGAE AND DUCKWEED. THE LOWEST VALUES FOR THE ACCUMULATION COEFFICIENTS OF THE SEPARATE ISOTOPE MIXTURES, THE CONCENTRATION BEING 1.0 X 10-8 CI/1 IN WATER, WERE FOUND IN THE MUSCLES OF FISH. AXIS SKELOTON ACCUMULATED MAINLY 89SR AND 140BA. (HOUSER-ORNL)

FIELD 05C

ACCESSION NO. W72-10694

THE INFLUENCE OF LIVING AND DEAD CELLS OF CHLORELLA VULGARIS AND SCENEDESMUS OBLIQUUS ON AQUATIC MICROORGANISMS,

WYZSZA SZKOLA ROLNICZA, OLSZTYN-KORTOWA (POLAND).

S. NIEWOLAK.

POLSKIE ARCHIWUM HYDROBIOLOGII, VOL 18, NO 1, P 43-54, 1971. 8 FIG, 38 REF.

DESCRIPTORS:
*CHLORELLA, *SCENEDESMUS, ALGAE, MICROORGANISMS, ENVIRONMENTAL EFFECTS, GROWTH RATES, E. COLI, AZOTOBACTER, PSEUDOMONAS, WATER POLLUTION EFFECTS.

IDENTIFIERS:
*MICROBIAL GROWTH, MICROCOCCUS, BACILLUS, VIBRIO, RHODOTORULA, AEROMONAS.

ABSTRACT:
THE STUDY CONCERNED THE INFLUENCE OF LIVING AND DEAD CELLS OF CHLORELLA VULGARIS AND SCENEDESMUS OBLIQUUS UPON SOME SPECIES OF SAPROPHYTIC BACTERIA SAMPLED RANDOMLY FROM AMONG A FEW HUNDRED SPECIES ISOLATED FROM ILAWA LAKES WATER. FOUR DIFFERENT PATTERNS OF BEHAVIOR OF MICROORGANISMS HAVE BEEN FOUND IN THE PRESENCE OF LIVING AND DEAD CELLS OF THESE TWO ALGAL SPECIES: (1) DYING OF MICROORGANISMS IN LIVING AND DEAD CULTURES OF ALGAE (MICROCOCCUS UREAE); (2) DYING OF MICROORGANISMS IN LIVING CULTURES OF ALGAE AND DEVELOPMENT IN DEAD CULTURES OF ALGAE (BACILLUS MYCOIDES, ESCHERICHIA COLI, VIBRIO SP.); (3) DYING OF MICROORGANISMS IN LIVING AND DEAD CULTURES OF ALGAE IN THE INITIAL PERIOD OF JOINT CULTIVATION AND DEVELOPMENT IN THE LATER PERIOD (AZOTOBACTER SP., PSEUDOMONAS FLUORESCENS, RHODOTORULA SP.); (4) INTENSIVE DEVELOPMENT OF MICROORGANISMS IN KILLED CULTURES OF ALGAE AND DYING IN LIVING CULTURES IN THE INITIAL PERIOD OF JOINT CULTIVATION, FOLLOWED BY DEVELOPMENT IN THE LATER PERIOD (AEROMONAS SP.). (LEGORE-WASHINGTON)

FIELD 05C

ACCESSION NO. W72-10861

GREAT LAKES ALGAE MONITORING PROGRAM, 1969.

MICHIGAN WATER RESOURCES COMMISSION, LANSING. DEPT. OF NATURAL RESOURCES.

REPORT, FEBRUARY 1970. 16 P, 7 FIG, 5 TAB, 2 REF.

DESCRIPTORS:
*ALGAE, *GREAT LAKES, *MONITORING, *BIOINDICATORS, THERMAL POLLUTION, MICROSCOPY, DIATOMS, NUTRIENTS, PHYTOPLANKTON, WATER QUALITY, MICHIGAN, SURFACE WATERS, COASTS, SAMPLING, BIOLOGICAL COMMUNITIES, LAKE MICHIGAN, LAKE ERIE, LAKE HURON, LAKE SUPERIOR, WATER TEMPERATURE, HYDROGEN ION CONCENTRATION, HARDNESS, ALKALINITY, CONDUCTIVITY, SUSPENDED SOLIDS, NITRATES, NITRITES, NITROGEN, PHOSPHORUS, CHLORIDE, CYANOPHYTA, CURRENTS(WATER), CHLOROPHYTA, SCENEDESMUS, *EUTROPHICATION, WATER ANALYSIS.

IDENTIFIERS:
SAMPLE PRESERVATION, FORMALIN, COUNTING, CYCLOTELLA, OSCILLATORIA, APHANOTHECE, ANABAENA, APHANIZOMENON, NAVICULA, GLENODINIUM, MICROSPORA, SYNEDRA, DINOBRYON, OEDOGONIUM, FRAGILARIA, DIATOMA, MELOSIRA, TABELLARIA, ACTINASTRUM, PHYTOCONIS, CYMBELLA, STEPHANODISCUS, ANKISTRODESMUS, TRACHELOMONAS, SAGINAW BAY, ORGANIC NITROGEN.

ABSTRACT:
WATER SAMPLES CONTAINING ALGAE WERE COLLECTED DURING 1969 FROM 49 STATIONS IN THE GREAT LAKES TO ATTEMPT TO CORRELATE THE ALGAL SPECIES WITH TROPHIC CONDITIONS IN EACH LAKE. EACH SAMPLE WAS PRESERVED WITH FORMALIN AND SENT TO THE LANSING LABORATORY OF THE WATER RESOURCES COMMISSION FOR MICROSCOPIC SORTING AND COUNTING. WATER TEMPERATURE, SUSPENDED SOLIDS, NITRATE-N, AMMONIA-N, ORGANIC N, PHOSPHATES, CHLORINE, SULFATES, PH, HARDNESS, CARBONATES, ALKALINITY, AND CONDUCTIVITY WERE ALSO DETERMINED AT THE SAMPLING SITES. IN GENERAL, THE ALGAL COMPOSITION OF LAKE SUPERIOR, LAKE MICHIGAN, AND LAKE HURON (WITH THE EXCEPTION OF SAGINAW BAY) SHOW LOW AVERAGE ALGAE COUNTS WITH GENERA CHARACTERISTIC OF OLIGOTROPHIC CLEAN WATER CONDITIONS. LAKE SUPERIOR AND LAKE HURON BOTH SUPPORT A CENTRIC DIATOM POPULATION WITH THE MOST COMMON GENUS BEING CYCLOTELLA. IN LAKE MICHIGAN ALGAL SAMPLES, PENNATE DIATOMS PREDOMINATED WITH SYNEDRA BEING THE MOST COMMON GENUS. BLUE-GREEN ALGAE WERE OBSERVED IN THE SOUTHERN LAKE MICHIGAN SAMPLES WHICH MAY INDICATE HIGH NUTRIENT LEVELS. LAKE ERIE AND SAGINAW BAY, ON THE OTHER HAND, HAVE HIGH ALGAL POPULATIONS WITH GENERA ASSOCIATED WITH EUTROPHIC CONDITIONS. IT IS CONCLUDED THAT ALGAE OF CERTAIN GENERA, ALONG WITH OTHER BIOLOGICAL ORGANISMS, ARE INDICATORS OF WATER QUALITY AND ARE USEFUL IN DETERMINING THE QUALITY OF MICHIGAN'S INSHORE WATERS. (MORTLAND-BATTELLE)28(U)

FIELD 05A, 05C

ACCESSION NO. W72-10875

RECENT DEVELOPMENTS IN THE MEASUREMENT OF THE RESPONSE OF PLANKTON AND PERIPHYTON TO CHANGES IN THEIR ENVIRONMENT,

NATIONAL ENVIRONMENTAL RESEARCH CENTER, CINCINNATI, OHIO. ANALYTICAL QUALITY CONTROL LAB.

C. I. WEBER.

PAPER PRESENTED AT THE SYMPOSIUM ON BIOASSAY TECHNIQUES IN ENVIRONMENTAL CHEMISTRY 162ND NATIONAL MEETING, WASHINGTON, D.C., SEPTEMBER 15, 1972. 29 P, 3 FIG, 7 TAB, 74 REF.

DESCRIPTORS:
*ANALYTICAL TECHNIQUES, CHEMICAL ANALYSIS, WATER ANALYSIS, EVALUATION, WATER QUALITY CONTROL, *BIOASSAY, *BIOMASS, CHLOROPHYLL, NITROGEN FIXATION, *PERIPHYTON, *PHYTOPLANKTON, BIOLOGICAL COMMUNITIES, DOMINANT ORGANISMS, REDUCTION(CHEMICAL), TROPHIC LEVELS, PROTOZOA, ALGAE, BACTERIA, DIATOMS, OCEANS, ACTIVATED SLUDGE, SOIL BACTERIA, PHOTOSYNTHESIS, FLUOROMETRY, ABSORPTION, CYANOPHYTA, CHLOROPHYTA, YEASTS.

IDENTIFIERS:
*ADENOSINE TRIPHOSPHATE, *AUTOTROPHIC INDEX, MACROINVERTEBRATES, ACETYLENE, ETHYLENE, NITROGEN RADIOISOTOPES, N-15, LUCIFERIN, LUCIFERASE, CHLOROPHYLL A, OPTICAL DENSITY, ACETYLENE REDUCTION, SCHIZOTHRIX CALCIOCOLA, CLOSTERIUM MONILIFERUM, MELOSIRA VARIANS, NAVICULA TRIPUNCTATA, GOMPHONEMA PARVULUM, BACILLARI PARADOXA, SPHAEROTILUS NATANS, EPILIMNON, AMPHIDINIUM CARTERI, CHLORELLA VULGARIS, CYCLOTELLA NANA, DITYLUM BRIGHTWELLII.

ABSTRACT:
WATER QUALITY IS REFLECTED IN BIOASSAY ANALYSIS BASED UPON BIOMASS, POPULATION DENSITY, AND SPECIES COMPOSITION AND DIVERSITY OF AQUATIC ORGANISMS, E.G. PLANKTON, PERIPHYTON, MACROINVERTEBRATES, FISH. OF THE MORE RECENTLY DEVELOPED METHODS OF DETERMINING PLANKTON AND PERIPHYTON BIOMASS AND CONDITION, SPECIAL INTEREST HAS BEEN FOCUSED ON CHLOROPHYLL A, ATP, AND NITROGEN FIXATION AS INDEXES OF PRODUCTIVITY, RESPIRATION, AND SUCCESS IN THE ENVIRONMENT. THE PRIMARY PHOTOSYNTHETIC PIGMENT, CHLOROPHYLL A, HAS BEEN USED IN DEVELOPMENT OF AN 'AUTOTROPHIC INDEX' RELATIONSHIP WHICH HAS PROVEN VALUABLE IN DETERMINING THE DOMINANCE AND TROPHIC LEVELS WITHIN THE PLANKTON AND PERIPHYTON COMMUNITIES OF WATERWAYS. ADENOSINE TRIPHOSPHATE (ATP) OFFERS CONSIDERABLE PROMISE AS AN INDEX OF TOTAL VIABLE PLANKTON BIOMASS AND ALSO AS AN INDEX OF TOXIC SUBSTANCES AND THEIR EFFECTS UPON THE AQUATIC SYSTEM. SINCE NITROGEN IS A MAJOR CELL COMPONENT, ANALYSIS OF NITROGEN FIXATION, AS INDICATED BY ACETYLENE TO ETHYLENE REDUCTION, HAS LENT ITSELF TO RAPID, ACCURATE EVALUATIONS OF NITROGEN BUDGETS OF AQUATIC ORGANISMS AND PLANKTON POPULATION DYNAMICS. ORGANISMS AND SPECIFIC METHODOLOGY ARE INCLUDED IN THE DISCUSSION OF THESE PARAMETERS OF ANAYSIS AS WELL AS DATA COLLECTED TO SUBSTANTIATE THEIR VALUE AS ANALYTICAL TECHNIQUES. (MACKAN-BATTELLE)

FIELD 05A, 05C

ACCESSION NO. W72-10883

RADIOECOLOGY AND CHEMOECOLOGY IN THE SERVICE OF THE PROTECTION OF NATURE,

INSTITUTE OF BIOLOGY OF THE SOUTHERN SEAS, SEVASTOPOL (USSR).

G. G. POLIKARPOV, L. G. KULEBAKINA, V. G. TSITSUGINA, AND V. V. ANDRUDTSCHENKO.

PRESENTED AT THE COMMISSION OF THE EUROPEAN COMMUNITIES INTERNATIONAL SYMPOSIUM, RADIOECOLOGY APPLIED TO THE PROTECTION OF MAN AND HIS ENVIRONMENT, ROME, SEPT. 1971. 6 P, 1 TAB, 14 REF.

DESCRIPTORS:
*RADIOISOTOPES, *ABSORPTION, *ENVIRONMENTAL EFFECTS, NUCLEAR WASTES, PATH OF POLLUTANTS, MARINE ALGAE, CHROMOSOMES, EMBRYONIC GROWTH STAGE, CRUSTACEANS, MONITORING, WATER POLLUTION EFFECTS, MOLLUSKS.

IDENTIFIERS:
BLACK SEA.

ABSTRACT:
CONCENTRATION FACTORS FOR SR-90 UPTAKE IN THE BLACK SEA ARE 60-250 FOR BROWN ALGAE AND 70-300 FOR MOLLUSCS AND CRUSTACEA. INCUBATION OF MARINE ORGANISMS IN RADIOACTIVE SOLUTIONS INCREASES CHROMOSOMAL ABERRATIONS. LETHAL DDT CONCENTRATIONS IN SEAWATER ARE 0.1 MG/LITER FOR AN ALGA AND 0.01 MG/LITER FOR A CRUSTACEAN. (BOPP-ORNL)

FIELD 05B, 05C

ACCESSION NO. W72-10955

ZINC AND COBALT UPTAKE BY THE BROWN SEAWEED FUCUS SPIRALIS (L.),

REACTOR CENTRUM NEDERLAND, PETTEN.

A. W. VAN WEERS,

PRESENTED AT THE COMMISSION OF THE EUROPEAN COMMUNITIES INTERNATIONAL SYMPOSIUM, RADIOECOLOGY APPLIED TO THE PROTECTION OF MAN AND HIS ENVIRONMENT, ROME, SEPT. 1971. 11 P, 5 FIG, 5 REF.

DESCRIPTORS:
*ZINC RADIOISOTOPES, *COBALT RADIOISOTOPES, *MARINE ALGAE, HEAVY METALS, ABSORPTION, BIOINDICATORS, PATH OF POLLUTANTS, FOOD CHAINS, NUCLEAR WASTES, ENVIRONMENTAL EFFECTS, EFFLUENTS, MONITORING, CHELATION.

ABSTRACT:
HAZARDS FROM UPTAKE OF BOTH RADIOACTIVE AND NON-RADIOACTIVE, HEAVY-METAL POLLUTANTS WERE STUDIED; AND A BIOLOGICAL INDICATOR OF POLLUTION WAS EVALUATED. ADDITION OF 500 PPB OF STABLE CO REDUCED THE CONCENTRATION FACTOR FOR CO-60 UPTAKE (MEASURED AFTER 6 DAYS) FROM ABOUT 350 TO 60; 500 PPB OF STABLE ZN REDUCED THE CONCENTRATION FACTOR FOR ZN-65 FROM ABOUT 1000 TO 300. A CHELATING AGENT (EDTA) REDUCED UPTAKE OF ZN-65 TO 10% OF THE CONTROL AT AN EDTA CONCENTRATION OF 1 MICROMOLE/LITER; A SIMILAR REDUCTION OF CO-60 UPTAKE OCCURRED AT AN EDTA CONCENTRATION OF 30 MICROMOLES/LITER. A SEASONAL VARIATION IN THE ZN CONTENT OF SEAWEED OBSERVED OVER A PERIOD OF SEVERAL YEARS IS PROBABLY RELATED TO ITS GROWTH CYCLE. (BOPP-ORNL)

FIELD 05B, 05C

ACCESSION NO. W72-10957

INDICATOR BASED SURVEILLANCE PROGRAM (MARINE) AT A NUCLEAR SITE,

BHABHA ATOMIC RESEARCH CENTRE, BOMBAY (INDIA). HEALTH PHYSICS DIV.

I. S. BHAT, AND P. R. KAMATH.

PRESENTED AT THE COMMISSION OF THE EUROPEAN COMMUNITIES INTERNATIONAL SYMPOSIUM, RADIOECOLOGY APPLIED TO THE PROTECTION OF MAN AND HIS ENVIRONMENT, ROME, SEPT. 1971. 13 P, 4 TAB, 6 REF.

DESCRIPTORS:
*IODINE RADIOISOTOPES, *CESIUM RADIOISOTOPES, *COBALT RADIOISOTOPES, *BIOINDICATORS, ANALYTICAL TECHNIQUES, POLLUTANT IDENTIFICATION, NUCLEAR POWERPLANTS, EFFLUENTS, NUCLEAR WASTES, PATH OF POLLUTANTS, ASIA, MARINE ALGAE, SEDIMENTS, ESTUARINE ENVIRONMENT, MONITORING, CRABS, SHRIMP.

IDENTIFIERS:
INDIA, CESIUM RADIOISOTOPES.

ABSTRACT:
EFFLUENTS FROM THE TARAPUR (BWR REACTOR) POWERPLANT CONTAIN IODINE, CESIUM, AND COBALT RADIOISOTOPES. ENVIRONMENTAL MONITORING IS PARTICULARLY USEFUL FOR DETECTING AN OCCASIONAL LARGE DISCHARGE MADE UNWITTINGLY. RESEARCH WAS CONDUCTED TO OPTIMIZE TECHNIQUES FOR SAMPLING AND ANALYSIS. TRACE ELEMENT CONTENT WAS MEASURED FOR SEVERAL ORGANISMS (LOBSTER, OYSTER, CLAM, CRAB, BOMBAY DUCK, SINGHALA, AND SEA WEED). PRAWNS AND CRABS ARE SUITABLE INDICATORS FOR CS-137, AND SEA WEED (FOR IODINE AND COBALT RADIOISOTOPES) AND SEA SILT ARE VERSATILE INDICATORS. THE HIGH ISOTOPIC DILUTION OF SR-90 IN SEA WATER RESULTS IN LOW SENSITIVITY OF INDICATORS IN THIS CASE.
(BOPP-ORNL)

FIELD 05B, 05A

ACCESSION NO. W72-10958

QUANTITATIVE CHARACTERISTICS OF THE MEANS OF PENETRATION OF SR-90 INTO THE BODIES OF GASTROPOD MOLLUSKS,

V. M. B. NYANISHKENE, AND G. G. POLIKARPOV.

AVAILABLE FROM NTIS AS AEC-TR-7225, $3.00 IN PAPER COPY, $0.95 MICROFICHE. RADIOBIOLOGY, VOL. 10, NO. 6, P. 198-202, 1970. 3 TAB, 5 REF. TRANSLATION FROM RADIOBIOLOGIYA, VOL. 10, P. 928-930, 1970.

DESCRIPTORS:
*STRONTIUM RADIOISOTOPES, *ABSORPTION, *SNAILS, GASTROPODS, PATH OF POLLUTANTS, NUCLEAR WASTES, WATER POLLUTION EFFECTS, FOOD CHAINS, PUBLIC HEALTH, AQUATIC LIFE, AQUATIC ALGAE, CALCIUM, RADIOACTIVITY EFFECTS.

ABSTRACT:
FOOD-INTAKE AND ABSORPTION-FROM-WATER PATHWAYS OF SR-90 UPTAKE WERE COMPARED IN ABOUT ONE-MONTH EXPERIMENTS. BOTH THE SNAILS AND THEIR FOOD ORGANISMS (ELODEA) WERE KEPT IN WATER CONTAINING 10 MICROCI/LITER OF SR-90 AND 50 MG/LITER OF CA. EXPERIMENTS WERE CONDUCTED IN THE SPRING, SUMMER, AND AUTUMN PERIODS UNDER LABORATORY CONDITIONS IN TWO REPETITIONS AND FOUR VARIATIONS. IN VARIATION I THE SNAILS WERE KEPT IN WATER WITHOUT SR-90 AND WERE FED ELODEA THAT HAD ACCUMULATED SR-90; IN VARIATION II THEY WERE KEPT IN WATER WITH SR-90 AND WERE FED ELODEA THAT HAD ACCUMULATED SR-90; IN VARIATION III THEY WERE KEPT IN WATER WITH SR-90 AND WERE FED ELODEA THAT DID NOT CONTAIN SR-90; IN VARIATION IV THEY WERE KEPT IN WATER WITH SR-90 WITHOUT FEEDING. UPTAKE FROM THE WATER WAS ONE (FOR THE FLESH) AND TWO (FOR THE SHELL) ORDERS OF MAGNITUDE GREATER THAN FROM FOOD. (BOPP-ORNL)

FIELD 05C

ACCESSION NO. W72-10978

A SYMBIOTIC SYSTEM FOR ADVANCED TREATMENT OF WASTEWATER,

OHIO STATE UNIV., COLUMBUS. DEPT. OF CIVIL ENGINEERING.

F. J. HUMENIK, AND G. P. HANNA, JR.

WATER AND SEWAGE WORKS, VOL 117, NO 6, JUNE 1970, P 198-202. 6 FIG, 11 REF.

DESCRIPTORS:
*WASTE WATER TREATMENT, *NUTRIENTS, *ALGAE, *BACTERIA, *SYMBIOSIS, ACTIVATED SLUDGE, NITROGEN, PHOSPHORUS, CHEMICAL OXYGEN DEMAND, CHLORELLA, PERFORMANCE, EFFICIENCIES, LABORATORY TESTS, LABORATORY EQUIPMENT, AERATION, DISSOLVED OXYGEN, AEROBIC CONDITIONS, BIOMASS, RESPIRATION, PHOTOSYNTHETIC OXYGEN, *TERTIARY TREATMENT.

IDENTIFIERS:
ORGANIC NITROGEN, SETTLING CHARACTERISTICS, SOLIDS RECYCLE, SOLIDS CONTROL.

ABSTRACT:
UNDER CONTROLLED ENVIRONMENTAL CONDITIONS A SYMBIOTIC ALGAL-BACTERIAL CULTURE WAS DEVELOPED AND MAINTAINED WHICH WOULD EFFICIENTLY REMOVE NUTRIENTS FROM WASTEWATER. THE CULTURE OF THE BIOMASS GROWTH SYSTEM WAS A NATURAL MIXTURE OF ALGAE, PREDOMINANTLY CHLORELLA, AND ACTIVATED SLUDGE IN WHICH THE ALGAE BECAME ENMESHED WITHIN THE BACTERIAL MATRIX. THE ALGAL-BACTERIAL FLOC EXHIBITED GOOD SETTLING CHARACTERISTICS UNDER QUIESCENT CONDITIONS. MAXIMUM AND MOST CONSISTENT REMOVAL OF INFLUENT COD AND ORGANIC NITROGEN WAS OBTAINED DURING UNAERATED OPERATION WITH SOLIDS CONTROL AND THROUGH DAILY HARVEST OF EXCESS BIOMASS. COD REMOVAL FOR THE RECLARIFIED SUPERNATANT AVERAGED 82.5 PERCENT WITH 88.2 PERCENT AS THE MAXIMUM. ORGANIC NITROGEN REMOVAL AVERAGED 85.3 PERCENT. HOWEVER, NO APPRECIABLE REMOVAL OF PHOSPHORUS WAS RECORDED. THE SYMBIOTIC ALGAL-BACTERIAL SYSTEM REMAINED AEROBIC DURING UNAERATED PERIODS AS A RESULT OF A BALANCE BETWEEN RESPIRATION AND PHOTOSYNTHETIC OXYGENATION. AN AVERAGE VALUE OF 2 MG/L DISSOLVED OXYGEN WAS REPORTED. SUPPLEMENTAL AERATION DID NOT RESULT IN INCREASED NUTRIENT REMOVAL, AND WAS THEREFORE CONSIDERED AS A WASTED ENERGY INPUT. (GALWARDI-TEXAS)

FIELD 05D

ACCESSION NO. W72-11100

PHOSPHATES AND PHOSPHATE SUBSTITUTES IN DETERGENTS (PART 2).

COMMITTEE ON GOVERNMENT OPERATIONS (U.S. HOUSE).

HEARINGS--COMM. ON GOVERNMENT OPERATIONS, U.S. HOUSE OF REPRESENTATIVES, 92D CONG, 1ST SESS, OCTOBER 29, 1971. 394 P, 3 FIG, 6 ILLUS, 1 PHOTO, 38 TAB, 5 APPEND.

DESCRIPTORS:
*PHOSPHATES, *WATER POLLUTION SOURCES, *EUTROPHICATION, *DETERGENTS, WATER QUALITY CONTROL, DOMESTIC WASTES, RUNOFF, WATER SOFTENING, ENVIRONMENTAL EFFECTS, AQUATIC ALGAE, ALGAL CONTROL, ALGAL POISONING, AQUATIC ENVIRONMENT, AQUATIC FUNGI, ACQUIFER CHARACTERISTICS, GENETICS, DISSOLVED OXYGEN.

ABSTRACT:
TESTIMONY IS PRESENTED TO THE SUBCOMMITTEE ON NATURAL RESOURCES OF THE HOUSE GOVERNMENT OPERATIONS COMMITTEE CONCERNING THE EFFECTS OF USING PHOSPHATES IN DETERGENTS. TESTIMONY CONCERNS CONSUMER MARKETING ATTEMPTS TO ELIMINATE PHOSPHATE DETERGENTS BOTH VOLUNTARILY AND BY PROHIBITING THEIR USE THROUGH LAW. TESTIMONY IS ALSO PRESENTED CONCERNING THE MANNER BY WHICH PHOSPHATE DETERGENT DISCHARGES SPEED THE NORMAL EUTROPHICATION OF A WATER BODY. THE COMPLETE BAN ON PHOSPHATE DETERGENT SALE IN DADE COUNTY, FLORIDA, IS DISCUSSED ALONG WITH THE UNIQUE ECOLOGICAL CONDITONS PRESENT IN THAT AREA. EFFORTS ON THE PART OF THE ENVIRONMENTAL PROTECTION AGENCY TO SPEED UP THE REMOVAL OF PHOSPHATE DISCHARGES IN WATER BODIES ARE DESCRIBED. THE DANGEROUS HEALTH ASPECTS OF PHOSPHATE SUBSTITUTES ARE ALSO DISCUSSED. OPTICAL BRIGHTENERS USED IN DETERGENTS ARE EXAMINED AND A SCIENTIFIC PAPER IS APPENDED. ONE PROBLEM WITH BRIGHTENERS IS THAT NOT ENOUGH IS KNOWN ABOUT THEIR EFFECT ON THE ENVIRONMENT. ONE SCIENTIST BELIEVES THEY MAY CAUSE GENETIC MUTATIONS IN SOME ORGANISMS. APPENDICES CONTAIN INFORMATION ON EUTROPHICATION, DETERGENT CONTENT AND ANALYSIS OF THEIR EFFECTS ON THE ENVIRONMENT, AND SURVEY STUDIES ON POLLUTION AND EUTROPHICATION IN THE NATION'S LAKES. (GRANT-FLORIDA)

FIELD 05C, 05G, 06E

ACCESSION NO. W72-11186

BENTHIC MARINE ALGAE FROM WATERS ADJACENT TO THE CRYSTAL RIVER ELECTRIC POWER PLANT (1969 AND 1970),

FLORIDA DEPT. OF NATURAL RESOURCES, ST. PETERSBURG.

K. A. STEIDINGER, AND J. F. VAN BREEDVELD.

PROFESSIONAL PAPERS SERIES, NO 16, JUNE 1971. 46 P, 1 FIG, 22 TAB, 14 REF.

DESCRIPTORS:
*BENTHIC FLORA, *MARINE ALGAE, *GULF OF MEXICO, *ELECTRIC POWERPLANTS, *THERMAL POLLUTION, *AQUATIC ENVIRONMENT, FLORIDA, ELECTRIC GENERATORS, WATER TEMPERATURE, RHODOPHYTA, PHAEOPHYTA, CHLOROPHYTA, COOLING WATER, HYDROLOGIC DATA.

IDENTIFIERS:
*SPECIES DIVERSITY, CRYSTAL RIVER.

ABSTRACT:
ONE HUNDRED SIX TAXA OF MARINE ALGAE WERE IDENTIFIED FROM GULF OF MEXICO WATERS ADJACENT TO THE FLORIDA POWER CORPORATION ELECTRICAL GENERATING PLANT AT CRYSTAL RIVER, FLORIDA IN 1969 AND 1970. OF THE 106 TAXA, 19 BELONG TO CHLOROPHYTA, 24 TO PHAEOPHYTA, AND 63 TO RHODOPHYTA. REDUCTIONS IN SPECIES DIVERSITY, AS WELL AS INCIDENCE OF OCCURRENCE, WERE NOTED IN 1970. CAUSES OF THESE REDUCTIONS ARE NOT KNOWN. HOWEVER, INCREASED TEMPERATURE DOES NOT APPEAR TO BE A FACTOR SINCE DEEPER WATER STATIONS, WHICH WERE ALSO AFFECTED, ARE NOT EXPOSED TO THE THERMAL PLUME. THE STUDY AREA IS SEMI-TROPICAL OR WARM-TEMPERATE, WITH A PRIMARILY SEASONAL FLORA. WINTER APPEARED TO BE THE SEASON OF LOWEST SPECIES DIVERSITY. SEVERAL NEW DISTRIBUTION RECORDS ARE REPORTED. (SVENSSON-WASHINGTON)

FIELD 05C

ACCESSION NO. W72-11252

MICROBIOLOGICAL STUDIES ON NITROGEN FIXATION IN AQUATIC ENVIRONMENTS--VII. SOME ECCLOGICAL ASPECTS OF NITROGEN FIXING BACTERIA,

KYOTO UNIV. (JAPAN). RESEARCH INST. OF FOOD SCIENCE; AND MIE PREFECTURAL UNIV. TSU (JAPAN).

A. KAWAI, AND I. SUGAHARA.

BULLETIN OF THE JAPANESE SOCIETY OF SCIENTIFIC FISHERIES, VOL 38, NO 3, P 291-297, 1972. 3 FIG, 1 TAB, 7 REF.

DESCRIPTORS:
*MICROBIOLOGY, *NITROGEN FIXING BACTERIA, *NITROGEN CYCLE, ENVIRONMENTAL EFFECTS, ANAEROBIC CONDITICNS, NITRATES, DENITRIFICATION, BOTTOM SEDIMENTS, SEA WATER, FRESHWATER, AMMONIA, EUTROPHICATION, OXIDATION, SYMBIOSIS, AZOTCBACTER, MARINE ALGAE, CRGANIC MATTER.

IDENTIFIERS:
NITROSOMONAS, NITROBACTER, AZOTOBACTER.

ABSTRACT:
THE OCCURRENCE AND ABUNDANCE OF BACTERIA RESPONSIBLE FOR NITROGEN METABOLISM FORM A PECULIAR PATTERN FOR EACH WATER REGION. THE NUMBER OF EACH BACTERIA GROUP HAVING THE ABILITY TO PERFORM THE FOLLOWING PROCESSES IN THE NITROGEN CYCLE WAS COUNTED IN THE WATER AND BOTTOM SEDIMENTS OF VARIOUS REGIONS: AMMONIFICATION, NITRATE REDUCTION, DENITRIFICATION, NITRIFICATION. THE RELATICNSHIP IS CLOSE BETWEEN THE NUMBER OF NITROGEN FIXING BACTERIA OCCURRING IN THE WATER AND BOTTOM SEDIMENTS OF VARIOUS FRESH WATER REGIONS AND THAT CF NITRATE REDUCERS AND DENITRIFIERS. THE GROWTH OF THE THREE BACTERIA GROUPS MAY BE CCNTROLLED BY THE SAME ENVIRCNMENTAL FACTORS, PRESUMABLY CCNCENTRATION OF AVAILABLE ORGANIC MATTER. THIS RELATIONSHIP WAS NCT OBTAINED IN SEA WATER REGIONS WHERE AMOUNT OF ORGANIC MATTER IS CCNSIDERABLY POORER. BACTERIAL ADHESION ON THE SUSPENDED MATTER PRESENT IN WATER REGIONS WAS STUDIED; WATER SAMPLES WERE TREATED WITH SONIC OSCILLATION AND THE BACTERIA COUNTED. IT MAY BE THAT MANY NITROGEN BACTERIA CCCUR, TOGETHER WITH THE OTHER HETERCTROPHS, ON OR IN THE SUSPENDED PARTICLES AND UTILIZE THE PARTICLE COMPONENTS AS NUTRITIONAL SOURCES. IN GENERAL, A DENSE BACTERIAL BIOMASS IS FCUND IN FRESH WATER ENVIRONMENTS, THE POPULATION DECREASING GRADUALLY FRCM COASTAL TO OFFSHORE REGIONS. (JONES-WISCONSIN)

FIELD 05C

ACCESSION NO. W72-11563

A SHCCK-WAVE TECHNIQUE TO COLLAPSE THE VACUOLES OF BLUE-GREEN ALGAE,

UNIVERSITY OF WALES INST. CF SCIENCE AND TECH., CARDIFF.

D. C. MENDAY, AND A. A. BUCK.

WATER RESEARCH, VOL 6, NO 3, P 279-284, 1972. 5 FIG, 18 REF.

DESCRIPTORS:
*ALGAL CONTROL, *CYANOPHYTA, EXPLOSIVES, COSTS, FISH, FISHKILL, RESERVOIRS, FISHERIES, PRESSURE.

IDENTIFIERS:
*SHOCK-WAVES, *GAS-VACUOLE DEFLATION, SOUTH WALES, MICROCYSTIS AERUGINOSA, ANABAENA.

ABSTRACT:
A FIELD SYSTEM DESIGNED TO CCNTROL BLUE-GREEN ALGAE BY PRODUCING PRESSURE WAVES TO BURST THEIR GAS-VACUOLES, CAUSING THEM TO SINK, IS DESCRIBED. THE PRESSURES REQUIRED TO BURST GAS-VACUOLES OF MICROCYSTIS AERUGINOSA, THE DESIGN CF AN EXPLOSIVE SYSTEM, TCGETHER WITH METHODS TO RECORD PRESSURE DISTRIBUTION, AND THE ADVERSE EFFECTS OF SUCH PRESSURES ON FISH ARE CONSIDERED. THE PRIMARY CONCERN IS THE POSSIBLE APPLICATION IN AN INDUSTRIAL WATER-SUPPLY RESERVOIR IN SCUTH WALWES WHERE THE PRINCIPAL ALGAE IS MICROCYSTIS AERUGINOSA AND WHERE BROWN TROUT (SALMO TRUTTA) HAVE BEEN INTRODUCED FOR ANGLING. IT WAS FCUND THAT A PRESSURE OF ABOUT 4.5 KG/SQ CM WAS REQUIRED TO BURST VACUOLES IN ALL CELLS OF M. AERUGINOSA. THE TECHNIQUE DEVELOPEC PROVIDED A QUICK AND CHEAP CONTROL METHOD WHEN A BLUE-GREEN ALGAL BLOOM IS SERIOUSLY AFFECTING A WATER SUPPLY. IT IS NOT SUGGESTED THAT EXPLCSIVES SHOULC BE USED AS A REGULAR MEANS OF ALGAL CONTROL. IT SITUATICNS WHERE ALGAE ACCUMULATE NEAR ABSTRACTION POINTS, THIS TECHNIQUE HAS PRACTICAL ADVANTAGES WITH LOW OPERATIONAL COSTS AND AVOIDANCE OF CAPITAL CHARGES, PARTICULARLY APPLICABLE WHERE BLOOMS ARE INFREQUENT OR OF SHORT DURATION. (JONES-WISCONSIN)

FIELD 05G, 05C

ACCESSION NO. W72-11564

THE EFFECT OF CUPRIC AND FLUORIDE IONS IN THE RESPIRATION OF CHLORELLA,

UNIVERSITY OF WESTERN ONTARIO, LONDON (ENGLAND). DEPT. OF BIOPHYSICS.

D. F. SARGENT, AND C. P. S. TAYLOR.

CANADIAN JOURNAL OF BOTANY, VOL. 50, P 905-907, 1972. 2 TAB, 7 REF.

DESCRIPTORS:
*ALGAL CONTROL, *RESPIRATION, *CHLORELLA, *INHIBITION, COPPER, FLUORIDES, IONS, CYTOLOGICAL STUDIES, WATER POLLUTION EFFECTS.

IDENTIFIERS:
CHLORELLA PYRENOIDOSA.

ABSTRACT:
SIMULATANEOUS APPLICATION OF CUPRIC AND FLUORIDE IONS VIRTUALLY STOPS RESPIRATION OF CHLORELLA VULGARIS, ALTHOUGH EITHER CUPRIC OR FLUORIDE IONS ALONE CAUSE NO INHIBITION. TESTS WITH VARIOUS COMBINATIONS OF CUPRIC AND FLUORIDE IONS WERE MADE FOR ENDOGENOUSLY RESPIRING CHLORELLA PYRENOIDOSA. THE SUM OF THE INHIBITIONS CAUSED BY CUPRIC AND FLUORIDE IONS INDIVIDUALLY IS MUCH LESS THAN THE INHIBITION FOUND WHEN BOTH AGENTS ARE PRESENT. THIS WORK INDICATES THAT CUPRIC IONS ALONE DO NOT BLOCK THE ALTERNATE DISSIMILATION PATHWAY, BUT THAT BOTH CUPRIC AND FLUORIDE IONS ARE REQUIRED TO INHIBIT COMPLETELY. THE RESPONSE DEPENDS ON THE ORDER OF ADDITION AND THE TIME BETWEEN ADDITIONS OF THE TWO POISONS. THE RESPONSE TO ADDED CUPRIC DECREASES AS THE TIME SINCE THE ADDITION OF FLUORIDE IONS INCREASES. IN CHLORELLA POISONED BY DICHLOROPHENYLDIMETHYLUREA THERE IS A LIGHT-INDUCED RESPIRATORY INHIBITION WHICH DEPENDS UPON THE FUNCTIONING OF GLYCOLYSIS AND WHICH IS SUPPRESSED BY SUCH GLYCOLYTIC INHIBITORS AS FLUORIDE IONS AND IODOACETATE. WHEN FLUORIDE IONS ARE USED, THE LIGHT-INDUCED INHIBITION GRADUALLY REAPPEARS, UNTIL, AFTER AN 80-MINUTE INCUBATION WITH FLUORIDE, THE EFFECT IS AS GREAT AS BEFORE FLUORIDE WAS ADDED. CLEARLY GLYCOLYSIS IS NO LONGER INHIBITED. (JONES-WISCONSIN)

FIELD 05C

ACCESSION NO. W72-11572

CHARACTERIZATION OF TASTE AND ODORS IN WATER SUPPLIES,

CONNECTICUT UNIV., STORRS. REGULATORY BIOLOGY SECTION.

R. P. COLLINS.

COPY AVAILABLE FROM GPO SUP DOC EP2.10:16040 DGH 08/71, $0.35; MICROFICHE FROM NTIS AS PB-211 025, $0.95. ENVIRONMENTAL PROTECTION AGENCY WATER POLLUTION CONTROL RESEARCH SERIES. AUGUST, 1971, 21 P, 1 FIG, 8 REF. EPA PROGRAM 16040 DGH 08/71.

DESCRIPTORS:
*TASTE, *ODOR, *ORGANIC COMPOUNDS, ANALYTICAL TECHNIQUES, SEPARATION TECHNIQUES, GAS CHROMATOGRAPHY, SEPCTROSCOPY, ALGAE, ACTINOMYCETES, FUNGI, LABORATORY TESTS, WATER QUALITY CONTROL.

IDENTIFIERS:
*ODOR-PRODUCING ALGAE.

ABSTRACT:
CULTURED SAMPLES OF STREPTOMYCES ODORIFER, SYNURA PETERSENII, AND TRICHODERMA VIRIDE, AN ACTINOMYCETE, AN ALGAE, AND A FUNGUS, RESPECTIVELY, WERE STEAM DISTILLED AND THE AQUEOUS DISTILLATES WERE THEN EXTRACTED WITH AN ORGANIC SOLVENT AND REDUCED TO A SMALL VOLUME FOR ANALYSIS TO DETERMINE THEIR ODOROUS CONSTITUENTS. ANALYSIS OF THE EXTRACTS WAS BY GAS CHROMATOGRAPHY, INFRARED, MASS, AND NUCLEAR MAGNETIC SPECTROSCOPY. THE MAJOR ODOROUS CONSTITUENTS OF STREPTOMYCES ODORIFER WERE IDENTIFIED AS TRANS-1, 10-DIMETHYL-TRANS-9-DECALOL (GEOSIM), 2-EXO-HYDROXY-2-METHYLBORNANE, AND CADIN-4-ENE-1-01. A LARGE NUMBER OF ODORUS COMPOUNDS WERE IDENTIFIED FROM SYNURA PETERSENII, WITH HEPTANOL CONTAINING THE ODOR NORMALLY ASSOCIATED WITH SYNURA PETERSENII. THE MAJOR ODOROUS COMPOUND PRODUCED BY TRICHODERMA VIRIDE WAS 6-PENTYL-2-PYRONE. (LOWRY-TEXAS)

FIELD 05A, 05D

ACCESSION NO. W72-11604

AEROBIC LAKE MUDS FOR THE REMOVAL OF PHOSPHORUS FROM LAKE WATERS,

WISCONSIN UNIV., MADISON. WATER CHEMISTRY LAB.

G. P. FITZGERALD.

LIMNOLOGY AND OCEANOGRAPHY, VOL 15, NO 4, 1970, P 550-555. 5 TAB, 14 REF.

DESCRIPTORS:
*PHOSPHORUS, *LAKE BEDS, *ALGAE, ALGAL CONTROL, NUTRIENT REQUIREMENTS, SORPTION, EUTROPHICATION, DEFICIENT ELEMENTS, ESSENTIAL NUTRIENTS, *AEROBIC CONDITIONS, WATER POLLUTION CONTROL.

IDENTIFIERS:
LAKE MUD, PHOSPHORUS SORPTION, PHOSPHORUS REMOVAL, RATE OF SORPTION.

ABSTRACT:
BIOASSAYS OF ALGAE INDICATED THAT THEY BECAME PHOSPHORUS OR NITROGEN-LIMITED DESPITE THE FACT THAT THE ALGAE WERE CLOSE TO MUDS CONTAINING BOTH PHOSPHORUS AND NITROGEN COMPOUNDS. LABORATORY TESTS WERE CONDUCTED TO DETERMINE IF THE PHOSPHORUS OF LAKE MUDS IS READILY AVAILABLE TO ALGAE UNDER AEROBIC CONDITIONS. PHOSPHORUS-LIMITED SELENASTRUM AND CLADOPHORA WILL RESPOND BY GROWTH OR CHANGES IN EXTRACTABLE PO4-P TO AS LITTLE AS 0.02 MG PO4-P IN SOLUTION; THESE SPECIES DID NOT RESPOND WHEN EXPOSED FOR A PERIOD OF 1 OR 2 WEEKS TO AS MUCH AS 2 MG OF PHOSPHORUS AS LAKE MUDS UNDER AEROBIC CONDITIONS. STUDIES OF THE RATE OF SORPTION OF PHOSPHORUS BY LAKE MUDS UNDER AEROBIC CONDITIONS FROM TWO LAKES, AND FROM THREE DEPTHS IN ONE LAKE, INDICATED THAT AS LITTLE AS 0.4 G (DRY WEIGHT) OF MUD COULD SORB ABOUT 0.05 MG PO4-P IN LESS THAN 30 MIN. THESE FINDINGS SUGGEST THAT THE SORPTION OF PHOSPHORUS BY LAKE MUDS UNDER AEROBIC CONDITIONS CAN BE USED TO REMOVE PHOSPHORUS FROM LAKE WATER. (MORGAN-TEXAS)

FIELD 05G

ACCESSION NO. W72-11617

THE DAN REGION LARGE SCALE OXIDATION PONDS,

WATER PLANNING FOR ISRAEL LTD., TEL-AVIV. DAN REGION SEWAGE RECLAMATION PROJECT.

Y. FOLKMAN, M. KREMER, AND P. G. J. MEIRING.

PREPRINT, PRESENTED AT SIXTH INTERNATIONAL WATER POLLUTION RESEARCH CONFERENCE, SESSION 7, HALL C, PAPER NO 15, JUNE 21, 1972. 12 P, 4 FIG, 2 TAB, 14 REF.

DESCRIPTORS:
*OXIDATION LAGOONS, *MUNICIPAL WASTES, *WATER REUSE, ALGAE, PHOTOSYNTHESIS, DISSOLVED OXYGEN, AEROBIC CONDITIONS, ANAEROBIC CONDITIONS, ORGANIC LOADING, ODOR, SLUDGE, COST ANALYSIS, NITROGEN, CLAYS, *WASTE WATER TREATMENT.

IDENTIFIERS:
*RECIRCULATION RATIO, *ISRAEL(DAN REGION).

ABSTRACT:
ALTHOUGH MOST DESIGNERS RELEGATE OXIDATION LAGOONS TO ONLY THE SMALLEST INSTALLATIONS, LARGE SCALE OXIDATION LAGOONS DESIGNED TO HANDLE 22,000 CU. M. PER DAY OF RAW SEWAGE HAVE BEEN GIVING SATISFACTORY PERFORMANCE WITH EXCELLENT REDUCTIONS OF ALL INDICATORS OF POLLUTION. THE PONDS ARE PRESENTLY OPERATING AT THEIR OVERALL DESIGN LOAD OF 230 KG BOD/DAY/HECTARE OF SURFACE AREA, BUT LOADING IS TO BE INCREASED QUITE GRADUALLY. THE PERFORMANCE OF THE PRIMARY PONDS DEPENDS ALMOST ENTIRELY ON THE 1.5 TO 1 RECIRCULATION RATE OF EFFLUENT FROM THE LAST POND IN THE RECIRCULATION POND SYSTEM BACK TO THE PRIMARY POND. NO ODOR PROBLEMS OR SLUDGE BUILD-UP PROBLEMS HAVE BEEN ENCOUNTERED. COST OF SEWAGE TREATMENT IN THE SECONDARY PONDS WAS INCREASED TO 2.6 U. S. CENTS PER CUBIC METER TREATED BECAUSE OF THE QUANTITIES OF SAND DUNES TO BE MOVED, AS WELL AS THE NECESSITY OF LINING THE BOTTOMS. HOWEVER, A 65 TO 70% REDUCTION OF AN AVERAGE 90 MG/L TOTAL NITROGEN INFLUENT IS ACCOMPLISHED AT THIS PRICE, FAR LESS THAN THE COST OF REMOVING NITROGEN BY ANY OTHER BIOLOGICAL MEANS. (LOWRY-TEXAS)

FIELD 05D

ACCESSION NO. W72-11642

MANMADE POLLUTION AND AMERICA'S 100,000 LAKES,

CALIFORNIA UNIV., BERKELEY. SANITARY ENGINEERING RESEARCH LAB.

P. H. MCGAUHEY, E. J. MIDDLEBROOKS, AND C. B. PORCELLA.

PUBLIC WORKS, VOL. 103, NO. 3, P 87-88, MARCH 1972.

DESCRIPTORS:
*SEWAGE TREATMENT, *EFFICIENCIES, *EUTROPHICATION, *LAKES, *WATER POLLUTION EFFECTS, RECLAIMED WATER, SEWAGE EFFLUENTS, RESEARCH AND DEVELOPMENT, NUTRIENTS, NITROGEN, PHOSPHORUS, ALGAE, PLANT GROWTH, WATER POLLUTION TREATMENT, EVALUATION, WATER QUALITY.

IDENTIFIERS:
FATE OF POLLUTANTS.

ABSTRACT:
TO COMBAT THE PROBLEM OF POLLUTION OF OUR LAKES, IT HAS BEEN ADVISED THAT RESEARCH BE DONE TO DETERMINE SEVERAL FACTORS IN RELATION TO SEWAGE EFFLUENTS AND THEIR EFFECTS ON THE EUTROPHICATION PROCESS: (1) THE CONCENTRATION AT WHICH NITROGEN AND PHOSPHORUS WILL TRIGGER OR SUPPORT SUBSTANTIAL ALGAL GROWTH, (2) THE GROWTH STIMULATING EFFECT OF SEWAGE THAT HAS HAD ITS P AND N PARTIALLY REMOVED, AND (3) THE ABILITY OF NUTRIENT REMOVAL PROCESSES TO REDUCE N AND P TO LEVELS BELOW THAT CRITICAL TO ALGAL GROWTH. IN ADDITION, IT IS FELT THAT SINCE INCREASING POPULATIONS TEND TO ACCELERATE POLLUTION IN LAKE BASINS, THE NEED FOR POLLUTION RESEARCH TAKES ON OTHER ASPECTS: (1) THE RESIDUAL ABILITY OF TERTIARY SEWAGE EFFLUENTS TO STIMULATE ALGAL GROWTH; (2) THE ULTIMATE FATE OF NUTRIENTS REMOVED FROM SEWAGE IN A BASIN; AND (3) THE EFFECT OF MAN'S NEAR-SHORE AND SHORELINE MODIFICATIONS ON THE BEAUTY AND BIOLOGY OF THE SYSTEM. LAKE TAHOE IS CITED AS AN EXAMPLE OF EFFICIENT WASTEWATER RECLAMATION AND TREATMENT. (MACKAN-BATTELLE)

FIELD 05C, 02H

ACCESSION NO. W72-11702

OSCILLATORY VARIATION OF A PHYTOPLANKTON POPULATION IN A TROPICAL OCEAN,

MCGILL UNIV., MONTREAL (QUEBEC). MARINE SCIENCES CENTRE.

D. M. STEVEN, AND R. GLOMBITZA.

NATURE, VOL 237, NO 5350, P 105-107, MAY 12, 1972. 4 FIG, 1 TAB, 9 REF.

DESCRIPTORS:
*VARIABILITY, *PHYTOPLANKTON, *DEPTH, ON-SITE INVESTIGATIONS, *POPULATION, TEMPORAL DISTRIBUTION, CHLOROPHYLL, SAMPLING, MATHEMATICAL STUDIES, ATLANTIC OCEAN, *TROPICAL REGIONS, MARINE ALGAE, SPATIAL DISTRIBUTION, FOURIER ANALYSIS, BIORHYTHMS, CORRELATION ANALYSIS, PRIMARY PRODUCTIVITY.

IDENTIFIERS:
*PERIODICITY, TRICHODESMIUM THIEBAUDII, BARBADOS.

ABSTRACT:
TRICHODESMIUM THIEBAUDII, A TROPICAL PHYTOPLANKTER, EXHIBITED OSCILLATORY VARIATIONS IN PRODUCTIVITY WHILE UNDER OBSERVATION IN THE TROPICAL WESTERN ATLANTIC. SIXTY-FIVE COLLECTIONS WERE MADE AT REGULAR INTERVALS (AUGUST 29, 1967-JUNE 4, 1970) AT AN OCEANOGRAPHIC STATION ABOUT 9 KM WEST OF SPEIGHTSTOWN, BARBADOS, AT A DEPTH OF 460 M. CHLOROPHYLL CONCENTRATIONS WERE MEASURED AT 5, 25, 50, 75 AND 100 M IN 2 LITER SAMPLES THE NUMBERS OF PHYTOPLANKTON WERE ESTIMATED AT 5, 50, AND 100 M IN 0.5 LITER SAMPLES OF WATER FIXED WITH ONE PERCENT NEUTRALIZED FORMALIN. THE NUMBER OF FILAMENTS OF TRICHODESMIUM AND OTHER COMMON PHYTOPLANKTON WERE COUNTED. RAW DATA INDICATE THAT CHLOROPHYLL CONCENTRATIONS AND THE NUMBER OF FILAMENTS OF TRICHODESMIUM AT 5 M FLUCTUATED REGULARLY WITH A PERIODICITY OF BETWEEN 3 AND 4 MONTHS. OTHER PHYTOPLANKTON WAS USUALLY LESS NUMEROUS AND DID NOT SHOW SUCH VARIATION. THE ASSOCIATION BETWEEN CHLOROPHYLL AND THE TRICHODESMIUM WAS STUDIED BY HARMONIC ANALYSIS AND BY AUTO- AND CROSS-CORRELATIONS. IT WAS SHOWN THAT CHLOROPHYLL AND TRICHODESMIUM OSCILLATE IN PHASE WITH A PERIODICITY OF ABOUT 105-120 DAYS. A SIMILAR ANALYSIS APPLIED TO THE COMBINED DATA FROM ALL DEPTHS TO 100 M REVEALED CHLOROPHYLL CONCENTRATION INCREASES WITH DEPTH. THIS INCREASE IS DEEMED TO BE ASSOCIATED WITH OTHER ORGANISMS, NOT TRICHODESMIUM. (SNYDER-BATTELLE)

FIELD 05C

ACCESSION NO. W72-11719

DIURNAL PH PATTERNS AS PREDICTORS OF CARBON LIMITATION IN ALGAL GROWTH,

CLEMSON UNIV., S.C. DEPT. OF ENVIRONMENTAL SYSTEMS ENGINEERING.

L. G. RICH, J. F. ANDREWS, AND T. M. KEINATH.

WATER AND SEWAGE WORKS, VOL 119, NO 5, P 126-130, MAY 1972. 5 FIG, 1 TAB, 12 REF.

DESCRIPTORS:
*HYDROGEN ION CONCENTRATION, *ALGAE, *GROWTH RATES, *LIMITING FACTORS, *DIURNAL DISTRIBUTION, CARBON DIOXIDE, NITROGEN, PHOSPHORUS, LIGHT, NUTRIENTS, MATHEMATICAL STUDIES, EQUATIONS, AQUATIC ENVIRONMENT, WATER PROPERTIES, ALGAL CONTROL, WATER ANALYSIS, CHEMCONTROL, WATER POLLUTION EFFECTS, WATER QUALITY, CARBON.

ABSTRACT:
A MATHEMATICAL STUDY OF DIURNAL PH PATTERNS IS PRESENTED AS THEY RELATE TO ALGAL GROWTH LIMITING FACTORS, WHETHER THEY ARE CARBON DIOXIDE, NITROGEN, PHOSPHORUS, OR LIGHT. SUCH INFORMATION MIGHT BE OF USE IN ANALYZING DIURNAL PH PATTERNS IN NATURE, THEREBY ELUCIDATING THAT PARTICULAR FACTOR CONTROLLING LOCAL ALGAL GROWTH. DATA FROM COMPUTATIONS CONCLUDE THAT FAIRLY SIGNIFICANT CHANGES IN PH MAGNITUDE IN A DIURNAL PERIOD (OF THE ORDER OF SEVERAL TENTHS OF A PH UNIT) CAN OCCUR ONLY WHEN CO2 IS THE LIMITING FACTOR FOR AT LEAST A PORTION OF THE DIURNAL CYCLE. HOWEVER, IT MUST BE REMEMBERED THAT TIDAL ACTION AND FLOW PHENOMENA MAY ALSO DISTORT THE PH PATTERNS. (MACKAN-BATTELLE)

FIELD 05C

ACCESSION NO. W72-11724

TOXICITY OF MERCURY TO PHYTOPLANKTON,

NEW YORK OCEAN SCIENCE LAB., MONTAUK.

R. NUZZI.

NATURE, VOL 237, NO 5349, P 38-39, MAY 5, 1972. 3 FIG.

DESCRIPTORS:
*PHYTOPLANKTON, *MERCURY, *TOXICITY, CULTURES, *CHLORELLA, AQUATIC ALGAE, CHLOROPHYTA, HEAVY METALS, *CHLAMYDOMONAS, INHIBITORS, *LETHAL LIMIT, WATER POLLUTION EFFECTS, WATER POLLUTION SOURCES, INHIBITION.

IDENTIFIERS:
PHAEODACTYLUM TRICORNUTUM, PHENYLMERCURIC ACETATE, MERCURY CHLORIDE, PHENYLACETATE, MERCURY COMPOUNDS.

ABSTRACT:
EXPERIMENTS WERE PERFORMED TO DETERMINE THE TOXICITY OF MERCURY TO THREE PHYTOPLANKTON SPECIES AT CONCENTRATIONS AS LOW AS 0.06 MICROGRAMS/LITER. PHAEODACTYLUM TRICORNUTUM, CHLORELLA SP., AND CHLAMYDOMONAS SP. WERE GROWN AND TESTED AXENICALLY IN CHEMICALLY DEFINED MEDIA. EACH WAS INOCULATED INTO TWO SERIES OF TUBES CONTAINING HGCL2 AT CONCENTRATIONS RANGING FROM 0.74-66.6 MICROGRAMS HG/LITER AND PHENYLMERCURIC ACETATE (PMA) CONCENTRATIONS RANGING FROM 0.66-15.0 MICROGRAMS HG/LITER. P. TRICORNUTUM WAS ALSO TESTED AGAINST PHENYLACETATE CONCENTRATIONS EQUIVALENT TO THE PHENYLACETATE CONCENTRATION IN THE PMA SERIES OF TUBES. CELLS WERE COUNTED AFTER 16 DAYS INCUBATION AT 18 C AND IT WAS FOUND THAT ALL THE ORGANISMS HAD BEEN INHIBITED BY THE MERCURY COMPOUNDS. PMA WAS MORE TOXIC THAN MERCURY CHLORIDE, CAUSING INHIBITION AT CONCENTRATIONS AS LOW AS 0.06 MICROGRAMS HG PER LITER. THE INHIBITION OF P. TRICORNUTUM BY PMA AT 3 MICROGRAMS HG/L WAS ALMOST COMPLETELY REVERSED BY THE ADDITION OF 5 PERCENT GLUTATHIONE. CELLS GROWN AT SUBLETHAL CONCENTRATIONS SHOWED INCREASED INCIDENCE OF MORPHOLOGICAL ABNORMALITIES WITH AN INCREASE IN MERCURY CONTENT. (SNYDER-BATTELLE)

FIELD 05C

ACCESSION NO. W72-11727

A NEW MOUNTING MEDIUM FOR DIATOMS,

BEMIDJI STATE COLL., MINN. DEPT. OF BIOLOGY.

D. B. CZARNECKI, AND H. D. WILLIAMS.

TRANSACTIONS AMERICAN MICROSCOPICAL SOCIETY, VOL. 91, NO. 1, P 73, JANUARY 1972. 2 REF.

DESCRIPTORS:
*DIATOMS, *MICROSCOPY, CHRYSOPHYTA, AQUATIC ALGAE.

IDENTIFIERS:
*MOUNTING MEDIA, METHYLENE IODIDE, TOLUENE, POLYSTYRENE, SLIDE PREPARATION, SAMPLE PREPARATION.

ABSTRACT:
AN IMPROVED INEXPENSIVE MOUNTING MEDIUM, CONSISTING OF A MIXTURE OF POLYSTYRENE AND METHYLENE IODIDE IN TOLUENE, HAS BEEN USED IN PREPARING DIATOM SLIDES. THE MEDIUM HAS A REFRACTIVE INDEX OF 1.75 AT 24 C AND PROVIDES FOR DETAIL RESOLUTION EXCEEDING THAT OF HYRAX MOUNTS. FOR MIXING THE MEDIUM: ADD 15.75 G OF POLYSTYRENE IN SOME TOLUENE, ALLOW TO STAND UNTIL CLEAR, THEN MIX IN 200 G OF METHYLENE IODIDE. (SYNDER-BATTELLE)

FIELD 05A, 07B

ACCESSION NO. W72-11738

MARINE ALGAE OF THE SMITHSONIAN-BREDIN EXPEDITION TO YUCATAN-1960,

MICHIGEN UNIV., ANN ARBOR. DEPT. OF BOTANY.

W. R. TAYLOR.

BULLETIN OF MARINE SCIENCE, VOL. 28, NO. 1, P 34-44, MARCH 1972. 19 REF.

DESCRIPTORS:
*SYSTEMATICS, *MARINE ALGAE, *CYANOPHYTA, *CHLOROPHYTA, PHAEOPHYTA, RHODOPHYTA, CLADOPHORA.

IDENTIFIERS:
*YUCATAN, LYNGBYA CONFERVOIDES, LYNGBYA MAJUSCULA, GOMONTIA POLYRHIZA, ULVA RIGIDA, CLADOPHORA FULIGINOSA, BATOPHORA OERSTEDI, ACETABULARIA CRENULATA, VALONIA VENTRICOSA, VALONIA OCELLATA, ERNODESMIS VERTICILLATA, DICTYOSPHAERIA VANBOSSEAE, DICTYOSPHAERIA CAVERNOSA, CLADOPHOROPSIS MEMBRANACEA, BOODLEA COMPOSITA, ANADYOMENE STELLATA, BRYOPSIS PENNATA, CAULERPA SPECIES, AVRAINVILLA SPECIES, CLADOCEPHALUS LUTEOFUSCUS, UDOTEA FLABELLUM, PENICILLUS SPECIES, HALIMEDA SPECIES, DILOPHUS SPECIES, DICTYOTA SPECIES, POCOCKIELLA VARIEGATA, STYPODIUM ZONALE, PADINA SANCTAE-CRUCIS, SARGASSUM FILIPENDULA, TURBINARIA TURBINATA, GALAXAURA LAPIDESCENS, FOSLIELLA FARINOSA, GONIOLITHON STRICTUM, AMPHIRCA FRAGILISSIMA, AMPHIROA BRASILIANA, CORALLINA CUBENSIS, CORALLINA SUBULATA, JANIA CAPILLACEA, JANIA ADHERENS, HYPNEA CERVICORNIS, HYPNEA MUSCIFORMIS, CRYPTARACHNE PLANIFRONS, WRANGELIA PENICILLATA, GRIFFITHSIA SCHCUSBOEI, CERAMIUM BYSSOIDEUM, CENTROCERAS CLAVULATUM, SPYRIDIA FILAMENTOSA, SPYRIDIA ACULEATA, HETEROSIPHONIA WURDEMANNI, HETEROSIPHONIA GIBBESII, POLYSIPHONIA SUBTILISSIMA, BRYOTHAMNION TRIQUETRUM, DIGENIA SIMPLEX, LOPHOCLADIA TRICHOCLADOS, WRIGHTIELLA TUMANOWICZII, BOSTRYCHIA MONTAGNEI, HERPOSIPHONIA TENELLA, ACANTHOPHORA MUSCOIDES, ACANTHOPHORA SPICIFERA, LAURENCIA SPECIES.

ABSTRACT:
OVER 80 SPECIES OF MARINE ALGAE THAT WERE COLLECTED BY THE FOURTH SMITHSONIAN-BREDIN EXPEDITION OF 1960 CAME FROM THE TERRITORY OF QUINTANNA ROO ON THE EAST SIDE OF THE PENINSULA OF YUCATAN, MEXICO, AND SUBSTANIALLY ADD TO THE RECORDS OF MEXICAN MARINE ALGAE. THE PRINCIPAL SET OF SPECIMENS HAS BEEN DEPOSITED IN THE U. S. NATIONAL HERBARIUM. (HOLOMAN-BATTELLE)

FIELD 05A, 05C

ACCESSION NO. W72-11800

THE FLOCCULATION OF ALGAE WITH SYNTHETIC POLYMERIC FLOCCULANTS,

CONNECTICUT UNIV., HARTFORD. SCHOOL OF MEDICINE.

R. C. TILTON, J. MURPHY, AND J. K. DIXON.

WATER RESEARCH, VOL 6, NO 2, P 155-164, FEBRUARY 1972. 3 FIG, 26 REF.

DESCRIPTORS:
*ALGAE, *FLOCCULATION, WATER POLLUTION TREATMENT, COLLOIDS, *POLYMERS, ANIONS, CATIONS, ORGANIC COMPOUNDS, SEPARATION TECHNIQUES, CALCIUM, MAGNESIUM, E. COLI, SILICA, *FILTRATION, LIGHT PENETRATION, PHYSICAL PROPERTIES, ELECTROPHORESIS, CHEMICAL PROPERTIES, HYDROGEN ION CONCENTRATION, *WASTE WATER TREATMENT.

IDENTIFIERS:
POLYACRYLAMIDES, POLYSTYRENE SULFONATE, POLYETHYLENEIMINE, *CHLORELLA ELLIPSOIDIA.

ABSTRACT:
SAMPLES OF A PURE ALGAL CULTURE OF CHLORELLA ELLIPSOIDIA, AT CONCENTRATIONS OF 50-3000 MG/L, WERE SUBJECTED TO POLYMER CONCENTRATIONS OF 0.01-1000 MG/L AT PH 4-7. A NUMBER OF ANALYTICAL METHODS WERE USED IN ORDER TO MEASURE THE FLOCCULATION OF THE ALGAL CULTURE BY THE SYNTHETIC POLYMERIC FLOCCULANTS. BY EMPLOYING A 9.6 SQ CM MILLIPORE MEMBRANE FILTER, THE UPPER LIMIT OF POLYMER CONCENTRATION ABOVE WHICH FILTER BLOCKAGE OCCURRED WAS DETERMINED. IN ANOTHER TEST THE DEGREE OF FLOCCULATION WAS MEASURED BY MEASURING THE LIGHT TRANSMITTED BY ALGAL DISPERSION AFTER POLYMER ADDITION AND SETTLING. THE FLOCCULATION EFFICIENCY WAS FURTHER ANALYZED BY MEASUREMENT OF THE ELECTROPHORETIC MOBILITIES OF ALGAL DISPERSIONS IN A MODIFIED BRIGGS-TYPE CELL. A BRIEF DISCUSSION WAS ALSO PRESENTED ON A COMPARISON OF RESULTS ON ALGAE WITH THOSE OF OTHER COLLOIDS, IT BEING ESPECIALLY SIGNIFICANT THAT AT THE SAME CONCENTRATION OF BIOCOLLOID THE ALGAE NEED ABOUT 100 TIMES HIGHER CONCENTRATION OF POLYMER THAN DOES E. COLI. (MACKAN-BATTELLE)

FIELD 05D, 05A

ACCESSION NO. W72-11833

RESPONSE OF THE ALGA CHLORELLA SOROKINIANA TO CO-60 GAMMA RADIATION,

GRUMMAN AEROSPACE CORP., BETHPAGE, N.Y. RESEARCH DEPT.

W. F. KUNZ.

NATURE, VOL 236, NO 5343, P 178-179, MARCH 24, 1972. 2 FIG, 11 REF.

DESCRIPTORS:
*COBALT RADIOISOTOPES, *RADIOSENSITIVITY, *RADIOACTIVITY EFFECTS, *ALGAE, CHLORELLA, IRRADIATION, RADIOACTIVITY TECHNIQUES, RESISTANCE, CULTURES, *CHLOROPHYTA.

IDENTIFIERS:
*CHLORELLA SOROKINIANA, CO-60, GAMMA RADIATION, SURVIVAL, CHLORELLA PYRENOIDOSA.

ABSTRACT:
THE RESPONSE AND SURVIVAL CHARACTERISTICS OF CHLORELLA SOROKINIANA TO IONIZING RADIATION HAD NOT BEEN DETERMINED, ALTHOUGH THE THERMOPHILIC, BLUE-GREEN ALGAE HAD PREVIOUSLY BEEN USED IN BIOREGENERATION STUDIES IN RADIATION ENVIRONMENTS. THE SURVIVAL OF THE CELLS TO CO-60 GAMMA RADIATION WAS THEREFORE MEASURED IN TERMS OF COLONY-FORMING ABILITY AFTER EXPOSING ALIQUOTS OF THE STOCK CULTURE TO CO-60 RADIATION AT AN ABSORBED DOSE RATE OF 15 KRAD/MIN IN FULLY AEROBIC CONDITIONS. THE EXPOSED CELLS WERE PLATED AND INCUBATED ON GLUCOSE-SUPPLEMENTED KNOP'S AGAR, AND SCORED 6 DAYS LATER FOR VISIBLE CULTURES. IN ADDITION, THE SURVIVAL-ABSORPTION RELATIONSHIP WAS DETERMINED FOR 1-HR. OLD CELLS BY EXPOSURE TO DIFFERENT SINGLE DOSES AND MEASURING THE SURVIVING FRACTIONS. SUBLETHAL DAMAGE REPAIR, REPAIR RATE, AND THE INFLUENCE OF SUBLETHAL DAMAGE REPAIR ON AGE AND RESPONSE FUNCTIONS WERE ALL TABULATED. FROM THE RESULTS IT IS MOST OBVIOUS THAT SURVIVAL WITH REPAIR IS ALWAYS HIGHER THAN THAT WITHOUT REPAIR, REGARDLESS OF THE CELL'S AGE. SURVIVAL CURVES ARE INCLUDED FOR ALL THE TESTS. (MACKAN-BATTELLE)

FIELD 05C

ACCESSION NO. W72-11836

ECOLOGICAL STUDIES ON MACROINVERTEBRATE POPULATIONS ASSOCIATED WITH POLLUTED KELP FORESTS IN THE NORTH SEA,

DURHAM UNIV. (ENGLAND). DEPT. OF BOTANY.

D. J. JONES.

HELGOLAENDER WISSENSCHAFTLICHE MEERESUNTERSUCHUNGEN, VOL 22, P 417-441, 1971. 10 FIG, 7 TAB, 44 REF.

DESCRIPTORS:
WATER POLLUTION EFFECTS, ECOLOGY, ECOSYSTEMS, *BIOLOGICAL COMMUNITIES, *SUCCESSION, AQUATIC ENVIRONMENT, *ECOLOGICAL DISTRIBUTION, *ESTUARINE ENVIRONMENT, HABITATS, AQUATIC LIFE, HEAVY METALS, INDUSTRIAL WASTES, INVERTEBRATES, ALGAE, *KELP.

IDENTIFIERS:
*NORTH SEA, *LAMINAREA SPP., SEWAGE POLLUTION, EPIFAUNA.

ABSTRACT:
TWO GRADIENTS OF POLLUTION, ONE ESTUARINE AND ONE OFF THE OPEN COAST, ARE DESCRIBED. THE INTERVENING SEACOAST HAS LITTLE OR NO POLLUTION. A COMPARATIVE METHOD OF POLLUTION SURVEYING IS PRESENTED. ECOLOGICAL COMPARISON IS MADE OF THE COMMUNITY DEVELOPMENT DESCRIBED FOR CLEAN AND POLLUTED STATIONS. TWO ECOLOGICAL BARRIERS TO NORMAL COMMUNITY DEVELOPMENT IN THE POLLUTED ENVIRONMENT ARE POSTULATED. (SVENSSON-WASHINGTON)

FIELD 05C, 02L

ACCESSION NO. W72-11854

ENVIRONMENTAL CHANGES ASSOCIATED WITH A FLORIDA POWER PLANT,

ROSENSTIEL SCHOOL OF MARINE AND ATMOSPHERIC SCIENCES, MIAMI, FLA.

M. A. ROESSLER.

MARINE POLLUTION BULLETIN, VOL 2, NO 6, P 87-90, JUNE 1971. 3 FIG, 1 TAB, 9 REF.

DESCRIPTORS:
BAYS, *MARINE ALGAE, *MARINE ANIMALS, *THERMAL POLLUTION, *THERMAL POWER, ATLANTIC OCEAN, MARINE BIOLOGY, BENTHIC FLORA, BIOASSAY, CYANOPHYTA, *FLORIDA, CORAL, SALINITY, CURRENTS(WATER), MOLLUSKS, ENVIRONMENTAL EFFECTS.

IDENTIFIERS:
*BISCAYNE BAY(FLORIDA), HEATED EFFLUENT, TURKEY POINT(FLORIDA), TURTLE GRASS, THALASSIA SPP., UDOTEA SPP., PENICILLUS SPP., LAURENCIA SPP., SPONGES.

ABSTRACT:
DAMAGE TO THE BIOTA OF BISCAYNE BAY BY THE HEATED EFFLUENT OF A POWER PLANT IS DEMONSTRATED QUANTITATIVELY AND QUALITATIVELY. ALGAE AND GRASSES ARE REPLACED BY BLUE-GREEN FILAMENTOUS ALGAL MATS; SEASONAL RECOVERY IS SLOW AND THE AFFECTED AREAS CONTAIN FEWER KINDS AND SMALLER NUMBERS OF ANIMALS. INCREASED TEMPERATURE IS THE CHIEF CAUSE OF THE CHANGES. (KATZ-WASHINGTON)

FIELD 05C, 02L

ACCESSION NO. W72-11876

A QUANTITATIVE STUDY OF FACTORS AFFECTING ALGAL DIVISION SYNCHRONY MEASUREMENTS,

SAINT LOUIS UNIV., MO. DEPT. OF BIOLOGY.

D. W. ROONEY.

MATHEMATICAL BIOSCIENCES, VOL. 13, NO. 3/4, P 205-211, APRIL 1972. 2 FIG, 1 TAB, 3 REF.

DESCRIPTORS:
*ALGAE, *MODEL STUDIES, MATHEMATICAL STUDIES, DIGITAL COMPUTERS, DATA ANALYSIS, EQUATIONS, REPRODUCTION, BIORHYTHMS, MEASUREMENT, POPULATIONS, CYTOLOGICAL STUDIES, STATISTICAL METHODS, CULTURES.

IDENTIFIERS:
*ERRORS, *SYNCHRONOUS CULTURES, AUTOSPORES, VARIABILITY, BIAS, SYNCHRONY INDEX, COUNTING.

ABSTRACT:
A DIGITAL COMPUTER WAS USED TO SIMULATE ERRORS IN ALGAL CELL NUMBER MEASUREMENTS USED IN SYNCHRONY INDEX COMPUTATIONS. THE INDEX S SUB R IS BASED ON CELL COUNTS OVER A TIME PERIOD IN WHICH THE NUMBER OF CELLS PER UNIT VOLUME INCREASES. REPEATED SIMULATIONS DEMONSTRATED THAT THE RESULTING DEGREES OF BIAS AND VARIABILITY IN THE MEASURED SYNCHRONY INDEX S SUB R DEPEND NOT ONLY ON THE MAGNITUDE OF ERROR IN CELL NUMBER MEASUREMENTS, BUT ALSO ON THE NUMBER OF SUCH MEASUREMENTS AND THE AUTOSPORE YIELD AT DIVISION. THE SIMULATIONS DEMONSTRATED THAT BIAS AND VARIABILITY IN MEASURED S SUB R DECREASE WITH INCREASES IN AUTOSPORE YIELD. THE VARIABILITY IN MEASURED S SUB R INCREASES WITH INCREASING FREQUENCY OF PERIODIC CELL COUNTS. MINIMAL BIAS OF S SUB R OCCURS AT INTERMEDIATE CELL COUNT FREQUENCIES. (SNYDER-BATTELLE)

FIELD 05C, 05A

ACCESSION NO. W72-11920

EYES BENEATH THE WAVES,

DURHAM UNIV. (ENGLAND). DEPT. OF BOTANY.

D. BELLAMY.

NEW SCIENTIST, VOL. 54, NO. 791, P 76-68, APRIL 13, 1972. 3 FIG.

DESCRIPTORS:
*ON-SITE INVESTIGATIONS, *SCUBA DIVING, *SEA WATER, *DATA COLLECTIONS, COPPER, LEAD, HEAVY METALS, MUSSELS, *KELPS, STARFISH, CRUSTACEANS, FOOD CHAINS, FOOD WEBS, WATER POLLUTION EFFECTS, AQUATIC ENVIRONMENT, TURBIDITY, WATER POLLUTION, PHAEOPHYTA, MOLLUSKS, MARINE ALGAE, MARINE ANIMALS.

IDENTIFIERS:
ENGLAND, BIOACCUMULATION, BIOLOGICAL MAGNIFICATION, FATE OF POLLUTANTS.

ABSTRACT:
DATA COLLECTED BY AMATEUR DIVERS ALONG THE EAST COAST OF BRITAIN SHOWS THAT IN POLLUTED SITES KELP FAILS TO GROW AT DEPTHS BELOW 4 M, WHEREAS IN UNPOLLUTED WATERS IT NORMALLY GROWS TO AT LEAST TWICE THAT DEPTH. SINCE KELP DEPTH RANGES CAN BE CONTROLLED BY AT LEAST 2 MAIN FACTORS, LIGHT AND BARE ROCK AREA, IT IS PRESUMED THAT INCREASED TURBIDITY IS THE CONTROLLING FACTOR IN POLLUTED WATER. STUDIES IN NATURALLY TURBID WATER SUPPORT THIS CONTENTION. IN ANOTHER STUDY, THE BACKGROUND COUNT OF HEAVY METALS OF AN 'INSHORE FOOD WEB' WAS ASCERTAINED BY THE COLLECTION OF ALGAE, MUSSELS AND STARFISH FOR ANALYSIS OF LEAD AND COPPER IN THEIR TISSUES. AT THE TWO UNPOLLUTED SITES TESTED, BOTH METALS APPEARED TO BE BIOLOGICALLY CONCENTRATED ALONG THE FOOD CHAIN, WHILE AT THE 2 POLLUTED SITES NO OVERALL BIOLOGICAL CONCENTRATION WAS INDICATED. THE ANIMALS FROM THE POLLUTED AREAS DISPLAY CONCENTRATION OF THE METALS AT 1-2 TIMES THE CONCENTRATION IN UNPOLLUTED SAMPLES. (MACKAN-BATTELLE)

FIELD 05C, 05A

ACCESSION NO. W72-11925

CHEMICAL CHARACTERISTICS OF HUMIC COMPOUNDS ISOLATED FROM SOME DECOMPOSED MARINE ALGAE,

BEDFORD INST., DARTMOUTH (NOVA SCOTIA). ATLANTIC OCEANOGRAPHIC LAB.

M. A. RASHID, AND A. PRAKASH.

JOURNAL OF FISHERIES RESEARCH BOARD OF CANADA, VOL. 29, NO. 1, P 55-60, JANUARY 1972. 3 TAB, 30 REF.

DESCRIPTORS:
*CHEMICAL ANALYSIS, *DECOMPOSING ORGANIC MATTER, *MARINE ALGAE, HUMUS, ORGANIC COMPOUNDS, POLLUTANT IDENTIFICATION, AQUEOUS SOLUTIONS, ION EXCHANGE, METHODOLOGY, ORGANIC ACIDS, FULVIC ACIDS, KELPS, WATER ANALYSIS, EXUDATION, WATER POLLUTION, PHYTOPLANKTON, WATER POLLUTION SOURCES, PHAEOPHYTA, OXIDATION.

IDENTIFIERS:
*HUMIC COMPOUNDS, COLUMN CHROMATOGRAPHY, FUCUS VESICULOSUS, LAMINARIA DIGITATA, ST. MARGARET'S BAY, SCOUDOUC RIVER, HUMIC ACIDS, OXIMATION, SAMPLE PREPARATION.

ABSTRACT:
THE CHEMICAL CHARACTERISTICS OF HUMIC COMPOUNDS ISOLATED FROM DECOMPOSED MARINE ALGAE WERE DETERMINED AND COMPARED WITH THOSE EXTRACTED FROM RIVER WATER. THE CHEMICAL NATURE OF HUMIC COMPOUNDS ISOLATED FROM THE EXUDATES OF LIVE, INTACT PLANTS OF FUCUS VESICULOSUS AND LAMINARIA DIGITATA WAS DETERMINED BY ANALYSES OF THEIR FUNCTIONAL GROUPS, MOLECULAR WEIGHT, ELEMENTAL COMPOSITION, AND CERTAIN SPECTRAL PROPERTIES. WATER SAMPLES WITH A HIGH, DISSOLVED HUMIC CONTENT WERE OBTAINED FROM ST. MARGARET'S BAY, NOVA SCOTIA AND SCOUDOUC RIVER, NEW BRUNSWICK. AFTER THE IMPURITIES WERE REMOVED, THE MOLECULAR WEIGHT WAS DETERMINED BY A COLUMN CHROMATOGRAPHIC TECHNIQUE USING SEPHADEX GELS AND TOTAL ACIDITY MEASURED BY BARIUM HYDROXIDE SOLUTION IN N2. THE CARBOXYLS WERE DETERMINED BY ION-EXCHANGE WITH CALCIUM ACETATE SOLUTION, AND THE CARBONYLS BY OXIMATION. THE DATA FOR OXYGEN-CONTAINING FUNCTIONAL GROUPS INDICATED THAT AS COMPARED TO HUMIC ACIDS, THE FULVIC ACIDS SHOWED RELATIVELY LARGE VALUES OF TOTAL ACIDITY AND 1.5 - 2.5 TIMES HIGHER CARBOXYL CONTENT. THE PHENOLIC HYDROXYL GROUP CONTENT WAS LOW (0.0-1.0 MILLIEQUIVALENTS PER GRAM) IN ALL SAMPLES AND APPEARED TO BE CHARACTERISTIC OF HUMIC COMPOUNDS ORIGINATING ORIGINATING IN THE MARINE ENVIRONMENT. EXCEPT FOR LAMINARIA HUMIC ACIDS, CARBONYL GROUPS WERE HIGH IN ALL SAMPLES, PARTICULARLY THE FULVIC ACIDS ISOLATED FROM RIVER WATER AND LAMINARIA EXUDATE. IN GENERAL, THE MOLECULAR WEIGHT PROPERTIES OF HUMIC AND FULVIC ACIDS WERE SIMILAR TO THE CORRESPONDING FRACTIONS PREVIOUSLY OBTAINED FROM MARINE SEDIMENTS. OPTICAL DENSITY TESTS INDICATED A HIGHER DEGREE OF CONDENSATION FOR MARINE AS COMPARED WITH FRESHWATER HUMIC COMPOUNDS. (BYRD-BATTELLE)

FIELD 05B, 05A

ACCESSION NO. W72-12166

USEFULNESS OF CULTURES IN THE TAXONOMY OF BLUE-GREEN ALGAE, (BENUTZUNG VON KULTUREN IN DER BLAUALGENTAXONOMIE),

CESKOSLOVENSKA AKADEMIE VED, TREBON. BOTANICKY USTAV.

J. KOMAREK.

SCHWEIZERISCHE ZEITSCHRIFT FUR HYDROLOGIE, VOL. 33, NO. 2, P 553-565, DECEMBER 1971. 3 FIG, 2 TAB, 17 REF.

DESCRIPTORS:
*CYANOPHYTA, *CULTURES, *SYSTEMATICS, CLASSIFICATION, *AQUATIC ALGAE.

IDENTIFIERS:
CULTUTURING TECHNIQUES, CULTURE MEDIA, ALGOLOGY, TAXON.

ABSTRACT:
CULTIVATION, BEHAVIOR PATTERN AND OTHER DIFFICULTIES ARISING FROM THE STUDY OF BLUE-GREEN ALGAE IN CULTURES ARE STUDIED. SOME OF THE PROPERTIES OF THE CYANOPHYTA WHICH CAUSE COMPLICATIONS IN TAXONOMY INCLUDE THE FOLLOWING: (1) SPECIAL NUTRIENT SOLUTIONS ARE NEEDED FOR MANY STRAINS; (2) THE GREAT VARIETY OF ECOLOGICAL NEEDS PLACES PARTICULAR DEMANDS ON CULTIVATION AND PRESUPPOSES TECHNICAL EQUIPMENT WHICH CANNOT BE STANDARDIZED; (3) ATYPICAL FORMS THAT DO NOT CORRESPOND TO ECOTYPES IN NATURE RESULT WHEN MANY OF THE STRAINS ARE CULTIVATED; (4) IT IS DIFFICULT TO DRAW GENERAL CONCLUSIONS FROM THE RELATIVELY SMALL AMOUNTS OF COMPARABLE DATA OBTAINED FROM PARALLEL EXPERIMENTS. HOWEVER, CULTIVATION IS RECOMMENDED FOR THE FOLLOWING CASES: (A) STUDY OF POTENTIAL VARIABILITY IN THE CLONAL POPULATION; (B) COMPARISON OF INDUCED VARIABILITY OF ONE FEATURE IN SEVERAL STRAINS; (C) DEFINITION OF QUANTITATIVE OR QUALITATIVE DIFFERENCES AMONG SOME STRAINS OR VERIFICATION OF THEIR TAXONOMICAL IDENTITY; (D) DETERMINATION OF ULTRASTRUCTURAL, PHYSIOLOGICAL AND BIOCHEMICAL PROPERTIES OF A TAXON. (SNYDER-BATTELLE)

FIELD 05C, 05A

ACCESSION NO. W72-12172

OCEANOGRAPHY OF THE NEARSHORE COASTAL WATERS OF THE PACIFIC NORTHWEST RELATING TO POSSIBLE POLLUTION.

OREGON STATE UNIV., CORVALLIS. DEPT. OF OCEANOGRAPHY.

COPY AVAILABLE FROM GPO SUP DOC EP2.10:16070 EOK 07/71 SN5501-0140 VOL I, $5.25, VOL II, $6.00; MICROFICHE FROM NTIS AS PB-211 275, $0.95/VOL. 1. ENVIRONMENTAL PROTECTION AGENCY, WATER POLLUTION CONTROL RESEARCH SERIES, JULY 1971, VOL I. 615 P; VOL II. 744 P. EPA PROGRAM 16070 EOK 07/71.

DESCRIPTORS:
*OCEANOGRAPHY, *PACIFIC NORTHWEST, *COAST REVIEW, *BIBLIOGRAPHIES, WATER POLLUTION SOURCES, THERMAL POLLUTION, REVIEWS, *DATA COLLECTIONS, TOXICITY, PATH OF POLLUTANTS, *METALS, TRACE ELEMENTS, PESTICIDES, SALINITY, NUTRIENTS, ALGAE, AQUATIC ANIMALS, RADIONUCLIDES.

IDENTIFIERS:
*LITERATURE REVIEWS.

ABSTRACT:
THIS STUDY IS LIMITED TO THE COASTAL ZONE OF THE PACIFIC NORTHWEST FROM HIGH TIDE TO TEN KILOMETERS FROM SHORE, AND DOES NOT INCLUDE ESTUARIES AND BAYS. THE LITERATURE HAS BEEN REVIEWED IN 21 CHAPTERS INCLUDING CHAPTERS ON GEOLOGY, HYDROLOGY, WINDS, TEMPERATURE AND SALINITY, HEAT BUDGET, WAVES, COASTAL CURRENTS, CARBON DIOXIDE AND PH, OXYGEN, NUTRIENTS, AND BIOLOGY. SPECIAL CHAPTERS DEAL WITH FIELD STUDIES ON THERMAL DISCHARGES, HEAT DISPERSION MODELS, PULP AND PAPER INDUSTRIAL WASTES, TRACE METALS, RADIOCHEMISTRY, PESTICIDES AND CHLORINE, THERMAL ECOLOGY, AND BIOLOGY OF 20 SELECTED SPECIES. A SUMMARY CHAPTER IS ENTITLED 'THE NEARSHORE COASTAL ECOSYSTEM: AN OVERVIEW.' THE BIBLIOGRAPHY CONTAINS MORE THAN 3100 ENTRIES, MOST FROM THE OPEN LITERATURE, BUT SOME FROM UNPUBLISHED REPORTS. A SEPARATE VOLUME (VOL II), INCLUDES THE FOLLOWING APPENDICES: (1) WIND DATA; (2) TEMPERATURE AND SALINITY DATA; (3) WAVE DATA; (4) TRACE METALS (INCLUDING TRACE METAL TOXICITIES); (5) PESTICIDE TOXICITIES; (6) OXYGEN, NUTRIENT, AND PH DATA; (7) RADIONUCLIDES; AND (8) AN ANNOTATED CHECKLIST OF PLANTS AND ANIMALS (INCLUDING MORE THAN 4400 SPECIES).

FIELD 05B, 05A, 10F

ACCESSION NO. W72-12190

LAKE SUPERIOR PERIPHYTON IN RELATION TO WATER QUALITY,

MINNESOTA UNIV., MINNEAPOLIS.

T. A. OLSON, AND T. O. ODLAUG.

COPY AVAILABLE FROM GPO SUP DOC AS EPA 18050 DBM 02/72 FOR $2.00; MICROFICHE FROM NTIS AS PB-211 185, $0.95. ENVIRONMENTAL PROTECTION AGENCY, WATER POLLUTION CONTROL RESEARCH SERIES, FEBRUARY 1972. 253 P, 171 FIG, 41 TAB, 7 REF, 4 APPEND. EPA PROGRAM 18050 DBM 02/72 (FORMERLY WP-00828).

DESCRIPTORS:
*CHLOROPHYLL, *PLANT PIGMENTS, AQUATIC PRODUCTIVITY, *PRIMARY PRODUCTIVITY, *ALGAE, *DIATOMS, *PERIPHYTON, *PHYTOPLANKTON, *SESSILE ALGAE, *LAKE SUPERIOR, LIMNOLOGY, FERTILITY, PHOTCSYNTHESIS, RESPIRATION, EUTROPHICATION, PLANT POPULATION, CLASSIFICATION, SAMPLING, CHLOROPHYTA, *CHRYSOPHYTA, CYANOPHYTA.

ABSTRACT:
LABORATORY AND FIELD STUDIES WERE CONDUCTED TO EVALUATE THE IMPORTANCE OF PERIPHYTON IN WESTERN LAKE SUPERIOR WITH SPECIAL REFERENCE TO THE MAKE-UP AND DISTRIBUTION OF THE PERIPHYTON GROWTHS AND TO THE OVERALL IMPORTANCE OF PRODUCTIVE CAPACITY OF THIS ASSEMBLAGE OF ORGANISMS. THE TAXONOMIC PORTION OF THE INVESTIGATION INDICATED THAT OVER 90% OF THE TOTAL NUMBER OF ORGANISMS WERE DIATOMS AND THAT THE PHYLA TO WHICH THESE DIATOMS BELONGED WERE THE CHRYSOPHYTA, THE CHLOROPHYTA, AND THE CYANOPHYTA. PREDOMINANT GENERA WERE SYNEDRA, ACHNANTHES, NAVICULA, CYMBELLA, AND GOMPHONEMA. IN MANY RESPECTS, THE PERIPHYTON OF LAKE SUPERIOR WAS SIMILAR TO THAT FOUND IN STREAMS AND THERE WAS EVIDENCE THAT THE INTERRELATED FACTORS THAT AFFECTED PERIPHYTON GROWTHS WERE TEMPERATURE, LIGHT INTENSITY, DEPTH OF WATER, WATER MOVEMENTS, NUTRIENT LEVELS, AND THE TYPE OF SUBSTRATE. ARTIFICIALLY DENUDED ROCKS DEMONSTRATED DEFINITE RE-GROWTH BUT AFTER 46 DAYS THIS GROWTH LEVEL WAS ONLY 18% OF THAT OCCURRING NATURALLY. THE MEAN TOTAL COUNTS OF ORGANISMS IN THE PRIMARY SAMPLING AREA RANGED FROM 497,000 TO 1,470,000 PER SQUARE CENTIMETER OF ROCK SURFACE. STUDIES OF THE PIGMENT CONCENTRATIONS SHOWED THAT THE BIOMASS OF PERIPHYTON ALONG THE NORTH SHORE OF LAKE SUPERIOR RESEMBLE THOSE OF OTHER OLIGOTROPHIC BODIES OF WATER AND RANGE FROM 0.338 TO 3.59 MG OF TOTAL PIGMENT PER 100 SQUARE CENTIMETERS OF ROCK SURFACE. THE AVERAGE WAS 1.36 MG PER 100 SQUARE CENTIMETERS OF ROCK SURFACE. PIGMENT RATIOS INDICATED THAT THE LAKE SUPERIOR PERIPHYTON WAS DOMINATED BY THE CHRYSOPHYTA.

FIELD 05C, 02H

ACCESSION NO. W72-12192

FIXATION OF SR, ZN AND CE RADIONUCLIDES BY SODIUM ALGINATE AND ALGINIC ACID FROM SEA WATER,

INSTITUTE OF BIOLOGY OF THE SOUTHERN SEAS, SEVASTOPOL (USSR).

G. E. LAZORENKO.

PAPER PRESENTED AT THE COMMISSION OF THE EUROPEAN COMMUNITIES INTERNATIONAL SYMPOSIUM, RADIOECOLOGY APPLIED TO THE PROTECTION OF MAN AND HIS ENVIRONMENT, ROME, SEPT. 1971. 8 P, 5 FIG, 5 REF.

DESCRIPTORS:
*MARINE ALGAE, *STRONTIUM RADIOISOTOPES, *ZINC RADIOISOTOPES, ABSORPTION, SEDIMENTATION, ANALYTICAL TECHNIQUES, MOLECULAR STRUCTURE, CATION EXCHANGE, RADIOECOLOGY, *PATH OF POLLUTANTS.

IDENTIFIERS:
*CERIUM RADIOISOTOPES.

ABSTRACT:
SEDIMENTATION AND LIGHT SCATTERING SHOWED THAT SODIUM ALGINATE OBTAINED FROM BLACK SEA CYSTOSEIRA HAS TWO MOLECULAR WEIGHT FRACTIONS, 800,000 AND 37,000. ION EXCHANGE STUDIES SHOWED THAT ZN IS BOUND BY ONLY THE LOWER, CE IS BOUND TO A GREATER EXTENT BY THE HIGHER, AND SR IS BOUND BY BOTH. (BOPP-ORNL)

FIELD 05B

ACCESSION NO. W72-12203

EXPERIMENTAL STUDY OF TH ISOTOPES ACCUMULATION BY MARINE ORGANISMS,

INSTITUTE OF BIOLOGY OF THE SOUTHERN SEAS, SEVASTOPOL (USSR).

A. B. NAZAROV, AND A. YA. ZESENKO.

PAPER PRESENTED AT THE COMMISSION OF THE EUROPEAN COMMUNITIES INTERNATIONAL SYMPOSIUM, RADIOECOLOGY APPLIED TO THE PROTECTION OF MAN AND HIS ENVIRONMENT, ROME, SEPT. 1971. 9 P, 4 FIG, 11 REF.

DESCRIPTORS:
*RADIOISOTOPES, *MARINE ALGAE, *ABSORPTION, *MINE WASTES, RADIOECOLOGY, *PATH OF POLLUTANTS, PHYSICOCHEMICAL PROPERTIES.

IDENTIFIERS:
THORIUM RADIOISOTOPES.

ABSTRACT:
TO DETERMINE THE ROLE OF MARINE ORGANISMS IN MIGRATION OF TH IN MINE WASTES, UPTAKE WAS STUDIED BY FOUR SPECIES OF ALGAE FROM SEAWATER CONTAINING OXALIC AND HYDROCHLORIC TH COMPOUNDS. POSSIBLE REASONS FOR DIFFERENCES BETWEEN THE CONCENTRATION FACTORS FOUND AND LITERATURE VALUES ARE: (1) UNDER NATURAL CONDITIONS THE TIME OF ALGAE CONTACT WITH SEA WATER IS CONSIDERABLY MORE THAN IN THE PRESENT EXPERIMENT, (2) THE POSSIBLE EXISTENCE OF REGIONS OF THE OCEAN WITH HIGH TH CONCENTRATION, AND (3) DIFFERENCE IN THE PHYSICOCHEMICAL STATE. (BOPP-ORNL)

FIELD 05B

ACCESSION NO. W72-12204

NON-BIOLOGICAL UPTAKE OF ZINC-65 FROM A MARINE ALGAL NUTRIENT MEDIUM,

OREGON STATE UNIV., CORVALLIS.

R. D. TOMLINSON.

AVAILABLE FROM NTIS, SPRINGFIELD, VA. AS RLO-1750-83, $3.00 PAPER COPY; $0.95 MICROFICHE. RLO-1750-83, SEPT. 1970, 73 P.

DESCRIPTORS:
*ASSAY, *ZINC, *ADSORPTION, *SURFACES, LABORATORY EQUIPMENT, *LABORATORY TESTS, *MEASUREMENT, *MARINE ALGAE TRACERS, CHEMICAL PRECIPITATION, HYDROGEN ION CONCENTRATION.

IDENTIFIERS:
NON-BIOLOGICAL UPTAKE.

ABSTRACT:
BOTH QUALITATIVE AND QUANTITATIVE EVALUATION WERE MADE OF THE NON-BIOLOGICAL ADSORPTION OF ZINC IN A LABORATORY SYSTEM DESIGNED TO MEASURE ZINC UPTAKE BY A MARINE ALGA. CARRIER-FREE 65ZN WAS USED AS A ZINC RADIOTRACER. FOUR GENERAL AREAS OF INVESTIGATION RECEIVED SPECIAL EMPHASIS. THESE WERE: (1) THE PHYSIOCHEMICAL NATURE OF THE NUTRIENT MEDIUM PRECIPITATE, (2) ADSORPTION OF 65ZN BY THE NUTRIENT MEDIUM PRECIPITATE, (3) ADSORPTION OF 65ZN BY LABORATORY GLASSWARE, AND, AS A FACTOR POTENTIALLY INFLUENCING ADSORPTION, (4) PH LEVELS IN ALGAL CULTURE CONDITIONS. (HOUSER-ORNL)

FIELD 05A, 05B

ACCESSION NO. W72-12210

EXPERIMENTAL-ECOLOGICAL INVESTIGATIONS ON PHAEOCYSTIS POUCHETI (HAPTOPHYCEAE): CULTIVATION AND WASTE WATER TEST,

BIOLOGISCHE ANSTALT HELGOLAND (WEST GERMANY).

H. KAYSER.

HELGOLAENDER WISSENSCHAFTLICHE MEERESUNTERSUCHUNG, VOL 20, P 195-212, 1970. 13 FIG, 6 REF.

DESCRIPTORS:
*MARINE ALGAE, *LABORATORY TESTS, *CHEMICAL WASTE, *GROWTH RATES, *SEAWATER, *TOXICITY, *NUTRIENTS, LETHAL LIMITS, METHODOLOGY, INDUSTRIAL WASTES, DOMESTIC WASTES, BIOASSAY, SULFATES, IRON COMPOUNDS, SEWAGE, NITRATES, PHOSPHATES.

IDENTIFIERS:
SILT-WATER-PLANT RELATIONSHIPS, NORTH SEA, PHAEOCYSTIS SPP., SOIL CULTURES, TITANIUM DIOXIDE WASTES, HELGOLAND.

ABSTRACT:
THE INFLUENCE OF LIGHT, TEMPERATURE, NUTRIENTS, INDUSTRIAL AND DOMESTIC WASTES ON THE ALGA, PHAEOCYSTIS POUCHETI, UNDER LABORATORY CONDITIONS WAS OBSERVED. THE MULTIPLICATION RATES OF THE VARIOUS STAGES IN SOIL, NITRATE AND PHOSPHATE SOLUTIONS, AND SEAWATER SOLUTIONS UNDER VARIOUS CULTURAL METHODS WAS DETERMINED. INDUSTRIAL WASTE WATER (CONSISTING PRIMARILY OF H2SO4 AND FESO4) FROM A TITANIUM DIOXIDE FACTORY FAVORS THE GROWTH OF THE COLONY STAGE OF P. POUCHETI IN A DILUTION OF 1 PART WASTE WATER TO 100,000 PARTS OF NUTRIENT MEDIUM. A DILUTION OF 1:4000 SIGNIFICANTLY REDUCES THE MULTIPLICATION RATES OF COLONIES. A DILUTION OF 1:2250 IS LETHAL. UNFILTERED DOMESTIC SEWAGE IN CONCENTRATIONS OF 1-5 PARTS OF SEWAGE TO 1000 PARTS OF SEA WATER RESULTS IN A VIGOROUS DEVELOPMENT OF THE COLONY STAGE, WHICH IS FOLLOWED BY DAMAGE TO THE COLONIES. THE SINGLE CELL STAGE SHOWS SLIGHTLY INCREASED MULTIPLICATION RATES ALL OF THE TIME. TEN PARTS OF SEWAGE TO 1000 PARTS SEAWATER RESULTS IN TOXICITY TO BOTH STAGES. (KATZ-WASHINGTON)

FIELD 05C

ACCESSION NO. W72-12239

THE INFLUENCE OF ALCOHOL EXTRACTS OF SOME ALGAE (CHLORELLA AND SCENEDESMUS) ON AQUATIC MICROORGANISMS,

WYZSZA SZKOLA ROLNICZA, OLSZTYN-KORTOWA (POLAND).

S. NIEWOLAK.

POLSKIE ARCHIWUM HYDROBIOLOGII, VOL 18, NO 2, P 31-42, 1971. 1 FIG, 3 TAB, 25 REF.

DESCRIPTORS:
*ALGAL TOXINS, *INHIBITION, *CHLORELLA, *SCENEDESMUS, AQUATIC MICROORGANISMS, AQUATIC BACTERIA, AQUATIC ALGAE, ALGAE, CHLOROPHYTA, PSEUDOMONAS, ENTERIC BACTERIA.

IDENTIFIERS:
BACILLUS, MICROCOCCUS, GRAM NEGATIVE BACTERIA.

ABSTRACT:
ALCOHOL EXTRACTS OF ALGAE OF THE GENERAL CHLORELLA AND SCENEDESMUS INHIBIT THE DEVELOPMENT OF ABOUT 20% OF WATER MICROORGANISMS. PARTICULARLY VULNERABLE ARE GRAM-POSITIVE BACTERIA OF THE GENERA MICROCOCCUS AND BACILLUS. AMONG THE GRAM-NEGATIVE BACTERIA, THE GREATEST NUMBER OF VULNERABLE STRAINS CAN BE FOUND IN THE PSEUDOMONAS GROUP AND IN THE ENTEROBACTERIACEAE FAMILY. CHLORELLA EXTRACTS HAVE MORE EXTENSIVE SCOPE OF ANTAGONISTIC ACTION THAN HAVE SCENEDESMUS EXTRACTS. THE LATTER, HOWEVER, ARE MORE ACTIVE AND THEY INHIBIT THE GROWTH OF BACTERIA IN LARGER AREAS. (LEGORE-WASHINGTON)

FIELD 05C

ACCESSION NO. W72-12255

TOXICITY OF MERCURY COMPOUNDS TO AQUATIC ORGANISMS AND ACCUMULATION OF THE COMPOUNDS BY THE ORGANISMS,

FRESHWATER FISHERIES RESEARCH LAB., TOKYO (JAPAN).

Y. MATIDA, H. KUMADA, S. KIMURA, Y. SAIGA, AND T. NOSE.

BULLETIN OF THE FRESHWATER FISHERIES RESEARCH LABORATORY, VOL 21, NO 2, P 197-227, 1971. 4 FIG, 11 TAB, 33 REF.

DESCRIPTORS:
*MERCURY, *PUBLIC HEALTH, *SHELLFISH, WATER POLLUTION EFFECTS, TOXICITY, WATER POLLUTION, FOOD CHAINS, PATH OF POLLUTANTS, BIOASSAY, FISH, RAINBOW TROUT, AQUATIC LIFE, AQUATIC ENVIRONMENT, FISH PHYSIOLOGY, ALGAE, DAPHNIA.

IDENTIFIERS:
*MERCURY COMPOUNDS, *MINAMATA DISEASE, BIOLOGICAL MAGNIFICATION, MINAMATA BAY(JAPAN), ORGANIC MERCURY COMPOUNDS, BIOSYNTHESIS, GUPPIES.

ABSTRACT:
THIS STUDY CONFIRMED THAT FISH FEEDING ON MERCURY-CONTAMINATED ORGANISMS FROM MINAMATA BAY, JAPAN, SUFFER FROM MERCURY POISONING. THE RESULTS OF STUDIES OF CHRONIC TOXICITY TO FISH OF TOXIC SHELLFISH CONTAINING SOME METHYL MERCURY COMPOUNDS AND OF VARIOUS MERCURY COMPOUNDS ARE DISCUSSED. IN ADDITION, THE RESULTS OF SOME EXPERIMENTS ON BIOLOGICAL MAGNIFICATION OF MERCURY COMPOUNDS THROUGH THE FOOD CHAIN FROM PHYTOPLANKTON TO FISH, AND ON BIOSYNTHESIS OF ORGANIC MERCURY FROM INORGANIC MERCURY ARE CONSIDERED. (SVENSSON-WASHINGTON)

FIELD 05C

ACCESSION NO. W72-12257

PREDICTION OF PHOTOSYNTHETIC BIOMASS PRODUCTION IN ACCELERATED ALGAL-BACTERIAL WASTEWATER TREATMENT SYSTEMS,

HEBREW UNIV., JERUSALEM (ISRAEL). DEPT. OF MEDICAL ECOLOGY.

G. SHELEF, M. SCHWARZ, AND H. SCHECHTER.

PREPRINT, PRESENTED AT 6TH INTERNATIONAL WATER POLLUTION RESEARCH CONFERENCE, SESSION 5, HALL A, PAPER NO 9, JUNE 20, 1972. P A/5/9/1-A/5/9/10, 6 FIG, 2 TAB, 23 REF.

DESCRIPTORS:
*OXIDATION LAGOONS, *ALGAE, *WASTE WATER TREATMENT, *MATHEMATICAL MODELS, GROWTH RATES, SOLAR RADIATION, FLOTATION, FLOCCULATION, PERFORMANCE, EFFICIENCIES, *BIOMASS, ORGANIC MATTER, NUTRIENTS, PHOTOSYNTHESIS.

IDENTIFIERS:
ALUM, FERRIC CHLORIDE, *JERUSALEM(ISRAEL).

ABSTRACT:
THE USE OF SPECIALLY DESIGNED PONDS AS A CONTROLLED WASTE WATER TREATMENT WITH MAXIMIZED PHOTOSYNTHETIC ACTIVITY OF ALGAE WAS DEMONSTRATED. THE KINETICS OF ALGAE PRODUCTION WITH RESPECT TO SOLAR INCIDENT IRRADIANCE SERVED FOR THE CONSTRUCTION OF A MATHEMATICAL MODEL FOR PREDICTING ALGAE BIOMASS PRODUCTION AND THE CONCENTRATION OF ALGAE AS A FUNCTION OF HYDRAULIC DILUTION RATE UNDER GIVEN LEVELS OF SOLAR IRRADIANCE. THE PREDICTED LEVELS OF ALGAE PRODUCTION AND CONCENTRATIONS WERE CLOSE TO ACTUAL LEVELS, ALTHOUGH IN DETERMINING THE OPTIMAL DILUTION RATE, SOME CORRECTIONS HAD TO BE MADE. THE REMOVAL EFFICIENCY OF ORGANIC MATTER AND NUTRIENTS IN THE POND SYSTEM WAS RELATIVELY HIGH, PROVIDED THE ALGAE BIOMASS WAS SEPARATED FROM THE EFFLUENT. A 1,750 GAL/HR RECTANGULAR FLOTATOR WAS USED FOR SEPARATING THE ALGAL BIOMASS FROM THE POND EFFLUENT AFTER TREATMENT WITH ALUM OR FERRIC CHLORIDE. PERFORMANCE DATA OF THE ACCELERATED PHOTOSYNTHETIC POND FOR MONTHS JUNE AND DECEMBER WERE ARRANGED IN TABULAR FORM. (GALWARDI-TEXAS)

FIELD 05D

ACCESSION NO. W72-12289

EFFECT OF SPECTRAL COMPOSITICN ON PHOTOSYNTHESIS IN TURBIC RESERVOIRS--PHOTOSYNTHETIC PRODUCTION IN A TURBIC RESERVOIR II. DETAILS OF AN INCUBATION MODEL AND COMMENTS ON THE EFFECT OF LIGHT QUALITY ON PHOTOSYNTHESIS,

KANSAS WATER RESOURCES RESEARCH INST., MANHATTAN.

J. A. OSBORNE, AND G. R. MARZOLF.

AVAILABLE FROM THE NATIONAL TECHNICAL INFORMATION SERVICE AS PB-211 368, $3.00 IN PAPER COPY, $0.95 IN MICROFICHE. CONTRIBUTION NC 106, JUNE 1972. 102 P, 8 FIG, 11 TAB, 69 REF, 2 APPEND. OWRR A-047-KAN(1).

DESCRIPTORS:
*RESERVOIRS, *PRIMARY PRODUCTIVITY, *TURBIDITY, *PHOTOSYNTHESIS, LIMNOLOGY, *EUTROPHICATION, *KANSAS, MODEL STUDIES, ECOSYSTEMS, ALGAE, *LIGHT QUALITY, EUTROPHIC ZONE, CHLOROPHYLL, PHOSPHATES, NITRATES, SEASONAL, REGRESSION ANALYSIS, CORRELATION ANALYSIS.

IDENTIFIERS:
*PHOTOSYNTHETIC MODEL, *TUTTLE CREEK RESERVOIR(KAN), GREAT PLAINS, AQUATIC ECOSYSTEM.

ABSTRACT:
A NUMERICAL MODEL, BASED ON SOME SIMPLIFYING ASSUMPTIONS, WAS USED TO EVALUATE SPATIAL AND SEASONAL PATTERNS IN PRIMARY PRODUCTION IN TUTTLE CREEK RESERVOIR, KANSAS, FROM JUNE 1970 THROUGH MAY 1971. THE ASSUMPTIONS OF THE MODEL BEST FIT TURBID, HOMOTHERMAL LAKES, SUCH AS THOSE IN KANSAS AND THE GREAT PLAINS PROVINCE. THE MODEL DESCRIBES THE PHOTOSYNTHESIS-DEPTH PROFILE ON THE BASIS OF LIGHT ATTENUATION WITH DEPTH AND ALLOWS INTEGRAL PHCTOSYNTHESIS TO BE OBTAINED BY A SIMPLE ANALYTICAL CALCULATION. THE USE OF THE MODEL HAD ITS GREATEST ADVANTAGE IN ALLOWING PELAGIC ALGAL PHOTOSYNTHESIS TO BE MONITORED IN SHIPBOARD INCUBATORS. VARIOUS CHEMICAL, PHYSICAL, AND BIOLOGICAL FEATURES WERE MEASURED SIMULTANEOUSLY WITH PRIMARY PROCUCTION TO DESCRIBE THE DIVERSITY OF THE RESERVOIR, TO EVALUATE HORIZONTAL PRODUCTION PATTERNS, AND TO VALIDATE THE ASSUMPTIONS OF THE MODEL.

FIELD 05C, 02H, 02J

ACCESSION NO. W72-12393

COLOR INFRARED (CIR) PHOTOGRAPHY: A TOOL FOR ENVIRONMENTAL ANALYSIS,

DARTMOUTH COLL., HANOVER, N.H. DEPT. OF GEOGRAPHY.

D. T. LINDGREN.

AVAILABLE FROM THE NATIONAL TECHNICAL INFORMATION SERVICE AS PB-204 472, $3.00 IN PAPER COPY, $0.95 IN MICROFICHE. FINAL REPORT, AUGUST 1971. 42 P, 8 REF. CONTRACT NO. DI-14-08-0001-12958.

DESCRIPTORS:
*REMOTE SENSING, *POLLUTANT IDENTIFICATION, WATER POLLUTION, *MONITORING, AERIAL PHOTOGRAPHY, *INFRARED RADIATION, WATER QUALITY, ALGAE, WATER QUALITY CONTROL, EUTROPHICATION, WASTE WATER TREATMENT, SEDIMENTATION RATES, OIL SPILLS, WATER POLLUTION SOURCES.

IDENTIFIERS:
*COLOR INFRARED PHOTOGRAPHY.

ABSTRACT:
THE NATURE OF COLOR INFRARED FILM (CIR), ITS CAPABILITIES AND LIMITATIONS, ARE DESCRIBED AND ITS POTENTIAL AS A TOOL IN ENVIRONMENTAL STUDIES IS DISCUSSED. INCLUDED IS A SECTION ON THE USE OF CIR IN WATER QUALITY ANALYSIS. THE GROWTH OF ALGAE IS EASILY DETECTED BY THIS TECHNIQUE, WHICH CAN BE USED AS AN INDICATOR OF UNDESIRABLE CONDITIONS. IN TREATMENT FACILITIES, CIR HAS BEEN USED TO IDENTIFY TRICKLING FILTERS THAT ARE OVERLOADED SINCE THEY REFLECT GREATER AMOUNTS OF INFRARED THAN THOSE THAT ARE OPERATING EFFICIENTLY. CIR CAN ALSO BE USED TO DETERMINE THE DEGREE OF SEDIMENTATION OF BODIES OF WATER. FOR DETECTION OF OIL SPILLS, CIR HAS NOT PROVEN AS RELIABLE AS OTHER TYPES OF FILM BECAUSE OF ITS SENSITIVITY TO SUN ANGLES. FINALLY, CIR CAN BE USED FOR DEAD-FISH COUNTS WHERE SOME TOXIC SUBSTANCE HAS BEEN RELEASED IN THE WATER. (MORTLAND-BATTELLE)

FIELD 05A, 07B

ACCESSION NO. W72-12487

CAPACITY OF DESERT ALGAL CRUSTS TO FIX ATMOSPHERIC NITROGEN,

ARIZONA UNIV., TUCSON. DEPT. OF AGRICULTURAL CHEMISTRY AND SOILS.

A. N. MACGREGOR, AND D. E. JOHNSON.

SOIL SCIENCE SOCIETY OF AMERICA PROCEEDINGS, VOL 35, NC 5, P 843-844, SEPTEMBER-OCTOBER 1971. 7 REF.

DESCRIPTORS:
*ALGAE, *NITROGEN FIXATION, *NITROGEN FIXING BACTERIA, *RAINFALL, *ARID LANDS, *ARIZONA, GRASSLANDS, ON-SITE DATA COLLECTIONS, PLANT GROWTH.

ABSTRACT:
IN ADDITION TO THEIR ROLE IN INCREASING NITROGEN, ALGAL CRUSTS, WHICH CONTAIN A VARIETY OF MICROORGANISMS, ARE REPORTED TO ENHANCE SOIL WATER INFILTRATION AND TO PROVIDE A LAYER OF ORGANIC MATTER SUITABLE FOR SEEC GERMINATION. A STUDY WAS UNDERTAKEN TO CETERMINE HOW SCON ALGAL CRUSTS FROM A SEMIARID SOIL ARE ABLE TO FIX N AFTER BEING MOISTENED AND THE RATE OF N FIXATION PER UNIT AREA OF MOISTENEC CRUSTS. THE STUDY WAS CONDUCTED IN A GRASSLAND AREA IN THE TUCSON BASIN OF SOUTHERN ARIZONA WHERE APPROXIMATELY 4% OF THE SOIL SURFACE POSSESSED ALGAL CRUST FORMATIONS. THREE HOURS AFTER BEING MOISTENED, ALGAL CRUSTS PRODUCED DETECTABLE LEVELS OF N FIXATICN AS MEASURED BY THE ACETYLENE-ETHYLENE METHOD. IT WAS ESTIMATED THAT 1 HA OF CESERT GRASSLAND RECEIVES A N INPUT OF 3-4 G/HR FOLLOWING A RAINFALL. THIS IS FORTUNATE IN VIEW OF THE BURST OF HERBACECUS PLANT GROWTH ANC SUBSEQUENT PLANT N REQUIREMENTS DURING THE RAINY SEASON IN THE SONORAN DESERT. (CASEY-ARIZONA)

FIELD 05B, 02G

ACCESSION NO. W72-12505

INTERRELATIONS AMONG PLANKTON, ATTACHED ALGAE, AND THE PHCSPHORUS CYCLE IN ARTIFICIAL OPEN SYSTEMS,

ITHACA COLL., N.Y. DEPT. OF BIOLCGY.

J. L, CONFER.

ECCLOGICAL MONOGRAPHS, VOL 42, NO 1, P 1-22, 1972. 10 FIG, 9 TAB, 32 REF.

DESCRIPTORS:
*PLANKTON, *ALGAE, *PHOSPHORUS, *MODEL STUCIES, LAKES, CYCLING NUTRIENTS, TRACERS, LITTORAL, KINETICS, PHOSPHATES.

IDENTIFIERS:
*PHOSPHORUS CIRCULATION, SMALL LAKES.

ABSTRACT:
PHOSPHORUS CIRCULATICN DURING SUMMER STRATIFICATICN WAS STUDIED IN 200-LITER AQUARIA, CONTINUOUSLY SUPPLIED WITH TAP WATER, BY MEANS OF ANALYTICAL ANC TRACER METHODS. THIS MODEL OF PHOSPHORUS CIRCULATION IS BELIEVED TO APPLY TO SMALL LAKES WITH EXTENSIVE LITTORAL VEGETATION, THE PHOSPHORUS INFLUX TO VARIOUS BIOLOGICAL COMPARTMENTS BEING EQUALED BY A CORRESPONDING OUTFLCW. A MAJOR MEANS BY WHICH PHCSPHORUS WAS REMOVED FROM THE OPEN WATER WAS BY TRAPPING OF PARTICLES BY THE COMMUNITY ASSOCIATED WITH THE AQUARIA SICES. AFTER THE FIRST RUN THE PONDS WERE INOCULATED WITH WATER FROM THE PREVIOUS PONDS. A BIOTIC SUCCESSION DEVELOPED FROM A PURELY PLANKTONIC COMMUNITY TO A MORE COMPLEX, TWO-COMMUNITY SYSTEM OF ATTACHEC ORGANISMS CN THE SIDES AND THE PLANKTON. THE P REMOVAL RATE VARIED WIDELY, DEPENDING ON EXTENT OF 'LITTORAL' GROWTH AND NATURE OF THE PARTICLES. CCNSIDERING SUCCESSION TO BE THE TOTAL CHANGE IN PHYSICAL AND BIOLOGICAL CONDITIONS OVER PROLONGED TIME, THIS OPEN-SYSTEM DESIGN DEVELOPEC FRCM A ONE-COMMUNITY, FEW-SPECIES SYSTEM INTO A TWO-COMMUNITY, SEVERAL-SPECIES SYSTEM. NUTRIENT CIRCULATION RATE GREATLY INCREASED WITH TIME AND DEVELOPED INTO THIS STEADY STATE, NOT AN EQUILIBRIUM SYSTEM WHICH DEPENDED ON A CONTINUAL PHOSPHORUS INFLUX TO MAINTAIN CONCENTRATIONS AND CIRCULATION RATES. (JONES-WISCONSIN)

FIELD 05C, 02H

ACCESSION NO. W72-12543

NUTRIENTS LIMITING ALGAL GROWTH,

DSM'S CENTRAL LAB., GELEEN (NETHERLANDS). BIOLOGICAL DEPT.

J. W. WOLDENDORP.

STRIKSTOF, DUTCH NITROGENOUS FERTILIZER REVIEW, NO 15, P 16-27, 1972. 3 FIG, 12 PHOTOS, 26 REF.

DESCRIPTORS:
*EUTROPHICATION, *NUTRIENTS, *ALGAE, *LIMITING FACTORS, PHOSPHATES, CHLOROPHYTA, CYANOPHYTA, NITROGEN, CARBON, DIATOMS, BACTERIA, CARBON DIOXIDE, CHLAMYDOMONAS, SCENEDESMUS, CHLORELLA, SILICA, LIGHT INTENSITY, TEMPERATURE, MUD, GROWTH RATES, ANABAENA, NITROGEN FIXATION, DENITRIFICATION, WATER POLLUTION EFFECTS, WATER POLLUTION CONTROL.

IDENTIFIERS:
NETHERLANDS, LIMITING NUTRIENTS.

ABSTRACT:
LITERATURE ON EXCESSIVE ALGAL GROWTH IS REVIEWED. ALTHOUGH LIMNOLOGISTS ARGUE THAT ALGAL GROWTH IS GENERALLY LIMITED BY PHOSPHORUS, NITROGEN CAN ALSO BE LIMITING IN SOME CASES; IT IS ALSO ARGUED THAT CARBON, NOT PHOSPHORUS, IS THE LIMITING NUTRIENT. IF A CORRELATION IS FOUND BETWEEN PHOSPHORUS CONCENTRATION AND ALGAL YIELD, IT IS NOT PROPER TO CONCLUDE THAT PHOSPHORUS IS THE LIMITING NUTRIENT. IT WILL TAKE A CONSIDERABLE TIME BEFORE ANY EFFECT OF PHOSPHATE REMOVAL IS OBSERVED, CONSIDERING THE VAST QUANTITIES OF PHOSPHATES STORED IN THE MUD. MICROBIAL PROCESSES IN TOTO PROBABLY HAVE A REGULATING EFFECT ON NITROGEN CONCENTRATION OF SURFACE WATERS. IT IS UNLIKELY THAT CARBON WOULD BE THE LIMITING ELEMENT IN VIEW OF HIGH BICARBONATE CONTENT OF MOST WATERS. WHETHER BICARBONATES OR CARBON DIOXIDE ARE UTILIZED BY ALGAL POPULATIONS IS DISCUSSED. LITERATURE CONCERNING THE INFLUENCE OF N, P AND C ON ALGAL GROWTH IN SURFACE WATERS DOES NOT INDICATE CLEARLY WHICH ELEMENT HAS LIMITING EFFECT UNDER THE PREVAILING NETHERLAND CONDITIONS. ALL FORMS OF POLLUTION BY HEAT, CHEMICALS, PATHOGENIC MICROORGANISMS, ETC., MUST BE DIMINISHED IF THE WATER IS TO CONTINUE TO SUPPORT LIFE. (JONES-WISCONSIN)

FIELD 05C

ACCESSION NO. W72-12544

SYNOPTIC STUDY OF ACCELERATED EUTROPHICATION IN LAKE TAHOE--AN ALPINE LAKE,

CALIFORNIA UNIV., DAVIS. INST. OF ECOLOGY.

C. R. GOLDMAN, G. MOSHIRI, AND E. DE AMEZAGA.

IN: INTERNATIONAL SYMPOSIUM ON WATER POLLUTION CONTROL IN COLD CLIMATES, JULY 22-24, 1970, UNIVERSITY OF ALASKA, COLLEGE, P 1-21. 9 FIG, 2 TAB, 28 REF.

DESCRIPTORS:
*EUTROPHICATION, *LAKES, *ALPINE, *SYNOPTIC ANALYSIS, COLD REGIONS, NUTRIENTS, BIOLOGICAL COMMUNITIES, ALGAE, PERIPHYTON, PHYTOPLANKTON, PHOTOSYNTHESIS, INVERTEBRATES, BIOMASS, BENTHIC FAUNA, BENTHIC FLORA, HUMAN POPULATION, PRIMARY PRODUCTIVITY, WATER CHEMISTRY, *CALIFORNIA, *NEVADA.

IDENTIFIERS:
*LAKE TAHOE(CALIF-NEV).

ABSTRACT:
LAKE TAHOE WAS STUDIED BY THE SYNOPTIC APPROACH WHICH PROVIDES A NEARLY INSTANTANEOUS EVALUATION OF CONDITIONS EXISTING ON A GIVEN DAY, ALLOWING NUTRIENT SOURCES TO BE LOCATED ACCURATELY. INCREASED FERTILITY WAS EVIDENT AT THE SOUTH SHORE UNDER THE INFLUENCE OF THE TRUCKEE RIVER AND HIGH RESIDENT POPULATION, IN CRYSTAL BAY WHICH CONTAINED HIGHLY DISTURBED LAND DRAINAGE, AND NEAR THE LAKE OUTFLOW WHERE THERE WERE HIGH RESIDENT POPULATION AND FAIRLY EXTENSIVE SHALLOW WATER AREAS. ALTHOUGH OCCASIONAL HIGH PERIPHYTON VALUES WERE ENCOUNTERED NEAR TRIBUTARIES, THERE WAS LESS CORRELATION WITH TRIBUTARIES THAN WAS FOUND FOR PHYTOPLANKTON PRODUCTIVITY AND BIOMASS; DISTRIBUTION OF PERIPHYTON WAS FAIRLY UNIFORM AROUND THE LAKE. THE ABUNDANCE AND DIVERSITY OF BENTHIC ORGANISMS MAY NOT BE FUNCTIONS OF THE SAME ENVIRONMENTAL PROPERTY AS THE ABUNDANCE, PRODUCTIVITY, AND DIVERSITY OF THE PHYTOPLANKTON. THE PRIMARY PRODUCERS MUST BE VIEWED AS A MORE SENSITIVE INDICATION OF INCREASED FERTILITY THAN CHEMICAL PARAMETERS, SINCE ANY ADDITIONAL NUTRIENTS APPEAR TO MOVE RAPIDLY INTO THE PHYTOPLANKTON. LITTLE OR NO MEASURABLE CHANGE IN WATER CHEMISTRY WAS FOUND WHILE PHYTOPLANKTON PHOTOSYNTHESIS SHOWED A VERY SIGNIFICANT CHANGE. (JONES-WISCONSIN)

FIELD 05C, 02H

ACCESSION NO. W72-12549

MICROBIOLOGIC INDICATORS OF THE EFFICIENCY OF AN AERATED, CONTINUOUS-DISCHARGE, SEWAGE LAGOON IN NORTHERN CLIMATES,

NORTH DAKOTA UNIV., GRAND FORKS. SCHOOL OF MEDICINE.

J. W. VENNES, AND O. O. OLSON.

IN: INTERNATIONAL SYMPOSIUM ON WATER POLLUTION CONTROL IN COLD CLIMATES, JULY 22-24, 1970, UNIVERSITY OF ALASKA, COLLEGE, P 286-311. 4 TAB, 5 REF, APPEND.

DESCRIPTORS:
*SEWAGE BACTERIA, *WASTE WATER TREATMENT, *SEWAGE LAGOONS, *COLD REGIONS, AERATED LAGOONS, *NORTH DAKOTA, BIOCHEMICAL OXYGEN DEMAND, BACTERIA, TEMPERATURE, COLIFORMS, ENTERIC BACTERIA, NITROGEN, VITAMINS, SUSPENDED SOLIDS, LACTOBACILLUS, CHLORELLA, SULFUR BACTERIA, ALGAE, SALMONELLA, E. COLI.

IDENTIFIERS:
*CONTINUOUS DISCHARGE LAGOON.

ABSTRACT:
THE AERATED LAGOON, A DEVELOPMENT IN BIOLOGICAL WASTE TREATMENT WAS STUDIED IN HARVEY, NORTH DAKOTA. COLIFORM, FECAL COLIFORM, AND ENTEROCOCCI WERE DETERMINED AS WELL AS BOD, NITROGEN, PH, TOTAL AND SUSPENDED SOLIDS; AND TOTAL BACTERIAL POPULATIONS ENUMERATED. LAGOON EFFICIENCY DEPENDS ON TEMPERATURE AND OXYGEN; THEIR EFFECT ON BIOLOGIC STABILIZATION IS DETERMINED AND THE FINDINGS ARE REFLECTED IN RELATIVE ABUNDANCE OF SEVERAL MICROBIAL SPECIES AND IN BOD. COLIFORM, FECAL COLIFORM, AND ENTEROCOCCAL NUMBERS IN THE SECONDARY LAGOON DURING ZERO CENTIGRADE TEMPERATURES WERE DIRECTLY RELATED TO BOD AND TOTAL NITROGEN. DURING SUMMER TEMPERATURES, LITTLE CORRELATION BETWEEN THESE ORGANISMS AND BOD AND TOTAL NITROGEN WAS NOTED. THERE WAS A CORRELATION BETWEEN THE TOTAL MICROBIAL POPULATION AND BOD AT SUMMER TEMPERATURES. SINCE ONLY 1% OR LESS OF THE TOTAL MICROBIAL POPULATION IS REPRESENTED BY THE ENTERIC ORGANISMS STUDIED, IT IS APPARENT THAT OTHER ORGANISMS MUST BE STUDIED TO DEFINE BETTER THE ROLE OF MICROBIOLOGIC INDICATORS IN THE EFFICIENCY OF THIS SEWAGE TREATMENT SYSTEM. A SYSTEM CONCERNED WITH PRODUCTION AND UTILIZATION OF THE VITAMIN B, BIOTIN, IS BEING STUDIED, SINCE IT RELATES TO SEVERAL ORGANISMS THAT THRIVE IN SEWAGE OXIDATION LAGOONS. (JONES-WISCONSIN)

FIELD 05D, 05C, 02C

ACCESSION NO. W72-12565

STRUCTURE AND FUNCTIONING OF ESTUARINE ECOSYSTEMS EXPOSED TO TREATED SEWAGE WASTES.

NORTH CAROLINA UNIV., CHAPEL HILL. INST. OF MARINE SCIENCES.

AVAILABLE FROM THE NATIONAL TECHNICAL INFORMATION SERVICE AS COM-71-00688, $6.00 IN PAPER COPY, $0.95 IN MICROFICHE. ANNUAL REPORT FOR 1970-1971, FEBRUARY 1971. 345 P, 62 FIG, 50 TAB, 110 REF. NOAA GH 103.

DESCRIPTORS:
*ECOSYSTEMS, *PRODUCTIVITY, *ESTUARINE ENVIRONMENT, *DOMESTIC WASTES, *SALT MARSHES, *CARBON CYCLE, *PHOSPHORUS COMPOUNDS, *NITROGEN CYCLE, *CRABS, *FISHES, ECOLOGY, FOOD CHAINS, ESTUARINE FISHERIES, MARINE FISHERIES, SEWAGE DISPOSAL, SEWAGE LAGOONS, BRACKISH WATER FISH, BIOLOGICAL TREATMENT, OXIDATION LAGOONS, PHYTOPLANKTON, AQUATIC INSECTS, WATERFOWL, MARINE ALGAE, AQUICULTURE.

IDENTIFIERS:
MEIOFAUNA, FORAMINIFERA.

ABSTRACT:
THIS IS THE THIRD ANNUAL REPORT FROM AN INVESTIGATION OF THE ECOLOGICAL SYSTEMS WHICH DEVELOP WHEN ESTUARINE WATERS ARE ENRICHED WITH SEWAGE WASTES. VARIOUS PHASES OF COMMUNITY STRUCTURE AND METABOLISM OF SIX EXPERIMENTAL BRACKISH-WATER PONDS, THREE OF WHICH RECEIVED TREATED SEWAGE WASTES, AND OF A SMALL TIDAL CREEK AND ITS SALT MARSHES WERE STUDIED. INCLUDED ARE CHAPTERS ON PRODUCTIVITY, CARBON METABOLISM, THE PHOSPHOROUS BUDGET, NITROGEN, AND BACTERIAL HETEROTROPHY; ON THE STANDING CROPS OF PHYTOPLANKTON, DECAPOD CRUSTACEANS, FISHES, MEIOFAUNA, FORAMINIFERA, INSECTS, MOLLUSKS, AND BIRDS; ON CALCIUM ANALYSIS; AND ON GROWTH AND REPRODUCTION OF ALGAE. THE WASTES PONDS HAVE DEVELOPED INTO PRODUCTIVE, WELL INTEGRATED, BUT SLIGHTLY UNSTABLE SYSTEMS. THEY PERFORM SOME OF THE FUNCTIONS OF TERTIARY TREATMENT AND HOLD PROMISE FOR PRODUCTION OF HARVESTABLE SEAFOOD PROTEIN. (KATZ-WASHINGTON)

FIELD 05C, 02L

ACCESSION NO. W72-12567

THE EFFECT OF MARINE POLLUTANTS ON LAMINARIA HYPERBOREA,

MARINE BIOLOGICAL STATION, PORT ERIN, ISLE OF MAN (ENGLAND).

R. HOPKINS, AND J. M. KAIN.

MARINE POLLUTION BULLETIN, VOL. 2, NO. 5, P 75-77, MAY 1971. 1 TAB, 3 REF.

DESCRIPTORS:
*ALGAE, *BIOASSAY, *PLANT PHYSIOLOGY, POLLUTANTS, *TOXICITY, WATER POLLUTION EFFECTS, RESPIRATION, *HEAVY METALS, CHEMICAL WASTES, MERCURY, COPPER, ZINC, PESTICIDES, PESTICIDE TOXICITY, *HERBICIDES, DALAPON, 2-4-D, INSECTICIDES, PHENOLS, *DETERGENTS, *LETHAL LIMIT, ALKYLBENZENE SULFONATES.

IDENTIFIERS:
*LAMINARIA, SUBLITTORAL REGION, CULTURE EXPERIMENTS, ATRAZINE, MCPA, ENDOSULFAN, FAIRY LIQUID, BLUSYL, SODIUM LAURYL, ETHER SULFATE, SODIUM DODECYL, BENZENE SULFONATE, COCONUT FATTY ACID, PLURONICS, DIETHANOLAMIDE.

ABSTRACT:
A STUDY HAS BEEN MADE OF THE TOXIC CONCENTRATIONS OF 15 CHEMICAL POLLUTANTS, INCLUDING EXAMPLES OF HEAVY METALS, DETERGENTS, AND HERBICIDES, USING LAMINARIA HYPERBOREA AS THE TEST ORGANISM. THIS PLANT IS ECOLOGICALLY THE MOST IMPORTANT IN THE SUBLITTORAL REGION AROUND MUCH OF THE COAST OF BRITAIN. TWO DIFFERENT BIOASSAY TECHNIQUES WERE USED, ONE OF A CULTURE EXPERIMENT TYPE, AND THE OTHER UTILIZING RESPIRATION MEASUREMENTS. (SVENSSON-WASHINGTON)

FIELD 05C

ACCESSION NO. W72-12576

FIVE NEW SPECIES OF CHLAMYDOMONAS,

WISCONSIN STATE UNIV., LA CROSSE. DEPT. OF BIOLOGY.

J. M. KING.

JOURNAL OF PHYCOLOGY, VOL. 8, NO. 1, P 120-126, MARCH 1972. 23 FIG, 24 REF.

DESCRIPTORS:
*SYSTEMATICS, *CHLAMYDOMONAS, CULTURES, CHLOROPHYTA, PROTOZOA, *ALGAE.

IDENTIFIERS:
CHLAMYDOMONAS ISABELIENSIS, CHLAMYDOMONAS PALLIDOSTIGMATICA, CHLAMYDOMONAS FOTTII, CHLAMYDOMONAS TEXENSIS, CHLAMYDOMONAS PSEUDOMICROSPHAERA, CULTURE MEDIA, CULTURING TECHNIQUES, MORPHOLOGY, FLAGELLATES.

ABSTRACT:
FIVE NEW SPECIES OF CHLAMYDOMONAS, C. ISABELIENSIS, C. PALLIDOSTIGMATICA, C. FOTTII, C. PSEUDOMICROSPHAERA AND C. TEXENSIS WERE ISOLATED INTO AXENIC CULTURE DURING AN INVESTIGATION OF GREEN MICROALGAE. THE AXENIC CULTURES WERE MAINTAINED ON BOLDO BASAL MEDIUM (BBM) SOLIDIFIED WITH 1.5 PERCENT AGAR AND THE MORPHOLOGY OF THE ORGANISMS STUDIED WITH WET MOUNTS AND HANGING DROP SLIDES FROM ISOLATES ON BBM AGAR OR BBM LIQUID MEDIUM. EXPERIMENTS WERE PERFORMED IN A CONTROLLED ENVIRONMENT CULTURE ROOM WITH TEMPERATURE AT 22 C AND INCIDENT LIGHT PROVIDED BY FLUORESCENT BULBS. SUPPLEMENTARY ATTRIBUTES OF THE ALGAE MAY SERVE AS POSSIBLE TAXONOMIC AIDS. MORPHOLOGICAL DESCRIPTIONS AND CULTURING CHARACTERISTICS ARE GIVEN FOR EACH OF THE NEW SPECIES. (SNYDER-BATTELLE)

FIELD 05A

ACCESSION NO. W72-12632

STUDIES ON THE BIOLOGY OF BROWN ALGAE ON THE ATLANTIC COAST OF VIRGINIA. I. PORTERINEMA FLUVIATILE (PORTER) WAERN,

KENT STATE UNIV., OHIO. DEPT. OF BIOLOGICAL SCIENCES.

R. G. RHODES.

JOURNAL OF PHYCOLOGY, VOL. 8, NO. 1, P 117-119, MARCH 1972. 11 FIG, 8 REF.

DESCRIPTORS:
*PHAEOPHYTA, *TIDAL MARSHES, *MARINE ALGAE, CULTURES, SALINITY, SALT MARSHES, SYSTEMATICS, *VIRGINIA, *ATLANTIC OCEAN, NORTH AMERICA.

IDENTIFIERS:
*PORTERINEMA FLUVIATILE, CULTURE MEDIA, HUMMOCK CHANNEL, CHLOROPLASTS, MORPHOLOGY, EPIPHYTES.

ABSTRACT:
PORTERINEMA FLUVIATILE FOUND AS A MICROSCOPIC FILAMENT ON A CULM OF SPARTINA ON THE ATLANTIC COAST OF VIRGINIA, REPRESENTS THE FIRST REPORT OF THIS MARINE ALGA IN NORTH AMERICA. THE COLLECTION WAS MADE FROM A TIDAL MARSH WITH WATER TEMPERATURE OF 19.5 C AND SALINITY OF ADJACENT WATER BEING 2.98 PERCENT. SINGLE FILAMENTS WERE ISOLATED AND PLACED INTO CULTURES CONTAINING MODIFIED SCHREIBER'S SOLUTION. CULTURES WERE MAINTAINED AT 20 PLUS OR MINUS 1 DEGREE C WITH CONTINUOUS ILLUMINATION FROM FLUORESCENT LAMPS. PHOTOGRAPHS WERE TAKEN WITH BRIGHT FIELD ILLUMINATION AND CHLOROPLAST MORPHOLOGY AND NUMBER WERE EXAMINED IN LIGHT PASSING THROUGH A WRATTEN FILTER NO. 48 (EASTMAN KODAK). (SNYDER-BATTELLE)

FIELD 05A

ACCESSION NO. W72-12633

RADIOISOTOPIC STUDY OF CALCIFICATION IN THE ARTICULATED CORALLINE ALGA BOSSIELLA ORBIGNIANA,

CALIFORNIA INST. OF TECH., PASADENA. DIV. OF GEOLOGICAL AND PLANETARY SCIENCES.

V. B. PEARSE.

JOURNAL OF PHYCOLOGY, VOL. 8, NO. 1, P 88-97, MARCH 1972. 10 FIG, 39 REF.

DESCRIPTORS:
*PHOTOSYNTHESIS, CALCIUM CHLORIDE, RADIOACTIVITY TECHNIQUES, METABOLISM, TRACERS, RADIOACTIVITY, CARBON DIOXIDE, SALTS, MARINE ALGAE.

IDENTIFIERS:
*CALCIFICATION, *CORALLINE ALGAE, CALCIUM RADIOISOTOPES, *BOSSIELLA ORBIGNIANA, PAPER CHROMATOGRAPHY, SCINTILLATION COUNTING, CA-45.

ABSTRACT:
IN AN ATTEMPT TO LINK CALCIFICATION TO PHOTOSYNTHETIC PROCESSES IN PLANTS, 2-3 CM BRANCHES OF BOSSIELLA ORBIGNIANA WERE SEPARATED FROM THE PLANTS AND INCUBATED IN LIGHT OR DARK IN MILLIPORE FILTERED SEAWATER CONTAINING 2 MICROCURIES/ML CA-45 AS CALCIUM CHLORIDE. AFTER 32 HR INCUBATION, THE BRANCHES WERE WASHED, POST-INCUBATED IN MEDIA FREE OF CA-45, AND THE MEDIA PERIODICALLY ASSAYED FOR CA-45 ACTIVITY BY METHODS OF DECALCIFICATION, SCINTILLATION COUNTING, AND PAPER CHROMATOGRAPHY. DATA INDICATE THAT CALCIFICATION IN BOSSIELLA ORBIGNIANA (CORALLINACEAE) IS CHARACTERIZED BY A SERIES OF FEATURES WHICH VARY ACCORDING TO THE AGE OF THE SEGMENT (INTERGENICULUM). FROM THE YOUNGEST SEGMENT AT THE BRANCH TIP TO OLDER, MORE BASAL SEGMENTS: (1) WEIGHT AND DEGREE OF MINERALIZATION INCREASE, WHILE RATE OF WEIGHT GAIN DECREASES; (2) RATE AND STABILITY OF CA-45 LABELING DECREASE; (3) EFFECT OF LIGHT ON CA-45 LABELING RATE DECREASES, WHILE EFFECT OF LIGHT ON STABILITY OF LABEL INCREASES; HOWEVER, FOR ALL SEGMENTS, CA-45 LABELING IS MORE RAPID AND MORE STABLE IN THE LIGHT THAN IN THE DARK; (4) RATE OF CA-45 LABELING IN KILLED CONTROLS DECREASES. COMPARATIVE STUDIES WITH THE TROPICAL CORALLINE ALGA AMPHIROA YIELDED SIMILAR RESULTS. CHARACTERISTICS OF THE GRADIENTS IN CALCIFICATION AND EFFECT OF LIGHT IN THESE CORALLINE ALGAE ARE VERY SIMILAR TO THOSE FOUND IN A REEF-BUILDING CORAL CONTAINING SYMBIOTIC ALGAE, AND THE DATA SUGGEST THAT ORGANIC PRODUCTS OF PHOTOSYNTHESIS MAY BE OF GENERAL IMPORTANCE TO CALCIFICATION. (MACKAN-BATTELLE)

FIELD 05C, 05A

ACCESSION NO. W72-12634

CHARACTERIZATION OF NATURALLY OCCURRING DISSOLVED ORGANOPHOSPHORUS COMPOUNDS,

WASHINGTON UNIV., SEATTLE. DEPT. OF CIVIL ENGINEERING.

R. A. MINEAR.

ENVIRONMENTAL SCIENCE AND TECHNOLOGY, VOL. 6, NO. 5, P 431-437, MAY 1972. 11 FIG, 6 TAB, 30 REF.

DESCRIPTORS:
*ALGAE, *PHOSPHORUS, *LAKES, CULTURES, PHOSPHATES, NUTRIENTS, CHLOROPHYTA, *ORGANOPHOSPHORUS COMPOUNDS, WASHINGTON, SEPARATION TECHNIQUES, DISTILLATION, SPECTROPHOTOMETRY, CHLOROPHYLL, HYDROLYSIS, PHYTOPLANKTON, FLUORESCENCE, PHOSPHORUS COMPOUNDS, CHLAMYDOMONAS, WATER ANALYSIS, FLUOROMETRY, PIGMENTS.

IDENTIFIERS:
*ORTHOPHOSPHATES, CHLAMYDOMONAS REINHARDTII, MOSES LAKE, PINE LAKE, LAKE WASHINGTON, CHLOROPHYLL A, *DNA, *DISSOLVED ORGANOPHOSPHORUS COMPOUNDS, NATURAL ORGANICS.

ABSTRACT:
TWO SOURCES OF DISSOLVED ORGANIC PHOSPHORUS COMPOUNDS (DOP) WERE USED IN A STUDY CONCERNED WITH THE FORMATION AND MECHANISMS OF RELEASE OF DOP COMPOUNDS IN NATURAL WATERS: (1) PURE CULTURES OF THE GREEN ALGA, CHLAMYDOMAS REINHARDTII, GROWN IN A CARBON-DIOXIDE ENRICHED ATMOSPHERE; AND (2) NATURAL WATER SAMPLES TAKEN FROM THREE SEPARATE LAKES. SPECTROPHOTOMETRY WAS USED FOR ORTHOPHOSPHATE DETERMINATION AND DNA ANALYSES. CHLOROPHYLL A AND TOTAL PIGMENT WERE DETERMINED FLUORIMETRICALLY. IN THE PURE ALGAL CULTURES, HIGH LEVELS OF SOLUBLE ORGANIC PHOSPHORUS WERE OBTAINED WITH ACCOMPANYING LOW RESIDUAL ORTHOPHOSPHATE. MOLECULAR SIEVE SAMPLES OF THE NATURAL LAKE WATER SAMPLES SHOWED THAT UP TO 20 PERCENT OF THE RECOVERABLE ORGANIC PHOSPHORUS IS HIGH-MOLECULAR-WEIGHT MATERIAL. A SIZABLE PERCENTAGE (UP TO 50 PERCENT) OF THIS MATERIAL APPEARED, IN MOST CASES, TO BE DNA OR ITS FRAGMENTS. THE NATURAL ORIGIN OF THE DOP, ESPECIALLY, DNA, AND OTHER HIGH-MOLECULAR-WEIGHT COMPONENTS WAS SUBSTANTIATED IN SPECIAL CULTURE STUDIES WHICH UTILIZED DUAL CHAMBER VESSELS DIVIDED BY 0.22 MICRON MEMBRANES. THE PRESENCE OF DNA IN THE PURE ALGAL CULTURES AND THE CORRELATION WITH CHLOROPHYLL AND TOTAL PIGMENT FOR NATURAL WATER SAMPLES SUGGEST THAT THIS MATERIAL ORIGINATES FROM THE PHYTOPLANKTON. (MORTLAND-BATTELLE)

FIELD 05C, 05B

ACCESSION NO. W72-12637

NATURAL ABUNDANCE OF THE STABLE ISOTOPES OF CARBON IN BIOLOGICAL SYSTEMS,

TEXAS UNIV., AUSTIN. DEPT. OF BOTANY.

B. N. SMITH.

BIOSCIENCE, VOL 22, NO 4, P 226-231, APRIL 1972. 4 FIG, 1 TAB, 64 REF.

DESCRIPTORS:
*ISOTOPE STUDIES, *STABLE ISOTOPES, *CARBON RADIOISOTOPES, ENVIRONMENTAL EFFECTS, METHODOLOGY, ALGAE, AQUATIC PLANTS, ISOTOPE FRACTIONATION, CARBON DIOXIDE, RESPIRATION, MARINE PLANTS, BICARBONATES, MASS SPECTROMETRY, DISTRIBUTION, SEPARATION TECHNIQUES.

IDENTIFIERS:
*C-13, *C-12, BIOLOGICAL SYSTEMS, BELEMNITELLA AMERICANA, TERRESTRIAL PLANTS, COMBUSTION, ISOTOPE RATIOS, SPECTROMETRY, LICHENS, MOSSES, GYMNOSPERMS, ANGIOSPERMS.

ABSTRACT:
THE GASEOUS CO2 USED IN MEASURING THE STABLE CARBON ISOTOPES, C-12 AND C-13, IS COLLECTED WITH SPECIFIC APPARATUS FROM THE COMBUSTION OF ORGANIC MATERIAL AT 800-900 DEGREES C IN EXCESS OXYGEN OR FROM RESPIRATION OF LIVING TISSUE. AFTER COLLECTION, THE GASEOUS CO2 IS ANALYZED ON AN ISOTOPE RATIO MASS SPECTROMETER. THE C-13/C-12 RATIO IN ANY GIVEN SAMPLE IS THEN COMPARED WITH A STANDARD - CO2 FROM THE FOSSIL CARBONATE SKELETON OF BELEMNITELLA AMERICANA (PDB SUB L) ACCORDING TO THE FUNCTION: DELTA C-13 PER MIL IS EQUAL TO 1000 TIMES THE C-13/C-12 OF THE SAMPLE MINUS THE C-13/C-12 OF THE STANDARD DIVIDED BY C-13/C-12 OF THE STANDARD. THE PRECISION OF MEASURING DELTA C-13 WITH THE MASS SPECTROMETER IS PLUS OR MINUS 0.1 PER MIL. RESULTS OF THESE METHODS HAVE SHOWN MARINE AND FRESHWATER PLANTS TO HAVE RELATIVELY MORE C-13 THAN MOST TERRESTRIAL PLANTS. THE PRESUMED DIFFERENCE ALLOWED FOR DISTINGUISHING BETWEEN MARINE AND FRESHWATER SEDIMENTS, PLANTS AND PETROLEUM, AND HAS SUGGESTED THAT ENVIRONMENTAL EFFECTS (E.G., TEMPERATURE) MAY ACCOUNT FOR SOME OF THE ISOTOPIC FRACTIONATIONS IN ORGANISMS AND PLANTS. (MACKAN-BATTELLE)

FIELD 05B, 05A

ACCESSION NO. W72-12709

SKEWED ALGAL DIVISION PATTERNS: EFFECTS OF AUTOSPORE YIELD ON COMPUTED SYNCHRONY INDICES,

SAINT LOUIS UNIV., MO. DEPT. OF BIOLOGY.

D. W. ROONEY.

MATHEMATICAL BIOSCIENCIES, VOL 12, NO 3/4, P 367-373, DECEMBER 1971. 2 FIG, 3 REF.

DESCRIPTORS:
*MATHEMATICAL STUDIES, *ALGAE, EQUATIONS, BIOLOGICAL PROPERTIES, REPRODUCTION, ANALYTICAL TECHNIQUES, NUMERICAL ANALYSIS, MODEL STUDIES, COMPUTERS, CULTURES.

IDENTIFIERS:
*DIVISION PATTERNS, *AUTOSPORES, *SYNCHRONOUS CULTURES, SYNCHRONY INDEX.

ABSTRACT:
REPORTED ARE ALGEBRAIC AND NUMERICAL ANALYSES OF THE ALGAL DIVISION SYNCHRONY INDEX S SUB T WHICH REFLECTS THE DEGREE OF NONLINEARITY OF PLOTS OF LOG CELL NUMBER VERSUS TIME. IT HAS BEEN FOUND THAT FOR SKEWED DISTRIBUTIONS OF DIVISION TIMES, S SUB T IS AFFECTED BY THE NUMBER OF AUTOSPORES PRODUCED PER DIVISION. A NEW ALGAL DIVISICN SYNCHRONY INDEX S SUB R IS PRESENTED. FOR A CULTURE OF ANY GIVEN AUTCSPORE YIELD, S SUB R REFLECTS THE NONLINEARITY OF A SEMILOGARITHMIC PLOT VERSUS TIME OF CELL NUMBER IN A CULTURE OF IDENTICAL DIVISION TIME PATTERN YIELDING TWO AUTOSPORES PER DIVISION. THE INDEX S SUB R, INDEPENDENT OF AUTOSPORE YIELD FOR ANY DIVISION TIME DISTRIBUTICN, ASSUMES HIGHER VALUES FOR DISTRIBUTIONS SKEWED TOWARD LONGER TIMES THAN FOR OPPOSITELY SKEWED DISTRIBUTIONS. (MACKAN-BATTELLE)

FIELD 05A, 05C

ACCESSION NC. W72-12724

THE GRADUAL DESTRUCTION OF SWEDEN'S LAKES,

NATIONAL SWEDISH ENVIRCNMENT PROTECTION BOARD, STOCKHOLM.

T. WILLEN.

AMBIO, VOL 1, NO 1, P 6-14, FEBRUARY 1972. 9 FIG, 3 REF.

DESCRIPTORS:
*LAKES, *WATER POLLUTION EFFECTS, *LIMNOLOGY, WATER POLLUTION, WATER POLLUTION SOURCES, SEWAGE, INDUSTRIAL WASTES, BIOINDICATCRS, CHEMICAL PROPERTIES, PHYSICAL PROPERTIES, PHYTOPLANKTON, ZOOPLANKTON, CHLOROPHYTA, DIATOMS, CYANCPHYTA, BENTHIC FAUNA, AQUATIC ANIMALS, INVERTEBRATES, AQUATIC ALGAE, AERIAL PHOTOGRAPHY, PHOSPHORUS, NITRATES, NITROGEN, *EUTROPHICATION, HYDROGEN ION CONCENTRATION, NUTRIENTS, CHLOROPHYLL, PIKES, PERCHES, PHYSICOCHEMICAL PROPERTIES, TROUT, SALMON, FRESHWATER FISH, ANABAENA, *TURBIDITY, CHRYSOPHYTA, AQUATIC PLANTS, MIDGES, OLIGOCHAETES, ANNELIDS, OLIGOTROPHY, BACTERIA.

IDENTIFIERS:
*CHLOROPHYLL A, MACRCPHYTES, *SWEDEN, MICROCYSIS, CSCILLATORIA, SYNURA, AROGLENA, CRYPTOPHYCEAE, PONTOPOREIA AFFINIS, ESCX LUCIUS, PERCA FLUVIATILIS, SALVELINUS ALPINUS, SALMO TRUTTA, VENDACE, SALMON TROUT, LOTA LOTA, LAKE MALOGEN, SWEDEN, CHAR, LUCIOPERCA LUCISPERCA, BURBOT, COREGONUS LAVARETUS, COREGONUS ALBULA, WHITEFISH, BRANCHIURA SOWERBYI, PIKE PERCH, LAKE HJALMAREN, LAKE VATTERN, LAKE VANERN.

ABSTRACT:
DETRIMENTAL CHANGES ARE TAKING PLACE IN A GRCUP CF 4 SWEDISH LAKES CAUSED BY INCREASING SEWAGE DISPOSAL AND INDUSTRIAL POLLUTION OF THEIR WATERS. WATER SAMPLES COLLECTED 6 TIMES A YEAR AND BIOLOGICAL SAMPLES (BACTERIA, PLANKTON, CHLOROPHYLL) COLLECTED MONTHLY WERE TRESTED BY PHYSICO-CHEMICAL METHODS FOR PH, ORGANIC MATTER, OXYGEN CONTENT, SILICA, TURBIDITY, AND CONDUCTIVITY. THE FLORA ANC FAUNA WERE ALSO STUDIED. LONG TERM STUDIES REVEAL INCREASING DETERIORATION AROUND HIGHLY POPULATED OR INDUSTRIAL AREAS. LAKE MALAREN, TAKEN AS A REPRESENTATIVE LAKE, SHOWS INCREASING TURBIDITY, HIGH CONCENTRATIONS OF PLANT NUTRIENTS RESULTING IN RELATIVELY HIGH PHYTOPLANKTON LEVELS, AND HIGH VOLUMES OF ZOOPLANKTON FEEDING OR BOTH. ALTHCUGH ORGANISM QUANTITY APPEARS TO BE INCREASING, QUALITY AND COMPOSITION ARE NOT. CHLOROPHYLL A STUDIES SHOW CONSIDERABLE YEARLY VARIATION BUT APPEAR TO INVERSELY CORRELATE WITH TURBIDITY LEVELS. (MACKAN-BATTELLE)

FIELD 05C, 05B, 02H

ACCESSION NO. W72-12729

AUTOANTAGONISM, HETEROANTAGONISM AND OTHER CONSEQUENCES OF THE EXCRETIONS OF ALGAE FROM FRESH OR THERMAL WATER, (AUTO., HETEROANTAGONISME ET AUTRES CONSEQUENCES DES EXCRETIONS D'ALGUES D'EAU DOUCE OU THERMALE),

CENTRE NATIONAL DE LA RECHERCHE SCIENTIFIQUE, PARIS (FRANCE). LABORATOIRE DE CYTOLOGIE ET DE CYTOPHYSIOLOGIE DE LA PHOTOSYNTHESE.

M. TASSIGNY, AND M. LEFEVRE.

INTERNATIONALE VEREINIGUNG FUR THEORETISCHE UND ANGEWANDTE LIMNOLOGIE, MITTEILUNGEN NO 19, P 26-38, NOVEMBER 1971. 6 FIG, 4 TAB, 39 REF.

DESCRIPTORS:
*AQUATIC ALGAE, *COMPETITION, *NUTRIENTS, *GROWTH RATES, *INHIBITION, FUNGI, THERMAL SPRINGS, CYANOPHYTA, BACTERIA, ECOLOGY, WATER POLLUTION EFFECTS.

IDENTIFIERS:
*EXCRETORY PRODUCTS, *AUTOANTAGONISM, *HETEROANTAGONISM, AXENIC CULTURES, EXCRETION, APHANIZOMENON GRACILE.

ABSTRACT:
DIFFERENT SPECIES OF ALGAE COMPETE DIRECTLY FOR NUTRIENTS, OR INDIRECTLY BY THE PRODUCTION OF SUBSTANCES THAT INHIBIT THE GROWTH OF COMPETITORS (HETEROANTAGONISM). LEFEVRE ALSO RECOGNIZES AUTOANTAGONISM, THE INHIBITION OF THE GROWTH OF A SPECIES BY ITS OWN EXCRETORY PRODUCTS. CERTAIN ALGAE IN CULTURE REACH A STAGE WHERE GROWTH CEASES ALTHOUGH THERE APPEARS TO BE NO LACK OF NUTRIENTS AND GROWTH CANNOT BE RESTARTED BY THE ADDITION OF NUTRIENTS. IT IS OBSERVED IN NATURE THAT, AS ONE SPECIES BECOMES PARTICULARLY ABUNDANT, OTHER SPECIES BECOME SCARCE. WHEN THE ABUNDANT SPECIES DECLINES, THE OTHERS RESUME ACTIVE MULTIPLICATION. EXPERIMENTALLY IT WAS SHOWN THAT SPECIES CULTURED IN THE WATER FEEDING A CERTAIN CANAL FLOURISHED, WHEREAS THOSE CULTURED IN FILTERED WATER FROM THE CANAL, WHERE THERE WAS A LARGE POPULATION OF APHANIZOMENON GRACILE, DID NOT. BACTERIA AND FUNGI WERE ELIMINATED IN A DENSE POPULATION OF AN ALGA, AND HETEROANTAGONISM WAS DEMONSTRATED IN THE LABORATORY WITH STRAINS CULTURED FREE OF BACTERIA AND FUNGI. THE SUBSTANCES PRODUCED BY SOME SPECIES OCCASIONALLY STIMULATE THE GROWTH OF OTHERS. THE EFFECT OF ONE SPECIES UPON ANOTHER VARIES ACCORDING TO THE CONDITIONS OF THE MEDIUM. (HOLOMAN-BATTELLE)

FIELD 05C

ACCESSION NO. W72-12734

ACTION OF CALCIUM ON THE GROWTH OF AXENIC DESMIDS, (ACTION DU CALCIUM SUR LA CROISSANCE DE DESMIDIEES AXENIQUES),

CENTRE NATIONAL DE LA RECHERCHE SCIENTIFIQUE, PARIS (FRANCE). LABORATOIRE DE CYTOLOGIE ET DE CYTOPHYSIOLOGIE DE LA PHOTOSYNTHESE.

M. TASSIGNY.

INTERNATIONALE VEREINIGUNG FUR THEORETISCHE UND ANGEWANDTE LIMNOLOGIE, MITTEILUNGEN NO 19, P 292-313, NOVEMBER 1971. 9 FIG, 3 TAB, 31 REF.

DESCRIPTORS:
*CALCIUM, *GROWTH RATES, *CULTURES, *AQUATIC ALGAE, CHLOROPHYTA, INHIBITORS, WATER POLLUTION EFFECTS, ENVIRONMENTAL EFFECTS, COMPETITION, AQUATIC POPULATIONS, NUTRIENTS, NUTRIENT REQUIREMENTS.

IDENTIFIERS:
*DESMIDS, *AXENIC CULTURES, CULTURING TECHNIQUES, STAURASTRUM PARADOXUM, STAURASTRUM SEBALDII VAR. ORNATUM, MICRASTERIAS CRUX-MELITENSIS, CLOSTERIUM STRIGOSUM, CULTURE MEDIA.

ABSTRACT:
IN ORDER TO STUDY THE EFFECT OF CALCIUM ON THE GROWTH OF PURE CULTURES, EXPERIMENTS WERE PERFORMED IN A MEDIUM BUFFERED WITH SODIUM PHOSPHATE TO PREVENT EXCESSIVE CHANGES IN PH AND 300 MG/L POTASSIUM NITRATE WAS ADDED TO THE CULTURE MEDIUM TO MAINTAIN GROWTH. STAURASTRUM PARADOXUM, WHICH WAS TAKEN ORIGINALLY FROM A CALCIUM-RICH POLLUTED POND, GREW EQUALLY WELL IN ALL FOUR CONCENTRATIONS OF CALCIUM USED. POPULATIONS OF THREE OTHER SPECIES TESTED INCREASED MORE SLOWLY THE HIGHER THE CONCENTRATION OF CALCIUM IN THE CULTURE MEDIUM, THOUGH THE EFFECT ON EACH WAS NOT IDENTICAL. S. SEBALDII VAR. ORNATUM GREW SLOWLY IN ALL MEDIA; THE RATE AT WHICH THE POPULATION INCREASED WAS DEPRESSED BY CALCIUM, THOUGH THE TOTAL NUMBER ATTAINED VARIED LESS THAN DID THAT OF SOME OTHER SPECIES. CALCIUM NOT ONLY DEPRESSES THE TIME BETWEEN DIVISIONS OF MICRASTERIAS CRUX-MELITENSIS BUT REDUCES MARKEDLY THE FINAL NUMBERS ATTAINED. CLOSTERIUM STRIGOSUM DIFFERS FROM MICRASTERIAS MAINLY IN THAT 13.5 MG/L IS THE CONCENTRATION BELOW WHICH EFFECT ON THE FINAL SIZE OF THE POPULATION IS SLIGHT. WHEN M. CRUS-MELITENSIS AND S. SEBALDII VAR. ORNATUM ARE CULTURED TOGETHER IN 17.1 MG/L CA, THE PROPORTION OF STAURASTRUM FALLS AT FIRST BECAUSE IT TAKES LONGER TO ADAPT TO THE NEW CONDITIONS OF THE CULTURE MEDIUM. AT CA CONCENTRATIONS OF LESS THAN 17.1 MG/L THE RISE IN THE PROPORTION OF STAURASTRUM TAKES PLACE MORE SLOWLY BECAUSE COMPETITION WITH M. CRUX-MELITENSIS IS MAXIMUM. THERE IS NO HETEROANTAGONISTIC REACTION OF EITHER TO THE OTHER, THOUGH THIS APPEARS TO BE THE FACTOR INHIBITING THE GROWTH OF M. CRUX-MELITENSIS IN CULTURE WITH CLOSTERIUM STRIGOSUM.
(HOLOMAN-BATTELLE)

FIELD 05C

ACCESSION NO. W72-12736

PHYTOPLANKTON OF TWO DANISH LAKES, WITH SPECIAL REFERENCE TO SEASONAL CYCLES OF THE NANNOPLANKTON,

COPENHAGEN UNIV. (DENMARK). INST. OF PLANT ANATOMY AND CYTOLOGY.

J. KRISTIANSEN,

INTERNATIONALE VEREINIGUNG FUER THEORETISCHE UND ANGEWANDTE LIMNOLOGIE, MITTEILUNGEN NO 19, P 253-265, NOVEMBER 1971. 12 FIG, 15 REF.

DESCRIPTORS:
*PHYTOPLANKTON, *NANNOPLANKTON, LAKES, *SEASONAL, *PRIMARY PRODUCTIVITY, AQUATIC ALGAE, ZOOPLANKTON, CHRYSOPHYTA, CHLOROPHYTA, BIOMASS, PYRROPHYTA, SPRING, SUMMER, AUTUMN, SAMPLING, PLANKTON NETS, DIATOMS, FOOD CHAINS, ANABAENA, SCENEDESMUS, SEASONAL, DINOFLAGELLATES.

IDENTIFIERS:
LAKE ESROM, TYSTRUP LAKE, BAVELSE LAKE, CRYPTOPHYCEAE, *DENMARK, CYCLOTELLA COMTA, STEPHANODISCUS HANTZSCHII, APHANIZOMENON FLOS-AQUAE, CHROOMONAS ACUTA, MELOSIRA GRANULATA, CERATIUM, MICROCYSTIS, RHODOMONAS MINUTA, ASTERIONELLA FORMOSA, STEPHANODISCUS ASTREA, ANKISTRODESMUS, CHRYSOCOCCUS MINUTUS, DICTYOSPHAERIUM, LAGERHEIMIA.

ABSTRACT:
THE PHYTOPLANKTON OF TYSTRUP-BAVELSE LAKES AND LAKE ESROM (DENMARK) WAS EXAMINED IN RELATIONSHIP TO COMPOSITION, QUANTITY, AND SEASONAL CYCLES OF NANNOPLANKTON. FOR MOST OF THE YEAR, ABOUT 80 PERCENT OF THE TOTAL CELL NUMBER WAS NANNOPLANKTON. IN THE TYSTRUP-BAVELSE LAKES DIATOMS WERE DOMINANT IN SPRING, CHLOROPHYCEAE IN SUMMER, AND CRYPTOPHYCEAE IN AUTUMN AND WINTER. IN LAKE ESROM, CHRYSOPHYCEAE AND CHLOROPHYCEAE WERE DOMINANT IN WINTER AND SPRING, CRYPTOPHYCEAE IN SUMMER AND AUTUMN. NANNOPLANKTON CONSTITUTED 25 PERCENT OF THE PHYTOPLANKTON BIOMASS IN BOTH LAKES WITH A VERNAL MAXIMUM (50-80 PERCENT). THE SEASONAL CYCLES OF NANNOPLANKTON BIOMASS WERE DUE PRIMARILY TO DIATOMS. BIOMASS VALUES WERE CALCULATED FROM VOLUME ESTIMATIONS OF EACH SPECIES CONCERNED. RESULTS WERE EXPRESSED IN ONE MILLION CUBIC MICRONS/ML, WITH VALUES BEING TRANSFORMED INTO WEIGHT VALUES, MG/L. (SNYDER-BATTELLE)

FIELD 05C, 02H

ACCESSION NO. W72-12738

MINERAL NUTRITION OF PLANKTONIC ALGAE: SOME CONSIDERATIONS, SOME EXPERIMENTS,

KOHLENSTOFFBIOLOGISCHE FORSCHUNGSSTATION, DORTMUND (WEST GERMANY).

C. J. SOEDER, H. MULLER, H. D. PAYER, AND H. SCHULLE.

INTERNATIONALE VEREINIGUNG FUR THEORETISCHE UND ANGEWANDTE LIMNOLOGIE, MITTEILUNGEN NO 19, P 39-58, NOVEMBER 1971. 6 FIG, 5 TAB, 53 REF.

DESCRIPTORS:
*MINERALOGY, *NUTRIENT REQUIREMENTS, *PHYTOPLANKTON, *AQUATIC ALGAE, CULTURES, PHOSPHATES, NITRATES, NUTRIENTS, SALTS, TRACE ELEMENTS, LIGHT INTENSITY, SCENEDESMUS, CHLORELLA, GROWTH RATES, RESISTANCE, PHOSPHORUS, IRON, MANGANESE, ZINC, NICKEL, COBALT, COPPER SULFATE, HEAVY METALS, METALS, SULFATES.

IDENTIFIERS:
*CHLORELLA FUSCA, *SCENEDEMUS ACUTUS VAR ALTERNANS, *NITZSCHIA ACTINASTROIDES, *STAURASTRUM PINGUE, SCENEDESMUS QUADRICAUDA, CHLORELLA PYRENOIDOSA, LITHIUM CHLORIDE, LITHIUM, ASTERIONELLA FORMOSA, ASTERIONELLA JAPONICA, STAURASTRUM PARADOXIUM, CULTURE MEDIA, ZINC SULFATE, MANGANESE CHLORIDE, NICKEL SULFATE, COBALT SULFATE, POTASSIUM BROMIDE.

ABSTRACT:
SEVERAL ASPECTS OF MINERAL NUTRITION IN PLANKTONIC ALGAE ARE COMPARED BASED ON BOTH A LITERATURE REVIEW AND SEVERAL EXPERIMENTS. UPPER LIMITS OF NUTRIENT CONCENTRATIONS REPRESENTING THE NUTRIENT CONCENTRATION TOLERANCE WERE ESTABLISHED FOR EACH OF FOUR ORGANISMS: CHLORELLA FUSCA, SCENEDESMUS ACUTUS VAR. ALTERNANS, NITZSCHIA ACTINASTROIDES, AND STAURASTRUM PINGUE. FROM THESE EXPERIMENTS IT WAS TENTATIVELY CONCLUDED THAT THE RELATIVE SALT TOLERANCE OF FRESHWATER ALGAE IS INVERSELY PROPORTIONAL TO THE GROWTH RATE WHICH IS ATTAINED UNLESS THE SALT CONCENTRATION BECOMES SUPRAOPTIMAL. THE RELATIVITY OF THE OPTIMUM OF NUTRIENT CONCENTRATIONS WAS ALSO EVIDENT IN STUDIES ON TRACE ELEMENT DOSAGE. SUSPENSION DENSITY WAS FOUND TO INFLUENCE OPTIMAL TRACE ELEMENT CONCENTRATIONS, THE AVERAGE REQUIREMENTS OF THE DENSE CULTURES BEING HIGHER THAN THESE OF THE MORE DILUTE SYNCHRONOUS CULTURES. A LENGTHY DISCUSSION ON THE ROLE OF PHOSPHATE IS INCLUDED. (MORTLAND-BATTELLE)

FIELD 05C

ACCESSION NO. W72-12739

DICHCTOMOSIPHON IN FLORIDA SPRINGS,

FLCRIDA UNIV., GAINESVILLE. DEPT. CF BOTANY.

J. S. DAVIS, AND W. F. GWOREK.

JOURNAL OF PHYCOLOGY, VOL 8, NO 1, P 130-131, MARCH 1972. 6 REF.

DESCRIPTORS:
*CHLOROPHYTA, *SPRINGS, DIATCMS, BIOTA, PROTOZOA, *AMPHIPODA, *FLORIDA, STREAMS, ISOPODS, SCENEDESMUS, AQUATIC HABITATS, CLASSIFICATION, AQUATIC ALGAE.

IDENTIFIERS:
*DICHOTOMOSIPHON TUBEROSUS, *EPIPHYTES, GROWTH MEDIA, COCCONEIS PLACENTULA, CLOSTERIUM, GCMPHONEMA LCNGICEPS, CYMBELLA, VAUCHERIA.

ABSTRACT:
DICHOTOMOSIPHON AND ITS ASSOCIATED BIOTA WERE OBTAINED FROM SEVERAL FLORIDA SPRINGS. COLLECTIONS WERE MADE SEVERAL TIMES BY DIVING WITH SCUBA EQUIPMENT AND WADING. TO AID IN IDENTIFICATION, PIECES OF THE DICHOTOMOSIPHON MAT WERE PLACED IN 3 DIFFERENT GRCWTH MEDIA: SOIL AND WATER, SPRING WATER, AND A DEFINED MINERAL MEDIUM. THEY WERE INCUBATED AT 70F UNDER FLORESCENT LIGHT (350 FT-C INTENSITY WITH A 16 HR-LIGHT, 8 HR-DARK CYCLE). AFTER 12-15 DAYS, THE TUBES IN SPRING WATER PRCDUCED HIGHLY CHARACTERISTIC SEX ORGANS AND AKINETES. IN ITS NATURAL HABITAT THIS TUBULAR ALGA PRODUCES EXTENSIVE BRIGHT GREEN MATS WHICH HARBOR EPIPHYTES (COCCONEIS PLACENTULA, GCMPHONEMA LONGICEPS), UNICELLULAR ALGAE (CLOSTERIUM, SCENEDESMUS, AND CYMBELLA), CILIATES, AND AMPHIPODS. (SNYDER-BATTELLE)

FIELD 05A

ACCESSION NO. W72-12744

ACCUMULATION AND PERSISTENCE OF DDT IN A LCTIC ECOSYSTEM,

MAINE UNIV., ORONO. DEPT. OF ENTCMOLOGY; AND MAINE UNIV., CRONO. DEPT. OF BIOCHEMISTRY.

J. B. DIMOND, A. S. GETCHELL, AND J. A. BLEASE.

JOURNAL OF FISHERIES RESEARCH BOARD OF CANADA, VOL 28, NO 12, P 1877-1882, DECEMBER 1971. 3 TAB, 20 REF.

DESCRIPTORS:
*PESTICIDE RESIDUES, *LOTIC ENVIRONMENT, WATERSHEDS(BASINS), *DDT, *SMALL WATERSHEDS, ECOSYSTEMS, PESTICIDE KINETICS, *MAINE, MUDS, INVERTEBRATES, FISH, BIRDS, STREAMS, AQUATIC PLANTS, AQUATIC INSECTS, CRUSTACEANS, MUSSELS, COMMON MERGANSER DUCK, BIOASSAY, WATER POLLUTION EFFECTS, GAS CHROMATOGRAPHY, BROOK TROUT, SAMPLING, SEPARATION TECHNIQUES, ALGAE, CRAYFISH, FOOD CHAINS, PERSISTENCE, RUNOFF, SEDIMENTS.

IDENTIFIERS:
CHUBS, KINGFISHES, SALVELINUS FONTINALIS, SEMOTILUS ATROMACULATUS, CLEANUP, SAMPLE PRESERVATION, CAMBARUS BARTONI, BIOLCGICAL SAMPLES, MERGUS MERGANSER, MEGACERYLE ALCYCN, BIOACCUMULATION.

ABSTRACT:
SMALL WATERSHEDS IN MAINE THAT HAD RECEIVED VARIOUS NUMBERS OD DDT TREATMENTS STARTING IN 1958 WERE STUDIED FCR PERSISTENCE OF DDT RESIDUES WITH ADJACENT UNTREATED WATERSHEDS BEING USED AS CONTROLS. MUDS, PLANTS, INVERTEBRATES, AND FISH WERE COLLECTED IN 1967 AND 1968 AND BIRDS IN 1969. SAMPLES WERE FROZEN WITHIN A FEW HOURS AFTER COLLECTION AND HELD AT MINUS 15 C UNTIL EXTRACTICN OF PESTICIDES. VARIOUS COLLECTIONS AND EXTRACTICN TECHNIQUES WERE USED, DEPENDING UPON THE TYPE OF SAMPLE. DDT WAS FOUND TO PERSIST IN THE STREAMS FOR AT LEAST 10 YEARS FOLLOWING LIGHT APPLICATION TO THE FOREST; HOWEVER, RESIDUES DID DECLINE SHARPLY AFTER 2 - 3 YEARS. THE PROLONGED PERSISTENCE LED TO CUMULATIVE LEVELS IN STREAMS SPRAYED MORE THAN ONCE AND RESIDUE CONCENTRATION THROUGHOUT THE FOOD CHAIN WAS EVIDENT. (MORTLAND-BATTELLE)

FIELD 05B

ACCESSION NO. W72-12930

PHYSIOLOGICAL INVESTIGATIONS ON THE TOLERANCE OF FUCUS VIRSOIDES (DON) J. AG., (PHYSIOLOGISCHE UNTERSUCHUNGEN UBER DIE TOLERANZ VON FUCUS VIRSOIDES (DON) J. AG),

ZAVOD ZA BIOLOGIJU MORA, ROVINJ (YUGOSLAVIA).

F. GESSNER, AND L. HAMMER.

INTERNATIONALE REVIE DER GESAMTEN HYDROBIOLOGIE, VOL 56, NO 4, P 581-597, 1971. 9 FIG, 7 TAB, 16 REF.

DESCRIPTORS:
*MARINE ALGAE, *KELPS, *PHYSIOLOGICAL ECOLOGY, PHAEOPHYTA, STANDING CROPS, RESISTANCE, PRIMARY PRODUCTIVITY, BIOMASS, ANAEROBIC CONDITIONS, WATER TEMPERATURE, CHLORIDES, PHOTOSYNTHESIS, LITTORAL, RESPIRATION, PLANT PHYSIOLOGY, BICARBONATES, WATER POLLUTION EFFECTS.

IDENTIFIERS:
*FUCUS VIRSOIDES, FUCUS SPIRALIS, MEDITERRANEAN SEA.

ABSTRACT:
THE WORK OF LINARDIC IS REPORTED, WHO PROVED THAT FUCUS VIRSOIDES, THROUGH ISOLATION HAS BECOME A SEPARATE SPECIES, ENDEMIC TO THE MEDITERRANEAN; IT IS POSSIBLY A DESCENDANT OF FUCUS SPIRALIS. THE ECOLOGICAL HABITAT CONDITIONS ARE ANALYZED AND IT IS NOTED THAT FUCUS VIRSOIDES, AVOIDS VERY EXPOSED COASTS. ACCORDING TO LINARDIC, THE MAIN GROWTH PERIODS ARE WINTER AND SPRING; WHEN COMPARED TO NORTHERN LITTORAL ALGAL POPULATIONS, 'STANDING CROP' AND PRIMARY PRODUCTION SEEM TO BE VERY LIMITED. THE TEMPERATURE LIMIT FOR PHOTOSYNTHESIS AT 12 HOURS EXPOSURE IS ABOUT 34C. AFTER EXPOSURE TO TEMPERATURES A FEW DEGREES ABOVE ZERO, PHOTOSYNTHESIS REMAINS NORMAL, BUT FUCUS DOES NOT TOLERATE FREEZING. IN RESPECT TO TEMPERATURE, RESPIRATION PROVES TO BE MORE TOLERANT THAN PHOTOSYNTHESIS; HOWEVER, EXPOSURE TO SUN THROUGHOUT THE DAY DOES NOT LEAD TO A REDUCTION IN PHOTOSYNTHESIS BUT A RISE IN THE BICARBONATE CONTENT CAUSES A MAINFOLD INCREASE IN RATE. AIR-DRY CONDITIONS CAN BE TOLERATED FOR 2-3 DAYS WITHOUT DAMAGE; HOWEVER, SUBSEQUENT PROCESSES IN FUCUS EXPOSED TO DRYNESS LEAD TO THE DEATH OF THE ALGAE. THIS WAS PROVED BY THE INCREASE OF CL-LOSS AND THE WINKLER-TITRATION. FUCUS VIRSOIDES CAN TOLERATE 1 WEEK IN DISTILLED WATER ALMOST WITHOUT DAMAGE AND IS AMAZINGLY TOLERANT TO OXYGEN-FREE CONDITIONS (ANAEROBIOSIS). PHOTOSYNTHESIS DOES NOT DECREASE UNTIL SEVERAL DAYS IN ANAEROBIC CONDITIONS; IN CONTRAST, RESPIRATION REMAINS UNAFFECTED. FUCUS VIRSOIDES IS NOT AN INDICATOR FOR MINIMAL EFFECTS OF POLLUTED WATER; HOWEVER, IT DISAPPEARS RAPIDLY AFTER STRONG POLLUTION OF COASTAL AREAS. (IN GERMAN) (HOLOMAN-BATTELLE)

FIELD 05C

ACCESSION NO. W72-12940

INTERDISCIPLINARY MONITORING OF THE NEW YORK BIGHT,

GRUMMAN AEROSPACE CORP., BETHPAGE, N.Y. RESEARCH DEPT.

W. G. EGAN, J. M. CASSIN, AND M. E. HAIR.

AVAILABLE FROM THE NATIONAL TECHNICAL INFORMATION SERVICE AS AD-737 506, $3.00 IN PAPER COPY, $0.95 IN MICROFICHE. GRUMMAN RESEARCH DEPARTMENT MEMORANDUM NO RM-534J, JANUARY 1972. 15 P, 5 FIG, 7 REF.

DESCRIPTORS:
*INSTRUMENTATION, *MONITORING, *WATER PROPERTIES, *ON-SITE DATA COLLECTIONS, NEW YORK, BIOLUMINESCENCE, BIOLOGICAL PROPERTIES, CHLOROPHYLL, TURBIDITY, SAMPLING, MATHEMATICAL MODELS, ESTUARINE ENVIRONMENT, SALINITY, CHEMICAL ANALYSIS, BIOMASS, DISSOLVED OXYGEN, THERMOCLINE, HYDROGEN ION CONCENTRATION, FLOW, CHEMICAL PROPERTIES, PHYSICAL PROPERTIES, HUDSON RIVER, DEPTH, WATER TEMPERATURE, NUTRIENTS, ALGAE, NITROGEN, PHOSPHORUS, DINOFLAGELLATES, LABORATORY TESTS, WATER QUALITY CONTROL, ON-SITE INVESTIGATIONS, PHYSIOCHEMICAL PROPERTIES, WATER QUALITY.

IDENTIFIERS:
*NEW YORK BIGHT, COLLABORATIVE STUDIES.

ABSTRACT:
INTERDISCIPLINARY PHYSICAL, CHEMICAL, AND BIOLOGICAL MEASUREMENTS WERE MADE IN THE NEW YORK BIGHT DURING 1969-70 USING NEW IN SITU ELECTRONIC INSTRUMENTATION, WITH ASSOCIATED MONITORING BY CONVENTIONAL PROCEDURES TO CHECK THE PERFORMANCE. THIS NEW EQUIPMENT WAS PARTLY AN ORIGINAL DESIGN AND PARTLY MODIFIED COMMERCIAL INSTRUMENTATION AND INCLUDED PHOTOMETRIC SENSORS AND A MODIFIED BECKMAN CONDUCTIVITY CELL, PHOTOVOLT FLOW TYPE ELECTRODE SENSOR, AND DELTA SCIENTIFIC OXYGEN PROBE. THIS EQUIPMENT WAS FOUND TO BE FEASIBLE FOR THE MEASUREMENT OF CHLOROPHYLL (AND THE RELATED BIOMASS), BIOLUMINESCENCE, GELBSTOFF, PH, DISSOLVED O2, SALINITY, AND THE LOCATION OF THE THERMOCLINE. IT APPEARS THAT THE IN SITU INSTRUMENTATION MAY BE ADAPTED TO CONTINUOUS SYNOPTIC MONITORING OF THE ESTUARINE AND OCEANOGRAPHIC PARAMETERS NECESSARY FOR MATHEMATICAL MODELING. (SNYDER-BATTELLE)

FIELD 05A, 02L

ACCESSION NO. W72-12941

EUTROPHICATION OF SURFACE WATERS--LAKE TAHOE.

LAKE TAHOE AREA COUNCIL, SOUTH LAKE TAHOE, CALIF.

COPY AVAILABLE FROM GPO SUP DOC EP2.10:16010 DJW 05/71, $1.25; MICROFICHE FROM NTIS AS PB-211 460, $0.95. ENVIRONMENTAL PROTECTION AGENCY, WATER POLLUTION CONTROL RESEARCH SERIES, MAY 1971, 154 P. 17 FIG., 45 TAB., 25 REF., 5 APPEND. EPA PROGRAM 16010 DJW 05/71.

DESCRIPTORS:
*WATER POLLUTION SOURCES, *EUTROPHICATION, *LAKES, SEWAGE EFFLUENTS, SEEPAGE, SEPTIC TANKS, BIOASSAY, CHEMICAL ANALYSES, NITROGEN, NUTRIENTS, SURFACE WATERS, ALGAE, CHEMICAL PROPERTIES, LANDFILLS, PERCOLATION, DRAINAGE, RUNOFF, PRECIPITATION(ATMOSPHERIC).

IDENTIFIERS:
*LAKE TAHOE(CALIF.).

ABSTRACT:
A SURVEY WAS MADE OF NUTRIENT AND OTHER CHEMICAL CONSTITUENTS OF SURFACE WATERS FROM DEVELOPED AND UNDEVELOPED LAND AREAS, SEWAGE EFFLUENTS, SEEPAGE FROM SEPTIC TANK PERCOLATION SYSTEM AND REFUSE FILLS, DRAINAGE FROM SWAMPS, PRECIPITATION, AND LAKE TAHOE WATER. ALGAL GROWTH STIMULATING POTENTIAL OF THE SAMPLES WERE BIOASSAYED WITH SELENASTRUM GRACILE AS A TEST ORGANISM. ALGAL RESPONSE TO NUTRIENTS WAS MEASURED BY MAXIMUM GROWTH RATE AND MAXIMUM CELL COUNT IN A 5-DAY GROWTH PERIOD. PONDS SIMULATING THE SHALLOW PORTIONS OF THE LAKE WERE USED FOR CONTINUOUS FLOW ASSAY OF THE BIOMASS OF INDIGENOUS LAKE ORGANISMS PRODUCED BY SEWAGE EFFLUENT. FLASK ASSAYS AND CHEMICAL ANALYSES WERE MADE OVER TWO YEARS ON THREE MAJOR CREEKS. TWENTY-EIGHT OTHER CREEKS AND PRECIPITATIONS WERE MONITORED BY CHEMICAL ANALYSIS. EVALUATING THE EUTROPHICATION POTENTIAL, LAKE TAHOE IS NITROGEN SENSITIVE AND RESPONDS TO IT IN PROPORTION TO ITS CONCENTRATION. CREEKS DRAINING DEVELOPED LAND CARRIED TWICE THE NITROGEN AS THOSE DRAINING RELATIVELY UNDISTURBED WATERSHEDS. HUMAN ACTIVITY DOUBLES NITROGEN INFLOW TO THE LAKE. EXPORTING ALL SEWAGE WOULD REMOVE 70% OF THE NITROGEN. (JONES-WISCONSIN)

FIELD 05C, 02H

ACCESSION NO. W72-12955

USE OF ALGAL ASSAYS IN STUDYING EUTROPHICATION PROBLEMS,

NATIONAL ENVIRONMENTAL RESEARCH CENTER, CORVALLIS, OREG. NATIONAL EUTROPHICATION RESEARCH PROGRAM.

T. E. MALONEY.

PREPRINT, PRESENTED AT 6TH INTERNATIONAL WATER POLLUTION RESEARCH CONFERENCE, SESSION 3, NO. 6, HALL C, JUNE 20, 1972, 10 P, 6 TAB, 3 REF.

DESCRIPTORS:
*EUTROPHICATION, *ASSAY, *ALGAE, ANALYTICAL TECHNIQUES, WASTE WATER TREATMENT, NUTRIENTS, NUTRIENT REQUIREMENTS, GROWTH RATES, PHOSPHORUS, NITROGEN, *REGRESSION ANALYSIS, VARIABILITY.

IDENTIFIERS:
DISSOLVED PHOSPHORUS, *ALGAL ASSAYS, ASSAY PROCEDURE.

ABSTRACT:
THE USE OF ALGAL ASSAYS TO ASSIST IN THE SOLUTION OF PRACTICAL EUTROPHICATION PROBLEMS WAS ILLUSTRATED. THE BOTTLE TEST ASSAY, AS DEVELOPED BY THE NATIONAL EUTROPHICATION RESEARCH PROGRAM WAS UTILIZED DURING THESE STUDIES. PRACTICAL APPLICATION OF THE ASSAY PROCEDURE INCLUDED ASSESSMENT OF AFFECTS OF CHANGES IN WASTE TREATMENT PROCESSES ON RECEIVING WATERS, IDENTIFICATION OF ALGAL GROWTH-LIMITING NUTRIENTS AND ASSESSMENT OF RECEIVING WATERS TO DETERMINE THEIR NUTRIENT STATUS AND SENSITIVITY TO CHANGE. A MULTIPLE REGRESSION ANALYSIS, USING THE VARIOUS CHEMICAL PARAMETERS AS THE INDEPENDENT VARIABLES AND THE MAXIMUM ALGAL YIELD AS THE DEPENDENT VARIABLE WAS RUN TO DETERMINE THE NUTRIENT HAVING THE MOST INFLUENCE ON ALGAL GROWTH. DISSOLVED PHOSPHORUS ALONE EXPLAINED 79.84 PERCENT OF THE VARIABILITY IN ALGAL GROWTH IN THE SAMPLES, AND THE SIX NUTRIENTS ONLY INCREASED THIS BY 6.29%. (GALWARDI-TEXAS)

FIELD 05C, 05D

ACCESSION NO. W72-12966

PROJECT HYPO,

DEPARTMENT OF ENERGY, MINES AND RESOURCES, BURLINGTON (ONTARIO). CANADA CENTRE FOR INLAND WATERS; AND ENVIRONMENTAL PROTECTION AGENCY, FAIRVIEW PARK, OHIO. OHIO DISTRICT BASIN OFFICE.

NOEL M. BURNS, AND C. ROSS.

CANADA CENTRE FOR INLAND WATERS PAPER NO 6, AND ENVIRONMENTAL PROTECTION AGENCY TECHNICAL REPORT TS-05-71-208-24, FEBRUARY 1972. 182 P, 6 APPEND.

DESCRIPTORS:
*LAKE ERIE, *EUTROPHICATION, *HYPOLIMNION, *DISSOLVED OXYGEN, *ANAEROBIC CONDITIONS, ALGAE, SEDIMENTS, OXYGEN DEMAND, PHOSPHORUS, NITROGEN, NUTRIENTS, ON-SITE DATA COLLECTIONS, LAKES.

ABSTRACT:
INTERDISCIPLINARY FINDINGS AND ESTIMATES RESULTING FROM PROJECT HYPO, AN INVESTIGATION OF THE CENTRAL BASIN OF LAKE ERIE, LEAD TO ONE DEFINITE CONCLUSION: PHOSPHORUS INPUT TO LAKE ERIE MUST BE REDUCED IMMEDIATELY; IF THIS IS DONE A RAPID IMPROVEMENT CAN BE EXPECTED; IF IT IS NOT DONE, THE RATE OF DETERIORATION WILL BE MUCH GREATER THAN IT HAS BEEN IN RECENT YEARS. CONTRIBUTIONS DISCUSS IN DETAIL THE EFFECTS OF ALGAE, CAUSES AND SITE OF OXYGEN DEPLETION, SEDIMENT OXYGEN DEMAND, BUDGET CALCULATIONS, HYPOLIMNION VOLUME INCREASE, UPWELLING OF WATER MASSES, PROXIMITY OF A PROCESS OF CONTINUAL SELF-FERTILIZATION, AND PHOSPHORUS AND NITROGEN ELIMINATION FROM THE LAKE SYSTEM. EVIDENCE IS PRESENTED WHICH SUGGESTS THAT 76% OF THE PHOSPHORUS AND 57% OF THE NITROGEN IN ALGAL MATERIAL WHICH SEDIMENTS TO THE LAKE FLOOR IS RETAINED THERE IF OXIC CONDITIONS ARE MAINTAINED IN THE OVERLYING WATER. APPROXIMATELY 80% OF THE PHOSPHORUS AND 56% OF THE NITROGEN LOADED INTO LAKE ERIE FROM EXTERNAL SOURCES WILL BE REMOVED FROM THE WATER IF OXYGENATED CONDITIONS ARE MAINTAINED. APPENDIX I DESCRIBES AN AUTOMATIC UNDERWATER CAMERA SYSTEM DESIGNED FOR USE DURING THIS PROJECT AND APPENDIX II DESCRIBES A SUBMERSIBLE AUTOMATIC DISSOLVED OXYGEN-TEMPERATURE MONITORING SYSTEM. (SEE W72-12992 THRU W72-12997) (AUEN-WISCONSIN)

FIELD 05C, 02H

ACCESSION NO. W72-12990

PHYSICAL PROCESSES AFFECTING THE HYPCLIMNION OF THE CENTRAL BASIN OF LAKE ERIE,

DEPARTMENT OF ENERGY, MINES AND RESOURCES, BURLINGTON (CNTARIO). CANADA CENTRE FOR INLAND WATERS; AND ENVIRONMENTAL PROTECTION AGENCY, FAIRVIEW PARK, OHIO.

J. O. BLANTON, AND A. R. WINKLHOFER.

IN: PROJECT HYPO: CANADA CENTRE FOR INLAND WATERS PAPER NO 6, AND ENVIRONMENTAL PROTECTION AGENCY TECHNICAL REPORT TS-05-71-208-24, FEBRUARY 1972, P 9-37. 22 FIG, 4 TAB, 9 REF.

DESCRIPTORS:
*HYPOLIMNION, *LAKE ERIE, *DISSOLVED OXYGEN, TEMPERATURE, CURRENTS(WATER), WINDS, THERMOCLINE, ALGAE, SEDIMENTS.

IDENTIFIERS:
*LAKE ERIE CENTRAL BASIN, POINTE PELEE(ONTARIO), CXYGEN DEPLETION.

ABSTRACT:
IN SITU AUTOMATIC MONITORING DEVICES, EXISTING LAND MONITORING STATIONS, AND CRUISE SAMPLING FURNISH DATA TO DETERMINE RELATIONSHIP BETWEEN DOMINANT WINDS, DOMINANT MOTIONS IN THE HYPOLIMNION, AND THERMOCLINE RESPONSE TO THESE MOTICNS. HYPOLIMNION VOLUME INCREASES WERE EXPLAINED BY LOW WIND ENERGY PERIODS FOLLOWED BY BRIEF PERIODS OF HIGH WIND ENERGY. THIS WIND ENERGY CYCLING DOES NOT NECESSARILY OCCUR ANNUALLY. HYPOLIMNION OXYGEN ENTRAINMENT FROM ABOVE WOULD BE MINIMAL RESULTING IN THE HYPCLIMNION BECOMING ANOXIC AT AN UNPRECEDENTED RATE UNDER CONDITIONS OF LONG DURATION HIGH ENERGY WINDS WITH FEW INTERVENING CALM PERIODS. THE NET HYPOLIMNION WATER MOVEMENT IS THE RESULT OF THE DOMINANT SOUTHWEST WINDS CAUSING THE CANADIAN NEARSHORE AREAS, EAST OF POINTE PELEE, TO BE A POTENTIAL STAGING AREA FOR PROFUSE ALGAL BLOOMS DUE TO ACCUMULATION OF NUTRIENT-RICH ANOXIC WATER IN UPWELLED AREAS. DISSOLVED OXYGEN DEPLETION RATES AS DETERMINED, CONFIRM THE LONGTERM TREND OF INCREASING DEPLETION RATE. MEASURED HYPOLIMNION DISSOLVED OXYGEN FLUCTUATIONS SOLELY ATTRIBUTABLE TO ALGAL PRESENCE WERE NOT DETECTED. SEDIMENT OXYGEN DEMAND MIGHT ACCOUNT FOR THE MEASURED HYPOLIMNION RATE; HOWEVER, DUE TO THE VARIETY AND COMPLEXITY OF THE LAKE SYSTEM, IT CANNOT BE ASSUMED THAT THIS SINGLE PARAMETER CAN DEFINE A SITUATION. (SEE ALSO W72-12990) (JONES-WISCONSIN)

FIELD 05C

ACCESSION NO. W72-12992

SEDIMENT OXYGEN DEMAND IN LAKE ERIE'S CENTRAL BASIN, 1970.

ENVIRONMENTAL PROTECTION AGENCY, CINCINNATI, OHIO. DIV. CF FIELD INVESTIGATIONS.

A. M. LUCAS, AND N. A. THOMAS.

IN: PROJECT HYPO: CANADA CENTRE FOR INLAND WATERS PAPER NO 6, AND ENVIRONMENTAL PROTECTION AGENCY TECHNICAL REPORT TS-05-71-208-24, FEBRUARY 1972, P 45-50. 3 FIG, 1 TAB, 2 REF.

DESCRIPTORS:
*SEDIMENTS, *LAKE ERIE, *DISSOLVED OXYGEN, OXYGEN DEMAND, HYPOLIMNION, AQUATIC PLANTS, ALGAE.

IDENTIFIERS:
LAKE ERIE CENTRAL BASIN.

ABSTRACT:
THE BIOCHEMICAL AND CHEMICAL OXYGEN DEMAND IN LAKE ERIE'S CENTRAL BASIN HYPOLIMNETIC WATERS WAS TOO SMALL TO EXPLAIN THE OXYGEN DEPLETION RATES; TESTS INDICATED THAT DISSOLVED OXYGEN DEPLETION WAS CAUSED PRIMARILY BY THE OXYGEN DEMAND OF BOTTOM SEDIMENTS. SEDIMENT OXYGEN DEMAND RATES WERE MEASURED AT FIVE LOCATIONS IN JUNE, AUGUST, AND SEPTEMBER 1970. THE RATES WERE DETERMINED FROM CHANGES IN THE DISSOLVED OXYGEN CONCENTRATION OF WATER SEALED AND CIRCULATED WITHIN BLACK AND CLEAR PLEXIGLASS CHAMBERS IMBEDDED IN THE LAKE BOTTOM. SEDIMENT OXYGEN DEMAND RATES RECORDED IN JUNE VARIED FROM 1.2 TO 2.2 GM OXYGEN/SQ M PER DAY AND WERE INDICATIVE OF EUTROPHIC CONDITIONS. RATES MEASURED IN AUGUST DURING DAYLIGHT HOURS WITH A CLEAR CHAMBER WERE LESS THAN THOSE MEASURED AT NIGHT WITH THE CLEAR CHAMBER OR DURING THE DAY WITH THE BLACK CHAMBER. OXYGEN PRODUCED BY ALGAL PHOTOSYNTHETIC ACTIVITY ON THE LAKE BOTTOM OFFSET THE SEDIMENT OXYGEN DEMAND DURING PART OF THE DAY RESULTING IN DAILY SOD RATES OF 0.4 TO 0.7 GM OXYGEN/SQ M PER DAY. RATES MEASURED IN SEPTEMBER WITH OXYGENATED SURFACE WATER TRAPPED AND CARRIED TO THE BOTTOM IN THE CHAMBERS RANGED FROM 1.0 TO 2.4 GM OXYGEN/SQ M PER DAY. (SEE ALSO W72-12990) (AUEN-WISCONSIN)

FIELD 05C

ACCESSION NO. W72-12994

BIOLOGICAL STUDIES RELATED TO OXYGEN DEPLETION AND NUTRIENT REGENERATION PROCESSES IN THE LAKE ERIE CENTRAL BASIN,

ENVIRONMENTAL PROTECTION AGENCY, FAIRVIEW PARK, OHIO.

T. BRAIDECH, P. GEHRING, AND C. KLEVENO.

IN: PROJECT HYPO: CANADA CENTRE FOR INLAND WATERS PAPER NO 6, AND ENVIRONMENTAL PROTECTION AGENCY, TECHNICAL REPORT TS-05-71-208-24, FEBRUARY 1972, P 51-70. 17 FIG, 3 PLATES, 4 REF.

DESCRIPTORS:
*OXYGEN DEMAND, *ANAEROBIC CONDITIONS, *LAKE ERIE, *ALGAE, *BENTHIC FAUNA, *BOTTOM SEDIMENTS, *HYPOLIMNION, PHYTOPLANKTON, LIGHT PENETRATION, DISSOLVED OXYGEN, BACTERIA, THERMOCLINE, SEDIMENT-WATER INTERFACES, CHRYSOPHYTA, CHLOROPHYTA, CYANOPHYTA, BIOCHEMICAL OXYGEN DEMAND.

IDENTIFIERS:
*LAKE ERIE CENTRAL BASIN, TRIBONEMA, OEDOGONIUM, ANACYSTIS, CERATIUM, COSMARIUM.

ABSTRACT:
A COMPREHENSIVE ATTEMPT WAS MADE TO DEFINE PHYTOPLANKTON CONDITIONS THROUGHOUT THE WATER COLUMN AND ON THE SEDIMENTS OF LAKE ERIE CENTRAL BASIN. SPECIAL TECHNIQUES WERE EMPLOYED TO EXPLAIN PRESENCE, ORIGIN, AND VIABILITY OF HERETOFORE UNOBSERVED, APPARENTLY METABOLIZING ALGAE. TRIBONEMA AND OEDOGONIUM, DEPOSITED ON THE BOTTOM, WERE THE MAJOR SOURCE OF ORGANIC CARBON UTILIZED IN CONSUMPTION OF HYPOLIMNETIC OXYGEN AS A RESULT OF BACTERIAL ACTIVITY AT THE WATER-SEDIMENT INTERFACE. THE ALGAE ON THE SEDIMENTS MAINTAIN GROWTH AFTER LIGHT BECOMES LIMITING FOR OTHER SEDIMENTAL FORMS. THE MIXING OF NUTRIENT-RICH HYPOLIMNION WATER INTO THE THERMOCLINE AND LOWER EPILIMNION STIMULATED ALGAL GROWTH, PRIMARILY ANACYSTIS, AT THESE LEVELS, PARTICULARLY AT STATIONS WHERE DISSOLVED OXYGEN DEPLETION IN THE HYPOLIMNION HAD BEEN RECORDED. FROM APPROXIMATELY JULY 21 TO AUGUST 10 THERE WAS MAXIMUM PHOTOSYNTHETIC ACTIVITY ON THE BOTTOM RESULTING IN A REDUCED SEDIMENT OXYGEN DEMAND. INCREASING PHYTOPLANKTON VOLUMES IN THE OVERLYING WATERS AND THE DECREASING PHOTOPERIOD REDUCED LIGHT ON THE BOTTOM TO BIOLOGICALLY LIMITING LEVELS IN MID-AUGUST. ALGAE DO CONTRIBUTE OXYGEN TO THE HYPOLIMNION FOR A PERIOD OF TIME; HOWEVER, IMPACT AND MAGNITUDE OF THIS CONTRIBUTION APPEARS TO BE MASKED BY A STRONGER OXYGEN DEMANDING PHYSICAL AND CHEMICAL PHENOMENA. (SEE ALSO W72-12990) (JONES-WISCONSIN)

FIELD 05C

ACCESSION NO. W72-12995

MICROBIOLOGICAL STUDIES RELATED TO OXYGEN DEPLETION AND NUTRIENT REGENERATION PROCESSES IN THE LAKE ERIE CENTRAL BASIN,

DEPARTMENT OF ENERGY, MINES AND RESOURCES, BURLINGTON (ONTARIO). CANADA CENTRE FOR INLAND WATERS; AND ENVIRONMENTAL PROTECTION AGENCY, FAIRVIEW PARK, OHIO. MICROBIOLOGY LAB.

A. S. MENON, C. V. MARION, AND A. N. MILLER.

IN: PROJECT HYPO: CANADA CENTRE FOR INLAND WATERS PAPER NO 6, AND ENVIRONMENTAL PROTECTION AGENCY TECHNICAL REPORT TS-05-71-208-24, FEBRUARY 1972, P 71-84. 10 FIG, 4 TAB, 17 REF.

DESCRIPTORS:
*DISSOLVED OXYGEN, *LAKE ERIE, *BACTERIA, *SEDIMENT-WATER INTERFACES, *HYPOLIMNION, NITRIFICATION, ALGAE, THERMOCLINE, ANAEROBIC BACTERIA, AEROBIC BACTERIA, SULFUR BACTERIA, IRON COMPOUNDS.

IDENTIFIERS:
*DEOXYGENATION, *NUTRIENT REGENERATION, LAKE ERIE CENTRAL BASIN.

ABSTRACT:
THIS BACTERIOLOGICAL STUDY AIMED TO EVALUATE DISTRIBUTION OF BACTERIAL DENSITIES AND BIOTYPES AT THE SEDIMENT-WATER INTERFACE AND OVERLYING WATERS IN RELATION TO TIME, CHEMICAL, PHYSICAL, AND BIOLOGICAL DATA AND TO ASSESS THEIR ROLE IN OVERALL OXYGEN DEPLETION AND NUTRIENT REGENERATION PROCESSES IN THE HYPOLIMNION OF LAKE ERIE'S CENTRAL BASIN. MAJOR SITE OF INTENSIVE BACTERIAL ACTIVITY WAS THE SEDIMENT-WATER INTERFACE. ORGANIC DEPOSITS FROM ALGAL RAINS AND OTHER SOURCES, ACCUMULATED AT THE BOTTOM, UNDERWENT BACTERIAL DECOMPOSITION RESULTING IN OXYGEN DEPLETION AND FORMATION OF REDUCED PRODUCTS OF LOW MOLECULAR WEIGHT. THE REDUCED PRODUCTS WERE SUBSEQUENTLY OXIDIZED BY CHEMOAUTOTROPHIC BACTERIA AT THE SEDIMENT-WATER INTERFACE, OR IN THE OVERLYING WATERS, RESULTING IN ADDITIONAL OXYGEN DEPLETION. THIS PROCESS REPEATED ITSELF AFTER EACH ALGAL RAIN CAUSING FURTHER OXYGENLOSS. PRECIPITATION OF PARTICULATE MATTER THROUGH THE HYPOLIMNION FROM INTERMITTENT ALGAL RAINS WAS PRIMARILY RESPONSIBLE FOR HIGH BACTERIAL DENSITIES IN THE HYPOLIMNION BECAUSE PHYTOPLANKTON CONSTITUTED A LOCUS FOR BACTERIAL ATTACHMENT AND PRODUCED SOLUBLE ORGANIC SUBSTRATES FOR BACTERIAL GROWTH. A SIGNIFICANT CORRELATION WAS OBTAINED BETWEEN DESULFOVIBRIO DENSITIES AND DEGREE OF OXYGEN DEPLETION. THESE FACTORS STRONGLY SUGGEST THAT HETEROTROPHIC AND CHEMOAUTOTROPHIC BACTERIA ARE THE PRINCIPAL FACTORS IN DEPLETING OXYGEN. (SEE ALSO W72-12990) (JONES-WISCONSIN)

FIELD 05C

ACCESSION NO. W72-12996

OXYGEN-NUTRIENT RELATIONSHIPS WITHIN THE CENTRAL BASIN OF LAKE ERIE,

DEPARTMENT OF ENERGY, MINES AND RESOURCES, BURLINGTON (ONTARIO). CANADA CENTRE FOR INLAND WATERS; AND ENVIRONMENTAL PROTECTION AGENCY, FAIRVIEW PARK, OHIO.

N. M. BURNS, AND C. ROSS.

IN: PROJECT HYPO: CANADA CENTRE FOR INLAND WATERS PAPER NO 6, AND ENVIRONMENTAL PROTECTION AGENCY TECHNICAL REPORT TS-05-71-208-24, FEBRUARY 1972, P 85-119. 19 FIG, 13 TAB, 3 PLATES, 10 REF.

DESCRIPTORS:
*DISSOLVED OXYGEN, *NUTRIENTS, *LAKE ERIE, *HYPOLIMNION, MATHEMATICAL MODELS, IRON, MANGANESE, PHOSPHORUS, ANAEROBIC CONDITIONS, CHLOROPHYLL, EPILIMNION, SEDIMENTS, DECOMPOSING ORGANIC MATTER, BACTERIA, CHEMICAL PROPERTIES, NITROGEN, HEAT TRANSFER, PHOSPHATES, ALKALINITY, OXYGEN DEMAND, HEAT BUDGET, ALGAE, EUTROPHICATION.

IDENTIFIERS:
*LAKE ERIE CENTRAL BASIN, NUTRIENT REGENERATION, DEOXYGENATION, ALGAL SEDIMENTATION.

ABSTRACT:
THE CHEMICAL BUDGET COMPONENTS IN LAKE ERIE'S CENTRAL BASIN WERE CALCULATED. A LARGE EROSION OF THE HYPOLIMNION DURING JULY 1970 SEEMED THE INITIAL FACTOR IN STARTING THE ALGAL GROWTH PERIOD LASTING TO OCTOBER; THE FIRST HEAVY BLOOM RESULTED IN ALGAL FALL-OUT ONTO THE LAKE FLOOR AT THE END OF JULY. FOR A WHILE THESE SEDIMENTED ALGAE WERE PHOTOSYNTHETIC, AMELIORATING OXYGEN DEPLETION BUT NOT PREVENTING NET OXYGEN DEPLETION CAUSED BY ORGANIC DECAY. AEROBIC HETEROTROPHIC AND SULFATE-REDUCING BACTERIA INCREASED STEADILY WHILE THE ALGAE ON THE BOTTOM DIED AND MATTED DOWN. LOSS OF OXYGEN FROM PHOTOSYNTHESIS PLUS ACTIVITY OF THE LARGE BACTERIAL POPULATIONS CAUSED HIGH OXYGEN DEPLETION RATE WITH ANOXIC CONDITIONS APPEARING ON AUGUST 12TH. ANOTHER PERIOD OF ALGAL RAINS, ABOUT AUGUST 17TH, AGAIN DIMINISHED THE OXYGEN DEPLETION RATE BUT WAS INSUFFICIENT TO PREVENT SPREAD OF ANOXIA WHICH, BY AUGUST 25TH, EXTENDED ACROSS THE HYPOLIMNION AREA. THE ANOXIC CONDITION CAUSED LARGE SCALE NUTRIENT REGENERATION BY DISSOLUTION OF INORGANIC FORMS. A MASSIVE BLOOM RESULTED WHEN THESE NUTRIENTS WERE MIXED WITH SURFACE WATER DURING SEPTEMBER. OXYGENATED CONDITIONS SHOULD BE MAINTAINED IN THE WATER AS A SIMPLE MECHANISM FOR ENSURING THAT LITTLE OF THE PHOSPHORUS IN THE SEDIMENTS RETURNS TO THE OVERLYING WATER. (SEE ALSO W72-12990) (JONES-WISCONSIN)

FIELD 05C

ACCESSION NO. W72-12997

UPTAKE AND METABOLISM OF 2,2-BIS-(P-CHLOROPHENYL)-1,1,1-TRICHLOROETHANE(DDT) BY MARINE PHYTOPLANKTON AND ITS EFFECT ON GROWTH AND CHLOROPLAST ELECTRON TRANSPORT,

SCRIPPS INSTITUTION OF OCEANOGRAPHY, LA JOLLA, CALIF.

G. W. BOWES.

PLANT PHYSIOLOGY, VOL 49, P 172-176, 1972. 3 FIG, 1 TAB, 31 REF.

DESCRIPTORS:
*OCEANS, *MATABOLISM, *MARINE ALGAE, *CHLORINATED HYDROCARBON PESTICIDES, *CYTOLOGICAL STUDIES, ABSORPTION, DDT, PHYTOPLANKTON, PLANT GROWTH, CHLOROPHYLL, DDE, INHIBITION.

IDENTIFIERS:
*ELECTRON TRANSPORT, DUNALIELLA TERTIOLECTA, CYCLOTELLA NANA, THALASSIOSIRA FLUVIATILIS, AMPHIDINIUM CARTERI, COCCOLITHUS HUXLEYI, PORPHYRIDIUM, SKELETONEMA COSTATUM.

ABSTRACT:
MARINE PHYTOPLANKTERS WERE STUDIED RELATIVE TO THEIR INTERACTION WITH DDT. AT 80 PPB DDT, GROWTH OF DUNALIELLA TERTIOLECTA WAS UNAFFECTED, AND THERE WAS SLIGHT, IF ANY, INFLUENCE ON DEVELOPMENT OF CYCLOTELLA NANA, THALASSIOSIRA FLUVIATILIS, AMPHIDINIUM CARTERI, COCCOLITHUS HUXLEYI, AND PORPHYRIDIUM. SKELETONEMA COSTATUM EXHIBITED A 9-DAY LAG BEFORE CELL DIVISION, GROWTH SUBSEQUENTLY BEING THE SAME AS THE CONTROL. ABILITY OF MARINE PHYTOPLANKTON TO METABOLIZE DDT VARIED. DDE WAS THE ONLY SIGNIFICANT HEXANE-SOLUBLE METABOLITE DETECTED. IT OCCURRED IN CELLS OF S. COSTATUM, C. NANA, T. FLUVIATILIS AND D. TERTIOLECTA. MAXIMUM DEGREE OF CONVERSION WAS 7.5% AND WAS BASED ON DDT FOUND IN CELL-WATER SYSTEM OF 9-DAY D. TERTIOLECTA CULTURES. THE TOTAL RECOVERED FROM CULTURES IN 2- TO 3-WEEK EXPERIMENTS RANGED FROM 63.5% FOR T. FLUVIATILIS TO 90.7% FOR S. COSTATUM. AMOUNT OF DDT FOUND ASSOCIATED WITH THE CELLS, COLLECTED BY CENTRIFUGATION, IN THE CELL-WATER SYSTEM RANGED FROM 70.8 TO 99.5%. NONCYCLIC ELECTRON FLOW, MEASURED BY FERRICYANIDE REDUCTION WAS INHIBITED BY DDT AND DDE, AND COULD EXPLAIN GROWTH INHIBITION. PHYTOPLANKTON SENSITIVITY TO TOXIC HYDROPHOBIC CHLORINATED HYDROCARBONS MAY BE DEPENDENT UPON PENETRATION OF THE MOLECULES TO ACTIVE SITES WITHIN MEMBRANES. (JONES-WISCONSIN)

FIELD 05C, 05B

ACCESSION NO. W72-12998

ATTACHED ALGAE ON ARTIFICIAL AND NATURAL SUBSTRATES IN LAKE WINNIPEG, MANITOBA,

FISHERIES RESEARCH BOARD OF CANADA, WINNIPEG (MANITOBA). FRESHWATER INST.

D. EVANS, AND J. G. STOCKNER.

JOURNAL OF FISHERIES RESEARCH BOARD OF CANADA, VOL 29, NO 1, P 31-44, JANUARY 1972. 7 FIG, 4 TAB, 21 REF.

DESCRIPTORS:
*BIOMASS, *AQUATIC ALGAE, *AQUATIC PRODUCTIVITY, LIGHT, TURBIDITY, NUTRIENTS, HETEROGENEITY, CHLOROPHYTA, CHRYSOPHYTA, BIOINDICATORS, PHYSICOCHEMICAL PROPERTIES, MICROSCOPY, LIMNOLOGY, WATER QUALITY, ECOLOGY, AQUATIC HABITATS, CLADOPHORA, LIMITING FACTORS, ECOLOGICAL DISTRIBUTION.

IDENTIFIERS:
*LAKE WINNIPEG, *SUBSTRATES, ACHNANTHES MINUTISSIMA, ACHNANTHES CF. BIRGIANI, AMPHIPLEURA PELLUCIDA, AMPHORA OVALIS, AMPHORA OVALIS V. PEDICULUS, COCCONEIS PEDICULUS, COCCONEIS PLACENTULA, CYCLOTELLA CF. SOCIALIS, CYCLOTELLA STELLIGERA, CYMBELLA SPP, DIATOMA SSP.

ABSTRACT:
THE COMPOSITION AND DISTRIBUTION OF ATTACHED ALGAE ON ARTIFICIAL (NAVIGATIONAL BUOYS) AND NATURAL SUBSTRATES IN LAKE WINNIPEG IN 1969-70, WAS STUDIED, IN THE FALL AFTER ABOUT 145 DAYS' GROWTH, WITH SPECIAL EMPHASIS ON THE DIATOMS, THE MAJOR COMPONENT OF THE ATTACHED ALGAL FLORA. ATTACHED ALGAE FROM KNOWN AREAS OF THE SURFACE OF REPRESENTATIVE SUBSTRATES WERE WEIGHED (DRY-WEIGHT) AND COUNTED. TOTAL DIATOM VOLUMES (CU MM/SQ CM) WERE CALCULATED USING AN EQUATION FOR THE GEOMETRICAL SHAPE THAT CLOSELY RESEMBLED INDIVIDUAL SPECIES, AND BY APPLYING SUITABLE CORRECTIONS TO EACH CALCULATION. THE PERCENT BY VOLUME OF OTHER ALGAL GROUPS WAS ESTIMATED BY MICROSCOPY. ON BOTH NATURAL AND ARTIFICIAL SUBSTRATES A DISTINCT ZONATION PATTERN RELATED TO WATER TRANSPARENCY OR AVAILABLE LIGHT WAS PREVALENT. PRELIMINARY RESULTS INDICATED THAT LIGHT WAS THE MOST IMPORTANT PHYSICAL FACTOR AFFECTING GROWTH AND SPECIFIC COMPOSITION OF ATTACHED ALGAE ON SUBSTRATES. BIOMASS VALUES (DRY WEIGHT) RANGED FROM 1.7 TO 29.1 MG/SQ CM, WITH THE GREATEST VALUES CONSISTENTLY OCCURRING BETWEEN A 10 AND 25 CM DEPTH ON BUOYS IN THE SOUTH BASIN. DIATOM DENSITY AND DIATOM VOLUME WERE ESTIMATED FROM BUOY SAMPLES AND VARIED RESPECTIVELY, WITHIN THE RANGE 200,000-6,650,000 DIATOMS/SQ CM AND 0.32-10.00 CU MM/SQ CM. IT APPEARS THAT NUTRIENTS MAY LIMIT GROWTH OF ATTACHED ALGAE IN THE NORTH BASIN, WHEREAS IN THE SOUTH BASIN, LIGHT IS THE MAJOR LIMITING FACTOR. THE SPECIFIC COMPOSITION OF THE ALGAL ASSEMBLAGES ON PARTICULAR BUOYS WAS MORE RELATED TO THE PHYSICAL-CHEMICAL FEATURES OF A MAJOR RIVER PLUME IN CLOSE PROXIMITY TO THE BUOY THAN TO A HOMOGENEOUS LAKE WINNIPEGWATER MASS. THIS FACTOR, TOGETHER WITH GREATER TURBIDITY IN THE SHALLOWER SOUTH BASIN, IS LARGELY RESPONSIBLE FOR THE OBSERVED HETEROGENEITY AMONG THE ATTACHED ALGAL ASSEMBLAGES IN LAKE WINNIPEG. (BYRD-BATTELLE)

FIELD 05A

ACCESSION NO. W72-13142

A PILOT PROGRAM TO DETERMINE THE EFFECT OF SELECTED NUTRIENTS (DISSOLVED ORGANICS, PHOSPHORUS, AND NITROGEN) ON NUISANCE ALGAL GROWTH IN AMERICAN FALLS RESERVOIR,

IDAHO UNIV., MOSCOW. WATER RESOURCES RESEARCH INST.

F. L. ROSE, AND G. W. MINSHALL.

AVAILABLE FROM THE NATIONAL TECHNICAL INFORMATION SERVICE AS PB-211 612, $3.00 IN PAPER COPY, $0.95 IN MICROFICHE. MOSCOW, COMPLETION REPORT, JUNE 1972, 37 P, 15 FIG, 5 TAB, 9 REF, APPEND. OWRR A-039IDA(1).

DESCRIPTORS:
*ALGAL CONTROL, *PHOSPHORUS, *NITROGEN, *IDAHO, *CHLOROPHYLL, *ANABAENA, NUTRIENTS, DISSOLVED SOLIDS, PHYTOPLANKCN, WATER POLLUTION EFFECTS.

IDENTIFIERS:
*GLUCOSE, *SNAKE RIVER(IDAHO), *DISSOLVED ORGANIC CARBON, *AMERICAN FALLS RESERVOIR, *APHANIZCMENON, ALGAL GROWTH.

ABSTRACT:
A FIELD INVESTIGATION WAS CONDUCTED TO DETERMINE THE FEASIBILITY OF USING POLYETHYLENE TUBES AS IN SITU CULTURE CHAMBERS IN ASSESSING THE INFLUENCE OF SELECTED NUTRIENT FACTORS ON ALGAL GROWTH. THE STUDY WAS CARRIED OUT IN AMERICAN FALLS RESERVOIR, A LARGE MULTIPURPOSE IMPOUNDMENT ON THE SNAKE RIVER IN S.E. IDAHO, DURING THE SUMMER OF 1971. THE TUBES CONTAINED ADDED AMOUNTS OF PHOSPHORUS OR PHOSPHORUS AND ORGANIC CARBON (GLUCOSE) AND WERE SUBSEQUENTLY 'SEEDED' WITH NATURALLY OCCURRING ALGAL SPECIES. PERIODIC WATER SAMPLES TAKEN FROM THE TUBES WERE ANALYZED FOR TOTAL AND SOLUBLE PHOSPHORUS, TOTAL AND SOLUBLE ORTHOPHOSPHORUS, AMMONIUM-N, NITRITE-N, NITRATE-N, PARTICULATE AND DISSOLVED ORGANIC CARBON, TOTAL DISSOLVED SOLIDS, CHLOROPHYLL A, AND ABUNDANCE OF MAJOR PHYTOPLANKTON SPECIES. RESULTS WERE COMPARED WITH THOSE FROM AMBIENT WATER COLLECTED BOTH AT THE LOCATION OF THE EXPERIMENTAL WORK AND AT OTHER SITES IN THE RESERVOIR. GROWTH OF PLANKTONIC ORGANISMS, ESPECIALLY APHANIZOMENON AND ANABAENA, REACHED LEVELS 10 TIMES MORE NUMEROUS IN THE EXPERIMENTAL TUBES THAN IN THE AMBIENT WATER. ALTHOUGH MODIFICATION OF SOME PARTS OF THE TECHNIQUE ARE SUGGESTED THE USE OF POLYETHYLENE TUBES APPEARS TC BE OF CONSIDERABLE PROMISE IN FUTURE INVESTIGATIONS.

FIELD 05C, 04A, 05A

ACCESSION NO. W72-13300

CONTAMINATION OF MARINE LITTCRAL AND SEA FOODS BY PESTICIDES, (CONTAMINATION DU LITTORAL MARITIME ET DES FRUITS DE MER PAR LES PESTICIDES),

POITIERS UNIV. (FRANCE). FACULTE DE MEDICINE ET DE PHARMACIE.

J. BRISOU.

MEDEDELINGEN FACULTEIT LANDBOUW-WETENSCHAPPEN GENT, VOL 35, NO 2, P 739-743, 1970. 3 TAB, 3 REF. ENGLISH SUMMARY.

DESCRIPTORS:
*PESTICIDE RESIDUES, *AQUATIC LIFE, *AQUATIC ENVIRONMENT, *WATER POLLUTION, *PATH OF POLLUTANTS, *PUBLIC HEALTH, ENZYMES, ANALYTICAL TECHNIQUES, PESTICIDES, FUNGICIDES, ORGANOPHOSPHOROUS PESTICIDES, CHLORINATED HYDROCARBON PESTICIDES, CARBAMATE PESTICIDES, ORGANIC PESTICIDES, HERBICIDES, WATER POLLUTION SOURCES, PLANKTON, OYSTERS, SHELLFISH, ALGAE, MUD, SEDIMENTS, GAS CHROMATOGRAPHY, COLORIMETRY.

IDENTIFIERS:
MERCURIAL PESTICIDES, THIN LAYER CHROMATOGRAPHY, GROWTH INHIBITION.

ABSTRACT:
A SYSTEMATIC INVESTIGATION OF THE FRENCH COASTS OF THE CHANNEL, THE ATLANTIC AND THE MEDITERRANEAN SEA SHOWS A SIGNIFICANT POLLUTION OF MUDS, SEDIMENTS, ALGAE, PLANKTON, AND SHELLFISH BY PESTICIDES. THE PER CENT OF CONTAMINATION BY ONE OR SEVERAL PRODUCTS IS BY APPROXIMATION: OYSTERS - 72%, OTHER SHELLFISH - 67%, ALGAE - 66%, PLANKTON- 100%, AND WATER - 0%. THE PLANKTON IS ONE OF THE MOST CONTAMINATED MATERIALS BECAUSE MERCURIAL FUNGICIDES ARE MORE PREVALENT THAN OTHER POLLUTANT PRODUCTS, AND THEY ARE CONCENTRATED BY THE PLANKTON. DATA HAVE BEEN OBTAINED USING SEVERAL TECHNIQUES: GROWTH INHIBITIONS, ANTICHOLINESTERASE ACTIVITY OF EXTRACTS, THIN LAYER CHROMATOGRAPHY, SPECTROCOLORIMETRY, AND IN A FEW INSTANCES GAS CHROMATOGRAPHY. (SVENSSON-WASHINGTON)

FIELD 05C, 05A

ACCESSION NO. W72-13347

PHOTOSYNTHETIC YIELDS AND BYPRODUCT RECOVERY FROM SEWAGE OXIDATION PONDS,

ASIAN INST. OF TECH., BANGKOK (THAILAND). DEPT. OF ENVIRONMENTAL ENGINEERING.

M. G. MCGARRY, C. D. LIN, AND J. L. MERTO.

PREPRINT, PRESENTED AT 6TH INTERNATIONAL WATER POLLUTION RESEARCH CONFERENCE, SESSION 4, HALL C, PAPER NO. 7, JUNE 20, 1972, P C/4/7/1-C/4/7/11. 6 FIG, 2 TAB, 17 REF.

DESCRIPTORS:
*OXIDATION PONDS, *ALGAE, *WASTE WATER TREATMENT, *BYPRODUCTS, MIXING, SOLAR RADIATION, ORGANIC LOADING, DIURNAL DISTRIBUTION, HYDROGEN ION CONCENTRATION, DISSOLVED OXYGEN, BIOCHEMICAL OXYGEN DEMAND, PERFORMANCE, FLOCCULATION, FLOTATION, ECONOMIC FEASIBILITY, COSTS, ESTIMATED COSTS.

IDENTIFIERS:
ALUM, RECYCLING, TREATMENT COSTS, ASIA.

ABSTRACT:
ALGAL GROWTH AND SEWAGE TREATMENT UNDER TROPICAL CONDITIONS AS AFFECTED BY POND MIXING, SOLAR RADIATION AND DIURNAL VARIATIONS IN POND LOADING WERE DESCRIBED. FOR MASS ALGAL CULTURE, PONDS SHOULD BE THOROUGHLY MIXED DAILY AND THE MIXING PERIOD LIMITED IN ORDER TO MINIMIZE TURBIDITY. DIFFERENT SHADING MATERIAL OVER POND SURFACES ALLOWED A PREDETERMINED PERCENTAGE OF TOTAL INCIDENT RADIATION TO REACH EACH SURFACE. ALGAL CONCENTRATIONS INCREASED IN A NEAR LINEAR FASHION WITH RADIATION AVAILABLE. EFFECTS OF DIURNAL VARIATIONS IN POND LOADING AND EFFLUENT REMOVAL REQUIRED ON POND CHARACTERISTICS WERE PRESENTED IN TABULAR FORM. NORMAL PH, D.O., AND ALKALINITY VARIATIONS WERE EXHIBITED BY ALL PONDS AND FILTERED BOD'S DID NOT EXCEED 20 MG/L. ALUM FLOCCULATION AND DOWNFLOW FLOTATION WERE USED TO CLARIFY THE POND EFFLUENT AND CONCENTRATE THE ALGAE. ALUMINUM RECOVERY AND RECYCLING WERE INVESTIGATED. THE POTENTIAL USE OF ALGAE AS A LIVESTOCK FEED WAS INVESTIGATED. ESTIMATES INDICATED THAT IF SEWAGE TREATMENT COSTS WERE SET AGAINST ALGAE PRODUCTION, ALGAE COULD BE PRODUCED FOR APPROXIMATELY 26 CENTS/KG. (GALWARDI-TEXAS)

FIELD 05D

ACCESSION NO. W72-13508

ESTURINE ECOSYSTEMS AND HIGH TEMPERATURES,

NORTH CAROLINA WATER RESOURCES RESEARCH INST., RALEIGH.

B. J. COPELAND, AND H. LEE DAVIS.

AVAILABLE FROM THE NATIONAL TECHNICAL INFORMATION SERVICE AS PB-211 808, $3.00 IN PAPER COPY, $0.95 IN MICROFICHE. NORTH CAROLINA WATER RESOURCES RESEARCH INSTITUTE, RALEIGH, REPORT NO 68, JUNE 1972 (UNC-WRRI-72-68). 90 P, 39 FIG, 9 TAB, 47 REF. OWRR A-041-NC(1).

DESCRIPTORS:
*WATER POLLUTION EFFECTS, *HEATED WATER, *WATER TEMPERATURE, *WATER QUALITY STANDARDS, *ESTURARIES, *THERMAL POLLUTION, *NORTH CAROLINA, NUTRIENTS, SEWAGE EFFLUENTS, EUTNOPHICATION, ALGAE, CYANOPHYTA, CHLOROPHYTA, METABOLISM, NEKTON, BENTHOS, BIOMASS.

IDENTIFIERS:
*PAMLICO RIVER ESTUARY, PHOTOSYNTHESIS/RESPIRATION RATIOS.

ABSTRACT:
RESPONSES OF ESTUARINE COMMUNITY STRUCTURE, RESPIRATION AND PRODUCTION TO ADDED HEAT, SEWAGE AND THEIR COMBINATION WERE INVESTIGATED. PLASTIC POOLS CONTAINING TRANSPLANTED ECOSYSTEMS FROM SOUTH CREEK ESTUARY, N. C. WERE USED. TEMPERATURE REPLICATION ACHIEVED. TEMPERATURE IN THE HEATED POOLS WAS REGULATED AT 5 C(9F) ABOVE THAT OF THE AMBIENT POOLS. THERMAL TREATMENT INCREASED NUTRIENT REGENERATION RATES, YIELDING SLIGHTLY HIGHER ALGAL BIOMASS; ALTHOUGH, SEASONAL DIFFERENCES WERE MORE SIGNIFICANT. GROSS COMMUNITY PRODUCTIVITY WAS REGULATED BY AMMONIA, LIGHT, AND TEMPERATURE LEVELS AND TOTAL RESPIRATION BY TEMPERATURE AND PRIMARY PRODUCTIVITY. SEWAGE ADDITION SUBSTANTIALLY INCREASED AMMONIA LEVELS, PARTICULARLY DURING THE WINTER. COMMUNITY METABOLISM RESPONDED POSITIVELY TO THERMAL TREATMENT, BUT NOT TO SEWAGE TREATMENT. THERMAL TREATMENT AND THE COMBINATION OF SEWAGE AND THERMAL TREATMENTS INCREASED THE PHOTOSYNTHESIS/RESPIRATION RATIOS (P/R) DURING SPRING AND SUMMER, BUT DECREASED THE P/R WHEN TEMPERATURE WAS LIMITING DURING WINTER. TEMPERATURE HAD VERY LITTLE EFFECT ON PHYTOPLANKTON COMPOSITION DURING THE SPRING. BLUE-GREEN ALGAE AND COCCOID GREEN ALGAE DOMINATED IN THE HEATED AND SEWAGE-TREATED POOLS DURING SUMMER. NEKTON AND BENTHIC INCREASED TO HIGHER BIOMASS IN HEATED POOLS DURING SPRING AND ACHIEVED A LOWER BIOMASS IN THE HEATED POOLS DURING SUMMER THAN IN THE AMBIENT POOLS. NO SIGNIFICANT DIFFERENCES WERE OBSERVED DURING WINTER AMONG HEATED AND AMBIENT POOLS. SEWAGE ADDITION DID NOT SUBSTANTIALLY ALTER PATTERNS BETWEEN HEATED AND AMBIENT SYSTEMS. OYSTERS, BAY CLAMS AND WIDGEON GRASS REACHED HIGHER BIOMASS IN HEATED POOLS DURING WINTER THAN IN AMBIENT POOLS. A FLOW-THROUGH EXPERIMENT WAS CONDUCTED DURING THE 1971 SUMMER TO TST MORE REALISTIC ESTUARINE CONDITIONS. RESULTS WERE SUBSTANTIALLY THE SAME AS OBTAINED UNDER QUIESCENT CONDITIONS.

FIELD 05C, 02L

ACCESSION NO. W72-13636

AVAILABILITY OF PHOSPHORUS FOR CLADOPHORA GROWTH IN LAKE MICHIGAN,

WISCONSIN UNIV., MILWAUKEE. DEPT. OF BOTANY; AND WISCONSIN UNIV., MILWAUKEE. CENTER FOR GREAT LAKES STUDIES.

CHANG-KWEI LIN.

IN: PROCEEDINGS 14TH CONFERENCE ON GREAT LAKES RESEARCH, UNIVERSITY OF TORONTO, ONTARIO, APRIL 19-21, 1971, P 39-43. 4 FIG, 15 REF.

DESCRIPTORS:
*PHOSPHORUS, *CLADOPHORA, *PLANT GROWTH, *LAKE MICHIGAN, ALGAE, PHOSPHATES, HYDROLYSIS, STORM WATER, RAINFALL, WASTE WATER(POLLUTION), EUTROPHICATION, WATER POLLUTION SOURCES, NUTRIENTS.

IDENTIFIERS:
CLADOPHORA GLOMERATA, MILWAUKEE HARBOR(WISCONSIN), POLYPHOSPHATES, ORTHOPHOSPHATES.

ABSTRACT:
THE PROFUSE GROWTH OF THE ALGA, CLADOPHORA, IN THE GREAT LAKES HAS SERIOUSLY CURTAILED SHORELINE RECREATION. SAMPLED FROM MILWAUKEE HARBOR AND FROM SITES ALONG LAKE MICHIGAN NORTH OF THE HARBOR, CLADOPHORA GLOMERATA WAS ANALYZED FOR EXTRACTABLE INORGANIC ORTHOPHOSPHATE AND FOR ITS ABILITY TO HYDROLYZE POLYPHOSPHATES. THOSE ALGAE GROWING IN THE HARBOR AREA SHOWED A CONSISTENTLY HIGH VALUE OF SURPLUS STORED PHOSPHORUS AND THE VALUES DECREASED AS THE DISTANCE FROM THE HARBOR INCREASED. LONGER, DENSER, AND GREENER CLADOPHORA WAS FOUND NEAR THE HARBOR THAN FROM MORE DISTANT SITES. SAMPLES FROM ALL LOCATIONS TAKEN AFTER HEAVY RAINFALLS CONTAINED SUBSTANTIAL INCREASES IN STORED PHOSPHORUS. STORM SEWER OVERFLOW AND INCREASED RIVER DISCHARGE DUE TO PRECIPITATION COULD SPORADICALLY PROVIDE EXCESSIVE SUPPLIES OF PHOSPHORUS. GREAT VARIATION IN ABILITY TO HYDROLYZE POLYPHOSPHATES WAS FOUND IN ALGAE SAMPLED FROM DIFFERENT LOCATIONS. THE YIELD OF ORTHOPHOSPHATES FROM POLYPHOSPHATES WAS INVERSELY PROPORTIONAL TO THE AMOUNT OF SURPLUS STORED PHOSPHORUS IN THE ALGAE. THE ALGAE GROWING NEAR THE HARBOR ACCUMULATED PHOSPHATE PHOSPHORUS WHILE THE ALGAE GROWING AT GREATER DISTANCE WERE CONCURRENTLY PHOSPHORUS STARVED, PERHAPS THEY ADAPTED TO OBTAIN THEIR PHOSPHORUS PRIMARILY FROM HYDROLYSIS OF POLYPHOSPHATE. (JONES-WISCONSIN)

FIELD 05C

ACCESSION NO. W72-13644

DEVELOPMENT OF A THEORETICAL SEASONAL GROWTH RESPONSE CURVE OF CLADOPHORA GLOMERATA TO TEMPERATURE AND PHOTOPERIOD,

STATE UNIVERSITY OF NEW YORK, BUFFALO, DEPT. OF BIOLOGY; AND STATE UNIV. COLL., BUFFALO. GREAT LAKES LAB.

J. R. STORR, AND R. A. SWEENEY.

IN: PROCEEDINGS 14TH CONFERENCE ON GREAT LAKES RESEARCH, UNIVERSITY OF TORONTO, ONTARIO, APRIL 19-21, 1971, P 119-127. 4 FIG, 4 TAB, 14 REF.

DESCRIPTORS:
*ALGAE, *SEASONAL, *CLADOPHORA, *TEMPERATURE, *PHOTOPERIODISM, GREAT LAKES, MODEL STUDIES, NUTRIENTS, PLANT GROWTH, ANALYTICAL TECHNIQUES, WATER TEMPERATURE.

IDENTIFIERS:
*CLADOPHORA GLOMERATA, GROWTH CURVES.

ABSTRACT:
LABORATORY AND FIELD INVESTIGATIONS ASCERTAINED EFFECTS OF HEAT ON CLADOPHORA GROWTH WHEN THIS PLANT IS CULTURED UNDER SPRING AND SUMMER PHOTOPERIODS. RESULTS INDICATED THAT AT THE LEVEL OF NUTRIENTS PRESENT, OPTIMUM GROWTH WAS AROUND 18C. LOW GROWTH LEVELS OCCURRED AT THE LOW TEMPERATURE RANGE (14.5C) BUT RESULTS WERE INCONCLUSIVE. IN THE UPPER TEMPERATURE RANGE, GROWTH CESSATION APPEARED AT ABOUT 25C. GROWTH RESPONSE TO PHOTOPERIOD LEVELS IN THE EXPERIMENTS APPEARED TO BE EXPONENTIAL IN CHARACTER WHILE THE GROWTH RESPONSE CURVES GENERATED BY THE TEMPERATURE EXPERIMENTS WERE VERY SIMILAR IN FORM AT THE DIFFERENT LEVELS OF PHOTOPERIOD. A FAMILY OF TEMPERATURE RELATED GROWTH CURVES WAS CONSTRUCTED FOR DIFFERENT PHOTOPERIODS THROUGHOUT THE MAY-OCTOBER PERIOD. RECORDED LAKE WATER TEMPERATURE DATA WAS PLOTTED ON THESE CURVES FOR THE SAME PERIOD AND A SEASONAL GROWTH RESPONSE CURVE WAS GENERATED RELATING SEASONAL PHOTOPERIOD AND TEMPERATURE. THIS RESULTANT PLOT APPEARED TO BE IN CLOSE AGREEMENT WITH OBSERVED CLADOPHORA GROWTH THROUGHOUT THE SEASON. A THEORETICAL GROWTH CURVE WAS ALSO PLOTTED USING THIS TECHNIQUE AND THE TEMPERATURE DATA FOR THE SAME AREA AND TIME PERIOD. THE THEORETICAL MODEL OF GROWTH RESPONSE RATE MAY BE USEFUL IN PREDICTING PROBABLE SEASONAL GROWTH AND APPROXIMATE STANDING CROP, ESPECIALLY IN LOCALIZED AREAS AROUND THERMAL DISCHARGES. (JONES-WISCONSIN)

FIELD 05C

ACCESSION NO. W72-13651

CHLOROPHYLL A - TOTAL PHOSPHORUS RELATIONSHIPS IN LAKE ERIE,

ONTARIO WATER RESOURCES COMMISSION, REXDALE (ONTARIO).

THOMAS G. BRYDGES.

IN: PROCEEDINGS 14TH CONFERENCE ON GREAT LAKES RESEARCH, UNIVERSITY OF TORONTO, ONTARIO, APRIL 19-21, 1971, P 185-190. 5 FIG, 4 REF.

DESCRIPTORS:
*CHLOROPHYLL, *PHOSPHORUS, *LAKE ERIE, NITROGEN, ALGAE, INORGANIC COMPOUNDS, STANDING CROPS, LIMITING FACTORS, EUTROPHICATION.

ABSTRACT:
SINCE 1966 AN EXTENSIVE WATER QUALITY MONITORING PROGRAM HAS BEEN CONDUCTED ON THE GREAT LAKES. SOME LAKE WATER SAMPLES HAVE BEEN ANALYZED FOR CHLOROPHYLL A AS A ROUTINE SINCE 1967 TO ESTIMATE ALGAL STANDING CROPS. EXAMINATION OF CHOROPHYLL RESULTS FOR 1967, 1968, AND 1969 IS PRESENTED WITH PARTICULAR REFERENCE TO THEIR RELATIONSHIPS WITH INORGANIC NITROGEN AND PHOSPHORUS CONCENTRATIONS. IN 1967 AVERAGE CHLOROPHYLL A AND TOTAL PHOSPHORUS CONCENTRATIONS AT 87 OPEN LAKE STATIONS (THREE TO FIVE MEASUREMENTS AT EACH) WERE DIRECTLY PROPORTIONAL. HIGH TOTAL PHOSPHORUS AND RELATIVELY LOWER CHLOROPHYLL A CONCENTRATIONS CHARACTERIZED THE DETROIT, RAISIN, AND MAUMEE RIVER STATIONS. LAKE STATIONS INCLUDED THE WESTERN BASIN AND A FIVE MILE WIDE BAND ALONG NORTH SHORE FROM POINT PELEE TO BUFFALO. THERE WERE NO APPARENT TRENDS BETWEEN CHLOROPHYLL A AND INORGANIC NITROGEN CONCENTRATIONS. THE CONCENTRATIONS OF CHLOROPHYLL A AND TOTAL PHOSPHORUS MEASURED IN 1968 AND 1969 WERE DIRECTLY PROPORTIONAL. EIGHT STATIONS IN WESTERN BASIN WERE SAMPLED FOR NINE SUCCESSIVE DAYS IN JULY 1968. IN 1969, 97 STATIONS (422 MEASUREMENTS) WERE SAMPLED OVER THE FULL LENGTH OF THE LAKE. THAT PHOSPHORUS IS AN ALGAL GROWTH LIMITING FACTOR IN LAKE ERIE IS CONCLUDED FROM THESE EXTENSIVE EMPIRICAL OBSERVATIONS. (JONES-WISCONSIN)

FIELD 05C

ACCESSION NO. W72-13653

THE EFFECT OF DDT AND DIELDRIN UPON C-14 UPTAKE BY IN SITU PHYTOPLANKTON IN LAKES ERIE AND ONTARIO,

DEPARTMENT OF ENERGY MINES AND RESOURCES, BURLINGTON (ONTARIO). CANADA CENTRE FOR INLAND WATERS.

WALTER A. GLOOSCHENKO.

IN: PROCEEDINGS 14TH CONFERENCE ON GREAT LAKES RESEARCH, UNIVERSITY OF TORONTO, ONTARIO, APRIL 19-21, 1971, P 219-223. 6 FIG, 12 REF.

DESCRIPTORS:
*PESTICIDES, *DDT, *DIELDRIN, *PHYTOPLANKTON, CARBON RADIOISOTOPES, LAKE ERIE, LAKE ONTARIO, ADSORPTION, AGRICULTURAL RUNOFF, ALGAE, FOOD CHAINS, CHLORINATED HYDROCARBON PESTICIDES, FISHERIES.

ABSTRACT:
SINCE PREVIOUS WORK WITH CHLORINATED HYDROCARBONS HAS BEEN MAINLY RESTRICTED TO MARINE SPECIES OF PHYTOPLANKTON, THIS STUDY INVESTIGATED THE EFFECTS OF DDT AND DIELDRIN UPON C-14 UPTAKE BY NATURAL PHYTOPLANKTON COMMUNITIES IN LAKES ONTARIO AND ERIE. IN SITU STUDIES WERE PERFORMED, BUT THE CONCENTRATIONS OF DDT AND DIELDRIN USED WERE HIGHER THAN NORMALLY FOUND IN THE GREAT LAKES. THE LOWEST CONCENTRATION USED, 1 PPB, IS AT LEAST 200 TIMES THAT FOUND IN SITU. IF DEPRESSION OF CARBON FIXATION BY PHYTOPLANKTON IN THE GREAT LAKES IS OCCURRING, IT WILL BE NEGLIGIBLE. POSSIBLY AGRICULTURAL RUNOFF IN LOCAL AREAS MAY BE MUCH HIGHER IN PESTICIDE CONCENTRATION WHICH MAY LEAD TO DDT OR DIELDRIN LIMITING CARBON FIXATION IN LOCALIZED INSHORE AREAS. THE ADAPTATION OF ALGAE TO PESTICIDES NEEDS CONSIDERATION. THE MOST SERIOUS HAZARD RESULTING FROM PESTICIDES IN WATER IS BIOLOGICAL CONCENTRATION; WHILE ALGAE THEMSELVES MAY NOT BE SERIOUSLY AFFECTED, THEY MAY CONCENTRATE HIGH AMOUNTS OF PESTICIDE BY ACTIVE UPTAKE ON SURFACE ADSORPTION AND TRANSFER IT TO HIGHER TROPHIC LEVELS. CONCENTRATION AND TRANSFER OF THESE TWO COMPOUNDS TO HIGHER TROPHIC LEVELS IS OF MAJOR CONCERN TO GREAT LAKES FISHERIES.
(JONES-WISCONSIN)

FIELD 05C

ACCESSION NO. W72-13657

HYPOTHESIS FOR DISSOLVED OXYGEN DEPLETION IN THE CENTRAL BASIN HYPOLIMNION OF LAKE ERIE,

LAKE ERIE BASIN OFFICE, FAIRVIEW PARK, OHIO.

CONRAD KLEVENO, O. BRAIDECH, E. THOMAS, AND PHILIP E. GEHRING.

IN: PROCEEDINGS 14TH CONFERENCE ON GREAT LAKES RESEARCH, UNIVERSITY OF TORONTO, ONTARIO, APRIL 19-21, 1971, P 252-255. 1 FIG, 1 TAB.

DESCRIPTORS:
*DISSOLVED OXYGEN, *HYPOLIMNION, *LAKE ERIE, BENTHIC FLORA, ALGAE, LIGHT PENETRATION, PLANKTON, NUTRIENTS, BIOCHEMICAL OXYGEN DEMAND, INORGANIC COMPOUNDS, THERMAL STRATIFICATION, DIATOMS, EPILIMNION, EUTROPHICATION.

IDENTIFIERS:
*LAKE ERIE CENTRAL BASIN, *OXYGEN DEPLETION.

ABSTRACT:
IN 1968 A RAPID DEPLETION OF DISSOLVED OXYGEN IN THE HYPOLIMNION OF THE CENTRAL BASIN OF LAKE ERIE WAS OBSERVED FROM 8.2 MGL ON AUGUST 6 TO 1.9 ON AUGUST 8. IN THE SUMMER OF 1969 A STUDY WAS INITIATED TO DETERMINE THE RATE AND DESCRIBE THE MECHANICS OF DISSOLVED OXYGEN DEPLETION IN THE HYPOLIMNION IN THE BASIN. THE DECLINE OF THE DISSOLVED OXYGEN WAS GRADUAL DURING EARLY SUMMER, BUT IN AUGUST THE RATE OF DEPLETION INCREASED RAPIDLY. THE DEATH AND DECOMPOSITION OF BENTHIC ALGAE, TRIBONEMA UTRICULOSUM AND OEDOGONIUM MAY EXPLAIN THIS RELATIVELY SUDDEN DECLINE. THESE ALGAE WERE FOUND GROWING IN PROFUSION AT THE BOTTOM DURING THE SUMMER OF 1969. IT IS POSTULATED THAT THEY ARE KILLED BY A REDUCTION IN LIGHT CAUSED BY AN INCREASE OF PLANKTON IN OVERLYING WATERS. THE INCREASED PLANKTON IS CAUSED BY AN INCREASED VERTICAL CIRCULATION OF NUTRIENTS WHEN THE LAKE BEGINS TO COOL. THE DEATH OF THE BENTHIC ALGAE RESULTS IN A TREMENDOUS INCREASE IN BIOCHEMICAL OXYGEN DEMAND IN HYPOLIMNION WATERS, THUS THE RAPID DEPLETION OF DISSOLVED OXYGEN. (JONES-WISCONSIN)

FIELD 05C

ACCESSION NO. W72-13659

A CASE OF NUTRIENT ENRICHMENT IN AN INSHORE AREA OF GEORGIAN BAY,

ONTARIO WATER RESOURCES COMMISSION, REXDALE (ONTARIO).

DENIS M. VEAL, AND M. F. P. MICHALSKI.

IN: PROCEEDINGS 14TH CONFERENCE ON GREAT LAKES RESEARCH, UNIVERSITY OF TORONTO, ONTARIO, APRIL 19-21, 1971, P 277-292. 4 FIG, 5 TAB, 25 REF.

DESCRIPTORS:
*EUTROPHICATION, *NUTRIENTS, LAKE HURON, EFFLUENTS, WATER QUALITY, PHOSPHORUS, PHYTOPLANKTON, SEWAGE TREATMENT, NITROGEN, PHYSOCHEMICAL PROPERTIES, DIATOMS, ALGAE, CLADOPHORA, PRODUCTIVITY, WATER POLLUTION, STANDING CROPS.

IDENTIFIERS:
*GEORGIAN BAY(ONTARIO).

ABSTRACT:
SURVEYS AND PUBLIC COMPLAINTS INDICATED THAT PHYTOPLANKTON POPULATIONS IN PENETANG HARBOR, GEORGIAN BAY, WERE REACHING EXCESSIVE LEVELS. A GRADATION OF WATER QUALITY CONDITIONS NOT UNLIKE THOSE OF LAKE ERIE WERE REVEALED IN A STUDY OF THE PENETANGUISHENE TO WAUBAUSHENE AREA IN 1969. MAJOR EMPHASIS WAS PLACED ON AN EVALUATION OF NITROGEN AND PHOSPHORUS CONCENTRATIONS, STANDING STOCKS OF PHYTOPLANKTON AND ALGAL PRODUCTIVITY AND AN ASSESSMENT OF NUTRIENT LOADING. MEAN TOTAL PHOSPHORUS CONCENTRATIONS RANGED FROM 0.02 MG/L (AS P) OVER MOST OF THE STUDY AREA TO 0.03 MG/L IN MIDLAND BAY AND 0.06 MG/L IN THE SOUTHERN END OF PENETANG HARBOR. PHYTOPLANKTON VALUES INCREASED WITH HIGHER TOTAL PHOSPHORUS CONCENTRATIONS. THE MIDLAND AND PENETANG SEWAGE TREATMENT PLANTS WERE RESPONSIBLE FOR APPROXIMATELY 53% OF THE NET PHOSPHORUS LEADING. THE FLUSHING RATE AS WELL AS EXHANGE OF HIGH QUALITY WATER FROM THE OPEN LAKE WILL PROMOTE RAPID RECOVERY WHEN SIGNIFICANT MUNICIPAL NUTRIENT SOURCES ARE ELIMINATED. IT IS EVIDENT THAT MUNICIPAL WASTE DISCHARGES TO SMALL BAYS ALONG THE UPPER GREAT LAKES WILL REQUIRE AN IMPROVED DEGREE OF TREATMENT, INCLUDING NUTRIENT REMOVAL, IN ORDER TO PRESERVE WATER QUALITY AND TO MAINTAIN HIGH RECREATIONAL POTENTIAL OF THE SHORELINES.
(JONES-WISCONSIN)

FIELD 05C

ACCESSION NO. W72-13662

A QUANTITATIVE COMPARISON OF PIGMENT EXTRACTION BY MEMBRANE AND GLASS-FIBER FILTERS,

KENT STATE UNIV., OHIO. DEPT. OF BIOLOGICAL SCIENCES.

E. B. LONG, AND G. D. COOKE.

LINMOLOGY AND OCEANOGRAPHY, VOL. 16, NO. 6, P 990-992, NOVEMBER 1971. 1 FIG, 1 TAB, 3 REF.

DESCRIPTORS:
*PHYTOPLANKTON, *CHLOROPHYLL, *EVALUATION, *SEPARATION TECHNIQUES, SPECTROPHOTOMETRY, LAKES, PLANT PIGMENTS, DETRITUS, CHLOROPHYTA, CHRYSOPHYTA, SOLVENT EXTRACTIONS, DIATOMS, AQUATIC ALGAE, FILTRATION, CYANOPHYTA, *EUTROPHICATION, *OHIO.

IDENTIFIERS:
*GLASS FIBER FILTERS, *MEMBRANE FILTERS, GRANGE LAKE, EAST TWIN LAKE, LAKE ROCKWELL, LAKE HODGSON, CLEARWATER QUARRY, DOLLAR LAKE, CHLOROPHYLL A, ABSORBANCE.

ABSTRACT:
WATER SAMPLES COLLECTED FROM SIX EUTROPHIC NORTHEASTERN OHIO LAKES WERE USED IN A STUDY CONCERNED WITH A QUANITATIVE COMPARISON OF PIGMENT EXTRACTION BY MEMBRANE AND GLASS-FIBER FILTERS. THE PHYTOPLANKTON-CONTAINING WATER SAMPLES WERE MEASURED INTO 500-ML SUBSAMPLES AND FILTERED AT 0.25-0.33 ATM. THE FILTER TYPES TESTED WERE 47-MM-DIAMETER TYPE HA MILLIPORE MEMBRANE FILTERS (0.45 MICRON PORE SIZE; CELLULOSE ACETATE) AND 42.5-MM-DIAMETER TYPES GF/C AND GF/A WHATMAN GLASS-FIBER FILTERS. THE PHYTOPLANKTON WAS EXTRACTED INTO ACETONE AND ABSORBANCES MEASURED SPECTROPHOTOMETRICALLY AT 665 MILLIMICRONS. DATA WERE ANALYZED BY 'STUDENT'S' T STATISTIC. ABSORBANCES OF THE ACETONE EXTRACTS FROM LAKE PHYTOPLANKTON COLLECTED ON GLASS-FIBER FILTERS, GROUND AND EXTRACTED 2 HR, WERE 6-18 PERCENT GREATER TAHN FROM PHYTOPLANKTON COLLECTEC ON MEMBRANE FILTERS AND PROCESSED IDENTICALLY. USE OF THE GLASS-FIBER FILTERS CUT FILTERING TIME BY A FACTOR OF 10 AND MATERIAL COST BY A FACTOR OF 4 OR MORE.
(SNYDER-BATTELLE)

FIELD 05A, 05C

ACCESSION NO. W72-13673

NUTRIENT STUDIES IN TEXAS IMPOUNDMENTS,

TEXAS UNIV., AUSTIN. DEPT. OF ENVIRONMENTAL HEALTH AND ENGINEERING.

V. H. HUANG, J. R. MASE, AND E. G. FRUH.

DECEMBER 1970. 39 P., 10 FIG, 11 TAB, 6 REF. OWRR B-040-TEX(4).

DESCRIPTORS:
*PHYTOPLANKTON, *ALGAE, *NUTRIENTS, *NITROGEN FIXATION, LAKES, NITROGEN, PHOSPHORUS, IRON, *TEXAS, LABORATORY TESTS, CARBON RADIOISOTOPES, *CYANOPHYTA, EUTROPHICATION, WATER POLLUTION EFFECTS.

IDENTIFIERS:
*LAKE TRAVIS, *LAKE LIVINGSTON, FIELD MEASUREMENTS, CARBON-14 MEASUREMENTS, ACETYLENE REDUCTION MEASUREMENTS, NUTRIENT ENRICHMENT TESTS.

ABSTRACT:
BLUE-GREEN ALGAE BECAME DOMINANT IN LATE SUMMER 1970 IN LAKE LIVINGSTON ON THE TRINITY RIVER. PRELIMINARY BATCH LABORATORY NUTRIENT ENRICHMENT TESTS INDICATED NITROGEN TO BE THE LIMITING NUTRIENT. LABORATORY AND IN-SITU CARBON- 14 MEASUREMENTS SUBSTANTIATED THE CONCLUSIONS THAT NITROGEN WAS THE LIMITING NUTRIENT. IN-SITU ACETYLENE REDUCTION MEASUREMENTS INDICATED THAT THE NITROGEN FIXATION RATE WAS UNMEASURABLE IN AUGUST BUT WAS 0.74 MILLIMICRO MOLES C2H4/MG PROTEIN/ MIN IN SEPTEMBER. THE RATE OF NITROGEN FIXATION SIGNIFICANTLY INCREASED WITH THE ADDITION OF PHOSPHORUS AFTER 2 WEEKS OF INCUBATION. BLUE-GREEN ALGAE WERE ALSO ONE OF THE DOMINANT GROUPS OF PHYTOPLANKTON IN LAKE TRAVIS ON THE COLORADO RIVER. PRELIMINARY BATCH LABORATORY NUTRIENT ENRICHMENT TESTS INDICATED THAT NITROGEN, PHOSPHORUS, AND IRON ALL COULD BE LIMITING PHYTOPLANKTON GROWTH. LABORATORY CARBON- 14 ANALYSES INDICATED IRON TO BE MOST CRITICAL BUT FIELD TESTS WERE INCONCLUSIVE. IN-SITU ACETYLENE REDUCTION MEASUREMENTS INDICATED THAT NO NITROGEN FIXATION WAS OCCURRING. FIXATION IN THE LABORATORY DID OCCUR AFTER 33 DAYS OF INCUBATION OF SAMPLES ENRICHED WITH PHOSPHORUS AND PHOSPHORUS AND IRON WITH THE LATTER SAMPLE SHOWING A HIGHER RATE. HOWEVER, THIS RATE WAS SIGNIFICANTLY LOWER THAN THAT FOUND IN LAKE LIVINGSTON SAMPLES. (GALWARDI-TEXAS)

FIELD 05C, 02H

ACCESSION NO. W72-13692

SALMONELLA.

DEFENSE DOCUMENTATION CENTER, ALEXANDRIA, VA.

AVAILABLE FROM THE NATIONAL TECHNICAL INFORMATION SERVICE AS AD 737 900. $3.00 IN PAPER COPY, $0.95 IN MICROFICHE. REPORT NO. DDC-TAS-71-62, FEBRUARY 1972. 274 P, 171 REF.

DESCRIPTORS:
*SALMONELLA, *BIBLIOGRAPHIES, BIOLOGY, EPIDEMIOLOGY, CULTURES, METHODOLOGY, DOCUMENTATION, ABSTRACTS, DATA COLLECTIONS, MICROBIOLOGY, SYSTEMATICS, POLLUTANTS, PATHOGENIC BACTERIA, COLIFORMS, TOXINS, ANTIBIOTICS(PESTICIDES), ENTERIC BACTERIA, DISEASES, E. COLI, SHIGELLA, VIRUSES, CHROMATOGRAPHY, GAS CHROMATOGRAPHY, ANALYTICAL TECHNIQUES, SEPARATION TECHNIQUES, TEMPERATURE, OXYGEN, ENVIRONMENT, ENZYMES, GEOGRAPHICAL REGIONS, ALGAE, MICROORGANISMS, IRON, RADIATION, METABOLISM, INSTRUMENTATION, EQUIPMENT, LABORATORY EQUIPMENT, MICROBIOLOGY, BACTERIOPHAGE, AEROSOLS, GENETICS, WATER POLLUTION, RADIOACTIVITY EFFECTS.

IDENTIFIERS:
DETECTION, FLUORESCENT ANTIBODY TECHNIQUES, SALMONELLA TYPHIMURIUM, AMPICILLIN, VIBRIOS, SALMONELLA TYPHOSA, TYPHOID, COLIPHAGE, IMMUNOLOGY, PASTEURELLA PSEUDOTUBERCULOSIS, PENICILLIN, DRUGS, CORTISONE, HORMONES, CHLORAMPHENICOL, SALMONELLA ENTEROCOLITIS, SHIGELLA DYSENTERIAE, SHIGELLA FLEXNERI, BIOCHEMICAL STUDIES, ARTHOPODS, STREPTOMYCIN, SALMONELLA PARATYPHI, STAPHYLOCOCCUS, NUCLEAR WARFARE, CHOLERA, PASTEURELLA PESTIS, CHARACTERIZATION, BIOSYNTHESIS.

ABSTRACT:
THIS BIBLIOGRAPHY IS A COLLECTION OF UNCLASSIFIED REFERENCES ON SALMONELLA. THE DISEASE, THE MODES OF TRANSMISSION AND THE VECTORS ARE PRESENTED. INDEXES FOR CORPORATE AUTHOR-MONITORING AGENCY, SUBJECT, TITLE, AND PERSONAL AUTHOR ARE INCLUDED. (LONG-BATTELLE)

FIELD 05C, 05A, 05B, 10C

ACCESSION NO. W72-13800

DDT: INHIBITION OF SODIUM CHLORIDE TOLERANCE BY THE BLUE-GREEN ALGA ANACYSTIS NIDULANS,

WISCONSIN UNIV., MADISON. DEPT. OF ENTOMOLOGY.

J. C. BATTERTON, G. M. BOUSH, AND F. MATSUMURA.

SCIENCE, VOL 176, P 1141-1143, JUNE 9, 1972. 2 TAB, 14 REF.

DESCRIPTORS:
*DDT, *SODIUM CHLORIDE, *INHIBITION, *SALT TOLERANCE, WATER POLLUTION EFFECTS, CHLORINATED HYDROCARBON PESTICIDES, *CYANOPHYTA, SODIUM, POTASSIUM, CULTURES, RESISTANCE, AQUATIC ALGAE, GROWTH RATES, BIOASSAY, LABORATORY TESTS, CALCIUM.

IDENTIFIERS:
*ANACYSTIS NIDULANS, *ADENOSINE TRIPHOSPHATASE, BIOCHEMICAL TESTS, OUABAIN, POTASSIUM CHLORIDE.

ABSTRACT:
THE EFFECTS OF DDT ON THE SODIUM CHLORIDE TOLERANCE OF THE BLUE-GREEN ALGA, ANACYSTIS NIDULANS WERE INVESTIGATED BY GROWTH-RATE STUDIES. ALGAL CULTURES WERE GROWN AT 30 DEGREES C ON PREVIOUSLY DEFINED MEDIA WITH ADDITIONS OF SODIUM CHLORIDE (1 PERCENT BY WEIGHT) AND/OR DDT (800 PPB) AND GROWTH RATE CONSTANTS WERE CALCULATED FOR EACH CONDITION. A. NIDULANS WAS ABLE TO TOLERATE SODIUM CHLORIDE AND DDT ALONE BUT NOT IN COMBINATION. WHEN NACL WAS REPLACED BY KCL, NO GROWTH WAS OBSERVED WITH KCL IN COMBINATION WITH DDT. GROWTH DID OCCUR IN CULTURES WITH LOWER CONCENTRATIONS OF EITHER KCL OR DDT. THAT DDT LOWERED ALGAL TOLERANCE TO SALTS BY THE INHIBITION OF ADENOSINE TRIPHOSPHATASE ACTIVITY AND THE SODIUM-PUMP WAS SUBSTANTIATED BY (1) ADDITIONS OF CALCIUM, A TRANSPORT INHIBITOR, TO THE GROWTH CULTURES (THE CALCIUM REDUCED THE PERMEABILITY OF THE CELLS TO NACL, THUS REDUCING NACL STRESS) AND (2) ENZYME ASSAYS. TWO TYPES OF ATPASES WERE USED IN A COMPARISON OF INHIBITION BY DDT AND OUABAIN. THE DEGREE OF INHIBITION BY DDT WAS FOUND TO EXCEED THAT DUE TO OUABAIN. GROWTH EXPERIMENTS CARRIED OUT WITH GLYCEROL INDICATED THAT THE NACL PLUS DDT INHIBITION IS PROBABLY DUE TO AN IONIC EFFECT AND NOT TO OSMOTIC STRESS. (LONG-BATTELLE)

FIELD 05C, 05A

ACCESSION NO. W72-13809

VARIATION IN THE TOXICITY OF ARSENIC COMPOUNDS TO MICROORGANISMS AND THE SUPPRESSION OF THE INHIBITORY EFFECTS BY PHOSPHATE,

COMMONWEALTH SCIENTIFIC AND INDUSTRIAL RESEARCH ORGANIZATION, MELBOURNE (AUSTRALIA). DIV. OF FOREST PRODUCTS.

E. W. B. DA COSTA.

APPLIED MICROBIOLOGY, VOL 23, NO 1, P 46-53, JANUARY 1972. 2 FIG, 5 TAB, 11 REF.

DESCRIPTORS:
*TOXICITY, *ARSENIC COMPOUNDS, *PHOSPHATES, *INHIBITIONS, *MICROORGANISMS, BACTERIA, FUNGI, METHODOLOGY, CULTURES, BIOASSAY, ENVIRONMENTAL EFFECTS, WATER POLLUTION EFFECTS, ALGAE, HEAVY METALS, GROWTH, PSEUDOMONAS, POTASSIUM COMPOUNDS, POLLUTANT IDENTIFICATION, CHLOROPHYTA, CHLORELLA.

IDENTIFIERS:
ACREMONIELLA, DIMETHYL SODIUM ARSONATE, ARSENATES, SODIUM CACODYLATE, CULTURE MEDIA, PORIA MONTICOLA, PORIA COCOS, PORIA VAILLANTII, CONIOPHORA OLIVACEA, LENZITES TRABEA, SCOPULARIOPSIS BREVICAULIS, CLADOSPORIUM HERBARUM, TRAMETES VERSICOLOR, FOMES ANNOSUS, FUSCOPORIA CONTIGUA, TRAMETES LILACINO-GILVA, XYLOBOLUS FRUSTULATUS, ASPERGILLUS NIGER, CHAETOMIUM GLOBOSUM, FUSARIUM SOLANI, MUCOR MICROSPORUS, PENICILLIUM SPINULOSUM, STREPTOMYCES GRISEUS, BACILLUS SUBTILIS, PSEUDOMONAS AERUGINOSA, CHLORELLA PYRENOIDOSA.

ABSTRACT:
TWO TESTS WERE USED TO MEASURE THE TOXICITY OF ARSENIC COMPOUNDS TO MICROORGANISMS AND THE SUPPRESSION OF THE INHIBITORY EFFECTS BY PHOSPHATE. THE TOXICITY PHIAL TEST MEASURED THE GROWTH OF FUNGI ON AGAR CONTAINING VARIOUS CONCENTRATIONS OF ARSENATES AND PHOSPHATES. THE SECOND, A SEEDED PLATE INHIBITION ZONE TEST, MEASURED THE EFFECTS OF PHOSPHATE ON THE ARSENATE GROWTH INHIBITION ZONES. THE TOXICITY OF POTASSIUM ARSENATE, AS MEASURED BY RETARDATION OR INHIBITION OF GROWTH ON SOLID NUTRIENT MEDIA, SHOWED WIDE VARIATION AMONG DIFFERENT FUNGI BUT WAS CONSISTENTLY REDUCED BY THE ADDITION OF LARGE AMOUNTS OF POTASSIUM PHOSPHATE, WITH BOTH ARSENIC-SENSITIVE AND ARSENIC-TOLERANT FUNGI. PORIA MONTICOLA WAS COMPLETELY INHIBITED BY 0.0025 M ARSENATE BUT WAS PROGRESSIVELY LESS INHIBITED AS THE PHOSPHATE CONTENT OF THE MEDIUM INCREASED AND GREW SLOWLY AT 0.04 M ARSENATE WHEN 0.16 M KH2PO4 WAS ADDED. CLADOSPORIUM HERBARUM SHOWED 36 PERCENT REDUCTION IN GROWTH AT 0.08 M ARSENATE IN A LOW-PHOSPHATE MEDIUM, BUT WHEN 0.01 M KH2PO4 WAS ADDED, ARSENATE CONCENTRATIONS UP TO 0.64 M (AT WHICH THE MEDIUM CONTAINS 4.8 PERCENT AS) CAUSED NO REDUCTION IN GROWTH RATE. ADDITION OF PHOSPHATE ALSO REDUCED THE TOXICITY OF POTASSIUM ARSENITE BUT NOT THAT OF DIMETHYL SODIUM ARSONATE (SODIUM CACODYLATE). THE COUNTERACTING EFFECT OFPHOSPHATE ON ARSENATE TOXICITY WAS FOUND TO OCCUR WITH EVERY ONE OF A WIDE VARIETY OF MICROORGANISMS TESTED. THE PRACTICAL IMPLICATIONS OF THE COUNTER-INHIBITION PHENOMENON IN LABORATORY INVESTIGATIONS AND STANDARD TESTS OF ARSENICAL FUNGICIDES, IN BIOCHEMICAL RESEARCH, AND IN THE COMMERCIAL USE OF ARSENICAL BIOCIDES ARE SET OUT. (LONG-BATTELLE)

FIELD 05C

ACCESSION NO. W72-13813

CHANGES IN PLANKTON SPECIES COMPOSITION AND DIVERSITY IN A CONTROLLED NUTRIENT ENRICHMENT STUDY,

KANSAS UNIV., LAWRENCE. DEPT. OF SYSTEMATICS AND ECOLOGY.

W. J. O'BRIEN.

TRANSACTIONS OF THE AMERICAN MICROSCOPICAL SOCIETY, VOL 91, NO 1, P 77-91, JANUARY 1972. 2 FIG, 3 REF.

DESCRIPTORS:
*PHYTOPLANKTON, *NUTRIENT REQUIREMENTS, NUTRIENTS, BIOINDICATORS, WATER POLLUTION EFFECTS, FERTILIZERS, NITROGEN, BIOASSAY, PHOSPHORUS, CYANOPHYTA, OLIGOTROPHY, CHLAMYDOMONAS, DEFICIENT ELEMENTS, MESOTROPHY, EUTROPHICATION, TROPHIC LEVEL, PROTOZOANS, AQUATIC ALGAE, ENVIRONMENTAL EFFECTS, CHLOROPHYTA, PRIMARY PRODUCTIVITY, CHRYSOPHYTA.

IDENTIFIERS:
ENRICHMENT, CHROMULINA, CRYPTOMONAS EROSA, ERKINIA SUBEQUICILIATA, UROGLENOPSIS AMERICANA, CRYPTOMONAS PULSILLA, MICROCYSTIS AERUGINOSA, CHROOMONAS CAUDATA, CHLAMYDOMONAS PERTUSA, PANDORINA MORUM.

ABSTRACT:
AN INVESTIGATION WAS MADE OF CHANGES IN PLANKTON SPECIES COMPOSITION AND DIVERSITY AS BIOLOGICAL INDICATORS BY CONTROLLING THE NUTRIENT CONTENT/DENSITY OF THEIR ENVIRONMENT. CONTROLLED EXPERIMENTATION WAS CARRIED OUT FOR TWO YEARS AT EIGHT PONDS WHICH WERE ORGANIZED INTO FOUR TREATMENT LEVELS (CONTROL, LOW, MEDIUM, HIGH) BY ADDING SPECIFIED AMOUNTS OF HIGH QUALITY INORGANIC FERTILIZER. THE CONTROL PONDS WERE OLIGOTROPHIC, THE LOW AND MEDIUM PONDS MESOTROPHIC, AND THE HIGH PONDS EUTROPHIC. IN THE FIRST YEAR THE PHYTOPLANKTON DENSITY WAS QUITE VARIABLE IN THE TREATED PONDS WITH THE HIGH TREATMENT LEVEL PONDS AT TIMES HAVING THE LOWEST PHYTOPLANKTON DENSITY AND AT OTHER TIMES THE HIGHEST. BY MID-SUMMER OF THE SECOND YEAR, HOWEVER, THE PHYTOPLANKTON DENSITY INCREASED PROPORTIONATELY WITH INCREASING TREATMENT LEVEL. IN 1969, THE PRIMARY PRODUCTIVITY WAS MEASURED AND ITS RESPONSE TO TREATMENT LEVEL WAS SIMILAR TO THAT OF THE PHYTOPLANKTON DENSITY. IN A LABORATORY STUDY, WATER FROM EACH TREATMENT LEVEL WAS COLLECTED AND PLACED IN FLASKS TO WHICH WAS ADDED DISSOLVED INORGANIC NITROGEN OR PHOSPHORUS OR NITROGEN AND PHOSPHORUS TOGETHER. THE TEST ORGANISM, PANDORINA MORUM, WAS INCUBATED IN THESE SEPARATE MEDIA FOR 2 WEEKS, AFTER WHICH THE YIELD WAS MEASURED. THE FINAL YIELD OF P. MORUM INCREASED WITH INCREASING TREATMENT LEVEL. THE ADDITION OF NITROGEN STIMULATED FINAL YIELD AT ALL TREATMENT LEVELS BUT WAS PROPORTIONATELY LESS STIMULATORY WITH INCREASING TREATMENT LEVEL. THE ADDITION OF PHOSPHORUS DID NOT STIMULATE FINAL YIELD AT ALL. (BYRD-BATTELLE)

FIELD 05C

ACCESSION NO. W72-13816

LINMCLOGY OF ONEIDA LAKE WITH EMPHASIS ON FACTORS CONTRIBUTING TC ALGAL BLOOMS,

GECLOGICAL SURVEY, ALBANY, N.Y.

P. E. GREESON.

NEW YORK STATE DEPT OF ENVIRONMENT CONSERVATION, ALBANY, (GEOLOGICAL SURVEY OPEN-GILE REPORT), 1971. 185 P, 51 FIG, 32 TAB, 145 REF, APPEND.

DESCRIPTORS:
*LIMNOLOGY, *LAKES, *EUSTROPHICATICN, *HYDROLOGY, *NEW YORK, WATER QUALITY, NUTRIENTS, ALGAE, INFLCW, DISCHARGE(WATER), WATER BALANCE, SEDIMENTS, CHEMICAL PROPERTIES, DISSOLVED SOLIDS, HYDROLOGIC DATA, DATA COLLECTIONS, LAKE MORPHOLOGY, ECOLOGY, BIOLOGICAL PROPERTIES.

IDENTIFIERS:
*ONEIDA LAKE(NY).

ABSTRACT:
ONEIDA LAKE, IN THE STATE OF NEW YORK, IS A NATURALLY EUTROPHIC LAKE THAT HAS EXISTED FOR ABOUT 10,500 YEARS. IT HAS BEEN IN A EUTROPHIC STATE FOR AT LEAST 350 YEARS, AND THE GEOCHEMICALLY DERIVED DISSOLVED MATERIALS ENTERING THE LAKE FROM THE DRAINAGE BASIN ARE OF SUFFICIENT QUANTITY (449,700 TONS PER YEAR) TO SUPPORT ANNUAL ALGAL BLOOMS. THE LAKE RETAINS 50,000 TONS OF DISSOLVED SOLIDS EACH YEAR. THESE MATERIALS BECOME INCORPORATED IN THE BOOTTOM SEDIMENTS. WATER IN ONEIDA LAKE REPRESENTS A HYDROLOGIC EQUILIBRIUM. WATER ENTERING THE LAKE IS FROM DIRECT PRECIPITATION (149,600 ACRE-FEET PER YEAR) AND FROM SURFACE WATER INFLOW (1,729,000 ACRE-FEET PER YEAR(. WATER LEAVES THE LAKE BY OUTFLOW THROUGH THE CNEIDA RIVER (1,730,400 ACRE-FEET PER YEAR) AND BY EVAPORATION (148,200 ACRE-FEET PER YEAR). SIXTY-SEVEN PERCENT OF ALL WATER ENTERING ONEIDA LAKE ORIGINATES IN THE NORTHERN PART OF THE DRAINAGE BASIN. THE FOUR MOST IMPORTANT FACTORS AFFECTING THE ECOLOGICAL PROCESSES IN THE LAKE ARE: (1) HIGH FERTILITY OF THE DRAINAGE BASIN, (2) PHYSICAL POSITION AND SHALLOWNESS OF THE LAKE, (3) MIXING AS CAUSED BY WIND, AND (4) FERTILITY OF THE BOTTOM SEDIMENTS. (WOODARD-USGS)

FIELD 05C, 02H

ACCESSION NO. W72-13851

DYNAMICS OF PHYTOPLANKTON PRIMARY PRODUCTION AND BIOMASS IN LOVIISA ARCHIPELAGO (GULF OF FINLAND),

INSTITUTE OF MARINE RESEARCH, HELSINKI (FINLAND).

P. BAGGE, AND A. NIEMI.

MERENTUTKIMUSLAIT. JULK./HAVSFORSKNINGSINST. SKR., NO 233, P 19-41, 1971. 7 FIG, 2 TAB, 35 REF, 1 APP.

DESCRIPTORS:
*PHYTOPLANKTON, *PRIMARY PRODUCTIVITY, *EUTROPHICATION, *BIOMASS, *HYDROGRAPHY, ALGAE, AQUATIC ALGAE, OLIGOTROPHY, DISSOLVED OXYGEN, WATER TEMPERATURE, HYDROGEN ION CONCENTRATION, STRATIFICATION.

IDENTIFIERS:
ORTHOPHOSPHATE CONCENTRATICNS, GULF OF FINLAND.

ABSTRACT:
PHYTOPLANKTON PRIMARY PRODUCTION AND BIOMASS AT TWO SAMPLING LOCALITIES SITUATED IN THE LOVIISA ARCHIPELAGO WERE STUDIED DURING 1967-1969. PRIMARY PRODUCTION WAS MEASURED MAINLY BY THE IN SITU CARBON-14 TECHNIQUE AND QUANTITATIVE ANALYSES OF PHYTOPLANKTON WERE USUALLY MADE ONCE A MONTH DURING THE GROWING SEASON. NO LINEAR CORRELATION WAS OBSERVED BETWEEN PRODUCTION AND BIOMASS VALUES. THE DATA CF PRIMARY PRODUCTION AND PHYTOPLANKTON OBTAINED IN THE LOVIISA AREA ARE COMPARED WITH THOSE AVAILABLE FROM CTHER AREAS SITUATED ON THE NORTHERN COAST OF THE GULF OF FINLAND. THE MAGNITUDE OF PRODUCTION (CA. 30-40 G C(ASS.)/SQUARE METER/YEAR) AND THE SUCCESSION AND BIOMASS OF PHYTOPLANKTON IN THE LOVIISA AREA WERE FOUND TO BE TYPICAL OF THE OLIGOTROPHIC WATERS OF THE SOUTH COAST OF FINLAND, HOWEVER THE OCCURRENCE OF SOME ALGAL SPECIES POINTS TO A WEAK EUTROPHICATION OF THE WATERS STUDIED. (SVENSSON-WASHINGTON)

FIELD 05C

ACCESSION NO. W72-13871

A STUDY OF PHOSPHATE INDUCED ALGAL GROWTH IN ORDER TO SUPPRESS OR ELIMINATE THIS PHENOMENON,

NEW MEXICO STATE UNIV., UNIVERSITY PARK. WATER RESOURCES RESEARCH INST.

N. E. VANDERBORGH, AND A. G. BUYERS.

AVAILABLE FROM THE NATIONAL TECHNICAL INFORMATION SERVICE AS PB-212 026, $3.00 IN PAPER COPY, $0.95 IN MICROFICHE. NEW MEXICO WATER RESOURCES RESEARCH INSTITUTE, LAS CRUCES, COMPLETION REPORT 003, JUNE 1972, 35 P, 9 FIG, 7 TAB, 22 REF. OWRR A-035-NMEX (1).

DESCRIPTORS:
*ALGAL CONTROL, *PHOSPHATES, *ALGAL NUTRIENTS, *CHLORELLA, ALGAE, *EUTROPHICATION, NEW MEXICO, BIODEGRADATION.

IDENTIFIERS:
*ALGAL GROWTH, *RADIOISOTOPE P32, SODIUM PYROPHOSPHATE, RIO GRANDE VALLEY.

ABSTRACT:
THE KINETICS OF THE RADIOACTIVE ISOTOPE P32 UPTAKE BY THE FRESH WATER ALGAE, CHLORELLA, WERE INVESTIGATED IN SOIL-WATER CULTURES, NATURAL SYSTEMS, AND SYNTHETIC CULTURE MEDIA. LABORATORY INVESTIGATIONS WERE CONDUCTED AT PH VALUES OF 4, 7 AND 10, AT A CONSTANT TEMPERATURE 20 PLUS OR MINUS 2C. PORTIONS OF REAGENT GRADE SODIUM PYROPHOSPHATE WERE IRRADIATED WITH A FAST NEUTRON FLUX CONVERTING A PORTION OF THE SAMPLE INTO THE RADIOISOTOPE P32. (P32 EMITS A 1.7 MEV BETA WITH A HALF-LIFE OF 14.31 DAYS.) TAGGING WAS DONE AT A RATE SUFFICIENT TO PRODUCE AN ACTIVITY OF 0.03 MICROCURIES/GRAM. SOLUTIONS WERE THEN PREPARED CONTAINING THE TAGGED PYROPHOSPHATE AND USED TO SPIKE CHLORELLA CULTURES. THE UPTAKE OF THE ISOTOPE WAS THEN MONITORED AS A FUNCTION OF TIME, AFTER VARIOUS TIMES THE CULTURES WERE SAMPLED, FILTERED AND COUNTED FOR P32 ACTIVITY. RESULTS INDICATED THAT THE KINETIC INTERPRETATION OF P32 UPTAKE MUST ALLOW FOR FIRST ORDER KINETICS WITH RESPECT TO P32 AT A PH OF 10, AND A SECOND ORDER KINETIC SCHEME AT PH 7 AND 4. THIS EVIDENCE INDICATED THAT ALGAL SYSTEMS ONLY UTILIZE PHOSPHORUS IN THE ORTHOPHOSPHATE FORM. PHOSPHORUS EXCHANGE RATES WITH SOIL AND INFORMATION ABOUT DEGRADATION PATHWAYS ARE INCLUDED. TWO EXPERIMENTS, ONE WITH AN OXYGEN RICH ENVIRONMENT AND THE SECOND WITH A CARBON DIOXIDE RICH ENVIRONMENT, WERE CARRIED OUT WITH SODIUM PYROPHOSPHATE-CHLORELLA, SOIL-WATER AND CHEMICAL CULTURES. IN THE OXYGEN EXPERIMENT P32 REMOVAL WAS ELIMINATED IN THE CHEMICAL CULTURES AND REMOVAL WAS SLOWED FOR THE SOIL-WATER CULTURES. CO2 ENRICHMENT GREATLY ACCELERATED THE REMOVAL. (CREEL-NMEX STATE)

FIELD 05A, 05G

ACCESSION NO. W72-13992

COMPARATIVE EFFECTS OF AQUATIC BIOTOXINS ON CARDIAC SYSTEMS,

NEW HAMPSHIRE UNIV., DURHAM. DEPT. OF ZOOLOGY.

F. P. THURBERG.

PHD THESIS, JANUARY 1972. 74 P, 9 FIG, 3 TAB, 100 REF. OWRR A-013-NH(6) AND A-021-NH(3).

DESCRIPTORS:
*ALGAL TOXINS, *POISONOUS PLANTS, *CYANOPHYTA, *NUISANCE ALGAE, *AQUATIC ALGAE, *CRUSTACEANS, *FISH TOXINS, *ALGAL POISONING, DINOFLAGELLATES, BEGGIATOA, PHYTOPLANKTON BLOOMS, PHYSIOLOGICAL ECOLOGY, MORTALITY, RED TIDE, TOXICITY.

IDENTIFIERS:
AQUATIC BIOTOXINS, *BIOTOXINS.

ABSTRACT:
TOXINS FROM 3 MARINE DINOFLAGELLATES, GYMNODIUM BREVE, AMPHIDINIUM CARTERI AND GONYAULAX CATENALLA, AND A FRESHWATER BLUE-GREEN ALGA, APHANIZOMENON FLOS-AQUAE WERE OBTAINED BY LABORATORY CULTURE, FIELD COLLECTION AND CORRESPONDENCE WITH OTHER INVESTIGATORS. IN VIVO AND ISOLATED HEARTS OF DECAPOD CRUSTACEANS, CANCER IRRORATUS AND CARCINUS MAENUS, THE BIVALVE MOLLUSCS, MYA ARENARIA AND MERCENARIA MERCENARIA AND THE GRASS FROG RANA PIPIENS WERE EXPOSED TO THESE TOXINS, AND MECHANICAL AND ELECTRICAL ACTIVITY WERE MEASURED. GYMNODINIUM BREVE TOXIN EXCITED CRUSTACEAN HEARTS, DEPRESSED FROG HEARTS AND HAD NO EFFECT ON MOLLUSCAN HEARTS. THESE OBSERVATIONS AND EXPERIMENTS WITH HUMAN BLOOD CHOLINESTERASE AND MAMMALIAN INTESTINE SUGGEST ANTICHOLINESTERASE-LIKE ACTIVITY AS ONE ACTION OF G. BREVE TOXIN. AMPHIDINIUM CARTERI TOXIN EXCITED CRUSTACEAN HEARTS AND DEPRESSED MOLLUSCAN AND FROG HEARTS. THIS CHOLINE-LIKE ACTION WAS FURTHER DEMONSTRATED WITH MAMMALIAN INTESTINE PREPARATIONS AND THE USE OF THE CHOLINE BLOCKING COMPOUND, MYTOLON CHLORIDE. APHANIZOMENON FLOS-AQUAE AND GONYAULAX CATENELLA TOXINS DEPRESSED FROG AND CRUSTACEAN HEARTS BUT HAD NO EFFECT ON MOLLUSCAN HEARTS. THIS EVIDENCE SUPPORTS REPORTED PHYSIOLOGICAL AND CHEMICAL SIMILARITIES OF THESE TWO TOXINS.

FIELD 05C, 05A

ACCESSION NO. W72-14056

LIMNOLOGICAL EFFECTS OF SIMULATED PUMPED-STORAGE OPERATICN AT YARDS CREEK,

DELAWARE RIVER BASIN COMMISSION, TRENTON, N.J.

C. F. BAREN.

DECEMBER 1971. 158 P, 36 FIG, 31 TAB, 103 REF.

DESCRIPTORS:
*WATER LEVEL FLUCTUATIONS, *PUMPED STORAGE, PHYSICOCHEMICAL PROPERTIES, *PLANKTON, *AQUATIC PLANTS, *AQUATIC ANIMALS, *MODEL STUDIES, LIMNOLOGY, BIOLOGICAL PROPERTIES, MINERALOGY, FISH, FISH BEHAVIOR, CHEMICAL ANALYSIS, SAMPLING, VOLUMETRIC ANALYSIS, ALGAE, RESERVOIR STORAGE, IMPOUNDMENTS, DISSOLVED OXYGEN, CHLORIDES, INVERTEBRATES, HARDNESS(WATER), TURBIDITY, ALKALINITY, CARBCN DIOXIDE, PHOSPHORUS, NITROGEN, SILICA, COPPER, SULFATES, IRON, MANGANESE, SPAWNING, FISH REPRODUCTION, LIFE HISTORY STUDIES, PHYTOPLANKTON, STANDING CROP, BIOMASS, SPORT FISH, ZOOPLANKTON, WATER TEMPERATURE, AIR TEMPERATURE, JUVENILE FISHES, FRY, PLANKTON NETS, NETS, SUNFISHES, BASS, ROCK BASS, DIATOMS, CHRYSCPHYTA, BULLHEADS, SHINERS, YELLOW PERCH, WALLEYE, MINNOWS, PROTOZOA, ROTIFERS, BROOK TROUT, RAINBOW TROUT, PIKES, STRIPED BASS, CHLOROPHYTA, EUGLENOPHYTA, PHRROPHYTA, EUGLENA, CYANOPHYTA, CRUSTACEANS, COPEPODS, NESTING, BROWN TROUT, SCENEDESMUS, DINOFLAGELLATES, LARVAE, DAPHNIA, MIDGES, CADDISFLIES, GASTROPODS, PONDWEEDS, NEW JERSEY, DELAWARE RIVER, NUTRIENTS, PRIMARY PRODUCTIVITY, SECONDARY PRODUCTIVITY.

IDENTIFIERS:
MACROINVERTEBRATES, *TOCKS ISLAND RESERVOIR(NJ), AMERICAN SHAD, EKMAN DREDGE, PUMPKINSEED, BLACK CRAPPIE, CHAIN PICKEREL, TENDIPES TENTANS, TRICHOPTERA, LIMNEPHILUS, GLCSSOSOMA, PHYSA GYRINA, KEMMERER SAMPLER, SECCHI *YARDS CREEK(NJ), RELATIVE CONDITION FACTOR.

ABSTRACT:
A PUMPED-STORAGE POWER GENERATING SYSTEMS WHICH WOULD ALTER WATER LEVELS HAS BEEN PROPOSED FOR INCLUSION IN THE TOCKS ISLAND RESERVOIR PROJECT AT BLAIRSTOWN, NEW JERSEY. RESULTS ARE REPORTED OF A TWO-YEAR STUDY TO DETERMINE WHAT EFFECTS CHANGING WATER LEVELS MIGHT HAVE ON AQUATIC BIOTA. FOUR T DISC,EST PONDS WERE CONSTRUCTED IN THE VICINITY OF THE PLANNED PROJECT TO INVESTIGATE: (1) THE EFFECTS OF DIURNALLY FLUCTUATING WATER LEVELS ON THE LIFE HISTORY OF SELECTED, NATIVE NEST-BUILDING AND NON-NEST-BUILDING FISH; (2) THE PHYSICAL AND CHEMICAL CHARACTERISTICS OF VARIOUSLY FLUCTUATED TEST PONDS; (3) THE RESULTING EFFECTS OF FLUCTUATING WATER LEVELS ON MACROINVERTEBRATES, PLANKTON, AND AQUATIC PLANTS; AND (4) THE CONSEQUENT COMBINED EFFECTS OF PHYSICOCHEMICAL CHARACTERISTICS AND CHANGING WATER LEVELS ON THE BEHAVIOR OF TH SELECTED FISHES. IT WAS FOUND THAT (1) FISH ADAPTED TO THE UNIFORM REGIME OF WATER LEVEL FLUCTUATIONS AND WERE SUCCESSFUL IN SPAWNING AND HATCHING EGGS AND (2) THE MEASURABLE EFFECTS OBSERVED IN PLANKTON AND MACROINVERTEBRATE PRODUCTIVITY, PLANT COLONIZATION, RELATIVE CONDITION FACTORS OF FISH, AND THE PHYSICAL AND CHEMICAL PROPERTIES OF THE TEST POND HAD NO MARKED EFFECT ON THE FISH POPULATIONS. THE RESULTS OF THE PHYSIOCHEMICAL ANALYSES AND PROFILES, THE PLANTS AND ANIMALS STUDIED, AND SAMPLING PROCEDURES UTILIZED IN THE STUDY ARE INCLUDED IN FIGURES, TABLES AND APPENDICES. (HOLOMAN-BATTELLE)

FIELD 05C

ACCESSION NO. W72-14280

CONFERENCE IN THE MATTER OF POLLUTION OF LAKE ERIE AND ITS TRIBUTARIES-INDIANA-MICHIGAN-NEW YORK-OHIO-PENNSYLVANIA: VOLUMES 1 AND 2.

FEDERAL WATER QUALITY ADMINISTRATION, WASHINGTON, D.C.

HELD IN DETROIT, MICHIGAN, JUNE 3 AND 4 1970, 740 P. 21 FIG, 25 TAB, 18 APPEND.

DESCRIPTORS:
*WATER POLLUTION, *LAKE ERIE, *MERCURY, *MUNICIPAL WASTES, *INDUSTRIAL WASTES, WATER POLLUTION CONTROL, WATER POLLUTION SOURCES, TRIBUTARIES, FRESHWATER FISH, YELLOW PERCH, SEWAGE EFFLUENTS, OIL SPILLS, THERMAL POLLUTION, NUTRIENTS, BIOASSAY, *WATER QUALITY, ALGAE, PESTICIDES, HEAVY METALS, WASTEWATER TREATMENT, SMELTS, MICHIGAN, OHIO, NEW YORK, PENNSYLVANIA, TOXICITY, DISCHARGE(WATER), PHENOLS, ELECTRIC POWERPLANTS, NUCLEAR WASTES, PHOSPHORUS, DISSOLVED OXYGEN, CLADOPHORA, ACIDS, ORGANIC WASTES, BIOCHEMICAL OXYGEN DEMAND, SALMON, CARP, WHITE BASS, CHANNEL CATFISH, SUCKERS, BULLHEADS, BUFFALO FISH, PIKE, SULFATE, CHLORIDE, SODIUM, POTASSIUM, CALCIUM, DRUMS, SAUGER, WALLEYE, WATER QUALITY STANDARDS, WATER ANALYSIS.

IDENTIFIERS:
MAUMEE RIVER, ST. CLAIR RIVER, LAKE ST. CLAIR, DETROIT RIVER, BLACK RIVER, PINE RIVER, CLINTON RIVER, HURON RIVER, BELL RIVER, SALT RIVER, MICROCYSTIS, APHANIZOMENON, CYANIDES, COHO SALMON, SHEEPSHEAD, CHUBS, NORTHERN PIKE, GIZZARD SHAD.

ABSTRACT:
A VERBATIM TRANSCRIPT OF THE PROCEEDINGS OF A CONFERENCE CONCERNED WITH THE WATER POLLUTION PROBLEMS OF LAKE ERIE AND ITS TRIBUTARIES AND SUMMARIES OF THE VARIOUS REPORTS PRESENTED HAVE BEEN COMPILED INTO A TWO-VOLUME REPORT. STATEMENTS OF COMPLIANCE WITH THE LAKE ERIE ENFORCEMENT CONFERENCE ABATEMENT SCHEDULES FOR MUNICIPALITIES AND INDUSTRIES AND DETAILED DISCUSSIONS ON THE SOURCES OF POLLUTION ON THE LAKES, INCLUDING INDUSTRIAL WASTES FROM MAJOR CITIES SUCH AS DETROIT AND CLEVELAND, AND MERCURY STANDARDS, SOURCES, AND LEVELS ARE ALSO INCLUDED. PARTICIPANTS IN THE CONFERENCE WERE FROM WATER POLLUTION CONTROL AGENCIES OF THE STATES OF MICHIGAN, INDIANA, OHIO, PENNSYLVANIA AND NEW YORK, AND OF THE U. S. DEPARTMENT OF THE INTERIOR. (MORTLAND-BATTELLE)

FIELD 05B, 05C, 05G

ACCESSION NO. W72-14282

RESEARCH ON AND CULTURE OF CALCAREOUS GREEN ALGAE,

OHIO STATE UNIV. RESEARCH FOUNDATION, COLUMBUS.

L. HILLIS-COLINVAUX.

FINAL REPORT, FEBRUARY 11, 1972. 30 P, 1 FIG, 3 TAB, 27 REF.

DESCRIPTORS:
*CULTURES, *PLANT GROWTH, *CHLOROPHYTA, *MARINE ALGAE, *BIOASSAY, REPRODUCTION, SEA WATER, MARINE PLANTS, PRIMARY PRODUCTIVITY, CORAL, REEFS, ANTIBIOTICS(PESTICIDES), CHLORINATED HYDROCARBON PESTICIDES, PESTICIDE TOXICITY, PLANT POPULATIONS, AQUARIA, LABORATORY EQUIPMENT, PLANT PHYSIOLOGY.

IDENTIFIERS:
EPIPHYTES, CULTURING VESSELS, PENICILLIN, LINDANE, DIADEMA ANTILLARUM, AANDOMENE, SIPHONALES, *HALIMEDA SPP., PENICILLUS CAPITATUS, PENICILLUS DUMENTOSUS, RHIPOCEPHALUS PHOENIX, UDOTEA FLABELLUM, BRYOPSIS, CAULERPA, ACROPORA PALMATA, PENICILLUS LAMOUROUXII.

ABSTRACT:
FOUR GENERA OF CORAL REEF SIPHONALES, HALIMEDA, PENICILLUS, RHIPOCEPHALUS AND UDOTEA WERE GROWN IN LABORATORY AQUARIA UNDER LIGHT INTENSITIES OF 650, 200-375 AND 125-200 FT-CANDLES. THE COMMERCIAL PREPARATION 'INSTANT OCEAN' WAS TESTED FOR USE WITH PLANTS. PENICILLIN AND LINDANE WERE TESTED FOR THEIR EFFECT ON EPIPHYTES AND OTHER NUISANCES. A 15 PPM SOLUTION OF THE INVERTEBRATE KILLER LINDANE DID NOT KILL HALIMEDAE WHICH WERE IMMERSED IN IT FOR 15 MINUTES. PENICILLIN AT A CONCENTRATION OF 3000 UNITS PER ML WAS INITIALLY EFFECTIVE IN CONTROLLING THE WEEDS, BUT WITHIN 2 MONTHS THEY WERE ABUNDANT AGAIN AND PENICILLIN WAS THEREAFTER INEFFECTIVE. THE GROWTH AND DEVELOPMENT OF ALL 4 GENERA FROM TINY PROTUBERANCES ABOVE THE SAND TO WHITE, DYING AND DISINTEGRATING INDIVIDUALS WAS FOLLOWED. ALL PRODUCED NEW INDIVIDUALS FROM RHIZOIDAL-LIKE FILAMENTS THAT EXTENDED OUTWARDS THROUGH THE SAND FROM THE HOLDFAST OF AN OLDER PLANT. IT WAS DEMONSTRATED THAT VEGETATIVE REPRODUCTION MIGHT OCCUR WITH PARTIAL BURYING OF INDIVIDUALS BY SHIFTING SANDS IN A REEF, OR FROM PORTIONS OF PLANTS BROKEN OFF BY GRAZING OR OTHER NATURAL ACTIVITIES. PLANTS PRODUCED VEGETATIVELY ARE NOT INITIALLY EPIPHYTIZED AND SO MAY BE A SOURCE OF CLEAN PLANTS FOR LABORATORY EXPERIMENT. THE FIRST SWARMER-PRODUCING PENICILLUS PLANTS SO FAR KNOWN WERE PRODUCED DURING THIS STUDY. THE RATHER UNDISTINGUISHED STRUCTURES ASSOCIATED WITH THEM, THE RELEASE OF SWARMERS FROM HALIMEDA, AND ASSOCIATED OBSERVATIONS ARE DESCRIBED. INSIGHT INTO THE ROLE OF CALCAREOUS PLANTS IN A REEF WAS OBTAINED BY A PRODUCTIVITY STUDY WHICH COMBINED A CENSUS OF SUCH PLANTS IN A REEF AND A LABORATORY MEASURE OF OXYGEN CHANGES IN AN ENTIRE AQUARIUM. SOME HALIMEDA REEF POPULATIONS APPEARED TO BE GRAZED BYTHE URCHIN DIADEMA. A CONSERVATIVE FIGURE FOR THE PRODUCTIVITY OF LABORATORY HALIMEDA PLANTS IS 2.5 MG C/PLANT/DAY (NET) OR 4.5 MG C/PLANT/DAY (GROSS). (LONG-BATTELLE)

FIELD 05C, 05A

ACCESSION NO. W72-14294

TECHNIQUE FOR MEASURING C-1402 UPTAKE BY SOIL MICROORGANISMS IN SITU,

WISCONSIN UNIV., MADISON. DEPT. OF BACTERIOLOGY.

D. W. SMITH, C. B. FLIERMANS, AND T. D. BROCK.

APPLIED MICROBIOLOGY, VOL. 23, NO. 3, P 595-600, MARCH 1972. 3 FIG, 1 TAB, 12 REF.

DESCRIPTORS:
*SOIL MICROORGANISMS, *ABSORPTION, *CARBON DIOXIDE, *RADIOACTIVITY TECHNIQUES, SOIL ALGAE, *METHODOLOGY, MEASUREMENT, OXIDATION, CARBON RADIOISOTOPES, *ON-SITE TESTS, SULFUR BACTERIA, PRODUCTIVITY, RADIOCHEMICAL ANALYSIS, THERMOPHILIC BACTERIA, ORGANIC MATTER, CULTURES, SOIL BACTERIA.

IDENTIFIERS:
CHEMICAL INTERFERENCE, SCINTILLATION COUNTING, CYANIDIUM CALDARIUM, THIOBACILLUS, SULFOLOBUS, GLUCOSE, URACIL.

ABSTRACT:
UPTAKE OF C-1402 IN SOILS DUE TO ALGAE OR SULFUR-OXIDIZING BACTERIA WAS EXAMINED BY INCUBATION OF SOIL SAMPLES WITH GASEOUS C-1402 AND SUBSEQUENT CHEMICAL OXIDATION OF THE BIOLOGICALLY FIXED RADIOACTIVE ISOTOPE TO C-1402 FOR DETECTION WITH A LIQUID SCINTILLATION COUNTING SYSTEM. THE C-1402 WAS ADDED TO THE SOIL IN THE GAS PHASE SO THAT NO ALTERATION OF THE MOISTURE OR IONIC STRENGTH OF THE SOIL OCCURRED. WET OXIDATION OF RADIOACTIVE ORGANIC MATTER WAS CARRIED OUT IN SEALED AMPOULES, AND THE C-1402 PRODUCED WAS TRANSFERRED TO A PHENETHYLAMINE-LIQUID SCINTILLATION COUNTING SYSTEM WITH A SIMPLY CONSTRUCTED APPARATUS. THE TECHNIQUE IS INEXPENSIVE AND EFFICIENT AND DOES NOT REQUIRE ELABORATE TRAPS SINCE SEVERAL POSSIBLE INTERFERING FACTORS WERE FOUND TO HAVE NO HARMFUL EFFECTS. THE EFFICIENCY OF THE TECHNIQUE IN OXIDIZING VARIOUS COMPOUNDS WAS EXAMINED USING VARYING CONCENTRATIONS OF UNIFORMLY LABELED GLUCOSE AND URACIL, AND LABELED CYANIDIUM CALDARIUM CELLS. IN ALL CASES, THE PROCEDURE WAS PERFORMED ON SAMPLES CONTAINING 0.5 G OF SOIL PER AMPOULE TO REPRODUCE THE TREATMENT OF NATURAL SAMPLES AS CLOSELY AS POSSIBLE. SIXTY-ONE TO EIGHTY-SEVEN PERCENT OF THE ADDED RADIOACTIVITY WAS RECOVERED DURING THE OXIDATION OF GLUCOSE AND URACIL, WHEREAS 98 PERCENT OF THE ADDED RADIOACTIVITY WAS RECOVERED DURING THE OXIDATION OF C. CALDARIUM. THE RECOVERY EFFICIENCY WAS INDEPENDENT OF THE AMOUNT OF RADIOACTIVITY PRESENT. EXPERIMENTS IN COAL MINE REGIONS AND IN GEOTHERMAL HABITATS HAVE DEMONSTRATED THE ECOLOGICAL APPLICABILITY OF THIS TECHNIQUE FOR MEASUREMENT OF CO2 FIXATION BY SULFUR OXIDIZING BACTERIA AND SOIL ALGAE. (LONG-BATTELLE)

FIELD 05A, 02G

ACCESSION NO. W72-14301

HEAVY METALS POLLUTE NATURE, MAY REDUCE PRODUCTIVITY,

LUND UNIV. (SWEDEN). DEPT. OF ECOLOGICAL BOTANY.

G. TYLER.

AMBIO, VOL. 1, NO. 2, P 52-59, APRIL 1972. 8 FIG, 1 TAB, 11 REF.

DESCRIPTORS:
*TERRESTRIAL HABITATS, *HEAVY METALS, *PRODUCTIVITY, *ECOSYSTEMS, POTASSIUM, MAGNESIUM, SODIUM, CALCIUM, HYDROGEN, CHEMICAL WASTES, LEAD, NICKEL, TOXICITY, BIOINDICATORS, BIODEGRADATION, BIOMASS, LICHENS, MOSSES, FUNGI, ALGAE, SOIL CONTAMINATION, TRACE ELEMENTS, ION EXCHANGE, POLLUTANT IDENTIFICATION, COPPER, CADMIUM, CHROMIUM, NICKEL, MANGANESE, IRON, ZINC, ORGANIC COMPOUNDS, ADSORPTION.

IDENTIFIERS:
VANADIUM, BRYOPHYTES, BIOACCUMULATION, METAL COMPLEXES, PICEA ABIES, VACCINIUM VITIS IDAEA, VACCINIUM MYRTILLUS, DESCHAMPSIA FLEXUOSA, PARMELIA PHYSODES, HYPNUM CUPRESSIFORME.

ABSTRACT:
THE DEPOSITION, ACCUMULATION, AND POSSIBLE EFFECTS OF PB, CD, CU, ZN, NI, FE, AND MN ON TERRESTRIAL SITES IN SCANDINAVIA ARE DISCUSSED. BRYOPHYTES HAVE BEEN CITED AS THE MOST SENSITIVE BIOLOGICAL INSTRUMENTS FOR MEASURING THE DEPOSITION OF HEAVY METALS. ADDITIONAL WORK HAS SHOWN DEAD ORGANIC MATTER, LITTER, AND HUMUS THAT WAS DERIVED FROM MOSSES AND LICHENS TO HAVE A LARGE CAPACITY FOR CAPTURING HEAVY METALS THROUGH PASSIVE ION EXCHANGE. THE GREAT CAPACITY OF BRYOPHYTES AND TO A LESSER EXTENT, LICHENS, TO CAPTURE HEAVY METALS THROUGH ION EXCHANGE, IS ATTRIBUTABLE TO THE GREAT STABILITY OF THE CHEMICAL COMPLEXES FORMED BETWEEN HEAVY METAL IONS AND NEGATIVELY CHARGED ORGANIC GROUPS. ACCUMULATION IN MOSSES, LICHENS, AND IN LITTER AND HUMUS LAYERS OF NATURAL AND SEMI-NATURAL ECOSYSTEMS WAS FOUND TO BE GOVERNED BY ION EXCHANGE EQUILIBRIA DEPENDENT ON THE ABSORPTION COMPLEX BETWEEN IONS FROM PRECIPITATION, FOLIAGE LEACHING, AND THE DRY DEPOSITION PRESENT IN THE WATER PASSING THROUGH THESE COMPONENTS. INCREASING DEPOSITION OF HEAVY METALS DUE TO HUMAN ACTIVITY ADVERSELY AFFECTS BIODECOMPOSITION RATES, AFFECTING A DECREASE IN PRODUCTIVITY. (BYRD-BATTELLE)

FIELD 05C, 05B

ACCESSION NO. W72-14303

RELATIONSHIPS OF CHLOROPHYLL MAXIMA TO DENSITY STRUCTURE IN THE ATLANTIC OCEAN AND GULF OF MEXICO,

VICTORIA UNIV. (BRITISH COLUMBIA).

L. A. HOBSON, AND C. J. LORENZEN.

DEEP-SEA RESEARCH, VOL. 19, NO. 4, P 297-306, APRIL 1972. 7 FIG, 2 TAB, 19 REF.

DESCRIPTORS:
*CHLOROPHYLL, *SPATIAL DISTRIBUTION, *DENSITY, *DEPTH, MARINE ANIMALS, MARINE ALGAE, CYANOPHYTA, PYRROPHYTA, LARVAE, SAMPLING, INVERTEBRATES, *ATLANTIC OCEAN, *GULF OF MEXICO, PHYTOPLANKTON, CYTOLOGICAL STUDIES, SYSTEMATICS, BIOMASS, FLUOROMETRY, MICROSCOPY, ZOOPLANKTON, NUTRIENTS, PHOTOSYNTHESIS, STANDING CROP, ECOLOGICAL DISTRIBUTION, DINOFLAGELLATES, EQUIPMENT, LIGHT PENETRATION, NITROGEN, SALINITY, AMMONIA, NITRATES, NITRITES, COPEPODS, SEDIMENTATION, CARBON, PROTOZOA, RADIATION, ENZYMES, AQUATIC LIFE.

IDENTIFIERS:
*PYCNOCLINE, CHLOROPHYLL A, PHEOPHYTIN, TUNICATES, FORAMINIFERA, RADIOLARIA, TRICHODESMIUM, FLAGELLATES, CILIATES, NAUPLII, COPEPODIDS.

ABSTRACT:
SPATIAL DISTRIBUTION OF CHLOROPHYLL MAXIMA IN RELATION TO THE DEPTHS OF PYCNOCLINES WERE STUDIED IN THE ATLANTIC OCEAN AND IN THE GULF OF MEXICO. THE DISTRIBUTION OF DISSOLVED INORGANIC NITROGEN AND BIOMASS OF MICROORGANISMS AND THEIR TAXONOMIC COMPOSITION IN CHLOROPHYLL MAXIMA AND SURROUNDING WATER OF THE GULF OF MEXICO ALSO WERE EXAMINED. WATER SAMPLES WERE TAKEN FROM STATIONS ALONG THE SOUTHWEST COAST OF AFRICA, THE NORTHEASTERN U. S. COAST, AND THE GULF OF MEXICO AT DEPTHS TO 200 M USING NISKIN 5-LITER NON-METALLIC SAMPLING BOTTLES EQUIPPED WITH REVERSING THERMOMETERS. SALINITY MEASUREMENTS WERE TAKEN WITH A HYTECH MODEL 6210 LABORATORY SALINOMETER. THE START OF THE PYCNOCLINE WAS DETERMINED BY THE DEPTH AT WHICH THE STABILITY OF THE WATER COLUMN EXCEEDED .0001/DZ. STANDARD METHODS WERE USED TO DETERMINE AMMONIA, NITRATE, AND NITRITE AND A FLUORCMETRIC TECHNIQUE WAS USED TO MEASURE CHLOROPHYLL A. MICROORGANISMS WERE COUNTED BY MICROSCOPIC TECHNIQUES AND CELL VOLUMES WERE CALCULATED USING FORMULAS OF GEOMETRICAL SOLIDS THAT APPROXIMATED CELL SHAPES. CONVERSION FACTORS REPORTED EARLIER WERE USED TO CONVERT THESE VOLUMES TO CARBON. THE SPATIAL DISTRIBUTIONS OF CHLOROPHYLL MAXIMA WERE PATCHY AND THE MAXIMUM DEPTHS TO WHICH THEY FOLLOW PYCNOCLINES WERE VARIABLE. THIS VARIABILITY APPARENTLY RELATES TO LIGHT ADAPTATION BY PHYTOPLANKTON CELLS WHICH MAY BE A FUNCTION OF THE TAXONOMIC COMPOSITION OF THE PHYTOPLANKTON CROP. (MORTLAND-BATTELLE)

FIELD 05C, 05B

ACCESSION NO. W72-14313

ON PRODUCTIVITY ESTIMATIONS OF RIVER WATER SAMPLES BY MEANS OF ALGAE TEST PROCEDURES,

JENA UNIV. (EAST GERMANY). BIOLOGY SECTION.

W. BRAUNE.

INTERNATIONALE REVUE DER GESAMTEN HYDROBIOLOGIE, VOL. 56, NO. 5, P 795-810, 1971. 4 FIG, 2 TAB, 30 REF.

DESCRIPTORS:
*PRIMARY PRODUCTIVITY, BACTERIA, SCENEDESMUS, GROWTH RATES, CULTURES, AQUATIC PRODUCTIVITY, METABOLISM, ELECTRODES, ELECTROCHEMISTRY, PHOTOSYNTHESIS, RIVERS, PLANT PHYSIOLOGY, TESTING PROCEDURES, AQUATIC ALGAE, CHLOROPHYTA, LABORATORY TESTS, OXYGEN, WATER POLLUTION EFFECTS.

IDENTIFIERS:
SAALE RIVER, GERMANY, SCENEDESMUS OBLIQUUS.

ABSTRACT:
THE POSSIBLE INFLUENCE OF BACTERIA ON THE RESULTS OF LONG-TERM ALGAE TEST PROCEDURES (CULTIVATING OF ALGAE IN ISOLATED UNSTERILE WATER SAMPLES) WAS INVESTIGATED AND DIFFERENCES IN THE GROWTH RATES BETWEEN NONAXENIC AND AXENIC SCENEDESMUS CULTURES WERE OBSERVED, SHOWING DIFFERENCES DEPENDENT ON EXPERIMENTAL CONDITIONS. THE NECESSITY OF A LONG EXPERIMENTAL TIME IN SUCH TEST METHODS PROVED TO BE DISADVANTAGEOUS BECAUSE THERE WAS ALWAYS THE POSSIBILITY THAT IN THE UNSTERILE WATER SAMPLE SECONDARY ALTERATIONS (CAUSED BY BACTERIA) MAY OCCUR, THE ACTION OF WHICH IS HARD TO ESTIMATE UNDER TEST CONDITIONS. TO DIMINISH THESE DISADVANTAGES A PHYSIOLOGICAL TEST PROCEDURE WAS TRIED OUT. COMPARED TO METHODS USED TILL NOW THE NEW METHOD ALLOWED SHORT-TERM ESTIMATIONS (IN A FEW HOURS) OF THE PRODUCTIVITY OF ALGAE (UNIALGAL AND MIXED MATERIAL) IN WATER SAMPLES BY CONTINUOUSLY RECORDING THE O2-METABOLISM (ELECTROCHEMICAL O2-DETERMINATION BY MEANS OF MEMBRANE COVERED PT-ELECTRODES). IN WATER SAMPLES TAKEN FROM A POINT BELOW THE TOWN OF JENA IN THE RIVER SAALE THERE WERE REGULARLY HIGHER RATES OF NET-PHOTOSYNTHESIS THAN IN SAMPLES TAKEN FROM ABOVE THE TOWN (TEST ORGANISM: SCENEDESMUS OBLIQUUS). THE RESULTS AGREED WITH FINDINGS OBTAINED BY OTHER IN VITRO-METHODS AND ARE IN CONTRADICTION TO TESTING RESULTS IN SITU. (IN GERMAN) (MORTLAND-BATTELLE)

FIELD 05C, 05A

ACCESSION NO. W72-14314

PRODUCTIVITY OF THE BENTHIC MICROFLORA OF SHOAL ESTUARINE ENVIRONMENTS IN SOUTHERN NEW ENGLAND,

RHODE ISLAND UNIV., KINGSTON. GRADUATE SCHOOL OF OCEANOGRAPHY.

N. MARSHALL, C. A. OVIATT, AND D. M. SKAUEN.

INTERNATIONALE REVUE DER GESAMTEN HYDROBIOLOGIE, VOL. 56, NO. 6, P 947-956, 1971. 4 FIG, 1 TAB, 13 REF.

DESCRIPTORS:
*ESTUARINE ENVIRONMENT, *PRIMARY PRODUCTIVITY, *SHOALS, *BENTHIC FLORA, SEA WATER, SEDIMENTS, LIGHT INTENSITY, DIATOMS, CARBON RADIOISOTOPES, NEW ENGLAND, AQUATIC LIFE, AQUATIC MICROORGANISMS, MARINE ALGAE, SALINITY, MIGRATION, TIDES, LOW WATER MARK, CHRYSOPHYTA.

IDENTIFIERS:
*C-14 PRODUCTIVITY, LIQUID SCINTILLATION, SAMPLE PREPARATION, ZOSTREA, ULVA.

ABSTRACT:
CARBON-14 PRODUCTIVITY BY BENTHIC MICROFLORA FROM SHOAL ESTUARINE ENVIRONMENTS IN SOUTHERN NEW ENGLAND WAS MEASURED BY A C-14 UPTAKE METHOD. REPRESENTATIVE AND INTACT SEDIMENT SAMPLES WERE COLLECTED AND THE C-14 COUNTED BY LIQUID SCINTILLATION AFTER BEING PASSED THROUGH A PULVERIZED FILTER AND SUSPENDED IN CAB-O-SIL. PRODUCTIVITY WAS MEASURED BY CALCULATING THE DIFFERENCE IN C-14 UPTAKE MEASURED BETWEEN LIGHT AND DARK FLASKS. SEDIMENT VARIABILITY CAUSED EXTREMELY HIGH DEVIATIONS IN PRODUCTIVITY VALUES, PROBABLY AS A RESULT OF IRREGULAR MICROFLORA DISTRIBUTION. RESULTS OBTAINED WITH THE CAB-O-SIL SUSPENSION PROCESSING TECHNIQUE WHEN COMPARED WITH THE WET OXIDATION METHOD PROVED TO BE ACHIEVED MORE EASILY, DIRECTLY, AND CONSISTENTLY; HOWEVER, SIMILAR RESULTS WERE OBTAINED FROM BOTH METHODS. THE AVERAGE YEARLY C-14 PRODUCTIVITY OF ALL SAMPLED ESTUARINE SHOALS EQUALLED 20.1 MG/SQ M/HR WITH HIGH PEAKS OF PRODUCTIVITY OCCURRING IN THE WARMER MONTHS. THE LIMITED TESTS FOR LIGHT EFFECTS SEEMED TO INDICATE THAT THE STRONGEST MID-SUMMER LIGHT PEAKS ARE IN EXCESS OF THE OPTIMUM AND THAT AN UPWARD MIGRATION OF DIATOMS AT LOW TIDE MAY RESULT IN INCREASED PRODUCTIVITY. (BYRD-BATTELLE)

FIELD 05C, 05A, 02L

ACCESSION NO. W72-14317

APPARATUS FOR CONTINUOUS MEASUREMENT OF ACTIVE UPTAKE OF RADIOACTIVE SUBSTANCES,

REGENSBURG UNIV. (WEST GERMANY).

E. LOOS,

ANALYTICAL BIOCHEMISTRY, VOL 47, NO 1, P 90-101, MAY 1972. 6 FIG, 1 TAB, 8 REF.

DESCRIPTORS:
*LABORATORY EQUIPMENT, *RADIOISOTOPES, *MEASUREMENT, *KINETICS, *ABSORPTION, INSTRUMENTATION, RADIOACTIVITY TECHNIQUES, CARBON RADIOISOTOPES, PRODUCTIVITY, METABOLISM, FUNGI, ORGANIC COMPOUNDS, CHLORELLA, TRACERS, AQUATIC ALGAE.

IDENTIFIERS:
C-14, *CHLORELLA VULGARIS, NEUROSPORA CRASSA, SCINTILLATION COUNTING, GLUCOSE, MEMBRANE FILTERS, MONOSACCHARIDES, MEMBRANE FILTRATION.

ABSTRACT:
AN APPARATUS CONSISTING OF A MODIFIED SCINTILLATION CHAMBER, SAMPLE CELL, AND AN END-ON PHOTOMULTIPLIER HAS BEEN USED TO STUDY THE UPTAKE KINETICS OF RADIOACTIVE COMPOUNDS (C-14-LABELLED 3-O-METHYL-D-GLUCOSE) BY CHLORELLA VULGARIS AND CONIDIA OF NEUROSPORA CRASSA. THE APPARATUS IS CONSTRUCTED SO THAT THE CELLS ARE HELD IN A THIN LAYER CLOSE TO THE SCINTILLATION CRYSTAL WITH THE NUTRIENT MEDIUM DIFFUSING FROM ABOVE THROUGH A MEMBRANE FILTER. THE MEDIUM IS CONSTANTLY STIRRED TO MINIMIZE CONCENTRATION GRADIENTS DEVELOPED DURING UPTAKE. UPTAKE RATES ARE SEEN AS AN INCREASE OF MEASURED RADIOACTIVITY IN THE CELLS. FOR CHLORELLA A K SUB M OF 0.6 MM WAS FOUND FOR THE UPTAKE. PARALLEL EXPERIMENTS WITH THE MEMBRANE FILTER TECHNIQUE SHOW GOOD AGREEMENT AND CAN BE USED TO CALIBRATE THE APPARATUS. A DIAGRAM OF THE APPARATUS ACCOMPANIES THE DESCRIPTION. (MACKAN-BATTELLE)

FIELD 05A, 05C

ACCESSION NO. W72-14326

ISOLATION AND CHARACTERIZATION OF ULTRAVIOLET LIGHT-SENSITIVE MUTANTS OF THE BLUE-GREEN ALGA ANACYSTIS NIDULANS,

NATIONAL AERONAUTICS AND SPACE ADMINISTRATION, MOFFETT FIELD, CALIF. AMES RESEARCH CENTER.

Y. ASATO.

JOURNAL OF BACTERIOLOGY, VOL 110, NO 3, P 1058-1064, JUNE 1972. 8 FIG, 20 REF.

DESCRIPTORS:
*ISOLATION, *ULTRAVIOLET RADIATION, SEPARATION TECHNIQUES, CYANOPHYTA, CULTURES, ANTIBIOTICS(PESTICIDES), PHOTOACTIVATION, AQUATIC ALGAE, ENVIRONMENTAL EFFECTS.

IDENTIFIERS:
*WASTE CHARACTERIZATION, *ANACYSTIS NIDULANS, *MUTANTS, *SENSITIVITY, CHLORAMPHENICOL, RECOVERY, SURVIVAL.

ABSTRACT:
THREE UV-SENSITIVE MUTANTS OF THE BLUE-GREEN ALGA, ANACYSTIS NIDULANS WERE ISOLATED AND CHARACTERIZED AS TO THEIR ABILITY TO REPAIR UV DAMAGE (PHOTOREACTIVATION). UV LIGHT-SENSITIVE MUTANTS, PREPARED BY SUBJECTING CELLS TO UV RADIATION AND REPLATING THE SURVIVORS, WERE ISOLATED FOR STUDY BY TREATING EXPONENTIAL CULTURES WITH N-METHYL-N'-NITRO-N-NITROSOGUANIDINE FOLLOWED BY CENTRIFUGATION. THE PHOTOREACTIVE ABILITY OF VARIOUS A. NIDULANS MUTANTS WAS DETERMINED UNDER A VARIETY OF CONDITIONS WHICH INCLUDED EXPOSING CELLS TO CHLORAMPHENICOL, CAFFEINE, BLACK AND WHITE FLUORESCENT LIGHTS, AND RED LIGHTS. STRAIN UVS-1 WAS MOST SENSITIVE TO UV IN THE ABSENCE OF PHOTOREACTIVATION. PRETREATMENT WITH CAFFEINE SUPPRESSED THE DARK-SURVIVAL CURVE OF STRAIN UVS-1, INDICATING THE PRESENCE OF EXCISION ENZYMES INVOLVED IN DARK REPAIR. UNDER 'BLACK' AND 'WHITE' ILLUMINATION, STRAIN UVS-1 DISPLAYED PHOTOREACTIVATION PROPERTIES NEARLY COMPARABLE TO WILD-TYPE CULTURE. MUTANTS UVS-35 AND UVS-88 APPEARED TO HAVE PARTIAL PHOTORECOVERY CAPACITIES. UPON PRETREATMENT WITH CHLORAMPHENICOL, PHOTOREACTIVATION PROPERTIES OF STRAINS UVS-1 AND UVS-88 WERE NOT EVIDENT ALTHOUGH THE PARTIAL PHOTOREACTIVATION CHARACTERISTICS OF STRAIN UVS-35 REMAINED THE SAME. DATA INDICATE THAT STRAINS UVS-1, UVS-35, AND UVS-88 ARE PROBABLY GENETICALLY DISTINCT UV-SENSITIVE MUTANTS. (LONG-BATTELLE)

FIELD 05A, 05C

ACCESSION NO. W72-14327

WATER SUPPLY AND WASTE DISPOSAL SERIES: VOLUME I. TERMINCLOGY STANDARDIZATION AND MICROBIOLOGY.

FEDERAL HIGHWAY ADMINISTRATION, WASHINGTON, DC, OFFICE OF RESEARCH; AND FEDERAL HIGHWAY ADMINISTRATION, WASHINGTON, D.C. OFFICE OF DEVELOPMENT.

DEPT OF TRANSPORTATION-FEDERAL HIGHWAY ADMINISTRATION STAFF REPORT 72R-106S-1, NOVEMBER 1971. 51 P, 16 FIG, 12 TAB, 141 REF. APPEND.

DESCRIPTORS:
*WATER SUPPLY, *WASTE DISPOSAL, *RECREATION WASTES, *MICROORGANISMS, *WATER POLLUTION SOURCES, REVIEWS, MICROBIAL DEGRADATION, CLASSIFICATION, SEWAGE, SOIL PROPERTIES, GROUNDWATER MOVEMENT, OUTFALL SERVERS, STREAMS, SEPTIC TANKS, FUNGI, ALGAE, WATER POLLUTION CONTROL.

IDENTIFIERS:
*ROADSIDE REST AREAS.

ABSTRACT:
THIS IS THE FIRST VOLUME OF A SERIES OF REPORTS ON WATER SUPPLY AND WASTE DISPOSAL PUBLISHED BY THE FEDERAL HIGHWAY ADMINISTRATION. THIS SERIES EMPHASIZES WATER SUPPLIES, WATER-CARRIAGE WASTE TREATMENT, AND SOLID WASTE DISPOSAL, PARTICULARLY AS THEY RELATE TO ROADSIDE REST AREAS. BECAUSE DEFINITIONS, CLASSIFICATIONS, AND TERMINOLOGY CONCERNING MICROORGANISMS AND BIOLOGICAL TERMS SOMETIMES LACK UNIFORMITY AMONG THE DISCIPLINES, THIS VOLUME DEFINES AND STANDARDIZES THE TERMINOLOGY THAT WILL BE USED IN ALL VOLUMES OF THE SERIES. INFORMATION ON DISEASES THAT MIGHT BE TRANSMITTED AT REST OR RECREATION AREAS IS PRESENTED. THE CAPABILITY OF INDIVIDUAL SPECIES OF MICROORGANISMS TO SURVIVE IN THE SOIL, ON THE SOIL, IN WATER, AND IN OTHER SITUATIONS IS INDICATED. A BIBLIOGRAPHY CONTAINING 14 REFERENCES IS INCLUDED. (WOODARD-USGS)

FIELD 05B, 05D

ACCESSION NO. W72-14363

PREDICTING QUALITY EFFECTS OF PUMPED STORAGE,

WATER RESOURCES ENGINEERS, INC., WALNUT CREEK, CALIF.

C. W. CHEN, AND G. T. ORLOB.

JOURNAL OF THE POWER DIVISION, PROCEEDINGS OF ASCE, VOL. 98, NO. PO1, P 65-75, JUNE 1972. 7 FIG, 2 TAB, 13 REF.

DESCRIPTORS:
*WATER QUALITY, *HYDROELECTRIC PLANTS, *MATHEMATICAL MODELS, THERMAL STRATIFICATION, TEMPERATURE, DISSOLVED OXYGEN, COMPUTERS, ALGAE, AQUATIC LIFE, ENVIRONMENTAL EFFECTS, ENVIRONMENTAL ENGINEERING, *VIRGINIA, PROJECTIONS, MODEL STUDIES.

IDENTIFIERS:
*PUMPED STORAGE OPERATION, SMITH MOUNTAIN RESERVOIR(VA).

ABSTRACT:
PUMPED STORAGE OPERATION (PSO) AFFECTS THE WATER QUALITY AND IN TURN THE RESIDENT BIOTA. A PREDICTIVE TECHNIQUE WAS ADAPTED FROM A RECENTLY DEVELOPED ECOLOGIC MODEL TO CALCULATE THE ENVIRONMENTAL CHANGES AND TO EVALUATE THE ALTERNATIVES OF PSO THAT MIGHT ENHANCE A NORMAL ECOLOGICAL SUCCESSION, FORESTALLING ADVERSE ENVIRONMENTAL PROBLEMS. THE MODEL, IN AN APPLICATION TO SMITH MOUNTAIN RESERVOIR, VIRGINIA, PREDICTED ACCURATELY FOR AN ENTIRE YEAR THE DAILY TEMPERATURE, DISSOLVED OXYGEN AND OTHER QUALITY CONSTITUENTS OF THE WATER AT VARYING DEPTHS. PSO, UNDER SIMULATED CONDITIONS, WAS FOUND TO IMPROVE WATER QUALITY BY DEEPENING THE EPILIMNION. A PROLONGED PUMPING SCHEDULE AT A REDUCED RATE, HOWEVER, WAS A BETTER ALTERNATIVE, BECAUSE IT WOULD PROVIDE A MORE STABLE ECOSYSTEM WITHOUT SUBSTANTIALLY ALTERING THE THERMAL STRATIFICATION. THERMAL STRATIFICATION IS NECESSARY TO REGULATE ENERGY LOSS, AND THUS CONTROL EVAPORATION, AND TO ENSURE A VARIED ENVIRONMENT SUITABLE FOR HOSTING A DIVERSE AND STABLE BIOTA. (EAGLE-VANDERBILT)

FIELD 05C, 05G, 04A

ACCESSION NO. W72-14405

APPARATUS AND PROCESS FOR TREATING SEWAGE,

P. J. GRESHAM.

U. S. PATENT NO. 3,565,797, 7 P, 6 FIG, 6 REF; OFFICIAL GAZETTE OF THE UNITED STATES PATENT OFFICE, VOL. 883, NO. 4, P 1617, FEBRUARY 23, 1971.

DESCRIPTORS:
*PATENTS, *WASTE WATER TREATMENT, *FILTRATION, *SEWAGE TREATMENT, ALGAE, *AERATION, SEDIMENTATION, *CHLORINATION, OXIDATION, POLLUTION ABATEMENT, WATER POLLUTION TREATMENT, WATER QUALITY, STREAMS, LAKES, BIOCHEMICAL OXYGEN DEMAND, *OXYGENATION, *ANAEROBIC BACTERIA, BIOLOGICAL TREATMENT.

ABSTRACT:
SEWAGE IS FED INTO CONTACT WITH A BED OF LIVING FILAMENTOUS MACROSCOPIC ALGAE. THE SEWAGE IS OXYGENATED AND THE GROWTH OF ANAEROBIC BACTERIA IS INHIBITED. THE SEWAGE IS FILTERED THROUGH THE ALGAE BED OR SCREEN. SOLIDS ARE COLLECTED ON THE SCREENS AS SLUDGE FOR SEPARATE DISPOSAL AND THE SEWAGE LIQUID UNDERGOES CHLORINATION BEFORE DISPOSAL. THE SLUDGE IS RECYCLED FOR REMOVAL OF REMAINING LIQUIDS. (SINHA-OEIS)

FIELD 05D

ACCESSION NO. W72-14469

CHEMICAL STUDIES ON TOXINS FROM GYMNODINIUM BREVE AND APHANIZOMENON FLOS-AQUAE,

NEW HAMPSHIRE UNIV., DURHAM. DEPT. OF BIOCHEMISTRY.

M. ALAM.

PHD THESIS, NOVEMBER 1972. 95 P, 25 FIG, 12 TAB, 30 REF. OWRR A-013-NH(5) AND A-021-NH(2).

DESCRIPTORS:
*ALGAL TOXINS, *RED TIDE, *BEGGIATOA, *PHYTOTOXICITY, *CYANOPHYTA, *NEW HAMPSHIRE, MARINE ALGAE, PHYTOPLANKTON, AQUATIC PLANTS, CHEMICAL PROPERTIES, WATER QUALITY.

IDENTIFIERS:
*KEZAR LAKE(NH).

ABSTRACT:
PART I: THIS INVESTIGATION CONCERNS THE ISOLATION AND CHEMICAL CHARACTERIZATION OF THE TOXIN (S) FROM GYMNODINIUM BREVE WHICH IS LETHAL TO MARINE ORGANISMS. UNIALGAL CULTURES OF G. BREVE WERE GROWN AND THE TOXIC MATERIAL ISOLATED. THE PARTIALLY PURIFIED TOXIN ON COLUMN CHROMATOGRAPHY GAVE A TOXIC PALE YELLOW GLASSY RESIDUE. THE INFRA-RED SPECTRUM INDICATED PRESENCE OF A CARBONYL GROUP AND ABSENCE OF HYDROXYL AND AROMATIC GROUPS. ULTRA-VIOLET SPECTRUM INDICATED TOXIN POSSIBLY CONTAMINATED WITH VERY SMALL AMT OF PIGMENT. MOLECULAR WEIGHT DETERMINED BY MASS SPECTROMETRY AND OSMOMETRY. ON BASIS OF PRESENT INFORMATION IT IS DIFFICULT TO ASSIGN A DEFINITE FORMULA TO THE TOXIN FROM G. BREVE. PART II: THE PURPOSE OF THIS INVESTIGATION WAS TO ISOLATE AND CHARACTERIZE THE TOXIC COMPOUND(S) FROM NATURAL BLOOMS OF A.FLOS-AQUAE. THE TOXIN WAS PURIFIED AND OBTAINED AS A PALE YELLOW POWDER IN A CHROMATOGRAPHICALLY PURE FORM. THE A.FLOS-AQUAE TOXIN WAS A VERY BASIC SUBSTANCE. IT DIFFERED FROM SAXITOXIN IN ITS RF VALUES AND ITS BEHAVIOR TO SEVERAL SPRAY REAGENTS. PARTIALLY HYDRATED FORM HAD DIFFERENT INFRA-RED SPECTRUM THAN DOES SAXITIXIN. ON BASIS OF THIS AND OTHER INFORMATION IT IS CONCLUDED THAT A.FLOS-AQUAE IS DIFFERENT FROM SAXITOXIN.

FIELD 05C, 05A

ACCESSION NO. W72-14542

STUDIES ON THE EFFECTS OF A STEAM ELECTRIC GENERATING PLANT ON THE MARINE ENVIRONMENT AT NORTHPORT, NEW YORK,

STATE UNIV. OF NEW YORK, STONY BROOK. MARINE SCIENCES RESEARCH CENTER.

G. C. WILLIAMS, J. B. MITTON, T. H. SUCHANEK, JR., N. GEBELEIN, AND C. GROSSMAN.

TECHNICAL REPORT SERIES NO 9, NOVEMBER 1971. 123 P, 44 FIG, 9 TAB, 59 REF.

DESCRIPTORS:
*NEW YORK, *POWER PLANTS, *STREAM, *THERMAL POLLUTION, *ENVIRONMENTAL EFFECTS, *ECOLOGY, *PLANKTON, *ALGAE, BIOLOGICAL COMMUNITIES, PRODUCTIVITY, LIFE CYCLES, GROWTH STAGES, DISTRIBUTION, BEHAVIOR, SEASONAL, WATER POLLUTION EFFECTS.

IDENTIFIERS:
*NORTHPORT PLANT(LONG ISLAND).

ABSTRACT:
THE RESULTS OF A CONTINUING GROUP OF INVESTIGATIONS ON EFFECTS OF THE POWER PLANT AT NORTHPORT ARE SUMMARIZED. THE EFFECTS ON THE PLANKTON COMMUNITY ORIGINALLY IN THE WATER USED FOR COOLING OF STEAM CONDENSERS AND DISCHARGED BACK INTO LONG ISLAND SOUND WERE INVESTIGATED BY MEASURING PRODUCTIVITY CHANGES AND BY A DIRECT SEASONAL OBSERVATIONAL STUDY OF THE LIFE CYCLE STAGES, DISTRIBUTION AND BEHAVIOR OF PLANKTON ALGAE BEFORE AND AFTER USE. RESULTS LEFT LITTLE DOUBT THAT WASTE HEAT IS THE PRIMARY ECOLOGICAL DISTURBANCE IN THE DISCHARGE AREA. (ENSIGN-PAI)

FIELD 05C, 05B

ACCESSION NO. W72-14659

DISCOLORATION OF ALGAL BLOOMS AND SEAWEEDS (VEGETATIONSFARBUNGEN UND WASSERBLUTEN),

OESTERREICHISCHE GESELLSCHAFT FUER METEOROLOGIE, VIENNA (AUSTRIA).

F. WAWRIK.

WETTER UND LEBEN, VOL 23, NO 9-10, P 203-210, 1971. 11 REF.

DESCRIPTORS:
WATER POLLUTION EFFECTS, *ORGANIC MATTER, *DECOMPOSING ORGANIC MATTER, *EUTROPHICATION, *PHYTOPLANKTON, *ALGAE, *AQUATIC WEEDS, *WEEDS, *RED TIDE, COASTS, LAKES, PONDS, WATER POLLUTION CONTROL, AIRCRAFT, COPPER SULFATE.

ABSTRACT:
THE COMPLEX PHENOMENA OF ORGANIC WATER POLLUTION, OBSERVED IN POOLS, ALPINE LAKES AND COASTAL AREAS, ARE DISCUSSED. CAUSES AND CONSEQUENCES OF ORGANIC POLLUTION AND CONTROL OF HARMFUL SEAWEEDS ARE INCLUDED. THE FIRST FEW BLOOMS OF THE RED TIDE ARE CONTROLLED BY THE USE OF CUSO4-CRYSTALS DISPERSED FROM AIRPLANES. (ENSIGN-PAI)

FIELD 05C, 05B, 05G

ACCESSION NO. W72-14673

ENVIRONMENTAL LIMITS OF PLANTS IN FLOWING WATERS,

DURHAM UNIV. (ENGLAND). DEPT. OF BOTANY.

B. A. WHITTON.

SYMPOSIUM OF THE ZOOLOGICAL SOCIETY OF LONDON, NO 29, P 3-19, 1972. 60 REF.

DESCRIPTORS:
*RUNNING WATERS, *AQUATIC PLANTS, *AQUATIC ENVIRONMENT, WATER POLLUTION EFFECTS, STREAMS, RIVERS, ROOTED AQUATIC PLANTS, AQUATIC ALGAE, WATER TEMPERATURE, ACID STREAMS, HYDROGEN ION CONCENTRATION, HEAVY METALS, DISSOLVED OXYGEN, *PLANT GROWTH, LIGHT INTENSITY, TOXICITY, PRIMARY PRODUCTIVITY, STANDING CROPS, *PHOTOSYNTHESIS.

IDENTIFIERS:
DIVERSITY INDICES.

ABSTRACT:
A REVIEW IS PRESENTED SUMMARIZING THE LITERATURE AND SOME UNPUBLISHED DATA ON ATTACHED PHOTOSYNTHETIC PLANTS IN CERTAIN EXTREME FLOWING WATER ENVIRONMENTS. WITH UPPER TEMPERATURE LIMITS, IT IS POSSIBLE TO COMPARE SITUATIONS OCCURRING NATURALLY AND THOSE ARISING AS A RESULT OF POLLUTION. SUCH COMPARISONS SHOW UP OBVIOUS ANOMALIES. THE LIMITS OF PLANTS UNDER LOW PH CONDITIONS AND HIGH LEVELS OF HEAVY METALS ARE ALSO REVIEWED, ALTHOUGH WITH THESE EXTREMES THE DATA ARE INSUFFICIENT FOR COMPARISON OF NATURAL AND POLLUTED CONDITIONS. SECTIONS OF A MORE SPECULATIVE NATURE ARE ADDED ON LIGHT, CLIMATE, AND DISSOLVED OXYGEN AS FACTORS WHICH AT HIGH AND LOW LEVELS MAY LIMIT THE GROWTH OF PLANTS. THE QUANTITATIVE METHODS WHICH HAVE BEEN USED FOR COMPARING THE EFFECTS OF EXTREME ENVIRONMENTS ON PLANTS FALL INTO FOUR MAIN CATEGORIES: USE OF TOXICITY TESTS; USE OF DIVERSITY INDICES; MEASUREMENT OF PRIMARY PRODUCTION; MEASUREMENT OF STANDING CROPS. THESE METHODS DO NOT NECESSARILY GIVE SIMILAR RESULTS. AN IMPROVEMENT IN METHODS OF QUANTIFYING THE EXTENT TO WHICH AN ENVIRONMENT IS EXTREME IS IMPORTANT IF THE SIGNIFICANCE OF ENVIRONMENTAL LIMITS OF PLANTS IN NATURAL AND POLLUTED WATERS IS TO BE EVALUATED. (SVENSSON-WASHINGTON)

FIELD 05C, 10F

ACCESSION NO. W72-14690

FILAMENTOUS ALGAE AS WEEDS,

DURHAM UNIV. (ENGLAND). DEPT. OF BOTANY.

B. A. WHITTON.

IN: 3RD SYMPOSIUM ON AQUATIC WEEDS, EUROPEAN WEED RESEARCH COUNCIL, OXFORD, 1971, P 249-263. 40 REF.

DESCRIPTORS:
*EUTROPHICATION, *AQUATIC ALGAE, *ALGAL CONTROL, *AQUATIC WEED CONTROL, BIOCHEMICAL OXYGEN DEMAND, LAKE STAGES, LIMNOLOGY, NUISANCE ALGAE, NUTRIENT REMOVAL, NUTRIENTS, WATER POLLUTION EFFECTS, WATER QUALITY, HERBICIDES, ODOR-PRODUCING ALGAE.

IDENTIFIERS:
*FILAMENTOUS ALGAE, *AQUATIC ANGIOSPERM WEEDS, CLADOPHORA.

ABSTRACT:
IT STILL REMAINS DIFFICULT TO EVALUATE THE NUISANCE VALUE OF FILAMENTOUS ALGAE ON A WORLD SCALE, AND IT WOULD AT PRESENT BE IMPOSSIBLE TO ESTIMATE EVEN ROUGHLY THEIR ECONOMIC SIGNIFICANCE, EITHER AS AN ABSOLUTE VALUE OR IN COMPARISON WITH THAT OF AQUATIC ANGIOSPERM WEEDS. IT DOES, HOWEVER, SEEM CLEAR THESE ALGAE ARE A SPECIAL NUISANCE IN AT LEAST TWO RATHER DIFFERENT TYPES OF SITUATION. THEY ARE A PROBLEM AT THE EDGES OF SEVERAL LARGE LAKES, LIKE ONTARIO AND ZURICH, WHICH HAVE A HIGH AMENITY VALUE, AND WHICH HAVE BEEN SUBJECT TO CONSIDERABLE EUTROPHICATION IN RECENT YEARS. THEY ARE ALSO A PARTICULAR PROBLEM IN WARM DRY CLIMATES IN CHANNELS USED FOR THE TRANSPORT OF WATER FOR IRRIGATION AND OTHER PURPOSES. THEY MAY ALSO BE A PROBLEM IN MANY OTHER TYPES OF SITUATION, AS FOR INSTANCE THE DRAINAGE DITCHES OF FLAT, LOW-LYING AREAS LIKE PARTS OF THE NETHERLANDS, BUT THEY THEN OFTEN OCCUR TOGETHER WITH AQUATIC ANGIOSPERM WEEDS, AND BOTH THESE LATTER AND THE FILAMENTOUS ALGAE CAN NOW BE TREATED SIMULTANEOUSLY BY CHEMICAL MEANS. (SVENSSON-WASHINGTON)

FIELD 05C, 04A

ACCESSION NO. W72-14692

TOXICITY OF HEAVY METALS TO FRESHWATER ALGAE: A REVIEW,

DURHAM UNIV. (ENGLAND). DEPT. OF BOTANY.

B. A. WHITTON.

PHYKOS, VOL 9, NO 2, P 116-125, 1970. 38 REF.

DESCRIPTORS:
*REVIEWS, *HEAVY METALS, *ALGAE, *AQUATIC ENVIRONMENT, *TOXICITY, *WATER POLLUTION EFFECTS, ZINC, LEAD, COPPER, CHROMIUM, SILVER, CHLOROPHYTA, CYANOPHYTA, ALGAL CONTROL, ALGICIDES, COPPER SULFATE, SALINITY, HARDNESS(WATER), CHELATION, MINE DRAINAGE, WATER PROPERTIES.

IDENTIFIERS:
*FRESHWATER ALGAE, EDTA.

ABSTRACT:
A REVIEW IS PRESENTED OF THE LITERATURE ON TOXICITY OF HEAVY METALS TO FRESHWATER ALGAE. METALS CONSIDERED INCLUDE ZINC, LEAD, COPPER, CHROMIUM, AND SILVER. THE ACCOUNT IS CONCERNED MAINLY WITH FIELD OBSERVATIONS AND THOSE LABORATORY STUDIES WHICH ARE PARTICULARLY HELPFUL IN INTERPRETING FIELD DATA. EFFECTS OF SALINITY, WATER HARDNESS, AND ORGANIC COMPLEXATION ON TOXICITY ARE DISCUSSED. (SVENSSON-WASHINGTON)

FIELD 05C, 10F

ACCESSION NO. W72-14694

THE EFFECT OF CCC AND DIETHYLAMINE HYDROCHLORIDE ON CERTAIN SPECIES OF ALGAE BELONGING TO CYANOPHYCEAE, CHLOROPHYCEAE, AND DIATOMEAE,

BIALYSTOK MEDICAL ACADEMY (POLAND). DEPT. OF BIOLOGY.

R. CZERPACK.

ACTA HYDROBIOLOGIA, VOL 12, NO 2-3, P 143-151, 1970. 5 FIG, 33 REF.

DESCRIPTORS:
*INHIBITION, *PLANT GROWTH REGULATORS, *CHLORIDES, *CYANOPHYTA, *CHLOROPHYTA, *CHRYSOPHYTA, TOXICITY, INHIBITORS, TOXINS, INORGANIC COMPOUNDS, ALGAE, DIATOMS, CHLAMYDOMONAS, SCENEDESMUS, BIO-ASSAY, RESISTANCE, LETHAL LIMIT, POISONS, WATER POLLUTION EFFECTS.

IDENTIFIERS:
*CHLOROCHOLINE CHLORIDE, *CCC, *DIETHYLAMINE HYDROCHLORIDE, MERISOMOPEDIA SPP., ANABAENA SPP., GOMPHONEMA SPP., NITZCHIA SPP., SENSITIVITY.

ABSTRACT:
THE EFFECTS OF CHLOROCHOLINE CHLORIDE (CCC) AND DITETHYLAMINE HYDROCHLORIDE ON THE CONCENTRATION OF CELLS, THEIR SIZE, TOTAL CHLOROPHYLL CONTENT AND THE DYNAMICS OF THE GROWTH OF NINE SPECIES OF ALGAE BELONGING TO CYANOPHYCEAE (MERISMOPEDIA GLAUCA, ANABAENA CYLINDRICA), CHLOROPHYCEAE (CHLAMYDOMONAS NIVALIS, SCENEDESMUS QUADRICAUDA, S. ACUMINATUS, S. BASILIENSIS, S. BIJUGATUS), AND DIATOMEAE (GOMPHONEMA PARVULUM, NITZCHIA PALEA) WERE INVESTIGATED. DIETHYLAMINE, THE CHEMICAL ANALOG OF CCC, HAD SIMILAR EFFECT--EVEN A SOMEWHAT STRONGER ONE--TO THAT OF CCC AS A TYPICAL GROWTH RETARDANT AT SIMILAR CONCENTRATIONS. BOTH GROWTH REGULATORS HAD AN INHIBITORY EFFECT ON THE CELL CONCENTRATION, THEIR GROWTH DYNAMICS, AND THE TOTAL CHLOROPHYLL CONTENT OF THE ALGAE. THE DIATOMEAE AND FILAMENTOUS CYANOPHYCEAE WERE THE MOST SENSITIVE TO THE REGULATORS INVESTIGATED. ON THE OTHER HAND, A SLIGHT STIMULATORY EFFECT WAS OBSERVED WITH CHLOROPHYCEAE WHEN THE CHEMICALS WERE APPLIED IN CONCENTRATIONS OF 10 TO THE (-5) POWER TO 10 TO THE (-7) POWER M. (LEGORE-WASHINGTON)

FIELD 05C

ACCESSION NO. W72-14706

THE USE OF PHOTOGRAPHY IN WATER QUALITY RESEARCH,

WASHINGTON STATE UNIV., PULLMAN.

W. H. FUNK, AND D. C. FLAHERTY.

PHOTOGRAPHIC APPLICATIONS IN SCIENCE, TECHNOLOGY AND MEDICINE, VOL 7, NO 5, P 20-22, SEPTEMBER 1972. 7 FIG, 1 REF.

DESCRIPTORS:
SURVEYS, *AERIAL PHOTOGRAPHY, WATER POLLUTION EFFECTS, EUTROPHICATION, *LAKES, ALGAE, INFRARED RADIATION, *REMOTE SENSING, *POLLUTANT IDENTIFICATION, AQUATIC WEEDS, *WASHINGTON, *CHLOROPHYLL, WATER TEMPERATURE, ALKALINITY, DISSOLVED OXYGEN, CONDUCTIVITY, HYDROGEN ION CONCENTRATION.

IDENTIFIERS:
*SNAKE RIVER(WASH).

ABSTRACT:
THE FIRST STEP IN AERIAL PHOTOGRAPHIC STUDIES OF LAKES HAS BEEN TO MAKE AN AERIAL PHOTOGRAPHIC SURVEY OF THE PARTICULAR LAKE OR RESERVOIR WHICH HAS BEEN SCHEDULED FOR INVESTIGATION. THIS HELPS TO GIVE THE INVESTIGATORS A COMPREHENSIVE PICTURE OF THE LAKE ENVIRONMENT IN TERMS OF ITS DRAINAGE BASIN, SURFACE OUTLETS, AQUATIC GROWTH, POSSIBLE NUTRIENT SOURCES AND SIZE IN RELATION TO OTHER HYDROGRAPHIC FEATURES OF THE AREA. THIS INITIAL AERIAL SURVEY IS FOLLOWED BY CR CONDUCTED SIMULTANEOUSLY WITH INFRARED PHOTOGRAPHY OF THE LAKE AND ITS ENVIRONS. ONE OF THE CHARACTERISTICS OF THIS MATERIAL IS THAT THE CHLOROPHYL OF LIVING ALGAE AND AQUATIC WEEDS IS CAPTURED ON THE FILM IN VIVID PINKISH RED HUES, WHEREAS DEAD ALGAE OR VEGETATION EXHIBIT A WHITISH OR YELLOWISH ORANGE APPEARANCE. AERIAL RECONNAISSANCE HAS MADE IT POSSIBLE TO SURVEY EIGHT TO TEN LAKES IN A TWO-STATE AREA IN A SINGLE DAY. ON SAMPLING TRIPS, LIGHT PENETRATION, WATER TEMPERATURE, DISSOLVED OXYGEN, PH, ALKALINITY, AND CONDUCTIVITY TESTS ARE MADE AT VARIOUS DEPTHS. TWO-LITER WATER SAMPLES ARE ALSO COLLECTED FOR PLANKTON AND CHEMICAL ANALYSIS. A CURRENT PROJECT INVOLVING WATER QUALITY SURVEILLANCE IS A STUDY OF 150 MILES OF THE LOWER SNAKE RIVER DRAINAGE AREA LOCATED IN SOUTHEASTERN WASHINGTON. AERIAL PHOTOGRAPHY SHOWS THAT THIS STRETCH OF THE RIVER IS SUBJECTED TO EXCESSIVE RUNOFF OF TOPSOIL SHORTLY AFTER THE SNOW MELTS IN THE SPRING, AND ALSO DURING PERIODS OF HIGH RAINFALL. (KNAPP-USGS)

FIELD 05A, 07B, 05C, 02H

ACCESSION NO. W72-14728

THE PATHOLOGY OF LAKE ERIE,

Q. DADISMAN.

THE NATION, VOL. 214, P. 492-496, APRIL 17, 1972. 1 PLATE.

DESCRIPTORS:
*LAKE ERIE, *EUTROPHICATION, *CHEMICAL WASTES, *POLITICAL ASPECTS, LAKES, GREAT LAKES, ALGAL BLOOMS, AGING(BIOLOGICAL), LAKE STAGES, MESOTROPHY, OLIGOTROPHY, WATER POLLUTION SOURCES, WATER POLLUTION EFFECTS, MERCURY, SEWAGE DISPOSAL, PHOSPHATES, PHOSPHORUS, CHEMICAL WASTES, LIMNOLOGY, INTERNATIONAL JOINT COMMISSION, CANADA, FEDERAL GOVERNMENT, STATE GOVERNMENTS.

IDENTIFIERS:
*MERCURY POLLUTION, *ENVIRONMENTAL PROTECTION AGENCY.

ABSTRACT:
ALTHOUGH MOST REPORTS INDICATE THAT LAKE ERIE IS 'DEAD', A RECENT STUDY SHOWS THAT THE LAKE IN NOT DEAD AND COULD BE REVIVED IN ONE GENERATION. ONE SERIOUS PROBLEM IS THE INPUT OF PHOSPHORUS INTO THE LAKE. SCIENTISTS BELIEVE THAT IF THIS INPUT WERE REDUCED A QUICK IMPROVEMENT IN THE CONDITION OF THE LAKE COULD BE EXPECTED. OTHER PROBLEMS INCLUDE THE LARGE AMOUNTS OF MERCURY DISCHARGED BY CHLORALKALI PLANTS NEAR DETROIT AND THE DISCHARGES OF SEWAGE ALL ALONG THE COAST. ONE OF SEVERAL FACTORS PREVENTING IMPROVEMENT OF THE LAKE'S CONDITION IS THE LACK OF AN INSTITUTIONAL MECHANISM FOR COOPERATION BETWEEN THE UNITED STATES AND CANADA. THIS CAN BE TRACED TO THREE CONDITIONS IN AMERICA: INFERIOR AMERICAN TECHNOLOGY IN SEWAGE AND POLLUTION TREATMENT, THE U.S. HESITANCE TO ELIMINATE THE PHOSPHATE CONTENT OF HOUSEHOLD DETERGENTS, AND THE POLITICAL INABILITY OF THE U.S. TO ACT QUICKLY. AS A RESULT, PHOSPHATES AND OTHER POLLUTANTS CONTINUE TO ENTER LAKE ERIE WITH LITTLE HOPE REMAINING FOR THE LAKE. THE U.S. BUREAUCRACY, IN PARTICULAR THE ENVIRONMENTAL PROTECTION AGENCY, FAILS TO RECOGNIZE THAT STRICTER POLLUTION CONTROL STANDARDS ARE NEEDED IN THE GREAT LAKES THAN FOR THE OCEAN. (NIELSEN-FLORIDA)

FIELD 05G, 02H, 06E

ACCESSION NO. W72-14782

CARBON AND NITROGEN AS REGULATORS OF ALGAL GROWTH IN TREATED SEWAGE,

KENTUCKY WATER RESOURCES INST., LEXINGTON.

G. FOREE, AND R. SCROGGIN.

AVAILABLE FROM THE NATIONAL TECHNICAL INFORMATION SERVICE AS PB-212 375, $3.00 IN PAPER COPY, $0.95 IN MICROFICHE. RESEARCH REPORT NO. 49, MARCH 1972. 61 P, 15 FIG, 7 TAB, 32 REF. OWRR A-023-KY (2).

DESCRIPTORS:
*CARBON, *NITROGEN, *ALGAE, PLANT GROWTH, SEWAGE EFFLUENTS, REGULATION, CYANOPHYTA, CHLOROPHYTA, HYDROGEN ION CONCENTRATION, CHEMICAL PORPERTIES, NITROGEN FIXATION, CULTURES, CARBON DIOXIDE, NUTRIENTS, LIMITING FACTORS, ALLOGENIC SUCCESSION, EUTROPHICATION.

IDENTIFIERS:
*ALGAL GROWTH, CONTINUOUS CULTURES, BATCH CULTURES.

ABSTRACT:
TO PROVIDE INFORMATION ON ALGAL GROWTH REGULATORS THE FIRST PHASE OF THIS INVESTIGATION WAS A CONTINUOUS FLOW CULTURE STUDY IN WHICH ALGAE WERE GROWN UNDER VARIOUS CONDITIONS. WHEN THE CULTURES REACHED STEADY STATE, THEY WERE ASSAYED TO DETERMINE THE EXTENT AND RATE OF GROWTH, ALGAL GENERA PRESENT, NUTRIENT UTILIZATION, AND ALGAL COMPOSITION. THE SECOND PHASE WAS A BATCH CULTURE STUDY WHICH PROVIDED MEANS FOR COMPARING THE TWO TYPES OF CULTURES (I. E., BATCH AND CONTINUOUS) AND VERIFICATION OF SOME CONCLUSIONS OF THE CONTINUOUS CULTURE STUDY. EVEN IN SEWAGE EFFLUENTS CONTAINING SIGNIFICANT CONCENTRATIONS OF BOTH ORGANIC AND INORGANIC CARBON, ALGAL GROWTH WAS LIMITED BY AVAILABILITY OF CARBON DIOXIDE. THE RESULTS INDICATED THAT IN NATURAL SITUATIONS WHERE EXCESS CARBON DIOXIDE MIGHT BE AVAILABLE, BUT NITROGEN IS DEFICIENT IN SOLUTION, ALGAL SUCCESSION MAY OCCUR WITH SHIFTS TO CERTAIN BLUE-GREEN FORMS CONTAINING ATMOSPHERIC NITROGEN FIXING CAPABILITIES. A PREVIOUSLY DEVELOPED KINETIC THEORY WHICH DESCRIBES ALGAL GROWTH AS BEING PROPORTIONAL TO THE CELLULAR CONCENTRATION OF THE GROWTH REGULATING NUTRIENT WAS CONFIRMED WHEN APPLIED TO THE CARBON DIOXIDE ENRICHED CONDITIONS WITH 10 MG/L AMMONIA NITROGEN IN THE FEED SOLUTION. (JONES-WISCONSIN)

FIELD 05C

ACCESSION NO. W72-14790

GRAZING BY THE CILIATED PROTOZOON LOXODES MAGNUS ON THE ALGA SCENEDEMUS IN A EUTROPHIC POND,

FRESHWATER BIOLOGICAL ASSOCIATION, AMBLESIDE, (ENGLAND).

R. GOULDER.

OIKOS, VOL. 23, NO. 1, P. 109-115, 1972. 4 FIG., 3 TAB., 12 REF.

DESCRIPTORS:
*GRAZING, *PROTOZOA, *ALGAE, SCENEDESMUS, EUTROPHICATION, PONDS, FEEDING RATES, BENTHOS, STANDING CROPS.

IDENTIFIERS:
*LOXODES MAGNUS.

ABSTRACT:
AN ASSESSMENT WAS MADE OF THE IMPORTANCE OF GRAZING BY LOXODES MAGNUS, ONE OF A GROUP OF BENTHIC CILIATED PROTOZOA FEEDING ON SCENEDESMUS DENTICULATUS, S. QUADRICAUDA, AND S. ACUMINATUS. THE RATE AT WHICH L. MAGNUS DIGESTS SCENEDESMUS WAS ESTIMATED BY MEANS OF A LABORATORY STARVATION PROCEDURE AND IT WAS ASSUMED THAT, UNDER NATURAL CONDITIONS, THE DIGESTION RATE EQUALS THE FEEDING RATE. NO CORRELATION WAS FOUND BETWEEN NUMBERS OF L. MAGNUS (AND ALSO THE CILIATE LOXODES STRIATUS) AND SCENEDESMUS CROPS. SCENEDESMUS CELLS PER SQUARE METER IN THE SEDIMENT DECREASED FROM S. QUADRICAUDA TO S. DENTICULATUS TO S. ACUMINATUS BUT INSIDE L. MAGNUS, NUMBERS OF S. DENTICULATUS WERE HIGHEST, DECREASING TO S. QUADRICAUDA AND FINALLY S. ACUMINATUS; THEREFORE, L. MAGNUS MAY BE ABLE TO DISTINGUISH BETWEEN SPECIES OF SCENEDESMUS IN ITS FEEDING. THE POPULATION FEEDING RATE OF L. MAGNUS REPRESENTED BETWEEN 0.003% AND 0.68% OF THE SCENEDESMUS CROP EATEN PER DAY; THEREFORE GRAZING BY L. MAGNUS PROBABLY HAD NO SIGNIFICANT EFFECT. OTHER CILIATES WERE EATING SCENEDESMUS, AND ALSO PERHAPS BENTHIC ROTIFERS AND CRUSTACEANS, CONSEQUENTLY GRAZING BY ALL INVERTEBRATES COULD BE SIGNIFICANT. L. MAGNUS UTILIZES OTHER FOOD THAN SCENEDESMUS--MEMBERS OF THIS GENUS ARE KNOWN TO FEED ON BOTH ALGAE AND BACTERIA. (JONES-WISCONSIN)

FIELD 05C

ACCESSION NO. W72-14796

QUANTITATIVE STUDIES ON THE PHYTOPLANKTON OF THE RIVERS THAMES AND KENNET AT READING,

READING UNIV. (ENGLAND). DEPT. OF ZOOLOGY.

T. J. LACK.

FRESHWATER BIOLOGY, VOL. 1, P. 213-224, 1971. 2 FIG., 2 TAB., 22 REF.

DESCRIPTORS:
*ALGAE, *PHYTOPLANKTON, *RIVERS, *DIATOMS, CHLORCPHYTA, BENTHIC FLORA, EUTROPHICATION, DOMINANT ORGANISMS, BIOLOGICAL COMMUNITIES, NUTRIENTS, SCENEDESMUS, CHRYSOPHYTA, SEASONAL, SILICA, DISCHARGE(WATER).

IDENTIFIERS:
*CENTRIC DIATOMS, *RIVER THAMES(ENGLAND), RIVER KENNET(ENGLAND), STEPHANODISCUS HANTZSCHII, CRYPTOMONAS, RHODOMONAS.

ABSTRACT:
SEASONAL CHANGES IN DENSITY AND SPECIES COMPOSITION OF RIVER THAMES PHYTOPLANKTON AND ITS TRIBUTARY, RIVER KENNET, AT READING (ENGLAND) WERE STUDIED. OBSERVATIONS EXTENDED FROM MAY 1966 TO MAY 1968. SAMPLES CONSISTING OF 250 ML OF SUB-SURFACE WATER WERE TAKEN WEEKLY FROM THE CENTER OF EACH RIVER APPROXIMATELY 200 M UPSTREAM OF THE CONFLUENCE. THESE WERE CONCENTRATED BY SEDIMENTATION AND COUNTED IN A HAEMOCYTOMETER. DATA ON DISCHARGE, TEMPERATURE, AND SILICA CONCENTRATION ARE FOLLOWED BY DESCRIPTIONS OF THE VARIATIONS IN NUMBER AND PHYTOPLANKTON COMPOSITION. IN BOTH RIVERS THERE WERE SPRING AND AUTUMN PEAKS OF THE CENTRIC DIATOM, STEPHANODISCUS HANTZSCHII. CHLOROPHYCEAE WERE MOST ABUNDANT DURING SUMMER. TWO CRYPTOPHYCEANS, CRYPTOMONAS AND RHODOMONAS, WERE SOMETIMES NUMEROUS. IN THE THAMES, POPULATION SIZE WAS CLOSELY CORRELATED WITH THE DISCHARGE, HIGHEST NUMBERS ALWAYS OCCURRING DURING LOW DISCHARGE PERIODS. IN THE KENNET, INCREASES IN DISCHARGE OFTEN BROUGHT ABOUT INCREASES IN CELL NUMBER DUE TO INFLUX OF BENTHIC FORMS. IT SEEMS HIGHLY PROBABLE THAT THE RIVERS MAINTAIN HIGH LEVELS OF NITRATES AND PHOSPHATES SINCE BOTH CARRY CONSIDERABLE QUANTITIES OF TREATED SEWAGE EFFLUENT. COMPARISONS WITH EARLIER STUDIES ON THE THAMES SHOWED DIFFERENCE IN PHYTOPLANKTON COMPOSITION WHICH WERE ALMOST CERTAINLY DUE TO EUTROPHICATION. (JONES-WISCONSIN)

FIELD 05C

ACCESSION NO. W72-14797

LIGHT-INDUCED FLUORESCENCE CHANGES IN CHLORELLA, AND THE PRIMARY PHOTOREACTIONS FOR THE PRODUCTION OF OXYGEN,

ROCKEFELLER UNIV., NEW YORK.

D. MAUZERALL.

PROCEEDINGS OF THE NATIONAL ACADEMY OF SCIENCES, VOL 69, NO 6, P 1358-1362, 1972. 3 FIG, 27 REF.

DESCRIPTORS:
*ALGAE, *FLUORESCENCE, *CHLORELLA, *PHOTOACTIVATION, *OXYGEN, LIGHT, PHOTOSYNTHESIS, OPTICAL PROPERTIES, ABSORPTION, MATHEMATICAL STUDIES, CHLOROPHYLL, OXIDATION REDUCTION POTENTIAL.

IDENTIFIERS:
CHLORELLA VULGARIS.

ABSTRACT:
THE INCREASE OF THE QUANTUM FLUORESCENCE YIELD IN GREEN PLANTS AND ALGAE WITH INCREASING ILLUMINATION HAS LONG BEEN AN INDICATOR OF THE STATE OF THE PHOTOREACTIVE CENTERS IN PHOTOSYNTHESIS. THERE ARE DIFFERENT PARTS OF THIS NONLINEAR FLUORESCENCE; SOME PARTS ARE FAIRLY SLOW (SECONDS TO MINUTES); OTHERS ARE FAST (MILLISECONDS). THE LIGHT-INDUCED INCREASES OF THE EFFECTIVE FLUORESCENCE YIELD IN CHLORELLA ARE TOO SLOW TO BE A PRIMARY PROCESS IN PHOTOSYNTHESIS. THE FAST TRANSIENT STATE IS ATTRIBUTED TO A PRIMING REACTION FOR THE PHOTOSYSTEM THAT MAKES OXYGEN. THE SLOWER CYCLICAL PROCESS IS ATTRIBUTED TO THE DARK REACTIONS THAT MAKE OXYGEN AFTER PHOTOEXCITATION OF THIS SYSTEM. THE SLOWER CYCLICAL PROCESS IS ALSO DISTINGUISHED BY A NARROWER EMISSION SPECTRUM THAT PEAKS AT A SHORTER WAVELENGTH THAN THE DARK ADAPTED OR FAST TRANSIENT STATE. A MINIMUM OF SIX DIFFERENT FLUORESCENT STATES ARE REQUIRED TO EXPLAIN THE DATA. IN ADDITION TO THE USUAL ASSUMPTION ABOUT CHANGING QUANTUM YIELD OF FLUORESCENCE IN THESE PROCESSES, THE DATA SUGGEST THAT CHANGES IN CROSS SECTION OF OPTICAL ABSORPTION MUST ALSO BE CONSIDERED. THE SLOWEST RELAXATION TIMES OBSERVED ARE WELL CORRELATED WITH THE SLOW STEPS DETECTED IN EVOLUTION OF OXYGEN. (JONES-WISCONSIN)

FIELD 05C

ACCESSION NO. W72-14800

SOME DATA ON THE ROLE OF FOOD IN THE BIOLOGY OF EUDIAPTOMUS ZACHARIASI POPPE,

INSTYTUT RYBACTWA SRODLADOWEGO, WARSAW (POLAND).

E. GRYGIEREK.

EKOLOGIA POLSKA, VOL 19, NO 21, P 277-292, 1971. 8 FIG, 38 REF.

DESCRIPTORS:
*CRUSTACEANS, *FOODS, *LIFE HISTORY STUDIES, COPEPODS, DIATOMS, CHLOROPHYTA, TEMPERATURE, SIZE, FECUNDITY, ALGAE.

IDENTIFIERS:
*EUDIAPTOMUS ZACHARIASI.

ABSTRACT:
THE ROLE OF FOOD IN THE DEVELOPMENT OF THE CRUSTACEAN, EUDIAPTOMUS ZACHARIASI, IN A NATURAL HABITAT WAS STUDIED. BY DAILY SAMPLING FROM SIX SIMILAR FRY PONDS, EXAMINATION WAS MADE OF THE RELATION BETWEEN DURATION OF DEVELOPMENT, SIZE, AND FECUNDITY AND THE AMOUNT OF ITS FOOD. ANALYSIS WAS MADE OF TOTAL ABUNDANCE, OF VARIATION IN AGE AND SEX STRUCTURE, IN LENGTH OF INDIVIDUALS AT DIFFERENT STAGES OF DEVELOPMENT, ABUNDANCE OF EGGS AND NUMBERS OF FEMALES WITH EGGS. EUDIAPTOMUS ZACHARIASI FED ON ROTIFERS, BUT WHEN THERE WERE FEW ROTIFERS, ITS FOOD WAS ALGAE, MEASURING UP TO 50 MICRONS, MORE DIFFICULT TO DIGEST THAN ROTIFERS. THE DEVELOPMENT CYCLE OF THE GENERATION OF E. ZACHARIASI HATCHED FROM RESTING EGGS LASTED FROM 13 TO 23 DAYS, AS DID THAT OF THE NEXT GENERATION. DIFFERENCES WERE NOT GREATER THAN A FEW DAYS AND MIGHT HAVE BEEN CAUSED BY AMOUNT OF FOOD DURING DEVELOPMENT PERIOD OF THE GENERATION. IN EXTREME CASES THE FOOD LEVEL LIMITS THE NUMBER OF GENERATIONS. THE SIZE OF INDIVIDUALS, SEX STRUCTURE AND PARTICULARLY INDIVIDUAL FECUNDITY OF E. ZACHARIASI DEPENDED ON AMOUNT OF FOOD. INDIVIDUAL FECUNDITY DID NOT ALWAYS DETERMINE THE FECUNDITY OF THE WHOLE POPULATION. (JONES-WISCONSIN)

FIELD 05C, 02H

ACCESSION NO. W72-14805

RELATIVE TOLERANCE OF NITROGEN-FIXING BLUE-GREEN ALGAE TO PESTICIDES,

INDIAN AGRICULTURAL RESEARCH INST., NEW DELHI.

G. S. VENKATARAMAN, AND B. RAJYALAKSHMI.

INDIAN JOURNAL OF AGRICULTURAL SCIENCES, VOL 42, NO 2, P 119-121, 1972. 1 TAB, 11 REF.

DESCRIPTORS:
*ALGAE, *CYANOPHYTA, *PESTICIDES, FUNGICIDES, HERBICIDES, IRRIGATION, AGRICULTURE, RICE, CROP RESPONSE.

IDENTIFIERS:
*TOLERANCE, ANABAENA, NOSTOC, AULOSIRA, TOLYPOTHRIX, ANACYSTIS NIDULANS.

ABSTRACT:
IN-VITRO TOLERANCE OF BLUE-GREEN ALGAE TO PESTICIDES WAS STUDIED. TWENTY-SEVEN STRAINS OF NITROGEN-FIXING CYANOPHYTA FROM FOUR GENERA WERE TESTED. COMMERCIAL PREPARATIONS OF CERESAN M, DITHANE, 2,4-D, DELAPON, PROPAZINE, COTORON, DIURON, AND LINURON WERE USED AT CONCENTRATIONS VARYING FROM 0.01 TO 2000 PPM. MOST NITROGEN-FIXING BLUE-GREEN ALGAE COULD TOLERATE HIGH PESTICIDE LEVELS ALTHOUGH THEY SHOWED WIDE VARIATION IN RELATIVE TOLERANCE. LEVELS OF DIFFERENT PESTICIDES ARE DESCRIBED IN RELATION TO SPECIFIC STRAINS; FOR EXAMPLE, THOUGH MOST OF THE STRAINS OF ANABAENA COULD TOLERATE 100 PPM OF CERESAN, STRAIN 310 WAS SENSITIVE TO A CONCENTRATION OF 0.01 PPM; THOUGH DITHANE WAS LETHAL TO SOME STRAINS OF ANABAENA AND NOSTOC AT THE LOWEST CONCENTRATION USED, SOME STRAINS COULD GROW WELL AT 50 PPM. HIGH CONCENTRATIONS OF CERESAN, DITHANE, 2,4-D AND DELAPON WERE TOLERATED BY TOLYPOTHRIX TENUIS AND AULOSIRA FERTILISSIMA. PROPAZINE, COTORON, AND LINURON WERE LETHAL, AND DIURON LETHAL TO BOTH ALGAE. SPECIFIC STRAINS SHOULD BE USED WITH DIFFERENT PESTICIDES AND A MIXTURE OF DIFFERENT STRAINS--NOT A SINGLE STRAIN--SHOULD BE USED AS SEEDING MATERIAL WHEN MORE THAN ONE PESTICIDE IS APPLIED TO THE CROP. (JONES-WISCONSIN)

FIELD 05C

ACCESSION NO. W72-14807

OBSERVATIONS ON THE CYTOLOGY AND ULTRASTRUCTURE OF THE NEW ALGAL CLASS, EUSTIGMATOPHYCEAE,

LEEDS UNIV. (ENGLAND), DEPT. OF BOTANY.

D. J. HIBBERD, AND G. F. LEEDALE.

ANNALS OF BOTANY, VOL 36, P 49-71, 1972. 2 FIG, 7 PLATES, 65 REF.

DESCRIPTORS:
*CYTOLOGICAL STUDIES, *SYSTEMATICS, *ALGAE, CHRYSOPHYTA, PHAEOPHYTA.

IDENTIFIERS:
*EUSTIGMATOPHYCEAE, XANTHOPHYCEAE, POLYEDRIELLA HELVETICA, PLEUROCHLORIS COMMUTATA, PLEUROCHLORIS MAGNA, VISCHERIA PUNCTATA, VISCHERIA STELLATA, ELLIPSOIDION ACUMINATUM.

ABSTRACT:
THE FIRST DETAILED ACCOUNT OF A NEW CLASS OF ALGAE, EUSTIGMATOPHYCEAE IS GIVEN. ITS UNIQUE FEATURES ARE EMPHASIZED IN COMPARISON WITH THOSE COCCOID FORMS WHICH MUST BE CONSIDERED AS MEMBERS OF THE XANTHOPHYCEAE SENSU STRICTO. THE ZOOSPORES, AND TO A LESSER EXTENT, THE VEGETATIVE CELLS OF SIX SPECIES OF EUSTIGMATOPHYCEAE SPECIES WERE STUDIED WITH LIGHT AND ELECTRON MICROSCOPY. THE FEATURES OF ORGANIZATION ARE COMPARED IN DETAIL WITH THE SAME COMPONENTS IN THE XANTHOPHYCEAE AND OTHER CLASSES OF ALGAE, WITH PARTICULAR REFERENCE TO THE MOTILE CELLS. IT IS CONCLUDED THAT EUSTIGMATOPHYCEAE IS LESS LIKE THE XANTHOPHYCEAE SENSU STRICTO THAN THE LATTER IS LIKE THE CHRYSOPHYCEAE AND PHAEOPHYCEAE, AND IN MOST CHARACTERISTICS SHOWS FUNDAMENTAL DIFFERENCES FROM ALL OTHER CLASSES OF ALGAE. THE TAXONOMIC DIFFICULTY RAISED BY EUSTIGMATOPHYCEAN POSSESSION OF AN ANTERIOR HAIRY FLAGELLUM OF TYPICALLY 'HETEROKONT' CONSTRUCTION IS DISCUSSED. THE TAXONOMIC AND NOMENCLATURAL PROBLEMS RESULTING FROM THE REMOVAL OF SEVERAL SPECIES FROM THE XANTHOPHYCEAE INTO THE EUSTIGMATOPHYCEAE ARE BRIEFLY CONSIDERED AND PREVIOUS OBSERVATIONS BY OTHER AUTHORS ON SPECIES NOW PLACED IN THE EUSTIGMATOPHYCEAE ARE REVIEWED. (JONES-WISCONSIN)

FIELD 05C

ACCESSION NO. W72-14808

THE INFLUENCE OF EXPERIMENTAL INCREASE OF BIOMASS OF THE BLUE-GREEN ALGAE GLOEOTRICHIA ECHINULATA (SMITH) RICHTER ON PHYTOPLANKTON PRODUCTION,

POLISH ACADEMY OF SCIENCES, WARSAW. INST. OF ECOLOGY.

I. SPODNIEWSKA.

EKOLOGIA POLSKA, VOL 19, NO 31, P 476-483, 1971. 2 FIG, 5 TAB, 4 REF.

DESCRIPTORS:
*ALGAE, *BIOMASS, *CYANOPHYTA, PHYTOPLANKTON, PRODUCTIVITY, PHOTOSYNTHESIS, EPILIMNION, LIGHT INTENSITY, RESPIRATION, LIMITING FACTORS.

IDENTIFIERS:
*GLOEOTRICHIA ECHINULATA, MIKOLAJSKIE LAKE(POLAND), CERATIUM HIRUNDINELLA.

ABSTRACT:
INVESTIGATIONS IN MIKOLAJSKIE LAKE, POLAND WERE MADE OF THE POSSIBILITIES OF INCREASING PHYTOPLANKTON PRODUCTION IN PARTICULAR TROPHIC CONDITIONS BY INCREASING BIOMASS OF THE PLANKTONIC BLUE-GREEN ALGA GLOEOTRICHIA ECHINULATA AND OF THE ESTIMATION OF NET PRODUCTION OF THIS SPECIES IN NATURAL CONDITIONS. THIS SPECIES WAS CHOSEN BECAUSE OF THE GOOD POSSIBILITY OF ISOLATING IT FROM OTHER PHYTOPLANKTON. SURFACE WATER PLANKTON WAS CONDENSED AND COLONIES SEPARATED. PRIMARY PRODUCTION WAS ESTIMATED. OF THREE EXPERIMENTAL SERIES, THE FIRST AND SECOND WERE JUST UNDER THE WATER SURFACE, AND THE THIRD AT VARIOUS DEPTHS OF THE TROPHOGENIC LAYER (0-6 M). PHYTOPLANKTON PRODUCTION OF AN UNCHANGED BIOMASS WAS COMPARED WITH PHYTOPLANKTON PRODUCTION TO WHICH A KNOWN NUMBER OF COLONIES OF G. ECHINULATA WAS ADDED. PRODUCTION WAS THEN CALCULATED FROM THE DIFFERENCE. IN EACH ONE OF THE EXPERIMENTAL SERIES, INDEPENDENT OF THE PERIOD THEY WERE CARRIED OUT, AND THUS INDEPENDENT OF THE POSSIBLE EXISTENCE OF DIFFERENCES IN PHYSIOLOGICAL ALGAL PROPERTIES AND/OR IN ENVIRONMENTAL CONDITIONS AT THE TIME, IT WAS FOUND THAT PRODUCTION INCREASES AS A RESULT OF AN INCREASE OF PHYTOPLANKTON DENSITY DUE TO ADDITION OF G. ECHINULATA COLONIES. (JONES-WISCONSIN)

FIELD 05C

ACCESSION NO. W72-14812

WATER QUALITY CONTROL WITH SYNTHETIC POLYMERIC FLOCCULANTS: EFFECT OF METAL IONS ON FLOCCULATION OF BIOCOLLOIDS,

CONNECTICUT UNIV., STORRS. INST. OF WATER RESOURCES.

J. K. DIXON, AND R. C. TILTON.

AVAILABLE FROM THE NATIONAL TECHNICAL INFORMATION SERVICE AS PB-212 364, $3.00 IN PAPER COPY, $0.95 IN MICROFICHE. COMPLETION REPORT 1972. 9 P. OWRR A-031-CONN(2).

DESCRIPTORS:
*FLOCCULATION, *ALGAE, *SILICA, *E. COLI, *IONS, *WASTEWATER TREATMENT, *POLYMERS, CATIONS, ANIONS, CHLORELLA, FILTRATION, ELECTROPHORESIS, TRACERS.

IDENTIFIERS:
*MULTIVALENT METALLIC CATIONS, *POLYACRYLAMIDE, *POLYETHYLENEMINE, *POLYSTYRENE SULFONATE, CHLORELLA ELLIPSOIDIA.

ABSTRACT:
EFFECT OF THE ADDITION OF MULTIVALENT METALLIC CATIONS UP TO 0.02M ON THE RATE OF FLOCCULATION OF PURE ALGAL CULTURES OF CHLORELLA ELLIPSOIDIA BY SYNTHETIC POLYMERIC CATIONIC POLYETHYLENEIMINES (PEI) WAS FOUND TO BE NEGLIGIBLE. A HUNDREDFOLD VARIATION IN THE MG (++) CONCENTRATION PRESENT IN NUTRIENT USED DURING THE GROWTH OF THESE ALGAE ALSO HAD NO EFFECT ON EFFICIENCY OF FLOCCULATION OF THE ALGAE BY THE PEI POLYMERS. ADDITION OF NON-IONIC POLYACRYLAMIDE TO THE PEI POLYMERS TO THE BACTERIA, E. COLI, HAD NO EFFECT ON THEIR GOOD PERFORMANCE. ADDITION OF ANIONIC POLYSTYRENE SULFONATE POLYMERS, WHICH WERE INEFFECTIVE ALONE, TO PEI POLYMERS DECREASED THE FLOCCULATION PERFORMANCE OF THE LATTER FOR E. COLI. EFFECTS OF THE MULTIVALENT METALLIC CATIONS AND/OR THE USE OF ANIONIC-CATIONIC POLYMER MIXTURES FOR FLOCCULATING THE COMPLEX DISPERSIONS PRESENT IN COMMERCIAL OPERATIONS ARE OFTEN SUCCESSFUL. THE RESULTS ON THE SIMPLE ALGAL AND BACTERIA SYSTEMS APPARENTLY CANNOT BE TRANSLATED INTO SOME GUIDING PRINCIPLES USEFUL DURING THE FLOCCULATION OF THE MORE COMPLEX COMMERCIAL SYSTEMS. ON THE OTHER HAND, THE RESULTS OF PREVIOUS STUDIES ON FLOCCULATION OF E. COLI, ALGAE AND SILICA WITH RADIOACTIVE TAGGED PEI POLYMERS SHOULD BE QUITE USEFUL FOR APPLICATION TO COMMERCIAL OPERATIONS.

FIELD 05G, 05D

ACCESSION NO. W72-14840

IRRIGATION AND FERTILIZATION WITH WASTEWATER,

IDAHO UNIV., MOSCOW. DEPT. OF CIVIL ENGINEERING; AND IDAHO UNIV., MOSCOW. DEPT. OF HYDROGEOLOGY.

D. D. EIER, A. T. WALLACE, AND R. E. WILLIAMS.

COMPOST SCIENCE JOURNAL OF WASTE RECYCLING, VOL 12, NO 3, MAY-JUNE 1971, P 26-29. 2 FIG, 1 REF.

DESCRIPTORS:
*IRRIGATION, *FERTILIZATION, *WASTE WATER(POLLUTION), *WASTE WATER DISPOSAL, *WATER POLLUTION CONTROL, ECONOMIC FEASIBILITY, STREAMS, SEWAGE EFFLUENTS, ALGAE, TOURISM, RECREATION, FISHERIES, HYDROELECTRIC POWER, NAVIGATION, CONSUMPTIVE USE, WATER POLLUTION, OXYGEN, SOIL PROPERTIES, IDAHO.

ABSTRACT:
AN ECONOMICALLY PALATABLE METHOD OF WATER POLLUTION CONTROL IS PRESENTED. THE PLAN PRESENTS THE ENTRY INTO STREAMS OF THE $18 WORTH OF NUTRIENTS PER ACRE FOOT OF SECONDARY TREATMENT PLANT EFFLUENT. DRIFTING ISLANDS OF ALGAE IN RIVERS, AND FERTILIZER DEMANDS CAN BE REDUCED. TOURISM, RECREATION, FISHERY PRODUCTION, HYDROELECTRIC POWER, NAVIGATION, IRRIGATION AGRICULTURE, AND DOMESTIC AND INDUSTRIAL USE OF THE SNAKE RIVER IN SOUTHERN IDAHO ARE AFFECTED BY WATER POLLUTION. PARTICULAR PROBLEMS INCLUDE TASTE, ODOR, BACTERIAL CONTAMINATION, AQUATIC GROWTH AND THERMAL EFFECTS. THE PRINCIPAL CAUSE OF THIS POLLUTION IS THE DISPOSAL OF UNTREATED OR INADEQUATELY TREATED WASTES. THE WASTES REDUCE AVAILABLE OXYGEN AND ENCOURAGE GROWTH OF UNDESIRABLE ALGAE. CONTROL OF NUTRIENT INPUT WOULD GO A LONG WAY IN REDUCING THIS POLLUTION. CROP IRRIGATION OR SURFACE APPLICATION OF DOMESTIC AND INDUSTRIAL WASTE WATER FOR SOIL RENOVATION IS SUGGESTED. INVESTIGATIONS AT THE UNIVERSITY OF IDAHO INDICATE IRRIGATION AND FERTILIZATION WITH WASTE WATER ARE FEASIBLE AND ECONOMICAL. (POPKIN-ARIZONA)

FIELD 03F, 05D

ACCESSION NO. W72-14878

MICROBIAL CRITERIA OF ENVIRONMENT QUALITIES,

COPENHAGEN UNIV. (DENMARK). INST. OF HYGIENE.

E. FJERDINGSTAD.

ANNUAL REVIEW OF MICROBIOLOGY, VOL. 25, P 563-582, 1971. 3 TAB, 58 REF.

DESCRIPTORS:
*EUTROPHICATION, *BIOLOGICAL PROPERTIES, *WATER QUALITY, *BIOINDICATORS, *BIOLOGICAL COMMUNITIES, AQUATIC ALGAE, AQUATIC PRODUCTIVITY, BIOCHEMICAL OXYGEN DEMAND, LAKE STAGES, NUTRIENTS, MESTROPHY, OLIGOTROPHY, OXYGEN SAG, AQUATIC LIFE, BIOASSAY, CLASSIFICATION.

IDENTIFIERS:
BIOTESTS.

ABSTRACT:
AFTER A BRIEF DISCUSSION OF EUTROPHICATION AND A SURVEY OF BIOTESTS, BIOLOGICAL ASSESSMENT OF EUTROPHICATION AND POLLUTION IS DISCUSSED. A NUMBER OF CLASSIFICATION SYSTEMS FOR DENOTING THE DEGREE TO WHICH WATER IS POLLUTED, SOME DEFINED IN TERMS OF BIOLOGICAL POPULATIONS, ARE GIVEN. THE SYSTEM MOST DISCUSSED RELATES SAPROBICITY LEVEL AND STRUCTURES OF THE COMMUNITIES OF ORGANISMS. (SVENSSON-WASHINGTON)

FIELD 05C, 05B

ACCESSION NO. W73-00002

USE OF TESTS FOR LIMITING OR SURPLUS NUTRIENTS TO EVALUATE SOURCES OF NITROGEN AND PHOSPHORUS FOR ALGAE AND AQUATIC WEEDS,

WISCONSIN UNIV., MADISON. WATER CHEMISTRY LAB.

G. P. FITZGERALD, AND G. F. LEE.

REPORT JULY 1, 1971. 35 P, 4 FIG, 3 TAB, 23 REF.

DESCRIPTORS:
*ALGAE, *AQUATIC WEEDS, *NUTRIENTS, BIOASSAY, LIMITING FACTORS, NITROGEN, PHOSPHORUS, RAIN, LAKES, CLADOPHORA, ANALYSIS, PHYTOPLANKTON, CHEMICAL ANALYSIS, DIATOMS, CYANOPHYTA, TURNOVERS, EUTROPHICATION.

IDENTIFIERS:
LAKE MENDOTA(WIS.), LAKE MONONA(WIS.), LAKE WINGRA(WIS.), LIMITING NUTRIENTS, SURPLUS NUTRIENTS, NUTRIENT SOURCES.

ABSTRACT:
SOURCES OF NITROGEN AND PHOSPHORUS AVAILABLE TO ALGAE AND AQUATIC WEEDS AND THEIR NUTRITIONAL STATUS WERE DETERMINED BY A SIMPLE BIOASSAY PROCEDURE. THE ALGAE AND AQUATIC WEEDS WERE COLLECTED FROM LAKE MENDOTA'S SOUTH SHORE, THE WEST BAY OF LAKE MONONA, OR FROM LAKE WINGRA, ALL IN MADISON, WISCONSIN WITHIN ABOUT ONE MILE OF EACH OTHER, ALTHOUGH VARYING WIDELY. RAIN COULD BE THE MAJOR SOURCE OF AVAILABLE NITROGEN TO CLADOPHORA IN LAKE MENDOTA DURING SUMMER; INCREASES IN PHOSPHORUS ASSOCIATED WITH RAINFALLS WERE DETECTED BUT WERE NOT AS DRAMATIC AS NITROGEN INCREASES. MIXED BLOOMS OF PLANKTONIC ALGAE DO NOT SHARE THEIR NUTRIENTS WITH OTHER ALGAE EVEN WHEN ONE SPECIES MAY HAVE A SURPLUS AND ANOTHER IS NUTRIENT-LIMITED. IN COMPARING SURFACE AND SUBSURFACE PHYTOPLANKTON, AT CERTAIN TIMES SURFACE PLANKTON CAN BE NUTRIENT LIMITED WHILE THE SAME SPECIES FROM SUBSURFACE SOURCES HAS ADEQUATE OR SURPLUS NUTRIENTS. FALL OVERTURN AS A NUTRIENT SOURCE DEMONSTRATES HOW SIMILAR CHANGES CAN TAKE PLACE IN THE NUTRITION OF DIFFERENT ALGAL TYPES. FOR ACCURATE EVALUATION OF THE NUTRIENT LEVEL OF LAKES, BIOASSAYS OF SURFACE WATERS SHOULD BE CONTRASTED WITH ANALYSES OF WATERS FROM THE THERMOCLINE OR HYPOLIMNION AND WITH WATERS OBTAINED IN THE SPRING BEFORE MAXIMUM PLANT PRODUCTION. (AUEN-WISCONSIN)

FIELD 05C, 02H

ACCESSION NO. W73-00232

THE IMPORTANCE OF ALGAL CULTURES FOR THE ASSESSMENT OF THE EUTROPHICATION OF THE OSLOFJORD,

NORWEGIAN INST. FOR WATER RESEARCH, OSLO.

O. M. SKULBERG.

HELGOLANDER WISSENSCHAFTLICHE MEERESUNTERSUCHUNGEN, VOL 20, P 111-125, 1970. 10 FIG, 3 TAB, 30 REF.

DESCRIPTORS:
*ALGAE, *CULTURES, *INDICATORS, *EUTROPHICATION, FJORDS, BIOASSAY, CHLORELLA, NUTRIENTS, WATER SAMPLING, ORGANIC MATTER, WATER POLLUTION EFFECTS, SEWAGE, NITROGEN, PHOSPHORUS, IRON, PHYTOPLANKTON, DIATOMS, DINOFLAGELLATES.

IDENTIFIERS:
*OSLOFJORD(NORWAY).

ABSTRACT:
THE RELATIVE SIGNIFICANCE IN EUTROPHICATION OF THE DIFFERENT CONTRIBUTING WATERS OF THE OSLOFJORD SEA WATER, RUNOFF WATER FROM CATCHMENT AREAS, BOTTOM WATERS OF THE FJORD, AND SEWAGE, AND OTHER POLLUTING MATTER OF URBAN ORIGIN, ARE EVALUATED BY APPLYING ALGAL CULTURE ASSAYS. IN 1962 TO 1965 UNIALGAL CULTURES OF TEST ALGAE, INCLUDING SELENASTRUM CAPRICORNUTUM, CHLORELLA OVALIS, SKELETONEMA COSTATUM, AND PHAEODACTYLUM TRICORNUTUM WERE USED FOR BIOASSAYS. THE AMOUNTS AND AVAILABILITY OF PLANT NUTRIENTS IN THE WATER ARE THUS MEASURED AND THE QUALITY OF THE WATER FOR ALGAL GROWTH DETERMINED. THE RESULTS INDICATED THAT THE WATER OF THE INNER PART OF THE FJORD WAS IN A CONDITION WHERE A SIGNIFICANT INCREASE IN SEWAGE LOAD GAVE A RELATIVELY SMALL INCREASE IN ALGAL GROWTH. CONVERSELY, IN WATER SAMPLES FROM THE OUTER OSLOFJORD, A SMALL ADDITION OF SEWAGE WOULD GIVE RELATIVELY HIGH INCREASE IN GROWTH. IT WAS SHOWN, IN ENRICHMENT EXPERIMENTS WITH ESSENTIAL NUTRIENTS, THAT THE SUPPLY OF COMPOUNDS OF NITROGEN, PHOSPHORUS, AND IRON WAS IMPORTANT FOR THE RESULTING YIELD OF ALGAE. THE MAIN FERTILIZING EFFECT COULD NOT BE ASCRIBED TO ANY SINGLE COMPONENT. (JONES-WISCONSIN)

FIELD 05C

ACCESSION NO. W73-00234

ECOLOGICAL STUDIES ON DISSOLVED OXYGEN AND BLOOM OF MICROCYSTIS IN LAKE SUWA--I. HORIZONTAL DISTRIBUTION OF DISSOLVED OXYGEN IN RELATION TO DRIFTING OF MICROCYSTIS BY WIND,

SHINSHU UNIV., SUWA (JAPAN). SUWA HYDROBIOLOGICAL STATION.

H. YAMAGISHI, AND K. AOYAMA.

BULLETIN OF THE JAPANESE SOCIETY OF SCIENTIFIC FISHERIES, VOL 38, NO 1, P 9-16, 1972. 6 FIG, 1 TAB, 7 REF.

DESCRIPTORS:
*CYANOPHYTA, *DISSOLVED OXYGEN, *EUTROPHICATION, *DISTRIBUTION, FLOATING, WINDS, ALGAE, SURFACE WATERS, HYDROGEN ION CONCENTRATION, LIGHT PENETRATION, PHOTOSYNTHESIS, PHYTOPLANKTON, STRATIFICATION, DIATOMS, CURRENTS(WATER), FISH FARMING.

IDENTIFIERS:
*MICROCYSTIS, *LAKE SUWA(JAPAN).

ABSTRACT:
THE SEVERE WATER-BLOOM OF BLUE-GREEN ALGAE DOMINATED BY MICROCYSTIS APPEARS ALMOST EVERY SUMMER AND SUMMER KILL OFTEN ATTACKS THE CARP CULTURED IN FLOATING NETS. OBSERVATIONS WERE MADE ON HORIZONTAL DISTRIBUTION OF DISSOLVED OXYGEN IN THE SURFACE WATER ACROSS LAKE SUWA, JAPAN IN RELATION TO WIND INDUCED DRIFT OF MICROCYSTIS. TEMPERATURE, PH, AND CELL NUMBER OF MICROCYSTIS IN THE SURFACE WATER TOGETHER WITH TRANSPARENCY AND WIND DIRECTION WERE RECORDED. THE STUDIES DEMONSTRATED THAT DISSOLVED OXYGEN IN THE LAKE INCREASED FROM THE WINDWARD TO THE LEEWARD WITH CONCOMITANT INCREASES IN THE DENSITY OF DRIFTING MICROCYSTIS AND PH VALUES. OXYGEN WAS MARKEDLY UNDERSATURATED ON THE WINDWARD (LESS THAN 50% SATURATION), BUT EXTREMELY SUPERSATURATED ON THE LEEWARD (MORE THAN 250% SATURATION). CELL NUMBERS OF MICROCYSTIS REACHED 28 MILLION/ML AT AREAS OF HIGHEST DENSITY. LABORATORY TESTS AND IN SITU OBSERVATIONS INDICATE THAT A VERY THICK LAYER OF MICROCYSTIS IN THE TOPMOST WATER OF THE LAKE INHIBIT PHOTOSYNTHESIS OF UNDERLYING ALGAE BY SHADING BUT THE REASON FOR THE CONSTANT PH VALUE BEFORE AND AFTER ILLUMINATION REMAINED UNKNOWN. (JONES-WISCONSIN)

FIELD 05C, 02H

ACCESSION NO. W73-00236

ALGAE AS INDICATORS OF PESTICIDE,

STATE UNIV., COLL., BUFFALC, N.Y. GREAT LAKES LAB.

R. A, SWEENEY.

SPECIAL REPORT NO. 4, FEBRUARY 1970. 10 P. 1 FIG, 28 REF. (PRESENTED AT 'ALGAE AS INDICATIONS' SYMPOSIUM OF THE PHYCOLOGICAL SOCIETY OF AMERICA, SEPTEMBER 4, 1968, OHIO STATE UNIVERSITY, COLUMBUS).

DESCRIPTORS:
*ALGAE, *INDICATORS, *PESTICIDES, CHLORINATED HYDROCARBON PESTICIDES, INSECTICIDES, RODENTICIDES, HERBICIDES, FUNGICIDES, ALGICIDES, CARBAMATE PESTICIDES, ORGANOPHOSPHORUS PESTICIDES, ACARICIDES, FOOD CHAINS, WATER POLLUTION SOURCES.

ABSTRACT:
THE NATURE AND EXTENT OF PESTICIDE CONTAMINATION OF AQUATIC HABITATS IN THE UNITED STATES ARE REVIEWED. A ROUTINE SAMPLING PROGRAM FOR PESTICIDES HAS BEEN INITIATED AT APPROXIMATELY 10C SURVEILLANCE STATIONS ON THE MAJOR RIVER BASINS WITHIN THE UNITED STATES. THIS IS CONDUCTED THROUGH COORDINATED AND COOPERATIVE EFFCRTS OF FEDERAL, STATE, LOCAL, AND PRIVATE AGENCIES. SPECIAL ATTENTION HAS BEEN GIVEN TO ALDRIN, BENZENE HEXACHLORIDE, DIELDRIN, ENDRIN, HEPTACHLOR, HEPTACHLOR EPOXIDE, TOXAPHENE, AND DDT AND ITS DERIVATIVES--DDE AND DDD, ALL CHARACTERIZEC BY LONG RESIDUAL ACTION. SINCE ALGAE RAPIDLY ACCUMULATE NUMEROUS PESTICIDES AND CCMPRISE THE BASE OF MCST AQUATIC TROPHIC PYRAMIDS, THE ROUTINE COLLECTION AND ANALYSIS OF ALGAE SHOULD BE PART OF OUR NATIONWIDE MONITORING SYSTEM AND SHOULD ENLIST THE ASSISTANCE OF ALGAL TAXONOMISTS. THIS WOULD GIVE A MORE MEANINGFUL INDICATION OF THE THREAT OF PESTICIDE CONTAMINATION OF THE HIGHER AQUATIC ORGANISMS AND MAN. BIOLOGICAL DETERMINATIONS OF PESTICIDE UPTAKE ARE NECESSARY AS A CHECK BECAUSE CHEMICAL ANALYSES, USING THIN LAYER AND GAS CHROMATOGRAPHY, ARE NOT ALWAYS RELIABLE. ALGAE WHICH ACCUMULATE PESTICIDES CAN BE EMPLOYED IN THE COLLECTION, IDENTIFICATION, AND QUANTIFICATION OF PESTICIDES. (JONES-WISCONSIN)

FIELD 05C, 05B

ACCESSION NO. W73-00238

THE EFFECTS OF INCREASING LIGHT AND TEMPERATURE CN THE STRUCTURE OF DIATOM COMMUNITIES,

ACADEMY OF NATURAL SCIENCES OF PHILADELPHIA, PA. DEPT. CF LIMNOLCGY.

R. PATRICK.

LIMNOLOGY AND OCEANOGRAPHY, VOL 16, NO 2, P 405-421, MARCH 1971. 1 FIG, 4 TAB, 17 REF.

DESCRIPTORS:
*LIGHT, *TEMPERATURE, *DIATOMS, BIOLOGICAL COMMUNITIES, SYSTEMATICS, CYANOPHYTA, CHEMICAL ANALYSIS, SEASONAL, BIOMASS, ON-SITE INVESTIGATIONS, ALGAE.

IDENTIFIERS:
GOMPHONEMA OLIVACEOIDES, WHITE CLAY CREEK(PA.).

ABSTRACT:
THE DIFFERENCES IN TEMPERATURE OR LIGHT REQUIREMENTS, OR BOTH, APPARENTLY INFLUENCE SEASONAL SUCCESSION OF SPECIES. THE PROBLEM ARISES WHETHER RAISING THE TEMPERATURE A FEW DEGREES NEAR THE ENDS OF THEIR RANGE OF TOLERANCE HAS A GREATER EFFECT ON DIATOM COMMUNITIES THAN RAISING THE TEMPERATURE ABCVE AMBIENT IN THE MIDDLE PCRTION OF THE TOLERANCE RANGE. EFFECTS ON COMMUNITY STRUCTURE OF VARIOUS DAY LENGTHS AND OF ARTIFICIAL TEMPERATURE INCREASE COMPARED WITH NATURAL TEMPERATURE INCREASE WERE STUDIED. SUMMARY OF CHEMICAL ANALYSIS OF THE STREAM, THE CHARACTERISTICS OF THE DIATOM COMMUNITIES AND THE PERCENTAGE OCCURRENCE OF THE MORE COMMON SPECIES ARE TABULATED. NATURALLY INCREASING DAY LENGTH WAS MORE FAVORABLE FOR COMMUNITY DEVELOPMENT THAN INCREASING DAY LENGTH BY ARTIFICIAL LIGHT. INCREASING TEMPERATURE IS MOST BENEFICIAL WHEN TEMPERATURES ARE NEAR OC. MOVING AWAY FROM THE TOLERANCE LIMITS AT EITHER END OF THE RANGE PRODUCED THE GREATEST CHANGES IN DIATCM COMMUNITY STRUCTURE. INTERMEDIATE CHANGES NEAR THE OPTIMUM RANGE PRODUCED LESS PREDICTABLE RESULTS. ONE NEW TAXON NAMED GOMPHONEMA OLIVACEOIDES VAR. HUTCHINSONIANA VAR. NOV., IS DESCRIBED. (JONES-WISCONSIN)

FIELD 05C

ACCESSION NO. W73-00242

COMPARATIVE STUDIES ON ALGAL TOXINS,

NEW HAMPSHIRE UNIV., DURHAM. DEPT. OF ZOOLOGY; AND NEW HAMPSHIRE UNIV., DURHAM. JACKSON ESTUARINE LAB.

J. J. SASNER, JR.

(1971), 98 P. 13 FIG., 3 TAB., 121 REF. OWRR A-021-NH (1) AND A-013-NH (4).

DESCRIPTORS:
*ANALYTICAL TECHNIQUES, *ALGAL TOXINS, AQUATIC PLANTS, TOXINS, RED TIDE, PLANT PHYSIOLOGY, BIOASSAY, CYTOLOGICAL STUDIES, DINOFLAGELLATES, CYANOPHYTA.

IDENTIFIERS:
*TOXICOLOGY, BIOTOXINS, PHARMACOLOGY.

ABSTRACT:
CHANGES WERE DEVELOPED IN THE THEORIES CONCERNING THE IONIC BASIS OF CELL ELECTROGENESIS AND JUNCTIONAL TRANSMISSION BETWEEN CELLS WHICH TOXICOLOGISTS HAVE USED TO STUDY SITES AND MODES OF ACTION OF AQUATIC POISONS. THE COMPARATIVE PHYSIOLOGIST MAY FIND BIOTOXINS USEFUL IN DETERMINING SIMILARITIES AND DIFFERENCES IN ELECTROGENIC AND TRANSMITTER MECHANISMS IN A WIDE VARIETY OF ORGANIC SYSTEMS. TETRODOTOXIN HAS ALREADY RECEIVED WIDE USAGE IN BLOCKING NA+ CONDUCTANCE ACROSS NERVE AND MUSCLE MEMBRANES; THE EFFECT OF THIS TOXIN IS MORE SPECIFIC THAN PROCAINE AND OTHER ANAESTHETICS WHICH ALSO AFFECT K+ CONDUCTANCE. TO DETERMINE WHETHER DINOFLAGELLATE TOXINS THAT CAUSE GENERAL MEMBRANE DEPOLIRIZATION PRODUCE THEIR EFFECTS BY ALTERING MEMBRANE PERMEABILITY OR BY INACTIVATING THE IONIC PUMP ACTIVE TRANSPORT MECHANISM AND WHETHER DEPOLARIZING TOXINS CAN ACT AS METABOLIC POISONS, OR CAN PROVEN DEPOLARIZING EFFECTS BE ANTAGONIZED BY TETRODOTOXIN SAXITOXIN OR THEIR ANALOGUES WHICH BLCC ICNIC CHANNELS WILL BE OF INTEREST. THE APPARENT IMMUNITY TO POTENT BIOTOXINS OF TRANSVECTORS WHO APPARENTLY ACT AS 'BIOLOGICAL STORAGE DEPOTS' FOR ORGANISMS HIGHER IN THE FOOD CHAIN IS OF EQUAL INTEREST. IT IS CLEAR THE FLAGELLATE AND BLUE-GREEN ALGAL TOXINS AFFECT IONIC MECHANISMS OF ELECTROGENESIS ASSOCIATED WITH THE MEMBRANE AND THE TRANSMITTER SYSTEM EXERTING CHEMICAL CONTROL BETWEEN CONTIGUOUS CELLS. (AUEN-WISCONSIN)

FIELD 05C

ACCESSION NO. W73-00243

APHANIZOMENON FLOS-AQUAE: INFECTION BY CYANOPHAGES,

LANTBRUKHOGSKOLAN, UPPSALA (SWEDEN). DEPT. OF MICROBIOLOGY.

U. GRANHALL.

PHYSIOLOGIA PLANTARUM, VOL 26, P 332-337, 1972. 9 FIG, 18 REF.

DESCRIPTORS:
*ALGAL CONTROL, *INFECTION, *VIRUSES, CYANOPHYTA.

IDENTIFIERS:
*CYANOPHAGES, *APHANIZOMENON FLOS-AQUAE, LAKE ERKEN(SWEDEN).

ABSTRACT:
NITROGEN-FIXING BLUE-GREEN ALGAE WERE STUDIED IN LAKE ERKEN, SWEDEN DURING THE SUMMER OF 1970. A LYTIC AGENT WAS PRESENT IN LAKE ERKEN AND ELECTRON MICROGRAPHS OF ALGAL SAMPLES SHOWED THAT CYANOPHAGE-LIKE PARTICLES OCCURRED IN LARGE NUMBERS, SPECIFICALLY ASSOCIATED WITH VEGETATIVE CELLS OF APHANIZOMENON FLOS-AQUAE. ELECTRON MICROSCOPY, STERILE FILTRATIONS (INCLUDING STEPS TO EXCLUDE BACTERIA) AND PLAQUE TECHNIQUES WERE USED TO CORRELATE THE VIRUS FLUCTUATIONS WITH THE WATER-BLOOM AND DEGRADATION OF APHANIZOMENON. ONLY VEGETATIVE CELLS OF APHANIZOMENON WERE LYSED, NOT HETEROCYSTS AND AKINETES. INFECTED CELLS OF APHANIZOMENON AND EXTRACELLULAR VIRIONS COULD NOT BE DETECTED BY ELECTRON MICROSCOPY IN LAKE SAMPLES UNTIL THE LATE STAGE OF THE WATER-BLOOM; THE HOST SHOWED NORMAL ULTRASTRUCTURE UP TO THE PERIOD OF MOST EXTENSIVE WATER-BLOOM. ALGAL SAMPLES, TAKEN SOME WEEKS LATER SHOWED QUITE DIFFERENT ULTRASTRUCTURES DUE TO CELL INFECTION BY VIRUS PARTICLES. IT IS SUGGESTED THAT A CYANOPHAGE NAMED AP-1 REGULATES TERMINATION OF THE A. FLOS-AQUAE BLOOM. IF THE ISOLATED VIRUSES COULD BE USED IN ENVIRONMENTAL CONTROL AFTER PROPAGATION IN THE LABORATORY IT WOULD CONSTITUTE A NEW APPROACH IN CONTROLLING APHANIZOMENON BLOOMS. (JONES-WISCONSIN)

FIELD 05C

ACCESSION NO. W73-00246

ENDOGENOUS RHYTHM OF THE PRODUCTIVITY IN CHLORELLA ANC THE INFLUENCE OF LIGHT (ENDOGENE RHYTHMIK DER PRODUKTION-SFAHIGKEIT BEI CHORELLA UND IHRE BEEINFLUSSUNG DURCH LICHT),

GOETTINGEN UNIV. (WEST GERMANY). INST. FOR PLANT PHYSIOLOGY.

M. HESSE.

ZEITSCHRIFT FUR PFLANZENPHYSIOLOGIE, VOL. 67, P 58-77, 1972. 12 FIG, 2 TAB, 32 REF. ENGLISH SUMMARY.

DESCRIPTORS:
*BIORTHYMS, *METABOLISM, *CHLORELLA, *LIGHT, ALGAE, GROWTH RATES, CHLOROPHYLL.

IDENTIFIERS:
*CHLORELLA PYRENOIDOSA.

ABSTRACT:
SINCE CHLORELLA IS UNIVERSALLY USED FOR ALGAL RESEARCH, CHARACTERISTIC PHYSIOLOGICAL CHANGES WERE STUDIED. UNDER 54 HOURS OF CONSTANT DARK CONDITIONS THE AUTOSPORES OF CHLORELLA PYRENOIDOSA STRAIN 211-8B WERE FOLLOWED BY MEASURING DRY-WEIGHT, CARBOHYDRATE, PROTEIN, CHLOROPHYLL, AND CELL NUMBER. THE CHANGES IN ACTIVITY ARE ENDOGENOUS AND OF A CIRCADIAN NATURE AND ARE INDEPENDENT OF THE DEVELOPMENTAL STAGE OF THE CELLS. AFTER REMAINING IN CONTINUOUS DIM WHITE LIGHT DURING 'WAITING' THE RHYTHM COULD NOT BE DEMONSTRATED ANC THE PHYSIOLOGICAL ACTIVITY WAS MINIMUM; TRANSFER FROM DIM LIGHT TO CONTINUOUS DARKNESS RESTORED THE ENDOGENOUS RHYTHM. THIS CHANGE INITIATES A RHYTHM IN THE INDIVIDUAL CELL AND IS NOT A SYNCHRONIZING FACTOR. WHEN THE DARK PERIOD OF THE LIGHT-DARK-CHANGE IS REPLACED BY DIM LIGHT, MINIMUM PRODUCTION FOLLOWS. THIS IS ONE REASON WHY THE PRODUCTIVITY PER LIGHT-HOUR IN CONTINUOUS LIGHT IS NOT AS HIGH AS UNDER LIGHT-DARK-CHANGE. APPEARANCE OF MAXIMUM AND MINIMUM PRODUCTICN AFTER DIFFERENT INTERVALS IS DUE TO THE DIFFERENCE IN GROWTH RATE WHICH OCCURS ABOUT 8 HOURS AFTER THE START OF LIGHT. SIMILARLY THERE ARE LARGE DIFFERENCES IN THE RATE OF INCREASE OF PROTEIN AND CARBOHYDRATE PRODUCTION AT THE 2ND AND 4TH HOURS AFTER START OF LIGHT DEPENDING ON THE CURATION OF 'WAITING' RECEIVED BY THE AUTOSPORES. (JONES-WISCONSIN)

FIELD 05C

ACCESSION NO. W73-00251

THE INTERLABORATORY PRECISION TEST. AN EIGHT LABORATCRY EVALUATION OF THE PROVISIONAL ALGAL ASSAY PROCEDURE BOTTLE TEST,

NORTH CAROLINA UNIV., CHAPEL HILL. DEPT. OF ENVIRONMENTAL SCIENCES AND ENGINEERING.

C. M. WEISS, AND R. W. HELMS.

ENVIRONMENTAL PROTECTION AGENCY, NATIONAL EUTROPHICATION RESEARCH PROGRAM, OCTOBER 1971. 70 P, 10 FIG, 16 TAB, 3 APPEND. EPA PROGRAM 16010 DQT.

DESCRIPTORS:
*ANALYTICAL TECHNIQUES, *ALGAE, EVALUATION, TESTING, ASSAY, NUTRIENTS, CULTURES, MEASUREMENT, LABORATORY TESTS, BIOASSAY.

IDENTIFIERS:
*PAAP TEST, *ALGAL GROWTH, BOTTLE TEST.

ABSTRACT:
A COMPARATIVE EXPERIMENT TO EVALUATE THE 'BOTTLE TEST' OF THE PROVISIONAL ALGAL ASSAY PROCEDURE BY EIGHT LABORATORIES REQUIRED ALL PARTICIPANTS TO FOLLOW A SET OF PRESCRIBED CONDITIONS AND PROCEDURES USING COMMON MATERIALS AND CULTURES WITH THEIR RESULTS BEING SUBJECTED TO STATISTICAL ANALYSIS FOR THE PURPOSES OF ANSWERING THE QUESTION OF WHAT IS THE INHERENT VARIABILITY THAT MIGHT BE EXPECTED IN AN ALGAL ASSAY BOTTLE TEST. ALL LABORATORIES PREPARED THEIR TEST MEDIA FROM A COMMON BATCH OF REAGENTS SUPPLIED BY ONE LABORATCRY AND ALL USED FRESH SUBCULTURES OF SELENASTRUM CAPRICORNUTUM OBTAINED FROM THE NATIONAL EUTROPHICATION PROGRAM. A REGRESSION ANALYSIS FOR EACH OF THE LABORATORIES, INDIVIDUALLY AND FOR ALL EIGHT COMBINED, WAS CARRIED OUT. IN THE FINAL ASSESSMENT THE RESULTS OF ALL EIGHT LABORATORIES CAN BE CONSIDERED AS A TEST UNIT AND THE DEGREE OF VARIATION OR COEFFICIENT OF VARIATION COMPUTED TO ESTABLISH THE EXPECTED ORDER OF PRECISION FOR THIS TYPE OF ASSAY. THE DATA ON THE AVERAGE VALUES AND COEFFICIENT OF VARIATION OF THE MEAN INDICATED A CERTAIN CEGREE OF CONSISTENCY AMONG VARIOUS MEASURING PARAMETERS AS WELL AS WHICH WOULD BE PREFERABLE. (JONES-WISCONSIN)

FIELD 05C

ACCESSION NO. W73-00255

ECOLCGY OF PHYTOPLANKTON OF THE VOLTA LAKE,

GHANA UNIV., LEGON. VOLTA BASIN RESEARCH PROJECT.

S. BISWAS.

HYDROBIOLOGIA, VOL 39, NO 2, P 277-288, 1972. 8 FIG, 6 TAB, 19 REF.

DESCRIPTORS:
*PHYTOPLANKTON, *ALGAE, *DISSOLVED OXYGEN, PHOTOSYNTHESIS, TURBIDITY, LIGHT PENETRATION, IMPOUNDMENTS, SEASONAL, COLOR, IRCN, PHYSICOCHEMICAL PROPERTIES, DIATOMS, DOMINANT ORGANISMS.

IDENTIFIERS:
*VOLTA LAKE(GHANA), ACTINASTRUM GRACILIMUM, CRYPTOMONAS EROSA, PERIDINIUM AFRICANUM, NITZCHIA ACICULARIS, SYNEDRA ACUS.

ABSTRACT:
THE CHANGES WHICH TOOK PLACE IN THE PHYSICOCHEMICAL CONDITIONS IN A NEW IMPOUNDMENT OF THE VOLTA RIVER, GHANA ARE EXAMINED IN RELATION TO THE PHYTOPLANKTON. A CORRELATION BETWEEN THE DISSOLVED OXYGEN AND PHYTOPLANKTON, NOT OBSERVED BEFORE, WAS FOUND. INITIALLY FILLED WITH OPALESCENT BROWN FLOOD WATER RICH IN IRON, A DECREASE IN TRANSPARENCY LOWERED THE PHOTOSYNTHETIC ACTIVITY OF THE PHYTOPLANKTON. DATA FROM THE LOWER REACHES OF THE VOLTA LAKE ON TRANSPARENCY, COLOR, TOTAL IRON, AND DISSOLVED OXYGEN REVEALED SIGNIFICANT CHANGES FROM 1965 TO 1966, BUT NOT FROM 1966 TO 1967 THOUGH A COMMON SEASCNAL PATTERN WAS EVIDENT. THESE FACTORS SHOWED CORRELATIONS OF SUCH SIGNIFICANCE THAT MANY OF THEM COULD BE ESTIMATED FROM A SIMPLE MEASUREMENT OF THE TRANSPARENCY ALONE. THE PHYTOPLANKTON DENSITY DID NOT CHANGE SIGNIFICANTLY EITHER FROM 1965 TO 1966 OR FROM 1966 TO 1967. THE CHIEF CONSTITUENTS, HOWEVER, PASSED THROUGH A SERIES OF CHANGES FROM A GREEN ALGA (ACTINASTRUM) IN EARLY 1965 TO THE FLAGELLATES (CRYPTOMONAS AND PERIDINIUM) DURING 1965 TO 1966, TO THE DIATOMS (NITZSCHIA AND SYNEDRA) DURING 1966 TO 1967. THE PHYTOPLANKTON DENSITY CCRRELATED SIGNIFICANTLY ONLY WITH THE DISSOLVED OXYGEN FROM WHICH IT COULD BE ROUGHLY ESTIMATED. (JONES-WISCONSIN)

FIELD 05C, 02H

ACCESSION NO. W73-00256

A SLIDING-CHAMBER PHYTOPLANKTON SETTLING TECHNIQUE FOR MAKING PERMANENT QUANTITATIVE SLIDES WITH APPLICATIONS IN FLUORESCENT MICROSCOPY AND AUTORADIOGRAPHY,

ALASKA UNIV., COLLEGE. INST. OF MARINE SCIENCE.

C. COULON, AND V. ALEXANDER.

LIMNOLOGY AND OCEANOGRAPHY, VOL 17, NO 1, P 149-152, JANUARY 1972. 3 FIG, 1 TAB, 4 REF.

DESCRIPTORS:
*PHYTOPLANKTON, LABORATORY EQUIPMENT, SEDIMENTATICN, METHCDOLOGY, POLLUTANT IDENTIFICATICN, PROTOZOA, CHRYSOPHYTA, CHLOROPHYTA, AQUATIC ALGAE, MARINE ALGAE, SESSILE ALGAE.

IDENTIFIERS:
*FLUORESCENT MICRCSCOPY, *AUTORACIOGRAPHY, GLUTARALDEHYDE, SAMPLE PREPARATION, COUNTING, SAMPLE PRESERVATION, FLAGELLATES, RHODOMONAS, IKROAVICH LAKE, UTERMCHL METHOD.

ABSTRACT:
A SYSTEM HAS BEEN DEVISED FOR PREPARING PERMANENT QUANTITATIVE PHYTOPLANKTON SLIDES USING SEDIMENTATION TECHNIQUES. THESE SLICES CAN BE USED WITH A CONVENTIONAL MICROSCOPE. A THREE-PART SETTLING CHAMBER ALLOWS SEPARATION OF THE SUPERNATANT FROM THE SETTLED MATERIAL. WATER-SOLUBLE MOUNTING MEDIUM ADDED TO THE SAMPLE BEFORE SETTLING PREVENTS DESICCATION AND DISTORTION OF THE ORGANISMS. THE PIGMENT-PRESERVING AND MOUNTING-MEDIUM QUALITIES OF GLUTARALCEHYDE COMBINED WITH THIS METHOD HAVE APPLICATIONS IN PLANKTON COUNTING AND AUTORADIOGRAPHY. (LONG-BATTELLE)

FIELD 05A

ACCESSION NO. W73-00273

RESISTANCE TO DDT OF A FRESHWATER ALGA,

OBERLIN COLL., OHIO. DEPT. OF BIOLOGY.

D. A. EGLOFF, AND R. PARTRIDGE.

THE OHIO JOURNAL OF SCIENCE, VOL 72, NO 1, P 6-10, JANUARY 6, 1972. 1 FIG, 1 TAB, 18 REF.

DESCRIPTORS:
DESCRIPTORS:*DDT, *RESISTANCE, *PLANT PHYSIOLOGY, CULTURES, PHOTOSYNTHESIS, AQUATIC ALGAE, CHLOROPHYLL, OXYGEN, RESPIRATION, CHLORINATED HYDROCARBON PESTICIDES, GROWTH RATES, ENVIRONMENTAL EFFECTS, PHYTOPLANKTON, CHLAMYDOMONAS, CHLOROPHYTA, WATER POLLUTION EFFECTS, ABSORPTION, PESTICIDE TOXICITY, LABORATORY TESTS.

IDENTIFIERS:
*CHLAMYDOMONAS REINHARDTII, CULTURE MEDIA, CHLORINATED HYDROCARBONS.

ABSTRACT:
LABORATORY CULTURES OF A FRESHWATER ALGA (CHLAMYDOMONAS REINHARDTII) WERE EXPOSED TO DDT(100 - 1,000 PPB) FOR 16 - 96 HOURS AT 18 - 22 C IN AN INORGANIC BASAL MEDIUM WITH AND WITHOUT ACETATE TO DETERMINE ITS EFFECTS ON GROWTH PHOTOSYNTHESIS, RESPIRATION, AND CHLOROPHYLL CONCENTRATION. EXPERIMENTAL DATA SHOWED THAT THE AMOUNT OF CHLOROPHYLL, OXYGEN EVOLUTION IN THE LIGHT, AND OXYGEN UPTAKE IN THE DARK WERE UNAFFECTED BY EXPOSURE TO DDT. GROWTH RATE AND FINAL CELL DENSITY WERE IDENTICAL IN CONTROL AND EXPERIMENTAL CULTURES EXPOSED TO 1000 PPB DDT IN THE GROWTH MEDIUM FOR NINE DAYS AT 22 C. (SNYDER-BATTELLE)

FIELD 05C

ACCESSION NO. W73-00281

NORTH CAROLINA MARINE ALGAE. I. THREE NEW SPECIES FROM THE CONTINENTAL SHELF,

DUKE UNIV., DURHAM, N.C. DEPT. OF BOTANY.

R. B. SEARLES.

PHYCOLOGIA, VOL 11, NO 1, P 19-24, MARCH 1972. 4 FIG, 10 REF.

DESCRIPTORS:
*DISTRIBUTION PATTERNS, *CHLOROPHYTA, *MARINE ALGAE, *RHODOPHYTA, *SYSTEMATICS, PHOTOGRAPHY, SEA WATER, *NORTH CAROLINA, MICROSCOPY, ATLANTIC OCEAN.

IDENTIFIERS:
TREMATOCARPUS PAPENFUSSII, GLOIODERMA ATLANTICA, CODIUM CAROLINIANUM, *ONSLOW BAY(NC).

ABSTRACT:
THREE NEW SPECIES OF MARINE ALGAE WERE IDENTIFIED FROM COLLECTIONS MADE FROM THE R/V EASTWARD USING A CERAME VIVAS ROCK DREDGE OR A CAPE TOWN DREDGE IN ONSLOW BAY, NORTH CAROLINA. WHOLE PLANTS WHICH HAD BEEN PRESERVED IN FORMALIN SEAWATER WERE PHOTOGRAPHED WHILE IMMERSED IN WATER. RED ALGAE WERE STAINED WITH AQUEOUS ANILINE BLUE DYE FOR MICROSCOPIC STUDY; DRAWINGS AND PHOTOGRAPHS OF THEM EMPHASIZE CYTOPLASMIC BOUNDARIES RATHER THAN CELL WALLS. DRAWINGS WERE MADE WITH THE AID OF A CAMERA LUCIDA. THE THREE SPECIES WERE: ONE GREEN ALGA, CODIUM CAROLINIANUM SP. NOV. OF THE CODIALES, AND TWO RED ALGAE, TREMATOCARPUS PAPENFUSSII SP. NOV. OF THE GIGARTINALES AND GLOIODERMA ATLANTICA SP. NOV. OF THE RHODYMENIALES. THE GENUS GLOIODERMA HAS NOT PREVIOUSLY BEEN REPORTED FROM THE ATLANTIC OCEAN. ALL THREE PLANTS GROW OFFSHORE ON THE NORTH CAROLINA CONTINENTAL SHELF. (LONG-BATTELLE)

FIELD 05A

ACCESSION NO. W73-00284

BORON IN SWEDISH AND NORWEGIAN FRESH WATERS,

UPPSALA UNIV. (SWEDEN). INST. OF LIMNOLOGY.

TH. AHL, AND E. JONSSON.

AMBIO, VOL 1, NO 2, P 66-70, APRIL 1972. 4 FIG, 6 TAB, 10 REF.

DESCRIPTORS:
FRESHWATER, *BORON, *PATH OF POLLUTANTS, METHODOLOGY, RIVERS, WATER POLLUTION SOURCES, ION EXCHANGE, *CHEMICAL ANALYSIS, WATER POLLUTION, WATER ANALYSIS, BIOASSAY, CYANOPHYTA, RESINS, AQUATIC PLANTS, AQUATIC ALGAE, NITRATES, NITRITES, EUTROPHICATION.

IDENTIFIERS:
SAMPLE PREPARATION, CHEMICAL INTERFERENCE, INTERLABORATORY STUDIES, PRECISION, *SWEDEN, LAKE MALAREN, MICROCYSTIS FLOS-AQUAE, PHRAGMITES COMMUNIS, *NORWAY, MACROPHYTES, FYRISAN RIVER, ARBOGAAN RIVER, BOTORPSSTROMMEN RIVER, ATRAN RIVER, MOTALA STROM RIVER, SPARGANIUM ERECTUM, SCHOENOPLECTUS LACUSTRIS, ACORUS CALAMUS, TYPHA ANGUSTIFOLIA.

ABSTRACT:
MONTHLY WATER SAMPLES WERE COLLECTED FROM SWEDISH AND NORWEGIAN RIVERS AND ANALYZED BY A MODIFIED CURCUMIN METHOD. SAMPLES WERE PASSED THROUGH A SODIUM-LOADED DOWEX 50-W-X4 COLUMN AND THEN CENTRIFUGED 5 MIN BEFORE ABSORPTION MEASUREMENT TO AVOID INTERFERENCE FROM PRECIPITATES. A LINEAR RELATIONSHIP WAS OBTAINED BETWEEN ABSORPTION AND CONCENTRATION IN THE RANGE 0-100 MICROGRAMS/L. TESTS SHOWED INTERFERENCE STILL OCCURRING FROM NITRATE AND NITRITE SO SAMPLES HIGH IN THESE COMPOUNDS WERE NO LONGER TESTED. THE BORON CONTENT IN THE RIVERS RANGED FROM 1 TO 1,046 MICROGRAMS B/L, WITH A MEDIAN VALUE FOR ALL RIVERS TESTED AT 13 MICROGRAMS/L. ANY REGIONAL DIFFERENCES WERE ATTRIBUTED TO DIFFERENCES IN GEOLOGY, LAND USE, AND POPULATION DENSITY. AQUATIC PLANTS WERE SUBJECTED TO ANALYSIS, AND THE RESULTS SHOWED THAT THE ENRICHMENT OF BORON IN AQUATIC PLANTS IS VERY SMALL COMPARED WITH THE ENRICHMENT OF PHOSPHORUS AND NITROGEN. SEWAGE WATER WAS ANALYZED AND SHOWED A FAR LOWER BORON CONTENT THAN ENGLISH SEWAGE WATER. THE DATA SHOW THAT BORON CONTENT IN THESE FRESHWATERS IS LOW, WITH A MEAN CONCENTRATION CLOSE TO THAT OF THE RIVERS OF THE WORLD. (MACKAN-BATTELLE)

FIELD 05B, 05C

ACCESSION NO. W73-00286

DISTRIBUTION AND ECOLOGY OF CERATIUM EGYPTIACUM HALIM AND ITS VALIDITY AS INDICATOR OF THE CURRENT REGIME IN THE SUEZ CANAL,

ALEXANDRIA UNIV. (EGYPT). DEPT. OF OCEANOGRAPHY.

N. M. DOWIDAR.

INTERNATIONALE REVUE DER GESAMTEN HYDROBIOLOGIE, VOL 56, NO 6, P 957-966, 1971. 3 TAB, 23 REF.

DESCRIPTORS:
*BIOINDICATORS, *ECOLOGICAL DISTRIBUTION, *CURRENTS(WATER), *DINOFLAGELLATES, ECOLOGY, DISTRIBUTION PATTERNS, MARINE ANIMALS, SALINITY, SAMPLING, SEA WATER, BODIES OF WATER, PYRROPHYTA, PROTOZOA, MARINE ALGAE, ZOOPLANKTON, PHYTOPLANKTON.

IDENTIFIERS:
*CERATIUM EGYPTIACUM, *SUEZ CANAL, FLAGELLATES, MEDITERRANEAN SEA, RED SEA, DRIFT ORGANISMS.

ABSTRACT:
CERATIUM EGYPTIACUM, AN ERYTHRAEAN DINOFLAGELLATE SPECIES, WAS RECORDED FOR THE FIRST TIME FROM THE MEDITERRANEAN WATERS IN 1966. THE SPECIES IS INDIGENOUS AND PERENNIAL TO THE RED SEA AND THE SOUTHERN PART OF THE SUEZ CANAL. ECOLOGICALLY THE SPECIES PROVED TO BE A STRICTLY NERITIC SURFACE WATER FORM AVOIDING THE OCEANIC AND DEEP WATERS. IT IS ALSO HIGHLY TOLERANT; ITS SALINITY AND TEMPERATURE RANGES RECORDED IN THIS INVESTIGATION ARE RESPECTIVELY 3.3-4.7 PERCENT AND 14.2 C - 33 C. THE OCCURRENCE OF THE SPECIES IN THE MEDITERRANEAN WATERS SUGGESTS A RECENT IMMIGRATION TO THAT SEA THROUGH THE SUEZ CANAL AND PROVIDES A GOOD EXAMPLE FOR STUDYING THE CURRENT REGIME IN THE CANAL, PARTICULARLY AFTER THE CONSTRUCTION OF THE ASWAN HIGH DAM. THE DISTRIBUTION OF THE SPECIES IN THE MEDITERRANEAN IS DETAILED. THE CURRENT IN THE CANAL IS NORTHWARD THROUGHOUT THE WHOLE YEAR AND IS NOT REVERSED IN AUGUST AND SEPTEMBER AS WAS PREVIOUSLY NORMAL. THE IMPORTANCE OF THE SUEZ CANAL AS A BIOLOGICAL LINK BETWEEN THE MEDITERRANEAN AND THE RED SEA, THE MIGRATION OF RED SEA PLANKTON ORGANISMS THROUGH THE CANAL AND THE FUTURE PROSPECTS OF THE NEW CONDITIONS ARE DISCUSSED. (LONG-BATTELLE)

FIELD 05B, 05C

ACCESSION NO. W73-00296

OCCURRENCE OF ELECTRA CRUSTULENTA (BRYOZOA) IN RELATION TC LIGHT,

STOCKHOLM UNIV. (SWEDEN). ASKO LAB.; AND STOCKHOLM UNIV. (SWEDEN). DEPT. OF ZOOLOGY.

L. SILEN, AND B-O. JANSSON.

OIKOS, VOL 23, NO 1, P 59-62, 1972. 2 FIG, 10 REF.

DESCRIPTORS:
*LIGHT, *GROWTH RATES, *ENVIRONMENTAL EFFECTS, LARVAE, ON-SITE TESTS, INVERTEBRATES, WATER POLLUTION EFFECTS, ECOLOGY, SEA WATER, LABORATORY EQUIPMENT, INHIBITION, ECOLOGICAL DISTRIBUTION, MARINE ALGAE, MARINE ANIMALS.

IDENTIFIERS:
*ELECTRA CRUSTULENTA, *BRYOZCA.

ABSTRACT:
EXPERIMENTS WERE PERFORMED TO RELATE THE OCCURRENCE OF NEW BRYOZOAN (ELECTRA CRUSTULENTA) COLONIES AND THE REACTIONS CF LARVAE TO LIGHT. TO TEST THE EFFECTS OF LIGHT CN BRYOZCAN COLONIES, VARICUS SETTLING EXPERIMENTS WERE CARRIED OUT AT DIFFERENT TIMES OF THE YEAR BY ALLOWING THE INVERTEBRATES TO SETTLE ON A SUBSTRATE HAVING BOTH AN ILLUMINATED AND SHADOWED SURFACE. A BLACK POLYETHENE TUBE WAS USED AS A SUBSTRATE FOR THESE SESSILE ORGANISMS, SINCE ILLUMINATED ORGANISMS LIVING ON THE TUBE'S SURFACE COULD BE DIRECTLY COMPARED TO ORGANISMS ON A PERMANENTLY SHADOWED INTERIOR. PHOTIC RESPONSES OF LARVAE, IN DIFFERENT STAGES OF DEVELOPMENT, WERE EXAMINED BY PLACING THE BRYOZCANS IN SEA WATER (15C) AND EXPOSING THEM TO LIGHT. RESULTS SHOWED THAT BRYOZOANS WERE ABLE TO GROW ON EITHER LIGHT OR DARK SURFACES; HOWEVER, LARVAE SHOWED A WEAK PREFERENCE FOR DARK SURFACES. DIRECT ILLUMINATION CF LARVAE APPEARED TO HAVE NO EFFECT. CONVERSELY, LIGHT COULD INDIRECTLY AFFECT THE SETTLING OF BRYOZOAN LARVAE OR THE GROWTH OF COLONIES BY RESULTING IN ALGAL GROWTH WHICH IS INHIBITORY TO THE BRYOZOANS. (LONG-BATTELLE)

FIELD 05C

ACCESSION NO. W73-00348

METABOLIC TRANSFORMATION OF DDT, DIELDRIN, ALDRIN, AND ENDRIN BY MARINE MICROORGANISMS,

WISCONSIN UNIV., MADISON. DEPT. OF ENTOMOLOGY.

K. C. PATIL, F. MATSUMURA, AND G. M. BOUSH.

ENVIRONMENTAL SCIENCE AND TECHNOLOGY, VOL 6, NO 7, P 629-632, JULY 1972. 2 FIG, 3 TAB, 16 REF.

DESCRIPTORS:
*CHLORINATED HYDROCARBON PESTICIDES, SEA WATER, *METABOLISM, BIODEGRADATION, *MARINE MICROORGANISMS, *MICROBIAL DEGRADATION, *DDT, *DIELDRIN, *ALDRIN, *ENDRIN, MARINE ALGAE, BOTTOM SEDIMENTS, PLANKTON, RADIOACTIVITY TECHNIQUES, MONITORING, SURFACE WATERS, PERSISTENCE, PATH OF POLLUTANTS, OCEANS, ESTUARIES, CULTURES, SAMPLING, HAWAII, SEPARATION TECHNIQUES, FILMS.

IDENTIFIERS:
*BIOTRANSFORMATION, BIOLOGICAL SAMPLES, FATE OF POLLUTANTS, METABOLITES, DUNALIELLA, AGMENELLUM QUADRAPLICATUM, C-14, THIN LAYER CHROMATOGRAPHY, CLEANUP, RADIOAUTOGRAPHY, SUBSTRATE UTILIZATION.

ABSTRACT:
A STUDY WAS MADE OF THE METABOLISM OF CHLORINATED HYDROCARBON PESTICIDES IN SEAWATER TO DETERMINE THEIR BIODEGRADATION IN A MARINE ENVIRONMENT. SAMPLES OF SEAWATER, BOTTOM SEDIMENTS, SURFACE FILMS, ALGAE, AND MARINE PLANKTON COLLECTED IN OAHU, HAWAII WERE TREATED WITH RADIOLABELED INSECTICIDES AT THE COLLECTION SITE AND CULTURED FOR 30 DAYS AT 23 C IN THE LABORATORY. MICROORGANISMS ISCLATED FROM THE SAMPLES WERE MONITORED FOR METABOLIC ACTIVITY ALCNG WITH LABORATORY CULTURES OF UNICELLULAR ALGAE. THESE MICROBIAL CULTURES WERE ALSO LABELLED AND KEPT FOR 30 DAYS IN AN INCUBATOR AT 30 C. THIN-LAYER CHROMATOGRAPHIC TECHNIQUES WERE USED TO IDENTIFY METABOLITES AFTER CHLOROFORM EXTRACTION OF WATER SAMPLES. THE MOST SIGNIFICANT RESULT OF THE STUDY WAS THAT THE INSECTICIDES WERE NOT METABCLIZED IN PLAIN SEAWATER; EVEN IN RELATIVELY POLLUTED WATERS THEY WERE NOT DEGRADED. SEA BOTTOM SEDIMENTS SHOWED ONLY SLIGHT DEGRADATICN ACTIVITY. MOST DEGRADATION ACTIVITY WAS ASSOCIATED WITH BIOLOGICAL SAMPLES SUCH AS ALGAE, PLANKTON, AND SURFACE FILMS. A NUMBER OF MICROORGANISMS IN PURE CULTURE ALSO SHOWED METABOLIC ACTIVITIES WITH PATTERNS SIMILAR TO THOSE OBSERVED IN FIELD COLLECTED SAMPLES. (MORTLAND-BATTELLE)

FIELD 05B, 05C

ACCESSION NO. W73-00361

ECOLOGICAL ENERGETICS OF THE SEA-WEED ZONE IN A MARINE BAY ON THE ATLANTIC COAST OF CANADA. II. PRODUCTIVITY OF THE SEAWEEDS,

BEDFORD INST., DARTMOUTH (NOVA SCOTIA).

K. H, MANN.

MARINE BIOLOGY, VOL 14, NO 3, P 199-209, JUNE 1972. 6 FIG, 5 TAB, 22 REF.

DESCRIPTORS:
*GROWTH RATES, *PRIMARY PRODUCTIVITY, ENERGY BUDGET, *MARINE ALGAE, MARINE PLANTS, BIOMASS, CANADA, COASTS, ATLANTIC OCEAN, SEASONAL, BAYS, SEA WATER, ORGANIC MATTER, MORTALITY, WATER TEMPERATURE, PHAEOPHYTA, SAMPLING, SCUBA DIVING, ECOLOGY, STANDING CROPS, PATH OF POLLUTANTS.

IDENTIFIERS:
*SEAWEEDS, LAMINARIA LONGICRURIS, LAMINARIA DIGITATA, AGARUM CRIBROSUM, *NOVA SCOTIA, *ST. MARGARET'S BAY.

ABSTRACT:
SCUBA DIVING TECHNIQUES WERE USED TO FOLLOW THE GROWTH OF SEAWEEDS IN ST. MARGARET'S BAY, NOVA SCOTIA OVER A 2-YEAR PERIOD. IT WAS FOUND THAT THE BLADES OF LAMINARIA LONGICRURIS, L. DIGITATA, AND AGARUM CRIBROSUM TURN OVER THEIR BIOMASS MANY TIMES A YEAR AND HAVE AMONG THE HIGHEST PRODUCTIVITIES OF ANY NATURAL COMMUNITY. THE SCUBA DIVERS PUNCHED HOLES OF ABOUT 5 MM DIAMETER AT A DISTANCE OF 10 CM FROM THE JUNCTION OF STRIPE AND BLADE. THE MOVEMENT OF THESE HOLES WAS FOLLOWED THROUGHOUT THE SEASONS AND NEW HOLES WERE PUNCHED WHEN THE OLD ONES APPROACHED THE TIPS OF THE BLADES. THE BLADES WERE FOUND TO BEHAVE LIKE MOVING BELTS OF TISSUE, GROWING AT THE BASE AND ERODING AT THE TIPS. ALL SIZES OF PLANTS GREW ABOUT 200 CM ANNUALLY WITH THE YEAR'S GROWTH AMOUNTING TO 1-5 TIMES THE INITIAL LENGTH. BIOMASS INCREASE WAS APPROXIMATELY PROPORTIONAL TO THE SQUARE OF THE LENGTH INCREMENT. MAXIMUM GROWTH RATES WERE ACHIEVED BETWEEN JANUARY AND APRIL WHEN WATER TEMPERATURE WAS CLOSE TO 0 C. MINIMUM RATES OCCURRED IN THE JULY-OCTOBER PERIOD. THESE PLANTS MAKE A CONSIDERABLE CONTRIBUTION TO THE PRODUCTIVITY OF COASTAL WATERS. SINCE THEIR MAXIMUM GROWTH IS AT LOWER TEMPERATURES, THIS CONTRIBUTION MAY BE MORE SIGNIFICANT IN NORTHERN WATERS. (MORTLAND-BATTELLE)

FIELD 05C, 05B

ACCESSION NO. W73-00366

ENVIRONMENTAL FACTORS AFFECTING THE STANDING CROP OF FORAMINIFERA IN SUBLITTORAL AND PSAMMOLITTORAL COMMUNITIES OF A LONG ISLAND SALT MARSH,

CITY COLL., NEW YORK. DEPT. OF BIOLOGY.

N. J. MATERA, AND J. J. LEE.

MARINE BIOLOGY, VOL 14, NO 2, P 89-103, MAY 1972. 9 FIG, 6 TAB, 38 REF.

DESCRIPTORS:
*SALT MARSHES, *PROTOZOA, *BIOLOGICAL COMMUNITIES, *ENVIRONMENTAL EFFECTS, *STANDING CROPS, TIDAL EFFECTS, MARINE ALGAE, SEA WATER, SALINITY, SAMPLING, CORES, DISTRIBUTION PATTERNS, CURRENTS(WATER), PARTICLE SIZE, SEDIMENTS, SEPARATION TECHNIQUES, SPATIAL DISTRIBUTION, ECOLOGICAL DISTRIBUTION, BENTHIC FAUNA, SIEVES, WATER TEMPERATURE, FLUORESCENT DYE, CHLOROPHYTA, PHAEOPHYTA.

IDENTIFIERS:
*FORAMINIFERA, SUBSTRATES, EPIPHYTES, SAMPLE PRESERVATION, SAMPLE PREPARATION, SUBLITTORAL, LONG ISLAND, ENTEROMORPHA INTESTINALIS, POLYSIPHONIA, ULVA LACTUCA, ZANICHELLIA PALUSTRIS, CODIUM, FUSCUS, MACROPHYTES, AMMOBACULITES DILATATUS, AMMONIA BECCARII, AMMOTIUM SALSUM, ELPHIDIUM ADVENUM, ELPHIDIUM CLAVATUM, ELPHIDIUM GALVESTONENSE, ELPHIDIUM GUNTERI, ELPHICIUM INCERTUM, ELPHIDIUM TRANSLUCENS, PROTELPHIDIUM TISBURYENSIS, QUINQUELOCUTINA LATA, QUINQUELOCUTINA SEMINULUM, TROCHAMMINA INFLATA, TROCHAMMINA MACRESCENS.

ABSTRACT:
FORAMINIFERA SAMPLES WERE COLLECTED WEEKLY IN THE TOWD POINT SALT MARSH ON LONG ISLAND 13 JUNE - 3 SEPTEMBER 1968. ALGAL SAMPLES WERE TAKEN AT LOW TIDE AND DEPOSITED IN TEST TUBES CONTAINING MILLIPORE FILTERED SEA WATER. ONE HOUR AFTER THE ADDITION OF ROSE BENGAL TO EACH TUBE, EACH SAMPLE WAS PRESERVED WITH A NEUTRAL FORMALIN SOLUTION. IN THE LABORATORY, A VORTEX MIXER WAS USED TO SEPARATE THE EPIPHYTIC COMMUNITY FROM THE MACROALGAL SUBSTRATE. THE EPIPHYTIC COMMUNITY WAS TRANSFERRED INTO ETHANOL AND FORAMINIFERA WERE ISOLATED WITH A PASTEUR PIPETTE AND PLACED ON SLIDES FOR STORAGE AND STUDY. CORE SEDIMENT SAMPLES WERE TAKEN AT BOTH HIGH AND LOW TIDES AND PREPAREC ANC PRESERVED IN THE SAME MANNER. IN THE LABORATORY, THESE SAMPLES WERE SIEVED AND FRACTIONS GREATER THAN 0.074 MM WERE TRANSFERRED TO ETHANOL. CURRENT MOVEMENT WITHIN THE SAMPLING AREA WAS STUDIED WITH A FLUORESCIN DYE. SOME MACROPHYTES PROVIDE BETTER SUBSTRATES FOR FORAMINIFERA THAN OTHERS. DOMINANT SPECIES AMONG FORAMINIFERA WERE ELPHIDIUM INCESTUM AND PROTELPHIDIUM TISBURYENSIS. OVERALL DISTRIBUTION PATTERNS WERE DETERMINED BY SMALL RIVULETS FLOWING THROUGH THE MARSH WHICH ALTERED WATER TEMPERATURE, SALINITY, AND SEDIMENT GRAIN SIZE. THERE WAS NO EVIDENCE TOSUGGEST THAT FORAMINIFERA MIGRATE IN THE SEDIMENTS AS A FUNCTION OF TIDAL CHANGES. (MORTLAND-BATTELLE)

FIELD 05C, 05B, 05A

ACCESSION NO. W73-00367

INFLUENCE OF ENVIRONMENTAL PARAMETERS ON INTRASPECIFIC VARIATION IN FUCUS VESICULOSUS,

MAINE UNIV., ORONO. DEPT. OF OCEANOGRAPHY; AND MAINE UNIV., ORONO. DEPT. OF ZOOLOGY.

A. J. JORDAN, AND R. L. VADAS.

MARINE BIOLOGY, VOL 14, NO 3, P 248-252, JUNE 1972. 1 FIG, 2 TAB, 12 REF.

DESCRIPTORS:
*PHAEOPHYTA, *SALINITY, *ENVIRONMENTAL EFFECTS, *PLANT MORPHOLOGY, *GENETICS, *VARIABILITY, WAVES(WATER), MAINE, STATISTICAL METHODS, SEA WATER, COASTS, MARINE ALGAE.

IDENTIFIERS:
*FUCUS VESICULOSUS, VESICULATION, SAMPLE PREPARATION, BRANCHING.

ABSTRACT:
THE VESICULATION AND BRANCHING OF THE MARINE BROWN ALGAE, FUCUS VESICULOSUS, WERE EXAMINED FOR SAMPLES TAKEN FROM THREE LOCATIONS ON THE MAINE COAST. THESE SITES WERE A NORMAL SALINITY EXPOSED AREA, A LOW SALINITY PROTECTED AREA, AND A NORMAL SALINITY PROTECTED AREA. AFTER COLLECTION, THE TOP AND BOTTOM OF EACH PLANT WERE CUT AND EACH SEGMENT WAS NUMBERED AND WEIGHED AND THE VESICLES AND DICHOTOMIES WERE COUNTED. THE VESICLES WERE MEASURED AND THEIR VOLUMES CALCULATED. A TOTAL OF 10 MEASUREMENTS OF MORPHOMETRIC CHARACTERISTICS AND RATIOS OF THESE MEASUREMENTS WERE COMPARED FOR EACH OF 300 PLANTS BY STATISTICAL ANALYSES ON A COMPUTER. THE RESULTS INDICATED THAT DECREASED SALINITY CORRELATED WITH INCREASED VESICULATION AND BRANCHING. INCREASED WAVE ACTION CORRELATED WITH DECREASED VESICULATION. (MORTLAND-BATTELLE)

FIELD 05C, 05A

ACCESSION NO. W73-00369

DENSITY-GRADIENT CENTRIFUGATION AS AN AID TO SORTING PLANKTONIC ORGANISMS. I. GRADIENT MATERIALS,

MARINE BIOLOGICAL LAB., WOODS HOLE, MASS.

R. A. BOWEN, J. M. ST. ONGE, J. B. COLTON, JR., AND C. A. PRICE.

MARINE BIOLOGY, VOL 14, NO 3, P 242-247, JUNE 1972. 5 FIG, 11 REF.

DESCRIPTORS:
*ZOOPLANKTON, *LARVAE, *FISH EGGS, *SEPARATION TECHNIQUES, *MARINE FISH, SILICA, MARINE ALGAE, INVERTEBRATES, ISOLATION, REFRACTIVITY, COPEPODS, MARINE ANIMALS, MOLLUSKS, GASTROPODS, PHAEOPHYTA.

IDENTIFIERS:
*DENSITY-GRADIENT CENTRIFUGATION, ISOPYCNIC CENTRIFUGATION, SAMPLE PRESERVATION, SALPA, LUDOX AM, SUCROSE, DENSITY GRADIENTS, TUNICATES, PHAEOCYSTIS, SORTING, SODIUM BROMIDE, DEXTRAN, OIKOPLEURA, SAGITTA, PTEROPODS, EUPHAUSIIDS, GADUS MORHUA, UROPHYCIS.

ABSTRACT:
DENSITY-GRADIENT CENTRIFUGATION WAS STUDIED AS AN AID TO THE SORTING OF ICHTHYOPLANKTON AND IT WAS SHOWN THAT FISH EGGS AND LARVAE CAN BE SEPARATED BY THIS METHOD USING GRADIENTS OF SUCROSE OR SILICA. PRESERVED SAMPLES WERE LAYERED OVER LINEAR GRADIENTS IN 100 CC SWINGING BUCKETS AND CENTRIFUGED FOR 1 HR AT 1000 RPM. IN THE SUCROSE GRADIENTS, ZOOPLANKTON WERE CONFINED TO TWO ENDS OF THE GRADIENT AND THE FISH EGGS TO AN INTERMEDIATE ZONE. USING SILICA GRADIENTS, IN THE FORM OF LUDOX AM, THE FISH EGGS BANDED IN ONE NARROW ZONE, THE FISH LARVAE IN ANOTHER, AND NEARLY ALL OF THE INVERTEBRATE PLANKTON WERE BETWEEN. BOTH SODIUM BROMIDE AND DEXTRAN GRADIENTS WERE ALSO TRIED BUT PROVED UNSUITABLE. SIX CLASSES OF ZOOPLANKTON WERE STUDIED AND ONLY SALPA OVERLAPPED APPRECIABLY WITH FISH EGGS. THERE WAS NO OVERLAP WITH LARVAE. THE LUDOX AM GRADIENT WAS FOUND TO OFFER THE MOST ADVANTAGES. (MORTLAND-BATTELLE)

FIELD 05A

ACCESSION NO. W73-00371

STUDIES OF MARINE FOULING AND BORING OFF KODIAK ISLAND, ALASKA,

NAVAL OCEANOGRAPHIC OFFICE, WASHINGTON, D. C.

E. R. LONG,

MARINE BIOLOGY, VOL 14, NO 1, P 52-57, MAY 1972. 3 FIG, 3 TAB, 15 REF.

DESCRIPTORS:
*FOULING, *ARTIFICIAL SUBSTRATES, DEPTH, *SAMPLING, *PERIPHYTON, EQUIPMENT, TUBIFICIDS, MOLLUSKS, WATER TEMPERATURE, MARINE ANIMALS, ALASKA, BIOMASS, SALINITY, BENTHIC FAUNA, ASBESTOS, INORGANIC COMPOUNDS, WORMS, INVERTEBRATES, ANNELIDS, SEASONAL, CRUSTACEANS, ISOPODS, MARINE ALGAE, *ALASKA.

IDENTIFIERS:
*WOOD BORERS, BARNACLES, BRYOZOAS, HYDROIDS, SPONGES, COELENTERATES, PORIFERA, POLYCHAETES, SERPULIDS.

ABSTRACT:
MARINE FOULING AND BORING OFF KODIAK ISLAND, ALASKA, WAS STUDIED AT DEPTHS OF 5, 15, AND 30 M BY EXPOSING WOOD/ASBESTOS TEST PANELS AND RETRIEVING THEM AT MONTHLY AND CUMULATIVELY LONGER INTERVALS. FOULING WAS MODERATE BETWEEN JUNE AND OCTOBER AND NEGLIGIBLE BETWEEN NOVEMBER AND MAY. THE FOULING COMMUNITIES AT 15 AND 30 M WERE QUITE DISSIMILAR, PROBABLY DUE TO A LARGE SURFACE-TO-BOTTOM SALINITY GRADIENT. THE COMMUNITY AT THE 30 M LEVEL WAS DOMINATED BY BALANUS CRENATUS AND PSEUDOCHITINOPOMA OCCIDENTALIS, THAT AT 15 M BY ALCYONIDIUM POLYOUM AND OBELIA BOREALIS. AT THE 5 M DEEP, PIERSIDE SITE, THE MOST COMMON ORGANISMS WERE B. CRENATUS, O. BOREALIS AND DESMACYSTIS SANDALIA. BORER ATTACK WAS A TRACE AT 5 AND 15 M AND SLIGHTLY MORE SEVERE AT 30 M. (SNYDER-BATTELLE)

FIELD 05A

ACCESSION NO. W73-00374

PRODUCTION OF DISSOLVED ORGANIC MATTER FROM DEAD GREEN ALGAL CELLS. I. AEROBIC MICROBIAL DECOMPOSITION,

TOKYO METROPOLITAN UNIV. (JAPAN). DEPT. OF CHEMISTRY.

A. OTSUKI, AND T. HANYA.

LIMNOLOGY AND OCEANOGRAPHY, VOL 17, NO 2, P 248-257, MARCH 1972. 12 FIG, 3 TAB, 28 REF.

DESCRIPTORS:
*AEROBIC BACTERIA, *AEROBIC CONDITIONS, *KINETICS, *SCENEDESMUS, *DISSOLVED SOLIDS, *MICROBIAL DEGRADATION, CARBON, ORGANIC MATTER, WATER POLLUTION SOURCES, CHLOROPHYTA, AQUATIC PRODUCTIVITY, AQUATIC ALGAE, NITROGEN, POLLUTANT IDENTIFICATION, ENVIRONMENTAL EFFECTS, LABORATORY EQUIPMENT, FREEZE DRYING, SEPARATION TECHNIQUES, SOLVENT EXTRACTION, MATHEMATICAL STUDIES, SYSTEMATICS, METABOLISM.

IDENTIFIERS:
FATE OF POLLUTANTS, SUBSTRATE UTILIZATION, ORGANIC NITROGEN, ORGANIC CARBON, CULTURE MEDIA, PAPER CHROMATOGRAPHY, INFRARED SPECTROSCOPY, ASSIMILATION, KJEDAHL PROCEDURE, SAMPLE PREPARATICN.

ABSTRACT:
THE PRODUCTICN OF DISSOLVEC ORGANIC MATTER (DOM) FROM DEAD GREEN ALGAL CELLS (SCENEDESMUS) BY AEROBIC BACTERIA WAS INVESTIGATED WITH PARTICULAR EMPHASIS ON THE PATTERNS AND KINETICS OF C AND N. DOM WAS PREPARED FOR EXPERIMENTAL USE BY FREEZE-DRYING AND ASHING APPROXIMATELY 99 PERCENT PURE CULTURES OF SCENEDESMUS. AEROBIC DECOMPOSITION OF THE DOM WAS INITIATED BY INOCULATING MICROFLORA INTO A SPECIALLY PREPARED DECOMPOSITION APPARATUS CONTAINING CULTURE MEDIA AND DOM AT PH 7. AFTER INCUBATION, DISSOLVED ORGANIC CARBCN (DOC) AND NITROGEN (DON) WERE MEASURED BY AN ELEMENTAL ANALYZER AND BY THE MICRC-KJELDAHL METHOD. ORGANIC MATERIALS, SUCH AS PROTEIN AND AMINO ACIDS, WERE EXTRACTED BY THIN-LAYER CHROMATOGRAPHY FOR IR-SPECTROSCOPIC ANALYSIS. THE DOM PRODUCTION, BY THE 30TH DAY OF INCUBATION, WAS ABCUT 7 PERCENT C AND 6 PERCENT N. AEROBIC MICROBIAL DECOMPOSITION APPROXIMATED A FIRST ORDER REACTION DURING THIS PERICD. KINETIC CONSICERATICNS OF THE DECOMPOSITION PATTERN OF THE CELL NITROGEN AND THE PRODUCTION OF DON SUGGEST THAT DEAD ALGAL CELL SUBSTANCE MAY BE DIVIDED INTO LABILE AND REFRACTORY CONSTITUENTS BY THEIR RELATIVE RESISTANCE TC THE ACTION OF BACTERIA. THE DISSOLVED ORGANIC NITRCGENOUS MATERIAL PRODUCED IS COMPOSED OF TWO MAJOR FRACTICNS: ONE IS PRODUCED WITH THE DECOMPOSITION OF ALGAL CELLS, AND THE OTHER IS PROBABLY EXCRETED BY BACTERIA THROUGH REASSIMILATION OF MINERALIZED NITROGEN. (LONG-BATTELLE)

FIELD 05C, 05B

ACCESSION NO. W73-00379

UTILIZATION OF UREA BY SOME MARINE PHYTOPLANKTERS,

WOODS HOLE OCEANOGRAPHIC INSTITUTION, MASS.

E. J. CARPENTER, C. C. REMSEN, AND S. W. WATSON.

LIMNOLOGY AND OCEANOGRAPHY, VOL 17, NO 2, P 265-269, MARCH 1972. 1 FIG, 3 TAB, 19 REF.

DESCRIPTORS:
*PHYTOPLANKTON, *MARINE ALGAE, CHEMICAL ANALYSIS, *NITRATES, *AMMONIA, DIATOMS, GROWTH RATES, NITROGEN, WATER POLLUTION EFFECTS, RADIOACTIVITY, METABOLISM, PLANT PHYSIOLOGY, CHRYSOPHYTA, CULTURES, SEA WATER, NUTRIENT REQUIREMENTS, WATER POLLUTION.

IDENTIFIERS:
*SUBSTRATE UTILIZATION, *SARGASSO SEA, COCCOLITHOPHORIDS, FLAGELLATES, FATE OF POLLUTANTS, STEPHANOPYXIS COSTATA, ASSIMILATION, EMILIANA(COCCOLITHUS) HUXLEYI, AMPHIPRORA ALATA, CHRYSOCHROMULINA, SKELETONEMA, CHAETOCEROS SIMPLEX.

ABSTRACT:
THE POSSIBLE EXISTENCE OF ADDITIONAL UREA-UTILIZING PHYTOPLANKTERS IN OFFSHORE AND INSHORE AREAS WAS INVESTIGATED AND THEIR GROWTH RATES ON UREA, NITRATE, AND AMMONIA WERE COMPARED. THE CONCEPT OF WHETHER THESE SPECIES CAN ASSIMULATE UREA AT SIGNIFICANT RATES AT THE CONCENTRATIONS PRESENT IN THEIR NORMAL HABITATS WAS ALSO CONSIDERED. THREE DIATOMS FROM THE SARGASSO SEA AND TWO DIATOMS AND A HAPTOPHYTE FLAGELLATE FROM AN INSHORE AREA NEAR WOODS HOLE WERE FOUND TO EXHIBIT SIMILAR GROWTH RATES ON UREA, NITRATE, AND AMMONIA. A COCCOLITHOPHORE EMILIANA (COCCOLITHUS) HUXLEYI FROM THE SARGASSO SEA DID NOT GROW ON UREA. THE HALF-SATURATION CONSTANT FOR UREA DETERMINED FOR ONE INSHORE DIATOM, STEPHANOPYXIS COSTATA (SKELETONEMA COSTATUM), WAS 8.5 MICROGRAMS-ATOM UREA-N/LITER. AT THE UREA CONCENTRATIONS OF ITS HABITATS, THE CALCULATED DIVISION RATE FOR THIS SPECIES, GROWING ON UREA AS THE SOLE NITROGEN SOURCE, WAS 2.2 DAYS, SIMITAR TO THAT OF DIATOMS GROWING IN INSHORE HABITATS. IT APPEARS, FROM DATA ON NATURAL UREA CONCENTRATIONS AND UREA UPTAKE RATES BY S. COSTATA, THAT UREA CAN BE A SIGNIFICANT N SOURCE FOR AT LEAST ONE COMMON INSHORE PHYTOPLANKTER. ACCORDING TO GROWTH RATE STUDIES, AT LEAST THREE INSHORE AND THREE OFFSHORE MARINE PHYTOPLANKTON CAN GROW AS RAPIDLY ON UREA AS ON NITRATE OR AMMONIA. (BYRD-BATTELLE)

FIELD 05C, 05B

ACCESSION NO. W73-00380

SIGNIFICANCE TO EUTROPHICATION OF SPATIAL DIFFERENCES IN NUTRIENTS AND DIATOMS IN LAKE MICHIGAN,

WISCONSIN UNIV., MILWAUKEE. CENTER FOR GREAT LAKES STUDIES.

R. E. HOLLAND, AND A. M. BEETON.

LIMNOLOGY AND OCEANOGRAPHY, VOL 17, NO 1, P 88-96, JANUARY 1972. 4 FIG, 1 TAB, 22 REF.

DESCRIPTORS:
*SAMPLING, *NUTRIENTS, *DIATOMS, *LAKE MICHIGAN, *EUTROPHICATION, *SPATIAL DISTRIBUTION, PHOSPHORUS, NITRATES, SILICA, PATH OF POLLUTANTS, ORGANIC COMPOUNDS, PHYTOPLANKTON, CARBON, WATER TEMPERATURE, GREAT LAKES REGION, WISCONSIN, MICHIGAN, CHRYSOPHYTA, PLANT PIGMENTS, AQUATIC ALGAE, NITROGEN.

IDENTIFIERS:
CHLOROPHYLL A, CHLOROPHYLL C, CAROTENOIDS, FRAGILARIA CROTONENSIS, TABELLARIA FLOCCULOSA, STEPHANODISCUS HANTZSCHII, CYCLOT MICHIGANIANA, RHIZOSOLENIA ERIENSIS, MELOSIRA ISLANDICA, MELOSIRA ITALICA, MELOSIRA AMBIGUA, SYNEDRA FILIFORMIS, FRAGILARIA CAPUCINA, CYCLOTELLA STELLIGERA, STEPHANODISCUS.

ABSTRACT:
WATER SAMPLES WERE TAKEN FROM A WATER-COOLING INTAKE AT A DEPTH OF ABOUT 4 M FROM A RAILROAD FERRY BETWEEN MILWAUKEE, WISCONSIN, AND LUDINGTON, MICHIGAN, 27 MAY 1970 - 6 JANUARY 1971, TO DETERMINE INSHORE-OFFSHORE DIFFERENCES IN SPECIES AND ABUNDANCE OF DIATOMS, AND CONCENTRATIONS OF PHOSPHORUS, SILICA, NITRATE, AND PIGMENTS. THE VARIOUS NUTRIENTS AND PIGMENTS WERE DETERMINED, AND THE DIATOMS IDENTIFIED BY PREVIOUSLY DESCRIBED METHODS. INSHORE WATERS (WITHIN 16 KM OF SHORE) HAD GREATER DIATOM POPULATIONS, DIFFERENT SPECIES COMPOSITION WITHIN THE DIATOM COMMUNITIES, AND DIFFERENT CONCENTRATIONS OF MAJOR NUTRIENTS AND PIGMENTS THAN OFFSHORE WATERS (GREATER THAN OR EQUAL TO 16 KM FROM SHORE). RESULTS INDICATE THAT DATA FROM WATER INTAKES WHICH HAVE BEEN USED IN THE PAST TO DOCUMENT CHANGES IN THE LAKE MAY HAVE REPRESENTED ONLY LOCAL CONDITIONS. (SNYDER-BATTELLE)

FIELD 05C, 05B

ACCESSION NO. W73-00384

DISTRIBUTION AND TAXONOMY OF LAMINARIA SINCLAIRII AND L. LONGIPES (PHAEOPHYCEAE, LAMINARIALES),

BRITISH COLUMBIA UNIV., VANCOUVER. DEPT. OF BOTANY.

J. W. MARKHAM.

PHYCOLOGIA, VOL 11, NO 2, P 147-157, JUNE 1972. 6 FIG, 3 TAB, 79 REF.

DESCRIPTORS:
*DISTRIBUTION PATTERNS, *SYSTEMATICS, *ECOLOGICAL DISTRIBUTION, *SPATIAL DISTRIBUTION, GEOGRAPHICAL REGIONS, PHAEOPHYTA, MARINE ALGAE, ENVIRONMENTAL EFFECTS, MARINE PLANTS, CANADA, WASHINGTON, OREGON, CALIFORNIA, ALASKA, SEA WATER.

IDENTIFIERS:
*LAMINARIA SINCLAIRII, *LAMINARIA LONGIPES, BRITISH COLUMBIA, RUSSIA, LAMINARIA RODRIGUEZII.

ABSTRACT:
LAMINARIA SINCLAIRII (HARV.) FARL., AND., AND EAT. OCCURS FROM CENTRAL BRITISH COLUMBIA TO SOUTHERN CALIFORNIA. L. LONGIPES BORY OCCURS FROM THE KURILE ISLANDS THROUGHOUT THE ALEUTIAN ISLANDS TO SOUTHEAST ALASKA. DETAILED DISTRIBUTION RECORDS ARE PRESENTED. BOTH SPECIES ARE DISTINGUISHED FROM OTHER SPECIES OF LAMINARIA IN HAVING MULTIPLE STIPES FROM AN EXPANDED, RHIZOME-LIKE HOLDFAST. THE TWO SPECIES ARE VERY SIMILAR IN APPEARANCE. MOST AUTHORS HAVE DISTINGUISHED THEM ON THE BASIS OF INTERNAL STIPE ANATOMY; L. SINCLAIRII HAS MUCILAGE DUCTS IN THE STIPE, WHEREAS L. LONGIPES DOES NOT. ECOLOGICAL STUDIES WERE CARRIED OUT ON BOTH SPECIES IN THE FIELD AND IN THE LABORATORY. THE RESULTS SHOW THE SPECIES ALSO DIFFER IN SEVERAL OTHER POINTS, INCLUDING LENGTH OF STIPES, WIDTH OF BLADES, WINTER LOSS OF BLADES, MORPHOLOGY OF GAMETOPHYTES, AND HABITAT. THE EVIDENCE CONFIRMS THAT THEY SHOULD BE RETAINED AS TWO SEPARATE SPECIES. (LONG-BATTELLE)

FIELD 05A, 05C

ACCESSION NO. W73-00428

PROCEDURES ADOPTED FOR THE LABORATORY CULTIVATION OF TRICHODESMIUM ERYTHRAEUM,

CENTRE FOR ADVANCED STUDY IN MARINE BIOLOGY, PORTO NOVA (INDIA).

V. D. RAMAMURTHY.

MARINE BIOLOGY, VOL 14, NO 3, P 232-234, JUNE 1972. 11 REF.

DESCRIPTORS:
*CULTURES, *CYANOPHYTA, *MARINE ALGAE, *ISOLATION, MARINE PLANTS, EUTROPHICATION, OCEANS, SAMPLING, SEPARATION TECHNIQUES, CULTIVATION, VITAMIN B, PLANT GROWTH REGULATORS, NITRATES, PHOSPHATES, NUTRIENT REQUIREMENTS, ESSENTIAL NUTRIENTS.

IDENTIFIERS:
*TRICHODESMIUM ERYTHRAEUM, *CULTURING TECHNIQUES, CULTURE MEDIA, *BAY OF BENGAL, INDIAN OCEAN, SOIL EXTRACTS, ERDSCHREIBER MEDIUM, ESTUARINE SOILS, AXENIC CULTURES, GIBBERELIC ACID.

ABSTRACT:
VARIOUS CULTURING TECHNIQUES WERE INVESTIGATED FOR THE AXENIC OR PURE CULTIVATION OF THE MARINE BLUE-GREEN ALGA TRICHODESMIUM ERYTHRAEUM ISOLATED A DEPTH OF 9-12 M IN THE BAY OF BENGAL DURING THE PRE-BLOOM PERIOD IN MARCH, 1965. MODIFIED, ENRICHED, ERDSCHREIBER MEDIUM, CONTAINING A DIFFERENTIAL CONCENTRATION OF STREPTOMYCIN, TETRACYCLINE, AND SULFADIAZINE FOR INHIBITING MICROBIAL GROWTH, WAS USED FOR CULTIVATING THE ALGAE. SUB-CULTURES WERE FORMED EVERY 7-9 DAYS BY INOCULATING FRESH MEDIUM WITH 10-15 CELLS PER 10 ML. THE INGREDIENTS IN THE CULTURE MEDIUM INCLUDE: NITRATE, PHOSPHATE, A 50 ML ESTUARINE SOIL EXTRACT, SEAWATER (3.3 PERCENT SALINITY), VITAMIN B12, GIBBERELLIC ACID, AND ANTIBIOTICS. INSTRUCTIONS FOR THE ISOLATION OF THE ALGAE AND THE PREPARATION OF THE CULTURE MEDIA ARE PRESENTED. (LONG-BATTELLE)

FIELD 05C, 05A

ACCESSION NO. W73-00430

PRIMARY PRODUCTION AT THE THERMOCLINE LEVEL IN THE NERITIC ZONE OF THE NORTH-OCCIDENTAL MEDITERRANEAN, (PRODUCTION PRIMAIRE AU NIVEAU DE LA THERMOCLINE EN ZONE NERITIQUE DE MEDITERRANEE NORD-OCCIDENTALE),

ARAGO LAB., BANYULS-SUR-MER (FRANCE).

G. CAHET, M. FIALA, G. JACQUES, AND M. PANOUSE.

MARINE BIOLOGY, VOL 14, NO 1, P 32-40, MAY 1972. 7 FIG, 3 TAB, 26 REF.

DESCRIPTORS:
*PHYTOPLANKTON, *THERMOCLINE, *PRIMARY PRODUCTIVITY, *NERITIC, NITRATES, RADIOACTIVITY TECHNIQUES, EUTROPHICATION, MARINE ALGAE, WATER POLLUTION SOURCES, SPATIAL DISTRIBUTION, BIOMASS, PHOTOSYNTHESIS, NUTRIENTS, LIMITING FACTORS, SALINITY, CARBON RADICISOTOPES, SESTON, ORGANIC MATTER, DISTRIBUTION PATTERNS, THERMAL PROPERTIES, THERMAL STRATIFICATION, CHLOROPHYLL, MAGNESIUM COMPOUNDS, PIGMENT, SURFACE WATERS, SEDIMENTATION, PHOSPHATES.

IDENTIFIERS:
NITZSCHIA DELICATISSIMA, NITZSCHIA PUNGENS, C-14, CHLOROPHYLL A, MEDITERRANEAN SEA, FRANCE, PYCNOCLINE.

ABSTRACT:
VERTICAL PROFILES OF PHYSICAL, CHEMICAL AND PHYTOPLANKTONIC PARAMETERS ARE DESCRIBED, AT THE LEVEL OF THE THERMOCLINE, IN THE AREA OF BANYULS-SUR MER, FRANCE. THE RESULTS SHOW THAT THE THERMOCLINE DIVIDES TWO MASSES OF WATER: (1) MEDITERRANEAN SURFACE WATER WITH LOW NUTRIENT CONCENTRATIONS AND A SALINITY BELOW 3.8 PERCENT; (2) DEEP, NUTRIENT-RICH UPWELLED WATER (N-NO3 GREATER THAN 3 MICROAT-G/L, P-PO4 GREATER THAN 0.3 MICROAT-G/L, GREATER THAN 3.83 PERCENT S), WHICH COMES FROM THE UPPER LIMIT OF THE MEDITERRANEAN INTERMEDIATE WATER, USUALLY LOCATED AT THE 200 M LEVEL. CONSEQUENTLY, CONDITIONS ARE SUITABLE FOR HIGH PRODUCTION RATES AT THE BOTTOM OF THE THERMOCLINE, WHERE CHL A IS ABOVE 0.5 MG/CU M; DOMINANT SPECIES ARE NITZSCHIA DELICATISSIM AND N. PUNGENS. A DIAGRAM IS PRESENTED EXPLAINING THE DIFFERENT EFFECTS OF THE PYCNOCLINES ON PRIMARY PRODUCTION: EUTROPHICATION AT THE PYCNOCLINE LEVELS IS THE RESULT OF PASSIVE ACCUMULATION OF PHYTOPLANKTON AND ORGANIC MATTER DURING SEDIMENTATION, AND/OR OF REDUCED DIFFUSION OF NUTRIENTS FROM DEEP WATERS TOWARDS THE SURFACE. (LONG-BATTELLE)

FIELD 05C, 05B

ACCESSION NO. W73-00431

STEPHANODISCUS BINDERANUS (KUTZ.) KRIEGER OR MELOSIRA BINDERANA KUTZ. (BACILLARIOPHYTA, CENTRALES),

BRISTOL UNIV. (ENGLAND). DEPT. OF BOTANY.

F. E. ROUND.

PHYCOLOGIA, VOL 11, NO 2, P 109-116, JUNE 1972. 27 FIG, 22 REF.

DESCRIPTORS:
*DIATOMS, *BIOINDICATORS, *EUTROPHICATION, *POLLUTANT IDENTIFICATION, *LAKE MICHIGAN, CHRYSOPHYTA, AQUATIC ALGAE, ELECTRON MICROSCOPY, PHOTOGRAPHY, WATER POLLUTION EFFECTS, WATER PURIFICATION, *SYSTEMATICS.

IDENTIFIERS:
*STEPHANODISCUS BINDERANUS, *MELOSIRA BINDERANA, SCANNING ELECTRON MICROSCOPY, TRANSMISSION ELECTRON MICROSCOPY, LIGHT MICROSCOPY.

ABSTRACT:
THE CENTRIC DIATOM WHICH HAS BEEN IDENTIFIED EITHER AS A SPECIES OF MELOSIRA C. A. AG. OR OF STEPHANODISCUS C. G. EHRENB IS AN INDICATOR OF INCREASING EUTROPHICATION IN LAKES, AND IS ALSO RESPONSIBLE FOR SERIOUS PROBLEMS IN WATER PURIFICATION PLANTS. WATER SAMPLES WERE OBTAINED FROM LAKE MICHIGAN TO DETERMINE THE GENUS TO WHICH THIS SPECIES BELONGS. THE ALGAE WERE INVESTIGATED BY LIGHT, SCANNING, AND TRANSMISSION MICROSCOPY. THIS REPRESENTS THE FIRST DETAILED STUDY OF THIS SPECIES AND CONFIRMED ITS POSITION IN STEPHANODISCUS. (LONG-BATTELLE)

FIELD 05C, 05A, 02H

ACCESSION NO. W73-00432

INTERREGULATION OF MARINE PLANKTONIC DIATOMS IN MONO-AND MIXED CULTURES, (IN RUSSIAN),

MOSCOW STATE UNIV. (USSR).

V. D. FEDOROV, AND N. G. KUSTENKO.

OKEANOLOGIYA, VOL 12, NO 1, P 111-122, JANUARY/FEBRUARY 1972. 10 FIG, 2 TAB, 14 REF.

DESCRIPTORS:
*DIATOMS, *CULTURES, *NUTRIENT REQUIREMENTS, SEA WATER, NITRATES, PHOSPHORUS, NUTRIENTS, CHRYSOPHYTA, COMPETITION, WATER POLLUTION SOURCES, POLLUTANT IDENTIFICATION, PHYTOPLANKTON, MARINE ALGAE, ESSENTIAL NUTRIENTS.

IDENTIFIERS:
*INTERREGULATION, SKELETONEMA COSTATUM, THALASSIONEMA NITZSCHIOIDES, MIXED CULTURES, METABOLITES, SUBSTRATE UTILIZATION, GROWTH.

ABSTRACT:
A METHOD OF NUTRIENT SALT ADDITIONS IS USED IN COMBINATION WITH THE DESIGN OF THE FULL FACTOR EXPERIMENTS TO SHOW THE PREVAILING DEPENDENCE OF THE GROWTH OF THE ALGOLOGICALLY PURE CULTURE OF SKELETONEMA COSTATUM ON NITRATE NITROGEN CONCENTRATIONS AND OF THALASSIONEMA NITZSCHIOIDES ON PHOSPHORUS CONCENTRATIONS. BY VARYING THE CONCENTRATIONS OF THESE TWO ELÉMENTS IN THE NUTRIENT MEDIUM (THE SEA WATER WITH NITROGEN AND PHOSPHORUS ADDITIONS), AS WELL AS THE SOWING TITER OF EACH DIATOM SPECIES, THE EXPERIMENTERS SUCCEEDED IN CHANGING THE DIRECTION OF THE EXCLUSION OF SOME SPECIES BY THE COMPETITIVE ONE WHEN THE SPECIES ARE CULTIVATED TOGETHER. THE MECHANISM OF THE QUANTITATIVE INTERREGULATION OF BOTH CULTURES IS SHOWN TO BE BASED ON THE ACTIVITY OF EXTRACELLULAR DIATOM METABOLITES THE ACCUMULATION AND ACTION OF WHICH IS VERY SPECIFIC (I.E., CONNECTED WITH A CERTAIN STAGE OF DEVELOPMENT) FOR EACH DIATOM. (LONG-BATTELLE)

FIELD 05C, 05B

ACCESSION NO. W73-00441

SEASONAL CHARACTERISTICS OF PHYTOPLANKTON IN THE AMURSKY BAY OF THE SEA OF JAPAN, (IN RUSSIAN),

AKADEMIYA NAUK SSSR, VLADIVOSTOK. INSTITUT BIOLOGII.

G. V. KONOVALOVA.

OKEANOLOGIYA, VOL 12, NO 1, P 123-128, JANUARY/FEBRUARY 1972. 2 FIG, 14 REF.

DESCRIPTORS:
*PHYTOPLANKTON, *ENVIRONMENTAL EFFECTS, *SYSTEMATICS, *SEASONAL, SEA WATER, SAMPLING, BIOMASS, DIATOMS, POLLUTANT IDENTIFICATION, MARINE ALGAE, CHRYSOPHYTA, AUTUMN, WINTER, SUMMER, PRIMARY PRODUCTIVITY, DINOFLAGELLATES, PYRROPHYTA.

IDENTIFIERS:
SEA OF JAPAN, *AMURSKY BAY, PERIDINIANS, EUCAMPIA ZOODIACUS, CHAETOCEROS AFFINIS, SKELETONEMA COSTATUM, CHAETOCEROS SOCIALIS.

ABSTRACT:
SEASONAL NUMBER AND BIOMASS DYNAMICS OF THE PHYTOPLANKTON, AS WELL AS OF ITS CONSTITUENT SYSTEMATIC GROUPS AND MASS SPECIES, WERE DETERMINED ON THE BASIS OF THE ANNUAL PHYTOPLANKTON COLLECTIONS IN THE AMURSKY BAY OF THE SEA OF JAPAN. THE SPECIFIC COMPOSITION OF ALGAE WAS DETERMINED IN THE WATER-BOTTLE COLLECTIONS. A MAJOR PART OF THE BIOMASS CALCULATIONS WAS BASED ON THE TABLE OF VOLUMES COMPILED FOR THE MASS PHYTOPLANKTON SPECIES. THREE MAXIMA OF ALGAL BIOMASS ARE ESTABLISHED: IN WINTER, SUMMER AND AUTUMN. DIATOMS PLAY A LEADING ROLE IN THE FORMATION PHYTOPLANKTON BIOMASS AND ITS NUMBERS. PERIDINIANS STAND SECOND IN BIOMASS. THE GREATEST BIOMASS WAS OBSERVED IN EUCAMPIA ZOODIACUS AND CHAETOCEROS AFFINIS, THE GREATEST NUMBERS IN SKELETONEMA COSTATUM, CHAETOCEROS AFFINIS AND CHAETOCEROS SOCIALIS. AT ALL THE STATIONS THE BIOMASS MAXIMUM WAS RECORDED AT THE END OF SEPTEMBER AND IN OCTOBER, WHEREAS THE MINIMUM WAS IN APRIL. THERE WERE DISTINCT CHANGES OF FORMS IN THE PLANKTONIC ALGAE. (LONG-BATTELLE)

FIELD 05C, 02L

ACCESSION NO. W73-00442

TOXIC SUBSTANCES CONTROL ACT OF 1971 AND AMENDMENT.

HEARINGS ON S.1478--SUBCOMM. ON THE ENVIRONMENT--COMM. CN COMMERCE, U.S. SENATE, 92ND CONGRESS, AUGUST 3, 4, 5; OCTOBER 4 AND NOVEMBER 5, 1971. 3 VOLUMES. 1251 P.

DESCRIPTORS:
*LEGISLATION, *TOXINS, *ENVIRONMENTAL EFFECTS, FEDERAL GOVERNMENT, AIR POLLUTION, WATER POLLUTION, TOXICITY, CHEMICALS, REGULATION, ALGAL TOXINS, MERCURY, DETERGENTS, WATER POLLUTION SOURCES, ASBESTOS, EUTROPHICATION, POLYCHLORINATED BIPHENYLS, PUBLIC HEALTH, ENVIRONMENT CONTROL, LEGAL ASPECTS, REGULATICN, PERMITS.

IDENTIFIERS:
*LEGISLATION(PROPOSED), ENVIRONMENTAL PROTECTION AGENCY.

ABSTRACT:
THE PURPOSE OF THE PROPOSED LEGISLATION IS TO REGULATE TOXIC SUBSTANCES, RECOGNIZING THAT GAPS EXIST IN THE REGULATORY FRAMEWORK THAT PERMIT PUBLIC HEALTH AND ENVIRONMENTAL HAZARDS TO OCCUR WITHOUT REGULATION. FOUR CHEMICAL SUBSTANCES WHICH WILL BE PRIME CANDIDATES FOR REGULATION ARE DISCUSSED IN GREAT DETAIL. THESE ARE MERCURY, POLYCHLORINATED BIPHENYLS (PCB'S), ASBESTOS, AND DETERGENTS. UNDER THE ACT THE ADMINISTRATOR OF THE ENVIRCNMENTAL PROTECTION AGENCY WILL BE EMPOWERED TO RESTRICT THE USE OR DISTRIBUTION, INCLUDING A TOTAL PROHIBITION, OF A CHEMICAL SUBSTANCE, IF SUCH RESTRICTIONS ARE NECESSARY TO PROTECT THE ENVIRONMENT. THE BILL PRCVIDES FOR SETTING TEST STANDARDS ON VARIOUS CLASSES AND USES OF NEW CHEMICALS. A NEW COMPOUND WHICH DOES NOT MEET THESE STANDARDS WILL BE KEPT OFF THE MARKET. THE AMENDMENT'S PRECLEARANCE CERTIFICATICN PROCEDURE IS CRITIZED AS BEING NOT ONLY TOO BURDENSOME UPON THE GOVERNMENT, BUT IMPOSING A SUBSTANTIAL IMPEDIMENT TO TECHNCLOGICAL INNOVATION. THE BILL'S IMPACT ON EXPORTS AND IMPORTS IS CONSIDERED. SETTING STANDARDS IN THIS COUNTRY WILL CONSTITUTE HIDDEN TRADE BARRIERS. IF ANY EXPORT HAS AN EFFECT ON THE GLOBAL ENVIRONMENT, IT WILL COME WITHIN THE PROPOSED LEGISLATION. INCLUDED ARE NUMEROUS DETAILED CHEMICAL RESEARCH REPORTS ON THE TOXIC SUBSTANCES UNDER DISCUSSION. (BEARDSLEY-FLORIDA)

FIELD 05G, 06E, 05C

ACCESSION NO. W73-00703

AMCHITKA BIOENVIRONMENTAL PROGRAM. RESEARCH PROGRAM ON MARINE ECOLOGY, AMCHITKA ISLAND, ALASKA. JULY 1, 1970-JUNE 30, 1971,

WASHINGTON UNIV., SEATTLE. FISHERIES RESEARCH INST.

R. L. BURGNER, AND R. E. NAKATANI.

AVAILABLE FROM NTIS, SPRINGFIELD, VA., AS BMI-171-144; $3.00 IN PAPER COPY; $0.95 IN MICROFICHE. REPCRT BMI-171-144, 1972. 79 P, 10 FIG, 18 TAB, 40 REF.

DESCRIPTORS:
*NUCLEAR EXPLOSIONS, *ALASKA, *MARINE ALGAE, *FISHERIES, FISH POPULATIONS, FISH BEHAVIOR, INVERTEBRATES, OTTERS, ROCKSLIDES, INTERTIDAL AREAS, ON-SITE INVESTIGATIONS, FOOD CHAINS, SPECIATION, REPRODUCTION.

IDENTIFIERS:
AMCHITKA ISLAND, MILROW TEST, CANNIKIN TEST.

ABSTRACT:
EFFECTS WERE ESTIMATED ON THE MARINE ENVIRONMENT OFF AMCHITKA ISLAND FROM THE MILROW TEST (1969), AND BASELINE DATA WERE OBTAINED PRIOR TO THE CANNIKIN TEST. SEA OTTER STOMACH CONTENTS, FISH, INVERTEBRATES, AND ALGAE WERE SAMPLED. NO MILROW EFFECTS ON NEARSHORE FISHES WERE OBSERVED; BUT OCEAN PERCH WERE NOT FOUND IN AREAS OF PREVIOUS ABUNDANCE, AND PREVICUSLY UNSEEN JUVENILE OFFSHORE FISHES WERE PREDOMINANT. MILROW-PRODUCED ROCK FALLS WERE COLONIZED WITHIN 1 YEAR. ADDITIONAL CONTINUING STUDIES INCLUDE ALGAL AND INVERTEBRATE SUCCESSION ON ROCKS DISPLACED BY MILROW, AND TAXONOMY AND REPRODUCTIVE CYCLES OF ALGAE. (BOPP-ORNL)

FIELD 05C

ACCESSION NO. W73-00764

PLUTCNIUM CONCENTRATION ALONG FRESH-WATER FOOD CHAINS OF THE GREAT LAKES. PROGRESS, 1971-72.

WOODS HOLE OCEANOGRAPHIC INSTITUTION, MASS.

AVAILABLE FROM NTIS, SPRINGFIELD, VA., AS COO-3568-1; $3.00 IN PAPER COPY, $0.95 IN MICROFICHE. REPORT COO-3568-1, 1972. 26 P, 5 FIG, 6 TAB, 7 REF.

DESCRIPTORS:
*GREAT LAKES, *STRONTIUM RADIOISOTOPES, *FALLOUT, *FOOD CHAINS, ABSORPTION, SEDIMENTS, CORES, ANALYTICAL TECHNIQUES, RADIOACTIVITY TECHNIQUES, RADIOECOLOGY, TRACERS, MONITORING, NUCLEAR POWERPLANTS, NUCLEAR WASTES, PATH OF POLLUTANTS, WATER POLLUTICN SOURCES, PUBLIC HEALTH, ON-SITE INVESTIGATIONS, WASTE ASSIMILATIVE CAPACITY, WASTE DILUTION, BOTTOM SEDIMENTS, DECOMPOSING ORGANIC MATTER, RADIOCHEMICAL ANALYSIS, ALGAE.

IDENTIFIERS:
*PLUTONIUM, CESIUM RADIOISOTOPES.

ABSTRACT:
LESS THAN 10% OF THE PU239 DEPOSITED BY FALLCUT IS PRESENTLY DETECTABLE IN LAKE ONTARIO WATER; AND ALTHOUGH ENVIRONMENTAL FACTORS DIFFER, THE PU CONTENT OF THE WATER IS SIMILAR IN LAKE MICHIGAN AND FALMOUTH RESERVOIR. THE HYPOTHESIS THAT BIOLOGICAL SEDIMENTATION IS CONTROLLING ACCOUNTS FOR INCREASED PU OBSERVED AT DEPTHS. AN EAST-WEST CONCENTRATION GRADIENT SUGGESTS THAT NIAGRA CUTFALL PLAYS A KEY ROLE. THERE WAS NO INDICATION OF INCREASED PU IN THE WATER OR ALGAE NEAR A NUCLEAR POWERPLANT, ALTHOUGH CS134 WAS DETECTABLE IN ALGAE NEAR THE POWERPLANT AND IN ALEWIFE FISH. THE CONCENTRATION FACTOR FOR PU239 UPTAKE BY ALGAE WAS 1-2 THOUSAND WITH A MEAN WATER CONCENTRATION OF 1.7 DPM/1000 KG. THE RATIO PU238/PU239 WAS HIGHER FOR THE WATER THAN FOR SEDIMENT (CORRESPONDING TO THE SNAP-9A BURNUP IN 1964). PRELIMINARY MEASUREMENTS INDICATED 50% MORE SR90 IN OPEN LAKE ONTARIO WATER THAN IN LAKE MICHIGAN. SR90 FOUND IN SEDIMENT WAS ONLY 1% OF THAT ESTIMATED FOR THE WATER COLUMN. (BOPP-ORNL)

FIELD 05B, 02H

ACCESSION NO. W73-00790

EFFECTS OF ENVIRONMENTAL STRESS ON THE COMMUNITY STRUCTURE AND PRODUCTIVITY OF SALT MARSH EPIPHYTIC COMMUNITIES. PROGRESS REPORT, SEPTEMBER 1, 1971-AUGUST 31, 1972,

CITY COLL., NEW YORK.

J. J. LEE.

AVAILABLE AS COO-3054-3(PT. 1) FROM NTIS, SPRINGFIELD, VA., $3.00 IN PAPER COPY, $0.95 IN MICROFICHE. REPORT COO-3054-3 (PT. 1), 1972. 76 P, 45 FIG, 9 TAB, 4 REF.

DESCRIPTORS:
*MARINE ALGAE, *SALT MARSHES, *PROTOZOA, *BACTERIA, FLORIDA, RADIATION SENSITIVITY, RADIOECOLOGY, ECOLOGY, CYCLING NUTRIENTS, THERMAL POLLUTION, TROPHIC LEVEL, TROPICAL REGIONS, ULTRAVIOLET RADIATION, FOOD CHAINS, GAMMA RAYS, RESISTANCE, BIOLOGICAL COMMUNITIES, TIDAL MARSHES, FOOD WEBS, INTERTIDAL AREAS, CONNATE WATER.

IDENTIFIERS:
FORAMINIFERA.

ABSTRACT:
BY MEANS OF STUDIES ON LABELED UPTAKE AND THE GROWTH OF ALGAE AND BACTERIA, AN AUXOTROPIC PROFILE OVER A GROWING SEASON WAS OBTAINED. RECYCLED ORGANIC NUTRIENTS PLAY A CENTRAL ROLE IN THE NUTRITION, REGULATION, AND SUCCESSION WITHIN THE ALGAL COMPARTMENT. SENSITIVITY TO IONIZING RADIATION AND THERMAL STRESS VARIES WIDELY. AT 40 C, SOME OF THE THERMALLY MORE RESISTANT SPECIES HAVE A MINIMAL LETHAL DOSE (MLD) GREATER THAN 135 MIN., AND ARE ALSO MORE RESISTANT TO ULTRAVIOLET (MLD GREATER THAN 45,000 ERGS) AND GAMMA (MLD ABOUT 10 MILLION ERGS) IRRADIATION. THIS COMBINATION OF RESISTANCE TO STRESSES SEEMS TO BE AN ADAPTATION TO GROWTH OR REPEATED EXPOSURE AT EBB TIDE IN HOT SHALLOW TIDE POOLS OR IN INTERSTITIAL WATERS. LIGHT QUALITY IS ALSO IMPORTANT IN SELECTIVELY REGULATING GROWTH AND SUCCESSION AMONG THE ALGAE. THE TROPHIC DYNAMICS, MINERAL CYCLING, AND NICHES OF SUBLITTORAL FORAMINIFERA WERE ALSO STUDIED. (BOPP-ORNL)

FIELD 05C, 02L

ACCESSION NO. W73-00796

DISTRIBUTION OF RADIOACTIVE AND STABLE ZINC IN AN EXPERIMENTAL MARINE ECOSYSTEM,

NATIONAL MARINE FISHERIES SERVICE, BEAUFORT, N.C. ATLANTIC COASTAL FISHERIES CENTER.

F. A. CROSS, J. N. WILLIS, AND J. P. BAPTIST.

JOURNAL FISHERIES RESEARCH BOARD OF CANADA, VOL 28, NO 11, P 1783-1788, 1971. 4 TAB, 7 REF.

DESCRIPTORS:
*ZINC RADIOISOTOPES, *ABSORPTION, *MARINE ALGAE, *PHYTOPLANKTON, BACTERIA, AQUARIA, DECOMPOSING ORGANIC MATTER, WASTE ASSIMILATIVE CAPACITY, EUTROPHICATION, NUTRIENT REMOVAL, *ZINC, CHELATION, FOOD CHAINS, PATH OF POLLUTANTS, PUBLIC HEALTH, WATER POLLUTION EFFECTS, RADIOECOLOGY, RADIOACTIVITY TECHNIQUES, ANALYTICAL TECHNIQUES.

ABSTRACT:
THE BEHAVIOR OF ZN65 TRACER WAS STUDIED IN A PLANKTONIC BLOOM IN 2000 LITERS OF SEAWATER IN AN AQUARIUM. THE ZN65 SPECIFIC ACTIVITY WAS SIGNIFICANTLY LESS FOR THE PARTICULATE FRACTION WHICH CONSISTED MAINLY OF CHLORELLA SP., NITZSCHIA CL., AND BACTERIA. POSSIBLY EXCHANGE OF ZN WAS RELATIVELY SLOW, OR ORGANICALLY-COMPLEXED ZN WAS PRESENT WHICH DID NOT EXCHANGE READILY WITH DISSOLVED INORGANIC ZN BUT WAS AVAILABLE TO PHYTOPLANKTON. CONCENTRATION FACTORS WERE 980 FOR ZN65 AND 1400 FOR TOTAL ZN (FOR THE MIXED PHYTOPLANKTON COMMUNITY ALONG WITH BACTERIA AND DETRITUS). (BOPP-ORNL)

FIELD 05B, 05C

ACCESSION NO. W73-00802

RADIOECOLOGY OF CERTAIN MOLLUSCS IN INDIAN COASTAL WATERS,

BHABHA ATOMIC RESEARCH CENTER, BOMBAY (INDIA).

B. PATEL, P. G. VALANJU, C. D. MULAY, M. C. BALANI, AND

AVAILABLE FROM UNIPUB INC., P.O. BOX 433, NEW YORK, N.Y. 10016. IN: SYMPOSIUM ON THE INTERACTION OF RADIOACTIVE CONTAMINANTS WITH THE CONSTITUENTS OF THE MARINE ENVIRONMENT, JULY 10-14, 1972, SEATTLE, WASHINGTON. PAPER IZEA/SM-158/13. 34 P, 5 FIG, 11 TAB, 61 REF.

DESCRIPTORS:
*REVIEWS, *NUCLEAR WASTES, *BIOINDICATORS, *MARINE ANIMALS, ASIA, MOLLUSKS, COBALT RADIOISOTOPES, IODINE RADIOISOTOPES, MARINE ALGAE, TRACE ELEMENTS, PATH OF POLLUTANTS, ABSORPTION, RADIOISOTOPES, STABLE ISOTOPES, ESTUARINE ENVIRONMENT, COASTS, ANIMAL PHYSIOLOGY, SEDIMENTS, ANIMAL METABOLISM.

IDENTIFIERS:
CESIUM RADIOISOTOPES.

ABSTRACT:
UPTAKE BY MARINE ORGANISMS, WHICH MAY SERVE AS BIOINDICATORS OF CERTAIN RADIONUCLIDES AND STABLE ELEMENTS, IS REVIEWED. MEASUREMENTS ARE REPORTED FOR THE MOLLUSC, APLYSIA BENEDICTI, WHICH MOVES SHOREWARDS IN COASTAL WATERS DURING COLDER MONTHS. HEPATOPANCREAS AND GUT CONTAINED MOST CO; THE HORNY INTERNAL SHELL AND THE SPAWN CHORDS, MOST I. UPTAKE OF CS137 WAS FOUND WHEN ALGAL FOOD WAS PRESENT. CONCENTRATION FACTORS WERE (IN UNITS OF TEN THOUSANDS): CO, 5-11, NI, 1-2; FE, 5-12; MN, 2-3; CU, 2-8; ZN, 2-4; SR, 0.001-0.002; CO60, 5; CO58, 2. THE DIFFERENCES BETWEEN THE CO ISOTOPES RESULT FROM DIFFERENCES IN HALF-LIVES. (BOPP-ORNL)

FIELD 05B, 02L

ACCESSION NO. W73-00811

BIOLOGICAL PHYSICOCHEMICAL ASPECTS OF RADIOACTIVE CONTAMINATION OF MARINE ORGANISMS AND OF MARINE SEDIMENTS (ASPECTS BIOLOGIQUES ET PHYSICO-CHIMIQUES DE LA CONTAMINATION RADIOACTIVE D'ESPECES ET DE SEDIMENTS MARINE),

COMMISSARIAT A L'ENERGIE ATOMIQUE, CHERBOURG (FRANCE). CENTRE DE LA HAGUE.

J. ANCELLIN.

AVAILABLE FROM UNIPUB INC., P.O. BOX 433, NEW YORK, N.Y. 10016. IN: SYMPOSIUM ON THE INTERACTION OF RADIOACTIVE CONTAMINANTS WITH THE CONSTITUENTS OF THE MARINE ENVIRONMENT, JULY 10-14, 1972, SEATTLE, WASHINGTON. PAPER IAEA/SM-158/19. 32 P, 4 FIG, 4 TAB, 60 REF.

DESCRIPTORS:
*REVIEWS, *RADIOISOTOPES, *MARINE ANIMALS, *MARINE ALGAE, SEDIMENTS, NUCLEAR WASTES, PATH OF POLLUTANTS, SEA WATER, WATER ANALYSIS, RADIOACTIVITY TECHNIQUES, RADIOECOLOGY, ENVIRONMENTAL EFFECTS, ABSORPTION, SORPTION, WASTE DILUTION, WASTE ASSIMILATIVE CAPACITY, EUROPE, ANIMAL PHYSIOLOGY, TRACE ELEMENTS, HEAVY METALS, PHYSICOCHEMICAL PROPERTIES, FOOD CHAINS, PUBLIC HEALTH.

ABSTRACT:
UPTAKE BY MARINE LIFE OF RADIONUCLIDES FROM SEA WATER AND SEDIMENTS IS REVIEWED, INCLUDING SOME ASPECTS OF FOOD CHAINS. THE CONCENTRATION FACTORS MAY DECREASE WITH INCREASING CONCENTRATION IN THE WATER, IN THE CASE OF TRACE ELEMENTS WHICH HAVE PHYSIOLOGICAL FUNCTIONS (ZN, CU, FE, MG, MN, CO, NI), BECAUSE OF METABOLIC REGULATION. HOWEVER, THE EFFECT MAY DIFFER BETWEEN SPECIES, AND BETWEEN ORGANS OF A GIVEN ORGANISM. MANY WORKS ARE CITED WHICH SHOW THAT ENVIRONMENTAL RADIONUCLIDES WERE MORE READILY ASSIMILATED THAN THE STABLE ISOTOPES OF THE SAME ELEMENT, SINCE THE LATTER ARE PRESENT IN A DIFFERENT PHYSICOCHEMICAL FORM, HAVING UNDERGONE MORE AGING. THE PERMANENCE OF RADIONUCLIDE FIXATION BY SEDIMENTS REQUIRES STUDY, ESPECIALLY WITH REGARD TO SEDIMENT-FEEDING ORGANISMS. (BOPP-ORNL)

FIELD 05B, 02J

ACCESSION NO. W73-00817

FALLOUT MN54 ACCUMULATED BY BAY SCALLOPS ARGOPECTEN IRRADIANS (LAMARCK) NEAR BEAUFORT, NORTH CAROLINA,

MICHIGAN UNIV., ANN ARBOR. GREAT LAKES RESEARCH DIV.

C. L. SCHELSKE.

AVAILABLE FROM UNIPUB INC., P.O. BOX 433, NEW YORK, N.Y. 10016. IN: SYMPOSIUM ON THE INTERACTION OF RADIOACTIVE CONTAMINANTS WITH THE CONSTITUENTS OF THE MARINE ENVIRONMENT, JULY 10-14, 1972, SEATTLE, WASHINGTON. PAPER IAEA/SM-158/20. 19 P, 3 FIG, 5 TAB, 16 REF.

DESCRIPTORS:
*BIOINDICATORS, *FALLOUT, *MOLLUSKS, *ATLANTIC COASTAL PLAIN, NORTH CAROLINA, ABSORPTION, SOUNDS, ESTUARINE ENVIRONMENT, PATH OF POLLUTANTS, FOOD CHAINS, PHYTOPLANKTON, MARINE ALGAE, ENVIRONMENTAL EFFECTS, TRACE ELEMENTS, COLLOIDS, NUCLEAR WASTES, RADIOECOLOGY.

IDENTIFIERS:
MANGANESE RADIOISOTOPES, BOGUE SOUND, CORE SOUND.

ABSTRACT:
FALLOUT MN54 IN SCALLOP KIDNEYS PEAKED IN 1964, AND DECREASED FROM 1964-1966 WITH AN EFFECTIVE HALF-LIFE (240 DAYS) GREATER THAN THE BIOLOGICAL HALF-LIFE (30-50 DAYS) MEASURED IN THE LABORATORY. THE DIFFERENCE BETWEEN THE TWO IS RELATED TO THE SUPPLY FROM ENVIRONMENTAL RESERVOIRS. EXPERIMENTS SHOWED THAT MN54 TAKEN UP BY INGESTION OF LABELLED PHYTOPLANKTON WAS RETAINED LONGER (21% REDUCTION IN 21 DAYS) THAN THAT TAKEN UP DIRECTLY FROM SEA WATER (50% REDUCTION IN THE SAME TIME). SCALLOPS SAMPLED FROM SEVERAL LOCATIONS IN BOGUE SOUND AND AT BELLS ISLAND IN CORE SOUND CONTAINED MORE MN54 THAN THOSE FROM OTHER LOCATIONS IN CORE SOUND. (BOPP-ORNL)

FIELD 05B, 02L

ACCESSION NO. W73-00818

RADICSTRONTIUM UPTAKE BY MARINE ORGANISMS (INCORPORACION DE RADIOESTRONCIO POR ORGANISMOS MARINOS),

COMISION NACIONAL DE ENERGIA ATOMICA, BUENOS AIRES (ARGENTINA).

D. CANCIO, J. A. LLAURO, N. R. CIALLELLA, AND D. J. BENINSCN.

AVAILABLE FROM UNIPUB INC., P.O. BOX 433, NEW YORK, N.Y. 10016. IN: SYMPOSIUM ON THE INTERACTION OF RADIOACTIVE CONTAMINANTS WITH THE CONSTITUENTS OF THE MARINE ENVIRONMENT, JULY 10-14, 1972, SEATTLE, WASHINGTON. PAPER IAEA/SM-158/21. 14 P, 8 TAB, 17 REF.

DESCRIPTORS:
*STRONTIUM RADIOISOTOPES, *MARINE ANIMALS, *MARINE ALGAE, *FOOD CHAINS, ABSORPTION, CALCIUM, PUBLIC HEALTH, SOUTH AMERICA, ATLANTIC COASTAL PLAIN, COASTS, CRUSTACEANS, MARINE FISH, MOLLUSKS, POLLUTANT IDENTIFICATION.

IDENTIFIERS:
ARGENTINA.

ABSTRACT:
UPTAKE OF SR AND CA BY ORGANISMS (FISH, CRUSTACEANS, MOLLUSKS, AND ALGAE) OF THE ARGENTINA ATLANTIC COAST WAS STUDIED IN THE LABORATORY, BOTH BY DOSING WITH CA AND SR AND BY A DOUBLE-TRACER TECHNIQUE. DISCRIMINATION FACTORS (ACTIVITY OF SR/G CA, IN THE ORGANISM)/(ACTIVITY OF SR/G CA, IN SEA WATER) WERE NEARLY THE SAME FOR ALL PARTS OF A SPECIES OF FISH (0.23-0.34) AND FOR THE EXCSKELETCN AND SOFT PARTS OF CRUSTACEANS (0.71-0.95), BUT DIFFERED BETWEEN THE SHELL (0.07-0.12) AND THE SOFT PARTS (0.36-0.54) OF MOLLUSKS. FIELD RESULTS GAVE REASONABLE AGREEMENT. CONCENTRATION FACTORS FOR UPTAKE OF SR WERE: FISH, 2-10, CRUSTACEANS AND MOLLUSKS, 13-70; ALGAE, 0.2-70. (BOPP-ORNL)

FIELD 05B, 05A

ACCESSION NO. W73-00819

THE STATE OF COBALT IN SEA WATER AND ITS UPTAKE BY MARINE CRGANISMS AND SEDIMENTS,

PUERTO RICO NUCLEAR CENTER, RIO PIEDRAS. RADIOECOLOGY DIV.

F. G. LOWMAN, AND R. Y. TING.

AVAILABLE FROM UNIPUB INC., P.O. BOX 433, NEW YORK, NY 10016. IN: SYMPOSIUM CN THE INTERACTION OF RADIOACTIVE CONTAMINANTS WITH THE CONSTITUENTS OF THE MARINE ENVIRONMENT, JULY 10-14, 1972, SEATTLE, WASHINGTON. PAPER IAEA/SM-158/23. 25 P, 10 FIG, 2 TAB, 6 REF.

DESCRIPTORS:
*COBALT RADIOISOTOPES, *COBALT, *ABSORPTION, *MARINE ANIMALS, MARINE ALGAE, SEDIMENTS, PUERTO RICO, ESTUARINE ENVIRONMENT, PATH OF POLLUTANTS, FALLOUT, NUCLEAR WASTES, RADIOECOLOGY, RADIOACTIVITY TECHNIQUES, ANALYTICAL TECHNIQUES, CYCLING NUTRIENTS, TRACE ELEMENTS, CHELATION, ORGANIC COMPOUNDS, FOOD CHAINS, PUBLIC HEALTH, PHYTOPLANKTON, ZOOPLANKTON, WATER ANALYSIS, CATION EXCHANGE, RADIOCHEMICAL ANALYSIS.

ABSTRACT:
ANALYTICAL TECHNIQUES WERE TESTED TO MEASURE IONIC AND COMPLEXED CO IN SEA WATER USING IONIC CO58 AND VITAMIN B12 TAGGED WITH CO57 AS TRACERS. SEA WATER COLLECTED NEAR PUERTO RICO, WHICH CONTAINED 0.043 MICROG/LITER OF IONIC AND 0.03 MICROG/LITER OF COMPLEXED CO AND THE ABOVE TRACERS ADDED IN AMOUNTS TO GIVE ROUGHLY THE SAME RADIOACTIVITIES FOR CO58 AND CO57, WAS USED IN RADIONUCLIDE UPTAKE EXPERIMENTS WITH MARINE ORGANISMS. AFTER ABOUT 17 DAYS, RATIOS OF UPTAKE OF ORGANIC TRACER TO IONIC TRACER WERE: PHYTOPLANKTON, 15; ZOOPLANKTON, 15; BRINE SHRIMP, 15; CRAB LARVAE, 5; CLAM SOFT PARTS, 4; PERIPHYTON, 4; PELAGIC MACRURAN CRUSTACEA, 0.9; ESTUARINE SEDIMENTS, 0.5; SPINY LOBSTER MOLT, 0.5; SPINY LOBSTER (WHOLE), 0.4; FRESH WATER PHYTOPLANKTON, 0.2; CRAB SHELL, 0.03. (BOPP-ORNL)

FIELD 05A, 05B

ACCESSION NO. W73-00821

THE KINETICS OF AND A PRELIMINARY MODEL FOR THE UPTAKE OF RADIOZINC BY PHAEODACTYLUM TRICORNUTUM IN CULTURE,

INTERNATIONAL LAB. OF MARINE RADIOACTIVITY, MONTE CARLC (MCNACO).

A. G, DAVIES.

AVAILABLE FROM UNIPUB INC., P.O. BCX 433, NEW YORK, N.Y. 10016. IN: SYMPOSIUM ON THE INTERACTION OF RADIOACTIVE CONTAMINANTS WITH THE CCNSTITUENTS OF THE MARINE ENVIRONMENT, JULY 10-14, 1972, SEATTLE, WASHINGTON. PAPER IAEA/SM-158/25. 20 P, 7 FIG, 4 TAB, 14 REF.

DESCRIPTORS:
*ZINC RADIOISOTOPES, *ABSORPTION, *PHYTOPLANKTON, *MARINE ALGAE, FOOD CHAINS, PUBLIC HEALTH, NUCLEAR WASTES, PATH CF POLLUTANTS, DIFFUSION, CYTOLOGICAL STUDIES, ADSORPTION, PENETRATION, CATION ADSORPTION, CHELATION, MATHEMATICAL MODELS.

ABSTRACT:
INTEGRATION OF A MATHEMATICAL MODEL BY COMPUTER GAVE SATISFACTORY AGREEMENT WITH EXPERIMENT. IT WAS ASSUMED THAT THE RATE OF UPTAKE BY INTRACELLULAR PROTEIN IS DIFFUSION-CONTROLLEC, AND THAT THE EQUILIBRIUM CONTENT OF INTRACELLULAR PROTEIN DECREASES WITH CEPLETION OF NUTRIENT SALTS OF P AND EXPECIALLY N. LANGMUIR ADSORPTICN ISOTHERMS WERE USED TO REPRESENT EQUILIBRIA. THE EVIDENCE INDICATES THAT THE UPTAKE OF ZINC BY PHAEDACTYLUM TRICORNUTUM IS LARGELY PASSIVE AND DIFFUSION CONTROLLED AND THAT THE ZINC CONTENT OF THE CELLS IS PROBABLY RELATED TO INTRA-CELLULAR PROTEIN LEVEL. (BOPP-ORNL)

FIELD 05B

ACCESSION NO. W73-00823

ACCUMULATION AND LCSS OF COBALT AND CESIUM BY THE MARINE CLAM, MYA ARENARIA, UNDER LABORATORY AND FIELD CONDITIONS,

CALIFCRNIA UNIV., LIVERMORE. LAWRENCE LIVERMCRE LAB.

F. L. HARRISON,

AVAILABLE FROM UNIPUB INC., P.O. BCX 433, NEW YORK, N.Y. 10016. IN: SYMPOSIUM ON THE INTERACTION OF RADIOACTIVE CONTAMINANTS WITH THE CONSTITUENTS OF THE MARINE ENVIRONMENT, JULY 10-14, 1972, SEATTLE, WASHINGTON. PAPER IAEA/SM-158/28. 35 P, 10 FIG, 6 TAB, 24 REF.

DESCRIPTORS:
*MARINE ANIMALS, *CLAMS, *FOOD CHAINS, *NUCLEAR WASTES, *COBALT RADIOISOTOPES, PATH OF POLLUTANTS, WASTE ASSIMILATIVE CAPACITY, *CALIFORNIA, RADICECOLOGY, MARINE ALGAE, ESTUARINE ENVIRCNMENT, ABSORPTION, ANIMAL POPULATIONS, ANIMAL PHYSIOLOGY, RADIOACTIVITY TECHNIQUES, CYCLING NUTRIENTS, TRACE ELEMENTS, PUBLIC HEALTH, RADIOECOLOGY.

IDENTIFIERS:
*HUMBOLDT BAY, *CESIUM RADIOISOTOPES.

ABSTRACT:
UPTAKE OF RADICNUCLIDES FROM DIET AND WATER ARE BEING COMPARED BETWEEN AQUARIA AND THE HUMBOLDT BAY, CALIFORNIA, REACTOR DISCHARGE CANAL. STEADY-STATE CONCENTRATION FACTORS IN EDIBLE TISSUES (ABOUT 5 FOR CS AND 200 FOR CO), BIOLCGICAL HALF-TIMES FOR UPTAKE (2-15 DAYS FOR CS AND 50-120 DAYS FOR CO), BIOLOGICAL LOSS-RATE CONSTANTS, AND TURNOVER RATES IN THE VARIOUS BODY PARTS ARE TABULATEC. LITTLE METABOLIC REGULATION OF CS UPTAKE WAS OBSERVED IN THE RANGE 0.5-12.5 MICRCGRAMS/LITER, BUT SOME REGULATION OF CO UPTAKE OCCURRED. FOOD IN THE FORM OF SUSPENSICNS OF RADIOACTIVE DIATOMS INCREASED UPTAKE OVER THAT FRCM WATER ALONE WITH CO60, BUT NOT WITH CS137. (BOPP-ORNL)

FIELD 05C, 02L

ACCESSION NO. W73-00826

LEVELS OF RADIOACTIVITY IN THE MARINE ENVIRONMENT AND THE DOSE COMMITMENT TO MARINE ORGANISMS,

MINISTRY OF AGRICULTURE, FISHERIES AND FOOD, LOWESTOFT (ENGLAND). FISHERIES RADIOBIOLOGICAL LAB.

D. S. WOODHEAD.

AVAILABLE FROM UNIPUB INC., P.O. BOX 433, NEW YORK, N.Y. 10016. IN: SYMPOSIUM ON THE INTERACTION OF RADIOACTIVE CONTAMINANTS WITH THE CONSTITUENTS OF THE MARINE ENVIRONMENT, JULY 10-14, 1972, SEATTLE, WASHINGTON. PAPER IAEA/SM-158/31. 34 P, 2 FIG, 15 TAB, 63 REF.

DESCRIPTORS:
*BACKGROUND RADIATION, *RADIOACTIVITY EFFECTS, *FALLOUT, *MARINE ANIMALS, MARINE PLANTS, MARINE ALGAE, NUCLEAR WASTES, PATH OF POLLUTANTS, RADIOECOLOGY, MATHEMATICAL MODELS, WASTE ASSIMILATIVE CAPACITY, ESTUARINE ENVIRONMENT, FOOD CHAINS, PUBLIC HEALTH, EUROPE, RADIOISOTOPES, EFFLUENTS.

IDENTIFIERS:
WINDSCALE PROCESSING PLANT.

ABSTRACT:
DATA ARE REVIEWED ON SEA-WATER AND SEA-BED RADIOACTIVITY; AND, BASED ON IDEALIZED MODELS, DOSE RATES (MICRORADS/HOUR) ARE CALCULATED FOR FIVE GROUPS OF MARINE BIOTA (1) PHYTOPLANKTON, (2) ZOOPLANKTON, (3) MOLLUSCA, (4) CRUSTACEA, AND (5) FISH. VALUES ARE OBTAINED FOR: (A) NATURAL SOURCES OF RADIOACTIVITY - (1) 2.8-8.2, (2) 3.3-16.4, (3) 9.5-31.5, (4) 10.0-38.0, (5) 3.3-20.8; (B) FALLOUT - (1) 0.26-24.5, (2) 1.4-147, (3) 0.10-8.0, (4) 0.36-0.46, (5) 0.14-42.7; (C) WASTE DISPOSAL OPERATIONS IN THE WINDSCALE PROCESSING PLANT VICINITY - (1) 200-2100, (2) 530-6900, (3) 51.8-3400, (4) 43.3-3410, (5) 0.6-3340. ALTHOUGH THE PRACTICE OF MARINE DISPOSAL LEADS TO LOCALLY HIGH DOSE RATES, IN POPULATION OR GLOBAL TERMS THE DOSE RATE CONTRIBUTION IS AND WILL PROBABLY REMAIN NEGLIGIBLE. IN THE CASE OF SR90, CS137, H3, C14, AND PU239, FOR WHICH THERE ARE EXTENSIVE DATA, APPROXIMATE TISSUE LEVELS CAN BE DERIVED USING THE APPROPRIATE CONCENTRATION FACTORS. (BOPP-ORNL)

FIELD 05B, 05C

ACCESSION NO. W73-00828

RADIONUCLIDE TRANSPORT STUDIES IN THE HUMBOLDT BAY MARINE ENVIRONMENT,

CALIFORNIA UNIV., LIVERMORE. LAWRENCE LIVERMORE LAB.

R. E. HEFT, W. A. PHILLIPS, H. R. RALSTON, AND W. A. STEEL.

AVAILABLE FROM UNIPUB INC., PO BOX 433, NEW YORK, NY 10016. IN: SYMPOSIUM ON THE INTERACTION OF RADIOACTIVE CONTAMINANTS WITH THE CONSTITUENTS OF THE MARINE ENVIRONMENT, JULY 10-14, 1972, SEATTLE, WASHINGTON. PAPER IAEA/SM-158/37. 25 P, 3 FIG, 10 TAB, 6 REF.

DESCRIPTORS:
*PATH OF POLLUTANTS, *NUCLEAR POWERPLANTS, *RADIOACTIVITY EFFECTS, *MATHEMATICAL MODELS, FORECASTING, COBALT RADIOISOTOPES, MARINE ALGAE, ABSORPTION, SEDIMENTS, WASTE DILUTION, WASTE ASSIMILATIVE CAPACITY, CALIFORNIA, TRITIUM, MARINE PLANTS, MONITORING, SAMPLING, WATER ANALYSIS, SEA WATER, STABLE ISOTOPES, RADIOISOTOPES, NUCLEAR WASTES, ESTUARINE ENVIRONMENT, ON-SITE INVESTIGATIONS, LABORATORY TESTS, TURNOVERS, WATER CIRCULATION.

IDENTIFIERS:
CESIUM RADIOISOTOPES.

ABSTRACT:
FROM JUNE 1971 TO MARCH 1972 AT 5-6 WEEK INTERVALS, SAMPLES OF BOTTOM SEDIMENT, WATER PLUS SUSPENDED SEDIMENT, EELGRASS, AND ENTEROMORPHA (ALGAE) WERE ANALYSED FOR GAMMA-EMITTING RADIONUCLIDES AND STABLE ELEMENTS. ABOUT 1/4 THE TOTAL ESTIMATED ANNUAL RELEASE FROM THE WASTE TREATMENT SYSTEM OF A 65-MW BOILING-WATER REACTOR WAS FOUND IN THE BAY FOR CO60 AND OTHER RADIONUCLIDES (EXCEPTING TRITIUM WHICH WAS NEAR OCEANIC BACKGROUND AND CS137 WHICH WAS LARGELY FROM FALLOUT). EXPERIMENTS ARE PLANNED IN WHICH THE RADIOACTIVITY IN THE WATER IN THE DISCHARGE CANAL WILL BE MONITORED, AND UPTAKE WILL BE DETERMINED BY VARIOUS BOTTOM SEDIMENTS AND ALGAL SPECIES. AUXILIARY EXPERIMENTS WILL BE CONDUCTED IN THE LABORATORY. THE LONG-RANGE OBJECTIVE IS TO INCORPORATE PARAMETERS INTO AN ADVECTION/DIFFUSION MODEL FOR THE BAY IN ORDER TO PREDICT COMPARTMENTAL RADIONCULIDE CONCENTRATION. (BOPP-ORNL)

FIELD 05B, 02L, 05C

ACCESSION NO. W73-00831

ACCUMULATION OF CERTAIN TRACE ELEMENTS IN MARINE ORGANISMS FROM THE SEA AROUND THE CAPE OF GOOD HOPE,

ATOMIC ENERGY BOARD, PELINDABA, PRETORIA (SOUTH AFRICA).

D. VAN AS, H. O. FOURIE, AND C. M. VLEGGAAR.

AVAILABLE FROM UNIPUB INC., P.O. BOX 433, NEW YORK, N.Y. 10016. IN: SYMPOSIUM ON THE INTERACTION OF RADIOACTIVE CONTAMINANTS WITH THE CONSTITUENTS OF THE MARINE ENVIRONMENT, JULY 10-14, 1972, SEATTLE, WASHINGTON. PAPER IAEA/SM-158-39. 12 P, 1 FIG, 3 TAB, 12 REF.

DESCRIPTORS:
*NUCLEAR WASTES, *MARINE ALGAE, *MARINE ANIMALS, *MARINE FISH, MUSSLES, LOBSTERS, ABSORPTION, COBALT RADIOISOTOPES, ZINC RADIOISOTOPES, MANGANESE, CHROMIUM, IRON, ANALYTICAL TECHNIQUES, FOOD CHAINS, AFRICA, EFFLUENTS, PATH OF POLLUTANTS, NEUTRON ACTIVATION ANALYSIS, SPECTROSCOPY, TRACE ELEMENTS, SAMPLING, WATER ANALYSIS, SEA WATER.

IDENTIFIERS:
ANTIMONY, CESIUM, *CAPE OF GOOD HOPE.

ABSTRACT:
IN CONNECTION WITH PREDICTING EFFECTS FROM THE RELEASE OF RADIOACTIVE EFFLUENTS TO THE SEA, EDIBLE MARINE LIFE WAS ANALYZED AND CONCENTRATION FACTORS WERE CALCULATED. PARTICULAR ATTENTION WAS PAID TO THE INFLUENCE OF CONTAINER MATERIAL ON LIQUID SAMPLES AND TO THE EFFECT OF VARIOUS METHODS OF SAMPLE TREATMENT. WITH ANALYSIS OF FE BY ATOMIC ABSORPTION, BETTER RESULTS WERE OBTAINED FOR AN ORGANIC EXTRACT THAN BY PERFORMING THE ANALYSIS DIRECTLY ON A SULFURIC ACID MEDIUM CONTAINING THE DIGESTED SAMPLE. THE CONCENTRATION FACTORS FOUND FELL IN THE RANGES: CRUSTACEANS (CR, 3-10 THOUSAND; FE AND ZN, 2-27 THOUSAND; CO, 6-50; MN, 300-2400; SB, 200-500; CS, 2-22), SEAWEEDS (CR, 600-12,000; FE AND ZN, 1-133 THOUSAND; CO, 8-94; MN, 400-9200; SB, 160-480); FISH (CR, 40-60,000; FE AND ZN, 2000-17000; CO, 2-15; MN, 100-1500; SB, 40-420, CS, 12-90). (BOPP-ORNL)

FIELD 05C

ACCESSION NO. W73-00832

CONTRIBUTIONS FROM THE ALPHA EMITTER, POLONIUM-210, TO THE NATURAL RADIATION ENVIRONMENT OF THE MARINE ORGANISMS,

SCRIPPS INSTITUTION OF OCEANOGRAPHY, LA JOLLA, CALIF.

T. R, FOLSOM, AND T. M. BEASLEY.

AVAILABLE FROM UNIPUB INC., P.O. BOX 433, NEW YORK, N.Y. 10016. IN: SYMPOSIUM ON THE INTERACTION OF RADIOACTIVE CONTAMINANTS WITH THE CONSTITUENTS OF THE MARINE ENVIRONMENT, JULY 10-14, 1972, SEATTLE, WASHINGTON. PAPER IAEA/SM-158/41. 10 P, 1 FIG, 5 TAB, 12 REF.

DESCRIPTORS:
*BACKGROUND RADIATION, *NUCLEAR WASTES, *ABSORPTION, *FOOD CHAINS, PUBLIC HEALTH, PATH OF POLLUTANTS, RADIOECOLOGY, RADIOACTIVITY EFFECTS, NUCLEAR POWERPLANTS, MARINE ALGAE, MARINE ANIMALS, MARINE PLANTS, HUMAN POPULATION, RADIOACTIVITY TECHNIQUES, RADIUM RADIOISOTOPES.

IDENTIFIERS:
*POLONIUM, PLUTONIUM.

ABSTRACT:
PO210 ACCUMULATES TO RELATIVELY HIGH LEVELS IN SOME MARINE ECOSYSTEMS WHICH ALSO ACCUMULATE PU EFFECTIVELY; THUS STUDY OF THE NATURAL RADIONUCLIDE MAY SERVE TO INDICATE THE PROBABLE FATE OF PU. THE DOSE FROM PO210 IN TISSUES IS ABOUT 12 MRADS/YEAR TO MAN (5-20% OF THE DOSE FROM NATURAL RADIATION), AND FROM 0.004 TO 800 MRADS/YEAR FOR THE TISSUES OF VARIOUS MARINE ORGANISMS. PREDICTION OF THE DOSE TO CERTAIN CRITICAL TISSUES IS DIFFICULT OWING TO THE EXTREMELY SHORT RANGE OF ALPHA PARTICLES. (BOPP-ORNL)

FIELD 05B

ACCESSION NO. W73-00833

HEAT TOLERANCE OF REEF ALGAE AT LA PARGUERA, PUERTO RICO,

GUSTAVUS ADOLPHUS CCLL., ST. PETER, MINN.

S. L. SCHWARTZ, AND L. R. ALMODOVAR.

NOVA HEDWIGIA, VOL. 21, NO. 1, P 231-240, 1971. 4 FIG, 1 TAB, 12 REF.

DESCRIPTORS:
*HEAT RESISTANCE, *MARINE ALGAE, LIGHT INTENSITY, *ECOLOGICAL DISTRIBUTION, SURFACE WATERS, LAGOCNS, CHLCROPHYTA, RHODOPHYTA, PHYSICAL PROPERTIES, WATER TEMPERATURE, CATA COLLECTIONS, MARINE PLANTS, ENVIRCNMENTAL EFFECTS, REEFS, LABORATORY TESTS, SAMPLING, PHAEOPHYTA.

IDENTIFIERS:
SURVIVAL, PENICILLUS, CHONDRIA, CLADOPHOROPSIS MEMBRANACEA, PENICILLUS CAPITATUS, CERAMIUM NITENS, WRANGELIA ARGUS, VALCNIA VENTRICOSA, BRYOPSIS PENNATA, SPYRIDIA FILAMENTOSA, DICTYOTA BARTAYRESII, LAURENCIA OBTUSA, LAURENCIA PAPILLOSA, COELOTHRIX IRREGULARIS, CENTROCERAS CLAVULATUM, HYPENA MUSCIFORMIS, CERAMILM BUSSOIDEUM, CHONDRIA LITTORALIS, HYPNEA SPINELLA.

ABSTRACT:
HEAT AND LIGHT TOLERANCES ARE DESCRIBED FOR REEF ALGAE COLLECTED FROM ENRIQUE REEF AT MAGUEYES ISLAND, LA PARGUERA. SIXTEEN SPECIES OF ALGAE WERE TESTED INCLUDING ELEVEN REDS, FOUR GREENS, AND ONE BROWN. ALGAE WERE COLLECTED IN GROUPS OF FOUR FROM 3 SPECIFIC AREAS: GROUP I - SHALLOW WATER AROUND BREAKING WAVES; GROUPS II AND III - SHALLCW, CALM WATER AT HIGH TIDE; AND GROUP IV - LAGOONS AT 6-8 FEET DEPTH. ALGAL SAMPLES WERE PLACED ON A SEAWATER-MOISTENED TOWEL CONTAINED IN A WHITE ENAMEL PAN; THE TOWEL WAS FREQUENTLY SPRAYED WITH SEAWATER TO AVOID DRYING. THE ALGAE WERE EXPOSED TO DIRECT SUNLIGHT AND THE LENGTH OF EXPOSURE, TEMPERATURE, AND RANGE FROM FIRST DEATH SIGNS TO DEATH WERE MEASURED FOR EACH. THE RESULTS SHOWED THAT MEMBERS OF RHODOPHYCOPHYTA, THE SHALLOW WATER ORGANISMS, SURVIVED EXTREME TEMPERATURES BETTER THAN THOSE MEMBERS OF CHLOROPHYCOPHYTA, DEEP LIVING ORGANISMS. FURTHERMORE, THE SPECIMENS ALLOWED TO DRY OUT IN THE SHADE SURVIVED LONGER THAT THOSE EXPOSED TO THE SUN. (LONG-BATTELLE)

FIELD 05C

ACCESSION NO. W73-00838

EUTROPHICATION,

NEW YORK STATE DEPT. OF ENVIRONMENTAL CONSERVATION, RONKCNKOMA.

J. FOEHRENBACH.

JOURNAL WATER POLLUTICN CONTROL FEDERATION, VOL 44, NO 6, P 1150-1159, JUNE 1972. 65 REF.

DESCRIPTORS:
*REVIEWS, *EUTROPHICATION, *NUTRIENTS, *WATER POLLUTICN CONTROL, ORGANIC WASTES, INDUSTRIAL WASTES, WASTE WATER(PCLLUTION), BIOASSAY, ECOSYSTEMS, BICDEGRADATION, ALGAE, WATER PCLLUTICN EFFECTS, HEAVY METALS, CYCLING NUTRIENTS, CARBON, NITROGEN, PHOSPHORUS, SULFUR, CALCIUM, MANGANESE, POTASSIUM, SODIUM, CARBON DICXIDE, CYANOPHYTA, FISH, OXYGEN, NITRATES, CHEMICAL PRECIPITATION, IRCN, PHOSPHATES, ANAEROBIC CONDITIONS, AMMONIA, MUD-WATER INTERFACES, POLLUTANTS, COPPER, TRACE ELEMENTS, WATER POLLUTION, OLIGOTRCPHY, COAGULATION, SEA WATER, FRESHWATER, BIOCHEMICAL OXYGEN DEMAND, FILTRATION, HYDROGEN ION CONCENTRATION.

IDENTIFIERS:
BAY OF QUINTE, SALT CREEK, CHARA, ELODEA, CALLITRICHE, ANABAENA FLOSAQUAE, SPIRULINA, GYMNODINIUM BREVE, PERIDINIUM, OSCILLATORIA RUBECENS, ANABAENA, CRUSSOSTREA, OSCILLATORIA.

ABSTRACT:
THE LITERATURE IS REVIEWED ON SEVERAL ASPECTS OF THE EUTROPHICATION PROCESS FOR SOME NATURAL BODIES OF WATER. SOME OF THE MORE PERTINENT ASPECTS OF THE EUTROPHICATION PROCESS INCLUDED ARE THE FOLLOWING: A REVIEW OF THE ROLE OF VARIOUS NUTRIENTS; ANALYSIS OF THE EFFECT OF DIFFERENT WASTE PRODUCTS ON SEVERAL BOCIES OF WATER; METHODS TO CONTROL THE RATE OF EUTROPHICATION OF A BODY OF WATER BY PROPER MANAGEMENT OF ITS WATERSHED; AND APPROACHES FOR THE REMOVAL OF NUTRIENTS FROM BODIES OF WATERS. (BYRD-BATTELLE)

FIELD 05C, 10F

ACCESSION NO. W73-00840

LABORATORY STUDIES OF ASSEBLAGES OF ATTACHED ESTUARINE DIATOMS,

OREGON STATE UNIV., CORVALLIS. DEPT. OF BOTANY.

B. L. WULFF, AND C. D. MCINTIRE.

LIMNOLOGY AND OCEANOGRAPHY, VOL 17, NO 2, P 200-214, MARCH 1972. 7 FIG, 8 TAB, 18 REF.

DESCRIPTORS:
*ESTUARIES, *DIATOMS, *SESSILE ALGAE, LIGHT INTENSITY, TIDAL EFFECTS, SALINITY, HEATED WATER, BIOMASS, THERMAL STRESS, GROWTH RATES, SPATIAL DISTRIBUTION, PRIMARY PRODUCTIVITY, CHRYSOPHYTA, BIOLOGICAL COMMUNITIES, ESTUARINE ENVIRONMENT, WATER POLLUTION SOURCES, BIODEGRADATION, ECOSYSTEMS, WATER POLLUTION.

IDENTIFIERS:
DESSICATION, SPECIES DIVERSITY, MELOSIRA NUMMULOIDES, CHLOROPHYLL A, *YAQUINA BAY, AMPHORA SPP, MELOSIRA SULCATA, NAVICULA SPP, SYNEDRA FASCICULATA, PLAGIOGRAMMA BROCKMANNI, AMPHIPLEURA RUTILANS, THALASSIOSIRA, THALASSIONEMA NITZSCHIOIDES, PLAGIOGRAMMA VANHEURCKII, PLEUROSIGMA, GYROSIGMA FASCIOLA, NITZSCHIA SPP, BACILLARIA PAXILLIFER, LICMOPHORA PARADOXA, FRAGILIARIA STRIATULA VAR CALIFCRNICA, COCCONEIS SCUTELLUM VAR PARVA, ACHNANTHESE TEMPEREI.

ABSTRACT:
EFFECTS OF LIGHT INTENSITY, EXPOSURE TO DESICCATION, SALINITY, AND HEATED WATER ON THE VERTICAL DISTRIBUTION AND GROWTH OF POPULATIONS OF ATTACHED ESTUARINE DIATOMS WERE STUDIED IN A LABORATCRY MODEL ECOSYSTEM. OF THE 35 MOST ABUNDANT DIATOM TAXA FOUND IN 36 SAMPLES FROM THE LABORATORY SYSTEM, ALL BUT 8 WERE ALSO ABUNDANT IN SAMPLES OBTAINED FROM YAQUINA BAY AND ESTUARY. VERTICAL DISTRIBUTION OF DIATOMS WAS MORE CLOSELY RELATED TO LIGHT INTENSITY AND PERIOD OF EXPOSURE TO DESICCATION IN THE SUMMER EXPERIMENTS THAN IN THE WINTER EXPERIMENTS. A SUDDEN, UNREASONABLE DECREASE IN SALINITY OR INCREASE IN WATER TEMPERATURE HAD A MUCH GREATER EFFECT ON DIATOM ASSEMBLAGES EXPOSED TO 12,270 LUX THAN ON THOSE THAT DEVELOPED AT EITHER 1,030 OR 4,710 LUX; THE MOST NOTICEABLE CHANGES IN COMMUNITY STRUCTURE INCLUDED A DECREASE IN SPECIES DIVERSITY AND A RAPID GROWTH OF A POPULATION OF MELOSIRA NUMMULOIDES, A FILAMENTOUS SPECIES. PRIMARY PRODUCTIVITY IN DIATOM ASSEMBLAGES EXPOSED TO PERIODS OF DESICCATION WAS LESS UNDER WINTER CONDITIONS THAN UNDER CORRESPONDING CONDITIONS IN SUMMER. PRIMARY PRODUCTIVITY IN ASSEMBLAGES NOT EXPOSED TO DESICCATION WAS STRONGLY AFFECTED BY LIGHT INTENSITY DURING BOTH SUMMER AND WINTER EXPERIMENTS, AND THE RATIO PRIMARY PRODUCTIVITY: CHLOROPHYLL A WAS GREATER DURING WINTER THAN SUMMER. (BYRD-BATTELLE)

FIELD 05C, 02L

ACCESSION NO. W73-00853

FARTHEST SOUTH ALGAE AND ASSCCIATED BACTERIA,

CALIFORNIA INST. OF TECH., PASADENA. BIOSCIENCE AND PLANETOLOGY SECTION.

R. E. CAMERON.

PHYCOLOGIA, VOL 11, NO 2, P 133-139, JUNE 1972. 3 FIG, 1 TAB, 18 REF.

DESCRIPTORS:
*ANTARCTIC, *DISTRIBUTION PATTERNS, *SYSTEMATICS, *AQUATIC ALGAE, *SOIL BACTERIA, SAMPLING, CULTURES, ECOLOGY, SOIL MICROBIOLOGY, CYANOPHYTA, FROZEN SOILS, IONS, NOSTOC, ENVIRONMENTAL EFFECTS, POLLUTANT IDENTIFICATION, PHYSICAL PROPERTIES, BULK DENSITY, POROSITY, COLOR, HYDROGEN ION CONCENTRATION, NITROGEN, CARBON, SODIUM, CALCIUM, POTASSIUM, MAGNESIUM, IRON, ALUMINUM, NITRATES, NITRITES, PHOSPHATES, BICARBONATES, SULFATES, CHLORIDES.

IDENTIFIERS:
NEOCHLORIS AQUATICA, SCHIZOTHRIX CALCICOLA, BORATES, ARTHROBACTER, NOSTOC COMMUNE, PORPHYROSIPHON NOTARISII, PSYCHROPHILIC, BIOTIC ASSOCIATIONS, LA GORCE MOUNTAINS, AMMONIUM IONS.

ABSTRACT:
A NEW RECORD IS REPORTED FOR ALGAE COLLECTED FROM THE HIGHEST LATITUDE, A FROZEN POND IN THE LA GORCE MOUNTAINS, ANTARCTICA (86 DEGREES 45 MINUTE SOUTH, 146 DEGREES ZERO MINUTES WEST). CULTURABLE ALGAE INCLUDED NEOCHLORIS AQUATICA STARR AND SCHIZOTHRIX CALCICOLA (AG.) GOM. PORPHYROSIPHON NOTARISII (MENEGH.) KUTZ. WAS NOT RECOVERABLE IN CULTURE. ASSOCIATED BACTERIA WERE SOIL DIPHTHEROIDS OF THE GENUS ARTHROBACTER. SOIL CONDITIONS WHICH SUPPORTED THE GROWTH OF ARTHROBACTER ARE PRESENTED. THE OCCURRENCE OF HIGH LATITUDE PHOTOSYNTHETIC ORGANISMS IS IMPORTANT IN THE SEARCH FOR POSSIBLE EXTRA-TERRESTRIAL LIFE BECAUSE ENVIRONMENTAL CONDITIONS, IN SOME ASPECTS, APPROACH THOSE OF MARS. (LONG-BATTELLE)

FIELD 05B, 05A

ACCESSION NO. W73-00854

HYDROBIOLOGICAL NOTES ON THE HIGH-SALINITY WATERS OF THE SINAI PENINSULA,

HEBREW UNIV., JERUSALEM (ISRAEL). DEPT. OF ZOOLOGY.

F. D. POR.

MARINE BIOLOGY, VOL 14, NO 2, P 111-119, MAY 1972. 2 FIG, 4 TAB, 42 REF.

DESCRIPTORS:
*SALINITY, *LAGOONS, *HYDROBIOLOGY, *HALOPHILIC ANIMALS, *HALOPHYTES, SEA WATER, MARINE ANIMALS, BIOLOGICAL COMMUNITIES, CHLORINE, MOLLUSKS, COPEPODS, DIPTERA, GULFS, GEOLOGIA HISTORY, PLEISTORENE EPOCH, ADAPTATION, ENVIRONMENTAL EFFECTS, CHLOROPHYTA, MARINE ALGAE , SALT TOLERANCE, CRUSTACEANS, MARINE FISH, MARSH PLANTS.

IDENTIFIERS:
*SINAI PENINSULA, CHLORINITY, METAHALINE, HYPERHALINE, EURYHALINE, RED SEA, GULF OF SUEZ, DECAPODS, COLEOPTERA, MEDITERRANEAN SEA, COELENTERATES, ECHINODERMS, POLYCHAETES, CHIRONOMIDS, BARNACLES, ARTHROPODS, SARGASSUM CRISPUM, DIGENIA SIMPLEX, HALOPHILA STIPULACEA, DIPLANTHERA UNINERVIS, CASSIOPEA ANDROMEDA, AKABARIA, STROMBUS TRICORNIS, FUSUS MARMORATUS, BERTHELLA CITRINA, MACTRA OLORINA.

ABSTRACT:
A 4-YEAR SURVEY WAS MADE OF THE HIGH-SALINITY SEAS SURROUNDING THE SINAI PENINSULA AND THE SALINE LAGOONS BORDERING ON IT. THESE HIGH-SALINITY BODIES OF WATER HAVE IONIC CONTENT SIMILAR TO THAT OF THE SEA. THE RED SEA HAS FORMED A METHALINE MARINE FAUNA BY SUCCESSFULLY ADAPTING TO SALINITIES OF ABOUT 45 PERCENT. THE SALINE LAGOONS HAVING SALINITIES OF 6-8 PERCENT ARE ALSO INHIBITED BY METABOLINE FAUNA. LAGOONS OF GREATER SALINITY ARE INHABITED BY A SMALL NUMBER OF EURYHALINE MARINE ORGANISMS. HYPERSALINE LAGOONS ARE THOSE ISOLATED FROM THE SEA OR WITH SALINITY VALUES GREATER THAN 100 PERCENT. HYPERSALINE CONTINENTAL FRESH-WATER ELEMENTS INHABIT THESE POOLS. THE SPECIFIC OBSERVATIONS MADE ARE ALSO DISCUSSED IN LIGHT OF THEIR POSSIBLE GENERAL APPLICATION TO OTHER NEARSHORE AND LAGOON AREAS. (MORTLAND-BATTELLE)

FIELD 05A, 05C

ACCESSION NO. W73-00855

CHLOROPHYLL AND THE MARGALEF PIGMENT RATIO IN A MOUNTAIN LAKE,

BRADLEY UNIV., PEORIA, ILL. DEPT. OF BIOLOGY.

B. J. MATHIS.

THE AMERICAN MIDLAND NATURALIST, VOL 88, NO 1, P 232-235, JULY 1972. 2 TAB, 13 REF.

DESCRIPTORS:
*LAKES, *PIGMENTS, *SUCCESSION, *EUTROPHICATION, CHLOROPHYLL, PLANKTON, BIOLOGICAL COMMUNITIES, ECOSYSTEMS, SPECTROPHOTOMETRY, VIRGINIA, SAMPLING, SEPARATION TECHNIQUES, CENTRIFUGATION, TURBIDITY, WATER TEMPERATURE, ALKALINITY, HYDROGEN ION CONCENTRATION, WATER QUALITY, NUTRIENTS, COLORIMETRY, DEFICIENT ELEMENTS, AQUATIC ALGAE, BIOINDICATORS, VIRGINIA, TROPHIC LEVEL.

IDENTIFIERS:
MARGALEF PIGMENT RATIO, PIGMENT DIVERSITY INDEX, CHLOROPHYLL A, CARTENOIDS, SAMPLE PREPARATION, MOUNTAIN LAKE.

ABSTRACT:
MARGALEF'S PIGMENT DIVERSITY INDEX WAS USED TO ESTIMATE THE SUCCESSIONAL STAGE OF A PLANKTON COMMUNITY IN A MOUNTAIN LAKE. IN THIS METHOD, OPTICAL DENSITY OF PLANT PIGMENTS IN 90 PERCENT ACETONE IS MEASURED WITH A SPECTROPHOTOMETER TO ESTABLISH A YELLOW/GREEN RATIO. CHLOROPHYLL IS LOST AS CELLS BECOME NITROGEN DEFICIENT AND THUS, ACCORDING TO MARGALEF, THE YELLOW/GREEN RATIO INCREASES, IN GENERAL, AS SUCCESSION ADVANCES. IN THIS STUDY, 1000-ML WATER SAMPLES WERE TAKEN AT A DEPTH OF 3.5 M FROM MOUNTAIN LAKE, VIRGINIA. PIGMENTS WERE EXTRACTED, THE SAMPLES WERE CENTRIFUGED, THEIR OPTICAL DENSITY WAS DETERMINED COLORIMETRICALLY, AND CHLOROPHYLL A DETERMINATIONS WERE MADE. WATER TEMPERATURE, ALKALINITY, PH, AND SPECIFIC CONDUCTANCE WERE DETERMINED IN THE FIELD AND TURBIDITY WAS MEASURED IN THE LABORATORY. THESE LATTER DATA INDICATED AN OLIGOTROPHIC LAKE WITH AN IMMATURE PLANKTON COMMUNITY. HOWEVER, THE MARGALEF RATIO SUGGESTED AN OLDER MORE STABLE COMMUNITY. IT IS SUGGESTED THAT MORE COMPARATIVE STUDIES ARE NEEDED BEFORE THE INDEX CAN BE VALUABLE IN ESTIMATING SUCCESSIONAL STAGES IN AQUATIC COMMUNITIES. (MORTLAND-BATTELLE)

FIELD 05C, 02H

ACCESSION NO. W73-00916

ECOLCGICAL STUDIES OF EUGLENINEAE IN CERTAIN POLLUTED AND UNPOLLUTED ENVIRONMENTS,

OSMANIA UNIV., HYDERABAD (INDIA).

M. MUNAWAR.

HYDROBIOLOGIA, VOL 39, NO 3, P 307-320, MARCH 1972. 3 FIG, 3 TAB, 28 REF.

DESCRIPTORS:
*EUGLENOPHYTA, *BIOINDICATORS, *PHYTOPLANKTON, *POLLUTANT IDENTIFICATION, PROTOZOA, PONDS, LIMNOLOGY, ECOLOGY, CHEMICAL ANALYSIS, AQUATIC ALGAE, NUTRIENTS, WATER QUALITY, WATER ANALYSIS, FREQUENCY, FLUCTUATIONS, BIOLOGICAL COMMUNITIES, PHYSIOCHEMICAL PROPERTIES, DISTRIBUTION PATTERNS, ECOLOGICAL DISTRIBUTION, WATER POLLUTION EFFECTS, FRESHWATER, NITRATES, AMMONIA, CARBON DIOXIDE, PHOSPHATES, SULFATES, SODIUM, POTASSIUM, MAGNESIUM, CHLORIDES, NITROGEN, ALKALINITY, SILICA, CALCIUM, DISSOLVED OXYGEN, ORGANIC MATTER, SULFIDES.

IDENTIFIERS:
*INDIA, ORGANIC NITROGEN, EUGLENA ACUS, EUGLENA CAUDATA, EUGLENA LIMNOPHYLA, EUGLENA OBLONGA, EUGLENA OXYURIS, EUGLENA POLYMORPHA, EUGLENA SPIROGYRA, LEPOCINCLIS OVUM, PHACUS CIRCUMFLEXUS, PHACUS ONYX, PHACUS PLEURONECTES, PHACUS PYRUM, TRACHELOMONAS ARMATA, TRACHELOMONAS HISPIDA VAR CRENULATOCOLLIS.

ABSTRACT:
VARIOUS PHYSICOCHEMICAL PROPERTIES OF THREE FRESHWATER PONDS IN INDIA WERE STUDIED IN RELATION TO THE DISTRIBUTION AND PERIODICITY OF EUGLENINEAE. TWO PONDS, SEWAGE AND GARDEN, WERE CONSIDERABLY POLLUTED AND THE THIRD, TYPHA, WAS COMPARATIVELY PURE. THE HIGHEST PERCENTAGE OF EUGLENINEAE WERE FOUND IN SEWAGE POND, WHICH CONTAINED 13 DIFFERENT SPECIES. GARDEN POND HARBORED JUST 2 SPECIES, BUT IN LARGE NUMBERS WHILE THE TRACHELOMONAS HISPIDA AND EUGLENA SP. FOUND IN TYPHA POND WERE TOTALLY ABSENT IN THE OTHER TWO. SEWAGE POND HAD HIGH AVERAGE CONCENTRATIONS OF FREE CARBON DIOXIDE, WHICH SEEMED TO FAVOR EUGLENOID GROWTH. NITRATE FLUCTUATIONS IN ALL THREE PONDS COINCIDED WITH FLUCTUATIONS IN EUGLENOID GROWTH SUGGESTING THAT INORGANIC SOURCES OF NITROGEN MIGHT BE IMPORTANT IN THE ECOLOGY OF THESE ALGAE. THE FLAGELLATES WERE ABUNDANT DURING PERIODS OF LOW TOTAL SULPHIDES AND CONCENTRATIONS GREATER THAN 2.0 PPM ADVERSELY AFFECTED THEIR DEVELOPMENT. A TEMPERATURE RANGE OF 27-39 C FAVORED THEIR GROWTH AND THEY USUALLY DEVELOPED AFTER RAINS OR DURING INTERMITTENT SHOWERS. IN SEWAGE POND, THERE WAS A DIRECT RELATIONSHIP BETWEEN EUGLENOID POPULATION AND HIGHER CONCENTRATIONS OF OXIDIZABLE ORGANIC MATTER. THIS SUGGESTS THE USE OF CERTAIN EUGLENOID SPECIES AS BIOINDICATORS. (MORTLAND-BATTELLE)

FIELD 05C, 05A

ACCESSION NO. W73-00918

SUBLITTORAL ECOLOGY OF THE KELP BEDS OFF DEL MONTE BEACH, MONTEREY, CALIFORNIA,

NAVAL POSTGRADUATE SCHOOL, MONTEREY, CALIF.

C. S. MINTER, III.

AVAILABLE FROM THE NATIONAL TECHNICAL INFORMATION SERVICE AS AD-738 875, $3.00 IN PAPER COPY, $0.95 IN MICROFICHE. MASTER'S THESIS, SEPTEMBER 1971. 181 P, 102 FIG, 5 APPEND, 31 REF.

DESCRIPTORS:
*KELPS, *ECOLOGICAL DISTRIBUTION, *MARINE ANIMALS, *SYSTEMATICS, *ECOLOGY, *MARINE ALGAE, ANNELIDS, MOLLUSKS, RHODOPHYTA, PHAEOPHYTA, PHOTOGRAPHY, SCUBA DIVING, SEA WATER, SAMPLING, ENVIRONMENTAL EFFECTS, MARINE FISH, PERIPHYTON, ON-SITE DATA COLLECTIONS, CRUSTACEANS, GASTROPODS, SNAILS, CRABS, BENTHIC FLORA, BENTHIC FAUNA, DISTRIBUTION PATTERNS, PACIFIC OCEAN, AQUATIC HABITATS, ISOPODS, CLAMS, CALIFORNIA, SHELLFISH.

IDENTIFIERS:
COELENTERATES, ARTHROPODS, NEMERTEANS, ECHINODERMS, SIPUNCULIDS, BRYOZOA, INTOPROCTA, TUNICATES, SPONGES, VERTEBRATES, MACROINVERTEBRATES, SPECIES DIVERSITY, PORIFERA, SEA CUCUMBERS, STARFISH, CHITONS, RHABDODERMELLA, LOPHOPANOPEUS, EUPENTACTA, CNEMIDOCARPA, ENHYDRA LUTRIS NEREIS, SEA OTTER, MACROCYSTIS PYRIFERA, CHORDATES, POLYCHAETES, BARNACLES, DECAPODS, SEA URCHINS, ACARNUS ERITHACUS, CRANIELLA, HYMENAMPHIASTRA CYANOCRYPTA.

ABSTRACT:
IN ORDER TO ASSESS THE ECOLOGICAL EFFECTS OF WAVE BARRIERS IN MONTEREY HARBOR, AN OVERALL STUDY OF THE AREA WAS CONDUCTED. THIS PART OF THE STUDY DEALT WITH THE IDENTIFICATION OF MACROSCOPIC ORGANISMS EXISTING IN THE KELP BEDS OFF DEL MONTE BEACH AND THE MAPPING AND COUNTING OF BENTHIC PLANTS AND ANIMALS LIVING WITHIN TWO CAREFULLY SELECTED AND PERMANENTLY MARKED STATIONS ON SHALE SUBSTRATE. THESE SPECIMENS WERE COLLECTED BY SCUBA DIVERS, AND MORE THAN 160 SPECIES WERE FOUND TO EXIST. COLLECTION METHODS AND TECHNIQUES UTILIZED BY DIVERS WERE DOCUMENTED AND NUMEROUS UNDERWATER PHOTOGRAPHS WERE TAKEN. THE PERMANENTLY MARKED AREAS WERE FOUND TO BE OF GENERALLY SIMILAR BIOLOGICAL POPULATION BUT OF MARKEDLY DIFFERENT SPECIES DISTRIBUTION AND RELATIVE ABUNDANCE. (LONG-BATTELLE)

FIELD 05C

ACCESSION NO. W73-00932

ECOLOGICAL ASPECTS OF PLANKTCN PRODUCTION,

CENTRE OF ADVANCED STUDY IN MARINE BIOLOGY, PORTO NOVO (INDIA).

R. C. SUBBARAJU, AND K. KRISHNAMURTHY.

MARINE BIOLOGY, VOL 14, NO 1, P 25-31, MAY 1972. 4 FIG, 4 TAB, 42 REF.

DESCRIPTORS:
*ECOLOGY, *AQUATIC PRODUCTIVITY, *ZOOPLANKTON, *PHYTOPLANKTON, NERITIC, ESTUARINE ENVIRCNMENT, SEA WATER, DIATOMS, CCPEPODS, DINOFLAGELLATES, CRUSTACEANS, ANNELIDS, WATER POLLUTION EFFECTS, STANDING CROPS, MOLLUSKS, LARVAE, CHRYSOPHYTA, CYANOPHYTA, PYRROPHYTA, MARINE ALGAE, GASTROPODS.

IDENTIFIERS:
SECONDARY PRODUCTIVITY, CEPHALOCHORDATES, HEMICHCRDATES, ARTHROPODS, *BAY OF BENGAL, COELENTERATES, CTENOPHORA, CHAETCGNATHS, CYPHONAUTES, ECHINODERMS, TUNICATES, NOCTILUCA, SIPHONOPHORES, MEDUSAE, PILIDIUM, CLADOCERA, LINGULA, ACTINOTROCHA, VELIGER, PTEROPCDS, HETEROPODS, OSCILLATORIA, STOMATOPODA, POLYCHAETES, AMPHIOXUS, SALPS, BALANOGLOSSUS, DOLIOLIDS, EUCALANUS ELONGATUS, EUCHAETA MARINA, PONTELLA PRINCEPS, PCNTELLOPSIS HERDMANNI, NANNOCALANUS MINOR, CANTHOCALANUS PAUPER, RHINCALANUS SPP, UNDINULA VULGARIS, OITHONA SPP, SAPPHRINA NIGROMACULATA, ACROCALANUS GRACILIS, PARACALANUS PARVUS, EUTERPINA ACUTIFRONS, PSEUDODIAPTOMUS SPP, ACARTIA SPP.

ABSTRACT:
A STUDY WAS CARRIED OUT IN THE NERITIC AND ESTUARINE WATERS OF PORTO NOVO, COROMANDEL COAST, BAY OF BENGAL, INDIA BETWEEN 1960 AND 1967 ON THE ECOLOGICAL ASPECTS OF PLANKTCN PRODUCTION. REPRESENTATIVE SAMPLES WERE COLLECTED FROM BOTH NEARSHORE AND ESTUARINE WATERS OF PORTO NOVO, ANC THE PLANKTCN VOLUMES RECORDED BY THE DISPLACEMENT METHOD TO THE NEAREST 0.10 CC AND EXPRESSED IN CC/CU M. THE AVERAGE DISPLACEMENT VOLUME OF PLANKTON USUALLY VARIED BETWEEN 2 AND 4 CC/CU M. DURING SUMMER, WITH A SEASON OF HIGH PLANKTON PRODUCTIVITY, THE AVERAGE PLANKTON DISPLACEMENT VOLUME ROSE TO 8 CC/CU M. GENERALLY SPEAKING, THE AVERAGE ZOOPLANKTON DENSITY (STANDING CROP) WAS USUALLY BETWEEN 80,000 AND 100,000 ORGANISMS/CU M, OF WHICH COPEPODS ALCNE COMPRISED USUALLY BETWEEN 70,000 AND 90,000 ORGANISMS/CU M. THE AVERAGE COPEPOD DENSITY PER SAMPLE VARIED FROM 30,000 TO 50,000 ORGANISMS/CU M. HOWEVER, IN THE SUMMER MONTHS, THE CCPEPOD DENSITY WAS USUALLY NOT LESS THAN 100,000 ORGANISMS/CU M; IN SOME YEARS THIS WAS EVEN HIGHER (FROM 125,000 TO 170,000 ORGANISMS/CU M). COPEPODS COMPRISEC BETWEEN 80 AND 95 PERCENT OF THE ZOOPLANKTON PCPULATION. THE MAXIMUM NCN-COPEPOD POPULATION IN THE ZOOPLANKTON SELDCM REACHED 30 PERCENT, WAS OFTEN BELOW 25 PERCENT, AND USUALLY LESS THAN 20 PERCET. DURING THE PERIOD MARCH TO OCTOBER (IN SOME YEARS AS EARLY AS FEBRUARY, AND IN SCME YEARS UP TO NOVEMBER), EITHER AN INCREASING OR A STEADY TREND OF PLANKTCN PRODUCTION WAS EVIDENT. IT WOULD APPEAR THAT SALINITY AND RAINFALL DETERMINE THE OCCURRENCE AND DISTRIBUTION OF PLANKTON IN PORTO NOVO. (BYRD-BATTELLE)

FIELD 05C

ACCESSION NO. W73-00935

EFFECT OF DREDGING ON THE NUTRIENT LEVELS AND BIOLOGICAL POPULATIONS OF A LAKE,

SOUTH DAKOTA STATE UNIV., BROOKINGS. WATER RESOURCES INST.

C. L. CHURCHILL, AND C. K. BRASHIER.

AVAILABLE FROM THE NATIONAL TECHNICAL INFORMATION SERVICE AS PB-212 718, $3.00 IN PAPER COPY, $0.95 IN MICROFICHE. COMPLETION REPORT, AUGUST 1972. 155 P, 15 FIG, 9 TAB, 16 REF, APPEND. OWRR B-013-SDAK(1).

DESCRIPTORS:
*NUTRIENTS, *DREDGING, DESILTING, SEDIMENTATION, *EUTROPHICATION, BIOLOGICAL COMMUNITIES, FOOD CHAINS, LAKE MORPHOLOGY, RESERVOIR SILTING, LAKES, *SOUTH DAKOTA, PLANKTON, *ALGAE, *PHOSPHORUS COMPOUNDS, CYANOPYTA, CHLOROPHYTA.

IDENTIFIERS:
PRAIRIE LAKES, *LAKE HERMAN(SO. DAKOTA), *ORTHOPHOSPHATES.

ABSTRACT:
LAKE HERMAN, A SHALLOW WARM WATER PRAIRIE LAKE, HAS BEEN MONITORED FOR CHEMICAL NUTRIENTS AND PLANKTONIC ALGAE BEFORE AND DURING A DREDGING PROJECT. THE MAJOR CHANGE IN THE LAKE WITH THE COMMENCEMENT OF DREDGING ACTIVITY WAS A 300% INCREASE IN BOTH ORTHO PHOSPHATE AND TOTAL PHOSPHORUS. CALCIUM AND TOTAL HARDNESS DECREASED SLIGHTLY AND SILICA AND TURBIDITY WERE JUDGED TO HAVE INCREASED SLIGHTLY. THERE WERE APPARENTLY NO ACCOMPANYING CHANGES IN PLANKTONIC ALGAE, EITHER IN POPULATION DENSITIES OR GENERA PRESENT. THE SILT AND WATER REMOVED WITH DREDGING WAS DEPOSITED IN A LOW LYING AREA NEAR THE LAKE, WHERE THE SILT SETTLED OUT AND THE WATER WAS RETURNED TO THE LAKE. THE WATER IN THIS SILT DEPOSIT AREA WAS ALSO MONITORED. IT WAS AS MUCH AS 2 PH UNITS CLOSER TO NEUTRALITY THAN THE LAKE WATER AND HAD A HIGHER CONDUCTIVITY (UP TO 10% HIGHER) THAN THE LAKE WATER. EARLY IN THE DREDGING SEASON, THIS WATER IN THE SILT DEPOSIT AREA HAD HIGH LEVELS OF ORTHO PHOSPHATE (0.90 MG PO4/1), BUT OVER A FOUR MONTH PERIOD, AS DREDGING PROCEEDED, THE ORTHO PHOSPHATE IN THE SILT DEPOSIT AREA WATER DECREASED UNTIL IT WAS ONLY ABOUT HALF AS GREAT (0.19 MG PO4/1) AS THE LEVEL IN THE LAKE BEFORE DREDGING BEGAN. (WIERSMA-SOUTH DAKOTA)

FIELD 05C, 02H

ACCESSION NO. W73-00938

I. CHEMORECEPTION IN MARINE BACTERIA. II. CHEMICAL DETECTICN OF MICROBIAL PREY BY BACTERIAL PREDATORS,

HARVARD UNIV., CAMBRIDGE, MASS. DIV. OF ENGINEERING AND APPLIED PHYSICS.

S. FOGEL, I. CHET, AND R. MITCHELL.

AVAILABLE FROM THE NATIONAL TECHNICAL INFORMATION SERVICE AS AD-736 982, $3.00 IN PAPER COPY, $0.95 IN MICROFICHE. TECHNICAL REPORT NO. 2, NR-306-025, SEPTEMBER 1971. 28 P, 10 TAB, 20 REF. CONTRACT NO. N00014-67A-0298-0026.

DESCRIPTORS:
*MARINE BACTERIA, *ATTRACTANTS, *PREDATION, ORGANIC COMPOUNDS, AMINO ACIDS, VITAMINS, CARBOHYDRATES, ALCOHOLS, POLLUTANT IDENTIFICATION, SYSTEMATICS, NUTRIENTS, PSEUDOMONAS, CULTURES, VITAMIN B, BIOASSAY, CELLULOSE, DIALYSIS, SEA WATER, ORGANIC ACIDS, CHRYSOPHYTA, INHIBITION, MARINE ALGAE, DIATOMS, BEHAVIOR.

IDENTIFIERS:
*CHEMORECEPTION, *CHEMOTAXIS, *PREY, PYTHIUM DEBARYANUM, SKELETONEMA COSTATUM, NUCLEOTIDES, ETHANOL, MONOSACCHARIDES, CULTURE MEDIA, DISACCHARIDES, THREONINE, PROLINE, GLUTAMATE, LEUCINE, ARGININE, SERINE, TYROSINE, TRYPTOPHAN, HISTIDINE, ASPARTATE, METHIONINE, PHENYLALANINE, ALANINE, ISOLEUCINE, VALINE, LYSINE, ASPARAGINE, CYSTINE, CYSTEINE, GLYCINE, GLYCEROL, RIBOSE, MALTOSE, FRUCTOSE, GLUCOSE 6-P, GALACTOSE, GLUCOSAMINE, ARABINOSE, GLUCOSE, RAFFINOSE, SUCROSE, INOSITOL, LACTOSE, ADENINE, URACIL, ACETATE, CITRATES, PYRUVATE, GLUCURONIC ACID, THIAMINE, NIACIN, BIOTIN, NUTRIENT BROTH.

ABSTRACT:
A WIDE VARIETY OF MOTILE MARINE BACTERIA WERE TESTED FOR THEIR ABILITY TO DETECT AND BE ATTRACTED TO ORGANIC CHEMICALS. ALL BACTERIA TESTED DISPLAYED CHEMOTACTIC RESPONSES TO MANY DIFFERENT MATERIALS. THESE RESPONSES WERE HIGHLY SPECIFIC FOR EACH BACTERIUM. THE ATTRACTANT WAS NOT NECESSARILY METABOLIZED. THE ECOLOGICAL SIGNIFICANCE OF THIS WIDESPREAD PHENOMENON OF CHEMORECEPTION AMONG MARINE BACTERIA IS DISCUSSED. A MOTILE, PREDACIOUS BACTERIUM WHICH DEGRADED PYTHIUM DEBARYANUM WAS STRONGLY ATTRACTED TO SUBSTANCES RELEASED INTO THE MEDIUM BY THE FUNGUS. A NONPREDACIOUS BACTERIUM WAS NOT ATTRACTED TO THESE SUBSTANCES. THE PREDATOR BACTERIUM WAS SPECIFICALLY ATTRACTED TO
THESE SUBSTANCES. THE PREDATOR BACTERIUM WAS SPECIFICALLY ATTRACTED TO
CELLULOSE AND ITS OLIGOMERS WHICH ARE KNOWN TO BE COMPONENTS OF THE CELL WALL OF PYTHIUM. ETHANOL INHIBITED CHEMOTAXIS OF THE BACTERIUM WITHOUT AFFECTING EITHER ITS MOTILITY OR ITS ABILITY TO DEGRADE CELLULOSE. A SECOND PREDACIOUS BACTERIUM WAS ISOLATED FOR THE ALGA, SKELETONEMA COSTATUM. THE ROLE OF CHEMORECEPTION IN THE DETECTION OF MICROBIAL PREY BY BACTERIAL PREDATORS IN NATURAL HABITATS IS DISCUSSED. (LONG-BATTELLE)

FIELD 05A, 05C

ACCESSION NO. W73-00942

PRODUCTION OF DISSOLVED ORGANIC MATTER FROM CEAD GREEN ALGAL CELLS. II. ANAEROBIC MICROBIAL DECOMPCSITION,

TCKYO METROPOLITAN UNIV., (JAPAN). DEPT. OF CHEMISTRY.

A. OTSUKI, AND T. HANYA.

LIMNOLOGY AND OCEANOGRAPHY, VOL 17, NO 2, P 258-264, MARCH 1972. 9 FIG, 2 TAB, 19 REF.

DESCRIPTORS:
*ANAEROBIC BACTERIA, *ANAEROBIC CONDITIONS, *KINETICS, *DISSOLVED SOLIDS, ORGANIC MATTER, NITROGEN, WATER POLLUTION SOURCES, ENVIRONMENTAL EFFECTS, AEROBIC CONDITIONS, CHLOROPHYTA, SCENEDESMUS, AQUATIC ALGAE, AQUATIC PRODUCTIVITY, POLLUTANT IDENTIFICATION, LABORATORY EQUIPMENT, AEROBIC BACTERIA, SOLVENT EXTRACTIONS, FREEZE DRYING, SYSTEMATICS, ORGANIC COMPOUNDS, WATER ANALYSIS.

IDENTIFIERS:
ORGANIC CARBON, ORGANIC NITROGEN, SUBSTRATE UTILIZATION, PARTICULATE MATTER, FATE OF POLLUTANTS, PAPER CHRCMATOGRAPHY, INFRARED SPECTROSCOPY, CULTURE MEDIA, FATTY ACIDS, SAMPLE PREPARATION, NATURAL ORGANICS.

ABSTRACT:
THE PRODUCTION OF DISSOLVED ORGANIC MATTER (DOM) FROM DEAD GREEN ALGAL CELLS (SCENEDESMUS) BY ANAEROBIC BACTERIA WAS INVESTIGATED AND COMPARED WITH THOSE PRODUCED UNDER AEROBIC CONDITIONS. DOM WAS PREPARED FOR EXPERIMENTAL USE BY FREEZE-DRYING AND ASHING APPROXIMATELY 99 PERCENT PURE CULTURES OF SCENEDESMUS. ANAEROBIC DECOMPOSITION OF THE DOM WAS INITIATED BY INOCULATING MICROFLORA INTO A SPECIALLY PREPARED DECOMPOSITION APPARATUS CONTAINING CULTURE MEDIA AND DOM AT PH 7. AFTER INCUBATION, DISSOLVED ORGANIC CARBCN (DOC) AND NITROGEN (DON) WERE MEASURED BY AN ELEMENTAL ANALYZER AND BY THE MICRC-KJELDAHL METHOD. ORGANIC ACIDS, PARTICULARLY FATTY ACIDS, WERE EXTRACTED BY THIN-LAYER CHROMATOGRAPHY FOR IR SPECTROPHOTOMETRIC DETERMINATICN. THIRTY PERCENT OF ADDED ALGAL CELL CARBON WAS TRANSFORMED INTO DCC AND 20 PERCENT MINERALIZED; 50 PERCENT REMAINED AS PARTICULATE MATTER. ON THE OTHER HAND, 8 PERCENT OF THE ADDED ALGAL CELL NITROGEN WAS TRANSFORMED INTO DON, 48 PERCENT WAS MINERALIZED, AND 44 PERCENT REMAINED IN PARTICULATE FORM. THE DISSOLVED ORGANIC COMPOUNDS CONSISTED MAINLY OF LOWER FATTY ACIDS AND YELLOWISH ACIDIC SUBSTANCES. SOME PROTEINACEOUS MATERIAL WAS FOUND. ANAEROBIC DECOMPOSITION PATTERNS AS COMPARED WITH THOSE UNDER AEROBIC CONDITIONS SUGGEST THE PRESENCE OF RELATIVELY HIGH CONCENTRATIONS CF DOM IN ANAEROBIC NATURAL ENVIRONMENTS. (SEE ALSO W73-00379) (LONG-BATTELLE)

FIELD 05C, 05B

ACCESSION NO. W73-01066

INVESTIGATIONS ON THE INFLUENCE OF OIL POLLUTIONS ON MARINE ALGAE. I. THE EFFECT OF CRUDE-OIL FILMS CN THE CO2 GAS EXCHANGE OUTSIDE THE WATER (UNTERSUCHUNGEN UBER DEN EINFLUSS VON OLVERSCHMUTZUNGEN AUF MEERESALG EN. I. DIE WIRKUNG VON ROHOFILMEN AUF DEN CO2 - GASWECHSEL AUSSERHALB DES WASSERS),

KIEL UNIV. (WEST GERMANY). INST. FOR HYDROGRAPHY.

W. SCHRAMM.

MARINE BIOLOGY, VOL 14, NO 3, P 189-198, JUNE 1972. 9 FIG, 1 TAB, 9 REF.

DESCRIPTORS:
*MARINE ALGAE, *WATER POLLUTION EFFECTS, *PHOTOSYNTHESIS, *OIL POLLUTION, *TOXICITY, CHEMICAL REACTIONS, CARBON DIOXIDE, WATER POLLUTION SOURCES, OIL SPILLS, SEA WATER, MARINE PLANTS, PHAEOPHYTA, OILY WATER, EVAPORATICN, DIFFUSION, RHODOPHYTA.

IDENTIFIERS:
FUCUS VESICULOSUS, LAMINARIA DIGITATA, PORPHYRA UMBILICALIS, SAMPLE PREPARATION, CRUDE OIL.

ABSTRACT:
OIL POLLUTION IN THE SEA IS GENERALLY RESTRICTED TC THIN OIL FILMS FLOATING ON THE WATER SURFACE. SUCH OIL FILMS TEND TC COAT LITTORAL PLANTS OR ANIMALS DURING LCW TIDE. THE EFFECTS OF COATING WITH CRUDE OIL ON THE CO2-UPTAKE OF VARIOUS MARINE ALGAE HAVE BEEN INVESTIGATED UNDER CONDITICNS OF EMERSION. IN EMERSED ALGAE, CO2-UPTAKE IS MORE OR LESS DEPRESSED, DEPENDING ON THE THICKNESS OF THE OIL FILM (0.1 TO 0.0001 MM) AND THE TYPE OF CRUDE OIL (IRAN, LIBYA, VENEZUELA). ON THE OTHER HAND, WATER LOSS DURING EXPOSURE IS RECUCED, SO THAT THE OIL-COVERED ALGAE ARE ABLE TC PHOTOSYNTHESIZE OVER A LCNGER PERIOD THAN ALGAE WITHOUT CIL COVER. AFTER RETRANSFER TO OIL-FREE SEA WATER, IN MOST CASES PHOTOSYNTHESIS RATES REMAINED DEPRESSED THRCUGHOUT THE PERIOD OF OBSERVATION. THERE ARE TWO EFFECTS WHICH PROBABLY INTERFERE WITH GAS EXCHANGE: (1) LCWERING OF DIFFUSION RATES OF PHOTOSYNTHETIC GASES AND OF WATER EVAPORATION BY THE OIL FILMS; (2) TCXIC EFFECTS OF CRUDE-OIL COMPONENTS. (LONG-BATTELLE)

FIELD 05C

ACCESSION NO. W73-01074

STUDIES OF A SIMPLE LABORATORY MICROSYSTEM: EFFECTS CF STRESS,

GECRGIA UNIV., ATHENS. INST. OF ECOLOGY.

M. C. FERENS, AND R. J. BEYERS.

REPORT AVAILABLE FROM THE NATIONAL TECHNICAL INFORMATION SERVICE AS SRO-310-1 (PT 1), $3.00 IN PAPER COPY, $0.95 IN MICROFICHE. IN: SAVANNAH RIVER ECOLOGY LABORATORY ANNUAL REPORT NO SRO-310-1 (PT 1), AUGUST 1971, P 8-24. 6 FIG, 1 TAB, 11 REF. AEC AT (38-1) 310.

DESCRIPTORS:
*STRESS, *PRIMARY PRCDUCTIVITY, *AQUATIC ALGAE, *CCMPARATIVE PRODUCTIVITY, *ECOSYSTEMS, *BIOLOGICAL COMMUNITIES, BIOMASS, PRODUCTIVITY, METABOLISM, CHLORELLA, RESPIRATION, CHLOROPHYLL, LABORATORY TESTS, IRRADIATION, CHLOROPHYTA, INHIBITION, GAMMA RAYS.

IDENTIFIERS:
CHLOROPHYLL A, SCHIZOTHRIX CALCICOLA.

ABSTRACT:
INOCULATICN FROM AN ESTABLISHED MICROCOSM WAS SUBJECTED TO ACUTE GAMMA RADIATION LEVELS OF 0.798 KILO- AND 79.8 MEGARADS. TWO SETS OF 240 FLASKS WERE PREPARED WITH ONE HALF RECEIVING IRRACIATED INOCULA AND THE ONE-HALF CONTROLLED INOCULA. THESE WERE SAMPLED CVER A 40-DAY PERIOD AND MEASURED FOR CHLOROPHYLL A CONTENT, DIURNAL COMMUNITY METABOLISM, AND PARTICULATE BIOMASS. BIOMASS AND CHLOROPHYLL SAMPLES WERE OBTAINED BY FILTRATION WITH WHATMAN GF/C GLASS FIBER FILTERS. SIGNIFICANT REDUCTION IN CHLOROPHYLL A WAS FOUND WHEN EACH TREATMENT LEVEL WAS COMPARED TO ITS CONTROL. BIOMASS WAS SIGNIFICANTLY REDUCED THROUGHOUT SUCCESSION FOR BOTH TREATMENT LEVELS EXCEPT FOR THE LAST SAMPLE OF 79.8 MEGARAD MICROCOSMS. NET DAYTIME PRODUCTIVITY AND NIGHTTIME RESPIRATION VARIED AND WERE REDUCED DURING EARLY SUCCESSION AT BCTH TREATMENT LEVELS. 0.798 KILORAD LEVELS SHOWED MORE STRESS THAN 7.98 MEGARAD LEVELS WHEN COMPARING RADIATION TREATMENT EFFECTS AT THE CCMMUNITY LEVEL AS RELATED TO OTHER FORMS OF STRESS WITHIN THE COMMUNITY. (SNYDER-BATTELLE)

FIELD 05C

ACCESSION NO. W73-01080

MARSH PRODUCTION: A SUMMARY OF THE LITERATURE,

MARYLAND UNIV., SOLOMONS. NATURAL RESOURCES INST.

C. W. KEEFE.

CONTRIBUTIONS IN MARINE SCIENCE, VOL 16, P 163-181, MARCH 1972. 1 FIG, 3 TAB, 86 REF.

DESCRIPTORS:
*PRIMARY PRODUCTIVITY, *FRESHWATER, *MARSHES, SALINE WATER, BIOMASS, NUTRIENTS, FOOD WEBS, DETRITUS, ECOSYSTEMS, ENERGY BUDGETS, NUTRIENTS, REVIEWS, BIOLOGICAL COMMUNITIES, AQUATIC PLANTS, LIMITING FACTORS, ROOTED AQUATIC PLANTS, CYCLING NUTRIENTS, RESPIRATION, MARSH PLANTS, CHEMICAL ANALYSIS, LEAVES, SAMPLING, PHYTOPLANKTON, IRON, MANGANESE, IONS, SALINITY, PHOSPHORUS, NITROGEN, ALGAE, ORGANIC MATTER, CARBON, CATTAILS.

IDENTIFIERS:
ENGLAND, SWEDEN, GERMANY, SPARTINA SPP, DISTICHLIS SPICATA, FIMBRISTYLIS, BORRICHIA, PHRAGMITES COMMUNIS, JUNCUS SPP, ATRIPLEX HASTATA, SCIRPUS SPP, TYPHA SPP, ZIZANIA AQUATICA, CAREX SPP, BUTOMUS UMBELLATUS, SPARGANIUM RAMOSUM, LEERSIA ORYZOIDES, NUPHAR ADVENA, EICHHORNIA CRASSIPES, JUSTICIA AMERICANA, ALTERNANTHERA PHILOXEROIDES, GLYCERIA MAXIMA, SAGITTARIA SPP, GLYCERIA FLUITANS, SPARGANIUM SPP, ELEOCHARIS PALUSTRIS, PHALARIS ARUNDINACEA, MOLINIA CAERULEA, MANYANTHES TRIFOLIATA, CALAMOGROSTIS LANCEOLATA, CLADIUM MARISCUS, JUSTICIA AMERICANA.

ABSTRACT:
STUDIES OF PRIMARY PRODUCTION IN BOTH SALT AND FRESHWATER MARSHES ARE REVIEWED. METHODS ARE DISCUSSED FOR MEASURING PRODUCTION WHICH RELY PRIMARILY ON THE AERIAL PORTIONS OF THE PLANT SINCE ROOT MATERIAL IS DIFFICULT TO SAMPLE. ONE METHOD IS TO CLIP AERIALS IN A UNIT AREA SELECTED AT RANDOM FROM A LARGER AREA AND TO SEPARATE LIVING AND DEAD COMPONENTS. THE TOTAL OF THE WEIGHT INCREASES IN MATERIAL PROVIDES A MEASURE OF PRODUCTION. IN A SECOND METHOD, ONLY LIVING MATERIAL IS CLIPPED AND THE WEIGHT OF AN AVERAGE MATURE LEAF IS DETERMINED. PRODUCTION IS CALCULATED FROM THE WEIGHT OF MATERIAL PRESENT PLUS MATERIAL ESTIMATED TO HAVE BEEN REMOVED BEFORE HARVEST. SEVERAL REASONS FOR THE HIGH PRODUCTIVITY OF MARSH COMMUNITIES AS COMPARED TO TERRESTRIAL COMMUNITIES ARE DISCUSSED. IT IS CONCLUDED THAT THE RELATIVE CONTRIBUTIONS OF PRIMARY PRODUCERS TO THE FOOD SUPPLY MAY AFFECT THE POPULATIONS OF CONSUMERS AND THE ENERGY FLOW PATTERNS OF THE ENTIRE ECOSYSTEM. (MORTLAND-BATTELLE)

FIELD 05C, 05A, 02L, 10F

ACCESSION NO. W73-01089

TAXONOMY OF AUSTRALIAN FRESHWATER ALGAE. 2. SOME PLANKTIC STAURASTRA FROM TASMANIA,

AQUATIC BIOLOGY INST., UPPSALA (SWEDEN)

K. THOMASSON, AND P. A. TYLER.

NOVA HEDWIGIA, VOL 21, NO 1, P 287-319, 1971. 12 FIG, 1 TAB, 57 REF.

DESCRIPTORS:
*PHYTOPLANKTON, *AQUATIC ALGAE, *SYSTEMATICS, *DISTRIBUTION PATTERNS, LAKES, AQUATIC PLANTS, AUSTRALIA, SAMPLING, MICROSCOPY, ECOLOGICAL DISTRIBUTION, PLANKTON NETS, CHLOROPHYTA, PLANKTON.

IDENTIFIERS:
*STAURASTRUM, DESMIDS, STAURASTRUM LONGIPES VAR CONTRACTUM, ARTHURS LAKE, LAKE SORELL, LAKE CRESCENT, TOOMS LAKE, WOODS LAKE, LAGOON OF ISLANDS, STAURASTRUM SPP.

ABSTRACT:
THE TAXONOMY AND DISTRIBUTION OF AUSTRALIAN FRESHWATER ALGAE OF THE GENUS, STAURASTRUM ARE DESCRIBED. PLANKTON SAMPLES WERE TAKEN WITH A 10-MICRON OR 60-MICRON PLANKTON NET FROM LAKES IN TASMANIA, AUSTRALIA, MICROSCOPICALLY IDENTIFIED AND DIAGRAMMED FOR REFERENCE. FIFTY-NINE TAXA OR VARIATIONS THEREOF WERE FOUND. (LONG-BATTELLE)

FIELD 05A, 02I

ACCESSION NO. W73-01094

AMINC ACID FLUX IN A NATURALLY EUTHROPHIC LAKE,

OREGON STATE UNIV., CORVALLIS.

B. K. BURNISON.

AVAILABLE FROM UNIV. MICROFILMS, INC., ANN ARBOR, MICH., 48106, ORDER NO. 71-27, 850. PH D DISSERTATION, 1971. 89 P.

DESCRIPTORS:
*AMINO ACIDS, *EUTROPHICATION, *PLANKTON, OREGON, ORGANIC COMPOUNDS, ABSORPTION, KINETICS, WATER TEMPERATURE, RESPIRATION, ALGAE, CYANOPHYTA, ANALYTICAL TECHNIQUES, WATER POLLUTICN EFFECTS, NUTRIENT REQUIREMENTS.

IDENTIFIERS:
UPPER KLAMATH LAKE, SUBSTRATE UTILIZATION.

ABSTRACT:
THE YEAR'S STUDY MADE ON THE IN SITU ACTIVITIES OF THE HETEROTROPHIC PLANKTON IN UPPER KLAMATH LAKE, OREGON USING THE WRIGHT-HOBBIE KINETIC APPROACH OF MEASURING THE UPTAKE OF ORGANIC COMPCUNDS WAS INITIATED TO TEST THE APPLICABILITY OF THE KINETIC APPROACH TO A HIGHLY EUTROPHIC SYSTEM. SIXTEEN AMINO ACIDS WERE USED AS THE ORGANIC SUBSTRATES AND IT WAS SHOWN THAT THE MAXIMUM VELOCITY, V SUB MAX, FOR THE UPTAKE OF ALL THE AMINO ACIDS IS PROPORTIONAL TO TEMPERATURE. THE TURNOVER TIME, T SUB T, AND THE SUM OF A TRANSPORT CONSTANT AND THE NATURAL SUBSTRATE CONCENTRATION, (K SUB T PLUS S SUB N), ARE DIFFICULT TO INTERPRET BECAUSE OF POSSIBLE COMPETITIVE INHIBITION EFFECTS AMONG THE AMINO ACIDS. THE V SUB MAX IS UNAFFECTED BY COMPETITIVE INHIBITION, BUT THE TURNOVER TIME AND SUBSTRATE CONCENTRATION VALUES ARE INCREASED. THESE LATTER VALUES THEN REFLECT THE TOTAL NATURAL CONCENTRATION AND AFFINITIES OF THE AMINO ACIDS TRANSPORTED BY A PARTICULAR TRANSPORT SYSTEM. THOSE AMINO ACIDS EXHIBITING THE HIGHEST V SUB MAX VALUES HAD THE HIGHEST PERCENT RESPIRED AS CO2. THE PERCENT RESPIRED APPEARS TO BE INVERSELY PROPORTIONAL TO TEMPERATURE. THE AMINO ACID CONCENTRATIONS DETERMINED FOR UPPER KLAMATH LAKE SURFACE WATERS PROBABLY ARE THE SUM OF THE FREE AND ADSORBED AMINO ACIDS. BIOCHEMICAL ANALYSES WERE MADE OF THE PREDOMINANT BLUE-GREEN ALGAE, FOUND IN THE NUISANCE ALGAL BLOOMS, TO BETTER UNDERSTAND THEIR CONTRIBUTION TO THE NUTRITION OF THE HETROTROPHIC PLANKTON. (MORTLAND-BATTELLE)

FIELD 05C, 05B

ACCESSION NO. W73-01269

AN INVESTIGATION INTO THE DEVELOPMENT OF ELECTROPHORETIC AND ELECTROCHEMICAL WATER PURIFICATION SYSTEMS,

TEXAS A AND M UNIV., COLLEGE STATION.

W. M. LYLE.

AVAILABLE FROM UNIVERSITY MICROFILMS, ANN ARBOR, MICHIGAN, ORDER NO. 71-17, 816. PH. D. DISSERTATION, 1970. 127 P.

DESCRIPTORS:
*SUSPENSION, *ELECTROPHORESIS, *ELECTROCHEMISTRY, *WATER PURIFICATION, CLAY, ALGAE, BACTERIA, WASTE WATER TREATMENT, WATER TREATMENT, MODEL STUDIES.

IDENTIFIERS:
*PARALLEL PLATE MODEL.

ABSTRACT:
ELECTROPHORETIC AND ELECTROCHEMICAL SYSTEMS APPEARED TO BE USEFUL IN THE REMOVAL OF ELECTRICALLY CHARGED SUSPENDED POLLUTANTS SUCH AS CLAY, ALGAE, AND BACTERIA. DESIGN CONCEPTS OF A PARALLEL PLATE MODEL (ENTIRELY FOR ELECTRCPHORETIC REMOVAL) AND POROUS FILLER AND ELECTRODE GRID MODELS (INCORPORATING BOTH ELECTROPHORETIC AND ELECTROCHEMICAL REMOVAL) WERE TESTED AS LABORATORY MODELS. THE PARALLEL PLATE MODEL ACHIEVED SUCCESSFUL WATER CLARIFICATION ONLY WHEN THE INFLUENT WATER HAD A VERY LOW ELECTRICAL CONDUCTIVITY. ELECTROCHEMICAL WATER PURIFICATION NOT ONLY OVERCAME THIS PROBLEM BUT WAS ALSO SUCCESSFUL BOTH ECONOMICALLY AND OPERATIONALLY. AN EXAMPLE DESIGN OF A SMALL SEMIAUTOMATED ELECTRCCHEMICAL WATER SYSTEM WHICH INCCRPORATED ELECTROCHEMICAL FLOCCULATION, SETTLING AND DISINFECTICN OPERATIONS IS PRESENTED. (ALBERT-TEXAS)

FIELD 05D, 05F

ACCESSION NO. W73-01362

PARAMETERS INFLUENCING PHOSPHORUS ELIMINATION BY ALGAE,

STUTTGART UNIV. (WEST GERMANY). INST. FOR SANITARY ENGINEERING.

K-H. HUNKEN, AND I. D. SEKCULOV.

WATER RESEARCH, VOL 6, P 1087-1096, 1972. 11 FIG, 7 REF.

DESCRIPTORS:
*PHOSPHORUS, *ALGAE, *WASTE WATER TREATMENT, HYDROGEN ION CONCENTRATION, CYANOPHYTA.

IDENTIFIERS:
*PHOSPHORUS ELIMINATION.

ABSTRACT:
THE PROCESS OF BIOPRECIPITATION OF PHOSPHORUS USES ALGAL PHOTOSYNTHESIS TO REMOVE CARBONATE CARBON TO ALTER THE PH. TO ENHANCE THE PHOTOSYNTHETIC ACTIVITY, ARTIFICIAL LIGHTS WERE USED, MAINTAINING THE PH ABOVE 9. USING THE ACCOMPANYING DIAGRAM, IT IS POSSIBLE TO CALCULATE DETENTION TIMES NEEDED TO REACH SUITABLE PH LEVELS, THUS DETERMINING TECHNICAL AND ECONOMIC FEASIBILITY. IN THE BATCH TESTS STUDIED, THE FINAL PH DETERMINED THE FINAL P CONCENTRATION. MIXING BY AERATION PREVENTED OVERSATURATION WITH CO2 AND STARTED PHOTOSYNTHESIS AT A PH OF 8.3. THE SUBMERGED ILLUMINATION MADE THE REACTION CHAMBER INDEPENDENT OF SURFACE AREA, ALLOWING MORE CONSTRUCTION FREEDOM. FOR AN ALGAE-FREE EFFLUENT, A FILTER AND A SETTLING BASIN WERE SATISFACTORY. (ANDERSON-TEXAS)

FIELD 05D

ACCESSION NO. W73-01367

RELATIONSHIP OF TRACE ELEMENTS TO ALGAE GROWTH,

WASHINGTON STATE UNIV., PULLMAN.

J. J. KOELLING.

AVAILABLE FROM UNIV. MICROFILMS, INC., ANN ARBOR, MICH., 48106, ORDER NO. 71-28,778. PH. D. THESIS, 1971, 142 P.

DESCRIPTORS:
*ALGAE, *GROWTH RATES, *PRIMARY PRODUCTIVITY, *HEAVY METALS, *TRACE ELEMENTS, NUTRIENTS, MERCURY, CHROMIUM, COBALT, FRESH WATER, LAKES, PHYSICAL PROPERTIES, CHEMICAL PROPERTIES, GOLD, IRON, CESIUM, ALKALI METALS, SEASONAL, ZINC, ABSORPTION, LIGHT PENETRATION, ALKALINITY, HALOGENS, HYDROGEN ION CONCENTRATION, PHOSPHATES, NITRATES, DISSOLVED OXYGEN, TEMPERATURE, SPECTROMETERS, TOXICITY, DIATOMS, CHRYSOPHYTA, CYANOPHYTA, METALS, IRRADIATION, AUTOMATION, EUTROPHICATION, NUCLEAR REACTORS, COMPUTERS, DATA PROCESSING, *WASHINGTON, SAMPLING.

IDENTIFIERS:
*WILLIAMS LAKE(WASH), URANIUM, THORIUM, ANTIMONY, MACRONUTRIENTS, MICRONUTRIENTS, SCANDIUM, RUBIDIUM, BROMINE, ACTINIDES, RADIOACTIVE DECAY, GAMMA-RAY SPECTROMETRY, COUNTING.

ABSTRACT:
RESEARCH DIRECTED TOWARD FINDING CORRELATIONS BETWEEN ALGAL POPULATIONS AND TRACE ELEMENTS IN A NATURAL AQUATIC ENVIRONMENT, WILLIAMS LAKE (WASHINGTON), BASICALLY INVOLVED THE IDENTIFICATION AND QUANTIFICATION OF ALGAL SPECIES AS WELL AS THE MEASUREMENT OF SOME 14 TRACE ELEMENTS PRESENT IN THE LAKE WATER. EIGHTY-TWO SAMPLES OF WATER WITH ALGAE WERE COLLECTED OVER A 13-MONTH PERIOD TO OBSERVE SEASONAL FLUCTUATIONS IN BOTH ALGAL SPECIES AND TRACE ELEMENTS. COLLECTION WAS MADE IN POLYETHYLENE BOTTLES FROM THREE DIFFERENT DEPTHS AT ONE LOCATION WHERE THE LAKE WAS THE DEEPEST. THESE SAMPLES ALONG WITH APPROPRIATE STANDARDS WERE IRRADIATED IN THE WASHINGTON STATE UNIVERSITY TRIGA III RESEARCH REACTOR FOR 12-15 HOURS. AFTER A DECAY PERIOD OF APPROXIMATELY 2 WEEKS, EACH SAMPLE AND STANDARD WAS COUNTED ON A NUCLEAR DATA 2200 GAMMA-RAY SPECTROMETER SYSTEM USING A HIGH RESOLUTION GE(LI) DETECTOR FOR APPROXIMATELY 17 HOURS TO DETERMINE TRACE ELEMENT CONCENTRATIONS. ALGAE CELL IDENTIFICATION AND QUANTITATIVE ESTIMATES WERE DETERMINED BY A MICROSCOPIC COUNT. DATA ANALYSIS AND A LINEAR CORRELATION STUDY BETWEEN ALGAL SPECIES AND TRACE ELEMENT CONCENTRATION WAS PERFORMED UTILIZING AN IBM 360/67 COMPUTER. SEVERAL PHYSICAL AND CHEMICAL MEASUREMENTS WERE ALSO MADE INCLUDING LIGHT TRANSMISSION, ALKALINITY, ACIDITY, PH, DISSOLVED OXYGEN, TEMPERATURE, PHOSPHATES, AND NITRATES. RESULTS INDICATE THAT: (1) URANIUM, THORIUM, AND CESIUM MAY BE PRESENT IN EXCESS OF GROWTH REQUIREMENTS OR MAY POSSIBLY HAVE NO AFECT ON THE GROWTH RATE OF ANY OF THE ALGAL FORMS STUDIED. (2) MERCURY, ALTHOUGH TOXIC IN LARGE QUANTITIES, MAY BE UTILIZED BY DIATOMS IN SMALL QUANTITIES. (3) CHROMIUM, GOLD, COBALT, AND ANTIMONY APPEAR TO BE TAKEN UP BY DIATOMS. (4) ZINC AND IRON APPEAR TO BE UTILIZED BY BLUE-GREEN ALGAE. (5) UPTAKE OF SCANDIUM, RUBIDIUM, AND BROMINE BY ALGAE IS NOT APPARENT, BUT MAY BE POSSIBLE. (HOLOMAN-BATTELLE)

FIELD 05C

ACCESSION NO. W73-01434

LIMNOLOGICAL STUDIES ON BIGHORN LAKE (YELLOWTAIL DAM) AND ITS TRIBUTARIES,

MONTANA STATE UNIV., BOZEMAN.

R. A. SOLTERO.

AVAILABLE FROM UNIV. MICROFILMS, INC., ANN ARBOR, MICH., 48106, ORDER NO. 71-28,873. PH. D. THESIS, 1971, 290 P.

DESCRIPTORS:
*LIMNOLOGY, *PHYTOPLANKTON, *WATER ANALYSIS, *STANDING CROP, *PRIMARY PRODUCTIVITY, LAKES, SAMPLING, RESERVOIRS, TRIBUTARIES, DAMS, CHEMICAL ANALYSIS, BIOMASS, ALGAE, PHYSICOCHEMICAL PROPERTIES, DIATOMS, SUCCESSION, NUTRIENTS, RIVERS, INFLUENT STREAMS, EFFLUENT STREAMS, LIGHT INTENSITY, NITROGEN, CHRYSOPHYTA, NITRATES, CHLOROPHYLL, PHOSPHATES, TURBIDITY, CONDUCTIVITY, OXYGEN, HYDROGEN ION CONCENTRATION, TEMPERATURE, FRESH WATER, WATER QUALITY, NUTRIENTS, *MONTANA.

IDENTIFIERS:
*BIGHORN LAKE(MONT), BIGHORN RIVER, SHOSHONE RIVER, YELLOWTAIL DAM, FRAGILARIA CROTONENSIS, CRYPTOMONAS OVATA, STEPHANODISCUS NIAGARAE, ASTERIONELLA FORMOSA, RHODOMONAS LACUSTRIS, COUNTING, ORTHOPHOSPHATES, CHLOROPHYLL A.

ABSTRACT:
THE PHYTOPLANKTON COMMUNITY OF BIGHORN LAKE IN RELATION TO ITS PHYSICAL AND CHEMICAL ENVIRONMENT WAS STUDIED DURING 1968-70. SAMPLES AND IN SITU MEASUREMENTS WERE TAKEN AT SIX PERMANENT SAMPLING STATIONS ON THE RESERVOIR DURING 56 CRUISES. CHARACTERIZATION OF THE INFLUENT AND EFFLUENT WATERS OF THE RESERVOIR REVEALED THAT BIGHORN LAKE WAS FERTILIZED BY THE BIGHORN AND SHOSHONE RIVERS. THE RELATIONSHIPS OF CONDUCTIVITY, TURBIDITY, NITROGEN, AND PHOSPHATE TO TRIBUTARY DISCHARGE WERE EXAMINED. UNDER AVERAGE CONDITIONS TOTAL VISIBLE LIGHT WAS REDUCED TO 1 PERCENT OF SURFACE INTENSITY AT A DEPTH OF APPROXIMATELY 10 METERS NEAR THE DAM, BUT ONLY 1 METER AT STATION 5 (80.5 KM UPSTREAM FROM THE DAM). RESULTS OF CHEMICAL ANALYSES SHOWED BIGHORN LAKE TO BE PREDOMINANTLY A CALCIUM-SODIUM-SULFATE-BICARBONATE WATER. NITRATE-NITROGEN AND ORTHOPHOSPHATE AS WELL AS THE OTHER CHEMICAL CONSTITUENTS DETERMINED WERE RELATIVELY HIGH AND APPEARED NOT TO BE LIMITING. STANDING CROPS OF THE PHYTOPLANKTON TAXA WERE DETERMINED BY DIRECT COUNT. FRAGILARIA CROTONENSIS, CRYPTOMONAS OVATA, STEPHANODISCUS NIAGARAE AND ASTERIONELLA FORMOSA REACHED THE LARGEST STANDING CROPS ACCORDING TO ABSOLUTE MEAN CELL VOLUMES, WHEREAS RHODOMONAS LACUSTRIS, CRYPTOMONAS OVATA, ASTERIONELLA FORMOSA AND FRAGILARIA CROTONENSIS WERE MORE IMPORTANT ON A PRESENCE BASIS. THE MEAN TOTAL ALGAL STANDING CROPS AND MEAN CHLOROPHYLL A FOR THE RESERVOIR WERE 2.3-23.2 CU MM/1 AND 0.6-10.4 MG/CU M, RESPECTIVELY. THE CHLOROPHYLL TO ALGAL CELL OLUME RATIO WAS 3.4 MICROGRAMS CHLOROPHYLL A TO 1.0 CU MM OF ALGAL CELLS. A DEFINITE ALGAL SUCCESSION WAS OBSERVED FOR ALL THREE YEARS. NET PRIMARY PRODUCTIVITY WAS CALCULATED AND A MEAN FOR ALL STATIONS SHOWED A RANGE OF 0.51 G C/SQ M/DAY - 1.42 G C/SQ M/DAY DURING 1968 AND 1970, RESPECTIVELY. (HOLOMAN-BATTELLE)

FIELD 05C, 02H

ACCESSION NO. W73-01435

THE ORGANIC GEOCHEMISTRY OF HYDROCARBONS IN COASTAL ENVIRONMENTS,

TEXAS UNIV., AUSTIN.

J. R. SEVER.

AVAILABLE FROM UNIV. MICROFILMS, INC., ANN ARBOR, MICH., 48106 ORDER NO. 72-2418. PH. D. THESIS, 1970. P 155.

DESCRIPTORS:
*COASTS, *ORGANIC COMPOUNDS, *GEOCHEMISTRY, CYANOPHYTA, BACTERIA, ALGAE, PLANKTON, PLANTS, SPECTROSCOPY, CHROMATOGRAPHY, GAS CHROMATOGRAPHY.

IDENTIFIERS:
NOSTOC MUSCORUM, LYNGBYA LAGERHAIMII, 7-METHYLHEPTADECANE, 8-METHYLHEPTADECANE, HYDROCARBONS, ALKANES, ALIPHATIC HYDROCARBONS, MASS SPECTROMETRY, INFRARED SPECTROSCOPY.

ABSTRACT:
SINCE HYDROCARBONS IN THE MARINE ENVIRONMENT ARE RELATIVELY BIOLOGICALLY AND CHEMICALLY INERT, A STUDY WAS CONDUCTED TO DETERMINE IF A PRECURSOR RELATIONSHIP COULD BE OBSERVED BETWEEN HYDROCARBONS IN THE SEDIMENTS AND HYDROCARBONS IN THE DOMINANT ORGANISMS WHICH CONTRIBUTE ORGANIC MATTER TO THE SEDIMENTS. BY EXAMINING THE HYDROCARBON DISTRIBUTIONS OF THE ORGANISMS AND ENVIRONMENTS OF THREE COASTAL AREAS - AN OPEN SHALLOW BAY, AN INTERTIDAL BLUE-GREEN ALGAL LAGOON, AND A SUPRATIDAL LAGOON, - IT WAS POSSIBLE TO OBSERVE DIRECT CORRELATIONS BETWEEN THE BIOLOGICAL AND GEOLOGICAL ALKANES. IDENTIFICATION OF THE HYDROCARBONS WAS DONE BY INFRARED SPECTROSCOPY, MASS SPECTROMETRY, AND GAS-LIQUID CHROMATOGRAPHY. FIVE CULTURED BLUE-GREEN ALGAE SHOWED A NARROW DISTRIBUTION OF NORMAL ALKANES WITH CHAIN LENGTHS BETWEEN C15 AND C19 WITH C17 PREDOMINATING. BACTERIAL ALKANE DISTRIBUTIONS WERE OF TWO TYPES: C14 - C18 RANGE AND BRANCHED HYDROCARBONS IN THE RANGE OF C15 - C20. ANALYSIS OF SEVEN DOMINANT HIGHER MARINE PLANTS REVEALED A DISTINCT ODD-CARBON NUMBER DISTRIBUTION IN THE C25 - C31 RANGE, WHILE A PLANKTON TOW YIELDED AN UNBIASED SMOOTH DISTRIBUTION OVER THE C14 - C28 RANGE. ANALYSIS OF EIGHT ANCIENT SHALES SHOWED THE UNIQUE 50:50 ALKANE MIXTURE OF 7-METHYL AND 8-METHYL HEPTADECANE WITH THE MAJORITY OF SAMPLES DISPLAYING THE C17 ALKANE AS ONE OF THE LARGEST HYDROCARBON COMPONENTS. THIS SUGGESTS LARGE DEPOSITS OF BLUE GREEN ALGAE AT THE TIME OF DEPOSITION. (MACKAN-BATTELLE)

FIELD 05B, 02L, 05C

ACCESSION NO. W73-01439

GROWTH AND PHOSPHATE REQUIREMENTS OF NITZSCHIA ACTINASTROIDES (LEMM.) V. GOOR IN BATCH AND CHEMOSTAT CULTURE UNDER PHOSPHATE LIMITATION, (WACHSTUM UND PHOSPHATBEDARF VON NITZSCHIA ACTINASTROICES (LEMM.) V GCCR IN STATISCHER UND HOMOKONTINUIERLICHER KULTUR UNTER PHOSPHATLIMITIERUNG),

FREIBURG UNIV. (WEST GERMANY). LIMNOLOGISCHES INSTITUT.

V. H. MULLER.

ARCHIV FUR HYDROBIOLOGIE, VOL 38, NO 4, P 399-484, MARCH 1972. 43 FIG, 14 TAB, 158 REF.

DESCRIPTORS:
*NUTRIENT REQUIREMENTS, *LIMITING FACTORS, *PHOSPHATES, DIATOMS, POLLUTANT IDENTIFICATION, GROWTH RATES, DEFICIENT ELEMENTS, WATER POLLUTION EFFECTS, AQUATIC ALGAE, CHRYSOPHYTA, LABORATORY EQUIPMENT, FRESHWATER, WATER POLLUTION SOURCES, INSTRUMENTATION, METHODOLOGY, NITROGEN, SILICCN, ABSORPTION, PHYTOPLANKTON.

IDENTIFIERS:
*NITZSCHIA ACTINASTRCIDES, *BATCH CULTURES, *CHEMOSTATS, ORTHOPHOSPHATES, SUBSTRATE UTILIZATION, MONOD EQUATION, PENICILLIN G, TERRAMYCIN, CHLORAMPHENICOL, ACTIDON, CULTURE MEDIA, CHLOROPHYLL A, TERATOLOGY, NITZSCHIA.

ABSTRACT:
THE CONSTRUCTION OF A CHEMOSTAT WITH A CAPACITY OF 2 OR 4 L IS DESCRIBED. IT WAS USED TO DETERMINE GROWTH RESPONSES OF THE PLANKTONIC FRESHWATER DIATOM NITZSCHIA ACTINASTROIDES (LEMM.) V. GOOR TO LIMITING CONCENTRATIONS OF ORTHOPHOSPHATE. THE MAXIMAL SPECIFIC GROWTH-RATE WAS CALCULATED FROM THE DATA OBTAINED IN THE CHEMOSTAT AT 23 C WITH CONTINUOUS ILLUMINATION. IT IS 0.087 /HR AND AGREES WELL WITH THE FIGURE OF 0.083 /HR OBTAINED IN BATCH CULTURES WITHOUT SUBSTRATE LIMITATION. MONOD'S EQUATION DESCRIBES THE CORRELATICN BETWEEN GROWTH RATE AND SUBSTRATE CONCENTRATION OF PHOSPHORUS P SUB L AT MEDIUM AND HIGH GROWTH-RATES. THE SATURATION CONSTANT LIES BETWEEN 0.40 AND 0.44 MICROGRAM/L P SUB L. AT MEDIUM GROWTH-RATES, THE GROWTH-LIMITING FACTOR IS THE RATE OF P UPTAKE, AS IS INDICATED BY COMPARING UPTAKE RATES WITH THE KINETICS OF P UPTAKE IN P-STARVED CELLS. THE DISCREPANCIES OBSERVED AT LOW GROWTH RATES ARE CHARACTERIZED BY INCREASING P SUB L, HIGH DEATH RATES, AND HIGH FRACTIONS OF TERATOLOGICAL CELLS. THIS INDICATES THE APPEARANCE OF FACTORS OTHER THAN PHOSPHORUS LIMITING GROWTH. THRESHOLD VALUES OF THE DILUTICN RATE EXIST, BELCW WHICH THE CULTURE IS WASHED OUT, THEY VARY WITH CONDITIONS IN THE CHEMOSTAT. THE CONTENTS IN P AND CHLOROPHYLL A OF CELLS WITH P-LIMITED GROWTH ARE DIRECTLY PROPORTIONAL TO P SUB L. P-STORAGE IN THE CELLS OCCURS IF GROWTH IS LIMITED BY FACTORS OTHER THAN P (E. G. CO2). THEREFORE THE YIELD COEFFICIENT IS CONTANT ONLY IF REFERRED TC THE CHLOROPHYLL A PRODUCED. REFERRED TO OTHER REFERENCE FIGURES, IT CECREASES WITH INCREASING GROWTH-RATE AND P-STORAGE, THE N-CONTENT OF THE CELLS INCREASES WITH INCREASING GROWTH RATE AND INCREASING STORAGE OF P. THE SI-CONTENT OF THE CELLS IS NOT INFLUENCED BY THESE FACTORS. THE RELIABILITY OF THE RESULTS AND THEIR TRANSFERABILITY TO NATURAL CONDITICNS ARE DISCUSSED. (LONG-BATTELLE)

FIELD 05C, 05A

ACCESSION NO. W73-01445

THE SPECIES COMPOSITION, SEASONAL SUCCESSION, REPRODUCTION AND DISTRIBUTION OF MARINE ALGAE FROM SCITUATE TO WOODS HOLE, MASSACHUSETTS,

NEW HAMPSHIRE UNIV., DURHAM.

D. C. COLEMAN.

AVAILABLE FROM UNIV. MICROFILMS, INC., ANN ARBOR, MICH., 48106 ORDER NO. 72-3735. PH. D. THESIS, 1971. P 143.

DESCRIPTORS:
*SEASONAL, *DISTRIBUTION PATTERNS, *MASSACHUSETTS, SYSTEMATICS, *REPRODUCTION, RHODOPHYTA, PHAEOPHYTA, WATER QUALITY, *MARINE ALGAE.

IDENTIFIERS:
*CAPE COD CANAL(MASS), BUZZARDS BAY, CAPE COD BAY, DATA INTERPRETATION, SPECIES DIVERSITY.

ABSTRACT:
BIMONTHLY COLLECTIONS WERE MADE FROM JANUARY TO DECEMBER, 1969 AND TOTAL OF 106 SPECIES WAS RECORDED. CONSPICUOUS DIFFERENCES IN SPECIES NUMBERS WERE EVIDENT AT EACH STATION. SCITUATE AND WOODS HOLE HAD THE HIGHEST NUMBERS, WHILE INTERMEDIATE VALUES WERE RECORDED IN THE CAPE COD CANAL. MOST OF THE SPECIES AT SCITUATE AND IN THE CANAL WERE PERENNIALS; ANNUALS WERE MOST ABUNDANT AT WINGS NECK AND WOODS HOLE. THE RHODOPHYCEAE ACCOUNTED FOR MOST OF THE SPRING AND SUMMER ANNUALS, WHILE THE BROWNS WERE THE MAJOR CONTRIBUTORS OF WINTER ANNUALS. SPRING AND EARLY SUMMER ANNUALS APPEARED SEVERAL WEEKS SOONER AT WINGS NECK AND WOODS HOLE THAN AT SCITUATE. SOME SPECIES APPEARED AS SPRING ANNUALS SOUTH OF THE CAPE, BUT AS SUMMER ANNUALS AT SCITUATE. SOME SPECIES REPRODUCED THROUGHOUT THE YEAR, WHILE OTHERS WERE RESTRICTED TO EITHER WARM OR COLD SEASONS. MOST SPECIES AT SCITUATE WERE FOUND IN THE INTERTIDAL AND/OR SUBTIDAL ZONE(S), WHILE THOSE AT ALL OTHER STATIONS WERE COLLECTED PRIMARILY IN THE SUBTIDAL ZONE. NUMEROUS FACTORS ARE RESPONSIBLE FOR THE VERTICAL DISPLACEMENT OF SPECIES RECORDED. AN INTERPRETATION OF THE DISTRIBUTIONAL PATTERNS IS GIVEN. (MORTLAND-BATTELLE)

FIELD 05C, 05B

ACCESSION NO. W73-01449

FIELD ASSESSMENT OF N2-FIXATION BY LEGUMES AND BLUE-GREEN ALGAE WITH THE ACETYLENE REDUCTION TECHNIQUE,

WISCONSIN UNIV., MADISON.

T. H. MAGUE.

AVAILABLE FROM UNIV. MICROFILMS, INC., ANN ARBOR, MICH., 48106 ORDER NO. 71-28,353. PH.D. THESIS, 1971, 200 P.

DESCRIPTORS:
*ALGAE, *LEGUMES, *NITROGEN FIXATICN, *CYANOPHYTA, PLANTS, NITROGEN, REDUCTION(CHEMICAL), GAS CHRCMATOGRAPHY, SEPARATION TECHNIQUES, *WISCONSIN, OLIGOTROPHY, EUTROPHICATION, LAKE SUPERICR, LAKE MICHIGAN, LAKE HURON, LAKE ERIE, GREAT LAKES, SOYBEANS, LIGHT INTENSITY, AIR TEMPERATURE, SAMPLING, AQUATIC ALGAE, PHYTOPLANKTCN, TROPHIC LEVELS.

IDENTIFIERS:
*ACETYLENE REDUCTION, RHIZCBIUM JAPONICUM, LAKE MENDOTA, CRYSTAL LAKE, TROUT LAKE, BIG ARBOR VITAE LAKE, LITTLE ARBOR VITAE LAKE, GREEN BAY, FOX RIVER, DETROIT RIVER, HETEROTROPHIC BACTERIA, ACETYLENE, ETHYLENE.

ABSTRACT:
FIELD STUDIES OF NITROGEN-FIXATION, BOTH IN AQUATIC AND TERRESTRIAL ENVIRONMENTS, UTILIZED A PROCEDURE WHICH CONSISTS OF (1) EXPOSING THE TEST MATERIAL (EXCISED AND INTACT SOYBEAN NODULES AND BLUE-GREEN ALGAE) TO APPROXIMATELY 0.1 ATM. OF ACETYLENE IN A CLOSED SYSTEM FOR 1/2 HOUR AND (2) QUANTITATING THE RESULTING ETHYLENE BY FLAME IONIZATION AFTER ITS GAS CHROMATOGRAPHIC SEPARATICN FROM ACETYLENE. THE EXPERIMENTS WITH EXCISED SOYBEAN NODULES SHOWED THAT ACETYLENE REDUCTION WAS LINEAR FOR AT LEAST 80 MINUTES BUT WAS UNDETECTABLE IN THE ABSENCE CF OXYGEN. TWENTY SAMPLINGS FROM LAKE MENDOTA (MAY-NOVEMBER, 1970) REVEALED TWO MAJOR PEAKS IN BLUE-GREEN ALGAE N2-FIXING ACTIVITY, BUT NO CORRELATION BETWEEN DISSOLVED INCRGANIC P OR N CONCENTRATIONS AND ACETYLENE REDUCTION RATES FOR THE SAME SAMPLES. CF THE FOUR WISCONSIN LAKES SAMPLED, (SUMMER, 1970) CRYSTAL AND TROUT LAKES (CLIGOTROPHIC) SUPPORTED LITTLE ACETYLENE REDUCTION/UNIT VOLUME; BIG AND LITTLE ARBOR VITAE LAKES (EUTROPHIC) SHOWED HIGH RATES OF ACETYLENE REDUCTION AND SUPPORTED HEAVY BLUE-GREEN ALGAL BLOOMS. SAMPLES (SEPTEMBER, 1970) FROM LAKES SUPERIOR, HURON, AND MICHIGAN SHOWED PRACTICALLY NC ACETYLENE REDUCTION. ACETYLENE REDUCTION BY SAMPLES TAKEN FRCM GREEN BAY OF LAKE MICHIGAN VARIEC AS A RESULT CF A NUTRIENT CONCENTRATION WHICH AT FIRST SUPPORTED GROWTH OF PHYTOPLANKTON REQUIRING COMBINED N, THEN WAS DILUTED SO THAT IT CCULD CNLY SUPPORT PRIMARILY THCSE SPECIES ABLE TO FIX N2, AND FINALLY WAS DEPLETED TO THE POINT WHERE ORGANISMS TYPICAL OF OLIGOTROPHIC WATERS PREDOMINATED. THESE EXPERIMENT DEMCNSTRATED THE SUITABILITY OF THE ACETYLENE REDUCTION ASSAY FCR FOLLOWING CHANGES IN RATES OF N2-FIXATION BY NODULATED LEGUMES IN THE FIELD AND BY BLUE-GREEN ALGAE IN LAKES. THE ASSAY ALSO SERVED TC DEFINE VARIOUS TROPHIC LEVELS IN LAKES AND TO LOCATE ZONES OF EUTROPHICATICN IN THE GREAT LAKES. (HOLOMAN-BATTELLE)

FIELD 05C, 05A, 02H

ACCESSION NO. W73-01456

SELECTED REFERENCES CONCERNING THE ALGAE OF LAKE ERIE. II,

STATE UNIV., COLL., BUFFALO, N.Y. GREAT LAKES LAB.

R. A. SWEENEY.

SPECIAL REPORT NO 6, MARCH, 1970. 11 P.

DESCRIPTORS:
*BIBLIOGRAPHIES, *ALGAE, *LAKE ERIE, PLANKTON, BENTHIC FLORA, DIATOMS, PHYTOPLANKTON, PHOTOSYNTHESIS, PRIMARY PRODUCTIVITY, GREAT LAKES, NEW YORK, WATER SUPPLY, SYSTEMATICS, LIMNOLOGY, ECOLOGY, FISH, WATER POLLUTION, PROTOZOA, OHIO, EUTROPHICATION, LAKE ONTARIO, LIGHT PENETRATION, PHYSICOCHEMICAL PROPERTIES, AQUATIC PLANTS, INDUSTRIAL WASTES, FISH FOOD ORGANISMS, STREAMS.

IDENTIFIERS:
DESMIDS.

ABSTRACT:
THIS BIBLIOGRAPHY OF 127 REFERENCES IS COMPILED TO ASSIST THOSE CONDUCTING AND/OR CONTEMPLATING RESEARCH AND INSTRUCTION DEALING WITH ALGAE IN LAKE ERIE. IT IS A COMPILATION OF REFERENCES ON PLANKTONIC AND BENTHIC ALGAE PREVIOUSLY PUBLISHED IN VARIOUS JOURNALS, REPORTS, ETC., TO REDUCE LITERATURE SEARCH TIME. THE CITATIONS, LISTED ALPHABETICALLY, INCLUDE 13 PAPERS PUBLISHED BEFORE 1900, WITH THE EARLIEST DATED 1872. FORTY TITLES AND REFERENCES, DATED FROM 1900 TO 1949, RELATE TO TAXONOMY AND OF THAT GROUP FOUR INCLUDE POLLUTION PROBLEMS IN THE TITLES. THE PROBLEM OF PHOSPHORUS APPEARS IN PAPERS DATED 1951; THE EVIDENCE FOR EUTROPHICATION OF LAKE ERIE FROM PHYTOPLANKTON RECORDS IN 1964; REFERENCE TO INDUSTRIAL POLLUTION IN 1953; SEVERAL PAPERS ARE ON CLADOPHORA; NINE REFERENCES CONCERN PHOTOSYNTHESIS; AND SIX CONCENTRATE ON DIATOMS. (JONES-WISCONSIN)

FIELD 02H, 05C

ACCESSION NO. W73-01615

MORPHOGENESIS IN THE RED ALGA, GRIFFITHSIA PACIFICA: REGENERATION FROM SINGLE CELLS,

WASHINGTON UNIV., SEATTLE. DEPT. OF BOTANY.

C. S. DUFFIELD, S. D. WAALAND, AND R. CLELAND.

PLANTA (BERL.), VOL 105, P 185-195, 1972. 7 FIG, 27 REF. AEC AT(45-1)2225.

DESCRIPTORS:
*ALGAE, *RHODOPHYTA, *PLANT MORPHOLOGY, *PLANT GROWTH, CYTOLOGICAL STUDIES.

IDENTIFIERS:
*GRIFFITHSIA PACIFICA, *REGENERATION.

ABSTRACT:
A SYSTEM FOR STUDYING FORM DEVELOPMENT IN A RED ALGA IS DESCRIBED. PLANTS OF THE MARINE GIANT-CELLED GRIFFITHSIA PACIFICA REGENERATE FROM A SINGLE, ISOLATED SHOOT CELLS FOLLOWING A REGULAR AND PREDICTABLE PATTERN. THE CULTURE FOR THIS STUDY WAS COLLECTED IN MEXICO IN 1966 AND HAD BEEN MAINTAINED IN UNIALGAL CONDITION. REGENERATION OF PLANTS FROM ISOLATED SHOOT CELLS MAKES A SUITABLE SYSTEM FOR STUDY OF MORPHOGENESIS CONTROL IN RED ALGAE. THE CELLS' LARGE SIZE FACILITATES BOTH EXPERIMENTAL MANIPULATIONS AND OBSERVATIONS. BECAUSE OF MORPHOLOGY AND CELL SIZE, GRIFFITHSIA SPECIES HAVE BEEN WIDELY USED IN BIOCHEMICAL, CYTOLOGICAL, AND ELECTROPHYSIOLOGICAL STUDIES. IT HAS THE ADVANTAGE OF MULTICELLULARITY WITHOUT COMPLEXITIES SUCH AS CORTICATION FOUND IN MANY RED ALGAE. REGENERATION CAN BE INITIATED FROM ANY CELL OF THE PLANT. THE DEVELOPMENT IS SUFFICIENTLY RAPID THAT A PLANT OF 30-40 SHOOT-CELLS IS PRODUCED FROM A SINGLE CELL WITHIN A WEEK. THIS INVESTIGATION IS APPARENTLY THE FIRST IN WHICH THE PATTERN OF DEVELOPMENT OF INDIVIDUAL PLANTS HAS BEEN FOLLOWED WITH TIME AND IN WHICH KINETICS OF CELL DIVISION AND CELL ELONGATION HAVE BEEN ELUCIDATED. (JONES-WISCONSIN)

FIELD 05C

ACCESSION NO. W73-01622

RESPIRATORY ACTIVITIES OF CHLORELLA ELLIPSOIDEA IN VARIOUS NITRIENT MEDIA,

TAIWAN PROVINCIAL CHUNG HSING UNIV., TAICHUNG. INST. OF BOTANY.

J. TSCHEN, AND S. T. LIU.

BOTANICAL BULLETIN OF ACADEMIA SINICA, VOL 12, P 50-56, 1971. 2 FIG, 4 TAB, 27 REF.

DESCRIPTORS:
*LABORATORY TESTS, *RESPIRATION, *CHLORELLA, *NUTRIENTS, CULTURES, NITROGEN, POTASSIUM, PHOSPHATES, METABOLISM, ALGAE, SYNTHESIS, VITAMINS, SULFATES, IONS.

IDENTIFIERS:
*CHLORELLA ELLIPSOIDEA, CULTURE MEDIA.

ABSTRACT:
POPULATION DENSITY OF FRESH-WATER ALGAE IN A BATCH SYSTEM IN SMALL VESSELS DEPENDS ON CULTURAL CONDITIONS. IN SELECTING OPTIMAL CULTURAL MEDIA FOR FRESH-WATER ALGAE, CHLORELLA ELLIPSOIDEA WAS CULTURED IN VARIOUS NUTRIENT MEDIA, AND THE RESPIRATORY ACTIVITY DETERMINED BY A WARBURG RESPIROMETER. THE FOLLOWING MEDIA WERE STUDIED: BOLD'S BASAL MEDIUM AND MODIFICATION, 5N-BOLD'S BASAL MEDIUM, KANTZ' MODIFIED BBM AND OTT'S. OXYGEN UPTAKE WAS FOUND TO BE HIGHEST IN BBM WITH DECREASING ORDER IN 5N-BBM, KBBM, AND OTT'S MEDIUM. HIGHER RESPIRATORY ACTIVITY IN BBM MAY BE ATTRIBUTED TO NITRATE COMPONENTS. DECREASED RESPIRATION IN KBBM IS RELATED TO LOWER POTASSIUM AND PHOSPHATE CONTENTS. NO GROWTH TOOK PLACE IN OTT'S MEDIUM. NITROGEN IS ESSENTIAL IN RESPIRATION, PHOTOSYNTHESIS, AND IN THE SYNTHESIS OF PURINES AND PYRIMIDINES OF RNA AND DNA. NITROGEN SOURCE IN BBM IS SODIUM NITRATE, WHILE IN 5N-BBM, UREA IS USED INSTEAD. UREASE WAS NOT DETECTED IN C. ELLIPSOIDEA. THIS EXPLAINS WHY BBM IS SUPERIOR TO 5N-BBM. POTASSIUM IS ESSENTIAL AS AN ACTIVATOR FOR ENZYMES IN CERTAIN PEPTIDE BOND SYNTHESIS AND SO MAY AFFECT RESPIRATION AND PHOTOSYNTHESIS. PHOSPHORUS, FOUND IN MANY COENZYMES, BECOMES IMPORTANT IN PHOTOSYNTHESIS, GLYCOLYSIS, TCA CYCLE AND FATTY ACID SYNTHESIS. (JONES-WISCONSIN)

FIELD 05C

ACCESSION NO. W73-01626

COMPOSITION AND STRUCTURE OF ALGAL COMMUNITIES IN A TRIBUTARY STREAM OF LAKE ONTARIO,

GUELPH UNIV. (ONTARIO). DEPT. OF ZOOLOGY.

J. W. MOORE.

CANADIAN JOURNAL OF BOTANY, VOL 50, P 1663-1674, 1972. 6 FIG, 2 TAB, 33 REF.

DESCRIPTORS:
*ALGAE, *BIOLOGICAL COMMUNITIES, *STREAMS, LAKE ONTARIO, PLANKTON, DIATOMS, SEDIMENTS, HARDNESS(WATER), BENTHOS, GRAZING, SYSTEMATICS, CHEMICAL ANALYSIS, PHYSICAL PROPERTIES, CHLOROPHYTA, EUGLENOPHYTA, CYANOPHYTA, CHRYSOPHYTA, SCENEDESMUS, CHLAMYDOMONAS.

IDENTIFIERS:
SHELTER VALLEY CREEK(ONTARIO).

ABSTRACT:
SEASONAL SUCCESSION, COMMUNITY STRUCTURE, AND THE EFFECT OF HERBIVOROUS GRAZING ON PLANKTONIC AND EPIPELIC ALGAE IN SHELTER VALLEY CREEK, ONTARIO ARE DESCRIBED. THE COMPOSITION AND STRUCTURE OF THE ALGAL COMMUNITIES WERE STUDIED FOR ONE YEAR. THE OVERALL ASSEMBLAGE CONSISTED OF 388 TAXA OF WHICH 321 WERE BACILLARIOPHYCEAE. ON THE SEDIMENTS, SEASONAL SUCCESSION AND COMMUNITY STRUCTURE WAS, WITH SOME EXCEPTIONS, TYPICAL OF HARD-WATER STREAMS IN NORTHERN TEMPERATE ZONES. THE PLANKTONIC COMMUNITY WAS, FOR THE MOST PART, DERIVED FROM THE BENTHOS. HERBIVOROUS GRAZING BY LARVAL SEA LAMPREY, PETROMYZON MARINUS DID NOT NOTICEABLY AFFECT ALGAL NUMBERS IN EITHER THE PLANKTON OR SEDIMENTS. THE ALGAE COLLECTED ARE LISTED. THE 388 TAXA RECORDED CONTAINED 321 BACILLARIOPHYCEAE, 32 CHLOROPHYTA, 20 EUGLENOPHYTA, 14 CYANOPHYTA, AND 1 CHRYSOPHYCEAE. SIXTY-SIX TAXA ARE REPORTED AS NEW OCCURRENCES IN SOUTHERN ONTARIO. DIATOMS ACCOUNTED FOR 93.0 TO 99.5% BY NUMBERS OF THE EPIPELIC COMMUNITY. TAXA THAT OCCURRED FREQUENTLY THROUGHOUT THE YEAR WERE ACHANTHES MINUTISSIMA, COCCONEIS PLACENTULA VAR. EUGLYPTA, GOMPHONEMA OLIVACEUM VAR. BALTICUM, AND NAVICULA TRIPUNCTATA. (JONES-WISCONSIN)

FIELD 05C

ACCESSION NO. W73-01627

GENERAL CABLE THEORY FOR CELLS OF ALGAE CHARACEAE,

AGROFIZICHESKII NAUCHNO-ISSLEDOVATELSKII INSTITUT, LENINGRAD (USSR).

G. A. VOLKOV.

BIOCHIMICA ET BIOPHYSICA ACTA, VOL 255, P 709-719, 1972. 3 FIG, 3 REF.

DESCRIPTORS:
*BIOLOGICAL MEMBRANES, *ALGAE, *CYTYLOGICAL STUDIES, PLANT MORPHOLOGY, EQUATIONS, RESISTANCE, ELECTRICAL PROPERTIES.

IDENTIFIERS:
*CABLE THEORY, ELECTRONIC POTENTIAL, CHARACEAE.

ABSTRACT:
TO INVESTIGATE THE BIOLOGICAL MEMBRANE, THE ELECTRICAL CHARACTERISTICS OF MEMBRANES AND PARTICULARLY OF THE OUTER CYTOPLASMIC MEMBRANE OF THE INTERNODAL CELL OF ALGAE OF THE CHARACEAE ARE STUDIED. THE SO-CALLED CABLE PROPERTIES ARE CONSIDERED REFLECTION OF A CERTAIN MORPHOLOGICAL STRUCTURE OF THAT CELL. SOLUTION OF THE CABLE EQUATION IS OBTAINED TAKING INTO ACCOUNT THE DEFINITE RESISTANCE OF CELL NODES FOR THE GENERAL CASE OF AN APPLIED CURRENT, THE STRENGTH OF WHICH ALTERS WITH TIME. TWO IMPORTANT CASES FOR THE FUNCTION ARE CONSIDERED IN DETAIL. PRACTICAL FORMULAE FOR DETERMINING THE FUNDAMENTAL CHARACTERISTICS OF THE CELL ARE OBTAINED: MEMBRANE RESISTANCE AND CAPACITY (OF PLASMALEMMA), NODAL RESISTANCE, AND THE CHARACTERISTIC LENGTH OF THE CELL. FOR DETERMINATION OF THE MEMBRANE TIME CONSTANT AND HENCE THE MEMBRANE CAPACITY, A SIMPLIFIED METHOD FOR MEASURING MEMBRANE RESISTANCE, SUGGESTED BY OTHER AUTHORS, CAN BE USED. TAKING INTO ACCOUNT THE NODAL RESISTANCE OF THE CELL GIVES AN IMPROVED VALUE FOR THE ELECTRODE POSITION IN TERMS OF THE CRITICAL COORDINATE. THE CURRENT ELECTRODE HAS TO BE INSERTED IN THE MIDDLE OF THE CELL. (JONES-WISCONSIN)

FIELD 05C

ACCESSION NO. W73-01630

THERMAL POLLUTION AND ALGAE, (VARMVATTENUTSLAPP OCH ALGER),

UPPSALA UNIV. (SWEDEN). INST. FOR PHYSIOLOGIC BOTANY.

G. ERIKSSON, AND C. FORSBERG.

VATTEN, VOL 27, NO 4, P 441-448, 1971. 3 FIG, 41 REF.

DESCRIPTORS:
*INDUSTRIAL WASTES, *THERMAL POLLUTION, *COASTS, WATER POLLUTION EFFECTS, AQUATIC LIFE, HABITATS, ECOLOGY, *ECOSYSTEM, *ALGAE, *PHYTOPLANKTON, OXYGEN, PHOTOSYNTHESIS, RESPIRATION, SUCCESSION, PRODUCTIVITY, GROWTH RATES.

IDENTIFIERS:
*SWEDEN, SPECIES COMPOSITION.

ABSTRACT:
NUCLEAR POWER PLANTS BEING BUILT ALONG THE SWEDISH COASTS HAVE INCREASED INTEREST IN THE EFFECTS OF THERMAL POLLUTION. INCREASED TEMPERATURE FROM INSTALLATION DISCHARGES WILL CHANGE THE HABITATS OF MANY ORGANISMS. AS ALGAE, ESPECIALLY PHYTOPLANKTON, HAVE AN IMPORTANT ROLE IN THE ECOSYSTEM BECAUSE OF THEIR ABILITY TO PRODUCE O2 IN PHOTOSYNTHESIS, IT IS ESPECIALLY IMPORTANT TO KNOW THEIR REACTIONS TO TEMPERATURE CHANGES. THE EFFECTS OF TEMPERATURE CHANGE ON PHYTOPLANKTON PHOTOSYNTHESIS, RESPIRATION AND NET OXYGEN PRODUCTION ARE SUMMARIZED. THERMOPHILIC ALGAE, SPECIES COMPOSITION CHANGES, SUCCESSION, GROWTH RATE AND PRODUCTION ARE DISCUSSED ALONG WITH THE SOLUBILITY OF O2 IN WATER OF VARYING TEMPERATURES. (ENSIGN-PAI)

FIELD 05C

ACCESSION NO. W73-01632

CONCURRENT GROWTH OF BACTERIA AND ALGAE IN A CLOSED VESSEL,

NEBRASKA UNIV., LINCOLN. DEPT. OF CHEMICAL ENGINEERING.

P. J. REILLY.

AVAILABLE FROM THE NATIONAL TECHNICAL INFORMATION SERVICE AS PB-212 996, $3.00 IN PAPER COPY, $0.95 IN MICROFICHE. NEBRASKA WATER RESOURCES RESEARCH INSTITUTE TECHNICAL RESEARCH PROJECT COMPLETION REPORT, JULY 1972. 34 P, 13 TAB, 10 REF. OWRR A-020-NEB(1). 14-31-0001-3227.

DESCRIPTORS:
*LABORATORY TESTS, *BACTERIA, *ALGAE, *CULTURES, CYANOPHYTA, HYDROGEN ION CONCENTRATION, LIGHT INTENSITY, NUTRIENTS, PSEUDOMONAS, SYMBIOSIS.

IDENTIFIERS:
MICROCYSTIS AERUGINOSA, PSEUDOMONAS AERUGINOSA, LIMITING NUTRIENTS.

ABSTRACT:
GROWTH RESPONSES OF PURE CULTURES OF MICROCYSTIS AERUGINOSA TO CARBON DIOXIDE, PHOSPHATE, AND NITRATE ARE STUDIED TO IDENTIFY THE LIMITING NUTRIENT. EXPERIMENTS WERE CONDUCTED ON GROWTH OF THE OBLIGATE AEROBIC BACTERIUM, PSEUDOMONAS AERUGINOSA FOR USE AS A MODEL BACTERIUM IN INVESTIGATION OF SYMBIOTIC TWO-SPECIES BACTERIAL/ALGAL CULTURES WHICH MAY BE SIGNIFICANT IN THE EUTROPHICATION PROCESS. IN ADDITION, EVIDENCE WAS FOUND THAT BACTERIA REMOVES ALGAL TOXIC AGENTS WHICH MAY EXPLAIN WHY ALGAE BLOOMS SOMETIMES APPEAR SO SUDDENLY AND UNEXPECTEDLY. ONE BACTERIAL AND ONE ALGAL SPECIE WERE GROWN IN THE SAME VESSEL UNDER CONTROLLED CONDITIONS AFTER THE ALGAE WERE PURIFIED OF CONTAMINATING BACTERIA. ITS RESPONSES TO GLUCOSE, PH, LIGHT INTENSITY, AND AERATION, AND AGITATION RATE WERE DETERMINED. DURING THE PURIFICATION OF THE ALGAE OTHER OTHER INTERACTIONS OCCURRED IN ADDITION TO THE TRANSFER OF OXYGEN FROM ALGAE TO BACTERIA AND THE REVERSE TRANSFER OF CARBON DIOXIDE FROM BACTERIA TO ALGAE: TWO THAT WERE VERY OBVIOUS WERE THE TRANSFER OF ORGANIC EXCRETIONS TO THE BACTERIA AND THE BREAKDOWN OF AGENTS TOXIC TO THE ALGAE. (JONES-WISCONSIN)

FIELD 05C

ACCESSION NO. W73-01657

SPECIAL LAKE WATER TREATMENT PROBLEMS,

CHICAGO DEPT. OF WATER AND SEWERS, ILL. BUREAU OF WATER.

J. C. VAUGHN.

JOURNAL OF THE AMERICAN WATER WORKS ASSOCIATION, VOL 64, NO 9, P 585-589, SEPTEMBER 1972. 10 FIG, 2 REF.

DESCRIPTORS:
*WATER TREATMENT, WATER POLLUTION EFFECTS, *TREATMENT FACILITIES, *INFLUENT STREAMS, *GREAT LAKES, AQUATIC ALGAE, ICE, FISH, MICROORGANISMS, ODOR, TASTE, PLANKTON, CHEMICALS, PHENOLS, CHLORINE, OXIDATION, ADSORPTION, ACTIVATED CARBON, SLUDGE TREATMENT, DIATOMS, COAGULATION, CALCIUM CARBONATE, HYDROGEN ION CONCENTRATION, FILTRATION, PLANKTON NETS, NUISANCE ALGAE, CLADOPHORA.

IDENTIFIERS:
*INTERFERENCE, ALEWIVES, DICHOTOMOSIPHON, FRAGILARIA, TABELLARIA, ASTERIONELLA, SYNEDRA, DINOBRYON, MELOSIRA.

ABSTRACT:
WATER TREATMENT PLANTS DEPENDING ON SURFACE WATER FOR FILTRATION TREATMENT ARE LIABLE TO MANY INTAKE DELIVERY PROBLEMS. ICING UP OF INTAKES CAN BE AVOIDED BY DESIGN TECHNIQUES, BACKFLUSHING, TEMPERATURE CONTROL, AND CLEANING BY BLASTING. BLOCKAGE OF INTAKES BY FISH (ALEWIVES) HAS BEEN REMEDIED BY PROTECTIVE NETS AND SCREENS. ALGAL COLLECTIONS ON SCREENS OR NETS REQUIRE POSITIVE PRESSURE FOR REMOVAL (DICHOTOMOSIPHON, CLADOPHORA) OR INSTALLATION OF REVOLVING SCREENS AND PULVERIZING EQUIPMENT. PLANKTON (FRAGILLARIA, TABELLARIA, ASTERIONELLA, AND SYNEDRA) CAN BEST BE GOTTON RID OF BY A GOOD SURFACE-WASH SYSTEM SO THAT BROKEN-UP MATS CAN BE REMOVED FROM FILTERS BY BACKWASHING. PROBLEMS WITH TASTES AND ODORS CAN RESULT FROM MICROORGANISMS (DINOBRYON) OR CHEMICALS. TREATMENT USUALLY INVOLVES OXIDATION BY EXCESS CHLORINE TREATMENT AND/OR ADSORPTION ONTO ACTIVATED CARBON. MICROSTRAINERS AND SLUDGE BLANKET DEVICES REQUIREC NORMAL MAINTENANCE AND SOME SUPPLEMENTATION. THE MOST DIFFICULT AND EXPENSIVE PROBLEM INVOLVES WINTERTIME DIATOMS (MELOSIRA). A SIDE ISSUE OF THEIR PROLIFERATION IS DEVELOPMENT OF COLLOICAL TURBIDITY; DOUBLING COAGULANT DOSAGE APPEARS SOMEWHAT REMEDIAL. CALCIUM CARBONATE AND PH CHANGES APPEAR RELATED TO MELOSIRA BLOOMS. (MACKAN-BATTELLE)

FIELD 05F, 02H

ACCESSION NO. W73-01669

PRELIMINARY RADIATION SURVEILLANCE OF AN AQUATIC SYSTEM NEAR THE NEVADA SITE, JUNE - JULY, 1967,

ENVIRONMENTAL PROTECTION AGENCY, LAS VEGAS, NEV. WESTERN ENVIRONMENTAL RESEARCH LAB.

W. L. KLEIN, AND R. A. BRECHBILL.

AVAILABLE FROM THE NATIONAL TECHNICAL INFORMATION SERVICE AS SWRHL-65-R, $3.00 IN PAPER COPY, $0.95 IN MICROFICHE. REPORT NO SWRHL-65R, FEBRUARY 1972. 23 P, 2 FIG, 3 TAB, 10 REF, 2 APPEND. MEMORANDUM OF UNDERSTANDING NO SF 54 373.

DESCRIPTORS:
*SURVEYS, *AQUATIC ENVIRONMENT, *RADIOECOLOGY, *RADIOCHEMICAL ANALYSIS, MEASUREMENT, *NEVADA, AQUATIC PLANTS, WATER POLLUTION SOURCES, SEDIMENTS, AQUATIC SOILS, PLANT TISSUES, FRESHWATER FISH, AQUATIC ALGAE, EVALUATION, SAMPLING, CARP, POTASSIUM RADIOISOTOPES, STRONTIUM RADIOISOTOPES, SOIL ANALYSIS, SOIL CONTAMINATION, CLADOPHORA, CHARA, FISH EGGS.

IDENTIFIERS:
*BIOLOGICAL SAMPLES, SR-90, K-40, CS-137, *PAHRANAGAT LAKE, GAMMA RAY SPECTROMETRY, ANIMAL TISSUES, MACROPHYTES, CYPRINUS CARPIO, MOSQUITO FISH, GAMBUSIA AFFINIS, CESIUM RADIOISOTOPES, CLADOPHORA SPP, PASPALUM DISTICHUM, POTAMOGETON PECTINATUS, ELEOCHARIS MONTEVIDENSES, CHARA ASPPRA, SCIRPUS SPP, GILLS, MUSCLE, VISCERA, EYES, SCALES(FISH), BONE.

ABSTRACT:
A THREE-MONTH PRELIMINARY RADIATION SURVEILLANCE STUDY WAS MADE OF AN AQUATIC SYSTEM IN UPPER PAHRANAGAT LAKE NEAR THE NEVADA TEST SITE. THE OBJECTIVES WERE TO DETERMINE THE CONCENTRATIONS OF FISSION PRODUCTS IN SELECTED SAMPLES AND TO ESTABLISH THE NECESSARY METHODOLOGY FOR RADIATION SURVEILLANCE IN AN AQUATIC ECOSYSTEM. BIOLOGICAL SAMPLES FROM A FRESHWATER LAKE NEAR THE NEVADA TEST SITE (NTS) WERE ANALYZED FOR THE PRESENCE OF SELECTED RADIONUCLIDES IN ORDER TO ESTABLISH A BASE LINE FOR THIS PARTICULAR SYSTEM AND TO DEVELOP METHODOLOGY NECESSARY FOR ANY FURTHER DEFINITIVE STUDIES OF THIS TYPE. RADIONUCLIDE CONCENTRATIONS WERE INSIGNIFICANT IN WATER, AQUATIC PLANT, AND FISH SAMPLES. SEDIMENT SAMPLES HAD DETECTABLE LEVELS OF CS-137, K-40, SR-90, AND U. STRONTIUM-90 LEVELS IN FISHBONE WERE LOW (2.38 PCI/G BONE ASH) COMPARED TO THOSE FOUND IN BOVINE FEMUR SAMPLES (6.9 PCI/G BONE ASH) COLLECTED DURING THE SAME PERIOD. (MACKAN-BATTELLE)

FIELD 05A, 02H

ACCESSION NO. W73-01673

SURVEY OF TECHNIQUES USED TO PRESERVE BIOLOGICAL MATERIALS,

STANFORD RESEARCH INSTITUTE, MENLO PARK, CALIF.

E. J. FEINLER, AND R. W. HUBBARD.

AVAILABLE FROM THE NATIONAL TECHNICAL INFORMATION SERVICE AS N72-18080, $6.00 IN PAPER COPY, $0.95 IN MICROFICHE. NASA CONTRACT REPORT NO. 114422, JANUARY 1972. 390 P, 78 TAB, 1350 REF. PROJECT NO. SRI LSU-8930. CONTRACT NO. NAS2-6201.

DESCRIPTORS:
*SURVEYS, *METHODOLOGY, FREEZE DRYING, REFRIGERATION, INCUBATION, DIALYSIS, CHEMICALS, DRYING, FREEZING, RADIATION, CYTOLOGICAL STUDIES, PLANT TISSUES, ALGAE, MICROORGANISMS, BACTERIA, INVERTEBRATES, FUNGI, ENZYMES, ANIMAL PARASITES, VIRUSES, MOLLUSKS, PROTOZOA, YEASTS, INSECTS, MILDEWS, E. COLI, FISH, FOODS, GRASSES, WORMS.

IDENTIFIERS:
*BIOLOGICAL MATERIALS, *BIOLOGICAL SAMPLES, *SAMPLE PRESERVATION, CHEMICAL PRESERVATION, HISTOCHEMISTRY, LYOPHILIZATION, ANIMAL TISSUES, HISTOLOGICAL STUDIES, BODY FLUIDS, VERTEBRATES, HEAT STERILIZATION, FIXATION, EMBEDDING, ASHING, BONE, ORGANS, BRAIN, BLOOD, KIDNEYS, LIVER, PLASMODIUM BERGHEI, BORRELIA KANSAS, BORRELIA ANSERINA.

ABSTRACT:
EXISTING TECHNIQUES USED TO PRESERVE BIOLOGICAL MATERIALS ARE DESCRIBED. THIS INFORMATION IS PRESENTED IN A HANDBOOK FORMAT THAT CATEGORIZES THE MOST IMPORTANT PRESERVATION TECHNIQUES AVAILABLE, AND INCLUDES A REPRESENTATIVE SAMPLING OF THOUSANDS OF APPLICATIONS OF THOSE TECHNIQUES TO BIOLOGICAL MATERIALS AND ORGANISMS. THE HANDBOOK IS DIVIDED INTO FOUR MAIN SECTIONS: (1) A REVIEW OF REVIEWS, (2) TABLES OF TECHNIQUES OF PRESERVATION, (3) INDEXES, AND (4) A COMPREHENSIVE BIBLIOGRAPHY. (HOLOMAN-BATTELLE)

FIELD 05A

ACCESSION NO. W73-01676

THE RESPONSES OF THE BIOTA OF LAKE WABAMUN, ALBERTA, TO THERMAL EFFLUENT,

ALBERTA UNIV., EDMONTON. DEPT. OF ZOOLOGY.

J. R. NURSALL, AND D. N. GALLUP.

IN: PROCEEDINGS OF THE INTERNATIONAL SYMPOSIUM ON THE IDENTIFICATION AND MEASUREMENT OF ENVIRONMENTAL POLLUTANTS, OTTAWA, P 295-304, JUNE 1971. 6 FIG, 2 TAB, 23 REF.

DESCRIPTORS:
*BIOTA, *TEMPERATURE, *CANADA, *LAKES, *THERMAL POLLUTION, WATER POLLUTION, COOLING, THERMAL POWERPLANTS, HEATED WATER, CIRCULATION, INFRARED RADIATION, HYDROGRAPHY, PLANKTON, ALGAE, ENVIRONMENTAL EFFECTS, BIOLOGICAL COMMUNITY, CANADA.

IDENTIFIERS:
*LAKE WABAMUN, BIOLOGICAL STUDIES, MACROPHYTES.

ABSTRACT:
TWO POWER-GENERATING PLANTS PUT WATER FROM CONDENSER-COOLING SYSTEMS INTO LAKE WABAMUN AT TEMPERATURES 8 - 14 C ABOVE AMBIENT. THE INPUT REPRESENTS ABOUT 0.5 PERCENT OF THE LAKE VOLUME PER DAY WHEN AVERAGED OVER THE YEAR. CIRCULATION RESTRICTS THE HEATED WATER TO THE EAST END OF THE LAKE, THEREBY LOCALIZING ITS EFFECTS. INFRA-RED IMAGERY HAS BEEN USED TO SUPPORT DETAILED DIRECT THERMOMETRY. DISCUSSED ARE THE EFFECTS OF HEATED EFFLUENT ON PLANKTONIC ALGAE, SEVERAL SPECIES OF MACROPHYTES, FIVE SPECIES OF ROTIFERS AND FOUR SPECIES OF PLANKTONIC CRUSTACEANS. NOTED IN PARTICULAR ARE THE WIDE DISTRIBUTION OF ELODEA CANADENSIS IN THE HEATED ZONE, CONCENTRATION OF POTAMOGETON PECTINATUS IN THE EFFLUENT CANAL, REDUCTION IN NUMBERS OF DIAPTOMUS OREGONENSIS IN THE HEATED ZONE, AND VARIATION IN NUMBERS AND EGG RATIOS FOR PLANKTONIC ROTIFERS BETWEEN THE HEATED AND UNHEATED REGIONS OF THE LAKE. (OLESZKIEWICZ-VANDERBILT)

FIELD 05C, 02H

ACCESSION NO. W73-01704

AN ECOSYSTEMATIC STUDY OF THE SOUTH RIVER, VIRGINIA,

VIRGINIA POLYTECHNIC INST. AND STATE UNIV., BLACKSBURG. WATER RESOURCES RESEARCHCENTER.

J. CAIRNS, JR., AND K. L. DICKSON.

AVAILABLE FROM THE NATIONAL TECHNICAL INFORMATION SERVICE AS PB-213 159, $3.00 IN PAPER COPY, $0.95 IN MICROFICHE. VIRGINIA WATER RESOURCES RESEARCH CENTER, BLACKSBURG, BULLETIN 54, JULY 1972. 104 P, 5 FIG, 11 TAB, 22 REF, APPEND. OWRR A-999-VA (15).

DESCRIPTORS:
*BIOLOGICAL PROPERTIES, *ECOSYSTEMS, *RIVERS, *VIRGINIA, *WATER POLLUTION EFFECTS, SURVEYS, SAMPLING, DATA COLLECTIONS, INDUSTRIAL WASTES, AGRICULTURAL RUNOFF, DOMESTIC WASTES, ECOLOGY, FISH, ALGAE, AQUATIC PLANTS, PROTOZOA, BACTERIA, STREAMFLOW, FLOW RATES, WATER TEMPERATURE, WATER CHEMISTRY, CHEMICAL ANALYSIS, NUTRIENTS.

IDENTIFIERS:
*SOUTH RIVER(VA).

ABSTRACT:
A BIOLOGICAL SURVEY OF THE FISH, MACROINVERTEBRATES, ALGAE, AQUATIC PLANTS, PROTOZOANS, AND BACTERIA WAS CONDUCTED ON THE SOUTH RIVER, VIRGINIA, IN SEPTEMBER 1970. THE PURPOSE OF THIS SURVEY WAS TO MEASURE THE ECOLOGICAL CONDITION OF A SYSTEM RECEIVING AGRICULTURAL, DOMESTIC, AND INDUSTRIAL WASTES. SAMPLES WERE COLLECTED AT 8 SITES TO DETERMINE THE EFFECTS OF VARIOUS WASTES ON THE BIOTA OF THE SOUTH RIVER. THE FAUNA AND FLORA AT EACH STATION WERE EXAMINED TO DETERMINE THE DIVERSITY, DENSITY, AND DISTRIBUTION OF AQUATIC LIFE IN RELATION TO PHYSICAL AND CHEMICAL WATER QUALITY. THE DISCHARGE OF DOMESTIC AND INDUSTRIAL WASTES IN WAYNESBORO, VIRGINIA (1) EXCEEDED THE WASTE ASSIMILATIVE CAPACITY OF THE RIVER AND CAUSED THE DISSOLVED OXYGEN TO BE ENTIRELY DEPLETED IN CERTAIN REACHES OF THE RIVER AT TIMES OF LOW FLOW AND HIGH TEMPERATURE; (2) ENRICHED THE SYSTEM BY ADDING NUTRIENTS SUCH AS CARBON, PHOSPHORUS, AND NITROGEN, CAUSING A DEFINITE SHIFT IN THE COMPOSITION OF THE FLORA; AND (3) DECREASED THE DIVERSITY OF FISH AND MACROINVERTEBRATES AND CAUSED QUALITATIVE SHIFTS IN ALGAE, HIGHER PLANTS, BACTERIA, AND PROTOZOANS WHEN COMPARED TO AREAS OF THE SOUTH RIVER UPSTREAM OF WAYNESBORO. THE BIOLOGICAL RECOVERY WAS NOT COMPLETE FOURTEEN MILES DOWNSTREAM OF WAYNESBORO AT HARRISTON, VIRGINIA. (WOODARD-USGS)

FIELD 05C, 05B

ACCESSION NO. W73-01972

WATER QUALITY CRITERIA DATA BOOK - VOLUME 3: EFFECTS OF CHEMICALS ON AQUATIC LIFE, SELECTED DATA FROM THE LITERATURE THROUGH 1968.

BATTELLE MEMORIAL INST., COLUMBUS, OHIO.

COPY AVAILABLE FROM GPO SUP DOC AS EP1.16:18050GWV 05/71/V3, $3.75; MICROFICHE FROM NTIS AS PB-213 210, $0.95. ENVIRONMENTAL PROTECTION AGENCY, WATER POLLUTION CONTROL RESEARCH SERIES, MAY 1971. 528 P, 1 FIG, 10 TAB, 961 REF, 4 APP. EPA PROGRAM 18050 GW V0 05/71 68-01-0007.

DESCRIPTORS:
*TOXICITY, REVIEWS, *BIOASSAY, *INDUSTRIAL WASTES, *PESTICIDES, *AQUATIC ORGANISMS, *PEST CONTROL, *CHEMCONTROL, *BIBLIOGRAPHIES, *DATA COLLECTIONS, *WATER POLLUTION EFFECTS, DOCUMENTATION, PUBLICATIONS, PESTICIDE TOXICITY, BIOINDICATORS, AGRICULTURAL CHEMICALS, FISH, CHEMICALS, CHEMICAL WASTES, BIOCHEMICAL OXYGEN DEMAND, BACTERIA, ALGAE, AQUATIC FUNGI, INVERTEBRATES, AQUATIC INSECTS, OYSTERS, SHRIMP.

IDENTIFIERS:
SUMMARIES.

ABSTRACT:
ORIGINAL DATA FROM MORE THAN 500 TECHNICAL PUBLICATIONS CONCERNING THE SPECIFIC EFFECTS OF CHEMICALS ON INDIVIDUALS SPECIES OF AQUATIC BIOTA WERE COLLECTED AND SUMMARIZED IN UNIFORM FORMAT. ALPHABETICAL ASSEMBLY OF THE DATA BY CHEMICAL ALLOWS RAPID ACCESS TO CONSIDERABLE DETAILED INFORMATION. A SPECIES INDEX FACILITATES SEARCH FOR INFORMATION ON THE TOXICITY OF CHEMICALS TO INDIVIDUAL AQUATIC SPECIES. THE DETAILS OF MAJOR PROCEDURES IN LABORATORY BIOASSAY AND FIELD ASSESSMENT OF CHEMICAL TOXICITY IN WATER ARE DISCUSSED. FRESHWATER AND MARINE PROCEDURES ARE INCLUDED. A TOTAL OF 961 REFERENCES WERE UTILIZED. RECOMMENDATIONS INCLUDE: (1) ESTABLISHMENT OF AN INFORMATION-ANALYSIS CENTER ON CHEMICAL WATER POLLUTION BASED TO SOME EXTENT ON THE REPORT PREPARED, (2) PREPARATION OF A LISTING OF CHEMICAL CONSTITUENTS OF EFFLUENTS AND CONTINUED UP-DATING OF THIS LIST, (3) DEVELOPMENT OF A PATTERN OF BIOASSAYS FOR EVALUATING THE EFFECTS OF A CHEMICAL ON AQUATIC LIFE . (DATA FROM THESE EVALUATIONS WOULD BE USED IN DEVELOPING MATHEMATICAL MODELS FOR PREDICTING CHEMICAL TOXICITY IN A WIDE RANGE OF ENVIRONMENTAL CIRCUMSTANCES). (4) DEVELOPMENT OF IN SITU BIOASSAY PROCEDURES FOR MORE REALISTIC ASSESSMENT OF CHEMICAL TOXICITY TO AQUATIC LIFE. (LEGORE-WASHINGTON)

FIELD 05C, 05A, 07C, 05B

ACCESSION NO. W73-01976

ECOLOGICAL EFFECTS OF OFFSHORE CONSTRUCTION,

MARINE SCIENCE INST., BAYOU LA BATRE, ALA.

G. A. ROUNSEFELL.

AVAILABLE FROM THE NATIONAL TECHNICAL INFORMATION SERVICE AS AD-739 704, $3.00 IN PAPER COPY, $0.95 IN MICROFICHE. JOURNAL OF MARINE SCIENCE, VOL 2, NO 1, 1972. 208 P, 1 TAB, 252 REF, 3 APPEND. CONTRACT NO DACW72-71-C-0002.

DESCRIPTORS:
*ECOLOGY, *ENVIRONMENTAL EFFECTS, *MARINE PLANTS, *MARINE ANIMALS, OFFSHORE PLATFORMS, NUCLEAR POWERPLANTS, WATER PROPERTIES, SALINITY, WATER TEMPERATURE, OXYGEN, TURBIDITY, BENTHIC FAUNA, WATER POLLUTION EFFECTS, STRIPED BASS, DISSOLVED OXYGEN, OIL, BRIDGES, BUOYS, HARBORS, TUNNELS, MARINE ALGAE, ANIMAL POPULATIONS, LOBSTERS, CRUSTACEANS, BIOMASS, SHRIMP, PINK SHRIMP, CLAMS, WATER POLLUTION, CHLOROPHYTA, ESTUARINE ENVIRONMENT, ISLANDS, GRASSES, INVERTEBRATES, MOLLUSKS, *REVIEWS.

IDENTIFIERS:
*OFFSHORE CONSTRUCTION, DELAWARE BAY, SCOLOPLOS ARMIGER, ALIMITOS BAY, LAKE PONTCHARTRAIN, PEARL RIVER, LAKE BORGNE, SUBSTRATES, MACROINVERTEBRATES, NEW YORK BIGHT, EELGRASS, ZOSTERA, PINAEUS DUORARUM, PINAEUS AZTECUS, PENAEUS SETIFERUS, MERCENARIA MERCENARIA, CLYMENELLA TORQUATA, CAPITELLA CAPITATA, MACOMA CALCAREA, ASTARTE BOREALIS, ENTEROMORPHA MINIMA, ULVA DACTYLIFERA, PANULIRUS INTERRUPTUS, MICROPOGON UNDULATUS, CROAKERS, CYNOSCION REGALIS, ROCCUS SAXATILIS, FUNDULUS HETEROCLITUS.

ABSTRACT:
AN EVALUATION OF CURRENT KNOWLEDGE OF THE PROBABLE ECOLOGICAL EFFECTS OF VARIOUS TYPES OF OFFSHORE CONSTRUCTION REVEALS SLIGHT DANGER FROM THE MAJORITY OF CONSTRUCTION PROGRAMS. THE GREATEST DANGERS LIE IN THE PLACEMENT OF ARTIFICIAL ISLANDS WITHIN OR TOO CLOSELY ADJACENT TO ESTUARIES WHERE THEY CAN SIGNIFICANTLY AFFECT WATER EXCHANGE, AND IN THE PROLIFERATION OF WATER COOLED NUCLEAR POWER PLANTS. PERHAPS THE MOST PRESSING NEED FOR ULTIMATE HUMAN SURVIVAL IS THE FURTHER DEVELOPMENT OF POWER FROM NATURAL FORCES TO REPLACE POWER FROM NUCLEAR AND FOSSIL FUEL SOURCES. (BYRD-BATTELLE)

FIELD 05C, 02L, 10F

ACCESSION NO. W73-02029

THE STRUCTURE AND FUNCTION OF FRESH-WATER MICROBIAL COMMUNITIES.

VIRGINIA POLYTECHNIC INST. AND STATE UNIV. BLACKSBURG.

RESEARCH DIVISION MONOGRAPH 3, 1971. J. CAIRNS, JR., EDITOR, 301 P.

DESCRIPTORS:
*AQUATIC PRODUCTIVITY, *ECOLOGY, *AQUATIC ENVIRONMENT, *BIOLOGICAL COMMUNITIES, *AQUATIC MICROORGANISMS, CYCLING NUTRIENTS, ECOLOGICAL DISTRIBUTION, ECOSYSTEMS, AQUATIC HABITATS, PRODUCTIVITY, AQUATIC BACTERIA, AQUATIC ALGAE, CARBON CYCLE, ON-SITE TESTS, AQUATIC ANIMALS, AQUATIC POPULATIONS, BIOASSAY, EUTROPHICATION.

IDENTIFIERS:
MINERAL CYCLE, CHARACTERIZATION, CHLAMYDOMONAS REINHARDTII, TETRAHYMENA VORAX, PSEUDOMONAS FLUORESCENS, AUTOTROPHIC BACTERIA, HETEROTROPHIC BACTERIA, GLUCOSE, ACETATES, BIOTIN, NIACIN, COBALAMINS, EUGLENA GRACILIS, COLONIZATION, CAULOBACTER, NAJAS FLEXILIS, SCIRPUS ACUTUS, C-14, DISSOLVED ORGANIC MATTER, MACROPHYTES, BIOLOGICAL SAMPLES.

ABSTRACT:
THE 1969 SYMPOSIUM ON 'THE STRUCTURE AND FUNCTION OF FRESH-WATER MICROBIAL COMMUNITIES' SPONSORED BY THE AMERICAN MICROSCOPICAL SOCIETY WAS DIRECTED TOWARD AN UNDERSTANDING OF THE RELATIONSHIPS AMONG LIVING THINGS IN THE AQUATIC ENVIRONMENT. THE TOPICS INCLUDED WERE: 'ADAPTATIONS FOR PHOTOREGENERATIVE CYCLING', 'CARBON FLOW IN THE AQUATIC SYSTEM'; 'AQUATIC LABORATORY MICROSYSTEMS AND COMMUNITIES'; 'THE ROLE OF LABORATORY EXPERIMENTATION IN ECOLOGICAL RESEARCH'; 'A CONTINUOUS GNOTOBIOTIC (SPECIES DEFINED) ECOSYSTEM'; 'COMMUNITY STRUCTURE OF PROTOZOANS AND ALGAE WITH PARTICULAR EMPHASIS ON RECENTLY COLONIZED BODIES OF WATER'; 'DIATOM COMMUNITIES'; 'MICROBIAL RELATIONSHIPS IN BIOLOGICAL WASTEWATER TREATMENT SYSTEMS'; 'HETEROTROPHIC BACTERIA IN AQUATIC ECOSYSTEMS; SOME RESULTS OF STUDIES WITH ORGANIC RADIOISOTOPES'; 'SEASONAL DISTRIBUTION OF COBALAMINS, BIOTIN AND NIACIN IN RAINWATER'; 'LACUSTRINE FUNGAL COMMUNITIES'; 'FACTORS AFFECTING THE NUMBER OF SPECIES IN FRESH-WATER PROTOZOAN COMMUNITIES'; 'THE INTERRELATIONSHIP BETWEEN FRESHWATER BACTERIA, ALGAE, AND ACTINOMYCETES IN SOUTHWESTERN RESERVORIS'; 'CHEMO-ORGANOTROPHY IN EPIPHYTIC BACTERIA WITH REFERENCE TO MACROPHYTIC RELEASE OF DISSOLVED ORGANIC MATTER'; AND 'BACTERIOLOGICAL PROFILES AND SOME CHEMICAL CHARACTERISTICS OF TWO PERMANENTLY FROZEN ANTARCTIC LAKES'. (LONG-BATTELLE)

FIELD 05C, 05B

ACCESSION NO. W73-02095

STUDIES ON ALGAL GROWTH, DEVELOPMENT, AND REPRODUCTION,

CALIFORNIA UNIV., IRVINE.

S. N. MURRAY.

AVAILABLE FROM UNIV. MICROFILMS, INC., ANN ARBOR, MICH. 48106. ORDER NO 72-14,661. PH D DISSERTATION, 1971. 119 P.

DESCRIPTORS:
*ALGAE, *GROWTH RATES, *SYSTEMATICS, *LIFE HISTORY STUDIES, *GROWTH STAGES, REPRODUCTION, CARBON DIOXIDE, CULTURES, LIMITING FACTORS, RHODOPHYTA, AQUATIC ALGAE, MICROORGANISMS, CHLOROPHYTA, BIOLOGY, POLLUTANT IDENTIFICATION, LIGHT INTENSITY, BIOASSAY.

IDENTIFIERS:
PORPHYROPSIS COCCINEA VAR. DAWSONII, CALLOPHYLLIS FIRMA, PLENOSPORIUM DASYOIDES, SELENASTRUM CAPRICORNUTUM, PAAP BATCH TEST, CULTURE MEDIA.

ABSTRACT:
THE LIFE HISTORY OF THE BANGIOPHYCEAN RED ALGAE PROPHYROPSIS COCCINEA VAR. DOWSONII HAS BEEN COMPLETED IN LABORATORY CULTURE. AN UNDESCRIBED FILAMENTOUS PHASE, DISTINCT IN CERTAIN RESPECTS FROM THE CONCHOCELIS OF BANGIA AND PORPHYRA, IS REPORTED. CALLOPHYLLIS FIRMA HAS BEEN SHOWN TO HAVE A POLYSIPHONIA-TYPE OF LIFE HISTORY IN LABORATORY CULTURE. THIS IS THE FIRST LABORATORY CONFIRMATION OF A TYPICAL POLYSIPHONIA-TYPE OF LIFE HISTORY IN THE CRYPTONEMIALES. THE RATES OF APICAL CELL DIVISION FOR THE MARINE RED ALGA PLEONOSPORIUM DASYOIDES WERE DETERMINED AT VARIOUS LIGHT INTENSITIES. RATES OF APICAL CELL DIVISION WERE FOUND TO BE CORRELATED WITH THE TOTAL AMOUNT OF ILLUMINATION RECEIVED PER 24 HOUR PERIOD. AN ANALYSIS AND DISCUSSION OF THE MECHANISMS OF APICAL CELL DIVISION IN FILAMENTS OF UNLIMITED AND LIMITED GROWTH OF P. DASYOIDES IS PRESENTED. THE PAAP BATCH TEST HAS BEEN EVALUATED IN TERMS IF ITS EFFEDTIVENESS. IT HAS BEEN DEMONSTRATED THAT THE PRESCRIBED METHOD OF MEDIUM PREPARATION RESULTS IN THE REMOVAL OF FE AND MN FROM THE MEDIUM DURING FILTER STERILIZATION AND CONSEQUENTLY REDUCES ALGAL YIELDS. CO2 LIMITS THE GROWTH OF THE GREEN ALGA SELENASTRUM CAPRICORNUTUM, THE PRESCRIBED TEST ORGANISM, AND THE CONCOMITANT INCREASE IN MEDIUM PH DURING THE PERIOD OF EXPERIMENTATION APPEARS TO HAVE AN EFFECT ON NUTRIENT AVAILABILITY IN THE TEST CULTURES. THE EFFECTS OF LIGHT INTENSITY AND THE METHOD OF AIR ADDITION ARE ALSO DISCUSSED. (LONG-BATTELLE)

FIELD 05C, 05A

ACCESSION NO. W73-02099

RELEASE OF DISSOLVED ORGANIC MATTER BY MARINE MACROPHYTES,

GEORGIA UNIV., ATHENS.

M, BRYLINSKY.

AVAILABLE FROM UNIV. MICROFILMS, INC., ANN ARBOR, MICH. 48106. ORDER NO. 72-10,923. PH D DISSERTATION, 1971. 125 P.

DESCRIPTORS:
*SEA WATER, *MARINE ALGAE, *CARBOHYDRATES, MARINE PLANTS, MARINE MICROORGANISMS, WATER ANALYSIS, BENTHIC FLORA, ORGANIC MATTER, DISSOLVED SOLIDS, BENTHOS, CARBON, PRODUCTIVITY, ORGANIC COMPOUNDS, CARBOHYDRATES, CYCLING NUTRIENTS, ALGAE.

IDENTIFIERS:
DISSOLVED ORGANIC CARBON, MACROPHYTES, SARGASSO SEA, SAMPLE PREPARATION, SPERMATOPHYTES, PHOTOASSIMILATION, GLYOXYLIC ACID, ASSIMILATION.

ABSTRACT:
THE AMOUNT OF PHOTOASSIMILATED CARBON RELAEASED AS DISSOLVED ORGANIC CARBON WAS INVESTIGATED FOR SIX SPECIES OF BENTHIC MARINE MACROPHYTES AND ONE SPECIES OF PELAGIC MARINE MACROPHYTE. RELEASE RATES RANGED BETWEEN 0.223 AND 1.805 MG C/GM HR. PERCENT RELEASE VALUES RANGED FROM 1.09 TO 3.82 PERCENT. SPERMATOPHYTES HAD SLIGHTLY HIGHER PERCENT RELEASE VALUES THAN ALGAE. THE RESULTS OF QUALITATIVE ANALYSIS PERFORMED ON THE SOLUBLE CARBOHYDRATES RELEASED SHOWED NEUTRAL CARBOHYDRATES TO BE LIBERATED IN THE LARGEST QUANTITY FOLLOWED BY LESSER AMOUNTS OF ACIDIC AND BASIC CARBOHYDRATES. GLYCOLIC ACID WAS NOT OBSERVED TO BE RELEASED IN SIGNIFICANT QUANTITIES. INVESTIGATIONS PERFORMED ON THE ABILITY OF RELEASE PRODUCTS TO BE UTILIZED BY HETEROTROPHIC ORGANISMS SHOWED THAT 20-30 PERCENT OF THE RELEASED CARBON WAS ASSIMILATED WITHIN ONE HOUR. PRELIMINARY CALCULATIONS ON THE CONTRIBUTION OF ORGANIC RELEASE PRODUCTS TO THE DISSOLVED ORGANIC CARBON POOL OF SEA WATER SHOW THIS TO BE RELATIVELY LOW IN TERMS OF ABSOLUTE AMOUNTS OF ORGANIC MATTER. A TECHNIQUE IS PRESENTED FOR THE CONCENTRATION AND MEASUREMENT OF DISSOLVED CARBOHYDRATE MATERIALS IN SEAWATER. ANALYSES OF SEAWATER SAMPLES COLLECTED IN VARIOUS INSHORE PLANT COMMUNITIES AND THE SARGASSO SEA INDICATE THAT THE BASIC PROCEDURE IS USEFUL IN OBTAINING DETAILED INFORMATION ON THE DISSOLVED CARBOHYDRATE MATERIALS PRESENT IN SEAWATER. (LONG-BATTELLE)

FIELD 05C, 05B

ACCESSION NO. W73-02100

DDT RESIDUES IN COASTAL MARINE PHYTOPLANKTON AND THEIR TRANSFER IN PELAGIC FOOD CHAINS,

STANFORD UNIV., CALIF.

J. L. COX,

AVAILABLE FROM UNIV. MICROFILMS, INC. ANN ARBOR, MICH. 48106. ORDER NO 72-5903. PH D DISSERTATICN, 1971. 149 P.

DESCRIPTORS:
*DDT, *MARINE FISH, *PHYTOPLANKTON, *SHRIMP, *BICASSAY, *ABSORPTION, PESTICIDE RESIDUES, WATER POLLUTION EFFECTS, *CALIFORNIA, SEA WATER, WATER ANALYSIS, MARINE ALGAE, CHEMICAL ANALYSIS, ADSORPTION, SEPARATION TECHNIQUES, FOOD CHAINS, CULTURES, FISH DIET, GAS CHRCMATOGRAPHY, PARTICULATE MATTER, DETRITUS, ORGANIC COMPCUNDS, PATH OF POLLUTANTS.

IDENTIFIERS:
*ELECTRON CAPTURE GAS CHRCMATOGRAPHY *BIOLOGICAL MAGNIFICATION, GAS LIQUID CHROMATOGRAPHY, ELECTRON CAPTURE DETECTOR, PRECONCENTRATION, BIOLOGICAL SAMPLES, LIQUID-LIQUID EXTRACTICN, TRIPHOTURUS MEXICANUS, EUPHAUSIA PACIFICA, ENGRAULIS MORDAX.

ABSTRACT:
STUDIES WERE CONDUCTED ON THE ENTRY AND TRANSFER CF DDT RESIDUES IN PELAGIC MARINE FOOD CHAINS. ANALYSES WERE DONE ON PHYTOPLANKTON AND DETRITAL MATERIAL COLLECTED BY A NET OR BY CONTINUCUS-FLOW CENTRIFUGATION. SEAWATER SAMPLES WERE EXTRACTED BY CONTINUOUS FLOW, LIQUID-LIQUID EXTRACTION. GAS-LIQUID CHROMATOGRAPHY WITH ELECTRON CAPTURE DETECTION (GLC-EC) WAS EMPLOYED FOR THE ANALYSES. GLC-EC ANALYSES WERE ALSO PERFORMED ON SAMPLES OF SURFACE AND MIDWATER FISHES AND ZOOPLANKTON. EXPERIMENTAL WCRK WITH C-14-DDT WAS DCNE WITH PURE CULTURES OF PHYTOPLANKTON AND WITH A COMMON EUPHAUSIID SHRIMP. PHYTOPLANKTON SAMPLES CCLLECTED IN MONTEREY BAY, CALIFCRNIA, FROM 1955 TO 1969 CONTAINED COMPOUNDS IDENTIFIED AS P,P'-DDT, P,P'-DDD, AND P,P'-DDE. TOTAL CONCENTRATIONS OF THESE COMPCUNDS WERE APPROXIMATELY THREE TIMES GREATER IN THE LATER SAMPLES. UPTAKE STUDIES WITH PURE CULTURES OF MARINE PHYTOPLANKTON SHOWEC THAT THE ALGAL CELLS, WHEN EXPOSED TC LOW PARTS PER TRILLION NOMINAL CONCENTRATIONS OF C-14-DDT IN THE MEDIUM, COULD CONCENTRATE THE LABELLED DDT BY FACTORS RANGING FROM 30,000 - 80,000. DDT RESIDUE CONCENTRATIONS IN WHCLE SEAWATER RANGED FROM 2.3 PG/ML OFF OREGON AND WASHINGTON, TO 5.6 PG/ML OFF SOUTHERN CALIFORNIA. DCT RESIDUE CONCENTRATIONS IN PARTICULATE MATERIAL RANGED FROM 1.2 TO 5.7 MICRCGRAMS/G C (WITH ONE EXCEPTICN). EXPERIMENTAL RESULTS ARE DESCRIBED WHICH IMPLICATE ADSORPTION AS THE UPTAKE MECHANISM FOR ALGAL CELLS; THESE EXPERIMENTS ALSC SUPPORT THE IDEA THAT SMALL PARTICLES CARRY MOST OF THE DDT RESIDUES IN WHCLE SEAWATER. EUPHAUSIA PACIFICA HENSEN CAN ACQUIRE SUFFICIENT DDT RESIDUES FROM ITS FOOD TO ACOUNT FOR AMCUNTS FCUND IN ITS TISSUES. DIRECT UPTAKE OF C-14-DDT FROM WATER IS PARTIALLY REVERSIBLE BY RETURNING ANIMALS TO UNLABELLED FLOWING SEAWATER. GLC-EC ANALYSES OF TRIPHOTURUS MEXICANUS, SHOWED THAT OLDER FISH HAD HIGHER DDT RESICUE CONCENTRATIONS, SUGGESTING THAT FISH ACCUMULATE DDT RESIDUES FROM THE ENVIRONMENT DURING THEIR LIFE SPAN. THE DCT RESIDUE CONTENT CF DIFFERENT SIZE CLASSES OF ENGRAULIS MORDAX GIRARD RANGES FROM 0.2 TC 2.8 PPM, WET WEIGHT. THESE FINDINGS ARE DISCUSSED IN THE CONTEXT OF A SIMPLE MODEL OF DDT RESIDUE ASSIMILATICN FOOD AND DCT RESIDUE LCSS VIA TRANSPORT IN THE REPROCUCTIVE MATERIALS. (SNYDER-BATTELLE)

FIELD 05C, 05B

ACCESSION NO. W73-02105

HEAVY METAL ION INTERACTION AND TRANSPORT WITH SYNTHETIC COMPLEXING AGENTS AND DETERGENT PHOSPHATE SUBSTITUTES IN AQUATIC SYSTEMS,

MISSOURI WATER RESOURCES RESEARCH CENTER, ROLLA.

S. E. MANAHAN, AND M. J. SMITH.

AVAILABLE FROM THE NATIONAL TECHNICAL INFORMATION SERVICE AS PB-213 252, $3.00 IN PAPER COPY, $0.95 IN MICROFICHE. MISSOURI WATER RESOURCES RESEARCH CENTER, COLUMBIA, COMPLETION REPORT, AUGUST 30, 1972. 181 P, 25 FIG, 14 TAB, 55 REF. OWRR A-049-MO.(1) 14-01-0001-3525.

DESCRIPTORS:
*COPPER, CULTURES, *CHELATION, *ION TRANSPORT, *HEAVY METALS, EUTROPHICATION, POLLUTANT IDENTIFICATION, TOXICITY, *IRON, *CHLORELLA, NUTRIENT REQUIREMENT, *ALGAE, DETERGENTS, PHOSPHATES.

IDENTIFIERS:
*ION-SELECTIVE ELECTRODES, EDTA, *POTENTIOMETRY, ELECTROANALYSIS, *ALGAL CULTURES, CHELATING AGENTS, OOCYSTIS MARSSONII.

ABSTRACT:
THE CHEMICAL ASPECTS OF THE COPPER MICRONUTRIENT REQUIREMENT FOR ALGAE HAVE BEEN INVESTIGATED. A REPRODUCIBLE COPPER REQUIREMENT FOR CHLORELLA VULGARIS AND OOCYSTIS MARSSONII WAS DEMONSTRATED. OPTIMAL GROWTH WAS OBSERVED ABOVE 40 MICROGRAMS/L FOR OOCYSTIS AND 30 MICROGRAMS/L FOR CHLORELLA. A STUDY OF THE EFFECTS OF EDTA ON THE TOXICITYOOF COPPER TO CHLORELLA SHOWED THAT COPPER IN CHELATED FORM WAS NOT TOXIC TO THESE ALGAE AT CONCENTRATIONS UP TO 46 MG/L COPPER. WHEN ONLY SUFFICIENT CHELATING AGENT WAS PRESENT TO KEEP THE IRON (III) IN SOLUTION, HOWEVER, THE TOXIC EFFECTS OF COPPER WERE EVIDENT AT 7.00 MG/L OF COPPER. A SECOND ASPECT OF THE PROJECT INVOLVED THE DEVELOPMENT OF A SIMPLE, DIRECT MULTIPLE STANDARD ADDITION METHOD FOR THE POTENTIOMETRIC ANALYSIS OF COPPER IN WATER WITH A SOLID-STATE COPPER ION-SELECTIVE ELECTRODE. THE TECHNIQUE IS MORE SENSITIVE THAN CONVENTIONAL ATOMIC ABSORPTION ANALYSIS, THOUGH NOT SO RAPID. MEASUREMENTS ARE MADE IN A COMPLEXING ANTIOXIDANT BUFFER MEDIUM CONTAINING ACETATE (TO COMPLEX COPPER), FLUORIDE (TO COMPLEX IRON), AND FORMALDEHYDE (TO PROVIDE A REDUCING MEDIUM).

FIELD 05A, 05C, 02K

ACCESSION NO. W73-02112

ARTIFICIAL DESTRATIFICATION IN RESERVOIRS.

AMERICAN WATER WORKS ASSOCIATION, NEW YORK. QUALITY CONTROL IN RESERVOIRS COMMITTEE.

JOURNAL OF THE AMERICAN WATER WORKS ASSOCIATION, VOL 63, NO 9, P 597-604, SEPTEMBER 1971. 4 FIG, 3 TAB, 19 REF.

DESCRIPTORS:
*THERMAL STRATIFICATION, *TASTE, *ODOR, DISSOLVED SOLIDS, ALGAE, RESERVOIR STORAGE, *HYPOLIMNION, WATER QUALITY CONTROL.

IDENTIFIERS:
*MIXERS.

ABSTRACT:
THE QUALITY CONTROL IN RESERVOIRS COMMITTEE OF THE AMERICAN WATER WORKS ASSOCIATION CONDUCTED A STUDY OF WATER PURVEYORS USING ARTIFICIAL DESTRATIFICATION TECHNIQUES TO OVERCOME TASTES AND ODORS; HIGH CONCENTRATIONS OF IRON, MANGANESE, AND HYDROGEN SULFIDE; AND ALGAE DIFFICULTIES ASSOCIATED WITH ANAEROBIC CONDITIONS IN THE HYPOLIMNION LAYER IN RESERVOIRS. A HIGH RATE OF SUCCESS FROM BOTH HOMEMADE AND COMMERCIAL MIXING APPARATUS WAS RECORDED WITH AN ESTIMATED INITIAL COST PER UNIT VOLUME OF $3/MIL GAL AND AN OPERATING COST OF $0.45/MIL GAL/YR FOR A 10,000 ACRE-FT. RESERVOIR. MORE STUDY IS NEEDED TO EVALUATE THE VARIOUS TYPES OF APPARATUS, THE ACTUAL CHANGES IN WATER QUALITY, AND THE INFLUENCE OF ARTIFICIAL DESTRATIFICATION ON INDICATOR ORGANISMS AND ENTERIC PATHOGENS. (WEIR-AWWARF)

FIELD 05G, 02H, 05F

ACCESSION NO. W73-02138

COMMUNITY WATER POLLUTION R AND D NEEDS.

AMERICAN WATER WORKS ASSOCIATION, NEW YORK. COMMITTEE ON POLLUTION PARAMETERS.

JOURNAL OF THE AMERICAN WATER WORKS ASSOCIATION, VOL 64, NO 4, P 211-215, APRIL 1972. 9 REF.

DESCRIPTORS:
*WATER PURIFICATION, *WATER POLLUTION TREATMENT, *PUBLIC HEALTH, ANALYTICAL TECHNIQUES, WASTE WATER TREATMENT, TASTE, ODOR, POTABLE WATER, ACTIVATED CARBON, COLIFORMS, VIRUSES, CRUSTACEANS, NEMATODES, EPIDEMIOLOGY, PHOSPHATES, HEAVY METALS, PESTICIDES, TASTE-PRODUCING ALGAE, ODOR-PRODUCING ALGAE, ACTINOMYCETES, PHOSPHATES.

IDENTIFIERS:
*WATERBORNE DISEASES, *TOXIC CHEMICALS, *PUBLIC HEALTH STANDARDS, TOXICOLOGY, NTA.

ABSTRACT:
THE POLLUTION PARAMETERS COMMITTEE OF THE AWWA HAS PREPARED A REPORT ON THE CURRENT STATE OF THE ART IN THE FIELD OF ANTIPOLLUTION MEASURES AND HAS SET FORTH ITS RESEARCH RECOMMENDATIONS. AMONG THE WATER QUALITY PARAMETERS DISCUSSED ARE COLIFORM BACTERIA, VIRUSES, TASTES AND ODORS, PHOSPHATES, NTA, AND OTHER ORGANIC AND INORGANIC CHEMICALS. THE RESEARCH NEEDS RECOMMENDED BY THE COMMITTEE INCLUDE (1) IDENTIFICATION OF ALL POLLUTANTS PRESENTING A POTENTIAL HEALTH HAZARD AS WELL AS THOSE RESPONSIBLE FOR OBNOXIOUS TASTES AND ODORS; (2) INCREASING THE SENSITIVITY, PRECISION, AND RAPIDITY OF BOTH MICROBIOLOGICAL AND CHEMICAL TECHNIQUES FOR EXAMINING WASTEWATER; (3) TOXICOLOGICAL RESEARCH AND EPIDEMIOLOGICAL STUDIES THAT CORRELATE MORBIDITY DATA WITH SPECIFIC CHARACTERISTICS OF THE WATER; (4) LABORATORY RESEARCH AND PILOT-PLANT STUDIES TO DEVELOP AND EVALUATE PROCESSES THAT WILL CONTINUOUSLY TREAT DIRECTLY REUSED MUNICIPAL WASTEWATERS TO PRODUCE 'SAFE AND SATISFACTORY' DRINKING WATER. (NICHOLS-AWWARF)

FIELD 05G, 05D, 05F

ACCESSION NO. W73-02144

WHAT'S IT ALL ABOUT. ALGAE,

BUCK, SEIFEIT AND JOST, MORRISVILLE, PA.

J. M. FOULDS.

WATER AND WASTES ENGINEERING, VOL 9, NO 8, P 45-46, AUGUST 1972, 3 TAB.

DESCRIPTORS:
*ALGAE, *NITRIFICATION, WASTE TREATMENT, *PHOSPHORUS, *NITROGEN, COST ANALYSIS, OZONE, DENITRIFICATION, POLLUTION ABATEMENT, WATER POLLUTION CONTROL, COAGULATION.

IDENTIFIERS:
*ORGANIC NUTRIENTS, ALUM COAGULATION.

ABSTRACT:
BY COMMERCIAL GROWTH OF ALGAE FROM NITRIFICATION-DENITRIFICATION PLANTS OR FROM RAW SEWAGE, FERTILIZER MIGHT BE PRODUCED WITH ORGANIC NITROGEN AND PHOSPHATE IN SLOW RELEASE, HARD TO LEACH FORMS. HIGH RATE ALGAE FORMS SHOW 70% NITROGEN REMOVAL AND 50% PHOSPHORUS REMOVAL, INCREASING TO NEAR 100% PHOSPHORUS REMOVAL WITH ALUM COAGULATION HARVESTING. THE ESTIMATED COST IS $20,000 PER ACRE, INCLUDING LAND. LABORATORY BATCH AND CONTINUOUS REACTOR STUDIES IN AN OZONATED SYSTEM WERE MADE TO PREPARE SUITABLE EFFLUENTS FOR A SOURCE OF ORGANIC NUTRIENTS. OZONE OXIDIZES METALS TO HIGHER STATES AND STERILIZES THE EFFLUENT. THE OPTIMUM OZONE-OXYGEN RATES WERE DETERMINED AND THE FROTH COMPOSITION AND VOLUME AT DIFFERENT GAS RATES WERE DETERMINED. IT APPEARS THAT THE PROCESS IS FEASIBLE WITH SINGLE OR COMBINATIONS OF TRANSITION METALS-NECESSARY TRACE METALS FOR LAWN FERTILIZER. (ANDERSON-TEXAS)

FIELD 05G, 05C

ACCESSION NO. W73-02187

METABOLISM OF DDT BY FRESH WATER DIATOMS,

MANITOBA UNIV., WINNIPEG. DEPT. OF ENTOMOLOGY.

S. MIYAZAKI, AND A. J. THORSTEINSON.

BULLETIN OF ENVIRONMENTAL CONTAMINATION AND TOXICOLOGY, VOL 8, NO 2, P 81-83, 1972. 1 TAB, 13 REF.

DESCRIPTORS:
*DDT, *DDE, *PESTICIDE KINETICS, *PERSISTENCE, *WATER POLLUTION EFFECTS, *PLANT PHYSIOLOGY, *DIATOMS, PESTICIDES, DDD, PESTICIDE TOXICITY, PATH OF POLLUTANTS, PESTICIDE RESIDUES, CHLORINATED HYDROCARBON PESTICIDES, PESTICIDE DRIFT, METABOLISM, ALGAE, AQUATIC ALGAE, CHRYSOPHYTA, PLANKTON, PHYTOPLANKTON.

ABSTRACT:
REDUCTIVE DECHLORINATION OF DDT TO TDE (DDD) UNDER ANAEROBIC CONDITIONS HAS BEEN DEMONSTRATED IN BAKER'S YEAST, A FUNGUS, SEVERAL ACTINOMYCETES, AND BACTERIA. DDE, THE DEHYDROCHLORINATION PRODUCT OF DDT, ALSO OCCURS IN SEVERAL BACTERIA AND IN A MARINE DIATOM. A POSSIBLE ROLE OF DIATOMS IN DETOXIFYING DDT IN AQUATIC ENVIRONMENTS HAS BEEN SUGGESTED. RADIOLABELLING AND THIN-LAYER CHROMATOGRAPHY WERE USED IN THIS STUDY OF DDT METABOLISM BY A NITZSCHIA SP. AND AN UNIDENTIFIED DIATOM SPECIES. RESULTS SUGGEST THAT SOME SPECIES OF FRESHWATER DIATOMS MAY BE SIGNIFICANT IN THE DEGRADATION OF DDT TO THE NON-INSECTICIDAL METABOLITE, DDE, IN NATURE. (LEGORE-WASHINGTON)

FIELD 05C

ACCESSION NO. W73-02280

THE IMPACT OF REDUCED LIGHT PENETRATION ON A EUTROPHIC FARM POND,

NEBRASKA UNIV., LINCOLN. DEPT. OF ZOOLOGY.

E. G. BUGLEWICZ.

M.S. THESIS, AUGUST 1972. 99 P, 20 FIG, 11 TAB, 24 REF. OWRR A-014-NEB(7) 14-31-0001-3527.

DESCRIPTORS:
*EUTROPHICATION, AQUATIC ECOSYSTEMS, WATER QUALITY, *ALGAE, *LIGHT PENETRATION, *MACROPHYTES, *AQUATIC WEED CONTROL, *LIGHT INTENSITY, *NEBRASKA, *CYANOPHYTA, ALGAL CONTROL, *PONDWEEDS, PONDS, *THERMAL STRATIFICATION, DIATOMS, *CHARA.

IDENTIFIERS:
*ANALINE DYE.

ABSTRACT:
AN EXPERIMENTAL APPROACH TO THE CONTROL OF EUTROPHICATION PROBLEMS USING ANILINE DYES WAS ATTEMPTED ON AN ENRICHED FARM POND IN EASTERN NEBRASKA. PRIMARY PRODUCTIVITY WAS SHOWN TO DECREASE NOT ONLY IN THE DYED BOXES, BUT ALSO IN POND SAMPLES SUSPENDED IN THE BOXES. POTAMOGETON SP. WAS ELIMINATED FROM BLUE AND BROWN-DYED BOXES, WHILE CHARA SP. WAS ONLY ELIMINATED FROM THE BLUE-DYED BOXES. A STRONG TEMPERATURE STRATIFICATION OCCURRED IN ALL DYED BOXES DUE TO ABSORPTION OF INFRARED ENERGY IN THE FIRST FEW CM OF THE WATER COLUMNS. ALGAL POPULATIONS AND PULSES WITHIN THE BOXES DIFFERED FROM BOX TO BOX AND DID NOT REFLECT THE SAME CHANGES OBSERVED IN THE POND, BUT DID RESEMBLE FALL AND SPRING DIATOM PULSES. BLUE-GREEN BLOOMS PRESENT IN EXPERIMENTAL BOXES 2, 4, 5 AND 6 BEFORE DYE ADDITION WERE REPLACED BY POPULATIONS OF DIATOMS WITH SOME GREEN AND A FEW BLUE-GREEN ALGAE. BOX 3, WHICH WAS TURBID BEFORE DYE ADDITION, NEVER EXPERIENCED A BLUE-GREEN ALGAL PULSE, BUT MAINTAINED A HIGH POPULATION OF DIATOMS. LACK OF MIXING OF THE WATER COLUMNS MAY HAVE HAD A PROFOUND EFFECT ON THE RESULTS OF THE EXPERIMENT BY CAUSING TEMPERATURE STRATIFICATION, EXCESSIVE PRIMARY PRODUCTIVITY, PROLONGED ANAEROBIC CONDITIONS BELOW THE SURFACE OF THE WATER, AND INCREASED NUTRIENT RELEASE FROM THE BOTTOM MUDS.

FIELD 05C, 02H

ACCESSION NO. W73-02349

MULTIVARIATE APPROACHES TO ALGAL STRATEGIES AND TACTICS IN THE SYSTEMS ANALYSIS OF PHYTOPLANKTON,

WISCONSIN UNIV., MADISON. DEPT. OF BOTANY.

T. F. H. ALLEN, AND J. F. KOONCE.

MEMO REPORT NO 72-24 (PREPRINT), EASTERN DECIDUOUS FOREST BIOME, MAY, 1972. 52P, 12 FIG, 2 TAB, 32 REF. 16010 EHR.

DESCRIPTORS:
*ANALYTICAL TECHNIQUES, *COMPUTER PROGRAMS, *ALGAE, *PHYTOPLANKTON, BIOLOGICAL COMMUNITIES, STANDING CROPS, WISCONSIN, PRIMARY PRODUCTIVITY, SYSTEMATICS, GROWTH RATES, BIOMASS, DATA COLLECTIONS.

IDENTIFIERS:
*LAKE WINGRA(WIS.), MULTIVARIATE ANALYSIS, ORDINATION ANALYSIS.

ABSTRACT:
TRADITIONALLY, PHYTOPLANKTON DATA ARE ORGANIZED ON CHRONOLOGICAL AXES; THE POWER OF MULTIVARIATE ANALYSES OF THE TYPE EMPLOYED HERE LIES IN THEIR ABILITY TO ORGANIZE SPECIES DATA INTO NATURAL GROUPS NOT NECESSARILY CONFORMING TO CHRONOLOGICAL GROUPS OR, INDEED, ANY OTHER IMPOSED STRUCTURE. A LOGARITHMIC TRANSFORMATION IS OFTEN APPROPRIATE TO PHYTOPLANKTON DATA BECAUSE THESE PLANTS CAN GROW EXPONENTIALLY. A PRESENCE/ABSENCE TRANSFORMATION IGNORES STANDING CROPS AND GIVES INFORMATION AS TO SPECIES TOLERANCES RATHER THAN SPECIES OPTIMA. CERTAIN TRANSFORMATIONS, IN WHICH DATA ARE RELATIVIZED, GIVE EQUAL WEIGHT TO RARE AND COMMON SPECIES WHILE PRESERVING MANY QUANTITATIVE ASPECTS. BY USING SEVERAL TRANSFORMATIONS AND ANALYSES INSIGHT HAS BEEN GAINED INTO DIFFERENT ECOLOGICAL ASPECTS OF PHYTOPLANKTON, AND THE BIOLOGICAL IMPLICATIONS OF CERTAIN DATA TRANSFORMATIONS. AT WEEKLY INTERVALS, MARCH 1970 UNTIL FEBRUARY 1971, WATER SAMPLES WERE COLLECTED FROM THREE DEPTHS AT ONE STATION IN LAKE WINGRA, WISCONSIN FOR ESTIMATION OF PRIMARY PRODUCTION, IDENTIFICATION AND ENUMERATION OF PHYTOPLANKTON SPECIES, AND DETERMINATION OF IMPORTANT WATER CHEMISTRY PARAMETERS. A DISTINCTION IS DRAWN BETWEEN ALGAL TACTICS, WHICH GIVE THE PLANT THE ABILITY TO EXIST IN A PARTICULAR ENVIRONMENTAL SITUATION, AND ALGAL STRATAGEMS, WHICH DEFINE THE ORGANISMS' PLACE IN THE COMMUNITY. (JONES-WISCONSIN)

FIELD 05C, 07B

ACCESSION NO. W73-02469

ALGAL NITROGEN FIXATION IN TEMPERATE REGIONS,

UPPSALA UNIV. (SWEDEN). INST. OF PHYSIOLOGICAL BOTANY.

E. HENRIKSSON.

PLANT AND SOIL, SPECIAL VOLUME, P 415-419, 1971. 1 TAB, 16 REF.

DESCRIPTORS:
*ALGAE, *NITROGEN FIXATION, *SOILS, *TEMPERATE, CLIMATIC ZONES, CYANOPHYTA, SYMBIOSIS, LICHENS, NITROGEN, LIGHT INTENSITY, TEMPERATURE, ANABAENA, BACTERIA, PHOTOSYNTHESIS.

IDENTIFIERS:
SWEDEN, COLLEMA, PELTIGERA, NOSTOC.

ABSTRACT:
A FEW EARLIER INVESTIGATIONS OF NITROGEN-FIXATION BY BLUE-GREEN ALGAE IN TEMPERATE SOILS ARE REVIEWED INCLUDING RECENT STUDIES ON THE OCCURRENCE OF POTENTIAL NITROGEN-FIXING ALGAE IN SEVERAL DIFFERENT SOIL TYPES ON VARIOUS CONTINENTS. IN THE SUMMER OF 1969 AN INVESTIGATION WAS MADE OF NITROGEN FIXATION BY ALGAE IN SWEDISH SOILS. MEASUREMENTS WERE MADE FROM APRIL TO OCTOBER AT VARIOUS TIMES OF THE DAY. GOOD NITROGEN FIXATION BY BLUE-GREEN ALGAE WAS OBSERVED DURING THE WHOLE EXPERIMENTAL TIME. FIXATION OCCURRED AT ABOUT THE SAME RATE DURING BOTH THE LIGHT AND DARK PERIODS OF EACH DAY ON WHICH IT WAS MEASURED. NEITHER LIGHT NOR TEMPERATURE CONDITIONS GREATLY INFLUENCED RATE OF NITROGEN FIXATION. THE LOW RATES OF FIXATION FREQUENTLY OBSERVED IN SOILS CONTAINING FREE-LIVING ALGAE MAY BE EXPLAINED BY THE FACT THAT THE CELLS OF THE SPECIES ANABAENA AND NOSTOC IN SOILS CAN VERY EASILY BE CONVERTED TO SPORES (AKINETES) WHICH DO NOT FIX NITROGEN. DATA SHOW THAT BLUE-GREEN ALGAE CAN CONTRIBUTE TO THE TOTAL NITROGEN ECONOMY OF SOILS IN THE NORTHERN TEMPERATE ZONE BOTH AS FREE-LIVING CELLS AND IN SYMBIOTIC ASSOCIATIONS IN LICHENS. (JONES-WISCONSIN)

FIELD 05C, 02G

ACCESSION NO. W73-02471

NITROGEN FIXATION IN LAKES,

LONDON UNIV. (ENGLAND); AND WESTFIELD COLL., LONDON (ENGLAND). DEPT. OF BOTANY.

G. E. FOGG.

PLANT AND SOIL, SPECIAL VOLUME, P 393-401, 1971. 2 TAB, 24 REF.

DESCRIPTORS:
*NITROGEN FIXATION, *LAKES, CYANOPHYTA, LIGHT INTENSITY, EUTROPHICATION, PHYTOPLANKTON, PRODUCTIVITY, ANABAENA, AMMONIA, NITRATES, LIMITING FACTORS, NITROGEN, PLANKTON, DEPTH, ALGAE.

IDENTIFIERS:
*WINDERMERE(ENGLAND), HETEROCYSTS, BENTHIC BLUE-GREEN ALGAE.

ABSTRACT:
FOR VALID ESTIMATE OF TOTAL NITROGEN FIXED IN A LAKE IN A YEAR, DETERMINATIONS SHOULD BE MADE WITH ENOUGH REPRESENTATIVE SAMPLES TO TAKE ACCOUNT OF HORIZONTAL AND VERTICAL VARIATIONS IN DENSITY OF BLUE-GREEN PLANKTONIC ALGAE AT FREQUENT INTERVALS WHEN THE ALGAE ARE ABUNDANT. DETERMINATIONS IN OPEN WATERS USING N-15 TRACERS SHOW THAT NITROGEN FIXATION IS GENERALLY ASSOCIATED WITH PRESENCE OF HETEROCYSTOUS BLUE-GREEN ALGAE AND IS LIGHT DEPENDENT. FIXATION ITSELF IS NOT NECESSARILY INHIBITED BY PRESENCE OF NITRATE OR AMMONIA IN THE WATER, ALTHOUGH NITROGEN-FIXING BLUE-GREEN ALGAE TEND TO BE ABUNDANT WHEN THESE SOURCES OF COMBINED NITROGEN ARE LOW. ACTIVITY OF THESE ALGAE SHOWS A DIRECT RELATIONSHIP TO DISSOLVED ORGANIC NITROGEN CONCENTRATION. NITROGEN FIXATION PER UNIT AREA OF LAKE SURFACE PER YEAR TENDS TO BE GREATEST AT AN EARLY EUTROPHICATION STAGE. CONTRIBUTION OF BIOLOGICAL NITROGEN FIXATION IN RELATION TO THE TOTAL NITROGEN BUDGET OF A LAKE IS PROBABLY ALWAYS SMALL, BUT AT CERTAIN TIMES AND IN PARTICULAR WATER STRATA IT MAY CONTRIBUTE A MAJOR PART OF THE NITROGEN ASSIMILATED BY PHYTOPLANKTON. NITROGEN FIXATION MAY FREQUENTLY BE IMPORTANT IN EUTROPHIC LAKES IN ENABLING HIGHER PRODUCTION RATES THAN WOULD OTHERWISE BE POSSIBLE. (JONES-WISCONSIN)

FIELD 05C, 02H

ACCESSION NO. W73-02472

ALGAL NITROGEN FIXATION IN THE TROPICS,

SEIJO UNIV., TOKYO (JAPAN). BIOLOGICAL LAB.

A. WATANABE, AND Y. YAMAMOTA.

PLANT AND SOIL, SPECIAL VOLUME, P 403-413, 1971. 2 FIG, 4 TAB, 25 REF.

DESCRIPTORS:
*ALGAE, *NITROGEN FIXATION, *TROPICAL REGIONS, RICE, CYANOPHYTA, FERTILIZATION, TEMPERATURE, CROP PRODUCTION, FUNGI, SOILS, SYMBIOSIS, AFRICA, ASIA, ANABAENA.

IDENTIFIERS:
*GREEN MANURE, INDIA, JAPAN, RICE FIELDS.

ABSTRACT:
SOME RECENT DATA OBTAINED ON NITROGEN FIXATION IN RICE FIELDS, IN TROPICAL WATERS, AND IN SYMBIOTIC BLUE-GREEN ALGAE ARE SUMMARIZED. NITROGEN-FIXING BLUE-GREEN ALGAE SEEM TO GROW MOST ABUNDANTLY IN TROPICAL AND SUBTROPICAL REGIONS AND TO A LESSER EXTENT IN TEMPERATE AND SUB-TEMPERATE REGIONS. OF BLUE-GREEN ALGAE TESTED FOR NITROGEN-FIXING CAPACITY, TOLYPOTHRIX TENUIS WAS THE MOST EFFICIENT. THE USE OF TOLYPOTHRIX TENUIS AS A SOURCE OF GREEN MANURE IN AGRICULTURAL PRACTICE WAS TESTED DURING 1951-1956 ON ABOUT 40 RICE FIELDS IN VARIOUS PARTS OF JAPAN. OVER-ALL PRODUCTION SHOWS AN AVERAGE INCREASE IN RICE YIELDS EVERY YEAR. CERTAIN SPECIES OF BLUE-GREEN ALGAE CAN WITHSTAND EXTREME ENVIRONMENTAL CONDITIONS SUCH AS HIGH TEMPERATURE AND SALINITY. THOSE FIXING NITROGEN CONTRIBUTE TO THE NITROGEN FERTILITY OF THE SEAS AND LAKES. IN JAPANESE HOT SPRINGS, 320 SPECIES OF BLUE-GREEN WERE FOUND 10 OF WHICH FIXED NITROGEN. CERTAIN BLUE-GREEN ALGAE SPECIES FORM ASSOCIATIONS WITH OTHER ORGANISMS SUCH AS FUNGI, LIVERWORTS, FERNS, AND SEED PLANTS. THE RELATIONSHIP BETWEEN THESE TWO ORGANISMS IS SOMETIMES COMMENSAL AND OTHER TIMES SYMBIOTIC. CERTAIN SYMBIOTIC BLUE-GREEN ALGAE ARE PROVIDED WITH THE ABILITY TO FIX THE ATMOSPHERIC NITROGEN. (JONES-WISCONSIN)

FIELD 05C

ACCESSION NO. W73-02473

ROLE OF PHOSPHORUS IN EUTROPHICATION AND DIFFUSE SOURCE CONTROL,

WISCONSIN UNIV., MADISON. WATER CHEMISTRY PROGRAM.

G. F. LEE.

(PREPRINT) IN: CONFERENCE ON PHOSPHORUS IN FRESH WATER AND MARINE ENVIRONMENT, APRIL 1972, INTERNATIONAL ASSOCIATION ON WATER POLLUTION RESEARCH, LONDON (ENGLAND). 28 P, 1 TAB, 22 REF.

DESCRIPTORS:
*WATER POLLUTION CONTROL, *PHOSPHORUS, *EUTROPHICATION, CHEMICAL ANALYSIS, WATER POLLUTION SOURCES, BIOASSAY, ALGAE, DOMESTIC WASTES, NUTRIENTS, SEDIMENTS, ANALYTICAL TECHNIQUES, FERTILIZATION, AQUATIC PLANTS, NITROGEN, NUTRIENT REMOVAL, WATER TREATMENT, ASSAY, BACTERIA.

IDENTIFIERS:
ALGAL ASSAY, ORTHOPHOSPHATE, PHOSPHORUS SOURCES.

ABSTRACT:
WHAT IS NEEDED TO DEVELOP THE MOST MEANINGFUL CONTROL PROGRAM FOR EXCESSIVE FERTILIZATION OF NATURAL WATERS IS A METHOD OF ASSESSING THE ROLE OF PHOSPHORUS IN EUTROPHICATION OF A PARTICULAR LAKE. THE CURRENT TECHNOLOGY WITH RESPECT TO THE TESTS USED IS REVIEWED: BIOASSAY OF NUTRIENT STATUS, DETERMINATION OF CRITICAL NUTRIENT CONCENTRATIONS; FACTORS INFLUENCING EXCHANGE OF NUTRIENTS BETWEEN SEDIMENTS AND WATER; METHODS OF PHOSPHORUS CONTROL. AT THIS TIME THERE IS INSUFFICIENT EVIDENCE TO PREDICT QUANTITATIVELY THE ROLE OF PHOSPHORUS IN EXCESSIVE FERTILIZATION OF A GIVEN LAKE, THUS RESEARCH IS NEEDED TO ASSESS THIS WITH VARIOUS CHEMICAL AND BIOLOGICAL TECHNIQUES. EMPHASIS SHOULD BE PLACED ON DETERMINING PHOSPHORUS SOURCES, FORMS, RATES OF TRANSFORMATION, AND AVAILABILITY FOR AQUATIC PLANT GROWTH IN NATURAL WATERS. RESULTS SHOULD BE FORMULATED IN MATHEMATICAL MODELS ENABLING PREDICTIONS OF RELATIONSHIPS BETWEEN PHOSPHORUS INPUT AND EXCESSIVE GROWTHS OF AQUATIC PLANTS. EFFORTS SHOULD BE INCREASED TO CONTROL PHOSPHORUS FROM DIFFUSE SOURCES SUCH AS URBAN STORM WATER AND AGRICULTURAL RUNOFF. DETERGENTS WITHOUT PHOSPHORUS, PROPERLY EVALUATED WITH RESPECT TO POTENTIAL ENVIRONMENTAL IMPACT AND PERSONAL SAFETY HAZARDS, SHOULD BE FORMULATED. DIRECT LAKE WATER TREATMENT WITH PHOSPHORUS-PRECIPITATING CHEMICALS SHOULD BE STUDIED ON A LARGE SCALE. (JONES-WISCONSIN)

FIELD 05C, 05B

ACCESSION NO. W73-02478

WAYS IN WHICH A RESIDENT OF THE MADISON LAKES' WATERSHED MAY HELP TO IMPROVE WATER QUALITY IN THE LAKES,

WISCONSIN UNIV., MADISON. DEPT. OF CIVIL AND ENVIRONMENTAL ENGINEERING.

G. F. LEE.

JULY 1972. 10 P.

DESCRIPTORS:
*EUTROPHICATION, *SOCIAL PARTICIPATION, *WATER POLLUTION CONTROL, WISCONSIN, LAKES, URBAN RUNOFF, AGRICULTURAL RUNOFF, FARM WASTES, ALGAE, AQUATIC WEEDS, FERTILIZATION, LEAVES, LEGISLATION, LAND USE, SHORES, MONITORING, MARSHES, SOIL EROSION, WASTE WATER TREATMENT.

IDENTIFIERS:
*MADISON(WIS), LAKE MENDOTA(WIS), LAWN MAINTENANCE.

ABSTRACT:
METHODS ARE RECOMMENDED BY WHICH MADISON RESIDENTS MAY PREVENT NUTRIENTS FROM ENTERING THE CITY'S LAKES, THEREBY REDUCING FREQUENCY AND SEVERITY OF EXCESSIVE ALGAL AND WEED GROWTH. A CENTRALIZED REGULATORY GOVERNMENTAL UNIT SHOULD BE DEVELOPED WITH RESPONSIBILITY FOR WATER QUALITY CONTROL IN THE COUNTY. BRIEFLY RECOMMENDATIONS ARE: FERTILIZER APPLICATION BY HOMEOWNERS SHOULD NOT EXCEED THE MINIMUM NEEDED FOR A HEALTHY LAWN; GRASS CLIPPINGS, LEAVES, ETC., SHOULD BE WORKED INTO SOIL OR PLACED IN IMPERVIOUS CONTAINERS FOR COLLECTION; DOMESTIC AND RURAL IRRIGATION RUNOFF SHOULD BE PREVENTED; FUNDS SHOULD BE APPROPRIATED FOR MORE FREQUENT STREET SWEEPING; DAIRY FARMERS, LIVESTOCK OWNERS, AND FEEDLOT OPERATORS SHOULD NOT SPREAD MANURE ON FROZEN GROUND; CLOSE SCRUTINY SHOULD ASCERTAIN IF ANY FERTILIZING PROCESSES TEND TO INCREASE PHOSPHORUS RUNOFF; ORDINANCES SHOULD BE ADOPTED FOR PENALTY ASSESSMENT; ALL NEW URBAN DEVELOPMENTS SHOULD BE REQUIRED TO USE THE LATEST TECHNIQUES TO MINIMIZE PHOSPHORUS TRANSPORT; FUNDS SHOULD BE APPROPRIATED FOR ENFORCEMENT OF THE ORDINANCE PROHIBITING DISCHARGE TO STORM SEWERS OF COOLING TOWER BLOWDOWN WATER; SHORELINE DEBRIS SHOULD BE COLLECTED; FUNDS FOR LONG-TERM MONITORING PROGRAMS SHOULD BE APPROPRIATED; MARSH DRAINAGE SHOULD BE PROHIBITED; SOIL EROSION PREVENTED; EXCESSIVE GROWTHS OF AQUATIC WEEDS HARVESTED. (JONES-WISCONSIN)

FIELD 05C

ACCESSION NO. W73-02479

SALINITY-RELATED POLYMORPHISM IN THE BRACKISH-WATER DIATOM CYCLOTELLA CRYPTICA,

CONNECTICUT UNIVERSITY, STORRS. BIOLOGICAL SCIENCES GROUP.

M. E. SCHULTZ.

CANADIAN JOURNAL OF BOTANY, VOL 49, NO 8, P 1285-1289, 1971. 1 FIG, 1 TAB, 9 REF. OWRR A-014-CONN(5), 14-01-0001-901.

DESCRIPTORS:
*DIATOMS, *SALINITY, *BIOINDICATORS, BRACKISH WATER, FRESHWATER, *ALGAE, WATER POLLUTION EFFECTS.

IDENTIFIERS:
*CYCLOTELLA CRYPTICA, *CYCLOTELLA MENEGHINIANA, *POLYMORPHISM, *VALVE MORPHOLOGY, *VALVE PATTERN CHARACTERISTICS, CLONES.

ABSTRACT:
THE BRACKISH-WATER DIATOM CYCLOTELLA CRYPTICA IS A POLYMORPHIC SPECIES. NINE CLONES ARE CAPABLE OF PRODUCING THE VALVE PATTERN CHARACTERISTIC OF THE SPECIES C. MENEGHINIANA, AS WELL AS THE C. CRYPTICA PATTERN. A STUDY OF THE EFFECTS OF SALINITY AND FRESHWATER CONDITIONS ON THE MORPHOLOGY OF THE VALVE SHOWS THAT THE CRYPTICA PATTERN IS PRODUCED IN SALINITIES OF ABOUT 4.3% TO FULL-STRENGTH SEAWATER, 28.7%. THE 'MENEGHINIANA' PATTERN IS THE FRESHWATER OR LOW SALINITY (1.4%) FORM. CHARACTERISTICS OF THE VALVE MORPHOLOGY AND LIFE HISTORY STAGES WHICH DISTINGUISH C. CRYPTICA FROM C. MENEGHINIANA AND CYCLOTELLA SP., CLONE 03A, ARE PRESENTED AND DISCUSSED.

FIELD 05C, 05A

ACCESSION NO. W73-02548

SIGNIFICANT DESCRIPTOR INDEX

The starred descriptors indexed here have been selected as best describing the content of each paper. If additional references to these or other terms are desired, consult the *General Index.* The numbers at the right are the accession numbers appearing at the end of the abstracts.

MENTS, *WATER POLLUTION SOURCES,	*AGRICULTURAL RUNOFF, *WATER POLL	W72-08401
RODUCTION, *IRRIGATION PROGRAMS,	*AGRICULTURE, *ECONOMICS, CYANOPH	W72-03327
NTRO/ *MANAGEMENT, *ENVIRONMENT,	*AGRICULTURE, *WATER POLLUTION CO	W72-00846
BOLISM, MOBILITY, PARTICULATE F/	*AGRICULTURAL DRAINAGE, LAKE META	W71-06443
RGY LOSSES, *WATER CONSERVATION,	*AIR CONDITIONING, *HEATING, *COO	W70-09193
STACEANS, ISOPODS, MARINE ALGAE,	*ALASKA.: /NNELIDS, SEASONAL, CRU	W73-00374
S, FISH PO/ *NUCLEAR EXPLOSIONS,	*ALASKA, *MARINE ALGAE, *FISHERIE	W73-00764
VITY, EUTROPHICATI/ *ECOSYSTEMS,	*ALASKA, *LAKES, ARCTIC, PRODUCTI	W71-09190
BENTHOS, SCUBA / *MARINE PLANTS,	*ALASKA, SAMPLING, *SYSTEMATICS,	W72-09120
ORINATED HYDROCARBON PESTICIDES,	*ALDRIN, *DIELDRIN, *ENDRIN, *GRO	W72-03545
AL DEGRADATION, *DDT, *DIELDRIN,	*ALDRIN, *ENDRIN, MARINE ALGAE, B	W73-00361
LUBLE ORGANIC SOLIDS.:	*ALEXANDRIA(EGYPT), PANDORINA, SO	W70-06619
CONTROL/ *WATER QUALITY CONTROL,	*ALGAE CONTROL, *FORAGE FISH, BIO	W71-09163
WEED STUDY FACILITY, PRODUCT E/	*ALGAE CULTURE TECHNIQUE, AQUATIC	W71-06102
	*ALGAE GROWTH, *CHEMOSTATS.:	W72-06612
	*ALGAE HARVESTING.:	W71-08951
GAL GROWTH, ALGAL HARVESTING, */	*ALGAE STRIPPING, SCENEDESMUS, AL	W72-02975
*PHYTOPLANKTON, *PACIFIC OCEAN,	*ALGAE.: *DIATOMS,	W71-11527
DUCTIVITY, *OIL WASTES, *ARCTIC,	*ALGAE.: *PRIMARY PRO	W72-07922
CULTURES, CHLOROPHYTA, PROTOZOA,	*ALGAE.: /ATICS, *CHLAMYDOMONAS,	W72-12632
ATER TREATMENT, *MICROORGANISMS,	*ALGAE, *ALGICIDES, MICROSTRAININ	W72-06536
S, BIOASSAY, LIMITING FACTORS, /	*ALGAE, *AQUATIC WEEDS, *NUTRIENT	W73-00232
XICITY/ *REVIEWS, *HEAVY METALS,	*ALGAE, *AQUATIC ENVIRONMENT, *TO	W72-14694
*EUTROPHICATION, *PHYTOPLANKTON,	*ALGAE, *AQUATIC WEEDS, *WEEDS, *	W72-14673
LABORATORY TESTS, C/ *ALGICIDES,	*ALGAE, *ALGAL CONTROL, TESTING,	W71-06189
TROL, GROWTH, *CULTURES, *ALGIC/	*ALGAE, *AQUATIC WEEDS, *WEED CON	W71-06102
ANALYTICAL TECH/ *RADIOISOTOPES,	*ALGAE, *ABSORPTION, MONITORING,	W71-08032
ITY, *LABORATORY TESTS, DETERGE/	*ALGAE, *ACTIVATED SLUDGE, *TOXIC	W70-08976
ES, CHE/ *ANALYTICAL TECHNIQUES,	*ALGAE, *ASSAY, CHLORELLA, NITRAT	W70-10403
ACTIVATED SLUDGE, W/ *SYMBIOSIS,	*ALGAE, *BACTERIA, *RESPIRATION,	W71-04079
D, *EUTROPHICATION, PHOSPHOROUS/	*ALGAE, *BIOCHEMICAL OXYGEN DEMAN	W72-04431
OTOSYNTHESIS, OXIDATION, RESPIR/	*ALGAE, *BACTERIA, *SYMBIOSIS, PH	W72-04983
RIOPHAG/ *SEPARATION TECHNIQUES,	*ALGAE, *BACTERIA, *FUNGI, *BACTE	W72-06124
YTOPLANKTON, PRODUCTIVITY, PHOT/	*ALGAE, *BIOMASS, *CYANOPHYTA, PH	W72-14812
RIVERS, BIOLOGICAL COMMUNITIES,/	*ALGAE, *BENTHIC FLORA, *OREGON,	W72-07901
*STREAMS, LAKE ONTARIO, PLANKTO/	*ALGAE, *BIOLOGICAL COMMUNITIES,	W73-01627
OGY, POLLUTANTS, *TOXICITY, WAT/	*ALGAE, *BIOASSAY, *PLANT PHYSIOL	W72-12576
NAEROBIC CONDITIONS, *LAKE ERIE,	*ALGAE, *BENTHIC FAUNA, *BOTTOM S	W72-12995
EFFECTS, *NITRATES, *NUTRIENTS,	*ALGAE, *BACTERIA, AESTHETICS, CO	W71-09958
POSING ORGANIC MATTER, AERATION,	*ALGAE, *BIODEGRADATION, LABORATO	W72-01881
ALYTICAL TECHNIQUES, *NUTRIENTS,	*ALGAE, *BIOASSAY, PHOSPHORUS COM	W72-10608
STE WATER TREATMENT, *NUTRIENTS,	*ALGAE, *BACTERIA, *SYMBIOSIS, AC	W72-11100
ER, *WYOMIN/ *NITROGEN FIXATION,	*ALGAE, *CYANOPHYTA, *THERMAL WAT	W72-08508
, *LAKES, RHODOPHYTA, SHORES.:	*ALGAE, *CHLOROPHYTA, *CYANOPHYTA	W71-12489
ATORY TESTS, CYTOLOGICAL STUDIE/	*ALGAE, *CHLORELLA, *BORON, LABOR	W71-12858
NT MORPH/ *BIOLOGICAL MEMBRANES,	*ALGAE, *CYTYLOGICAL STUDIES, PLA	W73-01630
D/ *LABORATORY TESTS, *BACTERIA,	*ALGAE, *CULTURES, CYANOPHYTA, HY	W73-01657
ACIDS, *PLANT PHYSIOLOGY, *GRO/	*ALGAE, *CARBOHYDRATES, *ORGANICS	W72-07710
ABITATS, *SAMPLING, *ECOSYSTEMS,	*ALGAE, *CHLORAPHYTA, *AQUATIC IN	W72-06526
, BIOMASS, LIGHT INTENSITY, GRO/	*ALGAE, *CULTURES, PHOTOSYNTHESIS	W72-06274
TA, *PYRROPHYTA, *SYSTEMATICS, /	*ALGAE, *CHLOROPHYTA, *EUGLENOPHY	W72-07683
EUTROPHICATION, FJORDS, BIOASSA/	*ALGAE, *CULTURES, *INDICATORS, *	W73-00234
FUNGICIDES, HERBICIDES, IRRIGA/	*ALGAE, *CYANOPHYTA, *PESTICIDES,	W72-14807
TION, *OXIDATION LAGOONS, CHLOR/	*ALGAE, *CHLOROPHYLL, *CLASSIFICA	W71-00139
RIA, *OXIDATION LAGOONS, *PATHO/	*ALGAE, *CULTURES, *ENTERIC BACTE	W70-08319
ER, NUTRIENTS,/ *EUTROPHICATION,	*ALGAE, *DECOMPOSING ORGANIC MATT	W70-10405
THERMAL SPRINGS, CYANOPHYTA, AL/	*ALGAE, *DISTRIBUTION PATTERNS, *	W72-05610
TRIBUTION PATTE/ *PHYTOPLANKTON,	*ALGAE, *DIATOMS, *CURRENTS, *DIS	W72-03543
YNTHESIS, TURBI/ *PHYTOPLANKTON,	*ALGAE, *DISSOLVED OXYGEN, PHOTOS	W73-00256
ACTIVATED SLUDGE, *NUTRIENTS, N/	*ALGAE, *DETERGENTS, *TOXICITY, *	W71-11972
UCTIVITY, *PRIMARY PRODUCTIVITY,	*ALGAE, *DIATOMS, *PERIPHYTON, *P	W72-12192
G FACTORS, PHOTOSYNTHE/ *CARBON,	*ALGAE, *EUTROPHICATION, *LIMITIN	W71-12068
AL, *SEWAGE, *INDUSTRIAL WASTES/	*ALGAE, *ESTUARIES, *WASTE DISPOS	W72-08804
TS, *PLANKTON, *RECREATION, *WA/	*ALGAE, *EUTROPHICATION, *NUTRIEN	W72-10431
RATES, CHLORELLA, CARB/ *CARBON,	*ALGAE, *EUTROPHICATION, *GROWTH	W72-09168
OTOPERIODISM, GROWTH RATES, TEM/	*ALGAE, *ECOLOGY, ENVIRONMENT, PH	W72-00377
GENIC BACTERIA, CULTURES, E. CO/	*ALGAE, *ENTERIC BACTERIA, *PATHO	W72-07664
RUS, *NITROGEN, EVA/ *NUTRIENTS,	*ALGAE, *EUTROPHICATION, *PHOSPHO	W72-04070
QUATIC/ *PHOSPHORUS, *SEDIMENTS,	*ALGAE, *EUTROPHICATION, LAKES, A	W72-03742
, *BOTTOM SEDIMENTS, *LIMNOLOGY,	*ALGAE, *EUTROPHICATION, LAKES, W	W70-08944
RICIDES, GROWTH STAGES, PATHOGE/	*ALGAE, *ENTERIC BACTERIA, *BACTE	W71-05155
WATER QUALITY CONTROL, *ECOLOGY,	*ALGAE, *EUTROPHICATION, NUTRIENT	W71-08775
L, PLANT/ *MOLECULAR STRUCTURES,	*ALGAE, *FLUID FRICTION, *LITTORA	W71-07878
N PATTERNS, BEHAVIOR.:	*ALGAE, *FRESH WATER, DISTRIBUTIO	W71-07360
COSTS, SEPARATION TECHNIQUES, /	*ALGAE, *FLOCCULATION, EQUIPMENT,	W70-08904
, *PHOTOACTIVATION, *OXYGEN, LI/	*ALGAE, *FLUORESCENCE, *CHLORELLA	W72-14800
LUTANTS, DIELD/ *COPPER SULFATE,	*ALGAE, *FLOODWATER, *PATH OF POL	W71-11498
, FLOATING PLANTS, S/ *CHORELLA,	*ALGAE, *FOULING, *NUISANCE ALGAE	W71-09261
UTION TREATMENT, COLLOIDS, *POL/	*ALGAE, *FLOCCULATION, WATER POLL	W72-11833
FA/ *HYDROGEN ION CONCENTRATION,	*ALGAE, *GROWTH RATES, *LIMITING	W72-11724
CS, *LIFE HISTORY STUDIES, *GRO/	*ALGAE, *GROWTH RATES, *SYSTEMATI	W73-02099
, *BICINDICATORS, THERMAL POLLU/	*ALGAE, *GREAT LAKES, *MONITORING	W72-10875
*FLUORESCENCE, / *HEAVY METALS,	*ALGAE, *GROWTH RATES, *TOXICITY,	W72-07660
, CULTURES, INHIBITION, OLIGOTR/	*ALGAE, *GROWTH RATES, *NUTRIENTS	W72-08376
RODUCTIVITY, *HEAVY METALS, *TR/	*ALGAE, *GROWTH RATES, *PRIMARY P	W73-01434
ITY, *PLANT GROWTH SUBSTANCES, /	*ALGAE, *GROWTH RATES, *PRODUCTIV	W71-08687
*PHOSPHORUS, *NUTRIENT REQUIRE/	*ALGAE, *GROWTH RATES, *KINETICS,	W72-04788
IZERS, DIATOMACEOUS EARTH, FOOD/	*ALGAE, *HUMAN POPULATION, FERTIL	W71-09157
COPPER, CHLOROPHY/ *INDICATORS,	*ALGAE, *HEAVY METALS, *TOXICITY,	W71-05991
CHLORELLA, RESP/ *CHLAMYDOMONAS,	*ALGAE, *HERBICIDES, PESTICIDES,	W70-07255

ON, *EUTROPHICATION/ *NUTRIENTS,	*DREDGING, DESILTING, SEDIMENTATI	W73-00938
*FLOCCULATION, *ALGAE, *SILICA,	*E. COLI, *IONS, *WASTEWATER TREA	W72-14840
RADOR), NEW BRUNSWI/ *CHECKLIST,	*EASTERN CANADA, CAPE CHIDLEY(LAB	W71-10066
HMOPLEA SPHAEROPHORA, MELANOSIP/	*EASTERN CANADA, NOVA SCOTIA, IST	W72-01370
C A/ *AERATION, *EUTROPHICATION,	*ECOLOGICAL DISTRIBUTION, *AQUATI	W72-09061
TS(WATER), *DIN/ *BIOINDICATORS,	*ECOLOGICAL DISTRIBUTION, *CURREN	W73-00296
RIBUTION PATTERNS, *SYSTEMATICS,	*ECOLOGICAL DISTRIBUTION, *SPATIA	W73-00428
ANIMALS, *SYSTEMATICS,/ *KELPS,	*ECOLOGICAL DISTRIBUTION, *MARINE	W73-00932
*MARINE ALGAE, LIGHT INTENSITY,	*ECOLOGICAL DISTRIBUTION, SURFACE	W73-00838
SUCCESSION, AQUATIC ENVIRONMENT,	*ECOLOGICAL DISTRIBUTION, *ESTUAR	W72-11854
, *MARINE ANIMALS, *SYSTEMATICS,	*ECOLOGY, *MARINE ALGAE, ANNELIDS	W73-00932
*ZOOPLANKTON, *PHYTOPLANKTON, N/	*ECOLOGY, *AQUATIC PRODUCTIVITY,	W73-00935
BIOLOGIC/ *AQUATIC PRODUCTIVITY,	*ECOLOGY, *AQUATIC ENVIRONMENT, *	W73-02095
*MARINE PLANTS, *MARINE ANIMAL/	*ECOLOGY, *ENVIRONMENTAL EFFECTS,	W73-02029
LLUTION, *ENVIRONMENTAL EFFECTS,	*ECOLOGY, *PLANKTON, *ALGAE, BIOL	W72-14659
TIC HABITATS, *SAMPLING, *ECOSY/	*ECOLOGY, *PREIMPOUNDMENTS, *AQUA	W72-06526
SIS, TEMPERATURE, SEASONAL, ION/	*ECOLOGY, *PONDS, *CHEMICAL ANALY	W72-01363
DISM, GROWTH RATES, TEM/ *ALGAE,	*ECOLOGY, ENVIRONMENT, PHOTOPERIO	W72-00377
ERS, *SEA WATER, *WATER QUALITY,	*ECOLOGY, *CALIFORNIA, HYDROLOGIC	W71-10577
RES, *ALGICIDES, BIBLIOGRAPHIES,	*ECOLOGY, AQUATIC LIFE, IRRIGATIO	W71-06102
UTRIENTS, WATER QUALITY CONTROL,	*ECOLOGY, *ALGAE, *EUTROPHICATION	W71-08775
S, *MARINE PLANTS, DEPTH, CALIF/	*ECOLOGY, *WASTES, *MARINE ANIMAL	W71-01475
QUALITY, *MODEL S/ *ECOSYSTEMS,	*ECOLOGY, *PUMPED STORAGE, *WATER	W72-05282
L/ *DETERGENTS, *EUTROPHICATION,	*ECONOMIC EFFICIENCY, *PUBLIC HEA	W72-04266
GE EFFLUENTS, *SEWAGE TREATMENT,	*ECONOMIC JUSTIFICATION, *COBALT,	W72-00383
RIGATION PROGRAMS, *AGRICULTURE,	*ECONOMICS, CYANOPHYTA, ALGAE, PL	W72-03327
RESOURCES, FOOD/ *MARINE ALGAE,	*ECONOMICS, *PRODUCTIVITY, ALGAE,	W71-13246
NG NUTR/ *RADIOACTIVITY EFFECTS,	*ECOSYSTEMS, *RAIN FORESTS, CYCLI	W72-00949
ORAGE, *WATER QUALITY, *MODEL S/	*ECOSYSTEMS, *ECOLOGY, *PUMPED ST	W72-05282
*WATER POLLUTION, *FOOD CHAINS,	*ECOSYSTEMS, ABSORPTION, FISH, IN	W72-05952
GEMENT, *HERBICIDES, *NUTRIENTS,	*ECOSYSTEMS, *WATERSHEDS(BASINS),	W71-01489
NVERTEBRATES, *MICROENVIRONMENT,	*ECOSYSTEMS, CESIUM, COBALT RADIO	W70-06668
TIC, PRODUCTIVITY, EUTROPHICATI/	*ECOSYSTEMS, *ALASKA, *LAKES, ARC	W71-09190
UCTIVITY, RADIOISOTOPES, STRONT/	*ECOSYSTEMS, *SALT MARSHES, *PROD	W71-09254
FECTS, *MODEL STUDIES, *STREAMS,	*ECOSYSTEMS, AQUATIC HABITATS, BI	W72-06340
NKTON, *PREDATION, *BIOASSAY, B/	*ECOSYSTEMS, FOOD CHAINS, *ZOOPLA	W72-07132
S, *AQUATIC HABITATS, *SAMPLING,	*ECOSYSTEMS, *ALGAE, *CHLORAPHYTA	W72-06526
S, *HEAVY METALS, *PRODUCTIVITY,	*ECOSYSTEMS, POTASSIUM, MAGNESIUM	W72-14303
AQUATIC LIFE, HABITATS, ECOLOGY,	*ECOSYSTEM, *ALGAE, *PHYTOPLANKTO	W73-01632
*WATER / *BIOLOGICAL PROPERTIES,	*ECOSYSTEMS, *RIVERS, *VIRGINIA,	W73-01972
LGAE, *COMPARATIVE PRODUCTIVITY,	*ECOSYSTEMS, *BIOLOGICAL COMMUNIT	W73-01080
ARINE ENVIRONMENT, *DOMESTIC WA/	*ECOSYSTEMS, *PRODUCTIVITY, *ESTU	W72-12567
LAKES, *WATE/ *SEWAGE TREATMENT,	*EFFICIENCIES, *EUTROPHICATION, *	W72-11702
	*ELECTRA CRUSTULENTA, *BRYOZOA.:	W73-00348
N-SITE DATA COLLECTIONS, RIVERS,	*ELECTRIC POWERPLANTS, WATER POLL	W72-07791
WERPLANTS, *DESALINATION PLANTS,	*ELECTRIC POWER PRODUCTION, *IRRI	W72-03327
*MARINE ALGAE, *GULF OF MEXICO,	*ELECTRIC POWERPLANTS, *THERMAL P	W72-11252
WATER PURIFICATION, *FILTRATION,	*ELECTROCHEMISTRY, *WATER POLLUTI	W71-12467
*SUSPENSION, *ELECTROPHORESIS,	*ELECTROCHEMISTRY, *WATER PURIFIC	W73-01362
UES, *WATER ANALYSIS, *NITRATES,	*ELECTRODES, IONS, FLUORIDES, ALG	W71-00474
	*ELECTROFILTRATION.:	W71-12467
ERTIOLECTA, CYCLOTELLA NANA, TH/	*ELECTRON TRANSPORT, DUNALIELLA T	W72-12998
PHY *BIOLOGICAL MAGNIFICATION, /	*ELECTRON CAPTURE GAS CHROMATOGRA	W73-02105
RY, *WATER PURIFIC/ *SUSPENSION,	*ELECTROPHORESIS, *ELECTROCHEMIST	W73-01362
IL WASTES, / BIOASSAY, TOXICITY,	*EMULSIFIERS, DETERGENTS, OIL, *O	W70-09429
PESTICIDES, *ALDRIN, *DIELDRIN,	*ENDRIN, *GROWTH RATES, *CYANOPHY	W72-03545
ATION, *DDT, *DIELDRIN, *ALDRIN,	*ENDRIN, MARINE ALGAE, BOTTOM SED	W73-00361
PRODUCTIVITY, *MIDGES, LAGOONS/	*ENERGY BUDGET, *SEWAGE, *AQUATIC	W72-03556
*STANDING CROP, *TROPHIC LEVEL,	*ENERGY BUDGET, PRODUCTIVITY, HER	W71-10064
IC FLORA, *PRIMARY PRODUCTIVITY,	*ENERGY CONVERSION, STANDING CROP	W72-06283
ERMAL POWERPLANTS, *POWERPLANTS,	*ENERGY LOSSES, *WATER CONSERVATI	W70-09193
	*ENGLAND, *LAMINARIA SACCHARINA.:	W71-09795
RHAM COUNTY RIVERS.:	*ENGLAND, *POLLUTED ESTUARIES, DU	W72-08804
, *DOMESTIC WASTES, ALGAE, INDU/	*ENTERIC BACTERIA, *WATER QUALITY	W72-09590
OONS, *PATHO/ *ALGAE, *CULTURES,	*ENTERIC BACTERIA, *OXIDATION LAG	W70-08319
GROWTH STAGES, PATHOGE/ *ALGAE,	*ENTERIC BACTERIA, *BACTERICIDES,	W71-05155
CTERIA, CULTURES, E. CO/ *ALGAE,	*ENTERIC BACTERIA, *PATHOGENIC BA	W72-07664
CA, *SARCOPTIFORMES, *ROCKPOOLS,	*ENTEROMORPHA, SWEDEN, TIGRIOPUS	W71-08026
), CONNECTICUT RIVER(CONNECTICU/	*ENTRAINMENT, WHITE RIVER(INDIANA	W71-11517
DROGEN ION CONCENTRATION, ALGAE,	*ENTROPHICATION, NEUTRALIZATION,	W70-10427
NT REQUIREMENTS, NUISANCE ALGAS,	*ENTROPHICATION, CYTOLOGICAL STUD	W72-03985
	*ENUMERATION, ANACYSTIS.:	W71-00139
T, WATER Q/ *FARM WASTES, *HOGS,	*ENVIRONMENT, *POLLUTION ABATEMEN	W71-08214
MINANT, STREAM QUALITY, WASTE M/	*ENVIRONMENTAL QUALITY, AIR CONTA	W71-08214
ANALYSIS, RADIOACTI/ *SAMPLING,	*ENVIRONMENT, *NEUTRON ACTIVATION	W71-11036
AS/ *RADIOISOTOPES, *ABSORPTION,	*ENVIRONMENTAL EFFECTS, NUCLEAR W	W72-10955
AQUATIC WEED CONTROL, *PARAQUAT,	*ENVIRONMENTAL EFFECTS, DIQUAT, A	W72-00155
R POLLUTION CONTRO/ *MANAGEMENT,	*ENVIRONMENT, *AGRICULTURE, *WATE	W72-00846
EXPLOSI/ *RADIOACTIVITY EFFECTS,	*ENVIRONMENTAL EFFECTS, *NUCLEAR	W72-00960
LAKES, *WATER POLLUTION CONTROL,	*ENVIRONMENTAL SANITATION, *LEGIS	W72-01660
*MEASUREMENT, *DATA COLLECTIONS,	*ENVIRONMENT, *RADIOACTIVITY, *RA	W72-07841
N-SITE T/ *LIGHT, *GROWTH RATES,	*ENVIRONMENTAL EFFECTS, LARVAE, O	W73-00348
: *MERCURY POLLUTION,	*ENVIRONMENTAL PROTECTION AGENCY.	W72-14782
TS, *STREAM, *THERMAL POLLUTION,	*ENVIRONMENTAL EFFECTS, *ECOLOGY,	W72-14659
OVERNMEN/ *LEGISLATION, *TOXINS,	*ENVIRONMENTAL EFFECTS, FEDERAL G	W73-00703
OTOZOA, *BIOLOGICAL COMMUNITIES,	*ENVIRONMENTAL EFFECTS, *STANDING	W73-00367
ICS, *SEASONAL,/ *PHYTOPLANKTON,	*ENVIRONMENTAL EFFECTS, *SYSTEMAT	W73-00442
RPHOLOG/ *PHAEOPHYTA, *SALINITY,	*ENVIRONMENTAL EFFECTS, *PLANT MO	W73-00369
LANTS, *MARINE ANIMAL/ *ECOLOGY,	*ENVIRONMENTAL EFFECTS, *MARINE P	W73-02029
BON DIOXIDE, NITRATES, NITRITES,	*ENZYMES.: /ON CONCENTRATION, CAR	W72-04309

*NUTRIENTS, ALGAE, LAKES, PHOSP/ *EUTROPHICATION, *ON-SITE TESTS, W72-09165
NS, *SYSTEMATICS, *SALINE LAKES, *EUTROPHICATION, WATER SAMPLING, W72-09111
CRUS, NITROGEN, ALGAE.: *EUTROPHICATION, *CARBON, *PHOSPH W72-09155
R QUALITY CONTROL, SEWAGE TREAT/ *EUTROPHICATION, *RECYCLING, WATE W72-09378
CONTROL, MANAGEMEN/ *ESTUARIES, *EUTROPHICATION, *WATER POLLUTION W72-09171
N, *WATER QUALI/ *ALGAL CONTROL, *EUTROPHICATION, *DESTRATIFICATIO W72-09304
HLORELLA, CARB/ *CARBON, *ALGAE, *EUTROPHICATION, *GROWTH RATES, C W72-09168
SMENT/ *DETERGENTS, *PHOSPHATES, *EUTROPHICATION, NUTRIENTS, ASSES W72-09169
DELS, *ESSENTIAL NUTRIENTS, *PR/ *EUTROPHICATION, *MATHEMATICAL MO W72-08986
RIBUTION, *AQUATIC A/ *AERATION, *EUTROPHICATION, *ECOLOGICAL DIST W72-09061
OLOGY, OREGON, *SED/ *VITAMIN B, *EUTROPHICATION, *AQUATIC MICROBI W72-08989
HATES, *WATER POLLUTION SOURCES, *EUTROPHICATION, *DETERGENTS, WAT W72-11186
AE, SEWAGE TREATMENT, WATER POL/ *EUTROPHICATION, *NUTRIENTS, *ALG W72-10625
SEWAGE TREATMENT, *EFFICIENCIES, *EUTROPHICATION, *LAKES, *WATER P W72-11702
NKTON, *RECREATION, *WA/ *ALGAE, *EUTROPHICATION, *NUTRIENTS, *PLA W72-10431
CTIVITY, *RESERVOIRS, *NEBRASKA, *EUTROPHICATION, PLANKTON, CHLORO W72-04270
*ALGAL CONTROL, *AQUATIC ALGAE, *EUTROPHICATION, MUNICIPAL WASTES W72-03970
ONTROL, *NON-STRUCTURAL/ *MAINE, *EUTROPHICATION, *WATER QUALITY C W72-04279
ENCY, *PUBLIC HEAL/ *DETERGENTS, *EUTROPHICATION, *ECONOMIC EFFICI W72-04266
*PHOSPHORUS, *SEDIMENTS, *ALGAE, *EUTROPHICATION, LAKES, AQUATIC P W72-03742
TROGEN, EVA/ *NUTRIENTS, *ALGAE, *EUTROPHICATION, *PHOSPHORUS, *NI W72-04070
POLLUTION CONTROL, WATER POLLUT/ *EUTROPHICATION, *CARBON, *WATER W72-04769
CONTROL, *WASTE WATER(POLLUTIO/ *EUTROPHICATION, *WATER POLLUTION W72-05473
EAT PLAINS, NUTRIENTS, RECREATI/ *EUTROPHICATION, *RESERVOIRS, *GR W72-04759
ARBON DIOXIDE, AERATION, MIXING, *EUTROPHICATION, *ANALYTICAL TECH W72-04433
GAE, *BIOCHEMICAL OXYGEN DEMAND, *EUTROPHICATION, PHOSPHORUS, NIT W72-04431
RECREATION, WATER SUPPLY, WATER/ *EUTROPHICATION, *LAKES, *MAINE, W72-04280
TERIA, *LAKE SOILS, *PHOSPHORUS, *EUTROPHICATION, LAKES, AQUATIC A W72-04292
TER POLLUTION CONTROL, *WASTE W/ *EUTROPHICATION, *PHOSPHATES, *WA W72-04734
ALITY CONTROL, *ECOLOGY, *ALGAE, *EUTROPHICATION, NUTRIENT REQUIRE W71-08775
, *NITROGEN, *PHOSPHORUS, ALGAE, *EUTROPHICATION, AQUATIC ENVIRONM W71-08953
NS, *SEWAGE LAGOONS, BIODEGRADA/ *EUTROPHICATION, *OXIDATION LAGOO W71-07973
PHOSPHATES, *PATH OF POLLUTANTS, *EUTROPHICATION, *LEACHING NUTRIE W71-04216
ROL, *FORMULATION, *SURFACTANTS, *EUTROPHICATION, LINEAR ALKYLATE W71-06247
CES DEVELOPMENT, *MASSACHUSETTS, *EUTROPHICATION, NUTRIENTS, ALGAE W71-05553
PHORUS, NUTRIENTS, ALGAE, NITRA/ *EUTROPHICATION, *NITROGEN, *PHOS W71-06443
SOURCES, *ESTUARIE/ *NUTRIENTS, *EUTROPHICATION, *WATER POLLUTION W70-06509
T REQUIREMENTS, CHLOROPHYTA, CY/ *EUTROPHICATION, *ALGAE, *NUTRIEN W70-05469
ONTROL, *NITROGEN C/ *NUTRIENTS, *EUTROPHICATION, *WATER QUALITY C W70-04080
DETERGENTS, *TERTIARY TREATMENT, *EUTROPHICATION, WASTE WATER TREA W70-09325
M SEDIMENTS, *LIMNOLOGY, *ALGAE, *EUTROPHICATION, LAKES, WATER QUA W70-08944
OSPHATES, *NITRATES, *NUTRIENTS, *EUTROPHICATION, ALGAE, SURFACTAN W70-09388
SLUDGE WORMS, *DISSOLVED OXYGEN, *EUTROPHICATION, *BIOCHEMICAL OXY W70-09189
RBON, BACTERIA, ALGAE, SYMBIOSI/ *EUTROPHICATION, *PHOSPHORUS, *CA W70-07283
ENVIRONMENT, CHEMICAL / *SEWAGE, *EUTROPHICATION, *LAKES, AQUATIC W71-00117
SING ORGANIC MATTER, NUTRIENTS,/ *EUTROPHICATION, *ALGAE, *DECOMPO W70-10405
YSTEMS ANALYSIS, *WATER QUALITY, *EUTROPHICATION, *MATHEMATICAL MO W71-03021
LANKTON/ *WATER QUALITY, *LAKES, *EUTROPHICATION, PHYTOPLANKTON, P W71-02880
OLLUTION, MINNESOTA, WATER QUAL/ *EUTROPHICATION, *LAKES, *WATER P W71-03012
CONTROL, WATER POLLUTION EFFEC/ *EUTROPHICATION, *WATER POLLUTION W71-01486
HURON, EFFLUENTS, WATER QUALIT/ *EUTROPHICATION, *NUTRIENTS, LAKE W72-13662
C ALGAE, FILTRATION, CYANOPHYTA, *EUTROPHICATION, *OHIO.: / AQUATI W72-13673
AE, *LIMITING FACTORS, PHOSPHAT/ *EUTROPHICATION, *NUTRIENTS, *ALG W72-12544
ITY, *PHOTOSYNTHESIS, LIMNOLOGY, *EUTROPHICATION, *KANSAS, MODEL S W72-12393
*SYNOPTIC ANALYSIS, COLD REGIO/ *EUTROPHICATION, *LAKES, *ALPINE, W72-12549
ISSOLVED OXYGEN, *A/ *LAKE ERIE, *EUTROPHICATION, *HYPOLIMNION, *D W72-12990
FFLUE/ *WATER POLLUTION SOURCES, *EUTROPHICATION, *LAKES, SEWAGE E W72-12955
ANALYTICAL TECHNIQUES, WASTE WA/ *EUTROPHICATION, *ASSAY, *ALGAE, W72-12966
POLLUTION CONTROL, *PHOSPHORUS, *EUTROPHICATION, CHEMICAL ANALYSI W73-02478
MS, WATER QUALITY, *ALGAE, *LIG/ *EUTROPHICATION, AQUATIC ECOSYSTE W73-02349
ATION, *WATER POLLUTION CONTROL/ *EUTROPHICATION, *SOCIAL PARTICIP W73-02479
E/ *PHYTOPLANKTON, *CHLOROPHYLL, *EVALUATION, *SEPARATION TECHNIQU W72-13673
*DISSOLVED ORGANIC PHOSPHORUS, *EXCRETION, ALKALINE PHOSPHATASE, W71-10083
CCHARIDES, OLIGOSACCHARIDES, CL/ *EXCRETION, DISACCHARIDES, MONOSA W72-07710
ISM, *HETEROANTAGONISM, AXENIC / *EXCRETORY PRODUCTS, *AUTOANTAGON W72-12734
ESTERN ONTARIO, EPILITHIC ALGAE/ *EXPERIMENTAL LAKES AREA, *NORTHW W71-11029
INATION, *ORIGIN, *SIGNIFICANCE, *EXTRACTION, LUCIFERIN-LUCIFERASE W72-04292
LAGOONS.: *FACULTATIVE LAGOONS, *ANAEROBIC W72-09386
*SOIL-WATER-PLANT RELATIONSHIPS, *FALLOUT, *CYCLING NUTRIENTS, ECO W72-03342
TIVITY EFFECTS, *NUCLEAR WASTES, *FALLOUT, *OCEANS, WATER POLLUTIO W72-04463
LIFE, R/ *RADIOACTIVITY EFFECTS, *FALLOUT, *FOOD CHAINS, *AQUATIC W72-04461
LAKES, *STRONTIUM RADIOISOTOPES, *FALLOUT, *FOOD CHAINS, ABSORPTIO W73-00790
DIATION, *RADIOACTIVITY EFFECTS, *FALLOUT, *MARINE ANIMALS, MARINE W73-00828
ASTAL PLAIN, NO/ *BIOINDICATORS, *FALLOUT, *MOLLUSKS, *ATLANTIC CO W73-00818
ING AND COOLING, *SPACE HEATING, *FARM USES OF WASTE HEAT, STEAM P W70-09193
IZATION, *WASTE WATER TREATMENT, *FARM WASTES.: /SYNTHESIS, STABIL W70-06619
GOONS, *WATER POLLUTION EFFECTS, *FARM WASTES, OKLAHOMA, CYANOPHYT W72-04773
, *POLLUTION ABATEMENT, WATER Q/ *FARM WASTES, *HOGS, *ENVIRONMENT W71-08214
AGOONS, *OXIDATION LAGOONS, AER/ *FARM WASTES, *WASTE DISPOSAL, *L W71-07551
CATTLE, DISSOLVED OXYGEN, DIVE/ *FARM WASTES, *RUNOFF, *FISHKILL, W72-08401
LAGOONS, PUMPING, AEROBIC TREAT/ *FARM WASTES, POULTRY, NITROGEN, W72-08396
LAGOONS, NUTRIENTS, AQUATIC AL/ *FARM WASTE, *POULTRY, *OXIDATION W72-02850
OGRAPHY, METHYL ESTERS, ESTERS,/ *FATTY ACIDS, *GAS LIQUID CHROMAT W72-09365
CONTROL, *WATER QUALITY CONTROL, *FEDERAL GOVERNMENT, *GRANTS, EUT W72-01659
MENTAL SANITATION, *LEGISLATION, *FEDERAL GOVERNMENT, EUTROPHICATI W72-01660
POLLUTION CONTROL, *PHOSPHATES, *FEDERAL GOVERNMENT, UNITED STATE W72-01093
ATER POLLUTION EFFECTS, *CATTLE, *FEED LOTS.: /CULTURAL RUNOFF, *W W72-08401
LANDICUS, COSCINODISCUS, COPEPO/ *FEEDING PATTERNS, *CALANUS HELGO W72-09248
*FEEDLOTS, LEMNA.: W72-04773
, POULTRY, HOGS, ANIMAL DISEASE/ *FEEDS, *RUMINANTS, ALGAE, CATTLE W71-07544

NTAL PROTECTION AGENCY.:	*LEGISLATION(PROPOSED), ENVIRONME	W73-00703
ANOPHYTA, PLANTS, NITRO/ *ALGAE,	*LEGUMES, *NITROGEN FIXATION, *CY	W73-01456
ECTICIDES, PHENOLS, *DETERGENTS,	*LETHAL LIMIT, ALKYLBENZENE SULFO	W72-12576
ALS, *CHLAMYDOMONAS, INHIBITORS,	*LETHAL LIMIT, WATER POLLUTION EF	W72-11727
*CHLOROPHYTA, *GROWTH CHAMBERS,	*LIFE CYCLES, CULTURES, WATER QUA	W72-07144
AE, *GROWTH RATES, *SYSTEMATICS,	*LIFE HISTORY STUDIES, *GROWTH ST	W73-02099
DIATOMS, / *CRUSTACEANS, *FOODS,	*LIFE HISTORY STUDIES, COPEPODS,	W72-14805
A, EPIPHYTOLOGY, ALGAE, HABITAT/	*LIFE HISTORY STUDIES, *PHAEOPHYT	W72-01370
TIVITY, *RADIOACTIVITY TECHNIQU/	*LIGHT INTENSITY, *PRIMARY PRODUC	W72-09102
ON, *ABSORPTION, *AQUATIC ALGAE,	*LIGHT INTENSITY, *PHOSPHORUS, PL	W72-07166
ROPHYTES, *AQUATIC WEED CONTROL,	*LIGHT INTENSITY, *NEBRASKA, *CYA	W73-02349
, P/ *ALGAE, *OXIDATION LAGOONS,	*LIGHT INTENSITY, PONDS, ROTIFERS	W71-07382
OSYSTEMS, WATER QUALITY, *ALGAE,	*LIGHT PENETRATION, *MACROPHYTES,	W73-02349
ODEL STUDIES, ECOSYSTEMS, ALGAE,	*LIGHT QUALITY, EUTROPHIC ZONE, C	W72-12393
NTAL EFFECTS, LARVAE, ON-SITE T/	*LIGHT, *GROWTH RATES, *ENVIRONME	W73-00348
TON, NATURAL RES/ *MARINE ALGAE,	*LIGHT, *GROWTH RATES, PHYTOPLANK	W71-13252
YTA, NITROGEN FIXATION, OXYGEN,/	*LIGHT, *PHOTOSYNTHESIS, *CYANOPH	W72-08584
ALGAE, CULTURES, VO/ *CHLORELLA,	*LIGHT, *PHOTOSYNTHETIC OXYGEN, *	W70-05547
IOLOGICAL COMMUNITIES, SYSTEMAT/	*LIGHT, *TEMPERATURE, *DIATOMS, B	W73-00242
RTHYMS, *METABOLISM, *CHLORELLA,	*LIGHT, ALGAE, GROWTH RATES, CHLO	W73-00251
COMMUNITIES, SEASONAL, DIATOMS,	*LIGHT, SOLAR RADIATION, *PHOTOPE	W72-07901
	*LIMITING FACTORS, SILICON.:	W72-07937
(WIS).:	*LIMITING NUTRIENTS, LAKE MENDOTA	W72-07933
IATOMS,/ *NUTRIENT REQUIREMENTS,	*LIMITING FACTORS, *PHOSPHATES, D	W73-01445
GTON(WASH).:	*LIMITING NUTRIENTS, *LAKE WASHIN	W72-09167
	*LIMITING NUTRIENTS.:	W72-09155
(MINN), WALDO LAKE(ORE).:	*LIMITING NUTRIENTS, SHAGAWA LAKE	W72-09165
	*LIMITING NUTRIENTS.:	W72-09162
	*LIMITING NUTRIENTS.:	W72-09164
LORIDE, SHRINER'S POND(GA).:	*LIMITING NUTRIENTS, POTASSIUM CH	W72-09159
*PHOSPHORUS, *PLANKTON, *ALGAE,	*LIMITING FACTORS, GROWTH RATES,	W72-09163
RBON SOURCES.:	*LIMITING NUTRIENTS, ANABAENA, CA	W72-09168
AKE(NY), LAKE GEORGE(NY).:	*LIMITING NUTRIENTS, CANADARAGO L	W72-09163
IVITY/ *CARBON, *SEWAGE LAGOONS,	*LIMITING FACTORS, ALGAE, PRODUCT	W72-09162
	*LIMITING NUTRIENTS, CANADA.:	W71-12857
CARBON, *ALGAE, *EUTROPHICATION,	*LIMITING FACTORS, PHOTOSYNTHESIS	W71-12068
ROPHICATION, *NUTRIENTS, *ALGAE,	*LIMITING FACTORS, PHOSPHATES, CH	W72-12544
NTRATION, *ALGAE, *GROWTH RATES,	*LIMITING FACTORS, *DIURNAL DISTR	W72-11724
LAKES, *WATER POLLUTION EFFECTS,	*LIMNOLOGY, WATER POLLUTION, WATE	W72-12729
R ANALYSIS, *STANDING CROP, *PR/	*LIMNOLOGY, *PHYTOPLANKTON, *WATE	W73-01435
ION, *HYDROLOGY, *NEW YORK, WAT/	*LIMNOLOGY, *LAKES, *EUSTROPHICAT	W72-13851
*NUTRIENTS, *BOTTOM SEDIMENTS,	*LIMNOLOGY, *ALGAE, *EUTROPHICATI	W70-08944
IS, *PHYTOPLANKTON, *PRODUCTIVI/	*LIMNOLOGY, *LAKES, *PHOTOSYNTHES	W71-08670
DS, SAMPLING, THERMAL STRATIFIC/	*LIMNOLOGY, *FLORIDA, *LAKES, PON	W72-04269
*DESMIDS, *EUPLANKTON,	*LIMNOPLANKTON.:	W71-13149
LABORATORY/ *ALGAE, *INHIBITION,	*LIPIDS, TOXICITY, HEAVY METALS,	W72-06037
PATENTS, *WASTE WATER TREATMENT,	*LIQUID WASTES, SEWAGE TREATMENT,	W72-05315
.: *ANIMAL MANURE,	*LITERATURE REVIEW, FEEDING VALUE	W71-07544
E TREATMENT, SECONDARY SEWAGE T/	*LITERATURE SURVEY, PRIMARY SEWAG	W70-09907
	*LITERATURE REVIEWS.:	W72-12190
CTURES, *ALGAE, *FLUID FRICTION,	*LITTORAL, PLANT GROWTH SUBSTANCE	W71-07878
THIC FLORA/ *PERIPHYTON, *ALGAE,	*LITTORAL, LIGHT PENETRATION, BEN	W71-11029
BASIN(ONTARIO), CANADA, BUFFER/	*LOADING GUIDELINES, *GRAND RIVER	W71-11017
ATION, *DETERGENTS, *PHOSPHATES,	*LOCAL GOVERNMENTS, *WATER POLLUT	W72-06638
Z CREEK, ARROYO GRANDE CREEK.:	*LOPEZ WATER SUPPLY PROJECT, LOPE	W72-05594
SINS), *DD/ *PESTICIDE RESIDUES,	*LOTIC ENVIRONMENT, WATERSHEDS(BA	W72-12930
	*LOXODES MAGNUS.:	W72-14796
USES, *PLECTONEMA BORYANIUM, SU/	*LPP VIRUS, *BLUE-GREEN ALGAL VIR	W72-04736
EM, PL/ *ADENOSINE TRIPHOSPHATE,	*LUCIFERIN-LUCIFERASE ENZYME SYST	W71-11916
ETABOLITES.:	*LYSIS, *CONTACT TIME, CELLULAR M	W72-00411
LPP-1, LPP-1D, PHYCOVIRUS, MITO/	*LYSOGENY, *PLECTONEMA BORYANUM,	W72-03188
EEN ALGAE, ANABAENA VARIABILIS,/	*LYSOZYME, *SPHEROPLASTS, BLUE-GR	W71-08683
6, CESIUM RADIOISOTOPES, SPIROG/	*MACROINVERTEBRATES, RUTHENIUM-10	W70-06668
*CHEMICAL EVOLUTION,	*MACROMOLECULES.:	W72-01473
YTIC ALGAE, *EPIPHYTIC BACTERIA,	*MACROPHYTE-EPIPHYTE METABOLISM,	W72-03216
OLAND), COLONIZATION, POTAMOGET/	*MACROPHYTES, *MIKOLAJSKIE LAKE(P	W71-11012
ISION, DEPTH SOUNDE/ *SEA WEEDS,	*MACROPHYTES, *SEA GRASSES, TELEV	W72-09120
ITY, *ALGAE, *LIGHT PENETRATION,	*MACROPHYTES, *AQUATIC WEED CONTR	W73-02349
LAWN MAINTENANCE.:	*MADISON(WIS), LAKE MENDOTA(WIS),	W73-02479
UALITY CONTROL, *NON-STRUCTURAL/	*MAINE, *EUTROPHICATION, *WATER Q	W72-04279
ECOSYSTEMS, PESTICIDE KINETICS,	*MAINE, MUDS, INVERTEBRATES, FISH	W72-12930
WATER/ *EUTROPHICATION, *LAKES,	*MAINE, RECREATION, WATER SUPPLY,	W72-04280
*PHO/ *SEA NETTLES, *POLLUTANTS,	*MAINTENANCE, *PLANT MORPHOLOGY,	W72-10162
HROMES, PHOSPHORYLATI/ *GLUCOSE,	*MALATE, CITRIC ACID CYCLE, CYTOC	W72-04309
ULTURE, *WATER POLLUTION CONTRO/	*MANAGEMENT, *ENVIRONMENT, *AGRIC	W72-00846
ER POLLUTION EFFECTS, *PROTOZOA,	*MARINE ALGAE, *SALT MARSHES, NEW	W72-00948
TES, PHYTOPLANKTON, NATURAL RES/	*MARINE ALGAE, *LIGHT, *GROWTH RA	W71-13252
CTIVITY, ALGAE, RESOURCES, FOOD/	*MARINE ALGAE, *ECONOMICS, *PRODU	W71-13246
HNIQUES, RATES, *PHOTOSYNTHESIS,	*MARINE ALGAE, DDT, *PESTICIDES,	W72-09653
ELECTRIC POWERP/ *BENTHIC FLORA,	*MARINE ALGAE, *GULF OF MEXICO, *	W72-11252
ISOTOPES, *COBALT RADIOISOTOPES,	*MARINE ALGAE, HEAVY METALS, ABSO	W72-10957
ANALYSIS,/ *TOXICITY, *AMMONIA,	*MARINE ALGAE, *DIATOMS, CHEMICAL	W71-11008
STS, BENTHIC FLORA, RHODOPHYTA,/	*MARINE ALGAE, *SYSTEMATICS, *COA	W71-10066
WATER POLLUTION EFFECTS, *ALGAE,	*MARINE ALGAE, *BIOINDICATORS, *P	W71-10553
HYTA, CHLOR/ *YEASTS, *CULTURES,	*MARINE ALGAE, RHODOPHYTA, PHAEOP	W72-04748
*AQUATIC PLANTS, *DISTRIBUTION,	*MARINE ALGAE, *ESTUARIES, PRODUC	W72-04766
ER, *SEDIMENTS, *MARINE ANIMALS,	*MARINE ALGAE, RADIOISOTOPES, FAL	W72-05965
*POLLUTANTS, *SEA WATER, *ALGAE,	*MARINE ALGAE, CULTURES, PHYTOPLA	W70-07280
DIOISOTOPES, *ORGANIC COMPOUNDS,	*MARINE ALGAE, PHYTOPLANKTON, HUM	W71-09188
ICAL ECOLOGY, PHAEOPHYTA, STAND/	*MARINE ALGAE, *KELPS, *PHYSIOLOG	W72-12940

THERMAL POLLUTION, *THERM/ BAYS,	*MARINE ALGAE, *MARINE ANIMALS, *	W72-11876
ROPHYTA, PHAEOPHY/ *SYSTEMATICS,	*MARINE ALGAE, *CYANOPHYTA, *CHLO	W72-11800
*CHEMICAL WASTE, *GROWTH RATES/	*MARINE ALGAE, *LABORATORY TESTS,	W72-12239
OTOPES, *ZINC RADIOISOTOPES, AB/	*MARINE ALGAE, *STRONTIUM RADIOIS	W72-12203
IS, *DECOMPOSING ORGANIC MATTER,	*MARINE ALGAE, HUMUS, ORGANIC COM	W72-12166
*LABORATORY TESTS, *MEASUREMENT,	*MARINE ALGAE TRACERS, CHEMICAL P	W72-12210
WASTES, RADIOE/ *RADIOISOTOPES,	*MARINE ALGAE, *ABSCRPTION, *MINE	W72-12204
, / *PHAEOPHYTA, *TIDAL MARSHES,	*MARINE ALGAE, CULTURES, SALINITY	W72-12633
CARBON PE/ *OCEANS, *MATABOLISM,	*MARINE ALGAE, *CHLORINATED HYDRO	W72-12998
RINE PLANTS, MARINE/ *SEA WATER,	*MARINE ALGAE, *CARBOHYDRATES, MA	W73-02100
ES, *PLANT GROWTH, *CHLOROPHYTA,	*MARINE ALGAE, *BIOASSAY, REPRODU	W72-14294
ARY PRODUCTIVITY, ENERGY BUDGET,	*MARINE ALGAE, MARINE PLANTS, BIO	W73-00366
RIBUTION PATTERNS, *CHLOROPHYTA,	*MARINE ALGAE, *RHODOPHYTA, *SYST	W73-00284
*NITRATES, *AM/ *PHYTOPLANKTON,	*MARINE ALGAE, CHEMICAL ANALYSIS,	W73-00380
PLANTS/ *CULTURES, *CYANOPHYTA,	*MARINE ALGAE, *ISOLATION, MARINE	W73-00430
RIMARY PRODUCTIVITY, *ENERGY CO/	*MARINE ALGAE, *BENTHIC FLORA, *P	W72-06283
ES, *REVIEWS, B/ *PHYTOPLANKTCN,	*MARINE ALGAE, *SEDIMENTATION RAT	W72-06313
HYTA, CHL/ *BACTERIA, *TOXICITY,	*MARINE ALGAE, SEA WATER, CHLOROP	W72-08051
IOACTIVITY EFFECTS, *MONITORING,	*MARINE ALGAE, RHODOPHYTA, PHAEOP	W72-07826
CHNIQUES, *CARBON, *AMINO ACIDS,	*MARINE ALGAE, PHYTOPLANKTON, NIT	W72-07497
HYTA, PHAEOPHYTA, WATER QUALITY,	*MARINE ALGAE.: /ODUCTION, RHODOP	W73-01449
FFECTS, *PHOTOSYNTHESIS, *OIL P/	*MARINE ALGAE, *WATER POLLUTION E	W73-01074
ECOLOGICAL DI/ *HEAT RESISTANCE,	*MARINE ALGAE, LIGHT INTENSITY, *	W73-00838
OTOZOA, *BACTERIA, FLORIDA, RAD/	*MARINE ALGAE, *SALT MARSHES, *PR	W73-00796
O/ *NUCLEAR EXPLOSIONS, *ALASKA,	*MARINE ALGAE, *FISHERIES, FISH P	W73-00764
ES, *ABSORPTION, *PHYTOPLANKTON,	*MARINE ALGAE, FOOD CHAINS, PUBLI	W73-00823
RADIOISOTOPES, *MARINE ANIMALS,	*MARINE ALGAE, *FOOD CHAINS, ABSO	W73-00819
MARINE FISH, M/ *NUCLEAR WASTES,	*MARINE ALGAE, *MARINE ANIMALS, *	W73-00832
ZINC RADIOISOTOPES, *ABSORPTION,	*MARINE ALGAE, *PHYTOPLANKTON, BA	W73-00802
*RADIOISOTOPES, *MARINE ANIMALS,	*MARINE ALGAE, SEDIMENTS, NUCLEAR	W73-00817
ANIMALS, *SYSTEMATICS, *ECOLOGY,	*MARINE ALGAE, ANNELIDS, MOLLUSKS	W73-00932
KELPS, *ECOLOGICAL DISTRIBUTION,	*MARINE ANIMALS, *SYSTEMATICS, *E	W73-00932
AINS, *NUCLEAR WASTES, *COBALT /	*MARINE ANIMALS, *CLAMS, *FOOD CH	W73-00826
FOOD / *STRONTIUM RADIOISOTOPES,	*MARINE ANIMALS, *MARINE ALGAE, *	W73-00819
RADIOACTIVITY EFFECTS, *FALLOUT,	*MARINE ANIMALS, MARINE PLANTS, M	W73-00828
EDIME/ *REVIEWS, *RADIOISOTOPES,	*MARINE ANIMALS, *MARINE ALGAE, S	W73-00817
*NUCLEAR WASTES, *MARINE ALGAE,	*MARINE ANIMALS, *MARINE FISH, MU	W73-00832
OISOTOPES, *COBALT, *ABSORPTICN,	*MARINE ANIMALS, MARINE ALGAE, SE	W73-00821
*NUCLEAR WASTES, *BIOINDICATORS,	*MARINE ANIMALS, ASIA, MOLLUSKS,	W73-00811
NMENTAL EFFECTS, *MARINE PLANTS,	*MARINE ANIMALS, OFFSHORE PLATFOR	W73-02029
ON, *THERM/ BAYS, *MARINE ALGAE,	*MARINE ANIMALS, *THERMAL POLLUTI	W72-11876
ARIA, MOLLUSKS,/ *RADIOISOTOPES,	*MARINE ANIMALS, *ABSCRPTION, AQU	W71-09182
, *DISTRIBUTION, *RADIOISOTOPES,	*MARINE ANIMALS, *MARINE PLANTS,	W71-09168
DEPTH, CALIF/ *ECOLOGY, *WASTES,	*MARINE ANIMALS, *MARINE PLANTS,	W71-01475
NALYSIS, *SEA WATER, *SEDIMENTS,	*MARINE ANIMALS, *MARINE ALGAE, R	W72-05965
PREDATION, ORGANIC COMPOUNDS, A/	*MARINE BACTERIA, *ATTRACTANTS, *	W73-00942
BENTHOS, CRABS, / *CRUSTACEANS,	*MARINE BACTERIA, *WATER QUALITY,	W72-08142
EDEN.:	*MARINE DISPOSAL, *GOTHENBRUG, SW	W70-09947
*MARINE ALGAE, *MARINE ANIMALS,	*MARINE FISH, MUSSLES, LOBSTERS,	W73-00832
SH EGGS, *SEPARATION TECHNIQUES,	*MARINE FISH, SILICA, MARINE ALGA	W73-00371
RIMP, *BIOASSAY, *ABSORPT/ *DDT,	*MARINE FISH, *PHYTOPLANKTON, *SH	W73-02105
SEPARATION TECHNIQUES, ATLANTI/	*MARINE FUNGI, *ISOLATICN, ALGAE,	W71-11851
ER, *METABOLISM, BIODEGRADATICN,	*MARINE MICROORGANISMS, *MICROBIA	W73-00361
SOTOPES, *ANALYTICAL TECHNIQUES,	*MARINE MICROORGANISMS, MARINE PL	W71-08042
*GLYCEROL, *PHOTOTROPHIC GROWTH,	*MARINE PHYTOPLANKTON(PRYMNESIUM	W70-07842
OLOGY, *WASTES, *MARINE ANIMALS,	*MARINE PLANTS, DEPTH, CALIFORNIA	W71-01475
*RADIOISOTOPES, *MARINE ANIMALS,	*MARINE PLANTS, AQUATIC ANIMALS,	W71-09168
R POLLUTION EFFECTS, *INTERTIDA/	*MARINE PLANTS, *OILY WATER, WATE	W71-08792
ECOLOGY, *ENVIRONMENTAL EFFECTS,	*MARINE PLANTS, *MARINE ANIMALS,	W73-02029
*ORGANIC COMPOUNDS, *PHOSPHORUS,	*MARINE PLANTS, *PHYTOPLANKTCN, A	W71-10083
, *SYSTEMATICS, BENTHOS, SCUBA /	*MARINE PLANTS, *ALASKA, SAMPLING	W72-09120
IMARY PRODUCTIVITY, *FRESHWATER,	*MARSHES, SALINE WATER, BIOMASS,	W73-01089
EASONAL, *DISTRIBUTION PATTERNS,	*MASSACHUSETTS, SYSTEMATICS, *REP	W73-01449
S, *WATER RESOURCES DEVELOPMENT,	*MASSACHUSETTS, *EUTROPHICATION,	W71-05553
RINATED HYDROCARBON PE/ *OCEANS,	*MATABOLISM, *MARINE ALGAE, *CHLO	W72-12998
UATIONS, BIOLOGICAL PROPERTIES,/	*MATHEMATICAL STUDIES, *ALGAE, EQ	W72-12724
*ALGAE, *WASTE WATER TREATMENT,	*MATHEMATICAL MODELS, GROWTH RATE	W72-12289
YCLE, *OXIDATION, AMMONIA, NITR/	*MATHEMATICAL MODELS, *NITROGEN C	W71-05390
IMULATION, / *OXIDATION LAGOONS,	*MATHEMATICAL MODELS, *COMPUTER S	W71-07084
ER RE-USE, / *OXIDATION LAGOONS,	*MATHEMATICAL MODELS, MIXING, WAT	W71-07099
PULATION, DYNAMICS, LAKES, DISS/	*MATHEMATICAL MODELS, *ALGAE, *PO	W71-03034
*WATER QUALITY, *EUTROPHICATION,	*MATHEMATICAL MODELS, SIMULATION,	W71-03021
RPLANTS, *RADIOACTIVITY EFFECTS,	*MATHEMATICAL MODELS, FORECASTING	W73-00831
QUALITY, *HYDROELECTRIC PLANTS,	*MATHEMATICAL MODELS, THERMAL STR	W72-14405
ROL, *RESERVOIR/ *PHYTOPLANKTCN,	*MATHEMATICAL MODELS, *ALGAL CONT	W72-08559
NUTRIENTS, *PR/ *EUTROPHICATION,	*MATHEMATICAL MODELS, *ESSENTIAL	W72-08986
TIFICATION, *AERIAL PHOTOGRAPHY,	*MATHEMATICAL METHODS, MUNICIPAL	W71-11685
A COLLECTIONS, *HYDROLOGIC DATA,	*MEASUREMENT, *ASSAY, ANALYTICAL	W72-10678
SOURCES, AREA / *EUTROPHICATION,	*MEASUREMENT, *SAMPLING, WATER RE	W72-03215
ATORY EQUIPMENT, *RADIOISOTOPES,	*MEASUREMENT, *KINETICS, *ABSORPT	W72-14326
*ENVIRON/ *MONITORING, *SURVEYS,	*MEASUREMENT, *DATA COLLECTIONS,	W72-07841
RY EQUIPMENT, *LABORATORY TESTS,	*MEASUREMENT, *MARINE ALGAE TRACE	W72-12210
OF ALGAE, *AQUATIC WEED CONTRCL,	*MECHANICAL CONTROL, *DEWATERING.	W71-06188
TENAN/ *CHRYSAORA QUINQUECIRRHA,	*MEDUSAE MAINTENANCE, *POLYP MAIN	W72-10162
CTR/ *STEPHANODISCUS BINDERANUS,	*MELOSIRA BINDERANA, SCANNING ELE	W73-00432
*ANAEROBIC BACTERIA, *ISOLATION,	*MEMBRANE PROCESSES, BACTERIA, SU	W72-04743
AST TWIN / *GLASS FIBER FILTERS,	*MEMBRANE FILTERS, GRANGE LAKE, E	W72-13673
*DISSOLVED ORGANIC CARBCN,	*MENAI STRAIT(WALES).:	W72-09108
EASE, BIOLOGICAL MAGNIFICATION,/	*MERCURY COMPOUNDS, *MINAMATA DIS	W72-12257
L PROTECTION AGENCY.:	*MERCURY POLLUTION, *ENVIRONMENTA	W72-14782

FLOS-AQUAE, PHRAGMITES COMMUNIS,	*NORWAY, MACROPHYTES, FYRISAN RIV	W73-00286
ARIA DIGITATA, AGARUM CRIBROSUM,	*NOVA SCOTIA, *ST. MARGARET'S BAY	W73-00366
RINE ALGAE, *FISHERIES, FISH PO/	*NUCLEAR EXPLOSIONS, *ALASKA, *MA	W73-00764
TION EFFECTS, *RADIOACTIVITY EF/	*NUCLEAR EXPLOSIONS, *WATER POLLU	W72-03347
EFFECTS, *ENVIRONMENTAL EFFECTS,	*NUCLEAR EXPLOSIONS, BIRDS, SALMO	W72-00960
ATOMACEOUS EARTH, *POROUS MEDIA/	*NUCLEAR EXPLOSION, *FILTERS, *DI	W72-00974
CHAINS, *RADIOACTIVITY EFFECTS,	*NUCLEAR POWERPLANTS, MONITORING,	W72-00959
ON PLANTS, *ELECTRIC POWER PROD/	*NUCLEAR POWERPLANTS, *DESALINATI	W72-03327
ITY EFFECT/ *PATH OF POLLUTANTS,	*NUCLEAR POWERPLANTS, *RADIOACTIV	W73-00831
ATER, *WATE/ *THERMAL POLLUTION,	*NUCLEAR POWER PLANTS, *COOLING W	W70-09612
ATER,/ *WATER POLLUTION CONTROL,	*NUCLEAR POWER PLANTS, *POTABLE W	W71-09018
S, WATE/ *RADIOACTIVITY EFFECTS,	*NUCLEAR WASTES, *FALLOUT, *OCEAN	W72-04463
MARINE ANIMALS, *MARINE FISH, M/	*NUCLEAR WASTES, *MARINE ALGAE, *	W73-00832
*MARINE ANIMALS, ASIA/ *REVIEWS,	*NUCLEAR WASTES, *BIOINDICATORS,	W73-00811
OD CHAIN/ *BACKGROUND RADIATION,	*NUCLEAR WASTES, *ABSORPTION, *FO	W73-00833
E ANIMALS, *CLAMS, *FOOD CHAINS,	*NUCLEAR WASTES, *COBALT RADIOISO	W73-00826
RADIOACTIVITY EFFECTS, *TRITIUM,	*NUCLEAR WASTES, WATER POLLUTION	W72-06342
CIFIC COAST REGION, TRITIUM, OC/	*NUCLEAR WASTES, *MONITORING, *PA	W72-03324
FFECTS, *THERMAL POLLUTION, CON/	*NUCLEAR WASTES, *RADIOACTIVITY E	W72-00939
ENTS, ALGAL CONTROL, *CYANPHYTA,	*NUISANCE ALGAE, PLANKTON, WASHIN	W72-00798
LUTION, WATER POLLUTION CONTROL,	*NUISANCE ALGAE, *EUTROPHICATION,	W72-00799
REVIEWS.: *CYANOPHYTA,	*NUISANCE ALGAE, *EUTROPHICATION,	W72-01358
GAL CONTROL, LABORATORY TESTS, /	*NUISANCE ALGAE, *CYANOPHYTA, *AL	W71-13184
*ALGAL CONTROL, CYANOPHYTA, CA/	*NUISANCE ALGAE, *CARBON DIOXIDE,	W72-10607
CONTROL, *DIFFUSION, *NUTRIENTS,	*NUISANCE ALGAE, WASHINGTON, PHYT	W72-08049
REVIEWS, TROPHI/ *ALGAL CONTROL,	*NUISANCE ALGAE, EUTROPHICATION,	W72-07937
*POISONOUS PLANTS, *CYANOPHYTA,	*NUISANCE ALGAE, *AQUATIC ALGAE,	W72-14056
S/ *CHORELLA, *ALGAE, *FOULING,	*NUISANCE ALGAE, FLOATING PLANTS,	W71-09261
OHIO).:	*NUTRIENT LEVELS, CUYAHOGA RIVER(	W72-04263
, CULTURES, *BIOMASS, NUTRIENTS,	*NUTRIENT REQUIREMENTS, NUISANCE	W72-03985
H RATES, *KINETICS, *PHOSPHORUS,	*NUTRIENT REQUIREMENTS, TEMPERATU	W72-04788
INETICS, *WASTE WATER TREATMENT,	*NUTRIENT REQUIREMENTS, TEMPERATU	W72-04789
*/ *ALGAE, *KINETICS, *NITROGEN,	*NUTRIENT REQUIREMENTS, AMMONIA,	W72-04787
, RESPIRATORY RATES, PHOTOSYNTH/	*NUTRIENT REMOVAL, OXYGEN DECLINE	W71-04079
TA, CY/ *EUTROPHICATION, *ALGAE,	*NUTRIENT REQUIREMENTS, CHLOROPHY	W70-05469
, NITRATES/ *DIATOMS, *CULTURES,	*NUTRIENT REQUIREMENTS, SEA WATER	W73-00441
FACTORS, *PHOSPHATES, DIATOMS,/	*NUTRIENT REQUIREMENTS, *LIMITING	W73-01445
CTION, LAKE MENDOTA(WIS), LAKE /	*NUTRIENT SOURCES, ACETYLENE REDU	W72-10608
(ONTARIO), NUTRIENT INPUTS, CHI/	*NUTRIENT BUDGETS, *BAY OF QUINTE	W71-11009
, BIOINDICATORS/ *PHYTOPLANKTON,	*NUTRIENT REQUIREMENTS, NUTRIENTS	W72-13816
NKTON, *AQUATIC AL/ *MINERALOGY,	*NUTRIENT REQUIREMENTS, *PHYTOPLA	W72-12739
CENTRAL BASIN./ *DEOXYGENATION,	*NUTRIENT REGENERATION, LAKE ERIE	W72-12996
I/ *AQUATIC ALGAE, *COMPETITION,	*NUTRIENTS, *GROWTH RATES, *INHIB	W72-12734
ION, MATHEMA/ *DISSOLVED OXYGEN,	*NUTRIENTS, *LAKE ERIE, *HYPOLIMN	W72-12997
, WATER QUALIT/ *EUTROPHICATION,	*NUTRIENTS, LAKE HURON, EFFLUENTS	W72-13662
AKES, N/ *PHYTOPLANKTON, *ALGAE,	*NUTRIENTS, *NITROGEN FIXATION, L	W72-13692
WTH RATES, *SEAWATER, *TOXICITY,	*NUTRIENTS, LETHAL LIMITS, METHOD	W72-12239
TORS, PHOSPHAT/ *EUTROPHICATION,	*NUTRIENTS, *ALGAE, *LIMITING FAC	W72-12544
PHICATION, LAKE ERIE, ALGAE, NI/	*NUTRIENTS, *LAKE ONTARIO, *EUTRO	W71-11009
BORATORY TESTS, *EUTROPHICATION,	*NUTRIENTS, *ALGAE, SEWAGE, PRODU	W71-12857
ER POLLUTION EFFECTS, *NITRATES,	*NUTRIENTS, *ALGAE, *BACTERIA, AE	W71-09958
S, *TOXICITY, *ACTIVATED SLUDGE,	*NUTRIENTS, NITROGEN, WATER POLLU	W71-11972
N, *WA/ *ALGAE, *EUTROPHICATION,	*NUTRIENTS, *PLANKTON, *RECREATIO	W72-10431
EPARATION TECHNIQUES, FERMENTAT/	*NUTRIENTS, *MUNICIPAL WASTES, *S	W72-10390
OSPHORU/ *ANALYTICAL TECHNIQUES,	*NUTRIENTS, *ALGAE, *BIOASSAY, PH	W72-10608
T IN/ *PHYTOPLANKTON, *NITROGEN,	*NUTRIENTS, *PHOTOSYNTHESIS, LIGH	W72-09103
ENT, WATER POL/ *EUTROPHICATION,	*NUTRIENTS, *ALGAE, SEWAGE TREATM	W72-10625
YMBIOSI/ *WASTE WATER TREATMENT,	*NUTRIENTS, *ALGAE, *BACTERIA, *S	W72-11100
POLLUTION EFFECTS, EUTROPHICAT/	*NUTRIENTS, *PHYTOPLANKTON, WATER	W72-09167
IOASSAY, OREGON, LAKES,/ *ALGAE,	*NUTRIENTS, *LABORATORY TESTS, *B	W72-09164
*EUTROPHICATION, *ON-SITE TESTS,	*NUTRIENTS, ALGAE, LAKES, PHOSPHO	W72-09165
LUTION EFFECTS, *ORGANIC WASTES,	*NUTRIENTS, *POLLUTANT IDENTIFICA	W72-01472
, *ALGAL BLOOMS, *PHYTOPLANKTON,	*NUTRIENTS, WATER POLLUTION EFFEC	W72-01329
, ANALYTICAL TECHNIQUES, *ALGAE,	*NUTRIENTS, LAKES, PLANT GROWTH,	W72-00801
EFFECTS, *EUTROPHICATION, LAKES,	*NUTRIENTS, ALGAL CONTROL, *CYANP	W72-00798
Y PRODUCTIVITY, *EUTROPHICATION,	*NUTRIENTS, *BIOLOGICAL COMMUNITI	W72-03216
MENT, PLANKTON, *EUTROPHICATION,	*NUTRIENTS, MICROORGANISMS, EFFLU	W71-13553
TESTS, *RESPIRATION, *CHLORELLA,	*NUTRIENTS, CULTURES, NITROGEN, P	W73-01626
ROL,/ *REVIEWS, *EUTROPHICATION,	*NUTRIENTS, *WATER POLLUTION CONT	W73-00840
SEDIMENTATION, *EUTROPHICATION/	*NUTRIENTS, *DREDGING, DESILTING,	W73-00938
GAN, *EUTROPHICATION/ *SAMPLING,	*NUTRIENTS, *DIATOMS, *LAKE MICHI	W73-00384
CTORS, / *ALGAE, *AQUATIC WEEDS,	*NUTRIENTS, BIOASSAY, LIMITING FA	W73-00232
ING/ *ALGAL CONTROL, *DIFFUSION,	*NUTRIENTS, *NUISANCE ALGAE, WASH	W72-08049
OLIGOTR/ *ALGAE, *GROWTH RATES,	*NUTRIENTS, CULTURES, INHIBITION,	W72-08376
ORGANISMS, *INORGANIC COMPOUNDS,	*NUTRIENTS, FRESH WATER, ALGAE, B	W72-07933
*PLANKTON, *BIOMASS, *SAMPLING,	*NUTRIENTS, *HYDROGRAPHY, *PACIFI	W72-07892
Y, *ST/ *EUTROPHICATION, *LAKES,	*NUTRIENTS, *FRESHWATER, *BIOASSA	W72-07648
ER POLLUTION SOURCES, *ESTUARIE/	*NUTRIENTS, *EUTROPHICATION, *WAT	W70-06509
ER QUALITY CONTROL, *NITROGEN C/	*NUTRIENTS, *EUTROPHICATION, *WAT	W70-04080
UNDED WATERS, STREAMS, FLOW, SE/	*NUTRIENTS, *WATER QUALITY, *IMPO	W71-03028
CUTTING MANAGEMENT, *HERBICIDES,	*NUTRIENTS, *ECOSYSTEMS, *WATERSH	W71-01489
EUTROPHICATION, *ALGAL BLOOMS,	*NUTRIENTS, *PHOSPHORUS, *NITROGE	W70-09669
EFFECTS, *PHOSPHATES, *NITRATES,	*NUTRIENTS, *EUTROPHICATION, ALGA	W70-09388
IMNOLOGY, *ALGAE, *EUTROPHICATI/	*NUTRIENTS, *BOTTOM SEDIMENTS, *L	W70-08944
N DIOXIDE, PHOSPHATES, NITRATES,	*NUTRIENTS, ALKALINITY, CHEMICAL	W72-04983
*SEDIM/ *TROPHIC LEVEL, *LAKES,	*NUTRIENTS, *CHEMICAL PROPERTIES,	W72-05469
ON, *PHOSPHORUS, *NITROGEN, EVA/	*NUTRIENTS, *ALGAE, *EUTROPHICATI	W72-04070
HATES, *GREAT LAKES, *SEDIMENTS,	*NUTRIENTS, EUTROPHICATION, NITRO	W72-04263
TION CONTROL, *SEA WATER, *OILS,	*NUTRIENTS, AIRCRAFT, SATELLITES(	W72-04742
HIC FLORA, *PHYTOPLANKTON, ALGA/	*NUTRIENTS, *BENTHIC FAUNA, *BENT	W72-06046

*SPECIES DIVERSITY,	*POTOMAC RIVER.:	W71-11583
	*POTOMAC RIVER ESTUARY.:	W71-08775
	*POTOMAC RIVER.:	W70-08248
*SURVEYS, *STATISTICAL METHODS,	*POTOMAC RIVER, DATA COLLECTIONS,	W72-07791
RIENTS, AQUATIC AL/ *FARM WASTE,	*POULTRY, *OXIDATION LAGCCNS, NUT	W72-02850
POLLUTION, *ENVIRONM/ *NEW YORK,	*POWER PLANTS, *STREAM, *THERMAL	W72-14659
, *THERMAL STRATIFICATION, HYDR/	*POWER PLANTS, *THERMAL POLLUTION	W70-09614
, *REVIEWS, WATER TEMPERATURE, /	*POWER PLANTS, *THERMAL PCLLUTION	W71-07417
TER CONSE/ *THERMAL POWERPLANTS,	*POWERPLANTS, *ENERGY LOSSES, *WA	W70-09193
SPECIFIC RESI/ *MICROSTRAINING,	*PRECOAT FILTRATION, BACKWASHING,	W72-09393
TEMS, FOOD CHAINS, *ZOOPLANKTON,	*PREDATION, *BIOASSAY, BIOLOGICAL	W72-07132
*MARINE BACTERIA, *ATTRACTANTS,	*PREDATION, ORGANIC COMPOUNDS, AM	W73-00942
TS, *SAMPLING, *ECOSY/ *ECOLOGY,	*PREIMPOUNDMENTS, *AQUATIC HABITA	W72-06526
O/ *CHEMORECEPTION, *CHEMOTAXIS,	*PREY, PYTHIUM DEBARYANUM, SKELET	W73-00942
ALS, *TR/ *ALGAE, *GROWTH RATES,	*PRIMARY PRODUCTIVITY, *HEAVY MET	W73-01434
*WATER ANALYSIS, *STANDING CROP,	*PRIMARY PRODUCTIVITY, LAKES, SAM	W73-01435
LGAE, *COMPARATIVE PRO/ *STRESS,	*PRIMARY PRODUCTIVITY, *AQUATIC A	W73-01080
R, *MARSHES, SALINE WATER, BIOM/	*PRIMARY PRODUCTIVITY, *FRESHWATE	W73-01089
*MARINE ALGAE, *BENTHIC.FLORA,	*PRIMARY PRODUCTIVITY, *ENERGY CO	W72-06283
S, *ARCTIC, *ALGAE.:	*PRIMARY PRODUCTIVITY, *OIL WASTE	W72-07922
TES, *PLANKT/ *ALGAE, *PROTOZOA,	*PRIMARY PRODUCTIVITY, INVERTEBRA	W72-07899
LORA, MICROORGANISMS, ALGAE, LA/	*PRIMARY PRODUCTIVITY, *BENTHIC F	W72-07501
N/ *PHYTOPLANKTON, *THERMOCLINE,	*PRIMARY PRODUCTIVITY, *NERITIC,	W73-00431
BENTHIC/ *ESTUARINE ENVIRONMENT,	*PRIMARY PRODUCTIVITY, *SHOALS, *	W72-14317
ATION, *BIOMASS/ *PHYTOPLANKTON,	*PRIMARY PRODUCTIVITY, *EUTROPHIC	W72-13871
SCENEDESMUS, GROWTH RATES, CULT/	*PRIMARY PRODUCTIVITY, BACTERIA,	W72-14314
GET, *MARINE ALG/ *GROWTH RATES,	*PRIMARY PRODUCTIVITY, ENERGY BUD	W73-00366
*MARINE ALGAE, DDT, *PESTICIDES,	*PRIMARY PRODUCTIVITY, WATER POLL	W72-09653
, CHLOR/ *ALGAE, *PHYTOPLANKTON,	*PRIMARY PRODUCTIVITY, MICROSCOPY	W72-09239
ORDS, SURFACE / *CARBON DIOXIDE,	*PRIMARY PRODUCTIVITY, ALASKA, FI	W72-09122
VITY TECHNIQU/ *LIGHT INTENSITY,	*PRIMARY PRODUCTIVITY, *RADIOACTI	W72-09102
AL MODELS, *ESSENTIAL NUTRIENTS,	*PRIMARY PRODUCTIVITY, WATER QUAL	W72-08986
ATION, *NUTRIENTS, *BIOLOGICAL /	*PRIMARY PRODUCTIVITY, *EUTROPHIC	W72-03216
KTON, *SEA WATER, COASTS, BIORT/	*PRIMARY PRODUCTIVITY, *PHYTOPLAN	W72-03228
D HYDROC/ *INSECTICIDES, *ALGAE,	*PRIMARY PRODUCTIVITY, CHLORINATE	W71-10096
CROP, *ALGAE, DIATOMS, MEASUREM/	*PRIMARY PRODUCTIVITY, *STANDING	W71-12855
*WATER QUALITY, *EUTROPHICATION,	*PRIMARY PRODUCTIVITY, *ALGAE, NI	W71-13183
PHYTOPLANKTON, *ESTUARIES, *STA/	*PRIMARY PRODUCTIVITY, *COASTS, *	W71-07875
GAE, *AQUATIC ANIMALS, *BIOMASS,	*PRIMARY PRODUCTIVITY, *SECONDARY	W72-05968
S, *NEBRASKA, *EUTROPHICATION, /	*PRIMARY PRODUCTIVITY, *RESERVOIR	W72-04270
, *PHOTOSYNTHESIS,/ *RESERVOIRS,	*PRIMARY PRODUCTIVITY, *TURBIDITY	W72-12393
PIGMENTS, AQUATIC PRODUCTIVITY,	*PRIMARY PRODUCTIVITY, *ALGAE, *D	W72-12192
NANNOPLANKTON, LAKES, *SEASONAL,	*PRIMARY PRODUCTIVITY, AQUATIC AL	W72-12738
MENT, *DOMESTIC WA/ *ECOSYSTEMS,	*PRODUCTIVITY, *ESTUARINE ENVIRON	W72-12567
*PHOTOSYNTHESIS, *PHYTOPLANKTON,	*PRODUCTIVITY, ALGAE, NITROGEN, P	W71-08670
TANCES, / *ALGAE, *GROWTH RATES,	*PRODUCTIVITY, *PLANT GROWTH SUBS	W71-08687
ONT/ *ECOSYSTEMS, *SALT MARSHES,	*PRODUCTIVITY, RADIOISOTOPES, STR	W71-09254
OMASS, / *PERIPHYTON, *PLANKTON,	*PRODUCTIVITY, PHOTOSYNTHESIS, BI	W71-09165
FOOD/ *MARINE ALGAE, *ECONOMICS,	*PRODUCTIVITY, ALGAE, RESOURCES,	W71-13246
ESTRIAL HABITATS, *HEAVY METALS,	*PRODUCTIVITY, *ECOSYSTEMS, POTAS	W72-14303
YTOPLANKTON, *LAKES, *TURBIDITY,	*PROFILES, PHOTOMETER, HYPOLIMNIO	W72-07145
OOCYSTIS, OSCILLATORIA, SPIRULI/	*PROTEASE ENZYMES, CHLOROGONIUM,	W72-07661
AL PROPERTIES, ANTICANCER, ANTI/	*PROTHROMBIN TIME, *PHARMACOLOGIC	W72-07360
INVERTEBRATES, *PLANKT/ *ALGAE,	*PROTOZOA, *PRIMARY PRODUCTIVITY,	W72-07899
S, *ENVIRONMENTA/ *SALT MARSHES,	*PROTOZOA, *BIOLOGICAL CCMMUNITIE	W73-00367
UTROPHICATION, PONDS,/ *GRAZING,	*PROTOZOA, *ALGAE, SCENEDESMUS, E	W72-14796
D/ *MARINE ALGAE, *SALT MARSHES,	*PROTOZOA, *BACTERIA, FLORIDA, RA	W73-00796
ARSHE/ *WATER POLLUTION EFFECTS,	*PROTOZOA, *MARINE ALGAE, *SALT M	W72-00948
WATER POLLUTION EFFECTS, *ALGAE,	*PROTOZOA, *BIOLOGICAL CCMMUNITIE	W72-08579
CHLOROPHYTA, HABITATS, AQUATIC /	*PROTOZOA, *SYSTEMATICS, *ALGAE,	W72-09119
TS, *DISPERSION, CADDIS/ *ALGAE,	*PROTOZOA, *FUNGI, *AQUATIC INSEC	W70-06655
PHICATION, *ECONOMIC EFFICIENCY,	*PUBLIC HEALTH, WATER POLLUTION,	W72-04266
QUES, / *AQUATIC PLANTS, *ALGAE,	*PUBLIC HEALTH, ANALYTICAL TECHNI	W72-04275
ATHOGENIC BACTERIA, CONFERENCES,	*PUBLIC HEALTH, RIVERS, *SEWAGE,	W72-09675
POLLUTION EFFECTS, T/ *MERCURY,	*PUBLIC HEALTH, *SHELLFISH, WATER	W72-12257
POLLUTION, *PATH OF POLLUTANTS,	*PUBLIC HEALTH, ENZYMES, ANALYTIC	W72-13347
ION, *WATER POLLUTION TREATMENT,	*PUBLIC HEALTH, ANALYTICAL TECHNI	W73-02144
ORNE DISEASES, *TOXIC CHEMICALS,	*PUBLIC HEALTH STANDARDS, TOXICOL	W73-02144
N, COAGULATION, BENTONITE, LIME,	*PULP WASTES, TEOTIARY TREATMENT,	W72-03753
*MODEL S/ *ECOSYSTEMS, *ECOLOGY,	*PUMPED STORAGE, *WATER QUALITY,	W72-05282
MOUNTAIN RESERVOIR(VA).:	*PUMPED STORAGE OPERATION, SMITH	W72-14405
PROP/ *WATER LEVEL FLUCTUATIONS,	*PUMPED STORAGE, PHYSICOCHEMICAL	W72-14280
HYTIN, TUNICATES, FORAMINIFERA,/	*PYCNOCLINE, CHLOROPHYLL A, PHEOP	W72-14313
AE, *CHLOROPHYTA, *EUGLENOPHYTA,	*PYRROPHYTA, *SYSTEMATICS, DINOFL	W72-07683
ECONOMIC JUSTIFICATION, *COBALT,	*RADIATION, COSTS, SEWAGE BACTERI	W72-00383
TON, PR/ *ANALYTICAL TECHNIQUES,	*RADIOACTIVITY TECHNIQUES, *PLANK	W72-00161
POWERPLANTS, MONI/ *FOOD CHAINS,	*RADIOACTIVITY EFFECTS, *NUCLEAR	W72-00959
MS, *RAIN FORESTS, CYCLING NUTR/	*RADIOACTIVITY EFFECTS, *ECOSYSTE	W72-00949
ENTAL EFFECTS, *NUCLEAR EXPLOSI/	*RADIOACTIVITY EFFECTS, *ENVIRONM	W72-00960
POLLUTION, CON/ *NUCLEAR WASTES,	*RADIOACTIVITY EFFECTS, *THERMAL	W72-00939
ANIMALS, *ALGAE, *INVERTEBRATES,	*RADIOACTIVITY EFFECTS, *RADIOISO	W71-13233
SIONS, *WATER POLLUTION EFFECTS,	*RADIOACTIVITY EFFECTS, UNDERGROU	W72-03347
NTENSITY, *PRIMARY PRODUCTIVITY,	*RADIOACTIVITY TECHNIQUES, *PHOTO	W72-09102
LOGY, *RESERVOI/ *RADIOISOTOPES,	*RADIOACTIVITY EFFECTS, *RADIOECO	W72-10694
IC FLORA, *GROWTH RATE, *AQUATI/	*RADIOACTIVITY TECHNIQUES, *BENTH	W71-11486
ODEGRADATION, *METABOLISM, *DDT,	*RADIOACTIVITY TECHNIQUES, PESTIC	W72-07703
NG, *MARINE ALGAE, RHODOPHYTA, /	*RADIOACTIVITY EFFECTS, *MONITORI	W72-07826
S, *ENVIRONMENT, *RADIOACTIVITY,	*RADIOACTIVITY EFFECTS, RIVERS, E	W72-07841
*DATA COLLECTIONS, *ENVIRONMENT,	*RADIOACTIVITY, *RADIOACTIVITY EF	W72-07841
*NUCLEAR WASTES,/ *FOOD CHAINS,	*RADIOACTIVITY EFFECTS, *TRITIUM,	W72-06342

, *ANACYSTIS NIDULANS, *MUTANTS,	*SENSITIVITY, CHLCRAMPHENICOL, RE	W72-14327
OOPLANKTON, *LARVAE, *FISH EGGS,	*SEPARATION TECHNIQUES, *MARINE F	W73-00371
BACTERIA, *FUNGI, *BACTERIOPHAG/	*SEPARATION TECHNIQUES, *ALGAE, *	W72-06124
N LAGOONS, *HARVESTING OF ALGAE,	*SEPARATION TECHNIQUES, CHEMICAL	W72-10237
*NUTRIENTS, *MUNICIPAL WASTES,	*SEPARATION TECHNIQUES, FERMENTAT	W72-10390
KTON, *CHLOROPHYLL, *EVALUATION,	*SEPARATION TECHNIQUES, SPECTROPH	W72-13673
MS, *PERIPHYTON, *PHYTOPLANKTON,	*SESSILE ALGAE, *LAKE SUPERIOR, L	W72-12192
T. LAWRENC/ *BENTHOS, *SAMPLING,	*SESSILE ALGAE, *INVERTEBRATES, S	W72-05432
TIDAL EFF/ *ESTUARIES, *DIATOMS,	*SESSILE ALGAE, LIGHT INTENSITY,	W73-00853
, *NITROGEN, PHOSPHORUS, CARBON,	*SEWAGE BACTERIA, SEWAGE, ALGAE,	W72-04309
EATMENT, *SEWAGE LAGOONS, *COLD/	*SEWAGE BACTERIA, *WASTE WATER TR	W72-12565
TION, BIOCHEMICAL OXYGEN DEMAND,	*SEWAGE EFFLUENTS, *SEWAGE TREATM	W72-06612
WATER TREATMENT, A/ *PHOSPHORUS,	*SEWAGE EFFLUENTS, *ALGAE, WASTE	W72-09176
ENT, *ECONOMIC JUSTIFICATION, */	*SEWAGE EFFLUENTS, *SEWAGE TREATM	W72-00383
OCLINE/ *EUTROPHICATION, *ALGAE,	*SEWAGE EFFLUENTS, GEORGIA, THERM	W71-11914
TANS, ZOOGLOEA/ *UNITED KINGDOM,	*SEWAGE FUNGUS, SPHAEROTCTILUS NA	W72-08573
, *OXIDATION LAGOONS, *DESIGN C/	*SEWAGE LAGOONS, *AERATED LAGOONS	W72-09386
S, ALGAE, PRODUCTIVITY/ *CARBON,	*SEWAGE LAGOONS, *LIMITING FACTOR	W72-09162
ERGENTS, *ALGAE, *PHYTOPLANKTON,	*SEWAGE LAGOONS, BIODEGRADATION,	W72-07661
ACTERIA, *WASTE WATER TREATMENT,	*SEWAGE LAGOONS, *COLD REGIONS, A	W72-12565
*METABOLISM, *SEWAGE TREATMENT,	*SEWAGE LAGOONS, OHIO, CXYGEN, DI	W72-05459
ROPHICATION, *OXIDATION LAGOONS,	*SEWAGE LAGOONS, BIODEGRADATION,	W71-07973
MENT, BIOCHEMI/ *SURFACE RUNOFF,	*SEWAGE TREATMENT, TERTIARY TREAT	W71-06445
S,/ *PHYTOPLANKTON, *METABOLISM,	*SEWAGE TREATMENT, *SEWAGE LAGOON	W72-05459
ES, CHEMIC/ *NITROGEN COMPOUNDS,	*SEWAGE TREATMENT, *DOMESTIC WAST	W70-09907
REATMENT, *TERTIARY TREATMENT, /	*SEWAGE TREATMENT, *WASTE WATER T	W70-10381
XYGEN DEMAND, *SEWAGE EFFLUENTS,	*SEWAGE TREATMENT, CULTURE, *WAST	W72-06612
TE WATER TREATMENT, *FILTRATION,	*SEWAGE TREATMENT, ALGAE, *AERATI	W72-14469
*EUTROPHICATION, *LAKES, *WATE/	*SEWAGE TREATMENT, *EFFICIENCIES,	W72-11702
EUTROPHICATION, *WATER POLLUTIO/	*SEWAGE TREATMENT, *PHOSPHATES, *	W71-11507
IFICATION, */ *SEWAGE EFFLUENTS,	*SEWAGE TREATMENT, *ECONOMIC JUST	W72-00383
ES, *EUTROPHICATION, *CHELATION,	*SEWAGE, *ORGANIC COMPOUNDS, WATE	W71-09674
AE, *ESTUARIES, *WASTE DISPOSAL,	*SEWAGE, *INDUSTRIAL WASTES, WATE	W72-08804
CAL COMMUNITIES, RIVERS, BACTER/	*SEWAGE, *FUNGI, *SLIME, *BIOLOGI	W72-08573
AQUATIC ENVIRONMENT, CHEMICAL /	*SEWAGE, *EUTROPHICATION, *LAKES,	W71-00117
MIDGES, LAGOONS/ *ENERGY BUDGET,	*SEWAGE, *AQUATIC PRODUCTIVITY, *	W72-03556
ERENCES, *PUBLIC HEALTH, RIVERS,	*SEWAGE, BIOCHEMICAL OXYGEN DEMAN	W72-09675
TS, T/ *MERCURY, *PUBLIC HEALTH,	*SHELLFISH, WATER POLLUTION EFFEC	W72-12257
IRONMENT, *PRIMARY PRODUCTIVITY,	*SHOALS, *BENTHIC FLORA, SEA WATE	W72-14317
ION, SOUTH WALES, MICROCYSTIS A/	*SHOCK-WAVES, *GAS-VACUOLE DEFLAT	W72-11564
T, *MARINE FISH, *PHYTOPLANKTON,	*SHRIMP, *BIOASSAY, *ABSORPTION,	W73-02105
STAGE, *DETERMINATION, *ORIGIN,	*SIGNIFICANCE, *EXTRACTION, LUCIF	W72-04292
ECTROPHORESIS, SETTLING, *ALGAE,	*SILICA, *BACTERIA, WATER POLLUTI	W71-01899
ATER TRE/ *FLOCCULATION, *ALGAE,	*SILICA, *E. COLI, *IONS, *WASTEW	W72-14840
*PLUTONIUM,	*SILVER RADIOISOTOPES.:	W72-07826
ARCINA, ANKISTRODESMUS BRAUNII,/	*SIMAZINE, *DETOXICATION, CHLOROS	W71-07675
*SOIL SURFACES, *SEDIMENT YIELD,	*SIMULATED RAINFALL, *SOIL TYPES,	W72-02218
AHALINE, HYPERHALINE, EURYHALIN/	*SINAI PENINSULA, CHLORINITY, MET	W73-00855
DISPERSANTS, *COLLECTING AGENTS,	*SINKING AGENTS, EFFECTIVENESS, T	W71-09789
DIFFUS/ *ANTIMICROBIAL ACTIVITY,	*SKELLYSOLVE, *FILTER PAPER DISC	W72-08586
RIVERS, BACTER/ *SEWAGE, *FUNGI,	*SLIME, *BIOLOGICAL COMMUNITIES,	W72-08573
	*SLIMICIDES.:	W72-08683
*EUTROPHICATION, *BIOCHEMICAL /	*SLUDGE WORMS, *DISSOLVED OXYGEN,	W70-09189
ROBIC CONDI/ *OXIDATION LAGOONS,	*SLUDGE, ANAEROBIC CONDITIONS, AE	W71-07102
NMENT, WATERSHEDS(BASINS), *DDT,	*SMALL WATERSHEDS, ECOSYSTEMS, PE	W72-12930
TION, WATER POLLUTION EFFECTS, /	*SNAILS, *FROGS, *TRITIUM, ABSORP	W72-03348
TIUM RADIOISOTOPES, *ABSORPTION,	*SNAILS, GASTROPODS, PATH OF POLL	W72-10978
RGANIC CARBON, *AMERI/ *GLUCOSE,	*SNAKE RIVER(IDAHO), *DISSOLVED O	W72-13300
	*SNAKE RIVER(WASH).:	W72-14728
LUTION CONTROL/ *EUTROPHICATION,	*SOCIAL PARTICIPATION, *WATER POL	W73-02479
ALT TOLERANCE, WATER POLL/ *DDT,	*SODIUM CHLORIDE, *INHIBITION, *S	W72-13809
IMENT YIELD, *SIMULATED RAINFAL/	*SOIL ALGAE, *SOIL SURFACES, *SED	W72-02218
S, *SYSTEMATICS, *AQUATIC ALGAE,	*SOIL BACTERIA, SAMPLING, CULTURE	W73-00854
SPHORUS, *EUTROPHICATION, LAKES/	*SOIL BACTERIA, *LAKE SOILS, *PHO	W72-04292
, *CARBON DIOXIDE, *RADIOACTIVI/	*SOIL MICROORGANISMS, *ABSORPTION	W72-14301
*SIMULATED RAINFAL/ *SOIL ALGAE,	*SOIL SURFACES, *SEDIMENT YIELD,	W72-02218
MENT YIELD, *SIMULATED RAINFALL,	*SOIL TYPES, SOIL TEXTURE, CLAYS,	W72-02218
*FALLOUT, *CYCLING NUTRIENTS, E/	*SOIL-WATER-PLANT RELATIONSHIPS,	W72-03342
S, / *ALGAE, *NITROGEN FIXATION,	*SOILS, *TEMPERATE, CLIMATIC ZONE	W73-02471
.:	*SOLAR INSOLATION, *STOICHIOMETRY	W72-02363
CULTURES.: *PHOSPHATES,	*SOLUBILITY, *CYANOPHYTA, ALGAE,	W71-10070
OLOGY, RESERVOIR SILTING, LAKES,	*SOUTH DAKOTA, PLANKTON, *ALGAE,	W73-00938
	*SOUTH RIVER(VA).:	W73-01972
A COSTATUM, TH/ *COASTAL WATERS,	*SOUTHERN LONG ISLAND, SKELETONEM	W71-07875
T, *CENTRAL HEATING AND COOLING,	*SPACE HEATING, *FARM USES OF WAS	W70-09193
*DEPTH, MARINE AN/ *CHLOROPHYLL,	*SPATIAL DISTRIBUTION, *DENSITY,	W72-14313
ATICS, *ECOLOGICAL DISTRIBUTION,	*SPATIAL DISTRIBUTION, GEOGRAPHIC	W73-00428
*LAKE MICHIGAN, *EUTROPHICATION,	*SPATIAL DISTRIBUTION, PHOSPHORUS	W73-00384
ULATION, *TEMPORAL DISTRIBUTION,	*SPATIAL DISTRIBUTION, ENVIRONMEN	W72-08560
.:	*SPECIES DIVERSITY, CRYSTAL RIVER	W72-11252
R.:	*SPECIES DIVERSITY, *POTOMAC RIVE	W71-11583
ENTIFIC/ *CALCIUM, *FOOD CHAINS,	*SPECTROPHOTOMETRY, *POLLUTANT ID	W71-11529
ANABAENA VARIABILIS,/ *LYSOZYME,	*SPHEROPLASTS, BLUE-GREEN ALGAE,	W71-08683
*MET/ *CLADOPHORA, *ACHNANTHES,	*SPIROGYRA, CLODOPHORA GLOMERATA,	W72-07901
A, *AMPHIPODA, *F/ *CHLOROPHYTA,	*SPRINGS, DIATOMS, BIOTA, PROTOZO	W72-12744
ON SOURCES, WATE/ *INSECTICIDES,	*ST LAWRENCE RIVER, WATER POLLUTI	W70-08098
OXYGEN DEMAND, HYDROGEN ION CON/	*ST. LAWRENCE RIVER, BIOCHEMICAL	W70-10427
AGARUM CRIBROSUM, *NOVA SCOTIA,	*ST. MARGARET'S BAY.: / DIGITATA,	W73-00366
REATMENT, B/ *OXIDATION LAGOONS,	*STABILIZATION, *ALGAE, AEROBIC T	W71-03896
ERRIC CHLORIDE, *FERRIC SULFATE,	*STABILIZATION PONDS, ALUM, POST-	W71-13356

OEIC BACTERIA, *TEXAS, / *ALGAE, *STABLE ISOTOPES, *CARBON, *ANAER W71-04873
OTOPES, ENVIR/ *ISOTOPE STUDIES, *STABLE ISOTOPES, *CARBON RADIOIS W72-12709
EASONAL, / *TEMPERATURE CONTROL, *STANDARDS, *WATER TEMPERATURE, S W70-09595
STS, *PHYTOPLANKTON, *ESTUARIES, *STANDING CROP, SEASONAL, DIATOMS W71-07875
ENERGY BUDGET, PRODUC/ *STREAMS, *STANDING CROP, *TROPHIC LEVEL, * W71-10064
MEASUREM/ *PRIMARY PRODUCTIVITY, *STANDING CROP, *ALGAE, DIATOMS, W71-12855
UNITIES, *ENVIRONMENTAL EFFECTS, *STANDING CROPS, TIDAL EFFECTS, M W73-00367
TRIENTS, *FRESHWATER, *BIOASSAY, *STANDING CROP, CARBON, NITROGEN, W72-07648
*PHYTOPLANKTON, *WATER ANALYSIS, *STANDING CROP, *PRIMARY PRODUCTI W73-01435
ION, POYELECTROLYTES.: *STARCH REMOVAL, STARCH SUBSTITUT W72-03753
VER, DATA COLLECTIONS/ *SURVEYS, *STATISTICAL METHODS, *POTOMAC RI W72-07791
OLOGICAL STUDIE/ *MODEL STUDIES, *STATISTICAL METHODS, *ALGAE, CYT W71-12868
M LONGIPES VAR CONTRACTUM, ARTH/ *STAURASTRUM, DESMIDS, STAURASTRU W73-01094
NANS, *NITZSCHIA ACTINASTROIDES, *STAURASTRUM PINGUE, SCENEDESMUS W72-12739
LLUTION CONTROL, *WATER COOLING, *STEAM TURBINES, *THERMAL POLLUTI W70-09193
SIRA BINDERANA, SCANNING ELECTR/ *STEPHANODISCUS BINDERANUS, *MELO W73-00432
*SOLAR INSOLATION, *STOICHIOMETRY.: W72-02363
, *RECREATION, *TREATMENT FACIL/ *STORM RUN-OFF, *SUSPENDED SOLIDS W72-00042
C BACTERIA, *POPULATION, *ALGAE, *STRATIFICATION, *ENZYMES, THERMO W72-03224
IRONM/ *NEW YORK, *POWER PLANTS, *STREAM, *THERMAL POLLUTION, *ENV W72-14659
*ALGAE, *BIOLOGICAL COMMUNITIES, *STREAMS, LAKE ONTARIO, PLANKTON, W73-01627
LLUTION EFFECTS, *MODEL STUDIES, *STREAMS, *ECOSYSTEMS, AQUATIC HA W72-06340
C LEVEL, *ENERGY BUDGET, PRODUC/ *STREAMS, *STANDING CROP, *TROPHI W71-10064
CREATI/ *SURFACE WATERS, *LAKES, *STREAMS, *REVIEWS, *ILLINOIS, RE W72-05869
LVE/ *SELF-PURIFICATION, *ALGAE, *STREAMS, LABORATORY TESTS, DISSO W71-04518
AQUATIC ALGAE, *COMPARATIVE PRO/ *STRESS, *PRIMARY PRODUCTIVITY, * W73-01080
ICROORGANISMS, *ACID MINE WATER, *STRIP MINES, *WATER POLLUTION EF W71-07942
POLLUTION EFFECTS, NUCLEAR WAST/ *STRONTIUM RADIOISOTOPES, *WATER W71-09225
T, *FOOD CHAINS, / *GREAT LAKES, *STRONTIUM RADIOISOTOPES, *FALLOU W73-00790
ANIMALS, *MARINE ALGAE, *FOOD / *STRONTIUM RADIOISOTOPES, *MARINE W73-00819
*FLAME EMISSION SPECTROPHOTOME/ *STRONTIUM, *SAMPLE PRESERVATION, W71-11529
TION, *SNAILS, GASTROPODS, PATH/ *STRONTIUM RADIOISOTOPES, *ABSORP W72-10978
ADIOISOTOPES, AB/ *MARINE ALGAE, *STRONTIUM RADIOISOTOPES, *ZINC R W72-12203
RIA DE ALDAN(SPAIN), LAMINARIA/ *SUBLITTORAL BENTHIC MACROPHYTES, W72-04766
ODES, ALGAE, OLIGO/ *PERIPHYTON, *SUBMERGED PLANTS, BIOMASS, NEMAT W71-11012
MA, ACHNANTHES / *LAKE WINNIPEG, *SUBSTRATES, ACHNANTHES MINUTISSI W72-13142
SEA, COCCOLITHOPHORIDS, FLAGEL/ *SUBSTRATE UTILIZATION, *SARGASSO W73-00380
OROPHYLL, PL/ *LAKES, *PIGMENTS, *SUCCESSION, *EUTROPHICATION, CHL W73-00916
YSTEMS, *BIOLOGICAL COMMUNITIES, *SUCCESSION, AQUATIC ENVIRONMENT, W72-11854
RANEAN SE/ *CERATIUM EGYPTIACUM, *SUEZ CANAL, FLAGELLATES, MEDITER W73-00296
ENT-WATER IN/ *AQUATIC BACTERIA, *SULFUR BACTERIA, *OXYGEN, *SEDIM W72-06690
SPENDED SOLIDS.: *SURFACE CHARGE, REGENERATION, SU W71-00138
T, TERTIARY TREATMENT, BIOCHEMI/ *SURFACE RUNOFF, *SEWAGE TREATMEN W71-06445
ES, MINERALIZATION, NITROGEN SO/ *SURFACE WATERS, GEOLOGICAL SOURC W71-06435
C WEED CONTROL, *DOMESTIC WATER, *SURFACE WATERS, WATER SUPPLY, EU W71-05083
T, *FILTRATION, *MICROORGANISMS, *SURFACE WATERS, *PLANKTON, *ALGA W72-06191
, *REVIEWS, *ILLINOIS, RECREATI/ *SURFACE WATERS, *LAKES, *STREAMS W72-05869
, PHOSPHATES, CLAY, TASTE, ODOR, *SURFACE WATERS, *GROUNDWATERS, S W72-06536
RARED RADIATION, *AQUATIC ALGAE, *SURFACE WATERS, ANALYTICAL TECHN W71-12034
POLLU/ *EUTROPHICATION, *ALGAE, *SURFACE WATERS, *REVIEWS, *WATER W71-10580
ER QUALITY, *ECOLOGY, *CALIFORN/ *SURFACE WATERS, *SEA WATER, *WAT W71-10577
*LA/ *ASSAY, *ZINC, *ADSORPTION, *SURFACES, LABORATORY EQUIPMENT, W72-12210
EANING, *DISPERSION, *CHEMICALS, *SURFACTANTS, *TOXICITY, *BIOASSA W71-09789
S, *ALGAL CONTROL, *FORMULATION, *SURFACTANTS, *EUTROPHICATION, LI W71-06247
ROLOGIC DATA, *MEASUREMENT, *AS/ *SURVEYS, *DATA COLLECTIONS, *HYD W72-10678
LECTIONS, *ENVIRON/ *MONITORING, *SURVEYS, *MEASUREMENT, *DATA COL W72-07841
POTOMAC RIVER, DATA COLLECTIONS/ *SURVEYS, *STATISTICAL METHODS, * W72-07791
RADIOECOLOGY, *RADIOCHEMICAL AN/ *SURVEYS, *AQUATIC ENVIRONMENT, * W73-01673
YING, REFRIGERATION, INCUBATION/ *SURVEYS, *METHODOLOGY, FREEZE DR W73-01676
TREATMENT FACIL/ *STORM RUN-OFF, *SUSPENDED SOLIDS, *RECREATION, * W72-00042
DE, *ROCK FILTERS.: *SUSPENDED SOLIDS, SODIUM HYDROXI W71-13341
MOUNT CLEMENS/ *COMBINED SEWERS, *SUSPENDED SOLIDS, BACKWASHING, * W72-00042
BLUE-GREEN ALGAE.: *SUSPENDED SEDIMENT PRODUCTION, * W72-02218
LECTROCHEMISTRY, *WATER PURIFIC/ *SUSPENSION, *ELECTROPHORESIS, *E W73-01362
TREATMENT, *FILTRATION, *ALGAE, *SUSPENSION, HARVESTING, PONDS, C W70-10174
ERLABORATORY STUDIES, PRECISION, *SWEDEN, LAKE MALAREN, MICROCYSTI W73-00286
, / *CHLOROPHYLL A, MACROPHYTES, *SWEDEN, MICROCYSIS, OSCILLATORIA W72-12729
*SWEDEN, SPECIES COMPOSITION.: W73-01632
ESPIRATION, ACTIVATED SLUDGE, W/ *SYMBIOSIS, *ALGAE, *BACTERIA, *R W71-04079
TION, RESPIR/ *ALGAE, *BACTERIA, *SYMBIOSIS, PHOTOSYNTHESIS, OXIDA W72-04983
, *NUTRIENTS, *ALGAE, *BACTERIA, *SYMBIOSIS, ACTIVATED SLUDGE, NIT W72-11100
S SOLI, SCENEDESMUS QUADRICAUDA/ *SYNCHRONOUS CULTURES, SCENEDESMU W72-07144
*DIVISION PATTERNS, *AUTOSPORES, *SYNCHRONOUS CULTURES, SYNCHRONY W72-12724
, VARIABILITY, BIAS, S/ *ERRORS, *SYNCHRONOUS CULTURES, AUTOSPORES W72-11920
EUTROPHICATION, *LAKES, *ALPINE, *SYNOPTIC ANALYSIS, COLD REGIONS, W72-12549
OXYGEN, BI/ *NITROGEN FIXATION, *SYNTHESIS, *CULTURES, *NITROGEN, W72-07941
THYL MERCURIC CHLORID/ *MERCURY, *SYNTHESIS, MERCURIC CHLORIDE, ME W72-06037
REPRODUCTION, LIFE CYCL/ *ALGAE, *SYSTEMATICS, *PLANT MORPHOLOGY, W72-07141
YTA, *EUGLENOPHYTA, *PYRROPHYTA, *SYSTEMATICS, DINOFLAGELLATES, SP W72-07683
, SEASONAL, FLU/ *PHYTOPLANKTON, *SYSTEMATICS, *DIATOMS, SEA WATER W72-07663
ES, *GRO/ *ALGAE, *GROWTH RATES, *SYSTEMATICS, *LIFE HISTORY STUDI W73-02099
*PHYTOPLANKTON, *AQUATIC ALGAE, *SYSTEMATICS, *DISTRIBUTION PATTE W73-01094
L DISTRIBUTION, *MARINE ANIMALS, *SYSTEMATICS, *ECOLOGY, *MARINE A W73-00932
TARCTIC, *DISTRIBUTION PATTERNS, *SYSTEMATICS, *AQUATIC ALGAE, *SO W73-00854
YTA, *MARINE ALGAE, *RHODOPHYTA, *SYSTEMATICS, PHOTOGRAPHY, SEA WA W73-00284
UTION, / *DISTRIBUTION PATTERNS, *SYSTEMATICS, *ECOLOGICAL DISTRIB W73-00428
LANKTON, *ENVIRONMENTAL EFFECTS, *SYSTEMATICS, *SEASONAL, SEA WATE W73-00442
ION EFFECTS, WATER PURIFICATION, *SYSTEMATICS.: /PHY, WATER POLLUT W73-00432
, PHAEOPH/ *CYTOLOGICAL STUDIES, *SYSTEMATICS, *ALGAE, CHRYSOPHYTA W72-14808
TURES, CHLOROPHYTA, PROTOZOA, */ *SYSTEMATICS, *CHLAMYDOMONAS, CUL W72-12632

CTIVITY EF/ *NUCLEAR EXPLOSIONS, *WATER POLLUTION EFFECTS, *RADIOA W72-03347
TROPHICATION, *LABORATORY TESTS, *WATER POLLUTION EFFECTS, *ALGAE, W72-03218
ENT, *ENVIRONMENT, *AGRICULTURE, *WATER POLLUTION CONTROL, SEDIMEN W72-00846
ATES, *DETERGETNTS, *PHOSPHATES, *WATER POLLUTION CONTROL, *WATER W72-01085
ATES, *FEDERAL GOV/ *DETERGENTS, *WATER POLLUTION CONTROL, *PHOSPH W72-01093
OA, *MARINE ALGAE, *SALT MARSHE/ *WATER POLLUTION EFFECTS, *PROTOZ W72-00948
HATES, *WATER POLLUTION CONTRCL, *WATER POLLUTION SOURCES, EUTROPH W72-01085
, *PATHOGENIC / *MICROORGANISMS, *WATER POLLUTION CONTROL, *WASTES W72-09675
QUATI/ *HYDROBIOLOGY, *TOXICITY, *WATER POLLUTION, *FRESHWATER, *A W72-09649
OXICITY, *WATER QUALITY CONTROL, *WATER POLLUTION EFFECTS, WATER P W72-09646
ABSORPTION, *PATH OF POLLUTANTS, *WATER POLLUTION, WATER POLLUTION W72-10686
ENCIES, *EUTROPHICATION, *LAKES, *WATER POLLUTION EFFECTS, RECLAIM W72-11702
HICATION, *DETERGE/ *PHOSPHATES, *WATER POLLUTION SOURCES, *EUTROP W72-11186
*PROT/ *AQUATIC MICROORGANISMS, *WATER POLLUTION EFFECTS, *ALGAE, W72-08579
RAPHIES, *ANALYTICAL TECHNIQUES, *WATER POLLUTION CONTROL, *INSTRU W72-08790
TS, CALIFOR/ *INDUSTRIAL WASTES, *WATER POLLUTION EFFECTS, EFFLUEN W72-09092
EN/ *ESTUARIES, *EUTROPHICATICN, *WATER POLLUTION CONTROL, MANAGEM W72-09171
LGAE, *SURFACE WATERS, *REVIEWS, *WATER POLLUTION SOURCES, NUTRIEN W71-10580
4-D, *DDT, *DIELDRIN, *DIAZINCN, *WATER POLLUTION EFFECTS, *PHOTOS W71-10566
*DELTAS, *BAYS, *WATER QUALITY, *WATER POLLUTION CONTROL, *DIVERS W71-09581
ES, *NUTRIENTS, *ALGAE, *BACTER/ *WATER POLLUTION EFFECTS, *NITRAT W71-09958
ICATION, *BIOASSAY, *INDICATORS, *WATER POLLUTION EFFECTS, *ALGAE, W71-10553
C WASTES, *EUTROPHICATION, *CHE/ *WATER POLLUTION EFFECTS, *ORGANI W71-09674
LUTION CONTROL, *EUTROPHICATION, *WATER POLLUTION SOURCES, POPULAT W71-12091
IDES, *DDT, *PESTICIDE TOXICITY, *WATER POLLUTION EFFECTS, PESTICI W71-12303
*FILTRATION, *ELECTROCHEMISTRY, *WATER POLLUTION TREATMENT, ANALY W71-12467
HICATION, *WATER POLLUTICN SCUR/ *WATER POLLUTION CONTROL, *EUTROP W71-12091
TORS, *SEA WATER, MARIN/ *ALGAE, *WATER POLLUTION EFFECTS, *INDICA W71-12867
T, *PHOSPHATES, *EUTROPHICATION, *WATER POLLUTION CONTROL, SEWAGE W71-11507
ATCHERIES, *SALMONIDS, WATER RE/ *WATER POLLUTION SOURCES, *FISH H W71-11006
UTRIEN/ *EUTROPHICATION, *LAKES, *WATER POLLUTION EFFECTS, OHIO, N W71-11949
S, WATER POLLUTION SOURCES, IND/ *WATER POLLUTION EFFECTS, *DIATOM W71-11583
HICATION, *DETERGENTS, *PHOSPHA/ *WATER POLLUTION CONTROL, *EUTROP W71-11516
DEMAND, *CHEMICAL OXYGEN DEMAND, *WATER POLLUTION SOURCES, *WATER W70-09189
EMAND, *WATER POLLUTION SOURCES, *WATER POLLUTION EFFECTS, *ORGANI W70-09189
ER SUPPLY, *POLLUTION ABATEMENT, *WATER POLLUTION CONTROL, *WATER W70-09193
ATES, *NITRATES, *NUTRIENTS, *E/ *WATER POLLUTION EFFECTS, *PHOSPH W70-09388
CT OF COLUMBIA, TID/ *ESTUARIES, *WATER POLLUTION EFFECTS, *DISTRI W70-08249
CT OF COLUMBIA, TID/ *ESTUARIES, *WATER POLLUTION EFFECTS, *DISTRI W70-08248
RMAL/ *RESEARCH AND DEVELOPMENT, *WATER POLLUTION, PHOSPHORUS, THE W70-09947
OLLUTION SOURCES, INDUSTRIAL WA/ *WATER POLLUTION EFFECTS, WATER P W71-03095
R QUAL/ *EUTROPHICATION, *LAKES, *WATER POLLUTION, MINNESOTA, WATE W71-03012
OLLUTION EFFEC/ *EUTROPHICATION, *WATER POLLUTION CONTROL, WATER P W71-01488
IE/ *NUTRIENTS, *EUTROPHICATICN, *WATER POLLUTION SOURCES, *ESTUAR W70-06509
WATER(POLLUTIO/ *EUTROPHICATION, *WATER POLLUTION CONTROL, *WASTE W72-05473
ECOSYSTEMS, ABSORPTION, FISH, I/ *WATER POLLUTION, *FOOD CHAINS, * W72-05952
LYCHLORINATED BIPHENYLS, *ALGAE, *WATER POLLUTION EFFECTS, *PESTIC W72-05614
TER, *OILS, *N/ *REMOTE SENSING, *WATER POLLUTION CONTROL, *SEA WA W72-04742
OLLUT/ *EUTROPHICATION, *CARBCN, *WATER POLLUTION CONTROL, WATER P W72-04769
W/ *EUTROPHICATION, *PHOSPHATES, *WATER POLLUTION CONTROL, *WASTE W72-04734
*COLORADO, *RE/ *WATER QUALITY, *WATER POLLUTION EFFECTS, *LAKES, W72-04791
ASTES, OKLA/ *OXIDATION LAGOONS, *WATER POLLUTION EFFECTS, *FARM W W72-04773
RESOURCES DEVELOPME/ *ESTUARIES, *WATER POLLUTION EFFECTS, *WATER W71-05553
AY, *TOXICITY, *CHLORINATED HYD/ *WATER POLLUTION EFFECTS, *BIOASS W71-07412
POLLU/ *PHOSPHATES, *DETERGENTS, *WATER POLLUTION EFFECTS, *WATER W71-08768
*ACID MINE WATER, *STRIP MINES, *WATER POLLUTION EFFECTS, *KENTUC W71-07942
GENTS, *WATER POLLUTION EFFECTS, *WATER POLLUTION CONTROL, *BIODEG W71-08768
WAST/ *STRONTIUM RADIOISOTOPES, *WATER POLLUTION EFFECTS, NUCLEAR W71-09225
R POWER PLANTS, *POTABLE WATER,/ *WATER POLLUTION CONTROL, *NUCLEA W71-09018
WATER, *WATER TEMPERATURE, *WA/ *WATER POLLUTION EFFECTS, *HEATED W72-13636
ATIC LIFE, *AQUATIC ENVIRONMENT, *WATER POLLUTION, *PATH CF POLLUT W72-13347
OGY, WATER POLLUTION, W/ *LAKES, *WATER POLLUTION EFFECTS, *LIMNOL W72-12729
HICATION, *LAKES, SEWAGE EFFLUE/ *WATER POLLUTION SOURCES, *EUTROP W72-12955
ESTICIDE KINETICS, *PERSISTENCE, *WATER POLLUTION EFFECTS, *PLANT W73-02280
IC HEALTH,/ *WATER PURIFICATION, *WATER POLLUTION TREATMENT, *PUBL W73-02144
HICATION, *SOCIAL PARTICIPATICN, *WATER POLLUTION CONTROL, WISCONS W73-02479
ORUS, *EUTROPHICATION, CHEMICAL/ *WATER POLLUTION CONTROL, *PHOSPH W73-02478
*INSTRUMENTATION, *MONITORING, *WATER PROPERTIES, *ON-SITE DATA W72-12941
, *CULTIVATED L/ *TILE DRAINAGE, *WATER PROPERTIES, *FERTILIZATION W71-12252
*ELECTROCHEMISTRY, *WATER POLL/ *WATER PURIFICATION, *FILTRATION, W71-12467
TION TREATMENT, *PUBLIC HEALTH,/ *WATER PURIFICATION, *WATER POLLU W73-02144
ESERVOIRS, RESERVOIRS, PHOTOSYN/ *WATER PURIFICATION, *DETENTION R W71-01491
CTROPHORESIS, *ELECTROCHEMISTRY, *WATER PURIFICATION, CLAY, ALGAE, W73-01362
*CRUSTACEANS, *MARINE BACTERIA, *WATER QUALITY, BENTHOS, CRABS, C W72-08142
VITY, *AQUATIC MICROORGANISMS, / *WATER QUALITY, *AQUATIC PRODUCTI W72-06532
ICATION, *BIOLOGICAL PROPERTIES, *WATER QUALITY, *BIOINDICATORS, * W73-00002
ANTS, *MATHEMATICAL MODELS, THE/ *WATER QUALITY, *HYDROELECTRIC PL W72-14405
POLLUTION, NUTRIENTS, BIOASSAY, *WATER QUALITY, ALGAE, PESTICIDES W72-14282
*MATHEMATICA/ *SYSTEMS ANALYSIS, *WATER QUALITY, *EUTROPHICATION, W71-03021
CATION, PHYTOPLANKTON, PLANKTCN/ *WATER QUALITY, *LAKES, *EUTROPHI W71-02880
, STREAMS, FLOW, SE/ *NUTRIENTS, *WATER QUALITY, *IMPOUNDED WATERS W71-03028
C/ *NUTRIENTS, *EUTROPHICATION, *WATER QUALITY CONTROL, *NITROGEN W70-04080
HORUS, PULP AND PAPER INDUSTRY,/ *WATER QUALITY, *WISCONSIN, PHOSP W70-05103
ERTIARY TREATMENT, *WATER REUSE, *WATER QUALITY, WATER SUPPLY, ALG W70-10381
NTROL, *FORAGE FISH, BIOCONTROL/ *WATER QUALITY CONTROL, *ALGAE CO W71-09163
ATION LAGOONS, *DESIGN CRITERIA, *WATER QUALITY, ALGAE, HARVESTING W71-07100
WEED CONTROL, *DOME/ *ALGICIDES, *WATER QUALITY CONTROL, *AQUATIC W71-05083
TEMS, *ECOLOGY, *PUMPED STORAGE, *WATER QUALITY, *MODEL STUDIES, Z W72-05282
EFFECTS, *LAKES, *COLORADO, *RE/ *WATER QUALITY, *WATER POLLUTION W72-04791
N, *ALGAE, TURBIDITY, *NEW YORK, *WATER QUALITY.: /ATERS, *PLANKTO W72-06191

GENERAL INDEX

The middle rank of this index lists in alphabetical order each word of each descriptor except those included in the *Significant Descriptor Index*, which should be consulted first. The numbers at the right are the accession numbers appearing at the end of the abstract.

RODUCTIVITY, *ENERGY CO/ *MARINE ALGAE, *BENTHIC FLORA, *PRIMARY P W72-06283
NT GROWTH, *CHLOROPHYTA, *MARINE ALGAE, *BIOASSAY, REPRODUCTION, S W72-14294
LLUTION EFFECTS, *ALGAE, *MARINE ALGAE, *BIOINDICATORS, *PHOTOSYNT W71-10553
NTROL, CYANOPHYTA, CA/ *NUISANCE ALGAE, *CARBON DIOXIDE, *ALGAL CO W72-10607
, *DEFICIENT NUTRIENTS, *AQUATIC ALGAE, *CARBON DIOXIDE, NUISANCE W72-07514
NTS, MARINE/ *SEA WATER, *MARINE ALGAE, *CARBOHYDRATES, MARINE PLA W73-02100
BERS, *LIFE CYCLES, CU/ *AQUATIC ALGAE, *CHLOROPHYTA, *GROWTH CHAM W72-07144
E/ *OCEANS, *MATABOLISM, *MARINE ALGAE, *CHLORINATED HYDROCARBON P W72-12998
*GROWTH RATES, *INHIBI/ *AQUATIC ALGAE, *COMPETITION, *NUTRIENTS, W72-12734
*PRIMARY PRODUCTIVITY, *AQUATIC ALGAE, *COMPARATIVE PRODUCTIVITY, W73-01080
PHYTA, *NUISANCE ALGAE, *AQUATIC ALGAE, *CRUSTACEANS, *FISH TOXINS W72-14056
PHAEOPHY/ *SYSTEMATICS, *MARINE ALGAE, *CYANOPHYTA, *CHLOROPHYTA, W72-11800
COLOGICAL DISTRIBUTION, *AQUATIC ALGAE, *CYANOPHYTA, *BEGGIATOA, P W72-09061
L, LABORATORY TESTS, / *NUISANCE ALGAE, *CYANOPHYTA, *ALGAL CONTRO W71-13184
ONS, BIODEGRADATION, *NUTRIENTS, ALGAE, *DEGRADATION(STREAM), WATE W71-07973
S,/ *TOXICITY, *AMMONIA, *MARINE ALGAE, *DIATOMS, CHEMICAL ANALYSI W71-11008
ALGAE, RESOURCES, FOOD/ *MARINE ALGAE, *ECONOMICS, *PRODUCTIVITY, W71-13246
AND, HYDROGEN ION CONCENTRATION, ALGAE, *ENTROPHICATION, NEUTRALIZ W70-10427
OPHYTE-EPIPHTYE META/ *EPIPHYTIC ALGAE, *EPIPHYTIC BACTERIA, *MACR W72-03216
ER(ENGLAND), AMPHOR/ *EPIPSAMMIC ALGAE, *EPIPELIC ALGAE, SHEAR WAT W71-12855
C PLANTS, *DISTRIBUTION, *MARINE ALGAE, *ESTUARIES, PRODUCTIVITY, W72-04766
ATMENT, *ALGAL CONTROL, *AQUATIC ALGAE, *EUTROPHICATION, MUNICIPAL W72-03970
TRIENTS, *NITROGEN, *PHOSPHORUS, ALGAE, *EUTROPHICATION, AQUATIC E W71-08953
*CYANOPHYTA, *NUISANCE ALGAE, *EUTROPHICATION, REVIEWS.: W72-01358
TER POLLUTION CONTROL, *NUISANCE ALGAE, *EUTROPHICATION, LAKES, WA W72-00799
S, *ALGAL NUTRIENTS, *CHLORELLA, ALGAE, *EUTROPHICATION, NEW MEXIC W72-13992
FAR EXPLOSIONS, *ALASKA, *MARINE ALGAE, *FISHERIES, FISH POPULATIO W73-00764
OTOPES, *MARINE ANIMALS, *MARINE ALGAE, *FOOD CHAINS, ABSORPTION, W73-00819
POWERP/ *BENTHIC FLORA, *MARINE ALGAE, *GULF OF MEXICO, *ELECTRIC W72-11252
RATES, GROWTH AUTOSTIMUL/ *GREEN ALGAE, *HORMOTILA BLENNISTA, FILT W71-08687
C/ *BENTHOS, *SAMPLING, *SESSILE ALGAE, *INVERTEBRATES, ST. LAWREN W72-05432
OLIDS, F/ *ALGAE, *HARVESTING OF ALGAE, *ION EXCHANGE, DISSOLVED S W71-00138
*CULTURES, *CYANOPHYTA, *MARINE ALGAE, *ISOLATION, MARINE PLANTS, W73-00430
NDUSTRIAL WASTES, INVERTEBRATES, ALGAE, *KELP.: /, HEAVY METALS, I W72-11854
LOGY, PHAEOPHYTA, STAND/ *MARINE ALGAE, *KELPS, *PHYSIOLOGICAL ECO W72-12940
AL WASTE, *GROWTH RATES/ *MARINE ALGAE, *LABORATORY TESTS, *CHEMIC W72-12239
PHYTON, *PHYTOPLANKTON, *SESSILE ALGAE, *LAKE SUPERIOR, LIMNOLOGY, W72-12192
T LAKES, *LAKES, EUTROPHICATION, ALGAE, *LAKE SUPERIOR, *LAKE MICH W72-01094
ENSITY STRATIFICATION,/ *AQUATIC ALGAE, *LAKES, *PHYTOPLANKTON, *D W71-09852
TOPLANKTON, NATURAL RES/ *MARINE ALGAE, *LIGHT, *GROWTH RATES, PHY W71-13252
NSAS, MODEL STUDIES, ECOSYSTEMS, ALGAE, *LIGHT QUALITY, EUTROPHIC W72-12393
OPLANKTON, *ABSORPTION, *AQUATIC ALGAE, *LIGHT INTENSITY, *PHOSPHO W72-07166
ISH, M/ *NUCLEAR WASTES, *MARINE ALGAE, *MARINE ANIMALS, *MARINE F W73-00832
POLLUTION, *THERM/ BAYS, *MARINE ALGAE, *MARINE ANIMALS, *THERMAL W72-11876
*RADIOACTIVITY TECHNIQUES, SOIL ALGAE, *METHODOLOGY, MEASUREMENT, W72-14301
IS, *TRACE ELEMENTS, C/ *AQUATIC ALGAE, *NEUTRON ACTIVATION ANALYS W71-11515
GAE, STATISTICAL METHODS, MARINE ALGAE, *OREGON, *SEASONAL, CHRYSO W72-08141
IOISOTOPES, *ABSORPTION, *MARINE ALGAE, *PHYTOPLANKTON, BACTERIA, W73-00802
*EUTRO/ *PHYTOPLANKTON, *AQUATIC ALGAE, *PHOSPHORUS, *CYANOPHYTA, W71-10645
S, *R/ *AQUATIC PLANTS, *AQUATIC ALGAE, *PHYTOPLANKTON, *RESERVOIR W72-04855
ONDS, WATER POLLUTION / *AQUATIC ALGAE, *RADIOACTIVITY EFFECTS, *P W71-09256
, TEMPERATURE, STRONTI/ *AQUATIC ALGAE, *RADIOISOTOPES, ABSORPTION W71-09250
PATTERNS, *CHLOROPHYTA, *MARINE ALGAE, *RHODOPHYTA, *SYSTEMATICS, W73-00284
*BACTERIA, FLORIDA, RAD/ *MARINE ALGAE, *SALT MARSHES, *PROTOZOA, W73-00796
TION EFFECTS, *PROTOZOA, *MARINE ALGAE, *SALT MARSHES, NEW YORK, N W72-00948
XIDATION LAGOONS, *HARVESTING OF ALGAE, *SEPARATION TECHNIQUES, CH W72-10237
IEWS, B/ *PHYTOPLANKTON, *MARINE ALGAE, *SEDIMENTATION RATES, *REV W72-06313
PATTERNS, *SYSTEMATICS, *AQUATIC ALGAE, *SOIL BACTERIA, SAMPLING, W73-00854
YIELD, *SIMULATED RAINFAL/ *SOIL ALGAE, *SOIL SURFACES, *SEDIMENT W72-02218
*ZINC RADIOISOTOPES, AB/ *MARINE ALGAE, *STRONTIUM RADIOISOTOPES, W72-12203
Y, *INFRARED RADIATION, *AQUATIC ALGAE, *SURFACE WATERS, ANALYTICA W71-12034
THIC FLORA, RHODOPHYTA,/ *MARINE ALGAE, *SYSTEMATICS, *COASTS, BEN W71-10066
N PATT/ *PHYTOPLANKTON, *AQUATIC ALGAE, *SYSTEMATICS, *DISTRIBUTIO W73-01094
*CYANOPHYTA, ALGAE, *VIRUSES.: W72-03188
*NITRATES, TREATMENT FACILITIES, ALGAE, *WASTE WATER TREATMENT, *A W72-02975
*PHOTOSYNTHESIS, *OIL P/ *MARINE ALGAE, *WATER POLLUTION EFFECTS, W73-01074
NG, COBALT RADIOISOTOPES, MARINE ALGAE, ABSORPTION, SEDIMENTS, WAS W73-00831
, CLAMS, CARP, CRAYFISH, AQUATIC ALGAE, ABSORPTION, AQUARIA, ANIMA W72-06342
*NORTHWESTERN ONTARIO, EPILITHIC ALGAE, ACHNANTHES MINUTISSIMA, AC W71-11029
AS CHROMATOGRAPHY, SEPCTROSCOPY, ALGAE, ACTINOMYCETES, FUNGI, LABO W72-11604
-PRODUCING ALGAE, ODOR-PRODUCING ALGAE, ACTINOMYCETES, PHOSPHATES. W73-02144
E, *WATER QUALITY, WATER SUPPLY, ALGAE, ACTIVATED CARBON, ADSORPTI W70-10381
ANIMALS, INVERTEBRATES, AQUATIC ALGAE, AERIAL PHOTOGRAPHY, PHOSPH W72-12729
EMICAL OXYGEN DEMAND, ECONOMICS, ALGAE, AGRICULTURAL: /NING, BIOCH W72-04734
AL, INVERTEBRATES, ALGAE, MARINE ALGAE, ALGAL POISONING, CHROMATOG W71-08597
NISMS, AQUATIC BACTERIA, AQUATIC ALGAE, ALGAE, CHLOROPHYTA, PSEUDO W72-12255
, ENVIRONMENTAL EFFECTS, AQUATIC ALGAE, ALGAL CONTROL, ALGAL POISO W72-11186
*ENVIRONMENTAL EFFECTS, DIQUAT, ALGAE, ALGAL CONTROL, DIATOMS, PE W72-00155
ON, WATER TEMPERATURE, BACTERIA, ALGAE, ALKALINITY, IRON, PHOSPHAT W71-05626
OZYME, *SPHEROPLASTS, BLUE-GREEN ALGAE, ANABAENA VARIABILIS, ANACY W71-08683
, RE-AERATION, DISSOLVED OXYGEN, ALGAE, ANAEROBIC CONDITIONS, AERO W71-07099
ATION, POLLUTANT IDENTIFICATION, ALGAE, ANALYTICAL TECHNIQUES, *UL W72-01780
BUOYS, HARBORS, TUNNELS, MARINE ALGAE, ANIMAL POPULATIONS, LOBSTE W73-02029
*SYSTEMATICS, *ECOLOGY, *MARINE ALGAE, ANNELIDS, MOLLUSKS, RHODOP W73-00932
CONTROL, ASSAY, REMOTE SENSING, ALGAE, ANTIBIOTICS, MERCURY, ARSE W72-10162
ION, PHOSPHORUS, EUTROPHICATION, ALGAE, AQUATIC WEEDS, PHYTOPLANKT W72-08459
ENTS, *POLLUTANT IDENTIFICATION, ALGAE, AQUATIC BACTERIA, SEWAGE B W72-01472
LGAE, DDT, PHYTOPLANKTON, MARINE ALGAE, AQUATIC ALGAE, COLORIMETRY W72-03545
S, PHOSPHORUS COMPOUNDS, POTASH, ALGAE, AQUATIC PLANTS, CHEMICAL A W71-12252
HLOROPHYTA, SCENEDESMUS, AQUATIC ALGAE, AQUATIC PRODUCTIVITY, POLL W73-01066

*AQUATIC MICROORGANISMS, SPORES, ALGAE, CYTOLOGICAL STUDIES, MYCOB W71-08027
IC ENVIRONMENT, FISH PHYSIOLOGY, ALGAE, DAPHNIA.: /TIC LIFE, AQUAT W72-12257
RATES, *PHOTOSYNTHESIS, *MARINE ALGAE, DDT, *PESTICIDES, *PRIMARY W72-09653
PESTICIDES, PESTICIDE RESIDUES, ALGAE, DDT, PHYTOPLANKTON, MARINE W72-03545
, TROPHIC LEVEL, EUTROPHICATION, ALGAE, DECOMPOSING ORGANIC MATTER W72-01784
ITY, *BIOASSAY, FISH, SHELLFISH, ALGAE, DEGRADATION.: /NTS, *TOXIC W71-09789
, DESALINATION, DRAIN/ NITRATES, ALGAE, DENTRIFICATION, FILTRATION W71-11698
, WATER POLLUTION, DESALINATION, ALGAE, DENITRIFICATION, CALIFORNI W71-08223
ERCHES, PIKES, SAMPLING, SNAILS, ALGAE, DIATOMS, CLAMS, SHELLFISH, W70-08098
RIVERS, WATER POLLUTION EFFECTS, ALGAE, DIATOMS.: /ALGAL CONTROL, W71-00099
CIAL DRIFT, NUTRIENTS, SALANITY, ALGAE, DIATOMS.: /OTA, LAKES, GLA W71-13149
RS, TOXINS, INORGANIC COMPOUNDS, ALGAE, DIATOMS, CHLAMYDOMONAS, SC W72-14706
CHRYSOPHYTA, INHIBITION, MARINE ALGAE, DIATOMS, BEHAVIOR.: /CIDS, W73-00942
AGOONS, BIOMASS, OREGON, SLUDGE, ALGAE, DIPTERA, DISSOLVED OXYGEN, W72-03556
ER POLLUTION SOURCES, WATER LAW, ALGAE, DISSOLVED OXYGEN.: /L, WAT W71-01192
TER RELATIONSHIPS, PERMEABILITY, ALGAE, DISSOLVED OXYGEN.: /OUNDWA W71-12410
TER POLLUTION SOURCES, BIOASSAY, ALGAE, DOMESTIC WASTES, NUTRIENTS W73-02478
SORPTION, NUTRIENTS, PESTICIDES, ALGAE, DREDGING, ESTUARINE ENVIRO W71-09581
HYSIOLOGY, LAKES, PONDS, AQUATIC ALGAE, ECOLOGY, BIOLOGICAL PROPER W71-10580
TION EFFECTS, *DATA COLLECTIONS, ALGAE, ECOTYPES, INDUSTRIAL PLANT W71-13746
ARVEST, PLANKTON, BENTHIC FAUNA, ALGAE, ECOLOGY, MISSOURI.: /ISH H W70-09596
GEMENT, WISCONSIN, PRODUCTIVITY, ALGAE, ECONOMICS, POLITICAL ASPEC W71-03012
*FRESHWATER ALGAE, EDTA.: W72-14694
HICATION, WASTE WATER TREATMENT, ALGAE, EFFLUENTS, NUTRIENTS, SEWA W70-09325
E MICHIGAN, CHRYSOPHYTA, AQUATIC ALGAE, ELECTRON MICROSCOPY, PHOTO W73-00432
ICAL REGIONS, PHAEOPHYTA, MARINE ALGAE, ENVIRONMENTAL EFFECTS, MAR W73-00428
CIDES), PHOTOACTIVATION, AQUATIC ALGAE, ENVIRONMENTAL EFFECTS.: /I W72-14327
ADIATION, HYDROGRAPHY, PLANKTON, ALGAE, ENVIRONMENTAL EFFECTS, BIO W73-01704
OD CHAINS, PHYTOPLANKTON, MARINE ALGAE, ENVIRONMENTAL EFFECTS, TRA W73-00818
OPHIC LEVEL, PROTOZOANS, AQUATIC ALGAE, ENVIRONMENTAL EFFECTS, CHL W72-13816
IFICATION, PHYTOPLANKTON, MARINE ALGAE, ESSENTIAL NUTRIENTS.: /ENT W73-00441
CALIFORNIA, RADIOECOLOGY, MARINE ALGAE, ESTUARINE ENVIRONMENT, ABS W73-00826
USTRIAL WASTES, FLORIDA, SESSILE ALGAE, ESTUARIES, CHEMICAL WASTES W71-07412
MS, ALGAE, MARINE ALGAE, AQUATIC ALGAE, EUGLENA, CHLAMYDOMONAS.: / W72-08436
ION EFFECTS, BIOMASS, NUTRIENTS, ALGAE, EUGLENOPHYTA, CHRYSOPHYTA, W72-09111
UTRIENTS, MEASUREMENT, BACTERIA, ALGAE, EUTROPHICATION, DATA COLLE W71-13794
A, *HUMIC ACIDS, ORGANIC MATTER, ALGAE, EUTROPHICATION, CHELATION, W72-01097
AL TURNOVER, ENERGY, CYANOPHYTA, ALGAE, EUTROPHICATION, SUMMER, EN W72-03228
, *WATER QUALITY CONTROL, *ODOR, ALGAE, EUTROPHICATION, TASTE, RES W71-13187
ROPHI/ *ALGAL CONTROL, *NUISANCE ALGAE, EUTROPHICATION, REVIEWS, T W72-07937
ALGAE, *CARBON DIOXIDE, NUISANCE ALGAE, EUTROPHICATION, PHOSPHATES W72-07514
ITRATES, PHOSPHORUS, PHOSPHATES, ALGAE, EUTROPHICATION, WATER QUAL W71-06445
ITY, OXYGEN DEMAND, HEAT BUDGET, ALGAE, EUTROPHICATION.: / ALKALIN W72-12997
NISTRATION, NON-CONSUMPTIVE USE, ALGAE, EVAPORATION, REASONABLE US W70-07100
ISSUES, FRESHWATER FISH, AQUATIC ALGAE, EVALUATION, SAMPLING, CARP W73-01673
ICATION, HYDRAULIC MODELS, FISH, ALGAE, FATHOMETERS, BENTHIC FAUNA W70-09614
, AMMONIA, NITROGEN, PHOSPHORUS, ALGAE, FEASIBILITY STUDIES, TECHN W71-09450
EPODS, CRUSTACEANS, ZOOPLANKTON, ALGAE, FEEDING RATES, *LARVAL GRO W72-09248
F RATES, AFTERGROWTH, BLUE-GREEN ALGAE, FIELD STUDIES, GREEN ALGAE W70-08319
NT EXTRACTIONS, DIATOMS, AQUATIC ALGAE, FILTRATION, CYANOPHYTA, *E W72-13673
QUALITY, TROPHIC LEVEL, AQUATIC ALGAE, FISH POPULATIONS, FLORIDA, W72-08986
OIRS, *FRESHWATER, HYDROBIOLOGY, ALGAE, FISH, ABSORPTION, BIOASSAY W72-10694
S, CHEMICALS, PLANKTON, BENTHOS, ALGAE, FISH, WATER POLLUTION, WAT W72-07841
, HEADLOSS, PRESSURE, TURBIDITY, ALGAE, FLOCCULATION, CLEANING, *W W72-09393
LLA, *ALGAE, *FOULING, *NUISANCE ALGAE, FLOATING PLANTS, SEAWATER, W71-09261
WATER POLLUTION SOURCES, AQUATIC ALGAE, FOOD CHAINS, MEASUREMENT, W72-10686
, PHYTOPLANKTON, MAMMALS, MARINE ALGAE, FOOD CHAINS, MARINE FISHER W72-00960
TAMIN B, PIGMENTS, GROWTH RATES, ALGAE, FOOD WEBS, COPEPODS, ORGAN W72-07892
ORPTION, *PHYTOPLANKTON, *MARINE ALGAE, FOOD CHAINS, PUBLIC HEALTH W73-00823
ADSORPTION, AGRICULTURAL RUNOFF, ALGAE, FOOD CHAINS, CHLORINATED H W72-13657
ICATION, NUTRIENTS, ASSESSMENTS, ALGAE, FORMULATION, WASTE TREATME W72-09169
ORUS, *PHOSPHORUS RADIOISOTOPES, ALGAE, FRESH WATER, FILTRATION, L W72-07143
TED WATER, CONDENSERS, PROTOZOA, ALGAE, FUNGI, WATER POLLUTION EFF W72-07526
BS, FOOD CHAINS, FROGS, LICHENS, ALGAE, FUNGI, SMALL ANIMALS(MAMMA W72-00949
, CYANOPHYTA, PYRROPHYTA, MARINE ALGAE, GASTROPODS.: / CHRYSOPHYTA W73-00935
NITRATES, MUSSELS, SNAILS, FISH, ALGAE, GASTROPODS, ADSORPTION, TE W71-09168
MICALS, WATER POLLUTION SOURCES, ALGAE, GEOLOGY, DATA COLLECTIONS. W72-05869
EALTH, COST COMPARISONS, AQUATIC ALGAE, GRANTS, MODEL STUDIES.: /H W71-09429
*TOXICITY, *CYANOPHYTA, SULFUR, ALGAE, GRANULES, INHIBITION, META W72-00153
TICAL MODELS, SELF-PURIFICATION, ALGAE, GREAT LAKES, NUTRIENTS, SE W71-12072
*BENTHIC FLORA, *PHYTOPLANKTON, ALGAE, GROWTH RATES, BIOASSAY, DI W72-06046
CULTURES, CHLOROPHYLL, PROTEINS, ALGAE, GROWTH RATES, TRACERS.: / W72-03557
RATURE, SORPTION, RADIOISOTOPES, ALGAE, GROWTH RATES, THERMAL POLL W71-02075
*METABOLISM, *CHLORELLA, *LIGHT, ALGAE, GROWTH RATES, CHLOROPHYLL. W73-00251
M, CULTURES, RESISTANCE, AQUATIC ALGAE, GROWTH RATES, BIOASSAY, LA W72-13809
DIES, *PHAEOPHYTA, EPIPHYTOLOGY, ALGAE, HABITATS, CULTURES.: / STU W72-01370
DESIGN CRITERIA, *WATER QUALITY, ALGAE, HARVESTING, DEPTH, TEMPERA W71-07100
, RADIOISOTOPES, TRACERS, MARINE ALGAE, HAWAII, WATER TEMPERATURE, W71-11486
, *COBALT RADIOISOTOPES, *MARINE ALGAE, HEAVY METALS, ABSORPTION, W72-10957
FFECTS, WATER POLLUTION EFFECTS, ALGAE, HEAVY METALS, GROWTH, PSEU W72-13813
OMPOSING ORGANIC MATTER, *MARINE ALGAE, HUMUS, ORGANIC COMPOUNDS, W72-12166
EROBIC CONDITIONS, PILOT PLANTS, ALGAE, HYDROGEN ION CONCENTRATION W71-07102
IES, FISH, MARINE PLANTS, MARINE ALGAE, HYDROLYSIS, CHELATION, SED W72-04463
ARBON, *SEWAGE BACTERIA, SEWAGE, ALGAE, HYDROGEN ION CONCENTRATION W72-04309
STORAGE, NUTRIENTS, TEMPERATURE, ALGAE, HYDROLOGIC DATA, DISSOLVED W72-05282
S, EQUATIONS, LIGHT PENETRATION, ALGAE, HYDROGEN ION CONCENTRATION W72-05459
T STREAMS, *GREAT LAKES, AQUATIC ALGAE, ICE, FISH, MICROORGANISMS, W73-01669
R, ATLANTIC OCEAN, GROWTH RATES, ALGAE, INCUBATION, PHYTOPLANKTON. W71-11562
WATER QUALITY, *DOMESTIC WASTES, ALGAE, INDUSTRIAL WASTES, METABOL W72-09590
EFFECTS, EUTROPHICATION, *LAKES, ALGAE, INFRARED RADIATION, *REMOT W72-14728

BSORPTION, MARINE PLANTS, MARINE ALGAE, PHAEPHYTA, AQUARIA, PARTIC W71-09193
CARBCN, ADSORPTION, FILTRATION, ALGAE, PHENOLS, DETERGENTS, PESTI W72-08357
TTS, *EUTROPHICATION, NUTRIENTS, ALGAE, PHCSPHATES, NITRATES, SURV W71-05553
RATURE, TURBULENCE, OXYGENATION, ALGAE, PHOTOSYNTHESIS, MUNICIPAL W71-06033
STRATIFICATION, SOLAR RADIATION, ALGAE, PHCTOSYNTHESIS, ANAEROBIC W71-07084
ISONS, *SEA WATER, *FRESH WATER, ALGAE, PHOTOSYNTHESIS, CRGANIC MA W71-03027
*MUNICIPAL WASTES, *WATER REUSE, ALGAE, PHOTOSYNTHESIS, DISSOLVED W72-11642
DESIGN CRITERIA, BIODEGRADATION, ALGAE, PHOTOSYNTHESIS, SUSPENDED W72-09386
S, EUTROPHICATION, PRODUCTIVITY, ALGAE, PHOSPHATES, NITROGEN, SEWA W72-09167
, LABORATORY TESTS, TEMPERATURE, ALGAE, PHOSPHCRUS, ION EXCHANGE, W72-01108
MENT, SURFACE WATERS, NUTRIENTS, ALGAE, PHOSPHORUS, NITROGEN, EUTR W72-00847
, *PLANT GROWTH, *LAKE MICHIGAN, ALGAE, PHOSPHATES, HYDROLYSIS, ST W72-13644
ING, SECONDARY TREATMENT, MARINE ALGAE, PHYTOPLANKTON, BICMASS, GR W72-09378
XICITY, *AQUATIC PLANTS, AQUATIC ALGAE, PHYTOPLANKTON, AQUATIC MIC W72-09650
UCTIVITY, SEASONAL, CHLOROPHYLL, ALGAE, PHYTOPLANKTON, WATER ANALY W72-09108
UTROPHICATION, RED TIDE, SILICA, ALGAE, PHYTOPLANKTON, FREEZING, T W71-09190
PES, *ORGANIC COMPOUNDS, *MARINE ALGAE, PHYTOPLANKTON, HUMIC ACIDS W71-09188
LY IMPOUNDMENT, BIOMASS, AQUATIC ALGAE, PHYTOPLANKTON, ZOCPLANKTON W72-04853
AL TECHNIQUES, MICHIGAN, SESTON, ALGAE, PHYTOPLANKTON, NANNOPLANKT W72-08048
, *CARBON, *AMINO ACIDS, *MARINE ALGAE, PHYTOPLANKTON, NITROGEN, A W72-07497
ANOPHYTA, *NEW HAMPSHIRE, MARINE ALGAE, PHYTOPLANKTON, AQUATIC PLA W72-14542
AMS, CHEMICAL ANALYSIS, BIOMASS, ALGAE, PHYSICOCHEMICAL PROPERTIES W73-01435
R TEMPERATURE, SAMPLING, AQUATIC ALGAE, PHYTOPLANKTON, TRCPHIC LEV W73-01456
INO-3,5,6-TR/ TORDON, FRESHWATER ALGAE, PICLORAM, PEDIASTRUM, 4-AM W72-08440
CHEMISTRY, CYANOPHYTA, BACTERIA, ALGAE, PLANKTCN, PLANTS, SPECTROS W73-01439
UTION), WATER POLLUTION SOURCES, ALGAE, PLANKTON, SEASONAL.: /POLL W72-05506
OTOPES, STRONTIUM RADIOISOTOPES, ALGAE, PLANT GROWTH, TEMPERATURE, W71-09254
NUTRIENTS, NITROGEN, PHOSPHORUS, ALGAE, PLANT GROWTH, WATER POLLUT W72-11702
NTHIC FAUNA, PERIPHYTON, AQUATIC ALGAE, PLANTS, ENVIRONMENTAL ENGI W72-00939
L CONTROL, *CYANPHYTA, *NUISANCE ALGAE, PLANKTCN, WASHINGTON, DIAT W72-00798
RESERVOIRS, TERTIARY TREATMENT, ALGAE, PLANTS, TROUT, AQUATIC ENV W72-02412
CULTURE, *ECONOMICS, CYANOPHYTA, ALGAE, PLANTS, PROTEINS, NUTRIENT W72-03327
INFILTRATION, INDUSTRIAL WASTES, ALGAE, PONDS, FARM WASTES.: /GE, W71-06435
MENT, WATER POLLUTION TREATMENT, ALGAE, POTABLE WATER, TASTE, AMMO W71-01491
ALGAE, BIOINDICATORS, BIOASSAY, ALGAE, POWERPLANTS, EFFLUENTS, DI W71-11583
PERATURE EFFEC/ *BP 1002, *GREEN ALGAE, PRASINOCLADUS MARINUS, TEM W70-09429
ON, SECCNDARY TREATMENT, AQUATIC ALGAE, PRIMARY PRODUCTIVITY, TRAC W71-09674
UTRIENTS, BIOMASS, HARVESTING OF ALGAE, PRCDUCTIVITY.: /RCTEINS, N W72-03327
DOLOGY, FOREIGN RESEARCH, MARINE ALGAE, PRODUCTIVITY, POLLUTANTS, W72-08804
TER POLLUTION, ACTIVATED SLUDGE, ALGAE, PROTOZOA, DIATOMS, EUGLENA W72-08573
WAGE LAGOONS, *LIMITING FACTORS, ALGAE, PRODUCTIVITY, PHOTOSYNTHES W72-09162
ONDS, *WATER REUSE, RECLAMATION, ALGAE, PROTEIN, *OXIDATION LAGOON W71-08951
XYGEN, BIOCHEMISTRY, INHIBITORS, ALGAE, PROTEINS.: /, *NITROGEN, O W72-07941
AL TECHNIQUES, DATA COLLECTIONS, ALGAE, PROTOZA, FUNGI, EQUIPMENT, W72-07225
MOLLUSKS, MARINE PLANTS, MARINE ALGAE, PUBLIC HEALTH, ESTUARINE E W72-04461
WATER POLLUTION SOURCES, AQUATIC ALGAE, PUBLIC HEALTH, BICDEGRADAT W71-11516
IMENTS, *MARINE ANIMALS, *MARINE ALGAE, RADIOISOTOPES, FALLOUT, CH W72-05965
S, CESIUM, COBALT RADIOISOTOPES, ALGAE, RADIOACTIVITY, TROPHIC LEV W70-06668
SEDIMENTS, EUTROPHICATION, FISH, ALGAE, RECREATION, ENVIRONMENTAL W71-10577
ONS, HYDROGEN ION CONCENTRATION, ALGAE, REPRODUCTION, SAMPLING, SE W71-08026
LGAE, *ECONOMICS, *PRODUCTIVITY, ALGAE, RESOURCES, FOOD CHAINS, FI W71-13246
, SAMPLING, VOLUMETRIC ANALYSIS, ALGAE, RESERVOIR STORAGE, IMPOUND W72-14280
*TASTE, *ODOR, DISSOLVED SOLIDS, ALGAE, RESERVOIR STORAGE, *HYPOLI W73-02138
TY EFFECTS, *MONITCRING, *MARINE ALGAE, RHODOPHYTA, PHAEPHYTA, *C W72-07826
LOR/ *YEASTS, *CULTURES, *MARINE ALGAE, RHODOPHYTA, PHAEOPHYTA, CH W72-04748
IDES, *FERTILIZATION, NUTRIENTS, ALGAE, RICE, NITROGEN, NITROGEN F W71-12071
*WATER ANALYSIS, *BIOINDICATORS, ALGAE, RIVERS, STREAMS, CELAWARE W72-04736
RVOIRS, WATER POLLUTION SOURCES, ALGAE, RIVER FLOW, PHOSPHCRUS, NI W72-04260
TIC WEED CONTROL, TILAPIA, CARP, ALGAE, ROOTED AQUATIC PLANTS, PLA W71-04528
IS, TEMPERATURE, SEASONAL, IONS, ALGAE, SALINITY, PHYTOPLANKTON, T W72-01363
, AQUATIC MICROORGANISMS, MARINE ALGAE, SALINITY, MIGRATICN, TIDES W72-14317
LUS, CHLORELLA, SULFUR BACTERIA, ALGAE, SALMONELLA, E. COLI.: /CIL W72-12565
COLIFORMS, MATHEMATICAL MODELS, ALGAE, SAMPLIN: /HATES, NITROGEN, W71-13356
ASSIUM, IRON, MANGANESE, COPPER, ALGAE, SAMPLING, CHLOROPHYTA.: /T W71-11001
DAPHNIA, AQUATIC PLANTS, AQUATIC ALGAE, SCENEDESMUS, RADIOISOTOPES W72-09649
FF, SALINITY, DEPTH, ABSORPTION, ALGAE, SEASONAL, WEATHER, WATER A W72-09122
ING CROPS, TIDAL EFFECTS, MARINE ALGAE, SEA WATER, SALINITY, SAMPL W73-00367
L/ *BACTERIA, *TOXICITY, *MARINE ALGAE, SEA WATER, CHLOROPHYTA, CH W72-08051
TS, SYNTHESIS, SLUDGE, EFFLUENT, ALGAE, SECONDARY TREATMENT, BIOCH W71-12188
PATH CF POLLUTANTS, ASIA, MARINE ALGAE, SEDIMENTS, ESTUARINE ENVIR W72-10958
OTOPES, *MARINE ANIMALS, *MARINE ALGAE, SEDIMENTS, NUCLEAR WASTES, W73-00817
ORPTION, *MARINE ANIMALS, MARINE ALGAE, SEDIMENTS, PUERTC RICO, ES W73-00821
ENTS(WATER), WINDS, THERMOCLINE, ALGAE, SEDIMENTS.: /ERATURE, CURR W72-12992
D OXYGEN, *ANAEROBIC CONDITIONS, ALGAE, SEDIMENTS, OXYGEN DEMAND, W72-12990
ANTI/ *MARINE FUNGI, *ISOLATION, ALGAE, SEPARATION TECHNIQUES, ATL W71-11851
LOROPHYTA, AQUATIC ALGAE, MARINE ALGAE, SESSILE ALGAE.: /PHYTA, CH W73-00273
ES, NITROGEN COMPOUNDS, AMMONIA, ALGAE, SEWAGE TREATMENT, TERTIARY W71-13939
OR/ *EPIPSAMMIC ALGAE, *EPIPELIC ALGAE, SHEAR WATER(ENGLAND), AMPH W71-12855
BENTHOS, CRABS, COPEPODS, MARINE ALGAE, SHRIMP, FISH EGGS, FISH, B W72-08142
ITROGEN, PHOSPHORUS, METABOLISM, ALGAE, SLUDGE, ANALYTICAL TECHNIQ W71-11793
ACTIVATED CARBON, FLCCULATION, ALGAE, SLUDGE, CHROMIUM, TCXICITY W71-09524
RE, NUTRIENTS, DISSOLVED OXYGEN, ALGAE, SLUDGE, OXYGENATICN, ANAER W71-07109
PS, MARINE ALGAE, MOSSES, CRABS, ALGAE, SNAILS, MUSSELS, ISOPODS, W72-03347
TY, FCOD CHAINS, FISH, BIOASSAY, ALGAE, SNAILS, MOSQUITOES, MODEL W72-07703
BIOMASS, LICHENS, MOSSES, FUNGI, ALGAE, SOIL CONTAMINATION, TRACE W72-14303
EN ION CONCENTRATION, ECONOMICS, ALGAE, SOLUBILITY, SEWAGE, BIOCHE W71-12200
MENTAL EFFECTS, ALGAL POISONING, ALGAE, SOLVENTS, SURFACTANTS, WAT W70-09429
RBONATES, DIATONICS, CYANOPHYTA, ALGAE, SOUTH CAKOTA, WATER POLLUT W71-02880
HYTA, RESERVOIRS, ODOR-PRODUCING ALGAE, SOUTHWEST U.S., TEMPERATUR W72-10613
OCEAN, *TROPICAL REGIONS, MARINE ALGAE, SPATIAL DISTRIBUTION, FOUR W72-11719

ERTEBRATES, ALGAE, MARINE ALGAE,	ALGAL POISONING, CHROMATOGRAPHY,	W71-08597
ONIA-NITROGEN, *GROWTH REACTORS,	ALGAL PONDS, ALGAL GROWTH RATES,	W72-04787
QUATIC ECOSYSTEMS, ALGAE GROWTH,	ALGAL POPULATIONS.: / NIDULANS, A	W71-00475
ENT REGENERATION, DEOXYGENATION,	ALGAL SEDIMENTATION.: /SIN, NUTRI	W72-12997
*CYANOPHYTA, *LABORATORY TESTS,	ALGAL TOXINS, HUMAN PATHOLOGY.:	W72-04259
PHYTA, CHLOROPHYTA, SCENEDESMUS,	ALGAL TOXINS, ANTIBIOTICS(PESTICI	W71-05155
AERATION, MIXING, ALGAL BLOOMS,	ALGAL TOXINS, WATER CIRCULATION,	W72-09304
ENVIRONMENT, TOXICITY, POISONS,	ALGAL TOXINS, FISH TOXINS, SEDIME	W72-10678
OZOA, PATHOGENS, MICROORGANISMS,	ALGAL TOXINS.: /ERIA, FUNGI, PROT	W72-08586
TOXICITY, CHEMICALS, REGULATION,	ALGAL TOXINS, MERCURY, DETERGENTS	W73-00703
IUM, SU/ *LPP VIRUS, *BLUE-GREEN	ALGAL VIRUSES, *PLECTONEMA BORYAN	W72-04736
ATION LAGOONS, SEWAGE TREATMENT,	ALGAL: /Y, DESIGN CRITERIA, *OXID	W72-04789
*NUTRIENT REQUIREMENTS, NUISANCE	ALGAS, *ENTROPHICATION, CYTOLOGIC	W72-03985
FISH PONDS, ALGISTATIC ACTIVITY,	ALGICIDAL ACTIVITY.: /SYNERGISM,	W71-06189
HYTA, CYANOPHYTA, ALGAL CONTROL,	ALGICIDES, COPPER SULFATE, SALINI	W72-14694
TICIDES, HERBICIDES, FUNGICIDES,	ALGICIDES, CARBAMATE PESTICIDES,	W73-00238
S, FISH, ALGAE, CHEMICAL WASTES,	ALGICIDES, BACTERICIDES, CHROMIUM	W72-08683
AENA, OSCILLATORIA, MICROCYSTIS,	ALGISTAT.: /LONIL, ULOTHRIX, ANAB	W72-07508
FICATION, SYNERGISM, FISH PONDS,	ALGISTATIC ACTIVITY, ALGICIDAL AC	W71-06189
URING TECHNIQUES, CULTURE MEDIA,	ALGOLOGY, TAXON.: CULTUT	W72-12172
SPRINGS, JACKSON HOT SPRINGS, /	ALHAMBRA HOT SPRINGS, BOULDER HOT	W72-05610
DELAWARE BAY, SCOLOPLOS ARMIGER,	ALIMITOS BAY, LAKE PONTCHARTRAIN,	W73-02029
TADECANE, HYDROCARBONS, ALKANES,	ALIPHATIC HYDROCARBONS, MASS SPEC	W73-01439
PROPERTIES, GOLD, IRON, CESIUM,	ALKALI METALS, SEASONAL, ZINC, AB	W73-01434
N, MICROORGANISMS, FERMENTATION,	ALKALINE HYDROLYSIS, ALGAE BLOOMS	W71-12223
ORGANIC PHOSPHORUS, *EXCRETION,	ALKALINE PHOSPHATASE, REASSIMILAT	W71-10083
*EPIPHYTOLOGY, *LAKES, *ALGAE,	ALKALINE WATER, MAGNESIUM, SODIUM	W71-04526
GEN, HYDROGEN ION CONCENTRATION,	ALKALINITY, RESPIRATION, PHOTOSYN	W71-04518
ER TEMPERATURE, BACTERIA, ALGAE,	ALKALINITY, IRON, PHOSPHATE, PHOS	W71-05626
MENTS, VOLCANOES, BENTHIC FLORA,	ALKALINITY, PHOTOSYNTHESIS, HYDRO	W71-09190
S, *THERMAL SPRINGS, CYANOPHYTA,	ALKALINITY, NUTRIENTS, WATER QUAL	W72-05610
E, MINERALOGY, TURBIDITY, COLOR,	ALKALINITY, VOLUMETRIC ANALYSIS,	W72-05594
ION, DETERGENTS, CARBON DIOXIDE,	ALKALINITY, BURNING, DOMESTIC WAS	W72-05473
H, CHEMICAL PROPERTIES, ACIDITY,	ALKALINITY, CONDUCTIVITY, CALCIUM	W72-04269
HOSPHATES, NITRATES, *NUTRIENTS,	ALKALINITY, CHEMICAL OXYGEN DEMAN	W72-04983
OXYGEN DEMAND, LABORATORY TESTS,	ALKALINITY, CARBONATES, BICARBONA	W72-04431
ON, CHLORELLA, SODIUM COMPOUNDS,	ALKALINITY, MINING, BIOCHEMICAL O	W72-04734
ITY, HYDROGEN ION CONCENTRATION,	ALKALINITY, CARBONATES, HARDNE: /	W71-00117
OIS, HYDROGEN ION CONCENTRATION,	ALKALINITY, HARDNESS(WATER), SAMP	W71-03028
HYLL, DIVERSION, CARBON DIOXIDE,	ALKALINITY, EPILIMNION, WASHINGTO	W71-02681
AN, MEASUREMENT, SUSPENDED LOAD,	ALKALINITY, HARDNESS(WATER), ANAL	W71-11033
ION, DECOMPOSING ORGANIC MATTER,	ALKALINITY, HYDROGEN ION CONCENTR	W71-12068
NTHESIS, *ALGAE, CARBON DIOXIDE,	ALKALINITY, NUTRIENTS, AMMONIA, H	W71-11519
DUCTIVITY, PHOSPHATES, NITRATES,	ALKALINITY, DIATOMS, CYANOPHYTA,	W72-08560
NTS, EUTROPHICATION, CYANOPHYTA,	ALKALINITY, DIURNAL, NITROGEN.: /	W72-09162
GEN ION CONCENTRATION, HARDNESS,	ALKALINITY, CONDUCTIVITY, S: /DRO	W72-10875
N, ANAEROBIC DIGESTION, AEROBIC,	ALKALINITY, ACIDITY, LABORATORY,	W72-02850
, ABSORPTION, LIGHT PENETRATION,	ALKALINITY, HALOGENS, HYDROGEN IO	W73-01434
N, TURBIDITY, WATER TEMPERATURE,	ALKALINITY, HYDROGEN ION CONCENTR	W73-00916
, BICARBONATES, HARDNESS(WATER),	ALKALINITY, ENZYMES, TEMPERATURE,	W72-07514
LOROPHYTA, DIATOMS, CHLOROPHYLL,	ALKALINITY, CARBON DIOXIDE, TEMPE	W72-08049
*CHLOROPHYLL, WATER TEMPERATURE,	ALKALINITY, DISSOLVED OXYGEN, CON	W72-14728
OGEN, HEAT TRANSFER, PHOSPHATES,	ALKALINITY, OXYGEN DEMAND, HEAT B	W72-12997
CTUANS, ANTINEOPLASTIC ACTIVITY,	ALKALOIDS, FLAVONOIDS, STEROIDS,	W72-04275
METHYLHEPTADECANE, HYDROCARBONS,	ALKANES, ALIPHATIC HYDROCARBONS,	W73-01439
ERCURIAL COMPOUNDS, ESOX LUCIUS,	ALKYL MERCURY, EAGLES, SALMO GAIR	W72-05952
ACTANTS, *EUTROPHICATION, LINEAR	ALKYLATE SULFONATES, CHELATION, P	W71-06247
SULFONATES, SURFACTANTS, LINEAR	ALKYLATE SULFONATES, SEWAGE, NUTR	W72-06837
*POTABLE WATER, EUTROPHICATION,	ALKYLBENZENE SULFONATES, SURFACTA	W72-06837
OLS, *DETERGENTS, *LETHAL LIMIT,	ALKYLBENZENE SULFONATES.: /, PHEN	W72-12576
NDARDS, WATER POLLUTION EFFECTS,	ALKYLBENZENE SULFONATES, WATER PO	W71-11516
MERS, TURBULANT FRICTIONAL DRAG,	ALKYLENE OXIDE, POLYMERIZATION, E	W70-06968
DE, NUTRIENTS, LIMITING FACTORS,	ALLOGENIC SUCCESSION, EUTROPHICAT	W72-14790
E, *WATER TREATMENT, *FILTATION,	ALLUVIA CHANNELS, SURFACE-GROUNDW	W71-12410
, ANACYSTIS NIDULANS, GLOEOCAPSA	ALPICOLA, OSCILLATORIA CHALYBIA,	W72-07664
S, PERCA FLUVIATILIS, SALVELINUS	ALPINUS, SALMO TRUTTA, VENDACE, S	W72-12729
LA FUSCA, *SCENEDEMUS ACUTUS VAR	ALTERNANS, *NITZSCHIA ACTINASTROI	W72-12739
A CRASSIPES, JUSTICIA AMERICANA,	ALTERNANTHERA PHILOXEROIDES, GLY:	W73-01089
QUALITY CONTROL, *NON-STRUCTURAL	ALTERNATIVES, AGING(BIOLOGICAL),	W72-04279
ION, MIXING, ALGAE, TEMPERATURE,	ALTITUDE, CLIMATES, METABOLISM, B	W72-06838
*ORGANIC NUTRIENTS,	ALUM COAGULATION.:	W73-02187
OPERATION, ACID INJECTION PUMP,	ALUM INJECTION PUMP, MANURE GRIND	W72-02850
(ISRAEL).:	ALUM, FERRIC CHLORIDE, *JERUSALEM	W72-12289
C SULFATE, *STABILIZATION PONDS,	ALUM, POST-TREATMENT.: /E, *FERRI	W71-13356
ASIA.:	ALUM, RECYCLING, TREATMENT COSTS,	W72-13508
NTS, ALGAE, BACTERIA, LAKE ERIE,	ALUMINUM, MANGANESE, SODIUM, CHRO	W72-01101
OISOTOPES, RADIUM RAIODISOTOPES,	ALUMINUM, IRON, COPPER, MAGNESIUM	W71-09188
AMPLE PRESERVATION, SALPA, LUDOX	AM, SUCROSE, DENSITY GRADIENTS, T	W73-00371
DICA, MELOSIRA ITALICA, MELOSIRA	AMBIGUA, SYNEDRA FILIFORMIS, FRAG	W73-00384
RUS NEBULOSUS, PERCA FLAVESCENS,	AMBLOPLITES RUPESTRIS, ESOX LUCIU	W70-08098
NIKIN TEST.:	AMCHITKA ISLAND, MILROW TEST, CAN	W73-00764
S CARPIO, CATOSTOMUS COMMERSONI,	AMEIURUS NEBULOSUS, PERCA FLAVESC	W70-08098
VIRGINIA, *ATLANTIC OCEAN, NORTH	AMERICA.: /ARSHES, SYSTEMATICS, *	W72-12633
N, CALCIUM, PUBLIC HEALTH, SOUTH	AMERICA, ATLANTIC COASTAL PLAIN,	W73-00819
ES, *TOCKS ISLAND RESERVOIR(NJ),	AMERICAN SHAD, EKMAN DREDGE, PUMP	W72-14280
, EICHHORNIA CRASSIPES, JUSTICIA	AMERICANA, ALTERNANTHERA PHILOXER	W73-01089
BIOLOGICAL SYSTEMS, BELEMNITELLA	AMERICANA, TERRESTRIAL PLANTS, CO	W72-12709
NIA SUBEQUICILIATA, UROGLENOPSIS	AMERICANA, CRYPTOMONAS PULSILLA,	W72-13816
ROSIN	AMINE D ACETATE(RADA).:	W71-05083
A, CADMIUM, DIMETHYLNITROSAMINE,	AMINES, CARCINOGENS, TORULOPSIS S	W72-09675

, MUSSELS, MARINE ALGAE, AQUATIC	ANIMALS, FRESH WATER, FRESHWATER	W71-09182
ANOPHYTA, BENTHIC FAUNA, AQUATIC	ANIMALS, INVERTEBRATES, AQUATIC A	W72-12729
ANIMALS, *MARINE PLANTS, AQUATIC	ANIMALS, LABORATORY TESTS, NITRAT	W71-09168
RROPHYTA, INVERTEBRATES, AQUATIC	ANIMALS, LITTORAL, WAVES(WATER).:	W71-11012
A, SAMPLING, MARINE FISH, MARINE	ANIMALS, MARINE ALGAE, TRITIUM, I	W71-09018
LOGY, BIRD TYPES, BIRDS, AQUATIC	ANIMALS, MOSSES, INVERTEBRATES, D	W72-05952
S, *COBALT, *ABSORPTION, *MARINE	ANIMALS, MARINE ALGAE, SEDIMENTS,	W73-00821
OWERPLANTS, MARINE ALGAE, MARINE	ANIMALS, MARINE PLANTS, HUMAN POP	W73-00833
IVITY EFFECTS, *FALLOUT, *MARINE	ANIMALS, MARINE PLANTS, MARINE AL	W73-00828
, REFRACTIVITY, COPEPODS, MARINE	ANIMALS, MOLLUSKS, GASTROPODS, PH	W73-00371
BUTION, *DENSITY, *DEPTH, MARINE	ANIMALS, MARINE ALGAE, CYANOPHYTA	W72-14313
AQUATIC MICROORGANISMS, AQUATIC	ANIMALS, NUTRIENTS, WATER POLLUTI	W72-07896
EFFECTS, *MARINE PLANTS, *MARINE	ANIMALS, OFFSHORE PLATFORMS, NUCL	W73-02029
INITY, NUTRIENTS, ALGAE, AQUATIC	ANIMALS, RADIONUCLIDES.: /ES, SAL	W72-12190
Y, DISTRIBUTION PATTERNS, MARINE	ANIMALS, SALINITY, SAMPLING, SEA	W73-00296
TREATMENT, COLLOIDS, *POLYMERS,	ANIONS, CATIONS, ORGANIC COMPOUND	W72-11833
R TREATMENT, *POLYMERS, CATIONS,	ANIONS, CHLORELLA, FILTRATION, EL	W72-14840
LATION EFFICIENCES, CRYPTOMONAS,	ANKISTRODESMUS, ELAKOTOTHRIX, OSC	W72-07940
FORMOSA, PHORMIDIUM FAVEOLARUM,	ANKISTRODESMUS BRAUNII, CHLORELLA	W72-07664
FORMOSA, STEPHANODISCUS ASTREA,	ANKISTRODESMUS, CHRYSOCOCCUS MINU	W72-12738
ORIDE, METHYL MERCURIC CHLORIDE,	ANKISTRODESMUS BRAUNII, EUGLENA G	W72-06037
E, *DETOXICATION, CHLOROSARCINA,	ANKISTRODESMUS BRAUNII, CHLORELLA	W71-07675
A.: MOUGEOTIA GENUFLEXA,	ANKISTRODESMUS FALCATUS, PODOHEDR	W71-07360
CONIS, CYMBELLA, STEPHANODISCUS,	ANKISTRODESMUS, TRACHELOMONAS, SA	W72-10875
R(ME), CULTURAL EUTROPHIC/ *LAKE	ANNABESSACOOK(ME), *KENNEBEC RIVE	W72-04280
US, CYANOPHYTA, MOLLUSKS, MITES,	ANNE: /AS, CHLOROPHYTA, SCENEDESM	W72-00155
EACHES, SANDS, INTERTIDAL AREAS,	ANNELIDS, COPEPODS, CRABS, AQUATI	W72-07911
COMPOUNDS, WORMS, INVERTEBRATES,	ANNELIDS, SEASONAL, CRUSTACEANS,	W73-00374
S, DINOFLAGELLATES, CRUSTACEANS,	ANNELIDS, WATER POLLUTION EFFECTS	W73-00935
MATICS, *ECOLOGY, *MARINE ALGAE,	ANNELIDS, MOLLUSKS, RHODOPHYTA, P	W73-00932
ARUM, TRAMETES VERSICOLOR, FOMES	ANNOSUS, FUSCOPORIA CONTIGUA, TRA	W72-13813
, ECONOMICS, FEASIBILITY, COSTS,	ANNUAL COSTS, PILOT PLANTS.: /TES	W70-10174
CIPAL WASTES, INDUSTRIAL WASTES,	ANNUAL TURNOVER, ENERGY, CYANOPHY	W72-03228
S HOFFMEISTERI, STICTOCHIRONOMUS	ANNULIORUS, PHYSA HETEROSTROPHA,	W70-06668
RS.: *CHELATORS,	ANODIC STRIPPING, ORGANIC CHELATO	W71-09674
RGHEI, BORRELIA KANSAS, BORRELIA	ANSERINA.: / LIVER, PLASMODIUM BE	W73-01676
IRONOMUS ATTENTUATUS, CHIRONOMUS	ANTHRACINUS, LIMNODRILUS HOFFMEIS	W71-11009
C CHLORIDE, MULTI-LAYER FILTERS,	ANTHRACITE, *WAHNBACH RESERVOIR(G	W72-06201
OPHYTA, CHLOROPHYTA, INHIBITORS,	ANTIBIOTICS, SEA WATER, RHODE ISL	W72-04748
ODDER, GLUE, AGAR, POLYSIPHONIA,	ANTIBIOTICS, IODINE, ATOMIC ENERG	W71-09157
HYTA, SCENEDESMUS, ALGAL TOXINS,	ANTIBIOTICS(PESTICIDES), DATA PRO	W71-05155
HATES, ECOSYSTEMS, AQUATIC LIFE,	ANTIBIOTICS(PESTICIDES), ECOLOGY,	W72-09668
L, ASSAY, REMOTE SENSING, ALGAE,	ANTIBIOTICS, MERCURY, ARSENIC COM	W72-10162
MARY PRODUCTIVITY, CORAL, REEFS,	ANTIBIOTICS(PESTICIDES), CHLORINA	W72-14294
ECHNIQUES, CYANOPHYTA, CULTURES,	ANTIBIOTICS(PESTICIDES), PHOTOACT	W72-14327
FISH, BACTERIA, FISH PARASITES,	ANTIBIOTICS(PESTICIDES), BACTERIC	W72-08142
, DNA, MOTILITY, CYST FORMATION,	ANTIBIOTIC SENSITIVITY, CYTOPHAGA	W72-07951
NIC BACTERIA, COLIFORMS, TOXINS,	ANTIBIOTICS(PESTICIDES), ENTERIC	W72-13800
YPHIMURI/ DETECTION, FLUORESCENT	ANTIBODY TECHNIQUES, SALMONELLA T	W72-13800
ME, *PHARMACOLOGICAL PROPERTIES,	ANTICANCER, ANTINEOPLASTICS.: /TI	W72-07360
NUPHAR JAPONICUM, NUPHAR LUTEN,	ANTICANCER POTENTIAL, ANTICOAGULA	W72-04275
HAR LUTEN, ANTICANCER POTENTIAL,	ANTICOAGULANT POTENTIAL, SPARGANI	W72-04275
HIDLEY(LABRADOR), NEW BRUNSWICK,	ANTICOSTI ISLAND(QUEBEC), MAGDALE	W71-10066
ON METHOD, *ZONES OF INHIBITION,	ANTIFUNGAL ACTIVITY.: /SC DIFFUSI	W72-08586
LS, PENICILLIN, LINDANE, DIADEMA	ANTILLARUM, AANDOMENE, SIPHONALES	W72-14294
SELENOMETHIONINE, SELENOPURINE,	ANTIMETABOLITES.: /ATE, SELENITE,	W72-00153
PROPERTIES, PHYTOCHEMICAL SCRE/	ANTIMICROBIOTICS, PHARMACOLOGICAL	W72-04275
S-AQUAE, *PHARMACOLOGICAL TESTS,	ANTIMICROBIOTICS, ESCHERICHIA COL	W72-04259
RPTION, ARSENIC, SELENIUM, ZINC,	ANTIMONY, NORWAY, ASHING, BIOLOGI	W71-11515
OPE.:	ANTIMONY, CESIUM, *CAPE OF GOOD H	W73-00832
MS LAKE(WASH), URANIUM, THORIUM,	ANTIMONY, MACRONUTRIENTS, MICRONU	W73-01434
OLOGICAL PROPERTIES, ANTICANCER,	ANTINEOPLASTICS.: /TIME, *PHARMAC	W72-07360
POTENTIAL, SPARGANIUM FLUCTUANS,	ANTINEOPLASTIC ACTIVITY, ALKALOID	W72-04275
NABAENA, MICROCYSTIS, TRIBONEMA,	APHANIZOMENON, OOCYSTIS, GLOEOCYS	W72-07940
DISC, NEWFOUND LAKE, SQUAM LAKE,	APHANIZOMENON FLOW-AQUAE, DESTRAT	W72-07890
ENA SPP, MICROCYSTIS AERUGINOSA,	APHANIZOMENON HOLSATICUM, ICTALUR	W72-07890
OCYSTIS AERU/ *KEZAR LAKE(N.H.),	APHANIZOMENON, ANABAENA SPP, MICR	W72-07890
RIVER, SALT RIVER, MICROCYSTIS,	APHANIZOMENON, CYANIDES, COHO SAL	W72-14282
S HOFFMEISTERI, TUBIFEX TUBIFEX,	APHANIZOMENON, CLADOPHORA.: /RILU	W71-11009
IA, STEPHANODISCUS, MICROCYSTIS,	APHANIZOMENON, ANABENA SPIROIDES.	W71-12855
M,/ DOLLAR LAKE(OHIO), SPHAGNUM,	APHANIZOMENON FLOS-AQUAE, CERATIU	W71-11949
CIRCINALIS, ANABAENA CYLINDRICA,	APHANIZOMENON FLOS-AQUAE.: /AENA	W72-10613
LLATORIA, APHANOTHECE, ANABAENA,	APHANIZOMENON, NAVICULA, GLENODIN	W72-10875
ISM, AXENIC CULTURES, EXCRETION,	APHANIZOMENON GRACILE.: /OANTAGON	W72-12734
OMTA, STEPHANODISCUS HANTZSCHII,	APHANIZOMENON FLOS-AQUAE, CHROOMO	W72-12738
UYAHOGA RIVER(OHIO), CLADOPHORA,	APHANOCHAETE, RHIZOCLONIUM, VAUCH	W71-05629
NTING, CYCLOTELLA, OSCILLATORIA,	APHANOTHECE, ANABAENA, APHANIZOME	W72-10875
SEAWEEDS,	APHIDS, SPONGES, CARBON-14.:	W72-06283
TOCOCCUS PYRIFORMIS NAUTOCOCCUS,	APIOCOCCUS, OSCILLATORIA, NAUTOCO	W72-07141
TRIA), CHAMAESIPHON SUBAEQUALIS,	APISTONEMA EXPANSUM, PORPHYRIDIUM	W71-04526
AMONAS SAUPAULENSIS, ENEIDAMONAS	APPLANATA, BICUDOMONAS CYANOPH: /	W72-09119
C TANKS, OXIDATION LAGOONS, LAND	APPLICAT: /DENITRIFICATION, SEPTI	W70-09907
OIRS, SWIMMING POOLS, CHELATION,	APPLICATION METHODS, COOLING TOWE	W71-06189
NISMS, AQUATIC ANIMALS, DAPHNIA,	APPLICATION METHODS, GRANULES, PL	W72-09650
QUALITY, RIPARIAN RIGHTS, PRIOR	APPROPRIATION, GROUNDWATER, WATER	W70-07100
EN INACTIVATION./ *ANABAENA FLOS-	AQUA, *NITROGENASE ACTIVITY, OXYG	W72-07941
ISH, MUSSELS, SHELLFISH FARMING,	AQUACULTURE, COMMERCIAL SHELLFISH	W72-09378
A CYLINDRICA, APHANIZOMENON FLOS-	AQUAE.: /AENA CIRCINALIS, ANABAEN	W72-10613
LINE PHOSPHATASE, *ANABAENA FLOS-	AQUAE, *CONTINUOUS CULTURE, CHEMO	W72-05476
EUTIC VALUE, *APHANIZOMENON FLOS-	AQUAE, *PHARMACOLOGICAL TESTS, AN	W72-04259

C ALGAE, AQUATIC MICROORGANISMS,	AQUATIC ANIMALS, NUTRIENTS, WATER	W72-07896
ABITATS, BIOMASS, AQUATIC ALGAE,	AQUATIC ANIMALS, BALANCE OF NATUR	W72-06340
ATIC ENVIRONMENT, AQUATIC ALGAE,	AQUATIC ANIMALS, AQUATIC INSECTS,	W72-07132
TER POLLUTION EFFECTS, CATTAILS,	AQUATIC ANIMALS, CLAMS, CARP, CRA	W72-06342
AE, CARBON CYCLE, ON-SITE TESTS,	AQUATIC ANIMALS, AQUATIC POPULATI	W73-02095
DES, ECOLOGY, BIRD TYPES, BIRDS,	AQUATIC ANIMALS, MOSSES, INVERTEB	W72-05952
, SNAILS, MUSSELS, MARINE ALGAE,	AQUATIC ANIMALS, FRESH WATER, FRE	W71-09182
*MARINE ANIMALS, *MARINE PLANTS,	AQUATIC ANIMALS, LABORATORY TESTS	W71-09168
CTERIA, PHOTOSYNTHETIC BACTERIA,	AQUATIC BACTERIA, LAKES, FILTERS,	W72-04743
AQUATIC HABITATS, PRODUCTIVITY,	AQUATIC BACTERIA, AQUATIC ALGAE,	W73-02095
POLLUTANT IDENTIFICATION, ALGAE,	AQUATIC BACTERIA, SEWAGE BACTERIA	W72-01472
EDESMUS, AQUATIC MICROORGANISMS,	AQUATIC BACTERIA, AQUATIC ALGAE,	W72-12255
	AQUATIC BIOTOXINS, *BIOTOXINS.:	W72-14056
EK RESERVOIR(KAN), GREAT PLAINS,	AQUATIC ECOSYSTEM.: / *TUTTLE CRE	W72-12393
, *ALGAE, *LIG/ *EUTROPHICATION,	AQUATIC ECOSYSTEMS, WATER QUALITY	W73-02349
H CAROLINA), ANACYSTIS NIDULANS,	AQUATIC ECOSYSTEMS, ALGAE GROWTH,	W71-00475
ARBON, *PHOSPHORUS, *CYANOPHYTA,	AQUATIC ENVIRONMENT, CARBON DIOXI	W71-00475
SEWAGE, *EUTROPHICATION, *LAKES,	AQUATIC ENVIRONMENT, CHEMICAL PRO	W71-00117
E ERIE, AQUATIC HABITATS, LAKES,	AQUATIC ENVIRONMENT.: /LGAE, *LAK	W71-09156
FFECTS, ALGAE, AQUATIC HABITATS,	AQUATIC ENVIRONMENT.: /OLLUTION E	W71-07337
SPHORUS, ALGAE, *EUTROPHICATION,	AQUATIC ENVIRONMENTS, GREAT LAKES	W71-08953
SH, RAINBOW TROUT, AQUATIC LIFE,	AQUATIC ENVIRONMENT, FISH PHYSIOL	W72-12257
OGICAL COMMUNITIES, *SUCCESSION,	AQUATIC ENVIRONMENT, *ECOLOGICAL	W72-11854
D WEBS, WATER POLLUTION EFFECTS,	AQUATIC ENVIRONMENT, TURBIDITY, W	W72-11925
TER QUALITY CONTROL, PHOSPHORUS,	AQUATIC ENVIRONMENT, WATER POLLUT	W72-06638
COMMUNITIES, DOMINANT ORGANISMS,	AQUATIC ENVIRONMENT, AQUATIC ALGA	W72-07132
TREATMENT, ALGAE, PLANTS, TROUT,	AQUATIC ENVIRONMENT, EUTROPHICATI	W72-02412
EN ION CONCENTRATION, NUTRIENTS,	AQUATIC ENVIRONMENT, CARBON DIOXI	W71-12068
ALGAL CONTROL, ALGAL POISONING,	AQUATIC ENVIRONMENT, AQUATIC FUNG	W72-11186
MATHEMATICAL STUDIES, EQUATIONS,	AQUATIC ENVIRONMENT, WATER PROPER	W72-11724
ARBON, *PHOSPHORUS, *POPULATION,	AQUATIC ENVIRONMENT, ECOSYSTEMS,	W72-09159
POISONING, AQUATIC ENVIRONMENT,	AQUATIC FUNGI, ACQUIFER CHARACTER	W72-11186
, *ALGAE, CHLOROPHYTA, HABITATS,	AQUATIC HABITATS, EUGLENOPHYTA, P	W72-09119
POWER PLANTS, *DISCHARGE(WATER),	AQUATIC HABITATS, ENVIRONMENTAL E	W71-11517
STUDIES, *STREAMS, *ECOSYSTEMS,	AQUATIC HABITATS, BIOMASS, AQUATI	W72-06340
OGICAL DISTRIBUTION, ECOSYSTEMS,	AQUATIC HABITATS, PRODUCTIVITY, A	W73-02095
MNOLOGY, WATER QUALITY, ECOLOGY,	AQUATIC HABITATS, CLADOPHORA, LIM	W72-13142
, STREAMS, ISOPODS, SCENEDESMUS,	AQUATIC HABITATS, CLASSIFICATION,	W72-12744
WATER POLLUTION EFFECTS, ALGAE,	AQUATIC HABITATS, AQUATIC ENVIRON	W71-07337
LAKE SHORES, *ALGAE, *LAKE ERIE,	AQUATIC HABITATS, LAKES, AQUATIC	W71-09156
XIDATION LAGOONS, PHYTOPLANKTON,	AQUATIC I: /OLOGICAL TREATMENT, O	W72-12567
BIRDS, STREAMS, AQUATIC PLANTS,	AQUATIC INSECTS, CRUSTACEANS, MUS	W72-12930
AQUATIC ALGAE, AQUATIC ANIMALS,	AQUATIC INSECTS, DAPHNIA, AQUATIC	W72-07132
URCES, *FISH, DIATOMS, PROTOZOA,	AQUATIC INSECTS, AQUATIC ANIMALS,	W72-07791
REAS, ANNELIDS, COPEPODS, CRABS,	AQUATIC LIFE, BACTERIA.: /TIDAL A	W72-07911
HELLFISH, CLAMS, MARINE ANIMALS,	AQUATIC LIFE, ECOSYSTEMS, SALMONI	W72-07907
COASTS, WATER POLLUTION EFFECTS,	AQUATIC LIFE, HABITATS, ECOLOGY,	W73-01632
SOLVED OXYGEN, COMPUTERS, ALGAE,	AQUATIC LIFE, ENVIRONMENTAL EFFEC	W72-14405
RBON RADIOISOTOPES, NEW ENGLAND,	AQUATIC LIFE, AQUATIC MICROORGANI	W72-14317
TROPHY, OLIGOTROPHY, OXYGEN SAG,	AQUATIC LIFE, BIOASSAY, CLASSIFIC	W73-00002
ESTUARINE ENVIRONMENT, HABITATS,	AQUATIC LIFE, HEAVY METALS, INDUS	W72-11854
, BIOASSAY, FISH, RAINBOW TROUT,	AQUATIC LIFE, AQUATIC ENVIRONMENT	W72-12257
CIDES, BIBLIOGRAPHIES, *ECOLOGY,	AQUATIC LIFE, IRRIGATION SYSTEMS,	W71-06102
TEMPERATURE, SEASONAL, MONTHLY,	AQUATIC LIFE, INTERSTATE RIVERS,	W70-09595
BIVORES, CARNIVORES, ECOSYSTEMS,	AQUATIC LIFE, SAMPLING, ALGAE, LA	W71-10064
DS, NUTRIENTS, DISSOLVED OXYGEN,	AQUATIC LIFE, MUNICIPAL WASTES, P	W71-11017
CTS, FOOD CHAINS, PUBLIC HEALTH,	AQUATIC LIFE, AQUATIC ALGAE, CALC	W72-10978
FFECTS, WATER POLLUTION SOURCES,	AQUATIC LIFE, AQUATIC ANIMALS, AQ	W72-09646
TRACERS, PHOSPHATES, ECOSYSTEMS,	AQUATIC LIFE, ANTIBIOTICS(PESTICI	W72-09668
BATEMENT, ENVIRONMENTAL EFFECTS,	AQUATIC LIFE, FISH POPULATIONS, P	W72-00939
, AQUATIC LIFE, AQUATIC ANIMALS,	AQUATIC MICROORGANISMS, AQUATIC P	W72-09646
S, AQUATIC ALGAE, PHYTOPLANKTON,	AQUATIC MICROORGANISMS, AQUATIC A	W72-09650
ITION, *CHLORELLA, *SCENEDESMUS,	AQUATIC MICROORGANISMS, AQUATIC B	W72-12255
OPES, NEW ENGLAND, AQUATIC LIFE,	AQUATIC MICROORGANISMS, MARINE AL	W72-14317
YSIS, POLLUTANTS, AQUATIC ALGAE,	AQUATIC MICROORGANISMS, AQUATIC A	W72-07896
UTROPHICATION, LARVAE, ROTIFERS,	AQUATIC MICROORGANISMS, DIPTERA,	W72-07132
ANCE, TRACERS, PLANT PHYSIOLOGY,	AQUATIC PLANTS, RADIOECOLOGY.: /L	W72-06342
EFFECTS, STREAMS, RIVERS, ROOTED	AQUATIC PLANTS, AQUATIC ALGAE, WA	W72-14690
RE, MARINE ALGAE, PHYTOPLANKTON,	AQUATIC PLANTS, CHEMICAL PROPERTI	W72-14542
S, BIOASSAY, CYANOPHYTA, RESINS,	AQUATIC PLANTS, AQUATIC ALGAE, NI	W73-00286
TICAL TECHNIQUES, *ALGAL TOXINS,	AQUATIC PLANTS, TOXINS, RED TIDE,	W73-00243
, *DISTRIBUTION PATTERNS, LAKES,	AQUATIC PLANTS, AUSTRALIA, SAMPLI	W73-01094
REVIEWS, BIOLOGICAL COMMUNITIES,	AQUATIC PLANTS, LIMITING FACTORS,	W73-01089
PLANTS, LIMITING FACTORS, ROOTED	AQUATIC PLANTS, CYCLING NUTRIENTS	W73-01089
ION, PHYSICOCHEMICAL PROPERTIES,	AQUATIC PLANTS, INDUSTRIAL: /TRAT	W73-01615
ANALYSIS, MEASUREMENT, *NEVADA,	AQUATIC PLANTS, WATER POLLUTION S	W73-01673
IC WASTES, ECOLOGY, FISH, ALGAE,	AQUATIC PLANTS, PROTOZOA, BACTERI	W73-01972
RTEBRATES, FISH, BIRDS, STREAMS,	AQUATIC PLANTS, AQUATIC INSECTS,	W72-12930
GEN, OXYGEN DEMAND, HYPOLIMNION,	AQUATIC PLANTS, ALGAE.: /LVED OXY	W72-12994
TAL EFFECTS, METHODOLOGY, ALGAE,	AQUATIC PLANTS, ISOTOPE FRACTIONA	W72-12709
TICAL TECHNIQUES, FERTILIZATION,	AQUATIC PLANTS, NITROGEN, NUTRIEN	W73-02478
AQUATIC ANIMALS, CARP, DAPHNIA,	AQUATIC PLANTS, AQUATIC ALGAE, SC	W72-09649
ANIMALS, AQUATIC MICROORGANISMS,	AQUATIC PLANTS, AQUATIC ALGAE, PE	W72-09646
ICATION, ONTARIO, ALGAE, MOSSES,	AQUATIC PLANTS.: /LLUTANT IDENTIF	W71-11529
*EUTROPHICATION, FERTILIZATION,	AQUATIC PLANTS, MATHEMATICAL MODE	W71-12072
ROL, RIVERS, AGRICULTURE, ROOTED	AQUATIC PLANTS, DRAINAGE, RUNOFF,	W71-13172
PHORUS COMPOUNDS, POTASH, ALGAE,	AQUATIC PLANTS, CHEMICAL ANALYSIS	W71-12252
OL, TILAPIA, CARP, ALGAE, ROOTED	AQUATIC PLANTS, PLANKTON, ARKANSA	W71-04528
DITY, RUNOFF, PHOSPHORUS, ALGAE,	AQUATIC PLANTS, CLAYS, ODOR, WATE	W72-04759
FISH, INDUSTRIAL WASTES, ALGAE,	AQUATIC PLANTS, SEA WATER, TRACE	W72-05952

*ALGAE, *EUTROPHICATION, LAKES,	AQUATIC PLANTS, IRON, MICROORGANI	W72-03742
ER, ICNS, NUTRIENT REQUIREMENTS,	AQUATIC POPULATIONS, PRIMARY PROD	W72-09103
IRONMENTAL EFFECTS, COMPETITICN,	AQUATIC POPULATIONS, NUTRIENTS, N	W72-12736
ON-SITE TESTS, AQUATIC ANIMALS,	AQUATIC POPULATICNS, BIOASSAY, EU	W73-02095
IMALS, AQUATIC INSECTS, DAPHNIA,	AQUATIC POPULATIONS, FISH DIETS,	W72-07132
RIVERS, EFFLUENT, RADIOECOLOGY,	AQUATIC POPULATIONS, HYDROLCGIC A	W72-07841
YTA, SCENEDESMUS, AQUATIC ALGAE,	AQUATIC PRODUCTIVITY, PCLLUTANT I	W73-01066
ICAL COMMUNITIES, AQUATIC ALGAE,	AQUATIC PRODUCTIVITY, BIOCHEMICAL	W73-00002
EDESMUS, GROWTH RATES, CULTURES,	AQUATIC PRODUCTIVITY, METABOLISM,	W72-14314
POLLUTION SOURCES, CHLOROPHYTA,	AQUATIC PRODUCTIVITY, AQUATIC ALG	W73-00379
*CHLOROPHYLL, *PLANT PIGMENTS,	AQUATIC PRODUCTIVITY, *PRIMARY PR	W72-12192
ENT-WATER INTERFACES, SEDIMENTS,	AQUATIC PRODUCTIVITY, PHYTOPLANKT	W72-08989
COMMUNITIES, DOMINANT ORGANISMS,	AQUATIC PRODUCTIVITY, BIOMASS, CA	W72-09239
ED OXYGEN, DIURNAL DISTRIBUTION,	AQUATIC PRODUCTIVITY, NITROGEN, P	W71-11519
GING(BIOLOGICAL), AQUATIC ALGAE,	AQUATIC PRODUCTIVITY, FISH, FISHK	W72-04279
ER POLLUTION SOURCES, SEDIMENTS,	AQUATIC SOILS, PLANT TISSUES, FRE	W73-01673
SING, *POLLUTANT IDENTIFICATION,	AQUATIC WEEDS, *WASHINGTCN, *CHLO	W72-14728
OSPHORUS, EUTROPHICATION, ALGAE,	AQUATIC WEEDS, PHYTOPLANKTON, OLI	W72-08459
UCT E/ *ALGAE CULTURE TECHNIQUE,	AQUATIC WEED STUDY FACILITY, PROD	W71-06102
PULP AND PAPER INDUSTRY, ALGAE,	AQUATIC WEEDS, RECREATION, GROUND	W70-05103
DEMAND, ECONOMIC JUSTIFICATION,	AQUATIC WEED CONTROL, RIVERS, AGR	W71-13172
E ALGAE, COMPREHENSIVE PLANNING,	AQUATIC WEEDS, WATER POLLUTICN SO	W72-09171
UTRIENTS, LAKES, MUD, WISCONSIN,	AQUATIC WEEDS, EUTROPHICATION.: /	W72-10608
URAL RUNOFF, FARM WASTES, ALGAE,	AQUATIC WEEDS, FERTILIZATION, LEA	W73-02479
MENTS(ALGAE), NUTRIENT CHEMISTRY(	AQUATIC).: /), PHOSPHORUS REQUIRE	W70-04080
SSIPES, TYPHA ANGUSTATA, IPOMOEA	AQUATICA, TOTAL SOLIDS.: /NIA CRA	W72-01363
BORATES, ARTHROBACTE/ NEOCHLORIS	AQUATICA, SCHIZOTHRIX CALCICOLA,	W73-00854
SCIRPUS SPP, TYPHA SPP, ZIZANIA	AQUATICA, CAREX SPP, BUTCMUS UMBE	W73-01089
MUSSELS, CLAMS, FISH, SHELLFISH,	AQUEOUS SOLUTIONS, PHAECPHYTA, AT	W72-05965
OUNDS, POLLUTANT IDENTIFICATICN,	AQUEOUS SOLUTIONS, ION EXCHANGE,	W72-12166
OWERPLANTS, DISTILLATION, ALGAE,	AQUICULTURE, FISH, SHELLFISH, CRU	W70-08832
ESCENT CULTURES, XYLOSE, RIBOSE,	ARABINOSE, GALACTCSE, HEXURONIC A	W72-07710
WAY, MACROPHYTES, FYRISAN RIVER,	ARBOGAAN RIVER, BOTORPSSTROMMEN R	W73-00286
KE, BIG ARBOR VITAE LAKE, LITTLE	ARBOR VITAE LAKE, GREEN BAY, FOX	W73-01456
A, CRYSTAL LAKE, TROUT LAKE, BIG	ARBOR VITAE LAKE, LITTLE ARBOR VI	W73-01456
LAKE MARY, LAKE JOSEPHINE, LAKE	ARCO, LAKE DEMING, LOWER LA SALLE	W72-07145
TOMONAS-SPP.: *CANADIAN	ARCTIC, *CHLAMYDOMONAS-SPP, *CRYP	W72-07922
I/ *ECOSYSTEMS, *ALASKA, *LAKES,	ARCTIC, PRODUCTIVITY, EUTROPHICAT	W71-09190
ENT, *SAMPLING, WATER RESOURCES,	AREA REDEVELOPMENT, BIOLOGICAL CH	W72-03215
ITHIC ALGAE/ *EXPERIMENTAL LAKES	AREA, *NORTHWESTERN ONTARIO, EPIL	W71-11029
*ROADSIDE REST	AREAS.:	W72-14363
R POLLUTION EFFECTS, *INTERTIDAL	AREAS, *INVERTEBRATES, *ALGAE, CA	W71-08792
IDES, BEACHES, SANDS, INTERTIDAL	AREAS, ANNELIDS, COPEPODS, CRABS,	W72-07911
L MARSHES, FOOD WEBS, INTERTIDAL	AREAS, CONNATE WATER.: /IES, TIDA	W73-00796
ESTICIDES, ESTUARIES, INTERTIDAL	AREAS, ESTUARINE FISHERIES.: /, P	W72-10678
, OTTERS, ROCKSLIDES, INTERTIDAL	AREAS, CN-SITE INVESTIGATIONS, FO	W73-00764
	ARGENTINA.:	W73-00819
NE, PROLINE, GLUTAMATE, LEUCINE,	ARGININE, SERINE, TYROSINE, TRYPT	W73-00942
ATUS, CERAMIUM NITENS, WRANGELIA	ARGUS, VALONIA VENTRICOSA, BRYOPS	W73-00838
OOMS, 2,4 DINITRO PHENOL, TUCSON(	ARIZ).: /ALCIUM RADIOISOTOPES, BL	W71-01474
LIF), LOS ANGELES(CALIF), TUCSON(	ARIZ).: /LING, IMPERIAL VALLEY(CA	W71-09157
R POLLUTION CONTROL, CALIFORNIA,	ARIZONA.: /OMIC FEASIBILITY, WATE	W71-09157
OGEN ICN CONCENTRATION, PHENOLS,	ARIZONA.: /, MICROORGANISMS, HYDR	W71-01474
ROOTED AQUATIC PLANTS, PLANKTON,	ARKANSAS, INSECTS, TURBIDITY, CHA	W71-04528
A, NEPHTYS, CAPITELLA, MAGELONA,	ARMANDIA, TELLINA BUTTONI, VENUS,	W72-09092
TES, PHACUS PYRUM, TRACHELOMONAS	ARMATA, TRACHELOMONAS HISPIDA VAR	W73-00918
RUCTICN, DELAWARE BAY, SCOLOPLOS	ARMIGER, ALIMITOS BAY, LAKE PONTC	W73-02029
E.:	ARNOLDIELLA-CONCHOPHILA, LAKE ERI	W71-12489
LGAL SPECIES, WESTERN LAKE ERIE,	ARNOLDIELLA CONCHCPHILA MILLER.: /	W71-09156
FIS/ *POLYCHLORINATED BIPHENYLS,	AROCLOR 1254, ESCAMBIA RIVER, PIN	W71-07412
ICROCYSIS, OSCILLATORIA, SYNURA,	AROGLENA, CRYPTOPHYCEAE, PONTOPCR	W72-12729
TER SUPPLY PROJECT, LOPEZ CREEK,	ARROYO GRANDE CREEK.: *LOPEZ WA	W72-05594
IELLA, DIMETHYL SODIUM ARSONATE,	ARSENATES, SODIUM CACODYLATE, CUL	W72-13813
NG, ALGAE, ANTIBIOTICS, MERCURY,	ARSENIC COMPOUNDS, CHESAPEAKE BAY	W72-10162
ILICATES, POLYCARBOXYLATES.:	ARSENIC, NITRILOTRIACETATE, POLYS	W72-06837
, NORWAY, A/ *ATOMIC ABSORPTION,	ARSENIC, SELENIUM, ZINC, ANTIMONY	W71-11515
Y/ ACREMONIELLA, DIMETHYL SODIUM	ARSONATE, ARSENATES, SODIUM CACOD	W72-13813
LA NANA.: CHELEX 100,	ARTEMIA SALINA, BIO-RAD, CYCLOTEL	W72-04708
A FLEXNERI, BIOCHEMICAL STUDIES,	ARTHOPODS, STREPTOM: /AE, SHIGELL	W72-13800
SCHIZOTHRIX CALCICOLA, BORATES,	ARTHROBACTER, NOSTOC COMMUNE, POR	W73-00854
CHAETES, CHIRONOMIDS, BARNACLES,	ARTHROPODS, SARGASSUM CRISPUM, DI	W73-00855
MS, SIPUNCULIDS,/ COELENTERATES,	ARTHROPODS, NEMERTEANS, ECHINODER	W73-00932
CEPHALOCHORDATES, HEMICHORDATES,	ARTHROPODS, *BAY OF BENGAL, COELE	W73-00935
HIC SPECIES, ABALONE, TUNICATES,	ARTHROPODS, COELENTERATES, BRYOZO	W71-01475
RASTRUM LONGIPES VAR CONTRACTUM,	ARTHURS LAKE, LAKE SORELL, LAKE C	W73-01094
KE, THOMAS L/ COOLING RESERVOIR,	ARTIFICIAL RESERVOIR, MONTROSE LA	W70-09596
*NUTRIENTS, AIRCRAFT, SATELLITES(	ARTIFICIAL), PHOSPHORUS, NITROGEN	W72-04742
APHY, REMOTE SENSING, SATELLITES(	ARTIFICIAL), ALGAE, CHLOROPHYTA,	W72-09120
RGENTS, WATER POLLUTION SOURCES,	ASBESTOS, EUTROPHICATION, POLYCHL	W73-00703
IOMASS, SALINITY, BENTHIC FAUNA,	ASBESTOS, INORGANIC COMPOUNDS, WO	W73-00374
DIGITATA, LAMINARIA HYPERBOREA,	ASCOPHYLLUM NODOSUM, GIGARTINA MA	W71-11515
RULA GLUTINIS, SARGASSUM NATANS,	ASCOPHYLLUM NODOSUM, FUCUS VESICU	W72-04748
COMPOU/ LAKE TAHOE(CALIF), ZINC,	ASCORBIC ACID, PYRIDOXINE, INDAL	W72-07514
ION, NEUROANATOMY, OXYGEN PLASMA	ASHING, ATOMIC ABSORPTION, TITRAT	W71-11036
ELENIUM, ZINC, ANTIMONY, NORWAY,	ASHING, BIOLOGICAL MAGNIFICATION,	W71-11515
PHOTOMETRY, PERCH LAKE(ONTARIO),	ASHING, BIOLOGICAL MAGNIFICATION.	W71-11529
RILIZATION, FIXATION, EMBEDDING,	ASHING, BONE, ORGANS, BRAIN, BLOO	W73-01676
IN, LOW-TEMPERATURE OVEN DRYING,	ASHING, EXTRACTIVE DIGESTION, NEU	W71-11036
ELLIPSOIDEA, RNA, CHLOROPHYLL B,	ASHING, LABORATORY TECHNIQUES.: /	W72-06274
LUM, RECYCLING, TREATMENT COSTS,	ASIA.: A	W72-13508

ALGAE, AQUATIC PLANTS, PROTOZOA, BACTERIA, STREAMFLOW, FLOW RATES, W73-01972
ER, DETROIT RIVER, HETEROTROPHIC BACTERIA, ACETYLENE, ETHYLENE.: / W73-01456
UNDS, *GEOCHEMISTRY, CYANOPHYTA, BACTERIA, ALGAE, PLANKTON, PLANTS W73-01439
WATER PURIFICATION, CLAY, ALGAE, BACTERIA, WASTE WATER TREATMENT, W73-01362
, *MARINE ALGAE, *PHYTOPLANKTON, BACTERIA, AQUARIA, DECOMPOSING OR W73-00802
NTENSITY, TEMPERATURE, ANABAENA, BACTERIA, PHOTOSYNTHESIS.: /GHT I W73-02471
AQUATIC MICROORGANISMS, AQUATIC BACTERIA, AQUATIC ALGAE, ALGAE, C W72-12255
N, ALGAE, THERMOCLINE, ANAEROBIC BACTERIA, AEROBIC BACTERIA, SULFU W72-12996
CTERIA, AEROBIC BACTERIA, SULFUR BACTERIA, IRON COMPOUNDS.: /IC BA W72-12996
NTS, DECOMPOSING ORGANIC MATTER, BACTERIA, CHEMICAL PROPERTIES, NI W72-12997
INE, ANAEROBIC BACTERIA, AEROBIC BACTERIA, SULFUR BACTERIA, IRON C W72-12996
T PENETRATION, DISSOLVED OXYGEN, BACTERIA, THERMOCLINE, SEDIMENT-W W72-12995
*SEWAGE LAGOONS, *COLD/ *SEWAGE BACTERIA, *WASTE WATER TREATMENT, W72-12565
KOTA, BIOCHEMICAL OXYGEN DEMAND, BACTERIA, TEMPERATURE, COLIFORMS, W72-12565
GI, THERMAL SPRINGS, CYANOPHYTA, BACTERIA, ECOLOGY, WATER POLLUTIO W72-12734
ROGEN FIXATION, *NITROGEN FIXING BACTERIA, *RAINFALL, *ARID LANDS, W72-12505
LACTOBACILLUS, CHLORELLA, SULFUR BACTERIA, ALGAE, SALMONELLA, E. C W72-12565
HYTA, NITROGEN, CARBON, DIATOMS, BACTERIA, CARBON DIOXIDE, CHLAMYD W72-12544
TEMPERATURE, COLIFORMS, ENTERIC BACTERIA, NITROGEN, VITAMINS, SUS W72-12565
, *INHIBITIONS, *MICROORGANISMS, BACTERIA, FUNGI, METHODOLOGY, CUL W72-13813
ANTIBIOTICS(PESTICIDES), ENTERIC BACTERIA, DISEASES, E. COLI, SHIG W72-13800
TEMATICS, POLLUTANTS, PATHOGENIC BACTERIA, COLIFORMS, TOXINS, ANTI W72-13800
EN/ *AQUATIC PLANTS, *MINNESOTA, BACTERIA, FUNGI, PROTOZOA, PATHOG W72-08586
LS, THERMAL POLLUTION, SAMPLING, BACTERIA, TOXICITY, BEHAVIOR, BIT W72-08790
*BIOLOGICAL COMMUNITIES, RIVERS, BACTERIA, WATER POLLUTION, ACTIVA W72-08573
IC WASTES, ALGAE, INDU/ *ENTERIC BACTERIA, *WATER QUALITY, *DOMEST W72-09590
FACTORS, GROWTH RATES, DIATOMS, BACTERIA, LAKES, NITROGEN, ANALYT W72-09163
FER, *PLASTICS, DOMESTIC WASTES, BACTERIA, ALGAE, NUTRIENTS, OXIDA W72-10401
DGE TREATMENT, DISSOLVED OXYGEN, BACTERIA, FUNGI, PROTOZOA, FERMEN W72-09675
TREATMENT, PHYTOPLANKTON, FUNGI, BACTERIA, ALGAE.: /IOLOGY, WATER W72-10076
ON CONTROL, *WASTES, *PATHOGENIC BACTERIA, CONFERENCES, *PUBLIC HE W72-09675
TROPHIC LEVELS, PROTOZOA, ALGAE, BACTERIA, DIATOMS, OCEANS, ACTIVA W72-10883
, OCEANS, ACTIVATED SLUDGE, SOIL BACTERIA, PHO: /BACTERIA, DIATOMS W72-10883
*MICROBIOLOGY, *NITROGEN FIXING BACTERIA, *NITROGEN CYCLE, ENVIRO W72-11563
UTROPHICATION, NUTRIENTS, ALGAE, BACTERIA, LAKE ERIE, ALUMINUM, MA W72-01101
ALGAE, AQUATIC BACTERIA, SEWAGE BACTERIA, ENVIRONMENTAL EFFECTS, W72-01472
T IDENTIFICATION, ALGAE, AQUATIC BACTERIA, SEWAGE BACTERIA, ENVIRO W72-01472
ALANCE, DISSOLVED OXYGEN, ALGAE, BACTERIA, ODOR, INFLOW, DISSOLVED W72-02274
SPHORUS, NUTRIENTS, MEASUREMENT, BACTERIA, ALGAE, EUTROPHICATION, W71-13794
OBALT, *RADIATION, COSTS, SEWAGE BACTERIA, ALGAE, COLIFORMS, INSEC W72-00383
L CONTROL, EUTROPHICATI/ *ALGAE, BACTERIA, *LAKES, SYMBIOSIS, ALGA W71-13257
XIDATION, ADSORPTION, COLIFORMS, BACTERIA, CENTRIFUGATION, *WASTE W71-13326
L CONTROL, DIATOMS, PERSISTENCE, BACTERIA, BIRDS, FISH, ON-SITE IN W72-00155
STRONTIUM RADIOISOTOPES, MARINE BACTERIA, COASTAL MARSHES, MICROO W72-00948
ACTERISTICS, ALGAE, ZOOPLANKTON, BACTERIA, NUTRIENTS, SEDIMENTS, E W72-03215
UNITIES, *AQUATIC PLANTS, ALGAE, BACTERIA, LITTORAL, LAKES, PERIPH W72-03216
TA/ *EPIPHYTIC ALGAE, *EPIPHYTIC BACTERIA, *MACROPHYTE-EPIPHTYE ME W72-03216
TRATIFICATION, *ENZYME/ *AQUATIC BACTERIA, *POPULATION, *ALGAE, *S W72-03224
RICTION, *TURBULENT FLOW, ALGAE, BACTERIA, MARINE PLANTS, ENGINEER W72-03220
, SOIL CHEMICAL PROPERTIES, SOIL BACTERIA, SOIL ALGAE.: /MPERATURE W72-03342
UATIC ANIMALS, DIATOMS, INSECTS, BACTERIA, FUNGI, TURNOVERS.: / AQ W71-10064
SOLVED OXYGEN, ALGAE, NUTRIENTS, BACTERIA, VIRUSES, HEAVY METALS, W71-09788
N CONCENTRATION, MARINE ANIMALS, BACTERIA, ABSORPTION, CULTURES.: / W71-10083
AE, *NITROGEN FIXATION, ENZYMES, BACTERIA, SPORES, NITROGEN, METAB W71-12070
, WEEDS, TASTE, ODOR, PATHOGENIC BACTERIA, ORGANIC WASTES, SOLID W W71-11006
TEMPERATURE, COPEPODS, NITROGEN, BACTERIA, OXIDATION-REDUCTION PO: W71-11008
E MICROORGANISMS, MARINE PLANTS, BACTERIA, POPULATION, HAWAII, ABS W71-08042
CES, PHOTOSYNTHESIS, METABOLISM, BACTERIA, HYDROGEN ION CONCENTRAT W71-08687
CEAN, SAPROPHYTES, HETEROTROPHIC BACTERIA, C-14.: / ISLAND, OPEN O W71-08042
UTROPHICATION, ALGAE, PATHOGENIC BACTERIA, EFFLUENT, IRRIGATION, S W71-08214
DROGEN ION CONCENTRATION, ALGAE, BACTERIA, WATER POLLUTION SOURCES W71-07942
UTION EFFECTS, *KENTUCKY, SULFUR BACTERIA, WATER QUALITY, SULFATES W71-07942
IVERS, CULTURES, ORGANIC MATTER, BACTERIA, WATER ANALYSIS.: /NS, R W71-08032
ONCENTRATION, WATER TEMPERATURE, BACTERIA, ALGAE, ALKALINITY, IRON W71-05626
LE ISOTOPES, *CARBON, *ANAEROBIC BACTERIA, *TEXAS, BIODEGRADATION, W71-04873
TAGES, PATHOGE/ *ALGAE, *ENTERIC BACTERIA, *BACTERICIDES, GROWTH S W71-05155
CIDES, GROWTH STAGES, PATHOGENIC BACTERIA, CYANOPHYTA, CHLOROPHYTA W71-05155
G, ODORS, PHOTOSYNTHESIS, ALGAE, BACTERIA, CHLORINATION, GROUNDWAT W71-07123
EN ION CONCENTRATION, COLIFORMS, BACTERIA, PHOTOSYNTHESIS, ORGANIC W71-07100
YGEN, SATURATION, ALGAE, METHANE BACTERIA, EUTROPHICATION, NITRATE W71-07106
NTACT TIME, SAMPLING, COLIFORMS, BACTERIA, OXIDATION, ORGANIC LOAD W71-07096
DEL STUDIES, ZOOPLANKTON, BIOTA, BACTERIA, BENTHIC FAUNA, FISH, EN W72-05282
TIC PLA/ *AQUATIC LIFE, *AQUATIC BACTERIA, *AQUATIC ANIMALS, *AQUA W72-04853
ATER, / *BIOCONTROL, *PATHOGENIC BACTERIA, *ALGAL CONTROL, FRESH W W72-05457
BITION, HOSTS, MYXOPHYCEA/ LYTIC BACTERIA, ENGLAND, METABOLIC INHI W72-05457
S, *EUTROPHICATION, LAKES/ *SOIL BACTERIA, *LAKE SOILS, *PHOSPHORU W72-04292
GEN, PHOSPHORUS, CARBON, *SEWAGE BACTERIA, SEWAGE, ALGAE, HYDROGEN W72-04309
ATION, NUTRIENTS, SLUDGE, ALGAE, BACTERIA, *WASTE WATER TREATMENT, W72-04086
PHICATION, LAKES, AQUATIC ALGAE, BACTERIA, BIOCHEMISTRY, WISCONSIN W72-04292
*ISOLATION, *MEMBRANE PROCESSES, BACTERIA, SULFUR BACTERIA, PHOTOS W72-04743
PHOTOSYNTHETIC BACTERIA, AQUATIC BACTERIA, LAKES, FILTERS, COLIFOR W72-04743
DUSTRIAL WASTES, VIRUSES, ENERIC BACTERIA, TRITIUM, RADIOACTIVE WA W72-04742
SULFUR BACTERIA, PHOTOSYNTHETIC BACTERIA, AQUATIC BACTERIA, LAKES W72-04743
RANE PROCESSES, BACTERIA, SULFUR BACTERIA, PHOTOSYNTHETIC BACTERIA W72-04743
ROCESSES,/ *CULTURES, *ANAEROBIC BACTERIA, *ISOLATION, *MEMBRANE P W72-04743
THO/ *ALGAE, *CULTURES, *ENTERIC BACTERIA, *OXIDATION LAGOONS, *PA W70-08319
PHICATION, *PHOSPHORUS, *CARBON, BACTERIA, ALGAE, SYMBIOSIS, CARBO W70-07283
XATION, PLANKTON, PHYTOPLANKTON, BACTERIA, ZOOPLANKTON, SALMONIDS, W71-00117
UDGE TREATMENT, TOXICITY, ALGAE, BACTERIA, FISH, SEWAGE SLUDGE, FL W70-06968
R), SAMPLING, DIATOMS, NITROGEN, BACTERIA, ALGAE, THERMAL STRATIFI W71-03028

ENT REQUIREMENTS, INVERTEBRATES,	CHLAMYDOMONAS, MATHEMATICAL STUDI	W72-09248
CHL/ CHLAMYDOMONAS ISABELIENSIS,	CHLAMYDOMONAS PALLIDOSTIGMATICA,	W72-12632
LAKE, PINE LA/ *ORTHOPHOSPHATES,	CHLAMYDOMONAS REINHARDTII, MOSES	W72-12637
ORESCENCE, PHOSPHORUS COMPOUNDS,	CHLAMYDOMONAS, WATER ANALYSIS, FL	W72-12637
ATOMS, BACTERIA, CARBON DIOXIDE,	CHLAMYDOMONAS, SCENEDESMUS, CHLOR	W72-12544
CHLAMYDOMONAS PALLIDOSTIGMATICA,	CHLAMYDOMONAS FOTTII, CHLAMYDOMON	W72-12632
TIGMATICA, CHLAMYDOMONAS FOTTII,	CHLAMYDOMONAS TEXENSIS, CHLAMYDOM	W72-12632
YDOMONAS PALLIDOSTIGMATICA, CHL/	CHLAMYDOMONAS ISABELIENSIS, CHLAM	W72-12632
FOTTII, CHLAMYDOMONAS TEXENSIS,	CHLAMYDOMONAS PSEUDOMICROSPHAERA,	W72-12632
PHORUS, CYANOPHYTA, OLIGOTROPHY,	CHLAMYDOMONAS, DEFICIENT ELEMENTS	W72-13816
AERUGINOSA, CHROOMONAS CAUDATA,	CHLAMYDOMONAS PERTUSA, PANDORINA	W72-13816
TASTE, ODOR, CORROSION, ENZYMES,	CHLEATION, ALGAE, WATER QUALITY,	W72-06837
IDULANS, *MUTANTS, *SENSITIVITY,	CHLORAMPHENICOL, RECOVERY, SURVIV	W72-14327
ATION, PENICILLIN G, TERRAMYCIN,	CHLORAMPHENICOL, ACTIDON, CULTURE	W73-01445
LIN, DRUGS, CORTISONE, HORMONES,	CHLORAMPHENICOL, SALMONELLA ENTER	W72-13800
O-60, GAMMA RADIATION, SURVIVAL,	CHLORELLA PYRENOIDOSA.: /NIANA, C	W72-11836
PINGUE, SCENEDESMUS QUADRICAUDA,	CHLORELLA PYRENOIDOSA, LITHIUM CH	W72-12739
ENEMINE, *POLYSTYRENE SULFONATE,	CHLORELLA ELLIPSOIDIA.: /OLYETHYL	W72-14840
	CHLORELLA VULGARIS.:	W72-14800
EOLARUM, ANKISTRODESMUS BRAUNII,	CHLORELLA PYRENOIDOSA, CHLORELLA	W72-07664
BRAUNII, CHLORELLA PYRENOIDOSA,	CHLORELLA VULGARIS, SCENEDESMUS O	W72-07664
YLUM TRI/ SILVER, CADMIUM, LEAD,	CHLORELLA PHYRENOIDOSA, PHAEODACT	W72-07660
TOGRAPHY, METHYL ESTERS, ESTERS,	CHLORELLA VULGARIS, SAMPLE PREPAR	W72-09365
	CHLORELLA PYRENOIDOSA.:	W72-11572
TASE, RIBONUCLEIC ACID.:	CHLORELLA VULGARIS, NITRATE REDUC	W72-03222
NAS, FONTINALIS, AS/ PHORMIDIUM,	CHLORELLA PYRENOIDOSA, CHLAMYDOMO	W71-11519
E ENZYME SYSTEM, PLANT EXTRACTS,	CHLORELLA SOROKINIANA.: /UCIFERAS	W71-11916
SARCINA, ANKISTRODESMUS BRAUNII,	CHLORELLA VULGARIS.: /ION, CHLORO	W71-07675
PB, MCPA, CHLAMYDOMONAS GLOBUSA,	CHLORELLA PYRENOIDOSA, STICHOCOCC	W70-07255
PALEA, SKELETONEMA COSTATUM.:	CHLORELLA PYRENOIDOSA, NITZSCHIA	W71-03027
CHLORELLA / DNA, CHLOROPHYLL A,	CHLORELLA PYRENOIDOSA, SPIROGIRA,	W72-06274
HLORELLA PYRENOIDOSA, SPIROGIRA,	CHLORELLA ELLIPSOIDEA, RNA, CHLOR	W72-06274
, SATURATION CONSTANT, DILUTION,	CHLORELLA PYRENOIDOSA, MONOD GROW	W72-04788
IC/ POTOMAC RIVER, ALAFIA RIVER,	CHLORELLA PYRENOIDOSA, SODIUM SIL	W72-04734
AERATION, MICROBIOLOGY, BIOMASS,	CHLORELLA.: /AEROBIC CONDITIONS,	W71-04079
ORUS, NITROGEN, AMMONIA, CARBON,	CHLORELLA.: /TION EFFECTS, PHOSPH	W71-12857
RITES, NITRATES, AMMONIA, ALGAE,	CHLORELLA.: /ON, HYDROGRAPHY, NIT	W72-08056
EGULATORS, ALGAE, AQUATIC ALGAE,	CHLORELLA.: /-4-D, PLANT GROWTH R	W72-08440
DRATES, PROTEINS, ACTINOMYCETES,	CHLORELLA.: /Y, PIGMENTS, CARBOHY	W72-07951
ANT IDENTIFICATION, CHLOROPHYTA,	CHLORELLA.: /UM COMPOUNDS, POLLUT	W72-13813
S, LIGHT INTENSITY, SCENEDESMUS,	CHLORELLA, GROWTH RATES, RESISTAN	W72-12739
*RADIOACTIVITY EFFECTS, *ALGAE,	CHLORELLA, IRRADIATION, RADIOACTI	W72-11836
IDE, CHLAMYDOMONAS, SCENEDESMUS,	CHLORELLA, SILICA, LIGHT INTENSIT	W72-12544
SUSPENDED SOLIDS, LACTOBACILLUS,	CHLORELLA, SULFUR BACTERIA, ALGAE	W72-12565
TION, EUTROPHICATION, NUTRIENTS,	CHLORELLA, CHLAMYDOMONAS, DIATOMS	W72-07508
ESCENCE, TRACE ELEMENTS, COPPER,	CHLORELLA, CULTURES, NUTRIENTS, R	W72-07660
UTROPHICATION, FJORDS, BIOASSAY,	CHLORELLA, NUTRIENTS, WATER SAMPL	W73-00234
ENT, *POLYMERS, CATIONS, ANIONS,	CHLORELLA, FILTRATION, ELECTROPHO	W72-14840
OLISM, FUNGI, ORGANIC COMPOUNDS,	CHLORELLA, TRACERS, AQUATIC ALGAE	W72-14326
OMASS, PRODUCTIVITY, METABOLISM,	CHLORELLA, RESPIRATION, CHLOROPHY	W73-01080
POLYSIPHONIA SP., GELIDIUM SP.,	CHLORELLA, AGARS.: /ADOPHORA SP.,	W71-11851
N ION CONCENTRATION, CYANOPHYTA,	CHLORELLA, CHLOROPHYTA, SCENEDESM	W71-11519
SALINITY, NUTRIENT REQUIREMENTS,	CHLORELLA, GROWTH RATES, TEMPERAT	W71-10083
RHODOPHYTA, DIATOMS, PYRROPHYTA,	CHLORELLA, PSEUDOMONAS, TESTING,	W72-03220
SPHORUS, CHEMICAL OXYGEN DEMAND,	CHLORELLA, PERFORMANCE, EFFICIENC	W72-11100
PHYTA, SCENEDESMUS, *CYANOPHYTA,	CHLORELLA, BIOLOGICAL COMMUNITIES	W72-09239
*EUTROPHICATION, *GROWTH RATES,	CHLORELLA, CARBON DIOXIDE, BICARB	W72-09168
GAL CONTROL, CYANOPHYTA, CARBON,	CHLORELLA, BICARBONATES, AERATION	W72-10607
CHELATION, CARBON RADIOISOTOPES,	CHLORELLA, GROWTH RATES, DIATOMS,	W71-03027
*ALGAE, *HERBICIDES, PESTICIDES,	CHLORELLA, RESPIRATION, PHOSPHORU	W70-07255
PHYTA, CHLOROPHYTA, CHRYSOPHYTA,	CHLORELLA, ROTIFERS, CHLAMYDOMONA	W70-06655
OSS, HYDROGEN ION CONCENTRATION,	CHLORELLA, SCENEDESMUS, WASTE WAT	W71-00138
ONDITIONS, ANAEROBIC CONDITIONS,	CHLORELLA, CHLAMYDOMONAS, SCENEDE	W70-10405
SSIFICATION, *OXIDATION LAGOONS,	CHLORELLA, OXYGEN DEMAND, PHOTOSY	W71-00139
ICAL TECHNIQUES, *ALGAE, *ASSAY,	CHLORELLA, NITRATES, CHEMICAL ANA	W70-10403
RM WASTES, OKLAHOMA, CYANOPHYTA,	CHLORELLA, IONS, PHYTOPLANKTON, N	W72-04773
BON, LAKE ERIE, RUNOFF, EROSION,	CHLORELLA, SODIUM COMPOUNDS, ALKA	W72-04734
NOPHYTA, YEASTS, CARBON DIOXIDE,	CHLORELLA, CLOSTRIDIUM, E. COLI,	W72-06124
HICATION, OLIGOTROPHY, PROTOZOA,	CHLORELLA, CHLAMYDOMONAS, CYANOPH	W72-05968
E RESIDUES, *ALGAE, CHLOROPHYTA,	CHLORELLA, TOXICITY, WATER ANALYS	W72-04134
, METABOLISM, HYDROGEN, EUGLENA,	CHLORELLA, SCENEDESMUS, CHLAMYDOM	W71-09172
ECOLOGY, RHODOPHYTA, PHAEOPHYTA,	CHLORELLA, PLANT PATHOLOGY, INDUS	W71-09157
OLUMNS, BOTTOM SEDIMENTS, ALGAE,	CHLORELLA, BENTHIC FLORA, MUD, MI	W71-08042
EMICAL TESTS, OUABAIN, POTASSIUM	CHLORIDE.: /TRIPHOSPHATASE, BIOCH	W72-13809
RANCE, WATER POLL/ *DDT, *SODIUM	CHLORIDE, *INHIBITION, *SALT TOLE	W72-13809
S, MET/ *PHOTOSYNTHESIS, CALCIUM	CHLORIDE, RADIOACTIVITY TECHNIQUE	W72-12634
ALUM, FERRIC	CHLORIDE, *JERUSALEM(ISRAEL).:	W72-12289
, CHLORELLA PYRENOIDOSA, LITHIUM	CHLORIDE, LITHIUM, ASTERIONELLA F	W72-12739
E MEDIA, ZINC SULFATE, MANGANESE	CHLORIDE, NICKEL SULFATE, COBALT	W72-12739
*MERCURY, *SYNTHESIS, MERCURIC	CHLORIDE, METHYL MERCURIC CHLORID	W72-06037
RCURIC CHLORIDE, METHYL MERCURIC	CHLORIDE, ANKISTRODESMUS BRAUNII,	W72-06037
THRACITE, *WAHNBACH RES/ *FERRIC	CHLORIDE, MULTI-LAYER FILTERS, AN	W72-06201
TRATE, CHEMICAL ANALYSIS, SODIUM	CHLORIDE, COLUMBIA RIVER, ORGANIC	W72-05421
*LIMITING NUTRIENTS, POTASSIUM	CHLORIDE, SHRINER'S POND(GA).:	W72-09159
S, EFFLUENTS, DETERGENTS, SODIUM	CHLORIDE, CHEMICAL PROPERTIES, SE	W72-09167
PHENYLMERCURIC ACETATE, MERCURY	CHLORIDE, PHENYLACETATE, MERCURY	W72-11727
LIZA/ *ALUMINUM SULFATE, *FERRIC	CHLORIDE, *FERRIC SULFATE, *STABI	W71-13356
ROCHLORIDE, MERI/ *CHLOROCHOLINE	CHLORIDE, *CCC, *DIETHYLAMINE HYD	W72-14706
TOCEROS GALVESTONENSIS, MERCURIC	CHLORIDE, DIMETHYL MERCURY, FLUOR	W72-07660
IMPOUNDMENTS, DISSOLVED OXYGEN,	CHLORIDES, INVERTEBRATES, HA: /E,	W72-14280

MPERATURE, PHOSPHORUS, NITRATES,	CHLCRIDES, CALCIUM, MAGNESIUM, SO	W71-11001
, POTASSIUM, CALCIUM, MAGNESIUM,	CHLCRIDES, HYDROGEN ION CONCENTRA	W72-05610
NITRATES, SULFATES, PHOSPHATES,	CHLCRIDES, BORATES, ABSORBANCE.: /	W72-06274
C CONDITIONS, WATER TEMPERATURE,	CHLCRIDES, PHOTOSYNTHESIS, LITTOR	W72-12940
SCILLATORIA TENUIS, OSCILLATORIA	CHLORINA, OSCILLATORI: /ANOENA, O	W72-07896
REEFS, ANTIBIOTICS(PESTICIDES),	CHLCRINATED HYDROCARBON PESTICIDE	W72-14294
ONAS REINHARDTII, CULTURE MEDIA,	CHLCRINATED HYDROCARBONS.: /MYDOM	W73-00281
HLOROPHYLL, OXYGEN, RESPIRATION,	CHLCRINATED HYDROCARBON PESTICIDE	W73-00281
ALGAE, *INDICATORS, *PESTICIDES,	CHLORINATED HYDROCARBON PESTICIDE	W73-00238
S, ORGANOPHOSPHOROUS PESTICIDES,	CHLCRINATED HYDROCARBON PESTICIDE	W72-13347
URAL RUNOFF, ALGAE, FOOD CHAINS,	CHLCRINATED HYDROCARBON PESTICIDE	W72-13657
ERANCE, WATER POLLUTION EFFECTS,	CHLCRINATED HYDROCARBON PESTICIDE	W72-13809
POLLUTANTS, PESTICIDE RESIDUES,	CHLCRINATED HYDROCARBON PESTICIDE	W73-02280
, *ALGAE, *PRIMARY PRODUCTIVITY,	CHLORINATED HYDROCARBON PESTICIDE	W71-10096
FILTRATION, ANAEROBIC DIGESTION,	CHLCRINATION, DISINFECTICN, ULTRA	W72-00042
MICAL WASTES, INDUSTRIAL WASTES,	CHLCRINATION, PHOSPHATES, NITRATE	W72-05315
ION CONCENTRATION, TEMPERATURE,	CHLCRINATION, FLOATATION, DEWATER	W71-09951
ATION, NUTRIENTS, SEDIMENTATION,	CHLCRINATION, SLOPES, BIOCHEMICAL	W71-07113
PHOTOSYNTHESIS, ALGAE, BACTERIA,	CHLCRINATION, GROUNDWATER, COST A	W71-07123
TRIENTS, TEMPERATURE, COLIFORMS,	CHLCRINATION, SEEPAGE, PERCOLATIO	W71-07124
CARBCN, ADSORPTION, FILTRATION,	CHLORINATION, CHEMICAL PRECIPITAT	W7C-10381
ONS, *DISINFECTION, DEGRADATION,	CHLORINE, CONTACT TIME, SAMPLING,	W71-07096
MENT, CHEMICALS, COPPER SULFATE,	CHLCRINE, WATER RESOURCES, METHOD	W71-05083
ANIMALS, BIOLOGICAL COMMUNITIES,	CHLCRINE, MOLLUSKS, COPEODS, DIP	W73-00855
E, PLANKTON, CHEMICALS, PHENOLS,	CHLORINE, OXICATION, ADSCRPTION,	W73-01669
NE, EURYHALIN/ *SINAI PENINSULA,	CHLORINITY, METAHALINE, HYPERHALI	W73-00855
MONAS GELATINOSA, THIORHODACEAE,	CHLCROBACTERIACEAE.: /RHODOPSEUDO	W72-04743
RA, EPHEMEROPTERA, NANNOCHLORIS,	CHLCROCOCCUM, MICROCYSTIS, FUSAR	W70-06655
ARTERIA EUGAMETOS, STICHOCOCCUS,	CHLCROCOCCUM, HYALOTHECA, LAKE LI	W72-09239
RIA, SPIRULI/ *PROTEASE ENZYMES,	CHLCROGONIUM, OOCYSTIS, OSCILLATO	W72-07661
TURES, NUTRIENTS, RADIOISOTOPES,	CHLOROPHYLL, PIGMENT, MAGNESIUM,	W72-C7660
THIAMINE, BIOTI/ ORGANIC CARBON,	CHLCROPHYLL A, ORGANIC NITROGEN,	W72-07892
URE, PHYTOPLANKTON, ZOOPLANKTON,	CHLOROPHYLL, SOLAR RADIATION, VIT	W72-07892
UCCESSION, CHLOROPHYTA, DIATOMS,	CHLCROPHYLL, ALKALINITY, CARBON D	W72-08049
S, *SUCCESSION, *EUTROPHICATION,	CHLCROPHYLL, PLANKTON, BIOLCGICAL	W73-00916
RATIO, PIGMENT DIVERSITY INDEX,	CHLCROPHYLL A, CARTENOICS, SAMPLE	W73-00916
ABOLISM, CHLORELLA, RESPIRATION,	CHLOROPHYLL, LABORATORY TESTS, IR	W73-01080
OLA.:	CHLOROPHYLL A, SCHIZOTHRIX CALCIC	W73-01080
TRIS, COUNTING, ORTHOPHOSPHATES,	CHLOROPHYLL A.: /RHODOMONAS LACUS	W73-01435
HENICOL, ACTIDON, CULTURE MEDIA,	CHLOROPHYLL A, TERATOLOGY, NITZSC	W73-01445
NITROGEN, CHRYSOPHYTA, NITRATES,	CHLCROPHYLL, PHO: /HT INTENSITY,	W73-01435
DIVERSITY, MELOSIRA NUMMULOIDES,	CHLOROPHYLL A, *YAQUINA BAY, AMPH	W73-00853
LA, *LIGHT, ALGAE, GROWTH RATES,	CHLCROPHYLL.: /TABOLISM, *CHLOREL	W73-00251
, PHOTOSYNTHESIS, AQUATIC ALGAE,	CHLOROPHYLL, OXYGEN, RESPIRATION,	W73-00281
TES, FORAMINIFERA,/ *PYCNOCLINE,	CHLOROPHYLL A, PHEOPHYTIN, TUNICA	W72-14313
BSORPTION, MATHEMATICAL STUDIES,	CHLCROPHYLL, OXIDATION REDUCTION	W72-14800
OTENOIDS, FRAGILARIA CROTONENSI/	CHLOROPHYLL A, CHLOROPHYLL C, CAR	W73-00384
ISSIMA, NITZSCHIA PUNGENS, C-14,	CHLOROPHYLL A, MEDITERRANEAN SEA,	W73-00431
LARIA CROTONENSI/ CHLOROPHYLL A,	CHLOROPHYLL C, CAROTENOIDS, FRAGI	W73-00384
ES, N-15, LUCIFERIN, LUCIFERASE,	CHLOROPHYLL A, OPTICAL DENSITY, A	W72-10883
PULATION, TEMPORAL DISTRIBUTION,	CHLCROPHYLL, SAMPLING, MATHEMATIC	W72-11719
TY CONTROL, *BIOASSAY, *BIOMASS,	CHLOROPHYLL, NITROGEN FIXATION, *	W72-10883
NALYSIS, PRODUCTIVITY, SEASONAL,	CHLCROPHYLL, ALGAE, PHYTOPLANKTON	W72-09108
ATION, RADIOACTIVITY TECHNIQUES,	CHLOROPHYLL, SEA WATER, IONS, NUT	W72-09103
CALIFCRNIA, SEDIMENTS, SAMPLING,	CHLOROPHYLL, SEDIMENT TRANSPORT,	W72-09092
, CULTURES, *FLORIDA, SEA WATER,	CHLOROPHYLL, NITRATES, NITRITES,	W72-09102
E(ENGLAND), EXOENZYME PRODUCERS,	CHLCROPHYLL A.: /, LAKE WINDERMER	W72-03224
YSERS, PHOTOSYNTHESIS, CULTURES,	CHLOROPHYLL, PROTEINS, ALGAE, GRO	W72-03557
S, MEASUREMENT, CURRENTS(WATER),	CHLOROPHYLL, FLUORESCENCE, BATHYM	W72-01095
TS, *INDICATORS, CHROMATOGRAPHY,	CHLCROPHYLL, TROPHIC LEVEL, EUTRO	W72-01784
ICATICN, BIOMASS, WATER QUALITY,	CHLOROPHYLL, NITROGEN, WISCONSIN,	W71-10645
ER TEMPERATURE, WATER CHEMISTRY,	CHLOROPHYLL, HYPSOMETRIC ANALYSIS	W71-11949
, METHODOLOGY, SEDIMENTS, LAKES,	CHLOROPHYLL, HYDROGEN ION CONCENT	W71-12855
L METHODS, NITROGEN, PHOSPHORUS,	CHLCROPHYLL, DISTRIBUTION, CARBON	W71-11029
N FIXATION, ALGAE, CONDUCTIVITY,	CHLOROPHYLL, PIGMENTS, DISPERSION	W72-04773
IRA, CHLORELLA ELLIPSOIDEA, RNA,	CHLOROPHYLL B, ASHING, LABORATORY	W72-06274
OSA, SPIROGIRA, CHLORELLA / DNA,	CHLCROPHYLL A, CHLORELLA PYRENOID	W72-06274
ABORATORY TESTS, PHOTOSYNTHESIS,	CHLCROPHYLL, ENZYMES, PERMEABILIT	W72-06037
ASKA, *EUTROPHICATION, PLANKTON,	CHLCROPHYLL, RECREATION, RUNOFF,	W72-04270
, RIVERS, IOWA, CORN, DISCHARGE,	CHLCROPHYLL.: /ION, WATER QUALITY	W71-06445
SONAL, DIATOMS, DINOFLAGELLATES,	CHLOROPHYLL, PHOTOSYNTHESIS, TEMP	W71-07875
IFERS, PROTOZOA, PHOTOSYNTHESIS,	CHLCROPHYLL, TEMPERATURE, HYDROGE	W71-07382
RIENTS, MEASUREMENT, POPULATION,	CHLOROPHYLL, WATER POLLUTION EFFE	W71-C8670
OPHYTA, OCHROMONAS, RESPIRATION,	CHLOROPHYLL, CARBON RADIOISOTOPES	W71-09172
CLEARWATER QUARRY, DOLLAR LAKE,	CHLOROPHYLL A, ABSORBANCE.: /SON,	W72-13673
HOSPHORUS, ANAEROBIC CONDITIONS,	CHLOROPHYLL, EPILIMNION, SEDIMENT	W72-12997
NESCENCE, BIOLOGICAL PROPERTIES,	CHLOROPHYLL, TURBIDITY, SAMPLING,	W72-12941
DT, PHYTOPLANKTON, PLANT GROWTH,	CHLCROPHYLL, DDE, INHIBITION.: /D	W72-12998
*LIGHT QUALITY, EUTROPHIC ZONE,	CHLOROPHYLL, PHOSPHATES, NITRATES	W72-12393
AKE, PINE LAKE, LAKE WASHINGTON,	CHLOROPHYLL A, *DNA, *DISSOLVED O	W72-12637
DISTILLATION, SPECTROPHOTOMETRY,	CHLOROPHYLL, HYDROLYSIS, PHYTOPLA	W72-12637
, *LIMITING FACTORS, PHOSPHATES,	CHLCROPHYTA, CYANOPHYTA, NITROGEN	W72-12544
IES, PHYTOPLANKTON, ZOOPLANKTON,	CHLOROPHYTA, DIATOMS, CYANOPHYTA,	W72-12729
ATICS, *CHLAMYDOMONAS, CULTURES,	CHLCROPHYTA, PROTOZOA, *ALGAE.: /	W72-12632
CULTURES, PHOSPHATES, NUTRIENTS,	CHLOROPHYTA, *ORGANOPHOSPHORUS CO	W72-12637
ALGAE, ZOOPLANKTON, CHRYSOPHYTA,	CHLOROPHYTA, BIOMASS, PYRROPHYTA,	W72-12738
ATES, *CULTURES, *AQUATIC ALGAE,	CHLCROPHYTA, INHIBITORS, WATER PO	W72-12736
BACTERIA, AQUATIC ALGAE, ALGAE,	CHLOROPHYTA, PSEUDOMONAS, ENTERIC	W72-12255
ATION, CLASSIFICATION, SAMPLING,	CHLCROPHYTA, *CHRYSOPHYTA, CYANOP	W72-12192
T-WATER INTERFACES, CHRYSOPHYTA,	CHLOROPHYTA, CYANOPHYTA, BIOCHEMI	W72-12995

E, CYTOCHROMES, PHOSPHORYLATICN,	CLYCOLYSIS.: /E, CITRIC ACID CYCL	W72-04309
ETIFERUS, MERCENARIA MERCENARIA,	CLYMENELLA TORQUATA, CAPITELLA CA	W73-02029
ELLA, LOPHOPANOPEUS, EUPENTACTA,	CNEMIDOCARPA, ENHYDRA LUTRIS NERE	W73-00932
CHLORE/ *CHLORELLA SOROKINIANA,	CO-60, GAMMA RACIATION, SURVIVAL,	W72-11836
*ORGANIC NUTRIENTS, ALUM	COAGULATION.:	W73-02187
TEMENT, WATER POLLUTION CONTRCL,	COAGULATION.: /ION, POLLUTION ABA	W73-02187
RBON, SLUDGE TREATMENT, DIATOMS,	COAGULATION, CALCIUM CARBONATE, H	W73-01669
C PLANTS, *MINNESOTA, *TOXICITY,	COAGULATICN, WATER POLLUTION EFFE	W72-07360
E SULFONATES, SEWAGE, NUTRIENTS,	COAGULATION, SEDIMENTATICN, TOXIC	W72-06837
, ALGAE, THERMAL STRATIFICATION,	COAGULATION, BENTONITE, LIME, *PU	W72-03753
, CLIMATIC ZONES, PRECIPITATICN,	COAGULATION, LIME, HYDRCGEN ION C	W71-08951
T, COSTS, SEPARATION TECHNIQUES,	CCAGULATION, WASTE WATER TREATMEN	W70-08904
WAGE SLUDGE, FLOW, FLOCCULATION,	COAGULATION, FILTRATION, VISCOSIT	W70-06968
HNIQUES, CHEMICAL PRECIPITATICN,	COAGULATION, SEDIMENTATICN, POLYE	W72-10237
, PHOSPHORUS, NITRATES, AMMONIA,	COAGULATION, FLOCCULATION, SEDIME	W71-13341
REATMENT, ANALYTICAL TECHNIQUES,	COALESCENCE, FLOCCULATICN, LIQUID	W71-13356
US VAR. NEAPOLITANA, CHAETOCEROS	COARCTATUS, CHAETOCEROS LCRENZIAN	W71-11527
AR WASTES, *MONITORING, *PACIFIC	COAST REGION, TRITIUM, OCEAN CURR	W72-03324
OF THANET, MEDWELL ESTUA/ *KENT	COAST, INCUSTRIAL EXPANSICN, ISLE	W71-13746
RADIOISOTOPES, MARINE BACTERIA,	COASTAL MARSHES, MICROORGANISMS.:	W72-00948
HEALTH, SOUTH AMERICA, ATLANTIC	COASTAL PLAIN, COASTS, CRUSTACEAN	W73-00819
, *FALLOUT, *MOLLUSKS, *ATLANTIC	CCASTAL PLAIN, NORTH CAROLINA, AB	W73-00818
*GULF OF FINLAND, DARK FIXATION,	COASTAL WATERS, TROPHOGENIC LAYER	W72-03228
ISOTOPES, ESTUARINE ENVIRONMENT,	COASTS, ANIMAL PHYSIOLOGY, SEDIME	W73-00811
MARINE PLANTS, BIOMASS, CANADA,	COASTS, ATLANTIC OCEAN, SEASONAL,	W73-00366
ITY, *PHYTOPLANKTON, *SEA WATER,	COASTS, BIORTHYMS, ON-SITE TESTS,	W72-03228
AMERICA, ATLANTIC COASTAL PLAIN,	COASTS, CRUSTACEANS, MARINE FISH,	W73-00819
QUATIC WEEDS, *WEEDS, *RED TIDE,	COASTS, LAKES, PONDS, WATER POLLU	W72-14673
STATISTICAL METHODS, SEA WATER,	COASTS, MARINE ALGAE.: /), MAINE,	W73-00369
ALITY, MICHIGAN, SURFACE WATERS,	COASTS, SAMPLING, BIOLOGICAL COMM	W72-10875
UCOSE, ACETATES, BIOTIN, NIACIN,	COBALAMINS, EUGLENA GRACILIS, COL	W73-02095
*MARINE ANIMALS, ASIA, MOLLUSKS,	COBALT RADIOISOTOPES, ICDINE RADI	W73-00811
ATHEMATICAL MODELS, FORECASTING,	COBALT RADIOISOTOPES, MARINE ALGA	W73-00831
, MUSSLES, LOBSTERS, ABSORPTICN,	COBALT RADIOISOTOPES, ZINC RADIOI	W73-00832
REGION, TRITIUM, OCEAN CURRENTS,	COBALT RADIOISOTOPES, NUCLEAR POW	W72-03324
, LINC RADIOISOTOPES, MANGANESE,	COBALT RADIOISOTOPES, CADMIUM RAD	W71-13243
NVIRONMENT, *ECOSYSTEMS, CESIUM,	COBALT RADIOISOTOPES, ALGAE, RADI	W70-06668
DIOISOTOPES, ZINC RADIOISOTOPES,	COBALT RADIOISOTOPES, ABSORPTION,	W71-09018
DIOISOTOPES, ZINC RADIOISOTOPES,	COBALT RACIOISOTOPES, WATER POLLU	W71-09250
, DYSTROPHY, ZINC RADIOISOTOPES,	COBALT RADIOISOTOPES, PHOSPHORUS	W71-09190
FISH, MOLLUSKS, MARINE ANIMALS,	COBALT RADIOISOTOPES, FCCD CHAINS	W71-09225
DIOISOTOPES, ZINC RADIOISOTOPES,	COBALT RADIOISOTOPES, URANIUM RAD	W71-09188
GANESE CHLORIDE, NICKEL SULFATE,	COBALT SULFATE, POTASSIUM BR: /AN	W72-12739
ADIATION SYSTEM, MODULAR SYSTEM,	COBALT-60.: *BIOLOGICAL--GAMMA-R	W72-00383
, IRON, MANGANESE, ZINC, NICKEL,	COBALT, COPPER SULFATE, HEAVY MET	W72-12739
S, NUTRIENTS, MERCURY, CHROMIUM,	CCBALT, FRESH WATER, LAKES, PHYSI	W73-01434
, COPPER, MOLYBDENUM, MANGANESE,	CCBALT, IRON, INDUSTRIAL WASTES,	W71-11515
LIS FIRMA, PLENOSP/ PORPHYROPSIS	COCCINEA VAR. DAWSONII, CALLOPHYL	W73-02099
RA,/ *GAMBUSIA, CYCLOPS NAUPLII,	CCCCOCHLORIS PENIOCYSTIS, SPIROGY	W72-07132
ES, SARGASSO SEA, SHELF WATERS,/	COCCOLITHOPHORES, SILICOFLAGELLAT	W72-07663
RATE UTILIZATION, *SARGASSO SEA,	COCCOLITHOPHORIDS, FLAGELLATES, F	W73-00380
COSTATA, ASSIMILATION, EMILIANA(	COCCOLITHUS) HUXLEYI, AMPHIPRORA	W73-00380
ECTA, SYNECHOCOCCUS, RHODOMONAS,	COCCOLITHUS HUXLEYI, AUTOLYSIS, P	W71-10083
LUVIATILIS, AMPHIDINIUM CARTERI,	COCCOLITHUS HUXLEYI, PORPHYRIDIUM	W72-12998
IS, AMPHORA OVALIS V. PEDICULUS,	COCCONEIS PEDICULUS, COCCONEIS PL	W72-13142
ROSUS, *EPIPHYTES, GROWTH MEDIA,	COCCONEIS PLACENTULA, CLOSTERIUM,	W72-12744
PEDICULUS, COCCONEIS PEDICULUS,	CCCCONEIS PLACENTULA, CYCLOTELLA	W72-13142
TY INDEX, FRAGILARIA CONSTRUENS,	COCCONEIS PLACENTULA, NAVICULA SR	W72-08141
ES, FRAG/ LAKE FURESO (DENMARK),	CCCCONEIS, RHOICOSPHENIA, ACHANTH	W72-07501
VER, COLUMBIA RIVER(WASHINGTON),	COCKLES, GOLDFISH, MERRIMACK RIVE	W71-11517
DIUM DODECYL, BENZENE SULFONATE,	COCONUT FATTY ACID, PLURCNICS, DI	W72-12576
RE MEDIA, PORIA MONTICOLA, PORIA	COCOS, PORIA VAILLANTII, CONICPHO	W72-13813
CANAL(MASS), BUZZARCS BAY, CAPE	COD BAY, DATA INTERPRETATION, SPE	W73-01449
PE COD BAY, DATA INTERPRE/ *CAPE	COD CANAL(MASS), BUZZARDS BAY, CA	W73-01449
TERS, PANAMA, CALLAO(PERU), CAPE	COD(MASS), CAPE MAY(N J), SARGASS	W72-08056
PHONIA LANOSA, CHONDRUS CRISPUS,	COD, FLCUNDER, CRAB EGGS, PELAGIC	W72-08142
PENFUSSII, GLOIODERMA ATLANTICA,	CODIUM CAROLINIANUM, *ONSLOW BAY(	W73-00284
LACTUCA, ZANICHELLIA PALUSTRIS,	CODIUM, FUSCUS, MACROPHYTES, AMMO	W73-00367
MIXING, EUPHOTIC ZONE, DIFFUSION	COEFFICIENT, LAKE SAMMAMISH(WASH)	W71-03034
N, EDDY DIFFUSIVITY,/ EXTINCTION	COEFFICIENT, LAMBERT-BEER EQUATIO	W72-05459
IZOMENON, OOCYSTIS, GLOEOCYSTIS,	COELASTRUM, ASTERIONELLA, CRYPTOM	W72-07940
ES, BRYOZOAS, HYDROIDS, SPONGES,	COELENTERATES, PORIFERA, POLYCHAE	W73-00374
, COLEOPTERA, MEDITERRANEAN SEA,	COELENTERATES, ECHINODERMS, POLYC	W73-00855
TES, ARTHROPODS, *BAY OF BENGAL,	COELENTERATES, CTENOPHORA, CHAETO	W73-00935
EANS, ECHINODERMS, SIPUNCULIDS,/	COELENTERATES, ARTHROPODS, NEMERT	W73-00932
ABALONE, TUNICATES, ARTHROPODS,	COELENTERATES, BRYOZOA, ECHINODER	W71-01475
ARI/ *CLEVELAND(OHIO), CERATIUM,	COELOSPHAERIUM, DINOBRYCN, FRAGIL	W72-10076
CIA OBTUSA, LAURENCIA PAPILLOSA,	COELOTHRIX IRREGULARIS, CENTROCER	W73-00838
CYSTIS, APHANIZOMENON, CYANIDES,	COHO SALMON, SHEEPSHEAD, CHUBS, N	W72-14282
SOUNDER, VERTICAL DISTRIBUTION,	COLD BAY, BERMERS BAY, ALARIA, NE	W72-09120
ES, *ALPINE, *SYNOPTIC ANALYSIS,	COLD REGIONS, NUTRIENTS, BIOLOGIC	W72-12549
RUTICOSA, NODAMASTIX SPIROGYRAE,	COLEMANIA V: /NA, PAVLOVIAMCNAS F	W72-07683
RED SEA, GULF OF SUEZ, DECAPODS,	COLEOPTERA, MEDITERRANEAN SEA, CO	W73-00855
BACTERIA, ALGAE, SALMONELLA, E.	COLI.: /CILLUS, CHLORELLA, SULFUR	W72-12565
OCCULATION, *ALGAE, *SILICA, *E.	COLI, *IONS, *WASTEWATER TREATMEN	W72-14840
SOLVED OXYGEN, BIOINDICATORS, E.	COLI, AQUATIC ALGAE, SUNFISHES, S	W71-07417
MENTAL EFFECTS, GROWTH RATES, E.	COLI, AZOTOBACTER, PSEUDCMONAS, W	W72-10861
PLANKTCN, WATER TEMPERATURE, E.	CCLI, BENTHIC FAUNA, CELLULCSE, E	W72-09668
S, ANTIMICROBIOTICS, ESCHERICHIA	COLI, BETA STREPTOCOCCUS, PSEUDOM	W72-04259
ATHOGENIC BACTERIA, CULTURES, E.	COLI, BIOINDICATORS, PONDS, ANAER	W72-07664

EUTROPHICATION, *WATER POLLUTION	CONTROL, *WASTE WATER(POLLUTION),	W72-05473
Y CONTROL, EUTROPHICATION, ALGAL	CONTROL, *CHLORELLA, *WASTE WATER	W72-04788
, *WASTE WATER TREATMENT, *ALGAL	CONTROL, *AQUATIC ALGAE, *EUTROPH	W72-03970
R QUALITY, RESER/ *WATER QUALITY	CONTROL, *FLOW AUGMENTATION, WATE	W72-04260
*EUTROPHICATION, *WATER QUALITY	CONTROL, *NON-STRUCTURAL ALTERNAT	W72-04279
*EUTROPHICATION, *WATER QUALITY	CONTROL, *NITROGEN COMPOUNDS, *PH	W70-04080
RATURE, SEASONAL, / *TEMPERATURE	CONTROL, *STANDARDS, *WATER TEMPE	W70-09595
TION ABATEMENT, *WATER POLLUTION	CONTROL, *WATER COOLING, *STEAM T	W70-09193
HICATION, *AQUATIC ALGAE, *ALGAL	CONTROL, *AQUATIC WEED CONTROL, B	W72-14692
ANOPHYTA.: *ALGAL	CONTROL, *INFECTION, *VIRUSES, CY	W73-00246
YSIS, LAKES, SAMP/ WATER QUALITY	CONTROL, *CYANOPHYTA, *WATER ANAL	W72-07890
ICATION, REVIEWS, TROPHI/ *ALGAL	CONTROL, *NUISANCE ALGAE, EUTROPH	W72-07937
ON, *MATHEMATICAL MODELS, *ALGAL	CONTROL, *RESERVOIRS, DESTRATIFIC	W72-08559
LABORATORY TESTS, WATER QUALITY	CONTROL, *KINETICS, *CHLOROPHYTA,	W72-08376
, *KINETICS, *CHLOROPHYTA, ALGAL	CONTROL, *NEW YORK.: /ITY CONTROL	W72-08376
*NUISANCE ALGAE, WASHING/ *ALGAL	CONTROL, *DIFFUSION, *NUTRIENTS,	W72-08049
EFICIENT NUTRIENTS, *AQU/ *ALGAL	CONTROL, *ESSENTIAL NUTRIENTS, *D	W72-07514
CIDES, *AQUATIC ORGANISMS, *PEST	CONTROL, *CHEMCONTROL, *BIBLIOGRA	W73-01976
, *INHIBITION, COPPER, F/ *ALGAL	CONTROL, *RESPIRATION, *CHLORELLA	W72-11572
LYSIS, EVALUATION, WATER QUALITY	CONTROL, *BIOASSAY, *BIOMASS, CHL	W72-10883
COSTS, FISH, FISHKILL, / *ALGAL	CONTROL, *CYANOPHYTA, EXPLOSIVES,	W72-11564
OLOGY, *TOXICITY, *WATER QUALITY	CONTROL, *WATER POLLUTION EFFECTS	W72-09646
E WATER TREATMENT, WATER QUALITY	CONTROL, *REVIEWS.: /NANCE, *WAST	W72-09386
TIFICATION, *WATER QUALI/ *ALGAL	CONTROL, *EUTROPHICATION, *DESTRA	W72-09304
DESTRATIFICATION, *WATER QUALITY	CONTROL, *LAKE MORPHOLOGY, AERATI	W72-09304
MICROORGANISMS, *WATER POLLUTION	CONTROL, *WASTES, *PATHOGENIC BAC	W72-09675
CAL TECHNIQUES, *WATER POLLUTION	CONTROL, *INSTRUMENTATION, MATHEM	W72-08790
POLLUTION SOUR/ *WATER POLLUTION	CONTROL, *EUTROPHICATION, *WATER	W71-12091
ENTS, *PHOSPHA/ *WATER POLLUTION	CONTROL, *EUTROPHICATION, *DETERG	W71-11516
Y, *CHLORINATION, *WATER QUALITY	CONTROL, *ODOR, ALGAE, EUTROPHICA	W71-13187
AQUATIC ALGAE, / *WATER QUALITY	CONTROL, *LAKES, *EUTROPHICATION,	W71-13256
CATION, LAKES, *NUTRIENTS, ALGAL	CONTROL, *CYANPHYTA, *NUISANCE AL	W72-00798
YTICAL TECHNIQUES, WATER QUALITY	CONTROL, *WASTE WATER TREATMENT.:	W72-00422
L EF/ *HERBICIDES, *AQUATIC WEED	CONTROL, *PARAQUAT, *ENVIRONMENTA	W72-00155
*FEDER/ *LAKES, *WATER POLLUTION	CONTROL, *WATER QUALITY CONTROL,	W72-01659
CULTURAL WASTES, WATER POLLUTION	CONTROL, *BIOLOGICAL TREATMENT, *	W72-02975
, *LEGISLATION/ *WATER POLLUTION	CONTROL, *PHOSPHATES, *DETERGENTS	W72-01685
N, *LE/ *LAKES, *WATER POLLUTION	CONTROL, *ENVIRONMENTAL SANITATIO	W72-01660
, *WASTE WATER TREATMENT, *ALGAL	CONTROL, *AQUATIC WEED CONTROL, C	W72-02975
OLLUTION CONTROL, *WATER QUALITY	CONTROL, *FEDERAL GOVERNMENT, *GR	W72-01659
V/ *DETERGENTS, *WATER POLLUTION	CONTROL, *PHOSPHATES, *FEDERAL GO	W72-01093
EFFECTS, *MICROORGANISMS, ALGAL	CONTROL, *POLLUTANT IDENTIFICATIO	W72-00801
*WASTE DILUTION, WATER POLLUTION	CONTROL, *NUISANCE ALGAE, *EUTROP	W72-00799
S, *PHOSPHATES, *WATER POLLUTION	CONTROL, *WATER POLLUTION SOURCES	W72-01085
IGN, NITROGEN, DIFFUSION, *ALGAL	CONTROL, AERATION, *WASTE WATER T	W72-00800
S, SEWAGE EFFLUENTS, TEMPERATURE	CONTROL, ASSAY, REMOTE SENSING, A	W72-10162
BON DIOXIDE, BICARBONATES, ALGAL	CONTROL, AERATION.: /LORELLA, CAR	W72-09168
AL EFFECTS, AQUATIC ALGAE, ALGAL	CONTROL, ALGAL POISONING, AQUATIC	W72-11186
ON, PUBLIC HEALTH, WATER QUALITY	CONTROL, ALGAL CONTROL, FEDERAL G	W72-06655
, CHLOROPHYTA, CYANOPHYTA, ALGAL	CONTROL, ALGICIDES, COPPER SULFAT	W72-14694
S, LAKES, PONDS, WATER POLLUTION	CONTROL, AIRCRAFT, COPPER SULFATE	W72-14673
RTIARY TREATMENT, *WATER QUALITY	CONTROL, AMMONIA, NITROGEN, PHOSP	W71-09450
E, *ALGAL CONTROL, *AQUATIC WEED	CONTROL, BIOCHEMICAL OXYGEN DEMAN	W72-14692
TIARY TREATMENT, WATER POLLUTION	CONTROL, BIODEGRADATION.: /T, TER	W71-13939
*OCEANS, *COASTS, *WATER QUALITY	CONTROL, BIOMASS, TOXICITY, EUTRO	W71-11793
M, NITROGEN, PHOSPHOROUS, *ALGAL	CONTROL, CALIFORNIA.: / METABOLIS	W71-13553
T, *ALGAL CONTROL, *AQUATIC WEED	CONTROL, CALIFORNIA.: /R TREATMEN	W72-02975
E ALGAE, *CARBON DIOXIDE, *ALGAL	CONTROL, CYANOPHYTA, CARBON, CHLO	W72-10607
MIC FEASIBILITY, WATER POLLUTION	CONTROL, CALIFORNIA, ARIZONA.: /O	W71-09157
WASTES, SAMPLING, WATER QUALITY	CONTROL, COST ANALYSIS, WASTE WAT	W71-08953
UTION ABATEMENT, WATER POLLUTION	CONTROL, COAGULATION.: /ION, POLL	W73-02187
TION, *DETERGENTS, WATER QUALITY	CONTROL, DOMESTIC WASTES, RUNOFF,	W72-11186
AL EFFECTS, DIQUAT, ALGAE, ALGAL	CONTROL, DIATOMS, PERSISTENCE, BA	W72-00155
CTERIA, *LAKES, SYMBIOSIS, ALGAL	CONTROL, EUTROPHICATION, WATER PO	W71-13257
*DENITRIFICATION, WATER QUALITY	CONTROL, EUTROPHICATION, ACTIVATE	W72-00800
LLUTION EFFECTS, WATER POLLUTION	CONTROL, ENERGY, DIFFUSION, THERM	W71-12092
ER QUALITY, ALGAE, WATER QUALITY	CONTROL, EUTROPHICATION, WASTE WA	W72-12487
ION, EFFICIENCIES, WATER QUALITY	CONTROL, EUTROPHICATION, ALGAL CO	W72-04788
S, *WASTE WATER TREATMENT, ALGAL	CONTROL, EUTROPHICATION, MODEL ST	W72-04787
WATER DISPOSAL, *WATER POLLUTION	CONTROL, ECONOMIC FEASIBILITY, ST	W72-14878
TH, WATER QUALITY CONTROL, ALGAL	CONTROL, FEDERAL GOVERNMENT, NUTR	W72-06655
OL, *PATHOGENIC BACTERIA, *ALGAL	CONTROL, FRESH WATER, CYANOPHYTA,	W72-05457
TOXICI/ *EUTROPHICATION, *ALGAL	CONTROL, FINANCING, PLANT GROWTH,	W71-13172
TER CIRCULATION, WATER POLLUTION	CONTROL, FISHKILL, NEW HAMPSHIRE,	W72-09304
C/ *ALGAE, *AQUATIC WEEDS, *WEED	CONTROL, GROWTH, *CULTURES, *ALGI	W71-06102
ITY, FALLOUT, BIOCONTROL, INSECT	CONTROL, IRRIGATION, PREDATION, P	W72-00846
SANCE ALGAE, *CYANOPHYTA, *ALGAL	CONTROL, LABORATORY TESTS, BUOYAN	W71-13184
NYLS, PUBLIC HEALTH, ENVIRONMENT	CONTROL, LEGAL ASPECTS, REGULATIO	W73-00703
LGAE, *PLANKTON, WATER POLLUTION	CONTROL, METHODOLOGY, MICROENVIRO	W72-03218
EUTROPHICATION, *WATER POLLUTION	CONTROL, MANAGEMENT, ANALYSIS, WA	W72-09171
N, EUTROPHICATION, WATER QUALITY	CONTROL, MASSACHUSETTS, *NUTRIENT	W72-06046
L OXYGEN DEMAND, WATER POLLUTION	CONTROL, NITRATES, EFFLUENTS, DET	W72-09167
HORUS, *LAKE BEDS, *ALGAE, ALGAL	CONTROL, NUTRIENT REQUIREMENTS, S	W72-11617
ATION, LAKES, WASHINGTON, *ALGAL	CONTROL, NUTRIENTS, BIOMASS, *CYA	W72-00799
ON, *NUTRIENTS, *WATER POLLUTION	CONTROL, ORGANIC WASTES, INDUSTRI	W73-00840
POLLUTION CONTROL, WATER QUALITY	CONTROL, PHOSPHORUS, AQUATIC ENVI	W72-06638
*CHLOROPHYTA, *INHIBITORS, ALGAL	CONTROL, PLANT GROWTH REGULATORS,	W71-13793
OMIC JUSTIFICATION, AQUATIC WEED	CONTROL, RIVERS, AGRICULTURE, ROO	W71-13172
ON, WILDLIFE CONSERVATION, FLOOD	CONTROL, RECREATION, HYDROLOGIC D	W72-05594
S, ENZYMES, COLOR, WATER QUALITY	CONTROL, RESEARCH AND DEVELOPMENT	W72-04070

GES, EARLY IMPOUNDMENT, BIOMASS,	DIATOMS, CHLOROPHYTA, CYANOPHYTA,	W72-04855
EFFECTS, *ALGAE, PHYTOPLANKTON,	DIATOMS, CHRYSOPHYTA, CHLOROPHYTA	W70-07842
PIKES, SAMPLING, SNAILS, ALGAE,	DIATOMS, CLAMS, SHELLFISH, AMPHIP	W70-08098
PHYTA, CHLOROPHYTA, CHRYSOPHYTA,	DIATOMS, DINOFLAGELLATES, SCENEDE	W72-05968
ARIES, *STANDING CROP, SEASONAL,	DIATOMS, DINOFLAGELLATES, CHLOROP	W71-07875
PHOSPHORUS, IRON, PHYTOPLANKTON,	DIATOMS, DINOFLAGELLATES,: /GEN,	W73-00234
RON, PHYSICOCHEMICAL PROPERTIES,	DIATOMS, DOMINANT ORGANISMS.: / I	W73-00256
TIVATED SLUDGE, ALGAE, PROTOZOA,	DIATOMS, EUGLENA, MATHEMATICAL ST	W72-08573
TY, ALGAE, BIOMASS, OLIGOTROPHY,	DIATOMS, EUGLENA, DAPHNIA, HABITA	W71-09171
MPOUNDS, THERMAL STRATIFICATION,	DIATOMS, EPILIMNION, EUTROPHICATI	W72-13659
*ALGAE, FRESHWATER, MEASUREMENT,	DIATOMS, EUTROPHICATION.: /SSAY,	W72-10619
AUTUMN, SAMPLING, PLANKTON NETS,	DIATOMS, FOOD CHAINS, ANABAENA, S	W72-12738
L ANALYSIS, *NITRATES, *AMMONIA,	DIATOMS, GROWTH RATES, NITROGEN,	W73-00380
HATES, PHOSPHORUS RADIOISOTOPES,	DIATOMS, HYDROGEN ION CONCENTRATI	W71-10083
RITUS, BENTHOS, AQUATIC ANIMALS,	DIATOMS, INSECTS, BACTERIA, FUNGI	W71-10064
LIFE CYCLES, SEASONAL, DENSITY,	DIATOMS, LAKES, PROTOZOA, ROTIFER	W71-11012
LOGICAL COMMUNITIES, FOOD CHAIN,	DIATOMS, LARVAE, DIETS.: /ES, BIO	W72-09248
OW WATER, PERIPHYTON, SEDIMENTS,	DIATOMS, LIGHT INTENSITY, PHOTOSY	W72-07501
ON EFFECTS, TEMPERATURE, CARBON,	DIATOMS, LIGHT, OXYGEN, MINNESOTA	W71-08670
CTIVITY, *STANDING CROP, *ALGAE,	DIATOMS, MEASUREMENT, DETRITUS, M	W71-12855
, THERMAL POLLUTION, MICROSCOPY,	DIATOMS, NUTRIENTS, PHYTOPLANKTON	W72-10875
NITY, HARDNESS(WATER), SAMPLING,	DIATOMS, NITROGEN, BACTERIA, ALGA	W71-03028
ETS, CENTRIFUGATION, MICROSCOPY,	DIATOMS, NUTRIENTS, BACTERIA, CHR	W72-07899
VELS, PROTOZOA, ALGAE, BACTERIA,	DIATOMS, OCEANS, ACTIVATED SLUDGE	W72-10883
CA, CHROMIUM, COPPER, *NEW YORK,	DIATOMS, PHOSP: /INDICATORS, SILI	W72-09111
S, DIQUAT, ALGAE, ALGAL CONTROL,	DIATOMS, PERSISTENCE, BACTERIA, B	W72-00155
ROPHYTA, PHAEOPHYTA, RHODOPHYTA,	DIATOMS, PYRROPHYTA, CHLORELLA, P	W72-03220
WATER POLLUTION SOURCES, *FISH,	DIATOMS, PROTOZOA, AQUATIC INSECT	W72-07791
DATION, SEWAGE TREATMENT PLANTS,	DIATOMS, PHYTOPLANKTON, HYDORGEN	W72-06837
L, SEA WATER, SAMPLING, BIOMASS,	DIATOMS, POLLUTANT IDENTIFICATION	W73-00442
*LIMITING FACTORS, *PHOSPHATES,	DIATOMS, POLLUTANT IDENTIFICATION	W73-01445
E ERIE, PLANKTON, BENTHIC FLORA,	DIATOMS, PHYTOPLANKTON, PHOTOSYNT	W73-01615
OTOPES, CHLORELLA, GROWTH RATES,	DIATOMS, PHYTOPLANKTON, TOXICITY,	W71-03027
POLLUTION EFFECTS, *PESTICIDES,	DIATOMS, PHYTOPLANKTON, TOXICITY,	W72-05614
ANIMALS, MOSSES, INVERTEBRATES,	DIATOMS, PHYTOPLANKTON, PHOTOSYNT	W72-05952
GAE, PHYSICOCHEMICAL PROPERTIES,	DIATOMS, SUCCESSION, NUTRIENTS, R	W73-01435
STREAMS, LAKE ONTARIO, PLANKTON,	DIATOMS, SEDIMENTS, HARDNESS(WATE	W73-01627
IENTS, CHLORELLA, CHLAMYDOMONAS,	DIATOMS, SILICATES, CYANOPHYTA, T	W72-07508
WATER ANALYSIS, LAKES, SAMPLING,	DIATOMS, TURBIDITY, WATER POLLUTI	W72-07890
CHEMICAL ANALYSIS, CHLOROPHYTA,	DIATOMS, TEMPERATURE, OPTICAL PRO	W70-10403
FATES, CARBONATES, BICARBONATES,	DIATONICS, CYANOPHYTA, ALGAE, SOU	W71-02880
5,6-TRICHLOROPICOLINIC ACID, 2,4-	DICHLOROPHENOXYACETIC ACID.: /-3,	W72-08440
LLARIA/ *INTERFERENCE, ALEWIVES,	DICHOTOMOSIPHON, FRAGILARIA, TABE	W73-01669
NOBRYON, OSCILLATORIA LIMNETICA,	DICTYOSPHAERIUM, CELL COUNTS, FRA	W71-03014
ELLATA, ERNODESMIS VERTICILLATA,	DICTYOSPHAERIA VANBOSSEAE, DICTYO	W72-11800
LATA, DICTYOSPHAERIA VANBOSSEAE,	DICTYOSPHAERIA CAVERNOSA, CLADOPH	W72-11800
TRODESMUS, CHRYSOCOCCUS MINUTUS,	DICTYOSPHAERIUM, LAGERHEIMIA.: /S	W72-12738
S PENNATA, SPYRIDIA FILAMENTOSA,	DICTYOTA BARTAYRESII, LAURENCIA O	W73-00838
HOLOGY, SEDIMENTATION, PLANKTON,	DIEL MIGRATION, PROTOZOA, HYDROGE	W71-05626
EAMS, PULP WASTES, PRODUCTIVITY,	DIELDRIN, PATH OF POLLUTANTS, REV	W72-06340
FLOODWATER, *PATH OF POLLUTANTS,	DIELDRIN, DDT, ORGANIC MATTER, AN	W71-11498
SOLVED OXYGEN, CULTRUES, ENDRIN,	DIELDRIN, OXYGENATION, CARBAMATE	W71-10096
*MIXED CULTURES,	DIEOFF RATES, AFTERGROWTH.:	W71-05155
CANAL ZONE, MACROINVERTEBRATES,	DIESEL FUEL, BUNKER C OIL, AVICEN	W72-07911
UES, FOOD CHAINS, CULTURES, FISH	DIET, GAS CHROMATOGRAPHY, PARTICU	W73-02105
, COCONUT FATTY ACID, PLURONICS,	DIETHANOLAMIDE.: /NZENE SULFONATE	W72-12576
ES, FOOD CHAIN, DIATOMS, LARVAE,	DIETS.: /ES, BIOLOGICAL COMMUNITI	W72-09248
PHNIA, AQUATIC POPULATIONS, FISH	DIETS, FRESHWATER FISH, PONDS, LA	W72-07132
*SKELLYSOLVE, *FILTER PAPER DISC	DIFFUSION METHOD, *ZONES OF INHIB	W72-08586
*VERTICAL MIXING, EUPHOTIC ZONE,	DIFFUSION COEFFICIENT, LAKE SAMMA	W71-03034
WATER POLLUTION CONTROL, ENERGY,	DIFFUSION, THERMODYNAMICS, SOLUBI	W71-12092
IVATED SLUDGE, DESIGN, NITROGEN,	DIFFUSION, *ALGAL CONTROL, AERATI	W72-00800
LEAR WASTES, PATH OF POLLUTANTS,	DIFFUSION, CYTOLOGICAL STUDIES, A	W73-00823
OPHYTA, OILY WATER, EVAPORATION,	DIFFUSION, RHODOPHYTA.: /TS, PHAE	W73-01074
H BEHAVIOR, FERTILITY, SEASONAL,	DIFFUSIVITY.: /TY, VISCOSITY, FIS	W70-09612
ENT, LAMBERT-BEER EQUATION, EDDY	DIFFUSIVITY, DESHLER(OHIO).: /ICI	W72-05459
OXYGEN, DISSOLVED OXYGEN, LIGHT,	DIFFUSIVITY, CARBON DIOXIDE, LIGH	W72-05459
, ARTHROPODS, SARGASSUM CRISPUM,	DIGENIA SIMPLEX, HALOPHILA STIPUL	W73-00855
ALKALINITY, ACIDITY, LABORATORY,	DIGESTION TANK, PONDAGE, PILOT PL	W72-02850
UATIC ALGAE, NITROGEN, ANAEROBIC	DIGESTION, AEROBIC, ALKALINITY, A	W72-02850
MENTATION, FILTRATION, ANAEROBIC	DIGESTION, CHLORINATION, DISINFEC	W72-00042
OVEN DRYING, ASHING, EXTRACTIVE	DIGESTION, NEUROANATOMY, OXYGEN P	W71-11036
L STUDIES, MATHEMATICAL STUDIES,	DIGITAL COMPUTERS, DATA ANALYSIS,	W72-11920
HY, FUCUS VESICULOSUS, LAMINARIA	DIGITATA, ST. MARGARET'S BAY, SCO	W72-12166
IS, FUCUS VESICULOSUS, LAMINARIA	DIGITATA, LAMINARIA HYPERBOREA, A	W71-11515
AM/ FUCUS VESICULOSUS, LAMINARIA	DIGITATA, PORPHYRA UMBILICALIS, S	W73-01074
LAMINARIA LONGICRURIS, LAMINARIA	DIGITATA, AGARUM CRIBROSUM, *NOVA	W73-00366
UM, FUCUS VESICULOSUS, LAMINARIA	DIGITATA, LAMINARIA AGARDHII, ULV	W72-04748
STE ASSIMILATIVE CAPACITY, WASTE	DIL: / ON-SITE INVESTIGATIONS, WA	W73-00790
SCUS, MACROPHYTES, AMMOBACULITES	DILATATUS, AMMONIA BECCARII, AMMO	W73-00367
AE, ABSORPTION, SEDIMENTS, WASTE	DILUTION, WASTE ASSIMILATIVE CAPA	W73-00831
CTS, ABSORPTION, SORPTION, WASTE	DILUTION, WASTE ASSIMILATIVE CAPA	W73-00817
AL / *MOSES LAKE(WASH), NUTRIENT	DILUTION, LIMITING NUTRIENTS, ALG	W72-08049
ATICAL MODELS, *BIOASSAY, *WASTE	DILUTION, LAGOONS, *WASTE WATER T	W72-04787
ROWTH RATE, SATURATION CONSTANT,	DILUTION, CHLORELLA PYRENOIDOSA,	W72-04788
, *NUISANCE ALGAE, *EUTR/ *WASTE	DILUTION, WATER POLLUTION CONTROL	W72-00799
*DETENTION RESERVOIRS, / *WATER	DILUTION, *NITROGEN, *PHOSPHORUS,	W72-02412
ETALS, HERBICIDES, TOXINS, WASTE	DILUTION, NITROGEN, PHOSPHORUS, E	W72-01472
LVESTONENSIS, MERCURIC CHLORIDE,	DIMETHYL MERCURY, FLUOROMICROPHOT	W72-07660
ES, SODIUM CACODY/ ACREMONIELLA,	DIMETHYL SODIUM ARSONATE, ARSENAT	W72-13813

TIGRIOPUS BREVICORNIS, GAMMARUS	DUEBENI, FUCUS SERRATUS, FUCUS VE	W71-08026
PENICILLUS CAPITATUS, PENICILLUS	DUMENTOSUS, RHIPOCEPHALUS PHOENIX	W72-14294
FATE CF POLLUTANTS, METABOLITES,	DUNALIELLA, AGMENELLUM QUADRAPLIC	W73-00361
ASIOSIRA SPP., SKELETONEMA SPP.,	DUNALIELLA SPP.: /TER ALGAE, THAL	W72-08436
SEUDONANA, SKELETONEMA COSTATUM,	DUNALIELLA TERTIOLECTA, EUGLENA G	W72-05614
A NANA, TH/ *ELECTRON TRANSPORT,	DUNALIELLA TERTIOLECTA, CYCLOTELL	W72-12998
TE, BAYTEX, BAYGONE, CYCLOTELLA,	DUNALIELLA, SKELETONEMA.: ABA	W71-10096
TICA, THALASSIOSIRA FLUVIATILIS,	DUNALIELLA TERTIOLECTA, SYNECHOCO	W71-10083
LUM/ CARBON-14, *FLORIDA STRAIT,	DUNALIELLA TERIOLECTA, PHAEODACTY	W72-09103
YLU/ *FLORIDA STRAIT, CARBON-14,	DUNALIELLA TERTIOLECTA, PHAEODACT	W72-09102
KELETCNEMA, ANABAENA CYLINDRICA,	DUNALIELLA.: /RIN, METABCLITES, S	W72-03545
IGHT, EELGRASS, ZOSTERA, PINAEUS	DUORARUM, PINAEUS AZTECUS, PENAEU	W73-02029
*ENGLAND, *POLLUTED ESTUARIES,	DURHAM COUNTY RIVERS.:	W72-08804
	DURHAM(NC).:	W71-07382
NTS, BICARBONATES, PLANT GROWTH,	DURNAL.: /DIOXIDE, CYCLING NUTRIE	W72-09159
*ANALINE	DYE.:	W73-02349
SON(WIS), LAKE WASHINGTON(WASH),	DYEING WATER.: /RVOIRS(NEB), MADI	W72-04759
CAL MODELS, *ALGAE, *POPULATION,	DYNAMICS, LAKES, DISSOLVED OXYGEN	W71-03034
LMONELLA ENTEROCOLITIS, SHIGELLA	DYSENTERIAE, SHIGELLA FLEXNERI, B	W72-13800
ION, HYDROGEN ION CONCENTRATION,	DYSTROPHY, ZINC RADIOISCTOPES, CC	W71-09190
FUR BACTERIA, ALGAE, SALMONELLA,	E. COLI.: /CILLUS, CHLORELLA, SUL	W72-12565
DISSOLVED OXYGEN, BIOINDICATORS,	E. COLI, AQUATIC ALGAE, SUNFISHES	W71-07417
RONMENTAL EFFECTS, GROWTH RATES,	E. COLI, AZOTOBACTER, PSEUDOMONAS	W72-10861
ON, PLANKTON, WATER TEMPERATURE,	E. COLI, BENTHIC FAUNA, CELLULOSE	W72-09668
*PATHOGENIC BACTERIA, CULTURES,	E. COLI, BIOINDICATORS, PONDS, AN	W72-07664
DIOXIDE, CHLORELLA, CLOSTRIDIUM,	E. COLI, EUGLENA, EUGLENOPHYTA, C	W72-06124
TOZOA, YEASTS, INSECTS, MILDEWS,	E. COLI, FISH, FOODS, GRASSES, WO	W73-01676
TECHNIQUES, CALCIUM, MAGNESIUM,	E. COLI, SILICA, *FILTRATION, LIG	W72-11833
ES), ENTERIC BACTERIA, DISEASES,	E. COLI, SHIGELLA, VIRUSES, CHROM	W72-13800
AVATUM, ELPHIDIUM GALVESTONENSE,	E: /PHIDIUM ADVENUM, ELPHIDIUM CL	W73-00367
NDS, ESOX LUCIUS, ALKYL MERCURY,	EAGLES, SALMO GAIRDNERII, IRIS PS	W72-05952
URINE, OEDOGONIUM, PHYSA, CULEX,	EAMBUSIA, FATE OF POLLUTANTS.: /	W72-07703
UATIC PLANTS, *RESERVOIR STAGES,	EARLY IMPOUNDMENT, BIOMASS, AQUAT	W72-04853
*RESERVOIRS, *RESERVOIR STAGES,	EARLY IMPOUNDMENT, BIOMASS, DIATO	W72-04855
PLOSION, *FILTERS, *DIATOMACEOUS	EARTH, *POROUS MEDIA, *WASTE WATE	W72-00974
ATION, FERTILIZERS, DIATOMACEOUS	EARTH, FOOD CHAINS, OXYGEN, ECOLO	W71-09157
*MEMBRANE FILTERS, GRANGE LAKE,	EAST TWIN LAKE, LAKE ROCKWELL, LA	W72-13673
HROPODS, COELENTERATES, BRYOZOA,	ECHINODERMS.: /NE, TUNICATES, ART	W71-01475
TERATES, ARTHROPODS, NEMERTEANS,	ECHINODERMS, SIPUNCULIDS, BRYOZOA	W73-00932
HORA, CHAETOGNATHS, CYPHONAUTES,	ECHINODERMS, TUNICATES, NOCTILUCA	W73-00935
EDITERRANEAN SEA, COELENTERATES,	ECHINODERMS, POLYCHAETES, CHIRONO	W73-00855
ND), CERATIUM HIR/ *GLOEOTRICHIA	ECHINULATA, MIKOLAJSKIE LAKE(POLA	W72-14812
USCIFORMIS, GRACILARIA VERUCOSA,	ECKLONIA RADIATA, GROWTH REGULATI	W72-07496
ECHNIQUES, SPATIAL DISTRIBUTION,	ECOLOGICAL DISTRIBUTION, BENTHIC	W73-00367
ABORATORY EQUIPMENT, INHIBITION,	ECOLOGICAL DISTRIBUTION, MARINE A	W73-00348
, PHOTOSYNTHESIS, STANDING CROP,	ECOLOGICAL DISTRIBUTION: /TRIENTS	W72-14313
OPERTIES, DISTRIBUTION PATTERNS,	ECOLOGICAL DISTRIBUTION, WATER PO	W73-00918
AUSTRALIA, SAMPLING, MICROSCOPY,	ECOLOGICAL DISTRIBUTION, PLANKTON	W73-01094
CROORGANISMS, CYCLING NUTRIENTS,	ECOLOGICAL DISTRIBUTION, ECOSYSTE	W73-02095
LAKE CAYUGA, CLADOPHORA,	ECOLOGICAL STUDIES.:	W70-09614
S, CLADOPHORA, LIMITING FACTORS,	ECOLOGICAL DISTRIBUTION.: /ABITAT	W72-13142
OTOZOA, *BIOLOGICAL COMMUNITIES,	ECOLOGICAL DISTRIBUTION, BIOINDIC	W72-08579
*SALTON SEA(CALIF),	ECOLOGICAL MANAGEMENT.:	W71-10577
HRYSOPHYTA, RHODOPHYTA, TESTING,	ECOLOGY.: /OPHYTA, CHLOROPHYTA, C	W70-07280
EFFECTS, AQUATIC LIFE, HABITATS,	ECOLOGY, *ECOSYSTEM, *ALGAE, *PHY	W73-01632
SCOPY, LIMNOLOGY, WATER QUALITY,	ECOLOGY, AQUATIC HABITATS, CLADOP	W72-13142
TA COLLECTIONS, LAKE MORPHOLOGY,	ECOLOGY, BIOLOGICAL PROPERTIES.: /	W72-13851
MOSOMES, FUNGICIDES, PESTICIDES,	ECOLOGY, BIRD TYPES, BIRDS, AQUAT	W72-05952
GY, LAKES, PONDS, AQUATIC ALGAE,	ECOLOGY, BIOLOGICAL PROPERTIES, F	W71-10580
PLANT PHYSIOLOGY, PHYSIOLOGICAL	ECOLOGY, BIOASSAY, MICROORGANISMS	W71-12303
C LIFE, ANTIBIOTICS(PESTICIDES),	ECOLOGY, BENTHOS, INVERTEBRATES,	W72-09668
ION, BIOCHEMISTRY, PHYSIOLOGICAL	ECOLOGY, CESIUM, STRONTIUM RADIOI	W71-13243
ION, PROTOZOA, PONDS, LIMNOLOGY,	ECOLOGY, CHEMICAL ANALYSIS, AQUAT	W73-00918
ATION SENSITIVITY, RADIOECOLOGY,	ECOLOGY, CYCLING NUTRIENTS, THERM	W73-00796
RRENTS(WATER), *DINOFLAGELLATES,	ECOLOGY, DISTRIBUTION PATTERNS, M	W73-00296
ATH OF POLLUTANTS, PHYSIOLOGICAL	ECOLOGY, ENVIRONMENTAL EFFECTS, P	W72-06313
METRY, PENNSYLVANIA, INHIBITORS,	ECOLOGY, ECOTYPES, FRESH WATER.: /	W72-04743
COMMUN/ WATER POLLUTION EFFECTS,	ECOLOGY, ECOSYSTEMS, *BIOLOGICAL	W72-11854
NITROGEN CYCLE, *CRABS, *FISHES,	ECOLOGY, FOOD CHAINS, ESTUARINE F	W72-12567
INS, *CHLOROPHYTA, *PHYTOTOXINS,	ECOLOGY, FISH FOOD ORGANISMS, PAT	W71-08597
SUPPLY, SYSTEMATICS, LIMNOLOGY,	ECOLOGY, FISH, WATER POLLUTION, P	W73-01615
ULTURAL RUNOFF, DOMESTIC WASTES,	ECOLOGY, FISH, ALGAE, AQUATIC PLA	W73-01972
PLANKTON, BENTHIC FAUNA, ALGAE,	ECOLOGY, MISSOURI.: /ISH HARVEST,	W70-09596
TOPLANKTON BLOOMS, PHYSIOLOGICAL	ECOLOGY, MORTALITY, RED TIDE, TOX	W72-14056
PLANT PHYSIOLOGY, PHYSIOLOGICAL	ECOLOGY, ORGANIC COMPOUNDS, ON-SI	W72-03216
ER, NUCLEAR POWER PLANTS, STEAM,	ECOLOGY, PHYTOPLANKTON, ZOOPLANKT	W71-11517
TOXINS, SEDIMENTS, SALT MARSHES,	ECOLOGY, PESTICIDES, ESTUARIES, I	W72-10678
NE ALGAE, *KELPS, *PHYSIOLOGICAL	ECOLOGY, PHAEOPHYTA, STANDING CRO	W72-12940
RE, CHEMICAL ANALYSIS, SAMPLING,	ECOLOGY, PHYTOP: /WATER TEMPERATU	W72-07710
MPOUNDMENT, NUCLEAR POWERPLANTS,	ECOLOGY, RADIOECOLOGY, MONITORING	W72-00939
MARY PRODUCTIVITY, PHYSIOLOGICAL	ECOLOGY, RADIOECOLOGY, RADIOACTIV	W72-00948
EOUS EARTH, FOOD CHAINS, OXYGEN,	ECOLOGY, RHODOPHYTA, PHAEOPHYTA,	W71-09157
IL BACTERIA, SAMPLING, CULTURES,	ECOLOGY, SOIL MICROBIOLOGY, CYANO	W73-00854
EOPHYTA, SAMPLING, SCUBA DIVING,	ECOLOGY, STANDING CROPS, PATH OF	W73-00366
BRATES, WATER POLLUTION EFFECTS,	ECOLOGY, SEA WATER, LABORATORY EQ	W73-00348
ATIC ANIMALS, BALANCE OF NATURE,	ECOLOGY, TROPHIC LEVEL, HERBIVORE	W72-06340
S, CITIES, PHOSPHATES, ILLINOIS,	ECOLOGY, TERTIARY TREATMENT, SEWA	W72-03970
S, *HYDROGRAPHY, *PACIFIC OCEAN,	ECOLOGY, WATER TEMPERATURE, PHYTO	W72-07892
RYSOPHYTA, PYRROPHYTA, SAMPLING,	ECOLOGY, WATER TEMPERATURE.: / CH	W72-03543
L SPRINGS, CYANOPHYTA, BACTERIA,	ECOLOGY, WATER POLLUTION EFFECTS.	W72-12734

RMANCE, FLOCCULATION, FLOTATION,	ECONOMIC FEASIBILITY, COSTS, ESTI	W72-13508
POSAL, *WATER POLLUTION CONTROL,	ECONOMIC FEASIBILITY, STREAMS, SE	W72-14878
SYSTEMS, CALIFORNIA, IRRIGATION,	ECONOMIC FEASIBILITY, WATER POLLU	W71-09157
STUDIES, TECHNICAL FEASIBILITY,	ECONOMIC FEASIBILITY, PILOT PLANT	W71-09450
TER SUPPLY, COSTS, WATER DEMAND,	ECONOMIC JUSTIFICATION, AQUATIC W	W71-13172
RUNOFF, FERTILIZERS, LIVESTOCK,	ECONOMICS,: /IC PLANTS, DRAINAGE,	W71-13172
UDGE, EFFLUENT, *ORGANIC WASTES,	ECONOMICS, CHEMICAL OXYGEN DEMAND	W71-12223
ATION, WATER SUPPLY, SUCCESSION,	ECONOMICS, NUTRIENTS, NITROGEN FI	W71-13183
IME, HYDROGEN ION CONCENTRATION,	ECONOMICS, ALGAE, SOLUBILITY, SEW	W71-12200
NING, BIOCHEMICAL OXYGEN DEMAND,	ECONOMICS, ALGAE, AGRICULTURAL: /	W72-04734
S, FILTERS, ENGINEERS ESTIMATES,	ECONOMICS, FEASIBILITY, COSTS, AN	W70-10174
WISCONSIN, PRODUCTIVITY, ALGAE,	ECONOMICS, POLITICAL ASPECTS, GEO	W71-03012
RIFFLE BENTHOS, WOODLANC STREAM,	ECOSYSTEM ANALYSIS, MACRCONSUMER	W71-10064
VOIR(KAN), GREAT PLAINS, AQUATIC	ECOSYSTEM.: / *TUTTLE CREEK RESER	W72-12393
ATER POLLUTION EFFECTS, ECOLOGY,	ECOSYSTEMS, *BIOLOGICAL CCMMUNITI	W72-11854
ICATICN, *KANSAS, MODEL STUDIES,	ECOSYSTEMS, ALGAE, *LIGHT QUALITY	W72-12393
ASINS), *DDT, *SMALL WATERSHEDS,	ECOSYSTEMS, PESTICIDE KINETICS, *	W72-12930
, *LIG/ *EUTROPHICATION, AQUATIC	ECOSYSTEMS, WATER QUALITY, *ALGAE	W73-02349
CTIVITY, HERBIVORES, CARNIVORES,	ECOSYSTEMS, AQUATIC LIFE, SAMPLIN	W71-10064
NS, FISH FOOD ORGANISMS, OXYGEN,	ECOSYSTEMS.: /ESOURCES, FCOD CHAI	W71-13246
S, *FALLOUT, *CYCLING NUTRIENTS,	ECOSYSTEMS, RAIN FORESTS, MATHEMA	W72-03342
ALGAE, INVERTEBRATES, RADIATION,	ECOSYSTEMS, ESTUARIES, CCRAL, REE	W71-13723
, CHORELLA, TRACERS, PHOSPHATES,	ECOSYSTEMS, AQUATIC LIFE, ANTIBIO	W72-09668
POPULATION, AQUATIC ENVIRONMENT,	ECOSYSTEMS, GEORGIA, BIORTHYTHMS,	W72-09159
TIC ENVIRONMENT, CARBON DIOXIDE,	ECOSYSTEMS, SOUTH CAROLINA, ALGAE	W71-00475
NA), ANACYSTIS NIDULANS, AQUATIC	ECOSYSTEMS, ALGAE GROWTH, ALGAL P	W71-00475
YTA, RADIOISOTOPES, TEMPERATURE,	ECOSYSTEMS, CALIFCRNIA, IRRIGATIO	W71-09157
S, MARINE ANIMALS, AQUATIC LIFE,	ECOSYSTEMS, SALMCNIDS.: /SH, CLAM	W72-07907
ACTIVITY TECHNIQUES, PESTICIDES,	ECOSYSTEMS, POLLUTICN SCURCES, PE	W72-07703
ASTE WATER(POLLUTION), BIOASSAY,	ECOSYSTEMS, BIODEGRADATICN, ALGAE	W73-00840
LLUTICN SOURCES, BIODEGRADATION,	ECOSYSTEMS, WATER POLLUTION.: /PO	W73-00853
RIENTS, ECOLOGICAL DISTRIBUTION,	ECOSYSTEMS, AQUATIC HABITATS, PRO	W73-02095
NUTRIENTS, FOOD WEBS, DETRITUS,	ECOSYSTEMS, ENERGY BUDGETS, NUTRI	W73-01089
LANKTON, BIOLOGICAL COMMUNITIES,	ECOSYSTEMS, SPECTROPHOTCMETRY, VI	W73-00916
NNSYLVANIA, INHIBITORS, ECOLOGY,	ECOTYPES, FRESH WATER.: /ETRY, PE	W72-04743
FECTS, *DATA COLLECTIONS, ALGAE,	ECOTYPES, INDUSTRIAL PLANTS, INDU	W71-13746
FFICIENT, LAMBERT-BEER EQUATION,	EDDY DIFFUSIVITY, DESHLER(OHIO).:	W72-05459
*FRESHWATER ALGAE,	EDTA.:	W72-14694
YSIS/ *ION-SELECTIVE ELECTRODES,	EDTA, *POTENTIOMETRY, ELECTROANAL	W73-02112
ROINVERTEBRATES, NEW YORK BIGHT,	EELGRASS, ZOSTERA, PINAEUS DUCRAR	W73-02029
QUATIC ALGAE, SUNFISHES, SNAILS,	EELS, CHINOOK SALMON, BASS, TROUT	W71-07417
TH MOUNTAIN PROJECT(VA), WEATHER	EFFECT.: SMI	W72-05282
CA, ACETYLENE REDUCTION, WARBURG	EFFECT.: /ION, *ANABAENA CYLINDRI	W72-08584
YCLE, ANACYSTIS NIDULAN/ WARBURG	EFFECT, EUGLENA GRACLIS, CALVIN C	W71-09172
ICAL PROPERTIES, WATER POLLUTION	EFFECT, WATER POLLUTION SCURCES,	W72-00974
LECTING AGENTS, *SINKING AGENTS,	EFFECTIVENESS, TESTING PROCEDURES	W71-09789
TY, CONVECTION, WATTER POLLUTICN	EFFECTS.: / BUOYANCY, WIND VELOCI	W71-09852
L WASTES, SILTS, WATER POLLUTION	EFFECTS.: /OAC, SEWAGE, INDUSTRIA	W71-09795
WASTE DISPOSAL, WATER POLLUTICN	EFFECTS.: / METALS, HEAVY METALS,	W71-11793
NKTON, CULTURES, WATER POLLUTION	EFFECTS.: /GROWTH REGULATCRS, PLA	W71-13793
TION, CORROSION, WATER POLLUTION	EFFECTS.: /HYDROGEN ION CONCENTRA	W72-08683
ERSITY, DIRECT EFFECTS, INDIRECT	EFFECTS.: /CREEK(PA), SPECIES DIV	W72-08579
EWAGE EFFLUENTS, WATER POLLUTION	EFFECTS.: /TS, CHEMICAL WASTES, S	W72-08804
HAMPSHIRE, MIXING, ENVIRONMENTAL	EFFECTS.: /, PHYTOPLANKTCN, *NEW	W72-09061
ER, PSEUDOMONAS, WATER POLLUTION	EFFECTS.: /ES, E. COLI, AZOTOBACT	W72-10861
IC ALGAE, CALCIUM, RADIOACTIVITY	EFFECTS.: /H, AQUATIC LIFE, AQUAT	W72-10978
LOGICAL STUDIES, WATER POLLUTION	EFFECTS.: / FLUORIDES, ICNS, CYTO	W72-11572
SIZE, COLLOIDS, WATER POLLUTION	EFFECTS.: /YTA, AQUARIA, PARTICLE	W71-09193
WATER TREATMENT, WATER POLLUTICN	EFFECTS.: / OXYGEN DEMAND, WASTE	W71-07098
WATER TREATMENT, WATER POLLUTION	EFFECTS.: /BIODEGRADATICN, WASTE	W71-07097
WATER TREATMENT, WATER POLLUTION	EFFECTS.: / CONCENTRATION, WASTE	W71-05155
COOLING TOWERS, WATER POLLUTION	EFFECTS.: /, APPLICATION METHODS,	W71-06189
QUALITY CONTROL, WATER POLLUTION	EFFECTS.: /, PERMEABILITY, WATER	W72-06037
LTURE MEDIA, DRY WEIGHT, THERMAL	EFFECTS.: /VANNAH RIVER PLANT, CU	W71-02075
ASINOCLADUS MARINUS, TEMPERATURE	EFFECTS.: /1002, *GREEN ALGAE, PR	W70-09429
LIMIT, POISONS, WATER POLLUTION	EFFECTS.: /AY, RESISTANCE, LETHAL	W72-14706
Y TESTS, OXYGEN, WATER POLLUTION	EFFECTS.: /CHLOROPHYTA, LABCRATOR	W72-14314
ON, AQUATIC ALGAE, ENVIRONMENTAL	EFFECTS.: /ICIDES), PHOTCACTIVATI	W72-14327
AVIOR, SEASONAL, WATER POLLUTION	EFFECTS.: /GES, DISTRIBUTION, BEH	W72-14659
A, ALGAE, FUNGI, WATER POLLUTION	EFFECTS.: /R, CONDENSERS, PROTOZO	W72-07526
OZOA, EQUIPMENT, WATER POLLUTION	EFFECTS.: /AS, EUGLENOPHYTA, PROT	W72-07661
SHWATER, *ALGAE, WATER POLLUTION	EFFECTS.: /S, BRACKISH WATER, FRE	W73-02548
Y, BICARBONATES, WATER POLLUTION	EFFECTS.: /ATION, PLANT PHYSIOLOG	W72-12940
S, PHYTOPLANKON, WATER POLLUTION	EFFECTS.: /IENTS, DISSOLVED SCLID	W72-13300
(WATER), MOLLUSKS, ENVIRONMENTAL	EFFECTS.: /AL, SALINITY, CURRENTS	W72-11876
CTERIA, ECOLOGY, WATER POLLUTION	EFFECTS.: /PRINGS, CYANOPHYTA, BA	W72-12734
EUTROPHICATION, WATER POLLUTION	EFFECTS.: /ISOTOPES, *CYANOPHYTA,	W72-13692
ION, W/ *LAKES, *WATER POLLUTION	EFFECTS, *LIMNOLOGY, WATER POLLUT	W72-12729
MPERATURE, *WA/ *WATER POLLUTION	EFFECTS, *HEATED WATER, *WATER TE	W72-13636
, *PERSISTENCE, *WATER POLLUTION	EFFECTS, *PLANT PHYSIOLOGY, *DIAT	W73-02280
AE, RHODOPHYTA, / *RADIOACTIVITY	EFFECTS, *MONITORING, *MARINE ALG	W72-07826
S, OILY WATER,/ *WATER POLLUTION	EFFECTS, *OIL WASTES, *CRUSTACEAN	W72-07911
TY, COAGULATION, WATER POLLUTION	EFFECTS, *ALGAL TOXINS.: /*TOXICI	W72-07360
RMAL POLLUTION, *WATER POLLUTION	EFFECTS, *AQUATIC MICROCRGANISMS,	W72-07225
MICROORGANISMS, *WATER POLLUTION	EFFECTS, *ALGAE, *PROTOZCA, *BIOL	W72-08579
ULTURAL RUNOFF, *WATER POLLUTION	EFFECTS, *CATTLE, *FEED LCTS.: /C	W72-08401
HERMAL POLLUTION, *ENVIRONMENTAL	EFFECTS, *ECOLOGY, *PLANKTCN, *AL	W72-14659
OSING ORGANIC M/ WATER POLLUTION	EFFECTS, *ORGANIC MATTER, *DECOMP	W72-14673
ICAL CCMMUNITIES, *ENVIRONMENTAL	EFFECTS, *STANDING CRCPS, TIDAL E	W73-00367
PHYTA, *SALINITY, *ENVIRONMENTAL	EFFECTS. *PLANT MORPHOLOGY, *GENE	W73-00369

GAE, AQUATIC LIFE, ENVIRONMENTAL	EFFECTS, ENVIRONMENTAL ENGINEERIN	W72-14405
IAL PHOTOGRAPHY, WATER POLLUTION	EFFECTS, EUTROPHICATION, *LAKES,	W72-14728
UTION TREATMENT, WATER POLLUTION	EFFECTS, ENVIRONMENTAL EFFECTS, A	W70-09429
*OIL, SEEPAGE, WATER POLLUTION	EFFECTS, ENVIRONMENTAL EFFECTS, *	W71-08793
YTA, INHIBITORS, WATER POLLUTION	EFFECTS, ENVIRONMENTAL EFFECTS, C	W72-12736
OLOGICAL COMMUN/ WATER POLLUTION	EFFECTS, ECOLOGY, ECOSYSTEMS, *BI	W72-11854
, BIOINDICATORS, WATER POLLUTION	EFFECTS, FERTILIZERS, NITROGEN, B	W72-13816
ITY, ADSORPTION, WATER POLLUTION	EFFECTS, FOOD CHAINS.: /, SOLUBIL	W71-09182
T RADIOISOTOPES, WATER POLLUTION	EFFECTS, FRESH WATER.: /ES, COBAL	W71-09250
ALLOUT, *OCEANS, WATER POLLUTION	EFFECTS, FOOD CHAINES, PUBLIC HEA	W72-04463
SLATION, *TOXINS, *ENVIRONMENTAL	EFFECTS, FEDERAL GOVERNMENT, AIR	W73-00703
AL DISTRIBUTION, WATER POLLUTION	EFFECTS, F: /N PATTERNS, ECOLOGIC	W73-00918
NUCLEAR WASTES, WATER POLLUTION	EFFECTS, FOOD CHAINS, PUBLIC HEAL	W72-10978
LLUTION SOURCES, WATER POLLUTION	EFFECTS, FLUCTUATION, COMPARATIVE	W72-03228
UNITY DEVELOPMENT, ENVIRONMENTAL	EFFECTS, GOVERNMENT FINANCE,: /MM	W72-01660
E, MICROORGANISMS, ENVIRONMENTAL	EFFECTS, GROWTH RATES, E. COLI, A	W72-10861
DUCK, BIOASSAY, WATER POLLUTION	EFFECTS, GAS CHROMATOGRAPHY, BROO	W72-12930
AQUATIC HABITATS, ENVIRONMENTAL	EFFECTS, HEATED WATER, NUCLEAR PO	W71-11517
RADATION, ALGAE, WATER POLLUTION	EFFECTS, HEAVY METALS, CYCLING NU	W73-00840
NSIN, NUTRIENTS, WATER POLLUTION	EFFECTS, INTERFACES, DISSOLVED OX	W71-10645
ALGAE, RECREATION, ENVIRONMENTAL	EFFECTS, INVERTEBRATES, WATER RES	W71-10577
K(PA), SPECIES DIVERSITY, DIRECT	EFFECTS, INDIRECT EFFECTS.: /CREE	W72-08579
OISOTOPES, CESIUM, RADIOACTIVITY	EFFECTS, LABORATORY TESTS.: /RADI	W72-09649
NT IDENTIFICATION, ENVIRONMENTAL	EFFECTS, LABORATORY EQUIPMENT, FR	W73-00379
T, *GROWTH RATES, *ENVIRONMENTAL	EFFECTS, LARVAE, ON-SITE TESTS, I	W73-00348
TS, WATER QUALITY, ENVIRONMENTAL	EFFECTS, LIMNOLOGY, PLANNING, EUT	W72-06526
EFFECTS, *STANDING CROPS, TIDAL	EFFECTS, MARINE ALGAE, SEA WATER,	W73-00367
YTA, MARINE ALGAE, ENVIRONMENTAL	EFFECTS, MARINE PLANTS, CANADA, W	W73-00428
LLUTION SOURCES, WATER POLLUTION	EFFECTS, MERCURY, SEWAGE DISPOSAL	W72-14782
A WATER, SAMPLING, ENVIRONMENTAL	EFFECTS, MARINE FISH, PERIPHYTON,	W73-00932
ANS, MONITORING, WATER POLLUTION	EFFECTS, MOLLUSKS.: /GE, CRUSTACE	W72-10955
BON RADIOISOTOPES, ENVIRONMENTAL	EFFECTS, METHODOLOGY, ALGAE, AQUA	W72-12709
NUCLEAR WASTES, WATER POLLUTION	EFFECTS, MOLLUSKS, MARINE PLANTS,	W72-04461
*WASTE HEAT, SUBLETHEL	EFFECTS, MIXING ZONES, MODELING.:	W70-09612
LLUTION SOURCES, WATER POLLUTION	EFFECTS, MUD, MOLLUSKS, FISH, INS	W70-08098
TS, *DISTRICT OF COLUMBIA, TIDAL	EFFECTS, NUTRIENTS, EUTROPHICATIO	W70-08248
TS, *DISTRICT OF COLUMBIA, TIDAL	EFFECTS, NUTRIENTS, EUTROPHICATIO	W70-08249
LLUTION CONTROL, WATER POLLUTION	EFFECTS, NUTRIENTS, ALGAE, CYANOP	W72-04769
UTION ABATEMENT, WATER POLLUTION	EFFECTS, NUTRIENTS, PHOSPHORUS,	W71-09429
OLISM, KINETICS, WATER POLLUTION	EFFECTS, NUCLEAR WASTES.: / METAB	W71-09168
RADIOISOTOPES, *WATER POLLUTION	EFFECTS, NUCLEAR WASTES, RIVERS,	W71-09225
PES, *ABSORPTION, *ENVIRONMENTAL	EFFECTS, NUCLEAR WASTES, PATH OF	W72-10955
ATIONS, FLORIDA, WATER POLLUTION	EFFECTS, NITROGEN, PHOSPHORUS.: /	W72-08986
, *WATER SUPPLY, WATER POLLUTION	EFFECTS, NUTRIENTS, RESERVOIR OPE	W71-13185
YGONUM, TYPHA, ELODE/ BIOLOGICAL	EFFECTS, NOTTINGHAM(ENGLAND), POL	W72-00155
NTS, RADIOECOLOGY, RADIOACTIVITY	EFFECTS, NUCLEAR POWERPLANTS, MAR	W73-00833
ICAL TECHNIQUES, WATER POLLUTION	EFFECTS, NUTRIENT REQUIREMENTS.: /	W73-01269
CATION, *LAKES, *WATER POLLUTION	EFFECTS, OHIO, NUTRIENTS, WATER P	W71-11949
E, PRODUCTIVITY, WATER POLLUTION	EFFECTS, PHOSPHORUS, NITROGEN, AM	W71-12857
ICIDE TOXICITY, *WATER POLLUTION	EFFECTS, PESTICIDES, AGRICULTURAL	W71-12303
ABOLISM, CULTURES, ENVIRONMENTAL	EFFECTS, PLANT GROWTH, SULFATE, L	W72-00153
ICAL TECHNIQUES, WATER POLLUTION	EFFECTS, POLLUTANT IDENTIFICATION	W71-13794
LATIVE CAPACITY, WATER POLLUTION	EFFECTS, POLLUTANTS,: /STE ASSIMI	W72-01685
ILS, IONS, NOSTOC, ENVIRONMENTAL	EFFECTS, POLLUTANT IDENTIFICATION	W73-00854
DES, GROWTH RATES, ENVIRONMENTAL	EFFECTS, PHYTOPLANKTON, CHLAMYDOM	W73-00281
IOLOGICAL ECOLOGY, ENVIRONMENTAL	EFFECTS, PRODUCTIVITY, RADIOECOLO	W72-06313
NTHIC FAUNA, FISH, ENVIRONMENTAL	EFFECTS, RESERVOIR STORAGE, NUTRI	W72-05282
EFFECTS, *PONDS, WATER POLLUTION	EFFECTS, RADIOISOTOPES, COMPETITI	W71-09256
RATES, NITROGEN, WATER POLLUTION	EFFECTS, RADIOACTIVITY, METABOLIS	W73-00380
NS, MARINE PLANTS, ENVIRONMENTAL	EFFECTS, REEFS, LABORATORY TESTS,	W73-00838
ATES, NUTRIENTS, WATER POLLUTION	EFFECTS, RADIOACTIVITY TECHNIQUES	W72-07143
WATER POLLUTION, WATER POLLUTION	EFFECTS, RESEARCH AND DEVELOPMENT	W72-06655
, *RADIOACTIVITY, *RADIOACTIVITY	EFFECTS, RIVERS, EFFLUENT, RADIOE	W72-07841
LLUTION SOURCES, WATER POLLUTION	EFFECTS, RECREATION DEMAND, POPUL	W72-01660
POLLUTION EFFECTS, RADIOACTIVITY	EFFECTS, RADIOACTIVITY TECHNIQUES	W72-03348
IUM, ABSORPTION, WATER POLLUTION	EFFECTS, RADIOACTIVITY EFFECTS, R	W72-03348
POLLUTION SOURCES, RADIOACTIVITY	EFFECTS, RADIOACTIVITY TECHNIQUES	W72-00940
CATION, *LAKES, *WATER POLLUTION	EFFECTS, RECLAIMED WATER, SEWAGE	W72-11702
ANTS, *TOXICITY, WATER POLLUTION	EFFECTS, RESPIRATION, *HEAVY META	W72-12576
PES, HEAVY METALS, AIR POLLUTION	EFFECTS, SURFACE WATERS, PATH OF	W72-00940
LLUTION SOURCES, WATER POLLUTION	EFFECTS, SAMPLING, TEMPERATURE, C	W71-11008
ANALYSIS, WATER ANALYSIS, TIDAL	EFFECTS, STABILITY, SOLAR RADIATI	W72-08141
LE ALGAE, LIGHT INTENSITY, TIDAL	EFFECTS, SALINITY, HEATED WATER,	W73-00853
CEANS, ANNELIDS, WATER POLLUTION	EFFECTS, STANDING CROPS, MOLLUSKS	W73-00935
, BENTHIC FAUNA, WATER POLLUTION	EFFECTS, STRIPED BASS, DISSOLVED	W73-02029
ERS, *VIRGINIA, *WATER POLLUTION	EFFECTS, SURVEYS, SAMPLING, DATA	W73-01972
ORGANIC MATTER, WATER POLLUTION	EFFECTS, SEWAGE, NITROGEN, PHOSPH	W73-00234
TIC ENVIRONMENT, WATER POLLUTION	EFFECTS, STREAMS, RIVERS, ROOTED	W72-14690
IES, *DELAWARE RIVER, VEGETATION	EFFECTS, SLUDGE, RIVER BEDS, ALGA	W70-09189
ON, CHLOROPHYLL, WATER POLLUTION	EFFECTS, TEMPERATURE, CARBON, DIA	W71-08670
TON, MARINE ALGAE, ENVIRONMENTAL	EFFECTS, TRACE ELEMENTS, COLLOIDS	W73-00818
RIENTS, SEDIMENTS, ENVIRONMENTAL	EFFECTS, TRACE ELEMENTS, ANALYTIC	W72-03215
LTH, *SHELLFISH, WATER POLLUTION	EFFECTS, TOXICITY, WATER POLLUTIO	W72-12257
OLLUTION EFFECTS, *RADIOACTIVITY	EFFECTS, UNDERGROUND, ALASKA, PON	W72-03347
RIO, *LAKE ERIE, WATER POLLUTION	EFFECTS, WATER POLLUTION SOURCES,	W72-01094
, AQUATIC ALGAE, WATER POLLUTION	EFFECTS, WATER QUALITY, CHEMICAL	W72-01085
ATH OF POLLUTANTS, AIR POLLUTION	EFFECTS, WATER POLLUTION EFFECTS,	W72-00959
TER TEMPERATURE, WATER POLLUTION	EFFECTS, WATER POLLUTION CONTROL,	W71-12092
IENTS, NITROGEN, WATER POLLUTION	EFFECTS, WATER POLLUTION SOURCES.	W71-11972

NIA, ANTIBIOTICS, IODINE, ATOMIC	ENERGY, REACTOR COOLING, IMPERIAL	W71-09157
, ALGAE, EUTROPHICATION, SUMMER,	ENERGY, SEASONAL.: /Y, CYANOPHYTA	W72-03228
DES, INDUSTRIAL WASTES, VIRUSES,	ENERIC BACTERIA, TRITIUM, RADIOAC	W72-04742
ALGAE, BACTERIA, MARINE PLANTS,	ENGINEERING, WATER PROPERTIES, PH	W72-03220
TIC ALGAE, PLANTS, ENVIRONMENTAL	ENGINEERING, COOLING TOWER: /AQUA	W72-00939
T, EUTROPHICATION, ENVIRONMENTAL	ENGINEERING, WATER TREATMENT, WAT	W72-01660
MENT, SEWAGE EFFLUENTS, SANITARY	ENGINEERING, SECONDARY TREATMENT,	W72-09378
RONMENTAL EFFECTS, ENVIRONMENTAL	ENGINEERING, *VIRGINIA, PROJECTIO	W72-14405
UM CARBONATE, PROTEINS, FILTERS,	ENGINEERS ESTIMATES, ECONOMICS, F	W70-10174
LELHAM TARN(ENGLAND), WINDERMERE(	ENGLAND).: /TERIONELLA FORMOSA, B	W72-10625
LEVEN(SCOTLAND), LAKE WINDERMERE(	ENGLAND).: / LUXURY UPTAKE, LOCH	W71-13183
THAMES RIVER(ENGLAND), LEE RIVER(	ENGLAND).: /, RESEARCH STRATEGY,	W71-13172
RESEARCH STRATEGY, THAMES RIVER(	ENGLAND), LEE RIVER(ENGLAND).: /,	W71-13172
*ULVA, *LAMINARIA, LIVERPOOL BAY(	ENGLAND), IRISH SEA.:	W71-12867
*UNITED KINGDOM, ESTHWAITE WATER(	ENGLAND), BLELHAM TARN(ENGLAND),	W71-13183
WATER(ENGLAND), LAKE WINDERMERE(	ENGLAND), GREAT OUSE(ENGLAND), AL	W71-13185
AE, *EPIPELIC ALGAE, SHEAR WATER(	ENGLAND), AMPHORA OVALIS, OPEPHOR	W71-12855
WINDERMERE(ENGLAND), GREAT OUSE(	ENGLAND), ALGAL GROWTH.: /), LAKE	W71-13185
*UNITED KINGDOM, ESTHWAITE WATER(	ENGLAND), LAKE WINDERMERE(ENGLAND	W71-13185
ITE WATER(ENGLAND), BLELHAM TARN(	ENGLAND), LUXURY UPTAKE, LOCH LEV	W71-13183
TERIONELLA FORMOSA, BLELHAM TARN(	ENGLAND), WINDERMERE(ENGLAND).: /	W72-10625
), EXCENZYME PR/ ESTHWAITE WATER(	ENGLAND), LAKE WINDERMERE(ENGLAND	W72-03224
WATER(ENGLAND), LAKE WINDERMERE(	ENGLAND), EXOENZYME PRODUCERS, CH	W72-03224
BIOLOGICAL EFFECTS, NOTTINGHAM(	ENGLAND), POLYGONUM, TYPHA, ELODE	W72-00155
ER THAMES(ENGLAND), RIVER KENNET(	ENGLAND), STEPHANODISCUS HANTZSCH	W72-14797
*CENTRIC DIATOMS, *RIVER THAMES(	ENGLAND), RIVER KENNET(ENGLAND),	W72-14797
UE-GREEN ALGAE.: *WINDERMERE(	ENGLAND), HETEROCYSTS, BENTHIC BL	W73-02472
ATOMS, CARBON RADIOISOTOPES, NEW	ENGLAND, AQUATIC LIFE, AQUATIC MI	W72-14317
CAL MAGNIFICATION, FATE OF POLL/	ENGLAND, BIOACCUMULATION, BIOLOGI	W72-11925
STS, MYXOPHYCEA/ LYTIC BACTERIA,	ENGLAND, METABOLIC INHIBITION, HO	W72-05457
A SPP, DISTICHLIS SPICATA, FIMB/	ENGLAND, SWEDEN, GERMANY, SPARTIN	W73-01089
STE WATER TREATMENT, CYANOPHYTE,	ENGLENOPHYTA, ORGANIC LOADING.: /	W71-07973
S MEXICANUS, EUPHAUSIA PACIFICA,	ENGRAULIS MORDAX.: /N, TRIPHOTURU	W73-02105
OPEUS, EUPENTACTA, CNEMIDOCARPA,	ENHYDRA LUTRIS NEREIS, SEA OTTER,	W73-00932
ANDS.: PHILIPPINES,	ENIWETOK, MICRONESIA, FANNING ISL	W71-11486
GROWTH KINETICS, BIOENERGETICS,	ENRICHMENT, CULTURE MEDIA, MACROI	W72-09378
URO ATOLL, M/ PLEOPODS, UROPODS,	ENRICHMENT, LEUCOTHRIX MUCOR, MAJ	W72-08142
HOMICROBIUM, VIBRIO ANGUILLARUM,	ENRICHMENT CULTURES, AGARS, CHEMO	W72-05421
REDUCTION MEASUREMENTS, NUTRIENT	ENRICHMENT TESTS.: /S, ACETYLENE	W72-13692
AS EROSA, ERKINIA SUBEQUICILIAT/	ENRICHMENT, CHROMULINA, CRYPTOMON	W72-13816
TOXINS, ANTIBIOTICS(PESTICIDES),	ENTERIC BACTERIA, DISEASES, E. CO	W72-13800
ALGAE, CHLOROPHYTA, PSEUDOMONAS,	ENTERIC BACTERIA.: /UATIC ALGAE,	W72-12255
ACTERIA, TEMPERATURE, COLIFORMS,	ENTERIC BACTERIA, NITROGEN, VITAM	W72-12565
OPERTIES, *ALGAE, *HYDROBIOLOGY,	ENTERIC BACTERIA, SAMPLING, ZOOPL	W72-07896
OBLIQUUS, ALCALIGENES FAECALIS,	ENTEROBACTER AEROGENES, PROTEUS V	W72-07664
NES, CHLORAMPHENICOL, SALMONELLA	ENTEROCOLITIS, SHIGELLA DYSENTERI	W72-13800
EUNOTIA PECTINALIS, LITHOPHYTES,	ENTEROMORPHA, INSOLATION, TEXTURE	W72-08141
ATION, SUBLITTORAL, LONG ISLAND,	ENTEROMORPHA INTESTINALIS, POLYSI	W73-00367
ROGEN, POLLUTANT IDENTIFICATION,	ENVIRONMENTAL EFFECTS, LABORATORY	W73-00379
CARBON PESTICIDES, GROWTH RATES,	ENVIRONMENTAL EFFECTS, PHYTOPLANK	W73-00281
TIC LIFE, ENVIRONMENTAL EFFECTS,	ENVIRONMENTAL ENGINEERING, *VIRGI	W72-14405
COMPUTERS, ALGAE, AQUATIC LIFE,	ENVIRONMENTAL EFFECTS, ENVIRONMEN	W72-14405
PHOTOACTIVATION, AQUATIC ALGAE,	ENVIRONMENTAL EFFECTS.: /ICIDES),	W72-14327
ATERS, *AQUATIC PLANTS, *AQUATIC	ENVIRONMENT, WATER POLLUTION EFFE	W72-14690
Y, *SHOALS, *BENTHIC/ *ESTUARINE	ENVIRONMENT, *PRIMARY PRODUCTIVIT	W72-14317
*HEAVY METALS, *ALGAE, *AQUATIC	ENVIRONMENT, *TOXICITY, *WATER PO	W72-14694
GIONS, PHAEOPHYTA, MARINE ALGAE,	ENVIRONMENTAL EFFECTS, MARINE PLA	W73-00428
*LEGISLATION(PROPOSED),	ENVIRONMENTAL PROTECTION AGENCY.:	W73-00703
OPES, STABLE ISOTOPES, ESTUARINE	ENVIRONMENT, COASTS, ANIMAL PHYSI	W73-00811
INATED BIPHENYLS, PUBLIC HEALTH,	ENVIRONMENT CONTROL, LEGAL ASPECT	W73-00703
SOPHYTA, SYSTEMATICS, *ESTUARINE	ENVIRONMENT, SEA WATER, SALINITY,	W72-08141
RIBUTION, *SPATIAL DISTRIBUTION,	ENVIRONMENT, SAMPLING, CHEMICAL A	W72-08560
NTS, NITRIF/ *NITROGEN, *AQUATIC	ENVIRONMENT, PRODUCTIVITY, SEDIME	W72-08459
*DIATOMS, *RESEARCH EQUIPMENT,	ENVIRONMENTAL EFFECTS, ESTUARIES,	W72-08141
S, PATH OF POLLUTANTS, ESTUARINE	ENVIRONMENT, RADIOACTIVITY TECHNI	W72-07826
ITY CONTROL, PHOSPHORUS, AQUATIC	ENVIRONMENT, WATER POLLUTION EFFE	W72-06638
IES, DOMINANT ORGANISMS, AQUATIC	ENVIRONMENT, AQUATIC ALGAE, AQUAT	W72-07132
UBA DIVING, SEA WATER, SAMPLING,	ENVIRONMENTAL EFFECTS, MARINE FIS	W73-00932
TROGEN, WATER POLLUTION SOURCES,	ENVIRONMENTAL EFFECTS, AEROBIC CO	W73-01066
HYTOPLANKTON, NERITIC, ESTUARINE	ENVIRONMENT, SEA WATER, DIATOMS,	W73-00935
N, HYDROGRAPHY, PLANKTON, ALGAE,	ENVIRONMENTAL EFFECTS, BIOLOGICAL	W73-01704
PRODUCTIVITY, *ECOLOGY, *AQUATIC	ENVIRONMENT, *BIOLOGICAL COMMUNIT	W73-02095
IOLOGICAL COMMUNITIES, ESTUARINE	ENVIRONMENT, WATER POLLUTION SOUR	W73-00853
ASSIMILATIVE CAPACITY, ESTUARINE	ENVIRONMENT, FOOD CHAINS, PUBLIC	W73-00828
A, ABSORPTION, SOUNDS, ESTUARINE	ENVIRONMENT, PATH OF POLLUTANTS,	W73-00818
, PLEISTORENE EPOCH, ADAPTATION,	ENVIRONMENTAL EFFECTS, CHLOROPHYT	W73-00855
NS, PHYTOPLANKTON, MARINE ALGAE,	ENVIRONMENTAL EFFECTS, TRACE ELEM	W73-00818
EDIMENTS, PUERTO RICO, ESTUARINE	ENVIRONMENT, PATH OF POLLUTANTS,	W73-00821
DATA COLLECTIONS, MARINE PLANTS,	ENVIRONMENTAL EFFECTS, REEFS, LAB	W73-00838
YTA, FROZEN SOILS, IONS, NOSTOC,	ENVIRONMENTAL EFFECTS, POLLUTANT	W73-00854
ECOLOGY, MARINE ALGAE, ESTUARINE	ENVIRONMENT, ABSORPTION, ANIMAL P	W73-00826
TIVITY TECHNIQUES, RADIOECOLOGY,	ENVIRONMENTAL EFFECTS, ABSORPTION	W73-00817
OCHEMICAL AN/ *SURVEYS, *AQUATIC	ENVIRONMENT, *RADIOECOLOGY, *RADI	W73-01673
EVEL, PROTOZOANS, AQUATIC ALGAE,	ENVIRONMENTAL EFFECTS, CHLOROPHYT	W72-13816
METHODOLOGY, CULTURES, BIOASSAY,	ENVIRONMENTAL EFFECTS, WATER POLL	W72-13813
ISOTOPES, *CARBON RADIOISOTOPES,	ENVIRONMENTAL EFFECTS, METHODOLOG	W72-12709
STEMS, *PRODUCTIVITY, *ESTUARINE	ENVIRONMENT, *DOMESTIC WASTES, *S	W72-12567
BITORS, WATER POLLUTION EFFECTS,	ENVIRONMENTAL EFFECTS, COMPETITIO	W72-12736
WATER POLLUTION EFFECTS, AQUATIC	ENVIRONMENT, TURBIDITY, WATER POL	W72-11925
BOW TROUT, AQUATIC LIFE, AQUATIC	ENVIRONMENT, FISH PHYSIOLOGY, ALG	W72-12257

A COSTATUM, CARBOY, OSCILLATORIA	ERYTHRAEA,: /CORNUTUM, SKELETCNEM	W72-09103
, CULTURE MEDIA,/ *TRICHODESMIUM	ERYTHRAEUM, *CULTURING TECHNIQUES	W73-00430
RINATED BIPHENYLS, AROCLOR 1254,	ESCAMBIA RIVER, PINFISH, FLCUNDER	W71-07412
LOGICAL TESTS, ANTIMICROBIOTICS,	ESCHERICHIA COLI, BETA STREPTOCOC	W72-04259
RAIN, ORGANOMERCURIAL COMPOUNDS,	ESOX LUCIUS, ALKYL MERCURY, EAGLE	W72-05952
AVESCENS, AMBLOPLITES RUPESTRIS,	ESOX LUCIUS, CAMPELCMA, PSIDIUM,	W70-08098
PTOPHYCEAE, PONTOPOREIA AFFINIS,	ESOX LUCIUS, PERCA FLUVIATILIS, S	W72-12729
, CRYPTOPHYCEAE, *DENMARK,/ LAKE	ESROM, TYSTRUP LAKE, BAVELSE LAKE	W72-12738
IENT REQUIREMENTS, PRODUCTIVITY,	ESSENTIAL NUTRIENTS, WATER QUALIT	W70-04080
ON, PHYTOPLANKTON, MARINE ALGAE,	ESSENTIAL NUTRIENTS.: /ENTIFICATI	W73-00441
OSPHATES, NUTRIENT REQUIREMENTS,	ESSENTIAL NUTRIENTS.: /TRATES, PH	W73-00430
ROPHICATION, DEFICIENT ELEMENTS,	ESSENTIAL NUTRIENTS, *AEROBIC CON	W72-11617
EN COMPOUNDS, CYCLING NUTRIENTS,	ESSENTIAL NUTRIENTS, LAKES, MUD,	W72-10608
D CHRCMATOGRAPHY, METHYL ESTERS,	ESTERS, CHLORELLA VULGARIS, SAMPL	W72-09365
AS LIQUID CHROMATOGRAPHY, METHYL	ESTERS, ESTERS, CHLORELLA VULGARI	W72-09365
NDERMERE(ENGLAND), EXOENZYME PR/	ESTHWAITE WATER(ENGLAND), LAKE WI	W72-03224
TARN(ENGLAND)/ *UNITED KINGDOM,	ESTHWAITE WATER(ENGLAND), BLELHAM	W71-13183
NDERMERE(ENGLA/ *UNITED KINGDOM,	ESTHWAITE WATER(ENGLAND), LAKE WI	W71-13185
ON, ECONOMIC FEASIBILITY, COSTS,	ESTIMATED COSTS.: /ATION, FLOTATI	W72-13508
TE, PROTEINS, FILTERS, ENGINEERS	ESTIMATES, ECONOMICS, FEASIBILITY	W70-10174
CARBON DIOXIDE, NITROGEN, LAKES,	ESTUARIES, NUTRIENTS, DETERGENTS,	W70-07283
WASTES, FLORIDA, SESSILE ALGAE,	ESTUARIES, CHEMICAL WASTES, GAS C	W71-07412
OXYGEN DEMAND, DISSOLVED OXYGEN,	ESTUARIES, WATER POLLUTICN EFFECT	W71-05390
ASSESSMENTS, MONITORING, SEWAGE,	ESTUARIES, GROWTH RATES, BIOLOGIC	W71-12867
ISSOLVED OXYGEN, *PHYTOPLANKTON,	ESTUARIES, WATER TEMPERATURE, TUR	W71-11876
, INDUSTRIAL WASTES, OIL WASTES,	ESTUARIES, MONITORING.: /L PLANTS	W71-13746
TEBRATES, RADIATION, ECOSYSTEMS,	ESTUARIES, CORAL, REEFS, WATER RE	W71-13723
LT MARSHES, ECOLOGY, PESTICIDES,	ESTUARIES, INTERTIDAL AREAS, ESTU	W72-10678
*ENGLAND, *POLLUTED	ESTUARIES, DURHAM COUNTY RIVERS.:	W72-08804
NCE, PATH OF POLLUTANTS, OCEANS,	ESTUARIES, CULTURES, SAMPLING, HA	W73-00361
QUIPMENT, ENVIRONMENTAL EFFECTS,	ESTUARIES, *DISTRIBUTION PATTERNS	W72-08141
LEAR WASTES, PATH OF POLLUTANTS,	ESTUARINE ENVIRONMENT, RADIOACTIV	W72-07826
L EXTRACTS, ERDSCHREIBER MEDIUM,	ESTUARINE SOILS, AXENIC CULTURES,	W73-00430
RADIOISOTOPES, STABLE ISOTOPES,	ESTUARINE ENVIRONMENT, COASTS, AN	W73-00811
ANKTON, *PHYTOPLANKTON, NERITIC,	ESTUARINE ENVIRONMENT, SEA WATER,	W73-00935
NIA, RADIOECOLOGY, MARINE ALGAE,	ESTUARINE ENVIRONMENT, ABSORPTION	W73-00826
E ALGAE, SEDIMENTS, PUERTO RICO,	ESTUARINE ENVIRONMENT, PATH OF PO	W73-00821
TH CAROLINA, ABSORPTION, SOUNDS,	ESTUARINE ENVIRONMENT, PATH OF PO	W73-00818
LS, WASTE ASSIMILATIVE CAPACITY,	ESTUARINE ENVIRONMENT, FOOD CHAIN	W73-00828
SOPHYTA, BIOLOGICAL COMMUNITIES,	ESTUARINE ENVIRONMENT, WATER POLL	W73-00853
ES, ESTUARIES, INTERTIDAL AREAS,	ESTUARINE FISHERIES.: /, PESTICID	W72-10678
TIVITY, WATER POLLUTION EFFECTS,	ESTUARINE ENVIRONMENT.: /Y PRODUC	W72-09653
, ASIA, MARINE ALGAE, SEDIMENTS,	ESTUARINE ENVIRCNMENT, MCNITCRING	W72-10958
IGATICNS, ENVIRONMENTAL EFFECTS,	ESTUARINE ENVIRONMENT, REGULATION	W72-00959
TS, PESTICIDES, ALGAE, DREDGING,	ESTUARINE ENVIRONMENT, CALIFORNIA	W71-09581
TS, FOOD CHAINES, PUBLIC HEALTH,	ESTUARINE ENVIRONMENT, ZINC RADIO	W72-04463
TS, MARINE ALGAE, PUBLIC HEALTH,	ESTUARINE ENVIRONMENT, ON-SITE IN	W72-04461
, SAMPLING, MATHEMATICAL MODELS,	ESTUARINE ENVIRONMENT, SALINITY,	W72-12941
, *FISHES, ECOLOGY, FOOD CHAINS,	ESTUARINE FISHERIES, MARINE FISHE	W72-12567
*POTOMAC RIVER	ESTUARY.:	W71-08775
PCTOMAC RIVER BASIN, POTOMAC	ESTUARY.:	W70-06509
*PAMLICO RIVER	ESTUARY.:	W72-01329
PANSICN, ISLE OF THANET, MEDWELL	ESTUARY.: /T COAST, INDUSTRIAL EX	W71-13746
*POTOMAC	ESTUARY.:	W72-09171
POTOMAC	ESTUARY.:	W71-09788
CALIF), SEARSVILLE LA/ SAN PABLO	ESTUARY(CALIF), SACRAMENTO RIVER(	W70-10405
A(CALIF),/ ALGAL CYCLE, DELAWARE	ESTUARY(DEL-NJ), SAN JOAQUIN DELT	W71-05390
LVED CXYGEN CONCENTRA/ *DELAWARE	ESTUARY, *BIOLOGICAL LIFE, *DISSO	W70-09189
CAYUGA LAKE(NEW YORK), PATUXENT	ESTUARY, BISCAYNE BAY(FLORIDA), H	W71-11517
ROAL/ *YAQUINA BAY(ORE), YAQUINA	ESTUARY, DATA INTERPRETATION, MAC	W72-08141
ON RATIOS.: *PAMLICO RIVER	ESTUARY, PHOTOSYNTHESIS/RESPIRATI	W72-13636
ROL, LACTATE, PYRUVATE, ACETATE,	ETHANOL, BENZOATE, PROPIANATE, PR	W72-05421
ELETONEMA COSTATUM, NUCLEOTIDES,	ETHANOL, MONOSACCHARIDES, CULTURE	W73-00942
Y LIQUID, BLUSYL, SODIUM LAURYL,	ETHER SULFATE, SOCIUM DODECYL, BE	W72-12576
S HAUCKII, STIGMAPHORA ROSTRATA,	ETHMODISCUS OOZES.: /IA, HEMIAULU	W72-03567
RNIA, HEMIAUL/ *ETHMODISCUS REX,	ETHMODISCUS GAZELLAE, BAJA CALIFO	W72-03567
ALKYLENE OXIDE, POLYMERIZATION,	ETHYLENE OXIDE, WATER CLARIFICATI	W70-06968
TEROTROPHIC BACTERIA, ACETYLENE,	ETHYLENE.: /ER, DETROIT RIVER, HE	W73-01456
, MACROINVERTEBRATES, ACETYLENE,	ETHYLENE, NITROGEN RADIOISOTOPES,	W72-10883
APAN, *AMURSKY BAY, PERIDINIANS,	EUCAMPIA ZOODIACUS, CHAETOCEROS A	W73-00442
BOSMINA, CERIODAPHNIA,	EUDIAPTOMUS.:	W72-00161
RINA MORUM, PLATYDORINA CAUDATA,	EUDORINA CALIFORNICA, EUDORINA IL	W71-13793
A CAUDATA, EUDORINA CALIFORNICA,	EUDORINA ILLINOISENSIS.: /TYDORIN	W71-13793
UPLII, LAMELLIBRANCH, SPIROGYRA,	EUDORINA, CYCLOPS, KERATELLA, ASP	W71-11876
MIPTERA, *TRICHOPTERA, CORIXIDS,	EUDORINA, NOTONECTIDAE, BODO, HYD	W70-06655
CCUM, HYALO/ CARBON-14, CARTERIA	EUGAMETOS, STICHOCOCCUS, CHLOROCO	W72-09239
GLENA/ *INDIA, ORGANIC NITROGEN,	EUGLENA ACUS, EUGLENA CAUDATA, EU	W73-00918
ORGANIC NITROGEN, EUGLENA ACUS,	EUGLENA CAUDATA, EUGLENA LIMNOPHY	W73-00918
TES, BIOTIN, NIACIN, COBALAMINS,	EUGLENA GRACILIS, COLONIZATION, C	W73-02095
HLORIDE, ANKISTRODESMUS BRAUNII,	EUGLENA GRACILIS.: /YL MERCURIC C	W72-06037
OSTATUM, DUNALIELLA TERTIOLECTA,	EUGLENA GRACILIS, CHLAMYDOMONAS R	W72-05614
ACYSTIS NIDULAN/ WARBURG EFFECT,	EUGLENA GRACLIS, CALVIN CYCLE, AN	W71-09172
, EUGLENA ACUS, EUGLENA CAUDATA,	EUGLENA LIMNOPHYLA, EUGLENA OBLON	W73-00918
ENA CAUDATA, EUGLENA LIMNOPHYLA,	EUGLENA OBLONGA, EUGLENA OXYURIS,	W73-00918
ENA LIMNOPHYLA, EUGLENA OBLONGA,	EUGLENA OXYURIS, EUGLENA POLYMORP	W73-00918
UGLENA OBLONGA, EUGLENA OXYURIS,	EUGLENA POLYMORPHA, EUGLENA SPIRO	W73-00918
ENA OXYURIS, EUGLENA POLYMORPHA,	EUGLENA SPIROGYRA, LEPOCINCLIS OV	W73-00918
AE, MARINE ALGAE, AQUATIC ALGAE,	EUGLENA, CHLAMYDOMONAS.: /MS, ALG	W72-08436
AY, SAMPLING, OXIDATION LAGOONS,	EUGLENA, CHLAMYDOMONAS, EUGLENOPH	W72-07661
, ENZYMES, METABOLISM, HYDROGEN,	EUGLENA, CHLORELLA, SCENEDESMUS,	W71-09172

ARA, SUBMERGED PLANTS, PROTOZOA,	EUGLENA, CHLAMYDOMONAS, CHLOROPHY	W72-00155
RELLA, CHLOROPHYTA, SCENEDESMUS,	EUGLENA, DISSOLVED OXYGEN, DIURNA	W71-11519
, BIOMASS, OLIGOTROPHY, DIATOMS,	EUGLENA, DAPHNIA, HABITATS, NICHE	W71-09171
CHLORELLA, CLOSTRIDIUM, E. COLI,	EUGLENA, EUGLENOPHYTA, CHLOROPHYT	W72-06124
LUDGE, ALGAE, PROTOZOA, DIATOMS,	EUGLENA, MATHEMATICAL STUDIES.: /	W72-08573
LAGOONS, EUGLENA, CHLAMYDOMONAS,	EUGLENOPHYTA, PROTOZOA, EQUIPMENT	W72-07661
C ALGAE, LIMNOLOGY, CHRYSOPHYTA,	EUGLENOPHYTA, PYRROPHYTA, BIOMASS	W72-07145
LLATES, CYANOPHYTA, CHLOROPHYTA,	EUGLENOPHYTA, GRAZING, WATER QUAL	W72-07899
OPHYTA, CYANOPHYTA, CHRYSOPHYTA,	EUGLENOPHYTA, DISSOLVED OXYGEN, C	W72-08048
HYSICAL PROPERTIES, CHLOROPHYTA,	EUGLENOPHYTA, CYANOPHYTA, CHRYSOP	W73-01627
, CLOSTRIDIUM, E. COLI, EUGLENA,	EUGLENOPHYTA, CHLOROPHYTA, FERROB	W72-06124
OPHYTA, PYRROPHYTA, CHRYSOPHYTA,	EUGLENOPHYTA, WATER TEMPERATURE,	W72-04855
SOPHYTA, RHODOPHYTA, CYANOPHYTA,	EUGLENOPHYTA.: /CHLOROPHYTA, CHRY	W71-05629
RNAL, FLUCTUATIONS, CHLOROPHYTA,	EUGLENOPHYTA, CYANOPHYTA, CHRYSOP	W72-00150
YTA, HABITATS, AQUATIC HABITATS,	EUGLENOPHYTA, PYRROPHYTA, FRESHWA	W72-09119
ECTS, BIOMASS, NUTRIENTS, ALGAE,	EUGLENOPHYTA, CHRYSOPHYTA, CHLORO	W72-09111
INUTISSIMA, ACHNANTHES FLEXELLA,	EUNOTIA PECTINALIS, DESMIDS.: / M	W71-11029
HORA OVALIS, SYNEDRA FASCIULATA,	EUNOTIA PECTINALIS, LITHOPHYTES,	W72-08141
, RHABDODERMELLA, LOPHOPANOPEUS,	EUPENTACTA, CNEMIDOCARPA, ENHYDRA	W73-00932
TRACTION, TRIPHOTURUS MEXICANUS,	EUPHAUSIA PACIFICA, ENGRAULIS MOR	W73-02105
OIKOPLEURA, SAGITTA, PTEROPODS,	EUPHAUSIIDS, GADUS MORHUA, UROPHY	W73-00371
ALGAL SINKING, *VERTICAL MIXING,	EUPHOTIC ZONE, DIFFUSION COEFFICI	W71-03034
ON, WASTE ASSIMILATIVE CAPACITY,	EUROPE, ANIMAL PHYSIOLOGY, TRACE	W73-00817
ENT, FOOD CHAINS, PUBLIC HEALTH,	EUROPE, RADIOISOTOPES, EFFLUENTS.	W73-00828
RINITY, METAHALINE, HYPERHALINE,	EURYHALINE, RED SEA, GULF OF SUEZ	W73-00855
ATER TEMPERATURE, PHYTOPLANKTON,	EUTHROPHICATION.: /R OPERATION, W	W72-05282
NA, NUTRIENTS, SEWAGE EFFLUENTS,	EUTNOPHICATION, ALGAE, CYANOPHYTA	W72-13636
IUM SPLENDENS, EXUVIELLA MARINA,	EUTREPTIA, NITZSCHIA SERIATA, THA	W71-07875
OSYSTEMS, ALGAE, *LIGHT QUALITY,	EUTROPHIC ZONE, CHLOROPHYLL, PHOS	W72-12393
TY, PHOTOSYNTHESIS, RESPIRATION,	EUTROPHICATION, PLANT POPULATION,	W72-12192
Y, ALGAE, WATER QUALITY CONTROL,	EUTROPHICATION, WASTE WATER TREAT	W72-12487
YGEN DEMAND, HEAT BUDGET, ALGAE,	EUTROPHICATION.: / ALKALINITY, OX	W72-12997
AINFALL, WASTE WATER(POLLUTION),	EUTROPHICATION, WATER POLLUTION S	W72-13644
TIFICATION, DIATOMS, EPILIMNION,	EUTROPHICATION.: /S, THERMAL STRA	W72-13659
RBON RADIOISOTOPES, *CYANOPHYTA,	EUTROPHICATION, WATER POLLUTION E	W72-13692
DEFICIENT ELEMENTS, MESOTROPHY,	EUTROPHICATION, TROPHIC LEVEL, PR	W72-13816
TANDING CROPS, LIMITING FACTORS,	EUTROPHICATION.: /IC COMPOUNDS, S	W72-13653
ES, CYANOPHYTA, LIGHT INTENSITY,	EUTROPHICATION, PHYTOPLANKTON, PR	W73-02472
RATION, ALGAE, METHANE BACTERIA,	EUTROPHICATION, NITRATES, PHOSPHA	W71-07106
LINE, NITRATES, LIGHT INTENSITY,	EUTROPHICATION, MARINE AL: /ERMOC	W71-07875
RCOLATION, EVAPORATION, INSECTS,	EUTROPHICATION, ALGAE.: /PAGE, PE	W71-07124
ONS, *ALGAE, *TRICKLING FILTERS,	EUTROPHICATION, PHOTOSYNTHESIS, D	W71-07097
, PHOSPHORUS, PHOSPHATES, ALGAE,	EUTROPHICATION, WATER QUALITY, RI	W71-06445
ALGAE, COMPARATIVE PRODUCTIVITY,	EUTROPHICATION.: /OMASS, OXYGEN,	W71-09165
EFFECTS, WATER POLLUTION SOURCES,	EUTROPHICATION, COLIFORMS, SPORES	W71-09173
A, *LAKES, ARCTIC, PRODUCTIVITY,	EUTROPHICATION, RED TIDE, SILICA,	W71-09190
NUTRIENTS, NITROGEN, PHOSPHORUS,	EUTROPHICATION, ALGAE, PATHOGENIC	W71-08214
TROL, *BIODEGRADATION, FISHKILL,	EUTROPHICATION, ALGAL BLOOMS, WAT	W71-08768
RE, THERMAL STRATIFICATION, ICE,	EUTROPHICATION, OXYGEN, SEASONAL.	W72-04855
DETRITUS, PLANKTON, ADSORPTION,	EUTROPHICATION, TEMPERATURE, LIGH	W72-05469
ITY, PHYTOPLANKTON, ZOOPLANKTON,	EUTROPHICATION, OLIGOTROPHY, PROT	W72-05968
PHOTOSYNTHESIS, CARBON FIXATION,	EUTROPHICATION, WATER QUALITY CON	W72-06046
AL EFFECTS, LIMNOLOGY, PLANNING,	EUTROPHICATION, EVALUATION, WATER	W72-06526
ACTERIA, *WASTE WATER TREATMENT,	EUTROPHICATION.: /LUDGE, ALGAE, B	W72-04086
T LAKES, *SEDIMENTS, *NUTRIENTS,	EUTROPHICATION, NITROGEN, LITTORA	W72-04263
), *KENNEBEC RIVER(ME), CULTURAL	EUTROPHICATION.: /NNABESSACOOK(ME	W72-04280
NESS(WATER), COLOR, OLIGOTROPHY,	EUTROPHICATION, ALGAE, CYANOPHYTA	W72-04269
CIENCIES, WATER QUALITY CONTROL,	EUTROPHICATION, ALGAL CONTROL, *C	W72-04788
WATER TREATMENT, ALGAL CONTROL,	EUTROPHICATION, MODEL STUDIES, MA	W72-04787
, *SURFACE WATERS, WATER SUPPLY,	EUTROPHICATION, ALGAE, NUTRIENTS,	W71-05083
UTRIENTS, *PHOSPHORUS, *NITROGE/	EUTROPHICATION, *ALGAL BLOOMS, *N	W70-09669
UMBIA, TIDAL EFFECTS, NUTRIENTS,	EUTROPHICATION, WATER QUALITY, WA	W70-08249
UMBIA, TIDAL EFFECTS, NUTRIENTS,	EUTROPHICATION, WATER QUALITY, WA	W70-08248
ON, ACTIVATED SLUDGE, TREATMENT,	EUTROPHICATION, SURFACTANTS, ENZY	W71-00116
IR POLLUTION, CHEMICAL ANALYSIS,	EUTROPHICATION, TRANSPIRATION, PR	W71-01489
, *ION TRANSPORT, *HEAVY METALS,	EUTROPHICATION, POLLUTANT IDENTIF	W73-02112
, AQUATIC POPULATIONS, BIOASSAY,	EUTROPHICATION.: /AQUATIC ANIMALS	W73-02095
WATER POLLUTION, PROTOZOA, OHIO,	EUTROPHICATION, LAKE ONTARIO, LIG	W73-01615
NIQUES, *WISCONSIN, OLIGOTROPHY,	EUTROPHICATION, LAKE SUPERIOR, LA	W73-01456
UATIC ALGAE, NITRATES, NITRITES,	EUTROPHICATION.: /ATIC PLANTS, AQ	W73-00286
LGAE, *ISOLATION, MARINE PLANTS,	EUTROPHICATION, OCEANS, SAMPLING,	W73-00430
TER POLLUTION SOURCES, ASBESTOS,	EUTROPHICATION, POLYCHLORINATED B	W73-00703
RATES, RADIOACTIVITY TECHNIQUES,	EUTROPHICATION, MARINE ALGAE, WAT	W73-00431
ER, WASTE ASSIMILATIVE CAPACITY,	EUTROPHICATION, NUTRIENT REMOVAL,	W73-00802
GRAPHY, WATER POLLUTION EFFECTS,	EUTROPHICATION, *LAKES, ALGAE, IN	W72-14728
DIATOMS, CYANOPHYTA, TURNOVERS,	EUTROPHICATION.: /MICAL ANALYSIS,	W73-00232
G FACTORS, ALLOGENIC SUCCESSION,	EUTROPHICATION.: /RIENTS, LIMITIN	W72-14790
OMS, CHLOROPHYTA, BENTHIC FLORA,	EUTROPHICATION, DOMINANT ORGANISM	W72-14797
*PROTOZOA, *ALGAE, SCENEDESMUS,	EUTROPHICATION, PONDS, FEEDING RA	W72-14796
SYNTHESIS, PRIMARY PRODUCTIVITY,	EUTROPHICATION, SOLAR RADIATION,	W72-08054
NKTON, STANDING CROP, LIMNOLOGY,	EUTROPHICATION.: /ODUCTIVITY, PLA	W72-08559
ARBON, DISTRIBUTION, PHOSPHORUS,	EUTROPHICATION, ALGAE, AQUATIC WE	W72-08459
ORGANIC MATTER, PHOTOSYNTHESIS,	EUTROPHICATION, PHOSPHORUS, CHLOR	W72-08048
*ALGAL CONTROL, *NUISANCE ALGAE,	EUTROPHICATION, REVIEWS, TROPHIC	W72-07937
DUCTION POTENTIAL, LAKES, SOILS,	EUTROPHICATION, AMMONIFICATION, N	W72-07933
TS, *PHOSPHATES, *POTABLE WATER,	EUTROPHICATION, ALKYLBENZENE SULF	W72-06837
*ALGAL CONTROL, WATER POLLUTION,	EUTROPHICATION, NUTRIENTS, CHLORE	W72-07508
A, MICROORGANISMS, ALGAE, LAKES,	EUTROPHICATION, LITTORAL, SANDS,	W72-07501
FISH, PONDS, LAKES, FOOD HABITS,	EUTROPHICATION, LARVAE, ROTIFERS,	W72-07132
N SOURCES, *POLLUTION ABATEMENT,	EUTROPHICATION, LEGISLATION, PUBL	W72-06655

T PROBABLE-NUMBER TEST, MEMBRANE FILTRATION, RHODOPSEUDOMONAS GELA W72-04743
ORINATION, PHOSPHATES, NITRATES, FILTRATION, POLLUTION ABATEMENT, W72-05315
LTERS, MONOSACCHARIDES, MEMBRANE FILTRATION.: /LUCOSE, MEMBRANE FI W72-14326
ERS, CATIONS, ANIONS, CHLORELLA, FILTRATION, ELECTROPHORESIS, TRAC W72-14840
ATE, HYDROGEN ION CONCENTRATION, FILTRATION, PLA: / CALCIUM CARBON W73-01669
DIOISOTOPES, ALGAE, FRESH WATER, FILTRATION, LEACHING, FUNGI, LIGH W72-07143
R, ACTIVATED CARBON, ADSORPTION, FILTRATION, ALGAE, PHENOLS, DETER W72-08357
ACTIONS, DIATOMS, AQUATIC ALGAE, FILTRATION, CYANOPHYTA, *EUTROPHI W72-13673
NTROLS, AERATION, SEDIMENTATION, FILTRATION, ANAEROBIC DIGESTION, W72-00042
MAND, BIOCHEMICAL OXYGEN DEMAND, FILTRATION, DISINFECTION, FLOCCUL W72-00411
ON, FLOCCULATION, SEDIMENTATION, FILTRATION, LIME, ROCKS, METABOLI W71-13341
ION, HYDROGEN ION CONCENTRATION, FILTRATION, COLLOIDS, CARBON DIOX W72-03613
IODEGRADATION, LABORATORY TESTS, FILTRATION, AMMONIA, HYDROGEN ION W72-01881
CTROLYTES, LIME, NEUTRALIZATION, FILTRATION, CENTRIFUGATION, VACUU W72-10237
D SOLIDS, SEPARATION TECHNIQUES, FILTRATION, MUNICIPAL WASTES, IND W72-09386
RESI/ *MICROSTRAINING, *PRECOAT FILTRATION, BACKWASHING, SPECIFIC W72-09393
NITRATES, ALGAE, DENTRIFICATION, FILTRATION, DESALINATION, DRAINAG W71-11698
KE ONTARIO, NUTRIENTS, NITROGEN, FILTRATION, MICROSCOPY, INDUSTRIA W71-11507
PARTINA SPP, DISTICHLIS SPICATA, FIMBRISTYLIS, BORRICHIA, PHRAGMIT W73-01089
ANTS, EUTROPHICATION, GOVERNMENT FINANCE, ENVIRONMENTAL SANITATION W72-01659
NVIRONMENTAL EFFECTS, GOVERNMENT FINANCE,: /MMUNITY DEVELOPMENT, E W72-01660
*EUTROPHICATION, *ALGAL CONTROL, FINANCING, PLANT GROWTH, TOXICITY W71-13172
SPHATES, *OLIGASAPROBIC STREAMS, FINGERNAIL CLAMS, UNIONIDS, ORCON W72-06526
HOSPHATE CONCENTRATIONS, GULF OF FINLAND.: ORTHOP W72-13871
FINLAND, *BOTHNIAN BAY, *GULF OF FINLAND, DARK FIXATION, COASTAL W W72-03228
, *PRIMARY PRODUCTIVITY, ALASKA, FIORDS, SURFACE WATERS, NITROGEN, W72-09122
, SANBORN FILTER, BACKWASHING.: FIREBAUGH(CALIF), VOLATILE SOLIDS W70-10174
INEA VAR. DAWSONII, CALLOPHYLLIS FIRMA, PLENOSPORIUM DASYOIDES, SE W73-02099
ATER, *FISHERIES, WATER QUALITY, FISH ATTRACTANTS, FISH HARVEST, P W70-09596
LING TOWERS, DENSITY, VISCOSITY, FISH BEHAVIOR, FERTILITY, SEASONA W70-09612
E, *FISHERIES, FISH POPULATIONS, FISH BEHAVIOR, INVERTEBRATES, OTT W73-00764
AL PROPERTIES, MINERALOGY, FISH, FISH BEHAVIOR, CHEMICAL ANALYSIS, W72-14280
CHNIQUES, FOOD CHAINS, CULTURES, FISH DIET, GAS CHROMATOGRAPHY, PA W73-02105
S, DAPHNIA, AQUATIC POPULATIONS, FISH DIETS, FRESHWATER FISH, POND W72-07132
COPEPODS, MARINE ALGAE, SHRIMP, FISH EGGS, FISH, BACTERIA, FISH P W72-08142
FERTILIZERS, FARM WASTES, FISH, FISH EGGS, TURBIDITY, NUTRIENTS, W71-06444
ATION, DIATOMS, CURRENTS(WATER), FISH FARMING.: /ANKTON, STRATIFIC W73-00236
RIVER(CALIF), RIFLE FALLS(COLO), FISH FECAL WASTES, RESIDUAL FOOD, W71-11006
, ALGAE, RESOURCES, FOOD CHAINS, FISH FOOD ORGANISMS, OXYGEN, ECOS W71-13246
OROPHYTA, *PHYTOTOXINS, ECOLOGY, FISH FOOD ORGANISMS, PATH OF POLL W71-08597
WATER QUALITY, FISH ATTRACTANTS, FISH HARVEST, PLANKTON, BENTHIC F W70-09596
, CONDENSERS, DESIGN, STANDARDS, FISH KILL, ALGAE, COOLING TOWERS, W70-09612
MEDIES, FOODS, FISH, HERBIVORES, FISH MANAGEMENT,: /BIOCONTROL, RE W71-09163
RIMP, FISH EGGS, FISH, BACTERIA, FISH PARASITES, ANTIBIOTICS(PESTI W72-08142
UATIC LIFE, AQUATIC ENVIRONMENT, FISH PHYSIOLOGY, ALGAE, DAPHNIA.: W72-12257
ALGI/ DETOXIFICATION, SYNERGISM, FISH PONDS, ALGISTATIC ACTIVITY, W71-06189
GAE, PHYTOPLANKTON, ZOOPLANKTON, FISH POPULATIONS, WATER TEMPERATU W72-04853
ASKA, *MARINE ALGAE, *FISHERIES, FISH POPULATIONS, FISH BEHAVIOR, W73-00764
Y, TROPHIC LEVEL, AQUATIC ALGAE, FISH POPULATIONS, FLORIDA, WATER W72-08986
RONMENTAL EFFECTS, AQUATIC LIFE, FISH POPULATIONS, PHYTOPLANKTON, W72-00939
SOURCES, *FISHERIES, OIL WASTES, FISH REPRODUCTION, WASTE DISPOSAL W71-13723
TOXICITY, POISONS, ALGAL TOXINS, FISH TOXINS, SEDIMENTS, SALT MARS W72-10678
AL CONTROL, SPAWNING, WARM-WATER FISH.: /ANAEROBIC CONDITIONS, ALG W70-09595
*CLUSTER ANALYSIS(FISH), *DICKERSON PLANT.: W72-07791
IOASSAY, *ABSORPT/ *DDT, *MARINE FISH, *PHYTOPLANKTON, *SHRIMP, *B W73-02105
FRESHWATER, HYDROBIOLOGY, ALGAE, FISH, ABSORPTION, BIOASSAY, METAB W72-10694
POLLUTANTS, MARINE ALGAE, MARINE FISH, ABSORPTION, WATER POLLUTION W72-00940
Y, *BIOASSAYS, LABORATORY TESTS, FISH, ALGAE, CHEMICAL WASTES, ALG W72-08683
NITY, SEDIMENTS, EUTROPHICATION, FISH, ALGAE, RECREATION, ENVIRONM W71-10577
UNOFF, DOMESTIC WASTES, ECOLOGY, FISH, ALGAE, AQUATIC PLANTS, PROT W73-01972
TRATIFICATION, HYDRAULIC MODELS, FISH, ALGAE, FATHOMETERS, BENTHIC W70-09614
ESTS, NITRATES, MUSSELS, SNAILS, FISH, ALGAE, GASTROPODS, ADSORPTI W71-09168
SOILS, PLANT TISSUES, FRESHWATER FISH, AQUATIC ALGAE, EVALUATION, W73-01673
MARINE ALGAE, SHRIMP, FISH EGGS, FISH, BACTERIA, FISH PARASITES, A W72-08142
DERGROUND, ALASKA, PONDS, WELLS, FISH, BAYS, AIR POLLUTION EFFECTS W72-03347
PESTICIDE TOXICITY, FOOD CHAINS, FISH, BIOASSAY, ALGAE, SNAILS, MO W72-07703
CONTROL, *ALGAE CONTROL, *FORAGE FISH, BIOCONTROL, REMEDIES, FOODS W71-09163
, SEWAGE LAGOONS, BRACKISH WATER FISH, BIOLOGICAL TREATMENT, OXIDA W72-12567
CS, *MAINE, MUDS, INVERTEBRATES, FISH, BIRDS, STREAMS, AQUATIC PLA W72-12930
ICATORS, AGRICULTURAL CHEMICALS, FISH, CHEMICALS, CHEMICAL WASTES, W73-01976
BIOTA, BACTERIA, BENTHIC FAUNA, FISH, ENVIRONMENTAL EFFECTS, RESE W72-05282
TIC ALGAE, AQUATIC PRODUCTIVITY, FISH, FISHKILL, LIMNOLOGY, NUTRIE W72-04279
OSION, FERTILIZERS, FARM WASTES, FISH, FISH EGGS, TURBIDITY, NUTRI W71-06444
OLOGICAL PROPERTIES, MINERALOGY, FISH, FISH BEHAVIOR, CHEMICAL ANA W72-14280
*CYANOPHYTA, EXPLOSIVES, COSTS, FISH, FISHKILL, RESERVOIRS, FISHE W72-11564
WEEDS, RECREATION, GROUNDWATER, FISH, FLOOD PLAINS, IRRIGATION, P W70-05103
MINOUS MATERIALS, ANIMAL WASTES, FISH, FLUORESCENCE, INFRARED RADI W72-08790
ASTS, INSECTS, MILDEWS, E. COLI, FISH, FOODS, GRASSES, WORMS.: /YE W73-01676
HYTES, CYPRINUS CARPIO, MOSQUITO FISH, GAMBUSIA AFFINIS, CESIUM RA W73-01673
SH, BIOCONTROL, REMEDIES, FOODS, FISH, HERBIVORES, FISH MANAGEMENT W71-09163
CHAINS, *ECOSYSTEMS, ABSORPTION, FISH, INDUSTRIAL WASTES, ALGAE, A W72-05952
OLLUTION EFFECTS, MUD, MOLLUSKS, FISH, INSECTS, DDT, SILTS, SUCKER W70-08098
TA, CARBON RADIOISOTOPES, MARINE FISH, INSTRUMENTATION, ABSORPTION W72-06283
ALKALINITY, VOLUMETRIC ANALYSIS, FISH, INTAKE STRUCTURES, CARBON D W72-05594
TER HYACINTH, NUTRIENTS, MIDGES, FISH, LAKES, CONDUCTIVITY, ALGAE, W71-01488
YSIS, RADIOACTIVITY, ADSORPTION, FISH, LAKE ERIE, SEDIMENTS, SILTS W71-11036
PES, PHYSICOCHEMICAL PROPERTIES, FISH, MARINE PLANTS, MARINE ALGAE W72-04463
WATER, *BIOTA, SAMPLING, MARINE FISH, MARINE ANIMALS, MARINE ALGA W71-09018
T TOLERANCE, CRUSTACEANS, MARINE FISH, MARSH PLANTS.: /ALGAE , SAL W73-00855
GREAT LAKES, AQUATIC ALGAE, ICE, FISH, MICROORGANISMS, ODOR, TASTE W73-01669

*HORMOTILA BLENNISTA, FILTRATES, GROWTH AUTOSTIMULATION, GROWTH RE W71-08687
*CLADOPHORA GLOMERATA, GROWTH CURVES.: W72-13651
IDES, THIN LAYER CHROMATOGRAPHY, GROWTH INHIBITION.: /URIAL PESTIC W72-13347
VARIABILIS, *ANACYSTIS NIDULANS, GROWTH INHIBITION, SELENATE, SELE W72-00153
NRICHMENT, CULTURE MEDIA, MACRO/ GROWTH KINETICS, BIOENERGETICS, E W72-09378
OMOSIPHON TUBEROSUS, *EPIPHYTES, GROWTH MEDIA, COCCONEIS PLACENTUL W72-12744
ON, CHLORELLA PYRENOIDOSA, MONOD GROWTH MODEL, ALGAL GROWTH RATES. W72-04788
ILL), DIAGNOSTIC CRITERIO/ ALGAL GROWTH POTENTIAL, ILLINOIS RIVER(W71-11033
*PLANT GROWTH, *EUTROPHICATION, GROWTH RATES, SAMPLING, LAKE MICH W71-11033
, MONITORING, SEWAGE, ESTUARIES, GROWTH RATES, BIOLOGICAL COMMUNIT W71-12867
UTRIENT REQUIREMENTS, CHLORELLA, GROWTH RATES, TEMPERATURE, SEA WA W71-10083
ON, VERTICAL MIGRATION, AQUARIA, GROWTH RATES, SEDIMENTATION, BUOY W71-09852
ICATION, *BIOINDICATORS, *ALGAE, GROWTH RATES, CULTURES, SUSPENDED W71-09795
ORGANIC MATTER, ATLANTIC OCEAN, GROWTH RATES, ALGAE, INCUBATION, W71-11562
, INDUSTRIAL WASTES, METABOLISM, GROWTH RATES, OXIDATION LAGOONS, W72-09590
S, NITROGEN, PHOSPHORUS, CARBON, GROWTH RATES, OLIGOTROPHY, EUTROP W72-09164
KTON, *ALGAE, *LIMITING FACTORS, GROWTH RATES, DIATOMS, BACTERIA, W72-09163
E ALGAE, PHYTOPLANKTON, BIOMASS, GROWTH RATES, PRODUCTIVITY, ANIMA W72-09378
RGANISMS, ENVIRONMENTAL EFFECTS, GROWTH RATES, E. COLI, AZOTOBACTE W72-10861
GY, ENVIRONMENT, PHOTOPERIODISM, GROWTH RATES, TEMPERATURE, BIOCHE W72-00377
UM COMPOUNDS, NITRATES, DENSITY, GROWTH RATES, LABORATORY TESTS, C W72-03217
S, CHLOROPHYLL, PROTEINS, ALGAE, GROWTH RATES, TRACERS.: / CULTURE W72-03557
ILIZATION, PRIMARY PRODUCTIVITY, GROWTH RATES, RESPIRATION.: /STAB W72-03556
ATION, EFFICIENCES, COMPETITION, GROWTH RATES, PLANT MORPHOLOGY, W W72-04766
NUOUS CULTURES, MAXIMUM SPECIFIC GROWTH RATE, SATURATION CONSTANT, W72-04788
WTH REACTORS, ALGAL PONDS, ALGAL GROWTH RATES, *POTAMOCYPRIS.: /RO W72-04787
IDOSA, MONOD GROWTH MODEL, ALGAL GROWTH RATES.: / CHLORELLA PYRENO W72-04788
ALGAL GROWTH RATES.: W72-04789
YSIOLOGY, PLANT GROWTH, BIOMASS, GROWTH RATES, LIGHT INTENSITY, CA W72-05476
IC FLORA, *PHYTOPLANKTON, ALGAE, GROWTH RATES, BIOASSAY, DISSOLVED W72-06046
ANKTON, TOXICITY, CULTURES, DDT, GROWTH RATES.: / DIATOMS, PHYTOPL W72-05614
HESIS, BIOMASS, LIGHT INTENSITY, GROWTH RATES, MATHEMATICAL MODELS W72-06274
TE WATER TREATMENT, CHLOROPHYTA, GROWTH RATES.: /LITY CONTROL, WAS W72-04431
BUTION PATTERNS, CHROMATOGRAPHY, GROWTH RATES, POLLUTANT IDENTIFIC W72-04134
SORPTION, RADIOISOTOPES, ALGAE, GROWTH RATES, THERMAL POLLUTION.: W71-02075
CARBON RADIOISOTOPES, CHLORELLA, GROWTH RATES, DIATOMS, PHYTOPLANK W71-03027
TEMPERATURE, OPTICAL PROPERTIES, GROWTH RATES.: /OPHYTA, DIATOMS, W70-10403
UTRIENTS, NUTRIENT REQUIREMENTS, GROWTH RATES, PHOSPHORUS, NITROGE W72-12966
URES, RESISTANCE, AQUATIC ALGAE, GROWTH RATES, BIOASSAY, LABORATOR W72-13809
GHT INTENSITY, TEMPERATURE, MUD, GROWTH RATES, ANABAENA, NITROGEN W72-12544
TREATMENT, *MATHEMATICAL MODELS, GROWTH RATES, SOLAR RADIATION, FL W72-12289
TENSITY, SCENEDESMUS, CHLORELLA, GROWTH RATES, RESISTANCE, PHOSPHO W72-12739
IMARY PRODUCTIVITY, SYSTEMATICS, GROWTH RATES, BIOMASS, DATA COLLE W73-02469
UCTIVITY, BACTERIA, *SCENEDESMUS, GROWTH RATES, CULTURES, AQUATIC P W72-14314
LISM, *CHLORELLA, *LIGHT, ALGAE, GROWTH RATES, CHLOROPHYLL.: /TABO W73-00251
S, *NITRATES, *AMMONIA, DIATOMS, GROWTH RATES, NITROGEN, WATER POL W73-00380
ORINATED HYDROCARBON PESTICIDES, GROWTH RATES, ENVIRONMENTAL EFFEC W73-00281
A, AEROBIC CONDITIONS, BIOASSAY, GROWTH RATES, MORTALITY.: /NOPHYT W72-07664
RADIATION, VITAMIN B, PIGMENTS, GROWTH RATES, ALGAE, FOOD WEBS, C W72-07892
ATION, SUCCESSION, PRODUCTIVITY, GROWTH RATES.: /SYNTHESIS, RESPIR W73-01632
ATOMS, POLLUTANT IDENTIFICATION, GROWTH RATES, DEFICIENT ELEMENTS, W73-01445
WATER, BIOMASS, THERMAL STRESS, GROWTH RATES, SPATIAL DISTRIBUTIO W73-00853
ICIDES, HERBICIDES, 2-4-D, PLANT GROWTH REGULATORS, ALGAE, AQUATIC W72-08440
ARIA VERUCOSA, ECKLONIA RADIATA, GROWTH REGULATION.: /RMIS, GRACIL W72-07496
YANOPHYTA, / *INHIBITION, *PLANT GROWTH REGULATORS, *CHLORIDES, *C W72-14706
S, CULTIVATION, VITAMIN B, PLANT GROWTH REGULATORS, NITRATES, PHOS W73-00430
, CULTURES, *GROWTH RATES, PLANT GROWTH REGULATORS, PLANT GROWTH S W71-03095
SIOLOGY, *PLANT PATHOLOGY, PLANT GROWTH REGULATORS, SALINITY, PLAN W70-09429
LTRATES, GROWTH AUTOSTIMULATION, GROWTH REGULATION, EXTRACELLULAR W71-08687
INHIBITORS, ALGAL CONTROL, PLANT GROWTH REGULATORS, PLANKTON, CULT W71-13793
ICATION METHODS, GRANULES, PLANT GROWTH REGULATORS, INHIBITORS, HE W72-09650
HESIS, PESTICIDE TOXICITY, PLANT GROWTH REGULATORS, INHIBITION, IN W71-10566
N, ALGAE, FEEDING RATES, *LARVAL GROWTH STAGE, PHYTOPLANKTON, NUTR W72-09248
NE ALGAE, CHROMOSOMES, EMBRYONIC GROWTH STAGE, CRUSTACEANS, MONITO W72-10955
ENTERIC BACTERIA, *BACTERICIDES, GROWTH STAGES, PATHOGENIC BACTERI W71-05155
TIES, PRODUCTIVITY, LIFE CYCLES, GROWTH STAGES, DISTRIBUTION, BEHA W72-14659
EOTRICHIA, MICROC/ *FULVIC ACID, GROWTH STIMULATION, ANABAENA, GLO W72-01097
, PLANT GROWTH REGULATORS, PLANT GROWTH SUBSTANCES, ALGAE, IRON, B W71-03095
PHYSIOLOGY, *PLANT GROWTH, PLANT GROWTH SUBSTANCES, PLANT MORPHOLO W70-07842
WTH RATES, *PRODUCTIVITY, *PLANT GROWTH SUBSTANCES, PHOTOSYNTHESIS W71-08687
FLUID FRICTION, *LITTORAL, PLANT GROWTH SUBSTANCES, PHYTOPLANKTON, W71-07878
GANIC CARBON, *ALGAL MASS, ALGAL GROWTH.: *OR W72-04431
*MICRO-NUTRIENTS, *ALGAL GROWTH.: W72-00801
ALGAL GROWTH.: W71-12868
AND), GREAT OUSE(ENGLAND), ALGAL GROWTH.: /), LAKE WINDERMERE(ENGL W71-13185
, UNDISSOCIATED MOLECULES, ALGAL GROWTH.: /STICHOCOCCUS BACILLARIS W70-07255
ABOLITES, SUBSTRATE UTILIZATION, GROWTH.: /ES, MIXED CULTURES, MET W73-00441
ON-SITE DATA COLLECTIONS, PLANT GROWTH.: /, *ARIZONA, GRASSLANDS, W72-12505
RESERVOIR, *APHANIZOMENON, ALGAL GROWTH.: /ARBON, *AMERICAN FALLS W72-13300
AE, *BIOASSAY/ *CULTURES, *PLANT GROWTH, *CHLOROPHYTA, *MARINE ALG W72-14294
*ALGAE GROWTH, *CHEMOSTATS.: W72-06612
, *AQUATIC WEEDS, *WEED CONTROL, GROWTH, *CULTURES, *ALGICIDES, BI W71-06102
AT/ *FORECASTING, *ALGAE, *PLANT GROWTH, *EUTROPHICATION, GROWTH R W71-11033
*ALGAL GROWTH, *LAKE TAHOE(CALIF).: W71-13553
ELENASTRUM CAPRICORNUTUM, *ALGAL GROWTH, *LAKE GEORGE(NY).: *S W72-08376
*PHOSPHORUS, *CLADOPHORA, *PLANT GROWTH, *LAKE MICHIGAN, ALGAE, PH W72-13644
*ALGAE, *NUTRIENTS, LAKES, PLANT GROWTH, *MOLYBDENUM, BIOASSAY, NE W72-00801
MNESIU/ *GLYCEROL, *PHOTOTROPHIC GROWTH, *MARINE PHYTOPLANKTON(PRY W70-07842
PYROPHOSPHATE, RIO GRAN/ *ALGAL GROWTH, *RADIOISOTOPE P32, SODIUM W72-13992
ULANS, AQUATIC ECOSYSTEMS, ALGAE GROWTH, ALGAL POPULATIONS.: / NID W71-00475

S, RIVERS, PHOSPHATES, NITRATES,	HUDSON RIVER, ALGAE, WATER QUALIT	W70-06509
GICAL TISSUES, MERCURY ISOTOPES,	HUMAN BRAIN, LOW-TEMPERATURE OVEN	W71-11036
E WASHINGTON, NITRILOTRIACETATE,	HUMAN FECES.: / METASILICATE, LAK	W72-04734
L PROPERTIES, MINNESOTA, LIPIDS,	HUMAN PATHOLOGY, DISEASES.: /MICA	W72-04275
*LABORATORY TESTS, ALGAL TOXINS,	HUMAN PATHOLOGY.: *CYANOPHYTA,	W72-04259
RE, ENZYMES, CELLULOSE, VIRUSES,	HUMAN PATHOLOGY, SYSTEMATICS.: /U	W72-05457
BACTERIA, VIRUSES, HEAVY METALS,	HUMAN POPULATION, WASTE WATER TRE	W71-09788
AL WASTES, TURNOVERS, SEDIMENTS,	HUMAN POPULATION, TOURISM, WATERS	W71-11009
, MARINE ANIMALS, MARINE PLANTS,	HUMAN POPULATION, RADIOACTIVITY T	W73-00833
S, BENTHIC FAUNA, BENTHIC FLORA,	HUMAN POPULATION, PRIMARY PRODUCT	W72-12549
IN RIVER, MORRO BAY POWER PLANT,	HUMBOLDT BAY, CONNETICUT YANKEE N	W72-07907
SAN JOAQUI/ BARNACLES, FLOUNDER,	HUMBOLDT BAY, CONNECTICUT RIVER,	W71-07417
ESTUARY, BISCAYNE BAY(FLORIDA),	HUMBOLT BAY(CALIFORNIA), POTOMAC	W71-11517
S, *MARINE ALGAE, PHYTOPLANKTON,	HUMIC ACIDS, FULVIC ACIDS, TRACE	W71-09188
MARGARET'S BAY, SCOUDOUC RIVER,	HUMIC ACIDS, OXIMATION, SAMPLE PR	W72-12166
INEMA FLUVIATILE, CULTURE MEDIA,	HUMMOCK CHANNEL, CHLOROPLASTS, MO	W72-12633
G ORGANIC MATTER, *MARINE ALGAE,	HUMUS, ORGANIC COMPOUNDS, POLLUTA	W72-12166
IVER, PINE RIVER, CLINTON RIVER,	HURON RIVER, BELL RIVER, SALT RIV	W72-14282
SUPERIOR, *LAKE MICHIGAN, *LAKE	HURON, *LAKE ONTARIO, *LAKE ERIE,	W72-01094
EUTROPHICATION, *NUTRIENTS, LAKE	HURON, EFFLUENTS, WATER QUALITY,	W72-13662
, LAKE MICHIGAN, LAKE ERIE, LAKE	HURON, LAKE SUPERIOR, WATER TEMPE	W72-10875
KE SUPERIOR, LAKE MICHIGAN, LAKE	HURON, LAKE ERIE, GREAT LAKES, SO	W73-01456
IMILATION, EMILIANA(COCCOLITHUS)	HUXLEYI, AMPHIPRORA ALATA, CHRYSO	W73-00380
OCOCCUS, RHODOMONAS, COCCOLITHUS	HUXLEYI, AUTOLYSIS, PHAEODACTYLUM	W71-10083
AMPHIDINIUM CARTERI, COCCOLITHUS	HUXLEYI, PORPHYRIDIUM, SKELETONEM	W72-12998
ITY, CALCIUM, SCENEDESMUS, WATER	HYACINTH, LIGHT PENETRATION, OXYG	W72-04269
PLANTS, HARVESTING, PONDS, WATER	HYACINTH, NUTRIENTS, MIDGES, FISH	W71-01488
TOS, STICHOCOCCUS, CHLOROCOCCUM,	HYALOTHECA, LAKE LIVINGSTON, AUTO	W72-09239
PLANTS, DIATOMS, PHYTOPLANKTON,	HYDORGEN ION TEMPERATURE,: /TMENT	W72-06837
*OXIDATION DITCH,	HYDRAULIC MANURE HANDLING.:	W72-08396
LUTION, *THERMAL STRATIFICATION,	HYDRAULIC MODELS, FISH, ALGAE, FA	W70-09614
LAKES, LAKE ERIE, LAKE ONTARIO,	HYDRO: /RS, PHOTOSYNTHESIS, GREAT	W72-07648
ERVOIR, ANGARA RIVER, OKA RIVER,	HYDROBIOLOGY, BACTERIOLOGY, SAPRO	W72-04853
OLOGY, *RESERVOIRS, *FRESHWATER,	HYDROBIOLOGY, ALGAE, FISH, ABSORP	W72-10694
RIMARY PRODUCTIVITY, CHLORINATED	HYDROCARBON PESTICIDES, DDT, ORGA	W71-10096
E PES/ *PESTICIDES, *CHLORINATED	HYDROCARBON PESTICIDES, *CARBAMAT	W71-10566
STICIDE TOXICITY, / *CHLORINATED	HYDROCARBON PESTICIDES, *DDT, *PE	W71-12303
*DIELDRIN, *ENDRIN/ *CHLORINATED	HYDROCARBON PESTICIDES, *ALDRIN,	W72-03545
E RESIDUES, *ALGAE/ *CHLORINATED	HYDROCARBON PESTICIDES, *PESTICID	W72-04134
IOASSAY, *TOXICITY, *CHLORINATED	HYDROCARBONS, *BIOINDICATORS, IND	W71-07412
LORINATED BIPHENYLS, *CHLORNATED	HYDROCARBON PESTICIDES, *PESTICID	W72-08436
CATORS, *PESTICIDES, CHLORINATED	HYDROCARBON PESTICIDES, INSECTICI	W73-00238
BIOTICS(PESTICIDES), CHLORINATED	HYDROCARBON PESTICIDES, PESTICIDE	W72-14294
, *METABOLISM, BIO/ *CHLORINATED	HYDROCARBON PESTICIDES, SEA WATER	W73-00361
DTII, CULTURE MEDIA, CHLORINATED	HYDROCARBONS.: /MYDOMONAS REINHAR	W73-00281
OXYGEN, RESPIRATION, CHLORINATED	HYDROCARBON PESTICIDES, GROWTH RA	W73-00281
EPTADECANE, 8-METHYLHEPTADECANE,	HYDROCARBONS, ALKANES, ALIPHATIC	W73-01439
HYDROCARBONS, ALKANES, ALIPHATIC	HYDROCARBONS, MASS SPECTROMETRY,	W73-01439
PESTICIDE RESIDUES, CHLORINATED	HYDROCARBON PESTICIDES, PESTICIDE	W73-02280
SPHORUS PESTICIDES, CHLORINATED	HYDROCARBON PESTICIDES, CARBAMATE	W72-13347
ISM, *MARINE ALGAE, *CHLORINATED	HYDROCARBON PESTICIDES, *CYTOLOGI	W72-12998
R POLLUTION EFFECTS, CHLORINATED	HYDROCARBON PESTICIDES, *CYANOPHY	W72-13809
ALGAE, FOOD CHAINS, CHLORINATED	HYDROCARBON PESTICIDES, FISHERIES	W72-13657
NE CHLORIDE, *CCC, *DIETHYLAMINE	HYDROCHLORIDE, MERISOMOPEDIA SPP.	W72-14706
*INORGANIC IONS, ZINC, VOLVOX,	HYDRODICTYON, SPONDYLOSIUM, SPIROG	W71-11001
TOURISM, RECREATION, FISHERIES,	HYDROELECTRIC POWER, NAVIGATION,	W72-14878
LATION, CYANOPHYTA, CHLOROPHYTA,	HYDROGEN ION CONCENTRATION, CHEMI	W72-14790
G, WINDS, ALGAE, SURFACE WATERS,	HYDROGEN ION CONCENTRATION, LIGHT	W73-00236
DISSOLVED OXYGEN, CONDUCTIVITY,	HYDROGEN ION CONCENTRATION.: /TY,	W72-14728
WATER TEMPERATURE, ACID STREAMS,	HYDROGEN ION CONCENTRATION, HEAVY	W72-14690
NETRATION, ALKALINITY, HALOGENS,	HYDROGEN ION CONCENTRATION, PHOSP	W73-01434
, *ALGAE, *CULTURES, CYANOPHYTA,	HYDROGEN ION CONCENTRATION, LIGHT	W73-01657
COAGULATION, CALCIUM CARBONATE,	HYDROGEN ION CONCENTRATION, FILTR	W73-01669
, BULK DENSITY, POROSITY, COLOR,	HYDROGEN ION CONCENTRATION, NITRO	W73-00854
, WATER TEMPERATURE, ALKALINITY,	HYDROGEN ION CONCENTRATION, WATER	W73-00916
*ALGAE, *WASTE WATER TREATMENT,	HYDROGEN ION CONCENTRATION, CYANO	W73-01367
AE, COLOR, COPPER SULFATE, NETS,	HYDROGEN ION CONCENTRATION, PHOSP	W72-07890
EMICAL OXYGE/ *ALGAE, *NITROGEN,	HYDROGEN ION CONCENTRATION, BIOCH	W72-06612
GLENOPHYTA, PYRROPHYTA, BIOMASS,	HYDROGEN ION CONCENTRATION, MINNE	W72-07145
, TEMPERATURE, DISSOLVED OXYGEN,	HYDROGEN ION CONCENTRATION, SCENE	W72-08049
OSPHORUS RADIOISOTOPES, DIATOMS,	HYDROGEN ION CONCENTRATION, MARIN	W71-10083
AE, *DIATOMS, CHEMICAL ANALYSIS,	HYDROGEN ION CONCENTRATION, SEA W	W71-11008
, SEDIMENTS, LAKES, CHLOROPHYLL,	HYDROGEN ION CONCENTRATION, CARBO	W71-12855
GEN, AERATION, DISSOLVED OXYGEN,	HYDROGEN ION CONCENTRATION, SETTL	W71-13334
ALKALINITY, NUTRIENTS, AMMONIA,	HYDROGEN ION CONCENTRATION, CYANO	W71-11519
TURE, TURBIDITY, PHOTOSYNTHESIS,	HYDROGEN ION CONCENTRATION, NITRI	W71-11876
EN DEMAND, AERATION, *OXIDATION,	HYDROGEN ION CONCENTRATION, MICRO	W71-12223
MENT, *FOAM FRACTIONATION, LIME,	HYDROGEN ION CONCENTRATION, ECONO	W71-12200
SING ORGANIC MATTER, ALKALINITY,	HYDROGEN ION CONCENTRATION, NUTRI	W71-12068
ATION, DISSOLVED OXYGEN, SEWAGE,	HYDROGEN ION CONCENTRATION.: /LIZ	W71-12188
EFFLUENTS, GEORGIA, THERMOCLINE,	HYDROGEN ION CONCENTRATION, WATER	W71-11914
MUD, PHYSICOCHEMICAL PROPERTIES,	HYDROGEN ION CONCENTRATION, TEMPE	W72-03224
*ALGAE, CHEMICAL PRECIPITATION,	HYDROGEN ION CONCENTRATION, FILTR	W72-03613
YTOPLANKTON, THERMAL PROPERTIES,	HYDROGEN ION CONCENTRATION, DISSO	W72-01363
, OXIDATION-REDUCTION POTENTIAL,	HYDROGEN ION CONCENTRATION.: /NGE	W72-01108
TORY TESTS, FILTRATION, AMMONIA,	HYDROGEN ION CONCENTRATION, TEMPE	W72-01881
, AMMONIA, METHANE, TEMPERATURE,	HYDROGEN ION CONCENTRATION, PROTE	W72-10390
AKE SUPERIOR, WATER TEMPERATURE,	HYDROGEN ION CONCENTRATION, HARDN	W72-10875
RATES, CARBON DIOXIDE, POLYMERS,	HYDROGEN ION CONCENTRATION, PROTO	W72-10233
ES, CHROMIUM, ZINC, TEMPERATURE,	HYDROGEN ION CONCENTRATION, CORRO	W72-08683

HORUS, PHOSPHATES, PRODUCTIVITY,	LEACHING.: / WATER QUALITY, PHOSP	W70-08944
L PRECIPITATION, IRON, BIOASSAY,	LEACHING, VOLUMETRIC ANALYSIS, SO	W72-05965
RUNOFF, BASE FLOW, PERCOLATION,	LEACHING, DRAINAGE, FARM WASTES,	W71-06443
ALGAE, FRESH WATER, FILTRATION,	LEACHING, FUNGI, LIGHT, PHOSPHATE	W72-07143
EODACTYLUM TRI/ SILVER, CADMIUM,	LEAD, CHLORELLA PYRENOIDOSA, PHA	W72-07660
*WATER POLLUTION EFFECTS, ZINC,	LEAD, COPPER, CHROMIUM, SILVER, C	W72-14694
ATER, *DATA COLLECTIONS, COPPER,	LEAD, HEAVY METALS, MUSSELS, *KEL	W72-11925
RADIOMETRY, VI/ SPACE PLATFORMS,	LEAD, MERCURY, LASERS, MICROWAVE	W72-04742
CIUM, HYDROGEN, CHEMICAL WASTES,	LEAD, NICKEL, TOXICITY, BIOINDICA	W72-14303
OTA, CYANOPHYTA, TREES, GRASSES,	LEAVES, FOREST SOILS, SAMPLING, M	W72-01784
E, AQUATIC WEEDS, FERTILIZATION,	LEAVES, LEGISLATION, LAND USE, SH	W73-02479
MARSH PLANTS, CHEMICAL ANALYSIS,	LEAVES, SAMPLING, PHYTOPLANKTON,	W73-01089
STRATEGY, THAMES RIVER(ENGLAND),	LEE RIVER(ENGLAND).: /, RESEARCH	W71-13172
UMBELLATUS, SPARGANIUM RAMOSUM,	LEERSIA ORYZOIDES, NUPHAR ADVENA,	W73-01089
LIC HEALTH, ENVIRONMENT CONTROL,	LEGAL ASPECTS, REGULATION, PERMIT	W73-00703
CONTROL, STANDARDS, LEGISLATION,	LEGAL ASPECTS, WATER POLLUTION, A	W72-01093
ES, EUTROPHICATION, LEGISLATION,	LEGAL ASPECTS, REGULATION, POLLUT	W72-01085
S, PHOSPHOROUS, WATER SOFTENING,	LEGAL ASPECTS, HARD WATER, PUBLIC	W71-09429
RECIRCULATED WATER, WATER REUSE,	LEGAL ASPECTS, JUDICIAL DECISIONS	W70-07100
GAL ASPECTS, JUDICIAL DECISIONS,	LEGISLATION, ADMINISTRATION, NON-	W70-07100
TATES, ALGAL CONTROL, STANDARDS,	LEGISLATION, LEGAL ASPECTS, WATER	W72-01093
LLUTION SOURCES, EUTROPHICATION,	LEGISLATION, LEGAL ASPECTS, REGUL	W72-01085
WASTE TREATMENT, AIR POLLUTION,	LEGISLATION, BENTHIC FLORA, F: /,	W71-13723
ATION, *DETERGENTS, *PHOSPHATES,	LEGISLATION, WATER LAW, FEDERAL G	W71-11516
D DEVELOPMENT, MUNICIPAL WASTES,	LEGISLATION, REGULATION,: /RCH AN	W72-06638
UTION ABATEMENT, EUTROPHICATION,	LEGISLATION, PUBLIC HEALTH, WATER	W72-06655
IC WEEDS, FERTILIZATION, LEAVES,	LEGISLATION, LAND USE, SHORES, MO	W73-02479
*FEEDLOTS,	LEMNA.:	W72-04773
INALIS, CHARALES, CERATOPHYLLUM,	LEMNA.: /DENSIS, PHRAGMITES, FONT	W71-11012
VAILLANTII, CONIOPHORA OLIVACEA,	LENZITES TRABEA, SCOPULARIOPSIS B	W72-13813
SPHAERICUS, CHYDORUS, MONOSTYLA,	LEPADELLA, TRICHOCERCA, HAEMATOCO	W72-07132
A POLYMORPHA, EUGLENA SPIROGYRA,	LEPOCINCLIS OVUM, PHACUS CIRCUMFL	W73-00918
HE, TRIAENODES INUUSTA, OECETIS,	LEPTOCELLA, ISONYCHIA, CAENIS, PE	W70-06655
INLET(ALASKA), CYCLOTELLA NANA,	LEPTOCYLINDRUS, PACIFIC OCEAN, AL	W71-11008
LA, DACTYLIOSOLEN MEDITERRANEUS,	LEPTOCYLINDRUS DANICUS, RHIZOSOLE	W72-03543
TRATI/ *MOSS LANDING, BARNACLES,	LETHAL DOSAGE, CALCIUM ION CONCEN	W72-09092
ILIS, CHLAMYDOMANAS REINHARDTII,	LETHAL DOSAGE.: /TA, EUGLENA GRAC	W72-05614
NEDESMUS, BIO-ASSAY, RESISTANCE,	LETHAL LIMIT, POISONS, WATER POLL	W72-14706
SEAWATER, *TOXICITY, *NUTRIENTS,	LETHAL LIMITS, METHODOLOGY, INDUS	W72-12239
, THREONINE, PROLINE, GLUTAMATE,	LEUCINE, ARGININE, SERINE, TYROSI	W73-00942
PLECPODS, UROPODS, ENRICHMENT,	LEUCOTHRIX MUCOR, MAJURO ATOLL, M	W72-08142
GE, PHYSICOCHEMICAL PROP/ *WATER	LEVEL FLUCTUATIONS, *PUMPED STORA	W72-14280
TY, NITROGEN, NUTRIENTS, TROPHIC	LEVEL.: / PHOSPHATES, WATER QUALI	W71-11033
, SOUTH CAROLINA, ALGAE, TROPHIC	LEVEL.: /RBON DIOXIDE, ECOSYSTEMS	W71-00475
TREAMS, *STANDING CROP, *TROPHIC	LEVEL, *ENERGY BUDGET, PRODUCTIVI	W71-10064
CAL PROPERTIES, *SEDIM/ *TROPHIC	LEVEL, *LAKES, *NUTRIENTS, *CHEMI	W72-05469
UCTIVITY, WATER QUALITY, TROPHIC	LEVEL, AQUATIC ALGAE, FISH POPULA	W72-08986
MATOGRAPHY, CHLOROPHYLL, TROPHIC	LEVEL, EUTROPHICATION, ALGAE, DEC	W72-01784
PROPERTIES, PESTICIDES, TROPHIC	LEVEL, GASES, MUDS, ADSORPTION, C	W72-00847
ANCE OF NATURE, ECOLOGY, TROPHIC	LEVEL, HERBIVORES, NATURAL STREAM	W72-06340
S, ALGAE, RADIOACTIVITY, TROPHIC	LEVEL, LIFE HISTORY STUDIES, TIME	W70-06668
EUTROPHICATION, REVIEWS, TROPHIC	LEVEL, PHOSPHORUS, PHOSPHATES, CA	W72-07937
OTROPHY, EUTROPHICATION, TROPHIC	LEVEL, PROTOZOANS, AQUATIC ALGAE,	W72-13816
ENTS, THERMAL POLLUTION, TROPHIC	LEVEL, TROPICAL REGIONS, ULTRAVIO	W73-00796
IC ALGAE, PHYTOPLANKTON, TROPHIC	LEVEL: /PERATURE, SAMPLING, AQUAT	W73-01456
DIOISOTOPES, METABOLISM, TROPHIC	LEVELS.: /DIOISOTOPES, CADMIUM RA	W71-13243
TRIENTS, MEASUREMENT, / *TROPHIC	LEVELS, *BIOASSAY, PHOSPHORUS, NU	W71-13794
*NUTRIENT	LEVELS, CUYAHOGA RIVER(OHIO).:	W72-04263
WATER POLLUTION CONTROL, TROPHIC	LEVELS, PHOSPHATES, DETERGENTS, B	W71-12072
MS, REDUCTION(CHEMICAL), TROPHIC	LEVELS, PROTOZOA, ALGAE, BACTERIA	W72-10883
IRA FERTILISSIMA, AN/ *TOLERANCE	LEVELS, TOLYPOTHRIX TENUIS, AULOS	W71-12071
ISM, WATERSHEDS(BASINS), TROPHIC	LEVELS, TURBIDITY, OXYGEN, FISH,	W71-11009
RN(ENGLAND), LUXURY UPTAKE, LOCH	LEVEN(SCOTLAND), LAKE WINDERMERE(	W71-13183
S, CERAMIUM BUSSOIDEUM, CHONDRIA	LI: /LAVULATUM, HYPENA MUSCIFORMI	W73-00838
ACETYLENE TEST, CYLINDROSPERMUM	LICHENEFORME, MASTIGOCLADUS LAMIN	W71-12070
, FOOD WEBS, FOOD CHAINS, FROGS,	LICHENS, ALGAE, FUNGI, SMALL ANIM	W72-00949
CATORS, BIODEGRADATION, BIOMASS,	LICHENS, MOSSES, FUNGI, ALGAE, SO	W72-14303
N, ISOTOPE RATIOS, SPECTROMETRY,	LICHENS, MOSSES, GYMNOSPERMS, ANG	W72-12709
IC ZONES, CYANOPHYTA, SYMBIOSIS,	LICHENS, NITROGEN, LIGHT INTENSIT	W73-02471
AIR POLLUTION, WATER POLLUTION,	LICHENS, SOIL-WATER-PLANT RELATIO	W72-03347
S BREVIPES, ACHNANTHES JAVANICA,	LICMOPHORA JURGENSII, AMPHORA OVA	W72-08141
*PLANT MORPHOLOGY, REPRODUCTION,	LIFE CYCLES, AQUATIC ALGAE, PERIP	W72-07141
GICAL COMMUNITIES, PRODUCTIVITY,	LIFE CYCLES, GROWTH STAGES, DISTR	W72-14659
NEMATODES, ALGAE, OLIGOCHAETES,	LIFE CYCLES, SEASONAL, DENSITY, D	W71-11012
TROPHICATION, COLIFORMS, SPORES,	LIFE CYCLES, ODOR, RECREATION, SY	W71-09173
E, RADIOACTIVITY, TROPHIC LEVEL,	LIFE HISTORY STUDIES, TIME, SNAIL	W70-06668
ANIMALS, *AQUATIC PLA/ *AQUATIC	LIFE, *AQUATIC BACTERIA, *AQUATIC	W72-04853
R/ *PESTICIDE RESIDUES, *AQUATIC	LIFE, *AQUATIC ENVIRONMENT, *WATE	W72-13347
*DELAWARE ESTUARY, *BIOLOGICAL	LIFE, *DISSOLVED OXYGEN CONCENTRA	W70-09189
PHOSPHATES, ECOSYSTEMS, AQUATIC	LIFE, ANTIBIOTICS(PESTICIDES), EC	W72-09668
WATER POLLUTION SOURCES, AQUATIC	LIFE, AQUATIC ANIMALS, AQUATIC MI	W72-09646
POLLUTION, *FRESHWATER, *AQUATIC	LIFE, AQUATIC ANIMALS, CARP, DAPH	W72-09649
D CHAINS, PUBLIC HEALTH, AQUATIC	LIFE, AQUATIC ALGAE, CALCIUM, RAD	W72-10978
AY, FISH, RAINBOW TROUT, AQUATIC	LIFE, AQUATIC ENVIRONMENT, FISH P	W72-12257
IOISOTOPES, NEW ENGLAND, AQUATIC	LIFE, AQUATIC MICROORGANISMS, MAR	W72-14317
NELIDS, COPEPODS, CRABS, AQUATIC	LIFE, BACTERIA.: /TIDAL AREAS, AN	W72-07911
OLIGOTROPHY, OXYGEN SAG, AQUATIC	LIFE, BIOASSAY, CLASSIFICATION.: /	W73-00002
, CLAMS, MARINE ANIMALS, AQUATIC	LIFE, ECOSYSTEMS, SALMONIDS.: /SH	W72-07907
XYGEN, COMPUTERS, ALGAE, AQUATIC	LIFE, ENVIRONMENTAL EFFECTS, ENVI	W72-14405
, ENVIRONMENTAL EFFECTS, AQUATIC	LIFE, FISH POPULATIONS, PHYTOPLAN	W72-00939

CC, *DIETHYLAMINE HYDROCHLORIDE, MERISOMOPEDIA SPP., ANABAENA SPP. W72-14706
(WASHINGTON), COCKLES, GOLDFISH, MERRIMACK RIVER(NEW HAMPSHIRE), D W71-11517
A, BOSTRYCHIETUM, FIDDLER CRABS, MESCFAUNA, TURBELLARIA, SPONGES, W72-07911
A, LYNGBYA.: MESOTROPHIC, MICROCYSTIS, ANABAEN W72-04269
AGING(BIOLOGICAL), LAKE STAGES, MESOTROPHY, OLIGOTROPHY, WATER PO W72-14782
LAMYDOMONAS, DEFICIENT ELEMENTS, MESOTROPHY, EUTROPHICATICN, TROPH W72-13816
TOCERCS LORENZIANUS, CHAETOCEROS MESSANENSIS, CHAETOCEROS PERUVIAN W71-11527
DEMAND, LAKE STAGES, NUTRIENTS, MESTROPHY, OLIGCTROPHY, CXYGEN SA W73-00002
PHYCEA/ LYTIC BACTERIA, ENGLAND, METABOLIC INHIBITION, HOSTS, MYXO W72-05457
NOPHYTA, MYXOBACTERIA, CULTURES, METABOLISM, SEWAGE, CARBOHYDRATES W72-05457
NSITY, MIXING, DISSOLVED OXYGEN, METABOLISM, CARBON DIOXIDE, PHOSP W72-04983
CONDITIONS, AEROBIC CONDITIONS, METABOLISM, SLUDGE, BIOCHEMICAL O W72-04789
ERIA, LAKES, FILTERS, COLIFORMS, METABOLISM, CARBON, SULFUR, SAMPL W72-04743
ESTICIDES, RESISTANCE, CULTURES, METABOLISM.: /, SOILS, TRIAGINE P W71-07675
F/ *AGRICULTURAL DRAINAGE, LAKE METABOLISM, MOBILITY, PARTICULATE W71-06443
HYLOGENY, CARBON CYCLE, ENZYMES, METABOLISM, HYDROGEN, EUGLENA, CH W71-09172
OWTH SUBSTANCES, PHOTOSYNTHESIS, METABOLISM, BACTERIA, HYDROGEN IO W71-08687
ROPODS, ADSORPTION, TEMPERATURE, METABOLISM, KINETICS, WATER POLLU W71-09168
AL PHYSIOLOGY, SEDIMENTS, ANIMAL METABOLISM.: /NMENT, COASTS, ANIM W73-00811
BON RADIOISOTOPES, PRODUCTIVITY, METABOLISM, FUNGI, ORGANIC COMPOU W72-14326
CULTURES, AQUATIC PRODUCTIVITY, METABOLISM, ELECTRODES, ELECTROCH W72-14314
OLLUTION EFFECTS, RADIOACTIVITY, METABOLISM, PLANT PHYSIOLOGY, CHR W73-00380
TROGEN, *PARTICULATE ORGA/ *LAKE METABOLISM, *DISSOLVED ORGANIC NI W72-08048
TEMPERATURE, ALTITUDE, CLIMATES, METABOLISM, BIOCHEMICAL OXYGEN DE W72-06838
MUNITIES, BIOMASS, PRODUCTIVITY, METABOLISM, CHLORELLA, RESPIRATIO W73-01080
BON PESTICIDES, PESTICIDE DRIFT, METABOLISM, ALGAE, AQUATIC ALGAE, W73-02280
NITROGEN, POTASSIUM, PHOSPHATES, METABOLISM, ALGAE, SYNTHESIS, VIT W73-01626
ELEMENTS, NITROGEN, PHOSPHORUS, METABOLISM, ALGAE, SLUDGE, ANALYT W71-11793
MES, BACTERIA, SPORES, NITROGEN, METABOLISM.: /OGEN FIXATION, ENZY W71-12070
ISOTOPES, CADMIUM RADIOISOTOPES, METABOLISM, TROPHIC LEVELS.: /DIO W71-13243
TATION, FILTRATION, LIME, ROCKS, METABOLISM, DESIGN CRITERIA, *WAS W71-13341
S, *BACTERIA, *ALGAE, OXIDATION, METABOLISM, RESPIRATION, LIPIDS, W72-10233
ANCE, *POLYP MAINTENANCE, *POLYP METABOLISM, TETRACYCLINE, SULFANI W72-10162
GAE, FISH, ABSORPTION, BIOASSAY, METABOLISM.: /R, HYDROBIOLOGY, AL W72-10694
ASTES, ALGAE, INDUSTRIAL WASTES, METABOLISM, GROWTH RATES, OXIDATI W72-09590
YGEN DEMAND, SATURATION, OXYGEN, METABOLISM, NITROGEN, PHOSPHOROUS W71-13553
DS, *HYDROGEN ION CONCENTRATION, METABOLISM, CARBON DIOXIDE, ALGAE W72-00838
UR, ALGAE, GRANULES, INHIBITION, METABOLISM, CULTURES, ENVIRONMENT W72-00153
C BACTERIA, *MACROPHYTE-EPIPHTYE METABOLISM, LAWRENCE LAKE(MICH), W72-03216
PHYTON, ORGANIC MATTER, BIOMASS, METABOLISM, PHYTOPLANKTON, PLANT W72-03216
ALGAE, CYANOPHYTA, CHLOROPHYTA, METABOLISM, NEKTON, BENTHOS, BIOM W72-13636
ORIDE, RADIOACTIVITY TECHNIQUES, METABOLISM, TRACERS, RADIOACTIVIT W72-12634
*LYSIS, *CONTACT TIME, CELLULAR METABOLITES.: W72-00411
N PR-6, PHOTOALDRIN, KETOENDRIN, METABOLITES, SKELETONEMA, ANABAEN W72-03545
CTS, *ETHOXYCHLOR, *METHYLCHLOR, METABOLITES, BIOLOGICAL MAGNIFICA W72-07703
CAL SAMPLES, FATE OF POLLUTANTS, METABOLITES, DUNALIELLA, AGMENELL W73-00361
A NITZSCHIOIDES, MIXED CULTURES, METABOLITES, SUBSTRATE UTILIZATIO W73-00441
AL MAGNIFICATION, CHLAMYDOMONAS, METABOLITES, HAEMATOCRIT TUBE.: / W72-04134
N/ *SINAI PENINSULA, CHLORINITY, METAHALINE, HYPERHALINE, EURYHALI W73-00855
UM, BRYOPHYTES, BIOACCUMULATION, METAL COMPLEXES, PICEA ABIES, VAC W72-14303
BRATSK RESERVOIR, *HYDROBIOLOGY, METALIMNION, BACILLARIOPHYTA, VOL W72-04855
ERIOLOGY, SAPROPHYTES, STURGEON, METALIMNION.: /HYDROBIOLOGY, BACT W72-04853
, *POLYETHYLENEMIN/ *MULTIVALENT METALLIC CATIONS, *POLYACRYLAMIDE W72-14840
WATER CONVEYANCE, CANALS, HEAVY METALS.: /ENVIRONMENT, CALIFORNIA W71-09581
ENT, *TOXICITY/ *REVIEWS, *HEAVY METALS, *ALGAE, *AQUATIC ENVIRONM W72-14694
OXICITY, *FLUORESCENCE, / *HEAVY METALS, *ALGAE, *GROWTH RATES, *T W72-07660
QUATIC ALGAE, CHLOROPHYTA, HEAVY METALS, *CHLAMYDOMONAS, INHIBITOR W72-11727
S/ *TERRESTRIAL HABITATS, *HEAVY METALS, *PRODUCTIVITY, *ECOSYSTEM W72-14303
S, *PRIMARY PRODUCTIVITY, *HEAVY METALS, *TRACE ELEMENTS, NUTRIENT W73-01434
PHY/ *INDICATORS, *ALGAE, *HEAVY METALS, *TOXICITY, COPPER, CHLORO W71-05991
CHELATION, CHROMATOGRAPHY, HEAVY METALS, *TRACE ELEMENTS, RADIOCHE W72-04708
IOISOTOPES, *MARINE ALGAE, HEAVY METALS, ABSORPTION, BIOINDICATORS W72-10957
ENTS, *LEAD RADIOISOTOPES, HEAVY METALS, AIR POLLUTION EFFECTS, SU W72-00940
INC, METALS, SECCHI DISKS, HEAVY METALS, BIOINDICATORS, SILICA, CH W72-09111
ION EFFECTS, RESPIRATION, *HEAVY METALS, CHEMICAL WASTES, MERCURY, W72-12576
, WATER POLLUTION EFFECTS, HEAVY METALS, CYCLING NUTRIENTS, CARBON W73-00840
YDROGEN ION CONCENTRATION, HEAVY METALS, DISSOLVED OXYGEN, *PLANT W72-14690
HELATION, *ION TRANSPORT, *HEAVY METALS, EUTROPHICATION, POLLUTANT W73-02112
CHEMISTRY, TRACE ELEMENTS, HEAVY METALS, ELECTRODES, NITRATES, ALG W72-01693
POLLUTION EFFECTS, ALGAE, HEAVY METALS, GROWTH, PSEUDOMONAS, POTA W72-13813
S, BIOLOGICAL COMMUNITIES, HEAVY METALS, HERBICIDES, TOXINS, WASTE W72-01472
STRIAL WASTES, MUNICIPAL WASTES, METALS, HEAVY METALS, WASTE DISPO W71-11793
RIENTS, BACTERIA, VIRUSES, HEAVY METALS, HUMAN POPULATION, WASTE W W71-09788
T, HABITATS, AQUATIC LIFE, HEAVY METALS, INDUSTRIAL WASTES, INVERT W72-11854
BITION, *LIPIDS, TOXICITY, HEAVY METALS, LABORATORY TESTS, PHOTOSY W72-06037
L, COBALT, COPPER SULFATE, HEAVY METALS, METALS, SULFATES.: /NICKE W72-12739
COLLECTIONS, COPPER, LEAD, HEAVY METALS, MUSSELS, *KELPS, STARFISH W72-11925
AL), PHOSPHORUS, NITROGEN, HEAVY METALS, ORGANOPHOSPHORUS PESTICID W72-04742
EPIDEMIOLOGY, PHOSPHATES, HEAVY METALS, PESTICIDES, TASTE-PRODUCI W73-02144
HYSIOLOGY, TRACE ELEMENTS, HEAVY METALS, PHYSICOCHEMICAL PR: /AL P W73-00817
HERBICIDES, PHOSPHORUS, PHENOLS, METALS, POLLUTANTS, INDUSTRIAL WA W72-09646
TA, CYANOPHYTA, MANGANESE, ZINC, METALS, SECCHI DISKS, HEAVY METAL W72-09111
TIES, GOLD, IRON, CESIUM, ALKALI METALS, SEASONAL, ZINC, ABSORPTIO W73-01434
T, COPPER SULFATE, HEAVY METALS, METALS, SULFATES.: /NICKEL, COBAL W72-12739
INSECTICIDES, HERBICIDES, HEAVY METALS, TASTE, ODOR, RECREATION, W71-12091
NTS, NUTRIENTS, PESTICIDES, OIL, METALS, THERMAL POLLUTION, SAMPLI W72-08790
MUNICIPAL WASTES, METALS, HEAVY METALS, WASTE DISPOSAL, WATER POL W71-11793
UALITY, ALGAE, PESTICIDES, HEAVY METALS, WASTEWATER TREATMENT, SME W72-14282
NOIDOSA, SODIUM SILICATE, SODIUM METASILICATE, LAKE WASHINGTON, NI W72-04734
POLYEDRA, PERIDINIUM DEPRESSUM, METAZOA, CALANUS HELGOLANDICUS(PA W72-07892

METHYLHEPTADECANE, 8-MET/ NOSTOC MUSCORUM, LYNGBYA LAGERHAIMII, 7- W73-01439
ONS, COPPER, LEAD, HEAVY METALS, MUSSELS, *KELPS, STARFISH, CRUSTA W72-11925
S, AQUATIC INSECTS, CRUSTACEANS, MUSSELS, COMMON MERGANSER DUCK, B W72-12930
SIS, SPECTROMETERS, ZOOPLANKTON, MUSSELS, CLAMS, FISH, SHELLFISH, W72-05965
E, MOSSES, CRABS, ALGAE, SNAILS, MUSSELS, ISOPODS, AMPHIPODA, RADI W72-03347
TION, AQUARIA, MOLLUSKS, SNAILS, MUSSELS, MARINE ALGAE, AQUATIC AN W71-09182
E DISPOSAL, MARINE ALGAE, CLAMS, MUSSELS, OYSTERS, EUTROPHICATION, W72-07715
ATER, ALGAE, NEMATODES, OYSTERS, MUSSELS, PROTOZOA, MANGROVE SWAMP W72-07911
T, REGULATION, FALLOUT, OYSTERS, MUSSELS, PLANKTON, MARINE ALG: /N W72-00959
LANT GROWTH, OYSTERS, SHELLFISH, MUSSELS, SHELLFISH FARMING, AQUAC W72-09378
ALS, LABORATORY TESTS, NITRATES, MUSSELS, SNAILS, FISH, ALGAE, GAS W71-09168
, *MARINE ANIMALS, *MARINE FISH, MUSSLES, LOBSTERS, ABSORPTION, CO W73-00832
RES, ALGAE, CYTOLOGICAL STUDIES, MYCOBACTERIUM, PSEUDOMONAS, BACTE W71-08027
OCOCCUS, PSEUDOMONAS AERUGINOSA, MYCOBACTERIUM FURTUITUM, STAPHYLO W72-04259
OLIATUS, POTAMOGETON PECTINATUS, MYRIOPHYLLUM SPECATUM, ELODEA CAN W71-11012
VACCINIUM VITIS IDAEA, VACCINIUM MYRTILLUS, DESCHAMPSIA FLEXUOSA, W72-14303
FO/ *G + C RATIOS, *NONFRUITING MYXOBACTERIA, DNA, MOTILITY, CYST W72-07951
ONTROL, FRESH WATER, CYANOPHYTA, MYXOBACTERIA, CULTURES, METABOLIS W72-05457
OPHAGA JOHNSONII, SPOROCYTOPHAGA MYXOCOCCOIDES.: /SENSITIVITY, CYT W72-07951
ND, METABOLIC INHIBITION, HOSTS, MYXOPHYCEAE, SCOTLAND.: /A, ENGLA W72-05457
A LONGIROSTRIS/ *WINNISQUAM LAKE(N H), ANABAENA CIRCINALIS, BOSMIN W71-03014
(PERU), CAPE COD(MASS), CAPE MAY(N J), SARGASSO SEA.: /AMA, CALLAO W72-08056
P, MICROCYSTIS AERU/ *KEZAR LAKE(N.H.), APHANIZOMENON, ANABAENA SP W72-07890
THYLENE, NITROGEN RADIOISOTOPES, N-15, LUCIFERIN, LUCIFERASE, CHLO W72-10883
ENCE LAKE(MICH), SCIRPUS ACUTUS, NAJAS FLEXILIS.: /ETABOLISM, LAWR W72-03216
ILIS, COLONIZATION, CAULOBACTER, NAJAS FLEXILIS, SCIRPUS ACUTUS, C W73-02095
EMIA SALINA, BIO-RAD, CYCLOTELLA NANA.: CHELEX 100, ART W72-04708
DACTYLUM TRICORNUTUM, CYCLOTELLA NANA, CHAETOCEROS GALVESTONENSIS, W72-07660
COOK INLET(ALASKA), CYCLOTELLA NANA, LEPTOCYLINDRUS, PACIFIC OCE W71-11008
NALIELLA TERTIOLECTA, CYCLOTELLA NANA, THALASSIOSIRA FLUVIATILIS, W72-12998
ERA, MAGALOPTERA, EPHEMEROPTERA, NANNOCHLORIS, CHLOROCOCCUM, MICR W70-06655
HYDROGEN SULFIDE, DISTRIBUTION, NANNOPLANKTON, TURBIDITY, HYDROGE W71-00117
N, SESTON, ALGAE, PHYTOPLANKTON, NANNOPLANKTON, AMMONIA, NITRATES, W72-08048
ERS, GULF STREAM, CAPE HATTERAS, NANSEN BOTTLE, SPECIES DIVERSITY, W72-07663
BACILLARI PARADOXA, SPHAEROTILUS NAT: /TATA, GOMPHONEMA PARVULUM, W72-10883
RHODOTORULA GLUTINIS, SARGASSUM NATANS, ASCOPHYLLUM NODOSUM, FUCU W72-04748
ROSOCYSTIS OCEANUS, SPHAEROTILUS NATANS, GLUCOSE, GALACTOSE, SUCRO W72-05421
MIA SUBSPHAERICA, TETRACULAMONAS NATANS, MARINIAMONAS SAUPAULENSIS W72-09119
, *SEWAGE FUNGUS, SPHAEROTOTILUS NATANS, ZOOGLOEAL BACTERIA.: /DOM W72-08573
GAE, MASTIGOCLADUS/ *YELLOWSTONE NATIONAL PARK(WYO), BLUE-GREEN AL W72-08508
PHYTA, *THERMAL WATER, *WYOMING, NATIONAL PARKS, HOT SPRINGS, WATE W72-08508
FATTY ACIDS, SAMPLE PREPARATION, NATURAL ORGANICS.: /LTURE MEDIA, W73-01066
LVED ORGANOPHOSPHORUS COMPOUNDS, NATURAL ORGANICS.: / *DNA, *DISSO W72-12637
T, *GROWTH RATES, PHYTOPLANKTON, NATURAL RESOURCES, DEPTH, DISTRIB W71-13252
LOGY, TROPHIC LEVEL, HERBIVORES, NATURAL STREAMS, PULP WASTES, PRO W72-06340
, MOSCOW RIVER, OKA RIVER, FOAM, NATURAL WATERS.: *USSR W72-05506
GAE, AQUATIC ANIMALS, BALANCE OF NATURE, ECOLOGY, TROPHIC LEVEL, H W72-06340
, SPIROGYRA,/ *GAMBUSIA, CYCLOPS NAUPLII, COCCOCHLORIS PENIOCYSTIS W72-07132
ODESMIUM, FLAGELLATES, CILIATES, NAUPLII, COPEPODIDS.: /RIA, TRICH W72-14313
CHIA, OSCILLATORIA, PLEUROSIGMA, NAUPLII, LAMELLIBRANCH, SPIROGYRA W71-11876
EUSTONT, NAUTOCOCCUS MAMMILATUS, NAUTOCOCCUS PYRIFORMIS NAUTOCOCCU W72-07141
UDATUS, NAUTOCOCCUS CONSTRICTUS, NAUTOCOCCUS EMERSUS.: /COCCUS CA W72-07141
OCCUS, APIOCOCCUS, OSCILLATORIA, NAUTOCOCCUS CAUDATUS, NAUTOCOCCUS W72-07141
ILLATORIA, NAUTOCOCCUS CAUDATUS, NAUTOCOCCUS CONSTRICTUS, NAUTOCOC W72-07141
MMILATUS, NAUTOCOCCUS PYRIFORMIS NAUTOCOCCUS, APIOCOCCUS, OSCILLAT W72-07141
US PYRIFORMIS NAUT/ EPINEUSTONT, NAUTOCOCCUS MAMMILATUS, NAUTOCOCC W72-07141
ONSTRUENS, COCCONEIS PLACENTULA, NAVICULA SRYTOCEPHALA, ACHNANTHES W72-08141
, AMPHORA SPP, MELOSIRA SULCATA, NAVICULA SPP, SYNEDRA FASCICULATA W73-00853
M MONILIFERUM, MELOSIRA VARIANS, NAVICULA TRIPUNCTATA, GOMPHONEMA W72-10883
OTHECE, ANABAENA, APHANIZOMENON, NAVICULA, GLENODINIUM, MICROSPORA W72-10875
PHACUS, CHAETOCEROS, MOUGEOTIA, NAVICULA, PINNULARIA, SURIRELLA, W72-09111
S, OPEPHORA MARTYI, CELL COUNTS, NAVICULA, NITZSCHIA, STEPHANODISC W71-12855
HANTHES, FRAGILARIA CONSTRUENTS, NAVICULA, NITZSCHIA, EPIPYTIC ALG W72-07501
FISHERIES, HYDROELECTRIC POWER, NAVIGATION, CONSUMPTIVE USE, WATE W72-14878
*PAMLICO RIVER(NC).: W71-07337
DURHAM(NC).: W71-07382
CODIUM CAROLINIANUM, *ONSLOW BAY(NC).: /II, GLOIODERMA ATLANTICA, W73-00284
SUM, CHAETOCEROS ATLANTICUS VAR. NEAPOLITANA, CHAETOCEROS COARCTAT W71-11527
ON(WASH)/ SALT VALLEY RESERVOIRS(NEB), MADISON(WIS), LAKE WASHINGT W72-04759
*SALT VALLEY RESERVOIRS(NEB), ULTRA PLANKTON, C-14.: W72-04270
RUNOFF, URBAN HYDROLOGY, LINCOLN(NEBR).: *URBAN W71-13466
, RECREATION, LIGHT PENETRATION, NEBRASKA, WASTEWATER TREATMENT, T W72-04759
CATOSTOMUS COMMERSONI, AMEIURUS NEBULOSUS, PERCA FLAVESCENS, AMBL W70-08098
ANIZOMENON HOLSATICUM, ICTALURUS NEBULOSUS, NOTEMIGONUS CRYSOLEUCA W72-07890
BACILLUS, MICROCOCCUS, GRAM NEGATIVE BACTERIA.: W72-12255
OPHYTA, CHLOROPHYTA, METABOLISM, NEKTON, BENTHOS, BIOMASS.: / CYAN W72-13636
*CRUSTACEANS, OILY WATER, ALGAE, NEMATODES, OYSTERS, MUSSELS, PROT W72-07911
COLIFORMS, VIRUSES, CRUSTACEANS, NEMATODES, EPIDEMIOLOGY, PHOSPHAT W73-02144
ATES, ST. LAWRENCE RIVER, ALGAE, NEMATODES, GASTROPODS, MOLLUSKS, W72-05432
ALGAE, *SALT MARSHES, NEW YORK, NEMATODES, EPIPHYTOLOGY, PRIMARY W72-00948
TON, *SUBMERGED PLANTS, BIOMASS, NEMATODES, ALGAE, OLIGOCHAETES, L W71-11012
IDS,/ COELENTERATES, ARTHROPODS, NEMERTEANS, ECHINODERMS, SIPUNCUL W73-00932
CALCICOLA, BORATES, ARTHROBACTE/ NEOCHLORIS AQUATICA, SCHIZOTHRIX W73-00854
IONOSPIO, LUMBRINEREIS, GLYCERA, NEPHTYS, CAPITELLA, MAGELONA, ARM W72-09092
TA, CNEMIDOCARPA, ENHYDRA LUTRIS NEREIS, SEA OTTER, MACROCYSTIS PY W73-00932
, COLD BAY, BERMERS BAY, ALARIA, NEREOCYSTIS, LAMINARIA, THALASSIO W72-09120
Y, *ZOOPLANKTON, *PHYTOPLANKTON, NERITIC, ESTUARINE ENVIRONMENT, S W73-00935
TEROLAMPRA HEPTACTIS, BAC/ JUDAY NET, ASTEROLAMPRA MARYLANDICA, AS W71-11527
RLANDS), BIESBOSCH RESERVOIR(THE NETHERLANDS), KINETIC EQUATION.: / W72-08559
HE NETHERLANDS), MEUSE RIVER(THE NETHERLANDS), BIESBOSCH RESERVOIR W72-08559

SPHORUS REQUIREMENTS(ALGAE), NU/ NITROGEN REQUIREMENTS(ALGAE), PHO W70-04080
ERTEBRATES, ACETYLENE, ETHYLENE, NITROGEN RADIOISOTOPES, N-15, LUC W72-10883
LOGICAL SOURCES, MINERALIZATION, NITROGEN SOURCES, WELL WATER, FEE W71-06435
IS, CARBON EXTRUSION, CARBON-14, NITROGEN SOURCE.: /NITZSCHIA OVAL W72-07497
, PLANT PIGMENTS, AQUATIC ALGAE, NITROGEN.: /MICHIGAN, CHRYSOPHYTA W73-00384
CHELOMONAS, SAGINAW BAY, ORGANIC NITROGEN.: /, ANKISTRODESMUS, TRA W72-10875
CYANOPHYTA, ALKALINITY, DIURNAL, NITROGEN.: /NTS, EUTROPHICATION, W72-09162
LGAE, LAKES, PHOSPHORUS, CARBON, NITROGEN.: / TESTS, *NUTRIENTS, A W72-09165
RATES, DIATOMS, BACTERIA, LAKES, NITROGEN, ANALYTICAL TECHNIQUES, W72-09163
LORIDA, WATER POLLUTION EFFECTS, NITROGEN, PHOSPHORUS.: /ATIONS, F W72-08986
PHICATION, *CARBON, *PHOSPHORUS, NITROGEN, ALGAE.: *EUTRO W72-09155
TESTS, *BIOASSAY, OREGON, LAKES, NITROGEN, PHOSPHORUS, CARBON, GRO W72-09164
ALASKA, FIORDS, SURFACE WATERS, NITROGEN, SEA WATER, WINDS, SAMPL W72-09122
TY, CYANOPHYTA, LAKES, MISSOURI, NITROGEN, PHOSPHORUS.: /TER QUALI W72-10431
RATION TECHNIQUES, FERMENTATION, NITROGEN, PHOSPHORUS, AMMONIA, ME W72-10390
AGRICULTURAL RUNOFF, PHOSPHORUS, NITROGEN, PHOSPHATES, DETERGENTS, W72-10625
PRODUCTIVITY, ALGAE, PHOSPHATES, NITROGEN, SEWAGE, CARBON DIOXIDE, W72-09167
ON SOURCES, MATHEMATICAL MODELS, NITROGEN, ABATEMENT, COSTS, ANALY W72-09171
ARCH AND DEVELOPMENT, NUTRIENTS, NITROGEN, PHOSPHORUS, ALGAE, PLAN W72-11702
AL DISTRIBUTION, CARBON DIOXIDE, NITROGEN, PHOSPHORUS, LIGHT, NUTR W72-11724
S, SOLIDS RECYCLE, SOLI/ ORGANIC NITROGEN, SETTLING CHARACTERISTIC W72-11100
A, *SYMBIOSIS, ACTIVATED SLUDGE, NITROGEN, PHOSPHORUS, CHEMICAL OX W72-11100
NG, ODOR, SLUDGE, COST ANALYSIS, NITROGEN, CLAYS, *WASTE WATER TRE W72-11642
OPHICATION, *PHOSPHOROUS, ALGAE, NITROGEN, ACTIVATED SLUDGE, CHEMI W71-13334
R POLLUTION EFFECTS, PHOSPHORUS, NITROGEN, AMMONIA, CARBON, CHLORE W71-12857
ICAL OXYGEN DEMAND, *PHOSPHATES, NITROGEN, COLIFORMS, MATHEMATICAL W71-13356
, *PRIMARY PRODUCTIVITY, *ALGAE, NITROGEN, PHOSPHORUS, THERMAL STR W71-13183
TER QUALITY, NITRATES, PLANNING, NITROGEN, PHOSPHORUS, RESERVOIRS, W71-13172
ING, *BIOCHEMICAL OXYGEN DEMAND, NITROGEN, PHOSPHOROUS, PHOTOSYNTH W71-13326
TION, ENZYMES, BACTERIA, SPORES, NITROGEN, METABOLISM.: /OGEN FIXA W71-12070
, *ACTIVATED SLUDGE, *NUTRIENTS, NITROGEN, WATER POLLUTION EFFECTS W71-11972
IZATION, NUTRIENTS, ALGAE, RICE, NITROGEN, NITROGEN FIXATION, 2-4- W71-12071
ASS, WATER QUALITY, CHLOROPHYLL, NITROGEN, WISCONSIN, NUTRIENTS, W W71-10645
SAMPLING, TEMPERATURE, COPEPODS, NITROGEN, BACTERIA, OXIDATION-RED W71-11008
UTROPHICATION, LAKE ERIE, ALGAE, NITROGEN, PHOSPHORUS, DRAINAGE, I W71-11009
E ERIE, LAKE ONTARIO, NUTRIENTS, NITROGEN, FILTRATION, MICROSCOPY, W71-11507
L ANALYSIS, STATISTICAL METHODS, NITROGEN, PHOSPHORUS, CHLOROPHYLL W71-11029
EUTROPHICATION, TRACE ELEMENTS, NITROGEN, PHOSPHORUS, METABOLISM, W71-11793
TRIBUTION, AQUATIC PRODUCTIVITY, NITROGEN, PHOSPHORUS, LAKES.: /IS W71-11519
ATES, PHOSPHATES, WATER QUALITY, NITROGEN, NUTRIENTS, TROPHIC LEVE W71-11033
SATURATION, OXYGEN, METABOLISM, NITROGEN, PHOSPHOROUS, *ALGAL CON W71-13553
S, NUTRIENTS, ALGAE, PHOSPHORUS, NITROGEN, EUTROPHICATION, SEDIMEN W72-00847
ATION, ACTIVATED SLUDGE, DESIGN, NITROGEN, DIFFUSION, *ALGAL CONTR W72-00800
MICAL OXYGEN DEMAND, PHOSPHORUS, NITROGEN, EUTROPHICATION, *AEROBI W72-01881
GOONS, NUTRIENTS, AQUATIC ALGAE, NITROGEN, ANAEROBIC DIGESTION, AE W72-02850
BICIDES, TOXINS, WASTE DILUTION, NITROGEN, PHOSPHORUS, EUTROPHICAT W72-01472
, SUBSTRATE UTILIZATION, ORGANIC NITROGEN, ORGANIC CARBON, CULTURE W73-00379
TIC PRODUCTIVITY, AQUATIC ALGAE, NITROGEN, POLLUTANT IDENTIFICATIO W73-00379
*AMMONIA, DIATOMS, GROWTH RATES, NITROGEN, WATER POLLUTION EFFECTS W73-00380
NTS, BIOASSAY, LIMITING FACTORS, NITROGEN, PHOSPHORUS, RAIN, LAKES W73-00232
WATER POLLUTION EFFECTS, SEWAGE, NITROGEN, PHOSPHORUS, IRON, PHYTO W73-00234
SCILLATO/ SECCHI DISC, DISSOLVED NITROGEN, DISSOLVED PHOSPHORUS, O W72-07648
IOASSAY, *STANDING CROP, CARBON, NITROGEN, PHOSPHORUS, PHYTOPLANKT W72-07648
S, *MARINE ALGAE, PHYTOPLANKTON, NITROGEN, ABSORPTION, SEA WATER.: W72-07497
C NITROGEN, *PARTICULATE ORGANIC NITROGEN, CALCAREOUS LAKES, ULTRA W72-08048
REOUS LAKES, ULTRA-VIOLET LABILE NITROGEN, ULTRA-VIOLET REFRACTORY W72-08048
, PHOSPHATES, CARBON, VITAMIN B, NITROGEN, CYANOPHYTA, LAKE MICHIG W72-07937
ITROGEN, ULTRA-VIOLET REFRACTORY NITROGEN, NETPLANKTON, LAWRENCE L W72-08048
ATERS, CONTINENTAL SHELF, DEPTH, NITROGEN, PHYTOPLANKTON, HYDROGRA W72-08056
E METABOLISM, *DISSOLVED ORGANIC NITROGEN, *PARTICULATE ORGANIC NI W72-08048
S, FRESH WATER, ALGAE, BACTERIA, NITROGEN, PHOSPHORUS, OXIDATION-R W72-07933
ATIFICATION, MIXING, PHOSPHORUS, NITROGEN, TURBULENCE, PHOTOSYNTHE W72-08559
BITION, OLIGOTROPHY, PHOSPHORUS, NITROGEN, *EUTROPHICATION, LABORA W72-08376
IC TREAT/ *FARM WASTES, POULTRY, NITROGEN, LAGOONS, PUMPING, AEROB W72-08396
N ION CONCENTRATION, PHOSPHATES, NITROGEN, HARDNESS(WATER), CONDUC W72-07890
C CARBON, CHLOROPHYLL A, ORGANIC NITROGEN, THIAMINE, BIOTIN, VITAM W72-07892
USSELS, OYSTERS, EUTROPHICATION, NITROGEN, PHOSPHOROUS, CA: /MS, M W72-07715
PIGMENT, MAGNESIUM, PHOSPHATES, NITROGEN, SEA WATER, FLUOROMETRY, W72-07660
PARTICU/ ORGANIC CARBON, ORGANIC NITROGEN, SUBSTRATE UTILIZATION, W73-01066
FLUENT STREAMS, LIGHT INTENSITY, NITROGEN, CHRYSOPHYTA, NITRATES, W73-01435
ISSOLVED SOLIDS, ORGANIC MATTER, NITROGEN, WATER POLLUTION SOURCES W73-01066
CHLORELLA, *NUTRIENTS, CULTURES, NITROGEN, POTASSIUM, PHOSPHATES, W73-01626
N FIXATION, *CYANOPHYTA, PLANTS, NITROGEN, REDUCTION(CHEMICAL), GA W73-01456
S, INSTRUMENTATION, METHODOLOGY, NITROGEN, SILICON, ABSORPTION, PH W73-01445
AUDATA, EUGLENA/ *INDIA, ORGANIC NITROGEN, EUGLENA ACUS, EUGLENA C W73-00918
LOR, HYDROGEN ION CONCENTRATION, NITROGEN, CARBON, SODIUM, CALCIUM W73-00854
TALS, CYCLING NUTRIENTS, CARBON, NITROGEN, PHOSPHORUS, SULFUR, CAL W73-00840
CYANOPHYTA, SYMBIOSIS, LICHENS, NITROGEN, LIGHT INTENSITY, TEMPER W73-02471
TY, NUTRIENTS, DISSOLVED OXYGEN, NITROGEN, PHOSPHORUS, WATER QUALI W71-06444
WASTES, POLLUTANTS, FERTILIZERS, NITROGEN, NITRATES, PHOSPHORUS, P W71-06445
ETRATION, NUTRIENTS, PHOSPHORUS, NITROGEN, TIDAL WATERS, CHEMICAL W71-07098
*WATER QUALITY CONTROL, AMMONIA, NITROGEN, PHOSPHORUS, ALGAE, FEAS W71-09450
PENS, ORGANIC MATTER, NUTRIENTS, NITROGEN, PHOSPHORUS, EUTROPHICAT W71-08214
OPLANKTON, *PRODUCTIVITY, ALGAE, NITROGEN, PHOSPHORUS, MATHEMATICA W71-08670
IC MATTER, VOLUME, DRYING, ODOR, NITROGEN, AERATION, HOGS, CATTLE, W71-07551
MICAL OXYGEN DEMAND, PHOSPHATES, NITROGEN, STABILIZATION.: /D, CHE W70-07838
LGAE, SYMBIOSIS, CARBON DIOXIDE, NITROGEN, LAKES, ESTUARIES, NUTRI W70-07283
, CYANOPHYTA, CYCLING NUTRIENTS, NITROGEN, PHOSPHORUS, CHEMICAL AN W70-05469
TTER, NUTRIENTS, SURFACE WATERS, NITROGEN, WATER QUALITY, PHOSPHOR W70-10405
LIMNION, WASHINGTON, PHOSPHORUS, NITROGEN, NUTRIENTS.: /INITY, EPI W71-02681

CHEMICAL PROPERTIES, PHOSPHORUS,	NITROGEN, POTASSIUM, DIVERSION, A	W71-00117
, PHOSPHATE, PHOSPHORUS, CARBON,	NITROGEN, CURRENTS(WATER), WINDS,	W71-05626
DNESS(WATER), SAMPLING, DIATOMS,	NITROGEN, BACTERIA, ALGAE, THERMA	W71-03028
CHLORELLA PYRENOIDOSA, SOURCE OF	NITROGEN, NTA.: /ILOTRIACETATE, *	W70-08976
TOR-PREY RELATIONSHIPS, *AMMONIA-	NITROGEN, *GROWTH REACTORS, ALGAL	W72-04787
ELLITES(ARTIFICIAL), PHOSPHORUS,	NITROGEN, HEAVY METALS, ORGANOPHO	W72-04742
N CONCENTRATION, CARBON DIOXIDE,	NITROGEN, CHLOROPHYTA, NITROGEN F	W72-04769
, MANGANESE, PHOSPHORUS, CARBON,	NITROGEN, OXIDATION, BIOCHEMICAL	W72-05506
ES, IRON, MANGANESE, PHOSPHORUS,	NITROGEN, ORGANIC MATTER, CALCIUM	W72-05469
LIGHT INTENSITY, CARBON DIOXIDE,	NITROGEN, LABORATORY TESTS, CYANO	W72-05476
POLLUTION), VIRGINIA, NUTRIENTS,	NITROGEN, PHOSPHORUS, SEWAGE EFFL	W72-05473
N CONCENTRATION, CARBON, OXYGEN,	NITROGEN, HYDROGEN, PRIMARY PRODU	W72-06274
(KENT), VASCULAR PLANTS, NITRATE	NITROGEN, ORTHOPHOSPHATES, *OLIGA	W72-06526
NTS, *NUTRIENTS, EUTROPHICATION,	NITROGEN, LITTORAL, HARBORS, SPAT	W72-04263
D, *EUTROPHICATION, PHOSPHOROUS,	NITROGEN, CHEMICAL OXYGEN DEMAND,	W72-04431
NALYSIS, DETERGENTS, PHOSPHORUS,	NITROGEN, NUTRIENTS, SEWAGE, WATE	W72-04734
RE, COLIFORMS, ENTERIC BACTERIA,	NITROGEN, VITAMINS, SUSPENDED SOL	W72-12565
PHATES, CHLOROPHYTA, CYANOPHYTA,	NITROGEN, CARBON, DIATOMS, BACTER	W72-12544
OPHYLL, *PHOSPHORUS, *LAKE ERIE,	NITROGEN, ALGAE, INORGANIC COMPOU	W72-13653
KS, BIOASSAY, CHEMICAL ANALYSES,	NITROGEN, NUTRIENTS, SURFACE WATE	W72-12955
MENTS, GROWTH RATES, PHOSPHORUS,	NITROGEN, *REGRESSION ANALYSIS, V	W72-12966
, BACTERIA, CHEMICAL PROPERTIES,	NITROGEN, HEAT TRANSFER, PHOSPHAT	W72-12997
ENTS, OXYGEN DEMAND, PHOSPHORUS,	NITROGEN, NUTRIENTS, ON-SITE DATA	W72-12990
ENTS, *NITROGEN FIXATION, LAKES,	NITROGEN, PHOSPHORUS, IRON, *TEXA	W72-13692
POLLUTION EFFECTS, FERTILIZERS,	NITROGEN, BIOASSAY, PHOSPHORUS, C	W72-13816
PHYTOPLANKTON, SEWAGE TREATMENT,	NITROGEN, PHYSOCHEMICAL PROPERTIE	W72-13662
NIA, NITRATES, LIMITING FACTORS,	NITROGEN, PLANKTON, DEPTH, ALGAE.	W73-02472
, FERTILIZATION, AQUATIC PLANTS,	NITROGEN, NUTRIENT REMOVAL, WATER	W73-02478
OCIATION, CARBOY TRIALS, CANADA,	NITROGEN: PHOSPHORUS RATIO.: /ASS	W70-07283
TATS, NITROSOMONAS, NITROBACTER,	NITROSOCYSTIS OCEANUS, SPHAEROTIL	W72-05421
ENT CULTURES, AGARS, CHEMOSTATS,	NITROSOMONAS, NITROBACTER, NITROS	W72-05421
NTAL FOREST, PARTICULATE MATTER,	NITROSOMONAS, NITROBACTER, THIOBA	W71-01489
ACTER.:	NITROSOMONAS, NITROBACTER, AZOTOB	W72-11563
NITZSCHIA SERIATA, THALASSIONEMA	NITZ: /VIELLA MARINA, EUTREPTIA,	W71-07875
NAS EROSA, PERIDINIUM AFRICANUM,	NITZCHIA ACICULARIS, SYNEDRA ACUS	W73-00256
ANABAENA SPP., GOMPHONEMA SPP.,	NITZCHIA SPP., SENSITIVITY.: /P.,	W72-14706
A PUNGENS, C-14, CHLOROPHYLL A,/	NITZSCHIA DELICATISSIMA, NITZSCHI	W73-00431
YLL A,/ NITZSCHIA DELICATISSIMA,	NITZSCHIA PUNGENS, C-14, CHLOROPH	W73-00431
PLEUROSIGMA, GYROSIGMA FASCIOLA,	NITZSCHIA SPP, BACILLARIA PAXILLI	W73-00853
N, CARBON-14, NITRO/ PLATYMONAS,	NITZSCHIA OVALIS, CARBON EXTRUSIO	W72-07497
NS, EXUVIELLA MARINA, EUTREPTIA,	NITZSCHIA SERIATA, THALASSIONEMA	W71-07875
ATUM.: CHLORELLA PYRENOIDOSA,	NITZSCHIA PALEA, SKELETONEMA COST	W71-03027
ALIF), SAN JOAQUIN RIVER(CALIF),	NITZSCHIA.: /), SEARSVILLE LAKE(C	W70-10405
EDIA, CHLOROPHYLL A, TERATOLOGY,	NITZSCHIA.: /, ACTIDON, CULTURE M	W73-01445
RAGILARIA CONSTRUENTS, NAVICULA,	NITZSCHIA, EPIPYTIC ALGAE.: /S, F	W72-07501
LACIALIS, DITYLUM, PROROCENTRUM,	NITZSCHIA, LAUDERIA, CHAETOCEROS.	W72-09248
GMA, NAUPLII, LA/ COSCINODISCUS,	NITZSCHIA, OSCILLATORIA, PLEUROSI	W71-11876
A MARTYI, CELL COUNTS, NAVICULA,	NITZSCHIA, STEPHANODISCUS, MICROC	W71-12855
NS, THALASSIOSIRA, THALASSIONEMA	NITZSCHIOIDES, PLAGIOGRAMMA VANHE	W73-00853
LETONEMA COSTATUM, THALASSIONEMA	NITZSCHIOIDES, MIXED CULTURES, ME	W73-00441
IATOA ALBA, BEGGIATOA, THIOTHRIX	NIVEA, THIOTHRIX TENUIS, MERISMOP	W72-07896
EBRATES, *TOCKS ISLAND RESERVOIR(	NJ), AMERICAN SHAD, EKMAN DREDGE,	W72-14280
RER SAMPLER, SECCHI *YARDS CREEK(	NJ), RELATIVE CONDITION FACTOR.: /	W72-14280
LGAL CYCLE, DELAWARE ESTUARY(DEL-	NJ), SAN JOAQUIN DELTA(CALIF), ST	W71-05390
TAL O/ FUSARIUM NO 83, BACTERIUM	NO 52, ENDOGENOUS RESPIRATION, TO	W71-12223
S RESPIRATION, TOTAL O/ FUSARIUM	NO 83, BACTERIUM NO 52, ENDOGENOU	W71-12223
IUM TRIPOS, CERATIUM MACROCEROS,	NOCTILUCA MILIARIS, GYMNODINIUM S	W71-07875
NA, LITHODESMIUM, HARPESTICOIDS,	NOCTILUCA, CENTROPYXIS, BIDDULPHI	W71-11876
ONAUTES, ECHINODERMS, TUNICATES,	NOCTILUCA, SIPHONOPHORES, MEDUSAE	W73-00935
ILIANA, PAVLOVIAMONAS FRUTICOSA,	NODAMASTIX SPIROGYRAE, COLEMANIA	W72-07683
S, SARGASSUM NATANS, ASCOPHYLLUM	NODOSUM, FUCUS VESICULOSUS, LAMIN	W72-04748
AMINARIA HYPERBOREA, ASCOPHYLLUM	NODOSUM, GIGARTINA MAMILLOSA, RHO	W71-11515
INEATUS-EXCENTRICUS, CLADOGRAMMA	NODULIFER, DACTYLIOSOLEN: /AMMA L	W71-11527
	NON-BIOLOGICAL UPTAKE.:	W72-12210
NS, LEGISLATION, ADMINISTRATION,	NON-CONSUMPTIVE USE, ALGAE, EVAPO	W70-07100
ASSIOSIRA GRAVIDA, THALASSIOSIRA	NORDENSKIOLDII, THALASSIOSIRA ROT	W72-03543
ICS, *VIRGINIA, *ATLANTIC OCEAN,	NORTH AMERICA.: /ARSHES, SYSTEMAT	W72-12633
OLLUTION EFFECTS, AQUATIC ALGAE,	NORTH CAROLINA, *ALGAE, DIATOMS,	W72-01329
AL SHELF, VERTICAL DISTRIBUTION,	NORTH CAROLINA, SOUTH CAROLINA, S	W72-07663
LLUSKS, *ATLANTIC COASTAL PLAIN,	NORTH CAROLINA, ABSORPTION, SOUND	W73-00818
SILT-WATER-PLANT RELATIONSHIPS,	NORTH SEA, PHAEOCYSTIS SPP., SOIL	W72-12239
COHO SALMON, SHEEPSHEAD, CHUBS,	NORTHERN PIKE, GIZZARD SHAD.: /S,	W72-14282
GRAPHIE/ *OCEANOGRAPHY, *PACIFIC	NORTHWEST, *COAST REVIEW, *BIBLIO	W72-12190
*OSLOFJORD(	NORWAY).:	W73-00234
SENIC, SELENIUM, ZINC, ANTIMONY,	NORWAY, ASHING, BIOLOGICAL MAGNIF	W71-11515
ALCICOLA, BORATES, ARTHROBACTER,	NOSTOC COMMUNE, PORPHYROSIPHON NO	W73-00854
MII, 7-METHYLHEPTADECANE, 8-MET/	NOSTOC MUSCORUM, LYNGBYA LAGERHAI	W73-01439
HOT SPRINGS, WATER TEMPERATURES,	NOSTOC STREAMS, DATA COLLECTIONS,	W72-08508
SWEDEN, COLLEMA, PELTIGERA,	NOSTOC.:	W73-02471
ACYSTIS N/ *TOLERANCE, ANABAENA,	NOSTOC, AULOSIRA, TOLYPOTHRIX, AN	W72-14807
	NOSTOC, CYLINDROTHECA.:	W71-12858
CYANOPHYTA, FROZEN SOILS, IONS,	NOSTOC, ENVIRONMENTAL EFFECTS, PO	W73-00854
AENA, GLOEOTRICHIA, MICROCYSTIS,	NOSTOC, SLAKED LIME.: /TION, ANAB	W72-01097
TRICALCIUM PHOSPHATES, ANABAENA,	NOSTOC, TOLYPOTHRIX TENUIS, AULOS	W71-10070
, NOSTOC COMMUNE, PORPHYROSIPHON	NOTARISII, PSYCHROPHILIC, BIOTIC	W73-00854
HOLSATICUM, ICTALURUS NEBULOSUS,	NOTEMIGONUS CRYSOLEUCAS, CATOSTOM	W72-07890
TRICHOPTERA, CORIXIDS, EUDORINA,	NOTONECTIDAE, BODO, HYDROPSYCHIDA	W70-06655
YPHA, ELODE/ BIOLOGICAL EFFECTS,	NOTTINGHAM(ENGLAND), POLYGONUM, T	W72-00155
ORA, MELANOSIP/ *EASTERN CANADA,	NOVA SCOTIA, ISTHMOPLEA SPHAEROPH	W72-01370
LEN ISLAND(QUEBEC), SABLE ISLAND(	NOVA SCOTIA), ST PIERRE(CANADA),	W71-10066

MENT, *WATER RE-USE, BIOCHEMICAL OXYGEN DEMAND, ORGANIC LOADING, O W71-07123
TERTIARY TREATMENT, BIOCHEMICAL OXYGEN DEMAND, WASTE WATER TREATM W71-07382
OCNS, *NORTH DAKOTA, BIOCHEMICAL OXYGEN DEMAND, BACTERIA, TEMPERAT W72-12565
PLANKTON, NUTRIENTS, BIOCHEMICAL OXYGEN DEMAND, INORGANIC COMPOUND W72-13659
IC CONDITIONS, ALGAE, SEDIMENTS, OXYGEN DEMAND, PHOSPHORUS, NITROG W72-12990
, *LAKE ERIE, *DISSOLVED OXYGEN, OXYGEN DEMAND, HYPOLIMNION, AQUAT W72-12994
ROPHYTA, CYANOPHYTA, BIOCHEMICAL OXYGEN DEMAND.: /FRYSOPHYTA, CHLO W72-12995
N, DISSOLVED OXYGEN, BIOCHEMICAL OXYGEN DEMAND, PERFORMANCE, FLOCC W72-13508
RANSFER, PHOSPHATES, ALKALINITY, OXYGEN DEMAND, HEAT BUDGET, ALGAE W72-12997
LS, CHEMICAL WASTES, BIOCHEMICAL OXYGEN DEMAND, BACTERIA, ALG: /CA W73-01976
ITY, STREAMS, LAKES, BIOCHEMICAL OXYGEN DEMAND, *OXYGENATION, *ANA W72-14469
QUATIC WEED CONTROL, BIOCHEMICAL OXYGEN DEMAND, LAKE STAGES, LIMNO W72-14692
QUATIC PRODUCTIVITY, BIOCHEMICAL OXYGEN DEMAND, LAKE STAGES, NUTRI W73-00002
CHEMICAL OXYGEN DEMAND, CHEMICAL OXYGEN DEMAND, NUTRIENTS, *WASTE W72-08396
, AEROBIC TREATMENT, BIOCHEMICAL OXYGEN DEMAND, CHEMICAL OXYGEN DE W72-08396
URES, SEDIMENTATION, BIOCHEMICAL OXYGEN DEMAND, ALGAE, CONFINEMENT W72-08401
LIMATES, METABOLISM, BIOCHEMICAL OXYGEN DEMAND, PILOT PLANTS, *WAS W72-06838
CHEMICAL OXYGEN DEMAND, CHEMICAL OXYGEN DEMAND, PHOSPHORUS, NITROG W72-01881
OXIDATION, STORAGE, BIOCHEMICAL OXYGEN DEMAND, CHEMICAL OXYGEN DE W72-01881
EATMENT, *TURBIDITY, BIOCHEMICAL OXYGEN DEMAND, DISSOLVED OXYGEN, W72-03753
REDUCTION POTENTIAL, BIOCHEMICAL OXYGEN DEMAND, NITRIFICATION, NUT W72-04086
ORGANISMS, EFFLUENT, BIOCHEMICAL OXYGEN DEMAND, SATURATION, OXYGEN W71-13553
ALGAE, OXYGENATION, BIOCHEMICAL OXYGEN DEMAND, WATER SPORTS, PARK W72-00042
MICAL OXYGEN DEMAND, BIOCHEMICAL OXYGEN DEMAND, FILTRATION, DISINF W72-00411
OSYNTHESIS, SOLUBILITY, CHEMICAL OXYGEN DEMAND, BIOCHEMICAL OXYGEN W72-00411
, NITROGEN, PHOSPHORUS, CHEMICAL OXYGEN DEMAND, CHLORELLA, PERFORM W72-11100
TH, RIVERS, *SEWAGE, BIOCHEMICAL OXYGEN DEMAND, SOLID WASTES, AMMO W72-09675
R POLLUTION SOURCES, BIOCHEMICAL OXYGEN DEMAND, WATER POLLUTION CO W72-09167
OGEN, ACTIVATED SLUDGE, CHEMICAL OXYGEN DEMAND, OXYGEN, AERATION, W71-13334
CHEMICAL OXYGEN DEMAND, CHEMICAL OXYGEN DEMAND, PHOSPHORUS, NITRAT W71-13341
S, ORGANIC LOADING, *BIOCHEMICAL OXYGEN DEMAND, NITROGEN, PHOSPHOR W71-13326
CHEMICAL OXYGEN DEMAND, CHEMICAL OXYGEN DEMAND, *PHOSPHATES, NITRO W71-13356
ON, WATER POLLUTION, BIOCHEMICAL OXYGEN DEMAND, ORGANIC MATTER, PH W71-13257
ES, MUNICIPAL WASTE, BIOCHEMICAL OXYGEN DEMAND, CHEMICAL OXYGEN DE W71-13356
ILTRATION, *FILTERS, BIOCHEMICAL OXYGEN DEMAND, CHEMICAL OXYGEN DE W71-13341
ON CONTROL, WASHINGTON, CHEMICAL OXYGEN DEMAND, BIOCHEMICAL OXYGEN W71-11006
MICAL OXYGEN DEMAND, BIOCHEMICAL OXYGEN DEMAND, DISSOLVED OXYGEN: / W71-11006
2, ENDOGENOUS RESPIRATION, TOTAL OXYGEN DEMAND, GLYCOLS.: /UM NO 5 W71-12223
SECONDARY TREATMENT, BIOCHEMICAL OXYGEN DEMAND, CHEMICAL OXYGEN DE W71-12188
ALGAE, ZOOPLANKTON, BIOCHEMICAL OXYGEN DEMAND, COPEPODS,: /QUATIC W71-11876
ANIC WASTES, ECONOMICS, CHEMICAL OXYGEN DEMAND, BIOCHEMICAL OXYGEN W71-12223
CHEMICAL OXYGEN DEMAND, CHEMICAL OXYGEN DEMAND, CARBON, FERTILIZAT W71-12188
MICAL OXYGEN DEMAND, BIOCHEMICAL OXYGEN DEMAND, AERATION, *OXIDATI W71-12223
SOLUBILITY, SEWAGE, BIOCHEMICAL OXYGEN DEMAND, ION EXCHANGE, CHEM W71-12200
N DEMAND, ION EXCHANGE, CHEMICAL OXYGEN DEMAND, FERTILIZER, TURBID W71-12200
AL BASIN, POINTE PELEE(ONTARIO), OXYGEN DEPLETION.: /KE ERIE CENTR W72-12992
LOS-AQUA, *NITROGENASE ACTIVITY, OXYGEN INACTIVATION.: /ANABAENA F W72-07941
RACTIVE DIGESTION, NEUROANATOMY, OXYGEN PLASMA ASHING, ATOMIC ABSO W71-11036
Y, CULT/ *CHLORELLA PYRENOIDOSA, OXYGEN PRODUCTION, LIGHT INTENSIT W70-05547
L, LIMNOLOGY, NUTRIENTS, OXYGEN, OXYGEN REQUIREMENTS, OXYGEN SAG, W72-04279
POLLUTION TREATMENT, NUTRIENTS, OXYGEN REQUIREMENTS, AGING(BIOLOG W72-01685
STE DISPOSAL, PESTICIDE REMOVAL, OXYGEN SAG, DREDGING, STATE GOVER W72-01659
TS, OXYGEN, OXYGEN REQUIREMENTS, OXYGEN SAG, WASTE ASSIMILATIVE CA W72-04279
CATION, FOULING, TOXICITY, ODOR, OXYGEN SAG, TURBIDITY, WATER QUAL W71-09958
TRIENTS, MESTROPHY, OLIGOTROPHY, OXYGEN SAG, AQUATIC LIFE, BIOASSA W73-00002
TA, WATER TEMPERATURE, DISSOLVED OXYGEN.: / CONCENTRATION, MINNESO W72-07145
CES, WATER LAW, ALGAE, DISSOLVED OXYGEN.: /L, WATER POLLUTION SOUR W71-01192
SIS, ALGAE, COLIFORMS, DISSOLVED OXYGEN.: /, SAMPLING, WATER ANALY W70-08248
SIS, ALGAE, COLIFORMS, DISSOLVED OXYGEN.: /, SAMPLING, WATER ANALY W70-08249
, PERMEABILITY, ALGAE, DISSOLVED OXYGEN.: /OUNDWATER RELATIONSHIPS W71-12410
ACTERISTICS, GENETICS, DISSOLVED OXYGEN.: /IC FUNGI, ACQUIFER CHAR W72-11186
LORELLA, *LIGHT, *PHOTOSYNTHETIC OXYGEN, *ALGAE, CULTURES, VOLUME, W70-05547
CATION, *HYPOLIMNION, *DISSOLVED OXYGEN, *ANAEROBIC CONDITIONS, AL W72-12990
ICAL / *SLUDGE WORMS, *DISSOLVED OXYGEN, *EUTROPHICATION, *BIOCHEM W70-09189
UTION, / *CYANOPHYTA, *DISSOLVED OXYGEN, *EUTROPHICATION, *DISTRIB W73-00236
BENTHIC FLORA, ALGA/ *DISSOLVED OXYGEN, *HYPOLIMNION, *LAKE ERIE, W72-13659
EDIMENT-WATER INTERF/ *DISSOLVED OXYGEN, *LAKE ERIE, *BACTERIA, *S W72-12996
HYPOLIMNION, MATHEMA/ *DISSOLVED OXYGEN, *NUTRIENTS, *LAKE ERIE, * W72-12997
TRATION, HEAVY METALS, DISSOLVED OXYGEN, *PLANT GROWTH, LIGHT INTE W72-14690
STES, *WATER QUALITY, *DISSOLVED OXYGEN, *PHYTOPLANKTON, ESTUARIES W71-11876
NTATION/ *ALGAE, *PHOTOSYNTHETIC OXYGEN, *PHOTOSYNTHESIS, INSTRUME W71-10791
ASS, RESPIRATION, PHOTOSYNTHETIC OXYGEN, *TERTIARY TREATMENT.: /OM W72-11100
GAE, *OXIDATION, PHOTOSYNTHESIS, OXYGEN, *THERMODYNAMICS, ANALYTIC W72-02363
ALGAE, PHOTOSYNTHESIS, DISSOLVED OXYGEN, AEROBIC CONDITIONS, ANAER W72-11642
Y EQUIPMENT, AERATION, DISSOLVED OXYGEN, AEROBIC CONDITIONS, BIOMA W72-11100
SLUDGE, CHEMICAL OXYGEN DEMAND, OXYGEN, AERATION, DISSOLVED OXYGE W71-13334
OLIFORMS, TEMPERATURE, DISSOLVED OXYGEN, AERATION, OXYGENATION, NU W71-07113
LECTRICAL CONDUCTANCE, DISSOLVED OXYGEN, AIR POLLUTION, CHEMICAL A W71-01489
, WASTE WATER, RUNOFF, DISSOLVED OXYGEN, ALGAE, NUTRIENTS, BACTERI W71-09788
PORATION, RE-AERATION, DISSOLVED OXYGEN, ALGAE, ANAEROBIC CONDITIO W71-07099
EMPERATURE, NUTRIENTS, DISSOLVED OXYGEN, ALGAE, SLUDGE, OXYGENATIO W71-07109
TIVITY, PHOTOSYNTHESIS, BIOMASS, OXYGEN, ALGAE, COMPARATIVE PRODUC W71-09165
ERATURE, SALT BALANCE, DISSOLVED OXYGEN, ALGAE, BACTERIA, ODOR, IN W72-02274
HEMICAL OXYGEN DEMAND, DISSOLVED OXYGEN, ALGAE, THERMAL STRATIFICA W72-03753
ANIC LOADS, NUTRIENTS, DISSOLVED OXYGEN, AQUATIC LIFE, MUNICIPAL W W71-11017
TES, SLUDGE TREATMENT, DISSOLVED OXYGEN, BACTERIA, FUNGI, PROTOZOA W72-09675
ON, LIGHT PENETRATION, DISSOLVED OXYGEN, BACTERIA, THERMOCLINE, SE W72-12995
GEN ION CONCENTRATION, DISSOLVED OXYGEN, BIOCHEMICAL OXYGEN DEMAND W72-13508
WS, WATER TEMPERATURE, DISSOLVED OXYGEN, BIOINDICATORS, E. COLI, A W71-07417
SYNTHESIS, *CULTURES, *NITROGEN, OXYGEN, BIOCHEMISTRY, INHIBITORS, W72-07941

*CHLORELLA PYRENOIDOSA, LINDANE,	P P'-DDE, CLUPHEN A 50, ELECTRON	W72-04134
HLORELLA PYRENOIDOSA, LINDANE, P	P'-DDE, CLOPHEN A 50, ELECTRON CA	W72-04134
A OLIVACEOIDES, WHITE CLAY CREEK(	PA.).: GOMPHONEM	W73-00242
E, WHIPPLE DAM, STONE VALLEY DAM(	PA.), ATRAZINE, BIOTYPES, PCUR PL	W72-04743
ES, DETRIVORES, LINESVILLE CREEK(	PA).: /AUTUMNAL COMMUNITY, CALORI	W71-10064
VERSITY, DIRECT EF/ RIDLEY CREEK(	PA), LITITZ CREEK(PA), SPECIES DI	W72-08579
RIDLEY CREEK(PA), LITITZ CREEK(	PA), SPECIES DIVERSITY, DIRECT EF	W72-08579
IDES, SELENASTRUM CAPRICORNUTUM,	PAAP BATCH TEST, CULTURE MEDIA.: /	W73-02099
RIVER(CALIF), SEARSVILLE LA/ SAN	PABLO ESTUARY(CALIF), SACRAMENTO	W70-10405
ITRITE, SPHAEROTILUS, NUTRIENTS,	PACIFIC OCEAN, NITRATE, CHEMICAL	W72-05421
I, ABSORPTION, TROPICAL REGIONS,	PACIFIC OCEAN, COLUMNS, BOTTOM SE	W71-08042
CYCLOTELLA NANA, LEPTOCYLINDRUS,	PACIFIC OCEAN, ALEUTIAN ISLANDS,	W71-11008
TION TECHNIQUES, ATLANTIC OCEAN,	PACIFIC OCEAN, GULF OF MEXICO.: /	W71-11851
S, MAPS, MAPPING, PHYTOPLANKTON,	PACIFIC OCEAN, BOTTOM SEDIMENTS,	W72-03567
PODA, RADIOISOTOPES, ABSORPTION,	PACIFIC OCEAN, FOOD CHA: /, AMPHI	W72-03347
BUTION PATTERNS, ATLANTIC OCEAN,	PACIFIC OCEAN, TROPICAL REGIONS,	W72-03543
TRIPHOTURUS MEXICANUS, EUPHAUSIA	PACIFICA, ENGRAULIS MORDAX.: /N,	W73-02105
*GRIFFITHSIA	PACIFICA, *REGENERATION.:	W73-01622
, METAZOA, CALANUS HELGOLANDICUS(	PACIFICUS), THERMISTORS, DATA INT	W72-07892
TER POLLUTION, DES MOINES RIVER,	PACKING PLANTS.: INDUSTRIAL WA	W71-06445
CHLORELLA PYRENOIDOSA, NITZSCHIA	PALEA, SKELETONEMA COSTATUM.:	W71-03027
ISOTOPES, ANALYTICAL TECHNIQUES,	PALEOZOIC ERA, WATER CHEMISTRY.: /	W72-02073
ONAS ISABELIENSIS, CHLAMYDOMONAS	PALLIDOSTIGMATICA, CHLAMYDOMONAS	W72-12632
PERIDINIUM DEPRESSUM, PERIDINIUM	PALLIDUM, CERATIUM MASSILENSE, CE	W71-07875
, PROPIANATE, PROPANOL, LAURATO,	PALM: /ACETATE, ETHANOL, BENZOATE	W72-05421
GIGARTINA MAMILLOSA, RHODYMENIA	PALMATA.: /, ASCOPHYLLUM NODOSUM,	W71-11515
ARDHII, ULVA LACTUCA, RHODYMENIS	PALMATA, POLYSIPHONIA LANOSA, POL	W72-04748
UM, BRYOPSIS, CAULERPA, ACROPORA	PALMATA, PENICILLUS LAMOUROUXII.:	W72-14294
DINIUM SINICU/ KOLBEANA OVOIDEA,	PALMERIAMONAS PLANCTONICA, MANCHU	W72-07683
RPA SP., CAULERPICIN, CAULERPIN,	PALMITIC ACID, BETA-SITOSTEROL, T	W71-08597
HONIA, ULVA LACTUCA, ZANICHELLIA	PALUSTRIS, CODIUM, FUSCUS, MACROP	W73-00367
SS), CAPE MAY(/ *COASTAL WATERS,	PANAMA, CALLAO(PERU), CAPE COD(MA	W72-08056
CAUDATA, CHLAMYDOMONAS PERTUSA,	PANDORINA MORUM.: /SA, CHROOMONAS	W72-13816
RS, AUTOINHIBITORS, VOLVOCACEAE,	PANDORINA MORUM, PLATYDORINA CAUD	W71-13793
DOMONAS SPP, CARTERIA FRITSCHII,	PANDORINA MORUM, SPHAEROCYSTIS SC	W72-09111
CYSTIS, OSCILLATORIA, SPIRULINA,	PANDORINA, PHOCUS, STAURASTRUM, T	W72-07661
.: *ALEXANDRIA(EGYPT),	PANDORINA, SOLUBLE ORGANIC SOLIDS	W70-06619
, CODIUM CAROLINI/ TREMATOCARPUS	PAPENFUSSII, GLOIODERMA ATLANTICA	W73-00284
, ORGANIC CARBON, CULTURE MEDIA,	PAPER CHROMATOGRAPHY, INFRARED SP	W73-00379
LATE MATTER, FATE OF POLLUTANTS,	PAPER CHROMATOGRAPHY, INFRARED SP	W73-01066
ISOTOPES, *BOSSIELLA ORBIGNIANA,	PAPER CHROMATOGRAPHY, SCINTILLATI	W72-12634
ACTIVITY, *SKELLYSOLVE, *FILTER	PAPER DISC DIFFUSION METHOD, *ZON	W72-08586
*WISCONSIN, PHOSPHORUS, PULP AND	PAPER INDUSTRY, ALGAE, AQUATIC WE	W70-05103
SII, LAURENCIA OBTUSA, LAURENCIA	PAPILLOSA, COELOTHRIX IRREGULARIS	W73-00838
, GOMPHONEMA PARVULUM, BACILLARI	PARADOXA, SPHAEROTILUS NAT: /TATA	W72-10883
TERIONELLA JAPONICA, STAURASTRUM	PARADOXIUM, CULTURE MEDIA, ZINC S	W72-12739
ULTURING TECHNIQUES, STAURASTRUM	PARADOXUM, STAURASTRUM SEBALDII V	W72-12736
HODO/ *AGARS, EPIPHYTES, CANDIDA	PARAPSILOSIS, CHONDRUS CRISPUS, R	W72-04748
ABLE SOLIDS, HATCHERY EFFLUENTS,	PARASITES, JORDAN RIVER(MICH), SA	W71-11006
TEBRATES, FUNGI, ENZYMES, ANIMAL	PARASITES, VIRUSES, MOLLUSKS, PRO	W73-01676
FISH EGGS, FISH, BACTERIA, FISH	PARASITES, ANTIBIOTICS(PESTICIDES	W72-08142
CONTROL, IRRIGATION, PREDATION,	PARASITISM, INSECT RESISTANCE: /T	W72-00846
IGOCLADUS/ *YELLOWSTONE NATIONAL	PARK(WYO), BLUE-GREEN ALGAE, MAST	W72-08508
A, PHORMIDIUM, MAST/ YELLOWSTONE	PARK, SYNECHOCOCCUS, FLEXIBACTERI	W72-03557
CAL OXYGEN DEMAND, WATER SPORTS,	PARKS, COST ANALYSIS, MICHIGAN, W	W72-00042
HERMAL WATER, *WYOMING, NATIONAL	PARKS, HOT SPRINGS, WATER TEMPERA	W72-08508
MYRTILLUS, DESCHAMPSIA FLEXUOSA,	PARMELIA PHYSODES, HYPNUM CUPRESS	W72-14303
ONTROL/ *EUTROPHICATION, *SOCIAL	PARTICIPATION, *WATER POLLUTION C	W73-02479
UTION PATTERNS, CURRENTS(WATER),	PARTICLE SIZE, SEDIMENTS, SEPARAT	W73-00367
RIFICATION, CALIFORNIA, FILTERS,	PARTICLE SIZE, ANAEROBIC CONDITIO	W71-08223
ARINE ALGAE, PHAEPHYTA, AQUARIA,	PARTICLE SIZE, COLLOIDS, WATER PO	W71-09193
INTERGREEN LAKE(MICH), CARBONATE	PARTICLES, PHOSPHORUS CYCLE.: / W	W72-08048
NITROGEN, SUBSTRATE UTILIZATION,	PARTICULATE MATTER, FATE OF POLLU	W73-01066
, FISH DIET, GAS CHROMATOGRAPHY,	PARTICULATE MATTER, DETRITUS, ORG	W73-02105
NAGE, LAKE METABOLISM, MOBILITY,	PARTICULATE FORM, FEEDLOTS, NUTRI	W71-06443
BBARD BROOK EXPERIMENTAL FOREST,	PARTICULATE MATTER, NITROSOMONAS,	W71-01489
LUE HERON, STERNA HIRUNDO, RATS,	PARTRIDGE, DDE, KIDNEYS, LIVER, B	W72-05952
NAVICULA TRIPUNCTATA, GOMPHONEMA	PARVULUM, BACILLARI PARADOXA, SPH	W72-10883
AN, ALEUTIAN ISLANDS, PRYMNESIUM	PARVUM.: /OCYLINDRUS, PACIFIC OCE	W71-11008
*MARINE PHYTOPLANKTON(PRYMNESIUM	PARVUM)(CHROOMONAS SALINA), GLYCE	W70-07842
EA, BACCILARIOPHYCEA, PRYMNESIUM	PARVUM, PHOTOAUTOTROPHS, CHROOMON	W70-07280
M RADIOISOTOPES, CLADOPHORA SPP,	PASPALUM DISTICHUM, POTAMOGETON P	W73-01673
TYPHOID, COLIPHAGE, IMMUNOLOGY,	PASTEURELLA PSEUDOTUBERCULOSIS, P	W72-13800
, CASTLE LAKE(CALIF), DIVERSITY,	PATCHES.: /IUM, LAKE TAHOE(CALIF)	W71-09171
S, ECOLOGY, FISH FOOD ORGANISMS,	PATH OF POLLUTANTS, BIOASSAY, TOX	W71-08597
EYS, PESTICIDES, WASTE DISPOSAL,	PATH OF POLLUTANTS, WATER QUALITY	W71-05553
ONTROL, SURVEYS, INVESTIGATIONS,	PATH OF POLLUTANTS, SAMPLING, WAT	W70-08249
ONTROL, SURVEYS, INVESTIGATIONS,	PATH OF POLLUTANTS, SAMPLING, WAT	W70-08248
WASTES, PRODUCTIVITY, DIELDRIN,	PATH OF POLLUTANTS, REVIEWS, FOOD	W72-06340
RY, GEOCHEMISTRY, RADIOISOTOPES,	PATH OF POLLUTANTS, PHYSIOLOGICAL	W72-06313
SON RIVER, INDICATORS, DELAWARE,	PATH OF POLLUTANTS, HOSTS, PENNSY	W72-04736
WS, *DATA COLLECTIONS, TOXICITY,	PATH OF POLLUTANTS, *METALS, TRAC	W72-12190
Y, WATER POLLUTION, FOOD CHAINS,	PATH OF POLLUTANTS, BIOASSAY, FIS	W72-12257
ER, DETRITUS, ORGANIC COMPOUNDS,	PATH OF POLL: /, PARTICULATE MATT	W73-02105
ICIDES, DDD, PESTICIDE TOXICITY,	PATH OF POLLUTANTS, PESTICIDE RES	W73-02280
S, MARINE ALGAE, NUCLEAR WASTES,	PATH OF POLLUTANTS, RADIOECOLOGY,	W73-00828
FOOD CHAINS, AFRICA, EFFLUENTS,	PATH OF POLLUTANTS, NEUTRON ACTIV	W73-00832
R WASTES, *COBALT RADIOISOTOPES,	PATH OF POLLUTANTS, WASTE ASSIMIL	W73-00826
ON, *FOOD CHAINS, PUBLIC HEALTH,	PATH OF POLLUTANTS, RADIOECOLOGY,	W73-00833
T RADIOISOTOPES, NUCLEAR WASTES,	PATH OF POLLUTANTS, ESTUARINE ENV	W72-07826

LGAE, *TOXICITY, SOILS, TRIAGINE	PESTICIDES, RESISTANCE, CULTURES,	W71-07675
CULATION, ABSORPTION, NUTRIENTS,	PESTICIDES, ALGAE, DREDGING, ESTU	W71-09581
XICITY, CHROMOSOMES, FUNGICIDES,	PESTICIDES, ECOLOGY, BIRD TYPES,	W72-05952
, HEAVY METALS, ORGANOPHOSPHORUS	PESTICIDES, INDUSTRIAL WASTES, VI	W72-04742
AMYDOMONAS, *ALGAE, *HERBICIDES,	PESTICIDES, CHLORELLA, RESPIRATIO	W70-07255
FISH, FLOOD PLAINS, IRRIGATION,	PESTICIDES.: /ATION, GROUNDWATER,	W70-05103
, PHOSPHATES, NITRATES, SURVEYS,	PESTICIDES, WASTE DISPOSAL, PATH	W71-05553
ESMUS, ALGAL TOXINS, ANTIBIOTICS(	PESTICIDES), DATA PROCESSING, COL	W71-05155
L WASTES, MERCURY, COPPER, ZINC,	PESTICIDES, PESTICIDE TOXICITY, *	W72-12576
IVITY, CORAL, REEFS, ANTIBIOTICS(	PESTICIDES), CHLORINATED HYDROCAR	W72-14294
, COLIFORMS, TOXINS, ANTIBIOTICS(	PESTICIDES), ENTERIC BACTERIA, DI	W72-13800
BIOASSAY, *WATER QUALITY, ALGAE,	PESTICIDES, HEAVY METALS, WASTEWA	W72-14282
ICIDES), CHLORINATED HYDROCARBON	PESTICIDES, PESTICIDE TOXICITY, P	W72-14294
EFFECTS, CHLORINATED HYDROCARBON	PESTICIDES, *CYANOPHYTA, SODIUM,	W72-13809
UTANTS, *METALS, TRACE ELEMENTS,	PESTICIDES, SALINITY, NUTRIENTS,	W72-12190
YDROCARBON PESTICIDES, CARBAMATE	PESTICIDES, ORGANIC PESTICIDES, H	W72-13347
ENZYMES, ANALYTICAL TECHNIQUES,	PESTICIDES, FUNGICIDES, ORGANOPHO	W72-13347
ALGAE, *CHLORINATED HYDROCARBON	PESTICIDES, *CYTOLOGICAL STUDIES,	W72-12998
TICIDES, CHLORINATED HYDROCARBON	PESTICIDES, CARBAMATE PESTICIDES,	W72-13347
CHAINS, CHLORINATED HYDROCARBON	PESTICIDES, FISHERIES.: /AE, FOOD	W72-13657
APHY, GROWTH INHIBITI/ MERCURIAL	PESTICIDES, THIN LAYER CHROMATOGR	W72-13347
S, CARBAMATE PESTICIDES, ORGANIC	PESTICIDES, HERBICIDES, WATER POL	W72-13347
S, FUNGICIDES, ORGANOPHOSPHORCUS	PESTICIDES, CHLORINATED HYDROCARB	W72-13347
TS, *PLANT PHYSIOLOGY, *DIATOMS,	PESTICIDES, DDD, PESTICIDE TOXICI	W73-02280
OLOGY, PHOSPHATES, HEAVY METALS,	PESTICIDES, TASTE-PRODUCING ALGAE	W73-02144
ESIDUES, CHLORINATED HYDROCARBON	PESTICIDES, PESTICIDE DRIFT, META	W73-02280
IRATION, CHLORINATED HYDROCARBON	PESTICIDES, GROWTH RATES, ENVIRON	W73-00281
ATE PESTICIDES, ORGANOPHOSPHORUS	PESTICIDES, ACARICIDES, FOOD CHAI	W73-00238
TICIDES, CHLORINATED HYDROCARBON	PESTICIDES, INSECTICIDES, RODENTI	W73-00238
FUNGICIDES, ALGICIDES, CARBAMATE	PESTICIDES, ORGANOPHOSPHORUS PEST	W73-00238
M, BIO/ *CHLORINATED HYDROCARBON	PESTICIDES, SEA WATER, *METABOLIS	W73-00361
YANOPHYTA, CULTURES, ANTIBIOTICS(	PESTICIDES), PHOTOACTIVATION, AQU	W72-14327
*DDT, *RADIOACTIVITY TECHNIQUES,	PESTICIDES, ECOSYSTEMS, POLLUTION	W72-07703
ION, ALGAE, PHENOLS, DETERGENTS,	PESTICIDES, OPERATION AND MAINTEN	W72-08357
IDES, *PESTICIDE TOXICITY, *DDT,	PESTICIDES, PHYTOPLANKTON, DIATOM	W72-08436
RIA, FISH PARASITES, ANTIBIOTICS(	PESTICIDES), BACTERICIDES, ATOLLS	W72-08142
TER POLLUTION EFFECTS, BIOASSAY,	PESTICIDES, HERBICIDES, 2-4-D, PL	W72-08440
PHENYLS, *CHLORNATED HYDROCARBON	PESTICIDES, *PESTICIDE TOXICITY,	W72-08436
FICATION, DETERGENTS, NUTRIENTS,	PESTICIDES, OIL, METALS, THERMAL	W72-08790
H RATES, *CYANOPHYTA, *TOXICITY,	PESTICIDES, PESTICIDE RESIDUES, A	W72-03545
*ALGAE/ *CHLORINATED HYDROCARBON	PESTICIDES, *PESTICIDE RESIDUES,	W72-04134
CHROMATOGRAPHY, CENTRIFUGATION,	PESTICIDES, BIOASSAY, DISTRIBUTIO	W72-04134
ENDRIN/ *CHLORINATED HYDROCARBON	PESTICIDES, *ALDRIN, *DIELDRIN, *	W72-03545
NITRATES, WATER REUSE, SALINITY,	PESTICIDES, LIVESTOCK, WASTE DISP	W72-00846
ION, PHYSICOCHEMICAL PROPERTIES,	PESTICIDES, TROPHIC LEVEL, GASES,	W72-00847
DROCARBON PESTICIDES, *CARBAMATE	PESTICIDES, *ORGANOPHOSPHOROUS PE	W71-10566
E PESTICIDES, *ORGANOPHOSPHOROUS	PESTICIDES, *2-4-D, *DDT, *DIELDR	W71-10566
ESTICIDES, DDT, ORGANOPHOSPHORUS	PESTICIDES, PHYTOPLANKTON, TOXICI	W71-10096
DIELDRIN, OXYGENATION, CARBAMATE	PESTICIDES.: / CULTRUES, ENDRIN,	W71-10096
CTIVITY, CHLORINATED HYDROCARBON	PESTICIDES, DDT, ORGANOPHOSPHORUS	W71-10096
ICIDES, *CHLORINATED HYDROCARBON	PESTICIDES, *CARBAMATE PESTICIDES	W71-10566
ICITY, *WATER POLLUTION EFFECTS,	PESTICIDES, AGRICULTURAL CHEMICAL	W71-12303
ODYNAMICS, SOLUBILITY, SORPTION,	PESTICIDES, TOXICITY, DISSOLVED O	W71-12092
CITY, / *CHLORINATED HYDROCARBON	PESTICIDES, *DDT, *PESTICIDE TOXI	W71-12303
EDIMENTS, SALT MARSHES, ECOLOGY,	PESTICIDES, ESTUARIES, INTERTIDAL	W72-10678
, AQUATIC PLANTS, AQUATIC ALGAE,	PESTICIDES, INSECTICIDES, HERBICI	W72-09646
STEMS, AQUATIC LIFE, ANTIBIOTICS(	PESTICIDES), ECOLOGY, BENTHOS, IN	W72-09668
OMONAS, MALLOMONAS, STAURASTRUM,	PHACOMYXA, SPONDYLOSIUM, LYNGBYA,	W72-07648
ENA SPIROGYRA, LEPOCINCLIS OVUM,	PHACUS CIRCUMFLEXUS, PHACUS ONYX,	W73-00918
NCLIS OVUM, PHACUS CIRCUMFLEXUS,	PHACUS ONYX, PHACUS PLEURONECTES,	W73-00918
HACUS CIRCUMFLEXUS, PHACUS ONYX,	PHACUS PLEURONECTES, PHACUS PYRUM	W73-00918
HACUS ONYX, PHACUS PLEURONECTES,	PHACUS PYRUM, TRACHELOMONAS ARMAT	W73-00918
GONIUM, RHIZOCLONIUM, COSMARIUM,	PHACUS, CHAETOCEROS, MOUGEOTIA, N	W72-09111
RIDES, OLIGOSACCHARIDES, CLONES,	PHAEOCYSTIS POUCHETI, GLUCOSE, MA	W72-07710
E, DENSITY GRADIENTS, TUNICATES,	PHAEOCYSTIS, SORTING, SODIUM BROM	W73-00371
-PLANT RELATIONSHIPS, NORTH SEA,	PHAEOCYSTIS SPP., SOIL CULTURES,	W72-12239
M, LEAD, CHLORELLA PHYRENOIDOSA,	PHAEODACTYLUM TRICORNUTUM, CYCLOT	W72-07660
LAX POLYEDRA, PLATYMONAS VIRIDI/	PHAEODACTYLUM TRICORNUTUM, GONYAU	W72-07166
RBON-14, DUNALIELLA TERTIOLECTA,	PHAEODACTYLUM TRICORNUTUM.: /, CA	W72-09102
A STRAIT, DUNALIELLA TERIOLECTA,	PHAEODACTYLUM TRICORNUTUM, SKELET	W72-09103
MERCURIC ACETATE, MERCURY CHLOR/	PHAEODACTYLUM TRICORNUTUM, PHENYL	W72-11727
COCCOLITHUS HUXLEYI, AUTOLYSIS,	PHAEODACTYLUM.: /CUS, RHODOMONAS,	W71-10083
ASTS, BENTHIC FLORA, RHODOPHYTA,	PHAEOPHYTA, CHLOROPHYTA, MAINE, S	W71-10066
, WATER TEMPERATURE, RHODOPHYTA,	PHAEOPHYTA, CHLOROPHYTA, COOLING	W72-11252
ENT, TURBIDITY, WATER POLLUTION,	PHAEOPHYTA, MOLLUSKS, MARINE ALGA	W72-11925
LGAE, *CYANOPHYTA, *CHLOROPHYTA,	PHAEOPHYTA, RHODOPHYTA, CLADOPHOR	W72-11800
ARTIFICIAL), ALGAE, CHLOROPHYTA,	PHAEOPHYTA, RHODOPHYTA.: /LLITES(	W72-09120
IES, PHYTOPLANKTON, CHLOROPHYTA,	PHAEOPHYTA, RHODOPHYTA, DIATOMS,	W72-03220
RING, *MARINE ALGAE, RHODOPHYTA,	PHAEOPHYTA, *COBALT RADIOISOTOPES	W72-07826
AMINO ACIDS, GAS CHROMATOGRAPHY,	PHAEOPHYTA, EUTROPHICATION, PLANK	W72-07710
R, MORTALITY, WATER TEMPERATURE,	PHAEOPHYTA, SAMPLING, SCUBA DIVIN	W73-00366
E ANIMALS, MOLLUSKS, GASTROPODS,	PHAEOPHYTA.: /TY, COPEPODS, MARIN	W73-00371
TRIBUTION, GEOGRAPHICAL REGIONS,	PHAEOPHYTA, MARINE ALGAE, ENVIRON	W73-00428
YSTEMATICS, *ALGAE, CHRYSOPHYTA,	PHAEOPHYTA.: /LOGICAL STUDIES, *S	W72-14808
ANNELIDS, MOLLUSKS, RHODOPHYTA,	PHAEOPHYTA, PHOTOGRAPHY, SCUBA DI	W73-00932
EFS, LABORATORY TESTS, SAMPLING,	PHAEOPHYTA.: /NMENTAL EFFECTS, RE	W73-00838
TICS, *REPRODUCTION, RHODOPHYTA,	PHAEOPHYTA, WATER QUALITY, *MARIN	W73-01449
PILLS, SEA WATER, MARINE PLANTS,	PHAEOPHYTA, OILY WATER, EVAPORATI	W73-01074
ANKTON, WATER POLLUTION SOURCES,	PHAEOPHYTA, OXIDATION.: / PHYTOPL	W72-12166
*KELPS, *PHYSIOLOGICAL ECOLOGY,	PHAEOPHYTA, STANDING CROPS, RESIS	W72-12940

PLANTS, TOXINS, RED TIDE, PLANT PHYSIOLOGY, BIOASSAY, CYTOLOGICAL W73-00243
RADIOACTIVITY, METABOLISM, PLANT PHYSIOLOGY, CHRYSOPHYTA, CULTURES W73-00380
Y, PHOTOSYNTHESIS, RIVERS, PLANT PHYSIOLOGY, TESTING PROCEDURES, A W72-14314
YDRATES, *ORGANICS ACIDS, *PLANT PHYSIOLOGY, *GROWTH RATES, PLANT W72-07710
TON, SEWAGE TREATMENT, NITROGEN, PHYSOCHEMICAL PROPERTIES, DIATOMS W72-13662
, DESCHAMPSIA FLEXUOSA, PARMELIA PHYSODES, HYPNUM CUPRESSIFORME.: / W72-14303
TION, INHIBITORS, PHYTOPLANKTON, PHYTO-TOXICITY, HERBICIDES, INSEC W71-10566
ICS, PHARMACOLOGICAL PROPERTIES, PHYTOCHEMICAL SCREENINGS, NYMPHAE W72-04275
LOSIRA, TABELLARIA, ACTINASTRUM, PHYTOCONIS, CYMBELLA, STEPHANODIS W72-10875
CAL ANALYSIS, SAMPLING, ECOLOGY, PHYTOP: /WATER TEMPERATURE, CHEMI W72-07710
EAN, ECOLOGY, WATER TEMPERATURE, PHYTOPLANKTON, ZOOPLANKTON, CHLOR W72-07892
BACTERIA, SAMPLING, ZOOPLANKTON, PHYTOPLANKTON, WATER ANALYSIS, PO W72-07896
NDS, SHELLFISH, INSTRUMENTATION, PHYTOPLANKTON, AUTOMATION, EQUIPM W72-07715
ON, *AMINO ACIDS, *MARINE ALGAE, PHYTOPLANKTON, NITROGEN, ABSORPTI W72-07497
P, CARBON, NITROGEN, PHOSPHORUS, PHYTOPLANKTON, CARBON DIOXIDE, SE W72-07648
EWAGE TREATMENT PLANTS, DIATOMS, PHYTOPLANKTON, HYDORGEN ION TEMPE W72-06837
PHICATION, ALGAE, AQUATIC WEEDS, PHYTOPLANKTON, OLIGOTROPHY, HARDN W72-08459
CIDE TOXICITY, *DDT, PESTICIDES, PHYTOPLANKTON, DIATOMS, ALGAE, MA W72-08436
UTRIENTS, BACTERIA, CHRYSOPHYTA, PHYTOPLANKTON, ZOOPLANKTON, PYRRO W72-07899
NIQUES, MICHIGAN, SESTON, ALGAE, PHYTOPLANKTON, NANNOPLANKTON, AMM W72-08048
TS, *NUISANCE ALGAE, WASHINGTON, PHYTOPLANKTON, CYANOPHYTA, NITRAT W72-08049
*ALGAE, MEASUREMENT, CYANOPHYTA, PHYTOPLANKTON, PHOTOSYNTHESIS, PR W72-08054
TINENTAL SHELF, DEPTH, NITROGEN, PHYTOPLANKTON, HYDROGRAPHY, NITRI W72-08056
EMENTS, CYANOPHYTA, CHLOROPHYTA, PHYTOPLANKTON, GRAZING.: / REQUIR W72-07940
A, *NEW HAMPSHIRE, MARINE ALGAE, PHYTOPLANKTON, AQUATIC PLANTS, CH W72-14542
ATLANTIC OCEAN, *GULF OF MEXICO, PHYTOPLANKTON, CYTOLOGICAL STUDIE W72-14313
F POLLUTANTS, ORGANIC COMPOUNDS, PHYTOPLANKTON, CARBON, WATER TEMP W73-00384
OZOA, MARINE ALGAE, ZOOPLANKTON, PHYTOPLANKTON.: /PYRROPHYTA, PROT W73-00296
TH RATES, ENVIRONMENTAL EFFECTS, PHYTOPLANKTON, CHLAMYDOMONAS, CHL W73-00281
AGE, NITROGEN, PHOSPHORUS, IRON, PHYTOPLANKTON, DIATOMS, DINOFLAGE W73-00234
*ALGAE, *BIOMASS, *CYANOPHYTA, PHYTOPLANKTON, PRODUCTIVITY, PHOT W72-14812
GHT PENETRATION, PHOTOSYNTHESIS, PHYTOPLANKTON, STRATIFICATION, DI W73-00236
IN, LAKES, CLADOPHORA, ANALYSIS, PHYTOPLANKTON, CHEMICAL ANALYSIS, W73-00232
URCES, POLLUTANT IDENTIFICATION, PHYTOPLANKTON, MARINE ALGAE, ESSE W73-00441
PATH OF POLLUTANTS, FOOD CHAINS, PHYTOPLANKTON, MARINE ALGAE, ENVI W73-00818
NDS, FOOD CHAINS, PUBLIC HEALTH, PHYTOPLANKTON: /N, ORGANIC COMPOU W73-00821
IC ALGAE, CHRYSOPHYTA, PLANKTON, PHYTOPLANKTON.: /SM, ALGAE, AQUAT W73-02280
LIGHT INTENSITY, EUTROPHICATION, PHYTOPLANKTON, PRODUCTIVITY, ANAB W73-02472
ICAL ANALYSIS, LEAVES, SAMPLING, PHYTOPLANKTON, IRON, MAN: /, CHEM W73-01089
LANKTON, BENTHIC FLORA, DIATOMS, PHYTOPLANKTON, PHOTOSYNTHESIS, PR W73-01615
RATURE, SAMPLING, AQUATIC ALGAE, PHYTOPLANKTON, TROPHIC LEVEL: /PE W73-01456
, NITROGEN, SILICON, ABSORPTION, PHYTOPLANKTON.: /ION, METHODOLOGY W73-01445
MICROSCOPY, DIATOMS, NUTRIENTS, PHYTOPLANKTON, WATER QUALITY, MIC W72-10875
, MICROBIOLOGY, WATER TREATMENT, PHYTOPLANKTON, FUNGI, BACTERIA, A W72-10076
ING RATES, *LARVAL GROWTH STAGE, PHYTOPLANKTON, NUTRIENT REQUIREME W72-09248
*AQUATIC PLANTS, AQUATIC ALGAE, PHYTOPLANKTON, AQUATIC MICROORGAN W72-09650
ECOLOGY, BENTHOS, INVERTEBRATES, PHYTOPLANKTON, PLANKTON, WATER TE W72-09668
CONDARY TREATMENT, MARINE ALGAE, PHYTOPLANKTON, BIOMASS, GROWTH RA W72-09378
SEDIMENTS, AQUATIC PRODUCTIVITY, PHYTOPLANKTON, ALGAE.: /ERFACES, W72-08989
Y, SEASONAL, CHLOROPHYLL, ALGAE, PHYTOPLANKTON, WATER ANALYSIS, CA W72-09108
*PHOTOSYNTHESIS, CARBON DIOXIDE, PHYTOPLANKTON, ALGAE, CULTURES, * W72-09102
ALGAE, *CYANOPHYTA, *BEGGIATOA, PHYTOPLANKTON, *NEW HAMPSHIRE, MI W72-09061
ULATORS, INHIBITION, INHIBITORS, PHYTOPLANKTON, PHYTO-TOXICITY, HE W71-10566
TION, SEA WATER, PHOTOSYNTHESIS, PHYTOPLANKTON, AMMONIUM COMPOUNDS W71-11008
DT, ORGANOPHOSPHORUS PESTICIDES, PHYTOPLANKTON, TOXICITY, DISSOLVE W71-10096
GELLATES, SEASONAL, ZOOPLANKTON, PHYTOPLANKTON.: /NOPHYTA, DINOFLO W71-11949
NE ALGAE, *LIGHT, *GROWTH RATES, PHYTOPLANKTON, NATURAL RESOURCES, W71-13252
GROWTH RATES, ALGAE, INCUBATION, PHYTOPLANKTON.: /ATLANTIC OCEAN, W71-11562
AR POWER PLANTS, STEAM, ECOLOGY, PHYTOPLANKTON, ZOOPLANKTON, LARVA W71-11517
PESTICIDE RESIDUES, ALGAE, DDT, PHYTOPLANKTON, MARINE ALGAE, AQUA W72-03545
ATIC POPULATIONS, MAPS, MAPPING, PHYTOPLANKTON, PACIFIC OCEAN, BOT W72-03567
AQUATIC LIFE, FISH POPULATIONS, PHYTOPLANKTON, BENTHIC FAUNA, PER W72-00939
NIC MATTER, BIOMASS, METABOLISM, PHYTOPLANKTON, PLANT PHYSIOLOGY, W72-03216
IATION, DISSOLVED OXYGEN, LAKES, PHYTOPLANKTON, NUTRIENTS.: /T RAD W72-01780
, ENGINEERING, WATER PROPERTIES, PHYTOPLANKTON, CHLOROPHYTA, PHAEO W72-03220
EN, EUTROPHICATION, OLIGOTROPHY, PHYTOPLANKTON, ANALYTICAL TECHNIQ W72-03224
ES, FOREST SOILS, SAMPLING, MUD, PHYTOPLANKTON, FLUORESCENCE, PHOT W72-01784
CKS, CRABS, OTTERS, ZOOPLANKTON, PHYTOPLANKTON, MAMMALS, MARINE AL W72-00960
ION, DECOMPOSING ORGANIC MATTER, PHYTOPLANKTON, PRODUCTIVITY, IRON W72-01097
SEASONAL, IONS, ALGAE, SALINITY, PHYTOPLANKTON, THERMAL PROPERTIES W72-01363
TROPHICATION, ALGAE, CYANOPHYTA, PHYTOPLANKTON, LAKE MORPHOMETRY, W72-04269
GHT INTENSITY, DISSOLVED SOLIDS, PHYTOPLANKTON, ALGAE, NUTRIENTS.: W72-04270
IR OPERATION, WATER TEMPERATURE, PHYTOPLANKTON, EUTROPHICATION.: / W72-05282
MOSSES, INVERTEBRATES, DIATOMS, PHYTOPLANKTON, PHOTOSYNTHESIS, WA W72-05952
N EFFECTS, *PESTICIDES, DIATOMS, PHYTOPLANKTON, TOXICITY, CULTURES W72-05614
EMPERATURE, LIGHT, WAVES(WATER), PHYTOPLANKTON: /MPLING, OXYGEN, T W72-06283
TIVITY, *SECONDARY PRODUCTIVITY, PHYTOPLANKTON, ZOOPLANKTON, EUTRO W72-05968
MA, CYANOPHYTA, CHLORELLA, IONS, PHYTOPLANKTON, NITRATES, NITROGEN W72-04773
UNDMENT, BIOMASS, AQUATIC ALGAE, PHYTOPLANKTON, ZOOPLANKTON, FISH W72-04853
*ALGAE, *MARINE ALGAE, CULTURES, PHYTOPLANKTON, CYTOLOGICAL STUDIE W70-07280
L, *PHOTOTROPHIC GROWTH, *MARINE PHYTOPLANKTON(PRYMNESIUM PARVUM)(W70-07842
WATER POLLUTION EFFECTS, *ALGAE, PHYTOPLANKTON, DIATOMS, CHRYSOPHY W70-07842
ORELLA, ROTIFERS, CHLAMYDOMONAS, PHYTOPLANKTON, ZOOPLANKTON, PERIP W70-06655
UALITY, *LAKES, *EUTROPHICATION, PHYTOPLANKTON, PLANKTON, PRODUCTI W71-02880
AE, NITROGEN FIXATION, PLANKTON, PHYTOPLANKTON, BACTERIA, ZOOPLANK W71-00117
TION, AMMONIA, NITRITES, OXYGEN, PHYTOPLANKTON, BIOCHEMICAL OXYGEN W71-05390
HLORELLA, GROWTH RATES, DIATOMS, PHYTOPLANKTON, TOXICITY, IRON, LI W71-03027
, NUTRIENTS, CYANOPHYTA, MIXING, PHYTOPLANKTON.: /DISSOLVED OXYGEN W71-03034
RGANIC COMPOUNDS, *MARINE ALGAE, PHYTOPLANKTON, HUMIC ACIDS, FULVI W71-09188
CATION, RED TIDE, SILICA, ALGAE, PHYTOPLANKTON, FREEZING, TEMPERAT W71-09190

TTORAL, PLANT GROWTH SUBSTANCES,	PHYTOPLANKTON, FLUID MECHANICS, E	W71-07878
NA, NUTRIENTS, DISSOLVED SOLIDS,	PHYTOPLANKON, WATER POLLUTION EFF	W72-13300
*BOTTOM SEDIMENTS, *HYPOLIMNION,	PHYTOPLANKTON, LIGHT PENETRATION,	W72-12995
OGICAL STUDIES, ABSORPTION, DDT,	PHYTOPLANKTON, PLANT GROWTH, CHLO	W72-12998
ENTS, WATER QUALITY, PHOSPHORUS,	PHYTOPLANKTON, SEWAGE TREATMENT,	W72-13662
PROPERTIES, PHYSICAL PROPERTIES,	PHYTOPLANKTON, ZOOPLANKTON, CHLOR	W72-12729
OMETRY, CHLOROPHYLL, HYDROLYSIS,	PHYTOPLANKTON, FLUORESCENCE, PHOS	W72-12637
COMMUNITIES, ALGAE, PERIPHYTON,	PHYTOPLANKTON, PHOTOSYNTHESIS, IN	W72-12549
SIS, EXUDATION, WATER POLLUTION,	PHYTOPLANKTON, WATER POLLUTION SO	W72-12166
AL TREATMENT, OXIDATION LAGOONS,	PHYTOPLANKTON, AQUATIC I: /OLOGIC	W72-12567
ING, DINOFLAGELLATES, BEGGIATOA,	PHYTOPLANKTON BLOOMS, PHYSIOLOGIC	W72-14056
IOACCUMULATION, METAL COMPLEXES,	PICEA ABIES, VACCINIUM VITIS IDAE	W72-14303
UMPKINSEED, BLACK CRAPPIE, CHAIN	PICKEREL, TENDIPES TENTANS, TRICH	W72-14280
,6-TR/ TORDON, FRESHWATER ALGAE,	PICLORAM, PEDIASTRUM, 4-AMINO-3,5	W72-08440
), SABLE ISLAND(NOVA SCOTIA), ST	PIERRE(CANADA), MIQUELON(CANADA),	W71-10066
YLL A, / MARGALEF PIGMENT RATIO,	PIGMENT DIVERSITY INDEX, CHLOROPH	W73-00916
INDEX, CHLOROPHYLL A, / MARGALEF	PIGMENT RATIO, PIGMENT DIVERSITY	W73-00916
NTS, RADIOISOTOPES, CHLOROPHYLL,	PIGMENT, MAGNESIUM, PHOSPHATES, N	W72-07660
HOTOSYNTHESIS, PLANT PHYSIOLOGY,	PIGMENTS.: /TERNS, MEASUREMENT, P	W71-13252
AS, WATER ANALYSIS, FLUOROMETRY,	PIGMENTS.: /OMPOUNDS, CHLAMYDOMON	W72-12637
PRIMARY PR/ *CHLOROPHYLL, *PLANT	PIGMENTS, AQUATIC PRODUCTIVITY, *	W72-12192
SPECTROPHOTOMETRY, LAKES, PLANT	PIGMENTS, DETRITUS, CHLOROPHYTA,	W72-13673
C BACTERIA, ELECTRON MICROSCOPY,	PIGMENTS, CARBOHYDRATES, PROTEINS	W72-07951
YLL, SOLAR RADIATION, VITAMIN B,	PIGMENTS, GROWTH RATES, ALGAE, FO	W72-07892
IN, MICHIGAN, CHRYSOPHYTA, PLANT	PIGMENTS, AQUATIC ALGAE, NITROGEN	W73-00384
CTANTS, WATER TEMPERATURE, PLANT	PIGMENTS, PLANT PHYSIOLOGY, *PLAN	W70-09429
LGAE, CONDUCTIVITY, CHLOROPHYLL,	PIGMENTS, DISPERSION, RUNOFF.: /A	W72-04773
MON, SHEEPSHEAD, CHUBS, NORTHERN	PIKE, GIZZARD SHAD.: /S, COHO SAL	W72-14282
CKERS, CATFISHES, BASS, PERCHES,	PIKES, SAMPLING, SNAILS, ALGAE, D	W70-08098
CTILUCA, SIPHONOPHORES, MEDUSAE,	PILIDIUM, CLADOCERA, LINGULA, ACT	W73-00935
LISM, BIOCHEMICAL OXYGEN DEMAND,	PILOT PLANTS, *WASTE WATER TREATM	W72-06838
EASIBILITY, COSTS, ANNUAL COSTS,	PILOT PLANTS.: /TES, ECONOMICS, F	W70-10174
ASIBILITY, ECONOMIC FEASIBILITY,	PILOT PLANTS, ION EXCHANGE, ELECT	W71-09450
CONDITIONS, AEROBIC CONDITIONS,	PILOT PLANTS, ALGAE, HYDROGEN ION	W71-07102
RATORY, DIGESTION TANK, PONDAGE,	PILOT PLANT, SETTLING BASIN,: /BO	W72-02850
RASS, ZOSTERA, PINAEUS DUORARUM,	PINAEUS AZTECUS, PENAEUS SETIFERU	W73-02029
W YORK BIGHT, EELGRASS, ZOSTERA,	PINAEUS DUORARUM, PINAEUS AZTECUS	W73-02029
DOMONAS REINHARDTII, MOSES LAKE,	PINE LAKE, LAKE WASHINGTON, CHLOR	W72-12637
AIR, DETROIT RIVER, BLACK RIVER,	PINE RIVER, CLINTON RIVER, HURON	W72-14282
S, AROCLOR 1254, ESCAMBIA RIVER,	PINFISH, FLOUNDER, CROAKER, MENHA	W71-07412
HIA ACTINASTROIDES, *STAURASTRUM	PINGUE, SCENEDESMUS QUADRICAUDA,	W72-12739
HAETOCEROS, MOUGEOTIA, NAVICULA,	PINNULARIA, SURIRELLA, PERIDINIUM	W72-09111
N HOT SPRINGS, LOLO HOT SPRINGS,	PIPESTONE HOT SPRINGS, SLEEPING C	W72-05610
YAMAMONAS TERRESTRIS, BRASILOBIA	PISCIFORMIS, PROWSEMONAS TROPICA,	W72-07683
TIA, ANEMONE, MOLGULA, TUNICATE,	PISMO CLAM, TIVELA STULTORUM, OST	W72-07907
N ION CONCENTRATION, FILTRATION,	PLA: / CALCIUM CARBONATE, HYDROGE	W73-01669
FRAGILARIA CONSTRUENS, COCCONEIS	PLACENTULA, NAVICULA SRYTOCEPHALA	W72-08141
IPHYTES, GROWTH MEDIA, COCCONEIS	PLACENTULA, CLOSTERIUM, GOMPHONEM	W72-12744
, COCCONEIS PEDICULUS, COCCONEIS	PLACENTULA, CYCLOTELLA CF. SOCIAL	W72-13142
RA, THALASSIONEMA NITZSCHIOIDES,	PLAGIOGRAMMA VANHEURCKII, PLEUROS	W73-00853
VICULA SPP, SYNEDRA FASCICULATA,	PLAGIOGRAMMA BROCKMANNI, AMPHIPLE	W73-00853
SOUTH AMERICA, ATLANTIC COASTAL	PLAIN, COASTS, CRUSTACEANS, MARIN	W73-00819
UT, *MOLLUSKS, *ATLANTIC COASTAL	PLAIN, NORTH CAROLINA, ABSORPTION	W73-00818
TTLE CREEK RESERVOIR(KAN), GREAT	PLAINS, AQUATIC ECOSYSTEM.: / *TU	W72-12393
EATION, GROUNDWATER, FISH, FLOOD	PLAINS, IRRIGATION, PESTICIDES.: /	W70-05103
ROPHICATION, *RESERVOIRS, *GREAT	PLAINS, NUTRIENTS, RECREATION, LI	W72-04759
KOLBEANA OVOIDEA, PALMERIAMONAS	PLANCTONICA, MANCHUDINIUM SINICUM	W72-07683
OSCOPY, ECOLOGICAL DISTRIBUTION,	PLANKTON NETS, CHLOROPHYTA, PLANK	W73-01094
PRING, SUMMER, AUTUMN, SAMPLING,	PLANKTON NETS, DIATOMS, FOOD CHAI	W72-12738
ION, PLANKTON NETS, CHLOROPHYTA,	PLANKTON.: / ECOLOGICAL DISTRIBUT	W73-01094
MASS, *CYANOPHYTA, PLANT GROWTH,	PLANKTON.: /NTROL, NUTRIENTS, BIO	W72-00799
OL, *CYANPHYTA, *NUISANCE ALGAE,	PLANKTON, WASHINGTON, DIATOMS, CH	W72-00798
ONTROL, PLANT GROWTH REGULATORS,	PLANKTON, CULTURES, WATER POLLUTI	W71-13793
ANT IDENTIFICATION, / *NITROGEN,	PLANKTON, *EUTROPHICATION, POLLUT	W72-01780
TION, FALLOUT, OYSTERS, MUSSELS,	PLANKTON, MARINE ALG: /NT, REGULA	W72-00959
DIMENT TRANSPORT, BENTHIC FAUNA,	PLANKTON, BIOINDICATORS, CURRENTS	W72-09092
S, INVERTEBRATES, PHYTOPLANKTON,	PLANKTON, WATER TEMPERATURE, E. C	W72-09668
KINETICS, *BIOLOGICAL TREATMENT,	PLANKTON, *EUTROPHICATION, *NUTRI	W71-13553
ONAL, ACTINOMYCETES, CYANOPHYTA,	PLANKTON, ON-SITE TESTS, LABORATO	W71-13187
SH, LAKE ERIE, SEDIMENTS, SILTS,	PLANKTON, ALGAE, TRACE ELEMENTS,	W71-11036
RY, CYANOPHYTA, BACTERIA, ALGAE,	PLANKTON, PLANTS, SPECTROSCOPY, C	W73-01439
R SILTING, LAKES, *SOUTH DAKOTA,	PLANKTON, *ALGAE, *PHOSPHORUS COM	W73-00938
N, *EUTROPHICATION, CHLOROPHYLL,	PLANKTON, BIOLOGICAL COMMUNITIES,	W73-00916
UNITIES, *STREAMS, LAKE ONTARIO,	PLANKTON, DIATOMS, SEDIMENTS, HAR	W73-01627
INFRARED RADIATION, HYDROGRAPHY,	PLANKTON, ALGAE, ENVIRONMENTAL EF	W73-01704
LIOGRAPHIES, *ALGAE, *LAKE ERIE,	PLANKTON, BENTHIC FLORA, DIATOMS,	W73-01615
SH, MICROORGANISMS, ODOR, TASTE,	PLANKTON, CHEMICALS, PHENOLS, CHL	W73-01669
TES, LIMITING FACTORS, NITROGEN,	PLANKTON, DEPTH, ALGAE.: /, NITRA	W73-02472
GAE, AQUATIC ALGAE, CHRYSOPHYTA,	PLANKTON, PHYTOPLANKTON.: /SM, AL	W73-02280
ATER TEMPERATURE, WATER QUALITY,	PLANKTON, ANALYTICAL TECHNIQUES,	W72-07663
PHY, PHAEOPHYTA, EUTROPHICATION,	PLANKTON, ATLANTIC OCEAN, SEA WAT	W72-07710
C ASPECTS, NUTRIENTS, CHEMICALS,	PLANKTON, BENTHOS, ALGAE, FISH, W	W72-07841
, *LIGHT INTENSITY, *PHOSPHORUS,	PLANKTON, CHLOROPHYTA, PYRROPHYTA	W72-07166
RIENTS, TURBIDITY, PRODUCTIVITY,	PLANKTON, STANDING CROP, LIMNOLOG	W72-08559
NT, SAMPLING, CHEMICAL ANALYSIS,	PLANKTON, COMPUTER PROGRAMS, BIOM	W72-08560
MARINE ALGAE, BOTTOM SEDIMENTS,	PLANKTON, RADIOACTIVITY TECHNIQUE	W73-00361
FLORA, ALGAE, LIGHT PENETRATION,	PLANKTON, NUTRIENTS, BIOCHEMICAL	W72-13659
ICIDES, WATER POLLUTION SOURCES,	PLANKTON, OYSTERS, SHEL: /S, HERB	W72-13347
WATER POLLUTION SOURCES, ALGAE,	PLANKTON, SEASONAL.: /POLLUTION),	W72-05506
ER, CALCIUM CARBONATE, DETRITUS,	PLANKTON, ADSORPTION, EUTROPHICAT	W72-05469

NALYTICAL TECHNIQUES, / *AQUATIC PLANTS, *ALGAE, *PUBLIC HEALTH, A W72-04275
PROPERTIES, *PLANKTON, *AQUATIC PLANTS, *AQUATIC ANIMALS, *MODEL W72-14280
GAE, / *ALGAL TOXINS, *POISONOUS PLANTS, *CYANOPHYTA, *NUISANCE AL W72-14056
HERMAL POLLUTION, *NUCLEAR POWER PLANTS, *COOLING WATER, *WATER TE W70-09612
GAE, *ESTUARIES, PRODU/ *AQUATIC PLANTS, *DISTRIBUTION, *MARINE AL W72-04766
ICAL COMMUNITIES, *THERMAL POWER PLANTS, *DISCHARGE(WATER), AQUATI W71-11517
CLEAR POWERPLANTS, *DESALINATION PLANTS, *ELECTRIC POWER PRODUCTIO W72-03327
*WATER QUALITY, *HYDROELECTRIC PLANTS, *MATHEMATICAL MODELS, THE W72-14405
*ENVIRONMENTAL EFFECTS, *MARINE PLANTS, *MARINE ANIMALS, OFFSHORE W73-02029
AGULATION, WATER POLLU/ *AQUATIC PLANTS, *MINNESOTA, *TOXICITY, CO W72-07360
GI, PROTOZCA, PATHOGEN/ *AQUATIC PLANTS, *MINNESOTA, BACTERIA, FUN W72-08586
ION EFFECTS, *INTERTIDA/ *MARINE PLANTS, *OILY WATER, WATER POLLUT W71-08792
OLLUTION CONTROL, *NUCLEAR POWER PLANTS, *POTABLE WATER, *BIOTA, S W71-09018
COMPOUNDS, *PHOSPHORUS, *MARINE PLANTS, *PHYTOPLANKTON, ALGAE, LI W71-10083
ERIA, *AQUATIC ANIMALS, *AQUATIC PLANTS, *RESERVOIR STAGES, EARLY W72-04853
ON, *ENVIRONM/ *NEW YORK, *POWER PLANTS, *STREAM, *THERMAL POLLUTI W72-14659
EWS, WATER TEMPERATURE, / *POWER PLANTS, *THERMAL POLLUTICN, *REVI W71-07417
MAL STRATIFICATION, HYDR/ *POWER PLANTS, *THERMAL POLLUTICN, *THER W70-09614
BIOCHEMICAL OXYGEN DEMAND, PILOT PLANTS, *WASTE WATER TREATMENT, W W72-06838
STRY, *OXIDATION LAGOONS, *PILOT PLANTS, ALGAE, BIOCHEMICAL OXYGEN W70-06619
TIONS, AEROBIC CONDITIONS, PILOT PLANTS, ALGAE, HYDROGEN ION CONCE W71-07102
BIOLOGICAL COMMUNITIES, *AQUATIC PLANTS, ALGAE, BACTERIA, LITTORAL W72-03216
GEN DEMAND, HYPOLIMNION, AQUATIC PLANTS, ALGAE.: /LVED OXYGEN, OXY W72-12994
S, FISH, BIRDS, STREAMS, AQUATIC PLANTS, AQUATIC INSECTS, CRUSTACE W72-12930
S, *PESTICIDE TOXICITY, *AQUATIC PLANTS, AQUATIC ALGAE, PHYTCPLANK W72-09650
ANIMALS, CARP, DAPHNIA, AQUATIC PLANTS, AQUATIC ALGAE, SCENEDESMU W72-09649
AQUATIC MICROORGANISMS, AQUATIC PLANTS, AQUATIC ALGAE, PESTICIDES W72-09646
OTOPES, *MARINE ANIMALS, *MARINE PLANTS, AQUATIC ANIMALS, LABORATO W71-09168
STREAMS, RIVERS, ROOTED AQUATIC PLANTS, AQUATIC ALGAE, WATER TEMP W72-14690
SAY, CYANOPHYTA, RESINS, AQUATIC PLANTS, AQUATIC ALGAE, NITRATES, W73-00286
*INDICATORS, *SEA WATER, MARINE PLANTS, ASSESSMENTS, MONITORING, W71-12867
IBUTICN PATTERNS, LAKES, AQUATIC PLANTS, AUSTRALIA, SAMPLING, MICR W73-01094
, *MARINE MICROORGANISMS, MARINE PLANTS, BACTERIA, POPULATION, HAW W71-08042
GY BUDGET, *MARINE ALGAE, MARINE PLANTS, BIOMASS, CANADA, COASTS, W73-00366
, OLIGO/ *PERIPHYTON, *SUBMERGED PLANTS, BIOMASS, NEMATODES, ALGAE W71-11012
BON DIOXIDE, RESPIRATION, MARINE PLANTS, BICARBONATES, MASS SPECTR W72-12709
E, ENVIRONMENTAL EFFECTS, MARINE PLANTS, CANADA, WASHINGTCN, OREGO W73-00428
NE ALGAE, PHYTOPLANKTON, AQUATIC PLANTS, CHEMICAL PROPERTIES, WATE W72-14542
NG NUTRIENTS, RESPIRATION, MARSH PLANTS, CHEMICAL ANALYSIS, LEAVES W73-01089
OMPOUNDS, POTASH, ALGAE, AQUATIC PLANTS, CHEMICAL ANALYSIS, WATER W71-12252
NOFF, PHOSPHORUS, ALGAE, AQUATIC PLANTS, CLAYS, ODOR, WATER QUALIT W72-04759
SSAY, TOXINS, POISONS, POISONOUS PLANTS, CORAL, INVERTEBRATES, ALG W71-08597
EMNITELLA AMERICANA, TERRESTRIAL PLANTS, COMBUSTION, ISOTOPE RATIO W72-12709
LIMITING FACTORS, ROOTED AQUATIC PLANTS, CYCLING NUTRIENTS, RESPIR W73-01089
WASTES, *MARINE ANIMALS, *MARINE PLANTS, DEPTH, CALIFORNIA, SANDS, W71-01475
BIODEGRADATION, SEWAGE TREATMENT PLANTS, DIATOMS, PHYTOPLANKTON, H W72-06837
ERS, AGRICULTURE, ROOTED AQUATIC PLANTS, DRAINAGE, RUNCFF, FERTILI W71-13172
ISH, HEATED WATER, NUCLEAR POWER PLANTS, ELECTRICAL POWER PLANTS, W72-07907
RATURE, DATA COLLECTIONS, MARINE PLANTS, ENVIRONMENTAL EFFECTS, RE W73-00838
NT FLOW, ALGAE, BACTERIA, MARINE PLANTS, ENGINEERING, WATER PROPER W72-03220
AUNA, PERIPHYTON, AQUATIC ALGAE, PLANTS, ENVIRONMENTAL ENGINEERING W72-00939
MARINE ALGAE, *ISOLATION, MARINE PLANTS, EUTROPHICATICN, CCEANS, S W73-00430
ATER POLLUTION EFFECTS, *AQUATIC PLANTS, HARVESTING, PONDS, WATER W71-01488
R POWER PLANTS, ELECTRICAL POWER PLANTS, HEAT TRANSFER, WATER POLL W72-07907
NE ALGAE, MARINE ANIMALS, MARINE PLANTS, HUMAN POPULATION, RADIOAC W73-00833
SICOCHEMICAL PROPERTIES, AQUATIC PLANTS, INDUSTRIAL: /TRATION, PHY W73-01615
ONS, ALGAE, ECOTYPES, INDUSTRIAL PLANTS, INDUSTRIAL WASTES, OIL WA W71-13746
ITY, ECONOMIC FEASIBILITY, PILOT PLANTS, ION EXCHANGE, ELECTROLYSI W71-09450
*EUTROPHICATION, LAKES, AQUATIC PLANTS, IRON, MICROORGANISMS, COL W72-03742
CTS, METHODOLOGY, ALGAE, AQUATIC PLANTS, ISOTOPE FRACTIONATICN, CA W72-12709
, EFFLUENTS, SEWAGE, DETERGENTS, PLANTS, LA: /IC MATTER, LAKE ERIE W71-11017
BIOLOGICAL COMMUNITIES, AQUATIC PLANTS, LIMITING FACTORS, ROOTED W73-01089
FALLOUT, *MARINE ANIMALS, MARINE PLANTS, MARINE ALGAE, NUCLEAR WAS W73-00828
NE ALGAE, *CARBOHYDRATES, MARINE PLANTS, MARINE MICROORGANISMS, WA W73-02100
HICATION, FERTILIZATION, AQUATIC PLANTS, MATHEMATICAL MODELS, SELF W71-12072
*METABOLISM, *ABSORPTION, MARINE PLANTS, MARINE ALGAE, PHAEPHYTA, W71-09193
LUTION EFFECTS, MOLLUSKS, MARINE PLANTS, MARINE ALGAE, PUBLIC HEAL W72-04461
HEMICAL PROPERTIES, FISH, MARINE PLANTS, MARINE ALGAE, HYDROLYSIS, W72-04463
ITY, CALIFORNIA, TRITIUM, MARINE PLANTS, MONITORING, SAMPLING, WAT W73-00831
*NITROGEN FIXATION, *CYANOPHYTA, PLANTS, NITROGEN, REDUCTION(CHEMI W73-01456
OSP/ *SALT RIVER(KENT), VASCULAR PLANTS, NITRATE NITROGEN, ORTHOPH W72-06526
CHNIQUES, FERTILIZATION, AQUATIC PLANTS, NITROGEN, NUTRIENT REMOVA W73-02478
, MOSQUITOES, BACTERIA, FLOATING PLANTS, ON-SITE INVESTIGATIONS, T W71-01488
S, OIL SPILLS, SEA WATER, MARINE PLANTS, PHAEOPHYTA, OILY WATER, E W73-01074
PIA, CARP, ALGAE, ROOTED AQUATIC PLANTS, PLANKTON, ARKANSAS, INSEC W71-04528
S, ECOLOGY, FISH, ALGAE, AQUATIC PLANTS, PROTOZOA, BACTERIA, STREA W73-01972
REPRODUCTION, SEA WATER, MARINE PLANTS, PRIMARY PRODUCTIVITY, COR W72-14294
, *ECONOMICS, CYANOPHYTA, ALGAE, PLANTS, PROTEINS, NUTRIENTS, BICM W72-03327
IANS, TOXICITY, CHARA, SUBMERGED PLANTS, PROTOZOA, EUGLENA, CHLAMY W72-00155
ACERS, PLANT PHYSIOLOGY, AQUATIC PLANTS, RADIOECOLOGY.: /LANCE, TR W72-06342
NDUSTRIAL WASTES, ALGAE, AQUATIC PLANTS, SEA WATER, TRACE ELEMENTS W72-05952
ULING, *NUISANCE ALGAE, FLOATING PLANTS, SEAWATER, PLANKTON, BIODE W71-09261
HYTA, BACTERIA, ALGAE, PLANKTCN, PLANTS, SPECTROSCOPY, CHRCMATOGRA W73-01439
CTS, HEATED WATER, NUCLEAR POWER PLANTS, STEAM, ECOLOGY, PHYTOPLAN W71-11517
CHNIQUES, *ALGAL TOXINS, AQUATIC PLANTS, TOXINS, RED TIDE, PLANT P W73-00243
OIRS, TERTIARY TREATMENT, ALGAE, PLANTS, TROUT, AQUATIC ENVIRONMEN W72-02412
S, MEASUREMENT, *NEVADA, AQUATIC PLANTS, WATER POLLUTION SOURCES, W73-01673
QUATIC ANIMALS, *BIOTA, *AQUATIC PLANTS, WATER QUALITY, ENVIRONMEN W72-06526
DIGESTION, NEUROANATOMY, OXYGEN PLASMA ASHING, ATOMIC ABSORPTION, W71-11036
S, BRAIN, BLOOD, KIDNEYS, LIVER, PLASMODIUM BERGHEI, BORRELIA KANS W73-01676

CHELATION, FOOD CHAINS, PATH OF	POLLUTANTS, PUBLIC HEALTH, WATER	W73-00802
, *COBALT RADIOISOTOPES, PATH OF	POLLUTANTS, WASTE ASSIMILATIVE CA	W73-00826
, ESTUARINE ENVIRONMENT, PATH OF	POLLUTANTS, FALLOUT, NUCLEAR WAST	W73-00821
HEALTH, NUCLEAR WASTES, PATH OF	POLLUTANTS, DIFFUSION, CYTOLOGICA	W73-00823
, ESTUARINE ENVIRONMENT, PATH OF	POLLUTANTS, FOOD CHAINS, PHYTOPLA	W73-00818
E ALGAE, TRACE ELEMENTS, PATH OF	POLLUTANTS, ABSORPTION, RADIOISOT	W73-00811
ION, BIOLOGICAL SAMPLES, FATE OF	POLLUTANTS, METABOLITES, DUNALIEL	W73-00361
HORUS, NITRATES, SILICA, PATH OF	POLLUTANTS, ORGANIC COMPOUNDS, PH	W73-00384
ACE WATERS, PERSISTENCE, PATH OF	POLLUTANTS, OCEANS, ESTUARIES, CU	W73-00361
THOPHORIDS, FLAGELLATES, FATE OF	POLLUTANTS, STEPHANOPYXIS COSTATA	W73-00380
, ORGANIC NITROGEN, ORG/ FATE OF	POLLUTANTS, SUBSTRATE UTILIZATION	W73-00379
ECOLOGY, STANDING CROPS, PATH OF	POLLUTANTS.: /ING, SCUBA DIVING,	W73-00366
WA/ FRESHWATER, *BORON, *PATH OF	POLLUTANTS, METHODOLOGY, RIVERS,	W73-00286
ION, PARTICULATE MATTER, FATE OF	POLLUTANTS, PAPER CHROMATOGRAPHY,	W73-01066
AINS, AFRICA, EFFLUENTS, PATH OF	POLLUTANTS, NEUTRON ACTIVATION AN	W73-00832
E ALGAE, NUCLEAR WASTES, PATH OF	POLLUTANTS, RADIOECOLOGY, MATHEMA	W73-00828
D CHAINS, PUBLIC HEALTH, PATH OF	POLLUTANTS, RADIOECOLOGY, RADIOAC	W73-00833
*RADIOACTIVITY EFFECT/ *PATH OF	POLLUTANTS, *NUCLEAR POWERPLANTS,	W73-00831
DDD, PESTICIDE TOXICITY, PATH OF	POLLUTANTS, PESTICIDE RESIDUES, C	W73-02280
PACITY, WATER POLLUTION EFFECTS,	POLLUTANTS,: /STE ASSIMILATIVE CA	W72-01685
AE, MATHEMATICAL MODELS, PATH OF	POLLUTANTS, TRACERS, ANIMAL PHYSI	W72-03348
POWERPLANTS, MONITORING, PATH OF	POLLUTANTS, AIR POLLUTION EFFECTS	W72-00959
ITY TECHNIQUES, TRACERS, PATH OF	POLLUTANTS, ON-SITE INVESTIGATION	W72-00949
EFFECTS, SURFACE WATERS, PATH OF	POLLUTANTS, MARINE ALGAE, MARINE	W72-00940
COLORADO, CATTLE, URINE, PATH OF	POLLUTANTS, FARM WASTES.: /RCES,	W71-11496
E, *ALGAE, *FLOODWATER, *PATH OF	POLLUTANTS, DIELDRIN, DDT, ORGANI	W71-11498
E, WASTE WATER DISPOSAL, PATH OF	POLLUTANTS, STORM DRAINS, DETERGE	W71-13466
OTOSYNTHESIS, *PLANT PHYSIOLOGY,	POLLUTANTS, ANALYTICAL TECHNIQUES	W71-10553
SORPTION, BIOINDICATORS, PATH OF	POLLUTANTS, FOOD CHAINS, NUCLEAR	W72-10957
ON, *SNAILS, GASTROPODS, PATH OF	POLLUTANTS, NUCLEAR WASTES, WATER	W72-10978
EFFECTS, NUCLEAR WASTES, PATH OF	POLLUTANTS, MARINE ALGAE, CHROMOS	W72-10955
FLUENTS, NUCLEAR WASTES, PATH OF	POLLUTANTS, ASIA, MARINE ALGAE, S	W72-10958
, *RADIUM, *ABSORPTION, *PATH OF	POLLUTANTS, *WATER POLLUTION, WAT	W72-10686
BIFICIDS, *FOOD CHAINS, *PATH OF	POLLUTANTS, *ABSORPTION, ALGAE, C	W72-09668
ES, PHOSPHORUS, PHENOLS, METALS,	POLLUTANTS, INDUSTRIAL WASTES, CO	W72-09646
IOLOGICAL MAGNIFICATION, FATE OF	POLLUTANTS.: / BIOACCUMULATION, B	W72-11925
FATE OF	POLLUTANTS.:	W72-11702
ABS, ALGAE, INVERTEBRATES, FISH,	POLLUTANTS, TURBIDITY, ORGANIC LO	W71-01475
SURVEYS, INVESTIGATIONS, PATH OF	POLLUTANTS, SAMPLING, WATER ANALY	W70-08249
SURVEYS, INVESTIGATIONS, PATH OF	POLLUTANTS, SAMPLING, WATER ANALY	W70-08248
TICIDES, WASTE DISPOSAL, PATH OF	POLLUTANTS, WATER QUALITY.: / PES	W71-05553
CHING NUT/ *PHOSPHATES, *PATH OF	POLLUTANTS, *EUTROPHICATION, *LEA	W71-04216
R, INDICATORS, DELAWARE, PATH OF	POLLUTANTS, HOSTS, PENNSYLVANIA,	W72-04736
HEMISTRY, RADIOISOTOPES, PATH OF	POLLUTANTS, PHYSIOLOGICAL ECOLOGY	W72-06313
PRODUCTIVITY, DIELDRIN, PATH OF	POLLUTANTS, REVIEWS, FOOD CHAINS,	W72-06340
GY, FISH FOOD ORGANISMS, PATH OF	POLLUTANTS, BIOASSAY, TOXINS, POI	W71-08597
INDUSTRIAL WASTES, FARM WASTES,	POLLUTANTS, FERTILIZERS, NITROGEN	W71-06445
ND, WASTE WATER TREATMENT, WATER	POLLUTION EFFECTS.: / OXYGEN DEMA	W71-07098
N METHODS, COOLING TOWERS, WATER	POLLUTION EFFECTS.: /, APPLICATIO	W71-06189
ON, WASTE WATER TREATMENT, WATER	POLLUTION EFFECTS.: /BIODEGRADATI	W71-07097
PHATES, ORGANIC COMPOUNDS, WATER	POLLUTION CONTROL.: /LATION, PHOS	W71-06247
*WATER POLLUTION EFFECTS, *WATER	POLLUTION CONTROL, *BIODEGRADATIO	W71-08768
*PHOSPHATES, *DETERGENTS, *WATER	POLLUTION EFFECTS, *WATER POLLUTI	W71-08768
ARINE PLANTS, *OILY WATER, WATER	POLLUTION EFFECTS, *INTERTIDAL AR	W71-08792
, WATER POLLUTION EFFECTS, WATER	POLLUTION SOURCES, EUTROPHICATION	W71-09173
GEORGIA, SEWAGE EFFLUENTS, WATER	POLLUTION EFFECTS, WATER POLLUTIO	W71-09173
EFFECTS, */ *OIL, SEEPAGE, WATER	POLLUTION EFFECTS, ENVIRONMENTAL	W71-08793
ION, ECONOMIC FEASIBILITY, WATER	POLLUTION CONTROL, CALIFORNIA, AR	W71-09157
PLANTS, *POTABLE WATER,/ *WATER	POLLUTION CONTROL, *NUCLEAR POWER	W71-09018
, POPULATION, CHLOROPHYLL, WATER	POLLUTION EFFECTS, TEMPERATURE, C	W71-08670
URE, METABOLISM, KINETICS, WATER	POLLUTION EFFECTS, NUCLEAR WASTES	W71-09168
S, SOLUBILITY, ADSORPTION, WATER	POLLUTION EFFECTS, FOOD CHAINS.: /	W71-09182
RADIOISOTOPES, ABSORPTION, WATER	POLLUTION SOURCES.: /PES, COBALT	W71-09018
MINE WATER, *STRIP MINES, *WATER	POLLUTION EFFECTS, *KENTUCKY, SUL	W71-07942
GAE, *DEGRADATION(STREAM), WATER	POLLUTION EFFECTS, WASTE WATER TR	W71-07973
VARIABILITY, PROBABILITY, WATER	POLLUTION EFFECTS, ALGAE, AQUATIC	W71-07337
S, OYSTERS, SHRIMP, CRABS, WATER	POLLUTION SOURCES, SEA WATER, TRO	W71-07412
NTRATION, ALGAE, BACTERIA, WATER	POLLUTION SOURCES.: /EN ION CONCE	W71-07942
XICITY, *CHLORINATED HYD/ *WATER	POLLUTION EFFECTS, *BIOASSAY, *TO	W71-07412
OACTIVITY EFFECTS, *PONDS, WATER	POLLUTION EFFECTS, RADIOISOTOPES,	W71-09256
EWAGE, *ORGANIC COMPOUNDS, WATER	POLLUTION TREATMENT, WATER POLLUT	W71-09674
S, *BAYS, *WATER QUALITY, *WATER	POLLUTION CONTROL, *DIVERSION, SE	W71-09581
, PARTICLE SIZE, COLLOIDS, WATER	POLLUTION EFFECTS.: /YTA, AQUARIA	W71-09193
INDUSTRIAL WASTES, SILTS, WATER	POLLUTION EFFECTS.: /CAD, SEWAGE,	W71-09795
WATER POLLUTION TREATMENT, WATER	POLLUTION SOURCES, ENVIRONMENTAL	W71-09674
NTS, *POLLUTION ABATEMENT, WATER	POLLUTION EFFECTS, NUTRIENTS, PHO	W71-09429
TRIENTS, *ALGAE, *BACTER/ *WATER	POLLUTION EFFECTS, *NITRATES, *NU	W71-09958
S, *EUTROPHICATION, *CHE/ *WATER	POLLUTION EFFECTS, *ORGANIC WASTE	W71-09674
PES, COBALT RADIOISOTOPES, WATER	POLLUTION EFFECTS, FRESH WATER.: /	W71-09250
IND VELOCITY, CONVECTION, WATTER	POLLUTION EFFECTS.: / BUOYANCY, W	W71-09852
*STRONTIUM RADIOISOTOPES, *WATER	POLLUTION EFFECTS, NUCLEAR WASTES	W71-09225
*DETERGENTS, *PHOSPHATES, *WATER	POLLUTION SOURCES, *POLLUTION ABA	W72-06655
*TRITIUM, *NUCLEAR WASTES, WATER	POLLUTION EFFECTS, CATTAILS, AQUA	W72-06342
EFFECTS, ALGAE, NUTRIENTS, WATER	POLLUTION SOURCES, SEWAGE TREATME	W72-06638
TY, WATER QUALITY CONTROL, WATER	POLLUTION EFFECTS.: /, PERMEABILI	W72-06037
, *STREAMS, *ECOSYSTEMS,/ *WATER	POLLUTION EFFECTS, *MODEL STUDIES	W72-06340
ATES, *LOCAL GOVERNMENTS, *WATER	POLLUTION CONTROL, WATER QUALITY	W72-06638
REATMENT, WATER POLLUTION, WATER	POLLUTION EFFECTS, RESEARCH AND D	W72-06655
ORUS, AQUATIC ENVIRONMENT, WATER	POLLUTION EFFECTS, ALGAE, NUTRIEN	W72-06638
, WATER POLLUTION EFFECTS, WATER	POLLUTION SOURCES, ALGAE, NUTRIEN	W72-04280

RECREATION, WATER SUPPLY, WATER POLLUTION EFFECTS, WATER POLLUTIO W72-04280
ROPHICATION, *PHOSPHATES, *WATER POLLUTION CONTROL, *WASTE WATER T W72-04734
, WATER POLLUTION SOURCES, WATER POLLUTION EFFECTS, WATER POL: /ON W72-04279
WATER QUALITY, RESERVOIRS, WATER POLLUTION SOURCES, ALGAE, RIVER F W72-04260
WASTES, *FALLOUT, *OCEANS, WATER POLLUTION EFFECTS, FOOD CHAINES, W72-04463
BSORPTION, NUCLEAR WASTES, WATER POLLUTION EFFECTS, MOLLUSKS, MARI W72-04461
OPERTIES, WATER POLLUTION, WATER POLLUTION SOURCES, WATER POLLUTIO W72-04279
ADO, *RE/ *WATER QUALITY, *WATER POLLUTION EFFECTS, *LAKES, *COLOR W72-04791
OPPER SULFATE, HERBICIDES, WATER POLLUTION EFFECTS, WATER POLLUTIO W72-04759
, WATER POLLUTION SOURCES, WATER POLLUTION CONTROL.: /TION EFFECTS W72-04759
OKLA/ *OXIDATION LAGOONS, *WATER POLLUTION EFFECTS, *FARM WASTES, W72-04773
*WATER POLLUTION CONTROL, WATER POLLUTION EFFECTS, NUTRIENTS, ALG W72-04769
QUALITY, WATER CHEMISTRY, WATER POLLUTION SOURCES, METEOROLOGICAL W72-04853
, WATER POLLUTION EFFECTS, WATER POLLUTION SOURCES, WATER POLLUTIO W72-04759
E, CYANOPHYTA, PHOSPHORUS, WATER ROLLUTION SOURCES, HYDROGEN ION C W72-04769
*EUTROPHICATION, *CARBON, *WATER POLLUTION CONTROL, WATER POLLUTIO W72-04769
ILS, *N/ *REMOTE SENSING, *WATER POLLUTION CONTROL, *SEA WATER, *O W72-04742
S, WASTE WATER(POLLUTION), WATER POLLUTION SOURCES, ALGAE, PLANKTO W72-05506
OLLUTIO/ *EUTROPHICATION, *WATER POLLUTION CONTROL, *WASTE WATER(P W72-05473
, SEWAGE TREATMENT, ALGAE, WATER POLLUTION SOURCES, CHLOROPHYTA, C W72-05473
INATED BIPHENYLS, *ALGAE, *WATER POLLUTION EFFECTS, *PESTICIDES, D W72-05614
HOSPHATES, NITRATES, FILTRATION, POLLUTION ABATEMENT, ALGAE, NUTRI W72-05315
, WATER POLLUTION SOURCES, WATER POLLUTION EFFECTS,: /OTOSYNTHESIS W72-05952
OPLANKTON, PHOTOSYNTHESIS, WATER POLLUTION SOURCES, WATER POLLUTIO W72-05952
G, AGRICULTURAL CHEMICALS, WATER POLLUTION SOURCES, ALGAE, GEOLOGY W72-05869
ON, WASTE WATER TREATMENT, WATER POLLUTION EFFECTS.: / CONCENTRATI W71-05155
, WATER POLLUTION EFFECTS, WATER POLLUTION SOURCES, NITRIFICATION, W71-05390
N SOURCES, INDUSTRIAL WA/ *WATER POLLUTION EFFECTS, WATER POLLUTIO W71-03095
NTS, ALGAE, LAKE SUPERIOR, WATER POLLUTION SOURCES, MINNESOTA, WAS W71-04216
ES DEVELOPME/ *ESTUARIES, *WATER POLLUTION EFFECTS, *WATER RESOURC W71-05553
SSOLVED OXYGEN, ESTUARIES, WATER POLLUTION EFFECTS, WATER POLLUTIO W71-05390
*WATER POLLUTION EFFECTS, WATER POLLUTION SOURCES, INDUSTRIAL WAS W71-03095
TRIENTS, *EUTROPHICATION, *WATER POLLUTION SOURCES, *ESTUARIES, RI W70-06509
OLUMBIA, TID/ *ESTUARIES, *WATER POLLUTION EFFECTS, *DISTRICT OF C W70-08249
OLUMBIA, TID/ *ESTUARIES, *WATER POLLUTION EFFECTS, *DISTRICT OF C W70-08248
CIDES, *ST LAWRENCE RIVER, WATER POLLUTION SOURCES, WATER POLLUTIO W70-08098
LANKTON, / ORGANIC WASTES, WATER POLLUTION EFFECTS, *ALGAE, PHYTOP W70-07842
OPHICATION, WATER QUALITY, WATER POLLUTION CONTROL, SURVEYS, INVES W70-08248
, WATER POLLUTION SOURCES, WATER POLLUTION EFFECTS, MUD, MOLLUSKS, W70-08098
NITRATES, *NUTRIENTS, *E/ *WATER POLLUTION EFFECTS, *PHOSPHATES, * W70-09388
*WATER POLLUTION SOURCES, *WATER POLLUTION EFFECTS, *ORGANIC MATTE W70-09189
WATER POLLUTION TREATMENT, WATER POLLUTION EFFECTS, ENVIRONMENTAL W70-09429
TER, OIL-WATER INTERFACES, WATER POLLUTION TREATMENT, WATER POLLUT W70-09429
*CHEMICAL OXYGEN DEMAND, *WATER POLLUTION SOURCES, *WATER POLLUTI W70-09189
LY, *POLLUTION ABATEMENT, *WATER POLLUTION CONTROL, *WATER COOLING W70-09193
OPHICATION, WATER QUALITY, WATER POLLUTION CONTROL, SURVEYS, INVES W70-08249
ATION, ALGAE, SURFACTANTS, WATER POLLUTION SOURCES, SURFACE WATERS W70-09388
HYTA, ALGAE, SOUTH DAKOTA, WATER POLLUTION EFFECTS, WATER POLLUTIO W71-02880
N EFFEC/ *EUTROPHICATION, *WATER POLLUTION CONTROL, WATER POLLUTIO W71-01488
, WATER YIELD IMPROVEMENT, WATER POLLUTION TREATMENT, ALGAE, POTAB W71-01491
AERATION, EROSION CONTROL, WATER POLLUTION SOURCES, WATER LAW, ALG W71-01192
, WATER POLLUTION EFFECTS, WATER POLLUTION SOURCES, CULTIVATION.: / W71-02880
*WATER POLLUTION CONTROL, WATER POLLUTION EFFECTS, *AQUATIC PLANT W71-01488
ALGAE, *SILICA, *BACTERIA, WATER POLLUTION CONTROL, WASTE WATER TR W71-01899
N, *ALGAL CONTROL, RIVERS, WATER POLLUTION EFFECTS, ALGAE, DIATOMS W71-00099
*UNIVERSITIES, *ABSTRACTS, WATER POLLUTION TREATMENT, INSTRUMENTAT W71-01192
TER ANALYSIS, CHEMCONTROL, WATER POLLUTION EFFECTS, WATER QUALITY, W72-11724
ENTS, *AEROBIC CONDITIONS, WATER POLLUTION CONTROL.: /ENTIAL NUTRI W72-11617
IONS, CYTOLOGICAL STUDIES, WATER POLLUTION EFFECTS.: / FLUORIDES, W72-11572
*EUTROPHICATION, *LAKES, *WATER POLLUTION EFFECTS, RECLAIMED WATE W72-11702
STEMS, *BIOLOGICAL COMMUN/ WATER POLLUTION EFFECTS, ECOLOGY, ECOSY W72-11854
ORUS, ALGAE, PLANT GROWTH, WATER POLLUTION TREATMENT, EVALUATION, W72-11702
S, FOOD CHAINS, FOOD WEBS, WATER POLLUTION EFFECTS, AQUATIC ENVIRO W72-11925
OL/ *ALGAE, *FLOCCULATION, WATER POLLUTION TREATMENT, COLLOIDS, *P W72-11833
, WATER POLLUTION EFFECTS, WATER POLLUTION SOURCES, INHIBITION.: / W72-11727
INHIBITORS, *LETHAL LIMIT, WATER POLLUTION EFFECTS, WATER POLLUTIO W72-11727
TUARIES, *EUTROPHICATION, *WATER POLLUTION CONTROL, MANAGEMENT, AN W72-09171
, *WATER QUALITY CONTROL, *WATER POLLUTION EFFECTS, WATER POLLUTIO W72-09646
OGENIC / *MICROORGANISMS, *WATER POLLUTION CONTROL, *WASTES, *PATH W72-09675
, FISHKILL, NEW HAMPSHIRE, WATER POLLUTION EFFECTS, COPPER SULFATE W72-09304
TOXINS, WATER CIRCULATION, WATER POLLUTION CONTROL, FISHKILL, NEW W72-09304
E PLANNING, AQUATIC WEEDS, WATER POLLUTION SOURCES, MATHEMATICAL M W72-09171
*WATER POLLUTION EFFECTS, WATER POLLUTION SOURCES, AQUATIC LIFE, W72-09646
ES, *PRIMARY PRODUCTIVITY, WATER POLLUTION EFFECTS, ESTUARINE ENVI W72-09653
AZOTOBACTER, PSEUDOMONAS, WATER POLLUTION EFFECTS.: /ES, E. COLI, W72-10861
*ALGAE, SEWAGE TREATMENT, WATER POLLUTION SOURCES, WATER POLLUTIO W72-10625
LUTANTS, *WATER POLLUTION, WATER POLLUTION SOURCES, AQUATIC ALGAE, W72-10686
OLLUTANTS, NUCLEAR WASTES, WATER POLLUTION EFFECTS, FOOD CHAINS, P W72-10978
, WATER POLLUTION SOURCES, WATER POLLUTION EFFECTS, AGRICULTURAL R W72-10625
N, *DETERGE/ *PHOSPHATES, *WATER POLLUTION SOURCES, *EUTROPHICATIO W72-11186
, CRUSTACEANS, MONITORING, WATER POLLUTION EFFECTS, MOLLUSKS.: /GE W72-10955
NUTRIENTS, *PHYTOPLANKTON, WATER POLLUTION EFFECTS, EUTROPHICATION W72-09167
IFOR/ *INDUSTRIAL WASTES, *WATER POLLUTION EFFECTS, EFFLUENTS, CAL W72-09092
IDE, SEASONAL, PHOSPHORUS, WATER POLLUTION SOURCES, BIOCHEMICAL OX W72-09167
PHICATION, WATER SAMPLING, WATER POLLUTION EFFECTS, BIOMASS, NUTRI W72-09111
BIOCHEMICAL OXYGEN DEMAND, WATER POLLUTION CONTROL, NITRATES, EFFL W72-09167
GEN, WISCONSIN, NUTRIENTS, WATER POLLUTION EFFECTS, INTERFACES, DI W71-10645
TS, ANALYTICAL TECHNIQUES, WATER POLLUTION SOURCES, TOXINS, WASTES W71-10553
DT, *DIELDRIN, *DIAZINON, *WATER POLLUTION EFFECTS, *PHOTOSYNTHESI W71-10566
ES, *SALMONIDS, WATER RE/ *WATER POLLUTION SOURCES, *FISH HATCHERI W71-11006

Y, FERTILIZERS, EFFLUENTS, WATER	POLLUTION SOURCES, WATER POLLUTIO	W71-11008
, *BIOASSAY, *INDICATORS, *WATER	POLLUTION EFFECTS, *ALGAE, *MARIN	W71-10553
SURFACE WATERS, *REVIEWS, *WATER	POLLUTION SOURCES, NUTRIENTS, PLA	W71-10580
NIA, TUBIFICIDS, COLORADO, WATER	POLLUTION CONTROL, WASHINGTON, CH	W71-11006
, WATER POLLUTION SOURCES, WATER	POLLUTION EFFECTS, SAMPLING, TEMP	W71-11008
ALS, *ALGAE, *INVERTEBRA/ *WATER	POLLUTION EFFECTS, *PLANTS, *ANIM	W71-13233
GAE, SEWAGE, PRODUCTIVITY, WATER	POLLUTION EFFECTS, PHOSPHCRUS, NI	W71-12857
OPHICATION, *WATER SUPPLY, WATER	POLLUTION EFFECTS, NUTRIENTS, RES	W71-13185
SEA WATER, MARIN/ *ALGAE, *WATER	POLLUTION EFFECTS, *INDICATORS, *	W71-12867
NITRATES, *NEBRASKA, URB/ *WATER	POLLUTION SOURCES, *PHOSPHATES, *	W71-13466
ON/ *POLLUTION ABATEMENT, *WATER	POLLUTION EFFECTS, *WATER POLLUTI	W71-13723
ICATICN, *NITROGEN, ALGAE, WATER	POLLUTICN SOURCES, COLORADO, CATT	W71-11496
N, *DETERGENTS, *PHOSPHA/ *WATER	POLLUTION CONTROL, *EUTROPHICATIO	W71-11516
, ALKYLBENZENE SULFONATES, WATER	POLLUTICN SOURCES, AQUATIC ALGAE,	W71-11516
LLUTION EFFECTS, *DIATOMS, WATER	POLLUTICN SOURCES, INDUSTRIAL WAS	W71-11583
SPHATES, *EUTROPHICATION, *WATER	POLLUTICN CONTROL, SEWAGE EFFLUEN	W71-11507
ATER POLLUTION, STANDARDS, WATER	POLLUTION EFFECTS, ALKYLBENZENE S	W71-11516
R POLLUTION SOURCES, IND/ *WATER	POLLUTION EFFECTS, *DIATCMS, WATE	W71-11583
VY METALS, WASTE DISPOSAL, WATER	POLLUTICN EFFECTS.: / METALS, HEA	W71-11793
N, *WATER POLLUTION SOUR/ *WATER	POLLUTION CONTROL, *EUTROPHICATIO	W71-12091
PLANTS, CHEMICAL ANALYSIS, WATER	POLLUTICN SOURCES.: /AE, AQUATIC	W71-12252
CATION, WATER TEMPERATURE, WATER	POLLUTION EFFECTS, WATER POLLUTIO	W71-12092
, WATER POLLUTION EFFECTS, WATER	POLLUTICN CONTROL, ENERGY, DIFFUS	W71-12092
, WATER POLLUTION SOURCES, WATER	POLLUTICN CONTROL, TROPHIC LEVELS	W71-12072
EFFECTS, OHIO, NUTRIENTS, WATER	POLLUTICN SOURCES, DISSOLVED OXYG	W71-11949
DDT, *PESTICIDE TOXICITY, *WATER	POLLUTICN EFFECTS, PESTICIDES, AG	W71-12303
ATION, *ELECTROCHEMISTRY, *WATER	POLLUTION TREATMENT, ANALYTICAL T	W71-12467
DGE, *NUTRIENTS, NITROGEN, WATER	POLLUTION EFFECTS, WATER POLLUTIO	W71-11972
CONTROL, *EUTROPHICATION, *WATER	POLLUTION SOURCES, POPULATION, IN	W71-12091
, WATER POLLUTION EFFECTS, WATER	POLLUTION SOURCES.: /TS, NITROGEN	W71-11972
SEDIMENTS, LAKE MICHIGAN, WATER	POLLUTICN SOURCES, WATER POLLUTIO	W71-12072
*EUTROPHICATION, *LAKES, *WATER	POLLUTION EFFECTS, OHIO, NUTRIENT	W71-11949
INOFLAGELLATES, *RED TIDE, WATER	POLLUTICN SOURCES.: / DIATOMS, *D	W72-01329
UALITY CONTROL, STANDARDS, WATER	POLLUTION SOURCES, WATER POLLUTIO	W72-01660
S, *NUTRIENTS, *POLLUTAN/ *WATER	POLLUTION EFFECTS, *ORGANIC WASTE	W72-01472
*LAKE ONTARIO, *LAKE ERIE, WATER	POLLUTION EFFECTS, WATER POLLUTIO	W72-01094
, WATER POLLUTION SOURCES, WATER	POLLUTION EFFECTS, RECREATION DEM	W72-01660
ING, SOAPS, AQUATIC ALGAE, WATER	POLLUTION EFFECTS, WATER QUALITY,	W72-01085
RINE ALGAE, *SALT MARSHE/ *WATER	POLLUTION EFFECTS, *PROTOZOA, *MA	W72-00948
WATER	POLLUTICN CONTROL ACT.:	W72-01085
FEDERAL GOV/ *DETERGENTS, *WATER	POLLUTICN CONTROL, *PHOSPHATES, *	W72-01093
CONTROL, *FEDER/ *LAKES, *WATER	POLLUTION CONTROL, *WATER QUALITY	W72-01659
*WATER	POLLUTICN CONTROL ACT.:	W72-01093
TION, LEGAL ASPECTS, REGULATICN,	POLLUTION ABATEMENT, WASTE DISPOS	W72-01085
, WATER POLLUTION EFFECTS, WATER	POLLUTION SOURCES, LIMNOLOGY.: /E	W72-01094
DETERGETNTS, *PHOSPHATES, *WATER	POLLUTICN CONTROL, *WATER POLLUTI	W72-01085
SANITATION, *LE/ *LAKES, *WATER	POLLUTION CONTROL, *ENVIRONMENTAL	W72-01660
*WATER POLLUTION CONTROL, *WATER	POLLUTION SOURCES, EUTROPHICATION	W72-01085
ITORING, PATH OF POLLUTANTS, AIR	POLLUTICN EFFECTS, WATER POLLUTIO	W72-00959
S, WATER POLLUTION EFFECT, WATER	POLLUTION SOURCES, COST COMPARISO	W72-00974
DETERGENTS, *LEGISLATION/ *WATER	POLLUTION CONTROL, *PHOSPHATES, *	W72-01685
CHAINS, MARINE FISHERIES, LAKES,	POLLUTION ABATEMENT.: /GAE, FCOD	W72-00960
TS, AIR POLLUTION EFFECTS, WATER	POLLUTION EFFECTS, BAYS, WATERSHE	W72-00959
PHYTOPLANKTON, *NUTRIENTS, WATER	POLLUTION EFFECTS, AQUATIC ALGAE,	W72-01329
VERNMENTS, WASTE DISPOSAL, WATER	POLLUTICN SOURCES, SOAPS, TOXICIT	W72-01093
AE, BIOLOGICAL PROPERTIES, WATER	POLLUTION EFFECT, WATER POLLUTION	W72-00974
OGS, *TRITIUM, ABSORPTION, WATER	POLLUTION EFFECTS, RADIOACTIVITY	W72-03348
EF/ *NUCLEAR EXPLOSIONS, *WATER	POLLUTION EFFECTS, *RADIOACTIVITY	W72-03347
A, PONDS, WELLS, FISH, BAYS, AIR	POLLUTION EFFECTS, AIR PCLLUTION,	W72-03347
OOD CHAINS, PUBLIC HEALTH, WATER	POLLUTION SOURCES.: /ALIFORNIA, F	W72-03324
, WATER POLLUTION SOURCES, WATER	POLLUTION EFFECTS, FLUCTUATION, C	W72-03228
*WATER	POLLUTION CONTROL ACT.:	W72-01685
EATM/ AGRICULTURAL WASTES, WATER	POLLUTICN CONTROL, *BIOLCGICAL TR	W72-02975
BIORTHYMS, ON-SITE TESTS, WATER	POLLUTION SOURCES, WATER POLLUTIO	W72-03228
, WATER POLLUTION SOURCES, WATER	POLLUTION TREATMENT, NUTRIENTS, O	W72-01685
FFECTS, *ALGAE, *PLANKTON, WATER	POLLUTICN CONTROL, METHCDOLOGY, M	W72-03218
STE ASSIMILATIVE CAPACITY, WATER	POLLUTION EFFECTS, POLLUTANTS,: /	W72-01685
HICATION, WATER POLLUTION, WATER	POLLUTION SOURCES, WATER POLLUTIO	W72-01685
ATION, *LABORATORY TESTS, *WATER	POLLUTION EFFECTS, *ALGAE, *PLANK	W72-03218
NVIRCNMENT, *AGRICULTURE, *WATER	POLLUTION CONTROL, SEDIMENTS, FAR	W72-00846
*WATER POLLUTION EFFECTS, *WATER	POLLUTION SOURCES, *FISHERIES, OI	W71-13723
TMENT, TERTIARY TREATMENT, WATER	POLLUTICN CONTROL, BIODEGRADATION	W71-13939
S, ALGAL CCNTROL, *POLLUT/ WATER	POLLUTION EFFECTS, *MICROCRGANISM	W72-00801
N, LAKES, *NUTRIENTS, ALG/ WATER	POLLUTION EFFECTS, *EUTRCPHICATIO	W72-00798
RADIOISOTOPES, HEAVY METALS, AIR	POLLUTION EFFECTS, SURFACE WATERS	W72-00940
E, *EUTR/ *WASTE DILUTICN, WATER	POLLUTICN CONTROL, *NUISANCE ALGA	W72-00799
TS, ANALYTICAL TECHNIQUES, WATER	POLLUTION EFFECTS, POLLUTANT IDEN	W71-13794
OLOGY, RADIOECOLOGY, MONITORING,	POLLUTICN ABATEMENT, ENVIRONMENTA	W72-00939
ATORS, PLANKTON, CULTURES, WATER	POLLUTION EFFECTS.: /GRCWTH REGUL	W71-13793
, MARINE FISH, ABSORPTION, WATER	POLLUTION SOURCES, RADIOACTIVITY	W72-00940
ONS, ALGAE, ECOTYPES, IN/ *WATER	POLLUTION EFFECTS, *DATA COLLECTI	W71-13746
PTICN, PESTICIDE RESIDUES, WATER	POLLUTION EFFECTS, *CALIFCRNIA, S	W73-02105
TURBIDITY, BENTHIC FAUNA, WATER	POLLUTION EFFECTS, STRIPED BASS,	W73-02029
E KINETICS, *PERSISTENCE, *WATER	POLLUTION EFFECTS, *PLANT PHYSIOL	W73-02280
EUTROPHICATION, CHEMICAL/ *WATER	POLLUTICN CONTROL, *PHOSPHORUS, *	W73-02478
NALYSIS, OZONE, DENITRIFICATION,	POLLUTICN ABATEMENT, WATER PCLLUT	W73-02187
TION, POLLUTION ABATEMENT, WATER	POLLUTION CONTROL, COAGULATION.: /	W73-02187
CATION, CHEMICAL ANALYSIS, WATER	POLLUTICN SOURCES, BIOASSAY, ALGA	W73-02478
TH,/ *WATER PURIFICATION, *WATER	POLLUTION TREATMENT, *PUBLIC HEAL	W73-02144
ES, ESTUARINE ENVIRONMENT, WATER	POLLUTION SOURCES, BIODEGRADATION	W73-00853

MS, BIODEGRADATION, ALGAE, WATER	POLLUTION EFFECTS, HEAVY METALS,	W73-00840
TROPHICATION, *NUTRIENTS, *WATER	POLLUTION CONTROL, ORGANIC WASTES	W73-00840
, ECOLOGICAL DISTRIBUTION, WATER	POLLUTION EFFECTS, F: /N PATTERNS	W73-00918
REACTIONS, CARBON DIOXIDE, WATER	POLLUTION SOURCES, OIL SPILLS, SE	W73-01074
ES, CRUSTACEANS, ANNELIDS, WATER	POLLUTION EFFECTS, STANDING CROPS	W73-00935
ORGANIC MATTER, NITROGEN, WATER	POLLUTION SOURCES, ENVIRONMENTAL	W73-01066
TA, ANALYTICAL TECHNIQUES, WATER	POLLUTION EFFECTS, NUTRIENT REQUI	W73-01269
S, *OIL P/ *MARINE ALGAE, *WATER	POLLUTION EFFECTS, *PHOTOSYNTHESI	W73-01074
RATES, DEFICIENT ELEMENTS, WATER	POLLUTION EFFECTS, AQUATIC ALGAE,	W73-01445
ORY EQUIPMENT, FRESHWATER, WATER	POLLUTION SOURCES, INSTRUMENTATIO	W73-01445
, *NEVADA, AQUATIC PLANTS, WATER	POLLUTION SOURCES, SEDIMENTS, AQU	W73-01673
TEMS, *RIVERS, *VIRGINIA, *WATER	POLLUTION EFFECTS, SURVEYS, SAMPL	W73-01972
HERMAL POLLUTION, *COASTS, WATER	POLLUTION EFFECTS, AQUATIC LIFE,	W73-01632
PHIES, *DATA COLLECTIONS, *WATER	POLLUTION EFFECTS, DOCUMENTATION,	W73-01976
ILITIES/ *WATER TREATMENT, WATER	POLLUTION EFFECTS, *TREATMENT FAC	W73-01669
HLAMYDOMONAS, CHLOROPHYTA, WATER	POLLUTION EFFECTS, ABSORPTION, PE	W73-00281
, WATER POLLUTION SOURCES, WATER	POLLUTION EFFECTS, MERCURY, SEWAG	W72-14782
, ACARICIDES, FOOD CHAINS, WATER	POLLUTION SOURCES.: /S PESTICIDES	W73-00238
, MESOTROPHY, OLIGOTROPHY, WATER	POLLUTION SOURCES, WATER POLLUTIO	W72-14782
ANTS, METHODOLOGY, RIVERS, WATER	POLLUTION SOURCES, ION EXCHANGE,	W73-00286
), *WASTE WATER DISPOSAL, *WATER	POLLUTION CONTROL, ECONOMIC FEASI	W72-14878
SAMPLING, ORGANIC MATTER, WATER	POLLUTION EFFECTS, SEWAGE, NITROG	W73-00234
N, CARBON, ORGANIC MATTER, WATER	POLLUTION SOURCES, CHLOROPHYTA, A	W73-00379
S, GROWTH RATES, NITROGEN, WATER	POLLUTION EFFECTS, RADIOACTIVITY,	W73-00380
SITE TESTS, INVERTEBRATES, WATER	POLLUTION EFFECTS, ECOLOGY, SEA W	W73-00348
XINS, MERCURY, DETERGENTS, WATER	POLLUTION SOURCES, ASBESTOS, EUTR	W73-00703
CHRYSOPHYTA, COMPETITION, WATER	POLLUTION SOURCES, POLLUTANT IDEN	W73-00441
ROPHICATION, MARINE ALGAE, WATER	POLLUTION SOURCES, SPATIAL DISTRI	W73-00431
N MICROSCOPY, PHOTOGRAPHY, WATER	POLLUTION EFFECTS, WATER PURIFICA	W73-00432
POLLUTANTS, PUBLIC HEALTH, WATER	POLLUTION EFFECTS, RADIOECOLOGY,	W73-00802
ASTES, PATH OF POLLUTANTS, WATER	POLLUTION SOURCES, PUBLIC HEALTH,	W73-00790
LABORATORY TESTS, OXYGEN, WATER	POLLUTION EFFECTS.: /CHLOROPHYTA,	W72-14314
TRIENT REMOVAL, NUTRIENTS, WATER	POLLUTION EFFECTS, WATER QUALITY,	W72-14692
TION, POLLUTION ABATEMENT, WATER	POLLUTION TREATMENT, WATER QUALIT	W72-14469
R, *DECOMPOSING ORGANIC M/ WATER	POLLUTION EFFECTS, *ORGANIC MATTE	W72-14673
EPTIC TANKS, FUNGI, ALGAE, WATER	POLLUTION CONTROL.: /, STREAMS, S	W72-14363
NTS, *AQUATIC ENVIRONMENT, WATER	POLLUTION EFFECTS, STREAMS, RIVER	W72-14690
CE, LETHAL LIMIT, POISONS, WATER	POLLUTION EFFECTS.: /AY, RESISTAN	W72-14706
WASTES, *MICROORGANISMS, *WATER	POLLUTION SOURCES, REVIEWS, MICRO	W72-14363
IDE, COASTS, LAKES, PONDS, WATER	POLLUTION CONTROL, AIRCRAFT, COPP	W72-14673
VEYS, *AERIAL PHOTOGRAPHY, WATER	POLLUTION EFFECTS, EUTROPHICATION	W72-14728
C ENVIRONMENT, *TOXICITY, *WATER	POLLUTION EFFECTS, ZINC, LEAD, CO	W72-14694
ATION, *CHLORINATION, OXIDATION,	POLLUTION ABATEMENT, WATER POLLUT	W72-14469
UTION, BEHAVIOR, SEASONAL, WATER	POLLUTION EFFECTS.: /GES, DISTRIB	W72-14659
HYTA, PROTOZOA, EQUIPMENT, WATER	POLLUTION EFFECTS.: /AS, EUGLENOP	W72-07661
QUATIC ANIMALS, NUTRIENTS, WATER	POLLUTION EFFECTS, WATER QUALITY,	W72-07896
, WATER POLLUTION EFFECTS, WATER	POLLUTION SOURCES, *FISH, DIATOMS	W72-07791
PLING, DIATOMS, TURBIDITY, WATER	POLLUTION SOURCES, NUTRIENTS, SEW	W72-07890
AE, FISH, WATER POLLUTION, WATER	POLLUTION SOURCES.: /BENTHOS, ALG	W72-07841
RS, *ELECTRIC POWERPLANTS, WATER	POLLUTION EFFECTS, WATER POLLUTIO	W72-07791
HNIQUES, PESTICIDES, ECOSYSTEMS,	POLLUTION SOURCES, PESTICIDE TOXI	W72-07703
EMENT PENS, IMPOUNDMENTS, *WATER	POLLUTION SOURCES, *AGRICULTURAL	W72-08401
CONCENTRATION, CORROSION, WATER	POLLUTION EFFECTS.: /HYDROGEN ION	W72-08683
, *ANALYTICAL TECHNIQUES, *WATER	POLLUTION CONTROL, *INSTRUMENTATI	W72-08790
FISH POPULATIONS, FLORIDA, WATER	POLLUTION EFFECTS, NITROGEN, PHOS	W72-08986
ES, *AGRICULTURAL RUNOFF, *WATER	POLLUTION EFFECTS, *CATTLE, *FEED	W72-08401
STRIBUTION, BIOINDICATORS, WATER	POLLUTION SOURCES.: /COLOGICAL DI	W72-08579
EWAGE, *INDUSTRIAL WASTES, WATER	POLLUTION SOURCES, ON-SITE INVEST	W72-08804
*AQUATIC MICROORGANISMS, *WATER	POLLUTION EFFECTS, *ALGAE, *PROTO	W72-08579
ICI/ *PESTICIDE TOXICITY, *WATER	POLLUTION EFFECTS, BIOASSAY, PEST	W72-08440
WASTES, SEWAGE EFFLUENTS, WATER	POLLUTION EFFECTS.: /TS, CHEMICAL	W72-08804
WER PLANTS, HEAT TRANSFER, WATER	POLLUTION EFFECTS, WATER TEMPERAT	W72-07907
CRUSTACEANS, OILY WATER,/ *WATER	POLLUTION EFFECTS, *OIL WASTES, *	W72-07911
ORGA/ *THERMAL POLLUTION, *WATER	POLLUTION EFFECTS, *AQUATIC MICRO	W72-07225
HT, PHOSPHATES, NUTRIENTS, WATER	POLLUTION EFFECTS, RADIOACTIVITY	W72-07143
S, PROTOZOA, ALGAE, FUNGI, WATER	POLLUTION EFFECTS.: /R, CONDENSER	W72-07526
A, *TOXICITY, COAGULATION, WATER	POLLUTION EFFECTS, *ALGAL TOXINS.	W72-07360
N, *LAKES, SEWAGE EFFLUE/ *WATER	POLLUTION SOURCES, *EUTROPHICATIO	W72-12955
YANOPHYTA, EUTROPHICATION, WATER	POLLUTION EFFECTS.: /ISOTOPES, *C	W72-13692
*WATER TEMPERATURE, *WA/ *WATER	POLLUTION EFFECTS, *HEATED WATER,	W72-13636
IC PESTICIDES, HERBICIDES, WATER	POLLUTION SOURCES, PLANKTON, OYST	W72-13347
OLLUTION), EUTROPHICATION, WATER	POLLUTION SOURCES, NUTRIENTS.: /P	W72-13644
LVED SOLIDS, PHYTOPLANKON, WATER	POLLUTION EFFECTS.: /IENTS, DISSO	W72-13300
NUTRIENTS, BIOINDICATORS, WATER	POLLUTION EFFECTS, FERTILIZERS, N	W72-13816
AY, ENVIRONMENTAL EFFECTS, WATER	POLLUTION EFFECTS, ALGAE, HEAVY M	W72-13813
, WATER POLLUTION CONTROL, WATER	POLLUTION SOURCES, TRIBUTARIES, F	W72-14282
HIBITION, *SALT TOLERANCE, WATER	POLLUTION EFFECTS, CHLORINATED HY	W72-13809
ASTES, *INDUSTRIAL WASTES, WATER	POLLUTION CONTROL, WATER POLLUTIO	W72-14282
T REVIEW, *BIBLIOGRAPHIES, WATER	POLLUTION SOURCES, THERMAL POLLUT	W72-12190
GY, POLLUTANTS, *TOXICITY, WATER	POLLUTION EFFECTS, RESPIRATION, *	W72-12576
NTATION RATES, OIL SPILLS, WATER	POLLUTION SOURCES.: /MENT, SEDIME	W72-12487
POLLUTION, PHYTOPLANKTON, WATER	POLLUTION SOURCES, PHAEOPHYTA, OX	W72-12166
FIXATION, DENITRIFICATION, WATER	POLLUTION EFFECTS, WATER POLLUTIO	W72-12544
PUBLIC HEALTH, *SHELLFISH, WATER	POLLUTION EFFECTS, TOXICITY, WATE	W72-12257
, WATER POLLUTION EFFECTS, WATER	POLLUTION CONTROL.: /ITRIFICATION	W72-12544
OPHYTA, BACTERIA, ECOLOGY, WATER	POLLUTION EFFECTS.: /PRINGS, CYAN	W72-12734
MERGANSER DUCK, BIOASSAY, WATER	POLLUTION EFFECTS, GAS CHROMATOGR	W72-12930
TER POLLUTION, W/ *LAKES, *WATER	POLLUTION EFFECTS, *LIMNOLOGY, WA	W72-12729
PHYSIOLOGY, BICARBONATES, WATER	POLLUTION EFFECTS.: /ATION, PLANT	W72-12940
IMNOLOGY, WATER POLLUTION, WATER	POLLUTION SOURCES, SEWAGE, INDUST	W72-12729

ER, PATUXENT RIVER, CONTRA COSTA POWER PLANT, SAN JOAQUIN RIVER, M W72-07907
NUCLEAR POWER PLANTS, ELECTRICAL POWER PLANTS, HEAT TRANSFER, WATE W72-07907
NT, SAN JOAQUIN RIVER, MORRO BAY POWER PLANT, HUMBOLDT BAY, CONNET W72-07907
L EFFECTS, HEATED WATER, NUCLEAR POWER PLANTS, STEAM, ECOLOGY, PHY W71-11517
BIOLOGICAL COMMUNITIES, *THERMAL POWER PLANTS, *DISCHARGE(WATER), W71-11517
*DESALINATION PLANTS, *ELECTRIC POWER PRODUCTION, *IRRIGATION PRO W72-03327
NG, *HEATING, *COOLING, *THERMAL POWER, *MULTIPLE-PURPOSE PROJECTS W70-09193
LS, *THERMAL POLLUTION, *THERMAL POWER, ATLANTIC OCEAN, MARINE BIO W72-11876
EATION, FISHERIES, HYDROELECTRIC POWER, NAVIGATION, CONSUMPTIVE US W72-14878
*FARM USES OF WASTE HEAT, STEAM POWERED BUSES.: / *SPACE HEATING, W70-09193
*GREENHOUSES, WATER TEMPERATURE, POWERPLANTS, DISTILLATION, ALGAE, W70-08832
Y LOSSES, *WATER CONSE/ *THERMAL POWERPLANTS, *POWERPLANTS, *ENERG W70-09193
GY, TRACERS, MONITORING, NUCLEAR POWERPLANTS, NUCLEAR WASTES, PATH W73-00790
A COLLECTIONS, RIVERS, *ELECTRIC POWERPLANTS, WATER POLLUTION EFFE W72-07791
ATER POLLUTION, COOLING, THERMAL POWERPLANTS, HEATED WATER, CIRCUL W73-01704
T/ *PATH OF POLLUTANTS, *NUCLEAR POWERPLANTS, *RADIOACTIVITY EFFEC W73-00831
, RADIOACTIVITY EFFECTS, NUCLEAR POWERPLANTS, MARINE ALGAE, MARINE W73-00833
ALS, OFFSHORE PLATFORMS, NUCLEAR POWERPLANTS, WATER PROPERTIES, SA W73-02029
LGAE, *GULF OF MEXICO, *ELECTRIC POWERPLANTS, *THERMAL POLLUTION, W72-11252
OLLUTANT IDENTIFICATION, NUCLEAR POWERPLANTS, EFFLUENTS, NUCLEAR W W72-10958
, *ELECTRIC POWER PROD/ *NUCLEAR POWERPLANTS, *DESALINATION PLANTS W72-03327
S, COBALT RADIOISOTOPES, NUCLEAR POWERPLANTS, SITES, MOLLUSKS, MAR W72-03324
*RADIOACTIVITY EFFECTS, *NUCLEAR POWERPLANTS, MONITORING, PATH OF W72-00959
IVERS, POST-IMPOUNDMENT, NUCLEAR POWERPLANTS, ECOLOGY, RADIOECOLOG W72-00939
BIOINDICATORS, BIOASSAY, ALGAE, POWERPLANTS, EFFLUENTS, DISCHARGE W71-11583
CH REMOVAL, STARCH SUBSTITUTION, POLYELECTROLYTES.: *STAR W72-03753
AGMENELLUM QUADRUPLICATUM STRAIN PR-6, PHOTOALDRIN, KETOENDRIN, ME W72-03545
S, HEAVY METALS, PHYSICOCHEMICAL PR: /AL PHYSIOLOGY, TRACE ELEMENT W73-00817
AKOTA), *ORTHOPHOSPHATES.: PRAIRIE LAKES, *LAKE HERMAN(SO. D W73-00938
E EFFEC/ *BP 1002, *GREEN ALGAE, PRASINOCLADUS MARINUS, TEMPERATUR W70-09429
,/ *ANTIBIOSIS, OPTICAL DENSITY, PRASINOPHYCEAE, BACCILARIOPHYCEAE W72-08051
LTRATION, CHLORINATION, CHEMICAL PRECIPITATION, FLOCCULATION, FLOT W70-10381
, EUTROPHICATION, TRANSPIRATION, PRECIPITATI: /, CHEMICAL ANALYSIS W71-01489
LEACHING NUTRIENTS, FERTILIZERS, PRECIPITATION(ATMOSPHERIC), RUNOF W71-04216
IDATION LAGOONS, CLIMATIC ZONES, PRECIPITATION, COAGULATION, LIME, W71-08951
EXCHANGE, ELECTROLYSIS, CHEMICAL PRECIPITATION, NITRIFICATION, DEN W71-09450
ATURE, LIGHT INTENSITY, CHEMICAL PRECIPITATION, HYDROGEN ION CONCE W71-07100
*NITRATES, GROUNDWATER, AMMONIA, PRECIPITATION, SEDIMENTS, DENITRI W71-06435
RADIOISOTOPES, FALLOUT, CHEMICAL PRECIPITATION, IRON, BIOASSAY, LE W72-05965
PHATES, *CLAMS, *ALGAE, CHEMICAL PRECIPITATION, HYDROGEN ION CONCE W72-03613
*SEPARATION TECHNIQUES, CHEMICAL PRECIPITATION, COAGULATION, SEDIM W72-10237
, PERCOLATION, DRAINAGE, RUNOFF, PRECIPITATION(ATMOSPHERIC).: /LLS W72-12955
*MARINE ALGAE TRACERS, CHEMICAL PRECIPITATION, HYDROGEN ION CONCE W72-12210
ERENCE, INTERLABORATORY STUDIES, PRECISION, *SWEDEN, LAKE MALAREN, W73-00286
APHY, ELECTRON CAPTURE DETECTOR, PRECONCENTRATION, BIOLOGICAL SAMP W73-02105
ROL, INSECT CONTROL, IRRIGATION, PREDATION, PARASITISM, INSECT RES W72-00846
ONIA-NITROGEN, *GRO/ *OSTRACODS, PREDATOR-PREY RELATIONSHIPS, *AMM W72-04787
IMITS, CHEMICAL RECOVERY, SAMPLE PREPARATION.: / BONE, DETECTION L W72-05965
YRENE, SLIDE PREPARATION, SAMPLE PREPARATION.: /E, TOLUENE, POLYST W72-11738
IDE, TOLUENE, POLYSTYRENE, SLIDE PREPARATION, SAMPLE PREPARATION.: W72-11738
INGSTON, AUTORADIOGRAPHY, SAMPLE PREPARATION, SCINTILLATION COUNTI W72-09239
TERS, CHLORELLA VULGARIS, SAMPLE PREPARATION.: / METHYL ESTERS, ES W72-09365
ACROPHYTES, SARGASSO SEA, SAMPLE PREPARATION, SPERMATOPHYTES, PHOT W73-02100
TA, PORPHYRA UMBILICALIS, SAMPLE PREPARATION, CRUDE OIL.: / DIGITA W73-01074
LTURE MEDIA, FATTY ACIDS, SAMPLE PREPARATION, NATURAL ORGANICS.: / W73-01066
HLOROPHYLL A, CARTENOIDS, SAMPLE PREPARATION, MOUNTAIN LAKE.: /, C W73-00916
IOGRAPHY, GLUTARALDEHYDE, SAMPLE PREPARATION, COUNTING, SAMPLE PRE W73-00273
E, INTERLABORATORY STUDI/ SAMPLE PREPARATION, CHEMICAL INTERFERENC W73-00286
TES, SAMPLE PRESERVATION, SAMPLE PREPARATION, SUBLITTORAL, LONG IS W73-00367
ATION, KJEDAHL PROCEDURE, SAMPLE PREPARATION.: /CTROSCOPY, ASSIMIL W73-00379
ESICULOSUS, VESICULATION, SAMPLE PREPARATION, BRANCHING.: /FUCUS V W73-00369
TY, LIQUID SCINTILLATION, SAMPLE PREPARATION, ZOSTREA, ULVA.: /IVI W72-14317
, HUMIC ACIDS, OXIMATION, SAMPLE PREPARATION.: /AY, SCOUDOUC RIVER W72-12166
S ATROMACULATUS, CLEANUP, SAMPLE PRESERVATION, CAMBARUS BARTONI, B W72-12930
ISOPYCNIC CENTRIFUGATION, SAMPLE PRESERVATION, SALPA, LUDOX AM, SU W73-00371
A, SUBSTRATES, EPIPHYTES, SAMPLE PRESERVATION, SAMPLE PREPARATION, W73-00367
LE PREPARATION, COUNTING, SAMPLE PRESERVATION, FLAGELLATES, RHODOM W73-00273
, *SAMPLE PRESERVATION, CHEMICAL PRESERVATION, HISTOCHEMISTRY, LYO W73-01676
LS, *BIOLOGICAL SAMPLES, *SAMPLE PRESERVATION, CHEMICAL PRESERVATI W73-01676
*TEMPERATURE TOLERANCES, SAMPLE PRESERVATION, COUNTING CHAMBERS, W72-07899
OTTLE, SPECIES DIVERSITY, SAMPLE PRESERVATION, COUNTING.: /ANSEN B W72-07663
CYCLOTELLA, OSCILLATORI/ SAMPLE PRESERVATION, FORMALIN, COUNTING, W72-10875
CTROPHOTOME/ *STRONTIUM, *SAMPLE PRESERVATION, *FLAME EMISSION SPE W71-11529
, *DEOXYRIBONUCLEIC ACID, SAMPLE PRESERVATION, COULTER COUNTER.: / W71-11562
TA.: ENZYME PRESOAKS, DISHWATER DETERGENTS, N W71-08768
FISHKILL, RESERVOIRS, FISHERIES, PRESSURE.: /OSIVES, COSTS, FISH, W72-11564
ST ANALYSIS, *FILTERS, HEADLOSS, PRESSURE, TURBIDITY, ALGAE, FLOCC W72-09393
IS, INSTRUMENTATION, HYDROSTATIC PRESSURE, DIATOMS.: /PHOTOSYNTHES W71-10791
OGEN, *GRO/ *OSTRACODS, PREDATOR- PREY RELATIONSHIPS, *AMMONIA-NITR W72-04787
LAND MANAGEMENT, FERTILIZATION, PRIMARY PRODUCTIVITY.: /FIXATION, W72-04769
BON, OXYGEN, NITROGEN, HYDROGEN, PRIMARY PRODUCTIVITY, RESPIRATION W72-06274
NION, ZOOPLANKTON, DISTRIBUTION, PRIMARY PRODUCTIVITY, ALGAE, BIOM W71-09171
ONDARY TREATMENT, AQUATIC ALGAE, PRIMARY PRODUCTIVITY, TRACE ELEME W71-09674
PHOTOSYNTHESIS, CARBON DIOXIDE, PRIMARY PRODUCTIVITY, BIODEGRADAT W71-04518
PHYTA, CHRYSOPHYTA, CHLOROPHYTA, PRIMARY PRODUCTIVITY, CARBON RADI W71-03014
LANKTON, PLANKTON, PRODUCTIVITY, PRIMARY PRODUCTIVITY, ICED LAKES, W71-02880
, AMMONIUM COMPOUNDS, NUTRIENTS, PRIMARY PRODUCTIVITY, FERTILIZERS W71-11008
IORHYTHMS, CORRELATION ANALYSIS, PRIMARY PRODUCTIVITY.: /ALYSIS, B W72-11719
QUIREMENTS, AQUATIC POPULATIONS, PRIMARY PRODUCTIVITY.: /TRIENT RE W72-09103
OACTIVITY TECHNIQUES, *PLANKTON, PRIMARY PRODUCTIVITY, FOOD HABITS W72-00161

TA, EUGLENOPHYTA, GRAZING, WATER QUALITY.: / CYANOPHYTA, CHLCRCPHY W72-07899
ANTS, CHEMICAL PROPERTIES, WATER QUALITY.: /TOPLANKTON, AQUATIC PL W72-14542
POSAL, PATH OF POLLUTANTS, WATER QUALITY.: / PESTICIDES, WASTE DIS W71-05553
AE, TURBIDITY, *NEW YORK, *WATER QUALITY.: /ATERS, *PLANKTON, *ALG W72-06191
ROPHICATION, ALGAL BLOOMS, WATER QUALITY.: /ADATION, FISHKILL, EUT W71-08768
PHORUS, WATER QUALITY ACT, WATER QUALITY.: /OXYGEN, NITRCGEN, PHOS W71-06444
ION TREATMENT, EVALUATION, WATER QUALITY.: /T GROWTH, WATER POLLUT W72-11702
L PROPERTIES, FRESH WATER, WATER QUALITY.: /AE, ECOLOGY, BIOLOGICA W71-10580
RFACE WATERS, *SEA WATER, *WATER QUALITY, *ECOLOGY, *CALIFCRNIA, H W71-10577
ALITY CONTROL, *ODOR, AL/ *WATER QUALITY, *CHLCRINATION, *WATER QU W71-13187
Y PRODUCTIVITY, *ALGAE, / *WATER QUALITY, *EUTROPHICATION, *PRIMAR W71-13183
OPLA/ *INDUSTRIAL WASTES, *WATER QUALITY, *DISSOLVED OXYGEN, *PHYT W71-11876
ION, *CHEMICAL REACTIONS, *WATER QUALITY, *CHEMICAL PROPERTIES, EU W71-12092
INDU/ *ENTERIC BACTERIA, *WATER QUALITY, *DOMESTIC WASTES, ALGAE, W72-09590
RVOIRS, *HYDROLOGIC DATA/ *WATER QUALITY, *IMPOUNDED WATERS, *RESE W72-02274
, *DIVER/ *DELTAS, *BAYS, *WATER QUALITY, *WATER POLLUTICN CONTROL W71-09581
MENT, *WATER/ *ESTUARIES, *WATER QUALITY, *WATER RESOURCES DEVELOP W71-09788
AQUATIC MICROORGANISMS, / *WATER QUALITY, *AQUATIC PRODUCTIVITY, * W72-06532
ECOLOGY, *PUMPED STORAGE, *WATER QUALITY, *MODEL STUDIES, ZOOPLANK W72-05282
, *LAKES, *COLORADO, *RE/ *WATER QUALITY, *WATER POLLUTICN EFFECTS W72-04791
MS, FLOW, SE/ *NUTRIENTS, *WATER QUALITY, *IMPOUNDED WATERS, STREA W71-03028
ATICA/ *SYSTEMS ANALYSIS, *WATER QUALITY, *EUTROPHICATION, *MATHEM W71-03021
PULP AND PAPER INDUSTRY,/ *WATER QUALITY, *WISCONSIN, PHCSPHORUS, W70-05103
PHYTCPLANKTON, PLANKTON/ *WATER QUALITY, *LAKES, *EUTROPHICATION, W71-02880
ER, *OXYGEN SAG, *IMPAIRED WATER QUALITY, *ESTUARIES, *DELAWARE RI W70-09189
MATHEMATICAL MODELS, THE/ *WATER QUALITY, *HYDROELECTRIC PLANTS, * W72-14405
, *BICLOGICAL PROPERTIES, *WATER QUALITY, *BIOINDICATORS, *BIOLOGI W73-00002
ATION, AQUATIC ECOSYSTEMS, WATER QUALITY, *ALGAE, *LIGHT PENETRATI W73-02349
N, RHODOPHYTA, PHAEOPHYTA, WATER QUALITY, *MARINE ALGAE.: /ODUCTIO W73-01449
OURCES, NUTRIENTS, SEWAGE, WATER QUALITY, AERATICN, WATER PROPERTI W72-07890
AGOONS, *DESIGN CRITERIA, *WATER QUALITY, ALGAE, HARVESTING, DEPTH W71-07100
QUALITY, WASTE M/ *ENVIRONMENTAL QUALITY, AIR CONTAMINANT, STREAM W71-08214
APHY, *INFRARED RADIATION, WATER QUALITY, ALGAE, WATER QUALITY CON W72-12487
ION, NUTRIENTS, BIOASSAY, *WATER QUALITY, ALGAE, PESTICIDES, HEAVY W72-14282
ENZYMES, CHLEATION, ALGAE, WATER QUALITY, BIODEGRADATION, SEWAGE T W72-06837
ACEANS, *MARINE BACTERIA, *WATER QUALITY, BENTHOS, CRABS, COPEPODS W72-08142
MODELS, WATER TEMPERATURE, WATER QUALITY, CARBON DIOXIDE, HYDROGEN W72-06274
PPROPRIATION, GROUNDWATER, WATER QUALITY, COMPETING USES, DIVERSIO W70-07100
, WATER POLLUTION EFFECTS, WATER QUALITY, CHEMICAL DEGRADATION.: / W72-01085
MPARISONS, BENEFICIAL USE, WATER QUALITY, CALIFCRNIA, BIOCCNTRCL, W72-00974
ROL, MANAGEMENT, ANALYSIS, WATER QUALITY, CYANOPHYTA, NUISANCE ALG W72-09171
, WATER POLLUTION EFFECTS, WATER QUALITY, CARBON.: /S, CHEMCCNTROL W72-11724
, *PLANKTON, *RECREATION, *WATER QUALITY, CYANCPHYTA, LAKES, MISSC W72-10431
*EUTROPHICATION, BIOMASS, WATER QUALITY, CHLOROPHYLL, NITROGEN, W W71-10645
TA, ALKALINITY, NUTRIENTS, WATER QUALITY, CIATCMS, CHLORCPHYTA, TH W72-05610
, *BICTA, *AQUATIC PLANTS, WATER QUALITY, ENVIRONMENTAL EFFECTS, L W72-06526
UDIES, ECOSYSTEMS, ALGAE, *LIGHT QUALITY, EUTROPHIC ZONE, CHLOROPH W72-12393
ES, MICROSCOPY, LIMNOLOGY, WATER QUALITY, ECOLCGY, AQUATIC HABITAT W72-13142
UATIC PLANTS, CLAYS, ODOR, WATER QUALITY, FOOD CHAINS, DISSOLVED O W72-04759
COOLING WATER, *FISHERIES, WATER QUALITY, FISH ATTRACTANTS, FISH H W70-09596
*CULTIVATED LANDS, *IOWA, WATER QUALITY, FERTILIZERS, NUTRIENTS, W71-12252
, WATER POLLUTION EFFECTS, WATER QUALITY, HERBICIDES, ODCR-PRODUCI W72-14692
S, LITTORAL, AGE, BIOMASS, LIGHT QUALITY, LIGHT PENETRATICN, EFFIC W72-04766
NUTRIENTS, PHYTOPLANKTON, WATER QUALITY, MICHIGAN, SURFACE WATERS W72-10875
NIA, NITRATES, PHOSPHATES, WATER QUALITY, NITRCGEN, NUTRIENTS, TRO W71-11033
OWTH, TOXICITY, NUTRIENTS, WATER QUALITY, NITRATES, PLANNING, NITR W71-13172
ATER POLLUTION, MINNESOTA, WATER QUALITY, NUTRIENTS, SEDIMENTS, CO W71-03012
YDROGEN ION CONCENTRATION, WATER QUALITY, NUTRIENTS, COLCRIMETRY, W73-00916
ON, *HYDROLOGY, *NEW YORK, WATER QUALITY, NUTRIENTS, ALGAE, INFLOW W72-13851
TURE, SALINITY, SEA WATER, WATER QUALITY, CN-SITE TESTS, HYDROGEN W72-09092
TS, LAKE HURON, EFFLUENTS, WATER QUALITY, PHOSPHORUS, PHYTCPLANKTO W72-13662
TTERNS, WATER TEMPERATURE, WATER QUALITY, PLANKTON, ANALYTICAL TEC W72-07663
SURFACE WATERS, NITROGEN, WATER QUALITY, PHOSPHCRUS, AERCBIC COND W70-10405
E, *EUTROPHICATION, LAKES, WATER QUALITY, PHOSPHORUS, PHOSPHATES, W70-08944
TER UTILIZATION, *IMPAIRED WATER QUALITY, RIPARIAN RIGHTS, PRIOR A W70-07100
ES, ALGAE, EUTROPHICATION, WATER QUALITY, RIVERS, IOWA, CCRN, DISC W71-06445
DE, SESTON, FERTILIZATION, WATER QUALITY, RADON RADIOISOTCPES, PRI W72-07648
NTROL, *FLOW AUGMENTATION, WATER QUALITY, RESERVOIRS, WATER PCLLUT W72-04260
OTOSYNTHESIS, RESPIRATION, WATER QUALITY, STREAMFLOW, AGRICULTURE, W71-11017
S, *LIFE CYCLES, CULTURES, WATER QUALITY, SCENEDESMUS, LABORATORY W72-07144
, WATER POLLUTION EFFECTS, WATER QUALITY, SEASONAL, SUSPENDED LOAD W72-07896
WATER POLLUTICN TREATMENT, WATER QUALITY, STREAMS, LAKES, BIOCHEMI W72-14469
ENT, *POLLUTION ABATEMENT, WATER QUALITY, STANDARDS, WATER POLLUTI W71-08214
KENTUCKY, SULFUR BACTERIA, WATER QUALITY, SULFATES, HYDRCGEN ICN C W71-07942
ATES, HUDSON RIVER, ALGAE, WATER QUALITY, SEWAGE DISPOSAL, WASTE W W70-06509
G, BOATING, DOCUMENTATION, WATER QUALITY, SILTING, AGRICULTURAL CH W72-05869
TS, *PRIMARY PRODUCTIVITY, WATER QUALITY, TROPHIC LEVEL, AQUATIC A W72-08986
STE TREATMENT, INDUSTRIES, WATER QUALITY, TECHNOLOGY, SURFACTANTS, W72-09169
AQUATIC ALGAE, NUTRIENTS, WATER QUALITY, WATER ANALYSIS, FREQUENC W73-00918
T, *DISTRIBUTION SYSTEMS, *WATER QUALITY, WATER CONSERVATION, WILD W72-05594
OXIDATION, ORGANIC MATTER, WATER QUALITY, WATER CHEMISTRY, WATER P W72-04853
NUTRIENTS, EUTROPHICATION, WATER QUALITY, WATER POLLUTION CONTROL, W70-08248
VITY, ESSENTIAL NUTRIENTS, WATER QUALITY, WATER CHEMISTRY, CYCLING W70-04080
NUTRIENTS, EUTROPHICATION, WATER QUALITY, WATER POLLUTION CONTROL, W70-08249
TREATMENT, *WATER REUSE, *WATER QUALITY, WATER SUPPLY, ALGAE, ACT W70-10381
QUALITY, AIR CONTAMINANT, STREAM QUALITY, WASTE MANAGEMENT, MANURE W71-08214
OR, OXYGEN SAG, TURBIDITY, WATER QUALITY, WATER TREATMENT, NITROGE W71-09958
NS, PCNDS, FARM PONDS, MARYLAND, QUARRIES, OHIO RIVER, HUDSON RIVE W72-04736
CKWELL, LAKE HODGSON, CLEARWATER QUARRY, DOLLAR LAKE, CHLCROPHYLL W72-13673
EAR LAKE(CALIF), RICHELIEU RIVER(QUEBEC), GREEN BAY(WIS), MACKEREL W70-08098

NEW BRUNSWICK, ANTICOSTI ISLAND(	QUEBEC), MAGDALEN ISLAND(QUEBEC),	W71-10066
ISLAND(QUEBEC), MAGDALEN ISLAND(	QUEBEC), SABLE ISLAND(NOVA SCOTIA	W71-10066
CE, *POLYP MAINTENAN/ *CHRYSAORA	QUINQUECIRRHA, *MEDUSAE MAINTENAN	W72-10162
TA, TSUMURAIA NUMEROSA, HIROSEIA	QUINQUELOBATA, TROITSKIELLA TRIAN	W72-07683
CHI/ *NUTRIENT BUDGETS, *BAY OF	QUINTE(ONTARIO), NUTRIENT INPUTS,	W71-11009
BAY OF	QUINTE, *LAKE ONTARIO.:	W71-08953
, CALLITRICHE, ANABAENA / BAY OF	QUINTE, SALT CREEK, CHARA, ELODEA	W73-00840
CHELEX 100, ARTEMIA SALINA, BIO-	RAD, CYCLOTELLA NANA.:	W72-04708
ROSIN AMINE D ACETATE(	RADA).:	W71-05083
LUORESCENCE, INFRARED RADIATION,	RADAR, C: /TRAVIOLET RADIATION, F	W72-04742
S, GRACILARIA VERUCOSA, ECKLONIA	RADIATA, GROWTH REGULATION.: /RMI	W72-07496
CONCENTRATION, RADIATION SAFETY,	RADIATION CONTROL.:	W72-07841
OL.: CONCENTRATION,	RADIATION SAFETY, RADIATION CONTR	W72-07841
, *PROTOZOA, *BACTERIA, FLORIDA,	RADIATION SENSITIVITY, RADIOECOLO	W73-00796
COBALT-60.: *BIOLOGICAL--GAMMA-	RADIATION SYSTEM, MODULAR SYSTEM,	W72-00383
MENTATION, ALGAE, INVERTEBRATES,	RADIATION, ECOSYSTEMS, ESTUARIES,	W71-13723
BIDITY, HYDROLOGIC BUDGET, SOLAR	RADIATION, WATER TEMPERATURE, SAL	W72-02274
LYTICAL TECHNIQUES, *ULTRAVIOLET	RADIATION, DISSOLVED OXYGEN, LAKE	W72-01780
, *AERIAL PHOTOGRAPHY, *INFRARED	RADIATION, *AQUATIC ALGAE, *SURFA	W71-12034
ORELLA SOROKINIANA, CO-60, GAMMA	RADIATION, SURVIVAL, CHLORELLA PY	W72-11836
L, TROPICAL REGIONS, ULTRAVIOLET	RADIATION, FOOD CHAINS, GAMMA RAY	W73-00796
, *FALLOUT, *MARINE/ *BACKGROUND	RADIATION, *RADIOACTIVITY EFFECTS	W73-00828
ICATION, *LAKES, ALGAE, INFRARED	RADIATION, *REMOTE SENSING, *POLL	W72-14728
CYANO/ *ISOLATION, *ULTRAVIOLET	RADIATION, SEPARATION TECHNIQUES,	W72-14327
ZOOPLANKTON, CHLOROPHYLL, SOLAR	RADIATION, VITAMIN B, PIGMENTS, G	W72-07892
ES, FISH, FLUORESCENCE, INFRARED	RADIATION, IRRAD: /S, ANIMAL WAST	W72-08790
SEASONAL, DIATOMS, *LIGHT, SOLAR	RADIATION, *PHOTOPERIODISM, PERIP	W72-07901
DUCTIVITY, EUTROPHICATION, SOLAR	RADIATION, DISTRIBUTION, DEPTH, D	W72-08054
TIDAL EFFECTS, STABILITY, SOLAR	RADIATION, PHOTOPERI: / ANALYSIS,	W72-08141
RPTION, *FOOD CHAIN/ *BACKGROUND	RADIATION, *NUCLEAR WASTES, *ABSO	W73-00833
TED WATER, CIRCULATION, INFRARED	RADIATION, HYDROGRAPHY, PLANKTON,	W73-01704
IS, CHEMICALS, DRYING, FREEZING,	RADIATION, CYTOLOGICAL STUDIES, P	W73-01676
TORING, PHOTOGRAPHY, ULTRAVIOLET	RADIATION, FLUORESCENCE, INFRARED	W72-04742
ADIATION, FLUORESCENCE, INFRARED	RADIATION, RADAR, C: /TRAVIOLET R	W72-04742
BIOCHEMICAL OXYGEN DEMAND, SOLAR	RADIATION, NUTRIENTS, TEMPERATURE	W71-07124
ES, PHOSPHATES, SILICATES, SOLAR	RADIATION, DEPTH, THERMOCLINE, NI	W71-07875
MPERATURE, STRATIFICATION, SOLAR	RADIATION, ALGAE, PHOTOSYNTHESIS,	W71-07084
MENT, *BYPRODUCTS, MIXING, SOLAR	RADIATION, ORGANIC LOADING, DIURN	W72-13508
ICAL MODELS, GROWTH RATES, SOLAR	RADIATION, FLOTATION, FLOCCULATIO	W72-12289
G, AERIAL PHOTOGRAPHY, *INFRARED	RADIATION, WATER QUALITY, ALGAE,	W72-12487
TECHNIQUES, METABOLISM, TRACERS,	RADIOACTIVITY, CARBON DIOXIDE, SA	W72-12634
HOTOSYNTHESIS, CALCIUM CHLORIDE,	RADIOACTIVITY TECHNIQUES, METABOL	W72-12634
, MUD, MICROORGANISMS, PROTEINS,	RADIOACTIVITY TECHNIQUES, FOOD CH	W71-08042
RUSES, ENERIC BACTERIA, TRITIUM,	RADIOACTIVE WASTES, MONITORING, P	W72-04742
AL OXYGEN DEMAND, RADIOACTIVITY,	RADIOACTIVE WASTES, WASTE WATER(P	W72-05506
TION, BIOCHEMICAL OXYGEN DEMAND,	RADIOACTIVITY, RADIOACTIVE WASTES	W72-05506
EOPHYTA, ATLANTIC OCEAN, SHARKS,	RADIOACTIVITY TECHNIQUES.: /, PHA	W72-05965
UM, COBALT RADIOISOTOPES, ALGAE,	RADIOACTIVITY, TROPHIC LEVEL, LIF	W70-06668
E HISTORY STUDIES, TIME, SNAILS,	RADIOACTIVE WASTES, TENNESSEE, CR	W70-06668
HOSPHORUS RADIOISOTOPES, SEWAGE,	RADIOACTIVITY, PHOSPHATES, CHEMIC	W71-01474
MARINE PLANTS, HUMAN POPULATION,	RADIOACTIVITY TECHNIQUES, RADIUM	W73-00833
ATH OF POLLUTANTS, RADIOECOLOGY,	RADIOACTIVITY EFFECTS, NUCLEAR PO	W73-00833
M, RUBIDIUM, BROMINE, ACTINIDES,	RADIOACTIVE DECAY, GAMMA-RAY SPEC	W73-01434
LLUTANTS, ESTUARINE ENVIRONMENT,	RADIOACTIVITY TECHNIQUES, ABSORPT	W72-07826
RIENTS, WATER POLLUTION EFFECTS,	RADIOACTIVITY TECHNIQUES.: /, NUT	W72-07143
S, *ABSORPTION, INSTRUMENTATION,	RADIOACTIVITY TECHNIQUES, CARBON	W72-14326
S, CORES, ANALYTICAL TECHNIQUES,	RADIOACTIVITY TECHNIQUES, RADIOEC	W73-00790
ANTS, SEA WATER, WATER ANALYSIS,	RADIOACTIVITY TECHNIQUES, RADIOEC	W73-00817
POLLUTION EFFECTS, RADIOECOLOGY,	RADIOACTIVITY TECHNIQUES, ANALYTI	W73-00802
POPULATIONS, ANIMAL PHYSIOLOGY,	RADIOACTIVITY TECHNIQUES, CYCLING	W73-00826
T, NUCLEAR WASTES, RADIOECOLOGY,	RADIOACTIVITY TECHNIQUES, ANALYTI	W73-00821
RODUCTIVITY, *NERITIC, NITRATES,	RADIOACTIVITY TECHNIQUES, EUTROPH	W73-00431
GAE, BOTTOM SEDIMENTS, PLANKTON,	RADIOACTIVITY TECHNIQUES, MONITOR	W73-00361
TROGEN, WATER POLLUTION EFFECTS,	RADIOACTIVITY, METABOLISM, PLANT	W73-00380
*ALGAE, CHLORELLA, IRRADIATION,	RADIOACTIVITY TECHNIQUES, RESISTA	W72-11836
ARBON RADIOISOTOPES, SUCCESSION,	RADIOACTIVITY TECHNIQUES, CULTURE	W72-09239
HOSPHORUS RADIOISOTOPES, CESIUM,	RADIOACTIVITY EFFECTS, LABORATORY	W72-09649
EATHER, WATER ANALYSIS, TRACERS,	RADIOACTIVITY TECHNIQUES.: /AL, W	W72-09122
GAE, INCUBATION, CENTRIFUGATION,	RADIOACTIVITY TECHNIQUES, CHLOROP	W72-09103
IC LIFE, AQUATIC ALGAE, CALCIUM,	RADIOACTIVITY EFFECTS.: /H, AQUAT	W72-10978
T, *NEUTRON ACTIVATION ANALYSIS,	RADIOACTIVITY, ADSORPTION, FISH;	W71-11036
RPTION, WATER POLLUTION SOURCES,	RADIOACTIVITY EFFECTS, RADIOACTIV	W72-00940
ON LAGOONS, DEHYDRATION, RUNOFF,	RADIOACTIVITY, FALLOUT, BIOCONTRO	W72-00846
IOLOGICAL ECOLOGY, RADIOECOLOGY,	RADIOACTIVITY EFFECTS, ABSORPTION	W72-00948
SOURCES, RADIOACTIVITY EFFECTS,	RADIOACTIVITY TECHNIQUES.: /UTION	W72-00940
EFFECTS, RADIOACTIVITY EFFECTS,	RADIOACTIVITY TECHNIQUES, AQUARIA	W72-03348
RPTION, WATER POLLUTION EFFECTS,	RADIOACTIVITY EFFECTS, RADIOACTIV	W72-03348
HIPS, RADIOISOTOPES, ABSORPTION,	RADIOACTIVITY TECHNIQUES, TRACERS	W72-00949
N LAYER CHROMATOGRAPHY, CLEANUP,	RADIOAUTOGRAPHY, SUBSTRATE UTILIZ	W73-00361
, HEAVY METALS, *TRACE ELEMENTS,	RADIOCHEMICAL ANALYSIS, RESINS, *	W72-04708
, SULFUR BACTERIA, PRODUCTIVITY,	RADIOCHEMICAL ANALYSIS, THERMOPHI	W72-14301
LGAE, *ABSORPTION, *MINE WASTES,	RADIOECOLOGY, *PATH OF POLLUTANTS	W72-12204
ULAR STRUCTURE, CATION EXCHANGE,	RADIOECOLOGY, *PATH OF POLLUTANTS	W72-12203
LANT PHYSIOLOGY, AQUATIC PLANTS,	RADIOECOLOGY.: /LANCE, TRACERS, P	W72-06342
RONMENTAL EFFECTS, PRODUCTIVITY,	RADIOECOLOGY, OCEANOGRAPHY.: /NVI	W72-06313
IMILATIVE CAPACITY, *CALIFORNIA,	RADIOECOLOGY, MARINE ALGAE, ESTUA	W73-00826
LEAR WASTES, PATH OF POLLUTANTS,	RADIOECOLOGY, MATHEMATICAL MODELS	W73-00828
LYSIS, RADIOACTIVITY TECHNIQUES,	RADIOECOLOGY, ENVIRONMENTAL EFFEC	W73-00817
HEALTH, WATER POLLUTION EFFECTS,	RADIOECOLOGY, RADIOACTIVITY TECHN	W73-00802
UTANTS, FALLOUT, NUCLEAR WASTES,	RADIOECOLOGY, RADIOACTIVITY TECHN	W73-00821

FLORIDA, RADIATION SENSITIVITY,	RADIOECOLOGY, ECOLOGY, CYCLING NU	W73-00796
IQUES, RADIOACTIVITY TECHNIQUES,	RADIOECOLOGY, TRACERS, MONITORING	W73-00790
, TRACE ELEMENTS, PUBLIC HEALTH,	RADIOECOLOGY.: /CYCLING NUTRIENTS	W73-00826
MENTS, COLLOIDS, NUCLEAR WASTES,	RADIOECOLOGY.: /FFECTS, TRACE ELE	W73-00818
IVITY EFFECTS, RIVERS, EFFLUENT,	RADIOECOLOGY, AQUATIC POPULATIONS	W72-07841
BLIC HEALTH, PATH OF POLLUTANTS,	RADIOECOLOGY, RADIOACTIVITY EFFEC	W73-00833
UCTIVITY, PHYSIOLOGICAL ECOLOGY,	RADIOECOLOGY, RADIOACTIVITY EFFEC	W72-00948
T, NUCLEAR POWERPLANTS, ECOLOGY,	RADIOECOLOGY, MONITORING, POLLUTI	W72-00939
CELLULOSE, ENZYMES, PULP WASTES,	RADIOECOLOGY, WATER POLLUT: /NA,	W72-09668
ODES, LASERS, ORTH/ VOLTAMMETRY,	RADIOGRAPHY, ION SELECTIVE ELECTR	W72-08790
*PLUTONIUM, *SILVER	RADIOISOTOPES.:	W72-07826
CULTURES, PHOTOSYNTHESIS, CARBON	RADIOISOTOPES, CHROMATOGRAPHY, AM	W72-07710
RHODOPHYTA, PHAEOPHYTA, *COBALT	RADIOISOTOPES, NUCLEAR WASTES, PA	W72-07826
IVITY, CULTURES, BIOMASS, CARBON	RADIOISOTOPES, LIMITING FACTORS,	W72-07648
CHLORELLA, CULTURES, NUTRIENTS,	RADIOISOTOPES, CHLOROPHYLL, PIGME	W72-07660
PHOSPHORUS, OSCILLATORIA, RADON	RADIOISOTOPES, CHROMULINA, CRYPTO	W72-07648
TILIZATION, WATER QUALITY, RADON	RADIOISOTOPES, PRIMARY PRODUCTIVI	W72-07648
RPTION, *PHOSPHORUS, *PHOSPHORUS	RADIOISOTOPES, ALGAE, FRESH WATER	W72-07143
TION, COBALT RADIOISOTOPES, ZINC	RADIOISOTOPES, MANGANESE, CHROMIU	W73-00832
RADIOACTIVITY TECHNIQUES, RADIUM	RADIOISOTOPES.: /MAN POPULATION,	W73-00833
SIS, SEA WATER, STABLE ISOTOPES,	RADIOISOTOPES, NUCLEAR WASTES: /Y	W73-00831
ES, LOBSTERS, ABSORPTION, COBALT	RADIOISOTOPES, ZINC RADIOISOTOPES	W73-00832
CESIUM	RADIOISOTOPES.:	W73-00831
TASSIUM RADIOISOTOPES, STRONTIUM	RADIOISOTOPES, SOIL ANALYSIS, SOI	W73-01673
ATION, SAMPLING, CARP, POTASSIUM	RADIOISOTOPES, STRONTIUM RADIOISO	W73-01673
O FISH, GAMBUSIA AFFINIS, CESIUM	RADIOISOTOPES, CLADOPHORA SPP, PA	W73-01673
MARINE ALGAE, *FOOD / *STRONTIUM	RADIOISOTOPES, *MARINE ANIMALS, *	W73-00819
PATH OF POLLUTANTS, ABSORPTION,	RADIOISOTOPES, STABLE ISOTOPES, E	W73-00811
OPLANKTON, *MARINE ALGAE,/ *ZINC	RADIOISOTOPES, *ABSORPTION, *PHYT	W73-00823
MITING FACTORS, SALINITY, CARBON	RADIOISOTOPES, SESTON, ORGANIC MA	W73-00431
*HUMBOLDT BAY, *CESIUM	RADIOISOTOPES.:	W73-00826
ICAL MODELS, FORECASTING, COBALT	RADIOISOTOPES, MARINE ALGAE, ABSO	W73-00831
ON, *MARINE ANIMALS, MA/ *COBALT	RADIOISOTOPES, *COBALT, *ABSORPTI	W73-00821
NE ALGAE, *PHYTOPLANKTON,/ *ZINC	RADIOISOTOPES, *ABSORPTION, *MARI	W73-00802
ANIMALS, ASIA, MOLLUSKS, COBALT	RADIOISOTOPES, IODINE RADIOISOTOP	W73-00811
CHAINS, *NUCLEAR WASTES, *COBALT	RADIOISOTOPES, PATH OF POLLUTANTS	W73-00826
*PLUTONIUM, CESIUM	RADIOISOTOPES.:	W73-00790
SOUND.: MANGANESE	RADIOISOTOPES, BOGUE SOUND, CORE	W73-00818
D CHAINS, PUBLIC HEALTH, EUROPE,	RADIOISOTOPES, EFFLUENTS.: /, FOO	W73-00828
CESIUM	RADIOISOTOPES.:	W73-00811
AINS, / *GREAT LAKES, *STRONTIUM	RADIOISOTOPES, *FALLOUT, *FOOD CH	W73-00790
KS, COBALT RADIOISOTOPES, IODINE	RADIOISOTOPES, MARINE ALGAE, TRAC	W73-00811
LIGHT INTENSITY, DIATOMS, CARBON	RADIOISOTOPES, NEW ENGLAND, AQUAT	W72-14317
RADIOACTIVITY TECHNIQUES, CARBON	RADIOISOTOPES, PRODUCTIVITY, META	W72-14326
NTS, AQUATIC ALGAE, SCENEDESMUS,	RADIOISOTOPES, STRONTIUM RADIOISO	W72-09649
CHAINS, *PATH OF P/ *PHOSPHORUS	RADIOISOTOPES, *TUBIFICIDS, *FOOD	W72-09668
IC PRODUCTIVITY, BIOMASS, CARBON	RADIOISOTOPES, SUCCESSION, RADIOA	W72-09239
DESMUS, RADIOISOTOPES, STRONTIUM	RADIOISOTOPES, PHOSPHORUS RADIOIS	W72-09649
ONTIUM RADIOISOTOPES, PHOSPHORUS	RADIOISOTOPES, CESIUM, RADIOACTIV	W72-09649
INDIA, CESIUM	RADIOISOTOPES.:	W72-10958
PES, *COBALT RADIOISOTO/ *IODINE	RADIOISOTOPES, *CESIUM RADIOISOTO	W72-10958
*IODINE RADIOISOTOPES, *CESIUM	RADIOISOTOPES, *COBALT RADIOISOTO	W72-10958
PES, *MARINE ALGAE, HEAVY/ *ZINC	RADIOISOTOPES, *COBALT RADIOISOTO	W72-10957
S, ACETYLENE, ETHYLENE, NITROGEN	RADIOISOTOPES, N-15, LUCIFERIN, L	W72-10883
VY/ *ZINC RADIOISOTOPES, *COBALT	RADIOISOTOPES, *MARINE ALGAE, HEA	W72-10957
, *CESIUM RADIOISOTOPES, *COBALT	RADIOISOTOPES, *BIOINDICATORS, AN	W72-10958
LS, GASTROPODS, PATH/ *STRONTIUM	RADIOISOTOPES, *ABSORPTION, *SNAI	W72-10978
*RADIOACTIVITY EFFECTS/ *COBALT	RADIOISOTOPES, *RADIOSENSITIVITY,	W72-11836
D CHAINS, *TRACE ELEMENTS, *LEAD	RADIOISOTOPES, HEAVY METALS, AIR	W72-00940
ABSORPTION, FOOD WEBS, STRONTIUM	RADIOISOTOPES, MARINE BACTERIA, C	W72-00948
SOIL-WATER-PLANT RELATIONSHIPS,	RADIOISOTOPES, ABSORPTION, RADIOA	W72-00949
LS, MUSSELS, ISOPODS, AMPHIPODA,	RADIOISOTOPES, ABSORPTION, PACIFI	W72-03347
TRITIUM, OCEAN CURRENTS, COBALT	RADIOISOTOPES, NUCLEAR POWERPLANT	W72-03324
INES, *ALGAE, *OKLAHOMA, *IODINE	RADIOISOTOPES, URANIUM RADIOISOTO	W72-02073
, *IODINE RADIOISOTOPES, URANIUM	RADIOISOTOPES, ANALYTICAL TECHNIQ	W72-02073
WTH RATE, *AQUATIC PRODUCTIVITY,	RADIOISOTOPES, TRACERS, MARINE AL	W71-11486
EA WATER, PHOSPHATES, PHOSPHORUS	RADIOISOTOPES, DIATOMS, HYDROGEN	W71-10083
E, COBALT RADIOISOTOPES, CADMIUM	RADIOISOTOPES, METABOLISM, TROPHI	W71-13243
GICAL ECOLOGY, CESIUM, STRONTIUM	RADIOISOTOPES, LINC RADIOISOTOPES	W71-13243
RADIOISOTOPES, MANGANESE, COBALT	RADIOISOTOPES, CADMIUM RADIOISOTO	W71-13243
M, STRONTIUM RADIOISOTOPES, LINC	RADIOISOTOPES, MANGANESE, COBALT	W71-13243
, RHODOPHYTA, PHAEOPHYTA, CARBON	RADIOISOTOPES, MARINE FISH, INSTR	W72-06283
EWS, BIOCHEMISTRY, GEOCHEMISTRY,	RADIOISOTOPES, PATH OF POLLUTANTS	W72-06313
N, SEDIMENTS, SETTLING VELOCITY,	RADIOISOTOPES, ABSORPTION, REVIEW	W72-04463
UT, *FOOD CHAINS, *AQUATIC LIFE,	RADIOISOTOPES, ABSORPTION, NUCLEA	W72-04461
UTONIUM RADIOISOTOPES, RUTHENIUM	RADIOISOTOPES, IRON RADIOISOTOPES	W72-04463
TOPES, IRON RADIOISOT/ PLUTONIUM	RADIOISOTOPES, RUTHENIUM RADIOISO	W72-04463
S, RUTHENIUM RADIOISOTOPES, IRON	RADIOISOTOPES.: /IUM RADIOISOTOPE	W72-04463
LTH, ESTUARINE ENVIRONMENT, ZINC	RADIOISOTOPES, PHYSICOCHEMICAL PR	W72-04463
*MARINE ANIMALS, *MARINE ALGAE,	RADIOISOTOPES, FALLOUT, CHEMICAL	W72-05965
TA, PRIMARY PRODUCTIVITY, CARBON	RADIOISOTOPES, PHOTOSYNTHESIS.: /	W71-03014
O P/ *BIOLOGICAL UPTAKE, CALCIUM	RADIOISOTOPES, BLOOMS, 2,4 DINITR	W71-01474
S, THERM/ TEMPERATURE, SORPTION,	RADIOISOTOPES, ALGAE, GROWTH RATE	W71-02075
S, *ACTIVATED SLUDGE, PHOSPHORUS	RADIOISOTOPES, SEWAGE, RADIOACTIV	W71-01474
ENT, *ECOSYSTEMS, CESIUM, COBALT	RADIOISOTOPES, ALGAE, RADIOACTIVI	W70-06668
RTEBRATES, RUTHENIUM-106, CESIUM	RADIOISOTOPES, SPIROGYRA, LIMNODR	W70-06668
SPIRATION, PHOSPHORUS, TOXICITY,	RADIOISOTOPES, INHIBITION.: /, RE	W70-07255
TER, PLANKTON, CHELATION, CARBON	RADIOISOTOPES, CHLORELLA, GROWTH	W71-03027
ORATORY TESTS, CESIUM, STRONTIUM	RADIOISOTOPES, IONS, RIVERS, CULT	W71-08032
SOLVED OXYGEN, NUTRIENTS, CARBON	RADIOISOTOPES, PHOSPHATES, SILICA	W71-07875

BITION, OLIGOTR/ *ALGAE, *GROWTH	RATES, *NUTRIENTS, CULTURES, INHI	W72-08376
ION CONCEN/ *CYANOPHYTA, *GROWTH	RATES, *ORGANIC ACIDS, *HYDROGEN	W72-00838
LGAE, D/ *ANALYTICAL TECHNIQUES,	RATES, *PHOTOSYNTHESIS, *MARINE A	W72-09653
CTORS, ALGAL PONDS, ALGAL GROWTH	RATES, *POTAMOCYPRIS.: /ROWTH REA	W72-04787
TH SUBSTANCES, / *ALGAE, *GROWTH	RATES, *PRODUCTIVITY, *PLANT GROW	W71-08687
RGY BUDGET, *MARINE ALG/ *GROWTH	RATES, *PRIMARY PRODUCTIVITY, ENE	W73-00366
AVY METALS, *TR/ *ALGAE, *GROWTH	RATES, *PRIMARY PRODUCTIVITY, *HE	W73-01434
N, *MARINE ALGAE, *SEDIMENTATION	RATES, *REVIEWS, BIOCHEMISTRY, GE	W72-06313
TESTS, *CHEMICAL WASTE, *GROWTH	RATES, *SEAWATER, *TOXICITY, *NUT	W72-12239
Y STUDIES, *GRO/ *ALGAE, *GROWTH	RATES, *SYSTEMATICS, *LIFE HISTOR	W73-02099
*HEAVY METALS, *ALGAE, *GROWTH	RATES, *TOXICITY, *FLUORESCENCE,	W72-07660
T REQUIREMENTS, AMMONIA, *GROWTH	RATES, *WASTE ASSIMILATIVE CAPACI	W72-04787
FECTS, *AXENIC CULTURES, *DIEOFF	RATES, AFTERGROWTH, BLUE-GREEN AL	W70-08319
*MIXED CULTURES, DIEOFF	RATES, AFTERGROWTH.:	W71-05155
ION, VITAMIN B, PIGMENTS, GROWTH	RATES, ALGAE, FOOD WEBS, COPEPODS	W72-07892
C MATTER, ATLANTIC OCEAN, GROWTH	RATES, ALGAE, INCUBATION, PHYTOPL	W71-11562
ENSITY, TEMPERATURE, MUD, GROWTH	RATES, ANABAENA, NITROGEN FIXATIO	W72-12544
LIS, CYCLOPS STRENUUS, INGESTION	RATES, ASSIMILATION EFFICIENCES,	W72-07940
, EUTROPHICATION, PONDS, FEEDING	RATES, BENTHOS, STANDING CROPS.: /	W72-14796
RODUCTIVITY, SYSTEMATICS, GROWTH	RATES, BIOMASS, DATA COLLECTIONS.	W73-02469
ESISTANCE, AQUATIC ALGAE, GROWTH	RATES, BIOASSAY, LABORATORY TESTS	W72-13809
ORING, SEWAGE, ESTUARIES, GROWTH	RATES, BIOLOGICAL COMMUNITIES.: /	W71-12867
A, *PHYTOPLANKTON, ALGAE, GROWTH	RATES, BIOASSAY, DISSOLVED OXYGEN	W72-06046
*ALGAE, *EUTROPHICATION, *GROWTH	RATES, CHLORELLA, CARBON DIOXIDE,	W72-09168
CHLORELLA, *LIGHT, ALGAE, GROWTH	RATES, CHLOROPHYLL.: /TABOLISM, *	W73-00251
Y, BACTERIA, SCENEDESMUS, GROWTH	RATES, CULTURES, AQUATIC PRODUCTI	W72-14314
, *BIOINDICATORS, *ALGAE, GROWTH	RATES, CULTURES, SUSPENDED LOAD,	W71-09795
POLLUTANT IDENTIFICATION, GROWTH	RATES, DEFICIENT ELEMENTS, WATER	W73-01445
RADIOISOTOPES, CHLORELLA, GROWTH	RATES, DIATOMS, PHYTOPLANKTON, TO	W71-03027
ALGAE, *LIMITING FACTORS, GROWTH	RATES, DIATOMS, BACTERIA, LAKES,	W72-09163
S, ENVIRONMENTAL EFFECTS, GROWTH	RATES, E. COLI, AZOTOBACTER, PSEU	W72-10861
D HYDROCARBON PESTICIDES, GROWTH	RATES, ENVIRONMENTAL EFFECTS, PHY	W73-00281
S, BIO-INDICATORS, STORAGE, FLOW	RATES, INHIBITION, ANALYTICAL TEC	W72-04070
HRYSOPHYTA, CHLOROPHYTA, *GROWTH	RATES, INHIBITION, *PLANT PHYSIOL	W70-07842
OUNDS, NITRATES, DENSITY, GROWTH	RATES, LABORATORY TESTS, CYANOPHY	W72-03217
Y, PLANT GROWTH, BIOMASS, GROWTH	RATES, LIGHT INTENSITY, CARBON DI	W72-05476
BIOMASS, LIGHT INTENSITY, GROWTH	RATES, MATHEMATICAL MODELS, WATER	W72-06274
BIC CONDITIONS, BIOASSAY, GROWTH	RATES, MORTALITY.: /NOPHYTA, AERO	W72-07664
RATES, *AMMONIA, DIATOMS, GROWTH	RATES, NITROGEN, WATER POLLUTION	W73-00380
E WATER TREATMENT, SEDIMENTATION	RATES, OIL SPILLS, WATER POLLUTIO	W72-12487
OGEN, PHOSPHORUS, CARBON, GROWTH	RATES, OLIGOTROPHY, EUTROPHICATIO	W72-09164
TRIAL WASTES, METABOLISM, GROWTH	RATES, OXIDATION LAGOONS, COLIFOR	W72-09590
S, NUTRIENT REQUIREMENTS, GROWTH	RATES, PHOSPHORUS, NITROGEN, *REG	W72-12966
VAL, OXYGEN DECLINE, RESPIRATORY	RATES, PHOTOSYNTHETIC OXYGENATION	W71-04079
*MARINE ALGAE, *LIGHT, *GROWTH	RATES, PHYTOPLANKTON, NATURAL RES	W71-13252
ORATORY TESTS, CULTURES, *GROWTH	RATES, PLANT GROWTH REGULATORS, P	W71-03095
EFFICIENCES, COMPETITION, GROWTH	RATES, PLANT MORPHOLOGY, WATER CI	W72-04766
CIDS, *PLANT PHYSIOLOGY, *GROWTH	RATES, PLANT MORPHOLOGY, CULTURES	W72-07710
PATTERNS, CHROMATOGRAPHY, GROWTH	RATES, POLLUTANT IDENTIFICATION.:	W72-04134
, PHYTOPLANKTON, BIOMASS, GROWTH	RATES, PRODUCTIVITY, ANIMAL GROWT	W72-09378
ON, PRIMARY PRODUCTIVITY, GROWTH	RATES, RESPIRATION.: /STABILIZATI	W72-03556
, SCENEDESMUS, CHLORELLA, GROWTH	RATES, RESISTANCE, PHOSPHORUS, IR	W72-12739
GROWTH, *EUTROPHICATION, GROWTH	RATES, SAMPLING, LAKE MICHIGAN, M	W71-11033
TICAL MIGRATION, AQUARIA, GROWTH	RATES, SEDIMENTATION, BUOYANCY, W	W71-09852
NT, *MATHEMATICAL MODELS, GROWTH	RATES, SOLAR RADIATION, FLOTATION	W72-12289
BIOMASS, THERMAL STRESS, GROWTH	RATES, SPATIAL DISTRIBUTION, PRIM	W73-00853
REQUIREMENTS, CHLORELLA, GROWTH	RATES, TEMPERATURE, SEA WATER, PH	W71-10083
IRONMENT, PHOTOPERIODISM, GROWTH	RATES, TEMPERATURE, BIOCHEMISTRY,	W72-00377
ON, RADIOISOTOPES, ALGAE, GROWTH	RATES, THERMAL POLLUTION.: /ORPTI	W71-02075
ROPHYLL, PROTEINS, ALGAE, GROWTH	RATES, TRACERS.: / CULTURES, CHLO	W72-03557
OZOA, BACTERIA, STREAMFLOW, FLOW	RATES, WATER TEMPERATURE, WATER C	W73-01972
LS, CANADA, NITROGEN: PHOSPHORUS	RATIO.: /ASSOCIATION, CARBOY TRIA	W70-07283
*RECIRCULATION	RATIO, *ISRAEL(DAN REGION).:	W72-11642
N RESERVOI/ *NITROGEN:PHOSPHORUS	RATIO, LAKE WAUBESA(WIS), OCCOQUA	W72-05473
HLOROPHYLL A, / MARGALEF PIGMENT	RATIO, PIGMENT DIVERSITY INDEX, C	W73-00916
UARY, PHOTOSYNTHESIS/RESPIRATION	RATIOS.: *PAMLICO RIVER EST	W72-13636
, DNA, MOTILITY, CYST FO/ *G + C	RATIOS, *NONFRUITING MYXOBACTERIA	W72-07951
RIAL PLANTS, COMBUSTION, ISOTOPE	RATIOS, SPECTROMETRY, LICHENS, MO	W72-12709
RIA, BLUE HERON, STERNA HIRUNDO,	RATS, PARTRIDGE, DDE, KIDNEYS, LI	W72-05952
CS-137, *PAHRANAGAT LAKE, GAMMA	RAY SPECTROMETRY, ANIMAL TISSUES,	W73-01673
INIDES, RADIOACTIVE DECAY, GAMMA-	RAY SPECTROMETRY, COUNTING.: /ACT	W73-01434
, CHLOROPHYTA, INHIBITION, GAMMA	RAYS.: /RATORY TESTS, IRRADIATION	W73-01080
NNESSEE, CRAYFISH, MIDGES, GAMMA	RAYS, BLOODWORMS, SLUDGE WORMS.: /	W70-06668
TS, VOLATILITY, FREEZE DRYING, X-	RAYS, ELECTROLYSIS, COMPUTER PROG	W71-11036
OLYSIS, COMPUTER PROGRAMS, GAMMA	RAYS, POLLUTANT IDENTIFICATION, C	W71-11036
ET RADIATION, FOOD CHAINS, GAMMA	RAYS, RESISTANCE, BIOLOGICAL COMM	W73-00796
TH, TEMPERATURE, PHOTOSYNTHESIS,	RE-AERATION, LIGHT PENETRATION, N	W71-07098
E, LIGHT INTENSITY, EVAPORATION,	RE-AERATION, DISSOLVED OXYGEN, AL	W71-07099
, *WASTE WATER TREATMENT, *WATER	RE-USE, BIOCHEMICAL OXYGEN DEMAND	W71-07123
THEMATICAL MODELS, MIXING, WATER	RE-USE, LIGHT INTENSITY, EVAPORAT	W71-07099
MENT, LITTORAL, OXYGEN, CHEMICAL	REACTIONS, HYDROGEN ION CONCENTRA	W71-08026
L POLLUTION, *TOXICITY, CHEMICAL	REACTIONS, CARBON DIOXIDE, WATER	W73-01074
C/ *THERMAL POLLUTION, *CHEMICAL	REACTIONS, *WATER QUALITY, *CHEMI	W71-12092
IBIOTICS, IODINE, ATOMIC ENERGY,	REACTOR COOLING, IMPERIAL VALLEY(	W71-09157
HIPS, *AMMONIA-NITROGEN, *GROWTH	REACTORS, ALGAL PONDS, ALGAL GROW	W72-04787
UMPTIVE USE, ALGAE, EVAPORATION,	REASONABLE USE, BRACKISH WATER, D	W70-07100
EXCRETION, ALKALINE PHOSPHATASE,	REASSIMILATION, GREEN FLAGELLATE,	W71-10083
*PIT RECHARGE, *WAT/ *ARTIFICIAL	RECHARGE, *INDUCED INFILTRATION,	W71-12410
RGE, *INDUCED INFILTRATION, *PIT	RECHARGE, *WATER TREATMENT, *FILT	W71-12410
LITY, COMPETING USES, DIVERSION,	RECIRCULATED WATER, WATER REUSE,	W70-07100

ARBOR VITAE LAKE, GREEN BAY, FOX RIVER, DETROIT RIVER, HETEROTROPH W73-01456
*USSR, MOSCOW RIVER, OKA RIVER, FOAM, NATURAL WATERS.: W72-05506
UUS.: SAALE RIVER, GERMANY, SCENEDESMUS OBLIQ W72-14314
E, GREEN BAY, FOX RIVER, DETROIT RIVER, HETEROTROPHIC BACTERIA, AC W73-01456
SHIRE), DELAWARE RIVER, ILLINOIS RIVER, HOLSTON RIVER(TENNESSEE).: W71-11517
PONDS, MARYLAND, QUARRIES, OHIO RIVER, HUDSON RIVER, INDICATORS, W72-04736
TA, ST. MARGARET'S BAY, SCOUDOUC RIVER, HUMIC ACIDS, OXIMATION, SA W72-12166
BLACK RIVER, PINE RIVER, CLINTON RIVER, HURON RIVER, BELL RIVER, S W72-14282
TSK RESERVOIR, ANGARA RIVER, OKA RIVER, HYDROBIOLOGY, BACTERIOLOGY W72-04853
K RIVER(NEW HAMPSHIRE), DELAWARE RIVER, ILLINOIS RIVER, HOLSTON RI W71-11517
ND, QUARRIES, OHIO RIVER, HUDSON RIVER, INDICATORS, DELAWARE, PATH W72-04736
, CONNECTICUT, FLORIDA, DELAWARE RIVER, INDIANA.: /NIA, WASHINGTON W71-07417
ERIE, LAKE ONTARIO, ST LAWRENCE RIVER, INTERNATIONAL JOINT COMMIS W70-07283
VER, BL/ MAUMEE RIVER, ST. CLAIR RIVER, LAKE ST. CLAIR, DETROIT RI W72-14282
S BAY, LAKE PONTCHARTRAIN, PEARL RIVER, LAKE BORGNE, SUBSTRATES, M W73-02029
RIVER, CHRISTINA RIVER, BIG ELK RIVER, LITTLE ELK RIVER, OCTORARA W72-04736
R, HURON RIVER, BELL RIVER, SALT RIVER, MICROCYSTIS, APHANIZOMENON W72-14282
A COSTA POWER PLANT, SAN JOAQUIN RIVER, MORRO BAY POWER PLANT, HUM W72-07907
ER, BOTORPSSTROMMEN RIVER, ATRAN RIVER, MOTALA STROM RIVER, SPARGA W73-00286
NUTRIENT INTERCHANGE, HAW RIVER, NEW HOPE RIVER.: W70-09669
A, SURIRELLA, PERIDINIUM, SENECA RIVER, NINEMILE CREEK, VAN DORN S W72-09111
RIVER, BIG ELK RIVER, LITTLE ELK RIVER, OCTORARA RIVER, PLECTONEMA W72-04736
BLAST, *BRATSK RESERVOIR, ANGARA RIVER, OKA RIVER, HYDROBIOLOGY, B W72-04853
ATERS.: *USSR, MOSCOW RIVER, OKA RIVER, FOAM, NATURAL W W72-05506
LYSIS, SODIUM CHLORIDE, COLUMBIA RIVER, ORGANIC COMPOUNDS, VITAMIN W72-05421
ALGAE, RIVERS, STREAMS, DELAWARE RIVER, OXIDATION LAGOONS, PONDS, W72-04736
RIAL WATER POLLUTION, DES MOINES RIVER, PACKING PLANTS.: INDUST W71-06445
TA POWER / CHALK POINT, COLUMBIA RIVER, PATUXENT RIVER, CONTRA COS W72-07907
IPHENYLS, AROCLOR 1254, ESCAMBIA RIVER, PINFISH, FLOUNDER, CROAKER W71-07412
ST. CLAIR, DETROIT RIVER, BLACK RIVER, PINE RIVER, CLINTON RIVER, W72-14282
IVER, LITTLE ELK RIVER, OCTORARA RIVER, PLECTONEMA, LYNGBYA, PHORM W72-04736
M, SUSQUEHANNA RIVER, BRANDYWINE RIVER, RED CLAY RIVER, WHITE CLAY W72-04736
CLINTON RIVER, HURON RIVER, BELL RIVER, SALT RIVER, MICROCYSTIS, A W72-14282
UNDER, HUMBOLDT BAY, CONNECTICUT RIVER, SAN JOAQUIN RIVER(CALIFORN W71-07417
D/ *BIGHORN LAKE(MONT), BIGHORN RIVER, SHOSHONE RIVER, YELLOWTAIL W73-01435
RIVER, ATRAN RIVER, MOTALA STROM RIVER, SPARGANIUM ERECTUM, SCHOEN W73-00286
SPARENCY, ORTHOP/ GREEN BAY, FOX RIVER, SPECIFIC CONDUCTANCE, TRAN W71-10645
ER(MINNESOTA), LAKE WA/ ILLINOIS RIVER, ST. LOUIS RIVER, BLACK RIV W71-04216
CLAIR, DETROIT RIVER, BL/ MAUMEE RIVER, ST. CLAIR RIVER, LAKE ST. W72-14282
THIC FLORA, LAKE ONTARIO, HUDSON RIVER, STANDARDS, SURFACE WATERS, W70-09614
ANKEE NUCLEAR PLANT, CONNECTICUT RIVER, TURKEY POINT, BISCAYNE BAY W72-07907
R QUALITY, *ESTUARIES, *DELAWARE RIVER, VEGETATION EFFECTS, SLUDGE W70-09189
ANALYTICAL TECHNIQUES, COLUMBIA RIVER, WASHINGTON, OREGON.: /NTS, W72-03215
ATE/ *INSECTICIDES, *ST LAWRENCE RIVER, WATER POLLUTION SOURCES, W W70-08098
IVER, BRANDYWINE RIVER, RED CLAY RIVER, WHITE CLAY RIVER, CHRISTIN W72-04736
E(MONT), BIGHORN RIVER, SHOSHONE RIVER, YELLOWTAIL DAM, FRAGILARIA W73-01435
OLLUTED ESTUARIES, DURHAM COUNTY RIVERS.: *ENGLAND, *P W72-08804
ATION, ON-SITE DATA COLLECTIONS, RIVERS, *ELECTRIC POWERPLANTS, WA W72-07791
IA, CONFERENCES, *PUBLIC HEALTH, RIVERS, *SEWAGE, BIOCHEMICAL OXYG W72-09675
LLUTION EFFECTS, NUCLEAR WASTES, RIVERS, ABSORPTION, FISH, MOLLUSK W71-09225
IFICATION, AQUATIC WEED CONTROL, RIVERS, AGRICULTURE, ROOTED AQUAT W71-13172
*SLIME, *BIOLOGICAL COMMUNITIES, RIVERS, BACTERIA, WATER POLLUTION W72-08573
*ALGAE, *BENTHIC FLORA, *OREGON, RIVERS, BIOLOGICAL COMMUNITIES, S W72-07901
, STRONTIUM RADIOISOTOPES, IONS, RIVERS, CULTURES, ORGANIC MATTER, W71-08032
FISH, RECREATION, INVERTEBRATES, RIVERS, CYANOPHYTA, GEOLOGIC FORM W71-11009
CTIVITY, *RADIOACTIVITY EFFECTS, RIVERS, EFFLUENT, RADIOECOLOGY, A W72-07841
ONTHLY, AQUATIC LIFE, INTERSTATE RIVERS, HYPOLIMNION, VOLUME, MIXI W70-09595
DIATOMS, SUCCESSION, NUTRIENTS, RIVERS, INFLUENT STREAMS, EFFLUEN W73-01435
, EUTROPHICATION, WATER QUALITY, RIVERS, IOWA, CORN, DISCHARGE, CH W71-06445
YTA, WATER POLLUTION, NUTRIENTS, RIVERS, LAKES, FARM PONDS.: /ROPH W71-05991
R POLLUTION SOURCES, *ESTUARIES, RIVERS, PHOSPHATES, NITRATES, HUD W70-06509
LECTROCHEMISTRY, PHOTOSYNTHESIS, RIVERS, PLANT PHYSIOLOGY, TESTING W72-14314
*THERMAL POLLUTION, CONNECTICUT, RIVERS, POST-IMPOUNDMENT, NUCLEAR W72-00939
ATER POLLUTION EFFECTS, STREAMS, RIVERS, ROOTED AQUATIC PLANTS, AQ W72-14690
, CURRENTS(WATER), ALGAE, LAKES, RIVERS, SEWAGE DISPOSAL, SEWAGE T W70-09947
ANALYSIS, *BIOINDICATORS, ALGAE, RIVERS, STREAMS, DELAWARE RIVER, W72-04736
, *CHLORINATION, *ALGAL CONTROL, RIVERS, WATER POLLUTION EFFECTS, W71-00099
PATH OF POLLUTANTS, METHODOLOGY, RIVERS, WATER POLLUTION SOURCES, W73-00286
PIROGIRA, CHLORELLA ELLIPSOIDEA, RNA, CHLOROPHYLL B, ASHING, LABOR W72-06274
, *IODINE RADIOISO/ *SEDIMENTARY ROCKS, *BRINES, *ALGAE, *OKLAHOMA W72-02073
SEDIMENTATION, FILTRATION, LIME, ROCKS, METABOLISM, DESIGN CRITERI W71-13341
BEHAVIOR, INVERTEBRATES, OTTERS, ROCKSLIDES, INTERTIDAL AREAS, ON- W73-00764
RANGE LAKE, EAST TWIN LAKE, LAKE ROCKWELL, LAKE HODGSON, CLEARWATE W72-13673
CARBON PESTICIDES, INSECTICIDES, RODENTICIDES, HERBICIDES, FUNGICI W73-00238
TISH COLUMBIA, RUSSIA, LAMINARIA RODRIGUEZII.: /ARIA LONGIPES, BRI W73-00428
ATERIAL, FERTILITY, RHIZOSPHERE, ROOT SYSTEMS, SOIL FUNGI, SOIL EN W72-03342
ED CONTROL, RIVERS, AGRICULTURE, ROOTED AQUATIC PLANTS, DRAINAGE, W71-13172
LUTION EFFECTS, STREAMS, RIVERS, ROOTED AQUATIC PLANTS, AQUATIC AL W72-14690
QUATIC PLANTS, LIMITING FACTORS, ROOTED AQUATIC PLANTS, CYCLING NU W73-01089
D CONTROL, TILAPIA, CARP, ALGAE, ROOTED AQUATIC PLANTS, PLANKTON, W71-04528
IUM, MACROMONAS MO/ AMOEBOBACTER ROSEUS CHROMATIUM OKENII, CHROMAT W72-07896
ROSIN AMINE D ACETATE(RADA).: W71-05083
, HEMIAULUS HAUCKII, STIGMAPHORA ROSTRATA, ETHMODISCUS OCZES.: /IA W72-03567
NSITY, DIATOMS, LAKES, PROTOZOA, ROTIFERS, CRUSTACEANS, DIPTERA, M W71-11012
ROPHYTA, CHRYSOPHYTA, CHLORELLA, ROTIFERS, CHLAMYDOMONAS, PHYTOPLA W70-06655
S, DINOFLAGELLATES, SCENEDESMUS, ROTIFERS, CRUSTACEANS, COPEPODS, W72-05968
AGOONS, *LIGHT INTENSITY, PONDS, ROTIFERS, PROTOZOA, PHOTOSYNTHESI W71-07382
HABITS, EUTROPHICATION, LARVAE, ROTIFERS, AQUATIC MICROORGANISMS, W72-07132
AMIDOMONAS, BERENPLAAT RESERVOIR(ROTTERDAM), THE NETHERLANDS.: /HL W71-01491
RA NORDENSKIOLDII, THALASSIOSIRA ROTULA, SCHRODERELLA DELICATULA, W72-03543

YTA, BIOMASS, CHEMICAL ANALYSIS,	STATISTICAL METHODS, NITROGEN, PH	W71-11029
ARIABILITY, WAVES(WATER), MAINE,	STATISTICAL METHODS, SEA WATER, C	W73-00369
BUTION PATTERNS, *AQUATIC ALGAE,	STATISTICAL METHODS, MARINE ALGAE	W72-08141
A, SPIRULINA, PANDORINA, PHOCUS,	STAURASTRUM, TRACHELOMONAS.: /ORI	W72-07661
MULINA, CRYPTOMONAS, MALLOMONAS,	STAURASTRUM, PHACOMYXA, SPONDYLOS	W72-07648
UM, ARTH/ *STAURASTRUM, DESMIDS,	STAURASTRUM LONGIPES VAR CONTRACT	W73-01094
, WOODS LAKE, LAGOON OF ISLANDS,	STAURASTRUM SPP.: /NT, TOOMS LAKE	W73-01094
NOBRYON, FRAGILARIA, PEDIASTRUM,	STAURASTRUM, TABELLARIA, MOUGEOTI	W72-10076
CULTURES, CULTURING TECHNIQUES,	STAURASTRUM PARADOXUM, STAURASTRU	W72-12736
FORMOSA, ASTERIONELLA JAPONICA,	STAURASTRUM PARADOXIUM, CULTURE M	W72-12739
CHNIQUES, STAURASTRUM PARADOXUM,	STAURASTRUM SEBALDII VAR. ORNATUM	W72-12736
L-NJ), SAN JOAQUIN DELTA(CALIF),	STEADY-STATE MODELS.: /ESTUARY(DE	W71-05390
ATING, *FARM USES OF WASTE HEAT,	STEAM POWERED BUSES.: / *SPACE HE	W70-09193
TED WATER, NUCLEAR POWER PLANTS,	STEAM, ECOLOGY, PHYTOPLANKTON, ZO	W71-11517
A, BOODLEA COMPOSITA, ANADYOMENE	STELLATA, BR: /OROPSIS MEMBRANACE	W72-11800
A, VISCHERIA PUNCTATA, VISCHERIA	STELLATA, ELLIPSOIDION ACUMINATUM	W72-14808
FRAGILARIA CAPUCINA, CYCLOTELLA	STELLIGERA, STEPHANODISCUS.: /IS,	W73-00384
LOTELLA CF. SOCIALIS, CYCLOTELLA	STELLIGERA, CYMBELLA SPP, DIATOMA	W72-13142
AS MINUTA, ASTERIONELLA FORMOSA,	STEPHANODISCUS ASTREA, ANKISTRODE	W72-12738
EAE, *DENMARK, CYCLOTELLA COMTA,	STEPHANODISCUS HANTZSCHII, APHANI	W72-12738
TONENSIS, TABELLARIA FLOCCULOSA,	STEPHANODISCUS HANTZSCHII, CYCLOT	W73-00384
CAPUCINA, CYCLOTELLA STELLIGERA,	STEPHANODISCUS.: /IS, FRAGILARIA	W73-00384
ENGLAND), RIVER KENNET(ENGLAND),	STEPHANODISCUS HANTZSCHII, CRYPTO	W72-14797
CROTONENSIS, CRYPTOMONAS OVATA,	STEPHANODISCUS NIAGARAE, ASTERION	W73-01435
TINASTRUM, PHYTOCONIS, CYMBELLA,	STEPHANODISCUS, ANKISTRODESMUS, T	W72-10875
ELL COUNTS, NAVICULA, NITZSCHIA,	STEPHANODISCUS, MICROCYSTIS, APHA	W71-12855
FLAGELLATES, FATE OF POLLUTANTS,	STEPHANOPYXIS COSTATA, ASSIMILATI	W73-00380
A, MOUGEOTIA, ULOTHRIX AEQUALIS,	STEPHANOPYXIS TURRIS.: /UADRICAUD	W72-07144
, BODY FLUIDS, VERTEBRATES, HEAT	STERILIZATION, FIXATION, EMBEDDIN	W73-01676
SIMACHIA NUMMULARIA, BLUE HERON,	STERNA HIRUNDO, RATS, PARTRIDGE,	W72-05952
ACTIVITY, ALKALOIDS, FLAVONOIDS,	STEROIDS, TANNINS, SAPONINS.: /C	W72-04275
GLOBUSA, CHLORELLA PYRENOIDOSA,	STICHOCOCCUS BACILLARIS, UNDISSOC	W70-07255
CARBON-14, CARTERIA EUGAMETOS,	STICHOCOCCUS, CHLOROCOCCUM, HYALO	W72-09239
SMOPEDIA TROLLERI, CRYPTOMONADS,	STICHOGLOEA D: / CHROMATIUM, MERI	W72-07145
CLEAR EXPLOSIONS, BIRDS, SALMON,	STICKLEBACKS, CRABS, OTTERS, ZOOP	W72-00960
OGYRA, LIMNODRILUS HOFFMEISTERI,	STICTOCHIRONOMUS ANNULICRUS, PHYS	W70-06668
LOMERATA, ULOTRIC/ *ZINC, *LEAD,	STIGEOCLONIUM TENUE, CLADOPHORA G	W71-05991
A CALIFORNIA, HEMIAULUS HAUCKII,	STIGMAPHORA ROSTRATA, ETHMODISCUS	W72-03567
ISCIFORMIS, PROWSEMONAS TROPICA,	STIGMOBODO BRASILIANA, PAVLOVIAMO	W72-07683
IA, MICROC/ *FULVIC ACID, GROWTH	STIMULATION, ANABAENA, GLOEOTRICH	W72-01097
SPUM, DIGENIA SIMPLEX, HALOPHILA	STIPULACEA, DIPLANTHERA UNINERVIS	W73-00855
OPODS, HETEROPODS, OSCILLATORIA,	STOMATOPODA, POLYCHAETES, AMPHIOX	W73-00935
LAKE MINNETONKA(MINNESOTA), BIG	STONE LAKE(MINNESOTA), LAKE CRYST	W71-04216
B/ *ATHIORHODACEAE, WHIPPLE DAM,	STONE VALLEY DAM(PA.), ATRAZINE,	W72-04743
SECTS, *DISPERSION, CADDISFLIES,	STONEFLIES, DOBSONFLIES, MAYFLIES	W70-06655
RESERVOIR(VA).: *PUMPED	STORAGE OPERATION, SMITH MOUNTAIN	W72-14405
OND, RISING SUN POND, INDUSTRIAL	STORAGE T: /X CALCICOLA, ELKTON P	W72-04736
*ECOSYSTEMS, *ECOLOGY, *PUMPED	STORAGE, *WATER QUALITY, *MODEL S	W72-05282
SSOLVED SOLIDS, ALGAE, RESERVOIR	STORAGE, *HYPOLIMNION, WATER QUAL	W73-02138
NSITY, SEDIMANTATION, OXIDATION,	STORAGE, BIOCHEMICAL OXYGEN DEMAN	W72-01881
NOFF, BIOASSAYS, BIO-INDICATORS,	STORAGE, FLOW RATES, INHIBITION,	W72-04070
ETRIC ANALYSIS, ALGAE, RESERVOIR	STORAGE, IMPOUNDMENTS, DISSOLVED	W72-14280
ENVIRONMENTAL EFFECTS, RESERVOIR	STORAGE, NUTRIENTS, TEMPERATURE,	W72-05282
ATER LEVEL FLUCTUATIONS, *PUMPED	STORAGE, PHYSICOCHEMICAL PROPERTI	W72-14280
BACTERIA, EFFLUENT, IRRIGATION,	STORAGE, WASTE DISPOSAL.: /OGENIC	W71-08214
ER DISPOSAL, PATH OF POLLUTANTS,	STORM DRAINS, DETERGENTS, EUTROPH	W71-13466
*NEBRASKA, URBANIZATION, CITIES,	STORM RUNOFF, SEWAGE, WASTE WATER	W71-13466
SPHORUS, *NITROGEN, EVAPORATION,	STORM RUNOFF, BIOASSAYS, BIO-INDI	W72-04070
, ALGAE, PHOSPHATES, HYDROLYSIS,	STORM WATER, RAINFALL, WASTE WATE	W72-13644
ULANS, AGMENELLUM QUADRUPLICATUM	STRAIN PR-6, PHOTCALORIN, KETOEND	W72-03545
DISSOLVED ORGANIC CARBON, *MENAI	STRAIT(WALES).: *	W72-09108
TIOLECTA, PHAEODACTYLU/ *FLORIDA	STRAIT, CARBON-14, DUNALIELLA TER	W72-09102
AEODACTYLUM/ CARBON-14, *FLORIDA	STRAIT, DUNALIELLA TERIOLECTA, PH	W72-09103
URE TRENDS, CLADOPHORA, RESEARCH	STRATEGY, THAMES RIVER(ENGLAND),	W71-13172
E, NITROGEN, PHOSPHORUS, THERMAL	STRATIFICATION, WATER SUPPLY, SUC	W71-13183
*LAKES, PONDS, SAMPLING, THERMAL	STRATIFICATION, HARDNESS(WATER),	W72-04269
DISSOLVED OXYGEN, ALGAE, THERMAL	STRATIFICATION, COAGULATION, BENT	W72-03753
ND, INORGANIC COMPOUNDS, THERMAL	STRATIFICATION, DIATOMS, EPILIMNI	W72-13659
URE, HYDROGEN ION CONCENTRATION,	STRATIFICATION.: / WATER TEMPERAT	W72-13871
CONCENTRATION, DISSOLVED OXYGEN,	STRATIFICATION, SAMPLING, MICROOR	W71-07382
MIXING, TURBULENCE, TEMPERATURE,	STRATIFICATION, SOLAR RADIATION,	W71-07084
*LAKES, *PHYTOPLANKTON, *DENSITY	STRATIFICATION, SPECIATION, VERTI	W71-09852
HYTA, WATER TEMPERATURE, THERMAL	STRATIFICATION, ICE, EUTROPHICATI	W72-04855
IONS, WATER TEMPERATURE, THERMAL	STRATIFICATION, OXIDATION, ORGANI	W72-04853
TROGEN, BACTERIA, ALGAE, THERMAL	STRATIFICATION, DISSOLVED OXYGEN,	W71-03028
TS, *THERMAL POLLUTION, *THERMAL	STRATIFICATION, HYDRAULIC MODELS,	W70-09614
R POLLUTION, PHOSPHORUS, THERMAL	STRATIFICATION, CURRENTS(WATER),	W70-09947
WATERS, STREAMS, FLOW, SEASONAL,	STRATIFICATION, NITRATES, SILICA,	W71-03028
ROL, *PONDWEEDS, PONDS, *THERMAL	STRATIFICATION, DIATOMS, *CHARA.:	W73-02349
SSOLVED SOLIDS, ALGAE,/ *THERMAL	STRATIFICATION, *TASTE, *ODOR, DI	W73-02138
S, *MATHEMATICAL MODELS, THERMAL	STRATIFICATION, TEMPERATURE, DISS	W72-14405
, PHOTOSYNTHESIS, PHYTOPLANKTON,	STRATIFICATION, DIATOMS, CURRENTS	W73-00236
TA, NITRATES, PHOSPHATES, LAKES,	STRATIFICATION, MIXING, BOTTOM SE	W72-08049
TA, SEASONAL, NUTRIENTS, THERMAL	STRATIFICATION, CARBON, SEDIMENTS	W72-08560
YDROLOGIC DATA, WATER TREATMENT,	STREAM GAGES, GROUNDWATER, SAFE Y	W72-05594
MENTAL QUALITY, AIR CONTAMINANT,	STREAM QUALITY, WASTE MANAGEMENT,	W71-08214
*NUTRIENTS, ALGAE, *DEGRADATION(	STREAM), WATER POLLUTION EFFECTS,	W71-07973
SARGASSO SEA, SHELF WATERS, GULF	STREAM, CAPE HATTERAS, NANSEN BOT	W72-07663
CONSU/ *RIFFLE BENTHOS, WOODLAND	STREAM, ECOSYSTEM ANALYSIS, MACRO	W71-10064
ECOSYSTEMS, *WATERSHEDS(BASINS),	STREAMFLOW, IONS, NITRATES, SULFA	W71-01489

IGATICN SYSTEMS, WATER ANALYSIS,	TEMPERATURE, LIMNOLOGY, CYANCPHYT	W71-06102
EATMENT, *NUTRIENT REQUIREMENTS,	TEMPERATURE, *NITROGEN, *PHOSPHAT	W72-04789
S, RESERVOIR STORAGE, NUTRIENTS,	TEMPERATURE, ALGAE, HYDRCLOGIC DA	W72-05282
LANKTCN, FISH POPULATIONS, WATER	TEMPERATURE, THERMAL STRATIFICATI	W72-04853
O, *RECREATION FACILITIES, WATER	TEMPERATURE, CHEMICAL ANALYSIS, A	W72-04791
SPHORUS, *NUTRIENT REQUIREMENTS,	TEMPERATURE, HYDROGEN ICN CCNCENT	W72-04788
ATION, RESPIRATION, OXYGENATION,	TEMPERATURE, LIGHT INTENSITY, MIX	W72-04983
CHRYSOPHYTA, EUGLENOPHYTA, WATER	TEMPERATURE, THERMAL STRATIFICATI	W72-04855
THER, RESERVOIR OPERATION, WATER	TEMPERATURE, PHYTOPLANKTON, EUTHR	W72-05282
CONCENTRATION, PERIPHYTON, WATER	TEMPERATURE.: /DES, HYDROGEN ION	W72-05610
TES, HYDROGEN ION CONCENTRATION,	TEMPERATURE, ENZYMES, CELLULOSE,	W72-05457
TON, ADSORPTION, EUTROPHICATICN,	TEMPERATURE, LIGHT PENETRATION, A	W72-05469
TROGEN, NUTRIENTS, SEWAGE, WATER	TEMPERATURE, LIGHT PENETRATION, C	W72-04734
OADING, AERATION, MIXING, ALGAE,	TEMPERATURE, ALTITUDE, CLIMATES,	W72-06838
ATES, MATHEMATICAL MODELS, WATER	TEMPERATURE, WATER QUALITY, CARBO	W72-06274
DOLOGY, REEFS, SAMPLING, OXYGEN,	TEMPERATURE, LIGHT, WAVES(WATER),	W72-06283
OMS, PHYTOPLANKTON, HYDORGEN ION	TEMPERATURE,: /TMENT PLANTS, DIAT	W72-06837
ALUATION, WATER CHEMISTRY, WATER	TEMPERATURE, CIS: /OPHICATION, EV	W72-06526
ES, CHLOROPHYLL, PHOTOSYNTHESIS,	TEMPERATURE, SAMPLING, SALINITY,	W71-07875
OA, PHOTOSYNTHESIS, CHLOROPHYLL,	TEMPERATURE, HYDROGEN ICN CCNCENT	W71-07382
ERMAL POLLUTION, *REVIEWS, WATER	TEMPERATURE, DISSOLVED OXYGEN, BI	W71-07417
IOISOTOPES, ALGAE, PLANT GROWTH,	TEMPERATURE, PROTOZOA, CHLOROPHYT	W71-09254
ION, HYDROGEN ION CONCENTRATION,	TEMPERATURE, BIOCHEMICAL OXYGEN D	W71-09524
EMENTS, CHLORELLA, GROWTH RATES,	TEMPERATURE, SEA WATER, PHOSPHATE	W71-10083
PLANKTON, BIODEGRADATION, LIGHT,	TEMPERATURE.: /PLANTS, SEAWATER,	W71-09261
GAE, *RADIOISOTOPES, ABSORPTION,	TEMPERATURE, STRONTIUM RADIOISOTO	W71-09250
OPHYLL, WATER POLLUTION EFFECTS,	TEMPERATURE, CARBCN, DIATOMS, LIG	W71-08670
ALGAE, PHYTOPLANKTON, FREEZING,	TEMPERATURE, CHELATION, HYDROGEN	W71-09190
IONS, CYANOPHYTA, RADIOISOTOPES,	TEMPERATURE, ECOSYSTEMS, CALIFORN	W71-09157
IME, HYDROGEN ION CONCENTRATICN,	TEMPERATURE, CHLORINATICN, FLOATA	W71-08951
S, FRESH WATER, FRESHWATER FISH,	TEMPERATURE, KINETICS, NUCLEAR WA	W71-09182
, ALGAE, GASTROPODS, ADSORPTION,	TEMPERATURE, METABOLISM, KINETICS	W71-09168
NS, *OXYGENATION, *ALGAE, DEPTH,	TEMPERATURE, PHOTOSYNTHESIS, RE-A	W71-07098
*CLIMATIC ZONES, *COST ANALYSIS,	TEMPERATURE, NUTRIENTS, DISSOLVED	W71-07109
AND, SOLAR RADIATION, NUTRIENTS,	TEMPERATURE, COLIFORMS, CHLCRINAT	W71-07124
RGANIC LOADING, ODOR, COLIFORMS,	TEMPERATURE, DISSOLVED OXYGEN, AE	W71-07113
SIMULATION, MIXING, TURBULENCE,	TEMPERATURE, STRATIFICATION, SOLA	W71-07084
ALITY, ALGAE, HARVESTING, DEPTH,	TEMPERATURE, LIGHT INTENSITY, CHE	W71-07100
N EFFECTS, *HEATED WATER, *WATER	TEMPERATURE, *WATER QUALITY STAND	W72-13636
TH, ANALYTICAL TECHNIQUES, WATER	TEMPERATURE.: /RIENTS, PLANT GROW	W72-13651
OTROPHY, DISSOLVED OXYGEN, WATER	TEMPERATURE, HYDROGEN ICN CCNCENT	W72-13871
HEMICAL OXYGEN DEMAND, BACTERIA,	TEMPERATURE, COLIFORMS, ENTERIC B	W72-12565
ORELLA, SILICA, LIGHT INTENSITY,	TEMPERATURE, MUD, GROWTH RATES, A	W72-12544
, *LAKE ERIE, *DISSOLVED OXYGEN,	TEMPERATURE, CURRENTS(WATER), WIN	W72-12992
ASS, ANAEROBIC CONDITIONS, WATER	TEMPERATURE, CHLORIDES, PHOTOSYNT	W72-12940
TIONAL PARKS, HOT SPRINGS, WATER	TEMPERATURES, NOSTOC STREAMS, DAT	W72-08508
S, BACTERICIDES, CHROMIUM, ZINC,	TEMPERATURE, HYDROGEN ICN CONCENT	W72-08683
TON, COMPUTER PROGRAMS, BIOMASS,	TEMPERATURE, CONDUCTIVITY, PHOSPH	W72-08560
STACEANS, WORMS, MOLLUSKS, WATER	TEMPERATURE, SALINITY, SEA WATER,	W72-09092
YLL, ALKALINITY, CARBON DIOXIDE,	TEMPERATURE, DISSOLVED CXYGEN, HY	W72-08049
MENT, SEA WATER, SALINITY, WATER	TEMPERATURE, CHEMICAL ANALYSIS, W	W72-08141
ATER), CONDUCTIVITY, ZINC, WATER	TEMPERATURE, AL: /GEN, HARDNESS(W	W72-07890
, WATER POLLUTION EFFECTS, WATER	TEMPERATURE, ALGAE, BULLHEADS, SA	W72-07907
INVERTEBRATES, *PLANKTON, *WATER	TEMPERATURE, SYSTEMATICS, SAMPLIN	W72-07899
WATER, SALINITY, PHENOLS, WATER	TEMPERATURE, CHEMICAL ANALYSIS, S	W72-07710
, *PACIFIC OCEAN, ECOLCGY, WATER	TEMPERATURE, PHYTOPLANKTCN, ZOOPL	W72-07892
ESS(WATER), ALKALINITY, ENZYMES,	TEMPERATURE, PHOTOSYNT: /S, HARDN	W72-07514
, PHOTOSYNTHESIS, SESSILE ALGAE,	TEMPERATURE, DEPTH, LIGHT PENETRA	W72-07501
TA, DISTRIBUTION PATTERNS, WATER	TEMPERATURE, WATER QUALITY, PLANK	W72-07663
CONCENTRATION, MINNESOTA, WATER	TEMPERATURE, DISSOLVED OXYGEN.: /	W72-07145
RUMENTATION, FLUORESCENCE, WATER	TEMPERATURES, THERMAL STRESS.: /T	W72-07225
COPEPODS, DIATOMS, CHLOROPHYTA,	TEMPERATURE, SIZE, FECUNDITY, ALG	W72-14805
DS, PHYTOPLANKTON, CARBON, WATER	TEMPERATURE, GREAT LAKES REGION,	W73-00384
ENT, TUBIFICIDS, MOLLUSKS, WATER	TEMPERATURE, MARINE ANIMALS, ALAS	W73-00374
ORGANIC MATTER, MORTALITY, WATER	TEMPERATURE, PHAEOPHYTA, SAMPLING	W73-00366
ON, BENTHIC FAUNA, SIEVES, WATER	TEMPERATURE,: /LOGICAL DISTRIBUTI	W73-00367
MODELS, THERMAL STRATIFICATION,	TEMPERATURE, DISSOLVED CXYGEN, CO	W72-14405
TIC PLANTS, AQUATIC ALGAE, WATER	TEMPERATURE, ACID STREAMS, HYDROG	W72-14690
*WASHINGTON, *CHLOROPHYLL, WATER	TEMPERATURE, ALKALINITY, DISSOLVE	W72-14728
, SOYBEANS, LIGHT INTENSITY, AIR	TEMPERATURE, SAMPLING, AQUATIC AL	W73-01456
NDS, ABSORPTION, KINETICS, WATER	TEMPERATURE, RESPIRATION, ALGAE,	W73-01269
CENTRIFUGATION, TURBIDITY, WATER	TEMPERATURE, ALKALINITY, HYDROGEN	W73-00916
HYTA, PHYSICAL PROPERTIES, WATER	TEMPERATURE, DATA COLLECTIONS, MA	W73-00838
HENS, NITROGEN, LIGHT INTENSITY,	TEMPERATURE, ANABAENA, BACTERIA,	W73-02471
RICE, CYANOPHYTA, FERTILIZATION,	TEMPERATURE, CROP PRODUCTION, FUN	W73-02473
ATER PROPERTIES, SALINITY, WATER	TEMPERATURE, OXYGEN, TURBIDITY, B	W73-02029
A, STREAMFLOW, FLOW RATES, WATER	TEMPERATURE, WATER CHEMISTRY, CHE	W73-01972
ITE INVESTIGATIONS, *POPULATION,	TEMPORAL DISTRIBUTION, CHLOROPHYL	W72-11719
, BLACK CRAPPIE, CHAIN PICKEREL,	TENDIPES TENTANS, TRICHOPTERA, LI	W72-14280
OPHA, PROCLADIUS, WHITE OAK LAKE(	TENN), UPTAKE.: / PHYSA HETEROSTR	W70-06668
R, ILLINOIS RIVER, HOLSTON RIVER(	TENNESSEE).: /IRE), DELAWARE RIVE	W71-11517
IME, SNAILS, RADIOACTIVE WASTES,	TENNESSEE, CRAYFISH, MIDGES, GAMM	W70-06668
RAPPIE, CHAIN PICKEREL, TENDIPES	TENTANS, TRICHOPTERA, LIMNEPHILUS	W72-14280
RIC/ *ZINC, *LEAD, STIGEOCLONIUM	TENUE, CLADOPHORA GLOMERATA, ULOT	W71-05991
S, ANABAENA, NOSTOC, TOLYPOTHRIX	TENUIS, AULOSIRA FERTILISSIMA, AN	W71-10070
*TOLERANCE LEVELS, TOLYPOTHRIX	TENUIS, AULOSIRA FERTILISSIMA, AN	W71-12071
ATOA, THIOTHRIX NIVEA, THIOTHRIX	TENUIS, MERISMOPEDIA GLAUCA, DACT	W72-07896
SCILLATORIA AMOENA, OSCILLATORIA	TENUIS, OSCILLATORIA CHLCRINA, OS	W72-07896
, BENTONITE, LIME, *PULP WASTES,	TEOTIARY TREATMENT, EFFLUENTS.: /	W72-03753
N, CULTURE MEDIA, CHLOROPHYLL A,	TERATOLOGY, NITZSCHIA.: /, ACTIDO	W73-01445

L POLLUTION, SAMPLING, BACTERIA,	TOXICITY, BEHAVIOR, BITUMINOUS MA	W72-08790
MS, POLLUTION SOURCES, PESTICIDE	TOXICITY, FOOD CHAINS, FISH, BIOA	W72-07703
DIATOMS, SILICATES, CYANOPHYTA,	TOXICITY, COSTS, CARBON DIOXIDE.:	W72-07508
N EFFECTS, ABSORPTION, PESTICIDE	TOXICITY, LABORATORY TESTS.: /TIO	W73-00281
YTA, *CHLOROPHYTA, *CHRYSOPHYTA,	TOXICITY, INHIBITORS, TOXINS, INO	W72-14706
*PLANT GROWTH, LIGHT INTENSITY,	TOXICITY, PRIMARY PRODUCTIVITY, S	W72-14690
AIR POLLUTION, WATER POLLUTION,	TOXICITY, CHEMICALS, REGULATION,	W73-00703
NTATION, PUBLICATIONS, PESTICIDE	TOXICITY, BIOINDICATORS, AGRICULT	W73-01976
ATION, POLLUTANT IDENTIFICATION,	TOXICITY, *IRON, *CHLORELLA, NUTR	W73-02112
TOMS, PESTICIDES, DDD, PESTICIDE	TOXICITY, PATH OF POLLUTANTS, PES	W73-02280
LOGY, INVERTEBRATES, AMPHIBIANS,	TOXICITY, CHARA, SUBMERGED PLANTS	W72-00155
WATER POLLUTION SOURCES, SOAPS,	TOXICITY, PUBLIC HEALTH: /SPOSAL,	W72-01093
N, *AMMONIA, NITRATES, NITRITES,	TOXICITY, CORROSION, PUBLIC HEALT	W72-04086
*ALGAE, CHLOROPHYTA, CHLORELLA,	TOXICITY, WATER ANALYSIS, ABSORPT	W72-04134
ECTS, *PHOTOSYNTHESIS, PESTICIDE	TOXICITY, PLANT GROWTH REGULATORS	W71-10566
HORUS PESTICIDES, PHYTOPLANKTON,	TOXICITY, DISSOLVED OXYGEN, CULTR	W71-10096
INHIBITORS, PHYTOPLANKTON, PHYTO-	TOXICITY, HERBICIDES, INSECTICIDE	W71-10566
BON PESTICIDES, *DDT, *PESTICIDE	TOXICITY, *WATER POLLUTION EFFECT	W71-12303
OLUBILITY, SORPTION, PESTICIDES,	TOXICITY, DISSOLVED OXYGEN, LAKE	W71-12092
ONTROL, FINANCING, PLANT GROWTH,	TOXICITY, NUTRIENTS, WATER QUALIT	W71-13172
, PUBLIC HEALTH, BIODEGRADATION,	TOXICITY, CONTRACTS, GRANTS, RESE	W71-11516
*WATER QUALITY CONTROL, BIOMASS,	TOXICITY, EUTROPHICATION, TRACE E	W71-11793
*PESTICIDE RESIDUES, *PESTICIDE	TOXICITY, *AQUATIC PLANTS, AQUATI	W72-09650
ION, REVIEWS, *DATA COLLECTIONS,	TOXICITY, PATH OF POLLUTANTS, *ME	W72-12190
UALITY, TECHNOLOGY, SURFACTANTS,	TOXICITY, HAZARDS, BIODEGRADATION	W72-09169
SAMPLING, BIOASSAY, ENVIRONMENT,	TOXICITY, POISONS, ALGAL TOXINS,	W72-10678
ICALS, *PUBLIC HEALTH STANDARDS,	TOXICOLOGY, NTA.: /S, *TOXIC CHEM	W73-02144
WATER POLLUTION EFFECTS, *ALGAL	TOXINS.: /*TOXICITY, COAGULATION,	W72-07360
PATHOGENS, MICROORGANISMS, ALGAL	TOXINS.: /ERIA, FUNGI, PROTOZOA,	W72-08586
UTION SOURCES, SEA WATER, TROUT,	TOXINS.: /RIMP, CRABS, WATER POLL	W71-07412
UATIC ALGAE, *CRUSTACEANS, *FISH	TOXINS, *ALGAL POISONING, DINOFLA	W72-14056
*FOOD CHAINS, *TOXICITY, *ALGAL	TOXINS, *CHLOROPHYTA, *PHYTOTOXIN	W71-08597
*SCENEDESMUS, AQUATIC MI/ *ALGAL	TOXINS, *INHIBITION, *CHLORELLA,	W72-12255
A, CYANOPHYTA, SYSTEMATI/ *ALGAL	TOXINS, *MYXOBACTERIA, CHLOROPHYT	W72-07951
PHYTA, *NUISANCE ALGAE, / *ALGAL	TOXINS, *POISONOUS PLANTS, *CYANO	W72-14056
HYTOTOXICITY, *CYANOPHYT/ *ALGAL	TOXINS, *RED TIDE, *BEGGIATOA, *P	W72-14542
PATHOGENIC BACTERIA, COLIFORMS,	TOXINS, ANTIBIOTICS(PESTICIDES),	W72-13800
CHLOROPHYTA, SCENEDESMUS, ALGAL	TOXINS, ANTIBIOTICS(PESTICIDES),	W71-05155
*ANALYTICAL TECHNIQUES, *ALGAL	TOXINS, AQUATIC PLANTS, TOXINS, R	W73-00243
ONMENT, TOXICITY, POISONS, ALGAL	TOXINS, FISH TOXINS, SEDIMENTS, S	W72-10678
OPHYTA, *LABORATORY TESTS, ALGAL	TOXINS, HUMAN PATHOLOGY.: *CYAN	W72-04259
RYSOPHYTA, TOXICITY, INHIBITORS,	TOXINS, INORGANIC COMPOUNDS, ALGA	W72-14706
TY, CHEMICALS, REGULATION, ALGAL	TOXINS, MERCURY, DETERGENTS, WATE	W73-00703
S, PATH OF POLLUTANTS, BIOASSAY,	TOXINS, POISONS, POISONOUS PLANTS	W71-08597
, *ALGAL TOXINS, AQUATIC PLANTS,	TOXINS, RED TIDE, PLANT PHYSIOLOG	W73-00243
ITY, POISONS, ALGAL TOXINS, FISH	TOXINS, SEDIMENTS, SALT MARSHES,	W72-10678
ION, MIXING, ALGAL BLOOMS, ALGAL	TOXINS, WATER CIRCULATION, WATER	W72-09304
ITIES, HEAVY METALS, HERBICIDES,	TOXINS, WASTE DILUTION, NITROGEN,	W72-01472
NIQUES, WATER POLLUTION SOURCES,	TOXINS, WASTES, CHLOROPHYTA, SESS	W71-10553
ABAENA FLOW-AQUAE, HOMOGENIZING,	TRA: /AQUAE, DESTRATIFICATION, AN	W72-07890
I, CONIOPHORA OLIVACEA, LENZITES	TRABEA, SCOPULARIOPSIS BREVICAULI	W72-13813
UNGI, ALGAE, SOIL CONTAMINATION,	TRACE ELEMENTS, ION EXCHANGE, POL	W72-14303
TES, NITRATES, NUTRIENTS, SALTS,	TRACE ELEMENTS, LIGHT INTENSITY,	W72-12739
RATES, *TOXICITY, *FLUORESCENCE,	TRACE ELEMENTS, COPPER, CHLORELLA	W72-07660
TIVATION ANALYSIS, SPECTROSCOPY,	TRACE ELEMENTS, SAMPLING, WATER A	W73-00832
Y TECHNIQUES, CYCLING NUTRIENTS,	TRACE ELEMENTS, PUBLIC HEALTH, RA	W73-00826
CITY, EUROPE, ANIMAL PHYSIOLOGY,	TRACE ELEMENTS, HEAVY METALS, PHY	W73-00817
L TECHNIQUES, CYCLING NUTRIENTS,	TRACE ELEMENTS, CHELATION, ORGANI	W73-00821
INE RADIOISOTOPES, MARINE ALGAE,	TRACE ELEMENTS, PATH OF POLLUTANT	W73-00811
NE ALGAE, ENVIRONMENTAL EFFECTS,	TRACE ELEMENTS, COLLOIDS, NUCLEAR	W73-00818
DIMENTS, SILTS, PLANKTON, ALGAE,	TRACE ELEMENTS, VOLATILITY, FREEZ	W71-11036
OMASS, TOXICITY, EUTROPHICATION,	TRACE ELEMENTS, NITROGEN, PHOSPHO	W71-11793
ON, *NITROGEN, ELECTROCHEMISTRY,	TRACE ELEMENTS, HEAVY METALS, FLE	W72-01693
EDIMENTS, ENVIRONMENTAL EFFECTS,	TRACE ELEMENTS, ANALYTICAL TECHNI	W72-03215
TY, PATH OF POLLUTANTS, *METALS,	TRACE ELEMENTS, PESTICIDES, SALIN	W72-12190
TIC ALGAE, PRIMARY PRODUCTIVITY,	TRACE ELEMENTS, PHOSPHATES, NITRA	W71-09674
PHORUS RADIOISOTOPES, SEDIMENTS,	TRACE ELEMENTS, VOLCANOES, BENTHI	W71-09190
KTON, HUMIC ACIDS, FULVIC ACIDS,	TRACE ELEMENTS, CHEMICAL ANALYSIS	W71-09188
LGAE, AQUATIC PLANTS, SEA WATER,	TRACE ELEMENTS, TOXICITY, CHROMOS	W72-05952
TESTS, RESEARCH AND DEVELOPMENT,	TRACE ELEMENTS, ALGAE, NUTRIENTS,	W72-04433
, PROTEINS, ALGAE, GROWTH RATES,	TRACERS.: / CULTURES, CHLOROPHYLL	W72-03557
ANT IDENTIFICATION, COLORIMETRY,	TRACERS.: /MS, GAMMA RAYS, POLLUT	W71-11036
LA, FILTRATION, ELECTROPHORESIS,	TRACERS.: /TIONS, ANIONS, CHLOREL	W72-14840
I, ORGANIC COMPOUNDS, CHLORELLA,	TRACERS, AQUATIC ALGAE.: /M, FUNG	W72-14326
ICAL MODELS, PATH OF POLLUTANTS,	TRACERS, ANIMAL PHYSIOLOGY, NUCLE	W72-03348
STS, *MEASUREMENT, *MARINE ALGAE	TRACERS, CHEMICAL PRECIPITATION,	W72-12210
UDIES, LAKES, CYCLING NUTRIENTS,	TRACERS, LITTORAL, KINETICS, PHOS	W72-12543
ACTERIA, *TEXAS, BIODEGRADATION,	TRACERS, LAGOONS, CYANOPHYTA.: /B	W71-04873
TIC PRODUCTIVITY, RADIOISOTOPES,	TRACERS, MARINE ALGAE, HAWAII, WA	W71-11486
TIVITY TECHNIQUES, RADIOECOLOGY,	TRACERS, MONITORING, NUCLEAR POWE	W73-00790
PTION, RADIOACTIVITY TECHNIQUES,	TRACERS, PATH OF POLLUTANTS, ON-S	W73-00949
S, *ABSORPTION, ALGAE, CHORELLA,	TRACERS, PHOSPHATES, ECOSYSTEMS,	W72-09668
NIMAL PHYSIOLOGY, WATER BALANCE,	TRACERS, PLANT PHYSIOLOGY, AQUATI	W72-06342
ASONAL, WEATHER, WATER ANALYSIS,	TRACERS, RADIOACTIVITY TECHNIQUES	W72-09122
ACTIVITY TECHNIQUES, METABOLISM,	TRACERS, RADIOACTIVITY, CARBON DI	W72-12634
GAE, AQUATIC ALGAE, COLORIMETRY,	TRACES.: /HYTOPLANKTON, MARINE AL	W72-03545
STEPHANODISCUS, ANKISTRODESMUS,	TRACHELOMONAS, SAGINAW BAY, ORGAN	W72-10875
PANDORINA, PHOCUS, STAURASTRUM,	TRACHELOMONAS.: /ORIA, SPIRULINA,	W72-07661
CUS PYRUM, TRACHELOMONAS ARMATA,	TRACHELOMONAS HISPIDA VAR CRENULA	W73-00918
ACUS PLEURONECTES, PHACUS PYRUM,	TRACHELOMONAS ARMATA, TRACHELOMON	W73-00918

, CATFISH, WASTE TREATMENT, FARM	WASTES, DEHYDRATION, FEASIBILITY.	W71-07544
CES, INDUSTRIAL WASTES, DOMESTIC	WASTES, DINOFLAGELLATES, *LABORAT	W71-03095
ER POLLUTION SOURCES, INDUSTRIAL	WASTES, DOMESTIC WASTES, DINOFLAG	W71-03095
LIMITS, METHODOLOGY, INDUSTRIAL	WASTES, DOMESTIC WASTES, BIOASSAY	W72-12239
VATED SLUDGE, EFFLUENT, *ORGANIC	WASTES, ECONOMICS, CHEMICAL OXYGE	W71-12223
S, AGRICULTURAL RUNOFF, DOMESTIC	WASTES, ECOLOGY, FISH, ALGAE, AQU	W73-01972
POLLUTANTS, FOOD CHAINS, NUCLEAR	WASTES, ENVIRONMENTAL EFFECTS, EF	W72-10957
L PLANTS, INDUSTRIAL WASTES, OIL	WASTES, ESTUARIES, MONITORING.: /	W71-13746
MICAL WASTES, CLEANING, DOMESTIC	WASTES, EUTROPHICATION, WATER POL	W72-01685
TS, INDUSTRIAL WASTES, MUNICIPAL	WASTES, FARM WASTES, WASTE DISPOS	W72-01659
EMICAL OXYGEN DEMAND, INDUSTRIAL	WASTES, FARM WASTES, POLLUTANTS,	W71-06445
SOIL EROSION, FERTILIZERS, FARM	WASTES, FISH, FISH EGGS, TURBIDIT	W71-06444
LLUTION SOURCES, *FISHERIES, OIL	WASTES, FISH REPRODUCTION, WASTE	W71-13723
OR, BITUMINOUS MATERIALS, ANIMAL	WASTES, FISH, FLUORESCENCE, INFRA	W72-08790
BONS, *BIOINDICATORS, INDUSTRIAL	WASTES, FLORIDA, SESSILE ALGAE, E	W71-07412
SSILE ALGAE, ESTUARIES, CHEMICAL	WASTES, GAS CHROMATOGRAPHY, SEDIM	W71-07412
SOIL CULTURES, TITANIUM DIOXIDE	WASTES, HELGOLAND.: /CYSTIS SPP.,	W72-12239
NS), SURFACE DRAINAGE, MUNICIPAL	WASTES, INDUSTRIAL WASTES, SAMPLI	W71-08953
SEPARATION TECHNIQUES, *CHEMICAL	WASTES, INDUSTRIAL WASTES, CHLORI	W72-05315
E, ALKALINITY, BURNING, DOMESTIC	WASTES, INDUSTRIAL WAS: /N DIOXID	W72-05473
WATER POLLUTION CONTROL, ORGANIC	WASTES, INDUSTRIAL WASTES, WASTE	W73-00840
*MATHEMATICAL METHODS, MUNICIPAL	WASTES, INDUSTRIAL WASTES, NEW YO	W71-11685
PARATIVE PRODUCTIVITY, MUNICIPAL	WASTES, INDUSTRIAL WASTES, ANNUAL	W72-03228
ECHNIQUES, FILTRATION, MUNICIPAL	WASTES, INDUSTRIAL WASTES, OPERAT	W72-09386
C LIFE, HEAVY METALS, INDUSTRIAL	WASTES, INVERTEBRATES, ALGAE, *KE	W72-11854
EARCH AND DEVELOPMENT, MUNICIPAL	WASTES, LEGISLATION, REGULATION,:	W72-06638
IUM, CALCIUM, HYDROGEN, CHEMICAL	WASTES, LEAD, NICKEL, TOXICITY, B	W72-14303
PHOSPHATES, PHOSPHORUS, CHEMICAL	WASTES, LIMNOLOGY, INTERNATIONAL	W72-14782
IRATION, *HEAVY METALS, CHEMICAL	WASTES, MERCURY, COPPER, ZINC, PE	W72-12576
MESTIC WASTES, ALGAE, INDUSTRIAL	WASTES, METABOLISM, GROWTH RATES,	W72-09590
ES, INDUSTRIAL WASTES, MUNICIPAL	WASTES, METALS, HEAVY METALS, WAS	W71-11793
C BACTERIA, TRITIUM, RADIOACTIVE	WASTES, MONITORING, PHOTOGRAPHY,	W72-04742
.: *TEXTILE MILL	WASTES, MONTREAL, BALANCING TANKS	W70-10427
ATERS, WATER POLLUTION, DOMESTIC	WASTES, MUNICIPAL WASTES.: /ACE W	W70-09388
NALYTICAL TECHNIQUES, INDUSTRIAL	WASTES, MUNICIPAL WASTES, METALS,	W71-11793
GOONS, STABILIZATION, INDUSTRIAL	WASTES, MUNICIPAL WASTE, BIOCHEMI	W71-13356
ON, ALGAE, NUTRIENTS, INDUSTRIAL	WASTES, MUNICIPAL WASTES, FARM WA	W72-01659
DS, MUNICIPAL WASTES, INDUSTRIAL	WASTES, NEW YORK, ALGAE, WATER PO	W71-11685
LLUTION CONTROL, SEDIMENTS, FARM	WASTES, NUTRIENTS, PHOSPHORUS, AL	W72-00846
URCES, BIOASSAY, ALGAE, DOMESTIC	WASTES, NUTRIENTS, SEDIMENTS, ANA	W73-02478
S, INDUSTRIAL PLANTS, INDUSTRIAL	WASTES, OIL WASTES, ESTUARIES, MO	W71-13746
ULSIFIERS, DETERGENTS, OIL, *OIL	WASTES, OILY WATER, OIL-WATER INT	W70-09429
*WATER POLLUTION EFFECTS, *FARM	WASTES, OKLAHOMA, CYANOPHYTA, CHL	W72-04773
ON, MUNICIPAL WASTES, INDUSTRIAL	WASTES, OPERATION AND MAINTENANCE	W72-09386
*ENVIRONMENTAL EFFECTS, NUCLEAR	WASTES, PATH OF POLLUTANTS, MARIN	W72-10955
POWERPLANTS, EFFLUENTS, NUCLEAR	WASTES, PATH OF POLLUTANTS, ASIA,	W72-10958
DUES, RECYCLING FOOD, PROCESSING	WASTES, PATHOGENS, BIOENVIRONMENT	W72-00846
MARINE ALGAE, SEDIMENTS, NUCLEAR	WASTES, PATH OF POLLUTANTS, SEA W	W73-00817
NG, NUCLEAR POWERPLANTS, NUCLEAR	WASTES, PATH OF POLLUTANTS, WATER	W73-00790
NE PLANTS, MARINE ALGAE, NUCLEAR	WASTES, PATH OF POLLUTANTS, RADIO	W73-00828
D CHAINS, PUBLIC HEALTH, NUCLEAR	WASTES, PATH OF POLLUTANTS, DIFFU	W73-00823
, *COBALT RADIOISOTOPES, NUCLEAR	WASTES, PATH OF POLLUTANTS, ESTUA	W72-07826
OXYGEN, AQUATIC LIFE, MUNICIPAL	WASTES, PHOTOSYNTHESIS, RESPIRATI	W71-11017
S, PUMPING, AEROBIC TREAT/ *FARM	WASTES, POULTRY, NITROGEN, LAGOON	W72-08396
DEMAND, INDUSTRIAL WASTES, FARM	WASTES, POLLUTANTS, FERTILIZERS,	W71-06445
ERBIVORES, NATURAL STREAMS, PULP	WASTES, PRODUCTIVITY, DIELDRIN, P	W72-06340
RACE ELEMENTS, COLLOIDS, NUCLEAR	WASTES, RADIOECOLOGY.: /FFECTS, T	W73-00818
OF POLLUTANTS, FALLOUT, NUCLEAR	WASTES, RADIOECOLOGY, RADIOACTIVI	W73-00821
FAUNA, CELLULOSE, ENZYMES, PULP	WASTES, RADIOECOLOGY, WATER POLLU	W72-09668
MARINE ALGAE, *ABSORPTION, *MINE	WASTES, RADIOECOLOGY, *PATH OF PO	W72-12204
), RIFLE FALLS(COLO), FISH FECAL	WASTES, RESIDUAL FOOD, GREEN RIVE	W71-11006
WATER POLLUTION EFFECTS, NUCLEAR	WASTES, RIVERS, ABSORPTION, FISH,	W71-09225
WATER QUALITY CONTROL, DOMESTIC	WASTES, RUNOFF, WATER SOFTENING,	W72-11186
GE, MUNICIPAL WASTES, INDUSTRIAL	WASTES, SAMPLING, WATER QUALITY C	W71-08953
*WASTE WATER TREATMENT, *LIQUID	WASTES, SEWAGE TREATMENT, SEPARAT	W72-05315
OALESCENCE, FLOCCULATION, LIQUID	WASTES, SEPARATION TECHNIQUES, WA	W71-13356
GANESE, COBALT, IRON, INDUSTRIAL	WASTES, SEWAGE EFFLUENTS, POLLUTA	W71-11515
ODUCTIVITY, POLLUTANTS, CHEMICAL	WASTES, SEWAGE EFFLUENTS, WATER P	W72-08804
SPENDED LOAD, SEWAGE, INDUSTRIAL	WASTES, SILTS, WATER POLLUTION EF	W71-09795
, TEMPERATURE, KINETICS, NUCLEAR	WASTES, SOLUBILITY, ADSORPTION, W	W71-09182
LATION, LEACHING, DRAINAGE, FARM	WASTES, SOIL MANAGEMENT, GROUNDWA	W71-06443
OR, PATHOGENIC BACTERIA, ORGANIC	WASTES, SOLID WASTES, CHEMICALS,	W71-11006
ION(ATMOSPHERIC), RUNOFF, ANIMAL	WASTES, SURFACE WATERS, SOIL EROS	W71-04216
UDIES, TIME, SNAILS, RADIOACTIVE	WASTES, TENNESSEE, CRAYFISH, MIDG	W70-06668
GULATION, BENTONITE, LIME, *PULP	WASTES, TEOTIARY TREATMENT, EFFLU	W72-03753
DRAINAGE, INDUSTRIES, MUNICIPAL	WASTES, TURNOVERS, SEDIMENTS, HUM	W71-11009
HOSPHORUS PESTICIDES, INDUSTRIAL	WASTES, VIRUSES, ENERIC BACTERIA,	W72-04742
DIOISOTOPES, ABSORPTION, NUCLEAR	WASTES, WATER POLLUTION EFFECTS,	W72-04461
VITY EFFECTS, *TRITIUM, *NUCLEAR	WASTES, WATER POLLUTION EFFECTS,	W72-06342
MAND, RADIOACTIVITY, RADIOACTIVE	WASTES, WASTE WATER(POLLUTION), W	W72-05506
*ALGAE, PHYTOPLANKTON, / ORGANIC	WASTES, WATER POLLUTION EFFECTS,	W70-07842
ER, AGRICULTURAL CHEMICALS, FARM	WASTES, WASTE WATER TREATMENT, CA	W71-11698
L WASTES, MUNICIPAL WASTES, FARM	WASTES, WASTE DISPOSAL, PESTICIDE	W72-01659
*BIOLOGICAL TREATM/ AGRICULTURAL	WASTES, WATER POLLUTION CONTROL,	W72-02975
ODS, PATH OF POLLUTANTS, NUCLEAR	WASTES, WATER POLLUTION EFFECTS,	W72-10978
E DISPOSAL, *SEWAGE, *INDUSTRIAL	WASTES, WATER POLLUTION SOURCES,	W72-08804
TROL, ORGANIC WASTES, INDUSTRIAL	WASTES, WASTE WATER(POLLUTION), B	W73-00840
, *MUNICIPAL WASTES, *INDUSTRIAL	WASTES, WATER POLLUTION CONTROL,	W72-14282
ISOTOPES, RADIOISOTOPES, NUCLEAR	WASTES: /YSIS, SEA WATER, STABLE	W73-00831
ALGAE, PESTICIDES, HEAVY METALS,	WASTEWATER TREATMENT, SMELTS, MIC	W72-14282

S, *ALGAE, SEWAGE, PRODUCTIVITY, WATER PCLLUTION EFFECTS, PHOSPHOR W71-12857
TROPHICATION, WATER TEMPERATURE, WATER POLLUTION EFFECTS, WATER PO W71-12092
CHIGAN, WATER POLLUTION SOURCES, WATER POLLUTICN CONTROL, TROPHIC W71-12072
ED SLUDGE, *NUTRIENTS, NITROGEN, WATER POLLUTION EFFECTS, WATER PO W71-11972
RATURE, WATER POLLUTION EFFECTS, WATER POLLUTICN CONTROL, ENERGY, W71-12092
UATIC PLANTS, CHEMICAL ANALYSIS, WATER POLLUTICN SOURCES.: /AE, AQ W71-12252
IENTS, SEDIMENTS, LAKE MICHIGAN, WATER POLLUTION SOURCES, WATER PO W71-12072
TROGEN, WATER POLLUTION EFFECTS, WATER POLLUTICN SOURCES.: /TS, NI W71-11972
LLUTANTS, ANALYTICAL TECHNIQUES, WATER POLLUTION SOURCES, TOXINS, W71-10553
ALIFORNIA, TUBIFICIDS, COLORADO, WATER POLLUTICN CONTROL, WASHINGT W71-11006
LUENTS, WATER POLLUTION SOURCES, WATER POLLUTION EFFECTS, SAMPLING W71-11008
CTIVITY, FERTILIZERS, EFFLUENTS, WATER POLLUTION SOURCES, WATER PO W71-11008
NITROGEN, WISCONSIN, NUTRIENTS, WATER POLLUTICN EFFECTS, INTERFAC W71-10645
MS, *DINOFLAGELLATES, *RED TIDE, WATER POLLUTION SOURCES.: / DIATO W72-01329
SOFTENING, SOAPS, AQUATIC ALGAE, WATER POLLUTION EFFECTS, WATER QU W72-01085
ONS, ALGAE, CHEMICAL PROPERTIES, WATER POLLUTICN, PROTEIN, PRIMARY W72-01473
E ERIE, WATER POLLUTION EFFECTS, WATER POLLUTION SOURCES, LIMNOLOG W72-01094
LUTION, WATER POLLUTION SOURCES, WATER POLLUTICN TREATMENT, NUTRIE W72-01685
AE, WASTE ASSIMILATIVE CAPACITY, WATER POLLUTION EFFECTS, POLLUTAN W72-01685
WATER POLLUTICN CONTROL ACT.: W72-01085
PERTIES, WATER POLLUTION EFFECT, WATER PCLLUTION SOURCES, COST CCM W72-00974
IC ALGAE, BIOLOGICAL PROPERTIES, WATER POLLUTION EFFECT, WATER POL W72-00974
OMS, *PHYTOPLANKTON, *NUTRIENTS, WATER POLLUTICN EFFECTS, AQUATIC W72-01329
CAL GOVERNMENTS, WASTE DISPOSAL, WATER POLLUTICN SOURCES, SOAPS, T W72-01093
EUTROPHICATION, WATER POLLUTION, WATER POLLUTION SOURCES, WATER PO W72-01685
DOMESTIC WASTES, EUTROPHICATION, WATER POLLUTION, WATER PCLLUTION W72-01685
ATER QUALITY CONTROL, STANDARDS, WATER POLLUTION SOURCES, WATER PO W72-01660
RDS, LEGISLATION, LEGAL ASPECTS, WATER POLLUTICN, ADJUDICATION PRO W72-01093
URON, *LAKE ONTARIO, *LAKE ERIE, WATER POLLUTION EFFECTS, WATER PO W72-01094
NDARDS, WATER POLLUTION SOURCES, WATER POLLUTION EFFECTS, RECREATI W72-01660
TION, WATER QUALITY, RESERVOIRS, WATER PCLLUTION SOURCES, ALGAE, R W72-04260
S, *FROGS, *TRITIUM, ABSORPTION, WATER POLLUTION EFFECTS, RADICACT W72-03348
OMIC EFFICIENCY, *PUBLIC HEALTH, WATER POLLUTICN, HAZARDS, NUTRIEN W72-04266
TESTS, WATER POLLUTION SOURCES, WATER POLLUTION EFFECTS, FLUCTUAT W72-03228
CAL TREATM/ AGRICULTURAL WASTES, WATER POLLUTION CONTRCL, *BIOLOGI W72-02975
OASTS, BIORTHYMS, ON-SITE TESTS, WATER PCLLUTICN SOURCES, WATER PO W72-03228
NIA, FOOD CHAINS, PUBLIC HEALTH, WATER POLLUTICN SOURCES.: /ALIFOR W72-03324
TION EFFECTS, *ALGAE, *PLANKTON, WATER POLLUTION CONTROL, METHODOL W72-03218
OLLUTION EFFECTS, AIR POLLUTION, WATER POLLUTION, LICHENS, SOIL-WA W72-03347
ALGAE, MARINE FISH, ABSORPTION, WATER POLLUTICN SOURCES, RADICACT W72-00940
E ALGAE, *EUTR/ *WASTE DILUTICN, WATER POLLUTICN CONTROL, *NUISANC W72-00799
GANISMS, ALGAL CONTROL, *POLLUT/ WATER POLLUTICN EFFECTS, *MICROOR W72-00801
LLUTANTS, AIR POLLUTION EFFECTS, WATER POLLUTION EFFECTS, BAYS, WA W72-00959
ICATION, LAKES, *NUTRIENTS, ALG/ WATER POLLUTION EFFECTS, *EUTROPH W72-00798
NTROL, *WATER POLLUTION EFFECTS, WATER POLLUTICN SOURCES, AQUATIC W72-09646
YMES, PULP WASTES, RADIOECOLOGY, WATER POLLUT: /NA, CELLULCSE, ENZ W72-09668
ALGAL TOXINS, WATER CIRCULATION, WATER PCLLUTION CONTROL, FISHKILL W72-09304
STICIDES, *PRIMARY PRODUCTIVITY, WATER POLLUTION EFFECTS, ESTUARIN W72-09653
ONTROL, FISHKILL, NEW HAMPSHIRE, WATER POLLUTION EFFECTS, COPPER S W72-09304
*EUTROPHICATION, WATER SAMPLING, WATER PCLLUTICN EFFECTS, BIOMASS, W72-09111
HENSIVE PLANNING, AQUATIC WEEDS, WATER POLLUTION SOURCES, MATHEMAT W72-09171
CAT/ *NUTRIENTS, *PHYTOPLANKTCN, WATER POLLUTICN EFFECTS, EUTROPHI W72-09167
RCES, BIOCHEMICAL OXYGEN DEMAND, WATER POLLUTICN CONTROL, NITRATES W72-09167
N DIOXIDE, SEASONAL, PHOSPHORUS, WATER POLLUTION SOURCES, BICCHEMI W72-09167
DS, *POL/ *ALGAE, *FLOCCULATION, WATER PCLLUTICN TREATMENT, COLLOI W72-11833
LIMIT, WATER POLLUTION EFFECTS, WATER POLLUTION SOURCES, INHIBITI W72-11727
OL, WATER ANALYSIS, CHEMCONTROL, WATER POLLUTION EFFECTS, WATER QU W72-11724
ELPS, WATER ANALYSIS, EXUDATION, WATER POLLUTION, PHYTCPLANKTON, W W72-12166
WATER POLLUTION, PHYTOPLANKTCN, WATER POLLUTION SOURCES, PHAEOPHY W72-12166
*COAST REVIEW, *BIBLIOGRAPHIES, WATER POLLUTICN SCURCES, THERMAL W72-12190
NUTRIENTS, *AEROBIC CONDITIONS, WATER POLLUTION CONTROL.: /ENTIAL W72-11617
AQUATIC ENVIRONMENT, TURBIDITY, WATER PCLLUTICN, PHAEOPHYTA, MOLL W72-11925
ECOSYSTEMS, *BIOLOGICAL COMMUN/ WATER POLLUTION EFFECTS, ECOLOGY, W72-11854
TACEANS, FOOD CHAINS, FOOD WEBS, WATER POLLUTICN EFFECTS, AQUATIC W72-11925
PHOSPHORUS, ALGAE, PLANT GROWTH, WATER PCLLUTICN TREATMENT, EVALUA W72-11702
ONAS, INHIBITORS, *LETHAL LIMIT, WATER POLLUTION EFFECTS, WATER PO W72-11727
STAGE, CRUSTACEANS, MONITORING, WATER PCLLUTICN EFFECTS, MOLLUSKS W72-10955
IENTS, *ALGAE, SEWAGE TREATMENT, WATER POLLUTION SOURCES, WATER PO W72-10625
ALYTICAL TECHNIQUES, EVALUATION, WATER POLLUTICN, POLLUTANT IDENTI W72-10678
IDES, IONS, CYTOLOGICAL STUDIES, WATER POLLUTICN EFFECTS.: / FLUOR W72-11572
H OF POLLUTANTS, NUCLEAR WASTES, WATER POLLUTION EFFECTS, FOOD CHA W72-10978
OF POLLUTANTS, *WATER POLLUTICN, WATER PCLLUTICN SOURCES, AQUATIC W72-10686
ATMENT, WATER PCLLUTICN SOURCES, WATER POLLUTION EFFECTS, AGRICULT W72-10625
COLI, AZOTOBACTER, PSEUDOMONAS, WATER PCLLUTION EFFECTS.: /ES, E. W72-10861
LGAE, FISH POPULATIONS, FLORIDA, WATER PCLLUTICN EFFECTS, NITRCGEN W72-08986
CAL DISTRIBUTION, BIOINDICATORS, WATER POLLUTION SOURCES.: /COLOGI W72-08579
L COMMUNITIES, RIVERS, BACTERIA, WATER POLLUTION, ACTIVATED SLUDGE W72-08573
EN ION CONCENTRATION, CORROSION, WATER POLLUTICN EFFECTS.: /HYDROG W72-08683
AL, *SEWAGE, *INDUSTRIAL WASTES, WATER POLLUTION SOURCES, ON-SITE W72-08804
EMICAL WASTES, SEWAGE EFFLUENTS, WATER PCLLUTICN EFFECTS.: /TS, CH W72-08804
, RIVERS, *ELECTRIC POWERPLANTS, WATER POLLUTION EFFECTS, WATER PO W72-07791
SMS, AQUATIC ANIMALS, NUTRIENTS, WATER POLLUTICN EFFECTS, WATER QU W72-07896
S, SAMPLING, DIATOMS, TURBIDITY, WATER POLLUTION SOURCES, NUTRIENT W72-07890
PLANTS, WATER POLLUTION EFFECTS, WATER POLLUTION SOURCES, *FISH, D W72-07791
PLANKTON, BENTHOS, ALGAE, FISH, WATER POLLUTICN, WATER PCLLUTION W72-07841
, SUSPENDED LOAD, BIOINDICATORS, WATER POLLUTION, MICROB: /EASONAL W72-07896
CAL POWER PLANTS, HEAT TRANSFER, WATER PCLLUTICN EFFECTS, WATER TE W72-07907
S, ALGAE, FISH, WATER POLLUTION, WATER POLLUTION SOURCES.: /BENTHO W72-07841
GLENOPHYTA, PROTOZOA, EQUIPMENT, WATER PCLLUTION EFFECTS.: /AS, EU W72-07661
DENSERS, PROTOZOA, ALGAE, FUNGI, WATER PCLLUTICN EFFECTS.: /R, CCN W72-07526

OST COMPARISONS, BENEFICIAL USE,	WATER QUALITY, CALIFORNIA, BIOCON	W72-00974
AL ENGINEERING, WATER TREATMENT,	WATER QUALITY CONTROL, STANDARDS,	W72-01660
POLLUTION TREATMENT, EVALUATION,	WATER QUALITY.: /T GROWTH, WATER	W72-11702
ONTROL, WATER POLLUTION EFFECTS,	WATER QUALITY, CARBON.: /S, CHEMC	W72-11724
ATOMS, NUTRIENTS, PHYTOPLANKTON,	WATER QUALITY, MICHIGAN, SURFACE	W72-10875
SIS, WATER ANALYSIS, EVALUATION,	WATER QUALITY CONTROL, *BIOASSAY,	W72-10883
S, *EUTROPHICATION, *DETERGENTS,	WATER QUALITY CONTROL, DOMESTIC W	W72-11186
YCETES, FUNGI, LABORATORY TESTS,	WATER QUALITY CONTROL.: / ACTINOM	W72-11604
N CONTROL, MANAGEMENT, ANALYSIS,	WATER QUALITY, CYANOPHYTA, NUISAN	W72-09171
ON, WASTE TREATMENT, INDUSTRIES,	WATER QUALITY, TECHNOLOGY, SURFAC	W72-09169
TENANCE, *WASTE WATER TREATMENT,	WATER QUALITY CONTROL, *REVIEWS.:	W72-09386
AT/ *EUTROPHICATION, *RECYCLING,	WATER QUALITY CONTROL, SEWAGE TRE	W72-09378
ION CONCENTRATION, EFFICIENCIES,	WATER QUALITY CONTROL, EUTROPHICA	W72-04788
TION, OXIDATION, ORGANIC MATTER,	WATER QUALITY, WATER CHEMISTRY, W	W72-04853
AE, AQUATIC PLANTS, CLAYS, ODOR,	WATER QUALITY, FOOD CHAINS, DISSO	W72-04759
IDE, HYDROGEN ION CONCENTRATION,	WATER QUALITY CONTROL, WASTE WATE	W72-04431
ANOPHYTA, ALKALINITY, NUTRIENTS,	WATER QUALITY, DIATOMS, CHLOROPHY	W72-05610
OROPHYLL, ENZYMES, PERMEABILITY,	WATER QUALITY CONTROL, WATER POLL	W72-06037
WIMMING, BOATING, DOCUMENTATION,	WATER QUALITY, SILTING, AGRICULTU	W72-05869
ION, LEGISLATION, PUBLIC HEALTH,	WATER QUALITY CONTROL, ALGAL CONT	W72-06655
MENTS, *WATER POLLUTION CONTROL,	WATER QUALITY CONTROL, PHOSPHORUS	W72-06638
NIMALS, *BIOTA, *AQUATIC PLANTS,	WATER QUALITY, ENVIRONMENTAL EFFE	W72-06526
SION, ENZYMES, CHLEATION, ALGAE,	WATER QUALITY, BIODEGRADATION, SE	W72-06837
TICAL MODELS, WATER TEMPERATURE,	WATER QUALITY, CARBON DIOXIDE, HY	W72-06274
CARBON FIXATION, EUTROPHICATION,	WATER QUALITY CONTROL, MASSACHUSE	W72-06046
L, EUTROPHICATION, ALGAL BLOOMS,	WATER QUALITY.: /ADATION, FISHKIL	W71-08768
ES, INDUSTRIAL WASTES, SAMPLING,	WATER QUALITY CONTROL, COST ANALY	W71-08953
*ALGAE,/ *ESTUARIES, *NUTRIENTS,	WATER QUALITY CONTROL, *ECOLOGY,	W71-08775
, PHOSPHORUS, WATER QUALITY ACT,	WATER QUALITY.: /OXYGEN, NITROGEN	W71-06444
ED OXYGEN, NITROGEN, PHOSPHORUS,	WATER QUALITY ACT, WATER QUALITY.	W71-06444
OSPHATES, ALGAE, EUTROPHICATION,	WATER QUALITY, RIVERS, IOWA, CORN	W71-06445
VIRONMENT, *POLLUTION ABATEMENT,	WATER QUALITY, STANDARDS, WATER P	W71-08214
CTS, *KENTUCKY, SULFUR BACTERIA,	WATER QUALITY, SULFATES, HYDROGEN	W71-07942
TY, ODOR, OXYGEN SAG, TURBIDITY,	WATER QUALITY, WATER TREATMENT, N	W71-09958
PULATION, WASTE WATER TREATMENT,	WATER QUALITY CONTROL.: /HUMAN PO	W71-09788
TE DISPOSAL, PATH OF POLLUTANTS,	WATER QUALITY.: / PESTICIDES, WAS	W71-05553
USE, *COOLING WATER, *FISHERIES,	WATER QUALITY, FISH ATTRACTANTS,	W70-09596
IENTS, SURFACE WATERS, NITROGEN,	WATER QUALITY, PHOSPHORUS, AEROBI	W70-10405
ES, *WATER POLLUTION, MINNESOTA,	WATER QUALITY, NUTRIENTS, SEDIMEN	W71-03012
ODUCTIVITY, ESSENTIAL NUTRIENTS,	WATER QUALITY, WATER CHEMISTRY, C	W70-04080
ECTS, NUTRIENTS, EUTROPHICATION,	WATER QUALITY, WATER POLLUTION CO	W70-08249
*ALGAE, *EUTROPHICATION, LAKES,	WATER QUALITY, PHOSPHORUS, PHOSPH	W70-08944
, NITRATES, HUDSON RIVER, ALGAE,	WATER QUALITY, SEWAGE DISPOSAL, W	W70-06509
RIOR APPROPRIATION, GROUNDWATER,	WATER QUALITY, COMPETING USES, DI	W70-07100
S, *WATER UTILIZATION, *IMPAIRED	WATER QUALITY, RIPARIAN RIGHTS, P	W70-07100
ECTS, NUTRIENTS, EUTROPHICATION,	WATER QUALITY, WATER POLLUTION CO	W70-08248
C MATTER, *OXYGEN SAG, *IMPAIRED	WATER QUALITY, *ESTUARIES, *DELAW	W70-09189
RADIATION, WATER QUALITY, ALGAE,	WATER QUALITY CONTROL, EUTROPHICA	W72-12487
HOTOGRAPHY, *INFRARED RADIATION,	WATER QUALITY, ALGAE, WATER QUALI	W72-12487
OPERTIES, MICROSCOPY, LIMNOLOGY,	WATER QUALITY, ECOLOGY, AQUATIC H	W72-13142
UTRIENTS, LAKE HURON, EFFLUENTS,	WATER QUALITY, PHOSPHORUS, PHYTOP	W72-13662
HICATION, *HYDROLOGY, *NEW YORK,	WATER QUALITY, NUTRIENTS, ALGAE,	W72-13851
S, *MATHEMATICAL MODELS, MIXING,	WATER RE-USE, LIGHT INTENSITY, EV	W71-07099
S, *FISH HATCHERIES, *SALMONIDS,	WATER REQUIREMENTS, TEMPERATURE,	W71-11006
ONMENTAL EFFECTS, INVERTEBRATES,	WATER RESOURCES DEVELOPMENT.: /IR	W71-10577
YSTEMS, ESTUARIES, CORAL, REEFS,	WATER RESOURCES, SYSTEMS ANALYSIS	W71-13723
ATION, AQUATIC ALGAE, NUTRIENTS,	WATER RESOURCES, WASTE ASSIMILATI	W71-13256
CATION, *MEASUREMENT, *SAMPLING,	WATER RESOURCES, AREA REDEVELOPME	W72-03215
	WATER RESOURCES RESEARCH.:	W71-01192
ICALS, COPPER SULFATE, CHLORINE,	WATER RESOURCES, METHODOLOGY, EVA	W71-05083
, DIVERSION, RECIRCULATED WATER,	WATER REUSE, LEGAL ASPECTS, JUDIC	W70-07100
TER TREATMENT, *DENITRIFICATION,	WATER REUSE, *NITROGEN, PHOSPHORU	W72-04309
PORATION, RESERVOIRS, NUTRIENTS,	WATER REUSE.: /ION, RESERVOIR EVA	W72-02412
ARY TREATMENT, SEWAGE EFFLUENTS,	WATER REUSE.: /IS, ECOLOGY, TERTI	W72-03970
TS, PHOSPHORUS, ALGAE, NITRATES,	WATER REUSE, SALINITY, PESTICIDES	W72-00846
NUTRIENTS, RESERVOIR OPERATION,	WATER REUSE, POTABLE WATER, ALGAE	W71-13185
*SALINE LAKES, *EUTROPHICATION,	WATER SAMPLING, WATER POLLUTION E	W72-09111
BIOASSAY, CHLORELLA, NUTRIENTS,	WATER SAMPLING, ORGANIC MATTER, W	W73-00234
ONTROL, DOMESTIC WASTES, RUNOFF,	WATER SOFTENING, ENVIRONMENTAL EF	W72-11186
ARDS, PUBLIC HEALTH, INSPECTION,	WATER SOFTENING, SOAPS, AQUATIC A	W72-01085
EFFECTS, NUTRIENTS, PHOSPHOROUS,	WATER SOFTENING, LEGAL ASPECTS, H	W71-09429
TION, BIOCHEMICAL OXYGEN DEMAND,	WATER SPORTS, PARKS, COST ANALYSI	W72-00042
SPHORUS, THERMAL STRATIFICATION,	WATER SUPPLY, SUCCESSION, ECONOMI	W71-13183
OSPHORUS, RESERVOIRS, FISHERIES,	WATER SUPPLY, COSTS, WATER DEMAND	W71-13172
WATER, IRRIGATION, RETURN FLOW,	WATER SUPPLY, LIVESTOCK, SEWAGE,	W71-06435
AGEMENT, GROUNDWATER, WISCONSIN,	WATER SUPPLY.: / WASTES, SOIL MAN	W71-06443
ION, *LAKES, *MAINE, RECREATION,	WATER SUPPLY, WATER POLLUTION EFF	W72-04280
, ARROYO GRANDE CREEK.: *LOPEZ	WATER SUPPLY PROJECT, LOPEZ CREEK	W72-05594
DOMESTIC WATER, *SURFACE WATERS,	WATER SUPPLY, EUTROPHICATION, ALG	W71-05083
T, *WATER REUSE, *WATER QUALITY,	WATER SUPPLY, ALGAE, ACTIVATED CA	W70-10381
UCTIVITY, GREAT LAKES, NEW YORK,	WATER SUPPLY, SYSTEMATICS, LIMNOL	W73-01615
NTS, WATER PROPERTIES, SALINITY,	WATER TEMPERATURE, OXYGEN, TURBID	W73-02029
ACTERIA, STREAMFLOW, FLOW RATES,	WATER TEMPERATURE, WATER CHEMISTR	W73-01972
QUES, CENTRIFUGATION, TURBIDITY,	WATER TEMPERATURE, ALKALINITY, HY	W73-00916
RHODOPHYTA, PHYSICAL PROPERTIES,	WATER TEMPERATURE, DATA COLLECTIO	W73-00838
COMPOUNDS, ABSORPTION, KINETICS,	WATER TEMPERATURE, RESPIRATION, A	W73-01269
D AQUATIC PLANTS, AQUATIC ALGAE,	WATER TEMPERATURE, ACID STREAMS,	W72-14690
EEDS, *WASHINGTON, *CHLOROPHYLL,	WATER TEMPERATURE, ALKALINITY, DI	W72-14728
OMPOUNDS, PHYTOPLANKTON, CARBON,	WATER TEMPERATURE, GREAT LAKES RE	W73-00384
EQUIPMENT, TUBIFICIDS, MOLLUSKS,	WATER TEMPERATURE, MARINE ANIMALS	W73-00374

RIBUTION, BENTHIC FAUNA, SIEVES,	WATER TEMPERATURE,: /LOGICAL DIST	W73-00367
ATER, ORGANIC MATTER, MORTALITY,	WATER TEMPERATURE, PHAEOPHYTA, SA	W73-00366
NVIRONMENT, SEA WATER, SALINITY,	WATER TEMPERATURE, CHEMICAL ANALY	W72-08141
NG, NATIONAL PARKS, HOT SPRINGS,	WATER TEMPERATURES, NOSTOC STREAM	W72-08508
S, CRUSTACEANS, WORMS, MOLLUSKS,	WATER TEMPERATURE, SALINITY, SEA	W72-09092
RROPHYTA, DISTRIBUTION PATTERNS,	WATER TEMPERATURE, WATER QUALITY,	W72-07663
EN ION CONCENTRATION, MINNESOTA,	WATER TEMPERATURE, DISSOLVED OXYG	W72-07145
, INSTRUMENTATION, FLUORESCENCE,	WATER TEMPERATURES, THERMAL STRES	W72-07225
NESS(WATER), CONDUCTIVITY, ZINC,	WATER TEMPERATURE, AL: /GEN, HARD	W72-07890
N, SEA WATER, SALINITY, PHENOLS,	WATER TEMPERATURE, CHEMICAL ANALY	W72-07710
ANSFER, WATER POLLUTION EFFECTS,	WATER TEMPERATURE, ALGAE, BULLHEA	W72-07907
GRAPHY, *PACIFIC OCEAN, ECOLOGY,	WATER TEMPERATURE, PHYTOPLANKTON,	W72-07892
KE STRUCTURES, DISCHARGE(WATER),	WATER TEMPERATURE.: /IMNION, INTA	W70-09614
G, ALGAE, SOLVENTS, SURFACTANTS,	WATER TEMPERATURE, PLANT PIGMENTS	W70-09429
ZOA, HYDROGEN ION CONCENTRATION,	WATER TEMPERATURE, BACTERIA, ALGA	W71-05626
THERMAL POLLUTION, *GREENHOUSES,	WATER TEMPERATURE, POWERPLANTS, D	W70-08832
N ION CONCENTRATION, PERIPHYTON,	WATER TEMPERATURE.: /DES, HYDROGE	W72-05610
US, NITROGEN, NUTRIENTS, SEWAGE,	WATER TEMPERATURE, LIGHT PENETRAT	W72-04734
ON, EVALUATION, WATER CHEMISTRY,	WATER TEMPERATURE, DIS: /OPHICATI	W72-06526
OWTH RATES, MATHEMATICAL MODELS,	WATER TEMPERATURE, WATER QUALITY,	W72-06274
HYTA, CHRYSOPHYTA, EUGLENOPHYTA,	WATER TEMPERATURE, THERMAL STRATI	W72-04855
N, WEATHER, RESERVOIR OPERATION,	WATER TEMPERATURE, PHYTOPLANKTON,	W72-05282
OLORADO, *RECREATION FACILITIES,	WATER TEMPERATURE, CHEMICAL ANALY	W72-04791
, ZOOPLANKTON, FISH POPULATIONS,	WATER TEMPERATURE, THERMAL STRATI	W72-04853
S, *THERMAL POLLUTION, *REVIEWS,	WATER TEMPERATURE, DISSOLVED OXYG	W71-07417
ICAL PROPERTIES, EUTROPHICATION,	WATER TEMPERATURE, WATER POLLUTIO	W71-12092
RPHOLOGY, AQUATIC ALGAE, MOSSES,	WATER TEMPERATURE, WATER CHEMISTR	W71-11949
YGEN, *PHYTOPLANKTON, ESTUARIES,	WATER TEMPERATURE, TURBIDITY, PHO	W71-11876
INE, HYDROGEN ION CONCENTRATION,	WATER TEMPERATURE, DISSOLVED OXYG	W71-11914
, TRACERS, MARINE ALGAE, HAWAII,	WATER TEMPERATURE, WAVES, LIGHT I	W71-11486
, PYRROPHYTA, SAMPLING, ECOLOGY,	WATER TEMPERATURE.: / CHRYSOPHYTA	W72-03543
ROLOGIC BUDGET, SOLAR RADIATION,	WATER TEMPERATURE, SALT BALANCE,	W72-02274
T, FLORIDA, ELECTRIC GENERATORS,	WATER TEMPERATURE, RHODOPHYTA, PH	W72-11252
ERIE, LAKE HURON, LAKE SUPERIOR,	WATER TEMPERATURE, HYDROGEN ION C	W72-10875
BRATES, PHYTOPLANKTON, PLANKTON,	WATER TEMPERATURE, E. COLI, BENTH	W72-09668
, OLIGOTROPHY, DISSOLVED OXYGEN,	WATER TEMPERATURE, HYDROGEN ION C	W72-13871
T GROWTH, ANALYTICAL TECHNIQUES,	WATER TEMPERATURE.: /RIENTS, PLAN	W72-13651
, BIOMASS, ANAEROBIC CONDITIONS,	WATER TEMPERATURE, CHLORIDES, PHO	W72-12940
AE, ANALYTICAL TECHNIQUES, WASTE	WATER TREATMENT, NUTRIENTS, NUTRI	W72-12966
*OXIDATION PONDS, *ALGAE, *WASTE	WATER TREATMENT, *BYPRODUCTS, MIX	W72-13508
Y CONTROL, EUTROPHICATION, WASTE	WATER TREATMENT, SEDIMENTATION RA	W72-12487
*COLD/ *SEWAGE BACTERIA, *WASTE	WATER TREATMENT, *SEWAGE LAGOONS,	W72-12565
XIDATION LAGOONS, *ALGAE, *WASTE	WATER TREATMENT, *MATHEMATICAL MO	W72-12289
PERATION AND MAINTENANCE, *WASTE	WATER TREATMENT, WATER QUALITY CO	W72-09386
OPERATION AND MAINTENANCE, WASTE	WATER TREATMENT.: /NTIAL, ODORS,	W72-10401
MENT, TERTIARY TREATMENT, *WASTE	WATER TREATMENT.: /, *WATER TREAT	W72-09393
TECHNIQUES, MICROBIOLOGY, *WASTE	WATER TREATMENT.: /S, ANALYTICAL	W72-09590
E, *ODOR, SURVEYS, MICROBIOLOGY,	WATER TREATMENT, PHYTOPLANKTON, F	W72-10076
STS, COST ANALYSIS, ALGAE, WASTE	WATER TREATMENT.: / LABORATORY TE	W72-10390
DISSOLVED OXYGEN, SLUDGE, WASTE	WATER TREATMENT.: /ION, PROTOZOA,	W72-10233
AE, *BACTERIA, *SYMBIOSI/ *WASTE	WATER TREATMENT, *NUTRIENTS, *ALG	W72-11100
*SEWAGE EFFLUENTS, *ALGAE, WASTE	WATER TREATMENT, ACTIVATED SLUDGE	W72-09176
NALYSIS, NITROGEN, CLAYS, *WASTE	WATER TREATMENT.: /SLUDGE, COST A	W72-11642
DROGEN ION CONCENTRATION, *WASTE	WATER TREATMENT.: /PROPERTIES, HY	W72-11833
ATMENT FACILITIES, ALGAE, *WASTE	WATER TREATMENT, *ALGAL CONTROL,	W72-02975
ESIGN CRITERIA, KINETICS, *WASTE	WATER TREATMENT.: / TECHNIQUES, D	W72-02363
*AQUATIC AL/ *PHOSPHORUS, *WASTE	WATER TREATMENT, *ALGAL CONTROL,	W72-03970
UM, MAGNESIUM, NUTRIENTS, *WASTE	WATER TREATMENT, TERTIARY TREATME	W72-03613
HEMICAL OXYGEN DEMAND, D/ *WASTE	WATER TREATMENT, *TURBIDITY, BIOC	W72-03753
SLUDGE, ALGAE, BACTERIA, *WASTE	WATER TREATMENT, EUTROPHICATION.:	W72-04086
*ALGAL CONTROL, AERATION, *WASTE	WATER TREATMENT.: /N, DIFFUSION,	W72-00800
S, WATER QUALITY CONTROL, *WASTE	WATER TREATMENT.: /ICAL TECHNIQUE	W72-00422
SINFECTION, FLOCCULATION, *WASTE	WATER TREATMENT.: /FILTRATION, DI	W72-00411
, COST ANALYSIS, MICHIGAN, WASTE	WATER TREATMENT.: / SPORTS, PARKS	W72-00042
OUS EARTH, *POROUS MEDIA, *WASTE	WATER TREATMENT, *DENITRIFICATION	W72-00974
TION, ENVIRONMENTAL ENGINEERING,	WATER TREATMENT, WATER QUALITY CO	W72-01660
AL CHEMICALS, FARM WASTES, WASTE	WATER TREATMENT, CALIFORNIA.: /UR	W71-11698
ION, LIME, / *PHOSPHATES, *WASTE	WATER TREATMENT, *FOAM FRACTIONAT	W71-12200
TREATMENT, *PHOSPHOROUS, *WASTE	WATER TREATMENT, NUTRIENTS, SYNTH	W71-12188
TMENT, *ACTIVATED SLUDGE/ *WASTE	WATER TREATMENT, *BIOLOGICAL TREA	W71-12223
ABOLISM, DESIGN CRITERIA, *WASTE	WATER TREATMENT, EFFLUENTS.: /MET	W71-13341
N, CALCIUM, SEDIMENTATION, WASTE	WATER TREATMENT.: /MS, ACCLIMATIO	W71-13334
RESEARCH AND DEVELOPMENT, WASTE	WATER TREATMENT.: /TIVE CAPACITY,	W71-13256
BACTERIA, CENTRIFUGATION, *WASTE	WATER TREATMENT.: /N, COLIFORMS,	W71-13326
*DENITRIFICATION, *NITRA/ *WASTE	WATER TREATMENT, *NITRIFICATION,	W71-13939
IQUES, COA/ *COAGULATION, *WASTE	WATER TREATMENT, ANALYTICAL TECHN	W71-13356
, WATER POLLUTION EFFECTS, WASTE	WATER TREATMENT, CYANOPHYTE, ENGL	W71-07973
IZE, ANAEROBIC CONDITIONS, WASTE	WATER TREATMENT.: /RS, PARTICLE S	W71-08223
IOCH/ *OXIDATION LAGOONS, *WASTE	WATER TREATMENT, *WATER RE-USE, B	W71-07123
ORGANISMS, BIODEGRADATION, WASTE	WATER TREATMENT, WATER POLLUTION	W71-07097
EROBIC CONDITIONS, SLUDGE, WASTE	WATER TREATMENT, BIODEGRADATION,	W71-07084
N, ORGANIC LOADING, ALGAE, WASTE	WATER TREATMENT, BIOCHEMICAL OXYG	W71-07096
BIOCHEMICAL OXYGEN DEMAND, WASTE	WATER TREATMENT.: /Y, COLIFORMS,	W71-07106
BIO/ *OXIDATION LAGOONS, *WASTE	WATER TREATMENT, ORGANIC LOADING,	W71-07124
YDROGEN ION CONCENTRATION, WASTE	WATER TREATMENT, KANSAS.: /AND, H	W71-07113
BIOCHEMICAL OXYGEN DEMAND, WASTE	WATER TREATMENT.: /ST ANALYSIS, *	W71-07099
BIOCHEMICAL OXYGEN DEMAND, WASTE	WATER TREAT: /EROBIC CONDITIONS,	W71-07100
BIOCHEMICAL OXYGEN DEMAND, WASTE	WATER TREATMENT, WATER POLLUTION	W71-07098
BIOCHEMICAL OXYGEN DEMAND, WASTE	WATER TREATMENT, *NORTH CAROLINA.	W71-07382
BIOCHEMICAL OXYGEN DEMAND, WASTE	WATER TREATMENT.: /C CONDITIONS,	W71-07109

E, PHOSPHATES, HYDROLYSIS, STORM WATER, RAINFALL, WASTE WATER(POLL W72-13644
TA, INHIBITORS, ANTIBIOTICS, SEA WATER, RHODE ISLAND.: / CHLOROPHY W72-04748
EVELOPMENT, *WATER SUPPLY, WASTE WATER, RUNOFF, DISSOLVED OXYGEN, W71-09788
TIDAL EFFECTS, MARINE ALGAE, SEA WATER, SALINITY, SAMPLING, CORES, W73-00367
N, PLANKTON, ATLANTIC OCEAN, SEA WATER, SALINITY, PHENOLS, WATER T W72-07710
ICS, *ESTUARINE ENVIRONMENT, SEA WATER, SALINITY, WATER TEMPERATUR W72-08141
TS, *SYSTEMATICS, *SEASONAL, SEA WATER, SAMPLING, BIOMASS, DIATOMS W73-00442
, PHOTOGRAPHY, SCUBA DIVING, SEA WATER, SAMPLING, ENVIRONMENTAL EF W73-00932
TON, *SYSTEMATICS, *DIATOMS, SEA WATER, SEASONAL, FLUCTUATIONS, CO W72-07663
TY, *SHOALS, *BENTHIC FLORA, SEA WATER, SEDIMENTS, LIGHT INTENSITY W72-14317
TER POLLUTION EFFECTS, RECLAIMED WATER, SEWAGE EFFLUENTS, RESEARCH W72-11702
G, SAMPLING, WATER ANALYSIS, SEA WATER, STABLE ISOTOPES, RADIOISOT W73-00831
LUTION TREATMENT, ALGAE, POTABLE WATER, TASTE, AMMONIA, SILICA.: / W71-01491
STES, ALGAE, AQUATIC PLANTS, SEA WATER, TRACE ELEMENTS, TOXICITY, W72-05952
BS, WATER POLLUTION SOURCES, SEA WATER, TROUT, TOXINS.: /RIMP, CRA W71-07412
INTERTIDA/ *MARINE PLANTS, *OILY WATER, WATER POLLUTION EFFECTS, * W71-08792
NG USES, DIVERSION, RECIRCULATED WATER, WATER REUSE, LEGAL ASPECTS W70-07100
WASTES, PATH OF POLLUTANTS, SEA WATER, WATER ANALYSIS, RADIOACTIV W73-00817
WATER TEMPERATURE, SALINITY, SEA WATER, WATER QUALITY, ON-SITE TES W72-09092
LUTION EFFECTS, *CALIFORNIA, SEA WATER, WATER ANALYSIS, MARINE ALG W73-02105
*BIOMASS, *ORGANIC MATTER, *SEA WATER, WATER ANALYSIS, ALGAE, CHE W71-11234
GY, BIOLOGICAL PROPERTIES, FRESH WATER, WATER QUALITY.: /AE, ECOLO W71-10580
S, SURFACE WATERS, NITROGEN, SEA WATER, WINDS, SAMPLING, INCUBATIO W72-09122
RIVER, OKA RIVER, FOAM, NATURAL WATERS.: *USSR, MOSCOW W72-05506
SPHORUS, EUTROPHICATION, SURFACE WATERS.: /DILUTION, NITROGEN, PHO W72-01472
ENVIRONMENT, WATER PO/ *RUNNING WATERS, *AQUATIC PLANTS, *AQUATIC W72-14690
TES, CLAY, TASTE, ODOR, *SURFACE WATERS, *GROUNDWATERS, SURVEYS, * W72-06536
S, *ILLINOIS, RECREATI/ *SURFACE WATERS, *LAKES, *STREAMS, *REVIEW W72-05869
ATION, *MICROORGANISMS, *SURFACE WATERS, *PLANKTON, *ALGAE, TURBID W72-06191
DATA/ *WATER QUALITY, *IMPOUNDED WATERS, *RESERVOIRS, *HYDROLOGIC W72-02274
EUTROPHICATION, *ALGAE, *SURFACE WATERS, *REVIEWS, *WATER POLLUTIO W71-10580
Y, *ECOLOGY, *CALIFORN/ *SURFACE WATERS, *SEA WATER, *WATER QUALIT W71-10577
ELETONEMA COSTATUM, TH/ *COASTAL WATERS, *SOUTHERN LONG ISLAND, SK W71-07875
ES, NITROGEN, NUTRIENTS, SURFACE WATERS, ALGAE, CHEMICAL PROPERTIE W72-12955
IATION, *AQUATIC ALGAE, *SURFACE WATERS, ANALYTICAL TECHNIQUES, PH W71-12034
NTS, PHOSPHORUS, NITROGEN, TIDAL WATERS, CHEMICAL OXYGEN DEMAND, B W71-07098
WATER QUALITY, MICHIGAN, SURFACE WATERS, COASTS, SAMPLING, BIOLOGI W72-10875
UREAS, *COASTS, *OCEANS, SURFACE WATERS, CONTINENTAL SHELF, DEPTH, W72-08056
HUDSON RIVER, STANDARDS, SURFACE WATERS, EPILIMNION, INTAKE STRUCT W70-09614
ALIZATION, NITROGEN SO/ *SURFACE WATERS, GEOLOGICAL SOURCES, MINER W71-06435
FLAGELLATES, SARGASSO SEA, SHELF WATERS, GULF STREAM, CAPE HATTERA W72-07663
FLOATING, WINDS, ALGAE, SURFACE WATERS, HYDROGEN ION CONCENTRATIO W73-00236
ECOLOGICAL DISTRIBUTION, SURFACE WATERS, LAGOONS, CHLOROPHYTA, RHO W73-00838
GANIC MATTER, NUTRIENTS, SURFACE WATERS, NITROGEN, WATER QUALITY, W70-10405
CTIVITY, ALASKA, FIORDS, SURFACE WATERS, NITROGEN, SEA WATER, WIND W72-09122
S, *AQUATIC ENVIRONMENT, SURFACE WATERS, NUTRIENTS, ALGAE, PHOSPHO W72-00847
, AIR POLLUTION EFFECTS, SURFACE WATERS, PATH OF POLLUTANTS, MARIN W72-00940
E COD(MASS), CAPE MAY(/ *COASTAL WATERS, PANAMA, CALLAO(PERU), CAP W72-08056
TECHNIQUES, MONITORING, SURFACE WATERS, PERSISTENCE, PATH OF POLL W73-00361
*LAKE KINNERET(ISRAEL), BOTTOM WATERS, RIVER JORDAN(ISRAEL).: W72-05469
, RUNOFF, ANIMAL WASTES, SURFACE WATERS, SOIL EROSION, NUTRIENTS, W71-04216
ENTS, *WATER QUALITY, *IMPOUNDED WATERS, STREAMS, FLOW, SEASONAL, W71-03028
FINLAND, DARK FIXATION, COASTAL WATERS, TROPHOGENIC LAYER.: /F OF W72-03228
NTROL, *DOMESTIC WATER, *SURFACE WATERS, WATER SUPPLY, EUTROPHICAT W71-05083
WATER POLLUTION SOURCES, SURFACE WATERS, WATER POLLUTION, DOMESTIC W70-09388
ENVIRONMENTS, GREAT LAKES, BAYS, WATERSHEDS(BASINS), SURFACE DRAIN W71-08953
BAYS, WATERSHEDS(BASINS), SMALL WATERSHEDS, CROPS, EGGS, FISHERIE W72-00959
, WATER POLLUTION EFFECTS, BAYS, WATERSHEDS(BASINS), SMALL WATERSH W72-00959
ENTS, HUMAN POPULATION, TOURISM, WATERSHEDS(BASINS), TROPHIC LEVEL W71-11009
RE, INDUSTRIES, WASTE TREATMENT, WATERSHEDS(BASINS), PHOSPHORUS, A W71-11017
WATERSHEDS(BASINS), *DDT, *SMALL WATERSHEDS, ECOSYSTEMS, PESTICIDE W72-12930
DE RESIDUES, *LOTIC ENVIRONMENT, WATERSHEDS(BASINS), *DDT, *SMALL W72-12930
ANCY, WIND VELOCITY, CONVECTION, WATTER POLLUTION EFFECTS.: / BUOY W71-09852
*NITROGEN:PHOSPHORUS RATIO, LAKE WAUBESA(WIS), OCCOQUAN RESERVOIR(W72-05473
ING, OXYGEN, TEMPERATURE, LIGHT, WAVES(WATER), PHYTOPLANKTON: /MPL W72-06283
STANDING CROP, CURRENTS(WATER), WAVES(WATER), SAMPLING, BENTHOS, W72-04766
ATES, AQUATIC ANIMALS, LITTORAL, WAVES(WATER).: /OPHYTA, INVERTEBR W71-11012
HOLOGY, *GENETICS, *VARIABILITY, WAVES(WATER), MAINE, STATISTICAL W73-00369
UTH WALES, MICROCYSTIS A/ *SHOCK- WAVES, *GAS-VACUOLE DEFLATION, SO W72-11564
ENTS(WATER), TIDES, WIND, LIGHT, WAVES, CLAMS, CRUSTACEANS, WORMS, W72-09092
LGAE, HAWAII, WATER TEMPERATURE, WAVES, LIGHT INTENSITY.: /ARINE A W71-11486
SMITH MOUNTAIN PROJECT(VA), WEATHER EFFECT.: W72-05282
DROLOGIC DATA, DISSOLVED OXYGEN, WEATHER, RESERVOIR OPERATION, WAT W72-05282
TH, ABSORPTION, ALGAE, SEASONAL, WEATHER, WATER ANALYSIS, TRACERS, W72-09122
LGAL TOXINS, *CHLOROPHYTA/ *FOOD WEBS, *FOOD CHAINS, *TOXICITY, *A W71-08597
MENTS, GROWTH RATES, ALGAE, FOOD WEBS, COPEPODS, ORGANOPHOSPHORUS W72-07892
WATER, BIOMASS, NUTRIENTS, FOOD WEBS, DETRITUS, ECOSYSTEMS, ENERG W73-01089
TS, ON-SITE INVESTIGATIONS, FOOD WEBS, FOOD CHAINS, FROGS, LICHENS W72-00949
COMMUNITIES, TIDAL MARSHES, FOOD WEBS, INTERTIDAL AREAS, CONNATE W W73-00796
TIVITY EFFECTS, ABSORPTION, FOOD WEBS, STRONTIUM RADIOISOTOPES, MA W72-00948
, CRUSTACEANS, FOOD CHAINS, FOOD WEBS, WATER POLLUTION EFFECTS, AQ W72-11925
MENTAL EF/ *HERBICIDES, *AQUATIC WEED CONTROL, *PARAQUAT, *ENVIRON W72-00155
ATMENT, *ALGAL CONTROL, *AQUATIC WEED CONTROL, CALIFORNIA.: /R TRE W72-02975
ECONOMIC JUSTIFICATION, AQUATIC WEED CONTROL, RIVERS, AGRICULTURE W71-13172
ALGAE, *ALGAL CONTROL, *AQUATIC WEED CONTROL, BIOCHEMICAL OXYGEN W72-14692
ETRATION, *MACROPHYTES, *AQUATIC WEED CONTROL, *LIGHT INTENSITY, * W73-02349
E,/ *FISH, *HERBIVORES, *AQUATIC WEED CONTROL, TILAPIA, CARP, ALGA W71-04528
*WATER QUALITY CONTROL, *AQUATIC WEED CONTROL, *DOMESTIC WATER, *S W71-05083
*HARVESTING OF ALGAE, *AQUATIC WEED CONTROL, *MECHANICAL CONTROL W71-06188
ALGAE CULTURE TECHNIQUE, AQUATIC WEED STUDY FACILITY, PRODUCT EVAL W71-06102

AUTHOR INDEX

In order not to complicate the computer program needlessly, only initials and the word "and" have been excluded from the alphabetic listing. Both first and last names of authors appear in the index. The numbers at the right are the accession numbers appearing at the end of the abstracts.

C. E.	GIBSON.: /	W72-04070
JOHN J. ROOK, AND	GIJSBERT OSKAM.:	W71-01491
J. BIRCH, ROBERT E. HILLMAN, AND	GILBERT E. RAINES.: /VIN, THOMAS /	W71-07417
M. S.	GILES.: /	W72-00959
A.	GILLESPIE, AND Y. MORITA.: /	W72-08989
W. MARSHALL.: W. J.	GILLESPIE, C. A. MAZZOLA, AND D. /	W72-03753
FRED G. LEE, AND	GILMAN D. VEITH.: /	W71-12092
N ES.: G. E. BURNS, R. M.	GIRLING, A. R. PICK, AND D. W. VA/	W71-07109
R.	GLEN FULLER.: /	W72-00960
D. M. STEVEN, AND R.	GLOMBITZA.: /	W72-11719
WALTER A.	GLOOSCHENKO.: /	W72-13657
E. M. DAVIS, AND E. F.	GLOYNA.: /	W72-09590
ERNST DAVIS, AND EARNEST	GLOYNA.:	W70-08319
ERNST M. DAVIS, AND EARNEST F.	GLOYNA.:	W71-05155
E. F.	GLOYNA, AND J. AGUIRRE.: /	W71-07102
OY, M. MUELLER, S. STOKER, J. J.	GOERING, AND M. T. GOTTSCHALK.: //	W72-09120
MARVIN C.	GOLDBERG.: /	W71-06435
ICHARD ARMSTRONG, AND CHARLES R.	GOLDMAN.: PETER RICHERSON, R/	W71-09171
ON, H. GEE, G. SHELET, AND J. C.	GOLDMAN.: /. SOBSEY, J. E. HARRIS/	W72-04787
OE MIDDLEBROOKS, AND C/ JOEL C.	GOLDMAN, DONALD B. PORCELLA, E. J/	W71-12068
EZAGA.: C. R.	GOLDMAN, G. MOSHIRI, AND E. DE AM/	W72-12549
SHELEF, W. J. OSWALD, AND C. C.	GOLUEKE.: G.	W70-10403
, WILLIAM J. OSWALD, CLARENCE G.	GOLUEKE, AND HENRY GEE.: /. ZABAT/	W72-04788
C. G.	GOLUEKE, AND W. J. OSWALD.:	W71-00138
G. L. DUGAN, C. G.	GOLUEKE, AND W. J. OSWALD.: /	W72-08396
RIXFORD.: G. L. DUGAN, C. G.	GOLUEKE, W. J. OSWALD, AND C. E. /	W72-02850
W. L. STROMBERG, AND C. J.	GOODNIGHT.: /	W72-09668
JON E. SANGER, AND EVILLE	GORHAM.: /	W72-01784
STOKER, J. J. GOERING, AND M. T.	GOTTSCHALK.: /OY, M. MUELLER, S. /	W72-09120
ERR.: P. D.	GOULDEN, W. J. TRAVERSY, AND G. K	W70-09388
R.	GOULDER.: /	W72-14796
K. H.	GOULDING.: /	W72-07508
L, G. S. VENKATARAMAN, AND S. K.	GOYAL.: / BOSE, UJJAL SINGH NAGPA/	W71-10070
U.	GRANHALL.: /	W73-00246
U.	GRANHALL, AND A. LUNDGREN.: /	W72-08054
JAMES C.	GRATTEAU.:	W71-05083
DOUGLAS EDWIN	GREER.: /	W72-03613
P. E.	GREESON.: /	W72-13851
JOHN J. COCHRANE, CONSTANTINE J.	GREGORY, AND GERALD L. ARONSON.: /	W71-05553
P. J.	GRESHAM.: /	W72-14469
RICHARD W.	GRIGG, AND ROBERT S. KIWALA.:	W71-01475
CHANEK, JR., N. GEBELEIN, AND C.	GROSSMAN.: /. B. MITTON, T. H. SU/	W72-14659
R. D.	GRUNDY.: /	W72-04734
E.	GRYGIEREK.: /	W72-14805
R. R. L.	GUILLARD, AND J. A. HELLEBUST.: /	W72-07710
R. C.	GUMERMAN.: /	W72-01108
KORZHENEVSKAYA.: M. V.	GUSEV, K. A. NIKITINA, AND T. G. /	W71-08683
EDATELEV, B. N. ILYIN, A/ V. N.	GUSKOVA, A. N. BRAGINA, A. A. ZAS/	W72-10694
SKAYA/ S. M. VARDAPETYAN, B. L.	GUTELMACHER, AND N. G. OZERETSKOV/	W72-00161
JANET R. DUGLE, AND J. E.	GUTHRIE.: /	W71-09256
J. S. DAVIS, AND W. F.	GWOREK.: /	W72-12744
G. EGAN, J. M. CASSIN, AND M. E.	HAIR.: W. /	W72-12941
LEE SU, E. J. STABA AND Y. ABUL-	HAJJ.: K./	W72-08586
MILLARD W.	HALL.: /	W72-04280
T.: D. G. AHEARN, EDWARD T.	HALL, JR., AND DONALD J. REINHARD/	W71-09173
DONALD J. REINHARDT./ EDWARD T.	HALL, JR., DONALD G. AHEARN, AND /	W71-11914
BRUCE W.	HALSTEAD.: /	W72-04259
A. L.	HAMILTON.: /	W71-12091
F. GESSNER, AND L.	HAMMER.: /	W72-12940
MARK J.	HAMMER, AND GARY L. HERGENRADER.:/	W72-04759
R. H. SIDDIQI, AND B. K.	HANDA.: /	W71-06033
F. J. HUMENIK, AND G. P.	HANNA, JR.:	W71-04079
FRANK J. HUMENIK, AND GEORGE P.	HANNA, JR.: /	W72-04983
F. J. HUMENIK, AND G. P.	HANNA, JR.: /	W72-11100
P. J.	HANNAN, AND C. PATOUILLET.: /	W72-07660
R. E.	HANSEN.: /	W72-08357
OSMUND HOLM-	HANSEN.: /	W71-11234
A. OTSUKI, AND T.	HANYA.: /	W73-00379
A. OTSUKI, AND T.	HANYA.: /	W73-01066
	HAROLD L. ALLEN.: /	W72-03216
D. O.	HARRIS.: /	W71-13793
ENTYNE, W. E. JOHNSON, AND A. J.	HARRIS.: J. R. VALL/	W71-11507
D. E. ARMSTRONG, R. F.	HARRIS, AND J. K. SYERS.: /	W72-03742
MSTRONG.: C. C. LEE, R. F.	HARRIS, J. K. SYERS, AND D. E. AR/	W71-13794
CHRISTIE.: A. J.	HARRIS, K. J. ROBERTS, AND A. E. /	W72-06837
F. L.	HARRISON.: /	W73-00826
FLORENCE L.	HARRISON, AND JOHN J. KCRANDA.: /	W72-06342
J. C. GOLDMAN/ M. SOBSEY, J. E.	HARRISON, H. GEE, G. SHELET, AND /	W72-04787
WILLIAM R.	HARTLEY, AND CHARLES M. WEISS.: /	W71-07382
R. S.	HARVEY.:	W71-02075
R. S.	HARVEY.: /	W71-09250
GIBBS, ERWIN LATZKO, MICHAEL J.	HARVEY, ZVI PLAUT, AND: MARTIN/	W71-09172
J. P.	HARVILLE.: /	W72-09092
B.	HAVLIK.: /	W72-10686
R. C.	HAYNES.: /	W72-09061
RGAN, AND WALTER TH/ CHARLES R.	HAZEL, FRED KOPPERDAHL, NORMAN MO/	W71-09789
ON, AND W. A. STEEL.: R. E.	HEFT, W. A. PHILLIPS, H. R. RALST/	W73-00831
G. W.	HEINKE, AND J. D. NORMAN.:	W71-00116
E. E.	HELD.: /	W72-03347
R. R. L. GUILLARD, AND J. A.	HELLEBUST.: /	W72-07710

W. L.	KLEIN, AND R. A. BRECHBILL.: /	W73-01673
T. BRAIDECH, P. GEHRING, AND C.	KLEVENO.: /	W72-12995
AND PHILIP E. GEHRING.: CONRAD	KLEVENO, O. BRAIDECH, E. THOMAS, /	W72-13659
WART, G. M. ROSENTHAL, AND J. R.	KLINE.: M. L. STE/	W72-03348
J. W.	KLOCK.: /	W72-10401
. NEWMAN, N. W. MOORE, AND F. W.	KNAGGS.: J. M. WAY, J. F/	W72-00155
WILLIAM H. BOUGHTON, RICHARD C.	KNUDSEN, AND NORVAL A. SINCLAIR.:	W71-01474
J. J.	KOELLING.: /	W73-01434
J.	KOMAREK.: /	W72-12172
H. W. DE	KONING, AND D. C. MORTIMER.: /	W71-12303
G. V.	KONOVALOVA.: /	W73-00442
T. F. H. ALLEN, AND J. F.	KOONCE.: /	W73-02469
LTER TH/ CHARLES R. HAZEL, FRED	KOPPERDAHL, NORMAN MORGAN, AND WA/	W71-09789
LORENCE L. HARRISON, AND JOHN J.	KORANDA.: F/	W72-06342
GUSEV, K. A. NIKITINA, AND T. G.	KORZHENEVSKAYA.: M. V. /	W71-08683
V.	KOTHANDARAMAN, AND R. L. EVANS.: /	W72-10237
YEHUDA	KOTT.: /	W72-00422
O. M.	KOZHOVA.: /	W72-04855
O. M.	KOZHOVA.: /	W71-09165
DAVID B. BAKER, AND JACK W.	KRAMER.: /	W72-04260
. SCALF, W. R. DUFFER, AND R. D.	KREIS.: M. R/	W72-08401
Y. FOLKMAN, M.	KREMER, AND P. G. J. MEIRING.: /	W72-11642
R. C. SUBBARAJU, AND K.	KRISHNAMURTHY.: /	W73-00935
J.	KRISTIANSEN.: /	W72-12738
RAY B.	KRONE.: /	W71-09581
ROY B. EVANS, AND PAUL	KRUGER.: /	W72-00974
YU V.	KRUGLOV, AND L. N. PAROMENSKAYA.:/	W71-07675
LOUIS A.	KRUMHOLZ.: /	W72-06526
D. K.	KRUPATKINA.: /	W72-07166
L. E.	KUENTZEL.: /	W71-13257
EDWARD J.	KUENZLER.: /	W71-10083
WILLIAM J. JEWELL, AND S. A.	KULASOORIYA.: /	W71-12070
V. V./ G. G. POLIKARPOV, L. G.	KULEBAKINA, V. G. TSITSUGINA, AND/	W72-10955
R. G.	KULLBERG.: /	W72-05610
T. NOSE.: Y. MATIDA, H.	KUMADA, S. KIMURA, Y. SAIGA, AND /	W72-12257
: DONALD B. PORCELLA, JAMES S.	KUMAGAI, AND E. JOE MIDDLEBROOKS.	W70-08944
H. D.	KUMAR, AND G. PRAKASH.: /	W72-00153
W. F.	KUNZ.: /	W72-11836
EDATELEV, B. N. ILYIN, AND V. M.	KUPRIYANOVA.: /BRAGINA, A. A. ZAS/	W72-10694
V. D. FEDOROV, AND N. G.	KUSTENKO.: /	W73-00441
	KWEI LEE SU.: /	W72-04275
CHANG-	KWEI LIN.: /	W72-13644
T. J.	LACK.: /	W72-14797
E. BOURGET, AND G.	LACROIX.: /	W72-05432
ND JOHN SKOK.:	LANDY MCBRIDE, WILLIAM CHCRNEY, A/	W71-12858
WILLY	LANGE.: /	W72-01097
J. CAIRNS, AND G. R.	LANZA.: /	W72-07225
J. CAIRNS, JR., AND G. R.	LANZA.: /	W72-08579
SOLON.: KENNETH W. STEWART,	LARRY E. MILLIGER, AND BERNARD M.	W70-06655
A.	LATTERA, AND M. BERNHARD.: /	W71-13243
DONALD R. TIMMONS, AND JOSEPH J.	LATTERELL.: ROBERT F. HOLT,	W71-04216
AUT, AND: MARTIN GIBBS, ERWIN	LATZKO, MICHAEL J. HARVEY, ZVI PL/	W71-09172
GEORGE E. DREWRY, AND ROBERT J.	LAVIGNE.: RICHARD G. CLEMENTS,/	W72-00949
C. MCDONALD, R. D. SPEAR, R. J.	LAVIN, AND N. L. CLESCERI.: G./	W72-08376
. A. DRAHN.: C. H.	LAWRANCE, K. G. TRANBARGER, AND R/	W72-05594
G. E.	LAZORENKO.: /	W72-12203
LOVE JR.: GEORGE W. REID,	LEALE E. STREEBIN, AND OLIVER T. /	W71-07973
N. A. JAWORSKI, D. W.	LEAR, JR., AND O. WILLA, JR.: /	W72-09171
NORBERT A. JAWORSKI, DONALD W.	LEAR, JR., AND ORTERIO VILLA, JR./	W71-08775
RBERT A. JAWORSKI, AND DONALD W.	LEAR, JR.: JOHAN A. AALTO, NO	W70-08248
G. D.	LEBEDEVA.: /	W72-09649
MAXIMOVA, G. I. MELESHKO, E. K.	LEBEDEVA, AND T. B. GALKINA.: /V./	W71-08027
B. J. COPELAND, AND H.	LEE DAVIS.: /	W72-13636
KWEI	LEE SU.: /	W72-04275
K.	LEE SU, AND E. JOHN STABA.: /	W72-07360
AJJ.: K.	LEE SU, E. J. STABA AND Y. ABUL-H/	W72-08586
G. F.	LEE.: /	W73-02479
N. J. MATERA, AND J. J.	LEE.: /	W73-00367
G. F.	LEE.: /	W73-02478
J. J.	LEE.: /	W73-00796
G. P. FITZGERALD, AND G. F.	LEE.: /	W73-00232
G. FRED	LEE.: /	W71-12072
C. C.	LEE.: /	W72-04292
J. J.	LEE.: /	W71-09254
FRED G.	LEE, AND GILMAN D. VEITH.: /	W71-12092
J. STONE.: JOHN J.	LEE, JOHN H. TIETJEN, AND ROBERT /	W72-00948
ND D. E. ARMSTRONG.: C. C.	LEE, R. F. HARRIS, J. K. SYERS, A/	W71-13794
D. J. HIBBERD, AND G. F.	LEEDALE.: /	W72-14808
M. TASSIGNY, AND M.	LEFEVRE.: /	W72-12734
PAULI BAGGE, AND PASO O.	LEHMUSLUOTO.: /	W72-03228
, AND B. THOMAS JOHNSON.:	LELYN STADNYK, ROBERT S. CAMPBELL/	W71-10566
J. H. ELDER, C. A.	LEMBI, AND D. J. MORRE.: /	W72-08440
NER.: NORBERT A. JAWORSKI,	LEO J. CLARK, AND KENNETH D. FEIG/	W71-09788
NER.: NORBERT A. JAWORSKI,	LEO J. CLARK, AND KENNETH D. FEIG	W70-08249
NORBERT A. JAWORSKI, AND	LEO J. HETLING.:	W70-06509
	LEONARD W. HOM.: /	W71-07096
	LESLIE RALPH BERGER.: /	W71-10791
ETH M. MACKENTHUM, AND LOWELL E.	LEUP.: /RISH, CARL R. HIRTH, KENN/	W71-05626
, AND G. E. RAINES.: A. A.	LEVIN, T. J. BIRCH, R. E. HILLMAN/	W72-07907
HILLMAN, AND GILBER/ ARTHUR A.	LEVIN, THOMAS J. BIRCH, ROBERT E./	W71-07417

, AND M. T. GO/ C. P. MCROY, M. MUELLER, S. STOKER, J. J. GOERING/ W72-09120
B. PATEL, P. G. VALANJU, C. D. MULAY, M. C. BALANI, AND: / W73-00811
V. H. MULLER.: / W73-01445
LE.: C. J. SOEDER, H. MULLER, H. D. PAYER, AND H. SCHUL/ W72-12739
M. MUNAWAR.: / W73-00918
R. C. TILTON, J. MURPHY, AND J. K. DIXON.: / W72-11833
S. N. MURRAY.: / W73-02099
. DIXON.: STEVEN MURRAY, JAN SCHERFIG, AND PETER S/ W72-04433
ROBERT S. MATSON, GEORGE E. MUSTOE, AND S. B. CHANG.: / W72-06037
. K/ PAROMITA BOSE, UJJAL SINGH NAGPAL, G. S. VENKATARAMAN, AND S/ W71-10070
R. L. BURGNER, AND R. E. NAKATANI.: / W73-00764
KAR.: R. E. NAKATANI, D. MILLER, AND J. V. TO W70-09612
NAMAN E. OTTO.: / W71-06102
CIMBERG.: NANCY L. NICHOLSON, AND ROBERT L./ W71-08792
NAVA NARKIS.: W70-09907
R. H. WEAVER, AND H. D. NASH.: / W71-07942
K. V. NATARAJAN.: / W71-11008
NAVA NARKIS.: W70-09907
JOSEPH Y. CHENG, AND NAVAL J. ANTIA.: W70-07842
JOSEPH Y. CHENG, AND NAVAL J. ANTIA.: W70-07280
A. B. NAZAROV, AND A. YA. ZESENKO.: / W72-12204
B. MAGUIRE, JR., AND W. E. NEILL.: / W72-09239
G. NELSON.: / W72-01685
AGGS.: J. M. WAY, J. F. NEWMAN, N. W. MOORE, AND F. W. KN/ W72-00155
H. WAYNE NICHOLS.: / W72-00377
: NANCY L. NICHOLSON, AND ROBERT L. CIMBERG./ W71-08792
JOHN G. NICKUM.: W71-02880
B. NIEHOFF.: / W72-06638
E. STEEMANN NIELSEN, AND S. WIUM-ANDERSEN.: W71-03027
P. BAGGE, AND A. NIEMI.: / W72-13871
S. NIEWOLAK.: / W72-10861
S. NIEWOLAK.: / W72-12255
A.: M. V. GUSEV, K. A. NIKITINA, AND T. G. KORZHENEVSKAY/ W71-08683
AGAT.: NING-HSI TANG, AND SURINDER K. BH/ W72-04431
CHUTTER.: R. G. NOBLE, W. A, PRETORIUS, AND F. M./ W72-01472
NOEL M. BURNS, AND C. ROSS.: / W72-12990
. LEAR, JR.: JOHAN A. AALTO, NORBERT A. JAWORSKI, AND DONALD W W70-08248
ETLING.: NORBERT A. JAWORSKI, AND LEO J. H W70-06509
AR, JR., AND ORTERIO VILLA, JR/ NORBERT A. JAWORSKI, DONALD W. LE/ W71-08775
, AND KENNETH D. FEIGNER.: NORBERT A. JAWORSKI, LEO J. CLARK W70-08249
, AND KENNETH D. FEIGNER.: NORBERT A. JAWORSKI, LEO J. CLARK/ W71-09788
ARLES R. HAZEL, FRED KOPPERDAHL, NORMAN MORGAN, AND WALTER THOMSEN/ W71-09789
G. W. HEINKE, AND J. D. NORMAN.: W71-00116
C. G. STEPHENS, AND B. B. NORTH.: / W72-07497
OUGHTON, RICHARD C. KNUDSEN, AND NORVAL A. SINCLAIR.: /ILLIAM H. B W71-01474
ADA, S. KIMURA, Y. SAIGA, AND T. NOSE.: Y. MATIDA, H. KUM/ W72-12257
VICTOR E. NOSHKIN.: / W72-04461
E E. LIKENS, F. HERBERT BORMANN, NOYE M. JOHNSON, D. W. FISHER, AN W71-01489
J. R. NURSALL, AND D. N. GALLUP.: / W73-01704
R. NUZZI.: / W72-11727
.: V. M. B. NYANISHKENE, AND G. G. POLIKARPOV/ W72-10978
PARKER, G. L. SAMSEL, AND E. K. OBENG-ASAMOA.: B. C./ W72-07661
T. A. OLSON, AND T. O. ODLAUG.: / W72-12192
W. REID, LEALE E. STREEBIN, AND OLIVER T. LOVE JR.: GEORGE/ W71-07973
J. W. VENNES, AND C. O. OLSON.: / W72-12565
T. A. OLSON, AND T. O. ODLAUG.: / W72-12192
. PRICE/ R. A. BOWEN, J. M. ST. ONGE, J. B. COLTON, JR., AND C. A/ W73-00371
I. L. OPHEL, AND C. D. FRASER.: / W71-11529
E. J. ORDAL.: / W72-05421
C. W. CHEN, AND G. T. ORLOB.: / W72-05282
C. W. CHEN, AND G. T. ORLOB.: / W72-14405
WORSKI, DONALD W. LEAR, JR., AND ORTERIO VILLA, JR.: /ORBERT A. JA/ W71-08775
J. A. OSBORNE, AND G. R. MARZOLF.: / W72-12393
JOHN J. ROOK, AND GIJSBERT OSKAM.: W71-01491
F. OSKAM.: / W72-08559
OSMUND HOLM-HANSEN.: / W71-11234
C. G. GOLUEKE, AND W. J. OSWALD.: W71-00138
DUGAN, C. G. GOLUEKE, AND W. J. OSWALD.: G. L./ W72-08396
ZABAT.: WILLIAM J. OSWALD, AARON MERON, AND MARIO D./ W71-07100
G. SHELEF, W. J. OSWALD, AND C. C. GOLUEKE.: W70-10403
. L. DUGAN, C. G. GOLUEKE, W. J. OSWALD, AND C. E. RIXFORD.: G/ W72-02850
AARON MERON, WILLIAM J. OSWALD, AND HENRY GEE.: / W72-04789
HEN/ MARIO D. ZABAT, WILLIAM J. OSWALD, CLARENCE G. GOLUEKE, AND / W72-04788
A. OTSUKI, AND T. HANYA.: / W73-00379
A. OTSUKI, AND T. HANYA.: / W73-01066
NAMAN E. OTTO.: / W71-06102
C. D. MC INTIRE, AND W. S. OVERTON.: / W72-08141
N. MARSHALL, C. A. OVIATT, AND D. M. SKAUEN.: / W72-14317
M. G. JOHNSON, AND G. E. OWEN.: / W71-08953
M. G. JOHNSON, AND G. E. OWEN.: / W71-11009
AN, B. L. GUTELMACHER, AND N. G. OZERETSKOVSKAYA.: /. M. VARDAPETY/ W72-00161
K. BASU, B. B. GHOSH, AND R. N. PAL.: A./ W71-11876
ET, M. FIALA, G. JACQUES, AND M. PANOUSE.: G. CAH/ W73-00431
PAT C. KERR, DORIS F. PARIS, AND D. L. BROCKWAY.: W71-00475
. C. KERR, D. L. BROCKWAY, D. F. PARIS, AND J. T. BARNETT, JR.: /P/ W72-09159
BENG-ASAMOA.: B. C. PARKER, G. L. SAMSEL, AND E. K. O/ W72-07661
YU V. KRUGLOV, AND L. N. PAROMENSKAYA.: / W71-07675
, G. S. VENKATARAMAN, AND S. K/ PAROMITA BOSE, UJJAL SINGH NAGPAL/ W71-10070
. MAC/ GERALD D. MCKEE, LOYS P. PARRISH, CARL R. HIRTH, KENNETH M/ W71-05626
D. A. EGLOFF, AND R. PARTRIDGE.: / W73-00281

ER T. LOVE JR.: GEORGE W. REID, LEALE E. STREEBIN, AND OLIV/ W71-07973
P. J. REILLY.: / W73-01657
R. J. REIMOLD, AND C. J. DURANT.: / W72-10678
WARD T. HALL, JR., AND DONALD J. REINHARDT.: D. G. AHEARN, ED/ W71-09173
DONALD G. AHEARN, AND DONALD J. REINHARDT.: /EDWARD T. HALL, JR.,/ W71-11914
C. C. REMSEN.: / W72-08056
E. J. CARPENTER, C. C. REMSEN, AND S. W. WATSON.: / W73-00380
R. A. ENGBERG, AND T. O. RENSCHLER.: / W71-13466
R. G. RHODES.: / W72-12633
RUSSELL G. RHODES, AND ANTHONY J. TERZIS.: / W71-05629
INATH.: L. G. RICH, J. F. ANDREWS, AND T. M. KE/ W72-11724
GOLDMAN.: PETER RICHERSON, RICHARD ARMSTRONG, AND CHARLES R./ W71-09171
G. DANIEL HICKMAN, AND RICHARD B . MOORE.: / W72-01095
CARTY, JAMES A. ROBERTSON, AND/ RICHARD C. BAIN, JR., PERRY L. MC/ W71-07098
RVING YALL, WILLIAM H. BOUGHTON, RICHARD C. KNUDSEN, AND NORVAL A. W71-01474
RICHARD D. ROX.: / W71-13334
EWRY, AND ROBERT J. LAVIGNE.: RICHARD G. CLEMENTS, GEORGE E. DR/ W72-00949
RANEY.: WILLIAM H. FUNK, RICHARD J. CONDIT, AND WAYNE T. C/ W72-00801
RICHARD M. KLEIN.: / W71-09157
IWALA.: RICHARD W. GRIGG, AND ROBERT S. K W71-01475
RICHARD W. TERKELTOUB.: / W71-11001
CHARLES R. GOLDMAN.: PETER RICHERSON, RICHARD ARMSTRONG, AND/ W71-09171
GOLUEKE, W. J. OSWALD, AND C. E. RIXFORD.: G. L. DUGAN, C. G. / W72-02850
E. BENDER, WAYNE R. MATSON, AND ROBERT A. JORDAN.: MICHAEL/ W71-09674
EDWIN F. BARTH, AND ROBERT B. DEAN.: / W72-04086
LMAN.: ROBERT E. BAUMANN, AND SHELDON KE/ W71-06445
RTHUR A. LEVIN, THOMAS J. BIRCH, ROBERT E. HILLMAN, AND GILBERT E./ W71-07417
, AND JOSEPH J. LATTERELL.: ROBERT F. HOLT, DONALD R. TIMMONS W71-04216
CLEMENTS, GEORGE E. DREWRY, AND ROBERT J. LAVIGNE.: RICHARD G./ W72-00949
OHN J. LEE, JOHN H. TIETJEN, AND ROBERT J. STONE.: J/ W72-00948
ROBERT JACK FREEMAN, JR.: / W71-13326
FRANCIS M. MIDDLETOWN, AND ROBERT L. BUNCH.: / W71-07123
ROBERT L. BUNCH.: / W71-12188
NANCY L. NICHOLSON, AND ROBERT L. CIMBERG.: / W71-08792
G. D. COOKE, AND ROBERT L. KENNEDY.: / W71-11949
ROBERT O. MEGARD.: / W71-08670
JOHNSON.: LELYN STADNYK, ROBERT S. CAMPBELL, AND B. THOMAS/ W71-10566
WHITLEY.: ROBERT S. CAMPBELL, AND JAMES R. / W71-06444
JR., AND JAMES R. WHITLEY.: ROBERT S. CAMPBELL, ARTHUR WITT, W70-09596
RICHARD W. GRIGG, AND ROBERT S. KIWALA.: W71-01475
E, AND S. B. CHANG.: ROBERT S. MATSON, GEORGE E. MUSTO/ W72-06037
YE M. JOHNSON, D. W. FISHER, AND ROBERT S. PIERCE.: /T BORMANN, NO W71-01489
A. J. HARRIS, K. J. ROBERTS, AND A. E. CHRISTIE.: / W72-06837
D. E. ROBERTSON.: / W72-04463
JR., PERRY L. MCCARTY, JAMES A. ROBERTSON, AND WILLIAM H. PIERCE./ W71-07098
R. B. ROBIE.: W70-07100
G. ROBIN SOUTH, AND ANDRE CARDINAL.:/ W71-10066
ALBERT MASSEY, AND JOHN ROBINSON.: / W72-07937
. J. SMITH, D. ALEXANDER, AND P. ROBINSON.: S. E. MANAHAN, M/ W72-01693
JOSEPH F. ROESLER, AND HERBERT C. PREUL.: / W71-07099
M. A. ROESSLER.: / W72-11876
K.: JOHN CAIRNS, JR., ROGER L. KAESLER, AND RUTH PATRIC/ W71-11583
D. E. ARMSTRONG, AND G. A. ROHLICH.: / W71-06443
RONALD L. RASCHKE.: W70-07838
B. WELCH, JAMES A. BUCKLEY, AND RONALD M. BUSH.: EUGENE/ W72-00799
CH.: RONALD M. BUSH, AND EUGENE B. WEL/ W72-00798
J. RONALD MINER.: / W71-08214
JOHN J. ROOK, AND GIJSBERT OSKAM.: W71-01491
D. W. ROONEY.: / W72-11920
D. W. ROONEY.: / W72-12724
F. L. ROSE, AND G. W. MINSHALL.: / W72-13300
KARL A. ROSEMAN, AND WARNER M. LINFIELD.:/ W71-06247
M. L. STEWART, G. M. ROSENTHAL, AND J. R. KLINE.: / W72-03348
N. M. BURNS, AND C. ROSS.: / W72-12997
NOEL M. BURNS, AND C. ROSS.: / W72-12990
F. E. ROUND.: / W73-00432
M. HICKMAN, AND F. E. ROUND.: / W71-12855
G. A. ROUNSEFELL.: / W73-02029
RICHARD D. ROX.: / W71-13334
JAMES L. YOUNT, AND ROY A. CROSSMAN, JR.: W71-01488
ROY B. EVANS, AND PAUL KRUGER.: / W72-00974
SYLVIA B. DERBY, AND ERNEST RUBER.: / W71-10096
TERZIS.: RUSSELL G. RHODES, AND ANTHONY J./ W71-05629
IRNS, JR., ROGER L. KAESLER, AND RUTH PATRICK.: JOHN CA/ W71-11583
K. RUTZLER, AND W. STERRER.: / W72-07911
RE, AND J. E. HUGUENIN.: J. H. RYTHER, W. M. DUNSTAN, K. R. TENO/ W72-09378
PAUL E. SAGER.: / W71-10645
MATIDA, H. KUMADA, S. KIMURA, Y. SAIGA, AND T. NOSE.: Y. / W72-12257
SAM E. BEALL.: W70-08832
B. C. PARKER, G. L. SAMSEL, AND E. K. OBENG-ASAMOA.: / W72-07661
JON E. SANGER, AND EVILLE GORHAM.: / W72-01784
L S. DOTY, AND GERTRUDES AGUILAR- SANTOS.: MAXWEL/ W71-08597
D. F. SARGENT, AND C. P. S. TAYLOR.: / W72-11572
J. J. SASNER, JR.: / W73-00243
W. JOHNSON, J. MCN. SIEBURTH, A. SASTRY, C. R. ARNOLD, AND M. S. D/ W72-08142
N. DABES, C. R. WILKE, AND K. H. SAUER.: J. / W72-06274
CLAIR N. SAWYER.: / W72-04769
CLAIR N. SAWYER.: / W72-05473
PHILIP J. SAWYER.: W71-03014
EIS.: M. R. SCALF, W. R. DUFFER, AND R. D. KR/ W72-08401

GPSR Compliance
The European Union's (EU) General Product Safety Regulation (GPSR) is a set of rules that requires consumer products to be safe and our obligations to ensure this.

If you have any concerns about our products, you can contact us on

ProductSafety@springernature.com

In case Publisher is established outside the EU, the EU authorized representative is:

Springer Nature Customer Service Center GmbH
Europaplatz 3
69115 Heidelberg, Germany

www.ingramcontent.com/pod-product-compliance
Lightning Source LLC
LaVergne TN
LVHW082004200926
848481LV00010B/2801
* 9 7 8 1 4 7 5 7 0 4 2 2 8 *